Structural Inorganic Chemistry

Structural Inorganic Chemistry

A. F. WELLS

FOURTH EDITION

CLARENDON PRESS · OXFORD
1975

Oxford University Press, Ely House, London W.1

GLASGOW NEW YORK TORONTO MELBOURNE WELLINGTON
CAPE TOWN IBADAN NAIROBI DAR ES SALAAM LUSAKA ADDIS ABABA
DELHI BOMBAY CALCUTTA MADRAS KARACHI LAHORE DACCA
KUALA LUMPUR SINGAPORE HONG KONG TOKYO

ISBN 0 19 855354 4

© OXFORD UNIVERSITY PRESS 1975

PRINTED IN GREAT BRITAIN BY
WILLIAM CLOWES & SONS LIMITED
LONDON, COLCHESTER AND BECCLES

Preface

This book has been almost entirely rewritten, but its purpose and general organization remain the same as those of previous editions. The Introduction to the first (1945) edition included the following paragraph: 'The reasons for writing this book were, firstly, the conviction that the structural side of inorganic chemistry cannot be put on a sound basis until the knowledge gained from the study of the solid state has been incorporated into chemistry as an integral part of that subject, and secondly, the equally strong conviction that it is unsatisfactory merely to add information about the structures of solids to the descriptions of the elements and compounds as usually presented in a systematic treatment of inorganic chemistry.' Now, after a period of thirty years during which considerable advances have been made in solid state chemistry, it is still true to say that the structures and properties of solids receive very little attention in most treatments of inorganic chemistry, and this in spite of the fact that most elements and most inorganic compounds are solids at ordinary temperatures. This state of affairs would seem to be sufficient justification for the appearance of yet another edition of this book.

Since the results of structural studies of crystals are described in crystallographic language the first requirement is that these results be made available in a form intelligible to chemists. It was this challenge that first attracted the author, and it is hoped that this book will continue to provide teachers of chemistry with facts and ideas which can be incorporated into their teaching. However, while any addition of structural information to the conventional teaching of inorganic chemistry is to be welcomed the real need is a radical change of outlook and the recognition that not only is the structure of a substance in all states of aggregation an essential part of its full description (or characterization) but also that the structures and properties of solids form an integral part, perhaps the major part, of the subject.

The general plan of the book is as follows. Part I deals with a number of general topics and is intended as an introduction to the more detailed Part II, which forms the larger part of the book. In Part II the structural chemistry of the elements is described systematically, and the arrangement of material is based on the groups of the Periodic Table. The advances made during the past decade have necessitated considerable changes in these latter chapters, but the major structural changes have been made in the content of Part I.

Since a concise treatment of certain geometrical and topological topics is not readily available elsewhere more space has been devoted to these than in previous editions at the expense of subjects such as the experimental methods of structural chemistry, which at best can receive only a sketchy treatment in a volume such as

v

this. Many students find difficulty in appreciating the three-dimensional geometry of crystal structures from two-dimensional illustrations (even stereoscopic photographs). In order to acquire some facility in visualising the three-dimensional arrangements of atoms in crystals some acquaintance is necessary with symmetry, repeating patterns, sphere-packings, and related topics. Some of this material could be, and sometimes is, introduced into teaching at an early age. However, there is a tendency in some quarters to regard solid geometry as old-fashioned and to replace it in school curricula by more fashionable aspects of mathematics. This adds to the difficulties of those teachers of chemistry who wish to modernize their teaching by including information about the structures of solids. Unless the student has an adequate grounding in the topics noted above little is gained by adding diagrams of unit cells of crystal structures to conventional chemistry texts.

The educational value of building models representing the arrangements of atoms in crystals cannot be over-emphasized; and by this we mean that the student actually assembles the model and does not simply look at a ready-made model, however much more elegant the latter may be. Some very tentative suggestions for model building have been offered in the author's *Models in Structural Inorganic Chemistry*, to which the abbreviation MSIC in the present volume refers.

References. The present volume has never been intended as a reference work, though it may serve as a useful starting-point when information is required on a particular topic. As an essential part of the educational process the advanced student should be encouraged to adopt a critical attitude towards the written word (including the present text); he must learn where to find the original literature and to begin to form his own judgment of the validity of conclusions drawn from experimental data. It is becoming increasingly difficult to locate the original source of a particular item of information, and for this reason numerous references to the scientific literature are included in the systematic part of this book. These generally refer to the latest work, in which references to earlier work are usually included. To save space (and expense) the names of scientific journals have been abbreviated to the forms listed on pp. xxi–xxiii.

Indexes. There are two indexes. The arrangement of entries in the formula index is not entirely systematic for there is no wholly satisfactory way of indexing inorganic compounds which retains chemically acceptable groupings of atoms. The formulae have been arranged so as to emphasize the feature most likely to be of interest to the chemist. The subject index is largely restricted to names of minerals and organic compounds and to topics which are not readily located in the list of contents.

Acknowledgments. During the writing of this book, which of necessity owes much to the work and ideas of other workers in this and related fields, I have had the benefit of helpful discussions with a number of colleagues, of whom I would particularly mention Dr. B. C. Chamberland. I wish to thank Dr. B. G. Bagley and the editor of Nature (London) for permission to use Fig. 4.3, Dr. H. T. Evans and

John Wiley and Sons for Figs 5c, 7, 10, 11, and 12b in Chapter 11, and Drs. G. T. Kokotailo and W. M. Meier for Fig. 23.27. It gives me great pleasure to acknowledge the debt that I owe to my wife for her support and encouragement over a period of many years.

A. F. Wells

Department of Chemistry,
University of Connecticut,
Storrs, Connecticut,
U.S.A.
1974

Contents

Contents

x

Contents

Contents

Contents

Contents

Contents

Abbreviations

The following abbreviations are used in references to Journals throughout this book.

AANL	Atti dell'Accademia nazional dei Lincei
AC	Acta crystallographica
AcM	Acta Metallurgica
ACSc	Acta Chemica Scandinavica
ACSi	Acta Chimica Sinica
AJC	Australian Journal of Chemistry
AJSR	Australian Journal of Scientific Research
AK	Arkiv för Kemi
AKMG	Arkiv för Kemi, Mineralogi och Geologi
AlC	Analytical Chemistry
AM	American Mineralogist
AnC	Angewandte Chemie
AP	Annalen der Physik
APURSS	Acta Physicochimica URSS
ARPC	Annual Review of Physical Chemistry
ASR	Applied Scientific Research
B	Berichte
BB	Berichte der Bunsengesellschaft für physikalische Chemie
BCSJ	Bulletin of the Chemical Society of Japan
BSCB	Bulletin des Sociétés chimiques Belges
BSCF	Bulletin de la Société chimique de France
BSFMC	Bulletin de la Société française de minéralogie et de cristallographie
C	Chimia (Switzerland)
CB	Chemische Berichte
CC	Chemical Communications (Journal of The Chemical Society, Chemical Communications)
CJP	Canadian Journal of Physics
CR	Comptes rendus hebdomadaires des Séances de l'Académie des Sciences (Paris)
CRURSS	Comptes rendus de l'Académie des Sciences de l'URSS
DAN	Doklady Akademii Nauk SSSR
E	Experientia
FM	Fortschritte der Mineralogie
GCI	Gazzetta chimica italiana
HCA	Helvetica Chimica Acta
IC	Inorganic Chemistry
ICA	Inorganica Chimica Acta
IEC	Industrial and Engineering Chemistry
JACS	Journal of the American Chemical Society
JACeS	Journal of the American Ceramic Society
JAP	Journal of Applied Physics

JCG	Journal of Crystal Growth
JCP	Journal of Chemical Physics
JCS	Journal of the Chemical Society (London)
JES	Journal of the Electrochemical Society
JINC	Journal of Inorganic and Nuclear Chemistry
JLCM	Journal of the Less-common Metals
JM	Journal of Metals
JMS	Journal of Molecular Spectroscopy
JNM	Journal of Nuclear Materials
JOC	Journal of Organometallic Chemistry
JPC	Journal of Physical Chemistry
JPCS	Journal of the Physics and Chemistry of Solids
JPP	Journal de Physique (Paris)
JPSJ	Journal of the Physical Society of Japan
K	Kristallografiya
KDV	Kongelige Danske Videnskabernes Selkab Matematisk-fysiske Meddeleser
MH	Monatshefte für Chemie und verwandte Teile anderer Wissenschaften
MJ	Mineralogical Journal of Japan
MM	Mineralogical Magazine (and Journal of the Mineralogical Society)
MMJ	Mineralogical Magazine (Japan)
MRB	Materials Research Bulletin
MSCE	Mémorial des Services chimiques de l'état (Paris)
N	Nature
NBS	Journal of Research of the National Bureau of Standards
NF	Naturforschung
NJB	Neues Jahrbuch für Mineralogie
NPS	Nature (Physical Sciences)
NW	Naturwissenschaften
PCS	Proceedings of the Chemical Society
PKNAW	Proceedings koninklijke nederlandse Akademic van Wetenschappen
PM	Philosophical Magazine
PNAS	Proceedings of the National Academy of Sciences of the U.S.A.
PR	Physical Review
PRL	Physical Review Letters
PRP	Philips Research Reports
PSS	Physica Status Solidi
QRCS	Quarterly Reviews of The Chemical Society
RJIC	Russian Journal of Inorganic Chemistry
RMP	Reviews of Modern Physics
RPAC	Reviews of Pure and Applied Chemistry (Royal Australian Chemical Institute)
RS	Ricerca scientifica
RTC	Recueil des Travaux chimiques des Pays-Bas et de la Belgique
SA	Spectrochimica Acta
Sc	Science
SMPM	Schweizerische mineralogische und petrographische Mitteilungen
SPC	Soviet Physics: Crystallography
SR	Structure Reports
SSC	Solid State Communications
TAIME	Transactions of the American Institute of Mining and Metallurgical Engineers
TFS	Transactions of the Faraday Society
TKBM	Tidsskrift for Kjemi, Bergvesen og Metallurgi
ZaC	Zeitschrift für anorganische (und allgemeine) Chemie

ZE	Zeitschrift für Elektrochemie
ZFK	Zhurnal fizicheskoi Khimii
ZK	Zeitschrift für Kristallographie
ZN	Zeitschrift für Naturforschung
ZP	Zeitschrift für Physik
ZPC	Zeitschrift für physikalische Chemie
ZSK	Zhurnal strukturnoi Khimii

Part I

1

Introduction

In this introductory chapter we discuss in a general way a number of topics which are intended to indicate the scope of our subject and the reasons for the choice of topics which receive more detailed attention in subsequent chapters.

The number of elements known exceeds one hundred, so that if each one combined with each of the others to form a single binary compound there would be approximately five thousand such compounds. In fact not all elements combine with all the others, but on the other hand some combine to form more than one compound. This is true of many pairs of metals, and other examples, chosen at random, include:

$$YB_2, \ YB_4, \ YB_6, \ YB_{12}, \ and \ YB_{66};$$
$$CrF_2, CrF_3, \ CrF_4, \ CrF_5, \ CrF_6, \ and \ Cr_2F_5;$$
$$CrS, \ Cr_7S_8, Cr_5S_6, Cr_3S_4, \ and \ Cr_2S_3.$$

The number of binary compounds alone is evidently considerable, and there is an indefinitely large number of compounds built of atoms of three or more elements. It seems logical to concentrate our attention first on the simplest compounds such as binary halides, chalconides, etc., for it would appear unlikely that we could understand the structures of more complex compounds unless the structures of the simpler ones are known and understood. However, it should be noted that simplicity of chemical formula may be deceptive, for the structures of many compounds with simple chemical formulae present considerable problems in bonding, and indeed the structures of some elements are incomprehensibly complex (for example, B and red P). On the other hand, there are compounds with complex formulae which have structures based on an essentially simple pattern, as are the numerous structures described in Chapter 3 which are based on the diamond net, one of the simplest 3-dimensional frameworks. We shall make a point of looking for the simple underlying structural themes in the belief that Nature prefers simplicity to complexity and also because structures are most easily understood if reduced to their simplest terms.

The importance of the solid state

Since we shall devote most of the first part of this book to matters directly concerned with the solid state it is appropriate to note a few general points, to some of which we return later in this chapter.

Introduction

(i) Most of the elements (some 90 per cent) are solids at ordinary temperatures, and this is also true of the majority of inorganic compounds. It happens that most of the important reagents are liquids, gases, or solutions, but these constitute a small proportion of the more common inorganic compounds. Also, although it is true that chemical reactions are usually carried out in solution or in the vapour state, in most reactions the reactants or products, or both, are solids. Chemical reactions range from those between isolated atoms or discrete groups of atoms (molecules or complex ions), through those in which a solid is removed or produced, to processes such as the corrosion of metals where a solid product builds up on the surface of the solid reactant. In all cases where a crystalline material is formed or broken down, the process involves the lattice energy of the crystal. The familiar Born-Haber cycle for the reaction between solid sodium and gaseous chlorine to form solid NaCl provides a simple example of the interrelation of heats of dissociation, ionization energy and affinity, lattice energy, and heat of reaction.

(ii) Organic compounds (other than polymers) exist as finite molecules in all states of aggregation. This means, first, that the structural problem consists only in discovering the structure of the finite molecule, and second, that this could in principle be determined by studying its structure in the solid, liquid, or vapour state. Apart from possible geometrical changes such as rotation about single bonds and small dimensional changes due to temperature differences, the basic topology and geometry could be studied in any state of aggregation. Some inorganic compounds also exist as finite molecules in the solid, liquid, and gaseous states, for example, many simple molecules formed by non-metals (HCl, CO_2) and also some compounds of metals (SnI_4, $Cr(CO)_6$). Accurate information about the structures of simple molecules, both organic and inorganic, comes from spectroscopic and electron diffraction studies of the vapours, but these methods are not applicable to very complex molecules. Because crystalline solids are periodic structures they act as diffraction gratings for X-rays and neutrons, and in principle the structure of any molecule, however complex, can be determined by diffraction studies of the solid.

In contrast to organic compounds and the minority of inorganic compounds mentioned above, the great majority of solid inorganic compounds have structures in which there is linking of atoms into systems which extend indefinitely in one, two, or three dimensions. Such structures are characteristic only of the solid state and must necessarily break down when the crystal is dissolved, melted, or vaporized. The study of crystal structures has therefore extended the scope of structural chemistry far beyond that of the finite groups of atoms to which classical stereochemistry was restricted to include all the periodic arrangements of atoms found in crystalline solids.

Because the great majority of inorganic compounds are compounds of one or more metals with non-metals, and because most of them are solids under ordinary conditions, the greater part of structural inorganic chemistry is concerned with the structures of solids. The only compounds of metals which have any structural chemistry, apart from that of the crystalline compound, are those molecules or ions that can be studied in solution or the molecules of compounds that can be melted or vaporized without decomposition. It is unlikely that very much accurate

4

structural information will ever be obtained from liquids, whereas electron diffraction or spectroscopic studies can be made of molecules in the gas phase provided they are not too complex. It is important therefore to distinguish between solid compounds which can be vaporized without decomposition and those which can exist *only as solids*. By this we mean that their existence depends on types of bonding which are possible only in the solid state. Some simple halides and a few oxides of metals have been studied as vapours, and if the vapour species is not present in the crystal the information so obtained is complementary to that obtained by studying the solid. On the other hand, many simple compounds $M_m X_n$ are unlikely to exist in the vapour state because the particular ratio of metal to non-metal atoms is only realizable in an infinite array of atoms between which certain types of bonding can operate. Crystalline Cs_2O consists of infinite layers, but nevertheless we can envisage molecules of Cs_2O in the vapour. However, oxides such as Cs_3O and Cs_7O depend for their existence on extended systems of metal–metal bonds which would not be possible in a finite molecule.

Comparatively little is yet known about the high temperature chemistry of metal halides, oxides, etc.; for example, the structures of molecules such as $FeCl_3$ or of oxides M_2O_3, MO_2, M_2O_5, or indeed whether these species are formed or are unstable (like SiO_2). Certainly complex halides and oxides, can exist only in the crystalline state, and this is true also of other large and important groups of compounds such as salts containing oxy ions, 'acid' and 'basic' salts, and hydrates. One particularly important result of the study of crystal structures has been the recognition that *non-stoichiometric compounds* are not the rarities they were once thought to be. A non-stoichiometric compound may be very broadly defined as a solid phase which is stable over a range of composition. This definition covers at one extreme all cases of 'isomophous replacement' and all kinds of solid solution, the composition of which may cover the whole range from one pure component to the other. At the other extreme there are phosphors (luminescent ZnS or ZnS–Cu), which owe their properties to misplaced and/or impurity atoms which act as 'electron traps', and coloured halides (alkali or alkaline-earth) in which some of the halide-ion-sites are occupied by electrons (F-centres); these defects are present in very small concentration, often in the range 10^{-6}–10^{-4}. Of more interest to the inorganic chemist is the fact that many simple binary compounds exhibit ranges of composition, the range depending on the temperature and mode of preparation. The non-stoichiometry implies disorder in the structure and usually the presence of an element in more than one valence state, and can give rise to semiconductivity and catalytic activity. Examples of non-stoichiometric binary compounds include many oxides and sulphides, some hydrides, and interstitial solid solutions of C and N in metals. More complex examples include various complex oxides with layer and framework structures, such as the bronzes (p. 505). The existence of green and black NiO, with very different physical properties, the recent preparation for the first time of stoichiometric FeO, and the fact that Fe_6S_7 is not FeS containing excess S but FeS deficient in Fe (that is, $Fe_{1-x}S$) are matters of obvious importance to the inorganic chemist.

The compositions and properties and indeed the very existence of non-stoichiometric compounds can be understood only in terms of their structures. This is particularly evident in cases where the non-stoichiometry arises from the inclusion of foreign atoms or molecules in a crystalline structure. It can occur in crystals built of finite molecules or crystals containing large finite ions. For example, if $Pd_2Br_4[As(CH_3)_3]_2$ (p. 28) is crystallized from dioxane the crystals can retain non-stoichiometric amounts of the solvent in the tunnels between the molecules, and these molecules can be removed without disruption of the structure (Fig. 1.9(a)). In the mineral beryl (p. 815) the large cyclic $(Si_6O_{18})^{12-}$ ions are stacked in columns, and helium can be occluded in the tunnels. Some crystals with layer structures can take up material between the layers. Examples include the lamellar compounds of graphite (p. 734) and of clay minerals (p. 823). An unusual type of layer structure is that of $Ni(CN)_2 . NH_3$ which can take up molecules of $H_2O, C_6H_6, C_6H_5NH_2$, etc. between the layers (Fig. 1.9(b)). Structures of this kind are called 'clathrates', and examples are noted on p. 28.

(iii) The great wealth of information about atomic arrangement in crystals and in particular the detailed information about bond lengths and interbond angles provided by studies of crystal structures is the raw material for the theoretician interested in bonding and its relation to physical properties.

All elements and compounds can be solidified under appropriate conditions of temperature and pressure, and the properties and structures of solids show that we must recognize four extreme types of bonding:

(a) the polar (ionic) bond in crystalline salts such as NaCl or CaF_2,

(b) the covalent bond in non-ionizable molecules such as Cl_2, S_8, etc., which exist in both the crystalline elements and also in their vapours, and in crystals such as diamond in which the length of the C–C bond is the same as in molecules such as $H_3C–CH_3$,

(c) the metallic bond in metals and intermetallic compounds (alloys), which is responsible for their characteristic optical and electrical properties, and

(d) the much weaker van der Waals bond between chemically saturated molecules such as those just mentioned—witness the much larger distances between atoms of different molecules as compared with those within such molecules. In crystalline Cl_2 the bond length is 1·99 Å, but the shortest distance between Cl atoms of different molecules is 3·34 Å. The van der Waals bond is responsible for the cohesive forces in liquid or solid argon or chlorine and more generally between neutral molecules chains, and layers in numerous crystals whose structures will be described later.

Although it is convenient and customary to recognize these four extreme types of bonding it should be realized that bonds of these 'pure' types—if indeed the term 'pure' has any clear physical or chemical meaning—are probably rather rare, particularly in the case of the first two types. Bonds of an essentially ionic type occur in salts formed from the most electropositive combined with the most electronegative elements and between, for example, the cations and the O atoms of the complex ion in oxy-salts such as $NaNO_3$. Covalent bonds occur in the non-metallic elements and in compounds containing non-metals which do not differ greatly in 'electronegativity' (see p. 236). However, it would seem that the great majority of bonds in inorganic compounds must be regarded as intermediate in character between these extreme types. For example, most bonds between metals and non-metals have some ionic and some covalent character, and at present there is no entirely satisfactory way of describing such bonds.

Evidently many crystals contain bonds of two or more quite distinct types. In molecular crystals consisting of non-polar molecules the bonds within the molecule may be essentially covalent (e.g. S_6 or S_8) or of some intermediate ionic–covalent nature (e.g. SiF_4), and those between the molecules are van der Waals bonds. In a crystal containing complex ions the bonds within the complex ion may approximate to covalent bonds while those between the complex ion and the cations (or anions) are essentially ionic in character, as in the case of $NaNO_3$ already quoted. In other crystals there are additional interactions between certain of the atoms which are not so obviously essential as in these cases to the cohesion of the crystal. An example is the metal–metal bonding in dioxides with the rutile structure, a structure which in many cases is stable in the absence of such bonding.

It is also necessary to recognize certain other types of interactions which, although weaker than ionic or covalent bonds, are important in determining or influencing the structures of particular groups of crystalline compounds—for example, hydrogen bonds (bridges) and charge-transfer bonds. Hydrogen bonds are of rather widespread occurrence and are discussed in more detail in later chapters.

(iv) It is perhaps unnecessary to emphasize here that there is in general no direct relation between the chemical formula of a solid and its structure. For example, only the first member of the series

HI	AuI	CuI	NaI	CsI	AX
1	2	4	6	8	(C.N. of A by X)†

consists of discrete molecules A–X under ordinary conditions. All the other compounds are solids at ordinary temperatures and consist of infinite arrays of A and X atoms in which the metal atoms are bonded to, respectively, two, four, six, and eight X atoms. Figure 1.1 shows some simple examples of systems with the composition AX. Two are finite groups, (a) the dimer and (b) the tetramer; the remainder are infinite arrangements, (c) 1-dimensional, (d) and (e) 2-dimensional, and (f) 3-dimensional. The number of ways of realizing a particular ratio of atoms may be large; Fig. 1.2 shows some systems with the composition AX_3.

Examples of all the systems shown in Figs 1.1 and 1.2 will be found in later chapters. In the upper part of Table 1.1 we list seven different ways in which an F : M ratio of 5 : 1 is attained in crystalline pentahalides; the list could be extended if anions $(MX_5)^{n-}$ are included, as will be seen from the structures of complex halides in Chapter 10. Conversely, we may consider how *different formulae* MX_n arise with the *same coordination number of* M. For tetrahedral and octahedral coordination this problem is considered in some detail in Chapter 5; the examples of Table 1.1 may be of interest as examples of the less usual coordination number

† C.N. = Coordination number.

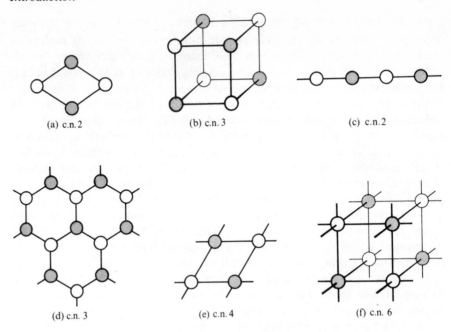

FIG. 1.1. Arrangements of equal numbers of atoms of two kinds.

nine. In order to gain a real understanding of the meaning of the formulae of inorganic compounds it is evidently necessary to think in three rather than two dimensions and in terms of infinite as well as finite groups of atoms.

(v) The chemist is familiar with *isomerism* (p. 47), which refers to differences in the structures of *finite* molecules or complex ions having a particular chemical composition. If infinite arrangements of atoms are permitted, in addition to finite groups, the probability of alternative atomic arrangements is greatly increased, as is evident from Figs 1.1 and 1.2.

An element or compound is described as *polymorphic* if it forms two or more crystalline phases differing in atomic arrangement. (The earlier term *allotropy* is still used to refer to different 'forms' of *elements*, but except for the special case of O_2 and O_3 allotropes are simply polymorphs.) Polymorphism of both elements and compounds is the rule rather than the exception, and the structural chemistry of any element or compound includes the structures of all its polymorphs just as that of a molecule includes the structures of its isomers. The differences between the structures of polymorphs range from minor difference such as the change from fixed to random orientation (or complete rotation) of a molecule or complex ion in the high temperature form of a substance (for example, crystalline HCl, salts containing NH_4^+, NO_3^-, CN^-, and other complex ions), or the α–β changes of the forms of SiO_2, to major differences involving reconstruction of the whole crystal (the polymorphs of C, P, SiO_2, etc.).

Originally the only variable in studies of polymorphism was the temperature, and substances are described as enantiotropic if the polymorphic change takes place

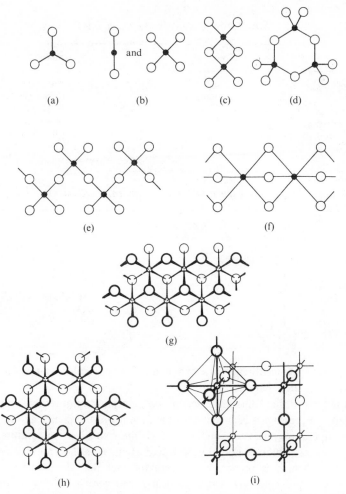

FIG. 1.2. Some ways of realizing a ratio of 3X:A in finite or infinite groupings of atoms: (a)–(d), finite groups AX_3, AX_2 and AX_4, A_2X_6 and A_3X_9, (e)–(g), infinite linear systems, (h) infinite two-dimensional system, (i) infinite three-dimensional complex.

at a definite transition temperature or monotropic if one form is stable at all temperatures under atmospheric pressure. Extensive work by Bridgman showed that many elements (and compounds such as ice) undergo structural changes when subjected to pressure, the changes being detected as discontinuities in physical properties such as resistivity or compressibility. In some cases the high pressure structure can be retained by quenching in liquid nitrogen and studied under atmospheric pressure by normal X-ray techniques. During the last decade the study of high pressure polymorphs has been greatly extended by the introduction of new apparatus (such as the *tetrahedral anvil*) which not only increase the range of pressures attainable but also permit the X-ray (and neutron) diffraction study of the phase while under pressure. Studies of halides and oxides, in addition to

9

TABLE 1.1

Structures of crystalline pentahalides

Halide	C.N. of M	Structural units in crystal
PBr_5	4	$(PBr_4)^+Br^-$
PCl_5	4 and 6	$(PCl_4)^+(PCl_6)^-$
$NbCl_5$	6	Nb_2Cl_{10}
MoF_5	6	Mo_4F_{20}
BiF_5	6	Chains $(BiF_5)_n$
$PaCl_5$	7	Chains $(PaCl_5)_n$
β-UF_5	7	(3D ionic structure)

Examples of tricapped trigonal prism coordination

AX_9 sharing	X : A	Examples
—	9	$[Nd(H_2O)_9](BrO_3)_3$, K_2ReH_9
2 edges	7	K_2PaF_7
2 faces	6	$[Sr(H_2O)_6]Cl_2$
2 edges, 4 vertices	5	$LiUF_5$
2 faces, 4 vertices	4	NH_4BiF_4, $NaNdF_4$
2 faces, 6 edges	3	UCl_3
2 faces, 12 edges	2	$PbCl_2$
(but see p. 221)		

elements, have produced many new examples of polymorphism and some of these are described in later chapters.

We noted above that some high-pressure polymorphs do not revert to the normal form when the pressure is reduced. Many high temperature phases do not revert to the low temperature phase on cooling through the transition temperature, witness the many high temperature polymorphs found as minerals. This non-reversibility of polymorphic changes is presumably due to the fact that the activation energies associated with processes involving a radical rearrangement of the atoms may be large, regardless of the difference between the lattice energies of the two polymorphs.

Members of families of closely related structures, the formation of which is dependent on the growth mechanism of the crystals, are termed *polytypes*. They are not normal polymorphs, and are formed only by compounds with certain types of structure. The best-known examples are SiC, CdI_2, ZnS, and certain complex oxides, notably ferrites, to which reference should be made for further details.

(vi) When atoms are bonded together to form either finite or infinite groupings complications can occur owing to the conflicting requirements of the various atoms due to their relative sizes or preferred interbond angles. It is well known that this problem arises in finite groups of atoms, as may be seen from scale models of molecules and complex ions. It is, however, less generally appreciated that geometrical and topological restrictions enter in much more subtle ways in 3D structures and may be directly relevant to problems which seem at first sight to be purely chemical in nature. As examples we may instance the relative stabilities of series of oxy-salts (for example, alkali-metal orthoborates and orthosilicates), the crystallization of salts from aqueous solution in the anhydrous state or as hydrates, and the behaviour of the nitrate ion as a bidentate or monodentate ligand. We return briefly to the subject later in this chapter and consider it in more detail in Chapter 7.

Structural formulae of inorganic compounds

Elemental analysis gives the relative numbers of atoms of different elements in a compound; it yields an 'empirical' formula. The simplest type of structural formula indicates how the atoms are linked together, and to this simple topological picture may be added information describing the geometry of the system. The nature of a structural formula depends on the extent of the linking of the atoms. If the compound consists of finite molecules it is necessary to know the molecular weight and then to determine the topology and geometry of the molecule:

$$HNO \longrightarrow H_2N_2O_2 \longrightarrow \overset{OH}{\underset{HO}{\diagdown N{=}N\diagup}} \longrightarrow \text{bond lengths and interbond angles}$$

elemental analysis	molecular weight	infrared and Raman spectroscopy indicate *trans* configuration	

If the atoms (in a solid) are linked to form a 1-, 2-, or 3-dimensional system the term molecular weight has no meaning, and the structural formula must describe some characteristic set of atoms which on repetition reproduces the arrangement found in the crystal. The repeat unit in an infinite 1-dimensional system is readily found by noting the points at which the pattern repeats itself:

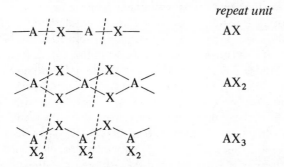

	repeat unit
—A—X—A—X—	AX
	AX$_2$
	AX$_3$

The complete description of the chain requires metrical information as in the case of a finite group. It should be noted that if the geometry of the chain is taken into account, that is, the actual spatial arrangement of the atoms in the crystal, then the (crystallographic) repeat unit may be larger than the simplest 'chemical' repeat unit. The crystallographic repeat unit is that set of atoms which reproduces the observed

11

structure when repeated *in the same orientation*, that is, by simple translations in one, two, or three directions. The chemical repeat unit is not concerned with orientation. This distinction is illustrated in Fig. 1.3(a) for the HgO chain. The chemical repeat unit consists of one Hg and one O atom whereas if we have regard to the geometrical configuration of the (planar) chain we must recognize a repeat unit containing 2 Hg + 2 O atoms. The various forms of AX_3 chains formed from tetrahedral AX_4 groups sharing two vertices (X atoms) provide further examples (p. 816); one is included in Fig. 1.3(b).

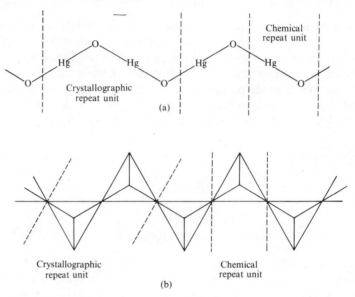

FIG. 1.3. Repeat units in chains.

Similar considerations apply to structures extending in two or three dimensions. The repeat unit of a 2D pattern is a *unit cell* which by translation in the directions of two (non-parallel) axes reproduces the infinite pattern. One crystalline form of As_2O_3 is built of infinite layers of the kind shown in Fig. 1.4(a), the unit cell being indicated by the broken lines. The pattern arises from AsO_3 groups sharing their O atoms with three similar groups, or alternatively, the repeat unit is $As(O_{1/2})_3$. These units are oriented in two ways to form the infinite layer, with the result that the crystallographic repeat unit—which must reproduce the pattern merely by translations in two directions—contains two of these $As(O_{1/2})_3$ units, or As_2O_3.

The crystallographic repeat unit of a 3D pattern is a parallelepiped containing a representative collection of atoms which on repetition in the directions of its edges forms the (potentially infinite) crystal. As in the case of a 1- or 2-dimensional pattern this unit cell may, and usually does, contain more than one basic 'chemical' unit (corresponding to the simplest chemical formula).

The following remarks may be helpful at this point; they are amplified in later chapters. There is no unique unit cell in a crystal structure, but if there are

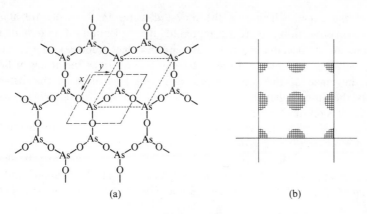

FIG. 1.4. (a) Alternative unit cells of layer structure of As_2O_3. (b) Projection of unit cell of a structure containing four atoms.

symmetry elements certain conventions are adopted about the choice of axes (directions of the edges of the unit cell). For example, crystalline NaCl has cubic symmetry (see Chapter 2) and the structure is therefore referred to a cubic unit cell. This cell contains 4 NaCl, but the structure may be described in terms of cells containing 2 NaCl or 1 NaCl; these alternative unit cells for the NaCl structure are illustrated in Fig. 6.3 (p. 197). It is sometimes convenient to choose a different origin, that is, to translate the cell in the directions of one or more of the axes, and the origin is not necessarily taken at an atom in the structure. For example, the unit cell of the projection of Fig. 1.4(a) does not have an atom at the origin but it is a more convenient cell than the one indicated by the dotted lines because it gives the coordinates $\pm (\frac{1}{3} \frac{2}{3})$ rather than (00) and $(\frac{2}{3} \frac{1}{3})$ for the two equivalent As atoms.

If there are atoms at the corners or on the edges or faces of a unit cell it may be difficult to reconcile the number of atoms shown in a diagram with the chemical formula—see, for example, the cell outlined by the broken lines in Fig. 1.4(a). It is only necessary to remember that the cell content includes all atoms whose centres lie within the cell and that atoms lying at the corner or on an edge or face count as follows:

    unit cell of 2D pattern:
        atom at corner belongs to four cells,
        atom on edge belongs to two cells
    unit cell of 3D pattern:
        atom at corner belongs to eight cells,
        atom on edge belongs to four cells
        atom in face belongs to two cells.

The cell content in each case could alternatively be shown by shading that portion of each atom which lies wholly within the cell (Fig. 14(b)).

Each atom shown in a *projection* repeats at a distance c above and below the

13

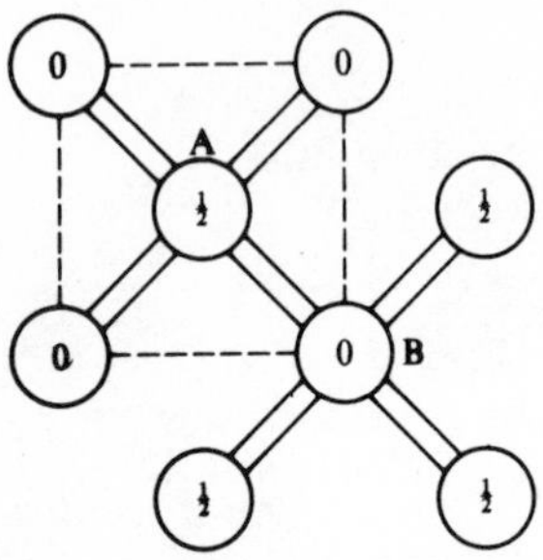

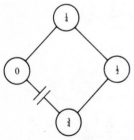

FIG. 1.5. Projection of body-centred structure showing 8-co-ordination of atoms.

plane of the paper, where c is the repeat distance in the structure along the direction of projection. Figure 1.5 represents the projection on its base of a cube containing an atom at its centre (body-centred cubic structure). The atom A has eight equidistant neighbours at the vertices of a cube, since the atoms at height 0 (i.e. in the plane of the paper) repeat at height 1 (in units of the distance c). Similarly the atom B has the same arrangement of eight nearest neighbours. (For some exercises on this topic see MSIC, p. 52.)

In order to simplify an illustration of a structure it is common practice to show a set of nearest neighbours (coordination group) as a polyhedral group. Thus the projection of the rutile structure of one of the forms of TiO_2 may be shown as either (a) or (b) in Fig. 1.6. In (a) the heavy lines indicate Ti–O bonds, and it may be deduced from the coordinates of the atoms that there is an octahedral

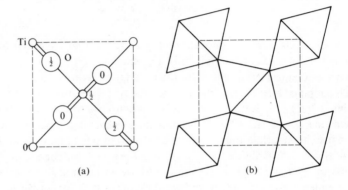

(a) (b)

FIG. 1.6. Projections of the structure of rutile (TiO_2): (a) showing atoms and their heights, (b) showing the octahedral TiO_6 coordination groups.

coordination group of six O atoms around each Ti atom. In (b) the lines represent the edges of the octahedral coordination groups. Since it is important that at least the two commonest coordination polyhedra should be recognized when viewed in a number of directions we illustrate several projections of the tetrahedron and octahedron at the beginning of Chapter 5.

Atoms arranged around 3-, 4-, or 6-fold helices project along the helical axis as triangle, square, or hexagon respectively. A pair of lines may be used to indicate that a number of atoms do not form a closed circuit but are arranged on a helix perpendicular to the plane of the paper (Fig. 1.7).

It is perhaps unnecessary to stress that a formula should correspond as closely as possible to the structure of the compound, that is, to the molecule or other grouping present, as, for example, $Na_3B_3O_6$ for sodium metaborate, which contains cyclic $B_3O_6^{3-}$ ions. Compounds containing metal atoms in two oxidation states are of interest in this connection. If the oxidation numbers differ by unity the formula does not reduce to a simpler form (for example, Fe_3O_4, Cr_2F_5), but if

FIG. 1.7. Projection of 4-fold helix along its axis. The atom at height $\frac{3}{4}$ is connected to the atom vertically above the one shown at height 0.

they differ by two the formula appears to correspond to an intermediate oxidation state:

Empirical formula	*Structural formula*
$GaCl_2$	$Ga^I(Ga^{III}Cl_4)$
PdF_3	$Pd^{II}(Pd^{IV}F_6)$
$CsAuCl_3$	$Cs_2(Au^ICl_2)(Au^{III}Cl_4)$
$(NH_4)_2SbCl_6$	$(NH_4)_4(Sb^{III}Cl_6)(Sb^VCl_6)$

Studies of crystal structures have led to the revision of many chemical formulae by regrouping the atoms to correspond to the actual groups present in the crystal. This is particularly true of compounds originally formulated as hydrates; some examples follow.

Hydrate	*Structural formula*
$HCl \cdot H_2O$	$(H_3O)^+Cl^-$
$NaBO_2 \cdot 2 H_2O$	$Na[B(OH)_4]$
$Na_2B_4O_7 \cdot 10 H_2O$	$Na_2[B_4O_5(OH)_4] \cdot 8 H_2O$
$FeCl_3 \cdot 6 H_2O$	$[FeCl_2(H_2O)_4] Cl \cdot 2 H_2O$
$ZrOCl_2 \cdot 8 H_2O$	$[Zr_4(OH)_8(H_2O)_{16}] Cl_8 \quad 12 H_2O$

The formulae of many inorganic compounds do not at first sight appear compatible with the normal valences of the atoms but are in fact readily interpretable in the light of the structure of the molecule or crystal. In organic chemistry we are familiar with the fact that the H : C ratios in *saturated* hydrocarbons, in all of which carbon is tetravalent, range from the maximum value four in CH_4 to two in $(CH_2)_n$ owing to the presence of C–C bonds. Similarly the unexpected formula, P_4S_3, of one of the sulphides of phosphorus arises from the presence of P–P bonds; the formation by P(III) of three and by S(II) of two bonds would give the formula P_4S_6 if all bonds were P–S bonds.

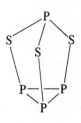

Bonding between atoms of the same element also occurs in many crystalline binary compounds and leads to formulae such as CdP_4, PdP_2, and PdS_2 which are not reconcilable with the normal oxidation numbers of Cd and Pd until their crystal structures are known. The structures of PdP_2 and PdS_2 are described shortly; for CdP_4 see p. 677.

Our final point relating to the structural formulae of solids is that in general crystallographers have not greatly concerned themselves with interpreting the structures of solids to chemists. As a result much of the structural chemistry of solids became segregated in yet another subdivision of chemistry (crystal chemistry, or more recently, solid-state chemistry), and many chemists still tend to make a mental distinction between the structures of solids and of the finite molecules and complex ions that can be studied in solution or in the gaseous state. The infinite layer structures of black and red phosphorus are manifestly only more complex examples of P forming three bonds as in the finite (tetrahedral) P_4 molecule of

15

white phosphorus, but equally the chain structure of crystalline $PdCl_2$ is simply the end-member of the series starting with the $PdCl_4^{2-}$ and $Pd_2Cl_6^{2-}$ ions:

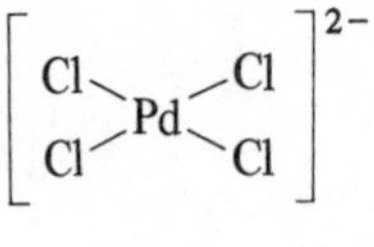

$$PdCl_4^{2-} \qquad\qquad Pd_2Cl_6^{2-} \qquad\qquad \text{crystalline } PdCl_2$$

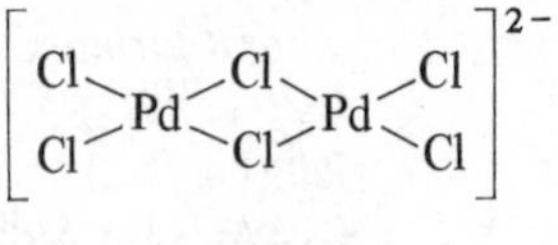

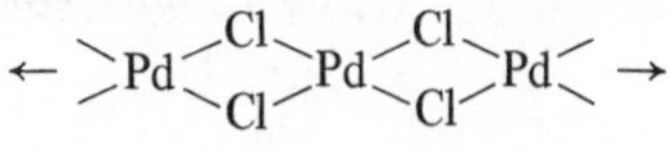

Diagrams purporting to show the origin of the electrons required for the various bonds are often given in elementary texts but only for (finite) molecules and complex ions—not for solids. It might help if 'Sidgwick-type' formulae were given for solids such as SnS (in which each atom forms three bonds), if only to show that the 'rules' which apply to finite systems also apply to some at least of the infinite arrays of atoms in crystals:

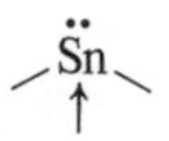

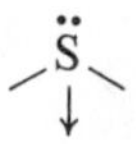

Many compounds of Pd(II) may be formulated in a consistent way so that the metal atom acquires a share in six additional electrons and forms planar dsp^2 bonds. Of the two simple possibilities (a) and (b) the former enables us to

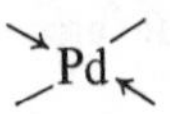

$$\text{(a)} \qquad\qquad \text{(b)} \qquad\qquad \text{(c)}$$

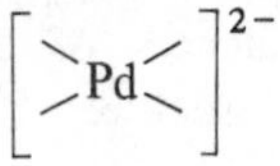

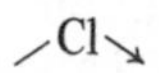

formulate the infinite chain of $PdCl_2$ and the $Pd_2Cl_6^{2-}$ ion (since a bridging Cl is represented as at (c)), while (b) represent the situation in $PdCl_4^{2-}$, though the actual state of the ion (e) is presumably intermediate between the 'ionic' picture (d) and the 'covalent' one (f):

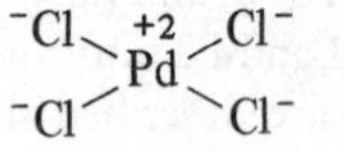

$$\text{(d)} \qquad\qquad \text{(e)} \qquad\qquad \text{(f)}$$

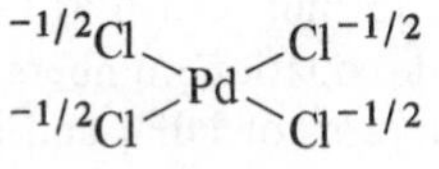

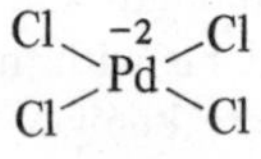

In crystalline PdO (and similarly for PdS and PtS) O forms four tetrahedral bonds and the metal forms four coplanar bonds, and we have the bond pictures

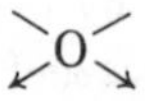

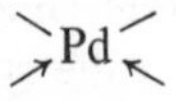

The compounds PdS_2 and PdP_2, which might not appear to be compounds of Pd(II), can be formulated in the following way. The disulphide consists of layers

16

(Fig. 1.8) in which Pd forms four coplanar bonds to S atoms which are bonded in pairs by covalent bonds of length 2·13 Å. We therefore have the bond pictures

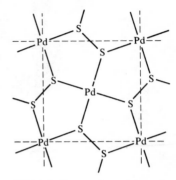

FIG. 1.8. Part of a layer in crystalline PdS_2.

Crystalline PdP_2 can be visualized as built from layers of the same general type, but each P atom forms a fourth bond to a P atom of an adjacent 'layer' so that there are continuous chains of P atoms. The structure is therefore not a layer structure like PdS_2 but a 3D framework in which the nearest neighbours are

$$\text{Pd—4 P (coplanar)} \qquad \text{and} \qquad \text{P} \begin{cases} 2\text{ P} \\ 2\text{ Pd} \end{cases} \text{(tetrahedral)}$$

and the bond pictures are

It is natural to enquire whether this somewhat naive treatment can be extended to related compounds. The compounds most closely related to PdS_2 are NiS_2 and PtS_2. All three compounds have different crystal structures and this, incidentally, is also true of the dichlorides. PtS_2 crystallizes with the CdI_2 structure in which Pt forms six octahedral bonds and S forms three pyramidal bonds, consistent with

The Pt atom thus acquires a share in eight additional electrons and forms octahedral d^2sp^3 bonds. NiS_2 crystallizes with the pyrites structure (p. 196) in which Ni is surrounded by six S atoms of S_2 groups, the Ni atoms and S_2 groups being arranged like the ions in the NaCl structure. For the pyrites structure of FeS_2, we write the bond pictures

so that with a share in ten additional electrons Fe acquires the Kr configuration. There are also compounds containing As—S, As—As, or P—P groups, instead of S—S, which adopt the pyrites or closely related structures. These groups supply nine or eight electrons instead of the ten supplied by S_2. In fact we find the pyrites (or a similar) structure for

$$FeS_2 \cdot \quad CoAsS \quad \underset{PtAs_2}{NiP_2} \text{ (high pressure)}$$

as expected, but we also find the same structure adopted by FeAsS and $FeAs_2$. The latter present no special problem, there being respectively one or two *fewer*

17

electrons to be accommodated in the d shell of the metal atom as compared with FeS_2. However, there are other compounds with the pyrites structure which have an *excess* of electrons, namely:

excess electrons

1	2	3	4
CoS_2	NiS_2	CuS_2†	ZnS_2†
NiAsS			

† high-pressure phases

which obviously do present a bonding problem. There is further discussion of the pyrites and related structures in Chapter 17.

Geometrical and topological limitations on the structures of molecules and crystals

Under this general heading we wish to draw attention to the importance of geometrical and topological factors which have a direct bearing not only on the details of molecular and crystal structures but also on the stability and indeed existence of some compounds. In spite of their relevance to structural problems, some of the factors we shall mention seem to have been completely ignored.

In any non-linear system of three atoms X—M—X (a), the distances M—X (bond length) and X—X (van der Waals contact) and the interbond angle X—M—X are necessarily related. In the regular tetrahedral molecule CCl_4 the C—Cl bond length (1·77 Å) implies a separation of only 2·9 Å between adjacent Cl atoms, a distance much less than the normal (van der Waals) distance between Cl atoms of different molecules (around 3·6 Å). If the X atoms are bonded to a second M atom, (b), there is formed a 4-ring, which we shall assume to be a planar parallelogram as it is in the examples we shall consider. The

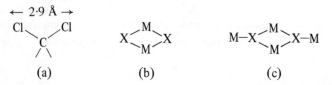

(a)   (b)   (c)

(non-bonded) distances M—M and X—X are related to the bond length M—X and the angle X—M—X. In the lithium chloride dimer these distances are 2·5 Å, 3·6 Å, and 2·2 Å, and the angle Cl—Li—Cl is approximately 110°. The Li—Li separation, which is shorter than the interatomic distance in the crystalline metal (3·1 Å) can obviously not be discussed in isolation since it is one of a number of interrelated quantities. Rings of type (b) also occur in molecules such as Fe_2Cl_6 and Nb_2Cl_{10} formed by the sharing of two X atoms between two tetrahedral MX_4 or two octahedral MX_6 groups.

Now suppose that each X atom is bonded to additional M atoms, as is the case in many crystals in which the X and M atoms form a 3D system. The bond angles around X are still related to those of the M atom. An example of (c) is the structure of rutile (one of the polymorphs of TiO_2) in which every Ti is surrounded octahedrally by six O atoms and each O by three coplanar Ti atoms. Evidently if the TiO_6 octahedra are regular the angle O–Ti–O is 90°, and hence one of the O bond angles would also be 90°. It follows that there cannot be regular octahedral coordination of M *and* the most symmetrical environment of X (three bond angles of 120°) in a compound MX_2 with the rutile structure. In the high-temperature form of BeO there is

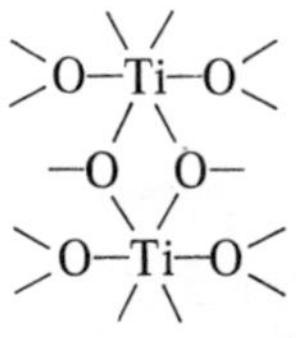

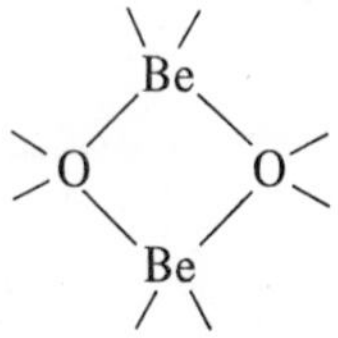

tetrahedral coordination of Be by four O atoms and the BeO_4 tetrahedra are linked in pairs by sharing an edge, these pairs being further linked by sharing vertices to form a 3D framework. The structure is illustrated in Fig. 12.3 (p. 446); the important feature in the present context is the presence of 4-rings. Here also regular coordination (in this case tetrahedral), of both O and Be atoms is not possible, and the distance Be–Be across the ring is only 2·3 Å, the same as in metallic beryllium.

The structure of crystalline PtS (see Fig. 17.3, p. 611) is a 3D structure which may be visualized as built up from two sets of planar chains

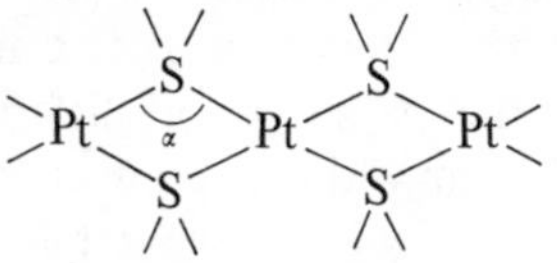

in which Pt(II) forms four coplanar bonds. The two sets of chains lie in planes which are perpendicular to one another and each S atom is common to a chain of each set. It is clearly impossible to have both *regular* tetrahedral coordination of S and the most symmetrical arrangement of four coplanar bonds around Pt, for the sulphur bond angle α is the supplement of the Pt bond angle. The actual sulphur bond angles represent a compromise between the values 90° and $109\frac{1}{2}$°; they are two of $97\frac{1}{2}$° and four of 115°.

In these examples we have progressed from the simple 4-ring molecule of Li_2Cl_2 to examples of 4-rings of which the X atoms are involved in further bonds to M atoms, and the last three examples are all crystalline solids in which the M–X bonding extends throughout the whole crystal (3D complex). A somewhat similar problem arises in *finite* molecules (or complex ions) if there are connections between the X atoms attached to the central atom. The 'ideal' stereochemistry of a metal atom forming six bonds might be expected when it is bonded to six identical atoms in a finite group MX_6. If two or more of the X atoms form parts of a polydentate ligand (that is, they are themselves bonded together in some way) restrictions have been introduced which may alter the angles X–M–X and possibly also the lengths of the M–X bonds. Many atypical stereochemistries of metal atoms have been produced in this way. For example, all four As atoms of the molecule (a)

19

can bond to the same metal atom, as in the ion (b) where the tetradentate ligand leads to a bond arrangement unusual for divalent Pt. On the other hand, if there is sufficient flexibility in the ligand this may suffer deformation rather than distort the bond arrangement around the metal atom. In the tetramethyldipyromethene derivative (c) it appears that the very short distances between the CH_3 groups in a model constructed with the usual coplanar bonds from Pd(II) might lead to a tetrahedral metal stereochemistry. In this case, however, the ring systems buckle in preference to distortion of the metal-bond arrangement.

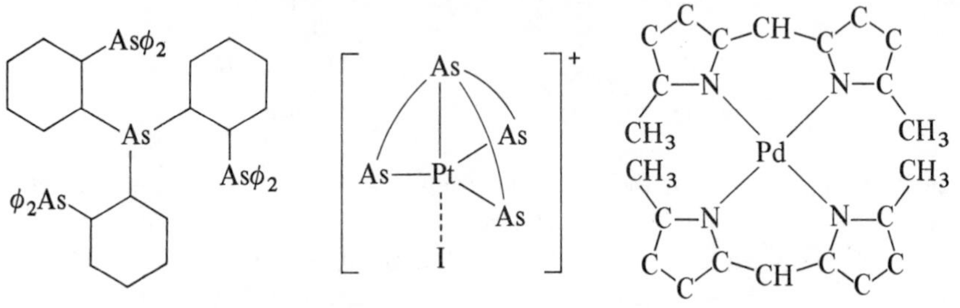

(a) (b) (c)

The angles subtended by the A atoms at X atoms (commonly oxygen or halogen) shared between two coordination groups have long been of interest as indications of the nature of the A–X bonds. Simple systems of this type include the 'pyro'-ions, in which A is Si, P, or S, and X is an O atom. If we make the reasonable assumption that the X atoms of different tetrahedra should not approach more closely than they do within a tetrahedral group (that is, the distance x should not be less than the edge-length of the tetrahedron) then a lower limit for the angle A–X–A can be calculated; the upper limit is $180°$. A similar calculation can be made for octahedral (or other) coordination groups meeting at a common X atom. Such simple geometrical considerations are obviously relevant to discussions of observed A–X–A bond angles, but even more interesting are the deductions that can be made concerning the sharing of edges between octahedra in complex oxide structures. These points are discussed in more detail in Chapter 5. It will also be evident that similar considerations limit the number of tetrahedral AX_4, octahedral AX_6, or other coordination groups AX_n that can meet at a point, that is, have a common vertex (X atom). These limits in turn have a bearing on the stability and indeed existence of crystalline compounds such as oxy-salts and nitrides, as we shall show in Chapter 7.

In contrast to these limitations on the structures of molecules and crystals which arise from metrical considerations there are others which may be described as topological in character. For example, the non-existence of compounds A_2X_3 (e.g. sesquioxides) with simple layer structures in which A is bonded to 6 X and X to 4 A is not a matter of crystal chemistry but of topology; it is concerned with the non-existence of the appropriate plane nets, as explained on p. 72. The non-existence of certain other structures for compounds A_mX_n, such as an AX_2 structure of $10:5$ coordination, may conceivably be due to the (topological) impossibility of constructing the appropriate 3D nets. On the other hand this problem may alternatively be regarded as a geometrical one, of the kind already mentioned, concerned with the numbers of coordination polyhedra of various kinds which can meet at a point, a matter which is further discussed on p. 265.

This interrelation of geometry (that is, metrical considerations) and topology (connectedness) seems to be a rather subtle one. It is well known that it is impossible to place five *equivalent* points on the surface of a sphere, if we exclude the trivial case when they form a pentagon around the equator, a fact obviously relevant to discussions of 5-coordination or the formation of five equivalent bonds. The most general (topological) proof of this theorem results from considering the linking of the points into a connected system of polygons (polyhedron) and showing that this cannot be done with the same number of connections to each point. Alternatively we may demonstrate that there is no *regular* solid with five vertices, when metrical factors are introduced. When we derive some of the possible 3D 4-connected nets in Chapter 3 simply as systems of connected points we find that the simplest (as judged by the number of points in the smallest repeat unit) is a system of 6-gons which in its most symmetrical configuration represents the structure of diamond. Although this net is derived as a 'topological entity', without reference to bond angles, it appears that it cannot be constructed with *any* arbitrary interbond angles, for example, with four coplanar bonds meeting at each point. Apparently geometrical limitations of a similar kind apply to other systems of connected points; this neglected field of 3D Euclidean geometry should repay study.

The complete structural chemistry of an element or compound

Having emphasized the importance of the solid state in inorganic chemistry we now examine what place this occupies in the complete structural chemistry of a substance—element or compound.

The complete structural chemistry of a substance could be summarized as in Chart 1.1. It includes not only the structures of the substance in the various states

CHART I.1

The complete structural chemistry of a substance

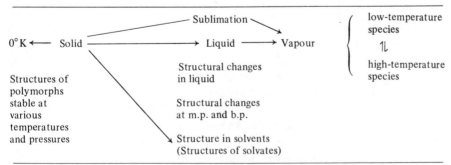

of aggregation but also the structural changes accompanying melting, vaporization of liquid or solid, or dissolution in a solvent, and those taking place *in* the solid, liquid, and vapour states. The complexity of the structural chemistry of an element or compound varies widely. At one extreme there are the noble gases which exist as discrete atoms in all states of aggregation. In these cases the only entries in the chart would be the arrangement of the atoms in the solid and the relatively small changes in structure of the simple atomic liquid with temperature. Next come gases such as H_2, N_2, O_2, and the halogens which persist as diatomic molecules from the solid through the liquid to the gaseous state and dissociate to single atoms only at higher temperatures. Sulphur, on the other hand, has an extremely complex structural chemistry in the elementary form which is summarized in Chapter 16. Unfortunately it is not possible to present anything like a complete picture of the structural chemistry of many compounds simply because the structural data are not available. Structural studies by particular investigators are usually confined to solids *or* gases (rarely liquids and rarely solids *and* gases); structural studies of a particular substance in more than one state are unusual. (We select $FeCl_3$ later to illustrate the structural chemistry of a relatively simple compound.) Here, therefore, we shall consider briefly the various entries in Chart I.1 and illustrate a number of points with examples.

Structure in the solid state

The structural chemistry of the solid includes the structures of its various crystalline forms (if it is polymorphic), that is, a knowledge of its structures over the temperature range $0°K$ to its melting point and also under pressure. In recent years studies of substances subjected to high pressures have widened the scope of structural chemistry in two ways. First, new polymorphs of many elements and known compounds have been produced in which the atoms are more closely packed, the higher density being achieved often, though not always, by increasing the coordination numbers of the atoms. For example, the 4 : 4 coordinated ZnO structure transforms to the 6 : 6 coordinated NaCl structure at 100 kbar, but higher density is achieved in coesite (a high-pressure form of SiO_2) without increasing coordination numbers. Second, new compounds have been produced which, though they cannot be made under atmospheric pressure, can persist under ordinary conditions once they have been made; examples include PbS_2, CuS_2, ZnS_2, and CdS_2. Stoichiometric FeO, a compound normally deficient in iron, has been made by heating the usual $Fe_{0.95}O$ with metallic Fe at $770°C$ under a pressure of 36 kbar. It is not always appreciated that our 'normal' chemistry and ideas on bonding and the structures of molecules and crystals strictly refer only to 'atmospheric pressure chemistry'. It has been possible to vary the temperature over a considerable range for a long time, and examples of polymorphism have been largely restricted to those resulting from varying the temperature. The large pressure range now readily attainable (accompanied unavoidably by an increase in temperature) may well prove more productive of structural changes than variation of temperature alone. The phase-diagram of water (Chapter 15) illustrates well the extension of the structural chemistry of a simple compound beyond that of the two forms (hexagonal and cubic ice) stable under atmospheric pressure.

Structural changes on melting

The structural changes taking place on melting range from the mere separation against van der Waals forces of atoms (noble gases) or molecules (molecular crystal built of non-polar molecules) to the catastrophic breakdown of infinite assemblies of atoms in the case of crystals containing chains, layers, or 3D frameworks. There is likely to be little difference in the immediate environment of an atom of a close-packed metal at temperatures slightly below and above its melting point, though there is a sudden disappearance of long-range order. On the other hand there is a larger structural change when metallic bismuth melts. Instead of the usual decrease in density on melting, which may be illustrated by the behaviour of a metal such as lead, there is an *increase* of two and a half percent due to the collapse of the rather 'open' structure of the solid to a more closely packed liquid:

	Bi (271°C)	Pb (328°C)	
Density of solid	9·75	11·35	g./c.c.
Density of liquid	10·00	10·68	

So little work has been done on the structures of liquids that it is not possible to give many examples of structural changes associated with melting. Crystals consisting of 3D complexes must break down into simpler units, while those built of molecules form a molecular melt. Simple ionic crystals such as NaCl melt to form a mixture of ions, but crystals such as SiO_2, which are built from SiO_4 groups sharing O atoms, do not break down completely. The tetrahedral groups remain linked together through the shared O atoms, accordingly the melt is highly viscous and readily forms a glass which is a supercooled liquid. Rearrangement of the tangled chains and rings of linked tetrahedra is a difficult process. The three crystalline forms of $ZnCl_2$ also consist of tetrahedral groups ($ZnCl_4$) linked to form layers or 3D frameworks, and this compound also forms a very viscous melt. The behaviour of $AlCl_3$ is intermediate between that of NaCl and $ZnCl_2$. The crystal consists of layers formed from octahedral $AlCl_6$ groups sharing three edges as shown diagrammatically for $FeCl_3$ on p. 25. These infinite layers break down on melting (or vaporization) to form dimeric molecules Al_2Cl_6, also of the same type as those in the vapour of $FeCl_3$. If crystalline $AlCl_3$ is heated the (ionic) conductivity, due to movement of Al^{3+} ions through the structure, increases as the melting point is approached and then drops suddenly to zero when the crystal collapses to a melt consisting of (non-conducting) Al_2Cl_6 molecules. The crystal structure of SbF_5 presumably consists of octahedral SbF_6 sharing a pair of vertices (F atoms), as in BiF_5, CrF_5, and other pentafluorides. The n.m.r. spectrum of the highly viscous liquid has been interpreted in terms of systems of SbF_6 groups sharing pairs of adjacent F atoms.

23

Introduction

Structural changes in the liquid state

With certain important exceptions, notably elementary S and H_2O, changes in structure taking place between the melting and boiling points have not been very much studied. The changes with temperature of the radial distribution curves of simple atomic liquids show only the variation in average number of neighbours at various distances. On the other hand, the structural changes in liquid sulphur are much more complex and have been the subject of a great deal of study. The complete structural chemistry of elementary sulphur is briefly described in Chapter 16; it provides an example of the opening up of cyclic molecules (S_6 or S_8) into chains followed by polymerization to longer chains. A note on the structure of water is included in Chapter 15.

Structural changes on boiling or sublimation

Since we know so little about the structures of liquids we are virtually restricted here to a comparison of structure in the solid and vapour states. In the prestructural era knowledge of the structures of vapours was confined to the molecular weight and its variation with temperature and pressure. The considerable amount of information now available relating to interatomic distances and interbond angles in vapour molecules has been obtained by electron diffraction or from spectroscopic studies of various kinds. This information is largely restricted to comparatively simple molecules, not only because it is impossible to determine the large number of parameters required to define the geometry of a more complex molecule from the limited experimental data, but also because the geometry of many molecules becomes indeterminate if the molecules are flexible. (In addition, certain methods of determining molecular structure are subject to special limitations; for example, in general only molecules with a permanent dipole moment give microwave spectra.) Information about the molecular weights of vapour species of certain compounds can be obtained by mass spectrometry, though the method has not as yet been widely applied to inorganic compounds—see, for example, the alkali halides and MoO_3. A less direct source of information is the study of the deflection of simple molecules in a magnetic field, which shows, for example, that some halide molecules AX_2 are linear and others angular (see Chapter 9). An interesting development in infrared spectroscopy is the trapping of species present in vapour at high temperatures in a (solid) noble gas matrix at very low temperatures, making possible the study of molecules which are too unstable to be studied by the more usual techniques.

In general we shall indicate what is known about the structural chemistry of a substance, in all states of aggregation, when it is described in the systematic part of this book. However, the following note on ferric chloride is included at this point to emphasize that even a compound as simple as $FeCl_3$ may have a very interesting structural chemistry of its own. In the vapour at low temperatures the compound is in the form of Fe_2Cl_6 molecules. When these condense to form a crystal a radical rearrangement takes place and instead of a finite molecule, in which each iron atom is attached to four chlorine atoms, there is an infinite 2-dimensional layer in which every Fe is bonded to six Cl atoms. When this crystal dissolves in a non-polar solvent such as CS_2 the Fe_2Cl_6 molecules reform, but in polar solvents such as ether simple tetrahedral molecules $(C_2H_5)_2O \rightarrow FeCl_3$ are formed. We are accustomed to regard a process like the dissolution of a crystal as a simple reversible process, in the sense that removal of the solvent leads to the recovery of the original solute. This is usually true for non-polar solvents but not for dissolution in a highly polar solvent like water. When a crystal of ferric chloride dissolves in water separation into Fe^{3+} and Cl^- ions occurs, and many of the properties of an aqueous solution of $FeCl_3$ are the properties of these individual ions rather than of ferric chloride. Thus the solution precipitates AgCl from a solution of $AgNO_3$ (property of Cl^- ion) and it oxidizes $SnCl_2$ to $SnCl_4$ or $TiCl_3$ to $TiCl_4$; these reactions are due to the transformation $Fe^{3+} + e \rightarrow Fe^{2+}$. In aqueous solution the ferric ion is hydrated and evaporation at ordinary temperatures leads to crystallization of the hexahydrate; the original anhydrous $FeCl_3$ cannot be recovered from the solution in this way. The 'hexahydrate' consists of octahedral $[FeCl_2(H_2O)_4]^+$ ions, Cl^- ions, and H_2O molecules which are not attached to metal atoms. We may set out the structural chemistry of $FeCl_3$ in the following way:

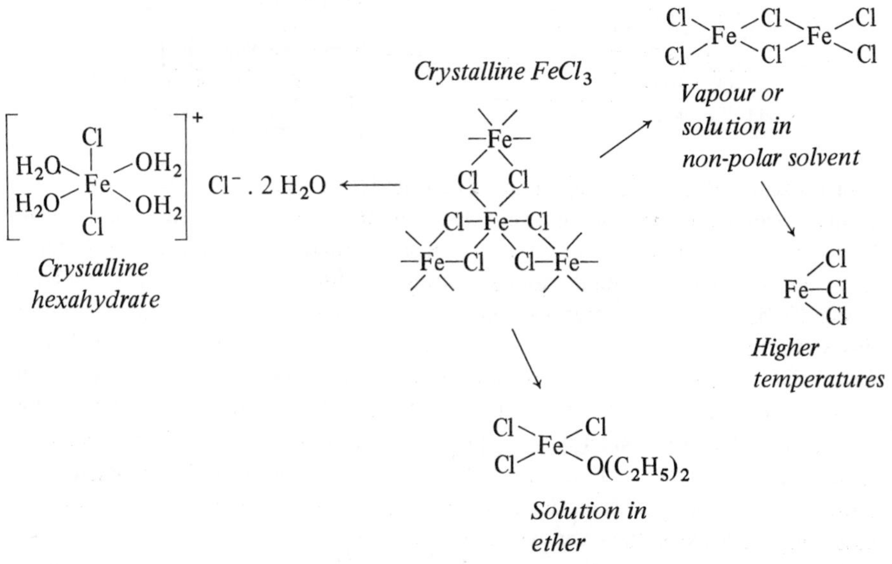

A classification of crystals

We have seen that in some crystals we may distinguish tightly-knit groups of atoms (*complexes*) within which the bonds are of a different kind from (and usually much shorter than) those between the complexes. The complexes may be finite or extend indefinitely in one, two, or three dimensions, and they are held together by ionic, van der Waals, or hydrogen bonds. The recognition of these types of complexes provides a basis for a broad geometrical classification of crystal structures, as shown in Chart 1.2. At first sight it might appear that the most obvious way of classifying

25

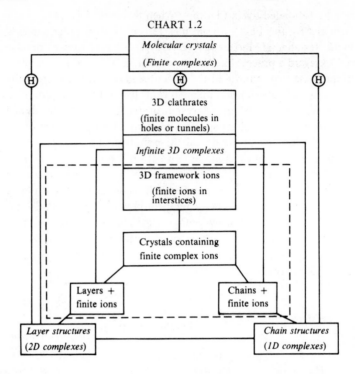

CHART 1.2

structures would be to group them according to the types of bonds between the atoms, recognizing the four extreme types, ionic, covalent, metallic, and van der Waals. A broad division into ionic, covalent, metallic, and molecular crystals is often made. However, bonds approximating to pure types (particularly ionic or covalent) are rare, and moreover in most crystals there are bonds of several different types. Numerous intermediate classes have to be recognized, and it is found that classifications based on bond types become complicated without being comprehensive. They also have the disadvantage that they over-emphasize the importance of 'pure' bond types, bonds of intermediate character being treated as departures from these extremes. It would seem preferable to be able to discuss the nature of the bonds in a particular crystal without having prejudged the issue by classifying a crystal as, for example, an ionic crystal.

We shall, of course, find structures that do not fall neatly into one of the main compartments of Chart 1.2. For example, in crystalline $HgBr_2$ and $CuCl_2$ the shortest bonds define finite molecules and chains respectively. These units are held together by weaker metal–halogen bonds to form layers, the bonds between which are the still weaker van der Waals bonds. Such structures find a place along one of the connecting lines of the chart; so also do structures in which molecules are linked by hydrogen bonds into chains, layers, or 3D complexes, at points marked H on the chart. It is inevitable that we refer in this section to many structures and topics which properly belong in later chapters; it may therefore be found profitable to refer back to this section at a later stage.

26

The four groups enclosed within the broken lines on the chart are closely related types of ionic structure, all of which involve two distinct kinds of charged unit. These are finite, 1-, 2-, or 3-dimensional ions in addition to discrete ions (usually monatomic). Examples of these four types of structure would include, for example, K_2SO_4, $KCuCl_3$, K_2CuF_4, and KB_5O_8, containing K^+ ions and respectively finite, chain, layer, and 3D complex ions. In each class of structure in the Chart there may be departures from strictly regular structure such as rotation or random orientation of complex ions or molecules, absent or misplaced atoms, and so on, as in the various types of 'defect' structure. We shall now discuss the four main structural types somewhat briefly as they will be considered in more detail in subsequent chapters.

Crystals consisting of infinite 3-dimensional complexes

The simplest structures of this group are those of

    (a) the noble gases,

    (b) metals and intermetallic compounds, and

    (c) many simple ionic and covalent crystals,

that is, all crystals in which no less extensive type of grouping is recognizable. It is immaterial whether crystals containing small molecules such as H_2, N_2, and H_2S are included here or with molecular crystals. Essentially covalent crystals of this class include diamond (and the isostructural Si, BN, etc.) and related compounds such as the various forms of SiC. Some of the simplest structure types for binary compounds in which the infinite 3D complex extends throughout the whole crystal are set out in Table 1.2.

TABLE 1.2

Simple structures for compounds A_mX_n

Coordination number of A	Coordination number of X				
	2	3	4	6	8
4	SiO_2		ZnS		
6	ReO_3	TiO_2	Al_2O_3	NaCl	
8			CaF_2		CsCl
9		LaF_3			

In crystals of groups (a), (b) and (c) all the atoms together form the 3D complex. Closely related to these are two other types of structure in which there is a 3D framework extending throughout the crystal but in addition discrete ions or molecules occupy interstices in the structure:

    (d) There are several kinds of 3-dimensional framework structures. In the zeolites (p. 827), a group of aluminosilicates, there are rigid frameworks built from (Si, Al)O_4 tetrahedra, and through these frameworks there run tunnels which are accessible from the surfaces of the crystals. Foreign ions and molecules can therefore enter or leave the crystal without disturbing the structure. The term

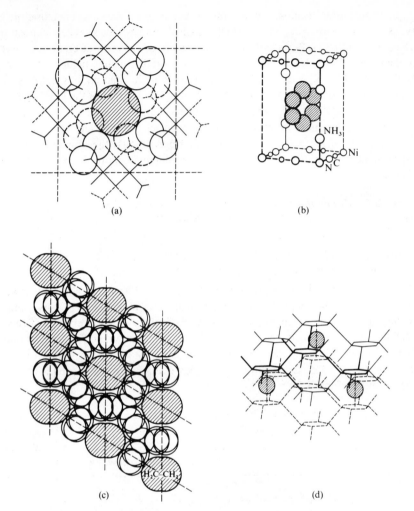

FIG. 1.9. Inclusion of foreign molecules in crystals of (a) $Pd_2Br_4[As(CH_3)_3]_2$, (b) $Ni(CN)_2NH_3 \cdot nC_6H_6$, (c) a urea–hydrocarbon complex, (d) β-quinol.

zeolitic is applied to crystals of this type. Some crystals, on the other hand, incorporate foreign molecules during growth in cavities from which they cannot escape until the crystal is dissolved or vaporized. These crystals are termed *clathrate* compounds. It would be more accurate to say that the framework of the 'host' forms around the included molecules, since the characteristic structure does not form in the absence of the 'guest' molecules. The clathrate structure is not stable unless a certain minimum proportion of the cavities is occupied; the structures of all the clathrate compounds of Fig. 1.9(b), (c), and (d), are different from those of the pure 'host' compounds. Clathrate compounds may have layer structures (for example, $Ni(CN)_2 \cdot NH_3 \cdot C_6H_6$, Fig. 1.9(b)) but more generally they have 3D structures, most of which are frameworks of hydrogen-bonded molecules. In the urea-hydrocarbon complexes (Fig. 1.9(c)) the urea molecules form a framework enclosing parallel tunnels around the linear hydrocarbon molecules, a much less dense packing of $CO(NH_2)_2$ molecules than in crystalline urea itself. The 'β-quinol' structure, Fig. 1.9(d), consists of two interpenetrating systems of hydrogen-bonded $HO . C_6H_4 . OH$ molecules which enclose molecules of water or certain gases if these are present during crystal growth (see also p. 97). The ice-like hydrates have frameworks of hydrogen-bonded water molecules which in the simplest cases form around neutral atoms or molecules such as those of the noble gases, chlorine, etc. They are described in Chapter 15 and 3.

(e) Crystals with 3D framework ions. These necessarily contain finite ions in the interstices to balance the charge on the framework.

Layer structures

It is conventional to describe as layer structures not only those in which all the atoms are incorporated in well-defined layers but also those in which there are ions or molecules between the layers. Since the layers may be electrically neutral, bonded together by van der Waals or hydrogen bonds, or charged, and since they may be identically constituted or have different structures, we may elaborate our Chart as shown in Fig. 1.10. Of the families of structures indicated in Fig. 1.10 by far the largest is (a_1), in which identical layers are held together by van der Waals bonds. The structure of the layer itself can range from the simplest possible planar network of atoms (graphite, white BN), through As_2O_3 (monoclinic), numerous layers formed from tetrahedral or octahedral coordination groups (halides AX_2 and AX_3), to the multiple layers in talc and related aluminosilicates. (The rarity of layer and chain structures among metal oxides and fluorides may be noted; these structures are more commonly adopted by the other chalconides and halides.) In the classes (a_2) and (c_2) identical layers are interleaved with molecules or ions, and there are many pairs of structures in which the layers are of the same or closely similar geometrical types but electrically neutral in (a_1) and charged in (c_2). Some examples are given in Table 1.3. For example, the $(Si_2O_5)_n^{2n-}$ layer in a number of alkali-metal silicates $M_2Si_2O_5$ is of the same general type (tetrahedral groups sharing three vertices) as the neutral layer in one polymorph of P_2O_5, and there is the same relation between the neutral layer of PbF_4 and the 2-dimensional ion in $K_2(NiF_4)$ or between talc, built of neutral layers $Mg_3(OH)_2Si_4O_{10}$, and the mica phlogopite, $KMg_3(OH)_2Si_3AlO_{10}$, in which the charged layers are held together by the K^+ ions. An interesting example of an (a_2) structure is the clathrate compound in which benzene or water molecules are trapped between neutral layers of composition $Ni(CN)_2 . NH_3$ (Fig. 1.9(b)).

Examples of layer structures of types (b_1) and (b_2) are much less numerous. The former, in which the layers are held together by hydrogen bonds, include $Al(OH)_3$, $HCrO_2$, γ-$FeO . OH$ and related compounds. Instead of direct hydrogen bonding between atoms of adjacent layers there is the possibility of hydrogen bonding through an intermediate molecule such as H_2O situated between the layers (class (b_2)). The structure of gypsum, $CaSO_4 . 2 H_2O$ (p. 560) provides an example of a structure of this type.

TABLE 1.3

Related structures containing layers of the same general type

(a_1)	(a_2) *or* (c_2)
Graphite	Interlamellar compounds of graphite, e.g. C_8K, $C_{16}K$ $CaSi_2$
P_2O_5	$Li_2(Si_2O_5)$
PbF_4	$K_2(NiF_4)$
Talc	Phlogopite ⎫
Pyrophyllite	Muscovite ⎭ (micas)

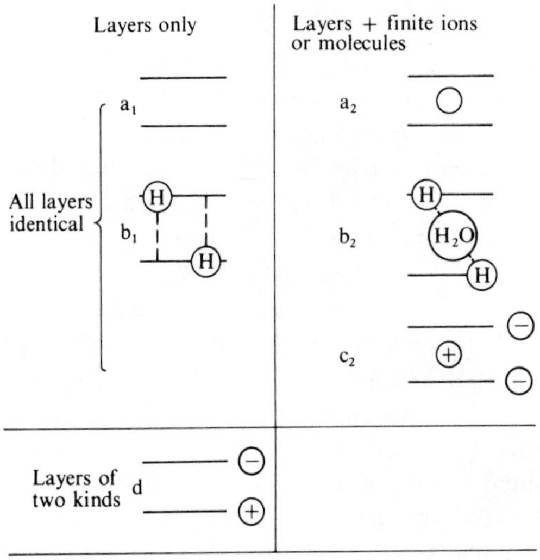

FIG. 1.10. Types of layer structure (diagrammatic).

The last group, (d), of structures in which positively and negatively charged layers alternate, is also a small one and includes the chlorite minerals (p. 824) and some hydroxyhalides such as $[Na_4Mg_2Cl_{12}]^{4-}$ $[Mg_7Al_4(OH)_{22}]^{4-}$ (p. 212).

In this very brief survey we have not been concerned with the detailed structure of the layers, various aspects of which are discussed in more detail later, in particular the basic 2D nets and structures based on the simplest 3- and 4-connected plane nets (Chapter 3), and layers formed from tetrahedral and octahedral coordination groups (Chapter 5). The CdI_2 layer and more complex structures derived from this layer are further discussed in Chapter 6. We shall see that there are corrugated as well as plane layers, and also composite layers consisting of two interwoven layers (red P, $Ag[C(CN)_3]$); these are included in the chapters just mentioned.

Between the four major classes our classification allows for structures of intermediate types. We can envisage, for example, structures in which the distinction between the bonds (and distances) between atoms within the layers and between those in different layers is not so clear-cut as we have supposed here. Alternatively, there may be well-defined layers but different types or strengths of

30

bonds between the units comprising the layers, so that the layer itself may be regarded as a rather loosely knit assembly of chains or finite molecules. We have already noted $CuCl_2$ and $HgBr_2$ as examples of structures of this kind, and other examples will be found in later chapters.

Chain structures

In principle the possible types of chain structure are similar to those for layer structures, and some are shown diagrammatically in Fig. 1.11, labelled in the same way. Examples are known of all the types of chain structure shown in the figure,

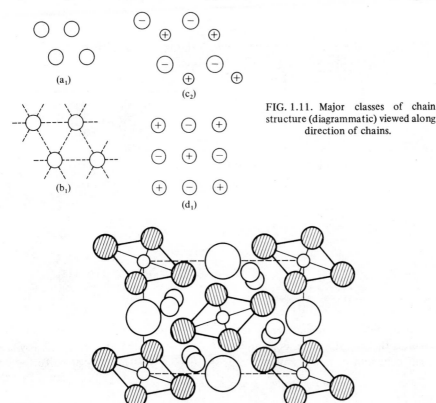

FIG. 1.11. Major classes of chain structure (diagrammatic) viewed along direction of chains.

FIG. 1.12. The structure of $K_2HgCl_4 . H_2O$ viewed along the direction of the infinite chains. The K^+ ions (pairs of overlapping circles) and H_2O molecules (large circles) lie, at various heights, between the chains.

the most numerous being those of class (c_2) which includes all oxy-salts containing infinite linear ions—the simplest of which are the linear meta-salts—and all complex halides containing chain anions. In some hydrated complex halides water molecules also are accommodated between infinite 1-dimensional anions, as in $K_2HgCl_4 . H_2O$ (Fig. 1.12)

   A simple example of class (b_1) is the structure of $LiOH . H_2O$, in which there are hydrogen bonds between OH groups and H_2O molecules of neighbouring chains (see Fig. 15.23, p. 561), and the structure of borax illustrates the rare (d_1) type. Here the Na^+ ions and water molecules form a linear $[Na(H_2O)_4]_n^{n+}$ chain (consisting of octahedral $Na(H_2O)_6$ groups sharing two edges) and the anion is an infinite chain formed from hydrogen-bonded $[B_4O_5(OH)_4]^{2-}$ ions, as described on p. 858. As in the case of the $(HCO_3)_n^{n-}$ chain in $NaHCO_3$ this borate anion is composed of sub-units which are hydrogen-bonded along the chain and is therefore not strictly a chain of the simplest type.

The simplest type of chain structure, (a_1), consists of identical chains which are

TABLE 1.4

Structures containing chains of the same general type

Type of chain	Neutral molecule	Infinite ion	
−A−X−A−X− X X	SeO_2	$(BO_2)_n^{n-}$ in CaB_2O_4	meta- ions
−A−X−A−X− (tetrahedral) X_2 X_2	SO_3	$(SiO_3)_n^{2n-}$, $(PO_3)_n^{n-}$ in meta-salts	
(planar)	AuF_3	$(CuCl_3)_n^{n-}$ in $CsCuCl_3$	
−A−X−A−X− X_4 X_4	BiF_5	$(AlF_5)_n^{2n-}$ in Tl_2AlF_5	
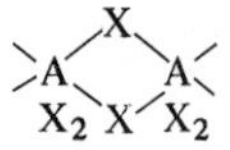 >A<X>A< X_2 X X_2	NbI_4	$(HgCl_4)_n^{2n-}$ in $K_2HgCl_4 \cdot H_2O$	
 >A−X−A< with bridging X	ZrI_3, Cs_3O	$(NiCl_3)_n^{n-}$ in $CsNiCl_3$	

packed parallel to one another. Since the basic requirement for the formation of a chain is that the structural unit can bond to two others, the simplest possibility is a chain consisting of similar atoms as in plastic sulphur or elementary Se or Te:

$$-S-S-S-$$

An obvious elaboration is the chain consisting of alternate atoms of different elements, as in HgO or AuI:

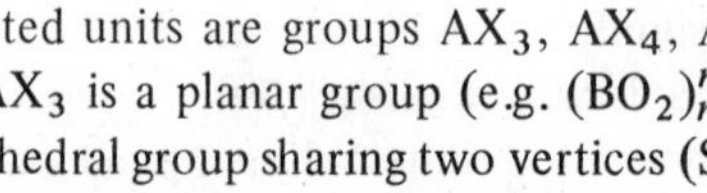

$$\_Hg\diagup^O\diagdown Hg\diagdown_O\diagup Hg\diagup^O\diagdown Hg\_$$

More complex 2-connected units are groups AX_3, AX_4, AX_6, etc., sharing one or more X atoms, where AX_3 is a planar group (e.g. $(BO_2)_n^{n-}$) or a pyramidal group (as in SeO_2), AX_4 a tetrahedral group sharing two vertices (SO_3) or two edges (SiS_2), AX_6 an octahedral group sharing two vertices (BiF_5) or two edges (NbI_4), or two faces (ZrI_3), and so on. These chains are described in more detail under tetrahedral

32

and octahedral structures in Chapter 5. The structure of $ReCl_4$ provides an example of a more complex chain in which the repeat unit is a pair of face-sharing octahedra which are further linked by sharing a pair of vertices:

$$\left[\begin{array}{ccc} Cl & Cl & Cl \\ Cl-Re-Cl-Re-Cl \\ Cl & Cl & Cl \end{array}\right]$$

As in the case of layer structures the same general type of structure can serve for infinite molecules as for infinite linear ions.

We do not propose to discuss here the geometry of chains, but we may note that whereas chains built of tetrahedral or octahedral groups sharing a pair of opposite edges or of octahedral groups sharing a pair of opposite faces are strictly linear, there are numerous configurations of chains built from tetrahedra sharing two vertices (p. 816) and two configurations are found for AX_5 chains in which octahedral AX_6 groups share two vertices, the latter being *cis* or *trans* to one another. A rotation combined with a translation produces a helix, the simplest form of which is the plane zigzag chain (as already illustrated for HgO) generated by a 2-fold screw axis. In a second polymorphic form of HgO the chain, like that in 'metallic' Se, is generated by a 3-fold screw axis, while the infinite chain molecule in crystalline AuF_3 results from the repetition of a planar AuF_4 group around a 6-fold screw axis.

Crystals containing finite complexes

Finite complexes comprise all molecules and finite complex ions. We are therefore concerned here with two main classes of Chart 1.2, namely, molecular crystals and crystals containing finite complex ions. Molecular crystals include many compounds of non-metals, some metallic halides AX_n and a number of polymeric halides A_2X_6 and A_2X_{10}, metal carbonyls and related compounds, and many coordination compounds built of neutral molecules (e.g. $(Co(NH_3)_3(NO_2)_3)$). In the simplest type of molecular crystal identical non-polar molecules are held together by van der Waals bonds. The structures of these crystals represent the most efficient packing of units of a given shape held together by undirected forces. When the molecules are roughly spherical the same structure types may occur as in crystals described as 3-dimensional complexes, a group of atoms replacing the single atom. As the shape of the molecule deviates from spherical so the structures bear more resemblance to those of crystals containing 1- or 2-dimensional complexes. More complex types of molecular crystal structure arise when there are molecules of more than one kind, as in $AsI_3 . 3 S_8$, and when there are hydrogen bonds between certain pairs of atoms of different molecules.

Finite complex ions include the numerous oxy-ions and complex halide ions, the aquo-ions in some hydrates (e.g. $Al(H_2O)_6^{3+}$), and all finite charged coordination complexes, in addition to the very simple ions such as CN^-, C_2^{2-}, O_2^-, O_2^{2-}, and many others. Crystals containing the smaller or more symmetrical complex ions often have structures similar to those of A_mX_n compounds; see, for example, the section on the NaCl structure in Chapter 6.

Relations between crystal structures

Our aim in the first part of this book is to set out the basic geometry and topology which is necessary for an understanding of the 3-dimensional systems of atoms which constitute molecules and crystals and to enable us to describe the more important structures in the simplest possible way. In view of the extraordinary variety of atomic arrangements found in crystals it is important to look for any principles which will simplify our task. We therefore note here some relations between crystal structures which will be found helpful.

(a) The same basic framework may be used for the structures of crystals with relatively complicated chemical formulae, as will be illustrated for the diamond net in Chapter 5.

(b) Structures may be related to a simple A_mX_n structure in one of the following ways:

  (I) Substitution structures
   (i) regular (superstructures),
   (ii) random.
 (II) Subtraction and addition structures
   (i) regular,
   (ii) random.

 (III) Complex groups replacing A and/or X.
 (IV) Distorted variants
   (i) minor distortion,
   (ii) major distortion.

> (A minor distortion implies that the topology of the structure has not radically altered, further distortion leading to changes in coordination number.)

We shall illustrate these relations by dealing in some detail with some of the simpler structures in Chapter 6. It will also be shown how the CdI_2 layer (Fig. 6.14) is used in building the structures of a number of 'basic salts', some with layer and others with 3D structures.

(c) The structures of some oxides and complex oxides are built of blocks or slices of simpler structures. Examples described in later chapters include oxides of Ti and V with formulae M_nO_{2n-1} built from blocks of rutile structure and oxides of Mo and W (for example M_nO_{3n-1}) related in a similar way to the structures of MoO_3 and WO_3 (shear structures). More complex examples include oxides $Sr_{n+1}Ti_nO_{3n+1}$ formed from slices of the perovskite (ABX_3) structure and the 'β-alumina' ($NaAl_{11}O_{17}$) and magnetoplumbite ($PbFe_{12}O_{19}$) structures in which slices of spinel-like (AB_2X_4) structure are held together by Na^+ and Pb^{2+} ions respectively.

34

2

Symmetry

Symmetry elements

Symmetry, in one or other of its aspects, is of interest in the arts, mathematics, and the sciences. The chemist is concerned with the symmetry of electron density distributions in atoms and molecules and hence with the symmetry of the molecules themselves. We shall be interested here in certain purely geometrical aspects of symmetry, namely, the symmetry of finite objects such as polyhedra and of repeating patterns. Inasmuch as these objects and patterns represent the arrangements of atoms in molecules or crystals they are an expression of the symmetries of the valence electron distributions of the component atoms. In the restricted sense in which we shall use the term, symmetry is concerned with the relations between the various parts of a body. If there is a particular relation between its parts the object is said to possess certain *elements of symmetry*.

The simplest symmetry elements are the centre, plane, and axes of symmetry. A cube, for example, is symmetrical about its body-centre, that is, every point (xyz) on its surface is matched by a point $(\bar{x}\bar{y}\bar{z})$. It is said to possess a centre of symmetry or to be centrosymmetrical; a tetrahedron does not possess this type of symmetry. Reflection of one-half of an object across a plane of symmetry (regarded as a mirror, hence the alternative name mirror plane) reproduces the other half. It can easily be checked that a cube has no fewer than nine planes of symmetry. The presence of an n-fold axis of symmetry implies that the appearance of an object is the same after rotation through $360°/n$; a cube has six 2-fold, four 3-fold, and three 4-fold axes of symmetry. We postpone further discussion of the symmetry of finite solid bodies because we shall adopt a more general approach to the symmetry of repeating patterns which will eventually bring us back to a consideration of the symmetry of finite groups of points.

Repeating patterns, unit cells, and lattices

A repeating pattern is formed by the repetition of some unit at regular intervals along one, two, or three (non-parallel) axes which, with the repeat distances, define the *lattice* on which the pattern is based. For a 1-dimensional pattern the lattice is a line, for a 2-dimensional pattern it is a plane network, and for a 3-dimensional pattern a third axis is introduced which is not coplanar with the first two (Fig. 2.1).

Symmetry

The parameters required to define the three types of lattice in their most general forms are

Lattice	Lattice translations (repeat distances)	Interaxial angles
1-dimensional	a	–
2-dimensional	a, b	γ
3-dimensional	a, b, c	γ, α, β

These parameters define the *unit cell* (repeat unit) of the pattern, which is accentuated in Fig. 2.1. Any point (or line) placed in one unit cell must occupy the same relative position in every unit cell, and therefore any pattern, whether 1-, 2-, or 3-dimensional, is completely described if the contents of one unit cell are specified.

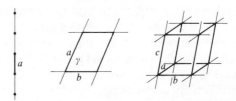

FIG. 2.1. One-, two-, and three-dimensional lattices.

Clearly, every lattice point has the same environment. A further property is that if along any line in the lattice there are lattice points distance x apart then there must be points at this separation (and no other lattice points) along this line when produced indefinitely in either direction. (We refer to this property of a lattice when we describe the closest packing of spheres in Chapter 4.) Note that the lattice has no physical reality; it does not form part of the pattern.

One- and two-dimensional lattices; point groups

The strictly 1-dimensional pattern is of somewhat academic interest since only points lying along a line are permitted. The only symmetry element possible is an inversion point (symmetry centre) and accordingly the pattern is either a set of points repeating at regular intervals a, without symmetry, or pairs of points related by inversion points (Fig. 2.2(a)). Although the inversion points (small black dots) were inserted only at the points of the lattice (that is, a distance a apart) further inversion points appear midway between the lattice points. This phenomenon, the

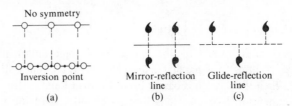

FIG. 2.2. Symmetry elements in one-dimensional patterns.

appearance of additional symmetry elements midway between those inserted at lattice points, is an important feature of 2- and 3-dimensional symmetry groups.

Two-dimensional patterns introduce two concepts that are of importance in the 3-dimensional patterns which are our main concern. First, the inversion (reflection) point is replaced by a reflection line (mirror line) (Fig. 2.2(b)) and two new types of symmetry element are possible, involving *translation* and *rotation*. The *glide-reflection line* combines in one operation reflection across a line with a translation of one-half the distance between lattice points (Fig. 2.2(c)). The translation is necessarily one-half because the point must repeat at intervals of the lattice translation. An *n*-fold *rotation* element produces sets of points related by rotation through $360°/n$, as, for example, the vertices of a regular *n*-gon. (When considering plane patterns we should imagine ourselves to be 2-dimensional beings capable of movement only in a plane and unaware of a third dimension. The operation which produces a set of *n* points arranged symmetrically around itself in the plane is strictly a 'rotation point'. It is, however, easier for three-dimensional beings to visualize such a rotation point as the point of intersection with the plane of an *n*-fold symmetry axis normal to the plane.) When illustrating symmetry elements it is convenient to use an asymmetric shape such as a comma so that the diagram shows only the intended symmetry. For example, Fig. 2.3(a) illustrates only 4-fold rotational symmetry, whereas (b) possesses also reflection symmetry. We shall see shortly that it is important to distinguish between the simple 4-fold symmetry (class 4) and the combination of this symmetry with reflection symmetry (class 4*mm*). These symbols are explained later in this Chapter.

We now come to the second point concerning plane patterns. An isolated object (for example, a polygon) can possess any kind of rotational symmetry but there is an important limitation on the types of rotational symmetry that a plane repeating pattern *as a whole* may possess. The possession of *n*-fold rotational symmetry would imply a pattern of *n*-fold rotation axes normal to the plane (or strictly a pattern of *n*-fold rotation points in the plane) since the pattern is a repeating one. In Fig. 2.4 let there be an axis of *n*-fold rotation normal to the plane of the paper at *P*, and at *Q* one of the nearest other axes of *n*-fold rotation. The rotation through $2\pi/n$ about *Q* transforms *P* into *P'* and the same kind of rotation about *P'* transforms *Q* into *Q'*. It may happen that *P* and *Q'* coincide, in which case $n = 6$. In all other cases *PQ'* must be equal to, or an integral multiple of, *PQ* (since *Q* was chosen as one of the nearest axes), i.e. $n \leqslant 4$. The permissible values of *n* are therefore 1, 2, 3, 4, and 6. Since a 3-dimensional lattice may be regarded as built of plane nets the same restriction on kinds of symmetry applies to the 3-dimensional lattices, and hence to the symmetry of crystals.

In Fig. 2.1 we illustrated the most general form of the 2-dimensional lattice, but it is clear that if a pattern has 3-, 4-, or 6-fold symmetry the lattice also will be more symmetrical than that of Fig. 2.1. It is found that there are five, and only five, 2-dimensional lattices consistent with the permissible symmetry of 2-dimensional patterns (Fig. 2.5).

We now have to find which combinations of symmetry elements are consistent with the five fundamental plane lattices. Let us consider the square lattice. We may

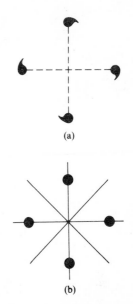

(a)

(b)

FIG. 2.3. Representation of symmetry elements (see text).

FIG. 2.4. Axial symmetry possible in plane patterns (see text).

37

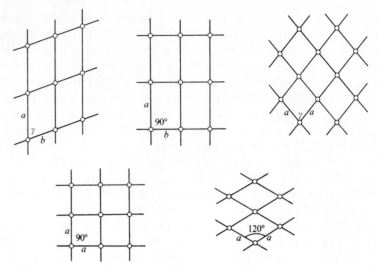

FIG. 2.5. The five plane lattices.

first have simply a 4-fold axis (□) at each point of the lattice (Fig. 2.6(a)). This automatically introduces a 4-fold axis at the centre of each square unit cell and 2-fold axes (●) at the mid-points of the sides. Any point placed in the plane is repeated in the manner shown as sets of four points (commas) arranged around each point of the lattice (unless it lies *on* the axis, when it will not be repeated). We could also make the pattern symmetrical across each side of the cell, as at (b), by making the edges of the cells mirror reflection lines. This combination of rotational and reflection symmetry turns a single point into a group of eight points around each lattice point. By finding all the permissible (different) combinations of symmetry with the various 2-dimensional lattices we arrive at a total of 17 *plane groups*, which form the bases of all 2-dimensional patterns.

Evidently different symmetries are associated with certain points in the unit cell. For example, in Fig. 2.6(a) the corners and centre of the unit cell have associated with them 4-fold symmetry, and the mid-points of the edges 2-fold symmetry.

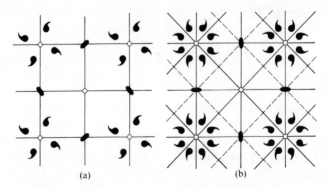

(a)   (b)

FIG. 2.6. 4-fold symmetry in plane patterns.

Similarly in (b) four reflection lines and 4-fold rotational symmetry are associated with the origin, but only two perpendicular reflection lines and 2-fold symmetry with the mid-points of the sides. The set or combination of symmetry elements associated with a point in a repeating pattern is called the *point group*. The symmetry elements all pass through the point and generate a set of symmetry-related (*equivalent*) points around it. There are ten different combinations (point groups) in two dimensions.

Three-dimensional lattices; space groups

As in the case of the 1- and 2-dimensional patterns we consider first the various possible lattices on which the patterns are based and then the possible combinations of symmetry elements which can be associated with the lattices.

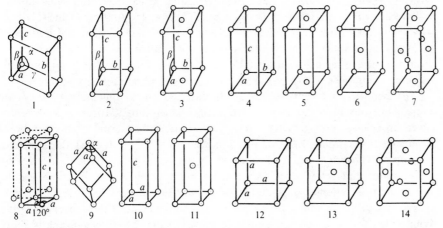

FIG. 2.7. The 14 Bravais lattices.

There are fourteen 3-dimensional lattices consistent with the types of rotational symmetry which a 3D repeating pattern may possess. These infinite 3D frameworks are the 14 Bravais lattices (Fig. 2.7) and Table 2.1. The repeat distances (unit translations) along the axes define the unit cell, and the full lines in Fig. 2.7 show one unit cell of each lattice.

TABLE 2.1
The 14 Bravais lattices

		Position in Fig. 2.7			Position in Fig. 2.7
Triclinic	P	1	Hexagonal	P	8
Monoclinic	P	2	Rhombohedral R		9
	C (or A)	3	Tetragonal	P	10
Orthorhombic	P	4		I	11
	C (B or A)	5	Cubic	P	12
	I	6		I	13
	F	7		F	14

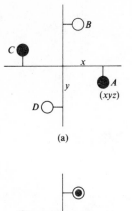

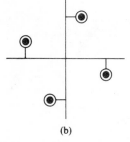

FIG. 2.8. Operation of a rotatory–inversion axis (see text).

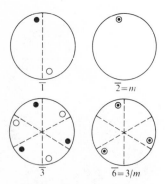

FIG. 2.9. The four kinds of rotatory–inversion axis.

In a simple or primitive (*P*) lattice there are lattice points only at the corners of the unit cell. If there is a lattice point at the centre of the unit cell the lattice is described as body-centred (*I* lattice), and if there are lattice points at the centres of some or all of the faces (in addition to those at the corners) the lattice is described as *A*- *B*-, or *C*-face-centred, or as an *F* lattice if it is centred on all faces. The unit cell of a centred lattice contains more than one lattice point (two for *I, A, B,* or *C,* and four for *F* lattices); a smaller cell containing only one lattice point could be found, but such cells correspond to lattices of lower symmetry already included in Fig. 2.7. It will be observed that Fig. 2.7 and Table 2.1 include only a limited number of centred lattices. There is, for example, no *B*-face-centred monoclinic lattice listed, because such a lattice may alternatively be described as a monoclinic *P* lattice with a different β angle. Similarly there is a *C*-face-centred orthorhombic lattice in Table 2.1 but no *C*-face-centred tetragonal lattice. The latter could equally well be described as a *P* lattice having *a* and *b* axes at 45° to those of the *C* lattice, whereas in the orthorhombic case the change to the smaller cell would lead to an interaxial angle not equal to 90°. It may easily be verified that a tetragonal *F* lattice reduces to the tetragonal *I* lattice. The only types of centred lattice consistent with cubic symmetry (see p. 43) are the body-centred and all-face-centred lattices.

The symmetry elements which can be inserted into 3-dimensional lattices are more numerous than those associated with 2-dimensional lattices. In addition to inversion (centre of symmetry), reflection (mirror plane), and simple rotational symmetry (axes of simple *n*-fold rotation, where *n* = 1, 2, 3, 4, or 6), there are *axes of rotatory inversion* and two kinds of operation involving translation, namely, *glide planes* and *screw axes*. An axis of rotatory inversion, $\bar{n}$, combines the operations of rotation through 360°/*n* with *simultaneous* inversion through a centre of symmetry. For example, the axis $\bar{4}$ (normal to the plane of the paper) converts a point (*xyz*) into a set of four points, as shown in Fig. 2.8, in which points above and below the plane of the paper are distinguished as full and open circles respectively. Rotated in a clockwise direction through 90° and then inverted *A* becomes *B* ($y\bar{x}\bar{z}$), *B* becomes *C* ($\bar{x}\bar{y}z$), and *C* becomes *D* ($\bar{y}x\bar{z}$). It should be emphasized that the two operations implied by an axis $\bar{n}$ are not separable, that is, $\bar{4}$ is not equivalent to a 4-fold rotation axis plus a centre of symmetry. This combination produces the set of eight points shown in Fig. 2.8(b) as compared with the four points produced by $\bar{4}$. It can easily be verified that $\bar{1}$ is a centre of symmetry, $\bar{2}$ is equivalent to a plane of symmetry (also written *m*), $\bar{3}$ to an ordinary 3-fold axis plus a centre of symmetry, and $\bar{6}$ to a 3-fold axis plus a plane of symmetry perpendicular to it (Fig. 2.9). The symbol 3/*m* is simply a convenient way of printing $\frac{3}{m}$ (3 over *m*) and indicates that the plane of symmetry is perpendicular to the 3-fold axis. The symbol 3*m* indicates that the plane is parallel to the symmetry axis. It is instructive to examine solid objects such as polyhedra and models of crystals which possess various types of symmetry, when it is found that $\bar{3}$ generates the six faces of a rhombohedron, $\bar{4}$ those of a tetrahedron, and $\bar{6}$ those of a trigonal bipyramid. When describing the symmetry of a polyhedron the symmetry element is operating on faces and not on points as in Figs. 2.8 and 2.9.

40

The *glide plane* is the 3-dimensional analogue of the glide-reflection line of the 2-dimensional patterns. As its name implies, it combines in one operation a movement with a reflection. If we imagine a point A (Fig. 2.10) on one side of a mirror moved first to A' and then reflected through the plane of the mirror to B, then A is converted into B by the operation of the glide plane. The same operation performed on B would bring it to C, the translation being always a constant amount $\frac{1}{2}a$, where a is the unit translation of the lattice. As we noted earlier for the

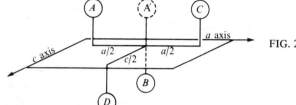

FIG. 2.10. Operation of a glide plane.

glide-reflection line the translation associated with a glide plane must be one-half of the lattice translation because the point must repeat at intervals equal to the lattice translation. A more complex type of glide plane transforms A into D, this involving translations of $\frac{1}{2}a + \frac{1}{2}c$, followed by reflection. If the unit translations a and c of the crystal lattice are not equivalent we clearly have three types of glide plane, with which are associated translations of $\frac{1}{2}a$, $\frac{1}{2}c$, and $\frac{1}{2}a + \frac{1}{2}c$ respectively; the corresponding symbols are $a, c,$ and d (for diagonal).

The *screw axis* derives its name from its relation to the screw thread. Rotation about an axis combined with simultaneous translation parallel to the axis traces out a helix, which is left- or right-handed according to the sense of the rotation. Instead of a continuous line on the surface of a cylinder there could be a series of discrete points, one marked after each rotation through $360°/n$. After n points we arrive back at one corresponding to the first but translated by x, the pitch of the helix, which in a 3-dimensional pattern corresponds to a lattice translation. The symbol for a screw axis indicates the value of n (rotation through $360°/n$) and, as a subscript, the translation in units of x/n where x is the pitch. The translation associated with each rotation of an n-fold screw axis may have any value from x/n to $(n-1)x/n$, and therefore the possible types of screw axis in periodic 3D patterns are the following:

$$2_1; 3_1, \text{ and } 3_2; 4_1, 4_2, \text{ and } 4_3; 6_1, 6_2, 6_3, 6_4; \text{ and } 6_5.$$

A convenient way of showing the sets of points generated by screw axes is to represent them by sets of figures giving the heights of the points above the plane of the paper in terms of x/n:

$6_1 \qquad 6_2 \qquad 6_3 \qquad 6_4 \qquad 6_5$

41

Starting in each case at the top of the diagram (height 0) and proceeding clockwise each point rises x/n for each rotation through $360°/n$. For 6_1 all the six points successively generated by the axis fall within one lattice translation normal to the plane of the paper, being at heights 0, 1/6, 2/6, 3/6, 4/6, and 5/6, but for the other 6-fold screw axes the points extend through 2, 3, 4, or 5 unit cells. For 6_4, for example, the heights of successive points are 0, 4, 8, 12, 16, and 20 units of $x/6$. However, since by definition each unit cell must contain the same arrangement of points we may deduct 6 or multiples of 6 from these values giving, 0, 4, 2, 0, 4, 2, a set of six points lying within one unit cell. Examination of the above diagrams shows that 6_1 and 6_5 are related as clockwise and anticlockwise helices, and similarly for 6_2 and 6_4.

The four types of symmetry element, axes of simple rotation or rotatory inversion, screw axes, and glide planes, have now to be inserted into the appropriate lattices. The only symmetry elements consistent with the first Bravais lattice of Fig. 2.7 (the triclinic lattice) are the axes 1 and $\bar{1}$, the former implying no symmetry and the latter a centre of symmetry. The highest rotational symmetry consistent with the lattices 2 and 3, which have two interaxial angles of 90° and one of β (hence the name monoclinic) is a 2-fold axis (2 or 2_1), the direction of which must correspond to the b axis. Alternatively, or in addition, one plane of symmetry is permissible, which must be perpendicular to the b axis. This may be a mirror plane (m or $\bar{2}$) or a glide plane. It is found that a total of 14 types of 3-dimensional symmetry (space-groups) can be associated with the two monoclinic lattices. It is interesting that the very considerable problem of determining the total number of space groups arising from all fourteen Bravais lattices was being studied independently during the same period (1885–94) by Fedorov in Russia, Schoenflies in Germany, and Barlow in England. It was established that there are 230 such space groups.

Point groups; crystal systems

We saw earlier that different symmetries are associated with particular positions in a plane pattern, and that the number of point groups in 2-dimensional patterns is ten. For 3D patterns the number of point groups is 32. These point groups are the possible symmetries of finite groups of points arranged around particular positions in a lattice; they cannot therefore include symmetry elements involving translations, namely, screw axes or glide planes, though they may include axes of rotatory inversion. We have approached the subject of symmetry from the standpoint of repeating patterns and lattices, leading to the enumeration of the symmetries of 3D patterns (the 230 space groups) and incidentally of the combinations of symmetry elements that can be associated with particular points in a lattice (the 32 point groups). Historically the development of symmetry theory was very different.

The science of crystallography began, in the seventeenth century, with the study of the shapes of crystals. It was observed that there is considerable variation in the overall shape of crystals of a particular substance (or of crystals of one form if it is polymorphic), but that however much a crystal departed from the 'ideal' shape, such as a cube or regular prism, the angles between corresponding pairs of faces were constant on all crystals. For this reason the crystallographer always works with face-normals, and the symmetry of a crystal shape (and hence its point symmetry) refers to the symmetry of the set of face-normals and not to the actual (often distorted) shape of a particular crystal. Variations in the relative developments of crystal faces are described as variations in *crystal habit*; they are not characteristic of the internal structure of the crystal but are attributable to variations in external conditions, for example, nearness to other crystals or presence of impurities in the solution or melt. Some examples are shown in Fig. 2.11.

Owing the regular internal structure of a crystal the symmetry of its external shape (the 'ideal' face development as defined above) is subject to the same limitations as regards types of rotational symmetry that apply to any 3D repeating pattern. Also the symmetry elements describing the shape of a crystal must all pass through a point and cannot involve translations, since they describe the arrangement of faces on a finite crystal. The 32 classes of crystal symmetry, which were derived as early as 1830, are therefore the 32 point groups to which we have already referred. They are grouped into seven crystal systems as shown in Table 2.2, which gives the Hermann–Mauguin symbols used by crystallographers and also the earlier Schoenflies symbols which are still favoured by spectroscopists.

The characteristic symmetries of the crystal systems and also the parameters required to define the unit cells are summarized in Table 2.3.

Comparison of Fig. 2.7 with Table 2.3 will show that whereas we have listed *trigonal* and hexagonal among the crystal systems, Fig. 2.7 includes *rhombohedral* and hexagonal lattices. The reason for this difference is the following. All crystals with a single axis of 3-fold or 6-fold symmetry can be referred to a hexagonal lattice (and unit cell), that is, all the points shown as open circles in Fig. 2.12 have the same environment. Some trigonal crystals also have the special property that the points marked as black circles have the same environment as those at the corners of the hexagonal unit cell. These points are vertices of a rhombohedron (with one-third the volume of the hexagonal cell), and the structure of such a crystal may therefore be referred to rhombohedral axes and a rhombohedral unit cell.

Attention should perhaps be drawn to the characteristic symmetry of the cubic system which is not, as might be supposed, the 4-fold (or 2-fold) axes of symmetry or planes of symmetry but four 3-fold axes parallel to the body-diagonals of the cubic unit cell. This combination of inclined 3-fold axes introduces either three 2-fold or three 4-fold axes which are mutually perpendicular and parallel to the cubic axes. Further axes and planes of symmetry may be present but are not essential to cubic symmetry and do not occur in all the cubic point groups or space groups.

Equivalent positions in space groups

We have stated that the arrangement of faces on a crystal (or more accurately the arrangement of face-normals) can be described by its point symmetry. The six

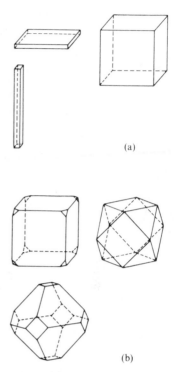

(a)

(b)

FIG. 2.11. Variation in habit of crystals. Different relative development of (a) cube faces, and (b) cube and octahedron faces.

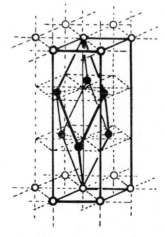

FIG. 2.12. Relation between hexagonal and rhombohedral unit cells.

TABLE 2.2

The thirty-two classes of crystal symmetry

Hermann–Mauguin	Schoenflies
Triclinic	
1	C_1
$\bar{1}$	C_i, S_2
Monoclinic	
2	C_2
m	C_s, C_{1h}
$2/m$	C_{2h}
Orthorhombic	
222	D_2, V
$mm2$	C_{2v}
mmm	D_{2h}, V_h
Tetragonal	
4	C_4
$\bar{4}$	S_4
$4/m$	C_{4h}
422	D_4
$4mm$	C_{4v}
$\bar{4}2m$	D_{2d}, V_d
$4/mmm$	D_{4h}
Trigonal	
3	C_3
$\bar{3}$	C_{3i}, S_6
32	D_3
$3m$	C_{3v}
$\bar{3}m$	D_{3d}
Hexagonal	
6	C_6
$\bar{6}$	C_{3h}
$6/m$	C_{6h}
622	D_6
$6mm$	C_{6v}
$\bar{6}m2$	D_{3h}
$6/mmm$	D_{6h}
Cubic	
23	T
$m3$	T_h
432	O
$\bar{4}3m$	T_d
$m3m$	O_h

TABLE 2.3

The crystal systems: unit cells and characteristic symmetry

System	Relations between edges and angles of unit cell	Lengths and angles to be specified	Characteristic symmetry
Triclinic	$a \neq b \neq c$ $\alpha \neq \beta \neq \gamma \neq 90°$	a, b, c α, β, γ	1-fold (identity or inversion) symmetry only
Monoclinic	$a \neq b \neq c$ $\alpha = \gamma = 90° \neq \beta$	a, b, c β	2-fold axis (2 or $\bar{2}$) in one direction only (y axis)
Orthorhombic	$a \neq b \neq c$ $\alpha = \beta = \gamma = 90°$	a, b, c	2-fold axes in three mutually perpendicular directions
Tetragonal	$a = b \neq c$ $\alpha = \beta = \gamma = 90°$	a, c	4-fold axis along z axis only
Trigonal† and Hexagonal	$a = b \neq c$ $\alpha = \beta = 90°$ $\gamma = 120°$	a, c	3-fold or 6-fold axis along z axis only
Cubic	$a = b = c$ $\alpha = \beta = \gamma = 90°$	a	Four 3-fold axes each inclined at 54°44' to cell axes (i.e. parallel to body-diagonals of unit cell)

† Certain trigonal crystals may also be referred to rhombohedral axes, the unit cell being a rhombohedron defined by cell edge a and interaxial angle α ($\neq 90°$).

triangular faces of a regular hexagonal pyramid are related by a 6-fold axis, but the basal face is not converted into a group of six faces because its normal is coincident with the symmetry axis. Similarly, six of the vertices are related by the symmetry axis but the seventh is a unique point, the difference being that the six basal vertices lie *off* the axis whereas the seventh lies *on* the axis. Similar considerations apply to the more complex systems of symmetry elements constituting the 230 space groups. The number of equivalent points (positions) in a unit cell depends not only on the types of symmetry present but also on the location of a point (atom) relative to the symmetry elements. Figure 2.13 represents the projection of a unit cell of a crystal. Planes of symmetry perpendicular to the paper intersect the plane of the projection in full lines. The planes repeat at intervals of the lattice spacing (unit cell edge) because we are dealing with a 3D lattice, and it will be observed that the planes through the origin also give rise to additional planes of symmetry midway between them. A point (atom) situated on the intersection of the symmetry planes is not operated upon by these symmetry elements; it is a 1-fold equivalent position. A point (shaded circle) situated on one of the planes is operated on only by the second plane(s)—2-fold position—while a point (small circle) which lies on neither plane is operated on by both—4-fold position. The

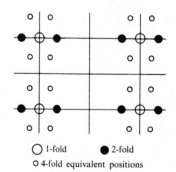

○ 1-fold   ● 2-fold
○ 4-fold equivalent positions

FIG. 2.13. Equivalent positions in a unit cell.

45

latter is called the *general* position and the 2- and 1-fold positions are called *special* positions.

Two points should be emphasized. First, according to classical structure theory, all the equivalent positions of a given set should be occupied and moreover they should all be occupied by atoms of the same kind. In later chapters we shall note examples of crystals in which one or both of these criteria are not satisfied; an obvious case is a solid solution in which atoms of different elements occupy at random one or more sets of equivalent positions. (The occupation of *different* sets of equivalent positions by atoms of the *same* kind occurs frequently and may lead to quite different environments of chemically similar atoms. Examples include the numerous crystals in which there is both tetrahedral and octahedral coordination of atoms of the same element—in the same oxidation state—as noted in Chapter 5, and crystals in which there is both coplanar and tetrahedral coordination of Cu(II), p. 890, or Ni(II), p. 965.) The second point for emphasis is if a molecule (or complex ion) is situated at one of the *special* positions it should possess the point symmetry of that position. A molecule lying on a plane of symmetry must itself possess a plane of symmetry, and one having its centre at the intersection of two planes of symmetry must itself possess two perpendicular planes of symmetry. If, therefore, it can be demonstrated that a molecule lies at such a position as, for example, would be the case if the unit cell of Fig. 2.13 contained only one molecule, (a fact deducible from the density of the crystal) this would constitute a proof of the symmetry of the molecule. Such a conclusion is not, of course, valid if there is any question of random orientation or free rotation of the molecules. Moreover, there is another reason for caution in applying this type of argument to inorganic crystals.

Suppose, for example, that we wish to know the arrangement of X atoms around B in a complex halide A_2BX_4 which is thought to contain planar or tetrahedral BX_4^{2-} ions. We shall further suppose that the X-ray diffraction data show that the B atoms are situated at centres of symmetry. If the BX_4^{2-} ions are finite they cannot be tetrahedral, since a tetrahedral group is not centrosymmetrical. However, the BX_4^{2-} anion could be infinite in extent, and an infinite linear ion consisting of octahedral BX_6 groups sharing opposite edges is also consistent with the observed point symmetry.

One final point about equivalent positions may be noted. There is no restriction on the symmetry of a molecule which occupies a *general* position in a space group (for example, the 4-fold position of Fig. 2.13); it may possess 8-fold, icosahedral, or any type of symmetry or no symmetry at all.

Examples of 'anomalous' symmetry

Special interest attaches to molecules or crystals that possess symmetry which is lower or higher than might be expected. This statement calls for a brief explanation. The packing of equal spheres is discussed in Chapter 4 where we shall see that the majority of metals adopt certain highly symmetrical structures. Some metals, however, crystallize with less symmetrical variants of these structures, and these are obviously of interest since the lower symmetry presumably indicates some peculiarity in the bonding in such crystals. Some crystals containing only highly symmetrical ions (for example FeO) exhibit lower symmetry when cooled. A particularly interesting case is that of metallic tin, the *higher* temperature form of which has the *lower* symmetry. There are a number of compounds which do not adopt the most symmetrical form of a particular structure but a distorted variant of the structure. In some cases a reasonable explanation can be given for the lower symmetry as, for example, some kind of interaction between metal atoms in NbI_4 or VO_2 or the asymmetry of the d^9 configuration of Cu^{2+} leading to the distorted rutile structure of CuF_2. In other cases there is as yet no obvious explanation for the lower symmetry, for example, of crystalline PdS as compared with PtS.

Conversely there are many cases where the symmetry is *higher* than might be expected, Many crystals containing complex ions have, either at room temperature or higher temperatures, structures indicating a symmetry for the complex ion higher than that corresponding to its known (or expected) structure (for example, spherical symmetry for OH^- or C_2^{2-}). In such cases the higher symmetry results from random orientation or free rotation of the non-spherical group. We have already mentioned the random occupation of sets of equivalent positions by atoms of more than one kind in solid solutions; the random arrangement of vacancies in a 'defect' structure can also lead to retention of the symmetry of the ideal structure.

It may appear surprising that molecules or ions presumably possessing symmetry do not always exhibit this symmetry in crystals, that is, they occupy positions of lower point symmetry. If the molecular symmetry is of a non-crystallographic type, for example, the 5-fold symmetry of a planar ring or an icosahedral group, clearly this cannot be exhibited in the crystal. The highest symmetry such a group could show would be, for example, a plane or 2-fold axis. The croconate ion in $(NH_4)_2C_5O_5$ has exact 5-fold symmetry to within the accuracy of the structure determination, but the ions must pack in a way consistent with one of the 230 space groups. Similarly, even if the molecules possess symmetry of a crystallographic type (e.g. 4- or 6-fold axis), the basic requirement is that they pack efficiently, and this may not be possible if they are arranged with their symmetry axes parallel, as would be necessary in a structure with tetragonal or hexagonal symmetry.

Isomerism

Isomerism is a rather comprehensive term embracing several types of structural differences between molecules (ions) having the same chemical composition. It is therefore closely related to polymorphism, for both are concerned with differences between the spatial arrangements of a given set of atoms which in the one case form a finite group and in the other an infinite array (crystal). The structural differences between isomers range from those between *topological isomers*, which represent different ways of connecting together the same set of atoms and are usually regarded as different chemical compounds, to those between a pair of *optical isomers*, which are structurally and chemically identical in every way except that they are related as object and mirror image. Because optical activity can be exhibited by both molecules and crystals and because molecules or ions having this

47

property do not necessarily exhibit any other type of isomerism, we shall discuss optical activity separately. Between these two extremes lie various types of *geometrical isomerism* or *stereoisomerism*. As the term implies, stereoisomerism includes all cases where a molecule (ion) can be obtained in two (or more) distinct forms which differ in the spatial arrangement of the constituent atoms but not in their topology. Thus in principle, one isomer could be converted into another simply by relative movements of certain of the atoms without breaking and remaking any of the σ bonds.

Between a pair of geometrical isomers there is an energy barrier which may be due to π-bonding (as in *cis–trans* isomerism of substituted ethylenes or the eclipsed and staggered forms of ferrocene) or to the fact that interconversion takes place through an intermediate with different geometry (for example, the rearrangement of a *cis* square-planar complex into the *trans* isomer via a tetrahedral intermediate). Geometrical isomers show differences in physical properties and in rare cases are recognized as different chemical compounds (for example, fumaric and maleic acids, the *trans* and *cis* forms of HOOC . CH=CH . COOH).

The subject of isomerism is closely concerned with symmetry. Consider, for example, the three molecules (i) O_2X_2, (ii) N_2X_2, and (iii) C_2X_2. In a molecule of type (i), formed by oxygen and sulphur, the two X atoms are not coplanar with the two O (S) atoms; the molecule is *enantiomorphic*, that is, it cannot be brought into coincidence with its mirror image; it exists in left- and right-handed forms.

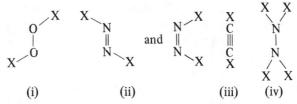

(i)   (ii)   (iii)   (iv)

Description of the molecular geometry requires a knowledge of the dihedral angle XOO/OOX. On the other hand all atoms of (ii) are coplanar, and such molecules exist in *cis* and *trans* forms. The four atoms of (iii) are collinear, it has cylindrical symmetry, and there are no isomers. A number of molecules of type (iv) are formed

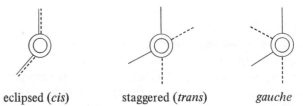

eclipsed (*cis*)   staggered (*trans*)   *gauche*

by nitrogen and phosphorus and they provide examples of two of the three extreme configurations possible for a molecule of this kind. The sketches show the molecule viewed along the N–N (P–P) axis. Examples include:

staggered: P_2H_4 (solid) P_2F_4 P_2Cl_4 P_2I_4
gauche: N_2H_4 P_2H_4 (gas)
staggered and *gauche*: N_2F_4

The dithionite ion ($S_2O_4^{2-}$) provides an example of a group of this kind with the eclipsed configuration.

The numbers and types of isomers of molecules such as Ma_2b_2 or Ma_2b_4 etc. depend on the spatial arrangement (symmetry) of the bonds around the central atom. Much of the 'classical' work on isomerism was expressly designed to distinguish between one or more plausible bond arrangements, such as the coplanar or tetrahedral arrangement of four bonds, or octahedral or trigonal prismatic arrangement of six bonds. Although determination of bond arrangements by enumeration of isomers and the resolution of optically active compounds has been superseded by direct structural studies, the older methods played an important part in the development of structural chemistry, and a few examples will be mentioned later.

Structural (topological) isomerism

As we have already remarked structural isomers are usually different chemical compounds, as in the case of NH_4NCO and $CO(NH_2)_2$, the earliest example of a pair of compounds to which the term 'isomers' was applied. This type of isomerism is extremely common in organic chemistry, simple examples being hydrocarbons such as normal and iso butane, *o*-, *m*- and *p*-substituted benzenes, and so on. The special case of two interlocked *n*-rings, isomeric with the 2*n*-ring, is mentioned in Chapter 3 as the simplest example of this kind of topological isomer. If the isomerism involves only movement of H atoms the term *tautomerism* is used, and this also is frequently encountered in organic chemistry, a simple example being the keto-enol tautomerism of acetylacetone:

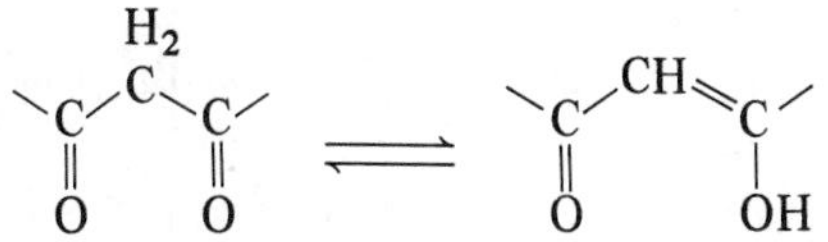

Thiocyanic acid (a), is tautomeric, existing essentially (95 per cent) in the iso form. Esters of both types, RSCN and SCNR, are known, and in some metal-coordination compounds the SCN ligand is attached to the metal through S and in others through N. On the other hand, the oxygen analogue cyanic acid exists entirely in the iso form and forms only one type of ester, RNCO. Both isomers (b)

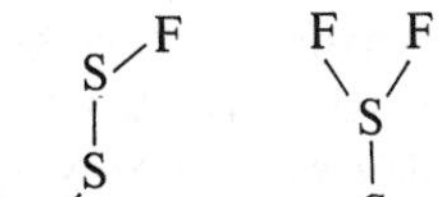

of S_2F_2 are formed when a mixture of dry AgF and S is heated in a glass vessel *in vacuo* to 120°C. The 'symmetrical' isomer FSSF has a dihedral structure similar to that of H_2O_2 (and is enantiomorphic), while SSF_2 is pyramidal and very similar geometrically to OSF_2. Although nitrous acid could be formulated as at (c) and (d)

49

Symmetry

there is no evidence for the existence of (c), the gas being a mixture of the *cis* and *trans* isomers (d).

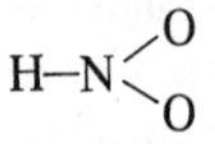

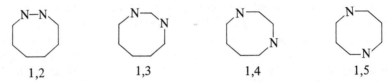

(c)

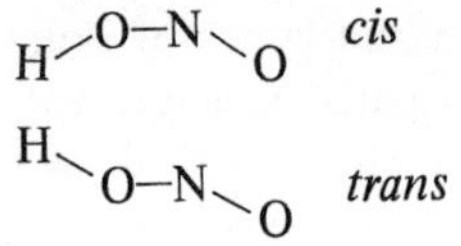

(d)

More complex examples of isomerism include that of carboranes, in which C atoms are introduced into polyhedral frameworks of B atoms, and of compounds such as $S_n(NH)_{8-n}$. These form cyclic molecules similar to the S_8 molecule of elementary sulphur but with some S atoms replaced by NH groups. Three of the four possible isomers of $S_6(NH)_2$ have been characterized, namely the 1,3, 1,4, and 1,5 isomers.

1,2   1,3   1,4   1,5

Coordination compounds provide numerous examples of structural isomerism, of which the following are a selection:

(i) 'Bond' isomerism
$[Co(NH_3)_5ONO] Cl_2$ and $[Co(NH_3)_5NO_2] Cl_2$.

(ii) 'Ionization' isomerism
$[Pt(NH_3)_4Cl_2] Br_2$ and $[Pt(NH_3)_4Br_2] Cl_2$.

(iii) 'Coordination' isomerism
$[Co(NH_3)_6] [Cr(CN)_6]$ and $[Co(CN)_6] [Cr(NH_3)_6]$

(iv) 'Polymerization' isomerism
$[Pt(NH_3)_4] [PtCl_4]$ and $Pt(NH_3)_2Cl_2$.

Geometrical isomerism

Two parts of a molecule which are connected by a single bond may be free to rotate relative to one another, but isomerism arises only if there are two (or more) configurations separated by energy barriers sufficiently large to prevent interconversion (see p. 641). The $S_2O_6^{2-}$ ion is found in crystalline $K_2S_2O_6$ with two configurations, the *trans* (centrosymmetrical) and an almost eclipsed configuration. The oxalate ion, $C_2O_4^{2-}$, is planar in the sodium and potassium salts but non-planar in the ammonium salt. These differences are due to interactions with adjacent ions in the crystal and are characteristic of the crystalline state, and are not retained in solution.

Some simple examples of *cis–trans* isomerism have already been noted. More complex examples include those of cyclic molecules such as chair-shaped 6-rings, (a), where a distinction must be made between equatorial 'e' and axial 'a' bonds,

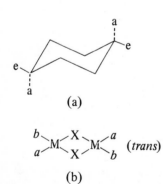

(a)

(b)

50

and bridged molecules (b); the latter could also exhibit 'structural' isomerism, in the unsymmetrical isomer $a_2MX_2Mb_2$.

Much attention has been paid in the past to the isomerism of mononuclear molecules and ions because of the relation between numbers of isomers and the bond arrangement around the central atom. Whereas a tetrahedral molecule Ma_2b_2 exists in only one form a planar molecule or ion of this type can in principle exist in *cis* and *trans* forms, and many pairs of Pt compounds of this type have been prepared. Before it became possible to determine crystal structures there was considerable interest in the numbers of isomers of complexes such as Coa_4b_2 which would be expected to have two isomers if octahedral but three if trigonal prismatic (or coplanar). Two points may be noted in connection with the deduction of bond arrangements from the numbers of isomers. First, failure to obtain the expected number of isomers may be due to large differences in stability or to interconversion in solution followed by crystallization of the less soluble isomer. For example, whereas many compounds Pta_2b_2 are known in both *cis* and *trans* forms, the corresponding Pd compounds are usually known only in the *trans* form (exceptions include $Pd(NH_3)_2Cl_2$ and $Pd(NH_3)_2(NO_2)_2$). In solutions of compounds $(R_3Sb)_2PdCl_2$ the two forms are in labile equilibrium with the *trans* form predominating, but the much less soluble *cis* isomer crystallizes out. Second, it is necessary to confirm that the various forms obtained are in fact geometrical isomers. For example, the isomers of the compound with the empirical composition $Pt(NH_3)_2Cl_2$ are structural isomers, one being a planar molecule $Pt(NH_3)_2Cl_2$ and the other the salt $[Pt(NH_3)_4][PtCl_4]$. A rather similar case, which for a time confused the stereochemistry of tellurium, concerned the existence of two forms of $Te(CH_3)_2I_2$, thought to be *cis* and *trans* isomers of a square planar complex. It has since been shown that one isomer is the salt $[Te(CH_3)_3]^+[TeCH_3I_4]^-$. Finally, a number of compounds of cobalt with the composition CoX_2R_2 exist in two forms, one consisting of finite tetrahedral complexes and the other of infinite chains in which the metal atom is octahedrally coordinated.

Optical activity

Optical activity, the property of rotating the plane of polarization of plane-polarized light, may be exhibited by matter in all states of aggregation. When an optically active compound is prepared there will normally be equal numbers of d- and l-molecules. They may generally be separated by forming a compound (e.g. a salt) with either the d- or l-form of a second active compound and utilizing differences between the physical properties of the resulting compounds $(d - A)(l - B)$ and $(l - A)(l - B)$—a process called *resolution*—though special methods (such as the hand-sorting of crystals) have been used in some cases. The resolution of tri-o-thymotide has been carried out[1] by forming a molecular compound with an *inactive* second component (benzene). This was possible because although the unsolvated material crystallizes from methanol as the racemate (see later) the molecular compound formed with benzene is optically active and large single crystals could be grown, each of which was either a d- or an l-crystal. The d- or l-forms of a compound exhibit optical activity in the crystalline, liquid, dissolved,

(1) JCS 1952 3747

and vapour states, though the activity may not persist indefinitely in the last three cases if interconversion of the two forms is possible; an inactive mixture of the two forms then results (*racemization*).

For a finite group of atoms the criterion for enantiomorphism is the absence of an axis of rotatory inversion. An axis $\bar{n}$ implies a centre of symmetry if n is odd, it introduces planes of symmetry if n is a multiple of 2 but not of 4, and if n is a multiple of 4 the system can be brought into coincidence with its mirror image. Of the simplest axes of these three types, $\bar{1}$ is synonymous with a centre of symmetry, and $\bar{2}$ with a plane of symmetry. Since axes of rotatory inversion $\overline{4n}$ are likely to occur very rarely in molecules, we may for practical purposes take as the criterion for enantiomorphism and for optical activity in a *finite* molecule or complex ion the absence of a *centre* or *plane of symmetry*.

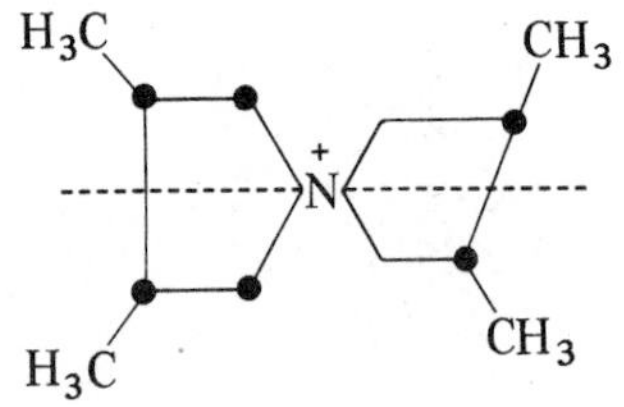

One isomer of the tetramethyl-*spiro*-bipyrrolidinium cation provides an example of a molecule with $\bar{4}$ symmetry. Each methyl group can project either above or below the plane of the ring to which it is attached, and the two rings lie in perpendicular planes. There are accordingly four forms of this molecule, which may be shown diagrammatically as set out below. The molecules are viewed along the direction of the dotted line and the thick lines indicate the rings:

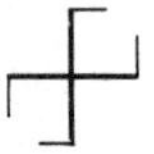

cis/cis	cis/trans	trans/trans	trans/trans
(active)	(active)	(active)	(meso)

All four forms have been prepared, the meso *trans/trans*-form being inactive[2], having $\bar{4}$ symmetry.

(Certain substituted cyclobutanes possess this unusual symmetry axis, e.g. that shown below (I). This molecule also, however, possesses diagonal planes of symmetry parallel to the $\bar{4}$ axis, so that its lack of enantiomorphism cannot be attributed to the presence of that axis. These planes of symmetry would disappear if groups C(XYZ) were substituted for the methyl groups, giving the molecule II(a). In the plan at II(b) only these C(XYZ) groups are shown to illustrate the absence of diagonal symmetry planes. This molecule would not be optically active since it is identical with its mirror image, though it contains no fewer than eight asymmetric carbon atoms.)

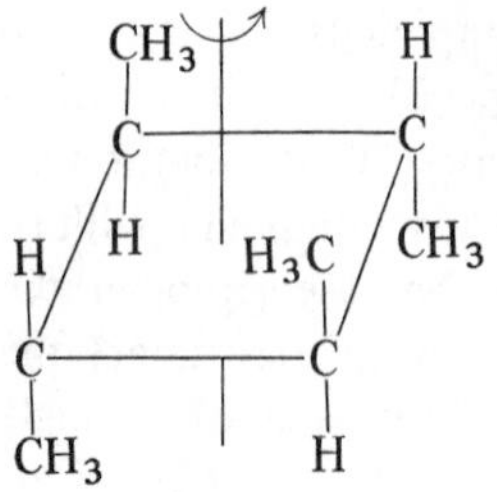

(2) JACS 1955 **77** 4688

I 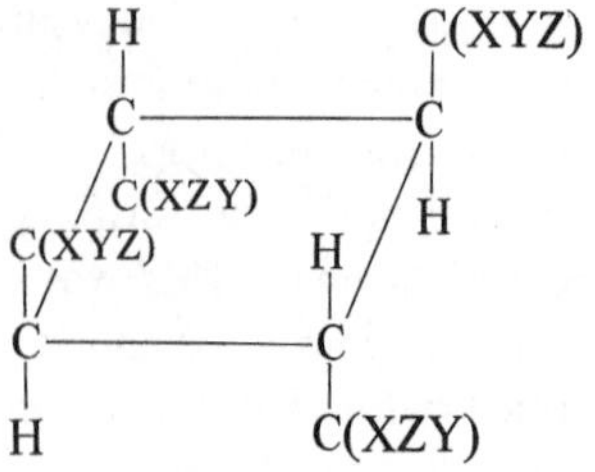   II(a) 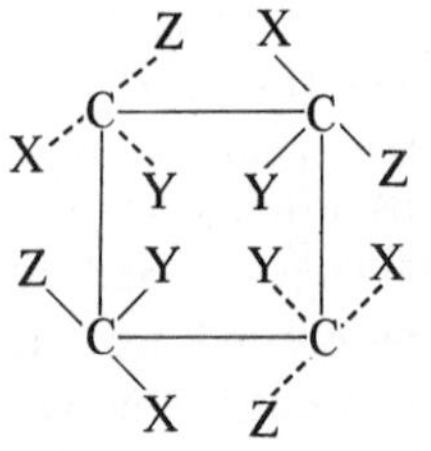   II(b)

Symmetry

The relation between optical activity and enantiomorphism is not quite so simple for a crystal. Of the thirty-two crystal classes eleven are enantiomorphic:

$$1, 2, 3, 4, 6, 23, 222, 32, 42, 62, \text{ and } 432.$$

A particular crystal having the symmetry of one of these classes is either left- or right-handed, and if suitable faces happen to develop when the crystal grows hand-sorting may be possible. Optical activity is also theoretically possible in four of the non-enantiomorphic classes:

$$\bar{2}\,(=m),\, mm,\, \bar{4},\, \text{and } \bar{4}2m.$$

In these classes directions of both left- and right-handed rotation of the plane of polarization must exist in the same crystal. An earlier claim that optical activity is exhibited by a crystal in class m (mesityl oxide oxalic methyl ester) has since been disproved, but it has been experimentally verified in both $\bar{4}$ ($CdGa_2S_4$)[3] and $\bar{4}2m$ ($AgGaS_2$).[4]

(3) AC 1969 **A25** 633
(4) AC 1968 **A24** 676

In general the optical activity exhibited by a crystal will persist in other states of aggregation only if it is due to the asymmetry of the finite molecule or complex ion. In this case it is also necessary that the energy of activation for racemization (change $d \rightleftharpoons l$) must exceed a certain value (~ 80 kJmol^{-1} at room temperature). The optical activity cannot, of course, be demonstrated unless there is a means of resolution (separation of d and l forms) or at least of altering the relative amounts of the two isomers. The optical activity of many crystals (e.g. quartz, cinnabar, $NaClO_3$) arises from the way in which the atoms are linked in the crystal; it is then a property of the crystalline material only.

Optical activity was first studied in compounds of carbon. In 1874 van't Hoff showed how optical activity could arise if the four bonds from a carbon atom were arranged tetrahedrally. Le Bel independently, and almost simultaneously, put forward somewhat similar views. A molecule $Cabcd$ in which a C atom is attached to four different atoms or groups should exist in two forms related as object and mirror image. The central C atom was described as an asymmetric C atom. A simple molecule of this sort is that of lactic acid, $CH(CH_3)(OH)COOH$, which exists in d- and l-forms. There is a third form of such a compound, the racemic form. A true racemate has a characteristic structure different from that of the active forms, there being equal numbers of d- and l-molecules in the unit cell of the crystal, which is optically inactive. We may note that inactive crystalline forms of optically active compounds are not necessarily racemates. For example, the inactive β-phenylglyceric acid (m.p. 141°C) is not a racemate but the crystals are built of submicroscopic lamellae which are alternately d- and l-rotatory. Slow recrystallization yields single crystals of the d- and l-forms (m.p. 164°C).[5]

(5) ACSc 1950 **4** 1020

A molecule containing two similar asymmetric C atoms, $cbaC-Cabc$, exists not only in d- and l-forms (which combine to form a racemate) but also in an inactive meso or 'internally compensated' form. The racemate is described as 'externally

53

compensated'. X-rays studies of crystalline tartaric acid show that the *d*- and *l*-forms of the molecule have the structures:

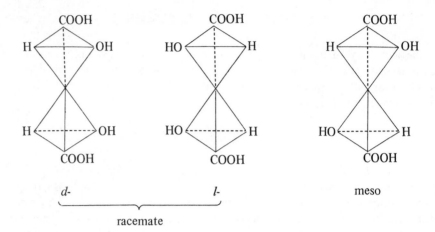

COOH
H
OH
H
OH
COOH
COOH
HO
H
HO
H
COOH
COOH
H
OH
HO
H
COOH

d- *l*- meso

racemate

We now know that (i) the presence of asymmetric carbon (or other) atoms does not necessarily lead to optical activity, and (ii) that optical activity may arise in the absence of asymmetric atoms. The criterion is the symmetry of the molecule as a whole. Some examples of (i) have already been given; another one is provided by dimethyl-diketopiperazine. This compound exists in *cis*- and *trans*-forms, each containing two asymmetric carbon atoms. The *trans*-form possesses a centre of symmetry and is therefore not optically active, while the *cis*-form is resolvable into optical antimers (*d*- and *l*-forms).

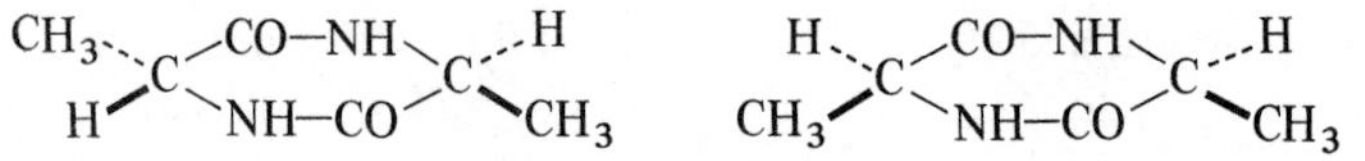

CH3
C
CO–NH
C
H
H
NH–CO
CH3
H
C
CO–NH
C
H
CH3
NH–CO
CH3

trans *cis*

The numerous examples of (ii) include molecules such as

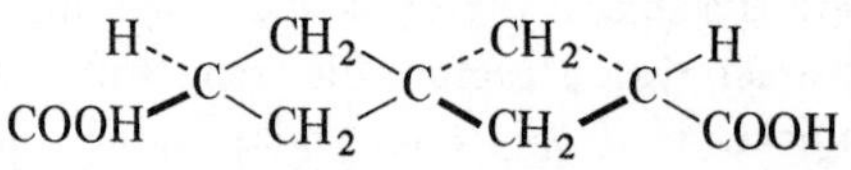

H
C
CH2
C
CH2
C
H
COOH
CH2
CH2
COOH

in which the bonds shown as dotted and heavy lines lie in a plane perpendicular to and respectively behind and in front of the plane of the paper, and many molecules which cannot adopt the most symmetrical configurations for steric reasons. In some cases the optical activity is due to restricted rotation around a single C–C bond. The presence of groups projecting from the aromatic rings in the molecules (1)–(3) leads to enantiomorphic forms. The replacement of SO_3^- in (3) by the smaller COO^- group destroys the optical activity.

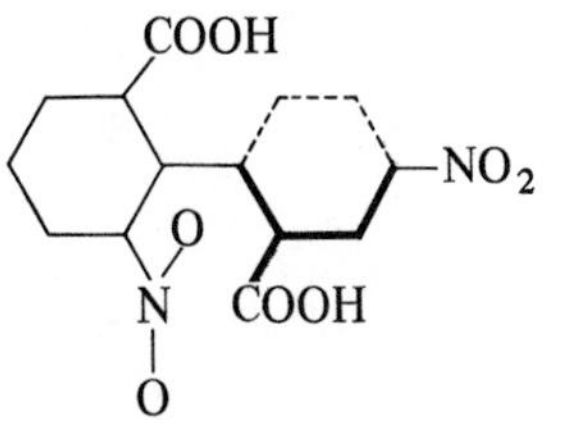

(1) 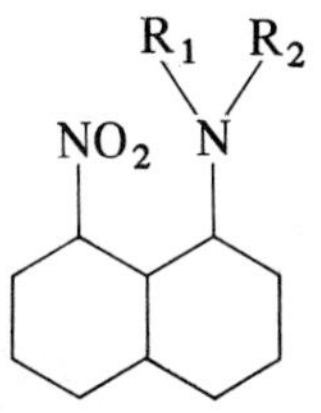   (2) 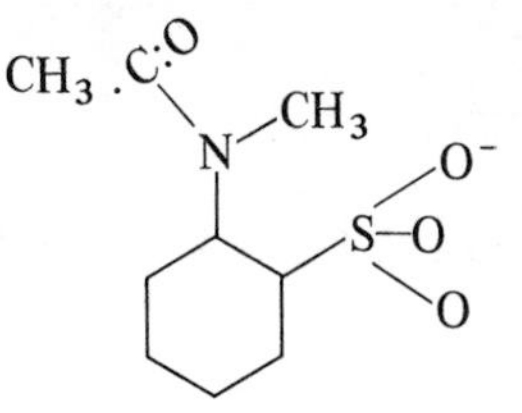   (3)

The molecules of many aromatic compounds would be highly symmetrical if planar, but if the planar configuration would imply impossibly short distances between certain atoms the molecules are forced into non-planar configurations which are often enantiomorphic. Examples of such 'overcrowded' molecules include:

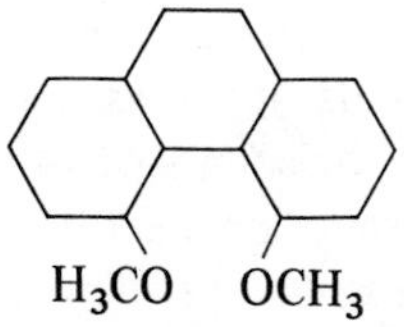

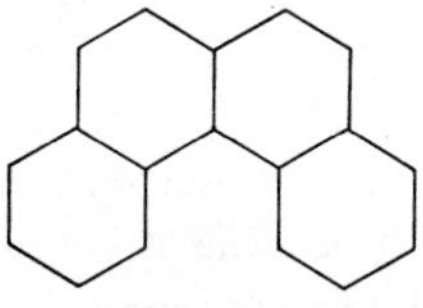

H_3CO OCH_3 and

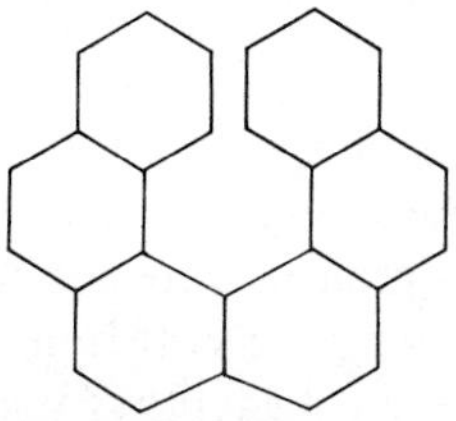

The compound on the right (hexahelicene) is also of interest because it was resolved by forming a crystalline molecular complex with a polynitro compound.[6]

(6) JACS 1956 78 4765

We conclude these remarks on optical activity by giving a few examples of molecules or ions the resolution of which demonstrates the general arrangement of the bonds formed by the central atom.

Three pyramidal bonds

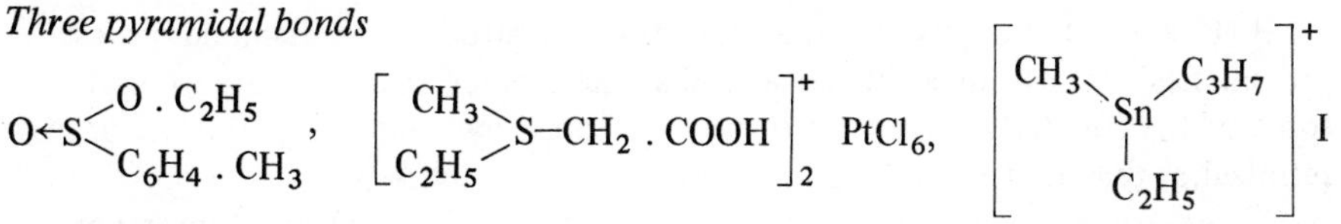

$$O \leftarrow S \begin{cases} O \cdot C_2H_5 \\ C_6H_4 \cdot CH_3 \end{cases}, \quad \left[\begin{array}{c} CH_3 \\ C_2H_5 \end{array} S - CH_2 \cdot COOH \right]_2^+ PtCl_6, \quad \left[\begin{array}{c} CH_3 \quad C_3H_7 \\ Sn \\ C_2H_5 \end{array} \right]^+ I$$

Four tetrahedral bonds

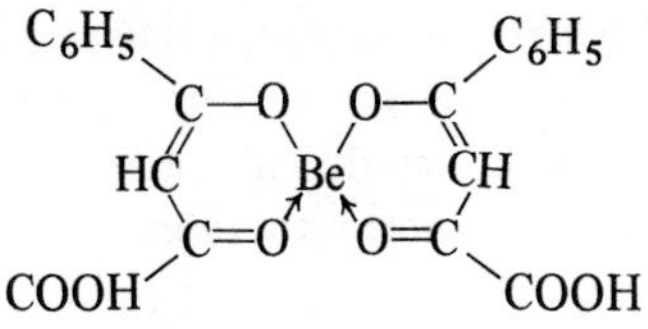

(also Cu^{II} and Zn)

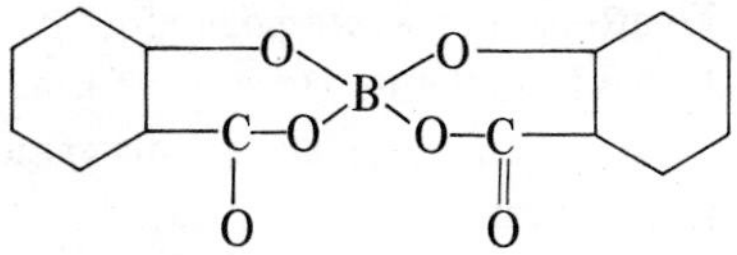

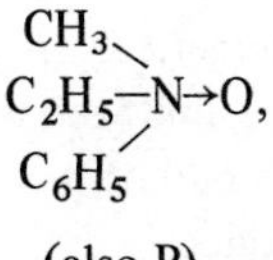

$$\begin{array}{c} CH_3 \\ C_2H_5 - N \rightarrow O, \\ C_6H_5 \end{array} \qquad C_6H_5 \cdot CH_2 - \underset{\underset{C_3H_7}{|}}{\overset{\overset{C_2H_5}{|}}{Si}} - CH_2 \cdot C_6H_4SO_3H,$$

(also P)

$$\left[\begin{array}{c} H \quad CH_2-CH_2 \quad CH_2\text{---}CH_2 \quad H \\ C \quad \quad \quad N \quad \quad \quad C \\ C_6H_5 \quad CH_2-CH_2 \quad CH_2-CH_2 \quad COOC_2H_5 \end{array} \right] Br$$

55

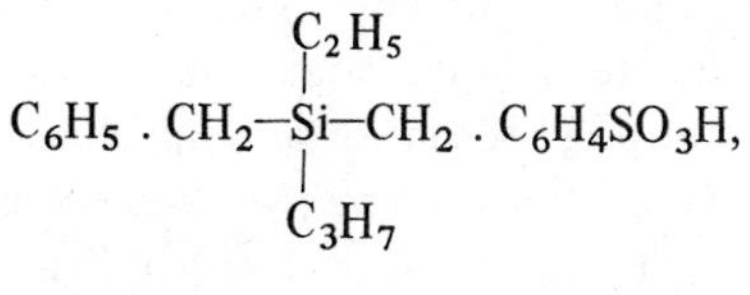

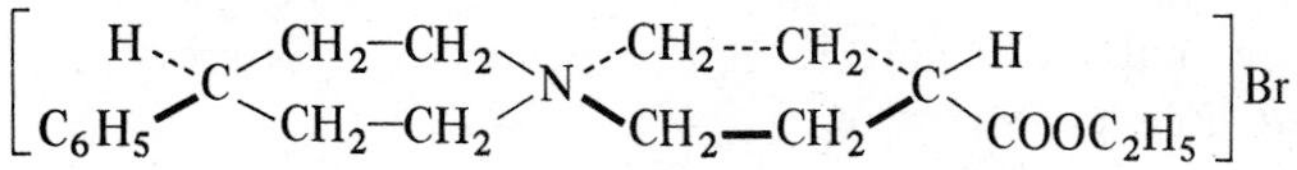

3

Polyhedra and Nets

Introduction

For a number of reasons it seems logical to postpone any discussion of structure and bonding until we have considered some basic geometry and topology:

(i) It is customary to discuss crystal and molecular structures in terms of other *known* structures. Ideally we should be able to set out the possible types of structure for a molecule or crystal composed of certain numbers of atoms with known bonding requirements so that the observed structures can be compared with all (geometrically and topologically) possible structures, up to some arbitrary limit of complexity. It is important to know why some apparently reasonable structures are never adopted by known molecules or crystals. Systematic studies of possible structure types have not been numerous; we shall survey here the simpler systems of connected points and the closest packing of equal spheres as essential parts of this basic geometry. A knowledge of the possible types of 3D nets throws at least some light on questions such as: why is diamond a system of rings of 6 carbon atoms, and why do certain crystalline forms of B_2O_3 and P_2O_5 consist of rings of 10 B (P) and 10 O atoms? We shall find that the structure of diamond represents the simplest 3D 4-connected net and that these oxide structures are examples of some of the simpler structures that can be formed by units (BO_3 triangles or PO_4 tetrahedra) that are to be joined to three other similar units.

(ii) We noted in Chapter 1 that many molecular and crystal structures represent a compromise between conflicting packing (or bonding) requirements of atoms of various kinds. Studies of the simpler types of structure for compounds A_mX_n in which X atoms are close-packed (Chapter 4) and of the linking together of tetrahedra and octahedra (Chapter 5) reveal interesting restrictions on bond angles which are obviously relevant to discussions of such simple structures as those of rutile and corundum.

(iii) Descriptions of the structures of ionic crystals usually start from a consideration of the environment of the individual ions (relation of coordination polyhedra to relative ionic sizes). However, the (energetically) preferred co-ordination polyhedron for an individual ion may be impossible (geometrically) in a 3D structure. For example, we find cubic coordination in two of the simplest ionic structures (CsCl and CaF_2) rather than antiprism or dodecahedral coordination. These two less symmetrical 8-coordination polyhedra are not possible in 3D AX or AX_2 structures but are important in more complex structures where the geometrical restrictions are less severe.

(iv) Crystal structures can be illustrated in various ways according to the feature it is wished to emphasize. For example, the structure of the P_4O_{10} molecule may be shown as a 'ball and spoke' model (Fig. 3.1 (a)), showing the atoms as spheres of arbitrary radii, to indicate the bond arrangement around the P and O atoms. Alternatively we may wish to emphasize the tetrahedral coordination of the P atoms, (b), or to focus attention on the group of ten close-packed O atoms, (c). There are similar types of representation of crystal structures, and it is important that the reader should be conversant with the types of illustration used by crystallographers.

(v) While some structures are associated more or less exclusively with bonding of a particular type (for example, the rutile structure of essentially ionic oxides and fluorides AX_2), other simple A_mX_n structures are not indicative of a particular type of bonding. For example, the NaCl structure is adopted not only by ionic oxides and halides but also by transition-metal nitrides and carbides and by some intermetallic compounds. The physical properties of compounds such as PbS, PbSe, and PbTe (NaCl structure) suggest more complex interactions than the essentially ionic bonding in the alkali halides; the oxides RuO_2 (rutile structure) and ReO_3 are metallic conductors. The zinc-blende structure, with tetrahedral coordination of both kinds of atom, is suitable for both ionic and covalent bonding. It is therefore preferable to describe the simple A_mX_n structures as geometrical entities rather than typical ionic, covalent, or metallic structures.

For these reasons three chapters are devoted to essentially topological and geometrical topics which are basic to an understanding of crystal structures. The present chapter is concerned with the ways in which points can be joined together to form finite or infinite systems, and includes some account of polyhedra and connected systems extending indefinitely in one, two, or three dimensions. In Chapter 4 we consider the packing of spheres, in particular the closest packing of equal spheres. Chapter 5 deals with the two most important coordination polyhedra in inorganic chemistry, the tetrahedron and the octahedron, and attempts a systematic account of the types of structure that can be built from these units by sharing vertices, edges, and /or faces.

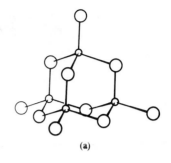

(a)

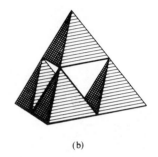

(b)

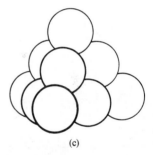

(c)

FIG. 3.1. Three representations of the P_4O_{10} molecule.

The basic systems of connected points

However complex the units that are to be joined together the problem may be reduced to the derivation of systems of points each of which is connected to some number (p) of others. In the simplest systems this number (the connectedness) is the same for all points:

$p = 1$: there is one solution only, a pair of connected points.

$p = 2$: the only possibilities are closed rings or an infinite chain.

$p \geqslant 3$: the possible systems now include finite groups (for example, polyhedra) and arrangements extending indefinitely in one, two, or three dimensions (p-connected nets).

57

It is not necessary to include singly-connected points ($p = 1$) in nets since they can play no part in extending the net. Also 2-connected points may be added along the links of any more highly connected net ($p > 2$) without altering the basic system of connected points. We therefore do not include either 1- or 2-connected points when deriving the basic nets though it may be necessary to add them to obtain the structures of actual compounds from the basic nets. For example, 2-connected points (representing $-O-$ atoms) are added along the edges of the 3-connected tetrahedral group of four P atoms to form the P_4O_6 molecule, and additional singly-connected points (representing $=O$ atoms) at the vertices to form P_4O_{10}.

In chain structures the repeating unit may be a single atom or a group of atoms:

	Repeat unit
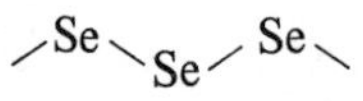	Se
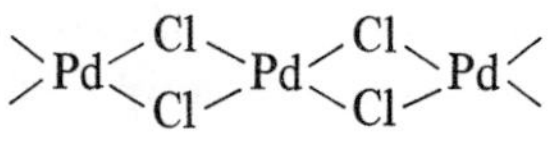	$PdCl_2$
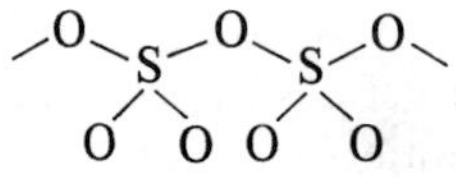	SO_3

but any simple chain may be reduced to the basic form $-A-A-$ (or ⌒A⌒A⌒A etc.), each unit being connected to two others.

Still retaining the condition that all points have the same connectedness we find that there are certain limitations on the types of system that can be formed. For example, there are no polyhedra with all vertices 6-connected, and there is only one 6-connected plane net. These limitations are summarized in Table 3.1, which may be considered in two ways:

TABLE 3.1

Systems of p-connected points

p	Polyhedra	Infinite periodic nets	
		Plane nets	3D nets
3			
4			
5			
6		(one only)	
7			

(i) Together with simple linear systems ($p = 2$) the *vertical* subdivisions correspond to the four main classes of crystal structure, namely, molecular crystals, chain, layer, and three-dimensional (macromolecular) structures. The division into polyhedra, plane, and 3D nets is also the logical one for the systematic derivation of these systems.

(ii) However, from the chemical standpoint we are more interested in the *horizontal* sections of Table 3.1. If we wish to discuss the structures which are possible for a particular type of unit, for example, an atom or group which is to be bonded to three others, we have to select the systems having $p = 3$.

We shall first say a little about the three vertical subdivisions of Table 3.1, confining our attention for the most part to systems in which $p = 3$ or 4. The coordination numbers from 6 to 12 are more conveniently considered under sphere packings, and the very high coordination numbers (13–16 in many transition-metal

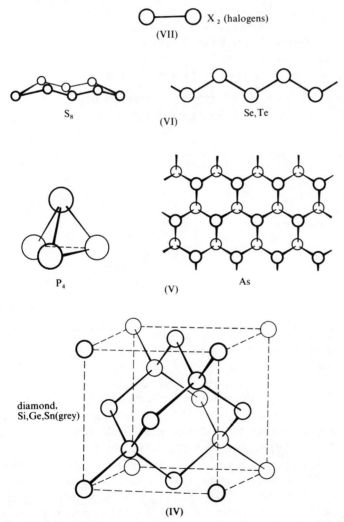

FIG. 3.2. The $8 - N$ rule illustrated for certain elements of the IVth, Vth, VIth, and VIIth Periodic Groups.

alloys, 22 and 24 in other intermetallic compounds, 24 in CaB_6, etc) will only be mentioned incidentally. We shall then give examples of polyhedral, cyclic, and linear systems and of structures based on 3- and 4-connected nets, dealing at some length with the three most important nets, namely, the plane hexagonal net, plane square net, and the diamond net.

The general topological relations already noted are well illustrated by the structures of the elements chlorine, sulphur, phosphorus, and silicon (Fig. 3.2). The number of bonds formed is $8 - N$, where N is the ordinal number of the Periodic Group, that is, 4 for Si, 5 for P, etc.

$p = 1$: Cl–Cl molecule in all states of aggregation.

$p = 2$: cyclic S_6, S_8, and S_{12} in three of the crystalline forms and in their vapours, and infinite chains in plastic S.

$p = 3$: tetrahedral P_4 molecule, 3-connected (buckled) 6-gon layer in black phosphorus, and complex 3-connected layer in red P.

$p = 4$: diamond-like structure of elementary silicon.

The same principles determine the structures of the oxides and oxy-ions of these elements in their highest oxidation states, the structural units being tetrahedral MO_4 groups:

p		*Oxy-ion*			*Oxide molecule*	
1	pyro	$Si_2O_7^{6-}$	$P_2O_7^{4-}$	$S_2O_7^{2-}$	Cl_2O_7	finite
2	meta	$(SiO_3)_n^{2n-}$	$(PO_3)_n^{n-}$	$(SO_3)_n$		cyclic or linear
3	infinite 2D	$(Si_2O_5)_n^{2n-}$	$(P_2O_5)_n$			polyhedral, 2D, or 3D
4	infinite 3D	$(SiO_2)_n$				infinite 3D

Polyhedra

Coordination polyhedra : polyhedral domains

Perhaps the most obvious connection of polyhedra with practical chemistry and crystallography is that crystals normally grow as convex polyhedra. The shapes of single crystals are subject to certain restrictions arising from the fact that only a limited number of types of axial symmetry are permissible in crystals, as explained in Chapter 2. We shall not be concerned here with the external shapes of crystals but with polyhedra which are of interest in relation to their internal structures and more generally to the structures of molecules and complex ions.

The nearest neighbours of an atom in a molecule, complex ion, or crystal, define a polyhedral *coordination group,* the number of vertices of which is the *coordination number* (c.n.) of the atom. It is convenient to describe the structures of many essentially ionic crystals in terms of the coordination groups around the cations. It is to be expected that coordination polyhedra with triangular faces will be prominent because these are the most compact arrangements of atoms in a polyhedral coordination shell. The tetrahedron and octahedron are the most important coordination polyhedra because their vertices represent the most symmetrical arrangements of four or six points around a central one. They are therefore stable arrangements of four or six ions around one of opposite charge in ionic crystals and also of these numbers of electron pairs in a covalent molecule or crystal. We shall see in Chapter 4 that the non-metal atoms in many crystalline halides and oxides are arranged in closest packing, the metal atoms occupying tetrahedral or octahedral interstices between the c.p. atoms. When a crystal is described in terms of the coordination polyhedra around the cations it becomes a system of polyhedra joined together by sharing vertices, edges, or (less frequently) faces. This approach is developed in Chapter 5.

An alternative is to describe the crystal structure in terms of the domains of the atoms, the *domain* being the polyhedron enclosed by planes drawn midway between the atom and each neighbour and perpendicular to the line of centres. Any atom for which the plane forms a face of this polyhedron is counted as belonging to the coordination sphere of the central atom. Each atom is then represented by a *polyhedral domain*, the number of *faces* of which is the coordination number of the atom, and the whole structure is a space-filling arrangement of such polyhedra. This concept is not especially useful for ionic crystals in which an ion has a well defined set of approximately equidistant nearest neighbours, though it does provide a more unambiguous definition of c.n. in crystals in which the environment of an atom is less regular. On the other hand, it is the logical way of describing structures built of atoms of comparable size, notably metals and intermetallic phases, where it is often difficult to define c.n.'s in terms of distances to nearest neighbours. A simple example is the body-centred cubic structure, in which an atom has eight nearest neighbours at a distance d and six more at a slightly greater distance ($1 \cdot 16\, d$) at the body-centres of the six adjacent unit cells. The domain of an atom in this structure is a truncated octahedron, with eight hexagonal and six square faces corresponding to the two sets of neighbours.

Space-filling arrangements of polyhedra, in which every polyhedron face is common to two polyhedra, are of interest in another connection. If we imagine atoms placed at the vertices of the polyhedra and bonded along the edges (either directly or through a 2-connected atom such as oxygen) then we have a rather open structure with polyhedral cavities. Such frameworks represent the structures of several groups of compounds which are described later.

The regular solids

The Greeks had a considerable knowledge of polyhedra, but only during the past two hundred years or so has a systematic study been made of their properties, following the publication in 1758 of Euler's *Elementa doctrinae solidorum*. From Euler's relation between the numbers of vertices (N_0), edges (N_1), and faces (N_2) of a simple convex polyhedron

$$N_0 + N_2 = N_1 + 2$$

equations may be derived relating to special types of polyhedra, as indicated in Appendix 1 of MSIC. If 3 edges meet at every vertex (3-connected polyhedron) and f_n is the number of faces with n edges (or vertices):

$$3f_3 + 2f_4 + f_5 \pm 0f_6 - f_7 - 2f_8 - \ldots = 12 \tag{1}$$

from which it follows that if all the faces are of the same kind they must be 3-gons, 4-gons, or 5-gons, and that the numbers of such faces must be four, six, and twelve respectively. The corresponding equations for 4- and 5-connected polyhedra are:

$$2f_3 \pm 0f_4 - 2f_5 - 4f_6 - \ldots = 16 \tag{2}$$

and

$$f_3 - 2f_4 - 5f_5 - 8f_6 - \ldots = 20, \tag{3}$$

of which the special solutions are: $f_3 = 8$ and $f_3 = 20$.

In the analogous equation for 6-connected polyhedra the coefficient of f_3 is zero and all the other coefficients have negative values. There is therefore no simple convex polyhedron having six edges meeting *at every vertex*. Polyhedra with more than six edges meeting at every vertex are impossible because all coefficients of f_n are negative.

The special solutions of equations (1)–(3), namely,

3-connected: $f_3 = 4; f_4 = 6; f_5 = 12$,
4-connected: $f_3 = 8$,
5-connected: $f_3 = 20$,

correspond to polyhedra having all vertices of the same kind and all faces of the same kind. This proof of the existence of only five such polyhedra is purely topological, the above equations making no reference to the regularity or otherwise of the faces.† The forms of these polyhedra when the faces are regular polygons are the five regular (Platonic) solids: tetrahedron, cube, pentagonal dodecahedron, octahedron, and icosahedron. The symbol (n, p) or (n^p) describes a polyhedron with p n-gon faces meeting at each vertex. The values of n and p (Table 3.2) shows that there is a special (reciprocal) relationship between certain pairs of these solids, the number of edges being the same but the numbers of faces and vertices are interchanged. The cube is reciprocal to the octahedron, the dodecahedron to the icosahedron; the tetrahedron is clearly reciprocal to itself.

All the regular solids are encountered in structural chemistry, but for reasons already given the tetrahedron and octahedron are of outstanding importance. The cube is the coordination polyhedron in certain simple structures (of both ions in CsCl and of the Ca^{2+} ion in CaF_2), but other arrangements of eight neighbours, which are described later, are more usual in finite molecules and ions and also in complex ionic crystals. The pentagonal dodecahedron, which has 20 vertices, is not of interest as a coordination polyhedron, but certain space-filling combinations of

---

† In addition to the five convex regular solids there are four stellated bodies, produced by extending outwards the faces of the convex solids until they meet.

TABLE 3.2

The regular (Platonic) solids

	n, p	Vertices (N_0)	Edges (N_1)	Faces (N_2)	Dihedral angle
Tetrahedron	3, 3	4	6	4	70° 32′
Octahedron	3, 4	6	} 12	8	109° 28′
Cube	4, 3	8		6	90°
Dodecahedron	5, 3	20	} 30	12	116° 34′
Icosahedron	3, 5	12		20	138° 12′

dodecahedra and related polyhedra are directly related to the structures of a family of hydrates, as described later. Icosahedral coordination is referred to under sphere packings (Chapter 4), it is found in numerous alloy structures and as the arrangement of the 12 oxygen atoms bonded to the metal in, for example, the $Ce(NO_3)_6^{3-}$ ion and of the carbonyl groups in $Fe_3(CO)_{12}$ and $Co_4(CO)_{12}$. There are icosahedral B_{12} groups in elementary boron and certain borides, and the boron skeleton in many boranes is composed of icosahedra or portions of icosahedra.

Semi-regular polyhedra

The regular solids have all vertices equivalent (isogonal) and all faces of the same kind (isohedral). If we retain the first condition but allow regular polygonal faces of more than one kind we find a set of semi-regular polyhedra, which are listed in Table 3.3. The symbols show the types of n-gon faces meeting at each vertex, the index being the number of such faces. These solids comprise three groups. The first

TABLE 3.3

The Archimedean semi-regular solids, prisms, and antiprisms

	Symbol	Name	Number of		
			faces	vertices	edges
1	$3, 6^2$	Truncated tetrahedron	8	12	18
2	$3, 8^2$	Truncated cube	14	24	36
3	$4, 6^2$	Truncated octahedron	14	24	36
4	$3^2 4^2$	Cuboctahedron	14	12	24
5	$4, 6, 8$	Truncated cuboctahedron	26	48	72
6	$3, 4^3$	Rhombicuboctahedron	26	24	48
7	$3^4, 4$	Snub cube	38	24	60
8	$3, 10^2$	Truncated dodecahedron	32	60	90
9	$3^2, 5^2$	Icosidodecahedron	32	30	60
10	$5, 6^2$	Truncated icosahedron	32	60	90
11	$4, 6, 10$	Truncated icosidodecahedron	62	120	180
12	$3, 4, 5, 4$	Rhombicosidodecahedron	62	60	120
13	$3^4, 5$	Snub dodecahedron	92	60	150
14	$n, 4^2$	Regular prisms	$n + 2$	$2n$	$3n$
15	$n, 3^3$	Regular antiprisms	$2n + 2$	$2n$	$4n$

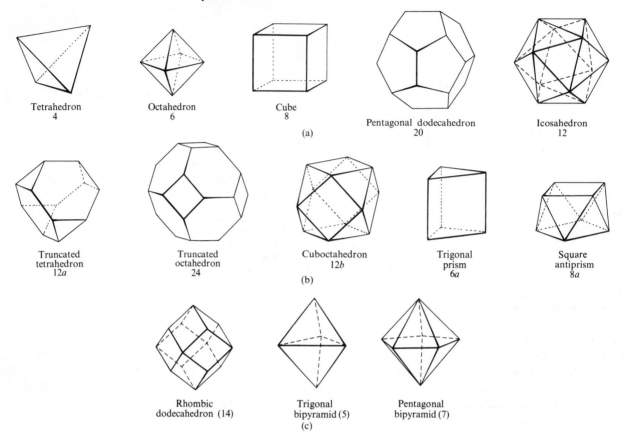

Tetrahedron
4

Octahedron
6

Cube
8

Pentagonal dodecahedron
20

Icosahedron
12

(a)

Truncated
tetrahedron
12*a*

Truncated
octahedron
24

Cuboctahedron
12*b*

Trigonal
prism
6*a*

Square
antiprism
8*a*

(b)

Rhombic
dodecahedron (14)

Trigonal
bipyramid (5)

Pentagonal
bipyramid (7)

(c)

FIG. 3.3. Polyhedra: (a) the regular solids, (b) some Archimedean semi-regular solids, (c) some Catalan semi-regular solids. (The numbers are the numbers of vertices.)

consists of 13 polyhedra (Archimedean solids), derivable from the regular solids by symmetrically shaving off their vertices (a process called truncation), of which only the first five are of importance in crystals. The other two groups are the prisms and antiprisms, both of which have, in their most regular forms, a pair of parallel regular *n*-gon faces at top and bottom and are completed by *n* square faces (regular prisms) or 2*n* equilateral triangular faces (antiprisms). The second prism, in its most symmetrical form, is the cube, and the first antiprism is the octahedron. The regular solids and some of the semi-regular solids are illustrated in Fig. 3.3.

The fact that the number of isogonal bodies is limited to the five regular solids and the semi-regular solids is of considerable importance in chemistry. There are many molecules and complex ions in which five atoms or groups surround a central atom. The fact that it is not possible to distribute five equivalent points uniformly over the surface of a sphere (apart from the trivial case when they form a regular pentagon) is obviously relevant to a discussion of the configuration of molecules or ions AX_5 or more generally of 5-coordination in crystals. Similar considerations apply to 7-, 9-, 10-, and 11-coordination.

64

Corresponding to the semi-regular polyhedra of Table 3.3 there are sets of reciprocal bodies named after Catalan, who first described them all in 1865. Of these we need note only the rhombic dodecahedron, which is the reciprocal of the cuboctahedron, and the family of bipyramids which are related in a similar way to the prisms.

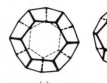

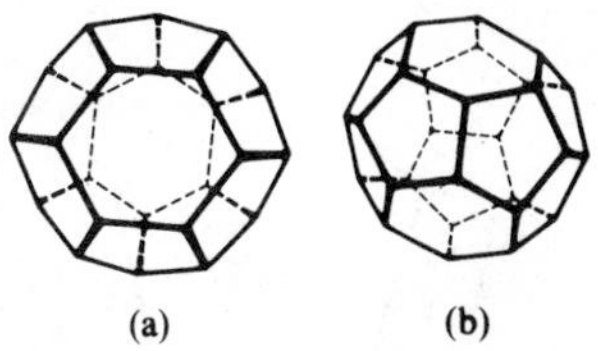

(a)   (b)

FIG. 3.4. (a) The tetrakaideca-hedron: $f_5 = 12$, $f_6 = 2$; (b) the hexakaidecahedron: $f_5 = 12$, $f_6 = 4$.

Polyhedra related to the pentagonal dodecahedron and icosahedron

In equation (1) for 3-connected polyhedra (p. 62) the coefficient of f_6 is zero, suggesting that polyhedra might be formed from simpler 3-connected polyhedra by adding any arbitrary number of 6-gon faces. Although such polyhedra would be consistent with equation (1) it does not follow that it is possible to construct them. The fact that a set of faces is consistent with one of the equations derived from Euler's relation does not necessarily mean that the corresponding convex polyhedron can be made. Three of the Archimedean solids are related in this way to three of the regular solids:

tetrahedron, $f_3 = 4$ truncated tetrahedron, $f_3 = 4$
$$f_6 = 4$$

cube, $f_4 = 6$ truncated octahedron, $f_4 = 6$
$$f_6 = 8$$

and

dodecahedron, $f_5 = 12$ truncated icosahedron, $f_5 = 12$
$$f_6 = 20$$

All the polyhedra intermediate between the dodecahedron and truncated icosa-hedron can be realized except $f_5 = 12$, $f_6 = 1$, and some in more than one form (different arrangements of the 5-gon and 6-gon faces). A number of these polyhedra are of interest in connection with the structures of clathrate hydrates (p. 543), because certain combinations of these solids with dodecahedra form space-filling assemblies in which four edges meet at every vertex. Two of these polyhedra, a tetrakaidecahedron and a hexakaidecahedron, are illustrated in Fig. 3.4.

The reciprocal polyhedra are triangulated polyhedra with twelve 5-connected vertices and two or more 6-connected vertices. Together with the icosahedron they are found as coordination polyhedra in numerous transition-metal alloys.

Some less-regular polyhedra

Other equations may be derived from Euler's relation which are relevant to polyhedra with, for example, a specified number of vertices or faces. The former are required in discussions of the coordination polyhedra possible for a particular number of neighbours; the latter are of interest in space-filling by polyhedra. For 8-coordination we need polyhedra with eight vertices. These must satisfy the equation $\Sigma(n - 2)f_n = 12$, that is,

$$f_3 + 2 f_4 + 3 f_5 + 4 f_6 + 5 f_7 = 12.$$

Solutions include

		N_1
$f_3 = 12$:	triangulated dodecahedron or bisdisphenoid	18
$f_3 = 8, f_4 = 2$:	square antiprism	16
$f_3 = 4, f_4 = 4$:		14
$f_4 = 6$:	cube	12

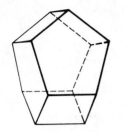

FIG. 3.5. The octahedron: $f_4 = f_5 = 4$.

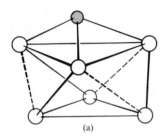

(a)

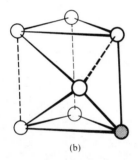

(b)

FIG. 3.6. 7-coordination poly-hedra: (a) monocapped octa-hedron, Ho(ϕCOCHCOϕ)$_3$. H$_2$O, (b) monocapped trigonal prism, Yb(acac)$_3$. H$_2$O. The broken lines show the edges spanned by the chelate ligands; the shaded circles represent H$_2$O molecules.

The triangulated dodecahedron (Fig. 3.7(a)) is the third dodecahedron we have encountered, the others being the pentagonal and rhombic dodecahedra (with 5-gon and 4-gon faces respectively).

On the other hand we may be interested in polyhedra with eight faces (8-hedra or octahedra), which must satisfy the equation

$$v_3 + 2v_4 + 3v_5 + 4v_6 + 5v_7 = 12,$$

where v_p is the number of p-connected vertices.

The regular octahedron has $v_4 = 6$, and in this book the term octahedron will normally refer to this solid. Among numerous other 8-hedra are the truncated tetrahedron, hexagonal prism and the polyhedron $v_3 = 12$ which has $f_4 = f_5 = 4$ (Fig. 3.5) and is the reciprocal of the triangulated dodecahedron mentioned above. This 8-hedron has the property of packing with a particular kind of 17-hedron to fill space, and the vertices of this space-filling arrangement are the positions of the water molecules in the hydrate (CH$_3$)$_3$CNH$_2$. $9\frac{3}{4}$ H$_2$O (p. 544). An 8-hedron of some interest as a 7-coordination polyhedron is the monocapped trigonal prism, illustrated in Fig. 3.6(b). The prefix 'capped' means that there is an atom above the (approximate) centre of a face of the simpler polyhedron. In the case of the trigonal prism (or the antiprism of Table 3.4) the capped face is a square (or rectangular) face; for an octahedron it is necessarily a triangular face.

Table 3.4 summarizes the polyhedra most frequently found as the arrangements of nearest neighbours in finite molecules (or ions) and crystals. The polyhedra listed at the left of the Table have only triangular faces; those to the right have some triangular and some 4-gon faces. Polyhedra on the same horizontal line have the same number of vertices, and are related in the following way. Buckling of a 4-gon face produces two triangular faces, so that the polyhedron on the right is converted into the one on the left. Such relationships are best appreciated by adding additional edges to 'outline' models of the former polyhedra to produce the triangulated polyhedra, and they are important when considering the geometry of the less symmetrical coordination groups. If there is appreciable departure from the most symmetrical form of one of the polyhedra of Table 3.4 (for example, trigonal bipyramid or square pyramid) the description of the coordination polyhedron may become somewhat arbitrary and of dubious value; a precise description of the geometry is then to be preferred.

Because of the outstanding importance of tetrahedral and octahedral coordination we devote the whole of Chapter 5 to systems built from these two polyhedra. The relatively few examples of 5-coordination are described in appropriate places.

66

TABLE 3.4
Polyhedral coordination groups

Vertices	Polyhedron	Faces			Polyhedron
		f_3	f_3	f_4	
4	Tetrahedron (4)	4			
5	Trigonal bipyramid (5)	6	4	1	Square pyramid
6	Octahedron (6)	8	2	3	Trigonal prism (6a)
7	Pentagonal bipyramid (7)	10			
	Monocapped octahedron	10	6	2	Monocapped trigonal prism
8	Dodecahedron	12	8	2	Square antiprism (8a)
			10	1	Bicapped trigonal prism
9	Tricapped trigonal prism	14	12	1	Monocapped antiprism
12	Icosahedron (12)	20	8	6	Cuboctahedron (12b)

Numbers in brackets refer to Fig. 3.3. Other polyhedra with seven, eight, or nine vertices are illustrated in Figs. 3.6, 3.7, and 3.8.

Trigonal prism coordination is very rare in finite complexes (for example, the chelates $M(S_2C_2R_2)_3$ of Mo, W, Re, V, and Cr, p. 940, and the Er chelate listed in Table 3.5) and not common in 3D structures other than NiAs, MoS_2, AlB_2 and related structures, and a few ionic crystals. Mono- or bicapped trigonal prism coordination is also not very common, and the following are examples in chemically related compounds:

monocapped trigonal prism:	EuO . OH	YO . OH	$Na_5Zr_2F_{13}$
bicapped:	$Eu(OH)_2 . H_2O$		$N_2H_6(ZrF_6)$
(tricapped:	$Eu(OH)_3$	$Y(OH)_3$	

We now comment briefly on 7-, 8- and 9-coordination.

7-coordination

Examples of the more well-defined coordination polyhedra include:

	Finite complexes	Infinite systems
Pentagonal bipyramid	K_3ZrF_7 K_3UF_7	$PaCl_5$
m.c. trigonal prism	K_2NbF_7 K_2TaF_7	
m.c. octahedron	chelates (Table 3.5)	A-La_2O_3

With bidentate ligands of the type R . CO . CH . CO . R yttrium and the smaller 4f ions form 7-coordinated complexes in which the seventh ligand is a water molecule. It is convenient to restrict the term monocapped octahedron to groups possessing exact or pseudo 3-fold symmetry with one ligand above one face of the octahedron, this face being somewhat enlarged. The complex $Ho(\phi CO . CH . CO\phi)_3 . H_2O$ has 3-fold symmetry with a propeller-like arrangement of the three rings. The 6 O atoms are situated at the vertices of a distorted octahedron, and the H_2O molecule caps one face (Fig. 3.6(a)). With this structure compare that of $Yb(acac)_3 . H_2O$, Fig. 3.6(b), in which the water molecule is not the capping ligand.

If we include metal–metal bonds there is 7-coordination of Nb in NbI_4 and of W in $K_3W_2Cl_9$, the seventh bond being perpendicular to an edge or face of the octahedral group of halogen atoms. The 7-coordination in $PbCl_2$ (p. 222) may be described as pseudo-9-coordination since there are ligands beyond the three rectangular faces of a trigonal prism, but the distances to two ligands are much greater than those to the other seven. Similarly the coordination of Ca^{2+} in $CaFe_2O_4$ is bicapped trigonal prismatic (8 + 1) rather than 9-coordination. The 7-coordination polyhedron of Zr^{4+} in monoclinic ZrO_2 (p. 449) may be described as related to either a capped trigonal prism or a capped octahedron. (See: AC 1970 **B26** 1129, and for a discussion of 7-coordination, CJC, 1963 **44** 1632.) For references see Table 3.5.

TABLE 3.5

Chelate molecules and ions containing acac *and related ligands*

C.N.	Coordination polyhedron	Complex	Reference
6	Trigonal prism	$Er[(CH_3)_3CCOCHCOC(CH_3)_3]_3$	AC 1971 **B27** 2335
7	Monocapped octahedron	$Ho(\phi COCHCO\phi)_3 . H_2O$	IC 1969 8 2680
	Monocapped octahedron	$Y(\phi COCHCOCH_3)_3 . H_2O$	IC 1968 7 1777
	Monocapped trigonal prism	$Yb(CH_3COCHCOCH_3)_3 . H_2O$	IC 1969 8 22
8	Dodecahedron	$[Y(CF_3COCHCOCF_3)_4]Cs$	IC 1968 7 1770
	Antiprism	$[Y(CH_3COCHCOCH_3)_3(H_2O)_2] . H_2O$	IC 1967 6 499

8-coordination

This is found in many molecules and crystals. Apart from the exceptional cubic 8-coordination in the body-centred cubic, CsCl, and CaF_2 structures (which are discussed in Chapter 7), the coordination polyhedron is usually either the Archimedean antiprism or the triangulated dodecahedron (bisdisphenoid). The bicapped trigonal prism is found in a few crystals, for example, $CaFe_2O_4$, $Sr(OH)_2 . H_2O$, and $N_2H_6(ZrF_6)$; it is closely related to the antiprism, as may be seen by joining the vertices *a* and *b* in Fig. 3.7(b). The dodecahedron of Fig. 3.7(a) is called a bisdisphenoid since it consists of two interpenetrating disphenoids (tetrahedra), one elongated (A) and the other flattened (B). Examples of the two kinds of coordination groups include:

Dodecahedral	Antiprism
$Na_4[Zr(C_2O_4)_4] . 3 H_2O$	$Zr(acac)_4$
K_2ZrF_6	$Na_3(TaF_8)$
$K_4[(Mo(CN)_8] . 2 H_2O$	$H_4[W(CN)_8] . 6 H_2O$
$Ti(NO_3)_4$	Ca^{2+} in $CaNa_2(CO_3)_2 . 2 H_2O$
$Sn(NO_3)_4$	Rb^+ in $RbLiF_2$

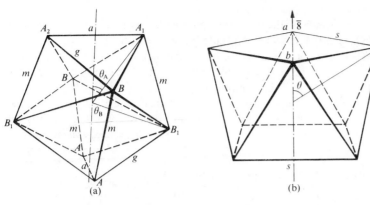

FIG. 3.7. 8-coordination polyhedra: (a) triangulated dodecahedron (bisdisphenoid), (b) antiprism.

Calculations of the ligand–ligand repulsion energies for 8-coordination polyhedra show that both of these polyhedra are more stable than the cube, for all values of n from 1 to 12 in a repulsion law of the type, force proportional to r^{-n}, and that the antiprism is marginally more stable than the dodecahedron. The very small difference in stability is shown by the adoption of the two configurations by chemically similar pairs of complexes such as those shown above and in Table 3.5.

Assuming incompressible atoms ('hard-sphere' model), that is, $a = m$ in Fig. 3.7(a) and $l = s$ in (b), the data for these polyhedra are:

	Radius ratio	A–X
Antiprism: $\theta = 59 \cdot 25°$	0·645	0·823l
Dodecahedron: $\theta_A = 36 \cdot 85°$, $\theta_B = 69 \cdot 46°$	0·668	0·834a

(The radius ratio (p. 261) is the ratio of the radius of the central atom A to that of the eight surrounding (equidistant) X atoms.) The ligand–ligand repulsion energy calculations show that more stable structures correspond to slight distortions of the hard sphere models, with $\theta = 57 \cdot 3°$ and $l : s = 1 \cdot 057$ for the antiprism, and θ_B approximately $72°$ for the dodecahedron, that is, a more nearly coplanar arrangement of the B ligands. The antiprismatic arrangement of eight covalent bonds using $d^4 sp^3$ hybrid orbitals has $\theta = 57 \cdot 6°$ and $l : s = 1 \cdot 049$. Since there is a delicate balance between the factors determining the choice of coordination polyhedron, and since the detailed geometry is dependent on the size and structure of the ligands and also on the interactions with more distant neighbours in the crystal, we shall not pursue this subject further but refer the reader to a number of discussions of 8-coordination: JCP 1950 **18** 746; IC 1963 **2** 235; JCS A 1967 345; IC 1968 **7** 1686. See also the section on the complex fluorides of zirconium on p. 396.

Polyhedra and Nets

9-coordination

Examples of tricapped trigonal prism coordination include the finite aquo complex in $Nd(BrO_3)_3 . 9 H_2O$ and the infinite linear one in $SrCl_2 . 6 H_2O$, (where these groups are stacked in columns in which they share basal faces), and numerous compounds of 4f and 5f elements, for example, YF_3 and compounds with the UCl_3 ($Y(OH)_3$) structure, and the complex fluorides of Th and U (Chapter 28). The less common monocapped antiprism occurs as the arrangement of Te around La atoms in $LaTe_2$ (Fe_2As, C38 structure); its relation to the tricapped trigonal prism may be seen by joining the vertices a and b in Fig. 3.8(a). The 'hard-sphere' model of the

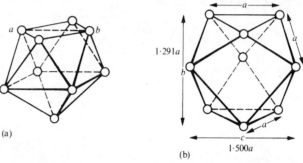

(a)

(b)

FIG. 3.8. 9-coordination polyhedra: (a) monocapped antiprism, (b) tricapped trigonal prism.

tricapped trigonal prism is shown in Fig. 3.8(b); the distance from the centre to the vertices is $0.866a$, corresponding to the radius ratio, 0.732.

We have not included 10-coordination in Table 3.4 because well-defined 10-coordination polyhedra are not common; examples include $U(CH_3COO)_4$, bicapped antiprism (AC 1964 **17** 758), and $La_2(CO_3)_3 . 8 H_2O$ (IC 1968 **7** 1340), where bidentate CO_3 groups occupy the positions B_1 in Fig. 3.7(a), producing two more edges parallel to the bottom edge AA. Further examples of high coordination will be found among complexes containing bidentate NO_3 groups (Chapter 18) and also in Chapter 28.

Plane nets

Derivation of plane nets

The division of an infinite plane into polygons is obviously related to the enumeration of polyhedra, which can be represented as tessellations of polygons on a simple closed surface such as a sphere. In fact we find equations somewhat similar to those for polyhedra except that instead of the *number* f_n of n-gon faces we have ϕ_n as the *fraction* of the polygons which are n-gons, since we are now dealing with an infinite repeating pattern. These equations, which are derived in Appendix 2, MSIC, are:

$$3\text{-connected: } 3\,\phi_3 + 4\,\phi_4 + 5\,\phi_5 + 6\,\phi_6 + 7\,\phi_7 + 8\,\phi_8 + \cdots n\phi_n = 6, \qquad (4)$$
$$4\text{-connected:} \qquad\qquad\qquad\qquad\qquad\qquad\qquad\qquad = 4, \qquad (5)$$
$$5\text{-connected:} \qquad\qquad\qquad\qquad\qquad\qquad\qquad\qquad = 10/3, \qquad (6)$$
$$6\text{-connected:} \qquad\qquad\qquad\qquad\qquad\qquad\qquad\qquad = 3.$$

There are three special solutions corresponding to plane nets in which all the polygons have the same number of edges (and the same number of lines meet at every point), namely:

$$3\text{-connected: } \phi_6 = 1,$$
$$4\text{-connected: } \phi_4 = 1,$$

and

$$6\text{-connected: } \phi_3 = 1.$$

They are illustrated in Fig. 3.9(a). The last is the only plane 6-connected net, and evidently plane nets with more than six lines meeting at every point are not

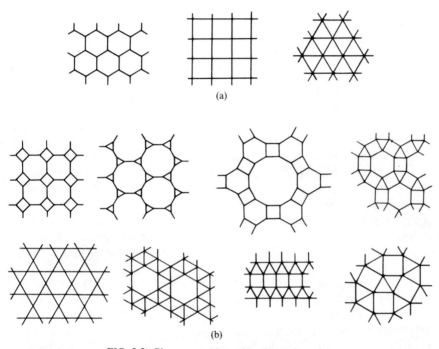

(a)

(b)

FIG. 3.9. Plane nets: (a) regular, (b) semi-regular.

possible. For 3-, 4-, and 5-connected nets there are other solutions corresponding to combinations of polygons of two or more kinds. For example, the next simplest solutions for 3-connected nets are: $\phi_5 = \phi_7 = \frac{1}{2}$, $\phi_4 = \phi_8 = \frac{1}{2}$, and $\phi_3 = \phi_9 = \frac{1}{2}$. The first of these is discussed later in connection with configurations of plane nets; the most symmetrical configuration of the second is illustrated in Fig. 3.9(b) as the first of the semi-regular plane nets. Examples of crystal structures based on these two nets are noted later (p. 93).

Corresponding to the five regular and thirteen semi-regular (Archimedean) solids, all of which have regular polygonal faces, there are three regular plane nets (the special solutions listed above) and eight semi-regular nets in which there are regular polygons of two or more kinds (Fig. 3.9(b)). The reciprocal relations between plane

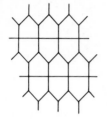

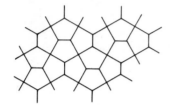

FIG. 3.10. Two plane nets consisting of pentagons.

nets are similar to those between pairs of polyhedra, for the nets (6, 3) and (3, 6) are related in this way while the reciprocal of (4, 4) is the same net (compare the tetrahedron). The reciprocals of the eight semi-regular nets of Fig. 3.9(b) may be drawn by joining the mid-points of adjacent (edge-sharing) polygons; they represent divisions of the plane into congruent polygons—compare the relation of the Catalan to the Archimedean solids. We illustrate (Fig. 3.10) only two of these reciprocal nets, those consisting of congruent pentagons, which are included in Table 3.7 (p. 79). These two nets, which are the reciprocals of the two at the bottom right-hand corner of Fig. 3.9(b), complete the series:

p	n-gons
3	6
3, 4	5
4	4
6	3

but, unlike the other three nets, cannot be realized with regular polygons.

In order to derive the general equation for nets containing both 3- and 4-connected points we must allow for variation in the proportions of the two kinds of points. If the ratio of 3- to 4-connected points is R it is readily shown that in a system of N points the number of links is $N(3R + 4)/2(R + 1)$. Using the same method as for deriving equations (4)–(6) it is found that $\Sigma n\phi_n = 2(3R + 4)/(R + 2)$. The value of $\Sigma n\phi_n$ ranges from 6, when $R = \infty$, to 4, when $R = 0$, and has the special value 5 if the ratio of 3- to 4-connected points is 2 : 1. The special solution of the equation $\Sigma n\phi_n = 5$ is $\phi_5 = 1$, corresponding to the 5-gon nets of Fig. 3.10. Although (3,4)-connected plane nets are not of much interest in structural chemistry the 3D nets of this type form the bases of a number of crystal structures (p. 77).

Plane nets in which points of two kinds (p- and q-connected) alternate are of interest in connection with layer structures of compounds $A_m X_n$ in which the coordination numbers of both A and X are 3 or more. For example, the simple CdI_2 layer may be represented as the plane (3, 6)-connected net. It is shown in Appendix 2 of MSIC that the only plane nets composed of alternate p- and q-connected points are those in which the values of p and q are 3 and 4, 3 and 5, or 3 and 6. The impossibility of constructing a plane net with alternate 4- and 6-connected points implies that a simple layer structure is not possible for a compound $A_2 X_3$ if A is to be 6-coordinated and X 4-coordinated. The non-existence of an octahedral layer structure for a sesquioxide is, therefore, not a matter of crystal chemistry in the sense in which this term is normally understood but receives a very simple topological explanation.

Configurations of plane nets
Two points call for a little further amplification. The equations (4)–(6) are concerned only with the *proportions* of polygons of different kinds and not at all with the *arrangement* of the polygons relative one to another. Moreover, all nets

72

have been illustrated as repeating patterns. For the special solutions $\phi_n = 1$ there is no question of different arrangements of the polygons since each net is entirely composed of polygons of only one kind. However, the *shape* of the hexagons in the net $\phi_6 = 1$ determines whether the net is a repeating pattern and if it is, the size of the unit cell. Figure 3.11 shows portions of periodic forms of this net; clearly a net in which the hexagons are distorted in a random way has no periodicity, or alternatively, it has an indefinitely large repeat unit (unit cell). The situation is more complex if the net contains polygons of more than one kind. Consider first the net $\phi_4 = \phi_8 = \frac{1}{2}$, which is one of the three solutions of equation (4) corresponding to nets consisting of equal numbers of polygons of only two kinds. If the 4-gons and 8-gons are regular there is a unique form of the net, that shown in Fig. 3.9(b). Let us drop the requirement that the polygons are regular but retain the same relative arrangement of 4-gons and 8-gons, that is, each 4-gon shares its edges with four 8-gons and each 8-gon shares alternate edges with 4-gons and 8-gons. If we make alternate 4-gons of different sizes (Fig. 3.12(a)) the content of the unit cell is doubled, to $Z = 8$. If we proceed a stage further, that is, we do not insist on the same relative arrangement of 4-gons and 8-gons but regard the net simply as a 3-connected system of equal numbers of the two kinds of polygon, then we find an indefinitely large number of nets (in which there is edge-sharing between 4-gons). An example is shown in Fig 3.12(b).

The nets $\phi_5 = \phi_7 = \frac{1}{2}$ and $\phi_3 = \phi_9 = \frac{1}{2}$ do not have configurations with regular polygons of both kinds, but here again there is an indefinitely large number of ways of arranging equal numbers of polygons of the two types. Four of the simplest ways of arranging equal numbers of 5-gons and 7-gons are shown in Fig. 3.13. Two of these, (a) and (d), are closely related, being built of the same sub-units, the strip *A* and its mirror-image *B*. It is of some interest that the form of this net adopted by ScB_2C_2 is not the simplest configuration (Fig. 3.13(a)) with eight points in the repeat unit but that of Fig. 3.13(d) with $Z = 16$.

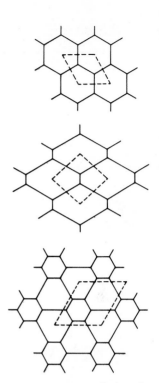

FIG. 3.11. Forms of the plane 6-gon net.

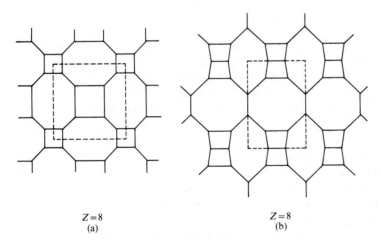

$Z=8$
(a)

$Z=8$
(b)

FIG. 3.12. Less regular forms of the 4:8 plane net (see text).

73

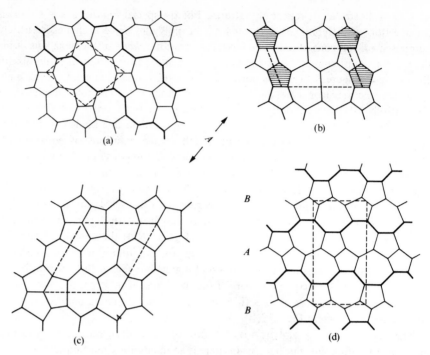

FIG. 3.13. Configurations of the plane net: $\phi_5 = \phi_7 = \frac{1}{2}$.

Three-dimensional nets

Derivation of 3D *nets*

No equations are known analogous to those for polyhedra and plane nets relating to the proportions of polygons (circuits) of different kinds. A different approach is therefore necessary if we wish to derive the basic 3D nets.

Any pattern that repeats regularly in one, two, or three dimensions consists of units that join together when repeated *in the same orientation*, that is, all units are identical and related only by translations. In order to form a 1-, 2-, or 3-dimensional pattern the unit must be capable of linking to two, four, or six others, because a one-dimensional pattern must repeat in both directions along a line, a two-dimensional pattern along two (non-parallel) lines, and a three-dimensional pattern along three (non-coplanar) lines (axes). The repeat unit may be a single point or a group of connected points, and it must have at least two, four, or six free links available for attachment to its neighbours. The requirement of a minimum of four free links for a 2D pattern may at first sight appear incompatible with the existence of 3-connected plane nets, but it can be seen (Fig. 3.11, p. 73) that even in the simplest of these nets ($\phi_6 = 1$) the repeat unit consists of a pair of connected points, this unit having the minimum number (4) of free links.

Evidently the simplest unit that can form a 3D pattern is a single point forming

74

six links, but for 4- and 3-connected 3D nets the units must contain respectively two and four points, as shown in Fig. 3.14. The series is obviously completed by the intermediate unit consisting of one 4- and two 3-connected points, which also has the necessary number (6) of free links. These values of Z, the number of points in the repeat unit, enable us to understand the nature of the simplest 3D nets. Similarly oriented units must be joined together through the free links, each one to six others. This implies that the six free links from each unit must form three pairs,

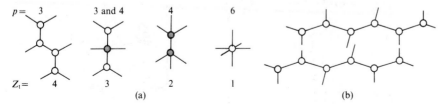

FIG. 3.14. Structural units for three-dimensional nets (see text).

one of each pair pointing in the opposite direction to the other. Identically oriented links repeat at intervals of $(Z + 1)$ points, so that circuits of $2(Z + 1)$ points are formed. We therefore expect to find the family of basic 3D nets listed in Table 3.6, where p is the number of links meeting at each point and n is the number of points in the smallest circuits. Note that the symbols for these nets are of the form (n, p); for example, $(10, 3)$ is a 3-connected net consisting of 10-gons.

TABLE 3.6

The basic 3-dimensional nets

Fig. 3.15	p	n	Z_t	Z_c	x	y
(a)	3	10	4	8	15	10
(b)	3, 4	8	3	6	$13\frac{1}{3}$†	8
(c)	4	6	2	8	12	6
(d)	6	4	1	1	12	4

† Weighted mean.

By analogy with the regular polyhedra and plane nets we might expect to find a set of regular 3D nets which have all their links equal in length and equivalent, all circuits (defined as the shortest paths including any two non-collinear links from any point) identical, and the most symmetrical arrangement of links around every point. Three of the nets of Table 3.6 satisfy all these criteria, namely, a 3-connected net, $(10, 3)$, the diamond net, $(6, 4)$, and the simple cubic framework (primitive cubic lattice), $(4, 6)$. These nets, all of which have cubic symmetry, are illustrated in Fig. 3.15(a), (c), and (d). There is a second 3-connected net $(10, 3)$, also with $Z = 4$ (Fig. 3.15(e)), and a second 4-connected net $(6, 4)$ with a less symmetrical (coplanar) arrangement of bonds from each point (Fig. 3.15(f)). It is convenient to refer to the nets (c), (e), and (f), as the diamond, $ThSi_2$, and NbO nets respectively because of their relation to the structures of these

75

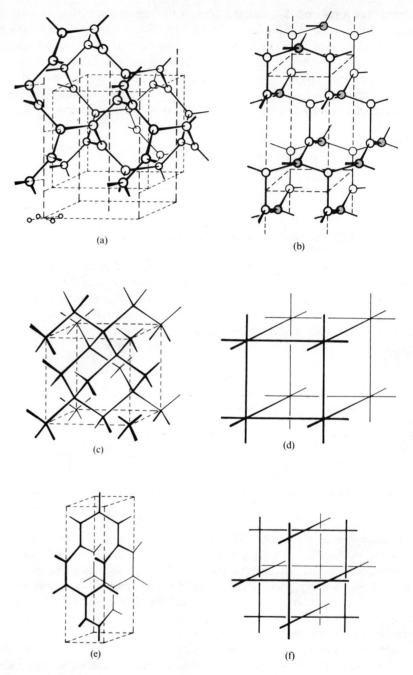

FIG. 3.15. Three-dimensional nets: (a)–(d), the four nets of Table 3.6. (e) $ThSi_2$ net, and (f) NbO net.

substances. The NbO net has cubic symmetry but the highest symmetry of the ThSi$_2$ net, and also of the net (b) is tetragonal.

A word of explanation is needed here concerning the values of Z_t and Z_c in Table 3.6. If the numbers of points in the unit cells of Fig. 3.15(a)–(d) are counted they will be found to be eight, six, eight and one respectively. In all cases except (d), which is a primitive lattice, these numbers Z_c are multiples of the values of Z_t in the Table. The reason for this is that the nets are illustrated in Fig. 3.15 in their most symmetrical configurations, and it is conventional to describe a structure in terms of a unit cell the edges of which are related to the symmetry elements of the structure. Such a unit cell is *usually* larger than the smallest one that could be chosen without relation to the symmetry; it contains Z_c points (atoms). Thus the cubic unit cell of diamond (c), contains 8 atoms, but the structure may also be described in terms of a tetragonal cell containing four atoms or a rhombohedral cell containing two atoms. The (10, 3) net of Fig. 3.30 (p. 96), on which the structure of B$_2$O$_3$ is based, is an example of a net for which the simplest topological unit has $Z_t = 6$ and this is also the value of Z_c for the most symmetrical (trigonal) form of the net.

No example appears to be known yet of a crystal structure based on the simplest 3D (3, 4)-connected net (Fig. 3.15(b)), but two more complex nets of this general type do represent crystal structures. A particularly interesting family of (3, 4)-connected nets includes those in which each 3-connected point is linked only to 4-connected points and vice versa. For such systems Z must be a multiple of 7, and the next two nets illustrated are of this type. Figure 3.16(a) represents the structure of Ge$_3$N$_4$, the open circles being N and the shaded circles Ge atoms. Essentially the same atomic arrangement is found in Be$_2$SiO$_4$ (phenacite), where 2 Be and 1 Si replace the Ge atoms in the nitride. This structure is suitable for atoms forming four tetrahedral bonds (Ge, Be, and Si) or three approximately coplanar bonds (N and O). The net of Fig. 3.16(b), on the other hand, is suited to a 4-connected atom forming four coplanar bonds (shaded circles), and represents the framework of Pt

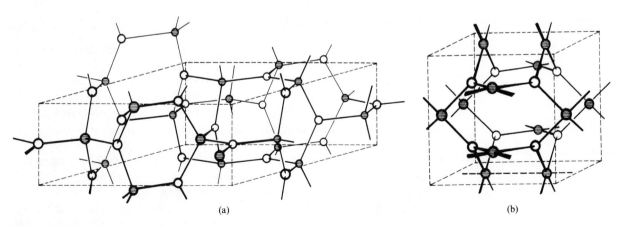

FIG. 3.16. Two (3, 4)-connected 3D nets representing (a) the structure of Ge$_3$N$_4$, (b) the arrangement of Pt and O atoms in Na$_x$Pt$_3$O$_4$ (shaded and open circles respectively).

77

and O atoms in $Na_xPt_3O_4$, a compound formed by oxidizing Pt wire in the presence of sodium. The Na atoms are situated in the interstices and are omitted from Fig. 3.16(b).

Uniform Nets. Certain nets have the property that the shortest path, starting from any point along any link and returning to that point along any other link, is a circuit of n points. Such nets may be called *uniform* nets. This condition is satisfied by the first three nets of Table 3.6, but it is not applicable in this simple form to nets with $p > 4$. However, it may be applied to 6-connected nets if we exclude circuits involving a pair of collinear bonds at any point. We may then include among uniform nets the (4, 6) net (primitive lattice) and also the pyrites structure as a net $(5, \frac{4}{6})$ containing 4-connected points (S atoms) and 6-connected points (Fe atoms) and consisting of 5-gon circuits. It is not yet known how many uniform nets there are, but uniform nets of all the following types have been derived and illustrated:

p	n			
3	12, 10, 9, 8, 7			
3, 4	9, 8, 7			
4		6		
4, 6			5	
6				4

The derivation and properties of certain families of 3D nets are described in papers by the author (AC 1972 **B28** 711 and earlier papers). The upper limit of n in 3-connected nets appears to be 12; the net earlier described as $(12^2, 14)$ (AC 1954 **7** 535) is in fact a uniform net (12, 3). Since no examples of this net are known it is not illustrated here or included in Table 3.7.

Further characterization of 3D nets

In addition to the two (10, 3) nets of Fig. 3.15(a) and (e) and the third (10, 3) net mentioned above there are also more complex 3-connected nets consisting of 10-gon circuits. These nets are not interconvertible without breaking and rejoining links, and therefore represent different ways of joining 3-connected points into 3D nets consisting of decagons. Evidently the symbol (n, p) is not adequate for distinguishing such nets which differ in *topological symmetry*. Unlike the crystallographic symmetry the topological symmetry does not involve reference to metrical properties of the nets, but only to the way in which the various polygonal circuits are related one to another. Two quantities that may be used as a measure of the topological symmetry of nets are x, the number of n-gons (here 10-gons) to which each *point* belongs, and y, the number of n-gons to which each *link* belongs. For a plane net x is equal to p, the connectedness, and y is always 2. In 3D nets x and y can attain quite high values, and for the very symmetrical nets of Table 3.6 y is equal to n. For these (regular) nets x and y are related: $x = py/2$, an expression which holds for the (3, 4)-connected net if weighted mean values of x and p are used. In the second 3-connected net ($ThSi_2$) there are two kinds of non-equivalent link, and the weighted mean value of y is $6\frac{2}{3}$, as compared with 10 in the regular (10, 3) net. The decagon net of B_2O_3 forms the third member of the series of 3-connected 10-gon nets, as shown by the values of x:

Z_t	Maximum symmetry	x
4	Cubic	15
4	Tetragonal	10
6	Hexagonal	5

For the second 4-connected net (6, 4), the NbO net, $y = 4$. The values of x and y are not of interest from the chemical standpoint, but it is recommended in MSIC that they be determined, if only to ensure that the models are examined carefully and not merely assembled and dismantled.

The basic systems of connected points, polyhedra, plane, and 3D nets are summarized in Table 3.7, which shows that the four plane and four 3D nets form series with $n = 6, 5, 4$, and 3, and with $n = 10, 8, 6$, and 4 respectively. All systems on the same horizontal line are composed of n-gon circuits, and all those in a

TABLE 3.7

Relation between polyhedra, plane, and 3-dimensional nets

n \ p	3	3 and 4	4	6
3	Tetrahedron	Trigonal bipyramid	Octahedron	Plane (3, 6)
4	Cubic	Rhombic dodecahedron	Plane (4, 4)	*P* lattice ①
5	Pentagonal dodecahedron	Plane (5, $\frac{3}{4}$)	R_4	
6	Plane (6, 3)	$R_{3,4}$	Diamond ②	
7	(7, 3)	(7, $\frac{3}{4}$)	(Value of Z_t)	
8	(8, 3)	(8, $\frac{3}{4}$) ③		
9	(9, 3) Z_t	(9, $\frac{3}{4}$)		
10	Cubic (10, 3) ④			

vertical column have the same value(s) of p. In the first column the 3-connected systems start with three of the regular solids, with $n = 3, 4,$ and 5, and the series continues through the plane net ($n = 6$) into the 3D nets with $n = 7, 8, 9,$ and 10. We shall not have occasion to refer to the nets with $n = 7, 8,$ or 9, since no examples of crystal structures based on these nets are known. In the second column the (3, 4)-connected 8-gon net (Fig. 3.15(b)) may be regarded as intermediate between the diamond (6, 4), and 3-connected (10, 3), nets. There are also (3, 4)-connected nets composed of 7-gons and 9-gons; these also are not illustrated since they are not known to occur in crystals. The remaining net $R_{3, 4}$ and also the net R_4 in the next column are apparently not realizable as periodic 3D nets but as radiating 3D nets built of 6-gons and 5-gons respectively. Although they are not of interest in relation to crystals, systems of this kind may well be relevant to the structures of partially ordered phases such as glasses and polymers. A short note on these is included in Appendix 3 of MSIC.

The fifth regular solid, the icosahedron, and other 5-connected systems are omitted from Table 3.7 because there are no 5-connected plane or 3D nets composed of polygons of one kind only. For coordination numbers greater than six only 3D nets are possible. Two aspects of 3D nets which are important in structural chemistry should be mentioned here; they will be discussed in more detail later.

Nets with polyhedral cavities

In certain 3D nets there are well-defined polyhedral cavities, and the links of the net may alternatively be described as the edges of a space-filling assembly of polyhedra. At least four links must meet at every point of such a net, and the most important nets of this kind are, in fact, 4-connected nets. Space-filling arrangements of polyhedra leading to such nets are therefore described after we have dealt with the simpler 4-connected nets.

Interpenetrating nets

We have supposed that in any system all the points together form one connected net, that is, it is possible to travel along links from any point to any other. There are some very interesting structures in which this is not possible, namely, those consisting of two or more interpenetrating (interlocking) structures (Fig. 3.17).

The simplest of these is a pair of linked n-rings, isomeric with a single $2n$-ring. It is likely that some molecules of this type are formed in many ring-closure reactions in which large rings are formed. No examples of intertwined linear systems are known in the inorganic field, but an example of two 'interwoven' layers is found in crystalline silver tricyanomethanide, $Ag[C(CN)_3]$ (p. 90). Several examples of crystals built of two or more identical interpenetrating 3D nets will be mentioned in connection with the diamond structure (p. 107).

We shall now give examples of molecular and crystal structures based on 2-, 3-, and 4-connected systems. Although logically the cyclic and chain systems (corresponding to $p = 2$) should precede the polyhedral ones ($p \geqslant 3$) we shall deal with the latter first so that we proceed from finite to infinite groups of atoms. This kind of treatment cuts right across the chemical classification of molecules and

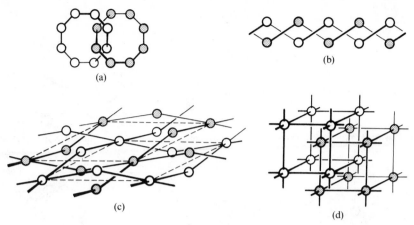

FIG. 3.17. Interpenetrating systems: (a) finite, (b)–(d), 1-, 2-, and 3-dimensional.

crystals and also is not concerned with details of structure, that is, it is topological rather than geometrical. It shows in a striking way how a small number of very simple patterns are utilized in the structures of a great variety of elements and compounds. The following are the main sub-divisions:

Polyhedral molecules and ions
Cyclic molecules and ions
Infinite linear molecules and ions
Structures based on 3-connected nets:
    the plane 6-gon net
    3D nets
Structures based on 4-connected nets:
    the plane 4-gon net
    the diamond net
        interpenetrating systems
    more complex nets
Polyhedral frameworks

Polyhedral molecules and ions

Structural studies have now been made of a number of polyhedral molecules and ions, and of these the largest class comprises the tetrahedral complexes.

Tetrahedral complexes

Most of these are of one of the four types shown in Fig. 3.18(a)–(d), the geometry of which is most easily appreciated if the tetrahedron is shown as four vertices of a cube. Disregarding the singly-attached ligands Y, there is first the simple tetrahedron A_4, (a). To this may be added either 4 X atoms situated above the centres of the faces, (b), or 6 X atoms bridging the edges, (c). These X atoms are, of course,

81

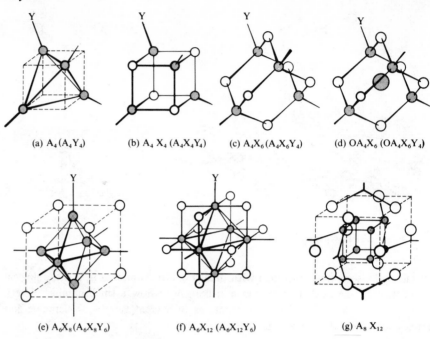

(a) $A_4 (A_4 Y_4)$ (b) $A_4 X_4 (A_4 X_4 Y_4)$ (c) $A_4 X_6 (A_4 X_6 Y_4)$ (d) $OA_4 X_6 (OA_4 X_6 Y_4)$

(e) $A_6 X_8 (A_6 X_8 Y_6)$ (f) $A_6 X_{12} (A_6 X_{12} Y_6)$ (g) $A_8 X_{12}$

FIG. 3.18. Polyhedral molecules and ions: (a)–(d), tetrahedral, (e) and (f), octahedral, (g) cubic.

wholly or largely responsible for holding the group of atoms together. In the most symmetrical form of (b) A and X form a cubic group, but some distortion is to be expected from this configuration, which would imply bond angles of $90°$ at A and X. In (c) the six X atoms delineate an octahedron; for the more familiar representation of the molecules of $P_4 O_6$ and $P_4 O_{10}$ see Fig. 19.7 (p. 685). In class (d) there is an oxygen atom at the centre of the tetrahedron, and in several molecules of this type X is a bidentate chelate group.

To all of the basic types of molecule A_4, $A_4 X_4$, $A_4 X_6$, and $OA_4 X_6$ there is the possibility of adding singly attached Y ligands to each A atom. There is usually one such atom, since A is already bonded to 3 X (in addition to any A–A interactions) and a fourth bond to a Y atom completes a tetrahedral group. Thus in $Cu_4 I_4 (AsEt_3)_4$ X = I and Y = $AsEt_3$. In another tetrameric cuprous molecule, $Cu_4 [S_2 P(i\text{-}C_3 H_7 O)_2]_4$, the chelate ligand behaves in a more complex way. One S atom coordinates to one Cu and the other to two Cu atoms, with the result that each Cu is bonded to 3 S, each $S_2 PR_2$ ligand providing three bonds to Cu as does I in $Cu_4 I_4 (AsEt_3)_4$. Alternatively, in (b), 3 Y atoms can complete an octahedral coordination group around A; this occurs in $Os_4 O_4 (CO)_{12}$ and $Pt_4 I_4 (CH_3)_{12}$ and similar molecules. In $Al_4 N_4 (C_6 H_5)_8$ and in 'cubane', $C_8 H_8$ (which is strictly a cubic rather than a tetrahedral molecule), a ligand ($C_6 H_5$ or H respectively) is attached to each A and to each X atom. In the less symmetrical tetrahedral $Co_4 (CO)_{12}$ molecule there are 3 CO bonded to the apical Co, 3 CO groups bridge the edges of the base, and two further CO are attached to each basal Co atom.

82

Octahedral molecules and ions

Compact groupings built around an octahedron of metal atoms are prominent in the halogen chemistry of the elements Nb, Ta, Mo, and W, where such groups are found as isolated ions or molecules and also as sub-units which are further linked through the Y atoms into layer and 3D structures. The two basic types of unit are shown (idealized) in Fig. 3.18(e) and (f). Examples of the simple A_6X_8 unit are not known, but a number of complexes of the general type $A_6X_8Y_6$ have been studied (Table 3.8). The A_6X_{12} unit of Fig. 3.18(f) represents the structure of the hexameric molecule in one polymorph of $PdCl_2$ and $PtCl_2$ and of the

TABLE 3.8

Polyhedral molecules and ions

Tetrahedral	Fig. 3.18		
	(a)	A_4	P_4, As_4
		A_4Y_4	B_4Cl_4
	(b)	$A_4X_4{}^a$	$Li_4(C_2H_5)_4$, $Tl_4(OAlk)_4$, $Pb_4(OH)_4^{4+}$
		$A_4X_4Y_4$	$Cu_4I_4(AsEt_3)_4$, $Al_4N_4\phi_8$, $Pt_4I_4(CH_3)_{12}$, $Co_4(CO)_{12}$
	(c)	A_4X_6	P_4O_6, As_4O_6, Sb_4O_6, $N_4(CH_2)_6$, $(CH)_4S_6$, $(CH)_4(CH_2)_6$, $(SiR)_4S_6$
		$A_4X_6Y_4$	P_4O_{10}, $P_4O_6[Ni(CO)_3]_4$
	(d)	OA_4X_6	$OBe_4(ac)_6$, $OZn_4(ac)_6$, $OZn_4(S_2PR_2)_6$, $OZn_4(OSPR_2)_6$, $OCo_4[C(CH_3)_3COO]_6$
		$OA_4X_6Y_4{}^b$	$OMg_4Br_6(C_4H_{10}O)_4$, $OCu_4Cl_6(pyr)_4$, $(OCu_4Cl_{10})^{4-}$
Octahedral		A_6Y_6	$B_6H_6^{2-}$
	(e)	A_6X_8	—
		$A_6X_8Y_6$	$Mo_6Cl_{14}^{2-}$, $Mo_6Br_{12}(H_2O)_2$
	(f)	A_6X_{12}	Pd_6Cl_{12}, Pt_6Cl_{12}, $Bi_6(OH)_{12}^{6+}$
		$A_6X_{12}Y_6$	$W_6Cl_{12}Cl_6$, $Nb_6Cl_{18}^{4-}$, $Ta_6Cl_{12}Cl_2(H_2O)_4$
Cubic	(g)	A_8X_{12}	$Cu_8[S_2C_2(CN)_2]_6$

For references see other chapters and also: *a* N 1970 **228** 648; *b* IC 1969 **8** 1982.

$Bi_6(OH)_{12}^{6+}$ ion in hydrolysed solutions of bismuth salts. Addition of a further six halogen atoms to A_6X_{12} gives the structure of the hexameric tungsten trichloride molecule, $W_6Cl_{12}Cl_6$ and the ion $Nb_6Cl_{18}^{4-}$. These complex halogen compounds are described in more detail in Chapter 9. The oxy-ions $M_6O_{19}^{8-}$ formed by V, Nb, and Ta could also be described as related to the A_6X_{18} complex, an additional O atom at the centre completing octahedral groups around the metal atoms. These and other complex oxy-ions of Groups VA and VIA metals are alternatively described as assemblies of MO_6 octahedra, as in Chapter 11.

The molecule of $Rh_6(CO)_{16}$ consists of an octahedral Rh_6 nucleus having 2 CO attached to each metal atom and four more CO situated above four faces of the octahedron, each of these bridging three Rh atoms; these four CO groups are arranged tetrahedrally.

Cubic molecules and ions

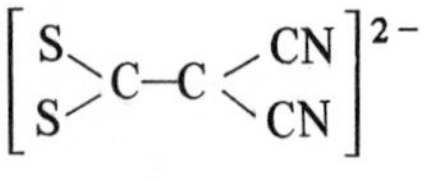

$$\left[\begin{array}{c} S \\ S \end{array} \!\!\! C\!\!-\!\!C \!\!\! \begin{array}{c} CN \\ CN \end{array} \right]^{2-}$$

Figure 3.18(g) shows an A_8X_{12} complex closely related to the A_6X_{12} complex of Fig. 3.18(f). There is the same cuboctahedral arrangement of 12 X atoms but with a cube of A atoms instead of an octahedral group. This structure is adopted by the anion $Cu_8^I L_6^{4-}$ in salts with the large cations $[N(C_6H_5)(CH_3)_3]^+$ or $[As(C_6H_5)_4]^+$, L being the ligand shown at the left. The 12 X atoms are made up of six pairs of S atoms.

Miscellaneous polyhedral complexes

These include a whole series of borohydride ions (p. 873), the B_8Cl_8 molecule (dodecahedral), the Bi_9^{5+} ion (tricapped trigonal prism) in $BiCl_{1\cdot167}$, and the boranes. For these complexes reference should be made to other chapters.

Cyclic molecules and ions

In Table 3.9 we distinguish between homocyclic and heterocyclic rings and give some examples. Homocyclic molecules and ions of which structural studies have been made are not as yet very numerous as compared with rings containing atoms of more than one element. The latter include many cyclic molecules and ions containing B, C, N, and P, the meta-ions (of Si, Ge, and P), the cyclic molecules S_3O_9 and Se_4O_{12}, and numerous cyclic oxyhalides and thiohalides of silicon, siloxanes, silthianes, and silazanes (Chapter 23). Not very much is yet known about the structures of very large rings or of doubly-bridged rings. Examples of the latter include the hydroxy-aquo cation (a) and the hexameric molecules of Ni and Pd mercaptides (b). A few heterocyclic rings contain atoms of more than two different kinds, of which (c) is one example. For details and references to the compounds mentioned other chapters should be consulted.

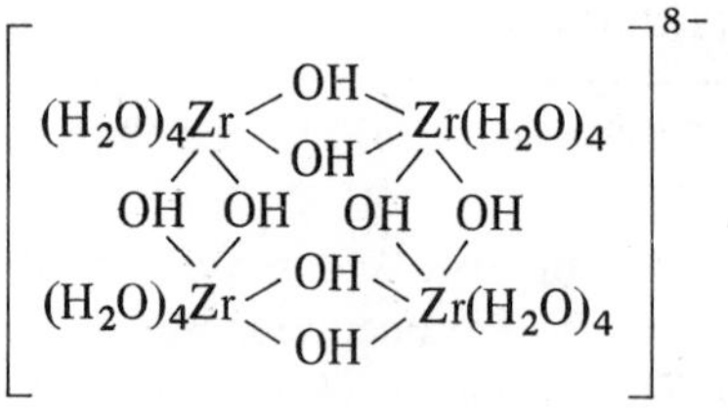

$$\left[\begin{array}{c} (H_2O)_4Zr \begin{array}{c} OH \\ OH \end{array} Zr(H_2O)_4 \\ OH \; OH \quad OH \; OH \\ (H_2O)_4Zr \begin{array}{c} OH \\ OH \end{array} Zr(H_2O)_4 \end{array} \right]^{8-}$$

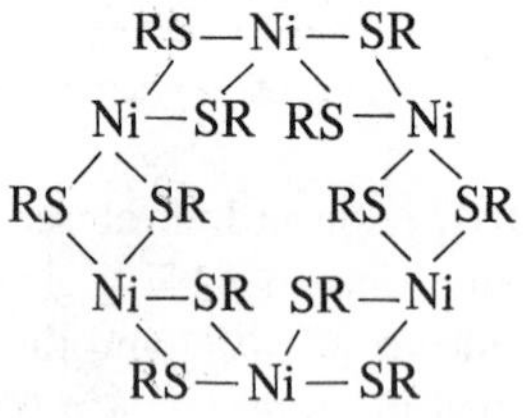

(a)

```
        RS—Ni—SR
       /   \   \
    Ni—SR   RS—Ni
   /   \       /   \
 RS    SR    RS    SR
   \   /       \   /
    Ni—SR   SR—Ni
       \   /   /
        RS—Ni—SR
```

(b)

```
   O       N       O
    \     / \     /
 Cl—S         S—Cl
     |         |
     N         N
      \       /
        P
       / \
     Cl   Cl
```

(c)

84

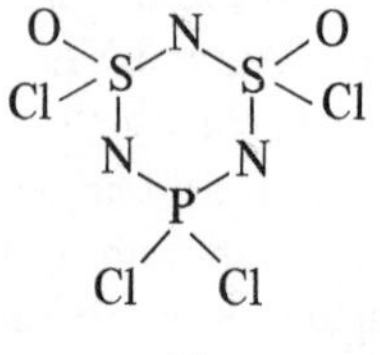

TABLE 3.9

Cyclic molecules and ions

Number of atoms in ring	Homocyclic	Heterocyclic A_nX_n with alternating A and X atoms
3	$Os_3(CO)_{12}$	—
4	$(PCF_3)_4$, $(AsCF_3)_4$, Se_4^{2+}	$(SSiCl_2)_2$
5	$(PCF_3)_5$, $(AsCH_3)_5$	—
6	$(AsC_6H_5)_6$, $(PC_6H_5)_6$, $(P_6O_{12})^{6-}$, S_6, $(Sn\phi_2)_6$	$(Si_3O_9)^{6-}$, $(P_3O_9)^{3-}$, S_3O_9, $B_3N_3H_6$, $(NSCl)_3$, $(PNCl_2)_3$, $Pd_3[S_3(C_2H_4)_2]_3$
8	S_8, S_8^{2+}, Se_8^{2+}	N_4S_4, $(Si_4O_{12})^{8-}$, $(P_4O_{12})^{4-}$, Se_4O_{12}, $(BNCl_2)_4$, $(PNCl_2)_4$, Mo_4F_{20}, $(OGaCH_3)_4$, $[(OH)Au(CH_3)_2]_4$
10		$(PNCl_2)_5$
12	S_{12}	$(Si_6O_{18})^{12-}$, $(P_6O_{18})^{6-}$, $[PN(NMe_2)_2]_6$
16		$[PN(OCH_3)_2]_8$

Infinite linear molecules and ions

Examples of the more important types of chain are given in Table 3.10. In the simplest chain, $-A-A-A-$, A may be a single atom as in plastic sulphur, or it may be a group of atoms, (a), as in the anion in $Ca[B_3O_4(OH)_3] \cdot H_2O$. In all the other chains of Table 3.10 the A atoms are linked through atoms X of a different element. In most of the examples the singly-attached X atoms are of the same kind as the bridging X atoms, but these may be ligands of two different kinds, as in oxyhalides, in a chain such as (b), and in the octahedral chains AX_2L_2 which are mentioned later.

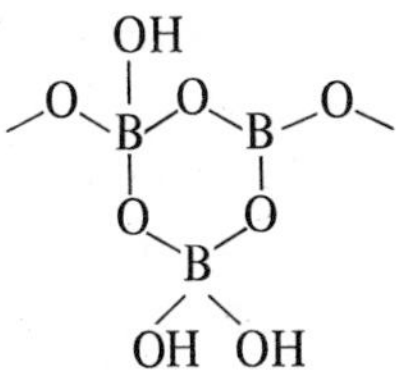

(a)   (b)

The linear molecule of $Zn[S_2P(OEt)_2]_2$, (c), is an example of a chain in which a ligand is behaving both as a bidentate and as a bridging ligand—contrast the chain in $Zn[O_2P\phi(n-C_4H_9)]_2$ described later. Since the chains in class (ii) of the table consist of coordination groups AX_n, which share two X atoms, they may be described as 'vertex-sharing' chains.

85

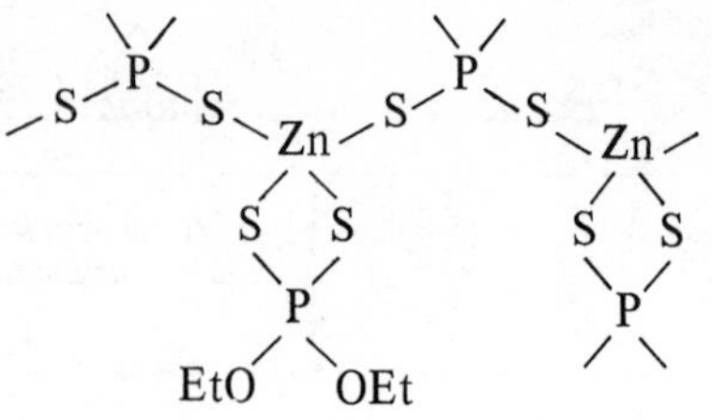

(c)

Chains in class (iii) may similarly be described as 'edge-sharing' chains. This class includes the important chain consisting of octahedral groups sharing two edges. These edges may be opposite (NbI_4) or not ($TcCl_4$), as described in Chapter 5, the

TABLE 3.10

Infinite linear molecules and ions

Type of chain	Formula	Molecules	Ions
(i) $-A-$	A	S (plastic), Se, Te	$[B_3O_4(OH)_3]_n^{2n-}$
(ii) $-A-X-$	AX	AuI, AuCN, HgO, In(C_5H_5), (ϕSeO_2)H	$(HCO_3)_n^{n-}$, $(HSO_4)_n^{n-}$
$\begin{array}{c}-A-X-\\X\end{array}$	AX_2	SeO_2, $Pb(C_5H_5)_2$	$(BO_2)_n^{n-}$, $[Cu(CN)_2]_n^{n-}$, $(AsO_2)_n^{n-}$
$\begin{array}{c}-A-X-\\X_2\end{array}$	AX_3	SO_3, CrO_3, AuF_3, $PNCl_2$, $SiOCl_2$	$(SiO_3)_n^{2n-}$ etc., $(Cu^{I}Cl_3)_n^{2n-}$ $(Cu^{II}Cl_3)_n^{n-}$
$\begin{array}{c}-A-X-\\X_4\end{array}$	AX_5	(*trans*): BiF_5, UF_5 (*cis*): CrF_5 etc., $MoOF_4$	$(AlF_5)_n^{2n-}$, $(PbF_5)_n^{n-}$ (in $SrPbF_6$)
(iii) 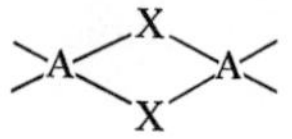	AX_2	(planar): $PdCl_2$, $CuCl_2$ (tetrahedral): $BeCl_2$, SiS_2	
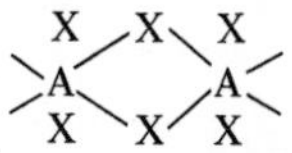	AX_4	NbI_4, $TcCl_4$	$(HgCl_4)_n^{2n-}$
	AX_5	$PaCl_5$	
Similarly	AX_6		$(ZrF_6)_n^{2n-}$ in K_2ZrF_6
	AX_7		$(PaF_7)_n^{2n-}$ in K_2PaF_7
(iv) 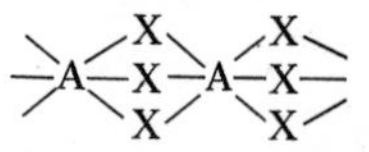	AX_3	ZrI_3 etc.	$(NiCl_3)_n^{n-}$ in $CsNiCl_3$
(v) 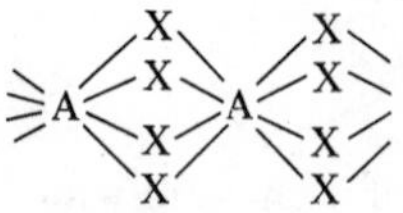	AX_4	U(ac)$_4$	
Hybrid chains		See text	
Multiple chains		Sb_2O_3, $NbOCl_3$	$(Si_4O_{11})_n^{6n-}$, $(CdCl_3)_n^{n-}$, $(Cu_2^{I}Cl_3)_n^{n-}$

86

former type of chain (*trans*) being more usual. As in class (ii) the unshared ligands may be different from the bridging ones, giving the composition AX_4 or AX_2L_2 where L is a ligand capable of forming only one bond to A, as in the example (d). Other examples are listed in Table 25.4 (p. 907).

More complex chains of this kind include the chain molecule $PaCl_5$ (pentagonal bipyramidal $PaCl_7$ groups sharing two equatorial edges), and the anions in K_2ZrF_6 (dodecahedral ZrF_8 groups sharing two edges), and K_2PaF_7 (tricapped trigonal prismatic groups PaF_9, sharing two edges).

In the much smaller classes, (iv) and (v), coordination groups share a pair of faces. For octahedral groups the formula is AX_3, as in ZrI_3 and $[Li(H_2O)_3]ClO_4$, and for tricapped trigonal prismatic groups it is AX_6, as in $[Sr(H_2O)_6]Cl_2$. In these chains triangular faces of coordination groups are shared; in class (v), $U(OOCCH_3)_4$, two opposite quadrilateral faces of antiprisms are shared, the bridges consisting of acetate groups, $-OC(CH_3)O-$.

The term 'hybrid' chain in Table 3.10 means a chain in which there are bridges of more than one kind. Examples include

$$Zn[O_2P\phi(n\text{-}C_4H_9)]_2:$$

$$\text{>Zn-O} \overset{\diagdown\diagup}{\underset{}{P}} \text{O-Zn} \overset{\diagup OPO \diagdown}{\underset{\diagdown OPO \diagup}{}} \text{Zn-}$$

and $ReCl_4$:

$$\text{-Cl-Re-Cl-Re-Cl-Re-Cl-}$$

which are examples of unexpectedly complex structures for compounds with formulae of the simple basic types AX_2 and AX_4. Under the heading 'multiple' chains in Table 3.10 we include the double chains in Sb_2O_3, and $NbOCl_3$, and silicate ions such as $(Si_4O_{11})_n^{6n-}$ which may be illustrated diagrammatically as in Fig. 3.19. An X atom is to be placed along each line joining a pair of A atoms; singly-attached X atoms are omitted. The double tetrahedral chain ion $(Cu_2Cl_3)_n^{n-}$ and the double octahedral ion $(CdCl_3)_n^{n-}$ are illustrated elsewhere.

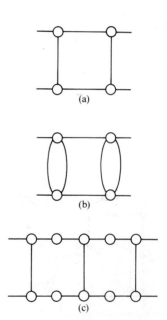

(d)

FIG. 3.19. Diagrammatic representations of the chains: (a) Sb_2O_3, (b), $NbOCl_3$, (c), $(Si_4O_{11})_n^{6n-}$.

Crystal structures based on 3-connected nets

Types of structural unit

The simplest units leading to structures of this kind are illustrated in Fig. 3.20. At (a) we have an atom forming three bonds which could be the structural unit in an element or in compounds in which all atoms are 3-connected. If atoms X are placed along the lines of any 3-connected net, as at (b), the formula is A_2X_3, and this arrangement is found in a number of oxides M_2O_3 and sulphides M_2S_3. At (c) we show a tetrahedral group AX_4 sharing three of its vertices with other similar groups.

Polyhedra and Nets

If this holds throughout the crystal the composition is A_2X_5. The most symmetrical way in which octahedral groups AX_6 can be joined to three others by sharing three edges is shown at (d). Since each X is common to two octahedra the formula is AX_3.

The four structural units (a)–(d) form the bases of the structures of numerous inorganic compounds in which the bonds are covalent or covalent–ionic. It is also convenient to represent diagrammatically, by means of 3-connected nets, the structures of certain crystals in which the structural units (which may be ions or molecules) form three hydrogen bonds. In the hydrates of some acids one H^+ is associated with H_2O to form H_3O^+ which is a unit of type (a). In certain hydroxy-compounds each OH group is involved in two hydrogen bonds, and accordingly a molecule of a dihydroxy-compound may be represented as at (e), it

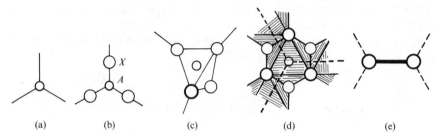

(a)   (b)   (c)   (d)   (e)

FIG. 3.20. Structural units forming 3-connected nets.

being necessary to indicate only the atoms involved in hydrogen bonding. The units of Fig. 3.20 may join together to form finite groups or arrangements extending indefinitely in one, two, or three dimensions. Finite (polyhedral) and infinite linear systems have already been mentioned, the only simple example of a chain being the double chain in the orthorhombic form of Sb_2O_3. The simplest plane 3-connected net is that consisting of hexagons. This is by far the most important plane 3-connected net in structural chemistry and will be dealt with in most detail. Examples will then be given of structures based on more complex plane nets before proceeding to 3D nets.

The plane hexagon net

Examples are known of structures based on this net incorporating all the five types of unit of Fig. 3.20. In its most symmetrical form this net is strictly planar and the hexagons are regular and of the same size. This form represents the structure of a layer of graphite or of B atoms in AlB_2. As the interbond angle decreases from $120°$ (in the plane regular hexagon) so the layer buckles. Crystalline As is built of buckled layers, and elementary Sb and Bi are structurally similar. The structures of these elements may be regarded as distorted versions of a simple cubic structure, in which each atom would have six equidistant neighbours arranged at the vertices of a regular octahedron. The relation between the As structure and the simple cubic structure is illustrated in Fig. 3.21(a). This is diagrammatic in the sense that the atoms are in the positions of the cubic structure (that is, interbond angles are shown as $90°$ and each As has six equidistant neighbours) but each atom is shown

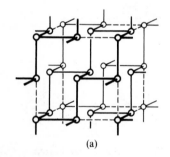

(a)

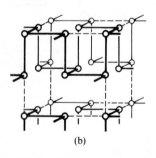

(b)

FIG. 3.21. Relation of the structures of (a) As and (b) black P to the simple cubic structure.

bonded to three others only, forming layers whose mean plane is perpendicular to a body-diagonal of the cubic unit cell. In the actual structure each atom has three close neighbours in its own layer and three more distant ones in the adjacent layer. In CaSi$_2$ the Si atoms form buckled layers very similar to those of As.

Figure 3.21(b) shows a second, more buckled, layer which is also derivable from the simple cubic structure and is idealized in the same way as the layer in (a); this more buckled layer represents the structure of a layer in black P.

If alternate atoms in the 6-gon layer are of different elements the composition is AB (Fig. 3.22), and we find hexagonal BN with plane layers and GeS (and SnS) with buckled layers similar to those in black P.

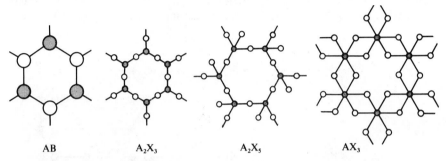

FIG. 3.22. The structures of binary compounds based on the plane 6-gon net.

Orthoboric acid is a trihydroxy-compound, B(OH)$_3$, and in the crystalline state each molecule is hydrogen-bonded to neighbouring molecules by six O–H···O bonds. These are in pairs to three adjacent molecules, an arrangement similar to that in the dimers of carboxylic acids:

and the molecules are arranged in layers based on the 6-gon net as shown in Fig. 24.11 (p. 853).

In the crystalline hydrates of some acids a proton is transferred to the water molecule forming the H$_3$O$^+$ ion, which can form three hydrogen bonds. Accordingly the structures of crystalline HCl . H$_2$O and HNO$_3$. H$_2$O consist of layers in which anions and H$_3$O$^+$ ions alternate in 6-rings, as in the AB layer of Fig. 3.22. In these two monohydrates the number of H atoms (3) is the number required · for each unit to be hydrogen-bonded to three neighbouring ones. A comparison with the structures of HClO$_4$. H$_2$O and H$_2$SO$_4$. H$_2$O illustrates how the structures of these hydrates are determined by the number of H atoms available for hydrogen bonding rather than by the structure of the anion. The low-temperature form of HClO$_4$. H$_2$O has a structure of exactly the same topological type as

89

that of $HNO_3 . H_2O$. As in $HNO_3 . H_2O$ there are three hydrogen bonds connecting each ion to its neighbours, and one O of each ClO_4^- ion is not involved in hydrogen bonding. In the monohydrate of sulphuric acid, on the other hand, there are sufficient H atoms for an average of four hydrogen bonds from each structural unit (H_3O^+ or HSO_4^-), and we find three hydrogen bonds from each H_3O^+ and five from each HSO_4^-; the resulting structure is no longer based on the hexagon net. These structures are described in more detail in Chapter 15.

Trithiane, $S_3(CH_2)_3$ a chair-shaped molecule like cyclohexane, forms many metal complexes. Two silver compounds have layer structures based on the 6-gon net. In both structures Ag is tetrahedrally coordinated, the fourth ligand (X in Fig. 3.23(a)) being H_2O in $Ag(trithiane)ClO_4 . H_2O$ and O of NO_3^- in $Ag(tri-thiane)NO_3 . H_2O$ (JCS A 1968 93). Silver tricyanmethanide, $Ag[C(CN)_3]$, has a

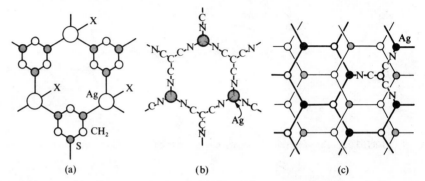

FIG. 3.23. Layers in the structures of (a) $Ag[S_3(CH_2)_3]NO_3 . H_2O$, (b) $Ag[C(CN)_3]$. (c) Two interwoven layers of type (b).

very interesting structure. The ligand $C(CN)_3$ forms with Ag (here forming only three bonds) a layer of the same basic type as in the trithiane complex (Fig. 3.23(b)), but pairs of layers are interwoven, as at (c). This is one of the two examples of interwoven 2D nets at present known, the other being the much more complex multiple layer of red P to which we refer later.

If A atoms at the points of a 6-gon net are joined through X atoms (Fig. 3.22) the result is a layer of composition A_2X_3. Crystalline As_2S_3 (the mineral orpiment) is built of layers of this kind (Fig. 3.24(a)), and one of the forms of the trioxide (claudetite) has a very similar layer structure. (The other form, arsenolite, is built of As_4O_6 molecules with the same type of structure as the P_4O_6 molecule.)

'Acid' (hydrogen) salts provide many interesting examples of hydrogen-bonded systems which are particularly simple if the ratio of H atoms to oxy-anions is that required for a 3-, 4-, or 6-connected net:

Salt	H : anion ratio	Type of net
$NaH_3(SeO_3)_2$	3 : 2	3-connected
KH_2PO_4	2 : 1	4-connected
$(NH_4)_2H_3IO_6$	3 : 1	6-connected

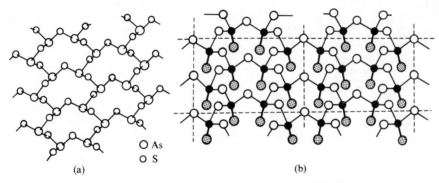

O As

O S

(a)   (b)

FIG. 3.24. Layers in (a) As_2O_3 (orpiment), (b) P_2O_5.

In $NaH_3(SeO_3)_2$ the SeO_3^{2-} ions are situated at the points of the plane 6-gon net and a H atom along each link, giving the required H : anion ratio, 3 : 2.

Layers formed by joining tetrahedral AX_4 groups through three of their vertices have the composition A_2X_5. Such a layer is electrically neutral if formed from PO_4 groups but the similar layer built from SiO_4 groups is a 2D ion, $(Si_2O_5)_n^{2n-}$. Figure 3.24(b) shows a projection of the atoms in a layer of one of the polymorphs of P_2O_5. Although at first sight this layer appears somewhat complex, removal of the shaded circles (the unshared O atoms) leaves a system of P and O atoms very similar to that of As and S in As_2S_3 (Fig. 3.24(a)). The charged layer in $Li_2Si_2O_5$ and other silicates of this kind is of the same topological type but has a very buckled configuration, presumably adjusting itself to accommodate the cations between the layers.

The fourth structural unit, (d) of Fig. 3.20, is an octahedral group AX_6 sharing three edges. The mid-points of these edges are coplanar with the central A atom, so that octahedra linked together in this way form a plane layer based on the 6-gon net. This layer is found in many compounds AX_3, the structures differing in the way in which the layers are superposed, that is, in the type of packing of the X atoms. In the BiI_3 structure there is hexagonal and in YCl_3 cubic closest packing of the halogen atoms, as described in Chapter 4; in $Al(OH)_3$ the more open packing of the layers is due to the formation of O–H$\cdots$O bonds between OH groups of adjacent layers.

We describe in Chapter 5 the formation of composite layers formed by the sharing of the remaining vertices of a tetrahedral A_2X_5 layer with certain of the vertices of an octahedral layer AX_3, both layers being based on the plane 6-gon net. These complex layers are the structural units in two important classes of minerals, the clay minerals (including kaolin, talc, and the bentonites) and the micas. One of these layers is illustrated as an assembly of tetrahedra and octahedra in Fig. 5.44 (p. 191).

A quite unexpected example of the use of the plane 6-gon net is found in crystalline ThI_4. There is 8-coordination of the Th atoms, and each antiprism ThI_8 shares one edge and two faces. In this way each I is bonded to two Th atoms, giving the composition ThI_4, as shown in Fig. 3.25.

91

TABLE 3.11

Layers based on the simple hexagon net

Layer type	Examples
A	C (graphite), As, Sb, Bi; P (black)
	$CaSi_2$, AlB_2
	$B(OH)_3$
AB	BN, GeS, SnS
	$(H_3O)^+Cl^-$, $(H_3O)^+NO_3^-$, $(H_3O)^+ClO_4^-$
	$Ag[S_3(CH_2)_3]ClO_4 . H_2O$, $Ag[S_3(CH_2)_3]NO_3 . H_2O$
	$Ag[C(CN)_3]$ (two interwoven layers)
A_2X_3	As_2O_3, As_2S_3
	$Na[H_3(SeO_3)_2]$
A_2X_5	P_2O_5
	$Li_2(Si_2O_5)$, $Rb(Be_2F_5)$
AX_3	YCl_3, BiI_3, $Al(OH)_3$
AX_4	ThI_4

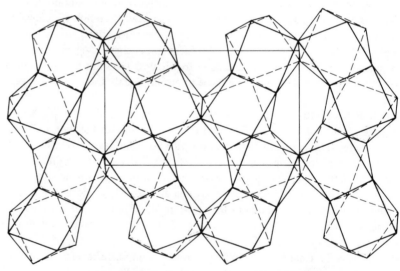

FIG. 3.25. Layer in crystalline ThI_4.

As an example of a unit of type (e) of Fig. 3.20 we illustrate diagrammatically the structure of one of the polymorphs of quinol, *p*-dihydroxybenzene. Each OH group can act as the donor and acceptor end of an O—H···O bond, and the simplest arrangement of molecules of this kind is the plane layer illustrated in Fig. 3.26. The structures we have described are summarized in Table 3.11.

This kind of topological representation of crystal structures may be extended to more complex compounds if we focus our attention on the limited number of stronger bonds that hold the structure together. Nylon is formed by condensing

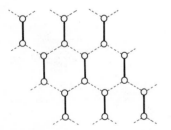

FIG. 3.26. Layer of molecules in crystalline γ-quinol (diagrammatic).

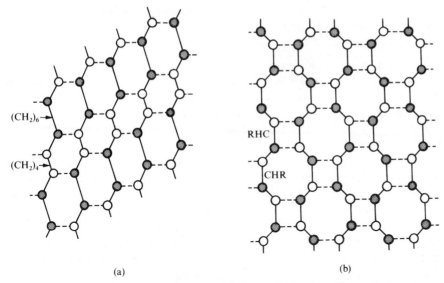

$(CH_2)_6 \rightarrow$

$(CH_2)_4 \rightarrow$

RHC

CHR

(a)   (b)

FIG. 3.27. Hydrogen bonding systems in layers of (a) nylon, (b) caprolactam.

hexamethylene diamine with adipic acid and forms long molecules which may be represented diagrammatically as in Fig. 3.27(a). The molecules are linked into layers by hydrogen bonds between the CO and NH groups of different chains, so that if we are interested primarily in the hydrogen bonding we may show only the CO and NH groups (open and shaded circles respectively) and omit the CH_2 groups. The hydrogen bonds are shown as broken lines. Reduced to this simple form the structure appears as the 6-gon net. The type of 3-connected net depends on the sequence of pairs of CO and pairs of NH groups along each chain molecule. If these groups alternate (–CO–NH–CO–NH–) as in caprolactam, the natural way for the chains to hydrogen-bond together is that shown in Fig. 3.27(b), which is one of the next simplest groups of plane 3-connected nets; the 4 : 8 net.

Structures based on other plane 3-connected nets

A few examples are known of crystal structures based on more complex nets. Borocarbides MB_2C_2 are formed by scandium and the rare-earth metals, and they consist of layers of composition B_2C_2 interleaved with metal atoms. In the compounds of the 4f metals the layer is the 4 : 8 layer, the pattern of atoms being similar to that of Fig. 3.27(b), the open and shaded circles now representing B and C atoms. The scandium compound is of special interest as the only example at present known of the layer consisting of equal numbers of 5-gons and 7-gons.

All plane 3-connected layers are, from the topological standpoint, possible structures for Si_2O_5 layers. The structures of several silicates are based on the 4 : 8 layer, shown as the first of the semi-regular nets in Fig. 3.9(b), for example, $BaFeSi_4O_{10}$ (gillespite) and $CaCuSi_4O_{10}$. The more complex $(Mn_{12}Fe_3Mg)$ $Si_{12}O_{30}(OH, Cl)_{20}$ is based on the 4 : 6 : 12 net illustrated as the third of the semi-regular nets in Fig. 3.9(b). See also Fig. 23.13 (p. 818).

93

Polyhedra and Nets

We mentioned earlier the structure of red phosphorus as a second example of interwoven layers. In this structure each individual layer is itself a multiple layer with a complex structure which is described on p. 675.

Structures based on 3D *3-connected nets*

There are very few examples as yet of simple inorganic compounds with structures based on 3-dimensional 3-connected nets. We might have expected to find examples among the crystalline compounds of boron, an element which forms three coplanar bonds in many simple molecules and ions. The structure of the normal form of B_2O_3 is in fact based on a simple 3-connected net; however, boron is 4-coordinated in many borates, and both triangular and tetrahedral coordination occur in many compounds. The complex crystalline forms of elementary boron are not simple covalent structures but electron-deficient systems of a quite special kind, the structures of which are briefly described in Chapter 24.

There are two 3-connected nets (10, 3) with four points in the simplest unit cell, but if the nets are constructed with equal bonds and interbond angles of 120° they have eight points in their unit cells and cubic and tetragonal symmetry respectively. The cubic net (Fig. 3.28) is clearly the 3-connected analogue of the diamond net. It

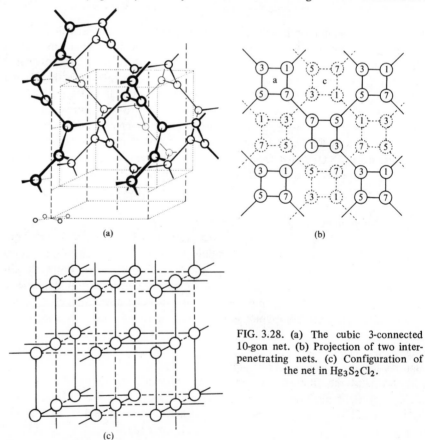

(a)

(b)

(c)

FIG. 3.28. (a) The cubic 3-connected 10-gon net. (b) Projection of two interpenetrating nets. (c) Configuration of the net in $Hg_3S_2Cl_2$.

represents the arrangement of Si atoms in SrSi$_2$ (p. 792). Although the symmetry (space group $P\,4_332$) is lower than that of the most symmetrical configuration (space group $I\,4_13$) with exactly coplanar bonds, it retains cubic symmetry. This net also represents the structure of crystalline H$_2$O$_2$. If the molecules are represented as at (a) and O atoms are placed at the points of the net, then two-thirds of the links represent hydrogen bonds between the molecules. The net is not in the most 'open' configuration of Fig. 3.28(a), which is drawn with equal coplanar bonds from each point, but is in the most compact form consistent with normal van der Waals contacts between non-hydrogen-bonded O atoms and with O—H···O distances of 2·70 Å and HO—OH (intramolecular) equal to 1·47 Å. An interesting property of this net is that it is enantiomorphic; accordingly, crystalline H$_2$O$_2$ is optically active.

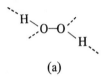

(a)

Examples of more complex 3-connected 3D nets formed by dihydroxy-compounds include the structures of α- and β-resorcinol (*m*-dihydroxybenzene) and of OH . Si(CH$_3$)$_2$. C$_6$H$_4$. Si(CH$_3$)$_2$. OH.

The projection of the cubic (10, 3) net on a face of the cubic unit cell, the full circles and lines of Fig. 3.28(b), shows that the net is built of 4-fold helices which are all anticlockwise upwards. The figures indicate the heights of the points in terms of $c/8$, where c is the length of the cell edge. A second net can be accommodated in the same volume, and if the second net is a mirror-image of the first in no case is the distance between points of different nets as short as the distance between (connected) points within a given net. In the second net of Fig. 3.28(b) (dotted circles and lines) the helices are clockwise. This type of structure, which would be a 3D racemate, is not yet known, but in view of its similarity to the β-quinol structure described later, there is no reason why it should not be adopted by some suitable compound.

We showed in Fig. 3.21 the relation of the layers of As and black P to the simple cubic structure, from which they may be derived by removing one-half of the links. We may derive 3D 3-connected nets in a similar way, by removing different sets of bonds. Figure 3.28(c) shows the cubic (10, 3) net drawn in this way with interbond angles of 90°; this configuration of the net is close to the arrangement of S atoms in one form of Hg$_3$S$_2$Cl$_2$. The Hg atoms are along the links of the net and the Cl ions are accommodated in the interstices of the Hg$_3$S$_2$ framework.

The second (10, 3) net is illustrated in Fig. 3.29(a) as the arrangement of Si atoms in ThSi$_2$, the large Th atoms being accommodated in the interstices of the framework. This net also represents the structure of the third crystalline form of P$_2$O$_5$, in which PO$_4$ tetrahedra are placed at all points of the net and joined by sharing three vertices (O atoms). As might be expected, this polymorph has the highest melting point of the three. It is interesting to find that a single compound, P$_2$O$_5$, has three crystalline forms which illustrate three of the four main types of crystal structure, namely, a finite (in this case polyhedral) group, a layer structure, and a 3D framework structure. A similar system of tetrahedra forms the framework in La$_2$Be$_2$O$_5$, in which the large La^{3+} ions occupy positions which are surrounded by irregular groups of 10 O atoms; contrast La—O, 2·42 Å with the mean Be—O, 1·64 Å, within the tetrahedra. As in the case of the cubic (10, 3) net in Fig. 3.28(c)

95

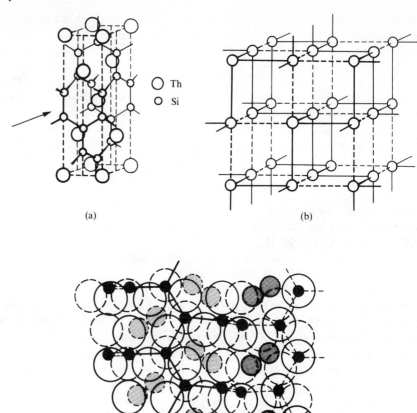

(a)   (b)

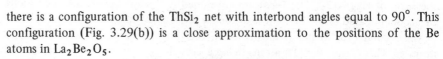

(c)

FIG. 3.29. (a) The tetragonal 3-connected 10-gon net in $ThSi_2$. (b) Configuration of the net in $La_2Be_2O_5$. (c) Part of the same net in $(Zn_2Cl_5)(H_5O_2)$.

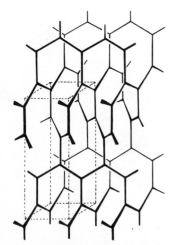

FIG. 3.30. The 10-gon net which is the basis of the structure of B_2O_3.

there is a configuration of the $ThSi_2$ net with interbond angles equal to 90°. This configuration (Fig. 3.29(b)) is a close approximation to the positions of the Be atoms in $La_2Be_2O_5$.

For a rather complex example of a structure in which there are two interpenetrating $ThSi_2$ nets see neptunite (p. 824).

A third 3-connected net, also consisting of 10-gons, forms the basis of the structure of the normal form of B_2O_3. This net has six points in the unit cell; it is illustrated in Fig. 3.30. It is related in a simple way to the $ThSi_2$ net, for both can be constructed from sets of zig-zag chains which are linked to form the 3D net. In the tetragonal net these sets are related by rotations through 90° along the direction of the c axis, whereas in the trigonal net successive sets of chains are rotated through 120°.

Of the indefinitely large number of more complex nets, we illustrate in Fig.

3.31 one consisting of 6-gons and 10-gons. This net is readily visualized in relation to a cube (or, more generally, a rhombohedron) with one body-diagonal vertical, the cube being outlined by the broken lines. A plane hexagon is placed with its centre at each vertex and its plane perpendicular to the body-diagonal of the cube. Each hexagon is connected to six others by the slanting lines. There is a large unoccupied volume at the centre of the cube. In fact a second, identical net can interpenetrate the first, displaced by one-half the vertical body-diagonal of the cube, so that a ring A of the second net occupies the position A'. This system of two interpenetrating nets forms the basis of the structure of β-quinol, in which the

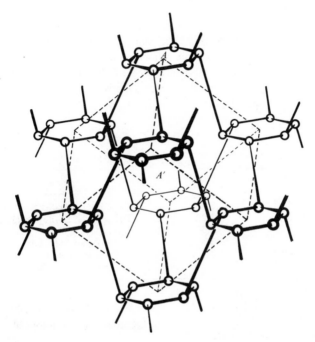

FIG. 3.31. 3D net of hydrogen-bonded molecules in β-quinol.

long slanting lines in Fig. 3.31 represent the molecules of $C_6H_4(OH)_2$ and the circles the terminal OH groups which are hydrogen-bonded into hexagonal rings. This structure is not a true polymorph of quinol, as was originally supposed, for it forms only in the presence of foreign atoms such as argon or molecules such as SO_2 or CH_3OH. These act as 'spacers' between rings such as A and A'; the structure is illustrated in Fig. 1.9 (p. 28) as an example of a 'clathrate' compound.

A *single* net of the type of Fig. 3.31 forms the framework of crystals of $N_4(CH_2)_6 \cdot 6\,H_2O$. The hexagons represent rings of six water molecules, and the $N_4(CH_2)_6$ molecules are suspended from three water molecules belonging to the eight hexagonal rings surrounding each large cavity, that is, they occupy the positions occupied by the rings of the second framework in β-quinol. This structure is illustrated in Fig. 15.8 (p. 547).

The obvious sequel to describing the topology of structures in terms of the basic 3-connected (or other) net is to enquire why a particular net is chosen from the indefinitely large number available. The clathrate β-quinol structure forms only in the presence of a certain minimum number of foreign atoms or molecules which must be of a size suitable for holding together the two frameworks. The choice of particular 3D framework in $N_4(CH_2)_6 . 6H_2O$ is obviously determined by the dimensions and hydrogen bonding requirements of the solute molecule. The growth of the crystal (which is synonymous with the formation of the hydrate) consists in the building of a suitable framework around the solute molecules which is compatible with the hydrogen bonding requirements of the H_2O molecules. Similarly, in the crystallization of $La_2Be_2O_5$ the Be_2O_5 framework must build up around the La^{3+} ions, this packing being the factor determining the choice of the particular net, since many 10-gon (or other) 3-connected nets could be built from tetrahedra sharing three vertices. The choice of the very simple 10-gon nets by H_2O_2, P_2O_5, and B_2O_3 is presumably also a matter of packing, that is, of forming a reasonably efficient packing of all the O atoms in the structure.

TABLE 3.12

Structures based on 3D 3-*connected nets*

Net	Fig.	Example
Cubic (10, 3)	3.28	$SrSi_2$, H_2O_2, $Hg_3S_2Cl_2$, $CsBe_2F_5$
Tetragonal (10, 3)	3.29	$ThSi_2$, P_2O_5, $La_2Be_2O_5$, $(Zn_2Cl_5)(H_5O_2)$
Trigonal (10, 3)	3.30	B_2O_3
	3.31	$N_4(CH_2)_6 . 6 H_2O$, β-quinol
More complex nets	—	α- and β- resorcinol
		$HO . Si(CH_3)_2 . C_6H_4 . Si(CH_3)_2OH$

Although we do not consider the closest packing of spheres until Chapter 4 it is convenient to illustrate at this point the synthesis of topological and packing requirements by a further example of a structure based on the $ThSi_2$ net of Fig. 3.15(e). From an acid aqueous solution of $ZnCl_2$ it is possible to crystallize a hydrate with the empirical composition $ZnCl_2 . \frac{1}{2} HCl . H_2O$. The crystals consist of a framework of $ZnCl_4$ tetrahedra, of the composition $(Zn_2Cl_5)_n^{n-}$, topologically similar to P_2O_5 or $(Be_2O_5)_n^{6n-}$ in $La_2Be_2O_5$. The framework encloses pairs of hydrogen-bonded water molecules, $(H_5O_2)^+$, so that the structural formula is $(Zn_2Cl_5)^-(H_5O_2)^+$. The Cl atoms are arranged in a distorted hexagonal closest packing, the distortion arising from the need to incorporate the $(H_5O_2)^+$ ions. Figure 3.29(c) illustrates the structure as close-packed layers, and shows that the growth of the crystalline hydrate involves the construction of a suitable (Zn_2Cl_5) framework around the aquo-ions; it also shows how the $ThSi_2$ net can be achieved for a compound A_2X_5 at the same time as closest packing of the X atoms, the structure now being viewed along the direction of the arrow in Fig. 3.29(a).

Table 3.12 summarizes the examples we have described of structures based on 3D 3-connected nets.

Crystal structures based on 4-connected nets

Types of structural unit

Some units that can form 4-connected nets are shown in Fig. 3.32, and of these t
simplest is a single atom capable of forming four bonds. There is no crystalli
element in which the atoms form four coplanar bonds, but we have noted that t
simplest 3D net in which each point is coplanar with its four neighbours (F
3.15(f)) represents the structure of NbO. More complex nets in which some of t

FIG. 3.32. Structural units forming 4-connected nets.

TABLE 3.13
Structures based on 4-connected nets

Plane 4-gon net

A *layers*
  $SO_2(OH)_2$, $SeO_2(OH)_2$, $[B_3O_5(OH)]Ca$, $[B_6O_9(OH)_2]Sr . 3 H_2O$

AX *layers* (4 : 4)
  PbO, LiOH, $Pd(S_2)$

AX_2 *layers* (4 : 2)
  HgI_2 (red), γ-$ZnCl_2$, $(ZnO_2)Sr$, $Zn(S_2COEt)_2$, $Cu(CN)(N_2H_4)$,

AX_2 *layers with additional ligands attached to* A
  AX_2Y: $[Ni(CN)_2 . NH_3]C_6H_6$
  AX_2Y_2: $SnF_2(CH_3)_2$, $UO_2(OH)_2$
  AX_4: SnF_4, $(NiF_4)K_2$, $(AlF_4)Tl$

3-dimensional nets

Tetrahedral	*Diamond* (see Table 3.14)	
	More complex nets	*Derived 4 : 2 structures*
	ZnS (wurtzite)	SiO_2 (tridymite), ice-I
	Ge (high pressure)	SiO_2 (keatite), ice-III
	β-BeO	
	SiC polytypes	
		GeS_2 β-$ZnCl_2$
		Other forms of SiO_2 and H_2O
		Aluminosilicates

Planar   NbO

Planar and
  tetrahedral   PtS, PdO, PdS, CuO; PdP_2

Tetrahedral
  Polyhedral frameworks
  Aluminosilicates
  Clathrate hydrates

points have four coplanar and the remainder four tetrahedral neighbours represent the structures of PtS and PdP$_2$. The simplest 3D 4-connected net in which all points have four tetrahedral neighbours is the diamond net, with which we shall deal in more detail shortly. Other units we may expect to find in 4-connected systems include tetrahedral AX$_4$ groups sharing all vertices (X atoms), giving the composition AX$_2$, octahedral AX$_6$ groups sharing four equatorial vertices to form a layer of composition AX$_4$, and molecules such as H$_2$O and H$_2$SO$_4$ (or O$_2$S(OH)$_2$) which have sufficient H atoms to form four hydrogen bonds to adjacent molecules.

Table 3.13 includes examples of structures based on 2D and 3D 4-connected nets. Only the simplest of the plane 4-connected nets is of importance in crystals as the basis of layer structures; in its most symmetrical configuration it is one of the three regular plane nets.

Structures based on the plane 4-gon net
The simplest types of layer based on this net are shown in Fig. 3.33:

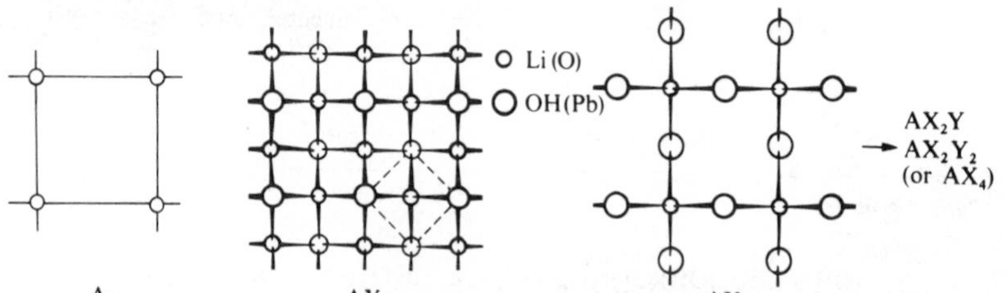

○ Li (O)
◯ OH(Pb)

AX$_2$Y
→ AX$_2$Y$_2$
(or AX$_4$)

A   AX   AX$_2$

A: all units of the same kind
AX: alternate units A and X, both 4-connected
AX$_2$: A (4-connected) linked through X (2-connected) which may be a single atom or a more complex ligand. In an AX$_2$ layer all the X atoms are not necessarily of the same kind (for example, Cu(CN)(N$_2$H$_4$)). Further (singly-connected) ligands Y may be attached to A; they may be the same as X (K$_2$NiF$_4$) or they may be different (SnF$_2$(CH$_3$)$_2$).

FIG. 3.33. Layers based on the plane 4-gon net.

Layers of type A. Simple examples include the layer structures of crystalline H$_2$SO$_4$ and H$_2$SeO$_4$ (p. 318) in which each structural unit is hydrogen-bonded to four others; contrast the topologically similar H$_2$PO$_4^-$ ion which in KH$_2$PO$_4$ forms a 3D structure based on the diamond net, to which we refer later. Two borates provide examples of more complex units forming layers of this kind. In CaB$_3$O$_5$(OH) the units of Fig. 3.34(a) are linked together by sharing the O atoms distinguished as shaded circles to form 2D ions; the layers are held together by the Ca^{2+} ions. In SrB$_6$O$_9$(OH)$_2$. 3 H$_2$O the primary structural unit of the anion is the tricyclic unit of Fig. 3.34(b), which is similarly joined to four others by sharing 4 O atoms.

Layers of type AX. Layers in which all A and X are coplanar are not known. Crystalline LiOH is built of the AX layers shown in Fig. 3.33. The small circles represent Li atoms in the plane of the paper and the larger circles OH groups lying in planes above (heavy) and below (light) that of the Li atoms. The layer is

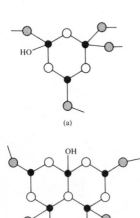

(a)

OH

OH

(b)

FIG. 3.34. Structural units in layers of (a) CaB$_3$O$_5$(OH), (b) SrB$_6$O$_9$(OH)$_2$. 3H$_2$O.

100

alternatively described as consisting of tetrahedral $Li(OH)_4$ groups each sharing four edges which are shown for one tetrahedron as broken lines. If Li is replaced by O and OH by Pb the layer is formed from tetrahedral OPb_4 groups, with Pb atoms on both the outer surfaces of the layer forming four pyramidal bonds to O atoms. This remarkable layer is the structural unit in the red form of PbO; the same structure is adopted by one polymorph of SnO.

A portion of the layer structure of PdS_2 was illustrated in Fig. 1.8 (p. 17); this may be regarded as an AX layer built from the 4-connected units

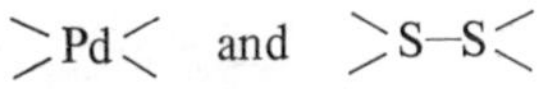

$$>Pd< \quad \text{and} \quad >S–S<$$

Layers of type AX_2. The simplest AX_2 layer based on the square net has A atoms at the points of the net joined through X atoms (2-connected) along the links of the net. The layer in which all A and X atoms are coplanar is not known. $Ni(CN)_2$ and $Pd(CN)_2$ might be expected to form layers of this type but their structures are not known. The clathrate compound $Ni(CN)_2 . NH_3 . C_6H_6$ is built of layers in which Ni atoms are joined through CN groups to form a square net, and NH_3 molecules complete octahedral coordination groups around *alternate* Ni atoms. Owing to the presence of the NH_3 molecules projecting from the layers there are cavities between the layers which can enclose molecules such as C_6H_6. The structure is illustrated in Fig. 1.9(b), p. 28. See also p. 753.

If the rows of X atoms lie alternately above and below the plane of the A atoms, as in Fig. 3.33, the layer consists of tetrahedral AX_4 groups sharing all their vertices. The layers in red HgI_2 (and the isostructural γ-$ZnCl_2$) and in $SrZnO_2$ are illustrated in this way in Fig. 5.6 on p. 162. Other examples of AX_2 layers include zinc ethyl xanthate and cuprous and argentous salts of nitrile complexes $M(nitrile)_2X$ ($X = NO_3, ClO_4$), in which the bridging ligands are (a) and (b), and

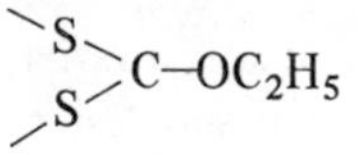

$$\begin{array}{c} {}^{\backslash}S{}_{\searrow} \\ {}_{\diagup}S{}^{\nearrow} \end{array} C–OC_2H_5$$

(a)

$$-NC–(CH_2)_n–CN-$$

(b)

$$\begin{array}{ccc} | & & | \\ -Cu–N–N–Cu-- \\ | & & | \\ C & & N \\ | & & | \\ N & & C \\ | & & | \\ -Cu–N–N–Cu- \\ | & & | \end{array}$$

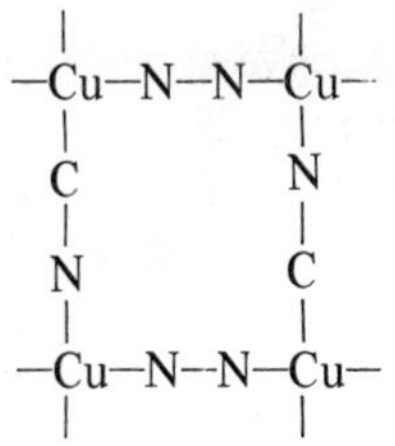

(c)

$CuCN(N_2H_4)$, (c), in which there are two different kinds of ligand X. The nitrile complexes, (b), provide examples of all three types of infinite complex AX_2:

	$n = 2$	$n = 3$	$n = 4$
$Cu(nitrile)_2NO_3$	chain	layer	3 D net

We refer later to the adiponitrile complex ($n = 4$).

101

Polyhedra and Nets

Layers of type AX_4. The plane layer consisting of octahedral groups sharing their four equatorial vertices occurs in a number of crystals. If formed from AX_6 groups the composition is AX_4, and this is the form of the infinite 2D molecules in crystalline SnF_4, PbF_4, and NbF_4, and of the 2D anions in $TlAlF_4$ and in the K_2NiF_4 structure. The latter structure is adopted by numerous complex fluorides and oxides and is illustrated in Fig. 5.16 (p. 172). If the unshared octahedron vertices are occupied by ligands Y different from the shared X atoms the composition is AX_2Y_2, as in the infinite 2D molecules of $SnF_2(CH_3)_2$ and $U(OH)_2O_2$. In $U(OH)_2O_2$ the two short U—O bonds characteristic of the uranyl ion are approximately perpendicular to the plane of the layer, the metal atoms being linked through the equatorial OH groups.

An interesting elaboration of the octahedral layer is found in the 'basic' salt $CuHg(OH)_2(NO_3)_2(H_2O)_2$. Distorted octahedral groups around Cu and Hg alternate in one direction, in which they share a pair of opposite edges, and are joined by sharing vertices (O atoms of NO_3^- ions) to form layers. This is one of the rare examples of the octahedral AX_3 layer of Fig. 5.26 (p. 179). The layers are held

$$
\begin{array}{c}
\text{>Hg}\diagup\!\!\overset{H_2O}{}\!\!\diagdown\text{Cu}\diagup\!\!\overset{OH}{}\!\!\diagdown\text{Hg<} \\[2pt]
\text{|}\quad\diagdown\text{OH}\quad\text{|}\quad H_2O\quad\text{|} \\[4pt]
ONO_2 \qquad ONO_2 \qquad ONO_2 \\[4pt]
\text{>Cu}\diagup\!\!\overset{H_2O}{}\!\!\diagdown\text{Hg}\diagup\!\!\overset{OH}{}\!\!\diagdown\text{Cu<} \\[2pt]
\text{|}\quad\diagdown\text{OH}\quad\text{|}\quad H_2O\diagup\quad\text{|}
\end{array}
$$

together by O—H...O bonds between unshared O atoms of NO_3^- ions and H_2O molecules.

Structures based on the diamond net

The diamond net (Fig. 3.35(a)) is well suited to showing how a very simple structural theme may serve as the basis of the structures of a variety of compounds of increasing complexity. We saw on p. 75 that the simplest three-dimensional framework in which every point is joined to four others is a system of puckered hexagons. Any atom or group which can form bonds to four similar atoms or

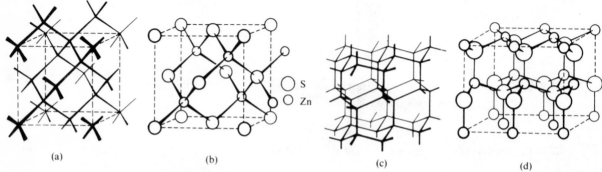

FIG. 3.35. 4-connected nets: (a) diamond, (b) zinc-blende, (c) hexagonal diamond, (d) wurtzite.

102

groups can be linked up in this way provided that the interbond angles permit a tetrahedral rather than coplanar arrangement of nearest neighbours.

The simplest structural unit which can form four tetrahedral bonds is an atom of the fourth Periodic Group, and accordingly we find this net as the structure of diamond, of the forms of silicon and germanium stable at atmospheric pressure, and of grey tin. The diamond-like structure of grey tin is stable at temperatures below $13 \cdot 2°C$, above which transformation into white tin takes place. The structure of white tin, the ordinary form of this metal, is related in an interesting way to that of the grey form. In Fig. 3.36 the grey tin structure is shown referred to a tetragonal unit cell (*a* axes at $45°$ to the axes of the conventional cubic unit cell). Compression along the *c* axis gives the white tin structure. In the diamond structure the six interbond angles at any atom are equal $(109\frac{1}{2}°)$. In white tin two of these angles, distinguished as θ_1 in Fig. 3.36, are enlarged to $149\frac{1}{2}°$ and the other four angles θ_2 are reduced to $94°$. In this process the nearest neighbours of an atom have changed from

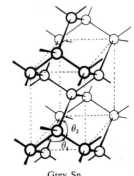

Grey Sn

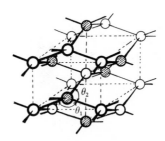

White Sn

FIG. 3.36. The structures of grey and white tin.

4 at $2 \cdot 80$ Å and 12 at $4 \cdot 59$ Å   to   4 at $3 \cdot 02$ Å, 2 at $3 \cdot 18$ Å and 4 at $3 \cdot 77$ Å

so that if we include the next two neighbours at $3 \cdot 18$ Å there are now six neighbours in white tin at approximately the same distance forming a distorted octahedral group. This structural change is remarkable, not only for the 26 per cent increase in density (from $5 \cdot 75$ to $7 \cdot 31$ g/cc) but also because white tin is the *high*-temperature form. A high-temperature polymorph is usually less dense than the form stable at lower temperatures. This anomalous behaviour is due to the fact that an electronic change takes place, the grey form consisting of $Sn(IV)$ atoms and the white form $Sn(II)$ atoms. (If white and grey tin are dissolved in hydrochloric acid the salts crystallizing from the solutions are respectively $SnCl_2 . 2 H_2O$ and $SnCl_4 . 5 H_2O$.) A more normal behaviour is that of Ge or InSb, which crystallize with the diamond structure under atmospheric pressure but can be converted into forms with the white tin structure under high pressure.

The total of eight electrons for the four bonds from each atom in the diamond structure need not be provided as two sets of four (as in diamond itself, elementary silicon, or SiC), but may be derived from pairs of atoms from Groups III and V, II and VI, or I and VII. Binary compounds with the same atomic arrangement as in the diamond structure therefore include not only SiC (carborundum) but also one form of BN and BP and also ZnS (cubic form, zinc-blende), BeS, CdS, HgS (which also has another form, cinnabar, with a quite different structure described on p. 923), cuprous halides, and AgI. The process of replacing atoms of one kind by equal numbers of atoms of two kinds (Zn and S), Fig. 3.35(b), goes a stage further in $CuFeS_2$ (the mineral chalcopyrite, or copper pyrites). One-half of the Zn atoms in zinc-blende are replaced by Cu and the remainder by Fe in a regular manner. Because of the regular replacement the repeat unit of the pattern is now doubled in one direction, since there must be atoms of the same kind at each corner of the unit cell. Further replacement of one-half of the Fe atoms by Sn gives Cu_2FeSnS_4.

Replacement of three-quarters of the metal atoms in ZnS by Cu and one-quarter by Sb gives Cu_3SbS_4. Other compounds have structures related in less simple ways

103

to the zinc-blende structure. For example, in Zn_3AsI_3 only three-quarters of the Zn positions are occupied, and 3 I + 1 As occupy the S positions at random. If only two-thirds of the metal positions of the ZnS structure are occupied the formula becomes $X_{8/3}Y_4$, or X_2Y_3, and this is the structure of one of the (disordered) high-temperature forms of Ga_2S_3. An alternative description of this sesquisulphide structure is that the metal atoms occupy at random two-thirds of the tetrahedral holes in a cubic close-packed assembly of S atoms. A further possibility is that some of the S positions of the ZnS structure are vacant. If one-quarter are unoccupied, as in Cu_3SbS_3 and Cu_3AsS_3, each Sb or As has three pyramidal instead of four tetrahedral S neighbours. These structures are described in more detail in Chapter 17 where a number of them are illustrated, as indicated in Table 3.14.

TABLE 3.14

Structures based on the diamond net

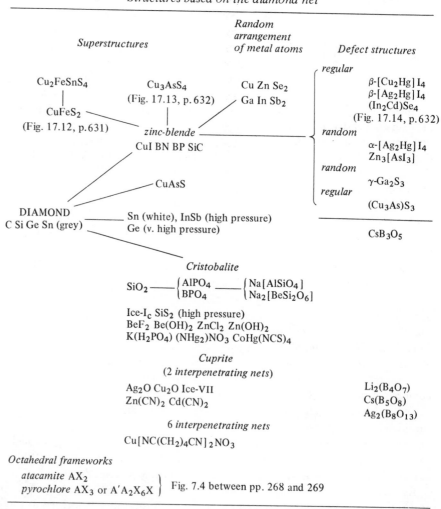

The structures of $CuFeS_2$ and Cu_3SbS_4 are derived from the zinc-blende structure simply by replacing one-half or one-quarter of the metal atoms by atoms of another kind. It is also possible to use the same basic structure for a compound consisting of *equal* numbers of atoms of *three* kinds, as in CuAsS. The unit cell of diamond (or ZnS) contains 8 atoms, but we now require a cell containing $3n$ atoms. The relation of the unit cell of CuAsS to the diamond cell is shown in Fig. 3.37, where it is seen that the dimensions of the CuAsS cell are $a' = 3a/\sqrt{2}$ and $b = a/\sqrt{2}$,

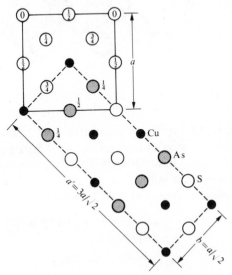

FIG. 3.37. Projection of the structure of CuAsS showing relation to the diamond structure (upper left).

and the third dimension (perpendicular to the plane of the projection) is the same as that of the diamond structure (a). The cell of CuAsS therefore contains twelve atoms, four each of Cu, As, and S.

AX_2 *structures.* The next way in which we may elaborate on the same basic pattern is to place atoms of a second kind along the links of the diamond net. The resulting AX_2 structure may alternatively be described as built of tetrahedral AX_4 groups sharing all their vertices (X atoms) with other similar groups. In order to simplify the diagram (Fig. 3.38(a)) the X atoms are shown lying on and at the mid-points of the links. In fact, in all known structures of this type, where X is O or Cl, the X atoms lie off the direct lines between A atoms because the preferred —X— bond angle is much less than $180°$. The structure of Fig. 3.38(a), but with —O— equal to $147°$, is that of cristobalite, one of the polymorphs of silica. It is also the structure of the high-pressure form of SiS_2, of one form of $ZnCl_2$ and of $Zn(OH)_2$ and $Be(OH)_2$, though for these dihydroxides it is somewhat distorted owing to the formation of hydrogen bonds between hydroxyl groups of different $M(OH)_4$ tetrahedra.

We noted that BP crystallizes with the zinc-blende structure. Replacement of alternate Si atoms in the cristobalite structure by B and P atoms gives the structure of BPO_4; in fact, BPO_4 crystallizes with all the structures of the normal

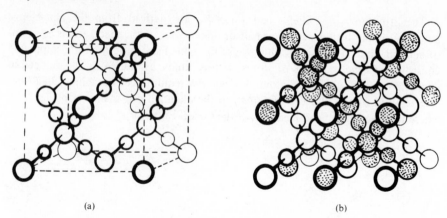

(a)   (b)

FIG. 3.38. (a) Structure of cristobalite (idealized), (b) two interpenetrating frameworks of
type (a) forming the structure of Cu_2O.

(atmospheric pressure) forms of SiO_2. Of the six known structures of crystalline
silica five represent topologically different ways of linking tetrahedral SiO_4 groups
into 3D 4-connected nets; the sixth form, stishovite, has the rutile structure in
which Si is 6-coordinated. Of the polymorphs normally encountered cristobalite is
the form stable over the highest temperature range and is the least dense:

$$\text{quartz} \xrightarrow{\quad 870°C \quad} \text{tridymite} \xleftrightarrow{\quad 1470°C \quad} \text{cristobalite} \xleftrightarrow{\quad 1710°C \text{ (m.p.)}}$$

$$2\cdot65 \qquad\qquad 2\cdot30 \qquad\qquad 2\cdot27 \text{ g/cc (density)}$$

Because of the relatively open structures of cristobalite and tridymite (see below),
both of which occur as minerals, these compounds usually—if not invariably—
contain appreciable concentrations of foreign ions, particularly of the alkali and
alkaline-earth metals. The holes in the structures of the high-temperature forms are
sufficiently large to accommodate these ions without much distortion of the
structure, but their inclusion is possible only if part of the Si is replaced by some
other tetrahedrally coordinated atom such as Al or Be. The framework then
acquires a negative charge which is balanced by those of the included ions. So we
find minerals such as carnegeite, $NaAlSiO_4$, with a cristobalite-like structure, and
kalsilite, $KAlSiO_4$, with a structure closely related to that of tridymite. In
$Na_2BeSi_2O_6$ one-third of the Si in the cristobalite structure is replaced by Be as
compared with the replacement of one-half of the Si by Al in $NaAlSiO_4$.

In the silica structures tetrahedral SiO_4 groups are linked together through their
common oxygen atoms. We noted earlier (Fig. 3.32) that the same topological
possibilities are presented by the H_2O molecule and by a molecule or ion of the
type $O_2M(OH)_2$. There are a number of forms of crystalline H_2O (ice). In addition
to those stable only under higher pressures, which are described on p. 537, there are
two forms of ice stable under atmospheric pressure. Ordinary ice (ice-I_h) has the
tridymite structure, but at temperatures around $-130°C$ water crystallizes with the
cristobalite structure. The tridymite structure is related to the net of Fig. 3.35(c),

which represents the structure of hexagonal diamond, in the same way that cristobalite is related to the cubic diamond net of Fig. 3.35(a). The two AX structures related to the nets of Fig. 3.35(a) and (c), with alternate A and X atoms, are the zinc-blende and wurtzite structures which are shown at (b) and (d). Both H_2O and SiO_2 crystallize with the cubic and hexagonal AX_2 structures:

Net	Fig. 3.35	ZnS	SiO_2	H_2O
Cubic diamond	(a)	zinc-blende	cristobalite	ice-I_c
Hexagonal diamond	(c)	wurtzite	tridymite	ice-I_h

We shall point out later other analogies between hydrate structures and silicates.

We mentioned earlier that crystalline H_2SO_4 has a layer structure, an arrangement of hydrogen-bonded molecules based on the simplest plane 4-

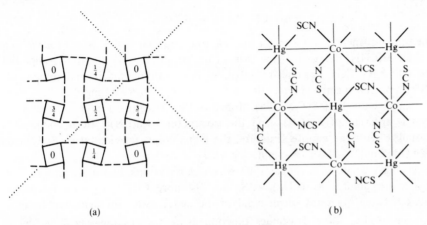

(a)   (b)

FIG. 3.39. Projections of the diamond-like structures of (a) the $PO_2(OH)_2^-$ anion in KH_2PO_4, (b) $CoHg(SCN)_4$.

connected net. In contrast to this layer structure the anion $O_2P(OH)_2^-$ in KH_2PO_4 is a 3D framework of hydrogen-bonded tetrahedral groups arranged at the points of the diamond net. A simplified projection of this anion is shown in Fig. 3.39, where the dotted lines indicate the directions of the cubic axes of the diamond structure. The K^+ ions, which are accommodated in the interstices of the framework, are omitted from Fig. 3.39(a). The structure of $CoHg(SCN)_4$ provides another example of a structure based on the diamond net (Fig. 3.39(b)). There is tetrahedral coordination of all the metal atoms, Co having 4 N and Hg 4 S nearest neighbours.

Structures based on systems of interpenetrating diamond nets. In the structures we have been describing it is possible to trace a path from any atom in the crystal to any other along bonds of the structure, that is, along C–C bonds in diamond, Si–O–Si bonds in cristobalite, etc. In Cu_2O (the mineral cuprite) each Cu atom forms two collinear bonds and each O atom four tetrahedral bonds, and these atoms are linked together in exactly the same way as the O and Si atoms

107

respectively in cristobalite. However, the distance O–Cu–O (3·7 Å) is appreciably greater than Si–O–Si (3·1 Å). As a result the Cu_2O framework has such a low density, or alternatively there is so much unoccupied space, that there is room for a second identical framework within the volume occupied by a single framework. The crystal consists of two identical interpenetrating frameworks which are not connected by any primary (Cu–O) bonds (Fig. 3.38(b)). Since ice-I_c has the cristobalite structure it is not surprising that one of the high-pressure forms of ice has the cuprite structure (see p. 538).

An even greater separation of the 4-connected points of the diamond net results if instead of a single atom X we use a longer connecting unit, that is, a group or molecule which can link at both ends to metal atoms. The CN group acts as a ligand of this kind in many cyanides of the less electropositive metals. The four atoms M–C–N–M are collinear. The simplest possibility is the linear chain in AgCN and AuCN:

$$-Au-CN-Au-CN-Au-CN-$$

The square plane net, for a metal forming four coplanar bonds, is not known in a cyanide $M(CN)_2$, though it is the basis of the structure of the clathrate compound $Ni(CN)_2 . NH_3 . C_6H_6$, as already noted. In Prussian blue and related compounds CN groups link together Fe or other transition-metal atoms arranged at the points of a simple cubic lattice (Fig. 22.5). More directly related to our present topic is the crystal structure of $Zn(CN)_2$ and the isostructural $Cd(CN)_2$. Both compounds crystallize with the cuprite structure, the metal atoms forming tetrahedral bonds (like O in Cu_2O) and –C–N– replacing –Cu–.

A longer ligand of a similar type which can also link two metal atoms is an organic dicyanide, $NC-(CH_2)_n-CN$; one or more CH_2 groups are interposed between two CN groups which bond to the metal atoms through their terminal nitrogen atoms. These dicyanides function as neutral coordinating molecules in salts of the type $[Cu(NC . R . CN)_2]NO_3$. In the glutaronitrile compound, $[Cu\{NC(CH_2)_3CN\}_2]NO_3$, the metal atoms are arranged at the points of a square plane net, but the next member of the series, the adiponitrile compound, $[Cu\{NC(CH_2)_4CN\}_2]NO_3$, has a quite remarkable structure. If metal atoms are placed at the points of the diamond net and joined through adiponitrile molecules

the distance between the metal atoms is approximately 8·8 Å, and the density of a single framework is so low that there is room for no fewer than six identical interpenetrating frameworks of composition Cu(adiponitrile)$_2$ in the same volume. The frameworks are, of course, positively charged, and the NO_3^- ions occupy interstices in the structure. It is of interest to see how this unique structure arises.

Figure 3.40(a) shows the simple diamond net, again referred to the tetragonal unit cell as in Fig. 3.36 (grey tin). The point B has coordinates of the type $(0, \frac{1}{2}, \frac{1}{4})$ and D is of the type $(\frac{1}{2}, 0, \frac{3}{4})$, while C is at the body-centre of the unit cell. It is clear that we can travel along lines of the net from A to B, to C, to D, to E, and in fact to any other point of the three-dimensional network of points. (The point F is the point at which a second net would start if we have two interpenetrating nets as in Cu_2O.) If, however, instead of joining the points in this way we take exactly the same set of points and join them as shown in Fig. 3.40(b), with a vertical component for each link of three-quarters of the height of the cell instead of one-quarter, then we find that the resulting system of lines and points has a remarkable property. The point A is no longer connected to the next point vertically above it (like the point E in (a)), and if in Fig. 3.40(b) we travel from $A \rightarrow B \rightarrow C$ etc. we do not arrive at a point vertically above A *which belongs to the same net as* A until we have travelled vertically the height of *three* unit cells. In other words, the points and lines of Fig. 3.40(b) form not one but three identical and interpenetrating frameworks. If, in addition, we add the points corresponding to the dotted circle (as F in (a)) and join these together in the same way, we should have a system of *six* identical interpenetrating nets. It should also be considered remarkable that this elaborate geometrical system can be so easily formed. It is necessary only to dissolve silver nitrate in warm adiponitrile, add copper powder (when metallic silver is deposited), and cool the solution, when the salt $[Cu\{NC(CH_2)_4CN\}_2]NO_3$ crystallizes out.

Further examples of structures based on two interpenetrating diamond nets are the anions in the borates $Li_2B_4O_7$, CsB_5O_8, and $Ag_2B_8O_{13}$ (Table 3.14) which are described in Chapter 24.

In the structures we have just been describing the interpenetrating nets are identical. We may also envisage structures consisting of two (or more) inter-penetrating nets which have different structures. In the pyrochlore structure (p. 209) adopted by certain complex oxides $A_2B_2O_7$, we may distinguish a 3D framework of octahedral BO_6 groups each of which shares its vertices with six others, giving the composition BO_3 (or B_2O_6). This framework (Fig. 7.4) is stable without the seventh O atom (as in $KSbO_3$) and may be constructed from octahedra arranged tetrahedrally around the points of the diamond net or alternatively from octahedra placed along each link of that net. (Another way of arranging groups of four octahedra tetrahedrally around the points of the diamond net is to make edge-sharing groups, and this gives the atacamite structure, also illustrated in Fig. 7.4.) The structure of $Hg_2Nb_2O_7$ may be described as a Nb_2O_6 framework of this kind through which penetrates a cuprite-like framework of composition Hg_2O with the same structure as one of the two interpenetrating nets of Cu_2O. The Hg(II) atoms have as nearest neighbours 2 O atoms of the Hg_2O framework and at a rather greater distance 6 O of the Nb_2O_6 framework.

Structures based on more complex 4-connected nets

Three classes of more complex 3D 4-connected nets are included in Table 3.13, namely, (i) nets in which the arrangement of bonds from each point is tetrahedral, these nets being suitable for the structures of the same types of elements and

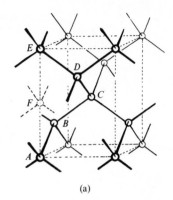

(a)

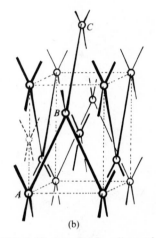

(b)

FIG. 3.40. (a) Single diamond net. (b) System of three interpenetrating diamond-like nets.

compounds as is the diamond net, (ii) nets in which the links from some points are tetrahedral and from others coplanar, these nets being suitable only for compounds $A_m B_n$, and (iii) nets in which there are polyhedral cavities. Inasmuch as most nets of class (iii) are tetrahedral nets these constitute a sub-group of class (i), but because of their special characteristics it is convenient to describe them separately. The sole example of a net in which all atoms form four coplanar bonds (NbO) has been illustrated in Fig. 3.15(f).

(i) *More complex tetrahedral nets.* All 3D 4-connected nets are more complex than the diamond net, which is the only one with the minimum number (2) of points in the (topological) unit cell. The closely related net of hexagonal diamond has already been mentioned. Like diamond it is an array of hexagons, but in its most symmetrical configuration it consists of both chair-shaped and boat-shaped rings, in contrast to those in diamond which are all of the former type. The positions of alternate points in these two nets (that is, the positions of the S (or Zn) atoms in zinc-blende and wurtzite) are related in the same way as cubic and hexagonal closest packing (Chapter 4), and accordingly there is an indefinite number of closely related structures which correspond to the more complex sequences of close-packed layers. Many of these structures have been found in crystals of SiC (see the polytypes of carborundum, p. 787). The structure of high-BeO (p. 445) is closely related to the wurtzite structure of the low-temperature form.

There are many AX_2 structures derived from more complex tetrahedral nets by placing an X atom along each link. They include those of quartz and other forms of SiO_2, the unexpectedly complex GeS_2, many aluminosilicates (some of which are mentioned later in class (iii)), and the high-pressure forms of Si, Ge, and of ice.

Among the high-pressure forms of elementary Si there is one with a body-centred cubic structure which is related to the diamond structure in an interesting way. The diamond net is shown in Fig. 3.41(a) viewed along the direction of a C–C bond, the 6-gon layers (distinguished as full and broken lines) being buckled. Alternate atoms of a 'layer' are joined to atoms of the layers above and below by bonds perpendicular to the plane of the paper, as indicated by the pair of circles. Suppose now that the latter bonds are broken and that alternate

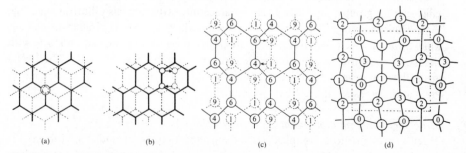

(a) (b) (c) (d)

FIG. 3.41. (a) Projection of diamond structure along C–C bond. (b) and (c), Structure of a high-pressure form of Si. (d) Structure of a high-pressure form of Ge.

layers are translated relative to the neighbouring ones, as in (b), and that connections between the layers are remade as shown by the arrows. An arrow indicates a bond rising to an atom of a layer above the layer from which it originates. The fourth bond from each atom is no longer normal to the plane of the paper, and the structure (which is that of a high-pressure form of elementary silicon) is therefore more dense than the diamond-like structure (a). The numerals in (c) are the heights of the atoms in units of $c/10$ where c is the repeat distance normal to the plane of the projection. Atoms such as 1, 4, 6, and 9 form a helical array, the bond 9 . . . 1 bringing the helix to a point at a height c above the original point 1.

Figure 3.41(d) shows the projection of the structure of a high-pressure form of Ge, the numerals here indicating the heights of the atoms above the plane of the paper in units of $c/4$. Note that the circuits 0-1-2-3 projecting as 4-gons represent 4-fold helices, and that the smallest circuits in this 3D net are 5-gons and 7-gons, in contrast to the 6-gon circuits in (c). The net of Fig. 3.41(d) is also the basis of the structures of the silica polymorph keatite and of ice-III.

The structures of the high-pressure forms of ice are described in Chapter 15, but it is worth noting here how some of the more dense structures arise. Polymorphs with different 3D H_2O frameworks include the following:

I_h	I_c	II	III	V	VI	VII
Density	0·92	1·17	1·16	1·23	1·31	1·50 g/cc

In contrast to ice-III, which has a structure based on the same net as keatite, ice-II and ice-V have structures not found as silica polymorphs, though that of ice-II has features in common with the structure of tridymite. All these three forms achieve their higher density by distortion of the tetrahedral arrangement of nearest neighbours and/or by the formation of more compact ring systems; there are 4-gon rings in ice-V and 5-gon rings in ice-III. As a result the next nearest neighbours are appreciably closer than in ice-I (4·5 Å). For example, the next nearest neighbours of a H_2O molecule in ice-V are at 3·28 Å and 3·64 Å. The even more dense forms VI and VII have structures consisting of two interpenetrating 4-connected frameworks, within each of which the H_2O molecules are hydrogen-bonded; there are no such bonds between molecules belonging to different frameworks. The arrangement of H_2O molecules in each net of ice-VI is similar to that of SiO_4 tetrahedra in the fibrous zeolite edingtonite (p. 828), while ice-VII consists of two frameworks of the cristobalite type which interpenetrate to form a pseudo-body-centred cubic structure (Fig. 15.2, p. 538). Each O atom is equidistant from *eight* others, but it is hydrogen-bonded to only *four*, each of the two interpenetrating nets being similar to the single net of cubic ice-I.

(ii) *Nets with planar and tetrahedral coordination.* Two simple compounds provide examples of nets of this type. In PtS (Fig. 3.42), equal numbers of atoms form coplanar and tetrahedral bonds; less symmetrical variants of this structure are adopted by PdS and CuO. Just as the diamond net may be dissected

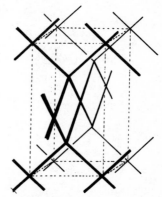

FIG. 3.42. The 4-connected net representing the structure of PtS, in which Pt forms 4 coplanar and S forms 4 tetrahedral bonds.

into buckled 6-gon (3-connected) layers, so the PdP_2 structure may be visualized in terms of puckered 5-gon (3,4-connected) layers (Fig. 3.43). The 4-connected points represent metal atoms forming four coplanar bonds and the 3-connected points represent the P atoms. The latter atoms of different layers are connected together to form the 3D structure, the interlayer bonds being directed to one or other side of the layer as indicated by the short vertical lines in the lower part of Fig. 3.43. In this structure each Pd is bonded to 4 P and each P to 2 Pd and 2 P, from which it

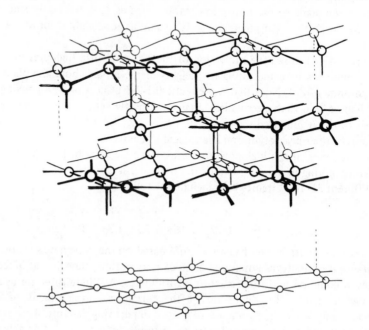

FIG. 3.43. The structure of PdP_2.

follows that continuous chains of linked P atoms can be distinguished in the 3D structure.

(iii) *Nets with polyhedral cavities or tunnels.* For the purposes of deriving nets and illustrating the basic topology of structures it is convenient to base the classification on the connectedness (p) of the points. There are, however, other general features of nets which are important in relation to both the chemical and physical properties of structures based on them. In any repeating pattern the number of points is necessarily the same in each unit cell, and on this scale there is a uniform distribution of points throughout space. However, the distribution of points *within* the unit cell may be such that there is an obvious concentration of points (and links) around points, around lines, or in the neighbourhood of planes. If there is a marked concentration of atoms around certain lines or sets of planes, with comparatively few links between the lines or planes, the crystal may be expected to imitate the properties of a chain or layer structure. In the fibrous zeolites the chains of (Si, Al)O_4 tetrahedra are cross-linked by relatively few bonds, with the

112

result that although the aluminosilicate framework is in fact a 3D one the crystals have a marked fibrillar cleavage.

In Chapter 1 we suggested a very simple classification of crystal structures based on recognition of the four main types of complex: finite (F), chain (C), layer (L), or 3D, hybrid systems including crystals containing complexes of more than one type (for example, chain ions and finite ions). If we arrange the four basic structure types at the vertices of a tetrahedron there are six families of intermediate structures, (1)–(6), corresponding to the six edges of the tetrahedron. Structures may be regarded as being intermediate between the four main types in one of two ways:

(a) Weaker (longer) bonds link the less extensive into the more extensive grouping. For example, finite molecules (ions) could be linked by hydrogen bonds into chains (class (1)), layers (class (2)), or 3D complexes (class (4)), or chains could be linked into layers (class (3)). This is the obvious chemical interpretation of the intermediate groups and could be represented on a diagram:

(b) A purely topological interpretation would be based not on differences in bond type (length) but simply on the relative numbers of bonds within and between the primary structural units. Thus in the intermediate group (1) relatively few bonds link the finite groups into chains, as in the hypothetical 4-connected chain of Fig. 3.44(a). This is a chain structure but one with a very uneven distribution of points along the chain, as in the $ReCl_4$ or $[B_3O_4(OH)_3]_n^{2n-}$ chains mentioned earlier. An example of (2) is the 2D anion in $Sr[B_6O_9(OH)_2]$ and of (3) the layer in red P. In the latter most of the P–P bonds link the atoms into tightly knit chains which are further linked by a very small number of bonds into the double layers illustrated in Fig. 19.2 (p. 000). Both (2) and (3) are layer structures but with obvious concentrations of atoms in finite groups and chains respectively.

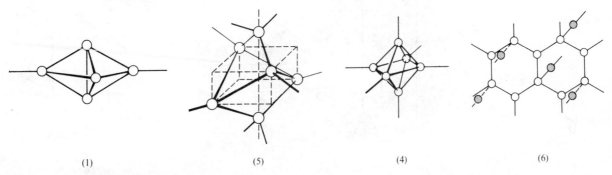

(1)   (5)   (4)   (6)

FIG. 3.44. Structures with uneven distributions of atoms in space (see text).

113

Similarly, (4), (5), and (6) are special types of 3D complexes. The intermediate classes correspond to structures of types different from those in (a). Examples are shown in Fig. 3.44; (4) is the octahedral B_6 group which is linked to six other similar groups at the points of a simple cubic lattice to form the boron framework in CaB_6, (5) is a portion of the infinite chain (Si atoms only indicated) in a fibrous zeolite, each chain being linked laterally to four others, and (6) is a portion of a layer of Si_2N_2O. This interesting compound is made by heating a mixture of Si and SiO_2 to 1450°C in a stream of argon containing nitrogen, and is intermediate between SiO_2 and Si_3N_4 not only as regards its composition but also structurally:

	SiO_2	Si_2N_2O	Si_3N_4
Coordination numbers	4 : 2	4 : 3 : 2	4 : 3

All the bonds from N and three-quarters of those from Si lie within the layers, which are held together only by the bonds through the 2-coordinated O atoms.

Instead of focussing attention on the regions of high density of points (atoms) we may consider the voids between them. These range from isolated cavities to complex 3D systems of tunnels. Crystals possessing these features are of interest as forming 'clathrate' compounds or exhibiting ion-exchange or molecular sieve properties. Three simple arrangements of tunnels found in crystals are shown in Fig. 3.45. In (a) and (b) the tunnels do not intersect, whereas in (c) they intersect at the points of a primitive cubic lattice. Two examples of (a) are illustrated in Fig. 1.9 (p. 28), the tunnels being formed in one case between stacks of molecules and in the other by the formation around hydrocarbon chains of a 3D net in which the hydrogen-bonded urea molecules form the walls of the tunnels. The actual tunnel system in (c) can be imagined to be formed by inflating the links of a 3D net so that they become tubes, the outer surfaces of which are covered by tessellations of bonded atoms; certain zeolite structures may be regarded in this way.

We shall conclude our remarks on nets by describing briefly two types of structure which are of particular interest in structural chemistry, namely, those in which there are well-defined polyhedral cavities or infinite tunnels. The B_6 groups in CaB_6 are preferably not described as polyhedral cavities because they are

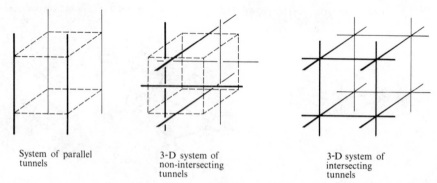

System of parallel tunnels
3-D system of non-intersecting tunnels
3-D system of intersecting tunnels

FIG. 3.45. Types of tunnel systems in crystals. The heavy lines represent the axes of the tunnels.

too small to accommodate another atom, but the large holes surrounded by 24 B atoms at the vertices of a truncated cube accommodate the Ca atoms. We shall suppose all the links in a net to be approximately equal in length, since the general nature of a net can obviously be changed if large alterations are permitted in the relative lengths of certain sets of bonds.

Space-filling arrangements of polyhedra

The division of space into polyhedral compartments is analogous to the division of a plane surface into polygons. One aspect of this subject was studied in 1904 by von Fedorov,† namely, the filling of space by identical polyhedra *all having the same orientation*. He noted that this is possible with five types of polyhedron, the most symmetrical forms of which are the cube, hexagonal prism, rhombic docecahedron, elongated dodecahedron, and truncated octahedron (Fig. 3.46). The third of von Fedorov's space-fillings is related to the closest packing of equal spheres, for the domain of a sphere in cubic closest packing is the rhombic dodecahedron; on uniform compression c.c.p. spheres are converted into rhombic dodecahedra. The space-filling by truncated octahedra (t.o.) corresponds to the

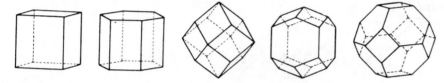

FIG. 3.46. The five space-filling polyhedra of Fedorov.

body-centred cubic packing of equal spheres, in which the t.o. is the polyhedral domain.

Another aspect of this subject was studied by Andreini‡ (1907), who considered the filling of space by regular and Archimedean semi-regular solids, either alone or in combination, but without von Fedorov's restriction on orientation. This restriction is not important in the present context, for evidently there may be polyhedra of the same kind in different orientations within the unit cell. Since a number of polyhedra meet around a common edge it is a matter of finding which combinations of dihedral angles add up to 2π. For example, neither regular tetrahedra nor regular octahedra alone fill space but a combination in the proportion of two tetrahedra to one octahedron does do so. (These polyhedra are the domains of the ions in the CaF_2 structure, which may therefore be represented in this way as a space-filling assembly of these two polyhedra.)

The polyhedra in the space-fillings of Andreini do not include the regular or semi-regular polyhedra with 5-fold symmetry. It is not possible to fill space with regular dodecahedra or icosahedra or the Archimedean solids derived from them (or with combinations of these polyhedra) owing to the unsuitable values of the

† ZK 1904 **38** 321.
‡ Mem. Soc. Ital. Sci. 1907 (3) **14** 75.

dihedral angles. However, there are space-filling assemblies of polyhedra including (irregular) pentagonal dodecahedra and polyhedra of the family $f_5 = 12$, $f_6 > 2$ which were mentioned on p. 65. In addition to these special families of space-filling polyhedra there are an indefinite number of ways of filling space with less regular polyhedra (of one or more kinds); an example is the packing of 8-hedra and 17-hedra noted on p. 66 as the basis of a hydrate structure.

In any space-filling arrangement of polyhedra at least four edges meet at each point (vertex). This number is not necessarily the same for all points; for example, in the space-filling by rhombic dodecahedra four edges meet at some points and eight at others. A special class of space-fillings comprises those in which the *same* number of edges (e.g. 4, 5, or 6) meet at each point, the edges forming a 3D 4-, 5-, or 6-connected net. The boron framework in CaB_6 (Fig. 3.47(c)) may be regarded as a space-filling by octahedra and truncated cubes in which 5 edges meet at every point. A more complex 5-connected net is that corresponding to the space-filling by truncated tetrahedra, truncated octahedra, and cuboctahedra (Fig. 3.47(d)), which represents the boron framework in UB_{12}. Here the uranium atoms are located in

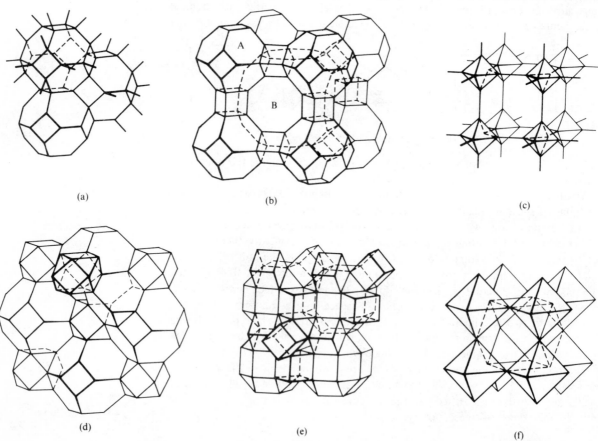

(a)

(b)

(c)

(d)

(e)

(f)

FIG. 3.47. Space-filling arrangements of regular and semi-regular polyhedra.

the largest (truncated octahedral) holes. Some space-filling arrangements of polyhedra are listed in Table 3.15.

TABLE 3.15

Space-filling arrangements of polyhedra

Fig. 3.47	p	Polyhedra
(a)	4	Truncated octahedra
(b)	4	Cube, truncated octahedron, truncated cuboctahedron
(c)	5	Octahedron, truncated cube
(d)	5	Truncated tetrahedron, truncated octahedron, cuboctahedron
(e)	6	Cube, cuboctahedron, rhombicuboctahedron
(f)	8	Octahedron, cuboctahedron

The space-fillings of greatest interest in structural chemistry are those in which four edges meet at each point. These structures are suitable for H_2O forming four hydrogen bonds and the topologically similar $(Si, Al)O_4$ tetrahedron linked to four others through its vertices; they belong to two families which we now consider in more detail.

Space-fillings of regular and Archimedean solids

The simplest representative of this group is von Fedorov's space-filling by truncated octahedra (t.o.) illustrated in Fig. 3.47(a). It represents the framework of Si (Al) atoms in ultramarine, $Na_8Al_6Si_6O_{24} . S_2$, in which Na^+ and S_2^{2-} ions occupy the t.o. interstices. At higher temperatures it is possible for ions to move from one cavity to another through the 6-rings of the framework. Accordingly, if the mineral sodalite (which has the ultramarine structure with 2 Cl^- replacing S_2^{2-}) is fused with sodium sulphate it is converted into $Na_8Al_6Si_6O_{24}. SO_4$, which also occurs as a mineral. The structure of the oxy-metaborate $Zn_4B_6O_{13}$ (or $OZn_4(BO_2)_6$) is also based on this framework. Tetrahedral BO_4 groups at the vertices of a t.o. framework form a 3D anion and in the interstices are situated tetrahedral OZn_4 groups. The same framework also represents the arrangement of water molecules in $HPF_6 . 6 H_2O$, the PF_6^- ions occupying the cavities, and in a modified form in $N(CH_3)_4OH . 5 H_2O$. Since the ratio of vertices to cavities is 6 : 1, the framework in the last compound is made up of $(5 H_2O + OH^-)$.

The space-filling of Fig. 3.47(b) results from inserting cubes between the square faces of the t.o.'s in the Fedorov packing, thereby producing much larger holes (truncated cuboctahedra, t.c.o.). The two kinds of larger polyhedron are distinguished as A and B in the figure. This framework represents the structure of the synthetic zeolite A, placing $Si(Al)O_4$ tetrahedra at all the points of the net. It is shown in Fig. 3.47(b) as a space-filling array of polyhedra of three kinds. Alternatively we could accentuate the framework formed by the t.o.'s and cubes, the space left being a 3D system of intersecting tunnels formed from t.c.o.'s sharing their 8-gon faces. It is the tunnel system that is the important feature as regards the behaviour of the crystal as a molecular sieve, the large tunnels (with free diameter

117

approximately $3 \cdot 8$ Å) allowing the passage of gas molecules. These aluminosilicate structures are more fully described in Chapter 23 p. 824, where we also illustrate the structure of the $(Si, Al)O_2$ framework in the mineral faujasite, $NaCa_{0.5}(Al_2Si_5O_{14}) \cdot 10\ H_2O$. This is a diamond-like arrangement of t.o.'s each of which is joined to four others through hexagonal prisms; this is not a space-filling arrangement of polyhedra.

Space-fillings of dodecahedra and related polyhedra

Our second family of polyhedral space-fillings comprises those formed from pentagonal dodecahedra in combination with one or more kinds of polyhedra of the type $f_5 = 12, f_6 \geqslant 2$. They represent the structures of hydrates of substances ranging from non-polar molecules of gases, such as chlorine and methane, and liquids such as chloroform, to amines and substituted ammonium and sulphonium salts. These hydrates may be described as expanded ice-like structures, since they are not stable at temperatures much above $0^{\pi}C$ and they consist of polyhedral frameworks of hydrogen-bonded water molecules built around the 'guest' molecules or ions. The latter almost invariably occupy the larger polyhedral cavities, from which they can escape only if the crystal is destroyed by dissolution or vaporization. The volumes of the cavities in these clathrate hydrates are: dodecahedra, 170 Å^3, 14-hedra, 220 Å^3, and 16-hedra, 240 Å^3, and the van der Waals diameter of the largest non-polar molecule forming a hydrate of this type is approximately 8 Å (n-propyl bromide).

The compositions of these hydrates indicate large numbers of molecules of water of crystallization. The ratio of the number of water molecules to 'guest' molecules is equal to that of the number of points (vertices) to the number of larger cavities in the framework, and is not necessarily integral. For example, in $CHCl_3 \cdot 17\ H_2O$ the molecules of chloroform occupy all the large (16-hedral) sites, and since there are 8 such sites in a unit cell containing $136\ H_2O$ the ratio of H_2O to $CHCl_3$ molecules is $136/8 = 17$. On the other hand, the unit cell of $(C_2H_5)_2NH \cdot 8\frac{2}{3}\ H_2O$ contains 104 H_2O and there are 12 large (18-hedral) cavities ($f_5 = 12, f_6 = 6$), in addition to some less regular ones. The ratio of water molecules to amine molecules is therefore $104/12$ or $8\frac{2}{3}$.

In hydrated salts such as $R_4NF \cdot mH_2O$ and $R_3SF \cdot nH_2O$ the positively charged N^+ or S^+ together with F^- ions replace some of the H_2O molecules in the framework. In such hydrates the bulky organic groups are accommodated in the appropriate number of cavities adjacent to the N or S atom to which they are bonded, that is, in the positions occupied by the guest molecules in the simpler hydrates. These structures, and also the structures of the hydrates mentioned in the previous section, are described in some detail in Chapter 15.

Sphere Packings

Periodic packings of equal spheres

Any arrangement of spheres in which each makes at least three contacts with other spheres may be described as a sphere packing. For example, equal spheres may be placed at the points of 3- or 4-connected nets. The densities of the resulting packings are low, density being defined as the fraction of the total space occupied by the spheres. If spheres are placed at the points of the diamond net and are in contact at the mid-points of the links the density is only 0·3401. Moreover, arrangements with small numbers of neighbours (low c.n.'s) are stable only if there are directed bonds between the spheres, and structures of this kind are therefore described under Nets. We consider here only those packings in which each sphere is in contact with six or more neighbours (Table 4.1).

TABLE 4.1
Densities of sphere packings

Coordination number	Name	Density
6	Simple cubic	0·5236
8	Simple hexagonal	0·6046
8	Body-centred cubic	0·6802
10	Body-centred tetragonal	0·6981
12	Closest packings	0·7405

Simple cubic packing

For 6-coordination the most symmetrical packing in three dimensions arises by placing spheres at the points of the simple cubic lattice; slightly more than one-half of the space is occupied by the spheres. Each sphere is in contact with six others situated at the vertices of an octahedron, the contacts being along the edges of the unit cube (Fig. 4.1(a)). There is no contact between spheres along face-diagonals or body-diagonals.

The low c.n. and low density of this structure make it unsuitable for most metals; this structure has been assigned to α-Po. Mercury does, however, crystallize with a closely related structure which can be derived from the simple cubic packing

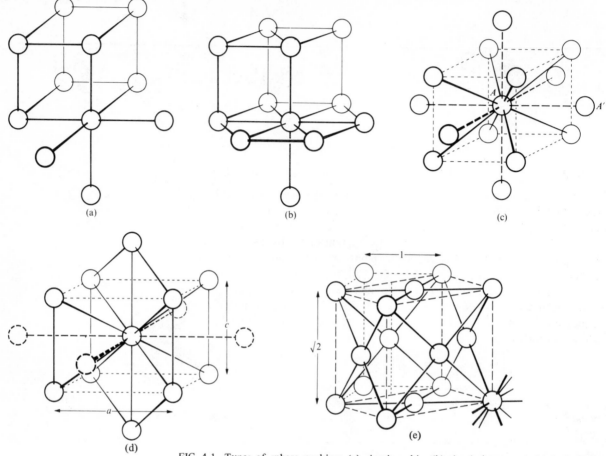

FIG. 4.1. Types of sphere packing: (a) simple cubic, (b) simple hexagonal, (c) body-centred cubic, (d) tetragonal (c.n. 10), (e) cubic closest packing.

by extension along one body-diagonal so that the cube becomes a rhombohedron (interaxial angle $70\frac{1}{2}°$ instead of $90°$). The atoms retain 6-coordination.

The structures of crystalline As (Sb and Bi) and of black P illustrate two different ways of distorting the simple cubic packing so that each atom has three nearest and three more distant neighbours. Both are layer structures which may alternatively be described as forms of the plane hexagon net. They are described and illustrated in the section on the plane hexagon net (p. 88).

The body-centred cubic packing

The simple hexagonal sphere packing (c.n. 8), Fig. 4.1(b), is not of great importance as a crystal structure; it is mentioned again later. A more dense arrangement with the same c.n. is the body-centred cubic packing illustrated in Fig. 4.1(c). Spheres are placed at the body-centre and corners of the cubic unit cell, and

120

are in contact only along the body-diagonals. This structure, with density 0·6801, is adopted by a number of metals (Table 29.3, p. 1015). Although each sphere is in contact with only eight others it has six next-nearest neighbours at a distance only 15 per cent greater than that to the eight nearest neighbours. For the atom A these six neighbours are those at the centres of neighbouring unit cells. If coordination number is defined in terms of the polyhedral domain (p. 61) of an atom it has the value 14 for this structure, the domain being the truncated octahedron.

The relatively small difference between the distances to the groups of eight and six neighbours suggests that structures with c.n.'s intermediate between eight and fourteen should be possible. The six more distant neighbours are arranged octahedrally. By compressing the structure in the vertical direction we may bring two of these neighbours to the same distance as the eight at the corners of the cell. Alternatively, by elongating the structure we may bring the four equatorial neighbours to this distance. If the height of the (tetragonal) cell is c and the edge of the square base is a, there are the two cases:

$$a^2/2 + c^2/4 = c^2, \text{ or } c : a = \sqrt{2}/\sqrt{3} \text{ for 10-coordination,}$$
$$\text{(Fig. 4.1 (d))}$$

or

$$a^2/2 + c^2/4 = a^2, \text{ or } c : a = \sqrt{2} \text{ for 12-coordination.}$$
$$\text{(Fig. 4.1 (e)).}$$

Only one metal (Pa) is known to crystallize with the 10-coordinated structure, but it is also the structure of $MoSi_2$ (p. 790). This packing has a density (0·6981) somewhat higher than the b.c. cubic packing, but the packing with 12-coordination, and density 0·7405, is of outstanding importance. Because the height of the cell ($\sqrt{2}a$) is equal to the diagonal of the square base there is an alternative unit cell (Fig. 4.1(e)) which is a cube with spheres at the corners and at the mid-points of all the faces, hence the name face-centred cubic (f.c.c.) structure. This is one of the forms of *closest* packing of equal spheres.

It is instructive to view a model of b.c.c. sphere packing along a direction parallel to a face-diagonal of the unit cell (Fig. 4.2(a)). The open circles represent spheres lying in the plane of the paper, and the shaded circles spheres in parallel planes above and below that of the paper. The lines connect spheres which are in contact. A small alteration in the structure of the layer and small shifts of the layers relative to one another converts (a) into (b), which is a projection of the (cubic) closest packing (f.c.c.). This close relationship between b.c.c. and f.c.c. sphere packings is of interest in view of the fact that both are important as the structures of many metals. Also of interest is the fact that although b.c.c. packing is less dense than f.c.c (or other closest packings) and therefore has more free space, we do not find 'interstitial' structures based on b.c.c. packing, in contrast to the numerous interstitial f.c.c. structures (p. 1051). The reason is that although the interstices in the b.c.c. structure are much more numerous than those in the f.c.c. structure they are not independent as are those in close-packed assemblies. They consist of distorted tetrahedral and distorted octahedral holes, and the former are

121

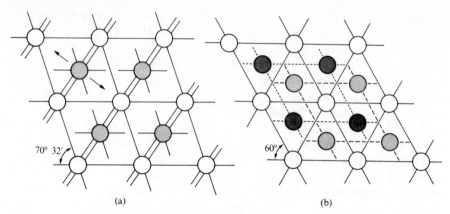

FIG. 4.2. (a) Body-centred cubic packing viewed along face-diagonal. (b) Projection of face-centred cubic packing.

arranged in groups of four *within* each 'octahedral' hole. Owing to the close proximity of the tetrahedral holes, which are *larger* than the octahedral holes, only one-quarter could be occupied simultaneously, corresponding to one interstitial atom placed off-centre in each face of the unit cell. We do not describe the interstices in close-packed structures until later, but the following figures relating to the interstices in an assembly of N spheres of unit radius in b.c.c. packing may be compared with those on p. 127 for close-packed structures:

Type of hole	Number	Maximum radius of interstitial sphere
'Tetrahedral'	6 N	0·291
'Octahedral'	3 N	0·155

The closest packing of equal spheres

It is usual to approach this subject by considering first the closest packing of spheres resting on a plane surface. The problem is evidently the same as that of arranging circles on a plane, the densest arrangement being that in which each sphere (circle) is in contact with six others. Layers of spheres of this kind may be superposed in various ways. The closest packing (maximum number of contacts made by each sphere) is achieved if each sphere touches three others in each adjacent layer, the total number of contacts then being 12. We shall pursue this approach in some detail shortly, but before doing so it is worth noting that this conventional treatment obscures an interesting fact about the closest packing of spheres. Although the packing in any layer is evidently the densest possible packing it may not be assumed that this is true of the space-filling arrangements resulting from stacking such layers. The reason for this uncertainty will be more evident if we look at the matter in a different way.

Three spheres are obviously most closely packed in a triangular arrangement, and a fourth sphere will make the maximum number of contacts (3) if it completes a tetrahedral group. We might then expect to obtain the densest packing by continuing to place spheres above the centres of triangular arrangements of three others. The lines joining the centres of spheres in such an array would outline a system of regular tetrahedra, each having faces in common with four other tetrahedra, that is, we should have a space-filling arrangement of regular tetrahedra. However, it is not possible to pack together regular tetrahedra to fill space because the dihedral angle of a regular tetrahedron $(70°32')$ is not an exact submultiple of $360°$.

Icosahedral sphere packings. Alternatively, suppose that we continue placing spheres around a central one, all the spheres having the same radius. The maximum number that can be placed in contact with the first sphere is 12. There is in fact more than sufficient room for 12 equal spheres in contact with a central sphere of the same radius but not sufficient for a thirteenth. There is therefore an infinite number of ways of arranging the twelve spheres, of which the *most symmetrical* is to place them at the vertices of a regular icosahedron—the only regular solid with 12 vertices. The length of an edge of a regular icosahedron is some five per cent greater than the distance from centre to vertex, so that a sphere of the outer shell of twelve makes contact only with the central sphere. (Conversely, if each sphere of an icosahedral group of twelve, all touching a central sphere, is in contact with its five neighbours, then the central sphere must have a radius of 0·902 if that of the outer ones is unity.) The icosahedral arrangement of nearest neighbours around every sphere does not lead to a periodic 3-dimensional array of spheres, but it is of sufficient interest to justify a brief description. Some other sphere packings which are not periodic in three dimensions but, like the icosahedral packing, can extend to fill space, are of interest from the structural standpoint. A number of these non-crystallographic sphere packings are described in MSIC, including an infinite packing with density (0·7236) approaching that of closest packing and having a unique axis of 5-fold symmetry.[1] A closely related packing with even higher density (0·7341) has also been described.[2]

First we note an interesting relation between the icosahedron and the cuboctahedron (Fig. 4.4(a)). If 24 rigid rods are jointed together at the 12 vertices of a cuboctahedron, rotation of the groups forming the triangular faces about their normals transforms the cuboctahedron into a regular icosahedron. In the course of the change each joint moves in towards the centre, so that the distance from centre to vertex contracts by some five per cent. If we imagine spheres of unit radius situated at the vertices and centre of the cuboctahedron the same transformation can be carried out, but now there is no radial contraction and in the icosahedral configuration the outer spheres are no longer in contact. If a close-packed arrangement is built starting with a cuboctahedral group of 12 spheres around a central sphere it will be found that the second shell of spheres packed over the first contains 42 spheres—in general $10 n^2 + 2$ for the nth layer. A second shell packed around an icosahedral group also contains 42 spheres, the spheres being in contact along the 5-fold axes, and further layers of $10 n^2 + 2$ spheres may be added as for the cuboctahedron. The whole assembly is not periodic in three dimensions but is a

(1) N 1965 **208** 674; JCG 1970 **6** 323
(2) N 1970 **225** 1040

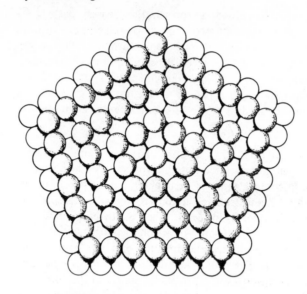

FIG. 4.3. A sphere-packing
with 5-fold symmetry.

(3) AC 1962 **15** 916

radiating structure, having a unique centre (Fig. 4.4(b)). It is less dense than the packing based on the cuboctahedron, the limiting density being 0·6882,[3] intermediate between that of body-centred cubic packing (0·6802) and the 'cuboctahedral' packing (0·7405), and it can be converted into the latter by the mechanism of Fig. 4.4(a).

Since the *most symmetrical* arrangement of 12 neighbours (the icosahedral coordination group) does not lead to the densest possible 3D packing of spheres we have to enquire which of the infinite number of arrangements of twelve neighbours lead to more dense packings and what is the maximum density that can be attained

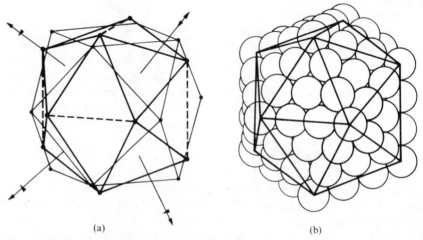

(a)

(b)

FIG. 4.4. (a) Relation between cuboctahedron and icosahedron. (b) Icosahedral packing of equal spheres, showing the third layer ($n = 3$). (AC 1962 **15** 916).

124

in an infinite sphere packing. It was shown by Barlow in 1883 that two coordination groups, alone or in combination, lead to infinite sphere packings which all have the same density (0·7405). These two coordination groups are the cuboctahedron and the related figure (the 'twinned cuboctahedron') obtained by reflecting one-half of a cuboctahedron cut parallel to a triangular face across the plane of section (Fig. 4.5). These are the arrangements of nearest neighbours in sphere packings formed by stacking in the closest possible way the close-packed layers mentioned at the beginning of this section. It is an interesting fact that it has yet to be proved that the density 0·7405 could not be exceeded in an infinite sphere packing of some (unknown) kind, though Minkowski showed that the packing based on the cuboctahedron (cubic closest packing) is the densest *lattice packing* of equal spheres. (A lattice packing has the following property. If on any straight line there are two spheres a distance a apart then there are spheres at all points along the line, extended in both directions, at this separation. Note that hexagonal closest packing (p. 130) is *not* a lattice packing–cubic closest packing is unique as a *closest lattice packing*.) The difficulty arises from the fact that whereas the closest packing of spheres in a plane is a unique arrangement, there is an infinite number of ways of placing 12 equal spheres in contact with a single sphere of the same radius.

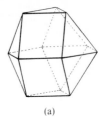

(a)

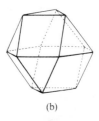

(b)

FIG. 4.5. Coordination polyhedra in (a) hexagonal, (b) cubic closest-packing.

Sphere packings based on closest-packed layers

A sphere in a closest-packed layer is in contact with six others (Fig. 4.6(a)). When such layers are stacked parallel to one another the number of additional contacts (on each side) will be 1, 2, or 3 if the centres of spheres in the adjacent layers fall above or below points such as A, D, or B (or C) respectively. We shall refer to such layers as A, D, B, or C layers. There are three D positions relative to a given sphere, D', D'', and D'''–the diametrically opposite point in each case is the position of another sphere in the same layer and is shown in parentheses in Fig. 4.6(b). If the same (vector) relationship is maintained between successive pairs of layers it is immaterial whether it is AD', AD'', or AD''', but if these translations are combined in a stacking sequence different structures result. The sequences $AD'AD' \ldots$,

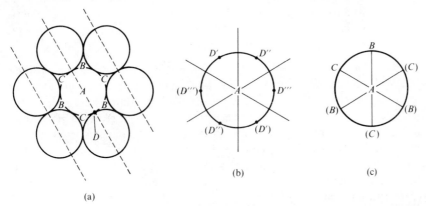

(a)

(b)

(c)

FIG. 4.6. Superposition of layers of c.p. spheres (see text).

125

$AD''AD''$ ..., and $AD'''AD'''$ .... correspond to the same structure, but $AD'AD''$ ... etc. represent different structures. There are only two different positions of the type B or C because all three B (or C) positions in Fig. 4.6(c) are positions of spheres in a given layer. The positions B and C are equivalent as regards the relation between two adjacent layers, so that the sequences $AB\,AB$ ... and $AC\,AC$ ... are the same structure, but $AB\,AC$ ... and other sequences involving mixtures of B and C layers represent different structures.

The total number of contacts (coordination number) can therefore have any value from eight to twelve, that is, $6 + 1 + 1$ to $6 + 3 + 3$, depending on the stacking sequence. Of these types of sphere packing only those with coordination numbers of 8, 10, and 12 are found in crystals.

c.n. 8   The layer sequence is $A\,A\,A$ .... This packing, in which the layers fall vertically above one another, is the *simple hexagonal* (lattice) *packing*.

c.n. 10   This can arise in two ways: (a) by a D-type contact with both adjacent layers $(6 + 2 + 2)$ or (b) by an A-type contact on one side and a B- (or C-) type contact on the other $(6 + 1 + 3)$.

(a) Layer sequence $A\,D\,A\,D$ .... There is an infinite number of structures because there is a choice between positions D', D'', and D''' at each layer junction. In the special case where all pairs of successive layers are related by a translation of the same kind, for example AD', there is also closest packing of the spheres in a second set of planes *perpendicular* to the original c.p. layers. These planes intersect the plane of the paper in the broken lines of Fig. 4.6(a). This is the body-centred tetragonal packing described on p. 121 and listed in Table 4.1.

(b) Layer sequence $\begin{smallmatrix}B\\C\end{smallmatrix}\,A\,A\,\begin{smallmatrix}B\\C\end{smallmatrix}$. Again there is an infinite number of possible structures because we must distinguish between B and C positions. For example, the simplest sequence is $\underline{AA\,BB}\,AA\,BB$ ... but any pair of B layers could be changed to a pair of C layers, and the sequence $\underline{AA\,BB\,AA\,CC}$ ... represents a different structure.

c.n. 12   The layer sequence may be any combination of A, B, and C provided that no two adjacent layers are of the same type. The two simplest possibilities are therefore $\underline{AB}\,AB$ ... and $\underline{ABC}\,ABC$ .... These closest-packed structures are considered in more detail later.

Of these four main classes of layer sequence only those corresponding to c.n. 10(a) and 12 are found as the structures of crystals *in which the only atoms are those of the c.p. layers*. Metals do not crystallize with the simple hexagonal packing $AAAA$ ... because a small relative displacement of the layers gives the more closely packed 10- or 12-coordinated structures. The $(6 + 2 + 2)$ structures are confined to a few disilicides ($MoSi_2$, $CrSi_2$, $TiSi_2$) and one metal (Pa), but many metals form one or other of the most closely packed sequences. In general we should not expect to find adjacent layers directly superposed unless there is some special reason for this less efficient packing. In fact it occurs only if there are either atoms in the (trigonal prism) holes between such layers or hydrogen atoms directly along the

$A \ldots A$ etc. contacts between the layers (see later). The reason for the great importance of the most closely packed structures is that in many halides, oxides, and sulphides the anions are appreciably larger than the metal atoms (ions) and are arranged in one of the types of closest packing. The smaller metal ions occupy the interstices between the c.p. anions. In another large group of compounds, the 'interstitial' borides, carbides, and nitrides, the non-metal atoms occupy interstices between c.p. metal atoms.

Interstices between close-packed layers

In the plane of a c.p. layer there are small holes surrounded by triangular groups of c.p. atoms, and between the layers there are holes surrounded by polyhedral groups

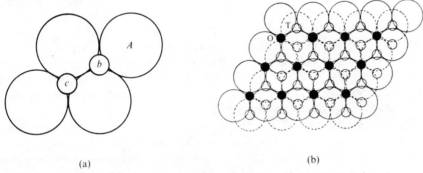

(a)   (b)

FIG. 4.7. (a) Trigonal prism holes between two directly superposed A layers. (b) Tetrahedral holes (small open circles) and octahedral holes (small black circles) between superposed A and B layers.

of c.p. atoms. These holes may be described as a, b, or c if they fall above positions A, B, or C (Fig. 4.7(a)). The types of hole are:

$$\begin{array}{cc} \underline{\quad A \quad} & \underline{\quad A \quad} \\ b \text{ or } c \text{ (trigonal prism)} & c \text{ (octahedral)} \\ \underline{\quad A \quad} & a \text{ or } b \text{ (tetrahedral)} \\ & \underline{\quad B \quad} \end{array}$$

and similarly for the a (octahedral) and b or c (tetrahedral) holes between B and C layers, and b (octahedral) and a or c (tetrahedral) holes between A and C layers. We

Sphere packing	Type of hole	Number	Maximum radius of interstitial sphere
Simple hexagonal Closest packing	Trigonal prism	$2N$	0·528
	Tetrahedral	$2N$	0·225
	Octahedral	N	0·414

give above the maximum radii of the spheres which can be accommodated in the interstices in packings of spheres of unit radius and also the numbers of these

127

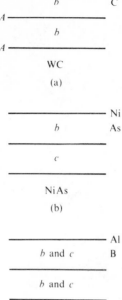

WC

(a)

NiAs

(b)

AlB$_2$

(c)

FIG. 4.8. Diagrammatic elevations of structures with trigonal prism coordination.

interstices in an assembly of N spheres. There are twice as many trigonal prism holes between a pair of AA layers as there are octahedral holes between a pair of AB layers.

Structures with some pairs of adjacent layers of type A

If all layers are directly superposed (for example, all of type A), all interstices are of type b or c surrounded by a trigonal prism of the larger spheres. Trigonal prism coordination is not to be expected in essentially ionic crystals, but may occur in other crystals for a number of special reasons: (i) an atom may have a preference for this arrangement of bonds, (ii) there may be some kind of interaction between the six atoms forming the trigonal prism coordination group which stabilizes this coordination group relative to the more usual octahedral one, or (iii) an interaction between the interstitial atoms themselves, in which case the coordination around these atoms may be incidental to satisfying other bonding or packing requirements. This is probably the case in AlB$_2$. It is obviously not easy (or perhaps possible) to distinguish between (i) and (ii). In the two structures shown as diagrammatic elevations in Fig. 4.8(a) and (b) there are metal–metal bonds between the atoms in the layers; NbS crystallizes with both these structures. For the alternative, more usual, description of the NiAs structure see p. 609.

Between a given pair of A layers all interstices b and c will not be occupied simultaneously because of their close proximity, unless the interstitial atoms are themselves bonded. This occurs in AlB$_2$ (Fig. 4.8(c)). The boron atoms form plane 6-gon nets, occupying all the b and c positions between directly superposed layers of Al atoms, as shown in a projection of the structure (Fig. 4.9(a)). In hexagonal MoS$_2$ metal atoms occupy one-half of the trigonal prism holes between alternate pairs of layers of S atoms (Fig. 4.9(b)).

Some further structures in which there are pairs of adjacent AA layers are shown diagrammatically as elevations in Fig. 4.10, where the type of coordination of the interlayer atoms is indicated. The simplest (layer) structure of the type $AbA\ BaB$, with trigonal prism coordination of all metal atoms, is that of the ordinary form of MoS$_2$ to which we have just referred. Because there are alternative b and c positions between two A layers and a and c positions between two B layers there are two

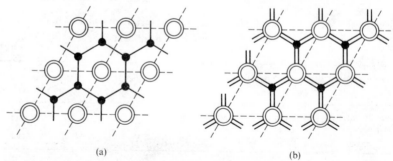

(a)   (b)

FIG. 4.9. Projections of atoms in trigonal prism holes (small circles) between two directly superposed c.p. layers (large circles): (a) AlB$_2$, showing linking of B atoms into plane 6-gon nets, (b) layer of the structure of hexagonal MoS$_2$.

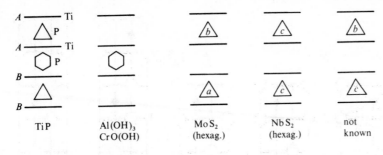

△ indicates trigonal prism hole between layers

⬡ indicates octahedral hole between layers

FIG. 4.10. Diagrammatic elevations of structures with trigonal prism and/or octahedral coordination.

other possible structures with 4-layer repeat units (*AABB*) namely, *AcA BcB* and *AbA BcB*. One is the structure of one form (hexagonal) of NbS_2; the anti-NbS_2 structure is adopted by Hf_2S. No example of the other structure is known. It may interest the reader to check that there are only two structures with a repeat unit consisting of three pairs of layers, namely,

AbA BcB CaC (rhombohedral forms of NbS_2 and MoS_2)

and

AbA BcB CbC (no example known),

but no fewer than ten structures with a repeat unit consisting of four pairs of layers, namely, *AA BB AA CC* (7 structures) and *AA BB AA BB* (3 structures). The three *AA BB* structures and the two *AA BB CC* structures are illustrated as elevations in Fig. 4.11.

There is an indefinite number of more complex layer sequences in which some pairs of adjacent layers are of the same kind (for example, *A A*), and others different (*A B*, etc.). An example is the structure of ThI_2, in which some Th atoms occupy trigonal prism and others octahedral holes in the following layer sequence (8-layer repeat):

$$...A \quad A \quad C \quad B \quad A \quad A \quad B \quad C \quad A \quad A...$$
$$\text{t. p.} \qquad \text{o} \qquad \text{t. p.} \qquad \text{o} \qquad \text{t. p.}$$

Such structures may be described as consisting of interleaved portions of CdI_2 and MoS_2 structures.

So far we have been concerned with the type of coordination of the (metal) atoms inserted between the c.p. layers. The way in which the double layers *AA* etc. are superposed also, of course, determines the coordination *around the c.p. atom*. This is not of special interest in MX_2 compounds since the contacts to one side are only of the van der Waals type. For example, in MoS_2, with the structure *AbA BaB*... each S has a trigonal prismatic arrangement of nearest neighbours (3 Mo + 3 S), but the S–S contacts are not of much chemical interest. The situation

129

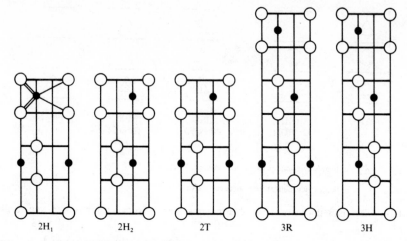

FIG. 4.11. Projections (along *a* axis) of the three *AA BB* and two *AA BB CC* structures for compounds MX_2 in which the coordination of M is trigonal prismatic.

is, however, different in the anti-MX_2 structures of this type, such as the anti-$2H_2$ structure (Fig. 4.11) of Hf_2S. Here the nearest neighbours of Hf are 3 Hf + 3 S, Hf being bonded to 3 Hf by metal–metal bonds. The coordination groups of the c.p. atoms in the three simplest structures of Fig. 4.11 are:

$2 H_1$	all trigonal prismatic
$2 H_2$	all octahedral
$2 T$	one-half trigonal prismatic
	one-half octahedral.

Hexagonal and cubic closest packing of equal spheres

We now consider the sphere packings in which close-packed plane layers are stacked in the closest possible way. If we label the positions of the spheres in one layer as *A* (Fig. 4.12) then an exactly similar layer can be placed above the first so that the centres of the spheres in the upper layer are vertically above the positions *B*. It is obviously immaterial whether we choose the positions *B* or the similar positions *C*, as may be seen by inverting Fig. 4.12. When the third layer is placed above the second (*B*) layer there are alternatives: the centres of the spheres may lie above either the *C* or *A* positions. The two simplest sequences of layers are evidently

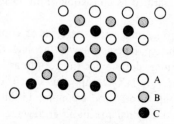

FIG. 4.12. Three successive layers of cubic closest packing.

130

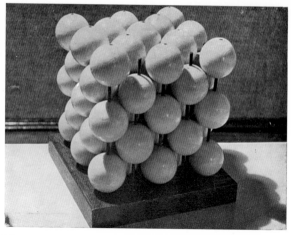

(a)

FIG. 4.14(a) Cubic close packing,
(b) the same with the corner atom removed,
(c) cubic close-packing with more atoms removed
to show a layer of close-packing atoms,
(d) hexagonal close-packing.

(b)

(d)

(c)

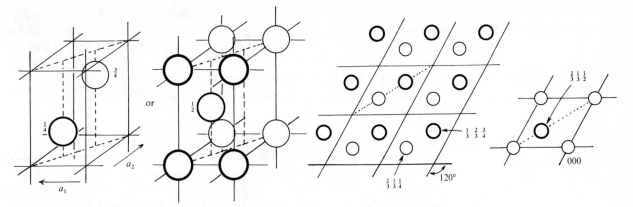

FIG. 4.13. Hexagonal close-packing of equal spheres: alternative unit cells and their projections.

ABAB... and *ABC ABC...*, but there is also an indefinite number of more complex sequences. In all such sphere packings the volume occupied per sphere is $5·66\,a^3$ for spheres of radius a. Alternatively, the mean density is $0·7405$ if that of the spheres is unity and that of the space between the spheres is zero.

The sequence *AB AB* ... is referable to a hexagonal unit cell (Fig. 4.13) and is called hexagonal closest packing (h.c.p.). In this arrangement, also illustrated in Fig. 4.14(d), the twelve neighbours of an atom are situated at the vertices of the coordination polyhedron of Fig. 4.5(a) ('twinned cuboctahedron').

The sequence *ABC ABC* ... possesses cubic symmetry, that is, 3-fold axes of symmetry in four directions parallel to the body-diagonals of a cube, and is therefore described as cubic closest packing (c.c.p.). A unit cell is shown in Fig. 4.1(e) (p. 120) and the packing is also illustrated in Fig. 4.14. The close-packed layers seen in plan in Fig. 4.12 are perpendicular to any body-diagonal of the cube. Since the atoms in c.c.p. are situated at the corners and mid-points of the faces of the cubic unit cell the alternative name 'face-centred cubic' (f.c.c.) is also applied to this packing. All atoms have their twelve nearest neighbours at the vertices of a cuboctahedron. (Fig. 4.5(b)).

More complex types of closest packing

In the sequences *AB* ... and *ABC* ... every sphere has the same arrangement of nearest neighbours. Also the arrangements of more distant neighbours are the same for all spheres in a given packing, h.c.p. or c.c.p. In more complex sequences this is not necessarily true, and it can be shown that the next family of closest packed structures, with two (and only two) types of non-equivalent sphere, comprises four sequences.

In a concise nomenclature for c.p. structures a layer is denoted by h if the two neighbouring layers are of the same type (i.e. both A, both B, or both C) or by c if they are of different types. Hexagonal close-packing is then denoted simply by h (i.e. $hhh...$) and cubic close-packing by c. The symbols for more complex sequences include both h and c. Now the spheres in h and c layers have different arrangements

131

of nearest neighbours (the coordination polyhedra of Fig. 4.5(a) and (b)) so that any more complex sequence necessarily has two types of non-equivalent sphere differing in arrangement of *nearest* neighbours. If we are to have *only* two kinds of non-equivalent sphere we must ensure that all spheres in *h* layers have identical arrangements of *more distant* neighbours, and similarly for spheres in *c* layers, that is, the sequence of *h* and *c* layers on each side of an *h* (or *c*) layer must always be the same.

Starting with *hc* we may add *c* or *h*. If we add *c*, giving *hcc*, the central *c* layer is surrounded by *h* and *c*. The second *c* layer must be surrounded in the same way, and therefore the next symbol added must be *h*. The addition of the third symbol *c* thus decides the choice of the fourth (*h*). We may then add either *h* or *c*. In the former case we have *hcchh* which fixes the environment of an *h* layer (fourth symbol) as *c* and *h*. This means that the next symbol added must be *c*, and since *c* must have *h* and *c* neighbours this in turn makes the next layer *c*. It is easily verified that the sequence becomes *hhcc* . . . . In a similar way we find that if we add *c* as the fifth symbol, instead of *h*, we should arrive at the sequence *hcc* . . . . Going back now to the original *hc* point, addition of *h* does not decide what the fourth symbol should be, and we find two possible sequences, *hc* and *hhc*. It can be shown in this way that only four sequences are possible for two types of non-equivalent sphere:

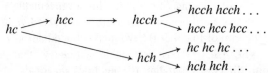

A symbol of the type *h*, *c*, *hc*, etc. obviously does not indicate the number of layers after which the sequence repeats, for this number is 2 for *h* (*AB* . . .), 3 for *c* (*ABC* . . .), and 4 for *hc* (*ABCB* . . .). The number of layers in the repeat unit of a particular stacking sequence may readily be found by deriving the latter from the symbol starting arbitrarily with *AB* . . . and continuing until the layer sequence repeats:

$$
\begin{array}{c c}
h\ c\ h\ c\ h\ c & h\ c\ c\ h\ c\ c\ h\ c\ c \\
(B)\ A\ B\ C\ B\ A\ B\ \ldots & \quad\text{or}\quad (B)\ A\ B\ C\ A\ C\ B\ A\ B\ \ldots \\
(4\ \text{layer repeat}) & (6\ \text{layer repeat})
\end{array}
$$

The alternative symbols in Table 4.2 indicate the numbers of layers in the repeat units; also the layer sequence may be derived directly from the symbol. The change from $A \to B \to C$ is represented as a unit vector $+1$ and in the reverse direction as -1. A succession of n positive (or negative) units is abbreviated to $n\,(\bar{n})$, so that $2\bar{2}$ stands for $11\bar{1}\bar{1}$ . . . and starting from A gives the sequence $\underline{A}\ B\ C\ B\ \underline{A}$.

The six c.p. sequences of Table 4.2 are those most frequently found in crystals, but the number of different sequences that repeat in 12 or fewer layers is very much larger if no account is taken of the number of kinds of non-equivalent sphere (Table 4.3). Only for 2, 3, 4, and 5 layers in the repeat unit is there a unique layer sequence.

132

TABLE 4.2

The closest packing of equal spheres

Number of kinds of non-equivalent sphere	Symbol		Number of layers in repeat unit
1	h	1 $\bar{1}$	2
	c	3	3
2	hc	2 $\bar{2}$	4
	hcc	3 $\bar{3}$	6
	chh	(1 $\bar{2}$)$_3$	9
	cchh	(1 $\bar{3}$)$_3$	12

TABLE 4.3

Numbers of different sequences of c.p. spheres

Number of layers in repeat unit	Number of different sequences
2	1
3	1
4	1
5	1
6	2
7	3
8	6
9	7
10	16
11	21
12	43

The simplest examples of the closest packing of identical units are the structures of crystalline metals or noble gases and of crystals built of molecules of one kind only. In the latter the molecules must either be approximately spherical in shape or become effectively spherical by rotation or random orientation. Metals provide many examples of hexagonal and cubic closest packing and a few examples of more complex sequences (for example, *hc*: La, Pr, Nd, Am, and *chh*: Sm). Close-packed molecular crystals include the noble gases, hydrogen, HCl, H_2S, and CH_4.

The great importance of closest packing in structural chemistry arises from the fact that the anions in many halides, oxides, and sulphides are close-packed (or approximately so) with the metal atoms occupying the interstices between them. As already noted, the polyhedral interstices between c.p. atoms are of two kinds, tetrahedral and octahedral (T and O in Fig. 4.7(b)). The number of tetrahedral holes is equal to twice the number of c.p. spheres; the number of octahedral holes is

133

TABLE 4.4
Examples of more complex sequences of c.p. layers

Number of layers in repeat unit	Symbol	Close-packed X layers	Close-packed AX$_3$ layers	Reference
4	hc	La, Pr, Nd, Am		
		HgBr$_2$		
		HgI$_2$ (yellow)		IC 1967 **6** 396
		Ti$_2$S$_3$		
		Cd(OH)Cl		
			high-BaMnO$_3$	AC 1962 **15** 179
			TiNi$_3$	
5	hhccc		Ba$_5$Ta$_4$O$_{15}$	AC 1970 **B26** 102
6	hcc		VCo$_3$	
			CsMnF$_3$	JCP 1962 **37** 697
			high-K$_2$(LiAl)F$_6$	AC 1954 **7** 33
			BaTiO$_3$	
			BaTi$_5$O$_{11}$[1] (see text)	AC 1969 **B25** 1444
7	chcccch		Ti(Pt$_{0.89}$Ni$_{0.11}$)$_3$	AC 1969 **B25** 996
8	ccch		Sr$_4$Re$_2$SrO$_{12}$	IC 1965 **4** 235
9	chh	Sm		
			BaPb$_3$	AC 1964 **17** 986
			high-YAl$_3$	AC 1967 **23** 729
		Mo$_2$S$_3$		JSSC 1970 **2** 188
			BaRuO$_3$	
			CsCoF$_3$	ZaC 1969 **369** 117
10	chhch	Ti$_4$S$_5$		
	cchhh		BaFe$_{12}$O$_{19}$[2]	ZK 1967 **125** 437
12	cchh	Fe$_3$S$_4$		AM 1957 **42** 309
		Ti$_5$S$_8$		RTC 1966 **85** 869
			BaCrO$_3$	IC 1969 **8** 286
			Ba$_4$Re$_2$MgO$_{12}$	IC 1965 **4** 235
			Ba$_6$Ti$_{17}$O$_{40}$[3]	AC 1970 **B26** 1645
14	hhhchhc		Ba(Pb$_{0.8}$Tl$_{0.2}$)$_3$	AC 1970 **B26** 653

(1) Two thirds of layers BaO$_7$−remainder O$_8$
(2) Every fifth layer BaO$_3$−remainder O$_4$
(3) Equal numbers of BaO$_{11}$ and Ba$_2$O$_9$ layers (O vacancy in latter).

equal to the number of c.p. spheres. By far the most important types of closest packing are the two simplest ones, namely, hexagonal and cubic, and a great variety of structures are derived by placing cations in various fractions of the tetrahedral and/or octahedral holes in such c.p. arrangements of anions. We shall consider these groups of structures in some detail shortly. As a matter of interest some examples of more complex layer sequences are given in Table 4.4. The sulphides of titanium have been included because they provide examples of a number of c.p. sequences (in addition to h in TiS and TiS_2); contrast the sulphides Cr_2S_3, Cr_5S_6, Cr_7S_8, and Cr_3S_4, which are all h.c.p. with different proportions and arrangements of metal atoms in octahedral holes.

Close-packed arrangements of atoms of two kinds

When discussing the crystal structure of KF (p. 263) we inquire why a 12-coordinated structure is not found, as this c.n. would be expected for spheres of the same radius. It is, however, evidently not possible to form a c.p. layer in which each ion is surrounded entirely by ions of the other kind, for if six X ions are placed around an A ion each X ion already has two X ions as nearest neighbours. This complication does not arise if all the c.p. atoms are anions, as in a hydroxyhalide. One of the forms of Cu(OH)Cl is built of c.p. layers of the type

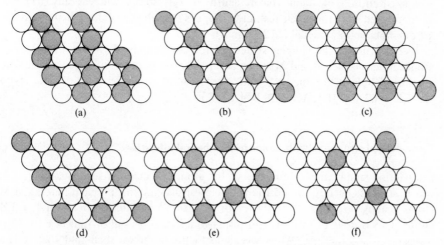

FIG. 4.15. Close-packed layers of spheres of two kinds: (a) AX, (b) AX_2, (c) and (d), AX_3, (e) AX_4, (f) AX_6.

illustrated in Fig. 4.15(a), and many compounds containing atoms of two kinds which are similar in size are built of c.p. layers of composition AX_2 or AX_3. The AX_4 layer of Fig. 4.15(e) is the basis of the structure of $MoNi_4$, and the AX_6 layer of Fig. 4.15(f) represents the arrangement of K^+ and F^- ions in $KOsF_6$.

The only c.p. AX_2 layer in which A atoms are not adjacent is that of Fig. 4.15(b). This type of layer is found in WAl_5, where alternate layers have the compositions Al_3 and WAl_2, and in a number of silicides with closely related structures (p. 790).

The two simplest c.p. AX_3 layers in which A atoms are not adjacent are shown in Fig. 4.15(c) and (d). They form the bases of the structures of three groups of compounds.

(i) Intermetallic compounds AX_3, for example, $TiAl_3$ and $ZrAl_3$, which are superstructures of the c.c.p. structure of Al, being built of layers (c) and layers (c) and (d) respectively. Both $SnNi_3$ and $TiNi_3$ are built of (d) layers, the layer sequences being h and hc respectively.

(ii) Complex halides and oxides $A_xB_yX_{3x}$. The formation of numerous complex halides containing K and F or Cs and Cl in the ratio 1 : 3 is related to the fact that these pairs of ions of similar sizes can form c.p. AX_3 layers. Similarly there are many complex oxides containing a large ion such as Ba^{2+} which is incorporated in close-packed MO_3 layers between which smaller ions of a second metal occupy octahedral holes between six O atoms. The two AX_3 layers illustrated in Fig. 4.15 differ in an important respect. Layers (c) cannot be superposed so as to produce octahedral holes surrounded by 6 X atoms without at the same time bringing A atoms into contact, and therefore (c) layers are not found in complex halides or oxides. Later in this chapter we survey briefly the structures of these compounds which are based on c.p. layers of type (d) in Fig. 4.15.

(iii) Hydroxyhalides $M_2X(OH)_3$. In these compounds the c.p. layers are composed entirely of anions (halide and hydroxyl), so that whereas only (d) layers are permitted in (ii) no such restriction applies to hydroxyhalides. Their structures are described in more detail in Chapter 10.

We now derive the simpler structures for compounds M_mX_n in which the X atoms are close-packed and M atoms occupy various fractions of tetrahedral and/or octahedral holes. (For the systematic derivation of structures built of AX_n layers see: ZK 1967 **124** 104; AC 1968 **B24** 1477.)

Close-packed structures with atoms in tetrahedral interstices

Considering simply the coordinates of the atoms we may list sets of structures in which X atoms occupy the positions of closest packing (h.c.p., c.c.p., etc.) and M atoms occupy positions of tetrahedral or octahedral coordination between 4 or 6 X atoms. This is done in Table 4.5 for tetrahedral structures and in Table 4.6 (p. 142) for octahedral structures. When we come to the latter we shall find that X is a halogen or chalconide and M is a metal atom, and this classification provides a realistic description of the structures. The situation is, however, rather different for the tetrahedral structures. There are twice as many tetrahedral interstices as c.p. atoms and the formula is therefore M_2X if all the tetrahedral positions are occupied. In the c.c.p. M_2X structure (*antifluorite structure*) X is surrounded by 8 M atoms at the vertices of a cube (Fig. 4.16(a)). If only the fraction $\frac{3}{4}$, $\frac{1}{2}$, or $\frac{1}{4}$ of the tetrahedral positions is occupied (in a regular manner) the coordination number of X falls to 6, 4, or 2 respectively. The corresponding coordination groups are derived by removing the appropriate number of M atoms from the cubic coordination group of Fig. 4.16(a) to give those shown at (b)–(f). These groups represent the arrangements of nearest neighbours of

(a) O^{2-} in Li_2O (antifluorite structure) or Ca^{2+} in CaF_2 (fluorite),

(b) and (c) two kinds of Mn^{3+} in O_3Mn_2,

(d) Zn or S in ZnS (zinc-blende),

(e) Pt in PtS, and

(f) Pb in PbO.

TABLE 4.5

Close-packed tetrahedral structures

Fraction of tetrahedral holes occupied	Sequence of c.p. layers		Formula	C.N.'s of M and X
	AB...	ABC...		
All	*	Li_2O (antifluorite)	M_2X	4 : 8
$\frac{3}{4}$	*	$O_3Bi_2\ O_6Sb_4\ O_3Mn_2$	M_3X_2	4 : 6
$\frac{1}{2}$	Wurtzite β-BeO	Zinc-blende PtS PbO	MX	4 : 4
$\frac{3}{8}$	Al_2ZnS_4	Al_2CdS_4 Cu_2HgI_4	M_3X_4	
$\frac{1}{3}$	β-Ga_2S_3	γ-Ga_2S_3	M_2X_3	
$\frac{1}{4}$	β-$ZnCl_2$	HgI_2 (γ-$ZnCl_2$) SiS_2 OCu_2 (α-$ZnCl_2$)	MX_2	4 : 2
$\frac{1}{6}$	Al_2Br_6	In_2I_6	MX_3	
$\frac{1}{8}$	$SnBr_4$	SnI_4 OsO_4	MX_4	4 : 1

This purely geometrical treatment regards Ca^{2+} as the c.p. ion in CaF_2, Mn^{3+} in Mn_2O_3, Pt in PtS, and Pb in PbO. As noted elsewhere, the description of the fluorite structure as a c.c.p. assembly of Ca^{2+} ions with F^- ions in the tetrahedral interstices is not a realistic description of the structure since F^- is appreciably larger than Ca^{2+}. It is preferable to place in Table 4.5 the antifluorite structure of compounds such as Li_2O. Similar considerations apply to the description of the

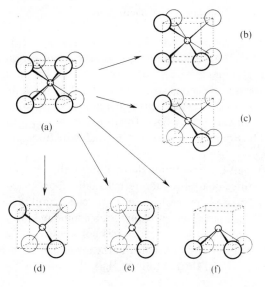

FIG. 4.16. Coordination groups in the fluorite family of structures.

C–Mn_2O_3 structure as a c.p. arrangement of Mn^{3+} ions with three-quarters of the tetrahedral holes occupied by O^{2-} ions. Although this structure is (geometrically) related in this simple way to the fluorite structure, the O^{2-} ions occupying three-quarters of the F^- positions of CaF_2, the tetrahedral arrangement of cations around O^{2-} is probably the factor determining the structure rather than the packing of the metal ions. Some other points relating to these tetrahedral structures should be noted.

The first is that in the cubic CaF_2 (and Li_2O), and ZnS structures the atomic positions are exactly those of cubic closest packing and certain of the tetrahedral interstices. In PtS and PbO this is not so. In the 'ideal' PtS (PdO) structure, with c.c.p. metal atoms, S would be in regular tetrahedral holes but the four S atoms

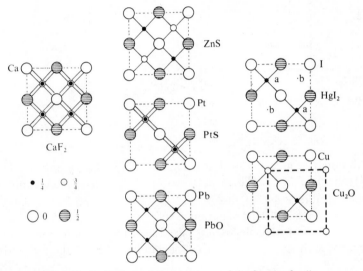

FIG. 4.17. Projections of the structures of the fluorite family.

would form a plane rectangular group around Pt, with edges in the ratio $\sqrt{2} : 1$. If the structure is elongated, to become tetragonal with $c : a = \sqrt{2}$, the coordination group around the metal would become a square. The actual structure is a compromise, with $c : a = 1\cdot24$. Similarly, the actual structure of PbO does not have c.c.p. Pb atoms but is tetragonal with $c : a = 1\cdot25$. This represents a compromise between the cubic structure ($c : a = 1$) and a tetragonal structure ($c : a = \sqrt{2}$) with body-centred cubic packing of Pb atoms. The PbO and related structures are discussed in more detail on p. 218.

The geometrical relations between the structures derived from the fluorite structure by removing one-half or three-quarters of the tetrahedrally coordinated atoms are shown in Fig. 4.17, referred in each case to a cubic unit cell corresponding to that of fluorite. The simplest layer structure for a compound MX_2 of this family, having atoms at the positions aa in each cell, is not known, but a variation with doubled c axis and the positions a occupied in the lower half of the cell and b in the upper half is the structure of the red form of HgI_2. The other MX_2

138

structure of Fig. 4.17, the cuprite (Cu_2O) structure, has collinear bonds from the Cu atoms (large circles). This structure is usually referred to the unit cell indicated by the heavier broken lines. It is unique among structures of inorganic compounds MX_2 for it consists of two identical interpenetrating networks which are not connected by any primary Cu–O bonds.

The second point regarding Table 4.5 is that the structures marked by an asterisk are not formed. The hypothetical structure in which all tetrahedral holes are occupied in h.c.p. is illustrated in Fig. 4.18(a). Placing the c.p. atoms at $\pm(\frac{2}{3}\ \frac{1}{3}\ \frac{1}{4})$ the coordinates of the 'interstitial' atoms would be $\pm(\frac{2}{3}\ \frac{1}{3}\ \frac{5}{8})$ and $\pm(\frac{1}{3}\ \frac{2}{3}\ \frac{1}{8})$, that is, these atoms would occur in pairs separated by only $c/4$, where $c/2$ is the perpendicular distance between c.p. layers. In other words there would be pairs of *face-sharing* tetrahedra. Such close proximity of interstitial atoms would also obtain in the corresponding $ABAC\ldots$ structure, but not in the c.c.p. antifluorite structure, which is shown in Fig. 4.18(b) referred to a hexagonal unit cell. There would be

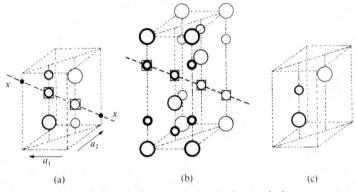

FIG. 4.18. Tetrahedral holes (smaller open circles) in close-packed structures (see text).

the same close approach of interstitial atoms in a h.c.p. structure M_3X_2 in which three-quarters of the tetrahedral holes are occupied, but this is no longer so if only one-half of these holes are filled, as in the wurtzite structure. However, it is possible to have an $ABAC\ldots$ structure for a compound M_3X_2 in which M atoms occupy three-quarters of the tetrahedral holes provided that no very close pairs ('face-sharing tetrahedral holes') are occupied simultaneously. Such a structure is adopted by high-Be_3N_2 (p. 670), in the unit cell of which two sets of tetrahedral sites are fully occupied and the third Be is disordered between two mutually exclusive tetrahedral sites.

A third point concerning the tetrahedral structures of Table 4.5 is that there are no known examples of *layer* structures in which tetrahedral interstices are occupied between alternate pairs of c.p. layers. Occupation of all tetrahedral sites between a pair of c.p. layers gives a layer of composition MX in which each MX_4 tetrahedron shares three edges with three other tetrahedra and also three vertices with a second set of six tetrahedra (Fig. 4.19(a)). The structure preferred by LiOH and OPb is the layer based on the square net in which each $Li(OH)_4$ (OPb_4) tetrahedron shares four edges (p. 100). Both layers are found as components of the La_2O_3 (Ce_2O_2S)

139

and U_2N_2Sb structures respectively (p. 1004). Occupation in the most symmetrical way of one-half of the tetrahedral holes between a pair of c.p. layers gives a layer of composition MX_2, the tetrahedral analogue of the CdI_2 layer. In this layer (Fig. 4.19(b)) each MX_4 tetrahedron shares vertices with three others and the fourth vertex is unshared, so that one-half of the X atoms have three equidistant M neighbours and the others only one M neighbour. While this is presumably the reason why halides and oxides do not have such a structure this type of non-equivalence of the c.p. atoms is inevitable in a molecular crystal such as Al_2Br_6. In this crystal pairs of adjacent tetrahedral holes indicated by the squares in Fig. 4.18(a) are occupied, leading to the formation of discrete Al_2Br_6 molecules consisting of two $AlBr_4$ groups sharing an edge. One-third of the c.p. atoms have 2 Al and two-thirds have 1 Al neighbour. A layer closely related to that of Fig. 4.19(b) is that of AlOCl (and GaOCl), but here there are not two c.p. layers. There is the same type of linking of the tetrahedra but equal numbers of the

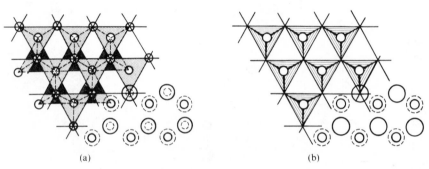

(a)   (b)

FIG. 4.19. (a) Pair of c.p. X layers with M atoms in all tetrahedral holes between them (composition MX). (b) Pair of c.p. X layers with M atoms in one-half of the tetrahedral holes (composition MX_2).

unshared vertices project on opposite sides of one layer of c.p. X atoms (see Fig. 10.16, p. 408). In the layer structure of red HgI_2 the planes containing the metal atoms are inclined to those of the c.p. halogen layers; see MSIC, p. 59.

Finally we may note that there is no h.c.p. analogue of the SiS_2 structure. Continued sharing of opposite edges of MX_4 tetrahedra to give a compound MX_2 is not possible in hexagonal closest packing because there are no tetrahedral sites at the points marked X in Fig. 4.18(a), but it can occur in cubic closest packing as may be seen by comparing Fig. 4.18(b) with (a). Occupation of one-quarter of the tetrahedral holes in a c.c.p. array of S atoms as at (b) gives chains formed from tetrahedra sharing opposite edges.

Close-packed structures with atoms in octahedral interstices

The most important structures of this kind are set out in Table 4.6. The relations between these structures may be illustrated in two ways: (a) we may consider the ways in which various proportions of the M atoms can be removed from the MX structure, in which all the octahedral holes are occupied, or (b) we may focus our attention on the patterns of sites occupied by the M atoms between the close-packed layers.

(a) The MX structures for hexagonal and cubic closest-packing are the NiAs and NaCl structures, in which the M atoms are arranged around X at the vertices of a trigonal prism or octahedron respectively. A feature of the NiAs structure is the close proximity of Ni atoms, suggesting interaction between the metal atoms in at least some compounds with this structure, which is not adopted by ionic compounds. In NaCl itself the Cl^- ions are not in contact, since Na^+ is rather too large for an octahedral hole between c.p. Cl^- ions. These ions are in contact in LiCl, compare

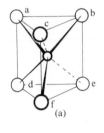

$$Cl–Cl \text{ in } NaCl \ 3·99 \ Å,$$
$$Cl–Cl \text{ in } LiCl \ \ 3·63 \ Å.$$

The c.c.p. structure would therefore more logically be called the LiCl structure, but since it has become generally known as the NaCl structure we keep that name here.

Removal of a proportion of the M atoms from these structures in a regular manner leaves X with fewer than six neighbours. For example, we may remove one-half of the Ni atoms from the NiAs structure so as to leave the close-packed (As) atoms with three neighbours arranged in one of the following ways (Fig. 4.20(a)):

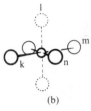

    (i) *abc*, a pyramidal arrangement with bond angles of 90°,

    (ii) *abf*, a nearly planar arrangement,

    (iii) *abd*, a pyramidal arrangement with bond angles 70°, 90°, and 132°.

FIG. 4.20. Trigonal prismatic and octahedral coordination groups.

Of these (iii) is not known; (i) corresponds to the CdI_2 (*C*6) structure. Figure 4.21(a) shows an elevation of the NiAs structure (viewed along an *a* axis) in which the small circles represent rows of Ni atoms perpendicular to the plane of the paper. Removal of those shown as dotted circles gives an MX_2 structure in which X has three nearly coplanar M neighbours. This is the $CaCl_2$ structure. A slight adjustment of the positions of the c.p. atoms gives the rutile structure (Fig. 4.21(b)), in which X has three coplanar neighbours. It is seen here viewed along the tetragonal *c* axis, that is, parallel to the chains of octahedra which are emphasized in (c).

Similarly, removal of one-half of the M atoms from the NaCl structure leaves X

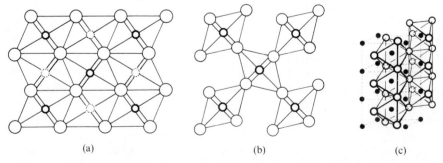

      (a)   (b)   (c)

FIG. 4.21. Relation between the NiAs and rutile structures (see text).

141

TABLE 4.6

Close-packed octahedral structures

Fraction of octahedral holes occupied	Sequence of c.p. layers		Formula	C.N.s of M and X
	AB...	ABC...		
All	NiAs[a]	NaCl	MX	6 : 6
$\frac{2}{3}$	α-Al$_2$O$_3$ LiSbO$_3$	*	M$_2$X$_3$	6 : 4
$\frac{1}{2}$[b]	*Layer structures* CdI$_2$	CdCl$_2$	MX$_2$	6 : 3
	Framework structures CaCl$_2$ Rutile NiWO$_4$ α-PbO$_2$ α-AlOOH	Atacamite Anatase		
$\frac{1}{3}$	*Chain structures* ZrI$_3$	—	MX$_3$	6 : 2
	Layer structures BiI$_3$	YCl$_3$		
	Framework structures RhF$_3$	—		
$\frac{1}{4}$	α-NbI$_4$ (chain)	NbF$_4$ (layer)	MX$_4$	6 : $\frac{2}{1}$
$\frac{1}{5}$	Nb$_2$Cl$_{10}$ Ru$_4$F$_{20}$ } (molecular)	U$_2$Cl$_{10}$ Mo$_4$F$_{20}$ } (molecular) UF$_5$ (chain)	MX$_5$	6 : $\frac{2}{1}$
$\frac{1}{6}$[c]	α-WCl$_6$	—	MX$_6$	6 : 1

[a] For structures intermediate between the NiAs and CdI$_2$ (C6) structures see sulphides of chromium (p. 621) and the NiAs structure (p. 609).

[b] For other structures in which one-half of the octahedral holes are occupied in a c.p. XY$_3$ assembly see hydroxyhalides M$_2$X(OH)$_3$ and in particular Table 10.15 (p. 411).

[c] For other structures in which $\frac{1}{4}$, $\frac{1}{6}$, or $\frac{1}{8}$ of the octahedral holes are occupied in a c.p. AX$_3$ assembly see 'Structures built from c.p. AX$_3$ layers' and in particular Table 4.10 (p. 155).

with three neighbours at three of the vertices of an octahedron, one of the combinations *lmn* or *klm* in Fig. 4.20(b). The two simplest ways of leaving X with the pyramidal arrangement of neighbours (*lmn*) are to remove alternate layers of X atoms parallel to the plane (111) giving the CdCl$_2$ (layer) structure (Fig. 4.22(a)) or to remove alternate rows of X atoms as in Fig. 4.22(b). It is an interesting fact that no dihalide crystallizes with this structure although the environments of M (octahedral) and X (pyramidal) are the same as in the CdCl$_2$ structure. Figure 4.22(b) does represent, however, the idealized structure of atacamite, Cu$_2$(OH)$_3$Cl, three-quarters of the X positions being occupied by OH groups.

The coplanar arrangement of neighbours *klm* in Fig. 4.20(b) arises by removing one-half of the Na atoms from the NaCl structure in the manner shown in Fig.

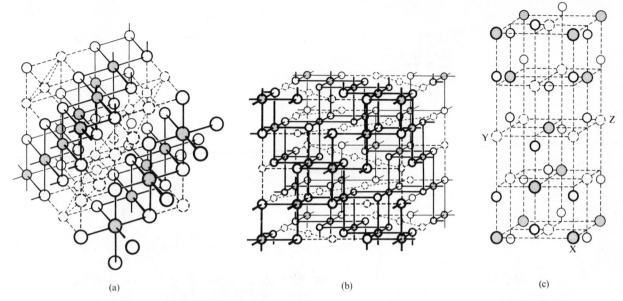

FIG. 4.22. Structures geometrically related to the NaCl structure: (a) $CdCl_2$, (b) atacamite, $Cu_2(OH)_3Cl$, (c) anatase, TiO_2. Metal ions removed from the NaCl structure are shown as dotted circles.

4.22(c); this necessitates doubling the c axis of the unit cell. The configuration klm of nearest neighbours is not suitable for an essentially ionic crystal, but by a slight adjustment of the atomic positions the coordination group around X becomes an approximately equilateral triangle. The hypothetical structure would have $c = 2a$ and $z = \frac{1}{4}$; in anatase, one of the polymorphs of TiO_2, $c : a = 2 \cdot 51$ and $z_0 = 0 \cdot 21$.

(b) Alternatively, we may consider the ways in which various proportions of the octahedral holes may be occupied in each type of closest packing. There are four variables:

(i) the sequence of c.p. layers,
(ii) the degree of occupancy of sites between successive pairs of layers,
(iii) there are different *patterns* of sites for a given fraction occupied, and
(iv) there is still the possibility of translating the sets of M atoms between successive pairs of layers relative to one another. (An exercise on this subject is included in MSIC.)

(i) The sequence of c.p. layers has already been discussed.

(ii) A particular composition implies the occupation of some fraction of the *total* number of octahedral sites, for example, one-half for MX_2. This may be achieved by filling all sites between *alternate* pairs of c.p. layers, one-half of the sites between each pair of layers, or alternately one-third and two-thirds or one-quarter and three-quarters of the sites between pairs of successive layers, and, of course, in more complex ways.

143

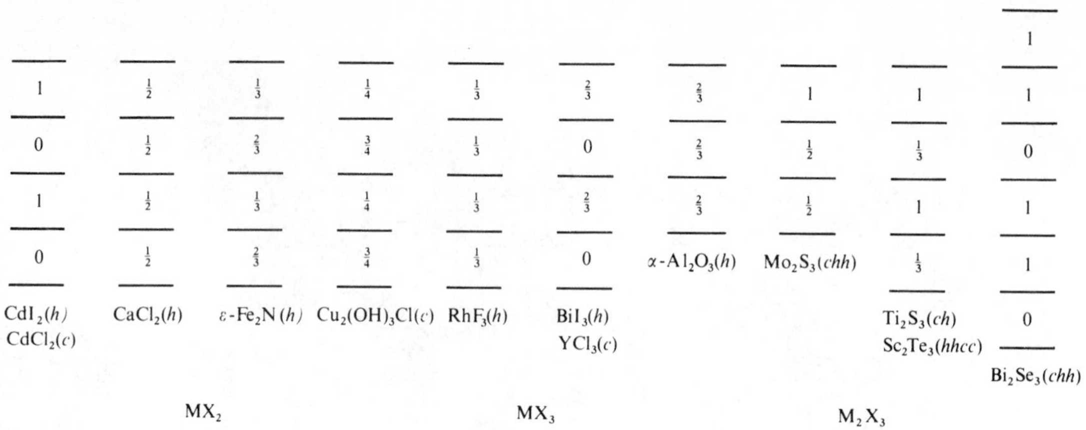

| | 1 |
| CdI₂ col | ... |

FIG. 4.23 diagram — fractions of octahedral sites occupied between c.p. layers:

$CdI_2(h)$ $CdCl_2(c)$	$CaCl_2(h)$	ε-$Fe_2N(h)$	$Cu_2(OH)_3Cl(c)$	$RhF_3(h)$	$BiI_3(h)$ $YCl_3(c)$	α-$Al_2O_3(h)$	$Mo_2S_3(chh)$	$Ti_2S_3(ch)$ $Sc_2Te_3(hhcc)$	$Bi_2Se_3(chh)$
									1
1	$\tfrac{1}{2}$	$\tfrac{1}{3}$	$\tfrac{1}{4}$	$\tfrac{1}{3}$	$\tfrac{2}{3}$	$\tfrac{2}{3}$	1	1	1
0	$\tfrac{1}{2}$	$\tfrac{2}{3}$	$\tfrac{3}{4}$	$\tfrac{1}{3}$	0	$\tfrac{2}{3}$	$\tfrac{1}{2}$	$\tfrac{1}{3}$	0
1	$\tfrac{1}{2}$	$\tfrac{1}{3}$	$\tfrac{1}{4}$	$\tfrac{1}{3}$	$\tfrac{2}{3}$	$\tfrac{2}{3}$	$\tfrac{1}{2}$	1	1
0	$\tfrac{1}{2}$	$\tfrac{2}{3}$	$\tfrac{3}{4}$	$\tfrac{1}{3}$	0			$\tfrac{1}{3}$	1
									0

MX₂   MX₃   M₂X₃

FIG. 4.23. Fractions of octahedral sites occupied between c.p. layers to give the compositions MX₂, MX₃, and M₂X₃. The c.p. sequence is shown after each formula.

These possibilities are shown diagrammatically in Fig. 4.23 together with some simple arrangements for compositions M_2X_3 and MX_3.

Two of the structures of Fig. 4.23 are *simple* layer structures, that is, metal atoms occur only between alternate pairs of c.p. X layers. Note that a simple layer structure is not possible for a c.p. octahedral M_xX_y structure if $x/y > \tfrac{1}{8}$ (as for M_3X_4 or M_2X_3) because all the octahedral holes between alternate pairs of c.p. layers are filled when x/y reaches the value $\tfrac{1}{2}$. More generally it can be shown that a periodic 2-dimensional system of composition M_2X_3 is not possible if M is to be 4-, and X 6-coordinated; a simple octahedral M_2X_3 layer is a special example of the more general topological limitation. Bi_2Se_3 provides an example of a more complex layer structure.

(iii) The small black (or open) circles in Fig. 4.24(a), (b), and (c) show three ways of filling one-half of the octahedral sites between a pair of c.p. layers, and examples of structures with these patterns of sites include

(a)   h.c.p.   c.c.p.
    $CaCl_2$ Co_2C
    (TiO_2, rutile)
(b) α-PbO_2 ʒ-V_2C ʒ-Fe_2N TiO_2 (anatase)

(c)   ξ-Nb_2C

(In the TiO_2 polymorphs there is some departure from ideal closest packing, as noted elsewhere.)

The most symmetrical ways of filling respectively three-quarters and two-thirds of the octahedral holes are shown by the small black circles in Fig. 4.24(d) and (e); the complementary positions (small open circles), which may be called d' and e', comprise one-quarter and one-third of the sites. As shown in Fig. 4.23 occupation

144

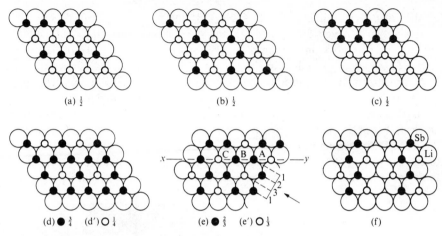

FIG. 4.24. Patterns of octahedral sites between pairs of c.p. layers (see text).

between alternate pairs of layers of (d) and (d′) or (e) and (e′) gives the composition MX_2, as in

(d) + (d′) $Cu_2(OH)_3Cl$ (atacamite)
(e) + (e′) ϵ-Fe_2N, ϵ-V_2C, etc.

Occupation of sites (d) or (e) between *alternate* pairs of layers gives layer structures:

(d) Nb_3Cl_8
(e) BiI_3 (h.c.p.), OTi_3 (h.c.p.), YCl_3 (c.c.p.).

Sites of type (e) are occupied between all successive pairs of layers in α-Al_2O_3 and related structures to which we refer shortly, while occupation of sites (e′) between all successive pairs of layers gives the h.c.p. MX_3 structure of trifluorides such as RhF_3. In this structure each MX_6 octahedron is linked to six others through shared vertices (X atoms); there is no c.c.p. structure of the same topological type, for in the ReO_3 structure the O atoms occupy only three-quarters of the positions of cubic closest packing (see MSIC, p. 61).

The pattern of sites of Fig. 4.24(f) is occupied in $LiSbO_3$.

(iv) Evidently a knowledge of the c.p. sequence and the site pattern is not sufficient to define a structure since the relative translations of the metal atoms are not defined in relation to some fixed frame of reference. We may illustrate this point by reference to a number of structures in which the sites of Fig. 4.24(e) are utilized between each pair of c.p. layers.

For a more complete description of a c.p. structure we may give elevations of the kind shown in Fig. 4.25, the structure being viewed in the direction of the arrow in Fig. 4.24(e). Note that the plane of the projection is not perpendicular to the arrow but is the plane intersecting the paper in the line xy in Fig. 4.24(e). In the corundum (α-Al_2O_3) structure the pattern of sites occupied by Al atoms is that

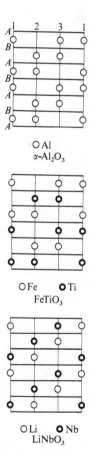

FIG. 4.25. Elevations of h.c.p. structures.

145

of the small black circles of Fig. 4.24(e). If the positions occupied between a particular pair of layers are ① and ② (and all equivalent ones, that is all the black circles) then the positions occupied between successive pairs of (h.c.p.) layers are ② and ③, ③ and ①, and so on. The elevation shown in Fig. 4.25 shows both the layer sequence and the pattern of sites occupied by the metal atoms. The other structures, $FeTiO_3$ and $LiNbO_3$, are superstructures of the corundum structure.

In Chapter 5 we show how certain structures may be constructed from octahedral coordination groups sharing vertices, edges, or faces (or combinations of these), and it is instructive to make models relating this way of describing structures to the description in terms of c.p. anions. For this reason we show in Fig. 4.26 both these types of representation of the anatase structure which should be compared

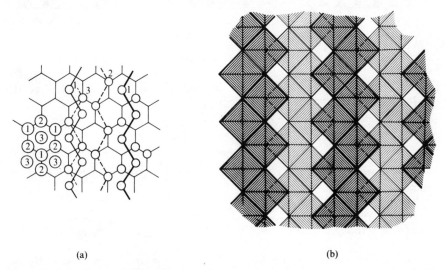

(a)   (b)

FIG. 4.26. The structure of anatase shown (a) as a c.p. structure, (b) as an assembly of edge-sharing octahedra. In (a) the numbers at the left indicate the positions of O atoms in the lowest (1) and succeeding layers, (2), (3), and the numbers against the Ti atoms (small circles) correspond to the c.p. layers on which they rest.

with Fig. 4.22(c), p. 143. When comparing Fig. 4.22(c) with Fig. 4.26 it is important to remember that the illustrations in the latter figure are projections in a direction perpendicular to the c.p. layers. These layers correspond to a plane such as XYZ in Fig. 4.22(c).

Whether vertices, edges, or faces are shared in the final structure depends not only on the patterns of sites occupied between pairs of layers but also on the relation between the sets of sites occupied between successive pairs of layers. For example, occupation of the sites of Fig. 4.24(e′) implies no sharing of X atoms between the octahedra formed around the e′ sites *in one layer*, but vertex-sharing occurs in the 3D structure (RhF_3) which results from the repetition of this pattern of site-filling. On the other hand, the occupation of more than one-third of the octahedral sites between a given pair of c.p. layers already implies edge-sharing between octahedra of one layer. Elevations of the type of Fig. 4.25 show at a

146

glance whether octahedral coordination groups are sharing vertices, edges, or faces with one another, as indicated in Fig. 4.27. Note that because of the 3-fold symmetry (a) and (b) in Fig. 4.27 imply sharing of *three* vertices or edges.

Some related MX_2, $MM'X_4$, and $M_2M'X_6$ structures

Any structure A_mX_n in which A atoms occupy some fraction of the octahedral holes in a c.p. assembly of X atoms is potentially a structure for more complex compounds, as we have already noted for $FeTiO_3$ and $LiNbO_3$, where atoms of two kinds occupy the Al sites in the corundum structure (Fig. 4.25). There are many complex halides and oxides containing cations of two or more kinds which have

Vertex-sharing
(a)

Edge-sharing
(b)

Face-sharing
(c)

FIG. 4.27. The sharing of octahedron vertices, edges, or faces, as indicated in the elevations of Fig. 4.25.

TABLE 4.7
Relations between some h.c.p. structures

Pattern of octahedral sites (*Fig.* 24.4) *occupied*	(a)	(b)	(e) (e')	–
Fractions of sites occupied between successive pairs of c.p. layers	$\frac{1}{2}$	$\frac{1}{2}$	$\frac{2}{3}$ $\frac{1}{3}$	$\frac{4}{9}$ $\frac{5}{9}$
MX_2	Rutile	α-PbO_2	ϵ-Fe_2N	–
$MM'X_4$	$MgUO_4$	$NiWO_4$	–	–
$M_2M'X_6$	Trirutile, $ZnSb_2O_6$	Columbite, (Fe, Mn)Nb_2O_6	Li_2ZrF_6 Li_2NbOF_5	Na_2SiF_6 NiU_2O_6

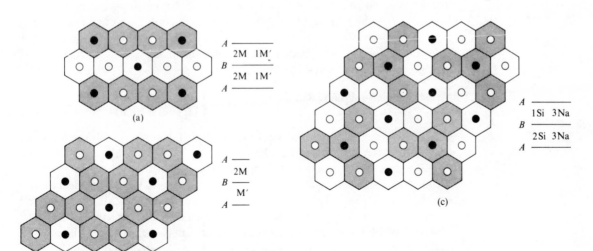

FIG. 4.28. The structures of compounds $M_2M'X_6$: (a) trirutile, (b) Li_2ZrF_6, (c) Na_2SiF_6. The open and filled circles represent M and M' atoms in octahedral holes in a h.c.p. assembly of X atoms. Clear and shaded octahedra contain cations at heights 0 and $c/2$. The sketches at the right show the types of atoms between successive pairs of c.p. layers.

147

similar radii and can therefore occupy, for example, octahedral sites in a c.p. structure. If the sites occupied by M and M' in MM'X$_4$ or M$_2$M'X$_6$ are those occupied by M in MX$_2$ the more complex structures may be described as superstructures of the simpler MX$_2$ structures. For example, MgUO$_4$ and ZnSb$_2$O$_6$ are related in this way to the rutile structure. In NiWO$_4$ the octahedral sites between alternate pairs of c.p. layers in the α-PbO$_2$ structure are occupied by Ni and W atoms, and in the columbite structure there is a more complex type of replacement:

$$\text{Pb; Pb; Pb; Pb.} \quad \text{Ni; W; Ni; W.} \quad \frac{\text{Fe}}{\text{Mn}}; \text{Nb; Nb;} \frac{\text{Fe}}{\text{Mn}}; \text{Nb; Nb.}$$

In all of the structures of Table 4.7 one-half of the octahedral sites are occupied in a h.c.p. array of anions. In Fig. 4.24 we showed some simple ways of occupying sites to give formulae MX$_2$, MX$_3$, and M$_2$X$_3$; the Na$_2$SiF$_6$ structure illustrates yet another way of filling one-half of the octahedral interstices. Three of the M$_2$M'X$_6$ structures are illustrated in Fig. 4.28.

Close-packed structures with atoms in tetrahedral and octahedral interstices

We have noted that all tetrahedral holes will not be occupied in hexagonal close-packing because they are too close together. If all octahedral and all tetrahedral holes are occupied in a cubic close-packed assembly the atomic positions are those of Fig. 9.7 (p. 357), where the small circles would be the c.p. atoms and the large open and shaded circles would represent the atoms in tetrahedral and octahedral holes respectively. It is the structure of BiLi$_3$ and other intermetallic phases. Now although one-third of the Li atoms are in octahedral holes and two-thirds in tetrahedral holes in a c.c.p. of Bi atoms and consequently have respectively six octahedral or four tetrahedral Bi neighbours, these are not the nearest (or the only equidistant) neighbours. Each Li has in fact a cubic arrangement of eight nearest neighbours:

$$\text{Li}_\text{I} \begin{cases} 8 \text{ Li}_\text{II} \text{ (cubic) at } \sqrt{3}a/4 \\ 6 \text{ Bi (octahedral) at } a/2 \end{cases} \qquad \text{Li}_\text{II} \begin{cases} 4 \text{ Bi (tetrahedral)} \\ 4 \text{ Li}_\text{I} \text{ (tetrahedral)} \end{cases} \text{at } \sqrt{3}a/4$$

This is therefore an 8-coordinated structure and it is grouped with the NaTl and related structures in Chapter 29 (p. 1035).

It is only in structures in which small proportions of the two types of hole are occupied that the nearest neighbours of an atom in a tetrahedral (octahedral) hole are *only* the nearest 4 (6) c.p. atoms. This is true, for example, in the spinel and olivine structures (Table 4.8). The Co$_9$S$_8$ structure is closely related to the spinel structure, which has 32 c.p. O atoms in the cubic unit cell. In Co$_9$S$_8$ there are 32 c.p. S atoms with 4 Co in octahedral and 32 Co in tetrahedral holes; compare Co$_3$S$_4$ with a slightly distorted spinel structure, also with 32 S in the cubic unit cell, but 16 Co in octahedral and 8 Co in tetrahedral holes. The spinel structure is discussed in detail on p. 490; for the olivine structure see p. 811. An expanded version of part of Table 4.8 will be found on p. 619, where the structures of a number

148

TABLE 4.8

Tetrahedral and octahedral coordination in close-packed structures

Fractions of holes occupied		c.c.p.	h.c.p.
Tetrahedral	Octahedral		
$\frac{1}{8}$	$\frac{1}{8}$		Al_2CoCl_8
$\frac{1}{8}$	$\frac{1}{6}$	Cr_5O_{12}	
$\frac{1}{8}$	$\frac{1}{2}$	$MgAl_2O_4$	Mg_2SiO_4
		(spinel)	(olivine)
$\frac{1}{8}$	$\frac{1}{4}$	$CrVO_4$	
$\frac{1}{2}$	$\frac{1}{8}$	Co_9S_8	
$\frac{1}{2}$	$\frac{1}{2}$	$\beta\text{-}Ga_2O_3$	

of metallic sulphides are described, and in Chapter 5 we list some structures (not necessarily close-packed) in which there is tetrahedral and octahedral coordination of the same element.

An alternative representation of close-packed structures

In the foregoing treatment of structures emphasis is placed on the coordination of the atoms occupying the interstices in the c.p. assemblies. Except for certain of the 'tetrahedral' structures of compounds M_2X and M_3X_2 of Table 4.5 these are metal atoms, the c.p. assembly being that of the anions. As we point out elsewhere the determining factor in some structures appears to be the environment of the anion rather than that of the cation, and this is emphasized in an alternative representation of c.p. structures.

Crystal structures may be described in terms of the coordination polyhedra MX_n of the atoms or in terms of their duals, that is, the polyhedra enclosed by planes drawn perpendicular to the lines M–X joining each atom to each of its neighbours at the mid-points of these lines. Each atom in the structure is then represented as a polyhedron (*polyhedral domain*), and the whole structure as a space-filling assembly of polyhedra of one or more kinds. We can visualize these domains as the shapes the atoms (ions) would assume if the structure were uniformly compressed. For example, h.c.p. and c.c.p. spheres would become the polyhedra shown in Fig. 4.29. These polyhedra are the duals of the coordination polyhedra illustrated in Fig. 4.5. These domains provide an alternative way of representing relatively simple c.p. structures (particularly of binary compounds) because the vertices of the domain are the positions of the interstices. The (8) vertices at which three edges meet are the tetrahedral interstices, and those (6) at which four edges meet are the octahedral interstices. Table 4.9 shows the octahedral positions occupied in some simple structures; c.p. structures in which tetrahedral or tetrahedral and octahedral sites are occupied may be represented in a similar way. (For examples see JSSC 1970 **1** 279.)

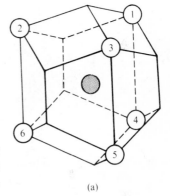

(a)

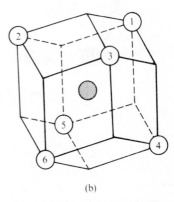

(b)

FIG. 4.29. The polyhedral domains for (a) hexagonal and (b) cubic closest packing, showing the six positions of octahedral coordination around a c.p. sphere. The remaining vertices of the domains are tetrahedral sites.

149

TABLE 4.9

Representation of c.p. structures by polyhedral domains

	C.N. of c.p. X	Vertices occupied	Structure MX_n
h.c.p. structures (Fig. 4.29 (a))	6	1 2 3 4 5 6	NiAs
	4	2 3 4 5	α-Al_2O_3
	3	2 4 5	$CaCl_2$
		4 5 6	CdI_2
	2	2 5	RhF_3
		2 6	ZrI_3
		5 6	BiI_3
c.c.p. structures (Fig. 4.29 (b))	6	1 2 3 4 5 6	NaCl
	4	2 4 5 6	Sc_2S_3
	3	4 5 6	$CdCl_2$
		1 4 6	TiO_2 (anatase)
	2	4 6	YCl_3
		1 6	ReO_3

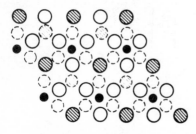

FIG. 4.30. Two close-packed AX_3 layers (full and dotted circles) showing the positions, midway between the layers, for metal atoms (small black circles) within octahedra of X atoms.

Structures built from close-packed AX_3 layers

We noted earlier that certain pairs of ions of similar size can together form c.p. layers AX_3 and that of the two possible AX_3 layers with non-adjacent A atoms only one is found in complex halides and oxides $A_xB_yX_{3x}$. This layer (Fig. 4.15(d)) can be stacked in closest packing to form octahedral X_6 holes between the layers without bringing A ions into contact. Figure 4.30 shows two such layers and the positions of octahedral coordination for the B atoms between the layers. The number of these X_6 holes is equal to the number of A atoms in the structure. All of these positions or a proportion of them may be occupied in a complex halide or oxide by cations B carrying a suitable charge to give an electrically neutral crystal $A_xB_yX_{3x}$, where A and X are, for example, K^+ and F^-, Cs^+ and Cl^-, Ba^{2+} and O^{2-}, etc. The formula depends on the proportion of B positions occupied:

all occupied	ABX_3
two-thirds	$A_3B_2X_9$
one-half	A_2BX_6

These fractions are, of course, respectively $\frac{1}{4}$, $\frac{1}{6}$, and $\frac{1}{8}$ of the total number of octahedral holes if we disregard the difference between the two kinds of c.p. atom A and X. In A_2BX_6 there are discrete octahedral BX_6 groups; in $A_3B_2X_9$ and ABX_3 the X : B ratios show that X atoms must be shared between BX_6 groups. A feature of the latter structures, which will be evident from Fig. 4.31, is that only *vertices* or *faces* of octahedral BX_6 groups are shared; this is in marked contrast to the edge-sharing found in many oxides and halides. This absence of edge-sharing is simply a consequence of the structure of the AX_3 layers, as may be seen by studying models. When a third layer is placed on a pair of layers there are two possible orientations of this layer, one leading to vertex-sharing and the other to face-sharing between BX_6 octahedra.

All the c.p. layer sequences are possible for AX_3 layers, and it is found that those most frequently adopted are the simplest ones having only one or two kinds of non-equivalent sphere, namely, those sequences which repeat after every 2, 3, 4, 6, 9, or 12 layers. Examples of structures based on c.p. AX_3 layers are included in the more general Table 4.4 (p. 134). Many compounds crystallize with more than one type of closest packing, for example, $BaMnO_3$ with the 2, 4, and 9 layer

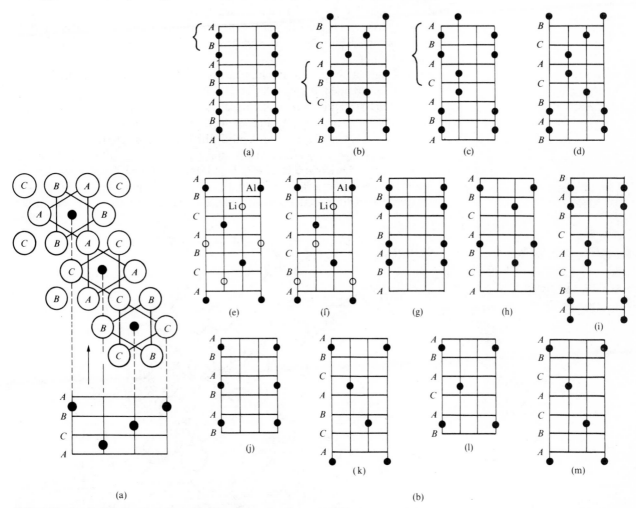

FIG. 4.31. At the left the plan shows the X atoms only of the c.p. AX_3 layers in the relative positions A, B and C, and the sites between pairs of layers AB, BC, and CA for B metal atoms (small black circles) surrounded octahedrally by six X atoms. The elevation shows the layers (horizontal lines) viewed in the direction of the arrow. (a)–(m), diagrammatic elevations of c.p. structures $A_xB_yX_{3x}$ in which there is octahedral coordination of B by 6 X atoms: (a)–(d) ABX_3, (e) and (f) low- and high-temperature forms of K_2LiAlF_6, (g)–(i) $A_3B_2X_9$, (j)–(m) A_2BX_6.

151

structures. In some cases this phenomenon is probably more accurately described as polytypism rather than polymorphism, as in the case of the very numerous polytypes of SiC and ZnS. For example, crystals of $BaCrO_3$, which is only formed under a pressure exceeding 3000 atmospheres, have been shown to have 4-, 6-, 9-, 14-, and 27-layer sequences of c.p. BaO_3 layers.

If all the octahedral X_6 holes are occupied by B ions there is only one possible structure (ABX_3) corresponding to each c.p. sequence, but if fewer are occupied there are various possible arrangements of the B ions in the available holes. This is exactly comparable to c.p. structures for binary compounds, there being one structure (NiAs, NaCl, etc.) for each type of closest-packing if all the octahedral holes are occupied but alternative arrangements of the same number of cations if only a fraction of the holes are filled.

We shall indicate here only structures with 2, 3, 4, and 6 layer sequences of c.p. layers, and for compounds $A_3B_2X_9$ and A_2BX_6 we shall illustrate first only the simplest of the structures for each c.p. sequence. The possible structures may be derived in the following way. By analogy with the nomenclature for c.p. layers of identical spheres we call layers *A*, *B*, and *C*, these positions referring to the larger cell of the AX_3 layer (Fig. 4.31). All A atoms of an *A* layer fall vertically above points such as *A*, those of a *B* layer over *B* and similarly for a *C* layer. A translation of *d* converts an *A* into a *B* layer, *B* into *C*, and *C* into *A*. In any sequence of these layers, stacked to form octahedral X_6 holes between the layers, the positions for

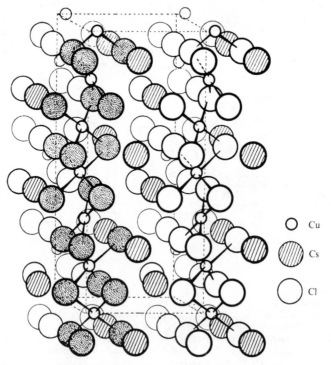

FIG. 4.32. The crystal structure of $CsCuCl_3$.

the B atoms in a compound ABX_3 also lie above the positions A, B, or C. Between A and B layers they lie above C, between B and C layers above A, and between A and C layers above B. The positions of the B atoms may therefore be shown in elevations of the same kind as those of binary c.p. structures given earlier.

ABX_3 structures

In these structures all the octahedral X_6 holes are occupied by B atoms. Figure 4.31(a)–(d) shows the structures based on the layer sequences *h, c, hc,* and *hcc.*

In the simple structure (a), with hexagonal closest packing of $A + 3X$, the octahedral BX_6 groups are stacked in columns sharing a pair of opposite faces. This is the structure of $CsNiCl_3$, $BaNiO_3$, and $LiI . 3H_2O$ (or $ILi(H_2O)_3$). A variant of this structure is adopted by $CsCuCl_3$, in crystals of which the hexagonal closest-packing is slightly distorted (by a small translation of the layers relative to one another) so as to give Cu only four (coplanar) nearest neighbours and two more at a considerably greater distance completing a distorted octahedral group. The Cu atoms are not vertically above one another, as implied by the elevation of Fig. 4.31(a) but are displaced a little off the axis of the chain of octahedra, actually in a helical array around a 6-fold screw axis (Fig. 4.32).

The structure (b), with cubic close-packed $A + 3X$ atoms, is the perovskite structure. This is illustrated in Fig. 4.33 with an A atom at the origin showing part of a c.p. AX_3 layer. It is more easily visualized as a 3D system of vertex-sharing BX_6 octahedra having B atoms at the corners of the cubic unit cell and X atoms midway along the edges, that is, as the ReO_3 structure of Fig. 5.18(a) (p. 173) with the A atom added at the body-centre of the cell. The ideal (cubic) perovskite structure (or a variant with lower symmetry) is adopted by many fluorides ABF_3 and complex oxides ABO_3; it is discussed in more detail in Chapter 13.

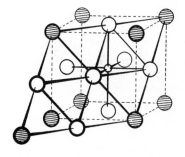

FIG. 4.33. The perovskite structure of $RbCaF_3$ showing a Ca^{2+} ion (small circle) surrounded by six F^- ions and a layer of close-packed Rb^+ and F^- ions (large shaded and open circles respectively).

The structure (c) is adopted by the high-temperature form of $BaMnO_3$. The MnO_6 octahedra are grouped in pairs with a face in common, the pairs are linked together by sharing vertices.

The 6-layer structure (d), with single and also face-sharing pairs of BX_6 octahedra, is that of hexagonal $BaTiO_3$ and $CsMnF_3$.

The basic structures (a)–(d) can be utilized by compounds with more complex formulae in a number of ways. The complex halide with empirical formula $CsAuCl_3$ is actually $Cs_2Au^IAu^{III}Cl_6$. It has a very distorted perovskite structure in which there are linear $(Cl-Au^I-Cl)^-$ and square planar $(Au^{III}Cl_4)^-$ ions; it is described in Chapter 10. The fluoride K_2LiAlF_6 has two enantiotropic forms. In the low-temperature form the Li^+ and Al^{3+} ions occupy in a regular manner the B positions of the perovskite structure, (b); this superstructure is the cryolite structure. The high-temperature form is a superstructure of the simple 6-layer structure (d). Elevations of these two superstructures are shown in Fig. 4.31(e) and (f). A much more complex variant of the 6-layer structure may be mentioned here though it is not built exclusively of c.p. AX_3 layers. In $BaTi_5O_{11}$ the layers are of two kinds, some consisting entirely of oxygen (O_8 in the unit cell) and others in which one-eighth of the O atoms are replaced by Ba (giving the composition BaO_7). There is room for 4 Ti between a BaO_7 and an O_8 layer but only for 2 octahedrally coordinated Ti atoms between BaO_7 layers, so that the structure may be represented diagrammatically

$$(BaO_7) \quad BaO_7 \quad O_8 \quad BaO_7 \quad BaO_7 \quad O_8 \quad BaO_7 \quad (BaO_7) \quad Ba_4Ti_{20}O_{44}$$
$$2\,Ti \quad\quad 4\,Ti \quad 4\,Ti \quad 2\,Ti \quad 4\,Ti \quad 4\,Ti \quad 2\,Ti \quad\quad = BaTi_5O_{11}$$

References to some of the structures mentioned are included in the summarizing Table 4.4 (p. 134); others are included in later chapters.

$A_3B_2X_9$ *structures*

In these structures B atoms occupy two-thirds of the X_6 holes. Elevations of structures with not more than six layers in the repeat unit are shown in Fig. 4.31,

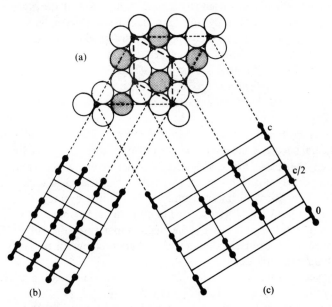

FIG. 4.34. The crystal structure of $Cs_3Tl_2Cl_9$ (see text).

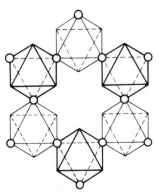

FIG. 4.35. Octahedral A_2X_9 layer.

(g)–(i). The structures (g) and (i) contain complex ions B_2X_9 consisting of two BX_6 octahedra sharing a face. Structure (i) is that of $K_2W_2Cl_9$. The simpler structure (g) is not known, but $Cs_3Tl_2Cl_9$ has a closely related structure (with hexagonal closest packing of A + 3 X) but a more uniform spatial distribution of the $Tl_2Cl_9^{3-}$ ions. In Fig. 4.34 the broken lines indicate the unit cell to which the structures of Fig. 4.31 can be referred. In structure (g) the $Tl_2Cl_9^{3-}$ ions would lie on lines (perpendicular to the plane of the paper) through all the small black circles and would be at the same heights on each line, as in the elevation (b). The more uniform distribution shown in the elevation (c) is that found in $Cs_3Tl_2Cl_9$. The structure (h) consists of corrugated layers, shown in plan in Fig. 4.35, formed from octahedra sharing three vertices with three other octahedra. No example of this structure is known, but a variant is adopted by $Cs_3As_2Cl_9$. The $AsCl_6$ groups are

154

distorted so that there are discrete $AsCl_3$ molecules embedded between Cs^+ and Cl^- ions rather than an $As_2Cl_9^{3-}$ ion formed from regular octahedral $AsCl_6$ groups.

A_2BX_6 structures

These arise by filling one-half of the available X_6 holes in the c.p. stacking of AX_3 layers. Elevations of the structures with *h*, *hc*, and *hcc* layer sequences are shown in Fig. 4.31 (j)–(m). In all these structures there are discrete BX_6 ions. Numerous examples of the first three structures, the trigonal K_2GeF_6 (Cs_2PuCl_6), the cubic K_2PtCl_6, and the hexagonal K_2MnF_6 structures, are given in Chapter 10, where the first two structures are illustrated. The structures of complex halides and oxides ABX_3, $A_3B_2X_9$, and A_2BX_6 are summarized in Table 4.10.

TABLE 4.10

Structures based on close-packed AX_3 *layers*

Layer sequence	Symbol	Fraction of octahedral X_6 holes occupied by B atoms		
		All ABX_3	$\frac{2}{3}$ $A_3B_2X_9$	$\frac{1}{2}$ A_2BX_6
AB . . .	*h*	$CsNiCl_3$ $BaNiO_3$ $CsCuCl_3$† (a)‡	$[Cs_3Tl_2Cl_9]$ (see text) (g)	K_2GeF_6 Cs_2PuCl_6 (j)
ABC . . .	*c*	Oxides ABO_3 $RbCaF_3$ $CsAuCl_3$† (b) (e)	$Cs_3As_2Cl_9$† (h)	K_2PtCl_6 (k)
ABAC . . .	*hc*	High-$BaMnO_3$ (c)	§	K_2MnF_6 (l)
ABCACB . . .	*hcc*	$BaTiO_3$ (hexag.) $CsMnF_3$ (d) (f)	$K_3W_2Cl_9$ (i)	— (m)

† Distorted variants of ideal c.p. structure.
‡ The inset letters refer to Fig. 4.31.
§ There is no $A_3B_2X_9$ structure with *ABAC* . . . packing.

This account of these structures has been based on the mode of packing of the c.p. layers. For the alternative description in terms of the way in which the octahedral BX_6 groups are joined together by sharing X atoms, see Chapter 5. The structures based on cubic closest-packing are normally illustrated and described in terms of the cubic unit cell. We noted earlier that the cryolite structure is a superstructure of perovskite; the relation between the perovskite, cryolite, and K_2PtCl_6 structures is described in Chapter 10 (p. 388).

5

Tetrahedral and Octahedral Structures

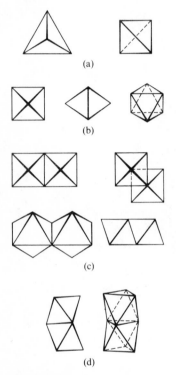

FIG. 5.2. (a) Tetrahedron viewed along line perpendicular to face or line joining mid-points of opposite edges. (b) Octahedron viewed along lines joining opposite vertices, mid-points of opposite edges, and mid-points of opposite faces. (c) Various projections of a pair of octahedra sharing an edge. (d) Projection of a pair of octahedra sharing a face.

Structures as assemblies of coordination polyhedra

Diagrams of crystal structures, particularly complex ionic structures, may be simplified by using a 'shorthand' notation analogous to be the organic chemist's use of the hexagon to represent a benzene ring. This may be illustrated by a 2D analogy. Suppose that in a compound AX_2 each A atom is bonded to four X atoms and that all the X atoms are equivalent. It follows that each X must be bonded to two A atoms, as in the simple examples of Fig. 5.1(a). Instead of showing all the A—X

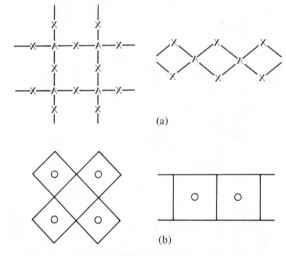

FIG. 5.1. Representation of AX_2 structures (a) as assemblies of AX_4 coordination groups (b).

bonds we may simplify the diagrams by representing the AX_4 groups as squares (Fig. 5.1(b)) which are linked by having edges or corners in common. The lines no longer represent chemical bonds. Square planar coordination groups are uncommon, and we are normally concerned with polyhedral groups AX_n, particularly tetrahedral AX_4 and octahedral AX_6 groups. In illustrations of crystal structures these will appear in various orientations and it is important that the reader should recognize these polyhedra when viewed in a number of directions. Figure 5.2 shows projections of a tetrahedron and octahedron and also projections of a pair of octahedra sharing one edge or one face.

In this chapter we shall describe the more important structures that may be built

156

from the two most important coordination polyhedra, the tetrahedron and octahedron, by sharing vertices, edges, or faces, or combinations of these polyhedral elements. With regard to the sharing of X atoms between AX_n coordination groups the convention adopted is that if an edge is shared its two vertices are not counted as shared vertices, and similarly the three edges and three vertices of a shared face are not counted as shared edges or vertices. The treatment will be essentially topological, that is, we shall be primarily concerned with the way in which the coordination polyhedra are connected together rather than with the detailed geometry of the systems. However, interatomic distances and interbond angles are of great interest to the structural chemist though they are usually discussed without reference to certain basic geometrical limitations which we shall examine first.

In descriptions of the structures of ionic crystals it is usual to point out that the sharing of edges and more particularly of faces of coordination polyhedra AX_n implies repulsions between the A atoms which lead, for example, to shortening of shared edges and to the virtual absence of face-sharing in essentially ionic structures. There are also, however, purely geometrical limitations on the interbond angles A–X–A at shared X atoms even if only vertices are shared, and these are clearly relevant to discussions of such angles in crystalline trifluorides with the ReO_3 or FeF_3 structures or in the cyclic tetramers M_4F_{20} of certain pentafluorides.

Limitations on bond angles at shared X atoms

It is convenient to consider first the angle A–X–A at an X atom shared between two regular tetrahedra or octahedra. This atom may be a shared vertex or it may belong to a shared edge or face, so that there are six cases to consider. If tetrahedra or octahedra share a face the system is invariant and the angle A–X–A has the value F in Fig. 5.3(a) or (b). For edge- or vertex-sharing there is in each case a *maximum* value of the angle A–X–A (points E and V for edge- and vertex-sharing

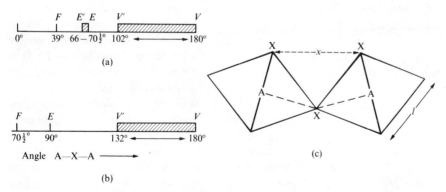

FIG. 5.3. Angles A–X–A for (a) tetrahedra AX_4 or (b) octahedra AX_6 sharing a face (F), an edge (E), or a vertex (V). (c) Restriction on distance between X atoms of different polyhedra (see text). The values given for certain of these angles are approximate. More precise values of those which are not directly derivable from the geometry of the polyhedra are: $E' = 65°58'$ ($\cos^{-1} 11/27$), $V'_{tetr.} = 102°16'$ [$\cos^{-1}(-17/81)$], and $V'_{oct.} = 131°48'$ ($2 \sin^{-1} \sqrt{5/6}$).

157

respectively) corresponding to the fully extended systems with maximum separation of the A atoms (centres of polyhedra). However, there is freedom to rotate about a shared edge or vertex which reduces this distance and also reduces the distance (x) between X atoms of different polyhedra (Fig. 5.3(c)). The *minimum* value of the angle A—X—A therefore depends on the lower limit placed on x. If we suppose that x will not be less than the X—X distance in AX_4 (or AX_6), that is, the edge-length l, we calculate the lower limits shown in Fig. 5.3 as E' and V'. In the tetrahedral case the range EE' is small and we shall neglect it.

From the A—X—A angles we may calculate the distance between the centres of the polyhedra (A—A) for face-, edge-, or vertex-sharing. These distances may be expressed in terms of either the edge-length X—X or the distance A—X from centre to vertex (bond length); both sets of values are given in Table 5.1. They illustrate the increasingly close approach of A atoms as edges or faces are shared; sharing of faces by tetrahedral coordination polyhedra does not occur, and sharing of faces by octahedral coordination groups is confined to a relatively small number of structures.

TABLE 5.1

The distance (A—A) *between centres of regular* AX_4 *or* AX_6 *groups sharing* X *atoms*

	In terms of X—X			In terms of A—X		
	Vertex†	*Edge†*	*Face*	*Vertex†*	*Edge†*	*Face*
Tetrahedron	1·22	0·71	0·41	2·00	1·16	0·67
Octahedron	1·41	1·00	0·82	2·00	1·41	1·16

† Maximum value

The angular range of A—X—A in Fig. 5.3(b) corresponds closely to the observed —F— bond angles in trifluorides of transition metals with ReO_3- or RhF_3-type structures and in cyclic M_4F_{20} molecules of transition-metal pentafluorides. In the MnF_4 layer of $BaMnF_4$ values of the angle Mn—F—Mn close to both the extreme values of Fig. 5.3(b), namely, 139° and 173°, are observed for the two kinds of non-equivalent F atom (Fig. 5.15(d), p. 170). Of more interest is the gap between 90° and 132°, from which the following conclusions may be drawn:

(i) In the rutile structure there cannot be both regular octahedral coordination of A and planar equilateral coordination of X, for the latter would require the angle A—X—A to be 120°.

(ii) In the corundum structure (α-Al_2O_3) there cannot be both regular octahedral coordination of Al and also regular tetrahedral coordination of O; the value $109\frac{1}{2}°$ occurs in the gap between E and V' (Fig. 5.3(b)). On this point see also p. 216.

(iii) The fact that the minimum value of A—X—A for vertex-sharing is greater that 120° implies that if three (or more) octahedra meet at a point at least one *edge* (or *face*) must be shared. We shall see that edge-sharing is a feature of many octahedral structures and arises in structures of compounds AX_n if $n \leqslant 2$ for this purely geometrical reason; it *may*, of course, occur if $n > 2$.

It should be emphasized that the validity of conclusion (iii) rests on our assumptions (a) that the distance of closest approach of X atoms of different vertex-sharing octahedra (x in Fig. 5.3(c)) may not be less than the distance X–X within an octahedron, and (b) that the octahedra are regular. Structures in which three octahedra share a common vertex but no edges or faces contravene (a) and/or (b).

(a) If x is less than the edge length X–X it is, of course, possible for three octahedra to meet at a point with only a vertex in common. This situation arises in the trimeric chromium oxyacetate ion in $[OCr_3(CH_3COO)_6(H_2O)_3]Cl \cdot 6\,H_2O$ (AC 1970 **B26** 673), in which the acetato groups bridge six pairs of vertices (Fig. 5.4).† The shortest O–O distances within the octahedral CrO_6 groups lie in the range 2·63–2·90 Å, but O–O in the acetato group is only 2·24 Å.

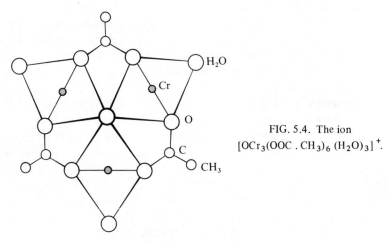

FIG. 5.4. The ion $[OCr_3(OOC \cdot CH_3)_6\,(H_2O)_3]^+$.

(b) It might have been expected that there would be a simple 3D AX_2 structure in which AX_6 octahedra share only vertices and each vertex is common to three octahedra—contrast the rutile structure, in which each AX_6 shares *edges* and *vertices*. Such a structure is not possible unless there is appreciable distortion of the octahedra and/or close contacts between vertices of different octahedra; in this connection see the discussion of the structures of AgF_2, PdS_2, and HgO_2 on p. 223.

The maximum number of polyhedra with a common vertex

There are two other geometrical theorems that we shall state without proof, namely, that the maximum number of *regular* tetrahedra that can meet at a point is eight, and the maximum number of *regular* octahedra that can meet at a point is six, assuming in each case that the distance between X atoms of different polyhedra is not less than the edge-length of the polyhedron. The numbers of these polyhedra

† In this chapter references to the literature are given only for compounds which are not described in other chapters.

are, of course, the numbers of tetrahedral and octahedral interstices surrounding a sphere in a closest packing of equal spheres. Accordingly there are two arrangements of six regular octahedra AX_6 (plus eight regular tetrahedra AX_4) meeting at a common vertex, corresponding to h.c.p. and c.c.p. X atoms. In the h.c.p. case each octahedron shares two edges and one face and in the c.c.p. case four edges with other octahedra of the group of six with the common vertex. In cubic closest packing the eight tetrahedral holes are separated by the six octahedral holes, but in hexagonal closest packing six of the tetrahedral holes form three face-sharing pairs. These three larger 'double tetrahedral' holes offer the possibility of inserting three more octahedra with the central X atom as common vertex, and hence the possibility of having up to nine (suitably distorted) octahedra meeting at a common vertex. Of these nine octahedra six would share three faces and two edges and three would share four faces with other octahedra of the group of nine octahedra.

These considerations are relevant to the existence of structures A_2X_3 of 9 : 6 coordination and structures A_3X_4 of 8 : 6 coordination, in which respectively nine and eight XA_6 groups would meet at each A atom (assuming equivalence of all A and of all X atoms). No structure of the former type is known, but Th_3P_4 is of the latter type. Evidently regular octahedral coordination of X is not possible, and moreover there must be considerable face-sharing between the XA_6 coordination groups. Accordingly we find that in Th_3P_4 one-half of the (distorted) octahedral PTh_6 groups share two faces (and three edges) and the others share three faces (and one edge) with other octahedra meeting at the common Th atom. It should be emphasized that the numbers of shared edges and faces given here are those shared with other octahedra belonging to the group of six, eight, or nine octahedra meeting at the common vertex; the numbers of edges and faces shared by each octahedron in the actual crystal structure will, of course, be greater. Some other consequences of these theorems are discussed in later chapters. They have a bearing on the stability (and therefore existence) of certain oxy-salts (p. 276), nitrides (p. 225), and cation-rich oxides (p. 278).

We now survey structures built from tetrahedra and octahedra. A survey of the kind attempted in this chapter not only emphasizes the great number and variety of structures arising from such simple building units as tetrahedra and octahedra but also draws attention to:

(a) the unexpected complexity of the structures of certain compounds for which there are geometrically simpler alternatives, for example, the unique layer structure of MoO_3 may be contrasted with the topologically simpler ReO_3 structure,

(b) the lack of examples of compounds crystallizing with certain relatively simple structures, which emphasizes the fact that the *immediate* environments of atoms are not the only factors determining their structures—note the non-existence of halides AX_2 with the atacamite structure,

(c) the large number of geometrically possible structures for a compound of given formula type (for example, AX_4), knowledge of which is necessary for a satisfactory discussion of any particular *observed* structure,

(d) the importance of the coordination group around the *anion*.

We may perhaps remind the reader at this point of the general types of connected system which are possible when any unit (atom or coordination group) is joined to a number (p) of others:

$p = 1$ dimer only,

$p = 2$ closed ring or infinite chain,

$p \geqslant 3$ finite (polyhedral), 1-, 2-, or 3D systems.

Tetrahedral structures

TABLE 5.2
Tetrahedral structures

	Formula	Type of complex	Examples
Number of shared vertices		*Vertices only shared* *Vertices common to two tetrahedra*	
1	A_2X_7	Finite molecule or pyro-ion	$Cl_2O_7, S_2O_7^{2-}$, etc.
2	$(AX_3)_n$	Cyclic molecule or meta-ion, infinite chain	$S_3O_9, Se_4O_{12} \cdot (PNCl_2)_n$ $(P_4O_{12})^{4-}, (Si_3O_9)^{6-},$ $(SO_3)_n, (PO_3)_n^{n-}$
3	$(A_2X_5)_n$	Finite polyhedral, double chain, layer or 3D structure	P_4O_{10} $Al[AlSiO_5]$ $P_2O_5, Li_2Si_2O_5$ $P_2O_5, La_2[Be_2O_5]$
4	$(AX_2)_n$	Layer, double layer, or 3D structure	HgI_2 (red) $CaSi_2Al_2O_8$ (hexag.) SiO_2 structures, GeS_2
		Vertices common to three tetrahedra	
3	$(AX_2)_n$	Infinite layer	AlOCl, GaOCl
		Edges only shared	
		Edges common to two tetrahedra	
Number of shared edges			
1	A_2X_6	Finite dimer	Al_2Cl_6, Fe_2Cl_6
2	$(AX_2)_n$	Infinite chain	$BeCl_2, SiS_2, Be(CH_3)_2$
3	$(A_2X_3)_n$	Infinite double chain	$Cs(Cu_2Cl_3)$
4	$(AX)_n$	Infinite layer	LiOH, PbO
6	$(A_2X)_n$	3D structure	Li_2O, F_2Ca
		Vertices and edges shared	
	$(AX)_n$	Double layer	$La_2O_3, Ce_2O_2S, U_2N_2Sb$
	$(AX)_n$	3D structure	β-BeO

Tetrahedral and Octahedral Structures

The most numerous and most important tetrahedral structures are those in which only vertices are shared. The sharing of faces of tetrahedral groups AX_4 would result in very close approach of the A atoms (to 0·67 AX or 0·41 XX, where XX is the length of the tetrahedron edge) and a very small A—X—A angle ($38°\,56'$ for regular tetrahedra) and for these reasons need not be considered.

Tetrahedra sharing vertices only

When a vertex is shared between two tetrahedra the maximum value of the A—A separation is 2 AX, for collinear A—X—A bonds, but a considerable range of A—X—A angles ($102°\,16'$–$180°$) is possible, consistent with the distance between X atoms of different tetrahedra being not less than the edge-length of the tetrahedron. Observed angles in oxy-compounds are mostly in the range 130–150°, though collinear O bonds apparently occur in ZrP_2O_7 and $Sc_2Si_2O_7$.

In most structures each X atom is common to only two tetrahedra, and if all the tetrahedra are topologically equivalent (that is, share the same number of vertices in the same way) the formulae are A_2X_7, AX_3, A_2X_5, and AX_2 according to whether 1, 2, 3, or 4 vertices are shared (Table 5.2). The first group includes Cl_2O_7 and the pyro-ions $S_2O_7^{2-}$ etc. and the second group contains cyclic and infinite linear molecules $(AX_3)_n$ and the meta-ions of the same types. The AX_3 chain is illustrated in Fig. 5.5(a). If three vertices of each tetrahedron are shared the possible structures include finite polyhedral groups such as P_4O_{10} (Fig. 3.1, p. 57) and infinite systems of which the simplest is the double chain of Fig. 5.5(b). Examples of 2D and 3D systems are included in Chapter 3 under 3-connected nets. The sharing of four vertices leads to layer, double layer, or 3D structures, and these

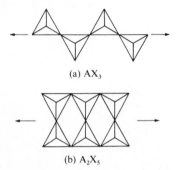

(a) AX_3

(b) A_2X_5

FIG. 5.5 Portions of infinite chains in which all tetrahedra share (a) two, (b) three vertices.

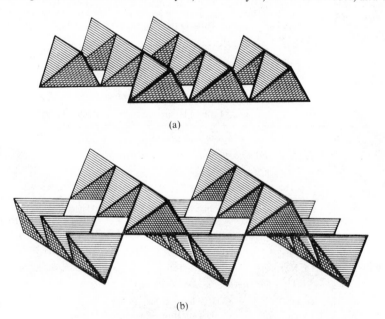

(a)

(b)

FIG. 5.6. Tetrahedral AX_2 layers: (a) HgI_2 (red), (b) $SrZnO_2$.

162

also are described in Chapter 3 under plane and 3D 4-connected nets. Two configurations of the simple AX_2 layer formed from tetrahedra sharing four vertices are shown in Fig. 5.6. In $SrZnO_2$ (Fig. 5.6(b)) the layer is the anion, and the Sr^{2+} ions are accommodated between the layers. A multiple unit consisting of four tetrahedra, each sharing three vertices, has the composition A_4X_{10} and represents the idealized structure of the P_4O_{10} molecule (Fig. 3.1(b)). Such a unit is topologically equivalent to a single tetrahedron since it can link to four others by sharing the four remaining vertices. Any structure that can be built of AX_4 tetrahedra sharing vertices with four other tetrahedra can therefore be built of these A_4X_{10} units, and the structure has the composition AX_2. A structure of this kind has been derived for an orange form of HgI_2, built of layers related in this way to the simple layers of red HgI_2.

For SiO_2 structures and aluminosilicates see Chapter 23.

The examples of Table 5.2 are restricted to systems in which all the tetrahedra are topologically equivalent. In finite linear systems consisting of more than two tetrahedra the terminal AX_4 groups share only one vertex but the intermediate tetrahedra share two vertices. In such hybrid systems the X : A ratio lies between $3\frac{1}{2}$ and 3, as in the $P_3O_{10}^{5-}$ and $S_3O_{10}^{2-}$ ions. Similarly there are chains in which some tetrahedra share two and others three vertices, as in the double-chain ions with X : A ratios between 3 and $2\frac{1}{2}$ found in some silicates (Fig. 5.7(a) and (b)). The X : A ratio falls to values between $2\frac{1}{2}$ and 2 if some of the tetrahedra share three and the remainder all their vertices, as in the layer of Fig. 5.7(c). This interesting layer is found in a number of compounds isostructural with melilite. $Ca_2MgSi_2O_7$ ($Ca_2SiAl_2O_7$, $Ca_2BeSi_2O_7$, $Y_2SiBe_2O_7$). In the A_3X_7 layer the tetrahedrally coordinated atom is Si, Al, Be, or Mg, and the larger Ca or Y ions are in positions of 8-coordination (distorted antiprism). This layer is based on one of the (3,4)-connected pentagon nets mentioned on p. 72. For a more complex A_3X_7 layer see $Na_2Si_3O_7$ (p. 819).

Because SiO_4 tetrahedra can share any number of vertices from none, in the orthosilicate ion $(SiO_4)^{4-}$ to four, in SiO_2, the extensive oxygen chemistry of silicon provides examples of all types of tetrahedral structures. Their variety is

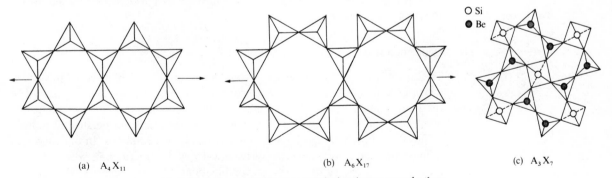

(a) A_4X_{11} (b) A_6X_{17} (c) A_3X_7

FIG. 5.7. (a) and (b) Portions of infinite chains in which some tetrahedra share two and others three vertices. (c) Portion of A_3X_7 (melilite) layer in which some tetrahedra share three and others four vertices.

considerably increased owing to the fact that Al may replace Si in some of the tetrahedra, leading to the aluminosilicates which account for most of the rock-forming minerals and soils. The only element rivalling silicon in this respect is germanium (the eka-silicon of Mendeleef) which has been shown to form oxy-salts analogous to all the families of silicates (or aluminosilicates).

In the structures we have considered above a shared vertex is common to two tetrahedra only. One structure in which each shared vertex is common to three tetrahedra is that of AlOCl (and the isostructural GaOCl); the layer is illustrated in Fig. 10.16 (p. 408).

Tetrahedra sharing edges only

When edges of tetrahedral AX_4 groups are shared the maximum A–A separation is 1·16 AX (or 0·71 XX), and a small variation in the A–X–A is possible ($66° - 70\frac{1}{2}°$) if we assume that X atoms of different tetrahedra may approach only as closely as within a tetrahedron. Edge-sharing between tetrahedral groups is not found in the more ionic crystals other than the fluorite and antifluorite structures, which may be described in terms of edge-sharing FCa_4 and LiO_4 groups. Here the edge-sharing is an unavoidable feature of the geometry of the 4 : 8 coordinated structure. Examples of edge-sharing tetrahedral structures are included in Table 5.2. They include the dimeric molecules A_2X_6 of certain halides (other than fluorides) in the vapour state and in some crystals, the corresponding infinite chain formed by sharing of opposite edges of each tetrahedron ($BeCl_2$ etc.), the double chain in a number of complex halides ($CsCu_2Cl_3$), and the layers in crystalline LiOH and PbO; the latter are described in Chapter 3 under plane 4-connected nets. We have included in Table 5.2 the infinite double layers of composition $(AX)_n$ in which each tetrahedral group shares 3 or 4 edges. They are known only as integral parts of the 3D La_2O_3, Ce_2O_2S, and U_2N_2Sb structures, and are illustrated in Fig. 28.9 (p. 1005).

Tetrahedra sharing edges and vertices

The sharing of edges *and* vertices of tetrahedral groups in a given structure is rare. An example is the 3D structure of the high-temperature form of BeO, in which pairs of edge-sharing tetrahedra are further linked into a 3D structure by sharing the four remaining vertices.

Octahedral structures

The greater part of this survey of octahedral structures will be devoted to systems which extend indefinitely in one, two, or three dimensions. However, in contrast to the limited number of types of finite complex built from tetrahedral groups the number of finite molecules and complex ions formed from octahedral units is sufficient to justify a separate note on this family of complexes. In our survey of infinite structures we shall deal systematically with structures in which there is sharing of vertices, edges, faces, or combinations of these elements. We shall not make these subdivisions for finite structures, which will simply be listed in order of increasing numbers of octahedra involved.

Some finite groups of octahedra

Some of the simpler octahedral complexes are illustrated in Fig. 5.8 and some examples are included in the more comprehensive Table 5.3. We comment here on only a few of these complexes.

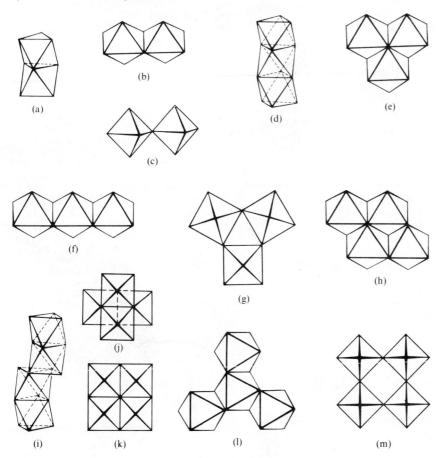

FIG. 5.8. Finite groups of octahedra of Table 5.3.

Further examples of the A_2X_{10} group of Fig. 5.8(b) include:

$$
\left[\begin{array}{c} \text{OH} \quad \text{OH} \\ | \qquad | \\ \text{O}-\text{O}-\text{O} \\ /\ \text{Te}\ /\ \text{Te}\ / \\ \text{O}-\text{O}-\text{O} \\ | \qquad | \\ \text{OH} \quad \text{OH} \end{array} \right]^{4-}
\qquad
\left[\begin{array}{c} \text{O} \quad \text{OH} \\ | \qquad | \\ \text{O}-\text{O}-\text{O} \\ \text{I}\quad \text{I} \\ \text{O}-\text{O}-\text{O} \\ | \qquad | \\ \text{OH} \quad \text{O} \end{array} \right]^{4-}
$$

The A_3X_{12} and A_4X_{16} complexes represent the structures of the trimeric Ni and tetrameric Co acetylacetonates (Fig. 5.9); the coordination groups around the metal

165

TABLE 5.3

Finite groups of octahedra

Fig. 5.8	Formula	Examples
(a)	A_2X_9	$Fe_2(CO)_9$, $W_2Cl_9^{3-}$, $Tl_2Cl_9^{3-}$, $I_2O_9^{4-}$
(b)	A_2X_{10}	Nb_2Cl_{10}, Mo_2Cl_{10}, U_2Cl_{10}
(c)	A_2X_{11}	$(Nb_2F_{11})^-$, $[(NH_3)_5Co \cdot NH_2 \cdot Co(NH_3)_5]^{5+}$
(d)	A_3X_{12}	$[Ni(acac)_2]_3$, Co_3L_6 (see text)
(e)	A_3X_{13}	
(f)	A_3X_{14}	
(g)	A_3X_{15}	
(h)	A_4X_{16}	$[Ti(OC_2H_5)_4]_4$
(i)	A_4X_{16}	$[Co(acac)_2]_4$
(j)	A_4X_{16}	
(k)	A_4X_{17}	
(l)	A_4X_{18}	$[Co_4(OH)_6(NH_3)_{12}]^{6+}$
(m)	A_4X_{20}	Mo_4F_{20}, $W_4O_4F_{16}$

More complex groups

Number of octahedra	Examples	Number of octahedra	Examples
6	$Nb_6O_{19}^{8-}$	12	$PW_{12}O_{40}^{3-}$, $H_4Co_2Mo_{10}O_{38}^{6-}$
7	$Mo_7O_{24}^{6-}$	13	$MnNb_{12}O_{38}^{12-}$, $CeMo_{12}O_{42}^{8-}$
8	$Mo_8O_{26}^{4-}$	18	$P_2W_{18}O_{62}^{6-}$
9	$Mn_9O_{32}^{6-}$		
10	$V_{10}O_{28}^{6-}$		

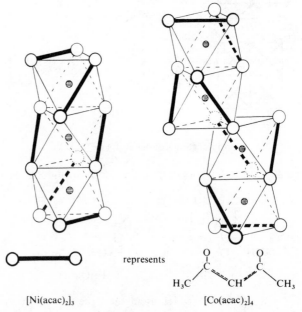

$[Ni(acac)_2]_3$ represents $[Co(acac)_2]_4$

FIG. 5.9. The molecules $[Ni(acac)_2]_3$ and $[Co(acac)_2]_4$.

atoms consist of O atoms of $CH_3CO.CH.CO.CH_3$ ligands. In the Co_3L_6 complex of Table 5.3[1] L represents the ligand $(C_2H_5O)_2PO.CH.CO.CH_3$ shown at the right. The A_4X_{18} complex of Table 5.3 consists of an octahedral $Co(OH)_6$ group sharing three edges with $Co(OH)_2(NH_3)_4$ groups. This ion is enantiomorphic, and is of special interest as the first purely inorganic coordination complex to be resolved into its optical antimers. Somewhat unexpectedly the ion $[Cr_4(OH)_6en_6]^{6+}$ (where en is ethylene diamine, $H_2N.CH_2.CH_2.NH_2$), which might have had the same structure as the Co ion $[Co_4(OH)_6(NH_3)_{12}]^{6+}$ has the quite different structure of Fig. 5.10[2] with edge- and vertex-sharing octahedra. It

(1) IC 1968 7 18

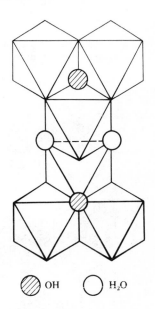

(2) JACS 1969 91 193

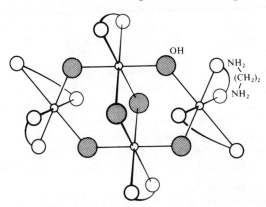

FIG. 5.10. The ion $[Cr_4(OH)_6(en)_6]^{6+}$.

is interesting that the simple A_4X_{16} unit of Fig. 5.8(j) is not known as a finite oxy-ion in solution or as a molecule. It has cubic ($\bar{4}3\,m$, T_d) symmetry and arises by adding a fourth octahedron below the centre of the unit (e). Each octahedron shares three edges, and the octahedra enclose a tetrahedral hole at the centre. This unit does, however, occur in a crystalline tungstate noted on p. 433.

Two 3-octahedra units of type (e) may be joined by sharing one more edge (the broken line in Fig. 5.11) to form the centrosymmetrical 6-octahedra unit which would have, in the simplest case, the composition A_6X_{24}. The vertices shown in Fig. 5.11 as shaded circles are common to three octahedra. These are OH groups, the open circles represent H_2O molecules, and the remaining twenty vertices are occupied by ten bidentate ligands $CF_3.CO.CH.CO.CH_3$ (L) in the complex $Ni_6L_{10}(OH)_2(H_2O)_2$[3]—an interesting example of a comparatively complicated formula arising from an essentially simple system of six edge-sharing octahedra.

Certain elements, notably V, Nb, Ta, Mo, and W, form complex oxy-ions built from larger numbers of octahedral coordination groups. Examples are included in Table 5.3. In the heteropolyacid ions such as $PW_{12}O_{40}^{3-}$ and $P_2W_{18}O_{62}^{6-}$ P atoms occupy the tetrahedral holes at the centres of the complexes. These more complex oxy-ions are described in Chapter 11.

FIG. 5.11. The molecule Ni_6-$(CF_3COCHCOCH_3)_{10}(OH)_2(H_2O)_2$.

(3) IC 1969 8 1304

Infinite systems of linked octahedra

Of the indefinitely large number of structures that could be built from octahedra sharing vertices, edges, and/or faces, a large number are already known, and a complete account of octahedral structures would cover much of the structural

167

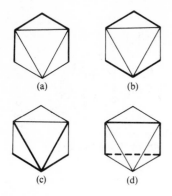

FIG. 5.12. Four ways of selecting four edges of an octahedron.

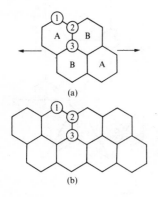

FIG. 5.13. Topological equivalence of octahedra (see text).

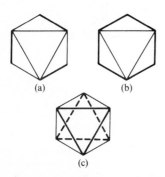

FIG. 5.14. Selection of (a) three, (b) and (c) six edges of an octahedron.

chemistry of halides and chalconides. The number of octahedral structures is large because (a) an octahedron has six vertices, eight faces, and twelve edges, various numbers of each of which can be shared, and (b) there are numerous ways of selecting, for example, a particular small number of edges from the twelve available. Figure 5.12 shows four ways of selecting four edges. Moreover, it is not necessary that all the octahedra in a structure are topologically equivalent, that is, share the same numbers and arrangements of vertices, edges, and/or faces. It may be assumed that all octahedra are topologically equivalent in the simpler systems we shall describe unless the contrary is stated. (Some comparatively simple octahedral complexes containing non-equivalent octahedra include the $[Na(H_2O)_4]_n^{n+}$ chain in borax, in which alternate $Na(H_2O)_6$ octahedra share a pair of opposite or non-opposite edges, and the Al_3F_{14} layer in $Na_5Al_3F_{14}$, in which some AlF_6 octahedra share four and the remainder two vertices.)

It is important to distinguish between the topological equivalence of the octahedra and the equivalence or otherwise of the X atoms (vertices). In the group of four octahedra of Fig. 5.13(a) there are two kinds of non-equivalent octahedron: A (sharing two edges), and B (sharing three edges). There are three kinds of non-equivalent X atom, belonging to 1, 2, and 3 octahedra respectively. If this group is extended indefinitely in the directions of the arrows to form the 'double chain' of Fig. 5.13(b) all the octahedra become equivalent but there are still three kinds of non-equivalent X atoms belonging, as before, to 1, 2, or 3 octahedra. In some of the simplest structures (for example, rutile, ReO_3) all the X atoms are in fact equivalent, but in MoO_3, for example, there are three kinds of non-equivalent oxygen atoms.

We may distinguish as a special set those structures in which all the octahedra are equivalent and each X atom belongs to the same number of octahedra, two in the AX_3 and three in the AX_2 structures. The simplest structures of this type are those in which each octahedron shares

(i)	6 vertices with 6 other octahedra	AX_3
(ii)	the 3 edges of Fig. 5.14 (a)	AX_3
(iii)	the 6 edges of Fig. 5.14 (b) or the 6 edges of Fig. 5.14 (c)	AX_2
(iv)	2 opposite faces	AX_3

We shall see later that (i) and (ii) correspond to families of related structures while (iii) and (iv) produce a single structure in each case. In addition to these structures in which octahedra share only vertices, edges, *or* faces, there are structures in which vertices *and* edges are shared in which all the octahedra and all the X atoms are equivalent. The rutile structure is a simple example—others are described later.

Octahedral structures may be classified according to the numbers and arrangements of shared vertices, edges, and/or faces (Table 5.4), and this systematic approach is adopted in MSIC. Here also it will be convenient to relate our treatment to the main classes of Table 5.4, dealing first with structures in which only vertices or only edges are shared and then proceeding to the more complex structures.

168

TABLE 5.4
Infinite structures built from octahedral AX_6 groups

Vertices only shared

2	AX_5 chains:
	cis: VF_5, CrF_5
	trans: BiF_5, $(CrF_5)^{2-}$, $\alpha\text{-}UF_5$
4	AX_4 layers:
	cis: $BaMnF_4$
	trans: SnF_4, K_2NiF_4
6	AX_3 frameworks:
	ReO_3, $Sc(OH)_3$,
	FeF_3 etc.,
	Perovskite,
	W bronzes,
	pyrochlore

Vertices and edges shared

AX_3, A_3X_8, A_2X_5 layers (V oxyhydroxides)
AX_2 frameworks
    Rutile structure

$\left.\begin{array}{l} \alpha\text{-}AlO . OH \\ Eu_3O_4 \\ CaTi_2O_4 \\ \alpha\text{-}MnO_2 \\ \hline BeY_2O_4 \end{array}\right\}$ (Table 5.5)

AX_3 layer: MoO_3
AX_3 framework: $CaTa_2O_6$

Edges only shared

2	AX_4 chains: $TcCl_4$, NbI_4
3	AX_3 layer: YCl_3, BiI_3
4	AX_3 double chain: NH_4CdCl_3
	AX_3 layer: NH_4HgCl_3
	A_3X_8 layer: Nb_3Cl_8
6	AX_2 layer: CdI_2, $CdCl_2$
	AX_2 double layer: $MOCl$, $\gamma\text{-}MO . OH$
	AX_2 framework: $Cu_2(OH)_3Cl$

Vertices, edges, and faces shared

$\alpha\text{-}Al_2O_3$ (corundum)
$\gamma\text{-}Cd(OH)_2$

Vertices and faces shared

ABO_3 structures: hexagonal $BaTiO_3$; high-$BaMnO_3$, $BaRuO_3$ (Table 5.6)

Faces and edges shared
Nb_3S_4

Faces only shared

2	AX_3 chain: ZrI_3, $Ba NiO_3$, $Cs NiCl_3$

Octahedra sharing only vertices

We noted earlier that the number of *regular* octahedra that can share a common vertex *without sharing edges or faces* is limited to two, assuming that the distance between any pair of non-bonded X atoms of different AX_6 groups is not less than the edge-length, X–X, taken to be the minimum van der Waals distance. If each vertex that is shared is common to two octahedra only, there is a simple relation between the formula of the structure and the number of shared vertices (X atoms):

Number of shared vertices	2	4	6
Formula	AX_5	AX_4	AX_3

The finite A_2X_{11} group, in which one vertex is shared, has been illustrated in Fig. 5.8(c). Examples are not known of structures in which all octahedra share three vertices; for a layer of this kind see Fig. 4.35, p. 154. The only example of octahedra sharing five vertices is the double layer of TiO_6 octahedra in $Sr_3Ti_2O_7$.

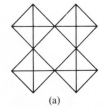 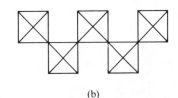 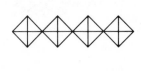 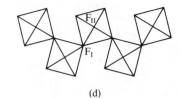

(a) (b) (c) (d)

FIG. 5.15. Octahedra sharing *cis* vertices to form (a) cyclic tetramer, (b) the *cis* chain. (c) Octahedra sharing *trans* vertices to form the *trans* (ReO_3) chain. (b) and (c) also represent end-on views (elevations) of the *cis* and *trans* layers. The actual configuration of the *cis* layer in $BaMnF_4$ is shown at (d), where F_I and F_{II} are the two non-equivalent F atoms referred to on p. 158. The bond angle $M–F_1–M$ is 139°; the bonds from F_2 are approximately perpendicular to the paper ($M–F_2–M$, 173°).

Two vertices of an octahedron are either adjacent (*cis*) or opposite (*trans*). Sharing of two *cis* vertices by each octahedron leads to cyclic molecules (ions) or zigzag chains (Fig. 5.15(a) and (b)). Sharing of *trans* vertices could lead to rings of eight or more octahedra (since the minimum value of the angle A–X–A is 132° for vertex-sharing octahedra) but such rings are not known. The simpler possibility (A–X–A = 180°) is the formation of linear chains (Fig. 5.15(c)).

The cyclic tetramers in crystals of a number of metal pentafluorides M_4F_{20} are of two kinds, with collinear or non-linear M–F–M bonds:

F bond angle 180°: M = Nb, Ta, Mo, W
132°: M = Ru, Os, Rh, Ir, Pt.

Other pentafluorides form one or other of the two kinds of chain shown in Fig. 5.15:

cis chain: VF_5, CrF_5, TcF_5, ReF_5; $MoOF_4$; $K_2(VO_2F_3)$
trans chain: BiF_5, α-UF_5; $WOCl_4$; $Ca(CrF_5)$, $Tl_2(AlF_5)$.

The factors determining the choice of cyclic tetramer or of one of the two kinds of chain are not understood, and this is true also of the more subtle difference (in

170

—F— bond angle) between the two kinds of tetrameric molecule, a difference which is similar to that between fluorides MF_3 to be noted shortly. The difference between $MoOF_4$ (chain) and WOF_4 (cyclic tetramer) is one of many examples of structural differences between compounds of these two elements (compare the structures of MoO_3 and WO_3, later). These two types (*cis* and *trans*) of AX_5 chain are also found in oxyhalides, and the *trans* chain in complex halides such as $Ca(CrF_5)$ and $(NH_4)_2MnF_5$. More complex types of chain with the same composition can be built by attaching additional octahedra (through shared vertices) to those of a simple AX_5 chain. An example of such a 'ramified' chain is found, together with simple *trans* chains, in $BaFeF_5$, and is illustrated in Fig. 10.2 (p. 383).

Corresponding to the two chains formed by sharing two *cis* or *trans* vertices there are layers formed by sharing four vertices, the *un*shared vertices being *cis* or *trans*. If each square in Fig. 5.15(b) and (c) represents a chain of vertex-sharing octahedra perpendicular to the plane of the paper these diagrams are also elevations of the *cis* and *trans* layers. Figure 5.15(d) shows the elevation of the *cis* layer which is the form of the anion in the isostructural salts $BaMF_4$ (M = Mg, Mn, Co, Ni, Zn) and in (triclinic) $BiNbO_4$. In the *trans* layer there is sharing of the four equatorial vertices of each octahedron. This AX_4 layer is alternatively derived by placing AX_6 octahedra at the points of the plane 4-gon net with X atoms at the mid-points of the links. Example of neutral molecules with this structure include SnF_4, PbF_4, $Sn(CH_3)_2F_2$, and $UO_2(OH)_2$. This layer also represents the structure of the 2D anion in $TlAlF_4$ and in the K_2NiF_4 structure (Fig. 5.16) which is adopted by numerous complex fluorides and oxides (see Chapters 10 and 13). Distorted variants of this structure are adopted by K_2CuF_4 and $(NH_4)_2CuCl_4$. In the former the two Cu—F bonds perpendicular to the plane of the layer are shorter than the four equatorial ones (Fig. 5.16(b)), while in the ammonium salt two of the equatorial bonds (broken lines in Fig. 5.16(c)) are longer than the other four Cu—Cl bonds. A fourth structure containing the *trans* layer is that of isostructural salts Ba_2MF_6 (M = Co, Ni, Cu, Zn) which also contain separate F^- ions, that is, the structural formula is $Ba_2(MF_4)F_2$ (Fig. 5.17).

The limit of vertex-sharing is reached when each vertex is shared with another octahedron, giving 3D structures of composition AX_3. This is the first of the very symmetrical octahedral structures listed on p. 168. Since each A atom is connected to six others (through X atoms) the A atoms lie at the points of a 3D 6-connected net, and there is therefore a family of such structures of which the simplest corresponds, in its most symmetrical configuration, to the primitive cubic lattice. A unit cell of the structure is illustrated in Fig. 5.18. A model built of rigid octahedra but with flexible joints at all vertices can adopt an indefinitely large number of configurations. The most symmetrical of these has cubic symmetry and represents the structure of crystalline ReO_3; WO_3 has this structure only at very high temperatures and adopts less symmetrical variants of the structure at lower temperatures.

In the ReO_3 structure the oxygen atoms occupy three-quarters of the positions of cubic closest packing, the position at the body-centre of the cube being

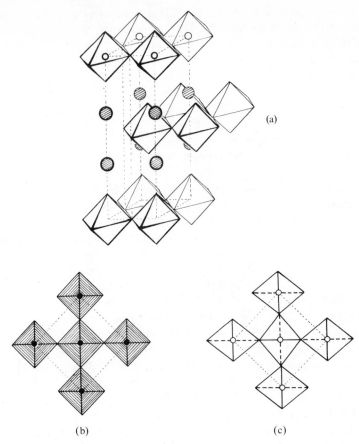

FIG. 5.16. The K_2NiF_4 structure: (a) portions of three layers, (b) plan of layer in K_2NiF_4 or K_2CuF_4, (c) plan of layer in $(NH_4)_2CuCl_4$.

unoccupied. (Occupation of this position by a large ion B comparable in size with O^{2-}, F^-, or Cl^- gives the perovskite structure for compounds ABX_3, in which the B and X ions together form the c.c.p. assembly.) For the fully-extended configuration of Fig. 5.18(a) the A–X–A angle is $180°$ but variants of the structure with smaller angles are also found. The most compact is that in which the X atoms are in the positions of hexagonal closest packing. This structure is adopted by a number of transition-metal trifluorides (see Table 9.16) (p. 355).

Just as α-$Zn(OH)_2$ and $Be(OH)_2$ crystallize with the simplest 3D framework structure possible for compound AX_2 with $4:2$ coordination (the cristobalite structure), distorted so as to bring together hydrogen-bonded OH groups of different coordination groups, so $Sc(OH)_3$ and $In(OH)_3$ have the simplest 3D framework structure of the AX_3 type (the ReO_3 structure), distorted so as to permit hydrogen bonding between OH groups of different $M(OH)_6$ octahedra. The nature of the distortion can be seen from Fig. 5.18(b). Instead of lying on the straight lines joining metal atoms the OH groups lie off these lines, each being

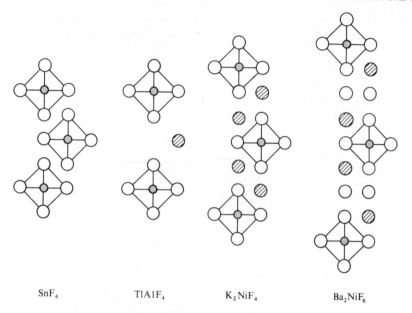

| SnF₄ | TlAlF₄ | K₂NiF₄ | Ba₂NiF₆ |

FIG. 5.17. Structures containing the *trans* AX₄ layer formed from octahedral AX₆ groups sharing their four equatorial vertices (diagrammatic elevations showing one octahedron of each layer). The larger shaded circles represent cations situated between the layers.

hydrogen-bonded to two others. The OH group A is bonded to the metal atom M and to a similar atom vertically above M and hydrogen-bonded to the OH groups B and C.

From the topological standpoint the ReO₃ structure is the simplest 3D framework structure for a compound AX₃ built of octahedral AX₆ groups, for it is based on the simplest 3D 6-connected net. More complicated structures of the same general type are known, that is, structures in which every octahedron is joined to six others through their vertices. The tungsten bronzes have structures of this kind

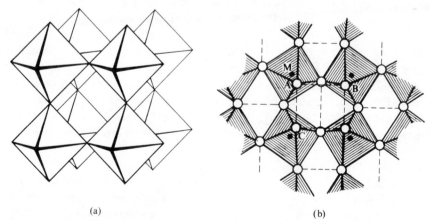

(a)   (b)

FIG. 5.18. The crystal structures of (a) ReO₃ and (b) Sc(OH)₃.

which are described in Chapter 13. A feature of the tungsten bronze structures is that there are tunnels parallel to the 4- or 6-fold axes, that is, in one direction only.

There is another framework structure built of octahedral groups each of which shares its vertices with six others. In the basic BX_3 framework of the pyrochlore structure octahedra which share all vertices are grouped tetrahedrally around the points of the diamond net. In this framework there are rather large holes, the centres of which are also arranged like the carbon atoms in diamond, and they can accommodate two larger cations and also one additional anion X for each $(BX_3)_2$ of the framework. The rigid octahedral framework is stable without either the large cations A or the additional X atoms, and it therefore serves as the basis of the structures of compounds of several types. If all the A and X positions are occupied the formula is $A_2B_2X_7$ (as in oxides such as $Hg_2Nb_2O_7$), but the positions for cations A may be only half occupied ($BiTa_2O_6F$) or they may be unoccupied as in $Al_2(OH,F)_6(H_2O)_{\frac{3}{4}}$ where, in addition, there is incomplete occupancy of the seventh X position. The pyrochlore structure is illustrated in Fig. 7.4 (facing p. 268) and the structure is further discussed in Chapters 6 and 13.

Octahedra sharing only edges

We noted earlier that an indefinitely large number of structures could be built from octahedra which share only edges because not only can the number of shared edges range from two up to a maximum of twelve, but there is the additional complication that there is more than one way of selecting a particular number of edges. It is not necessary to consider the sharing of more than six edges, for this is observed only in the NaCl structure, which may be represented as octahedral $NaCl_6$ or $ClNa_6$ coordination groups sharing all twelve edges. We shall describe here only some of the simpler structures; a somewhat more systematic treatment is adopted in MSIC.

There are four different ways of selecting two edges of an octahedron, namely, (i) *cis* edges (with a common vertex) inclined at $60°$, (ii) *cis* edges inclined at $90°$, (iii) 'skew' edges, and (iv) *trans* (opposite) edges. Of these (i) gives the finite group of three octahedra in Fig. 5.8(e) and (ii) gives the finite group of four octahedra of Fig. 5.8(k), or a zigzag chain (compare the elevation of the MoO_3 layer in Fig. 5.30(a), p. 181); no examples are known of this chain. The remaining possibilities, (iii) and (iv), are shown in Fig. 5.19(a) and (b), where each octahedron is shown resting on a face. The first, (a), leads to a ring of six octahedra or to an infinite chain. The former represents the arrangement of MoO_6 octahedra in the anion in $(NH_4)_6TeMo_6O_{24}$, though since the Te atom occupies the central hole (which is also octahedral) the ion may alternatively be described as a group of seven octahedra.

Crystalline $TcCl_4$ provides a simple example of the skew chain, which is also found as the form of the aquo-anion in the dihydrate of $LiCuCl_3$. One-half of the molecules of water of crystallization are incorporated into the chain and the remainder, together with the Li^+ ions, are accommodated between the chains. Other examples include the anion in $(C_5H_5NH)(SbCl_4)$ and the chain molecules

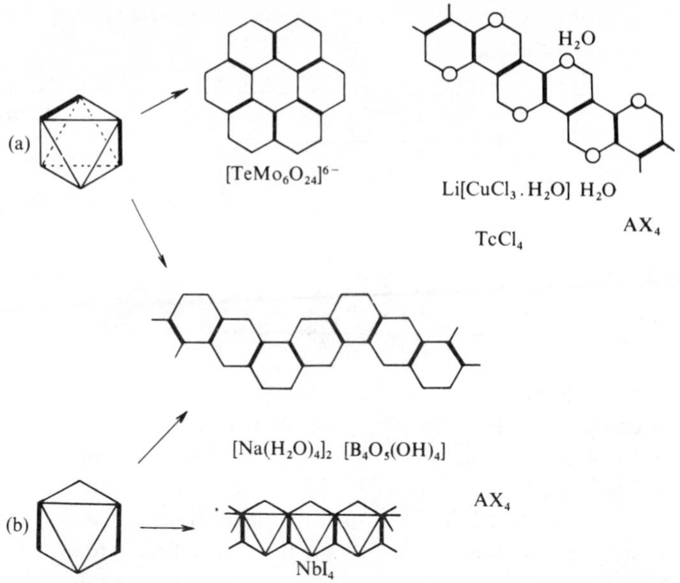

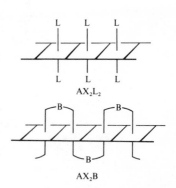

$[TeMo_6O_{24}]^{6-}$

$Li[CuCl_3 . H_2O] \, H_2O$

$TcCl_4$

AX_4

$[Na(H_2O)_4]_2 \; [B_4O_5(OH)_4]$

AX_4

NbI_4

FIG. 5.19. Octahedra sharing (a) 'skew' edges and (b) 'trans' edges.

$Mg(PO_2Cl_2)_2(POCl_3)_2$ and $Mn(PO_2Cl_2)_2(CH_3 . COO . C_2H_5)_2$. In the more complex chain which represents the cation–water complex in borax, $[Na(H_2O)_4]_2$ $[B_4O_5(OH)_4]$, alternate octahedra share a pair of opposite *or* skew edges. In this crystal the chain apparently adopts this configuration in order to pack satisfactorily with the hydrogen-bonded chains of anions, but it is less obvious why a similar chain is found in $Na_2SO_4.10\,H_2O$. In this hydrate only eight of the ten molecules of water of crystallization are associated with the cations to form a chain with the same ratio $4\,H_2O : Na$ as in borax.

Sharing of two opposite edges of each octahedron leads to the infinite AX_4 chain of Fig. 5.19(b) which represents the structure of the infinite molecules in crystalline NbI_4 and of the infinite anion in, for example, $K_2HgCl_4. H_2O$. In this AX_4 chain only 4 X atoms of each octahedral AX_6 group are acting as links between the A atoms. The other two are attached to one A atom only and may be ligands of a second kind, not necessarily capable of bridging two metal atoms (formula AX_2L_2) or they may be atoms forming part of a bidentate ligand, when the formula is AX_2B (Fig. 5.20). This simple octahedral chain thus serves for compounds of three types:

AX_4: NbI_4
AX_2L_2: $PbCl_2(C_6H_5)_2$, $CoCl_2 . 2\,H_2O$
AX_2B: $CuCl_2 . C_2N_3H_3$

For $CuCl_2. C_2N_3H_3$ and other examples see p. 903.

There is also a family of structures containing rutile-like chains which are held

AX_2L_2

AX_2B

FIG. 5.20. Octahedral chains AX_2L_2 and AX_2B.

175

together by ions of a second kind, the chain ions being arranged to give suitable coordination groups around these cations:

$M'_n MX_4$	C.N. of M'	Reference
Sr$_2$PbO$_4$, Ca$_2$PbO$_4$		NW 1965 **52** 492
Ca$_2$SnO$_4$, Cd$_2$SnO$_4$	6	NW 1967 **54** 17
Na$_2$MnCl$_4$		AC 1971 **B27** 1672
High-pressure Mn$_2$GeO$_4$		AC 1968 **B24** 740
Na$_2$CuF$_4$, Na$_2$CrF$_4$	7	ZaC 1965 **336** 200
Ca$_2$IrO$_4$	6, 7, 9	ZaC 1966 **347** 282

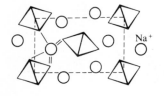

FIG. 5.21. Projection of the structure of Na$_2$MnCl$_4$ along the direction of the chain ions.

The same (orthorhombic) structure, Fig. 5.21, has been described several times for a number of compounds. It is interesting for the trigonal prismatic coordination of the M' atoms between the chains. The monoclinic Na$_2$CuF$_4$ structure is a version of the structure distorted to give Cu(II) (4 + 2)-coordination; Na$^+$ has 7 F neighbours in the range 2·27–2·66 Å. In the hexagonal Ca$_2$IrO$_4$ structure there is a somewhat different arrangement of the chains, but all three structures are basically of the same type. In K$_2$HgCl$_4$. H$_2$O there are H$_2$O molecules in addition to K$^+$ ions between the rutile-like (HgCl$_4$)$_2^{2n-}$ chains.

When resting on one set of octahedral faces the *trans* AX$_4$ chain has the appearance of Fig. 5.22(a), and when viewed along its length it appears as shown on the r.h.s. of the figure. Two such chains may be joined laterally to form the double chain of Fig. 5.22(b) with the composition AX$_3$, in which each octahedron shares four edges. This is the form of the anion in NH$_4$CdCl$_3$ and KCuCl$_3$. It is convenient to refer to the chains of Fig. 5.22(a) and (b) as the rutile and 'double rutile' chains. The end-on views will be used later in representations of more complex structures which result when these chains form a corrugated layer by sharing additional edges or 3D frameworks by sharing the projecting vertices.

We showed in Fig. 5.12 (p. 168) four ways of selecting 4 edges of an octahedron, and we have given examples of (a) and (b); sharing of the edges (c) is not observed in infinite 3D structures. The selection (d), the four equatorial edges of an octahedron, leads to a layer of composition AX$_3$ in which 4 X atoms of each AX$_6$ are common to 4 octahedra and 2 are unshared. This layer is found in NH$_4$(HgCl$_3$), though the octahedral coordination group is so distorted that the alternative description in terms of HgCl$_2$ molecules is to be preferred.

Continued lateral linking of simple AX$_4$ chains leads to the infinite layer in which the six edges of Fig. 5.22(c) of each octahedron are shared and each X atom is common to three octahedra. The composition of the layer is AX$_2$; this is the layer of the CdI$_2$ and CdCl$_2$ structures, for which see also Chapters 4 and 6. We have included in Fig. 5.22 a further choice of 4 edges giving the A$_3$X$_8$ layer found in crystalline Nb$_3$I$_8$.

We have seen that linear and zigzag chains result from the sharing of different pairs of edges. Similarly plane and corrugated layers arise from the sharing of different sets of 6 edges. Whereas the sharing of the very symmetrical arrangement

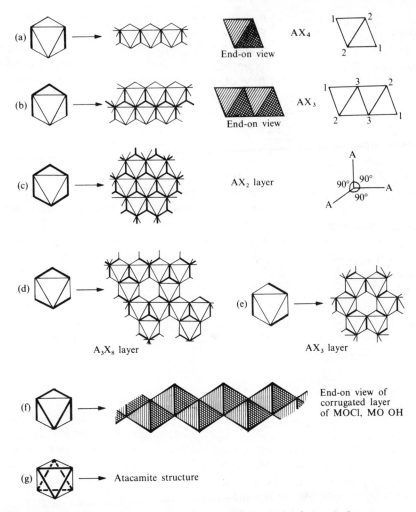

FIG. 5.22. Structures formed from octahedra sharing 2, 4 or 6 edges.

of 6 edges of Fig. 5.22(c) gives a plane layer, the sharing of the edges of Fig. 5.22(f) leads to a corrugated layer. This layer is the basis of the structures of a number of oxychlorides MOCl and oxyhydroxides MO(OH). The structures of pairs of compounds such as FeOCl and γ-FeO(OH) (the mineral lepidocrocite) differ in the way in which the layers are packed together, there being hydrogen-bonding between O atoms of different layers in the latter compound. It is interesting (and unexplained) that unlike the other 3d metal dihydroxides with the simple CdI_2 layer structure $Cu(OH)_2$ apparently crystallizes with the corrugated layer structure more characteristic of compounds MOCl and MO(OH). The corrugated layer of lepidocrocite is found also in the compound $Rb_x Mn_x Ti_{2-x} O_4$ (0.60 $< x <$ 0.80), where replacement of some Ti^{4+} in the octahedra by Mn^{3+} gives charged layers which are held together by the Rb^+ ions (Fig. 5.23). The same kind of layer is thus

177

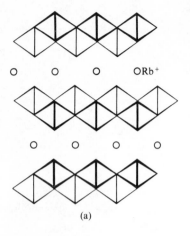

O   O   O   ORb$^+$

O   O   O   O

(a)

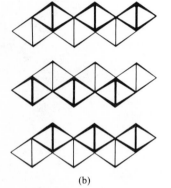

(b)

FIG. 5.23. Elevations of the structures of (a) $Rb_xMn_xTi_{2-x}O_4$, (b) γ-FeO . OH.

Tetrahedral and Octahedral Structures

found held together by van der Waals bonds in FeOCl, by O–H–O bonds in γ-FeO(OH), and by Rb$^+$ ions in the complex oxide.

The sharing of the six edges (g) of Fig. 5.22 with six other octahedra leads to an interesting 3D framework structure which is related to both the sodium chloride and diamond structures. It will be observed that the shared edges (g) are the six unshared ones of (c); sharing of all twelve gives the NaCl structure. Accordingly there are two very simple structures derivable from the NaCl structure by removing one-half of the metal ions. Removal of alternate *layers* of cations as in Fig. 4.22(a) (p. 143) gives the CdCl$_2$ (layer) structures, in which the octahedra share edges (c). On the other hand, if alternate *rows* of cations are removed as indicated by the dotted circles of Fig. 4.22(b) we obtain a structure in which each octahedron shares the edges (g) with its neighbours. This structure is illustrated as a system of octahedra in Fig. 7.4. Although the immediate environments of the A and of the X ions are exactly the same in both the structures of Fig. 4.22(a) and (b) it is an interesting fact that no dihalide crystallizes with the structure (b), in contrast to the considerable number that crystallize with the CdCl$_2$ structure. A distorted form of this structure is, however, adopted by the mineral atacamite, one of the polymorphs of $Cu_2(OH)_3Cl$, a second polymorph having a CdI$_2$ type of layer structure.

Before leaving the simple edge-sharing structures we have to mention the important AX$_3$ layer formed by octahedra sharing the three edges of Fig. 5.22(e). This is the structural unit in Al(OH)$_3$, in many trichlorides, and in some tribromides and triiodides.

Octahedra sharing edges and vertices

It is convenient to describe more complex octahedral structures as built of sub-units which may be, for example, finite groups or infinite chains which are linked by sharing additional vertices and/or edges to form the layer or 3D framework structure. While this device is often helpful in illustrating or constructing a model of a more complex structure it should be remembered that the sub-units have only been distinguished for this purpose and do not necessarily have any chemical or physical significance. We shall encounter structures built from edge-sharing and face-sharing pairs of octahedra, and in particular many structures may be described as built from chains, either the edge-sharing (rutile) AX$_4$ chain of Fig. 5.22(a) or the vertex-sharing (ReO$_3$) AX$_5$ chain of Fig. 5.15(c). Structures are often viewed along the direction of such a chain, and it is therefore important to note how the X atoms are shared in an isolated chain or multiple chain.

In the isolated AX$_4$ (rutile) chain the X atoms are of two kinds, as indicated by the numerals at the top right-hand corner of Fig. 5.22. Some belong to one octahedron only while those in shared edges belong to two octahedra. Similarly in the double (AX$_3$) chain of Fig. 5.22(b) X atoms belong to 1, 2, or 3 octahedra. We shall differentiate these different types of X atom as X_1, X_2, and X_3 respectively. There are obviously many ways in which such chains could be linked to form more complex structures by sharing X atoms. The most important group of these structures arises by sharing X_1 atoms of one chain with X_2 atoms of other chains.

The simplest structure of this kind is the rutile structure, shown in plan in Fig.

178

5.24. Because an X_1 atom of one chain is an X_2 atom of another chain each X is common to three octahedra, and the formula is accordingly AX_2. The rutile structure is described in more detail in Chapter 6. Corresponding to the rutile structure, which is built from single chains, there are 3D frameworks derived from the double chain of Fig. 5.22(b). Three of these are illustrated in Fig. 5.25. In all the structures of Fig. 5.25 each vertex is common to three octahedra; the formula is therefore AX_2. Figure 5.25(a) represents the structure of α-AlO(OH), the mineral diaspore. In the structures (b) and (c) the more open packing of the chains leaves room for additional atoms (cations), and these charged frameworks (or 3D ions)

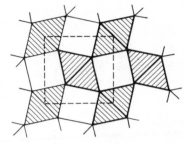

FIG. 5.24. The rutile structure.

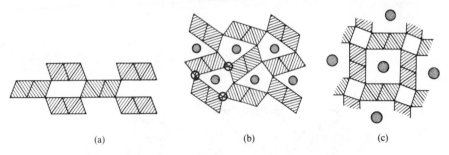

(a) (b) (c)

FIG. 5.25. Structures built from double rutile chains.

form around suitable ions indicated by the circles in (b) and (c). Examples of compounds with these structures are given in Table 5.5.

TABLE 5.5
Structures built from double octahedral chains

Nature of basic single chain	Type of structure built from double chain			
	Chain	Layer	3D framework	
	AX_3	AX_2	AX_2	
Edge-sharing AX_4	$NH_4(CdCl_3)$	γ-AlO . OH	α-AlO . OH	Eu_3O_4 $CaFe_2O_4$	α-MnO_2 (hollandite)
	AX_4	AX_3	AX_3	
Vertex-sharing AX_5	$NbOCl_3$	MoO_3	$CaTa_2O_6$	—	—

There is another family of possible structures in which the shared vertices are either X_1 atoms of both chains (which may be single or multiple) or X_2 atoms of both chains. The simplest structure of this kind is the AX_3 layer of Fig. 5.26(a) consisting of rutile chains (perpendicular to the plane of the paper, joined by sharing X_1 vertices. The analogous layer formed in the same way from double chains is shown at (c), and the intermediate possibility (from single and double chains) at (b). There are clearly two kinds of non-equivalent octahedra in (b).

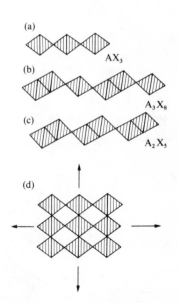

(a)

AX_3

(b)

A_3X_8

(c)

A_2X_5

(d)

FIG. 5.26. Layers built from single and double octahedral chains which are perpendicular to the plane of the paper: (a) AX_3, (b) A_3X_8, (c) A_2X_5, (d) projection of layer (a).

179

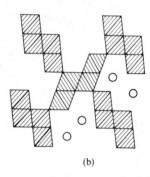

(a)

(b)

FIG. 5.27. Projections of the structures of (a) $CaTi_2O_4$, (b) $Na_xFe_xTi_{2-x}O_4$.

Vanadium oxyhydroxides (p. 471) provide examples of such layers. In Chapter 3 we gave the (layer) structure of the compound $CuHg(OH)_2(NO_3)_2(H_2O)_2$ as a more complex example of the plane 4-gon net. This is in fact the AX_3 layer of Fig. 5.26(a), a point more easily appreciated if this layer is shown in projection (on the plane of the layer) as in (d).

Three-dimensional framework structures, analogous to those of Fig. 5.25, are also possible. An example is the structure of $CaTi_2O_4$ (Fig. 5.27(a)) which should be compared with Fig. 5.25(b). Further sharing of edges between the double chains gives the structure of Fig. 5.27(b), adopted by $Na_xFe_xTi_{2-x}O_4$ ($0.90 > x > 0.75$). Yet another way of utilizing the double rutile chain is exemplified by the structure

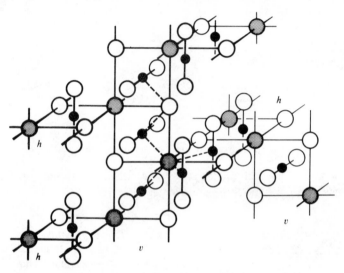

FIG. 5.28. The U_2O_7 framework in BaU_2O_7. The axes of the chains v and h lie respectively in and perpendicular to the plane of the paper. The UO_2 groups of chains v are therefore perpendicular to the paper and those of chains h parallel to the plane of the paper. The shaded circles represent O atoms bonded to 4 U and the open circles O atoms bonded to 2 U.

of $NaTi_2Al_5O_{12}$, where a framework built from double and single chains accommodates Al in tetrahedral and Na in octahedral holes (AC 1967 **23** 754).

Structures built from other types of multiple chain can also be visualized. The family of minerals related to MnO_2 provides examples of several structures of this general type, and the complex oxide BeY_2O_4 is built of quadruple rutile chains. These structures are described in Chapters 12 and 13 where further examples of compounds with the above structures are given.

All the chains are parallel in the structures we have described as built from rutile-like chains. In BaU_2O_7 such chains run in two perpendicular directions to form a 3D framework (Fig. 5.28). Alternate O atoms of shared edges in one chain are also the corresponding atoms of perpendicular chains, so that in each UO_6 octahedron two O atoms are bonded to 1 U, two to 2 U, and two to 4 U. The ratio of O : U atoms in the framework is therefore 7 : 2 $[2(1) + 2(\frac{1}{2}) + 2(\frac{1}{4}) = 3\frac{1}{2}]$.

We now come to structures built from the AX_5 chain composed of octahedral

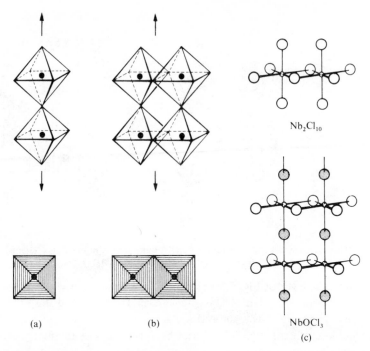

FIG. 5.29. (a) ReO_3 (AX_5) chain; (b) double ReO_3 (AX_4) chain, and (c) the Nb_2Cl_{10} molecule and $NbOCl_3$ chain molecule.

AX_6 groups sharing a pair of opposite vertices. This chain is conveniently represented by its end-on view, as also is the double chain formed from two single chains by edge-sharing (Fig. 5.29). This double chain is the infinite 'molecule' in crystalline $NbOCl_3$, the shared vertices being the O atoms; compare the dimeric Nb_2Cl_{10} molecule in the crystalline pentachloride, formed from two octahedra sharing one edge. Further *edge*-sharing between these double chains gives the corrugated layers seen end-on in Fig. 5.30(a) and (b). In each case three X atoms in each octahedron are bonded to three A atoms, two to 2 A and one to 1 A, so that the composition is AX_3; $(3 \times \frac{1}{3}) + (2 \times \frac{1}{2}) + 1 = 3$. Figure 5.30(a) represents a layer of crystalline MoO_3; the complexity of this structure may be compared with the simplicity of the ReO_3 structure, in which every oxygen atom is bonded to two

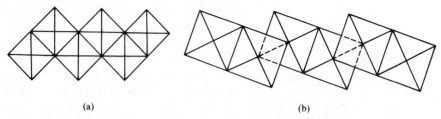

FIG. 5.30. Layers of composition AX_3 built from 'double ReO_3' chains: (a) MoO_3, (b) layer in $Th(Ti_2O_6)$.

181

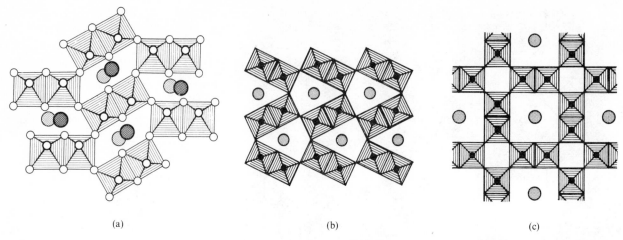

(a)   (b)   (c)

FIG. 5.31. Three frameworks of composition AX_3 built from double ReO_3 chains. (a) is a projection of the structure of $CaTa_2O_6$.

metal atoms. The layer of Fig. 5.30(b) is not known as a neutral AX_3 layer but it represents the arrangement of TiO_6 octahedra in $ThTi_2O_6$ (and the isostructural compounds UTi_2O_6, CdV_2O_6, and $NaVMoO_6$).

The double chains of Fig. 5.29(b) can be further linked to similar chains by vertex-sharing to form a family of 3D structures analogous to those of Fig. 5.25. In the frameworks of Fig. 5.31 each X atom is bonded to two A atoms; the formula is therefore AX_3. Examples of (b) and (c) are not known, but (a) represents the crystal structure of $CaTa_2O_6$ if the circles are the Ca^{2+} ions occupying the interstices in the framework.

We should not expect to find structures based on combinations of vertex-sharing AX_5 and edge-sharing AX_4 chains built from octahedra of the same type for the purely geometrical reason that the repeat distances along these two chains are not the same; they are $2(AX)$ and $\sqrt{2}(AX)$, respectively, if AX is the distance from centre to vertex of the octahedral AX_6 group. However, such structures are possible if the octahedra are of different sizes, as is the case if, for example, the atoms A are of different elements or of one element in different oxidation states. Chromous fluoride, CrF_2, has the rutile structure (distorted to give Cr^{II} a coordination group of four closer and two more distant neighbours), while CrF_3 has a ReO_3-type structure with only vertex-sharing between octahedral CrF_6 groups. Owing to the fact that the Cr^{III}–F bonds (1·89 Å) are shorter than the Cr^{II}–F bonds (four of 1·98 Å and two of 2·57 Å), the repeat distance is the same along the two types of chain. They can therefore fit together as shown diagrammatically in Fig. 5.32 to give the fluoride Cr_2F_5.

Many more structures are known in which vertices and edges of octahedral coordination groups are shared, and models of several of them are described in MSIC. A relatively simple example is the AX_3 framework formed from edge-sharing pairs of octahedra which are further linked to form the 3D framework of Fig. 5.33. This framework is the basis of the structure of one form of $KSbO_3$ and of $KBiO_3$,

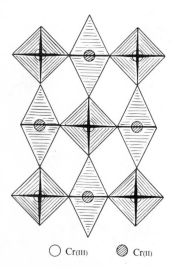

◯ Cr(III)   ◍ Cr(II)

FIG. 5.32. Projection of the structure of Cr_2F_5 showing the two kinds of octahedral chain linked by vertex-sharing into a 3D framework.

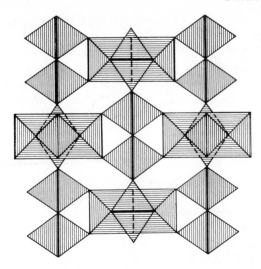

FIG. 5.33. Framework built from pairs of edge-sharing octahedra further linked by vertex-sharing.

where K^+ ions occupy the interstices. If only one-sixth of the positions occupied by K^+ in $KBiO_3$ are occupied by OLa_4 groups (consisting of an O atom surrounded by a tetrahedron of La^{3+} ions) in a framework built of ReO_6 octahedra the formula becomes $(OLa_4)Re_6O_{18}$ or $La_4Re_6O_{19}$. In this crystal there is interaction between the metal atoms within the edge-sharing pairs of octahedra (Re–Re = 2·42 Å) so that there is a physical basis for recognizing these sub-units in the structure.

The structures of two molybdenum bronzes differ from the tungsten bronzes in much the same way as does MoO_3 from WO_3; we have noted that there is edge-sharing as well as vertex-sharing in MoO_3 in contrast to the sharing of vertices only in WO_3. Similarly there is only vertex-sharing of WO_6 octahedra in the tungsten bronze structures, but two molybdenum bronzes with very similar compositions have layer structures which are built from edge-sharing sub-units containing respectively six and ten octahedra. These sub-units can be seen in Fig. 5.34, which shows how they are further linked into layers by sharing eight vertices

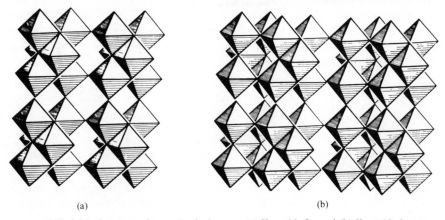

(a)   (b)

FIG. 5.34. Portions of layers in the bronzes (a) $K_{0.33}MoO_3$ and (b) $K_{0.30}MoO_3$.

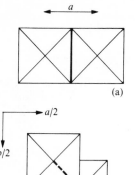

FIG. 5.35. Two ways of joining ReO_3 chains (perpendicular to paper) by edge-sharing.

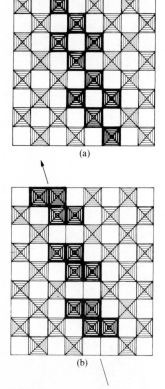

FIG. 5.36. Formation of shear structure (diagrammatic).

with other similar units. Both layers have the composition MoO_3 and they are held together by the potassium ions.

We comment elsewhere on the inadequacy of current bonding theory to account for the complexity of some binary systems in which solid phases appear with unexpected formulae and/or properties—for example, the oxides of caesium and the nitrides of calcium. Certain transition metals, notably Ti, V, Nb, Mo, and W, have a suprisingly complex oxide chemistry, and although it may be difficult to appreciate all the features of their structures from diagrams, these compounds are sufficiently important to justify mention here.

The structures in question are built from slices or blocks of the simpler rutile or ReO_3 structures which are displaced relative to one another to form structures with formulae that correspond in some cases to normal oxides (for example, Nb_2O_5) and in others to oxides with complex formulae implying non-integral (mean) oxidation numbers of the metal atoms. When the rutile structure is sheared along certain (regularly spaced) planes, sharing of *faces* of TiO_6 coordination groups occurs, giving a family of related structures with composition Ti_nO_{2n-1}. All members of this family have been prepared and characterized, in the titanium oxides for $n = 4$–10 inclusive, and in the vanadium oxides for $n = 4$–8. Their compositions are Ti_4O_7, Ti_5O_9 etc. Shearing the ReO_3 structure, in which there is only vertex-sharing, leads to sharing of *edges* of octahedral coordination groups and to homologous series of structures with formulae such as W_nO_{3n-2}. Since the ReO_3-type structures are more easily illustrated than those derived from the rutile structure we shall describe examples of the former. The structures will be shown as projections along the direction of the AX_5 (ReO_3) chains so that each square represents an infinite chain of vertex-sharing octahedra perpendicular to the plane of the paper. First we have to note that ReO_3 chains may be joined by edge-sharing in two essentially different ways, as shown in Fig. 5.35. In (a) the 'equatorial' edge (parallel to the plane of the paper) is shared, that is, the two chains are related by a translation a; in (b) the shared edge is inclined to the plane of the paper, one chain being displaced relative to the other by $a/2 + b/2 + c/\sqrt{2}$. Because there is now a translation in the direction of the length of the chain (i.e. perpendicular to the paper), we may refer to the chains in case (b) as being at different levels. Blocks of ReO_3 structure may be joined by sharing edges at the perimeters of the blocks in either or both of these two ways. As examples of these two types of edge-sharing we shall describe the structures of the following oxides:

Type of edge-sharing: (a) Mo_8O_{23}

(a) and (b) V_6O_{13},

(b) $WNb_{12}O_{33}$ and high-Nb_2O_5

Further examples are given in Chapters 12 and 13.

The first type of edge-sharing, (a), corresponds to shearing the ReO_3 structure so that the chains shown as black squares in Fig. 5.36 (a) are displaced as in (b), and if this shearing occurs at regular intervals the new structure is composed of infinite slabs of ReO_3 structure cemented together along the 'shear-planes' (arrow in Fig. 5.36(b)) by edge-sharing. The composition of the resulting oxide depends on the number of octahedra in the edge-sharing groups at the junctions and on the distance apart of the shear-planes. The examples of Fig. 5.37 show the structure of the oxide Mo_8O_{23} in the series M_nO_{3n-1} and of a hypothetical $Mo_{11}O_{31}$ in the series M_nO_{3n-2} (as a simpler example than the known oxide $W_{20}O_{58}$).

The structure of V_6O_{13} illustrates both the types of edge-sharing of Fig. 5.35. Blocks of ReO_3 structure consisting of 6 chains (3 x 2) are joined by sharing equatorial edges and then the second type of edge-sharing results in the 3D structure of Fig. 5.38.

The octahedra drawn with heavier lines are displaced both in the plane of the paper and also in a perpendicular direction with respect to those drawn with light lines. Each block extends indefinitely normal to the plane of the paper, but as noted earlier it is convenient to refer to the blocks as being at different levels. The formula depends on the size of the block, and for blocks of (3 x n) octahedra at each level the general formula is $M_{3n}O_{8n-3}$:

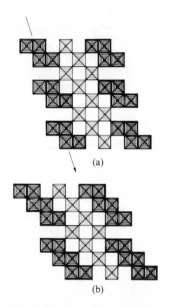

FIG. 5.37. Examples of shear structures: (a) Mo_8O_{23} in the M_nO_{3n-1} series, (b) hypothetical oxide $Mo_{11}O_{31}$ in M_nO_{3n-2} series.

n	Examples
2	V_6O_{13}
3	$TiNb_2O_7$ (M_9O_{21})
4	$Ti_2Nb_{10}O_{29})$ $(M_{12}O_{29})$

In Fig. 5.39 we illustrate the structure of $WNb_{12}O_{33}$ as an example of a structure in which there is edge-sharing of the second type only, that is, there is no edge-sharing between blocks at the same level. This particular structure is built of ReO_3 blocks consisting of (3 x 4) octahedra. This type of structure has the peculiarity that there are small numbers of *tetrahedral* holes at the points indicated by the black circles. The high-temperature form of Nb_2O_5 has a more complex structure of the same general type (see p. 505). There is one such tetrahedral hole for every 27 octahedra, so that the structural formula is $NbNb_{27}O_{70}$. The existence of structures of this kind is very relevant to discussions of bonding in transition-metal

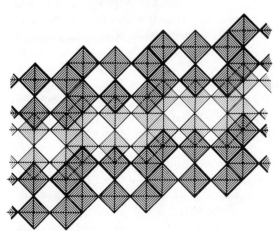

FIG. 5.38. The crystal structure of V_6O_{13}.

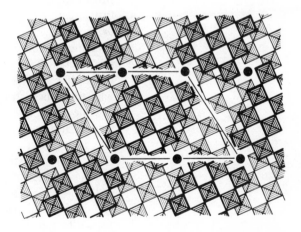

FIG. 5.39. The crystal structure of $WNb_{12}O_{33}$.

oxides. For example, $TiNb_{24}O_{62}$ is a normal valence compound, with one tetrahedral hole (occupied by Ti) to every 24 octahedral holes (occupied by Nb). However, the same structure is adopted by a niobium oxide, but the formula, $Nb_{25}O_{62}$, implies a fractional oxidation number or, presumably, one delocalized electron, $Nb_{25}^{V}O_{62}$ (1e).

Octahedra sharing faces only

As already noted only the sharing of a pair of opposite faces has to be considered; an infinite chain of composition AX_3 is then formed. A number of crystalline trihalides consist of infinite molecules of this type, and chain ions of the same kind exist in $BaNiO_3$ and $CsNiCl_3$. In the ZrI_3 structure the metal atoms occupy one-third of the octahedral holes in a close-packed assembly of halogen atoms to form infinite chains perpendicular to the planes of c.p. halogen atoms:

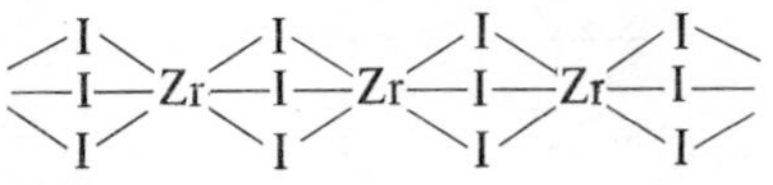

In $BaNiO_3$ there is close packing of Ba + 3 O, and the Ni atoms occupy octahedral holes between 6 O atoms to form an arrangement similar to that of the Zr and I atoms in ZrI_3.

Octahedra sharing faces and vertices

Although the sharing of faces of octahedral coordination groups in 3D (as opposed to chain) structures is not common it does occur in some close-packed structures. When considering structures built of c.p. AX_3 layers, between which smaller B ions occupy positions of octahedral coordination between 6 X atoms, we saw that only vertices and/or faces of BX_6 octahedra can be shared. The sharing of octahedral edges, which is a feature of so many oxide structures, cannot occur in $A_xB_yX_{3x}$ structures for purely geometrical reasons. The sharing of octahedron faces leads first to pairs (or larger groups) of octahedra which are then linked by vertex-sharing to form a family of structures which are related to the ReO_3 and perovskite structures. A linear group of face-sharing octahedra is topologically similar to a single octahedron because it has three vertices at each end through which it can be linked, like a single octahedron, to six other groups or to single octahedra. The simplest of these structures are listed in Table 5.6 and also in Chapter 4 as examples of some of the more complex sequences of c.p. layers, for in all of these structures the transition-metal atoms occupy octahedral holes between six X atoms of the c.p. AX_3 layers.

An interesting structure of a quite different type is that of $Cs_4Mg_3F_{10}$, in which groups of three face-sharing MgF_6 octahedra are linked by *four* of their terminal vertices to form layers (Fig. 5.40), between which Cs^+ ions occupy positions of 10- and 11-coordination. Inasmuch as all the bonds in this crystal are presumably essentially ionic in character, the description of the structure in terms of the coordination groups around the smaller cations is not intended to imply that this is

TABLE 5.6

Close-packed ABX_3 *structures in which octahedral coordination groups share faces and vertices*

Face-sharing groups	Structure	Other examples
Pairs + single	Hexagonal $BaTiO_3$	$CsMnF_3$, $RbNiF_3$
All pairs	High-$BaMnO_3$	Low-$BaNiO_3$
Groups of 3	$BaRuO_3$	$CsCoF_3$
Infinite chains	High-$BaNiO_3$	$CsNiCl_3$, $CsCuCl_3$
	Low-$BaMnO_3$	$BaTiS_3$, $BaVS_3$, $BaTaS_3$

For further examples of complex oxides with structures of these types see Table 4.4 (p. 134) and Table 13.5 (p. 481).

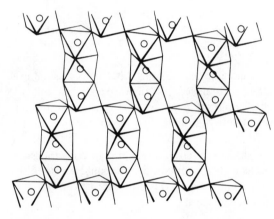

FIG. 5.40. Linking of MgF_6 octahedra into layers $(Mg_3F_{10})_n^{4n-}$ in $Cs_4Mg_3F_{10}$.

a layer structure. Isostructural crystals include the corresponding Co, Ni, and Zn compounds, and it is interesting that the groups of three face-sharing CoF_6 octahedra in $Cs_4Co_3F_{10}$ are also present in the structure of $CsCoF_3$ (Table 5.6).

We dealt separately at the beginning of this chapter with finite groups of octahedra (molecules and complex ions) but it seems justifiable to draw attention here to the anion $CeMo_{12}O_{42}^{8-}$ in view of its relation to some of the structures we have been describing. The four vertices of a pair of face-sharing octahedra which are marked A and B in Fig. 11.12(a) (p. 438) correspond to four vertices of an icosahedron. Six pairs of octahedra may therefore be joined together by sharing these vertices to form an icosahedral group of 12 octahedra. The Ce^{4+} ion occupies the position of 12-coordination at the centre of the complex ion.

Octahedra sharing faces and edges

An example of this rare phenomenon is provided by the remarkable structure of Nb_3S_4. Three rutile chains can coalesce into a triple chain by face-sharing provided that the octahedra are modified to make the dihedral angle α equal to $120°$. The shared faces in Fig. 5.41(a) are perpendicular to the plane of the paper and have a

187

common edge (also perpendicular to the paper) at the centre of the diagram. These triple chains may now join together by sharing all the edges projecting as *a, b,* and *c* to form the 3D structure shown in projection in Fig. 5.41(b). If built from MX_6 octahedra the structure has the composition M_3X_4. A model, which can be made from 'D-stix', shows that the pairs of vertices AA' etc. and D, E, and F outline a tricapped trigonal prism, so that the triple column of Fig. 5.41(a) may also be

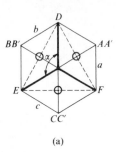 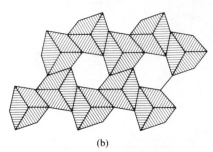

(a)   (b)

FIG. 5.41. The crystal structure of Nb_3S_4: (a) projection (idealized) of multiple chain formed from three rutile chains sharing the faces projecting as heavy lines, (b) projection of the Nb_3S_4 structure.

regarded as a column of tricapped trigonal prisms sharing basal faces (assuming all M and X atoms within the column to be removed). We then see a general similarity to the structure of UCl_3, which is built of columns of face-sharing tricapped trigonal prisms joined by edge-sharing into a structure of the same general form as Fig. 5.41(b). This will be evident from a comparison of Fig. 9.8 (p. 359) with Fig. 5.41(b). We have described the Nb_3S_4 structure here in terms of the NbS_6 coordination groups, that is, in terms of Nb—S bonds, but it should be noted that metal–metal bonding plays an important part in this structure (p. 623) as in a number of the other structures we have described.

Octahedra sharing faces, edges, and vertices

Structures of this type include the corundum (α-Al_2O_3) and γ-$Cd(OH)_2$ structures. The construction of models of both of these structures is described in MSIC. In the corundum structure there are pairs of octahedra with a common face, and in addition there is sharing of edges and vertices of coordination groups around the Al^{3+} ions. The complexity of this structure for a compound A_2X_3 appears to arise from the need to achieve a close packing of the anions and at the same time a reasonably symmetrical environment for these ions—see also the remarks on p. 216. In the structure of γ-$Cd(OH)_2$ (Fig. 5.42) pairs of rutile chains are joined through common faces and the double chains then share vertices.

Other structures in this class include the high-pressure form of Rh_2O_3 (p. 451) and $K_2Zr_2O_5$ (JSSC 1970 1 478).

Structures built from tetrahedra and octahedra

Structures which can be represented as assemblies of tetrahedral and octahedral coordination groups are very numerous. For example, they include all oxy-salts in

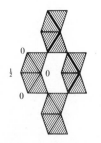

FIG. 5.42. The structure of γ-$Cd(OH)_2$.

188

which the cations occupy positions of octahedral coordination and the anion is a discrete BO_4 ion (as in Na_2SO_4 or $MgSO_4$) or a more complex ion built from such units (as in pyro- or meta- salts). In many such structures the oxygen atoms are close-packed, so that these structures, like those of many complex oxides, may be described as c.p. assemblies in which various proportions of tetrahedral and octahedral holes are occupied (Table 4.8, p. 149). Alternatively the structures may be described in terms of the way in which the tetrahedra and octahedra are joined together by sharing vertices, edges, or faces. It is not proposed to attempt an elaborate classification of this type but it is of interest to note a few simple examples of one class if only to show the relation between certain structures which are mentioned in later chapters.

We have already noted examples of structures in which tetrahedra *or* octahedra share various numbers of vertices. Structures in which tetrahedra *and* octahedra share only vertices form an intermediate group:

Tetrahedra sharing vertices only

$$\left\{ \begin{array}{ll} 4 & AX_2 \\ 3 & A_2X_5 \\ 2 & AX_3 \end{array} \right.$$

Tetrahedra and octahedra sharing vertices only:

$$\left\{ \begin{array}{l} A_3B_2X_{12} \\ ABX_5 \\ ABX_6 \\ ABX_7 \text{ etc.} \end{array} \right. \qquad \text{(Table 5.7)}$$

Octahedra sharing vertices only:

$$\left\{ \begin{array}{ll} 6 & BX_3 \\ 4 & BX_4 \\ 2 & BX_5 \end{array} \right.$$

Some of the simplest of this large family of structures are set out in Table 5.7; the structures are described in later chapters. The table shows the number of octahedra (o) with which each tetrahedron (t) shares vertices, the relative numbers of tetrahedra and octahedra, the contributions to the chemical formula made by each (a shared vertex counting as $\frac{1}{2}X$ and an unshared vertex as X), the formula, and examples of compounds with the structure.

In the garnet structure the $A_3B_2X_{12}$ system of linked tetrahedra and octahedra is a charged 3D framework which accommodates larger ions C, (in positions of 8-coordination) in the interstices. The general formula is $C_3A_3B_2X_{12}$ or $C_3B_2(AX_4)_3$ if we wish to distinguish the tetrahedral groups in, for example, an orthosilicate such as $Ca_3Al_2(SiO_4)_3$. In some garnets the same element occupies the positions of tetrahedral and octahedral coordination, when the formula reduces to $Y_3Al_5O_{12}$, for example. In all the last four examples of Table 5.7 there is both tetrahedral and octahedral coordination of one element in the same structure, and as a matter of interest we have collected together in Table 5.8 a more general set of

189

TABLE 5.7

Structures in which tetrahedral AX_4 *and octahedral* BX_6 *groups share only vertices*

			Formula			Example
All vertices of t and o shared	$\left.\begin{array}{l} t-4\,o \\ o-6\,t \end{array}\right\}$ t_3o_2		$(AX_2)_3(BX_3)_2 = A_3B_2X_{12}$			Garnet framework (see text)
	$\left.\begin{array}{l} t-4\,o \\ o-4\,t \\ \quad-2\,o \end{array}\right\}$ to		AX_2	BX_3	ABX_5	$NbOPO_4$ § $VOSO_4$ §
	$\left.\begin{array}{l} t-2\,o \\ \quad-2\,t \\ o-2\,t \\ \quad-4\,o \end{array}\right\}$ to		AX_2	BX_3	ABX_5	$Ca_2[Fe_2O_5]$ †
Each t and each o sharing 4 vertices	$\left.\begin{array}{l} t-4\,o \\ o-4\,t \end{array}\right\}$ to		AX_2	BX_4	ABX_6	$Na[PWO_6]$ §
	$\left.\begin{array}{l} t-2\,t \\ \quad-2\,o \\ o-4\,t \end{array}\right\}$ t_2o		$(AX_2)_2\,BX_4$	A_2BX_8		P_2MoO_8 §
t sharing 2 vertices o sharing 4 vertices	$\left.\begin{array}{l} t-2\,o \\ o-2\,t \\ \quad-2\,o \end{array}\right\}$ to		AX_3	BX_4	ABX_7	Re_2O_7 ‡ $Na_2[Mo_2O_7]$ ‡
	$\left.\begin{array}{l} t-2\,o \\ o-4\,t \end{array}\right\}$ t_2o		$(AX_3)_2\,BX_4$	A_2BX_{10}		$K_2[Mo_3O_{10}]$ ‡ $H_5[As_3O_{10}]$ ‡

† Square brackets enclose that part of the formula which corresponds to the t-o complex.
‡ Same element in t and o (A = B).
§ For these structures see p. 512.

TABLE 5.8

Structures in which an element exhibits both tetrahedral and
octahedral coordination

Structure	Element
γ-Fe_2O_3	Fe
β-Ga_2O_3, θ-Al_2O_3	Ga, Al
$X^{II}X_2^{III}O_4$ (regular spinel)	Mn, Co
$Y(XY)O_4$ (inverse spinel)	Fe in Fe_3O_4 and $Fe(MgFe)O_4$, Al in $Al(NiAl)O_4$, Co in $Co(SnCo)O_4$, Zn in $Zn(SnZn)O_4$
$Na_2Mo_2O_7$, $K_2Mo_3O_{10}$	Mo
$H_5As_3O_{10}$	As
Re_2O_7	Re
$M_2^{II}Mo_3O_8$	Mg, Mn, Fe, Co, Ni, Zn, Cd
$Zn_5(OH)_6(CO_3)_2$, $Zn_5(OH)_8Cl_2 . H_2O$	Zn
$(Mg_2Al)(OH)_4SiAlO_5$, $KAl_2(OH)_2Si_3AlO_{10}$	Al
$Na_4Ge_9O_{20}$, $K_3HGe_7O_{16}$	Ge
$Y_3Fe_2(FeO_4)_3$, $Y_3Al_5O_{12}$ (garnets)	Fe, Al
$Nb_{25}O_{62}$, high-Nb_2O_5	Nb
Cr_5O_{12}, KCr_3O_8	Cr(VI and III)

examples of this phenomenon, not restricted to vertex-sharing structures. It may also be of interest to set out the elements which exhibit both these coordination numbers when *in the same oxidation state*, in relation to the Periodic Table:

$$
\begin{array}{cccccccc}
 & & & & \text{Mg} & \text{Al} & & \\
 & \text{Mn}^{2+} & \text{Fe}^{3+} & \text{Co}^{2+} & \text{Ni}^{2+} & \text{Zn} & \text{Ga} & \text{Ge} & \text{As} \\
\text{Nb}^{5+} & \text{Mo}^{6+} & & & & \text{Cd} & & \\
 & & \text{Re}^{7+} & & & & &
\end{array}
$$

Referring again to the last four examples of Table 5.7 it may be noted that Re_2O_7 has a layer structure (p. 454), while the other three are chain structures. In the simple chain of Fig. 5.43(a), the topological repeat unit consists of one tetrahedron and one octahedron. This is the form of the anion in $Na_2Mo_2O_7$. The more complex chain of Fig. 5.43(b) is found in $K_2Mo_3O_{10}$, and a closely-related chain (Fig. 20.7) is the structural unit in $H_5As_3O_{10}$, an intermediate dehydration product of $As_2O_5 \cdot 7 H_2O$. (For $K_2Mo_3O_{10}$ see also p. 431.)

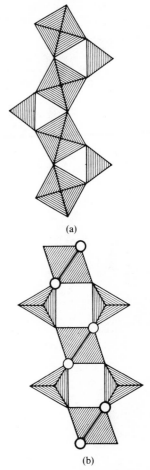

(a)

(b)

FIG. 5.43. Chains formed from tetrahedra and octahedra sharing only vertices.

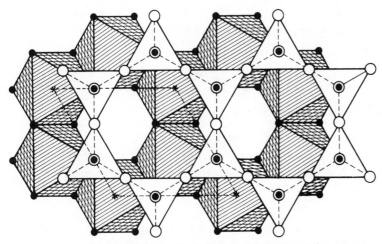

FIG. 5.44. The composite kaolin layer.

Of the structures listed in Table 5.8 the sesquioxide, spinel, and other complex oxide structures are described in Chapters 12 and 13. The structures of the two 'basic' zinc salts $Zn_5(OH)_6(CO_3)_2$ and $Zn_5(OH)_8Cl_2 \cdot H_2O$ are also described in more detail on p. 214. The composite layers with general formulae $Al_2(OH)_4Si_2O_5$ (or $Mg_3(OH)_4Si_2O_5$) and $Al_2(OH)_2Si_4O_{10}$ (or $Mg_3(OH)_2Si_4O_{10}$) which form the bases of the structures of many clay minerals and micas are described in Chapter 23 and the construction of an idealized model of the former (Fig. 5.44) in MSIC. Germanates provide examples of crystals containing Ge in positions of tetrahedral and octahedral coordination. A particularly elegant example, the 3D $Ge_7O_{16}^{4-}$ framework ion in $K_3HGe_7O_{16} \cdot 4 H_2O$, is shown in Fig. 5.45. We have referred earlier to oxides such as $Nb_{25}O_{62}$ and the high-temperature form of Nb_2O_5 in which a very small fraction of the metal atoms occupy tetrahedral holes in an essentially octahedral structure.

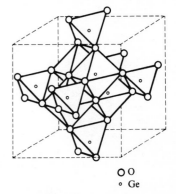

○ O
∘ Ge

FIG. 5.45. Framework of linked GeO_4 and GeO_6 groups in $K_3HGe_7O_{16} \cdot 4H_2O$.

6

Some Simple AX$_n$ Structures

We described in Chapter 3 a number of A$_m$X$_n$ structures in which the A atoms form three or four bonds. We now consider four simple 3D structures in which A has six or eight nearest neighbours and, as in the case of the diamond structure, we shall show how these structures can be adapted to suit the bonding requirements of various kinds of A and X atoms. The structures are:

AX   Sodium chloride (6 : 6); Caesium chloride (8 : 8).
AX$_2$ Rutile (6 : 3); Fluorite (8 : 4).

Then follow sections on

The CdI$_2$ and related structures,
The ReO$_3$ and related structures.

Next we consider some groups of structures:

The PbO and PH$_4$I structures,
The LiNiO$_2$, NaHF$_2$, and CsICl$_2$ structures,
The CrB, TlI (yellow), and related structures,
The PbCl$_2$ structure,
The PdS$_2$, AgF$_2$, and β-HgO$_2$ structures,

which are related in the following way. All structures in a particular group have the same symmetry (space group) and the atoms occupy the same sets of equivalent positions, but owing to the different values of one or more variable parameters, the similar analytical descriptions refer to atomic arrangements with quite different geometry (and/or topology). The chapter concludes with notes on the relations between the structures of some nitrides and oxy-compounds and on superstructures and other related structures.

The sodium chloride structure

In this structure (Fig. 6.1(a)) the A and X atoms alternate in a simple cubic sphere packing, each atom being surrounded by 6 others at the vertices of a regular octahedron. An alternative description, that the A (X) atoms occupy octahedral holes in a cubic closest packing of X (A) atoms is realistic only for compounds such as LiCl in which each Cl is actually in contact with 12 Cl atoms. This structure was derived by Barlow (1898) as a possible structure for crystals composed of

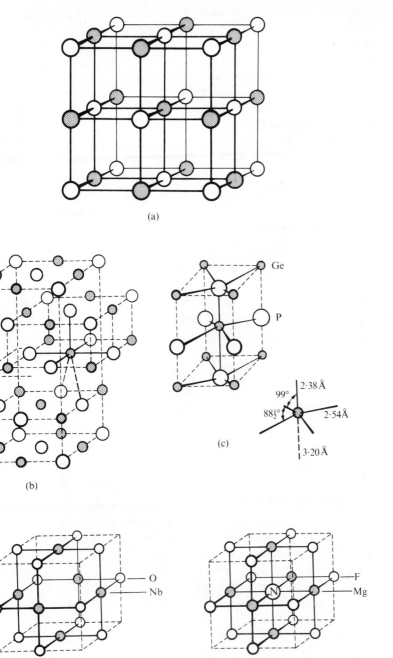

FIG. 6.1. The sodium chloride and related structures: (a) NaCl, (b) yellow TlI, (c) tetragonal GeP, (d) NbO, (e) Mg_3NF_3.

atoms (or ions) of suitable relative sizes; this aspect of the structure is discussed in Chapter 4. This structure is notable not only for the variety of types of chemical compound which crystallize in this way, but also for the large departures from the composition AX which it tolerates.

Compounds with the sodium chloride structure range from the essentially ionic halides and hydrides of the alkali metals and the monoxides and monosulphides of Mg and the alkaline-earths, through ionic–covalent compounds such as transition-metal monoxides to the semi-metallic compounds of B subgroup metals such as PbTe, InSb, and SnAs, and the interstitial carbides and nitrides (Table 6.1). Unique and different distorted forms of the structure are adopted by the Group IIIB

TABLE 6.1

Compounds AX *with the cubic* NaCl *structure*

Alkali halides and hydrides, AgF, AgCl, AgBr

Monoxides of

Mg								
Ca	Ti	V	–	Mn	Fe	Co	Ni	
Sr	Zr	Nb						Cd
Ba	Hf							

Eu; Th, Pa, U, Np, Pu, Am

Monosulphides of

Mg		
Ca	Mn	
Sr		
Ba		Pb

Ce, Sm, Eu; Th, U, Pu

Interstitial carbides and nitrides

	TiC,	VC,	UC
ScN,	LaN,	TiN,	UN

Phosphides etc.

InP, InAs, SnP (high pressure), SnAs

monohalides TlF and by the low-temperature form of InCl (see Chapter 9). The structure of the yellow form of TlI (Fig. 6.1(b)) is formed from slices of the NaCl structure which are displaced relative to one another with the result that the metal ion has five nearest neighbours at five of the vertices of an octahedron and then two pairs of nearly equidistant next-nearest neighbours (p. 349). The low-temperature form of NaOH has a similar structure.

When subjected to high pressure certain phosphides and arsenides of metals of Groups IIIB and IVB adopt either the cubic NaCl structure (InP, InAs) or a tetragonal variant of this structure (GeP, GeAs) shown in Fig. 6.1(c) in which there are bonds of three different lengths and effectively 5-coordination. The compounds of Ge and Sn, with a total of nine valence electrons, are metallic conductors. This property is not a characteristic only of the distorted NaCl structure, for SnP forms both the cubic and tetragonal structures, and both polymorphs exhibit metallic conduction. (IC 1970 **9** 335; JSSC 1970 **1** 143.)

194

Many oxides show a range of composition, the range depending on the temperature. As normally prepared, FeO is a typical non-stoichiometric compound which is deficient in Fe atoms, the range of composition being approximately $Fe_{0.95}O$–$Fe_{0.88}O$ at 1000°C. Some compounds MX formed by metals of Groups IVA and VA exhibit gross departures from the ideal composition; for example, the NaCl structure is stable for 'TiO' with O : Ti ratio from 0·7–1·25 at 1400°C. At the composition $TiO_{0.7}$ the Ti positions are almost fully occupied and one-third of the O positions vacant, while at the composition $TiO_{1.25}$ all the O positions are occupied but only about 80 per cent of the Ti positions. In the stoichiometric oxide about one-sixth of both metal and oxygen sites are vacant, and on cooling below 990°C the vacancies become ordered as shown in Fig. 12.17 (p. 466). In stoichiometric NbO the vacant sites are ordered, and the NbO structure (Fig. 6.1(d)) is preferably regarded as a distinct structure rather than as a defective NaCl structure ($Nb_{0.75}O_{0.75}$).

There are also ordered superstructures in nitrides and carbides, for example, Ti_2N, Th_4C_3, V_6C_5, and V_8C_7, the last two having helical arrays of vacancies (PM 1968 **18** 177; AC 1970 **B26** 1882). Other compounds with structures that may be described as defective NaCl structures include Mg_6MnO_8, $Li_2U_4O_7$, Mg_3NF_3, and Sc_2S_3, in which respectively $\frac{1}{8}$, $\frac{1}{7}$, $\frac{1}{4}$, and $\frac{1}{3}$ of the cation positions are vacant. The oxide structures are described in Chapter 12 and we refer to Sc_2S_3 later in this chapter. The very simple Mg_3NF_3 structure is illustrated in Fig. 6.1(e); its relationship to NaCl and NbO is obvious. Mg and N have octahedral arrangements of nearest neighbours (2 N + 4 F and 6 Mg respectively) while F has four coplanar Mg neighbours (p. 431).

The full cubic symmetry of the NaCl structure is retained in (a) high temperature defect structures with random distribution of vacancies, (b) solid solutions in which there is random arrangement of ions of two or more kinds in the anion and/or cation positions, and (c) crystals containing complex ions either if the complex ions have full cubic symmetry, as in $[Co(NH_3)_6] [TlCl_6]$, or if there is rotation or random orientation of less symmetrical groups, as in the high-temperature forms of alkaline-earth carbides, KSH, and KCN. In the latter cases the low-temperature forms have lower symmetry.

Three types of phase with NaCl-like structures may be illustrated by examples of oxides. In solid solutions where both ions have the same charge the relative numbers of the two kinds of ion may vary, the range of solid solution formation depending on the chemical nature of the ions and on their relative sizes. If the charges on the ions are different their proportions are fixed, but the ions of different kinds may be either randomly or regularly arranged:

Arrangement of cations		*Charges on ions*
Random	⎰ Solid solutions: (Mg, Ni)O	Same
	⎱ High temperature forms of $LiFeO_2$, Na_2SnO_3, Li_3TaO_4	
Ordered (superstructure)	$LiFeO_2$ (low), $LiNiO_2$, $LiInO_2$, Li_2TiO_3, Li_4UO_5	Different

In the random structures full cubic symmetry is retained; the ordered structures are superstructures of NaCl. That of $LiNiO_2$ is referable to the rhombohedral cell of Fig. 6.3(b), while the $LiInO_2$ structure is a tetragonal superstructure illustrated in Fig. 6.2(a).

We saw in Chapter 2 that the characteristic (minimum) symmetry of a cubic crystal is the set of 3-fold axes parallel to the body-diagonals of the cubic cell. These 3-fold axes also imply 2-fold axes parallel to the cube edges. The NaCl structure has the highest class of cubic symmetry, with 4-fold axes and planes of symmetry. An octahedral ion such as $(TlCl_6)^{3-}$ also has full cubic symmetry ($m3m$) can occupy the Cl^- positions in the normal NaCl structure. Groups such as S_2, $(F-H-F)^-$, and CN^- have lower symmetry and can form the fully symmetrical NaCl structure only if they are rotating or are randomly oriented with their centres

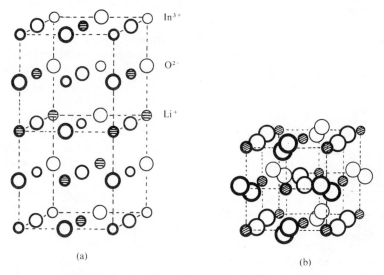

FIG. 6.2. The structures of (a) $LiInO_2$, (b) FeS_2 (pyrites).

at the anion positions of that structure, as in high-KCN. The S_2 group may, however, be arranged along the 3-fold axes of a lower class of cubic symmetry, as in the cubic pyrites (FeS_2) structure of Fig. 6.2(b). Here the 3-fold axes are directed along the body-diagonals of the octants of the cell, and do not intersect. This structure is adopted by numerous chalconides of 3d metals (Mn–Cu), in some cases only under pressure (see Chapter 17), and by the high-pressure form of SiP_2. The pyrites structure and the CaC_2 structure to be described shortly are also suitable for peroxides of the alkaline-earths (containing the O_2^{2-} ion) and for the superoxides of the alkali metals, which contain the O_2^- ion (Table 6.2). At higher temperatures the structures of some of these compounds (NaO_2, KO_2) change to the disordered pyrites structure, which is also the structure of the high-temperature form of KHF_2. In this structure there is random orientation of the $(F-H-F)^-$ ions along the directions of the four body-diagonals of the cube.

The structure of PdS_2 results from elongation of the pyrites structure in one

TABLE 6.2

Crystal structures of peroxides and superoxides

Pyrites structure	CaC$_2$ structure
MgO$_2$, ZnO$_2$, CdO$_2$ NaO$_2$	CaO$_2$, SrO$_2$, BaO$_2$ KO$_2$, RbO$_2$, CsO$_2$

direction so that Pd has only four (coplanar) nearest S neighbours as compared with the six octahedral neighbours of Fe in FeS$_2$; it is preferably regarded as a layer structure, and was so described in Chapter 1: for another way of describing the PdS$_2$ and pyrites structures see p. 223.

The two kinds of distortion of a cubic structure which preserve the highest axial symmetry are extension or compression parallel to a cube-edge or body-diagonal, leading to tetragonal or rhombohedral structures respectively. The structure of the tetragonal form of CaC$_2$ is an example of a NaCl-like structure in which the linear C_2^{2-} ions are aligned parallel to one of the cubic axes (Fig. 22.6, p. 757), while the structures of certain monoalkyl substituted ammonium halides (for example, NH$_3$CH$_3$I and NH$_3$C$_4$H$_9$I) illustrate a much more extreme extension of the structure along a 4-fold axis. In these halides there is presumably random orientation or rotation of the paraffin chains. Parallel alignment of the CN$^-$ ions in the low-temperature form of KCN (and the isostructural NaCN) leads to orthorhombic symmetry (Fig. 22.1, p. 750).

It is usually easy to see the relation of the unit cell of a tetragonal NaCl-like structure to the cubic unit cell of the NaCl structure, as in the case of CaC$_2$ or LiInO$_2$ (Fig. 6.2(a)) where one dimension is doubled. The relation of rhombohedral NaCl-like structures to the cubic NaCl structure is best appreciated from models. The cubic NaCl structure itself may be referred to rhombohedral unit cells containing one or two formula-weights, as compared with four for the normal cubic cell, as shown in Fig. 6.3(a) and (b):

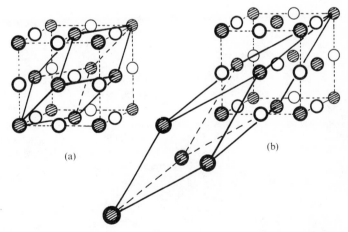

(a)

(b)

Unit cell	Edge	Angle α	$Z^\dagger$
Cubic	a	90°	4
Rhombohedral (a)	$a/\sqrt{2}$	60°	1
Rhombohedral (b)	$\sqrt{3}a/\sqrt{2}$	33° 34′	2

$^\dagger Z$ is the number of formula-units (NaCl) in the unit cell.

FIG. 6.3. Alternative rhombohedral unit cells in the NaCl structure.

197

Some simple AX_n Structures

The smallest rhombohedral distortion of the NaCl structure is found in crystals such as the low-temperature forms of FeO and SnTe, the high-temperature polymorphs of which have the normal NaCl structure. The rhombohedral structure of low-KSH is referable to a cell of the type of Fig. 6.3(a) with $\alpha = 68°$ instead of the ideal value, $60°$. Other rhombohedral NaCl-like structures include those of $LiNiO_2$ and $NaCrS_2$, which are superstructures of NaCl, and those of $NaHF_2$, NaN_3, and $CaCN_2$. These groups of structures are further discussed on p. 219. The planar anions in $NaNO_3$ and $CaCO_3$ possess 3-fold symmetry and in the calcite structure of these and other isostructural salts they are oriented with their planes perpendicular to one of the 3-fold axes of the original cubic structure; the cations and the centres of the anions are arranged in the same way as the Na^+ and Cl^- ions in NaCl. The structure of non-stoichiometric Sc_2S_3 is referable to the rhombohedral cell of Fig. 6.3(b) with α equal to the value $33°34'$, with 1 Sc in the position (000) but only 0·37 Sc in the position ($\frac{1}{2}$ $\frac{1}{2}$ $\frac{1}{2}$).

Apart from the defect structures and TlI, the structures we have mentioned are all essentially of the 6 : 6 coordination type. In the SnS (GeS) structure the distortion of the octahedral coordination groups is such as to bring three of the six original neighbours much closer than the other three, so that the structure could alternatively be described as consisting of very buckled 6-gon nets in which

CHART 6.1

The NaCl and related structures

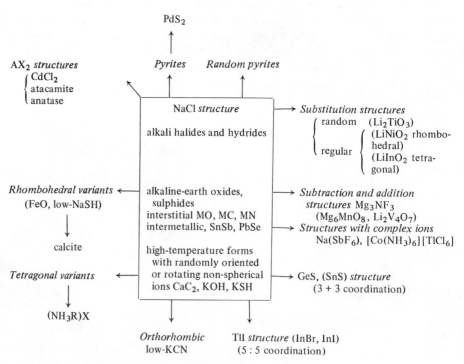

198

alternate atoms are Sn and S. The SnS structure has been described in this way in Chapter 3.

Structures related to the NaCl structure are summarized in Chart 6.1.

The caesium chloride structure

In this AX structure (Fig. 6.4) each atom (ion) has eight equidistant nearest neighbours arranged at the vertices of a cubic coordination group. Compared with the NaCl structure this is an unimportant structure. It is adopted by some intermetallic compounds, for example the ordered forms of β-brass (CuZn) and phases such as FeAl, TlSb, LiHg, LiTl, and MgTl, but not by any 'interstitial' compounds MC, MN, or MO. Only three of the alkali halides (and TlCl, TlBr, and TlI) have the CsCl structure at ordinary temperature and pressure, though some K and Rb halides adopt this structure under pressure; on the other hand those halides which do adopt the CsCl structure can be induced to crystallize with the NaCl structure on a suitable substrate. The importance of polarization of the Cs^+ or Tl^+ ions in this structure is emphasized in Chapter 7.

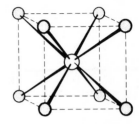

FIG. 6.4. The CsCl structure.

Of the alkali hydroxides, hydrosulphides, and cyanides only CsSH and CsCN have CsCl-like structures; the structure of CsOH is not known. Only one form of CsSH is known, and it has the cubic CsCl structure in which SH^- is behaving as a spherically symmetrical ion with the same radius as Br^-. The high-temperature form of CsCN has the cubic CsCl structure; the low-temperature form has the less symmetrical structure noted later. The bifluorides behave rather differently, the K and Rb salts showing a preference for CsCl-like structures. Ammonium salts differ from the corresponding alkali metal salts if there is the possibility of hydrogen bonding between cation and anion. For example, the structures of NH_4HF_2 and KHF_2 (low-temperature form) are similar in that both have CsCl-like arrangements of the ions (Fig. 8.6, p. 311) but owing to the different relative orientations of the $(F-H-F)^-$ ions in the two crystals NH_4^+ has only four nearest F^- neighbours while in KHF_2 K^+ has eight equidistant F^- neighbours. This breakdown of the coordination group of eight into two sets of four is due to the formation of N–H–F bonds; in NH_4CN the cation makes contact with only four anions, as described later. The random pyrites structure of the high-temperature form of KHF_2 has been noted under the sodium chloride structure.

As in the case of the NaCl structure, the two simplest types of distortion are those leading to rhombohedral or tetragonal structures. An example of the former is the low-temperature polymorph of CsCN, in which (with 1 CsCN in the unit cell) all the CN^- ions are oriented parallel to the triad axis (Fig. 22.2, p. 750). The edge of the tetragonal unit cell of the structure of NH_4CN (which is not known to exhibit polymorphism in the temperature range $-80°$ to $+35°C$) is doubled in one direction because the CN^- ions have two different orientations (Fig. 22.3(b), p. 751). Although each NH_4^+ ion is surrounded by 8 CN^- ions, owing to the orientations of these ions there are 4 at a distance of 3·02 Å and 4 more distant (at 3·56 Å). Other compounds with CsCl-like packing of ions include $KSbF_6$, $AgNbF_6$, and isostructural compounds, $[Be(H_2O)_4]SO_4$, and $[Ni(H_2O)_6]SnCl_6$.

The rutile structure

This tetragonal structure (Fig. 6.5(a)) is named after one of the polymorphs of TiO_2; it is also referred to as the cassiterite (SnO_2) structure. The coordinates of the atoms are:

$$Ti: (000), (\tfrac{1}{2}\tfrac{1}{2}\tfrac{1}{2}).$$
$$O: \pm (xx0), (\tfrac{1}{2} + x, \tfrac{1}{2} - x, \tfrac{1}{2}).$$

The structure consists of chains of TiO_6 octahedra, in which each octahedron shares a pair of opposite edges (Fig. 6.5(b)), which are further linked by sharing vertices to form a 3D structure of 6 : 3 coordination as shown in the projection of Fig. 6.5(c). With the above coordinates each O has three coplanar neighbours (2 at the distance d and 1 at e), Ti has six octahedral neighbours (4 at the distance d and 2 at e), and all Ti–Ti distances (between centres of octahedra along a chain) are

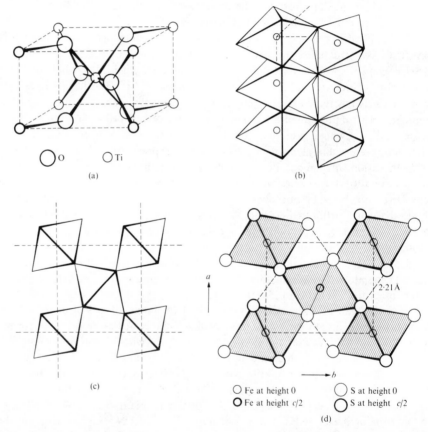

FIG. 6.5. The rutile structure: (a) unit cell, (b) parts of two columns of octahedral TiO_6 coordination groups, (c) projection of structure on base of unit cell, (d) corresponding projection of marcasite structure.

equal. Although there are two independent variables, $c : a$ and x, in the structure as described, this number reduces to one if all the Ti–O distances are made equal ($d = e$), when there is the following relation between x and c/a: $8x = 2 + (c/a)^2$. The special cases would be: (i) regular octahedral coordination of Ti ($c/a = 0.58$, $x = 0.29$), and (ii) equilateral triangular coordination of O ($c/a = 0.817$, $x = 0.33$). We saw in Chapter 5 that both of these highly symmetrical arrangements of nearest neighbours are not possible simultaneously, for (i) implies bond angles at O of $90°$ and $135°$ (two). In dioxides and difluorides with the tetragonal rutile structure there is usually very little difference between the two M–O or M–F distances, the structure being much closer to case (i) than to (ii), with x close to 0.30. Typical data are:

	$c : a$	M–O *or* M–F	
		4 of	2 of
TiO_2	0.645	1.94 Å	1.99 Å
ZnF_2	0.665	2.03 Å	2.04 Å
NiF_2	0.663	1.98 Å	2.04 Å

The normal rutile structure is restricted to the difluorides of Mg and certain 3d metals, a number of dioxides, some oxyfluorides (FeOF, TiOF, VOF), and MgH_2; it is *not* adopted by disulphides, by other dihalides, or by intermetallic phases. In

TABLE 6.3
Dioxides and difluorides with the rutile or fluorite structures

Dioxides   *Rutile structure*

Ti	V[a]	Cr	Mn			Si[b]
Zr	Nb	Mo	Tc	Ru	Rh[b]	Ge
Hf	Ta	W	Re	Os	Ir	Pt[b]	Pb	Po	Sn

Fluorite structure

| Ce | Pr | | Tb |
| Th | Pa | U | Np | Pu | Am | Cm |

[a] Structure modified by metal–metal bonding.
[b] When crystallized under pressure.

Difluorides   *Rutile structure*

Mg

| Ca | Cr[c] | Mn | Fe | Co | Ni | Cu[c] | Zn |
| Sr | | | | | | Cd |

Fluorite structure

| Ba | | | | | | Hg | Pb |

[c] Distorted (see text).

general, compounds MF_2 and MO_2 containing larger ions adopt the fluorite structure, and since in this sense the rutile and fluorite structures are complementary we group together in Table 6.3 compounds that crystallize with one or other of these structures. Some interesting modifications of the rutile structure show how it can be adapted to suit the special bonding requirements of certain atoms (ions).

The structure of CrF_2 (and the isostructural CuF_2) illustrates the distortion of the octahedral coordination groups to give four shorter and two longer bonds, the structure tending towards a layer structure. This can be seen from Fig. 6.6 by noting that the atom at the body-centre of the cell is connected through F atoms to four others by the stronger bonds (full lines), these five atoms defining a plane passing through the centre of the cell. The differences between the two sets of M—F distances are hardly sufficient to justify describing these as layer structures, though the difference is rather larger in CrF_2 than in CuF_2;

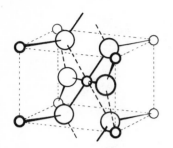

FIG. 6.6. Distorted rutile structure of CuF_2.

$$Cu—4\ F,\ 1\cdot93\ \text{Å};\qquad Cr—4\ F,\ 2\cdot00\ \text{Å}.$$
$$Cu—2\ F,\ 2\cdot27\ \text{Å};\qquad Cr—2\ F,\ 2\cdot43\ \text{Å}.$$

For a discussion of bond lengths in oxides and fluorides with the rutile structure (see: AC 1971 **B27** 2133).

A quite different type of distortion of the rutile structure is found in a number of transition-metal dioxides (of V, Nb, Mo, W, Tc, and Re). In contrast to the regular spacing of Ti atoms along the chains in TiO_2 (at $2\cdot96$ Å) the distances between metal atoms in these oxides alternate: $2\cdot65$ and $3\cdot12$ Å in VO_2, and $2\cdot80$ and $3\cdot20$ Å in NbO_2; the latter has a complex superstructure of the rutile type. The metal-metal interactions are considered responsible for the anomalously low paramagnetism of NbO_2, but it should be noted that RuO_2 (with Ru—Ru, $3\cdot11$ Å in a normal rutile structure) also has a very low paramagnetic susceptibility, and high electrical conductivity, in spite of the absence of direct metal–metal interactions.

In our survey of close-packed structures in Chapter 4 we noted that the packing of the O atoms in the rutile structure represents a considerable distortion of hexagonal closest packing. The packing of the anions in $CaCl_2$ is much closer to hexagonal closest packing, and in this structure Ca has 6 nearly equidistant Cl neighbours (4 at $2\cdot76$ Å and 2 at $2\cdot70$ Å) with Cl slightly out of the plane of its 3 nearest Ca neighbours. This change in the coordination of the anion is brought about by rotating the chains of octahedra relative to one another, and is more marked in $CrCl_2$ (Fig. 6.7). In this crystal there is also a pronounced elongation of the octahedral $CrCl_6$ groups (Cr—4 Cl, $2\cdot37$ Å, Cr—2 Cl, $2\cdot91$ Å). At the limit this type of distortion would lead to chains in which the metal atoms form four coplanar bonds, as in $PdCl_2$.

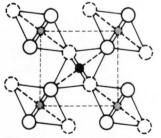

FIG. 6.7. Projection of the structure of $CrCl_2$.

When the edge-sharing chains in the rutile structure, seen in projection in Fig. 6.5(c), are rotated relative to one another the distance between pairs of X atoms of different chains is reduced. The angle of rotation in $CrCl_2$ is small ($20°$) but larger in InO . OH (and the isostructural CrO . OH (green) and CoO . OH), where it is due to hydrogen bonding, as shown in Fig. 14.11 (p. 525). Further rotation of the chains gives the marcasite structure of FeS_2 in which the close S–S contacts correspond to S–S bonds (2·21 Å) as shown in Fig. 6.5(d). Examples of compounds with the marcasite and the closely related löllingite and arsenopyrites structures are given in Chapter 17.

Compounds ABX_4, A_2BX_6, etc. with rutile-like structures
The cation sites in the rutile structure may be occupied by cations of two or more different kinds, either at random (random rutile structure) or in a regular manner (superstructure). In the former case the structure is referable to the normal rutile unit cell, but the unit cell of a superstructure is usually, though not necessarily, larger than that of the statistical structure. Phases with disordered structures arise if the charges on the cations are the same or not too different. If the charges are the same the composition may be variable, as in solid solutions $(Mn, Cr)O_2$, but it is necessarily fixed if the charges are different ($CrNbO_4$, $FeSbO_4$, $AlSbO_4$, etc.). Ordered arrangements of cations arise when either

(a) the ionic charges are very different, or
(b) special bonding requirements of particular atoms have to be satisfied.

    (a) A number of complex fluorides and oxides A_2BX_6 (X = F or O) crystallize with the *trirutile structure* (Fig. 6.8). Fluorides $A^{2+}B_2^{2+}F_6$ (e.g. $FeMg_2F_6$) adopt the disordered rutile structure but $Li_2^+Ti^{4+}F_6$ and one form of Li_2GeF_6 have the trirutile structure. There are three possible types of fluoride $ABCF_6$, namely, $A^+B^+C^{4+}F_6$, $A^{1+}B^{2+}C^{3+}F_6$, and $A^{2+}B^{2+}C^{2+}F_6$. The random rutile structure is adopted if all the ions are M^{2+} (as in $FeCoNiF_6$), but the trirutile structure is adopted by compounds of the first two classes. The arrangement of the different kinds of ion in the trirutile structure is such as to give maximum separation of the M^{4+} ions in the first group of compounds (e.g. Li_2TiF_6) or of the M^+ ions in the second group (e.g. $LiCuFeF_6$), that is, these ions occupy the Zn positions in Fig. 6.8. (ZN 1967 **22b** 1218; JSSC 1969 **1** 100). Oxides with the trirutile structure include $MgSb_2O_6$, $MgTa_2O_6$, Cr_2WO_6, $LiNbWO_6$, and VTa_2O_6 (JSSC 1970 **2** 295).

    (b) In $MgUO_4$ the U atom has, as in many uranyl compounds, two close neighbours (at approximately 1·9 Å) and four more distant O neighbours (at 2·2 Å). Within one rutile chain all metal atoms are U atoms. The environment of Mg^{2+} is similar to that of U, 2 O at 2·0 Å and 4 O at 2·2 Å, this evidently being a secondary result of the U coordination. In $CuUO_4$, on the other hand, Cu and U atoms alternate along each rutile chain, and here the rutile structure is modified to give U (2 + 4)–coordination (at the mean distances 1·9 Å and 2·2 Å close to those in $MgUO_4$) but Cu has (4 + 2)–coordination (at 1·96 Å and 2·59 Å).

Homologous series of oxides structurally related to the rutile structure. Vanadium and titanium form series of oxides M_nO_{2n-1} which have structures related to the rutile structure in the following way. Slabs of rutile-like structure, n octahedra thick but extending indefinitely in two dimensions, are connected across

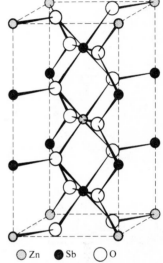

○ Zn   ● Sb   ○ O

FIG. 6.8. The trirutile structure of $ZnSb_2O_6$.

203

'shear planes'. These structures can be derived from the rutile structure by removing the O atoms in certain planes and then displacing the rutile slabs so as to restore the octahedral coordination of the metal atoms. This operation results in the sharing of faces between certain pairs of octahedra (in addition to the vertex- and edge-sharing in the rutile blocks) as in the corundum structure (p. 450)—compare the shear structures of Mo and W oxides derived in a similar way from the vertex-sharing ReO_3 structure to give structures in which some octahedron edges are shared. References to the Ti oxides ($4 \leqslant n \leqslant 9$) and the V oxides ($3 \leqslant n \leqslant 8$) are given in Chapter 12.

The fluorite (AX$_2$) and antifluorite (A$_2$X) structures

In the fluorite structure (Fig. 6.9) for a compound AX_2 the A atoms (ions) are surrounded by 8 X at the vertices of a cube and X by 4 A at the vertices of a regular tetrahedron. It is the structure of a number of difluorides and dioxides (Table 6.3) and also of some disilicides (see later), intermetallic compounds (for example, $GeMg_2$, $SnMg_2$, $PtAl_2$ etc.), and transition-metal dihydrides. The dihydrides of Ti, Zr, and Hf are typical defect structures, and in the systems Ti–H and Hf–H the cubic fluorite structure is not stable up to the composition MH_2. For example, in the Hf–H system the cubic structure occurs over the composition range $HfH_{1.7}$–$HfH_{1.8}$ (at room temperature), but at the composition HfH_2 the structure is less symmetrical (f.c. tetragonal). We refer later to the quite different behaviour of the 4f elements.

The positions of the Ca^{2+} ions in the fluorite structure are those of cubic closest packing and the positions of the F^- ions correspond to all the tetrahedral holes. However, the ions which are actually in contact are the F^- ions (F–F, 2·70 Å, twice the radius of F^-) whereas the shortest distance between Ca^{2+} ions is 3·8 Å, which may be compared with the radius of Ca^{2+} (1·0 Å). The structure is therefore

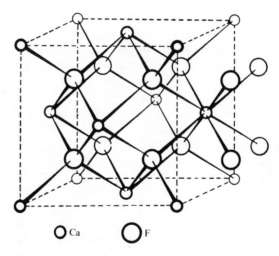

FIG. 6.9. The fluorite structure.

○ Ca　　◯ F

preferably described as a simple cubic packing of F^- ions in which alternate positions of cubic coordination are occupied by Ca^{2+} ions, that is, a CsCl structure from which one-half of the cations have been removed. The pattern of cation sites occupied is that which gives F^- four tetrahedral Ca^{2+} neighbours.

In the antifluorite structure the positions occupied by the anions and cations in the fluorite structure are interchanged. This structure is adopted by the oxides M_2O and sulphides M_2S of Li, Na, K, and Rb; Cs_2O has a layer structure, and the structure of Cs_2S is not known. The structure of Li_2O may be described as an approximately c.c.p. array of O^{2-} ions with Li^+ ions in the tetrahedral holes, since O^{2-} is much larger than Li^+, though O–O in Li_2O is appreciably larger (3·3 Å) than the 2·8 Å found in many c.p. oxide structures. In fact the antifluorite structure is also adopted by K_2O and Rb_2O, in which the cations are comparable in size with O^{2-}, yet 4-coordination of the cations persists. We return to this point in our discussion of ionic structures in Chapter 7. Other compounds with this structure include Mg_2Si, Mg_2Ge, etc.

A considerable number of fluorides, oxides, and oxyfluorides have structures which are related more or less closely to the fluorite structure. The relation is closest for compounds with cation: anion ratio equal to 2 : 1 with random arrangement of cations or anions, these structures having the normal cubic unit cell; examples include the high-temperature forms of $NaYF_4$ and K_2UF_6 and the oxyfluorides AcOF and HoOF. As in the case of the NaCl structure there are tetragonal and rhombohedral superstructures, which are described in Chapter 10. These are the structures of a number of oxyfluorides:

> Tetragonal: YOF, LaOF, PuOF,
> Rhombohedral: YOF, LaOF, SmOF, etc.

The fluorite superstructure of γ-Na_2UF_6 is illustrated in Fig. 28.3 (p. 995). There is slightly distorted cubic coordination of M^{5+} in Na_3UF_8 (and the isostructural Na_3PuF_8). Owing to the arrangement of cations in this structure (Fig. 6.10) the unit cell is doubled in one direction and is a b.c. tetragonal cell.

The ternary fluorides $SrCrF_4$ (and isostructural $CaCrF_4$, $CaCuF_4$, and $SrCuF_4$) and $SrCuF_6$ have structures which are related less closely to fluorite. The cell dimensions correspond to cells of the fluorite type doubled and tripled respectively in one direction, and the cation positions are close to those of the fluorite structure of, for example SrF_2, as shown in Fig. 6.10. There is, however, considerable movement of the F^- ions from the 'ideal' positions to give suitable environments for the transition-metal ions. The Sr^{2+} ions retain cubic 8-coordination but Cr^{2+} (Cu^{2+}) ions are surrounded by an elongated tetrahedron of anions (two angles of 94°, four of 118°). In a series of oxides MTe_3O_8 (M = Ti, Zr, Hf, Sn) the cation positions are close to those of the fluorite structure, as shown in the sub-cell of Fig. 6.10, but the actual unit cell has dimensions twice as large in each direction and the O atoms are displaced to give very distorted octahedral coordination of both M and Te (for example Te-4 O, 2·04 Å, Te-2 O, 2·67 Å).

An interesting distortion of the fluorite structure to give weak metal–metal interactions is reminiscent of the distorted rutile structures of certain dioxides.

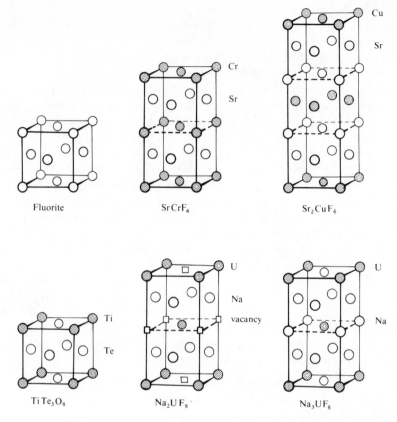

FIG. 6.10. Cation positions in structures related to fluorite.

Whereas $CoSi_2$ and $NiSi_2$ crystallize with the cubic fluorite structure, this structure is distorted in β-$FeSi_2$ to give Fe two Fe neighbours, the metal atoms being arranged in squares of side 2·97 Å (compare 2·52 Å in the 12-coordinated metal). The metal atom also has 8 Si neighbours at the vertices of a deformed cube. This distortion of the fluorite structure represents a partial transition towards the $CuAl_2$ structure (p. 1046), in which the coordination group of Cu is a square antiprism and there are chains of bonded Cu atoms. (AC 1971 **B27** 1209).

Addition of anions to the fluorite structure: the Fe_3Al *structure*
In a number of metal–non-metal systems the cubic fluorite-like phase is stable over a range of composition. For the hydrides of the earlier 4f metals (La–Nd) this phase is stable from about $MH_{1.9}$ to compositions approaching MH_3. (The range of stability of the 'dihydride' phase of the later 4f metals is more limited, and separate from that of the 'trihydride', which has a different (LaF_3) structure). The only sites available for the additional anions are at the mid-points of the edges and at the body-centre of the fluorite cell (the larger shaded circles in Fig. 6.11). The resulting AX_3 structure is sometimes described as the BiF_3 structure because it was

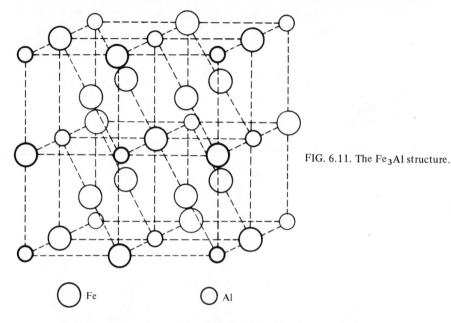

FIG. 6.11. The Fe_3Al structure.

○ Fe   ○ Al

once thought (erroneously) to be the structure of one form of that compound (see p. 357). It is preferably termed the Fe_3Al (or Li_3Bi) structure since it is the structure of the ordered form of a number of intermetallic phases.

In the Fe_3Al structure there is cubic 8-coordination of all the atoms, and writing the formula AX_2X' the nearest neighbours are:

$$A : 8\ X,\ X : 4\ A,\ and\ X' : 8\ X.$$
$$4\ X'$$

The environment of the (additional) X' atoms is therefore a group of 8 X at $0.433a$ (a = cell edge), and the nearest A neighbours are an octahedral group of 6 A at $0.5a$. Although satisfactory in an intermetallic compound this is an unlikely environment for an anion, and although fluorite-like phases certainly exist for the compounds MH_{2+x} and UO_{2+x} it seems likely that some rearrangement of the anions occurs in these structures. It seems improbable that any phases, other than intermetallic compounds form the ideal structure of Fig. 6.11 with its full complement of X' atoms. For further details of compounds with this type of structure see the discussion of the 4f hydrides (p. 295), Bi_2OF_4 (p. 358), U oxides UO_{2+x} (p. 998), and also Table 29.7 (p. 1036).

Defect fluorite structures

Fluorite-like structures deficient in cations seem to be rare; an example is Na_2UF_8 (earlier described as Na_3UF_9) in which there are discrete cubic UF_8 groups. The arrangement of the Na^+ ions is the same as that of two-thirds of those ions in Na_3UF_8, as shown in Fig. 6.10. On the other hand, there are a number of phases

Some simple AX_n Structures

(mostly oxides) with structures derivable from fluorite by removing a fraction of the *anions*. These structures are summarized in Table 6.4.

TABLE 6.4

Structures related to the fluorite structure

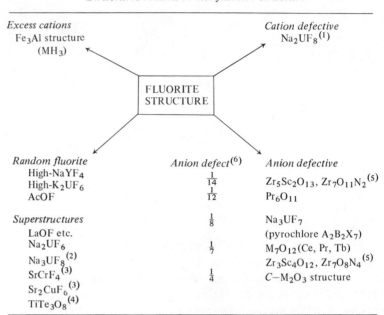

Excess cations		Cation defective
Fe$_3$Al structure (MH$_3$)		Na$_2$UF$_8$[1]

FLUORITE STRUCTURE

Random fluorite	Anion defect[6]	Anion defective
High-NaYF$_4$	$\frac{1}{14}$	Zr$_5$Sc$_2$O$_{13}$, Zr$_7$O$_{11}$N$_2$[5]
High-K$_2$UF$_6$	$\frac{1}{12}$	Pr$_6$O$_{11}$
AcOF		
Superstructures	$\frac{1}{8}$	Na$_3$UF$_7$
LaOF etc.		(pyrochlore A$_2$B$_2$X$_7$)
Na$_2$UF$_6$	$\frac{1}{7}$	M$_7$O$_{12}$(Ce, Pr, Tb)
Na$_3$UF$_8$[2]		Zr$_3$Sc$_4$O$_{12}$, Zr$_7$O$_8$N$_4$[5]
SrCrF$_4$[3]	$\frac{1}{4}$	C–M$_2$O$_3$ structure
Sr$_2$CuF$_6$[3]		
TiTe$_3$O$_8$[4]		

(1) IC 1966 5 130
(2) JCS 1969 A 1161
(3) JSSC 1970 2 262
(4) AC 1971 **B27** 602
(5) AC 1968 **B24** 1183; for other compounds see later chapters
(6) For a systematic treatment of the homologous series M$_n$O$_{2n-2}$ see: ZK 1969 **128** 55

The C-M$_2$O$_3$ structure (p. 451) may be derived from fluorite by removing $\frac{1}{4}$ of the anions and slightly rearranging the remainder. All the M atoms are octahedrally coordinated, the coordination being rather more regular for one-quarter of these atoms. Solid solutions UO$_{2+x}$–Y$_2$O$_3$ form defect f.c.c. structures with a degree of anion deficiency depending on the U content and on the partial pressure of oxygen. When the O deficiency becomes too large rearrangement takes place to form a rhombohedral phase M^{VI}M$_6^{III}$O$_{12}$ (M = Mo, W, U) in which M^{VI} is 6- and M^{III} 7-coordinated. This phase can also be formed by oxides M$_4^{III}$M$_3^{IV}$O$_{12}$ (e.g. Sc$_4$Zr$_3$O$_{12}$) and also as an oxide M$_7$O$_{12}$ by elements which can form ions M^{3+} and M^{4+} (e.g. Ce, Pr, Tb). Intermediate between this phase and a dioxide is the phase Zr$_5$Sc$_2$O$_{13}$, corresponding to removal of only $\frac{1}{14}$ of the O atoms, instead of $\frac{1}{7}$, from a fluorite structure. In these oxygen-deficient phases O retains its tetrahedral coordination, while the coordination of the M ions falls:

Oxygen deficiency:	0	$\frac{1}{14}$	$\frac{1}{7}$	$\frac{1}{4}$
Composition:	MO$_2$	M$_7$O$_{13}$	M$_7$O$_{12}$	M$_2$O$_3$
Coordination of M:	8	6, 7, 8	6, 7	6

Other defect fluorite structures with intermediate degrees of anion deficiency are noted elsewhere: Pr$_6$O$_{11}$ (p. 449) and Na$_3$UF$_7$ (p. 995).

208

The pyrochlore structure

This structure (Fig. 7.4 (between pp. 268 and 269), forms a link between the defect fluorite structures we have been describing and a topic discussed at the end of this chapter. With its complete complement of atoms this (cubic) structure contains eight $A_2B_2X_6X'$ in the unit cell, and there is only one variable parameter, the x parameter of the 48 X atoms in the position $(x, \frac{1}{8}, \frac{1}{8})$ etc. The structure may be described in two ways, according to the value of x:

(i) $x = 0.375$: coordination of A cubic and of B a very deformed octahedron (flattened along a 3-fold axis). This is a defect superstructure of fluorite, $A_2B_2X_7\square$ ($\square$ = vacancy).

(ii) $x = 0.3125$: coordination of A a very elongated cube (i.e. a puckered hexagon + 2), and coordination of B, a regular octahedron. This describes a framework of regular octahedra (each sharing vertices with six others) based on the diamond net, having large holes which contain the X' and 2 A atoms, which themselves form a cuprite-like net A_2X' interpenetrating the octahedral framework, as noted in Chapter 3. Because of the rigidity of the octahedral B_2X_6 framework this structure can tolerate the absence of some of the other atoms as illustrated for complex oxides in Chapter 13.

The coordination groups of the A atoms for (i) and (ii) are illustrated in Fig. 6.12. In fact no examples of this structure are known with x as high as 0.375, the maximum observed value being 0.355, but x is less than 0.3125 in some cases ($Cd_2Nb_2O_7$, $x = 0.305$), when the BX_6 octahedron is elongated. The relation of the limiting (unknown) structure (i) to fluorite is interesting because certain compounds show a transition from the pyrochlore structure to a defect fluorite structure.

The CdI_2 and related structures

The simple (C6) CdI_2 structure (also referred to as the brucite, $Mg(OH)_2$, structure) is illustrated in Fig. 6.13 as a 'ball and-spoke' model. It has been described in previous chapters in two other ways, as a hexagonal closest packing of I^- ions in which Cd^{2+} ions occupy all the octahedral holes between alternate pairs of c.p. layers (that is one-half of all the octahedral holes), and as built from layers of octahedral CdI_6 coordination groups each sharing an edge with each of six adjacent groups. These three ways of illustrating one layer of the structure are shown in Fig. 6.14. From the c.p. description of the structure it follows that there is an indefinite number of closely related structures, all built of the I–Cd–I layers of Fig. 6.14, but differing in the c.p. layer sequences. The three simplest structures of the family are often designated by their Strukturbericht symbols:

	c.p. sequence
C 6	h $(AB\ldots)$
C 19	c $(ABC\ldots)$
C 27	hc $(ABAC\ldots)$

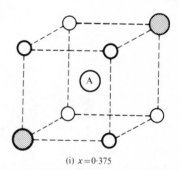

(i) $x = 0.375$

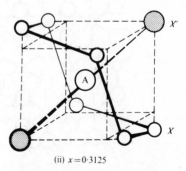

(ii) $x = 0.3125$

FIG. 6.12. Coordination of A atoms in the pyrochlore structure, $A_2B_2X_6X'$.

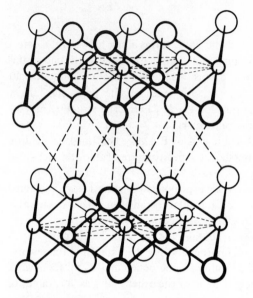

FIG. 6.13. Portions of two layers of the CdI$_2$ structure. The small circles represent metal atoms.

CdI$_2$ itself, mixed crystals CdBrI, PbI$_2$, etc. crystallize with more than one of these structures and also with structures with much more complex layer sequences; more than 80 polytypes of CdI$_2$ have been characterized. These are not of special interest here since their formation is obviously connected with the growth mechanism and they are all built of the same basic layer. We wish to show here how this very simple layer is utilized in the structures of a variety of compounds, and we shall not be particularly concerned with the mode of stacking of the layers. We shall describe five types of structure, starting with those in which the layer is of the simplest possible type, namely, in compounds MX$_2$ or M$_2$X in which all M atoms or all X atoms are of the same kind.

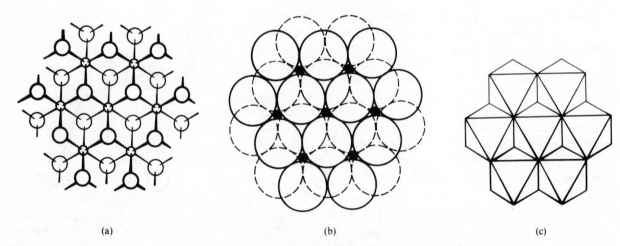

(a)   (b)   (c)

FIG. 6.14. Three representations of the CdI$_2$ layer.

210

(i) MX_2 and M_2X structures

The numerous compounds with these simple layer structures are described in other chapters; they include:

$C\,6$ structure: many dibromides and diiodides, dihydroxides, and disulphides,
$C\,19$ structure: many dichlorides,
$C\,27$ structure: β-TaS_2, some dihalides.

The following compounds crystallize with one or other of the anti-CdI_2 or anti-$CdCl_2$ structures i.e. structures in which non-metal atoms occupy octahedral holes between layers of metal atoms:

$$Ag_2F, Ag_2O \text{ (under pressure)}, Cs_2O, Ti_2O, Ca_2N, Tl_2S, \text{ and } W_2C.$$

In the rhombohedral form of $CrO \cdot OH$ (and the isostructural $CoO \cdot OH$) CdI_2-type layers are directly superposed and held together by short $O{-}H{-}O$ bonds.

The salts $Zr(HPO_4)_2 \cdot H_2O$ (IC 1969 **8** 431) and the γ form of $Zr(SO_4)_2 \cdot H_2O$ (AC 1970 **B26** 1125) have closely related structures based on $C\,6$ layers in which the 3-connected unit is a tetrahedral oxy-ion bridging three metal ions. In $Zr(HPO_4)_2 \cdot H_2O$ Zr^{4+} is 6-coordinated by O atoms of $O_3P(OH)^{2-}$ ions which lie alternately above and below the plane of the metal ions (Fig. 6.15). The P–OH bonds are perpendicular to this plane, and the H_2O molecules occupy cavities between the layers. The sulphate has a similar system of Zr^{4+} and bridging oxy-ions, but here the H_2O molecule is bonded to Zr^{4+}, which is thus 7-coordinated.

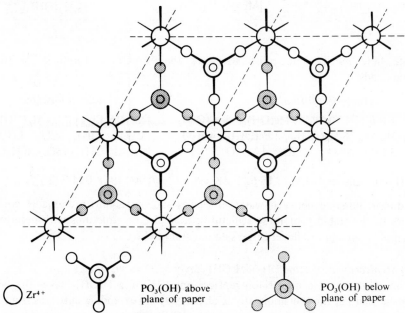

$\bigcirc\ Zr^{4+}$ $PO_3(OH)$ above plane of paper $PO_3(OH)$ below plane of paper

FIG. 6.15. Layer of the structure of $Zr(HPO_4)_2 \cdot H_2O$. The linking of anions and cations is similar in γ-$Zr(SO_4)_2 \cdot H_2O$.

(ii) *Anions or cations of more than one kind in each* MX_2 *layer*

Hydroxyhalides $M(OH)Cl$, $M_2(OH)_3Cl$, and $M(OH)_xCl_{2-x}$ are of a number of different kinds. In some there is random arrangement of OH and Cl (Br), in others a regular arrangement of OH and Cl in each c.p. layer ($Co(OH)Cl$, $Cu_2(OH)_3Cl$), and in $Cd(OH)Cl$ alternate c.p. layers consist entirely of OH^- or Cl^- ions.

In complex hydroxides such as $K_2Sn(OH)_6$ two-thirds of the Mg positions of the $Mg(OH)_2$ layer are replaced by K and the remainder by Sn; the structure contains discrete $Sn(OH)_6^{2-}$ ions.

(iii) *Replacement of cations to form charged layers*

If part of the Mg^{2+} in a (neutral) $MgCl_2$ layer is replaced by Na^+ the layer becomes negatively charged; conversely, if part of the Mg^{2+} in a $Mg(OH)_2$ layer is replaced by Al^{3+} the layer becomes positively charged. Layers of these two types can together form a structure consisting of alternate negatively and positively charged layers, as shown at (a). The layers extend throughout the crystal, the chemical formula simply representing the composition of the repeating unit of the layers. Alternatively, the charged layers may be interleaved with ions of opposite charge (together with water, if necessary, to fill any remaining space), as at (b) and (c):

$[Na_4Mg_2Cl_{12}]_n^{4n-}$	$[Mg_6Fe_2^{III}(OH)_{16}]_n^{2n+}$	$[Ca_4Al_2(OH)_{12}]_n^{2n+}$
	$CO_3^{2-}(H_2O)_4$	$SO_4^{2-}(H_2O)_6$
$[Mg_7Al_4(OH)_{22}]_n^{4n+}$	$[Mg_6Fe_2^{III}(OH)_{16}]_n^{2n+}$	$[Ca_4Al_2(OH)_{12}]_n^{2n+}$
(a)	(b)	(c)

Note how much more informative are the 'structural' formulae of these compounds:

Analytical formula	*Structural formula*
(a) $4 NaCl . 4 MgCl_2 . 5 Mg(OH)_2 . 4 Al(OH)_3$	$[Na_4Mg_2Cl_{12}][Mg_7Al_4(OH)_{22}]$
(b) $MgCO_3 . 5 Mg(OH)_2 . 2 Fe(OH)_3 . 4 H_2O$	$[Mg_6Fe_2(OH)_{16}]CO_3 . 4 H_2O$
(c) $3 CaO . Al_2O_3 . CaSO_4 . 12 H_2O$	$[Ca_4Al_2(OH)_{12}]SO_4 . 6 H_2O$

(For details see N 1967 **215** 622; ZK 1968 **126** 7; AC 1968 **B24** 972.)

Hydrogen bonding between water molecules and oxy-anions presumably contributes to the stability of these structures; examples of structures with negatively charged layers and interlayer cations do not seem to be known.

(iv) *Replacement of some* OH *in* $M(OH)_2$ *layer by* O *atoms of oxy-ions*

If one O atom of a nitrate ion replaces one OH^- in a $M(OH)_2$ layer the layer remains neutral, the remaining atoms of the NO_3^- ion extending outwards from the surface of the layer. The hydroxynitrate $Cu_2(OH)_3NO_3$ consists of neutral layers of this kind, O of NO_3^- replacing one-quarter of the OH^- ions in the layer, as shown diagrammatically in Fig. 6.16(a). There is hydrogen bonding between O of NO_3^- and OH of the adjacent layer. If O atoms of sulphate ions replace one-quarter of the OH^- ions in a $C6$ layer of composition $Cu(OH)_2$ the composite layer is charged, and it is necessary to interpolate cations; H_2O molecules occupy the remaining space (Fig. 6.16(b)). The structure of $CaCu_4(OH)_6(SO_4)_2 . 3 H_2O$ is of this type, in which the relative numbers of Cu, OH, and oxy-ions are the same as in $Cu_2(OH)_3NO_3$. (In the mineral serpierite (AC 1968 **B24** 1214) there is also replacement of about one-third of the Cu by Zn.)

A more complex arrangement is found in certain silicates and aluminosilicates. If one-third of the OH groups on one side of a $Mg(OH)_2$ layer are replaced by O of SiO_4 groups the other three O atoms of each SiO_4 can be shared with other SiO_4 groups (Fig. 6.16(c)). The formula of this composite layer, which is uncharged, is $Mg_3(OH)_4Si_2O_5$, as explained in Chapter 23. If there is a layer of linked SiO_4 groups on both sides of the $Mg(OH)_2$ layer the composition is $Mg_3(OH)_2Si_4O_{10}$. Crystals of chrysotile and talc consist of such neutral layers. Replacement of some of the Si by Al gives negatively charged layers which are the structural units in the micas, for example, $KMg_3(OH)_2[Si_3AlO_{10}]$. Corresponding to each of the Mg-containing layers there is an Al-containing layer in which the Mg^{2+} ions in the $Mg(OH)_2$ layer are replaced by two-thirds their number of Al^{3+} ions, as in $KAl_2(OH_2[Si_3AlO_{10}]$. In the chlorite minerals negatively charged mica-like layers are interleaved with positively charged brucite layers in which some Mg has been replaced by Al, as in $[Mg_3(OH)_2Si_3AlO_{10}]^-$ $[Mg_2Al(OH)_6]^+$; compare the layers in (iii) (a)–(c).

(v) Attachment of additional metal atoms to the surface of a layer

Since the X atoms on either side of a CdX_2 layer are close-packed any triangle of X atoms forms a suitable site for an additional metal atom (beyond the surface of the layer) which can be either tetrahedrally or octahedrally coordinated (Fig. 6.17(a)). The coordination group of this M ion would share three X atoms (face) with the octahedral coordination groups of the metal ion in the CdX_2 layer, but such close

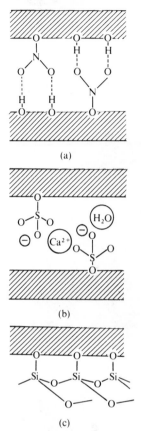

FIG. 6.16. Layers related to the CdI_2 layer: (a) $Cu_2(OH)NO_3$, (b) $[Cu_4(OH)_6(SO_4)_2]_n^{2n-}$, (c) $Mg_3(OH)_4Si_2O_5$.

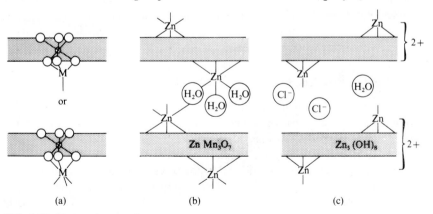
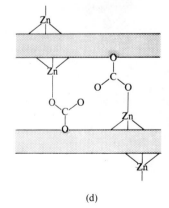

FIG. 6.17. (a) Attachment of metal ion to CdI_2-like layer. Diagrammatic representations of the structures of (b) $ZnMn_3O_7 . 3H_2O$, (c) $Zn_5(OH)_8Cl_2 . H_2O$, (d) $Zn_5(OH)_6(CO_3)_2$.

approach of metal ions can be avoided by removing the metal ion within the layer, the black circle of Fig. 6.17(a). The mineral chalcophanite, $ZnMn_3O_7 . 3 H_2O$, consists of layers of the $C6$ type built of Mn^{4+} and O^{2-} ions from which one-seventh of the metal ions have been removed, giving the composition Mn_3O_7 instead of MnO_2. Attached to each side of this layer, directly above and below the unoccupied Mn^{4+} positions, are Zn^{2+} ions, and between the (uncharged) layers are layers of water molecules. Three of these complete the octahedral coordination group around Zn^{2+} (Fig. 6.17(b)).

In $Zn_5(OH)_8Cl_2 . H_2O$ the basic structural unit is a charged layer. One-quarter of the Zn^{2+} ions are absent from a $Zn(OH)_2$ layer, and for every one removed $2 Zn^{2+}$ are added beyond the surfaces of the layer, the composition being therefore $Zn_5(OH)_8^{2+}$. The charge is balanced by Cl^- ions between the layers, with water molecules filling the remaining space (Fig. 6.17(c)). There is tetrahedral coordination of the Zn^{2+} ions on the outer surfaces of the layers, in contrast to the octahedral coordination within the layers. For another example of a layer structure of the same general type see $Zn_5(OH)_8(NO_3)_2 . 2 H_2O$ (p. 536).

Our last example is hydrozincite, $Zn_5(OH)_6(CO_3)_2$, a corrosion product of zinc which is also found accompanying zinc ores that have been subjected to weathering. Its structure arises from a combination of (iv) and (v). As in the previous example one-quarter of the Zn^{2+} ions are absent from a $Zn(OH)_2$ layer and replaced by twice their number of similar ions, equally distributed on both sides of the layer. In addition there is replacement of one-quarter of the OH^- ions by O atoms of CO_3^{2-} ions, so that the composition has changed from $Zn_4(OH)_8$ to $Zn_2[Zn_3(OH)_6](CO_3)_2$. A second O atom of each CO_3 is bonded to the 'added' Zn atoms, which are tetrahedrally coordinated as in the hydroxychloride. This is shown diagrammatically in Fig. 6.17(d). Although the structure as a whole is not a layer structure, for there is bonding between Zn and O atoms throughout, nevertheless the c.p. layers of O atoms around the octahedrally coordinated Zn atoms remain a prominent feature of the structure and may well act as a template in the growth of the crystal. It is not always appreciated that in the case of a compound of this kind, which exists only as a solid and can be formed on the surface of another solid (metallic Zn, ZnO, etc.), crystal growth is synonymous with the actual formation of a chemical compound.

The ReO_3 and related structures

The cubic ReO_3 structure has been described in Chapter 5 as the simplest 3D structure formed from vertex-sharing octahedral groups; the distorted variant adopted by $Sc(OH)_3$ was also illustrated. The structure is adopted by NbF_3 and several trioxides and oxyfluorides (Table 6.5), but is not possible for an ionic trinitride, since it would require cations M^{9+}; however, the anti-ReO_3 structure is found for Cu_3N. Similarly, the perovskite structure, derived from ReO_3 by adding a 12-coordinated ion at the body-centre of the unit cell of Fig. 13.3 (p. 483), is formed by complex fluorides and oxides (often with symmetry lower than cubic) and the anti-perovskite structure by a number of ternary nitrides and carbides, in which N or C occupy the positions of octahedral coordination. Alternatively, these nitrides and carbides may be described as c.c.p. metal systems in which N or C atoms occupy one-quarter of the octahedral interstices, a more obvious description if all the metal atoms are similar (as in Fe_4N, Mn_4N, and Ni_4N). Since the O atoms in ReO_3 occupy three-quarters of the positions of cubic closest packing, rearrangement to a more dense structure is possible, giving in the limit the h.c.p. structure of RhF_3 and certain other trifluorides (Chapter 9).

TABLE 6.5

The ReO_3 and related structures

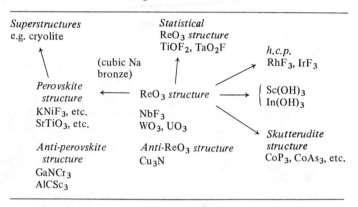

Superstructures
e.g. cryolite

↑

Perovskite structure
$KNiF_3$, etc.
$SrTiO_3$, etc.

Anti-perovskite structure
$GaNCr_3$
$AlCSc_3$

(cubic Na bronze)

Statistical
ReO_3 *structure*
$TiOF_2$, TaO_2F

↑

← ReO_3 *structure* →
NbF_3
WO_3, UO_3

Anti-ReO_3 structure
Cu_3N

h.c.p.
RhF_3, IrF_3

{ $Sc(OH)_3$
$In(OH)_3$

Skutterudite structure
CoP_3, $CoAs_3$, etc.

The relation between the ReO_3 and RhF_3 structures can easily be seen from a model consisting of two octahedra sharing a vertex. In Fig. 6.18(a) the vertices *a, b, c, d,* and *e* are coplanar, and the octahedra are related as in ReO_3, the angle AcB being 180°. Keeping these vertices coplanar, anticlockwise rotation through 60° of the upper octahedron B about the vertex *c* brings it to the position shown in (b), with the angle AcB equal to 132° and the X atoms in positions of hexagonal closest packing. This is the relation between adjacent octahedra in the h.c.p. RhF_3 structure, and it may be noted that the arrangement of RhF_6 octahedra is

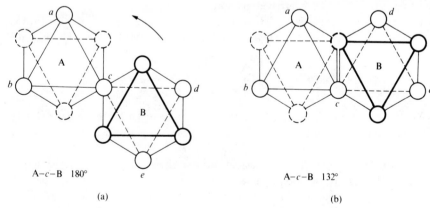

A–*c*–B 180°   A–*c*–B 132°

(a)   (b)

FIG. 6.18. Relation between vertex-sharing octahedra in (a) ReO_3, (b) RhF_3.

215

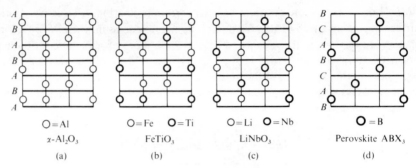

$O = Al$

χ-Al_2O_3

(a)

$O = Fe$ $O = Ti$

$FeTiO_3$

(b)

$O = Li$ $O = Nb$

$LiNbO_3$

(c)

$O = B$

Perovskite ABX_3

(d)

FIG. 6.19. Elevations of close-packed structures containing 3D systems of linked octahedra: (a) corundum, (b) $FeTiO_3$, (c) $LiNbO_3$, (d) perovskite.

essentially the same as that of the LiO_6 *or* the NbO_6 octahedra in $LiNbO_3$, a superstructure of corundum. This point is illustrated by the diagrammatic elevations of Fig. 6.19, where (a), (b), and (c) are (approximately) h.c.p. structures (M–O–M in the range $120°$–$140°$) and (d) represents the c.c.p. perovskite structure, with a similar system of vertex-sharing octahedra but M–O–M equal to $180°$. It is interesting to note that the corundum structure (and its superstructures) illustrate the limitations on the values of A–X–A angles for octahedra set out in Fig. 5.3, since in this structure octahedral MO_6 groups share vertices, edges, and one face:[1]

(1) JSSC 1973 6 469

	Octahedral element shared		
	Face	Edge	Vertex
Observed in α-Al_2O_3	$85°$	$94°$ (two)	$120°$ and $132°$ (two)
'Ideal' values for regular octahedra (p. 157)	$70\frac{1}{2}°$	$90°$ (two)	$132°$–$180°$ (three)

Since these are the bond angles at an O atom they also show the considerable departures from the regular tetrahedral coordination of that atom.

Although trinitrides MN_3 of transition metals do not occur, the analogous compounds with the heavier elements of Group VB are well known. They include

$$CoP_3 \quad NiP_3 \quad CoSb_3 \quad CoAs_3$$
$$RhP_3 \quad PdP_3 \quad RhSb_3 \quad RhAs_3$$
$$IrSb_3 \quad IrAs_3$$

which all crystallize with the skutterudite structure, named after the mineral $CoAs_3$ with that name. A point of special interest concerning this structure is that it contains well defined As_4 groups.

The skutterudite structure is related in a rather simple way to the ReO_3 structure. The non-metal atoms in the ReO_3 structure situated on four parallel edges of the unit cell are displaced into the cell to form a square group, as shown for two adjacent cells in Fig. 6.20(a). From the directions in which the various sets

216

of atoms are moved it follows that one-quarter of the original ReO$_3$ cells will not contain an X$_4$ group. Figure 6.20(b) shows a (cubic) unit cell of the CoAs$_3$ structure, which has dimensions corresponding to twice those of the ReO$_3$ structure in each direction; the bottom front right and upper top left octants are empty. (The unit cell contains 8 Co and 24 As atoms, the latter forming 6 As$_4$ groups.) Each As atom has a nearly regular tetrahedral arrangement of 2 Co + 2 As neighbours, and Co has a slightly distorted octahedral coordination group of 6 As neighbours. The planar As$_4$ groups are not quite square, the lengths of the sides of the rectangle being 2·46 Å and 2·57 Å.[2] The reason for the considerable reorientation of the octahedra which converts the ReO$_3$ to the CoAs$_3$ structure is that it makes possible the formation by each As atom of two bonds of the same length as those in the As$_4$ molecule. If CoAs$_3$ had the ReO$_3$ structure (with the

(2) AC 1971 **B27** 2288

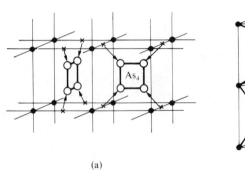 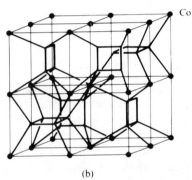

(a) (b)

FIG. 6.20. The skutterudite (CoAs$_3$) structure: (a) relation to the ReO$_3$ structure, (b) unit cell. In (b) only sufficient Co–As bonds are drawn to show that there is a square group of As atoms in only six of the eight octants of the cubic unit cell. The complete 6-coordination group of Co is shown only for the atom at the body-centre of the cell.

same Co–As distance, 2·33 Å) each As would have 8 equidistant As neighbours at about 3·3 Å. (The relatively minor distortions from perfectly square As$_4$ groups and exactly regular octahedra are due to the fact that in the highly symmetrical cubic CoAs$_3$ structure it is geometrically impossible to satisfy both of these requirements simultaneously, and the actual structure represents a compromise.) In the phosphides with this structure the difference between the two P–P bond lengths is smaller (for example, 2·23 Å and 2·31 Å), the shorter bonds corresponding to normal single P–P bonds.[3]

(3) AK 1968 **30** 103

The fact that a compound such as CoP$_3$ contains cyclic P$_4$ groups suggests that it might be of interest to try to formulate it in the way that was done for FeS$_2$, PdS$_2$, and PdP$_2$ in Chapter 1. There we saw that the S$_2$ groups bonded to six metal atoms in the pyrites structure can be regarded as a source of ten electrons, so that Fe in FeS$_2$ acquires the Kr configuration. Each P in a cyclic P$_4$ group requires an additional electron, which it could acquire either as an ionic charge, (a), or by forming a normal covalent bond, (b):

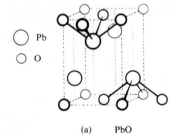

We should then formulate CoP_3 either as $(Co^{3+})_4(P_4^{4-})_3$ or as a covalent compound $Co_4(P_4)_3$; in the latter case each metal atom would acquire a share in nine electrons, since each P_4 group is a source of 12 electrons. The Co atom in CoP_3 would thus attain the Kr configuration like Fe in FeS_2. Just as CoS_2 with the pyrites structure has an excess of one electron per metal atom, so this would be true of Ni in NiP_3, accounting for the metallic conduction and Pauli paramagnetism of the latter.

Structures with similar analytical descriptions

We include here a note on a subject which sometimes presents difficulty when encountered in the crystallographic literature, namely, the fact that the same analytical description may apply to crystal structures which are quite different as regards their geometry and topology. A simple example is a structure referable to a rhombohedral unit cell with M at (000) and X at $(\frac{1}{2}\frac{1}{2}\frac{1}{2})$. This describes the CsCl structure (8-coordination of M and X) if $\alpha = 90°$ but the NaCl structure (6-coordination of M and X) if $\alpha = 60°$. This complication arises if there is (at least) one variable parameter, which may be either one affecting the shape of the unit cell (e.g. the angle α in a rhombohedral cell or the axial ratio of a hexagonal or tetragonal cell) or one defining the position of an atom in the unit cell. The following are further examples.

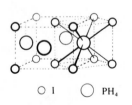

The PbO *and* PH$_4$I *structures*

A number of compounds MX have tetragonal structures in which two atoms of one kind occupy the positions (000) and $(\frac{1}{2}\frac{1}{2}0)$ and two atoms of a second kind the positions $(\frac{1}{2}0u)$ and $(0\frac{1}{2}\bar{u})$. The nature of the structure depends on the values of the two variable parameters, u and the axial ratio $c : a$. For $c : a = \sqrt{2}$ and $u = \frac{1}{4}$, the atoms in $(\frac{1}{2}0u)$ and $(0\frac{1}{2}\bar{u})$ are cubic close-packed with the atoms at (000) and $(\frac{1}{2}\frac{1}{2}0)$ in tetrahedral holes. For $c : a = 1$ and $u = \frac{1}{4}$ the packing of the former atoms would be body-centred cubic. The structures of PbO and LiOH are intermediate between these two extremes; they are layer structures with only Pb–Pb (OH–OH) contacts between the layers (Fig. 6.21(a)). If $c : a = \frac{1}{2}\sqrt{2}$ and $u = \frac{1}{2}$ the structure becomes the CsCl structure, in which each kind of ion is surrounded by eight of the other kind at the vertices of a cube; the structure of PH$_4$I (Fig. 6.21(b)) approximates to the CsCl structure. Some compounds with these structures are listed in Table 6.6; they fall into groups, with the exception of InBi ($c : a = 0.95$), with $c : a$ close to 1·25 or 0·70.

FIG. 6.21. The PbO and PH$_4$I structures.

218

TABLE 6.6

Compounds with the PbO and PH$_4$I (B 10) structures

	$c:a$	u	Atoms in	
			$(000), (\frac{1}{2}\frac{1}{2}0)$	$(\frac{1}{2}0u), (0\frac{1}{2}\bar{u})$
	1·414	0·25		
PbO	1·26	0·24	O	Pb
LiOH	1·22	0·20	Li	OH
	1·00	0·25		
InBi	0·95	0·38	In	Bi
N(CH$_3$)$_4$Cl, Br, I	0·71–0·72	0·37–0·39	N	Cl, Br, I
PH$_4$I	0·73	0·40	P	I
Ideal CsCl structure	0·707	0·50		
NH$_4$SH	0·667	0·34	N	SH

The LiNiO$_2$, NaHF$_2$, *and* CsICl$_2$ *structures*

Each of these three rhombohedral structures is described as having M at (000), X at
$(\frac{1}{2}\frac{1}{2}\frac{1}{2})$, and two Y and atoms at $\pm (uuu)$ situated along the body-diagonal of the cell.
If $u = \frac{1}{4}$ and $\alpha = 33° 34'$ this corresponds to the atomic positions of Fig. 6.3(b),
one-half of the Na ions having been replaced by M, the remainder by X, and the Cl
ions by Y. This is the type of structure adopted by a number of complex oxides
and sulphides MXO$_2$ and MXS$_2$ which are superstructures of the NaCl structure
(group (a) in Table 6.7). However, the same analytical description applies to two

TABLE 6.7

Data for some compounds MXY$_2$

	Compound	α	u
(a)	FeNaO$_2$ NiLiO$_2$ CrNaS$_2$ CrNaSe$_2$	*c.* 30°	*c.* 0·25
(b)	CrCuO$_2$ FeCuO$_2$	*c.* 30°	*c.* 0·40
	NaHF$_2$ NaN$_3$ CaCN$_2$	*c.* 40°	
(c)	CsICl$_2$	70°	0·31

other quite different structures, the NaHF$_2$ and CsICl$_2$ structures, (b) and (c) of
Table 6.7. All the compounds with the NaHF$_2$ structure have α in the range 30–40°
and u about 0·40 (or 0·10 if M and X are interchanged), while for CsICl$_2$ $\alpha = 70°$
and $u = 0·31$.

If these structures are projected along the trigonal axis of the rhombohedron,
that is, along the direction of the arrow in Fig. 6.22(a), all atoms fall on points of

the type A, B, or C of (b), and these are the positions of closest packing. The sequence of layers depends on the parameter u, and discrete groups XY_2 appear only if the X atoms and the pairs of nearest Y atoms in adjacent layers are of the same type, that is, all A, all B, or all C. The elevations shown at (c), (d), and (e) show how different are the layer sequences along the vertical axis of Fig. 6.22(a). In (c) the sequence of O layers is that of cubic close packing, and all the metal ions are in positions of octahedral coordination between the layers. In (d) and (e) linear HF_2^- and ICl_2^- ions can be distinguished, but whereas in $NaHF_2$ a Na^+ ion has six nearest F^- neighbours at the vertices of a distorted octahedron, in $CsICl_2$ a Cs^+ ion has six

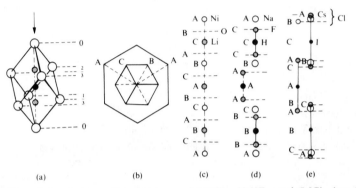

FIG. 6.22. The relation between the structures of $LiNiO_2$, $NaHF_2$, and $CsICl_2$ (see text). In (c)–(e) the horizontal dotted lines represent layers of oxygen or halogen atoms.

nearest Cl^- neighbours nearly coplanar with it and two more not much further away (at 3·85 Å as compared with 3·66 Å). Comparing the compounds $FeCuO_2$ and $FeNaO_2$, both with $\alpha \approx 30°$, we see that the different values of u (0·39 and 0·25 respectively) lead to entirely different layer sequences. In $FeCuO_2$ Cu has only two nearest oxygen neighbours, whereas in $FeNaO_2$ there is regular octahedral coordination of both Na and Fe.

On the basis of both interatomic distances and physical properties it would seem justifiable to subdivide the compounds with structure (b) of Table 6.7 (Fig. 6.22(d)) into two classes. There are numerous oxides with this structure, sometimes called the *delafossite* structure after the mineral $CuFeO_2$:

$Cu^IFe^{III}O_2$	$AgFeO_2$	$AgRhO_2$	$PdCoO_2$	$PtCoO_2$
Co	Co	In	Cr	
Cr	Cr	Tl	Rh	
Al	Al			
Ga	Ga			

Although normally a structure for oxides $M^IM^{III}O_2$ it is also adopted by a few oxides containing Pd or Pt, but whereas $CuFeO_2$ and $AgFeO_2$ are semiconductors the Pd and Pt compounds are metallic conductors. The conductivity is very anisotropic, being much greater perpendicular to the vertical axis of Fig. 6.22(a) (parallel to the layers of Pd(Pt) atoms), in which plane it is only slightly inferior to that of metallic copper. In this plane Pd(Pt) has six coplanar metal neighbours at 2·83 Å, a distance very similar to that in the metal itself; contrast Cu–Cu in $CuFeO_2$ (3·04 Å) with Cu–Cu in the metal, 2·56 Å. Counting these metal-metal contacts the coordination is 6 + 2 (hexagonal bipyramidal). For reference see p. 478.

The CrB, yellow TlI (B 33), and related structures
All the structures of Table 6.8 are described by the space group *Cmcm* with atoms in 4(*c*), $(0y\frac{1}{4})$, $(0\bar{y}\frac{3}{4})$, $(\frac{1}{2}, \frac{1}{2} + y, \frac{1}{4})$, and $(\frac{1}{2}, \frac{1}{2} - y, \frac{3}{4})$. They should not be described as the same structure because there are four variables (two axial ratios, y_A, and y_X), and the nature of the coordination group around the X atoms is very different in the various crystals. In CrB there is a trigonal prism of 6 Cr around B, but B also has 2 B neighbours at a distance clearly indicating B–B bonds; in CrB and CaSi

TABLE 6.8
The CrB, ThPt, and TlI (B 33) structures

AX	a	b	c	y_A	y_X	X–2 X in chain	A–X (shortest)
CrB	2·97 Å	7·86 Å	2·93 Å	0·15	0·44	1·74 Å	2·19 Å
CaSi	4·59	10·79	3·91	0·14	0·43	2·47	3·11
ThPt	3·90	11·09	4·45	0·14	0·41	2·99	2·99
NaOH (rh.)	3·40	11·38	3·40	0·16	0·37	3·49	2·30
TlI	4·57	12·92	5·24	0·11	0·37	4·32	3·36

For a more complete list see: AC 1965 **19** 214 (which includes older data for NaOH and KOH).

these –B–B– and –Si–Si– chains are an important feature of the structure. In ThPt and a number of other isostructural intermetallic phases this distinction between the X–X and A–X distances has gone, and the nearest neighbours of Pt in ThPt are 2 Pt and 1 Th at 2·99 Å, 4 Th at 3·01 Å, 2 Th at 3·21 Å, etc. In yellow TlI there is no question of I–I chains (the covalent I–I bond length is 2·76 Å), and the structure is described in terms of (5 + 4)-coordination of Tl by I and of I by Tl. In (rhombic) NaOH, with the TlI structure, the shortest (interlayer) OH–OH distance (3·49 Å) and the positions of the H atoms rule out the idea of preferential bonding (in this case hydrogen bonding) between the OH^- ions. The relative values of the X–X and A–X distances given in the last two columns of Table 6.8 shows that the same analytical description covers three quite different structures.

The $PbCl_2$ structure

We include a note on this structure to emphasize not only how little information about a structure is conveyed by its analytical description but how one structure type is utilized by compounds in which the bonds are of very different kinds—a point already noted in connection with the NaCl structure. The orthorhombic unit cell of the $PbCl_2$ structure contains three groups of four atoms, all in 4-fold positions $(x\frac{1}{4}z)$ etc. (space group *Pnma*). There are therefore eight variables, two axial ratios, three x, and three z parameters. More than one hundred compounds AX_2 or AXY are known to adopt a structure of this kind, and they may be divided into four groups which correspond to different values of the axial ratios (Table 6.9). The majority of the compounds fall into Class A, which includes the salt-like dihalides and dihydrides but also some sulphides, phosphides, silicides, and borides, and the intermetallic compounds Mg_2Pb and Ca_2Pb. Examples of compounds in the smaller Classes B and C, and the sole representative of Class D are given in the Table.

Although this is potentially a structure of $9:\frac{4}{5}$ coordination, with tricapped trigonal prismatic coordination of the large Pb^{2+} ion, there are never nine equidistant X neighbours. For example, in both PbF_2 and $PbCl_2$ the coordination group of the cation consists of seven close and two more distant anion neighbours, though in both structures there is appreciable spread of the interatomic distances:

PbF_2: 7 F at 2·41–2·69 Å (mean 2·55 Å); 2 at 3·03 Å.
$PbCl_2$: 7 Cl at 2·80–3·09 Å (mean 2·98 Å); 2 at 3·70 Å.

The two more distant neighbours are at two of the vertices of the trigonal prism, as shown in Fig. 6.23. This $(7 + 2)$-coordination is a feature also of the structures of $BaCl_2$, $BaBr_2$, and BaI_2, but in mixed halides $PbX'X''$ the A–X distances do not fall into well-defined groups of $7 + 2$ (for example, in PbBrI there is only one outstandingly large bond length). In the numerous ternary phases with the

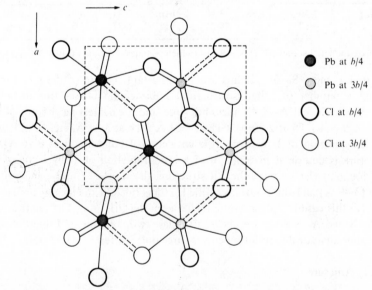

FIG. 6.23. Projection of the crystal structure of $PbCl_2$ on (010). Atoms at heights $b/4$ and $3b/4$ above the plane of the paper are distinguished as heavy and light circles. The broken lines indicate bonds from Pb^{2+} to its two more distant neighbours.

222

TABLE 6.9

Compounds with the PbCl$_2$ structure

Class	$a : c$	$(a+c)/b$	Examples
A	0·90 to 0·80	4·0 to 3·3	BaX$_2$, PbX$_2$, Pb(OH)Cl, EuCl$_2$, SmCl$_2$, CaH$_2$, etc., YbH$_2$ Mg$_2$Pb, ThS$_2$, US$_2$, Co$_2$P Ternary silicides, phosphides, etc.
B	0·75	4·3	TiP$_2$, ZrAs$_2$
C	0·74, 0·66	3·1, 3·3	Rh$_2$Si, Pd$_2$Al
D	0·55	5·3	Re$_2$P

For further examples and references see: AC 1968 **B24** 930.

'anti-PbCl$_2$' structure (e.g. TiNiSi, MoCoB, TiNiP, etc.) the $\frac{4}{5}$: 9 coordination is still recognizable but there are more neighbours within fairly close range, a general feature of intermetallic phases. For example, the following figures show the total numbers of neighbours up to $1·15(r_A + r_B)$, where r_A and r_B are the radii for 12-coordination (p. 1022):

	B	Mo	Co	Total
Mo	5	4	6	15
Co	4	6	2	12
B	–	5	4	9

The PdS$_2$, AgF$_2$, and β-HgO$_2$ structures and their relation to the pyrites structure

The structures of compounds AX$_2$ with octahedral coordination of A are of interest in connection with the 'theorem', stated on p. 158, that if three regular octahedral groups AX$_6$ have a common vertex at least one edge (or face) must be shared if reasonable distances are to be maintained between X atoms of different

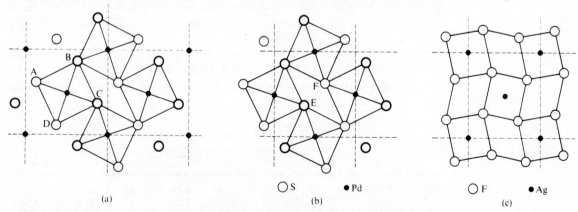

○ S   ● Pd   ○ F   ● Ag

(a)   (b)   (c)

FIG. 6.24. (a) and (b) Successive layers of the PdS$_2$ structure. (c) One layer of the structure of AgF$_2$.

AX$_6$ groups. We saw that there is one vertex common to three octahedral groups in OCr$_3$(OOC . CH$_3$)$_6$. 3 H$_2$O, this being possible because certain pairs of O atoms of different octahedral groups belong to bridging acetate groups, with O–O such shorter (2·24 Å) than the normal van der Waals distance. It is worth while to examine a 3D structure in which three octahedral AX$_6$ groups meet at each vertex, without edge-sharing.

Consider the buckled layer of Fig. 6.24(a) built of square planar coordination groups AX$_4$. This represents a layer of the structures of PdS$_2$. If layers (a) and the (translated) layers (b) of Fig. 6.24 alternate in a direction normal to the plane of the paper, an octahedral group ABCDEF around a metal atom can be completed by atoms E and F of layers (b) situated $c/2$ above and below the layer (a). In the resulting 3D structure each X atom is a vertex common to three octahedra (which share no edges). The structure with regular octahedra and normal van der Waals distances between X atoms of different octahedra is not possible for the reason stated above, but versions of this structure are adopted by PdS$_2$, AgF$_2$, and β-HgO$_2$. The geometrical difficulty is overcome in one or both of two ways, namely, (i) distortion of the AX$_6$ coordination groups, or (ii) bonding between pairs of X atoms (such as E and F) in each 'layer'. In AgF$_2$ there is moderate distortion of the octahedral groups (Ag–4 F, 2·07 Å and Ag–2 F, 2·58 Å) and normal F–F distances between atoms of different coordination groups (shortest F–F, 2·61 Å)–case (i). Figure 6.24(c) shows one 'layer' of the structure of AgF$_2$. In PdS$_2$ the S atoms are linked in pairs (S–S, 2·13 Å) and also the octahedral PdS$_6$ groups are so elongated that the structure is a layer structure (Pd–4 S, 2·30 Å, Pd–2 S, 3·28 Å). In β-HgO$_2$ the O atoms are bonded in pairs (O–O, 1·5 Å) and here the octahedra are compressed (Hg–2 O, 2·06 Å, Hg–4 O, 2·67 Å), so that instead of layers there are chains –Hg–O–O–Hg– in the direction of the c axis (normal to the plane of the layers in Fig. 6.24). In these structures, therefore, both factors (i) and (ii) operate. We thus have three closely related structures which are all derived from a hypothetical AX$_2$ structure built of regular octahedra which is not realizable in this 'ideal' form. All three structures have the same space group (*Pbca*) and the same equivalent positions are occupied, namely, (000) etc. for M and (*xyz*) etc. for X; there are rather similar parameters for X but very different c dimensions of the unit cell (Table 6.10).

We have discussed these structures at this point as examples of structures with similar analytical descriptions. We noted earlier that the PdS$_2$ structure can be described as a pyrites structure which has been elongated in one direction to such

TABLE 6.10

The PdS$_2$, AgF$_2$, and β-HgO$_2$ structures

	a	b	c	x	y	z	*Reference*
PdS$_2$	5·46 Å	5·54 Å	7·53 Å	0·11	0·11	0·43	AC 1957 **10** 329
AgF$_2$	5·07	5·53	5·81	0·18	0·19	0·37	JPCS 1971 **32** 543
β-HgO$_2$	6·08	6·01	4·80	0·08	0·06	0·40	AK 1959 **13** 515

an extent that it has become a layer structure. The pyrites structure may be described as a 3D assembly of FeS_6 octahedra, each vertex of which is common to three octahedra. Each vertex is also close to one vertex of another FeS_6 octahedron, these close contacts corresponding to the S–S groups and giving S its fourth (tetrahedral) neighbour. The pyrites structure is therefore one of a group of structures of the same topological type which represent compromises necessitated by the fact that a 3D AX_2 structure in which AX_6 groups share only vertices cannot be built with regular octahedra *and* normal van der Waals distances between X atoms of different octahedra:

Octahedral AX_2 structures in which each vertex is common to three octahedra and only vertices are shared

Regular octahedra and close X–X contacts	Distorted octahedra (4 + 2)-coordination: no close X–X contacts	Distorted octahedra (2 + 4)-coordination: close X–X contacts
Pyrites structure	*AgF_2 structure*	*β-HgO_2 structure*

↓

PdS_2 *structure*

Highly distorted
octahedra, giving
planar 4-coordination
and close X–X contacts

Relations between the structures of some nitrides and oxy-compounds

The close relation between structure and composition and simple geometrical considerations is nicely illustrated by some binary and ternary nitrides. In many of these compounds there is tetrahedral coordination of the smaller metal ions; in particular, there are many ternary nitrides containing tetrahedrally coordinated Li^+ ions. From the simple theorem that not more than eight regular tetrahedra can meet at a point (p. 159) it follows, for example, that there cannot be tetrahedral coordination of Li by N in Li_3N, a point of some interest in view of the unexpected (and unique) structure of this compound and the apparent non-existence of other alkali nitrides of type M_3N. It also follows that the limit of substitution of Be by Li in Be_3N_2 is reached at the composition BeLiN; for example, there cannot be a compound $BeLi_4N_2$ with tetrahedrally coordinated metal atoms.

These points are more readily appreciated if we formulate nitrides with 4 N atoms, for if all the metal atoms are 4-coordinated and all N atoms have the same coordination number this c.n. is equal to the total number of metal atoms in the

225

formula. This number (n_t) is the number of MN_4 tetrahedra meeting at each N atom, and it must not exceed eight:

n_t	Compound	Structure type
3	Ge_3N_4	Phenacite (Be_2SiO_4)
4	Al_4N_4 (AlN)	Zinc-blende or wurtzite
5	($Al_2Be_3N_4$)	—
6	Ca_6N_4 (Ca_3N_2)	Anti-Mn_2O_3
7	($Be_5Li_2N_4$)	—
8	$Be_4Li_4N_4$ (BeLiN)	Anti-CaF_2

Examples of structures in which five or seven tetrahedra meet at each point do not appear to be known, but the structures with n_t = 3, 4, 6, and 8 are well known. We see that replacement of Be by Li in Be_6N_4 must stop at $Be_4Li_4N_4$ (BeLiN) since further substitution would imply that more than eight tetrahedra must meet at each N atom. The end-member of the series would be $Li_{12}N_4$ (Li_3N) in which, assuming tetrahedral coordination of Li, every N would be common to 12 LiN_4 tetrahedra.

More generally we can derive the limits of substitution of a metal M^{m+} by tetrahedrally coordinated Li^+ in nitrides in the following way. Writing the formula of the ternary nitride $(Li_x M^m)_3 N_y$ we have

$$x + m = y \qquad \text{and} \qquad 3(x + 1)/y \geqslant 2.$$

The first equation corresponds to the balancing of charges while the second states that the c.n. of N must not exceed eight, assuming tetrahedral coordination of all metal atoms. It follows that x cannot exceed $2m - 3$, that is, the limiting values of Li : M in ternary nitrides are:

$$m = 2 \quad 3 \quad 4 \quad 5 \quad 6$$
$$\text{Li : M} \quad 1 \quad 3 \quad 5 \quad 7 \quad 9$$

corresponding to compounds such as

$$LiMgN, \ Li_3AlN_2, \ Li_5SiN_3, \ Li_7VN_4, \ and \ Li_9CrN_5,$$

all of which have been prepared and shown to crystallize with the antifluorite structure. There is, of course, no limit to the replacement of more highly-charged ions by ions carrying charges $\geqslant 2$ since the end-members are the binary nitrides Ca_3N_2, AlN, etc. The foregoing restrictions, which are purely geometrical in origin, apply only to substitution by Li^+ or other singly-charged ions.

The feature common to the structures of Ge_3N_4, AlN, Ca_3N_2 and the numerous ternary nitrides with the antifluorite structure is the tetrahedral coordination of the metal atoms, the c.n. of N being 3, 4, 6, and 8 respectively. It would seem that in these structures the determining factor is the type of coordination around the metal ions.

A table similar to that given for nitrides can be drawn up for oxides, but here there is no geometrical limitation of the proportion of Li^+ (assumed to be tetrahedrally coordinated) since the limiting case (Li_2O) corresponds to eight tetrahedral LiO_4 groups meeting at each O atom. A point of interest here is the fact that certain simple compounds require 'awkward' numbers of tetrahedra meeting at a point, notably five in Li_4SiO_4 and seven in Li_6BeO_4:

n_t		n_t	
3	Be_2SiO_4, $LiAlSiO_4$	6	Li_5AlO_4
4	$BeLi_2SiO_4$, Li_3PO_4	7	Li_6BeO_4
5	Li_4SiO_4	8	Li_2O

Restrictions on composition would arise in oxides in which there is octahedral coordination of all metal atoms, since not more than six regular octahedra may meet at a point, though of course this number may be increased if the octahedra are sufficiently distorted; see the note on the Th_3P_4 structure in Chapter 5. Octahedral structures may be listed in the same way as the tetrahedral structures:

n_o		*Known or probable structure type*
2	ReO_3	ReO_3
3	Al_2ReO_6	Rutile?
4	Mg_3ReO_6	Corundum, etc.
5	$Na_2Mg_2ReO_6$, Mg_4TiO_6	?
6	Na_4MgReO_6	NaCl?

Structures with $n_o = 6$ (e.g. Na_6ReO_6) would be impossible for *regular octahedral* coordination of all metal atoms, while that of Mg_4TiO_6 would involve 5-coordination of O (five octahedra meeting at a point).

It is perhaps worthwhile to list the simpler structures in order of increasing numbers of tetrahedra or octahedra meeting at each X atom (Table 6.11).

TABLE 6.11

Structures with tetrahedral and octahedral coordination

n	*Tetrahedral structures*		*Octahedral structures*	
2	AX_2	SiO_2 structures	AX_3	ReO_3
3	A_3X_4	Ge_3N_4 (phenacite)	AX_2	Rutile
4	AX	ZnS structures	A_2X_3	Corundum
6	A_3X_2	Anti-M_2O_3	AX	NaCl
8	A_2X	Anti-CaF_2	A_4X_3	See p. 160

Superstructures and other related structures

In a substitutional solid solution AA′ there is random arrangement of A and A′ atoms in equivalent positions in the crystal structure. If on suitable heat treatment the random solid solution rearranges into a structure in which the A and A′ atoms occupy the same set of positions but in a regular way, the structure is described as a *superstructure*. We use the term in this book to describe relations such as

$$AX \rightarrow (AA')X_2 \qquad \text{or} \qquad A_2X_3 \rightarrow (AA')X_3$$

regardless of whether or not the superstructure is formed from a random solid solution or whether such a solid solution exists. In the superstructure the positions occupied by A and A′ are, of course, no longer equivalent. Corresponding to the relation between the parent structure and superstructure:

Parent structure	*Superstructure*
All of a set of equivalent positions occupied by A atoms	The same set of positions occupied in a regular way by atoms of two or more kinds A and A′

there are pairs of structures related in the following way:

'Normal structure'	*'Degenerate structure'*
A and A′ occupy different sets of equivalent positions	Both sets of equivalent positions occupied by A atoms

As regards the formulae the relation is similar to that between the parent structure and superstructure except that we have

$$(AA')X_2 - A_2X_2 \text{ instead of } A_2X_2 - (AA')X_2$$

Some pairs of structures related in this way are listed in Table 6.12. Structures in column (1) are normal structures and examples are given only of compounds isostructural with a compound in column (2) or (3). The structures (2) are also to be expected inasmuch as ions of the same metal carrying different charges have different sizes and bonding requirements, and from the structural standpoint are similar to ions of different elements. The structures (3), however, may be described as 'degenerate' since chemically indistinguishable atoms occupy positions with different environments. To the chemist the gradation from (a) to (c) is of some interest. In (3)(a) there is a marked difference between the environments of atoms of the same element in the same oxidation state; in (3)(b) the difference is smaller and in (3)(c) smaller still. In (3)(c) there is, to a first approximation, no difference between the immediate environments of the chemically similar atoms although they occupy crystallographically non-equivalent positions. In BeY_2O_4 both sets of non-equivalent Y atoms occupy positions of 6-coordination in an octahedral framework which is built from quadruple strips of rutile-like chains (p. 497). Any such strip composed of three or more single chains contains (topologically or crystallographically) non-equivalent octahedra:

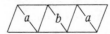   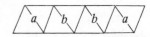

228

TABLE 6.12

Some related crystal structures

Difference between the various sets of equivalent positions	Different sets of equivalent positions occupied by:		
	(1) Atoms of *different* metals	(2) Atoms of *same* metal in *different* oxidation states	(3) Atoms of *same* metal in *same* oxidation state
(a) Different c.n.'s	GeV_3, NiV_3 Ba_2CaWO_6 K_2NaAlF_6 $Ca_3Al_2Si_3O_{12}$ $Na_3KAl_4Si_4O_{16}$ $CaFe_2O_4$ $ZnSb_2O_4$	Eu_3O_4 Pb_3O_4	β-W Ba_3WO_6 K_3AlF_6 $Y_3Al_2Al_3O_{12}$ $Na_4Al_4Si_4O_{16}$
(b) Same c.n.'s but different arrangements of nearest neighbours	$Cd_2Mn_3O_8$ $ScTiO_3$	Mn_5O_8	C-Mn_2O_3
(c) Same c.n.'s and essentially the same coordination group	Fe_2TiO_5 ULa_6O_{12} $BFeCoO_4$	Ti_3O_5 Tb_7O_{12}	BeY_2O_4

In the limit the process (a) → (b) → (c) in Table 6.12 would result in identical environments for all atoms of a given chemical species, that is, each would occupy its own set of equivalent positions. A structure in column (1) would then be a superstructure of the structure in column (3). Since any one of the actual structures in column (1) is a 'normal' structure, derived superstructures and statistical solid solutions are in principle possible. The following possibilities for oxides with the $B(FeCo)O_4$ and (BeY_2O_4) type of structure illustrate the relations between the types of structure we have been discussing:

Superstructure $\qquad$ *'Parent structure'* $\qquad$ *'Degenerate structure'*

$$M_2^{II}\begin{array}{|c|}\hline Fe \\\hline Ni \\\hline Ti_2 \\\hline\end{array}O_8 \qquad M^{II}\begin{array}{|c|}\hline Fe^{II} \\\hline Ti^{IV} \\\hline\end{array}O_4 \qquad M^{II}\begin{array}{|c|}\hline Y^{III} \\\hline Y^{III} \\\hline\end{array}O_4$$

Statistical solid solution

$$M^{II}\begin{array}{|c|}\hline Fe, Mg \\\hline Ti \\\hline\end{array}O_4$$

Symbols of metal atoms on different lines indicate sets of atoms in different equivalent positions. The above examples, except MY_2O_4, are purely hypothetical; it would be interesting to know the arrangement of metal ions in $B_2Mg_3TiO_8$ and related minerals, which form a related series containing B^{III} in place of M^{II}.

7

Bonds in Molecules and Crystals

Introduction

In conventional treatments of the 'chemical' bond it is usual for chemists to restrict themselves to the bonds in finite molecules and ions and for crystallographers and 'solid state chemists' to concern themselves primarily with the bonding in crystals. Moreover, effort is understandably concentrated on certain groups of compounds which are of special interest from the theoretical standpoint or on crystals which have physical properties leading to technological applications. As a result it becomes difficult for the student of structural chemistry to obtain a perspective view of this subject. The usual treatment of ionic and covalent bonds, with some reference to metallic, van der Waals, and other interactions, provides a very inadequate picture of bonding in many large groups of inorganic compounds. We shall therefore attempt to present a more general survey of the problem, the intention being to emphasize the complexity of the subject rather than to present an over-simplified approach which ignores many interesting facts. Clearly, a balanced review of bonding is not possible in the space available, and moreover would presuppose a knowledge of at least the essential structural features of (ideally) all finite groups of atoms and of all crystals. We shall therefore select a very limited number of topics and trust that the more detailed information provided in later chapters will provide further food for thought.

A glance at the Periodic Table will show the difficulty of making useful generalizations about bonding in many inorganic compounds. In Table 7.1 the full lines enclose elements which form ions stable in aqueous solution. At the extreme right the anions include only the halide ions; O^{2-} combines with H_2O to form (2)

TABLE 7.1

Cations +1 +2 +3																			Anions −3 −2 −1
H⁺																			H⁻
Li	Be													B	C	N	O		F
Na	Mg	Al												Si		P	S		Cl
K	Ca	Sc	Ti	V	Cr	Mn	Fe	Co	Ni	Cu	Zn	Ga	Ge			As	Se		Br
Rb	Sr	Y	Zr	Nb	Mo	Tc	Ru	Rh	Pd	Ag	Cd	In	Sn^{II}			Sb	Te		I
Cs	Ba	La	Hf	Ta	W	Re	Os	Ir	Pt	Au	Hg	Tl^{I}	Pb^{II}	Bi					

OH^-, and H^+ combines with H_2O to form H_3O^+. However, O^{2-} exists in many crystalline oxides, S^{2-}, etc., N^{3-} (and possibly also P^{3-}) in the appropriate compounds of the most electropositive metals. Next there are a few elements which do not form either cations or anions (B, C, Si: but note the formation of the finite C_2^{2-} ion in a number of carbides and the infinite 2D Si_n^{n-} ion in $CaSi_2$), a group which for most practical purposes also includes the neighbouring elements Ge, Sn(IV), As, and Sb. The Si—O bond is usually regarded as intermediate between covalent and ionic; Ge would be grouped with Si in view of the extraordinarily similar crystal chemistry of these two elements, though Ge—O bonds are probably closer to ionic than to covalent bonds.

The metals which form cations include the most electropositive elements at the left of the Table and a group of 3d metals in their lower oxidation states (usually 2 or 3), together with some of the earlier B subgroup metals, Tl(I) and Pb(II); note that Pb^{4+} is known only in crystalline oxides and oxy-compounds. In the lower centre of the Table is a group of metals which have no important aqueous ionic chemistry and probably not much tendency to form ions in the crystalline state—though the bonds in certain oxy- and fluoro-compounds may well have appreciable ionic character (for example, dioxides such as MoO_2, lower fluorides such as MoF_3, etc.).

The Group IVA metals have no important aqueous ionic chemistry but form ions in crystalline oxy-compounds. Bismuth can perhaps be grouped with Ti, Zr, and Hf; it has a strong tendency to form hydroxy-complexes, but salts such as $Bi(NO_3)_3 \cdot 5\,H_2O$, $M_3^{II}[Bi(NO_3)_6]_2 \cdot 24\,H_2O$, and $Bi_2(SO_4)_3$ presumably contain Bi^{3+} ions. It seems reasonable to put Ga(III), In(III), Tl(III), and Sn(II) in this group; the complex structural chemistry of these elements is outlined in Chapter 26.

The bonding in compounds of the non-metals, one with another, may be regarded as essentially covalent, but it is evident from the shortage of anions that the only large classes of essentially ionic binary compounds are those of the metals with O or F, sulphides etc. of the most electropositive elements (the I, II, and IIIA subgroups), the remaining monohalides of these metals and of Ag and Tl, and halides MX_2 or MX_3 of other metals enclosed within the full lines of Table 7.1. Clearly there remain large groups of compounds, even binary ones, which should be included in any comprehensive survey of bonding in inorganic compounds, and it must be admitted that a satisfactory and generally acceptable description of the bonding in many of these groups is not available. In the compounds of these metals with the more electronegative non-metals the bonding is probably intermediate between ionic and covalent, but as we proceed down the B subgroups the more metallic nature of the 'semi-metals' suggests bonding in their compounds of a kind intermediate between covalent and metallic. Even the structures of the crystalline B subgroup elements themselves present problems in bonding.

In Group IV there is a change from the essentially covalent 4-coordinated structure of diamond, Si, Ge, and grey Sn(IV) through white Sn(II) to Pb, with a c.p. structure characteristic of many metals. Group V begins with the normal molecular structure of N_2 and white P (P_4), but phosphorus also has the deeply

231

coloured black and red forms, both with layer structures in which P is 3-coordinated. The structure of red P is unique and inexplicably complex. Then follow As, Sb, and Bi, with structures that can be described either as simple cubic structures, distorted to give (3 + 3)-coordination, or as layer structures in which the distinction between the two sets of neighbours becomes less as the metallic character increases:

	Distances to nearest neighbours	
	3 at	3 at
As	2·51 Å	3·15 Å
Sb	2·91	3·36
Bi	3·10	3·47

In Group VI crystalline oxygen exists as O_2 molecules, and sulphur also forms normal molecular crystals containing S_6, S_8, and S_{12} molecules. However, this element also forms a fibrous polymorph built of chains with a unique configuration quite different from those in Se (metallic form) and Te. As regards its polymorphism Se occupies a position intermediate between S and Te. It has two red forms, both built of Se_8 molecules structurally similar to the S_8 molecule, and also a 'metallic' form built of helical chains. The only form of Te is isostructural with the metallic form of Se. In all these crystalline structures S, Se, and Te form the expected two bonds. However, the interatomic distances in Table 7.2 show that while rhombic S is adequately described as consisting of covalent S_8 molecules held together by van der Waals bonds this is an over-simplification in the cases of Se and Te. For example, there are many contacts between atoms of different Se_8 molecules in β-Se which are actually shorter than the shortest intermolecular contacts in rhombic S, and the shortest distances between atoms of different chains in Te are very little greater than the corresponding distances in metallic Se. A somewhat similar phenomenon is observed in Group VII. In crystalline Cl_2, Br_2, and I_2 the molecules are arranged in layers, the shortest intermolecular distances within the layers being appreciably less than those between the layers (Table 7.2); the latter are close to the expected van der Waals distances.

We have introduced the data of Table 7.2 at this point to emphasize that even the bonds in some crystalline elements cannot be described in a simple way. It appears that many bonds are intermediate in character between the four 'extreme' types, and of these the most important in inorganic chemistry are those which are intermediate in some way between ionic and covalent bonds.

A further complication is the presence of bonds of more than one kind in the same crystal or, less usually, in the same molecule. This is inevitable in any crystal containing complex ions, such as $NaNO_3$, the bonds between Na^+ and O atoms of NO_3^- ions being different in character from the N—O bonds within the NO_3^- ion; in some crystals bonds of three or four different kinds are recognizable. This complication also arises in much simpler crystals, for example, RuO_2, where there is electronic interaction leading to metallic conduction in addition to the metal–oxygen bonds which are the only kind of bonds we need to recognize in a

TABLE 7.2

Interatomic distances in crystalline S, Se, *and* Te

	Intramolecular	*Shortest intermolecular*	*Van der Waals distance* ($2 r_{s^{2-}}$ *etc.*)
S_8 (rhombic)	2·04 Å	3·49 Å[a]	*ca.* 3·6 Å
Se_8 (β)	2·34	3·58[a]	4·0
Se (metallic)	2·37	3·44	4·0
Te	2·84	3·50	4·4

(a) Mean of the shortest contacts (one for each of the non-equivalent S (Se) atoms); actual shortest contacts, S–S, 3·38 Å, Se–Se, 3·48 Å.

Interatomic distances in the crystalline halogens

	Intramolecular	*Intermolecular*		$2 r_{X^-}$
		In layer	*Between layers*	
Cl_2	1·98 Å	3·32 Å	3·74 Å	3·6 Å
Br_2	2·27	3·31	3·99	3·9
I_2	2·67	3·50	4·35	4·3

crystal such as TiO_2 which has a rather similar crystal structure. In numerous other cases we shall find it convenient to indicate that one (or more) of the valence electrons appears to be behaving differently from the others, when we shall use formulae such as $Cs_3^+O^{2-}(e)$ or $Th^{4+}I_2^-(e)_2$.

TABLE 7.3

Types of bonds

NO OVERLAP OF CHARGE CLOUDS	Valence electrons localized on particular atoms	*Van der Waals bond* charge-transfer bonds hydrogen bonds *Ionic bond* bonds between polarized ions Ionic-covalent bonds
OVERLAP OF CHARGE CLOUDS	Shared electrons localized in particular bonds Partial delocalization of some valence electrons	*Covalent bond*
Increasing delocalization of some or all of the valence electrons	Conjugated systems (–C=C–C=C– etc.) 'resonating' systems (CO_3^{2-}, C_6H_6, $P_3N_3X_3$) finite metal clusters crystalline semi-metals 'electron-excess' solids Complete delocalization of some proportion of valence electrons	Localized and delocalized bonds in the same system *Metallic bond*

We may set out the various types of bonding as in Table 7.3, making our first broad subdivision according to whether there is or is not sharing of electrons between the atoms. In systems without appreciable overlapping of electron density we have interactions ranging from those between ions, through ion : dipole, dipole : dipole, ion : induced-dipole, dipole : induced-dipole, to induced-dipole : induced-dipole (van der Waals) bonds. In systems where electrons are shared between atoms we have the various types of covalent bond, with various degrees of delocalization of some or all of the bonding electrons, leading in the limit to the metallic bond. By this we mean the bond in metals and intermetallic compounds which leads to the electronic properties which are characteristic, in varying degrees, of the metallic state. There would seem to be no good reason to regard the metal–metal bonds in many molecules (p. 250) as differing in any essential respect from other covalent bonds. Indeed, Pauling showed that for discussing certain features of the structures of metals it is feasible to extend the valence-bond theory to the metallic state, as noted in Chapter 29. Certain electron-deficient molecules and crystals have, in common with metals, the feature that the atoms have more available orbitals than valence electrons; it may prove useful also to recognize electron-excess systems, which have more valence electrons than are required for the primary bonding scheme. This general term would include a number of groups of compounds of quite different types, for example, 'sub-compounds' such as Cs_3O, Ca_2N, and ThI_2, 'inert-pair' ions and molecules (p. 245), and transition-metal compounds such as those to the right of the vertical lines:

$$
\begin{array}{c|cc}
WO_3 & ReO_3 & \\
\hline
MnO_4^- & MnO_4^{2-} & MnO_4^{3-} \\
\hline
& FeS_2 & CoS_2 \quad NiS_2 \\
\hline
& CoP_3 & NiP_3 \,.
\end{array}
$$

In such compounds there may be localized interactions between small numbers of metal atoms (as in MoO_2, NbI_4) or delocalization of a limited number of electrons leading to metallic conduction.

We now discuss some aspects of the covalent bond, metal–metal bonds, the van der Waals bond, and the ionic bond.

The lengths of covalent bonds

From the early days of structural chemistry there has been considerable interest in discussing bond lengths in terms of radii assigned to the elements, and it has become customary to do this in terms of three sets of radii, applicable to metallic, ionic, and covalent crystals. Distances between non-bonded atoms have been compared with sums of 'van der Waals radii', assumed to be close to ionic radii. The earliest 'covalent radii' for non-metals were taken as one-half of the M–M distances in molecules or crystals in which M forms $8 - N$ bonds (N being the number of the Periodic Group), that is, from molecules such as F_2, HO–OH, H_2N–NH_2, P_4, S_8, and the crystalline elements of Group IV with the diamond structure. This accounts for H and the sixteen elements in the block C–Sn–F–I. The origin of the covalent radii of metals was quite different owing to the lack of data from molecules containing M–M bonds. 'Tetrahedral radii' were derived from the lengths of bonds M–X in compounds MX with the ZnS structures, 'octahedral radii' were derived from crystals with the pyrites and related structures, assuming additivity of radii and using the $8 - N$ radii assigned to the non-metals. As Pauling remarked at the time, it is unlikely that a bond such as Zn–S is a covalent bond in the same sense as C–C or S–S, and it is obviously difficult to justify the later use of such radii in discussions of the ionic character of other bonds formed by Zn. It could be added that it is also not obvious why the S–O bond length in SO_4^{2-}, in which S forms four tetrahedral bonds and O one bond, should be compared with the sum of r_S (from S_8, in which S forms two bonds) and r_0 (from HO–OH, in which O forms two bonds).

The lengths of homonuclear bonds M–M are in general equal to or less than the standard single bond length in the molecules or crystals noted above. Exceptions include N–N in N_2O_3 (1·86 Å) and in N_2O_4, for which two determinations give 1·64 Å and 1·75 Å, both much longer than the bond in N_2H_4 (1·47 Å), S–S in $S_2O_5^{2-}$ (2·17 Å) and $S_2O_4^{2-}$ (2·39 Å), which are to be compared with the bond in S_8 (2·06 Å), and the bonds in I_3^- and other polyiodide ions which are discussed as a group in Chapter 9. Shorter bonds are regarded as having multiple-bond character. In some cases there are obvious standards for M=M and M≡M, as for

$$\begin{array}{ccccc} H_3C-CH_3 & H_2C=CH_2 & \text{and} & HC≡CH \\ 1·54\ \text{Å} & 1·35\ \text{Å} & & 1·20\ \text{Å}, \end{array}$$

and the numerous bonds of intermediate length are assigned non-integral bond orders. In other cases (for example, N=N) there has been less general agreement as to the precise values, owing to the absence of data from, or the non-existence of, suitable molecules or ions.

The situation with regard to heteronuclear bonds (M–X) is different. Shortening due to π-bonding is to be expected in many bonds involving O, S, N, P, etc. and is presumably the major reason for variations in the length of a particular bond such as S–O. This is consistent with the values of the stretching frequencies of the bonds:

Molecule	S–O *stretching frequency*	Length
	(cm^{-1})	
SO	1124	1·49 Å
Cl_2SO	1239	1·45
SO_2	1256	1·43
F_2SO	1312	1·41

However, the longest measured S–O bond has a length of 1·65 Å (excluding those in $S_3O_{10}^{2-}$ and $S_5O_{16}^{2-}$, for which high accuracy was not claimed), as compared with 1·77 Å, the sum of the covalent radii. In other cases (for example $SiCl_4$) where there is no reason for supposing appreciable amounts of π-bonding (though it cannot be excluded) bonds are much shorter (Si–Cl, 2·00 Å) than the sum of the covalent radii (2·16 Å). In many cases the discrepancies between the observed lengths of (presumably) single bonds and radius sums appeared to be greater the greater the difference between the electronegativities of the atoms concerned, and it is now generally assumed that 'ionic character' of bonds reduced their lengths as compared with those of hypothetical covalent bonds. The introduction of electronegativity coefficients is thus seen to be a consequence of assuming additivity of radii. It is not proposed to discuss here the derivation of electronegativity coefficients, for which there is no firm theoretical foundation, but since an equation due to Schomaker and Stevenson has been, and still is, widely used by those interested in relating bond lengths to sums of covalent radii, we give in Table 7.4 some electronegativity coefficients and covalent radii. The empirical equation has the form:

$$r_{AB} = r_A + r_B - 0.09 (x_A \sim x_B),$$

though a smaller numerical coefficient (0·06) has been suggested by some authors. This equation certainly removes some of the largest discrepancies, which arise for

TABLE 7.4

Normal covalent radii and electronegativity coefficients

Electronegativity coefficients					Normal covalent radii				
H	C	N	O	F	H	C	N	O	F
2·1	2·5	3·0	3·5	4·0	0·37	0·77	0·74	0·74	0·72 Å
	Si	P	S	Cl		Si	P	S	Cl
	1·8	2·1	2·5	3·0		1·17	1·10	1·04	0·99
	Ge	As	Se	Br		Ge	As	Se	Br
	1·7	2·0	2·4	2·8		1·22	1·21	1·17	1·14
	Sn	Sb	Te	I		Sn	Sb	Te	I
	1·7	1·8	2·1	2·4		1·40	1·41	1·37	1·33

bonds involving the elements N, O, and F, but the following figures show that it does not account even qualitatively for the differences between observed and estimated bond lengths for series of bonds such as C–Cl, Si–Cl, Ge–Cl, and Sn–Cl:

Bond	$r_{obs.}$	$r_M + r_{Cl}$	Correction required	S–S correction
C–Cl	1·76 Å	1·76 Å	0·00 Å	0·045 Å
Si–Cl	2·00	2·16	0·16	0·11
Ge–Cl	2·08	2·21	0·13	0·12
Sn–Cl	2·30	2·39	0·09	0·12

Moreover, the electronegativity correction is by no means sufficient for bonds such as those in the molecules SiF_4 and PF_3:

	Si–F	P–F
$r_{obs.}$	1·56 Å	1·54 Å
$r_M + r_F$	1·89	1·82
r_{corr} (S.–S.)	1·69	1·65

In this connection it is interesting to note that the difference between pairs of bond lengths M–F and M–Cl is approximately equal in many cases to the difference between the *ionic* radii (0·45 Å) rather than to the difference between the covalent radii (0·27 Å) of F and Cl. This is to be expected for ionic crystals and molecules (for example, gaseous alkali-halide molecules) but it is also true for the following molecules:

	BX_3	CX_4	SiX_4	PX_3	SX_2
$r_{M-Cl} - r_{M-F}$	0·45 Å	0·44 Å	0·44 Å	0·50 Å	0·41 Å,

though the difference becomes increasingly smaller for Br (0·38 Å), Cl (0·35 Å), O (0·28 Å), and F (0·21 Å), being finally equal to, or smaller than, the difference between the covalent radii.

It has long been recognized that an attractive alternative to the use of three sets of radii (metallic, ionic, and covalent) would be the adoption of one set of radii applicable to bonds in all types of molecules and crystals. This would obviate the need to prejudge the bond type. Such a set of radii was suggested by Bragg in 1920, and the idea has been revived by Slater (1964).[1] These radii agree closely in most cases with the calculated radii of maximum radial charge density of the largest shells in the atoms. The lengths of covalent or metallic bonds should therefore be equal to the sums of the radii since these bonds result from the overlapping of charge of the outer shells, and this overlap is a maximum when the maximum charge densities of the outer shells of the two atoms coincide. The radius of an ion with noble-gas configuration is approximately 0·85 Å greater or less than the atomic radius, depending on whether an anion or cation is formed. The length of an ionic bond is therefore not expected to differ appreciably from that of a covalent bond between the same pair of atoms. These radii have been shown to give the interatomic distances in some 1200 molecules and crystals with an average deviation of 0·12 Å. They were rounded off to 0·05 Å since for more precise discussions allowance would have to be made for coordination number and special factors such as crystal field effects. The agreement between observed bond lengths and radius sums is admittedly poor in some cases (bonds involving Ag, Tl, and the elements from Hg to Po), but some of these elements also present difficulty when more elaborate treatments are used. We do not give the Slater radii here because for metals they approximate to 'metallic radii' (Table 29.5, p. 1022) and for non-metals to Pauling's 'covalent radii' except for certain first-row elements, notably N (0·65 Å), O (0·60 Å), and F (0·50 Å). We have already noted that many M–F bond lengths suggest a radius for F much smaller than one-half the bond length in F_2 (0·72 Å); they agree much more closely with the Slater radius of 0·50 Å. Since bond length difficulties are most acute for certain bonds involving F, it should be noted that this element is abnormal in many ways, as shown by the data in Table 9.1 (p. 327). We have noted several examples of anomalously long bonds (N–N, S–S, I–I). If the bond in the F_2 molecule is also of this type (and possibly to some extent those in HO–OH and $H_2N–NH_2$) it may be necessary to re-examine the basis of the discussion of bond lengths and of the ionic–covalent character of bonds in terms of electronegativity coefficients.

(1) JCP 1964 **41** 3199

The shapes of simple molecules and ions of non-transition elements

The spatial arrangement of bonds in most molecules and ions AX_n formed by non-transition elements (and by transition elements in the states d^0, d^5, and d^{10}), where X represents a halogen, O, OH, NH_2, or CH_3, may be deduced from the total number of valence electrons in the system. If this number (V) is a multiple of eight the bond arrangement is one of the following highly symmetrical ones:

V = 16: 2 collinear bonds
  24: 3 coplanar bonds
  32: 4 tetrahedral bonds
  40: 5 trigonal bipyramidal bonds
  48: 6 octahedral bonds
  56: 7 pentagonal bipyramidal bonds

For intermediate values of V the configuration is found by expressing V in the form $8n + 2m$ (or $8n + 2m + 1$ if V is odd). The arrangement of the n ligands and m unshared electron pairs then corresponds to one of the symmetrical arrangements listed above, for example:

V	n	m	$n + m$	Bond arrangement
32	4	0 ⎫		Tetrahedral
26	3	1 ⎬ 4		Pyramidal
20	2	2 ⎭		Angular

Compounds of non-transition elements containing odd numbers of electrons are few in number, but they can be included in the present scheme since an odd electron, like an electron pair, occupies an orbital. Thus a 17-electron system has the same angular shape as an 18-electron one, as described later.

It will be appreciated that the term $8n$ implies completion of the octets of the ligands X (usually O or halogen) rather than that of A, which seems reasonable since X is usually more electronegative than A; witness the 16-electron, 18-electron, and 24-electron systems in which A has an incomplete octet of valence electrons, and the existence of O–S–O and S–S–O but non-existence of S–O–S and S–O–O. To each ligand there corresponds one electron pair in the valence group of A. If the ligand is a halogen, OH, NH_2, or CH_3, one electron for each bond is provided by X and one by A, but in the case of O two electrons are required from A (or from A and the ionic charge). The negative formal charge on O is reduced by use of some of its electron density to strengthen the A–O bond, which is invariably close to a double bond. The use of other electron pairs on O in this way does not affect the stereochemistry appreciably, and it is not necessary to distinguish =O from –X in what follows. For example, the 24-electron systems include not only the boron trihalides, but also the carbonyl and nitryl halides, the oxy-ions BO_3^{3-},

238

CO_3^{2-}, and NO_3^-, and the neutral SO_3 molecule, all of which are planar triangular molecules or ions. An *equilateral* triangular structure is, of course, only to be expected if all the ligands are of the same kind; deviations from bond angles of $120°$ occur in less symmetrical molecules such as $COCl_2$ or O_2NF.

In this section we shall not in general distinguish multiple from single bonds in the structural formulae of simple molecules and ions.

In the examples given earlier the stereochemistry follows directly from the value of V (20, 26, or 32), since there is only one possible arrangement of two or three bonds derivable from a regular tetrahedral arrangement of four pairs of electrons. However, for some of the larger numbers of electron pairs there are several ways of arranging a smaller number of bonds. For example, there are two ways of arranging two lone pairs at two of the apices of an octahedron (Fig. 7.1). Thus, although the planar structure of the ICl_4^- ion is consistent with the octahedral disposition of the electron pairs so also would be the structure (b) of Fig. 7.1. Similarly, the irregular tetrahedral shape of $TeCl_4$, the T-shape of ClF_3, and the linear configuration of ICl_2^- are not the only arrangements derivable from the trigonal bipyramid for one, two, or three lone pairs. The highly symmetrical arrangements found for various numbers of *shared* electron pairs are, as might be expected, the same as the arrangements of a number of similar ions around a particular ion; moreover, the general validity of the $8n + 2m$ 'rule' suggests that electron pairs, whether shared or unshared, tend to arrange themselves as far apart as possible. However, in order to account for the arrangement of ligands in cases where there is a choice of structures (as in ICl_4^-) and for the finer details of the structures of less symmetrical molecules such as $TeCl_4$ or ClF_3 it is necessary to elaborate the very simple treatment. A refinement is to suppose that the repulsions between the electron pairs in a valence shell decrease in the order

$$\text{lone pair}:\text{lone pair} \underset{\delta_1}{>} \text{lone pair}:\text{bond pair} \underset{\delta_2}{>} \text{bond pair}:\text{bond pair}$$

as is to be expected since lone pairs are closer to the nucleus than bonding pairs. The structure (a) of Fig. 7.1 then clearly has the minimum lone pair : lone pair repulsion and is to be preferred if $\delta_1 > \delta_2$. In $TeCl_4$, with only one lone pair, the repulsions between the lone pair and the bond pairs favour the structure in which the lone pair occupies an equatorial rather than an axial position. Similar arguments can be applied to ClF_3 and to other molecules.

We shall now review the ions and molecules having $V > 16$ and then comment briefly on those with 10–14 electrons and those with 17 or 19 electrons.

Linear 16-electron molecules and ions

This group includes linear molecules and ions of Ag and Au (for example, $Ag(NH_3)_2^+$, $Au(NH_3)_2^+$, $H_3N \cdot AuCl$, $AuCl_2^-$), mercuric halides, and a group of molecules and ions containing C, N, and O. In the latter all the bonds are very short compared with single bonds and the overall length of the molecule or ion is close to 2.3 Å except for CN_2^{2-}:

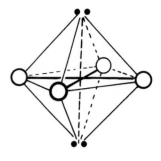

(a)

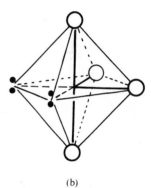

(b)

FIG. 7.1. The alternative ways of arranging two lone pairs at two of the apices of an octahedron.

TABLE 7.5

Bond lengths in linear molecules and ions

	Bond lengths		Lengths of single bonds	
O–C–O	1·16 Å	1·16 Å	C–C	1·54 Å
N–N–O	1·13	1·19	C–N	1·47
(O–N–O)$^+$	1·15	1·15	C–O	1·43
(N–C–O)$^-$	1·21	1·13	N–N	1·46
(N–N–N)$^-$	1·15	1·15	N–O	1·41
(N–C–N)$^{2-}$	1·22	1·22	O–O	1·47

If we wish to distribute the 16 electrons in the system A–B–C so that octets are maintained around all three atoms and all the bonds contain even numbers of electrons the possibilities are:

$$
\begin{array}{ccccc}
\text{A=B=C} & & & \text{A--B}\equiv\text{C} & \\
4\quad4\quad4\quad4 & \text{and} & \Bigg\{ & 6\quad2\quad6\quad2 & \\
& & & \text{A}\equiv\text{B--C} & \\
& & & 2\quad6\quad2\quad6 &
\end{array}
$$

The bond lengths show that all the above molecules and ions approximate to the symmetrical form A=B=C, though the extreme shortness of most of the bonds suggests that there is further interaction of the electron systems in these compact linear molecules.

The cyanogen halides NCX provide examples of the alternative structure A≡B–C. All the molecules NCCl, NCBr, and NCI have been shown to be linear, with N–C, 1·16 Å; the C–X bonds are uniformly about 0·14 Å shorter than the corresponding bonds in the carbon tetrahalides.

The gaseous molecules of the alkaline-earth dihalides are also 16-electron molecules, but are presumably ionic rather than covalent molecules; some are linear, as expected, others apparently non-linear (p. 373).

Triangular arrangement of 3 electron pairs

18 electrons   24 electrons

The 18-*electron group*. It is convenient to arrange the examples according to Periodic Groups; note that there are no compounds of C and that the only halogen species AX$_2$ are those of elements of Group IV. (A molecule such as HFC=O containing H does not count as an 18-electron but as a 24-electron molecule; if it is desired to include H it must be counted as contributing seven electrons like a

240

halogen.) The ClO_2^+ ion exists in ClO_2 (AsF_6), formed from ClO_2F and AsF_5, according to i.r. evidence.

Group IV	*Group* V	*Group* VI	*Group* VII

$$\underset{101°}{F-\overset{Si}{}-F}$$

$$\underset{115°}{O-\overset{N}{}-O^-}$$

$$\underset{117°}{O-\overset{O}{}-O}$$

$$O-\overset{Cl}{}\overset{+}{-}O$$

also
GeF$_2$ (94°)
TiF$_2$ (130°)
SnX$_2$
PbX$_2$

$$\underset{110°}{O-\overset{N}{}-F}$$

also
NOCl (116°)
NOBr (117°)
NO(OH) (111°)

$$\underset{117°}{N\equiv\overset{S}{}-F}$$

also
SSO (118°)
OSO (120°)

Note the remarkable similarity in interbond angles in all the compounds of N, O, and S despite great differences in multiplicity of bonds.

The 24-electron group. Although potentially a large group examples are not very numerous. In fact one reason for setting these ions and molecules out in this way, with oxy-compounds at the top and halogen compounds at the bottom, is to draw attention to the 'missing' compounds. For example, only one of the *ionic* species containing halogen or halogen and O ligands appears to be known. The $(NOF_2)^+$

BO_3^{3-}	CO_3^{2-}	NO_3^-	SO_3
–	–	O_2NX	–
–	OCX_2	$(NOF_2)^+$	–
BX_3	–	–	–

cation exists in salts of the type $(NOF_2)BF_4$ and $(NOF_2)AsF_6$ formed from NOF_3 and the appropriate halides. The i.r. spectrum shows that the ion is structurally similar to the planar isoelectronic COF_2 molecule.

The three oxy-ions all have the form of equilateral triangles, as also does SO_3. The neutral molecules, lying along the diagonal of the chart, range from SO_3 to BF_3, and include the nitryl halides (O_2NF and O_2NCl) with the angle O–N–O around 130° and the carbonyl halides, COF_2 (F–C–F, 112°), $COCl_2$ (Cl–C–Cl, 111°); all are planar molecules. No complete vertical family in the chart is known with O or halogen ligands, but with the isoelectronic NH_2 instead of halogen we have:

$$CO_3^{2-} \qquad \overset{O}{\underset{O}{>}}C{-}NH_2^- \qquad O{-}C\overset{NH_2}{\underset{NH_2}{<}} \quad \text{and} \quad H_2N{-}C\overset{NH_2^+}{\underset{NH_2}{<}}$$

carbonate ion   carbamate ion   carbamide   guanidinium ion,

all of which have been shown to be planar.

Tetrahedral arrangement of 4 electron pairs

20 electrons   26 electrons   32 electrons

The 20-*electron group.* This small group includes a number of ions and molecules formed by elements of Groups V, VI, and VII:

$$
\begin{array}{ccccc}
& \underset{H \quad H}{\overset{N^-}{\diagup\diagdown}} & \underset{\substack{F \quad F \\ 103^\circ}}{\overset{O}{\diagup\diagdown}} & \underset{\substack{S \quad S \\ 103^\circ}}{\overset{S^{2-}}{\diagup\diagdown}} & \underset{\substack{O \quad O \\ 111^\circ}}{\overset{Cl^-}{\diagup\diagdown}} \\[3ex]
& & \underset{\substack{Cl \quad Cl \\ 111^\circ}}{\overset{O}{\diagup\diagdown}} & \underset{\substack{Cl \quad Cl \\ 100^\circ}}{\overset{S}{\diagup\diagdown}} & \underset{\substack{F \quad F \\ 96^\circ}}{\overset{Cl^+}{\diagup\diagdown}} \\[3ex]
& & & \underset{\substack{F \quad F \\ 98^\circ}}{\overset{S}{\diagup\diagdown}} & \underset{\substack{F \quad F \\ 93\frac{1}{2}^\circ}}{\overset{Br^+}{\diagup\diagdown}}
\end{array}
$$

(Note the absence of SO_2^{2-}, $SOCl^-$, and $OClF$, for example.) SF_2 is an unstable species produced and studied by passing SF_6 through a radio frequency discharge, reacting the products with COS downstream from the discharge zone, and pumping them through a m.w. cell. The ClF_2^+ ion occurs in compounds such as $ClF_2 . AsF_6$ and the BrF_2^+ ion in $BrF_2 . SbF_6$. In the latter the $F-Br-F$ angle is $93\frac{1}{2}^\circ$, but there are also two weaker $Br-F$ bonds coplanar with the other two, suggesting a structure intermediate between BrF_2^+ and BrF_4^-. The FCl_2^+ ion is unsymmetrical, $(Cl-Cl-F)^+$. For halogen and interhalogen cations see p. 333.

The 26-*electron group.* Pyramidal AO_3 complexes include the SO_3^{2-}, SeO_3^{2-}, ClO_3^-, BrO_3^-, and IO_3^- ions and the XeO_3 molecule; note the absence of the pyramidal PO_3^{3-} (and NO_3^{3-}) ion. Less symmetrical pyramidal molecules include the thionyl and chloryl halides; iodyl fluoride, IO_2F, a white crystalline solid, which is stable in dry air, is probably structurally different from the very reactive ClO_2F:

$$
\begin{array}{lll}
& SO_3^{2-} \quad ClO_3^- \quad XeO_3 \\
& SO_2F^- \quad ClO_2F \\
& SOCl_2 \\
SnCl_3^- \quad NF_3, PX_3 & SX_3^+(?)
\end{array}
$$

It is possible that the ion SCl_3^+ may exist in SCl_4, and far i.r. studies of $SeCl_4$, $SeBr_4$, $TeBr_4$, and TeI_4, have suggested the formulation of these halides as $(SeX_3)X$ and $(TeX_3)X$. The pyramidal structure of $(SeF_3)^+$ has been established in $(SeF_3)(NbF_6)$ and $(SeF_3)(Nb_2F_{11})$, p. 600; for the structure of crystalline $TeCl_4$ see p. 577. The pyramidal structure of the isoelectronic ion $[(CH_3)_3Te]^+$ has been established in crystalline $[(CH_3)_3Te](CH_3TeI_4)$. We may include in this group the $SnCl_3^-$ ion in $K_2(SnCl_3)Cl . H_2O$, formerly thought to be $K_2SnCl_4 . H_2O$.

The 32-*electron group.* This very large group ranges from the tetrahedral oxy-ions of Si, P, S, and Cl—such ions are not formed by the first-row elements C, N, and F—to the halogeno-ions MX_4 formed by numerous non-transition elements, and also includes many intermediate oxy-halogen ions and molecules. Compounds of P are given below because they form a more complete series than the N analogues.

NF_4^+ and NOF_3 are known, but the highest oxyfluoride is NO_2F, though NS_3F has been prepared:

$$B(OH)_4^- \quad SiO_4^{4-} \quad PO_4^{3-} \quad SO_4^{2-} \quad ClO_4^- \quad XeO_4$$
$$PO_3X^{2-} \quad SO_3X^- \quad ClO_3F$$
$$PO_2X_2^- \quad SO_2X_2$$
$$POF_3$$
$$BF_4^- \quad SiX_4 \quad PX_4^+$$
$$BeCl_4^{2-} \quad AlCl_4^-$$

Note the complete vertical column of P complexes and the complete diagonal series of neutral molecules, from SiX_4 to XeO_4.

Trigonal bipyramidal arrangement of 5 electron pairs

| 22 electrons | 28 electrons | 34 electrons | 40 electrons |

Particular interest attaches to the 22, 28, and 34 electron systems because of the non-equivalence of the axial and equatorial positions in trigonal bipyramidal coordination, there being no spatial arrangement of five equivalent bonds around an atom other than the coplanar (pentagonal) one.

The 22-electron group. Examples are here confined to linear ions formed by the heavier halogens ($ClBr_2^-$, $BrCl_2^-$, numerous ions IX_2^- and $IX'X''$, Br_3^-, and I_3^-) and the KrF_2 and XeF_2 molecules.

The 28-electron group. The T-shaped configuration has been established for ClF_3 and BrF_3; ions such as SCl_3^- would presumably have the same type of structure.

The 34-electron group. The neutral molecules in this group are SF_4 (SCl_4, SeF_4, $TeCl_4$), IOF_3, and XeO_2F_2. The structure of IOF_3 is not known, nor is the detailed structure of XeO_2F_2, though i.r. and Raman data for this compound in an

$$IO_2F_2^- \quad XeO_2F_2$$
$$IOF_3$$
$$SF_4 \quad BrF_4^+$$

argon matrix suggest a structure similar to that of $IO_2F_2^-$. The expected structure, derived from a trigonal bipyramid has been confirmed for SF_4, SeF_4, $IO_2F_2^-$, BrF_4^+, and also for molecules such as $Se(C_6H_5)_2Br_2$ and $Te(CH_3)_2Cl_2$.

The 40-electron group. The only complexes in this group with the full symmetry of the trigonal bipyramid are Group V pentahalides and the $SnCl_5^-$ ion. In SOF_4 the O atom occupies an equatorial position and the structure is necessarily somewhat less symmetrical. It is interesting that the $InCl_5^{2-}$ ion, isoelectronic with $SnCl_5^-$, has the form of a tetragonal pyramid in which the metal atom is 0·6 Å above the base; the relation of this configuration to the trigonal bipyramid is discussed shortly. The $SnCl_3(CH_3)_2^-$ ion is trigonal bipyramidal like $SnCl_5^-$.

Octahedral arrangement of six electron pairs

| 36 electrons | 42 electrons | 48 electrons |

The 36-*electron group.*   The square planar configuration has been established for the ions BrF_4^- and ICl_4^- and for the XeF_4 molecule.

The 42-*electron group.*   The arrangement of five ligands at five of the vertices of an octahedron has been demonstrated by structural studies of the following molecules and ions:

$$XeOF_4$$
$$BrF_5$$
$$SbF_5^{2-} \quad TeF_5^- \qquad XeF_5^+$$
$$SbCl_5^{2-}$$

and a similar configuration may be presumed for ClF_5 and IF_5. It may be noted that in no case of tetragonal pyramidal coordination is the central atom A located *in* the base of the pyramid. This atom is either situated about 0·5 Å above the base, as in the 40 electron $InCl_5^{2-}$ ion with five shared electron pairs, and in a number of transition-metal complexes when the bond arrangement is an alternative to—and closely related to—the trigonal bipyramidal configuration, or it is situated some 0·3–0·4 Å *below* the base of the pyramid. This is the case in all the pentafluoro-ions, BrF_5, and $SbCl_5^{2-}$, and also in TeF_4, which is built from TeF_5 pyramids sharing 2 F atoms (but on this point see p. 578). For details and references see Table 7.6.

TABLE 7.6

Square pyramidal ions and molecules of non-transition elements

Ion or molecule	M–X (apical)	M–X (basal)	X_a–M–X_b	Reference
$(TeF_5)^-$	1·85 Å	1·96 Å	79°	IC 1970 9 2100
$(TeF_4)_n$	1·80	2·03†	82°	JCS A 1968 2977
$(XeF_5)^+$	1·81	1·88	79°	JCS A 1967 1190
$(SbF_5)^{2-}$	2·00	2·04	83°	IC 1970 9 2100
$(SbCl_5)^{2-}$	2·36	2·62	85°	ACSc 1955 9 122
BrF_5	1·68	1·81	84°	JCP 1957 27 982
$IF_5(XeF_2)$	1·88	1·88	81°	IC 1970 9 2264

† Mean of lengths 1·90–2·26 Å (unsymmetrical bridge, see p. 578).

244

The 48-*electron group.*   Octahedral molecules and ions of this group are more numerous, and include:

$$\text{Te(OH)}_6 \quad \text{IO(OH)}_5$$
$$\text{TeO}_6^{6-} \quad \text{IO}_6^{5-} \quad \text{XeO}_6^{4-}$$
$$\text{IOF}_5$$
$$\text{AlF}_6^{3-} \quad \text{SiF}_6^{2-} \quad \text{PF}_6^- \quad \text{SF}_6 \quad \text{IF}_6^+$$

and other ions of Te noted in Chapter 16.

The arrangement of 7 and 9 electron pairs

It is convenient to deal first with the 56 electron group since the only molecule AX_n in which there is a valence group consisting of 7 shared electron pairs is the IF_7 molecule. Its structure has caused a great deal of discussion, but it would seem that this molecule probably has the expected pentagonal bipyramidal configuration.

The 50-electron group comprises the following molecules and ions which arise by adding X^- to the 42 electron systems noted earlier:

$$\text{Sb}^{III}\text{X}_6^{3-} \quad \text{Te}^{IV}\text{X}_6^{2-} \quad \text{I}^{V}\text{F}_6^- \quad \text{and} \quad \text{Xe}^{VI}\text{F}_6.$$

The oxidation number of A is two less than the 'group valence', so that there is a lone pair of electrons in addition to the six bonding pairs. The structure of the IF_6^- ion is not yet known. Careful studies of the crystal structures of $(NH_4)_4(Sb^{III}Br_6)$ $(Sb^V Br_6)$ and of $(NH_4)_2TeCl_6$ and K_2TeBr_6 show that the ions under discussion form undistorted octahedra in spite of the presence of the seventh electron pair. Thus the latter does not occupy a bond position but is a 'stereochemically inert' pair.

No definite conclusion has yet been reached about the configuration of the XeF_6 molecule in the vapour state, except that it appears to be neither regular octahedral nor regular pentagonal bipyramidal. The structures of the crystalline forms are complex. The unique cyclic polymers can be described as built from tetragonal pyramidal XeF_5^+ groups and F^- ions, but there is no obvious simple description of the bonding.

Pauling commented many years ago on the abnormally long M—X bonds in $Se^{IV}X_6^{2-}$ and $Te^{IV}X_6^{2-}$ ions, comparing them with the sums of the tetrahedral covalent radii, which for Se and Te correspond to M(II). A more direct comparison could be made of the M—X bond lengths in the following pairs:

M—X			M—X
2·80 Å	$SbBr_6^{3-}$	$SbBr_6^-$	2·56 Å
?	TeF_6^{2-}	TeF_6	1·82
?	IF_6^-	IF_6^+	?

Unfortunately this comparison can be made at present only for Sb—X, for the only known hexahalide of Te is TeF_6 and the only accurate data are for the $TeCl_6^{2-}$ and $TeBr_6^{2-}$ ions; no data are available for IF_6^- or IF_6^+. It appears that although stereochemically inert the pair of 2s electrons increases the bond length, behaving as a spherically symmetrical shell resulting in an increase in the size of M(IV) as compared with M(VI).

The ion XeF_8^{2-} presents a somewhat similar problem, in that there are nine electron pairs, eight shared and one unshared, yet the shape is not far removed from a square antiprism. This is the sole representative of the 66-electron group.

The 10–14 electron groups

If we arrange from 10 to 14 electrons in a diatomic molecule (or ion) so as to maintain an octet of valence electrons on each atom the only possibilities are the symmetrical arrangements:

10	11	12	13	14
2 : 6 : 2	3 : 5 : 3	4 : 4 : 4	5 : 3 : 5	6 : 2 : 6

As more electrons are added to the system the number in the bond falls from six to two and the bond weakens and lengthens:

	10	11	12	13	14
	(N_2)	O_2^+	O_2	O_2^-	O_2^{2-}
Length	1·10	1·12	1·21	1·30	1·48 Å
Dissociation energy	941·4		493·7		154·8 kJ mol^{-1}
Stretching force constant	2290		1180		383 Nm^{-1}

(Since data are not available for the species O_2^{2+} the N_2 molecule is included as the representative of the 10-electron class.) The bond orders according to m.o. bond theory are 3, 2·5, 2, 1·5, and 1 respectively. The fact that the bond length increases on adding successive electrons to O_2^+ is due to the rearrangement of the electrons and corresponds to the change in the isoelectronic series:

10 electrons	12 electrons	14 electrons
HC≡CH	$H_2C=CH_2$	H_3C-CH_3
1·20 Å	1·35 Å	1·54 Å

Although this simple treatment accounts qualitatively for the bond lengths, it does not account for all of the facts. Not only are O_2^+ and O_2^- paramagnetic, as is to be expected for odd-electron ions, but so also is the O_2 molecule. This difficulty is overcome in m.o. theory by assigning separate (unpaired) electrons to two 'antibonding' orbitals.

Examples of the even-numbered members of this group include some important molecules and ions. The 10-electron group includes N_2, the high-temperature molecules PN and P_2, and the series:

C_2^{2-}	CN^-	CO	NO^+
1·20	1·17	1·13	1·06 Å.

The 12-electron group includes NO^-, O_2, SO, and S_2, and the 14-electron group, the molecules of the halogens and interhalogen compounds XX′, and halogen oxy-ions XO^-. There are few examples of the 11- and 13-electron groups:

11 electrons : NO, O_2^+; 13 electrons : O_2^-, Br_2^+.

Note that the simple arithmetic approach adopted here predicts facts such as the collinear arrangement of bonds M–C≡O or $M–(N≡O)^+$ for 10-electron ligands bonded to metal atoms as contrasted with the non-linear arrangement for a 12-electron ligand, $M–(N{=}O)^-$.

Odd electron systems AX_2 *and their dimers* $X_2A - AX_2$

The 17-electron and 19-electron systems are of special interest because they might be expected to dimerize to the 34- and 38-electron molecules and so get rid of the unpaired electrons.

17 electrons *19 electrons*

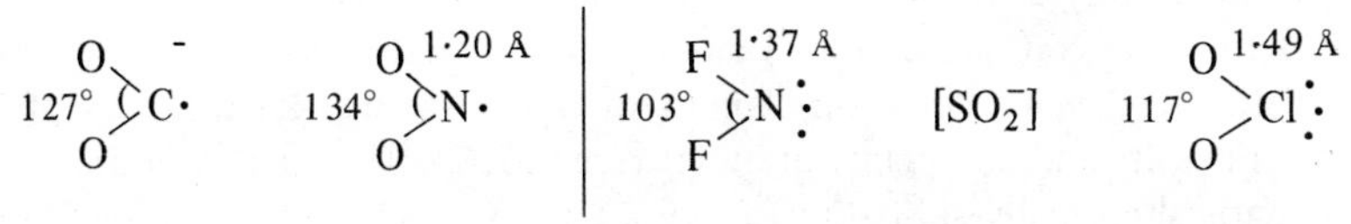

† in Xe matrix (JCP 1969 **51** 4710)

‡ in gas (PRS 1967 A **298** 145)

34 electrons *38 electrons*

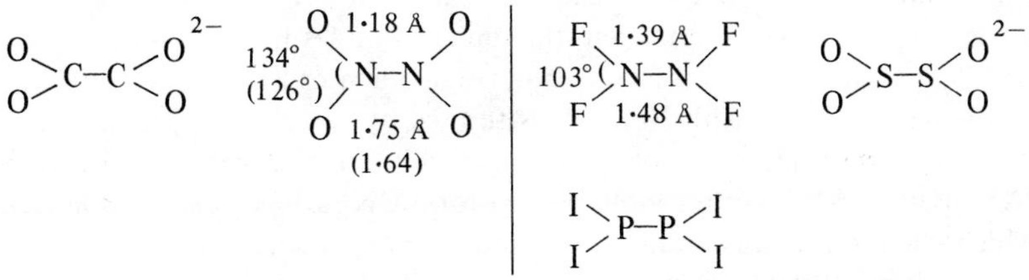

The effect on the stereochemistry of a system of adding first an odd electron and then a lone pair is nicely illustrated by the series: NO_2^+ (linear), NO_2, and NO_2^- (both angular), and it is noteworthy that the 17-electron system NO_2 is intermediate as regards both bond length and interbond angle between the 16- and 18-electron systems:

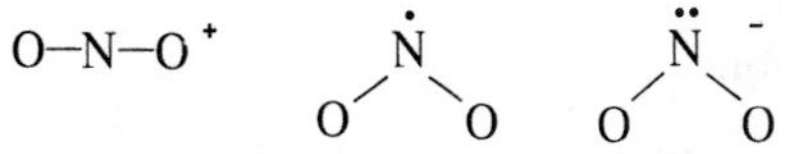

Angle O–N–O	180°	134°	115°
N–O	1·15 Å	1·20 Å	1·24 Å

The detailed structure of ClO_2^+ would be of great interest because, together with ClO_2 and ClO_2^- it forms another series including an odd-electron molecule (as also would SO_2^+, SO_2, and SO_2^-):

$$\ddot{C}l^+ \qquad \dot{\ddot{C}l} \quad 1\cdot49 \text{ Å} \qquad \ddot{\ddot{C}l}^- \quad 1\cdot57 \text{ Å}$$
$$O \qquad O \qquad O \quad 117° \quad O \qquad O \quad 111° \quad O$$

If eight electrons are assigned to each terminal O or X atom there are sufficient electrons for a single bond in the 34-electron systems while in the 38-electron systems there is a lone pair on each of the A atoms:

$$>A-A< \quad \text{and} \quad \ddot{A}-\ddot{A}$$

In the former there are therefore three coplanar bonds around A and in the latter three pyramidal bonds. In $C_2O_4^{2-}$ and B_2X_4 the central bonds are found to have the lengths expected for single bonds, but in N_2O_4 the N—N bond is abnormally long. Note the exactly similar structures of monomer and one-half of the dimer in the case of NO_2 and N_2O_4 and of NF_2 and N_2F_4. An interesting point about these systems is that N_2O_4 and B_2F_4 are planar (for B_2Cl_4 see p. 845) while $C_2O_4^{2-}$ is planar in some salts and non-planar in others (see p. 732). The oxalic acid molecule is planar, with abnormally short C=O (1·19 Å) and C—OH (1·29 Å) bonds. For a discussion of the bonding in molecules A_2X_4 formed by B, N, and P see: JACS 1969 **91** 1922 and IC 1969 **8** 2086.

Unlike NO_2 ClO_2 shows no tendency to dimerize, whereas the isoelectronic SO_2^- can only be isolated as $S_2O_4^{2-}$ (in dithionites), though there is some evidence for an equilibrium between monomer and dimer in solution. Only one of the 19-electron halogen species (NF_2) is known; the dimeric species include N_2F_4, P_2F_4, P_2Cl_4, and P_2I_4. The free radical NF_2 is quite stable and can exist indefinitely in the gas phase in equilibrium with N_2F_4. It is formed by reacting NF_3 with metallic Cu at 400°C and dimerizes on cooling. There is an interesting reaction between NF_2 and NO, another odd-electron molecule, to form $ON . NF_2$, a deep purple compound which at room temperature and atmospheric pressure rapidly dissociates to a colourless mixture of NO and NF_2.

The halogen compounds P_2X_4 apparently all exist in the centro-symmetrical *trans* form (in contrast to the *gauche* configuration of P_2H_4); the dithionite ion is notable for its eclipsed configuration and for the extraordinary length of the S—S bond (2·39 Å, as compared with 2·06 Å for the normal single bond in the various forms of elementary S).

The van der Waals bond

Under appropriate conditions of temperature and pressure it is possible to liquefy and solidify all the elements, including the noble gases, and all compounds, including those consisting of non-polar molecules such as CH_4, CCl_4, etc. The existence of a universal attraction between all atoms and molecules led van der

Waals to include a term a/V^2 in his equation of state. For molecules with a permanent dipole moment (μ) Keesom calculated the mean interaction energy

$$E_K = -\frac{2\mu^4}{3r^6 kT}$$

at a distance r, and to this expression Debye added the energy resulting from the interaction between the permanent dipole and the moments it induces in neighbouring molecules:

$$E_D = -\frac{2\alpha\mu^2}{r^6}$$

where α is the polarizability. The first expression, requiring the van der Waals factor a to be inversely proportional to the absolute temperature, is not consistent with observation, and moreover the sum of E_K and E_D is much too low (see Table 7.7).

TABLE 7.7
Lattice energy kJ *mol*$^{-1}$

	μ(D)	α (×10^{-24} cm^3)	E_K	E_D	E_L	*Total*
Ar	0·00	1·63	0·000	0·000	8·49	8·49
CO	0·12	1·99	0·000	0·008	8·74	8·74
HI	0·38	5·40	0·025	0·113	25·86	25·98
HBr	0·78	3·58	0·686	0·502	21·92	23·10
HCl	1·03	2·63	3·31	1·004	16·82	21·13
NH$_3$	1·50	2·21	13·31	1·26	14·73	29·58
H$_2$O	1·84	1·48	36·36	1·92	9·00	47·28

Neither expression, of course, accounts for the interaction between non-polar molecules, a type of interaction first calculated by London and hence called London, or dispersion, energy. This (wave-mechanical) calculation gives

$$E_L = -\frac{3 I\alpha^2}{4 r^6}$$

for similar particles with ionization potential I, or

$$-\frac{3 I_1 I_2 \alpha_1 \alpha_2}{2 r^6 (I_1 + I_2)}$$

for dissimilar particles.

The relative magnitudes of these three types of interaction can be seen from Table 7.7 for a few simple cases. For a non-polar molecule the London energy is necessarily the only contribution to the lattice energy; even for polar molecules such as NH$_3$ and H$_2$O for which E_K is appreciable, E_L forms an important part of the lattice energy. Note that these lattice energies are between one and two orders of magnitude smaller than for ionic crystals; for example, those of the 'permanent' gases are of the order of one per cent of the lattice energy of NaCl (approximately 753 kJ mol^{-1}).

As a contribution to the lattice energy of salts the London interaction can be as large as 20 per cent of the total in cases where the ions are highly polarizable (for example, TlI), and the adoption of the CsCl structure rather than the NaCl structure by CsCl, CsBr, and CsI is attributed to the van der Waals energy (see later).

Metal–metal bonding

The discussion of metal–metal (m–m) bonding by chemists is usually wholly or largely confined to finite molecules and complex ions. Here we shall attempt to place this aspect of the subject in perspective as part of a much more general phenomenon.

The formation of a chemical compound implies bonding between atoms of different elements, and for simplicity we may start by considering binary compounds. In some binary compounds $A_m X_n$ all the primary bonds (between nearest neighbours) are between A and X atoms (as in simple ionic crystals) but this is by no means generally true. Bonds between atoms of the same element (A–A or X–X) are to be expected in a compound containing a preponderance of atoms of one kind, whether it consists of molecules (for example, S_7NH) or is a crystalline compound such as a phosphorus-rich phosphide (CdP_4), a silicon-rich silicide ($CsSi_8$), a boron-rich boride (BeB_{12}) or at the other extreme a metal-rich compound such as $Li_{15}Si_4$, Cs_3O, or Ta_6S. Of special interest are compounds in which the greater part of the bonding is between A and X atoms, but where there is also some interaction between A and A or X and X atoms. Disregarding the other bonds formed by A and X atoms (which may form part of a 3D arrangement in a crystal) we may show this diagrammatically:

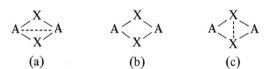

An example of (a) is the structure of NiAs, in which Ni has 6 As neighbours but is also bonded to 2 Ni; (b) represents a 'normal' binary compound with only A–X bonds, while an example of (c) is a boride such as FeB in which B is surrounded by 6 Fe but is also bonded to 2 B atoms.

Few elements other than carbon show much tendency to form molecules containing systems of bonded A atoms, a process sometimes called catenation, a term which should perhaps strictly refer only to chains. Diatomic systems include the singly-bonded molecules X_2 of the halogens, the multiply-bonded O_2, N_2, and the high-temperature species C_2, P_2, and S_2, and ions C_2^{2-}, O_2^{2-}, etc. Triatomic species include the multiply-bonded O_3, N_3^-, the singly-bonded S_3^-, and ions formed by the halogens such as I_3^-, in which the bonds are longer than the normal single bonds in the X_2 molecules. Molecules L_nX-XL_n are formed by a variety of elements,

250

CHART 7.1
Types of structure with metal-metal bonding

Class I
m–m bonds
no bridging atoms

finite: Cu_2,
    X—Hg—Hg—X, $Mn_2(CO)_{10}$
    $Cl_4Re\equiv ReCl_4$. etc.
    polyhedral Bi_9^{5+}
        $L_n L_n L_n$
[chains and layers, e.g. —M—M—M—]

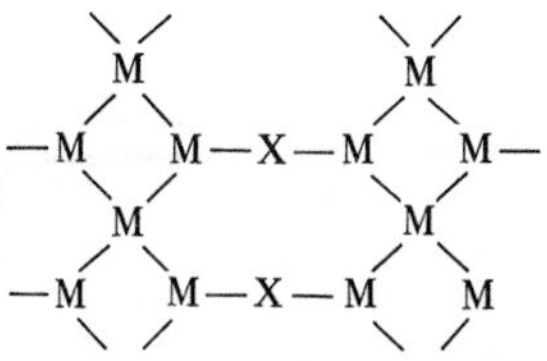

I–II hybrids, e.g.
$Co_4(CO)_{12}$

Class II
m–m bonds and bridging:

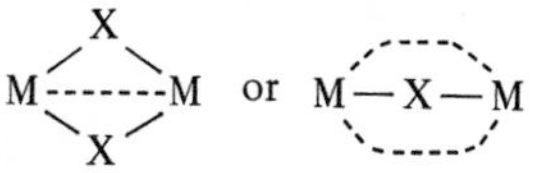

M⸱⸱⸱⸱⸱⸱⸱M   or   M—X—M
with bridging X (above and below)

(NiAs)   (ReO_3)

finite: metal cluster compounds of Cu, Ag, Nb, Ta, Mo, W
    carbonyls, $Fe_2(CO)_9$, $Os_3(CO)_{12}$, $Re_4(CO)_{16}^{2-}$
    carboxylates (Cr, Mo, Re, Cu)
1D AgCNO, AgNCO, RuO_2
3D ReO_3

Class Ia
m–m systems linked into more extensive system through
X atoms (no further m–m bonding)

m–m grouping
  finite =
  chain = Gd_2Cl_3, —M M—X—M M—
    Ta_2S, Ta_6S
  layer = —M M—X—M M

Class IIa
Units of class II further linked through M—X—M bridges to
more complex systems, e.g.

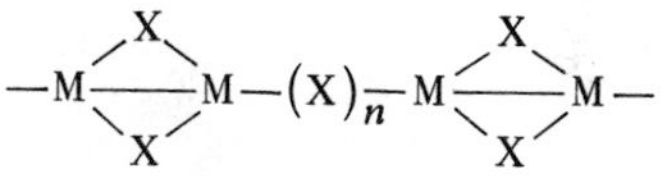

—M⟨X/X⟩M—$(X)_n$—M⟨X/X⟩M—

chains: NbI_4, $ReCl_4$, ReI_3, W_6Br_{16}
layers: $ReCl_3$, $MoCl_2$, WCl_2, Nb_6Cl_{14}, Ta_6I_{14}
3D: Nb_6I_{11}, Nb_6F_{15}, Ta_6Cl_{15}, MoO_2, WO_2, etc.

Class III
Finite units, chains and layers bonded together by m–m bonds:
'anti' chain and layer structures.

finite: Rb_9O_2, Rb_6O, Cs_7O

chains: Cs_3O

layers: Ag_2F, PbO, Ti_2O, Ti_3O

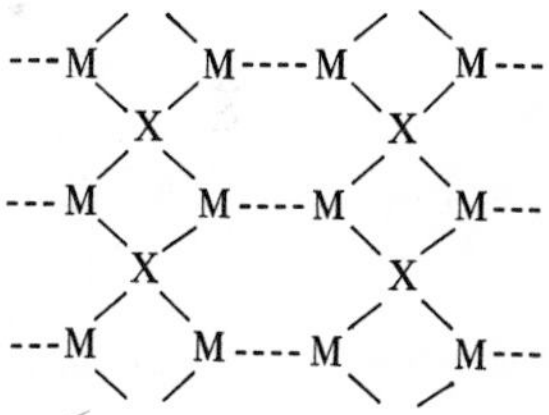

B, C, Si, Ge, Sn, Pb, N, P, S, Se, and Te, and a few molecules and ions containing chains of 3 X atoms are known (for example, $P_3(C_6H_5)_5$, $P_3O_8^{5-}$, XS_3X, etc.). Examples of homocyclic molecules are listed in Table 3.9 on p. 85. Chains containing more than three A atoms are formed by a very few elements (C, Si, Ge, S) and this is also true of polyhedral systems (P_4, Bi_9^{5+}, and the $B_nH_n^{2-}$ ions (p. 83)).

In Chart 7.1 are set out the main types of structure (molecular and crystal) in which there is m–m bonding, and we now comment briefly on the various classes.

Class I molecules (ions) containing directly bonded metal atoms without bridging ligands

The simplest molecules of this kind are the diatomic molecules in the vapours of a number of metals (of Groups I–III, Co and Ni). The bond dissociation energies vary

251

widely, from around 170 to 210 kJ mol^{-1} for Co_2, Ni_2, Cu_2, Ag_2 and Au_2 (very similar to those of Cl_2, Br_2, and I_2) and somewhat smaller values for the alkali metals (from 109 in Li_2 to 42 kJ mol^{-1} in Cs), to very much smaller values in Mg_2 (29 kJ mol^{-1}) and in Zn_2, Cd_2, and Hg_2 (around 4 to 8 kJ mol^{-1}). The next-simplest systems containing m–m bonds are those formed by a few B subgroup elements, and of these the most important are the mercurous compounds which contain the grouping X–Hg–Hg–X; compounds of Group IVB elements include Ge_2H_6, Sn_2H_6, and $(CH_3)_3Pb–Pb(CH_3)_3$.

Direct m–m bonds without any bridging ligands are formed by transition metals (in low oxidation states) only when special π-bonding ligands are attached to the metal, as in $(CO)_5Mn–Mn(CO)_5$ (and the isostructural Re and Tc compounds) and in the similarly constituted $[Co_2(NC \cdot CH_3)_{10}]^{4-}$ ion; the m–m bonds are rather long (2·9–3·0 Å), possibly due to repulsions between the ligands on different metal atoms. Intermediate between compounds of this type and the Group IVB molecules noted above are numerous molecules containing a number of directly bonded metal atoms, in which some are B subgroup metals (Cu, Au, Hg, In, Tl, Ge, Sn) and others are transition metals (Mo, W, Mn, Fe, Co, Ir, Pt). Examples include:

$(C_6H_5)_3Sn–Mn(CO)_5$

$(C_5H_5)Mo(CO)_3$

$(C_6H_5)_3Sn–Fe(CO)_2C_5H_5$

$C_5H_5(CO)_2Fe–Sn–Fe(CO)_2C_5H_5$

$[Cl_3Sn–Pt–SnCl_3][N(CH_3)_4]_2$
   $|$
   Cl_2

$(C_5H_5)Mo(CO)_3$

In a small group of ions and molecules there is multiple bonding between the metal atoms of a type possible only for transition metals. They include the ions $Re_2Cl_8^{2-}$, $Tc_2Cl_8^{3-}$, and $Mo_2Cl_8^{4-}$, in which the lengths of the m–m bonds are respectively 2·24, 2·13, and 2·14 Å. These bonds are described as quadruple bonds ($\sigma\pi^2\delta$), the δ component accounting for the eclipsed configuration of the ions and also of the similarly constituted molecule $Re_2Cl_6(PEt_3)_2$.

Class Ia. We can envisage systems of bonded metal atoms which are then cross-linked by ligands X, there being no metal–metal interactions between the different sub-units. Examples include Gd_2Cl_3, Ta_2S, and Ta_6S; further examples are likely to emerge from studies of sub-halides, sub-chalconides, etc.

Class II. *Molecules (ions) containing directly bonded metal atoms and bridging ligands*

This large class includes all compounds in which there is direct interaction between pairs of metal atoms which are also bridged by other ligands. These ligands may be halogen atoms, NH_2, CO, SR, PR_2, carboxylate groups, or other chelate organic ligands. Examples of finite systems with halogen bridges include those $M_2Cl_9^{3-}$ ions formed from a pair of face-sharing octahedra in which the M–M distance indicates interaction across the shared face, and the numerous metal cluster compounds formed by Nb, Ta, Mo, W, and Re, and also by Ag and Cu. For the polynuclear carbonyls in which some of the CO ligands act as bridges between metal atoms, and for compounds such as $[(CO)_3Fe(NH_2)]_2$ and $[(C_5H_5)Co(P\phi_3)]_2$ see Chapter 22. Two carboxylate bridges are present in compounds such as $Re_2(O_2CC_6H_5)_2I_4$ and four in $Mo_2(O_2CCH_3)_4$, molecules such as $Re_2(O_2CR)_4X_2$, and the monohydrates of Cr(II) and Cu(II) acetates. Examples of bridged molecules are given in Table 7.8.

TABLE 7.8

Some molecules containing metal–metal bonds*

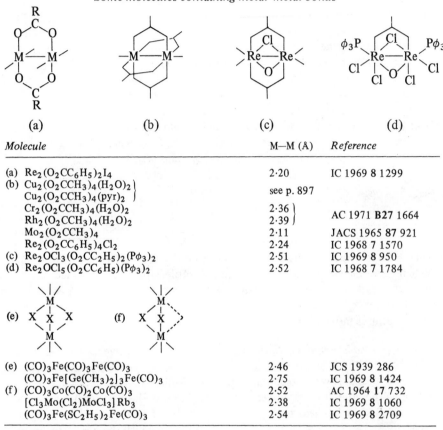

Molecule	M—M (Å)	Reference
(a) $Re_2(O_2CC_6H_5)_2I_4$	2·20	IC 1969 8 1299
(b) $Cu_2(O_2CCH_3)_4(H_2O)_2$	see p. 897	
$Cu_2(O_2CCH_3)_4(pyr)_2$		
$Cr_2(O_2CCH_3)_4(H_2O)_2$	2·36	AC 1971 **B27** 1664
$Rh_2(O_2CCH_3)_4(H_2O)_2$	2·39	
$Mo_2(O_2CCH_3)_4$	2·11	JACS 1965 87 921
$Re_2(O_2CC_6H_5)_4Cl_2$	2·24	IC 1968 7 1570
(c) $Re_2OCl_3(O_2CC_2H_5)_2(P\phi_3)_2$	2·51	IC 1969 8 950
(d) $Re_2OCl_5(O_2CC_6H_5)(P\phi_3)_2$	2·52	IC 1968 7 1784
(e) $(CO)_3Fe(CO)_3Fe(CO)_3$	2·46	JCS 1939 286
$(CO)_3Fe[Ge(CH_3)_2]_3Fe(CO)_3$	2·75	IC 1969 8 1424
(f) $(CO)_3Co(CO)_2Co(CO)_3$	2·52	AC 1964 17 732
$[Cl_3Mo(Cl_2)MoCl_3]Rb_3$	2·38	IC 1969 8 1060
$(CO)_3Fe(SC_2H_5)_2Fe(CO)_3$	2·54	IC 1969 8 2709

* For other examples see Chapters 9 and 22.

The structures of silver cyanate and fulminate are examples of simple chain structures in which there are m–m interactions along the chains:

In this class come oxides such as ReO_3 and RuO_2 in which there is not only the 'normal' bonding of the metal atoms through O atoms but also some kind of less direct interaction ('super-exchange') involving the oxygen orbitals. Thus RuO_2 has the normal rutile structure, with equidistant metal atoms in the chains, but the high electrical conductivity suggests an average oxidation state greater than four, that is, some delocalization of electrons through overlap of metal and oxygen orbitals.

Class IIa.   This class includes all structures in which units of class II are further linked, but without m–m interactions, into more extensive systems. Examples include the simple edge-sharing chain of NbI_4, with alternate short and long Nb–Nb distances, the chain molecules in crystalline $ReCl_4$ built from Re_2Cl_9 units, and the still more complex chains in ReI_3 (Re_3I_9 units) and in W_6Br_{16}. In the extraordinary structure of the latter compound $(W_6Br_8)Br_4$ clusters are linked into infinite chains by linear Br_4 groups. Infinite 2D structures in this class include $ReCl_3$, built from triangular Re_3Cl_9 units similar to those in Re_3I_9 but here further linked into layers rather than chains, and a number of halides built from the clusters of the two main types, M_6X_8 and M_6X_{12}. These units can also be connected into 3D systems, as in the following examples:

Type of metal cluster	2D structures	3D structures
M_6X_8	$MoCl_2$, WCl_2	Nb_6I_{11}
M_6X_{12}	Nb_6Cl_{14}, Ta_6I_{14}	Nb_6F_{15}, Ta_6Cl_{15}

These halide structures are described in more detail in Chapter 9. We may also include here those dioxides with distorted rutile structures in which there are alternate short and long m–m distances within each 'chain' (e.g. 2·5 Å and 3·1 Å in MoO_2 and WO_2), since the discrete pairs of close metal atoms are linked through oxygen bridges into a 3D structure.

Class III. *Crystals containing finite, 1-, or 2-dimensional complexes bonded through m–m bonds*

In Classes I and II we recognized units within which there is m–m bonding, with additional bonding through bridging X atoms in Class II. In classes Ia and IIa these units are further bonded, through X atoms, without further m–m bonding. It is necessary to recognize a further class of structures in which there is extensive m–m bonding *between* finite, 1-, or 2-dimensional sub-units; there may also be metal–metal bonding *within* the units, but this is not necessary. This class includes halides and chalconides (usually metal-rich 'sub-compounds') which have structures geometrically similar to those of normal halides or chalconides but with positions of metal and non-metal interchanged. Just as we find polarized ionic structures with only van der Waals bonds between the halogen atoms on the outer surfaces of the chains or layers, so we find the 'anti' chain and layer structures of the same geometrical types in which the bonds between the chains (layers) are bonds between the metal atoms on their outer surfaces. The metal-rich oxides of Rb and Cs (p. 443) provide examples of structures built entirely of finite units held together by m–m bonds (Rb_9O_2) and of structures which might be described as consisting of finite units inserted into the metal, there being metal atoms additional to those in the sub-units, as in Rb_6O ($Rb_9O_2 . Rb_3$) and Cs_7O ($Cs_{11}O_3 . Cs_{10}$).

The ionic bond

The ionic bond is the bond between charged atoms or groups of atoms (complex ions) and is the only one of the four main types of bond that can be satisfactorily described in classical (non-wave-mechanical) terms. Monatomic ions formed by elements of the earlier A subgroups and ions such as N^{3-}, O^{2-}, etc., F^- etc., have noble gas configurations, but many transition-metal ions and ions containing two s electrons (such as Tl^+ and Pb^{2+}) have less symmetrical structures. We shall not be concerned here with the numerous less stable ionic species formed in the gaseous phase.

The simplest systems containing ionic bonds are the gaseous molecules of alkali halides and oxides, the structures of which are noted in Chapters 9 and 12; we refer later to the halide molecules in connection with polarization. The importance of the ionic bond lies in the fact that it is responsible for the existence at ordinary temperatures, as stable solids, of numerous metallic oxides and halides (both simple and complex), of some sulphides and nitrides, and also of the very numerous crystalline compounds containing complex ions, particularly oxy-ions, which may be finite (CO_3^{2-}, NO_3^-, SO_4^{2-}, etc.) or infinite in one, two, or three dimensions.

The adequacy of a purely electrostatic picture of simple ionic crystals A_mX_n is demonstrated by the agreement between the values of the lattice energy resulting from direct calculation, from the Born–Haber cycle, and in a few cases from direct measurement.

The lattice energy of a simple ionic crystal

The lattice energy is the energy released when a mole of free gaseous ions are brought together from infinity to form the crystal; it is conveniently calculated by considering the reverse process, the complete break-up of the crystal. The major part of the lattice energy is the electrostatic or coulomb energy, and this is found by calculating the potential at the position of an ion in the crystal. For the NaCl structure this potential is the same for a cation as for an anion site (P_c and P_a respectively), and the energy required to remove an ion from the crystal to infinity is $-eP_c$ or $+eP_a$ respectively. Since each quantity takes account of the mutual interactions of the ion with all other ions in the crystal the sum of these terms for all ions would count the coulomb energy twice; its value is therefore $-\frac{1}{2}Ne(P_c - P_a)$ per mole.

In the NaCl structure an ion has six equidistant neighbours of the other kind at a distance r (Na–Cl), twelve of its own kind at $r\sqrt{2}$, eight of the other kind at $r\sqrt{3}$ and so on, so that the potential at an ion is found by summing the infinite series

$$-P_a = P_c = -\frac{6e}{r} + \frac{12e}{r\sqrt{2}} - \frac{8e}{r\sqrt{3}} + \frac{6e}{2r} - \ldots \text{ etc.}$$

If this summation is written $-A(e/r)$ then the coulomb energy is

$$U = -N \cdot \frac{Ae^2}{r}$$

since the coulomb attraction between two charges $\pm e$ is $-e^2/r$, or for a mole, $-Ne^2/r$. The constant A (Madelung constant) is a pure number characteristic only of the geometry of the structure, and it can therefore in principle be calculated for any structure. We give here the Madelung constants for the NaCl and CsCl structures since we shall refer to them later: they are 1·748 and 1·763 respectively.

Since crystals are not indefinitely compressible there is evidently a repulsion force which operates when the electron clouds of the ions begin to interpenetrate (without electron-sharing between the ions). This repulsion energy is not readily calculable, and Born represented it by B/r^n, a function which increases very rapidly with decreasing distance r if n is large, that is, it corresponds to the ions being 'hard' spheres. The expression for the lattice energy is now

$$U = -N \cdot \frac{Ae^2}{r} + \frac{B}{r^n}.$$

The value of the exponent n can be deduced from the compressibility of the crystal; values used in calculating lattice energies are 7, 9, 10, and 12 for ions with Ne, Ar, Kr, and Xe configurations. The value of B is calculated in the following way. At equilibrium the energy is a minimum, that is, the attractive and repulsive forces balance one another, therefore

$$\frac{dU}{dr} = 0 = N \cdot \frac{Ae^2}{r^2} - \frac{nB}{r^{n+1}}$$

from which B can be eliminated to give

$$U = -N \cdot \frac{Ae^2}{r_e} \left(1 - \frac{1}{n}\right)$$

where r_e is the equilibrium distance between the ions in the crystal. (An alternative expression for the repulsion energy, Be^{-kr}, gives values not very different from the simple Born formula.)

The calculation of lattice energy has been refined by including terms arising from the van der Waals (London) forces and from the zero-point energy. The former is important only if both ions are readily polarizable, as may be seen from the following figures:

Salt	Structure	Coulomb attraction	London attraction	Born repulsion	Zero-point energy	U (total) kJ mol^{-1}
NaCl	NaCl	−862	−13	+100	+8	−767
AgCl	NaCl	−875	−121	+146	+4	−846
CsI	CsCl	−619	−46	+ 63	+29	−573
TlCl	CsCl	−732	−117	+142	+4	−703

We shall see later that the van der Waals energy is apparently the critical factor responsible for the adoption of the CsCl structure by CsCl, CsBr, and CsI.

Indirect determinations of the lattice energies of many ionic crystals have been made using the Born–Haber cycle, which relates the following quantities:

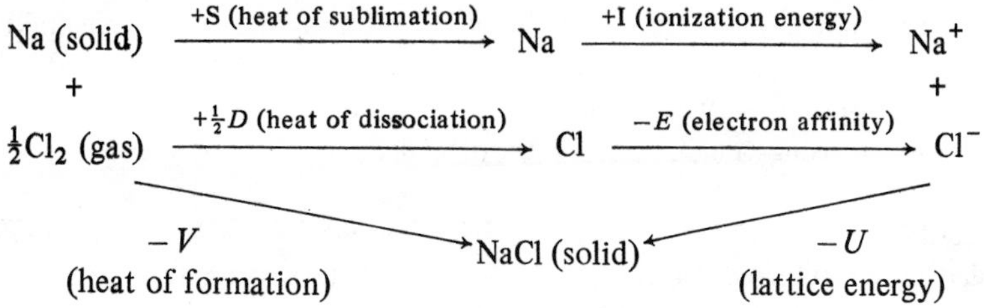

Na (solid) $\xrightarrow{\text{+S (heat of sublimation)}}$ Na $\xrightarrow{\text{+I (ionization energy)}}$ Na$^+$

$+$

$\frac{1}{2}$Cl$_2$ (gas) $\xrightarrow{\text{+}\frac{1}{2}D \text{ (heat of dissociation)}}$ Cl $\xrightarrow{-E \text{ (electron affinity)}}$ Cl$^-$

$-V$ → NaCl (solid) ← $-U$
(heat of formation)   (lattice energy)

From the relation: $V = U - I - S + E - \frac{1}{2}D$, the lattice energy U can be determined in cases where all the other quantities are known. Direct determinations of lattice energies have been made only in a few cases—by measurement of the equilibrium

$$MX \rightleftharpoons M^+ + X^-$$

combined with vapour pressure data for the solid salt. There is good agreement for a number of alkali-halides between the values of U calculated directly, indirectly from the Born–Haber cycle, and by direct measurement.

Ionic radii

To calculate the electrostatic contribution to the lattice energy of the NaCl (or any other structure) it is necessary to know only the relative positions of the atoms and the distances between them, both of which are directly determined from the diffraction data. No knowledge of the relative sizes of the ions is required. Barlow considered the packing of spheres of two different sizes and derived the NaCl and CsCl structures many years before they were confirmed by X-ray studies. If ions can be regarded as approximately incompressible spheres of various sizes having spherically symmetrical charge distributions we might expect to be able to relate the way in which they pack together to their relative sizes, that is, to relate the coordination numbers in different structures to the ionic radii. Since only *sums* of radii are measurable it is necessary to fix one ionic radius. Two early methods were:

(i) to start with the radii of the halide ions derived from the X–X distances in the lithium halides, assuming that in these crystals the anions are in contact (Landé, 1920), or to take the radius of O^{2-} as one-half the distance apart of these ions in silicates (c. 2·7 Å), making a similar assumption (Bragg, 1927). This method may be illustrated by the cation–anion distances in the following crystals, all of which have the NaCl structure:

	A–X		A–X
MgO	2·10 Å	MnO	2·24 Å
MgS	2·60	MnS	2·59
MgSe	2·73	MnSe	2·73

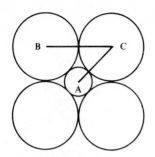

FIG. 7.2. Calculation of ionic radii from interatomic distances in crystals with the sodium chloride structure.

Bonds in Molecules and Crystals

From the constancy of the M–S and M–Se distances in the sulphides and selenides respectively it is reasonable to assume that in these crystals the anions are in contact with one another and to deduce the radii of S^{2-} and Se^{2-} ions as 1·84 $(2·60/\sqrt{2})$ Å and 1·93 $(2·73/\sqrt{2})$ Å respectively ($\frac{1}{2}$ CB, Fig. 7.2).

(2) Wasastjerna (1923) divided the interionic distances in the alkali halides and alkaline-earth oxides in the ratios of the molar refractivities of the ions, taking the molar refractivity of an ion to be roughly proportional to its volume. Goldschmidt adopted Wasastjerna's values for F^- (1·33 Å) and O^{2-} (1·32 Å) and extended the list of radii by using the observed interionic distances in many other essentially ionic crystals.

In addition to the above sets of radii based on O^{2-}, 1·35 Å or 1·32 Å, Pauling compiled a set referred to O^{2-}, 1·40 Å, and Zachariasen later used the value 1·46 Å. Since the effective radius of an ion depends on its coordination number (c.n.) it is usual to adopt the c.n. 6 as standard. The following relative values of radii for different c.n.'s are often quoted:

c.n.	4	6	8
radius	0·95	1·00	1·04

They were derived from the Born equation given earlier making certain assumptions about the relative values of the Madelung constants A and the Born repulsion coefficients B for the structures with different c.n.'s. More recently appreciably different ratios have been suggested for some ions, as implied by the figures given in Table 7.9 for 4-coordinated Li^+ and Be^{2+} and for 8-coordinated Y^{3+}.

Early compilations of 'ionic radii' were rather comprehensive, and included values for ions such as B^{3+}, Si^{4+}, N^{5+}, etc. derived from crystals containing complex ions in which the bonds would now usually be regarded as having appreciable covalent character. Only the most electropositive elements form essentially ionic crystals containing ions such as N^{3-} and those of the heavier chalcogens and halogens, and radii for many metals must be derived from oxides and fluorides. Accordingly, a recent list of ionic radii lists only F^- and O^{2-} as anions, and is admittedly applicable only to fluoride and oxide structures. Some of these radii are given in Table 7.9, which also includes some anion radii applicable to a limited number of compounds of the most electropositive elements. Even in the case of fluorides and oxides there are only a few structures in which the ions have symmetrical arrangements of *equidistant* nearest neighbours. For metals which do not form binary compounds with structures of this type it is necessary to take mean interionic distances from structures such as the C–Mn_2O_3 or LaF_3 structures, in which there are groups of 'nearest neighbours' at somewhat different distances, or from complex fluorides or oxides in which there are highly symmetrical coordination groups. In some structures the departures from regular coordination groups (for example, around Ti^{4+} or V^{5+}) may indicate partial covalent character of the bonds, so that the significance of 'ionic radii' derived from such structures is doubtful. Allowance has always been made for the c.n. of the cation; it seems reasonable (see Table 7.9) also to allow for the c.n. of the anion and probably also to use different radii for low- and high-spin states of the 3d metal ions.

258

TABLE 7.9

Ionic radii (Å)

−3	−2	−1	+1	+2	+3	+4	
			Li(a) 0·74	Be(b) 0·35			
N 1·50	O(c) 1·40	F(d) 1·33	Na 1·02	Mg 0·72	Al 0·53		
P 1·90	S 1·85	Cl 1·80	K 1·38	Ca 1·00	Sc 0·73	Ti 0·61	Zn²⁺ 0·75
As 2·20	Se 1·95	Br 1·95	Rb 1·49	Sr 1·16	Y(e) 0·89	Zr 0·72	Cd²⁺ 0·95
	Te 2·20	I 2·15	Cs 1·70	Ba(f) 1·36	La 1·06	Tl⁺ 1·50	Pb²⁺ 1·18
NH₄⁺ 1·50	OH⁻ 1·55	SH⁻ 2·00					

(a) c.n. 4, 0·59
(b) c.n. 4, 0·27
(c) c.n. 2, 1·35, c.n. 8, 1·42
(d) Note that M—F is at least 0·1 Å *greater* than M—O in ScOF and YOF (p. 404)
(e) c.n. 8, 1·02, c.n. 9, 1·10
(f) c.n. 8, 1·42, c.n. 12, 1·60 Å.

	Ti^{2+}	V^{2+}	Cr^{2+}	Mn^{2+}	Fe^{2+}	Co^{2+}	Ni^{2+}	Cu^{2+}
Low spin	0·86	0·79	0·73	0·67	0·61	0·65	0·70	0·73
High spin			0·82	0·82	0·77	0·74		

	Ti^{3+}	V^{3+}	Cr^{3+}	Mn^{3+}	Fe^{3+}	Co^{3+}	Ni^{3+}	(Ga^{3+})
Low spin	0·67	0·64	0·62	0·58	0·55	0·53	0·56	0·62
High spin				0·65	0·65	0·61	0·60	

The values are those of Shannon and Prewitt (AC 1969 **B25** 925) and are applicable to oxides and fluorides (c.n. 6, based on the value 1·40 Å for 6-coordinated O^{2-}) except those enclosed by the full line, which are values intended to give an approximate idea of the sizes of the ions.

We shall not be greatly concerned with differences between the various sets of radii since they do not have a precise physical meaning and moreover are of interest chiefly in detailed discussions of interatomic distances in crystals; the remarks on Slater's radii (p. 237) are relevant in this connection. However, and in spite of the various complications mentioned above, it is important to have some idea of the approximate relative sizes of ions if only as an aid to visualizing structures and understanding why, for example, pairs of ions such as K^+ and F^-, Cs^+ and Cl^-, or Ba^{2+} and O^{2-} form c.p. arrays in so many complex halides or oxides. As aids to structure determination or in discussing many features of ionic crystals sets of M—F or M—O distances would obviously serve the same purpose as sets of ionic radii. Thus, any one of the standard sets of radii serves to illustrate the following general points:

(a) A positive ion is appreciably smaller than the neutral atom of the same element owing to the excess of nuclear charge over that of the orbital electrons, whereas the radius of an anion is much larger than the covalent radius:

Li	1·5 Å		Cl^-	1·8 Å	
Li^+	0·7	compare	Cl	1·0	

(b) In a series of isoelectronic ions the radii decrease rapidly with increasing positive charge, but there is no comparable increase in size with increasing negative charge (see the values in a horizontal row of Table 7.9).

(c) *Most* metal ions are smaller than anions; a few pairs of cations and anions of comparable size lead to the special families of c.p. structures noted above, and we discuss later some structures containing the exceptionally large Cs^+, Tl^+, and Pb^{2+} ions. The very small Li^+ and Be^{2+} ions are typically found in tetrahedral coordination in halide and oxide structures, Al^{3+} in both 4- and 6-coordination, the numerous ions M^{2+} and M^{3+} with radii in the range 0·7–1·0 Å are usually octahedrally coordinated in such structures, while the largest ions are found in positions of higher coordination (up to 12-coordination in complex structures).

(d) In contrast to the steady increase of radii down a Periodic family the radii of the 4f M^{3+} ions show a steady decrease with increasing atomic number. For example, the M–O distances (for 6-coordination) decrease from La^{3+}–O, 2·44 Å, to Lu^{3+}–O, 2·23 Å. As a result of this 'lanthanide contraction' certain pairs of elements in the same Periodic Group have practically identical ionic (and atomic) radii, for example, Zr and Hf, Nb and Ta; the remarkable similarity in chemical behaviour of such pairs of elements is well known. An effect analogous to the lanthanide contraction is observed also in the 5f ions.

(e) The irregular variation of the radii of 3 d ions M^{2+} and M^{3+} with the numbers of e_g and t_{2g} electrons (p. 272). See reference in Table 7.9 and references quoted therein.

The structures of simple ionic crystals

Table 7.10 shows the c.n.'s of A in simple AX_n structures. The structures are divided into two groups which we shall describe as (a) normal and (b) polarized ionic structures. The former are typically those of fluorides and oxides (with the exception that the NaCl structure is adopted more generally by all halides of most of the alkali metals and by all chalconides of the alkaline-earths), while the

TABLE 7.10

The structures of simple ionic crystals

		(a)					(b)		
	4	6	8	9		4	6	8	9
AX	ZnO†	NaCl				ZnS‡	NaCl (distorted) TlI	CsCl	
AX_2	Cristobalite	Rutile	CaF_2			$BeCl_2$ HgI_2	CdI_2 $CdCl_2$	$[PbCl_2]$ (7 + 2)	
AX_3		ReO_3 RhF_3		LaF_3 YF_3			BiI_3 YCl_3		$LaCl_3$ UCl_3

† The hexagonal structure of ZnO (zincite) and also of ZnS (wurtzite).
‡ The cubic zinc-blende structure. As noted elsewhere the bonds in these, and possibly other, 4-coordinated structures, may have some covalent character.

structures (b) are typically those of chlorides, bromides, iodides, sulphides, and many hydroxides. The bonds in the tetrahedral structures, wurtzite (ZnO) and cristobalite, may have appreciable covalent character; this bond arrangement is to be expected for either type of bond. There are interesting differences between the anion coordinations in the $4:2$ (AX_2) and $6:2$ (AX_3) structures, namely, non-linear M–F–M or M–O–M bonds in crystals with the cristobalite structure, collinear M–F–M or M–O–M bonds in MoF_3 and ReO_3, but smaller M–F–M bond angles in a number of other trifluorides ($132°$ in the h.c.p. RhF_3 structure). The non-linear anion bonds in the cristobalite structure presumably result in a greater van der Waals contribution to the lattice energy of the more compact structure; in Cu_2O, which consists of two interpenetrating 'anticristobalite' nets, the collinear O–Cu–O bonds are usually attributed to covalent character, and here the better space-filling is attained by the interpenetration of the two nets. In ReO_3 the collinear Re–O–Re bonds appear to be due to a special type of interaction of the metal atoms via the O orbitals; the reason for the adoption of this structure by certain trifluorides, and not by others, is less obvious.

In any one of the structures of Table 7.10 the environment of all anions is the same, and all cations are surrounded in the same way by anions. In both (a) and (b) structures the maximum number of nearest neighbours of a cation does not exceed eight or nine, and except for the cases already noted the environment of the anions (and in some cases, for example, the TlI structure, also of the cations) is generally less symmetrical in the (b) than in the (a) structures. Departures from regularity of the octahedral coordination group in the rutile (and also in the α-Al_2O_3) structure, which are due to purely geometrical factors, have been discussed in Chapter 5.

It seems reasonable to suppose that the more symmetrical structures (a) are consistent with the efficient packing of the ions, regarded as incompressible spheres having spherical charge distributions. If the ions are to pack as closely as possible the determining factor will be the number of the larger ions that can pack around one of the smaller ions (usually the cation A). The c.n. of the larger ion follows from the fact that in A_mX_n the c.n.'s of A and X must be in the ratio $n:m$.

Radius ratio and shape of coordination group
Suppose that we have three X ions surrounding an A ion. The condition for stability is that each X ion is in contact with A, so that the limiting case arises when the X ions are also in contact with one another. The following relation exists between r_A and r_X, the radii of A and X respectively: $r_A/r_X = 0.155$. (If the radius of X is a we have

$$r_A = \tfrac{2}{3}\sqrt{3}a - a = 0.155a.)$$

If the radius ratio $r_A : r_X$ falls below this value, then the X ions can no longer all touch the central A ion and this arrangement becomes unstable. If r_A increases the X ions are no longer in contact with one another, and when $r_A : r_X$ reaches the value 0.225 it is possible to accommodate 4 X around A at the vertices of a regular tetrahedron. In general, for a symmetrical polyhedron, the critical minimum value of $r_A : r_X$ is equal to the distance from the centre of the polyhedron to a vertex less one-half of the edge length. The radius ratio ranges for certain highly symmetrical coordination polyhedra are:

$r_A:r_X$	0·155 ——	0·225 ——	0·414 ——	0·732–
c.n.	3	4	6	8
polyhedron	equilateral triangle	regular tetrahedron	regular octahedron	cube

If the packing of X ions around A is to be the densest possible we should expect the coordination polyhedra to be those having equilateral triangular faces and appropriate radius ratio. For example, for 8-coordination we should not expect the cube but the triangulated dodecahedron or even the intermediate polyhedron, the square antiprism, which for a finite AX_8 group is a more stable arrangement than the cube. Similarly, the icosahedron would be expected for 12-coordination in preference to the cuboctahedron, which is the coordination polyhedron found in a number of complex ionic crystals. For this reason we include a number of triangulated polyhedra in Table 7.11. For the geometry of 7- and 9-coordination

TABLE 7.11

Radius ratios

Coordination number	Minimum radius ratio	Coordination polyhedron
4	0·225	Tetrahedron
6	0·414	Octahedron
	0·528	Trigonal prism
7	0·592	Capped octahedron
8	0·645	Square antiprism
	0·668	Dodecahedron (bisdisphenoid)
	0·732	Cube
9	0·732	Tricapped trigonal prism
12	0·902	Icosahedron
	1·000	Cuboctahedron

polyhedra the reader is referred to Chapter 3, where examples of 7-, 8-, and 9-coordination are given. The radius ratio is not meaningful for 5-coordination, either trigonal bipyramidal or tetragonal pyramidal, since for the former the minimum radius ratio would be the same as for equilateral triangular coordination (0·155), that is, not a value intermediate between those for tetrahedral (0·225) and octahedral (0·414) coordination, while for tetragonal pyramidal coordination the radius ratio would be the same as for octahedral coordination.

Although there is a general increase in c.n.'s of cations with increase in ionic radius a detailed correspondence between c.n. and radius ratio is not observed for simple ionic crystals. For example, all the alkali halides at ordinary temperature and pressure except CsCl, CsBr, and CsI crystallize with the NaCl structure. For LiI and LiBr (and possibly LiCl) the radius ratio is probably less than 0·41, but the radius ratios for the lithium halides are somewhat doubtful because the interionic distances in these crystals are not consistent with constant (additive) radii:

Li—I	3·02 A	K—I	3·53 A	Rb—I	3·66 A
Li—F	2·01	K—F	2·67	Rb—F	2·82
Difference	1·01		0·86		0·84

(Plots of lattice energy of ionic crystals AX against radius ratio show that the cubic ZnS structure becomes more stable than the NaCl structure only at values of $r_A : r_X$ less than about 0·35 rather than 0·41.) Of the salts with radius ratio greater than 0·732 only CsCl, CsBr, and CsI normally adopt the 8-coordinated CsCl structure. We find the same persistence of the NaCl structure in monoxides, where the CsCl structure might have been expected for SrO, BaO, and PbO; in fact there is no monoxide with the CsCl structure. The c.n. six is not exceeded although the pairs of ions K^+ and F^-, Ba^{2+} and O^{2-}, have practically the same radius.

	Li	Na	K	Rb	Cs
F	(0·56)	0·77	0·96[a]	0·89[a]	0·78[a]
Cl	0·41	0·57	0·77	0·83	0·94
Br	0·38	0·52	0·71	0·76	0·87
I	0·34	0·47	0·64	0·69	0·79

[a] When r_A is larger than r_X we give $r_X : r_A$ since it is not meaningful to discuss the coordination of X^- around A^+ if the latter is the larger ion.

In the case of difluorides (and dioxides) the compounds containing the smaller cations have the rutile structure and those containing larger cations the fluorite structure, for example:

M	MF_2 (*rutile structure*)		MF_2 (*fluorite structure*)			
$r_{M^{2+}}$	Mg	Zn	Cd	Ca	Sr	Ba
	0·72	0·75	0·95	1·00	1·16	1·36 Å

However, for BaF_2 the radius ratio is very close to unity, yet the 8 : 4 structure is not replaced by one of higher coordination (e.g. 12 : 6). The radius of Na^+ is consistent with the 6–8 coordination observed in many oxy-salts, but Na_2O has the antifluorite structure in which this ion is 4-coordinated. Moreover the same structure is adopted by K_2O (radius ratio unity) and Rb_2O ($r_X : r_A = 0·94$) but not by Cs_2O, which has a quite different structure (see later). Clearly higher coordination of M^+ in M_2O would require higher coordination of O (e.g. 6 : 12 coordination).

The correspondence between the relative sizes of ions and their c.n.'s is in fact very much more satisfactory for complex ionic crystals (complex fluorides and oxides and crystals containing complex oxy- or fluoro-ions) than for simple ionic crystals; it is clearly not satisfactory for *simple* ionic crystals. The good agreement between calculated and experimentally determined lattice energies confirms the soundness of the electrostatic model of these structures, but it is worth noting that the calculations refer only to *observed* structures. They appear not to have been made either for certain structures which are not observed but which would seem to be perfectly satisfactory as regards their geometry (for example, those marked with asterisks in Table 7.13, p. 269) or for others such as the $AuCu_3$ structure which are adopted by numerous intermetallic compounds but not by ionic compounds. It therefore seems worthwhile to draw attention to the following points, some of which will be discussed in subsequent paragraphs.

(a) The occurrence of cubic rather than antiprism and of cuboctahedral rather than icosahedral coordination is due to the impossibility of building structures with particular types of coordination, and does not call for further discussion.

(b) Structures of very high coordination do not occur.

(c) The environment of the *anions* in certain of the structures of Table 7.10(b) is highly unsymmetrical.

(d) Emphasis on the coordination group of the cation has led to a tendency to underestimate the importance of the coordination of the anion.

We are not referring in (d) to the polarized ionic structures (p. 268) in which anions such as Cl^- have very unsymmetrical environments but to a number of oxide structures in which the attainment of a satisfactory environment of the anion seems to be at least as important as the coordination around the cation. It is usual to describe the structures of metal oxides in terms of the coordination polyhedra of the *cations*. This is convenient because their c.n.'s are usually greater than those of the anions, and a structure is more easily visualized in terms of the larger coordination polyhedra. However, this preoccupation with the coordination of the metal ions may lead us to underestimate the importance of the arrangement of ions around the *anions*, information which is regrettably omitted from many otherwise excellent descriptions of crystal structures. The formation of the unexpectedly complex corundum structure (with vertex-, edge-, and face-sharing of the octahedral AlO_6 groups) is presumably associated with the difficulty of building a c.p. M_2X_3 structure with octahedral coordination of M and an environment of the 4-coordinated X atom which approximates at all closely to a regular tetrahedral one. Similarly the two kinds of distorted octahedral coordination of Mn^{3+} ions in cubic Mn_2O_3 are probably incidental to attaining satisfactory 4-coordination around O^{2-} ions. The choice of tetrahedral holes occupied by the A atoms in a spinel AB_2O_4 (one-eighth of the total number) is that which gives a tetrahedral arrangement of 3 B + A around O^{2-}.

(e) Departures from regular tetrahedral or octahedral coordination around certain transition-metal ions (and also certain 'inert-pair' ions) are related to the asymmetry of the electron clouds of such ions. A short note on the *ligand field theory* is included later in this chapter.

Limitations on coordination numbers

We have commented on the absence of structures of ionic compounds A_mX_n with coordination numbers of A greater than eight or nine. If we derive 2-dimensional nets in which A has some number (p) of X atoms and X has some number (q) of A atoms as nearest neighbours we find that the only possible (p, q)-connected nets (p and $q \geqslant 3$) in which p- and q-connected points alternate are the (3, 4), (3, 5), and (3, 6) nets. Since there are upper limits to the values of p and q in plane nets, it is reasonable to assume that the same is true of 3D nets. We are not aware that this problem has been studied. However, the existence or otherwise of crystal structures with high c.n.'s is not simply a matter of topology (connectedness); we must take account of metrical factors.

In a structure $^aA_m{}^xX_n$, where a and x are the c.n.'s of A and X by X and A, the number a of coordination polyhedra XA_x meet at an A atom and x coordination polyhedra AX_a meet at an X atom (assuming that all atoms of each kind have the same environment). We saw in Chapter 5 that, allowing for the size of the X atoms, not more than six regular octahedral groups AX_6 can meet at a point, that is, have a common vertex (X atom), though this number can be increased to eight (in Th_3P_4) and possibly nine if the octahedra are suitably distorted (p. 278). Thus, we can construct a framework of composition AX_2 in which each A is connected to 12 X and each X to 6 A—the AlB_2 structure—but this structure cannot be built with octahedral coordination of B. It can be built with trigonal prism coordination of B because twelve trigonal prisms can meet at a point, though this brings certain of the B atoms very close together: B—B, 1·73 Å, compare Al—B, 2·37 Å, Al—Al, 3·01 Å and 3·26 Å. As c.n.'s increase and more coordination polyhedra meet at a given point, not only does this limit the types of coordination polyhedra (six regular octahedra, eight cubes or tetrahedra, or twelve trigonal prisms can meet at a point), but also the coordination polyhedra have to share more edges and then faces, with the result that distances between *like* atoms decrease. The AlB_2 structure is an extreme case, the very short B—B distances (covalent bonds) are those between the centres of BAl_6 trigonal prism groups across shared rectangular faces. Although this particular 12 : 6 structure exists, not all 12 : 6 structures are possible, as may be seen by studying one special set of structures.

The $AuCu_3$ structure is one of 12 : 4 coordination in which Au has 12 equidistant Cu neighbours and Cu has 4 Au and 8 Cu neighbours.† This is one of a family of cubic close-packed A_mX_n structures in which we shall assume that A has 12 X atoms as nearest neighbours. The neighbours of an X atom in these hypothetical structures would be $12m/n$ A atoms plus sufficient X atoms to complete the 12-coordination group:

	(AX)	A_3X_4	A_2X_3	AX_2	AX_3	AX_4	AX_6	(AX_{12})
Coordination group of X	(12 A)	9 A 3 X	8 A 4 X	6 A 6 X	4 A 8 X	3 A 9 X	2 A 10 X	(A) (11 X)

We may eliminate the first two structures immediately since X is obviously in contact with 4 X in the coordination shell around an A atom (cuboctahedron), but we can go further than this and show that all the structures to the left of the vertical line are impossible. In cubic close packing the coordination polyhedron of an atom is a cuboctahedral group, and in the above structures the coordination group of X is made up of certain numbers of A and X atoms. Since A is to be surrounded entirely by X atoms we cannot permit A atoms to occupy *adjacent* vertices of the cuboctahedral group around an X atom. The problem is therefore to find the *maximum* number of vertices of a cuboctahedron that may be occupied by

† For another (12, 4)-connected net see Fig. 23.19 (p. 825).

A atoms without allowing A atoms to occupy adjacent vertices. It is readily shown that this number is 4, and that there are two possible arrangements, shown in Fig. 7.3(a) and (b). (As a matter of interest we include in the Figure a solution for 3 A + 9 X, a coplanar arrangement of three A atoms which would be relevant to an AX_4 structure.) The arrangement (a) is that found in the $AuCu_3$ structure. Evidently c.p. structures for compounds such as KF (12 : 12) and BaF_2 (12 : 6) with 12-coordination of the cations are impossible. The $AuCu_3$ structure, with square planar coordination of Cu by 4 Au, is not found for the few trihalides in which M^{3+} ions are comparable in size with the halide ion (for example, the trifluorides of La^{3+} and the larger 4f and 5f ions) presumably because the observed structures have fewer X—X contacts and higher lattice energies; the nitrides Rb_3N and Cs_3N do not appear to be known. This structure is not adopted by any trioxides because there are no ions M^{6+} comparable in size with O^{2-}. It is, of course, not advisable to draw conclusions about the stability or otherwise of structures from the relative numbers of A—X, X—X, and A—A contacts. There are

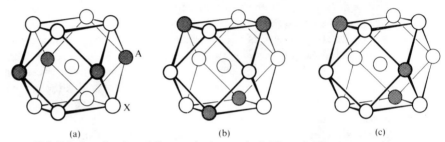

FIG. 7.3. Coordination of X atoms in close-packed AX_3 and AX_4 structures (see text).

anion–anion contacts in many ionic structures, for example, BeO (O–O, 2·70 Å), LiCl (Cl–Cl, 3·62 Å), and CaF_2, where F has 4 Ca neighbours but is also in contact with 6 F (F–F, 2·73 Å). In the CsCl structure the shortest anion–anion and cation–cation distances are only 15 per cent greater than Cs–Cl. Now that it is possible to determine the structures of high-pressure forms of elements and compounds, it may well prove profitable to include in theoretical studies and calculations of lattice energies structures which have hitherto been omitted from consideration because they are not stable under atmospheric pressure.

The polarizability of ions

We noted earlier that the magnitude of the London (van der Waals) attraction between non-bonded atoms is proportional to the product of the polarizabilities of the atoms (or ions) and that the London contribution to the lattice energy of an ionic crystal can cancel out the contribution of the Born repulsion term to the lattice energy. The polarizabilities of a number of ions are given in Table 7.12. We shall be particularly concerned with the highly polarizable ions to the right of the broken line. We may suppose that the polarizing power of a cation will be largest for small highly-charged ions, but whereas polarizability can be precisely defined as the dipole moment induced in a particle in a field of unit strength, ($\alpha = \mu/F$) and

266

TABLE 7.12

Polarizabilities of ions $(\times 10^{-24} \text{ cm}^3)$

+1	Li^+ 0·03	Na^+ 0·24		K^+ 1·00		Rb^+ 1·50		Ag^+ 1·9	Cs^+ 2·40		Tl^+ 3·9
+2	Be^{2+} 0·01	Mg^{2+} 0·10	Ca^{2+} 0·60	Sr^{2+} 0·90	Cd^{2+} 1·15		Ba^{2+} 1·70		Hg^{2+} 2·45	Pb^{2+} 3·6	
+3						La^{3+} 1·30					
−1				F^- 0·81			OH^- 1·89	Cl^- 2·98	Br^- 4·24	I^- 6·45	
−2								O^{2-} 3·0	S^{2-} 10·2	Se^{2-} 10·5	Te^{2-} 14·0

obtained from refractive index data (molar refractivity $R = (4\pi N\alpha)/3$), this is not true of 'polarizing power', to which we shall not assign numerical values.

The effect of polarization is evident in the gaseous alkali-halide molecules, being most pronounced for small M^+ and large X^- (LiI) and large M^+ and small X^- (CsF), and resulting in a dipole moment appreciably less than the product of the internuclear distance (d) and the electronic charge (e):

	$\mu(D)$	μ/ed
LiF	6·33	0·55
KCl	10·3	0·77
CsCl	10·4	0·71
CsF	7·85	0·63

We shall discuss here some general features of the structures of crystalline halides AX, AX_2, and AX_3, the structures of which are described in more detail in Chapter 9.

Monohalides. We have seen that the appearance of the CsCl structure cannot be explained in terms of the relative sizes of the ions A^+ and X^-. Although the Madelung constant is slightly larger for the CsCl than for the NaCl structure the electrostatic attraction is in fact greater for the latter structure, for the 3 per cent increase in interionic distance in the 8-coordinated structure more than compensates for the larger Madelung constant. The London interaction is proportional to $1/r^6$ and is therefore negligible except for nearest neighbour interactions and potentially greater for the CsCl than for the 6-coordinated NaCl structure. However, it becomes a determining factor only if the polarizabilities of *both* ions are large, since it is proportional to the product of the polarizabilities. This is presumably the reason for the adoption of the CsCl structure by CsCl, CsBr, and CsI. The only other halides with this structure at ordinary temperature and pressure are TlCl, TlBr, and the red form of TlI. The facts that thin layers of these Tl and Cs salts can be grown with the NaCl structure on suitable substrates, that CsCl transforms to the NaCl structure at $445°C$, and that RbCl, RbBr, and RbI adopt the CsCl structure under pressure, show that there is a fairly delicate balance between the various factors influencing the structures of these salts. Another structure in which polarization is important is that of the yellow form of TlI, which is a rather extraordinary layer structure; this structure is also that of NaOH, InBr, and InI. There would seem to be no simple explanation for the unique distorted forms of the NaCl structure adopted by TlF, and InCl; for details see Chapter 9.

Dihalides and trihalides. Here we comment only on those features of the structures which indicate the importance of polarization effects and the inadequacy of the simple 'hard sphere' treatment for halides other than fluorides.

We would first draw attention to the unexpected complexity of the crystal chemistry of the alkaline-earth halides, other than the fluorides, which all have the fluorite structure. The twelve compounds exhibit at least six different structures. Comparison of the Ca with the Cd halides is also of interest. Both CaF_2 and CdF_2 have the fluorite structure, with virtually the same interionic distance (2·36 and 2·38 Å respectively), but the interionic distances are appreciably different in the iodides (Ca–6 I, 3·12 Å, Cd–6 I, 2·99 Å) which have the same (layer) structure. Moreover, the chlorides and bromides have different structures, the Ca compounds having the $CaCl_2$ (3D h.c.p. structure) and the Cd compounds the $CdCl_2$ (layer) structure. Table 9.15 (p. 353) shows that the highly polarized CdI_2 structure occurs for the smallest cations in Group IIA and also for PbI_2 (both ions highly polarizable) and the fluorite structure for all the fluorides, but in addition the fluorides containing the largest ions (Sr^{2+}, Ba^{2+}, Pb^{2+}) also adopt the $PbCl_2$ structure. Between these extremes lies the complex crystal chemistry of the chlorides and bromides, and also the iodides of Sr and Ba. The fluorides of Mg and all the 3d metals from Cr to Cu have the rutile structure (distorted in the case of CrF_2 and CuF_2), but all the other dihalides of these metals (except $CrCl_2$, which has a distorted rutile-like structure) have the CdI_2 or $CdCl_2$ layer structures. Similarly, the trifluorides of Al, Ti, V, Cr, Mn, Fe, and Co have 3D structures, but the other trihalides have layer or chain structures, apart from $AlBr_3$ which consists of molecules Al_2Br_6. The BiI_3 and YCl_3 layer structures are very closely related to the CdI_2 and $CdCl_2$ structures, having respectively h.c.p. and c.c.p. halogen ions and, of course, one-third instead of one-half of the total number of octahedral holes occupied by metal ions.

The outstanding feature of these layer structures, as compared with the 3D rutile and ReO_3-type structures of the fluorides, is the very unsymmetrical environment of the anions, which have their cation neighbours (3 in AX_2, 2 in AX_3) all lying to one side. This unsymmetrical environment is not, however, peculiar to layer structures, and in Table 7.13 we set out the simplest structures for compounds AX_2 and AX_3 in which A is octahedrally coordinated by 6 X atoms. In each group of structures there is one in which X *may* have a rather symmetrical set of nearest A neighbours; in the rutile structure the coordination of X is plane triangular (albeit not equilateral if AX_6 is a regular octahedron), and in the ReO_3 structure Re–O–Re is linear. Each structure may be modified to give less symmetrical coordination of X, as in the $CaCl_2$ and RhF_3 structures. The simplest edge-sharing structures are of two types for both AX_2 and AX_3, namely, 3D structures (of which examples are shown in Fig. 7.4) and layer structures; for AX_3 there are also

FIG. 7.4(a). The 3D octahedral framework, of composition AX_2, is derived from the NaCl structure by removing alternate rows of metal ions as shown in Fig. 4.22(b). Although each AX_6 group shares the same set of six edges as in the $CdCl_2$ or CdI_2 layers, this structure is not adopted by any compound AX_2, but it represents the idealized structure of one form of $Cu_2(OH)_3Cl$, the mineral atacamite. Additional atoms (B) in positions of tetrahedral coordination give the spinel structure for compounds A_2BX_4. The positions of only a limited number of B atoms are indicated.

FIG. 7.4(b). The BX_6 octahedra grouped tetrahedrally around the points of the diamond net form the vertex-sharing BX_3 (B_2X_6) framework of the pyrochlore structure for compounds $A_2B_2X_6(X)$; the seventh X atom does not belong to the octahedral framework. In $Hg_2Nb_2O_7$ a framework of the cuprite (Cu_2O) type is formed by the seventh O atom and the Hg atoms, O forming tetrahedral bonds and Hg two collinear bonds.

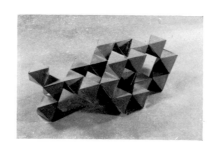

FIG. 7.4(c) and (d). Two of the (unknown) AX_3 structures in which octahedral AX_6 groups share three edges as in the $CrCl_3$ and BiI_3 layers, the A atoms being situated at the points of 3D 3-connected nets instead of the plane hexagon net as in the layer structures. In the structure (c), based on the cubic (10,3) net of Fig. 3.28, the X atoms occupy three-quarters of the positions of cubic closest packing—compare the ReO_3 structure. The spheres indicate the positions of the missing c.p. atoms. In the structure (d), based on the (10,3) net of Fig. 3.29, the X atoms occupy all the positions of cubic closest packing.

TABLE 7.13

The simplest structures for octahedral coordination of A *in compounds* AX_2 *and* AX_3

Type of structure AX_2	Octahedra sharing	Coordination of X		Octahedra sharing	Type of structure AX_3
3D: rutile CaCl$_2$	} 2 edges and } 6 vertices	Triangular Pyramidal	Linear Non-linear	6 vertices 6 vertices	3D: ReO$_3$ RhF$_3$
3D: atacamite* Layer (CdI$_2$, CdCl$_2$)	6 edges 6 edges	Pyramidal Pyramidal	Non-linear Non-linear	3 edges 3 edges	3D** Layer (BiI$_3$, YCl$_3$)
			Non-linear	2 faces	Chain (ZrI$_3$)

* No examples known for simple halides AX_2.
** Structures based on 3D 3-connected nets no examples known.

chain structures. No examples are known of simple halides with either of the two types of 3D structure marked (*) and (**) (Table 7.13), though the idealized structure of one polymorph of $Cu_2(OH)_3Cl$ (atacamite) has a structure of this kind. There are on the other hand many compounds with the layer structures. The point we wish to emphasize is that the unsymmetrical environment of X is not a feature only of layer (and chain) structures; it arises from the sharing of *edges* or *faces* of the coordination groups, and is found also in the $PbCl_2$ and UCl_3 structures, which are 3D structures for higher c.n.'s of A and correspond to the 3D structures of Table 7.13 which are marked (*) and (**). We do not understand why these 3D structures (for which lattice energy calculations do not appear to have been made) are not adopted by any halides AX_2 or AX_3, but it seems reasonable to suppose that the layer structures arise owing to the high polarizability of the larger halide ions. Layer structures are also adopted by many di- and tri-hydroxides, the polarizability of OH^- being intermediate between those of F^- and Cl^-. The positive charges of the anions will be localized close to the cations, with the result that the outer surfaces of the layers are electrically neutral (Fig. 7.5(a)). In the AX_3 chain

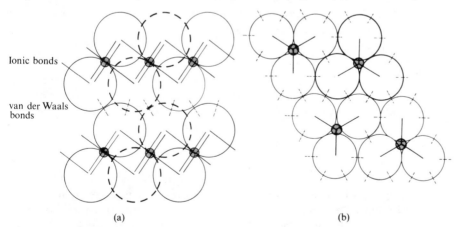

Ionic bonds

van der Waals bonds

(a)   (b)

FIG. 7.5. (a) Diagrammatic elevation of AX_3 layer structure (layer perpendicular to plane of paper), (b) AX_3 chain structure viewed along direction of chains.

269

structure of Table 7.13 the chains of face-sharing octahedra are perpendicular to the planes of c.p. X atoms (Fig. 7.5(b)). The large van der Waals contribution to the lattice energy arising from the close packing of the X atoms, giving the maximum number of X–X contacts between the layers or chains, explains why there is little difference in stability between the layer and chain structures for halides AX_3, a number of which crystallize with both types of structure ($TiCl_3$, $ZrCl_3$, $RuCl_3$).

The 'anti-layer' and 'anti-chain' structures

Just as we find Li_2O with the antifluorite structure so we find compounds adopting the layer and chain structures for compounds AX_2 and AX_3 with the positions of cations and anions interchanged:

Type of structure			C.N.'s
anti-$CdCl_2$ or anti-CdI_2	layer	Cs_2O, Ti_2O, Ag_2O,[a] Ag_2F, Ca_2N, Tl_2S	3 : 6
anti-ZrI_3	chain	Cs_3O	2 : 6
anti-MX_3	layer	Ti_3O	2 : 6

[a] formed from the normal form under pressure at 1400°C.

By analogy with the normal layer structures (large polarizable anions) we should expect the 'anti' structures to occur for the large polarizable cations such as Ag^+, Cs^+, Tl^+, and Pb^{2+}. However, the situation is more complicated because not all these compounds are 'normal valence compounds'; there is clearly some delocalization of electrons in a number of them. It is interesting to compare the oxides of Cs:

$$Cs_2O, Cs_3O, Cs_7O_2, Cs_4O, \text{ and } Cs_7O,$$

with the nitrides of Ca:

$$Ca_3N_4, Ca_{11}N_8, Ca_3N_2 \text{ (4 polymorphs), and } Ca_2N,$$

for it might not have been expected that these two groups of compounds would have much in common. However, Cs_2O and Ca_2N have similar layer structures— Cs_2O is the only alkali oxide M_2O which does not crystallize with the antifluorite structure—and both Cs_3O and Ca_2N form dark green crystals with metallic lustre. Both are apparently ionic crystals (for example Ca_2N reacts with water to give NH_3), so that they should presumably be formulated $(Cs^+)_3O^{2-}(e)$ and $(Ca^{2+})_2N^{3-}(e)$—compare $Ce^{3+}S^{2-}(e)$ and $Th^{4+}(I^-)_2(e)_2$. Evidently, reference only to the normal valence compounds Cs_2O and Ca_3N_2 gives little idea of the interest (and complexity) of these two groups of compounds.

Ligand field theory

We have seen that whereas in many ionic crystals there is a highly symmetrical arrangement of nearest neighbours around each type of ion, there are numerous structures with less symmetrical coordination groups. In the structures of some simple compounds, for example, La_2O_3 or monoclinic ZrO_2, the c.n. of the cation implies a much less symmetrical arrangement of the (seven) neighbours than is

270

possible for c.n.'s such as four or six. In other cases the coordination group of an ion is not far removed from a highly symmetrical one, and the lower symmetry has a purely geometrical explanation as, for example, in the rutile or corundum structures, though other factors such as partial covalent character cannot in all cases be excluded. In some crystals the lower symmetry results from polarization of some or all of the ions, though quantitative treatments of these effects have not been given (for example, TlF). We may also mention here the very small deviations from the cubic NaCl structure shown by MnO, FeO, CoO, and NiO below their transition temperatures, which are connected with magnetic ordering processes, and the distortions from cubic symmetry of $BaTiO_3$, $KNbO_3$, and other perovskite-type compounds, which are associated with ferroelectric effects. These are noted in Chapters 12 and 13. A further type of distorted coordination group is characteristic of certain transition-metal ions such as Cr^{2+}, Cu^{2+}, and Mn^{3+}.

It is to be expected that the behaviour of an ion when closely surrounded by others of opposite charge will depend not only on its size and polarizability but also on its outer electronic structure, and it is necessary to distinguish between ions with (a) closed shells of 8 or 18 electrons, or (b) incomplete outer shells with two s electrons (In^+, Tl^+, Pb^{2+}), or (c) transition-metal ions with incomplete d shells. The theoretical study of the interactions of transition-metal ions with the surrounding ions throws considerable light on the unsymmetrical coordination groups of certain of these ions. Features of many transition metals include the formation of more than one type of ion, many of which are coloured and paramagnetic, and the formation of covalent compounds in which a range of formal valences is exhibited, and a feature of which is the bonding of the metal atom to ligands such as CO, C_2H_4 and other unsaturated hydrocarbons, NH_3, PF_3, PR_3, etc.

The electrostatic or crystal field theory was originally developed by Bethe, Van Vleck, and others during the period 1929-35 to account for the magnetic properties of compounds of transition and rare-earth metals in which there are non-bonding d or f electrons. An alternative, molecular orbital, approach was also suggested as early as 1935 by Van Vleck. After a period of comparative neglect both theories have been widely applied since 1950 to the interpretation of the spectroscopic, thermodynamic, and stereochemical properties of finite complexes of transition metals and also to certain aspects of the structures of their crystalline compounds. The electrostatic and m.o. theories represent two approaches to the problem of dealing with the effect of the non-bonding d or f electrons on the behaviour of the transition (including 4f and 5f) elements, and they may be regarded as aspects of a general ligand field theory. In contrast to valence-bond theory, which is concerned primarily with the bonding electrons, ligand field theory considers the effect of the electric field due to the surrounding ligands on the energy levels of all the electrons in the outer shell of the central atom. Since an adequate account of ligand field theory could not be given in the space available and since we are not concerned in this book with such specialized subdivisions of physical chemistry as the optical spectra of transition-metal complexes, we shall simply note here some points relevant to the geometry of the structures of molecules (and complex ions) and crystals.

271

(a) *Preference for tetrahedral or octahedral coordination.* When a transition-metal ion is surrounded by a regular tetrahedral or octahedral group of ions (or dipoles) the five d orbitals no longer have the same energy but are split into two groups, a doublet e_g and a triplet t_{2g}. Ligands may be arranged in order of their capacity to cause d-orbital splitting (the spectrochemical series):

$$I^- < Br^- < Cl^- < F^- < OH^- < C_2O_4^{2-} \approx H_2O < {-}NCS < pyr \approx NH_3 < en < NO_2^- < CN^-.$$

The difference between the mean energies of the two groups e_g and t_{2g} (Δ) increases from left to right. If Δ is large (strong ligand field) as many electrons as possible occupy the orbitals of lower energy, which in an octahedral environment are the t_{2g} and in a tetrahedral environment are the e_g orbitals, while if Δ is small they are distributed so as to give the maximum number of parallel spins. This is shown in Table 7.14, from which low-spin (strong ligand field) tetrahedral

TABLE 7.14

Electronic structures of transition-metal ions

Number of d electrons	Octahedral environment				Tetrahedral environment		
	Weak ligand field		Strong ligand field		Weak field		
	t_{2g}	e_g	t_{2g}	e_g	e_g	t_{2g}	
1	↑	−	↑	−	↑	−	
2	↑ ↑	−	↑ ↑	−	↑ ↑	−	
3	↑ ↑ ↑	−	↑ ↑ ↑	−	↑ ↑	↑	*b*
4	↑ ↑ ↑	↑ *a*	↑↓ ↑ ↑	−	↑ ↑	↑ ↑	*c*
5	↑ ↑ ↑	↑ ↑	↑↓ ↑↓ ↑	−	↑ ↑	↑ ↑ ↑	
6	↑↓ ↑ ↑	↑ ↑	↑↓ ↑↓ ↑↓	−	↑↓ ↑	↑ ↑ ↑	
7	↑↓ ↑↓ ↑	↑ ↑	↑↓ ↑↓ ↑↓	↑	↑↓ ↑↓	↑ ↑ ↑	
8	↑↓ ↑↓ ↑↓	↑ ↑	↑↓ ↑↓ ↑↓	↑ ↑	↑↓ ↑↓	↑↓ ↑ ↑	*b*
9	↑↓ ↑↓ ↑↓	↑↓ ↑ *a*	↑↓ ↑↓ ↑↓	↑↓ ↑	↑↓ ↑↓	↑↓ ↑↓ ↑	*c*

a, b: Large tetragonal distortion (usually $c:a > 1$, but see p. 888).
c: Large tetragonal distortion ($c:a < 1$).

complexes are omitted since none is known, Accordingly cyanido- and nitrito-complexes are of the low-spin type but aquo- or halogen complexes are of the high-spin type. The main factors leading to spin-pairing appear to be high electronegativity of the metal (i.e. high atomic number and valence) and low electronegativity of the ligand, which should have a readily polarizable lone pair and be able to form $d_\pi - p_\pi$ or $d_\pi - d_\pi$ bonds by overlap of vacant p_π or d_π orbitals with filled t_{2g} orbitals of the metal.

In a regular octahedral field the energy of the e_g orbitals is higher (by $\Delta_{oct.}$) than that of the t_{2g} orbitals, and a simple electrostatic treatment gives for an ion with the configuration $(t_{2g})^m (e_g)^n$ the crystal field stabilization energy as $\Delta_{oct.}(4m - 6n)/10$. From the configurations for weak field (high-spin) of Table 7.14 it is readily found that the maximum value of this stabilization energy occurs for the following ions: d^3 (V^{2+}, Cr^{3+}, Mn^{4+}) and d^8 (Ni^{2+}). In a regular tetrahedral field the relative energies of the two sets of orbitals are interchanged, and the corresponding crystal field stabilization for a configuration $(e_g)^p(t_{2g})^q$ is $\Delta_{\text{tetr.}}(6p-4q)/10$, which is a maximum for d^2 (Ti^{2+}, V^{3+}) and d^7 (Co^{2+}). These figures indicate that essentially ionic complexes of Co(II) will tend to be tetrahedral and those of Ni(II) octahedral. In the normal spinel structure for complex oxides AB_2O_4 the A atoms occupy tetrahedral and the B atoms octahedral sites, but in some ('inverse') spinels the ions arrange themselves differently in the two kinds of site. These stabilization energies are obviously relevant to a discussion of the types of site occupied by various transition-metal ions in this structure (see Chapter 13).

(b) *Distorted coordination groups.* When there is unsymmetrical occupancy of the subgroups of orbitals, more particularly of the subgroup of higher energy, a more stable configuration results from distorting the regular octahedral or tetrahedral coordination group (Jahn–Teller distortion). From Table 7.14 it follows that for weak-field octahedral complexes this effect should be most pronounced for d^4 (Cr^{2+}, Mn^{3+}) and d^9 (Cu^{2+}). The result is a tetragonal distortion of the octahedron, usually an extension ($c:a > 1$) corresponding to the lengthening of the two bonds on either side of the equatorial plane—(4 + 2)-coordination. Examples include the distorted rutile structures of CuF_2 and CrF_2 (compare the regular octahedral coordination of Cr^{3+} in CrF_3) and the (4 + 2)-coordination found in many cupric compounds. Distortion of the octahedral group in the opposite sense, giving two short and four longer bonds, (2 + 4)-coordination, is also found in certain cupric compounds (e.g. K_2CuF_4), but examples are much fewer than those of (4 + 2)-coordination.

The structural chemistry of Cu(II) (and of isostructural Cr(II) compounds) is described in some detail in Chapter 25. Fewer data are available for Mn(III), but both types of octahedral distortion have been observed, in addition to regular octahedral coordination in $Mn(acac)_3$, a special case with three bidentate ligands attached to the metal atom:

$MnF_3\dagger$	2 F at 1·79 Å	2 F at 2·09 Å	AC 1957 **10** 345
	2 F 1·91		
$K_2MnF_5 . H_2O$ (p. 383)	4 F 1·83	2 F 2·07	JCS A 1971 2653
$K_2MnF_3(SO_4)$ (p. 586)	2 F 1·82	2 F 2·04	JCS A 1971 3074
		2 O 2·01	
$Mn(acac)_3$	6 O at 1·89 Å		IC 1968 **7** 1994

† From X-ray powder data; it may be significant that the mean of the four shorter bond lengths is close to the shorter Mn–F found in the other compounds.

In a tetrahedral environment the t_{2g} orbitals are those of higher energy, and large Jahn–Teller distortions are expected for d^3, d^4, d^8, and d^9 configurations. The distortion could take the form of an elongation or a flattening of the tetrahedral coordination group, as indicated in Table 7.14. The flattened tetrahedral $CuCl_4^{2-}$ ion is described in Chapter 25, but it is interesting to note that there is no distortion of the $NiCl_4^{2-}$ ion in $[(C_6H_5)_3CH_3As]_2 NiCl_4$ (p. 969). There are numerous crystal structures which are distorted variants of more symmetrical structures, the distortion being of the characteristic Jahn–Teller type. They include the following, the structures being distorted forms of those shown in parenthesis: CrF_2 and CuF_2 (rutile), $CuCl_2 . 2 H_2O$ ($CoCl_2 . 2 H_2O$), $CuCl_2$ (CdI_2), MnF_3 (VF_3), and γ-$MnO . OH$ (γ-$AlO . OH$). These are structures in which there is octahedral coordination of the metal ions. Examples of distorted tetrahedral coordination leading to lower symmetry include the spinels $CuCr_2O_4$ and $NiCr_2O_4$, containing respectively $Cu^{2+}(d^9)$ and $Ni^{2+}(d^8)$.

A further question arises if the six ligands in a distorted octahedral group are not identical: which will be at the normal distances from the metal ion and which at the greater distances in (4 + 2)-coordination? For example, in both $CuCl_2 . 2 H_2O$ and $K_2CuCl_4 . 2 H_2O$ we find

$$Cu \begin{cases} 2\,Cl & \text{at 2·3 Å} \\ 2\,H_2O & \text{at 2·0 Å} \end{cases} \quad \text{and} \quad 2\,Cl\ \text{at 2·95 Å}$$

rather than $4\,Cl$ or $4\,H_2O$ as nearest neighbours. Similar problems arise in the structures of hydrated transition-metal halides (e.g. the structure of $FeCl_3 . 6 H_2O$ is $[FeCl_2(H_2O)_4]Cl . 2 H_2O$), as noted in Chapter 15. It does not appear that satisfactory answers can yet be given to these more subtle structural questions.

The structures of complex ionic crystals

The term 'complex ionic crystal' is applied to solid phases of two kinds. In $MgAl_2O_4$ or $CaMgF_3$ the bonds between all pairs of neighbouring atoms are essentially ionic in character, so that such crystals are to be regarded as 3D assemblies of ions. The anions are O^{2-}, F^-, or less commonly, S^{2-} or Cl^-. The structures of many of these complex ('mixed') oxides or halides are closely related to those of simple oxides or halides, being derived from the simpler A_mX_n structure by regular or random replacement of A by ions of different metals (see, for example, Table 13.1, p. 477), though there are also structures characteristic of complex oxides or halides; these are described in Chapters 10 and 13. In a second large class of crystals we can distinguish tightly-knit groups of atoms within which there is some degree of electron-sharing, the whole group carrying a charge which is distributed over its peripheral atoms. Such *complex ions* may be finite or they may extend indefinitely in one, two, or three dimensions. The structures of many mononuclear complex ions are included in our earlier discussion of simple molecules and ions; the structures of polynuclear complex ions are described under the chemistry of the appropriate elements.

For simplicity we shall discuss complex oxides and complex oxy-salts, but the same principles apply to complex fluorides and ionic oxyfluorides. A complex oxide is an assembly of O^{2-} ions and cations of various kinds which have radii ranging from about one-half to values rather larger than the radius of O^{2-}. It is usual to mention in the present context some generalizations concerning the structures of complex ionic crystals which are often referred to as Pauling's 'rules'. The first relates the c.n. of M^{n+} to the radius ratio $r_M : r_0$. The general increase of c.n. with increasing radius ratio is too well known to call for further discussion here. We have seen that for *simple* ionic crystals the relation between c.n. and ionic size is complicated by factors such as the non-existence of alternative structures and the polarizability of ions, but in crystals containing complex ions there is a reasonable correlation between the c.n.'s of cations and their sizes. For example, a number of salts MXO_3 crystallize with one or other (or both) of two structures, the calcite structure, in which M is surrounded by 6 O atoms, and the aragonite structure, in which M is 9-coordinated. The choice of structure is determined by the size of the cation:

Calcite structure: $LiNO_3$, $NaNO_3$; $MgCO_3$, $CaCO_3$, $FeCO_3$; $InBO_3$, YBO_3
Aragonite structure: KNO_3 $CaCO_3$, $SrCO_3$; $LaBO_3$

(For salts such as $RbNO_3$ and $CsNO_3$ containing still larger cations neither of these structures is suitable.)

The second 'rule' states that as far as possible charges are neutralized locally, a principle which is put in a more precise form by defining the *electrostatic bond strength* (e.b.s.) in the following way. If a cation with charge $+ze$ is surrounded by n anions the strength of the bonds from the cation to its anion neighbours is z/n. For the Mg–O bond in a coordination group MgO_6 the e.b.s. is 1/3, for Al–O in AlO_6 it is $\frac{1}{2}$, and so on. If O^{2-} forms part of several coordination polyhedra the sum of the e.b.s.'s of the bonds meeting at O^{2-} would then be equal to 2, the numerical value of the charge on the anion. If a cation is surrounded by various numbers of non-equivalent anions at different distances it is necessary to assign different strengths to the bonds. In monoclinic ZrO_2 Zr^{4+} is 7-coordinated, by 3 O_I at 2·07 Å and 4 O_{II} at 2·21 Å, O_I and O_{II} having c.n.'s three and four respectively. Assigning bond strengths of $\frac{2}{3}$ and $\frac{1}{2}$ to the shorter and longer bonds, there is exact charge balance at all the ions.

In many essentially ionic crystals coordination polyhedra share vertices and/or edges and, less frequently, faces. Pauling's third rule states that the presence of shared edges and especially of shared faces of coordination polyhedra decreases the stability of a structure since the cations are thereby brought closer together, and this effect is large for cations of high charge and small coordination number. A corollary to this rule states that in a crystal containing cations of different kinds those with large charge and small c.n. tend not to share polyhedron elements with one another, that is, they tend to be as far apart as possible in the structure.

These generalizations originated in the empirical 'rules' developed during the early studies of (largely) mineral structures, particularly silicates. They are not relevant to the structures of *simple* ionic crystals, in which sharing of edges (and sometimes faces) of coordination groups is necessary for purely geometrical reasons. For example, the NaCl structure is an assembly of octahedral $NaCl_6$ (or $ClNa_6$) groups each sharing all twelve edges, and the CaF_2 structure is an assembly of FCa_4 tetrahedra each sharing all six edges. In the rutile structure each TiO_6 group shares two edges, and in $\alpha\text{-}Al_2O_3$ there is sharing of vertices, edges, and faces of AlO_6 coordination groups. We have shown in Chapter 5 that the sharing of edges (and faces) of octahedral coordination groups in these simple structures has a simple geometrical explanation. It is, however, a feature of many more complex

275

structures built from octahedral coordination groups, and in fact a characteristic feature of many complex oxides of transition metals such as V, Nb, Mo, and W is the widespread occurrence of rather compact groups of edge-sharing octahedra which are then linked, often by vertex-sharing, into 2- or 3-dimensional arrays, as described in Chapters 5 and 13. Even in some simple oxides such as MoO_3 there is considerable edge-sharing which is not present in the chemically similar WO_3. Since these compounds contain transition metals in high oxidation states it could be argued that this feature of their structures is an indication of partial covalent character of the bonds, and certainly there is often considerable distortion of the octahedral coordination. (It would be more accurate to say 'considerable range of M—O bond lengths', since the octahedra of O atoms may be almost exactly regular but with M displaced from their centres.)

We shall now show that a consideration of both the size factor and the principle of the local balancing of charges leads to some interesting conclusions about salts containing complex ions.

The structures and stabilities of anhydrous oxy-salts $M_m(XO_n)_p$

Much attention has been devoted over the years to that part of the electron density of complex ions which is concerned with the bonding within the ions, since this determines the detailed geometry of the ion. Here we shall focus our attention on that part of the charge which resides on the periphery of the ion, since this has a direct bearing on the possibility, or otherwise, of building crystals of salts containing these ions and therefore on such purely chemical matters as the stability (or existence) of anhydrous ortho-salts (as opposed to hydrated salts or 'hydrogen' salts) or the behaviour of the NO_3^- and other ions as bidentate ligands. In so doing we shall observe the importance of purely geometrical factors which, somewhat surprisingly, are never mentioned in connection with Pauling's 'rules' for complex ionic crystals.

Consider a crystal built of cations and oxy-ions XO_n, and let us assume for simplicity that there is symmetrical distribution of the anionic charge over all the O atoms in XO_n so that the charge on each is -1 in SiO_4^{4-}, $-\frac{3}{4}$ in PO_4^{3-}, $-\frac{1}{2}$ in SO_4^{2-}, $-\frac{1}{4}$ in ClO_4^-, and similarly for ions XO_3. These O atoms form the coordination groups around the cations, which must be arranged so that charges are neutralized locally. For example, in the calcite structure of $NaNO_3$, $CaCO_3$, and $InBO_3$ there is octahedral coordination of the cations and therefore electrostatic bond strengths of $+\frac{1}{6}$, $+\frac{1}{3}$, and $+\frac{1}{2}$ respectively for the M—O bonds. These are balanced by arranging that each O belongs to two MO_6 coordination groups:

In the following series of salts the charge on O increases as shown:

$$NaClO_4 \quad Na_2SO_4 \quad Na_3PO_4 \quad Na_4SiO_4$$

Charge on each O atom $\quad -\frac{1}{4} \qquad -\frac{1}{2} \qquad -\frac{3}{4} \qquad -1.$

If there is octahedral coordination of Na^+ the e.b.s. of a Na—O bond is $\frac{1}{6}$ and therefore in Na_4SiO_4 each O must belong to one tetrahedral (SiO_4) and to six octahedral (NaO_6) groups, or in other words, one tetrahedron and six octahedra must meet at the common vertex (O atom). There are limits to the number of polyhedra that may meet at a point, without bringing vertices of different polyhedra closer than the edge-length, as was pointed out for tetrahedra and octahedra in Chapter 5. No systematic study appears to have been made of the permissible combinations of polyhedra of different kinds which can share a common vertex (subject to the condition noted above), but it is reasonable to suppose that in general the numbers will decrease with increasing size of the polyhedra. In the present case the factors which increase this purely geometrical difficulty are those which increase the size and/or number of the polyhedral coordination groups which (in addition to the oxy-ion itself) must meet at each O atom. The problem is therefore most acute when

the charge on the anion is large (strictly, charge on O of anion),
the charge on the cation is small, and
the c.n. of the cation is large (large cation).

We have taken here as our example a tetrahedral oxy-ion; the same problem arises, of course, for other oxy-ions XO_3, XO_6, etc. The figures in the self-explanatory Table 7.15 are simply the numbers of M—O bonds required to balance the charge on O of the oxy-ion; non-integral values, which would be mean values for non-equivalent O atoms, are omitted. It seems likely that structures corresponding to entries below and to the right of the stepped lines are geometrically impossible. If we wish to apply this information to particular compounds we have to assume reasonable

TABLE 7.15

Number of coordination groups MO_x to which O must belong in alkali-metal salts (in addition to its XO_3, XO_4, or XO_6 ion)

	Charge on O atom of oxy-ion								
	$-\frac{1}{3}$	$-\frac{2}{3}$	-1	$-\frac{1}{4}$	$-\frac{1}{2}$	$-\frac{3}{4}$	-1	$-\frac{5}{6}$	-1
c.n. of M^+	NO_3^-	CO_3^{2-}	BO_3^{3-}	ClO_4^-	SO_4^{2-}	PO_4^{3-}	SiO_4^{4-}	IO_6^{5-}	TeO_6^{6-}
4			4	1	2	3	4		4
6	2	4	6		3		6	5	6
8			8	2	4	6	8		8
12	4	8	12	3	6	9	12	10	12

coordination numbers for the cations. For example, assuming tetrahedral co-ordination of Li^+ we see that none of the lithium salts presents any problem, nor do sodium salts (assuming 6-coordination) except Na_6TeO_6. Turning to the vertical columns, evidently the alkali nitrates and perchlorates present no problems, but this is not true of salts containing some of the more highly charged anions and larger cations. It will be appreciated that although the Table predicts the non-existence of, for example, orthoborates and orthosilicates of the larger alkali metals with their normal oxygen coordination numbers (in the range 8–12) it is possible that particular compounds will exist with abnormally low coordination numbers of the cations, for example, as a high-temperature phase. Lowering of the c.n. compensates for the increasing shortage of O atoms which occurs in a series of compounds such as MNO_3, M_2CO_3, and M_3BO_3. It is, of course, immaterial whether we regard a compound such as M_6TeO_6 as a salt containing TeO_6^{6-} ions or as a complex oxide $M_6^+Te^{6+}O_6^{2-}$; there is a charge of -1 on each O to be compensated by M–O bonds.

Systematic studies of cation-rich oxides $M_nM'O_6$ (M = alkali metal) have already produced interesting results. The compounds include Li_5ReO_6, Li_6TeO_6, Li_7SbO_6, and Li_8SnO_6. As regards accommodating the cations in (approximately regular) tetrahedral and/or octahedral interstices it is advantageous to have close-packed O atoms, when there are 12 tetrahedral and 6 octahedral positions per formula-weight. One of the latter is occupied by M', and therefore there is no structural problem for values of $n \leqslant 5$. In fact Li_5ReO_6 has a c.p. structure rather similar to that of α-$NaFeO_2$. All cations occupy octahedral interstices, Li replacing Na and a mixture of cations replacing Fe between alternate pairs of c.p. layers:

$$\begin{array}{ccc} Na & Fe & O_2 \\ Li & (Li_{\frac{2}{3}}Re_{\frac{1}{3}}) & O_2 = Li_5ReO_6. \end{array}$$

None of the Li compounds presents any difficulty because Li^+ can go into the tetrahedral holes, and this apparently happens in all these structures (which are all close-packed):

	Tetrahedral	Octahedral	Reference
Li_5ReO_6	–	Li_5Re	ZN 1968 **23b** 1603
Li_6TeO_6	Li_6	Te	ZN 1969 **24b** 647
Li_7SbO_6	Li_6	LiSb	ZN 1969 **24b** 252
Li_8SnO_6	Li_6	Li_2Sn	ZaC 1969 **368** 248

A number of Na compounds are known, but they are few in number compared with the Li compounds: Na_5ReO_6 (also I, Tc, Os), Na_6AmO_6 (also Te, W, Np, Pu), Na_7BiO_6, and Na_8PbO_6 (also Pt). For example, 11 compounds Li_8MO_6 are known, 2 Na compounds, but none containing K, Rb, or Cs. For the larger alkali metals there are insufficient octahedral holes if $n > 5$, and for the Na compounds the possibilities are therefore (a) tetrahedral coordination or (b) very distorted octahedral coordination in a non-close-packed structure (compare Th_3P_4, p. 160, in which eight *distorted* 6-coordination polyhedra meet at a point). Tetrahedral coordination of all the alkali metals except Cs is found in the oxides M_2O

(antifluorite structure), but the O^{2-} ions are by no means close-packed, even in Li_2O, as may be seen from the O–O distances:

Li_2O	Na_2O	K_2O	Rb_2O
3·27	3·92	4·55	4·76 Å

It will be interesting to see how the geometrical difficulties are overcome in these Na-rich oxides.

An alternative to the formation of the normal salt is to form 'hydrogen' salts. For example, assuming K^+ to be 8-coordinated, every O in K_3PO_4 would have to belong to 6 KO_8 coordination groups in addition to its own PO_4^{3-} ion. In KH_2PO_4, on the other hand, there is a charge of only $-\frac{1}{4}$ on each O, and moreover the H atoms can link the $PO_4H_2^-$ groups into a 3D network by hydrogen bonds. There is no difficulty in arranging that each O also has two K^+ neighbours situated in the interstices of the framework.

An entirely different problem arises if we *reduce* the charge on O of the oxy-ion, for example, by changing from $CaCO_3$ to $Ca(NO_3)_2$, and/or *increase* that on M, as in the series: $M(NO_3)_2$, $M(NO_3)_3$, and $M(NO_3)_4$. In calcite and aragonite, two polymorphs of $CaCO_3$, Ca^{2+} is 6- and 9-coordinated respectively. If there were octahedral coordination of this ion in $Ca(NO_3)_2$ the crystal would consist of CaO_6 groups linked through N atoms of NO_3^- ions, each O of which would be bonded to only one Ca^{2+} ion, as at (a). Structures of this kind, for example, a rutile-like

(a)   (b)

structure in which Ti is replaced by Ca and O by NO_3, would presumably have low density and stability. In fact, anhydrous $Ca(NO_3)_2$ adopts essentially the same structure as the corresponding Sr, Ba, and Pb salts, which is really a structure more suited to the larger 12-coordinated M^{2+} ions. In $Pb(NO_3)_2$ Pb^{2+} has 12 equidistant O neighbours (at 2·81 Å), while according to an early study of $Ca(NO_3)_2$ Ca^{2+} has 6 O at 2·50 Å and 6 O at 2·93 Å. By increasing the c.n. of M^{2+} to 12 we arrive at the same arrangement, as regards charge balance, as in $NaNO_3$, as shown at (b).

Now consider the extreme case of a compound such as $Ti(NO_3)_4$ containing a small ion Ti^{4+} which is normally coordinated octahedrally by 6 O. The Ti–O bond strength would be $\frac{2}{3}$, twice the charge on O of NO_3^-, so that it is impossible to achieve a charge balance. In order to reduce the e.b.s. of the Ti–O bond to $\frac{1}{3}$ the c.n. of Ti^{4+} would have to be increased to the impossibly high value 12. Clearly this difficulty is most pronounced when

the charge on the anion is small,
the charge on the cation is large, and
the c.n. of the cation is small.

To be more precise we may say that the charge on each O of the oxy-ion must be equal to or greater than the charge on the cation divided by its c.n. (the e.b.s. of the M–O bond). The values of this quantity (Table 7.16) show that this condition cannot be satisfied for a number of anhydrous salts containing ions XO_3^- and XO_4^- assuming that the ionic charge is distributed equally over the O atoms, that is, $-\frac{1}{3}$ or $-\frac{1}{4}$ on each respectively. It is probable that the only normal ionic nitrates that can be crystallized from aqueous solution in the anhydrous form are those of the alkali metals, the alkaline-earths, and Pb. (The structures of the anhydrous nitrates of Be, Mg, Zn, Cd, and Hg do not appear to be known.) A number of anhydrous chlorates, bromates, iodates, and perchlorates are known (for example, $Ca(ClO_3)_2$, $Ba(BrO_3)_2$, $Ca(IO_3)_2$, $Ba(IO_3)_2$, and $M(ClO_4)_2$) (M = Cd, Ca, Sr, Ba). These are likely to be normal ionic salts, but again there is little information about their

TABLE 7.16

Electrostatic bond strengths of bonds M–O

C.N. of M	M^+	M^{2+}	M^{3+}	M^{4+}
4	$\frac{1}{4}$	$\frac{1}{2}$		1
6	$\frac{1}{6}$	$\frac{1}{3}$	$\frac{1}{2}$	$\frac{2}{3}$
8	$\frac{1}{8}$	$\frac{1}{4}$		$\frac{1}{2}$
9			$\frac{1}{3}$	
12	$\frac{1}{12}$	$\frac{1}{6}$	$\frac{1}{4}$	$\frac{1}{3}$

Limit for normal ionic $M(XO_4)_x$ ⟶
$M(XO_3)_x$ ⟶

crystal structures. In the case of the corresponding salts containing ions M^{3+} or M^{4+} very few, if any, can be crystallized anhydrous from aqueous solution. Most of these compounds are highly hydrated, for example, $Al(ClO_4)_3$. 6, 9, and 15 H_2O, $Ga(ClO_4)_3$. 6 and 9 H_2O, and $Cr(NO_3)_3$. 9 H_2O, though in a few instances the anhydrous salts have been described ($Al(ClO_4)_3$, $In(IO_3)_3$). Anhydrous salts $M(XO_3)_4$ crystallizable from (acid) aqueous solution include $Ce(IO_3)_4$ and $Zr(IO_3)_4$, to which we refer shortly.

Clearly this charge-balance difficulty is overcome by the formation of hydrates when the salt is crystallized from aqueous solution. In a hydrate in which the cation is surrounded by a complete shell of water molecules the charge on the central ion M^{n+} is spread over the surface of the hydrated ion, and the aquo-complexes can be hydrogen-bonded to the oxy-ions. The structural problem is now entirely different, since there are no bonds between cations and O atoms of oxy-ions, and it is now a question of packing large $[M(H_2O)_x]^{n+}$ groups and anions as in salts such as $Nd(BrO_3)_3$. 9 H_2O and $Th(NO_3)_4$. 12 H_2O. There is, however, an alternative structure for an anhydrous compound $M(NO_3)_3$, $M(NO_3)_4$, or the corresponding perchlorates, etc., which calls for the rearrangement of the charge distribution on the XO_3^- or XO_4^- ion from

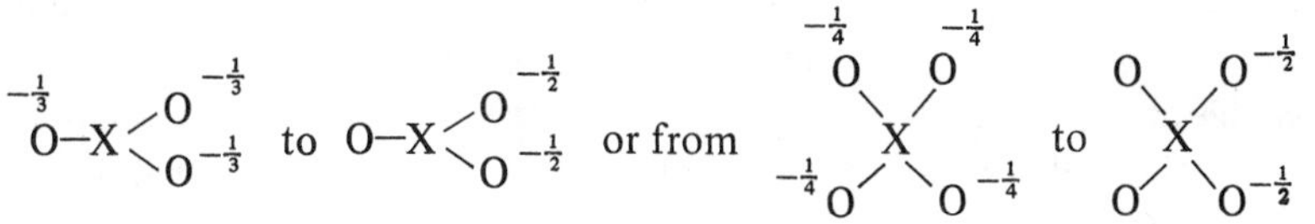

$$O-X{\overset{\overset{-\frac{1}{3}}{O}}{\underset{\underset{-\frac{1}{3}}{O}}{}}}^{-\frac{1}{3}} \quad \text{to} \quad O-X{\overset{\overset{-\frac{1}{2}}{O}}{\underset{\underset{-\frac{1}{2}}{O}}{}}}^{} \quad \text{or from} \quad {\overset{\overset{-\frac{1}{4}\ \ \ -\frac{1}{4}}{O\quad O}}{\underset{\underset{-\frac{1}{4}O\quad O-\frac{1}{4}}{}}{X}}} \quad \text{to} \quad {\overset{\overset{O\quad O^{-\frac{1}{2}}}{}}{\underset{\underset{O\quad O^{-\frac{1}{2}}}{}}{X}}}$$

Only two of the O atoms of, for example, a NO_3^- ion are then coordinated to cations, and the ion can function either as a bidentate ligand, (a), or as a bridging ligand, in which case there are the three possibilities (b), (c), and (d), corresponding to the linking of vertices, edges, or faces of different cation coordination groups by anions.

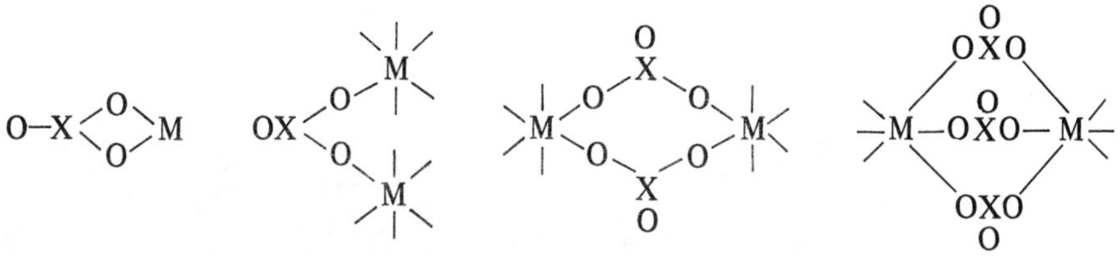

(a)	(b)	(c)	(d)

The use of new nitrating agents has led to the preparation of many anhydrous nitrates (nitrato compounds) which cannot be obtained from aqueous solution or by dehydrating hydrated salts. They include:

$M(NO_3)_2$: M = Be, Mg, Zn, Cd, Hg, Mn, Co, Ni, Cu, Pd
$M(NO_3)_3$: M = Al, Sc, Y, La, Cr, Fe, Au, In, Bi
$M(NO_3)_4$: M = Ti, Zr, Th, Sn.

Many of these compounds can be sublimed (under low pressure or *in vacuo*) and are clearly molecular, having bidentate NO_3 groups, as has been established for the vapour of $Cu(NO_3)_2$ and crystalline $Ti(NO_3)_4$ and $Sn(NO_3)_4$. Bidentate NO_3 groups are also present in the ion $[Co(NO_3)_4]^{2-}$, which has a structure very similar to that of the $Ti(NO_3)_4$ molecule, and also in $[Th(NO_3)_6]^{2-}$, $[Ce(NO_3)_6]^{2-}$, and $[Ce(NO_3)_6]^{3-}$. These compounds are discussed in Chapter 18.

This behaviour as a bidentate or bridging ligand should not be peculiar to NO_3^-; it is to be expected for ClO_3^-, BrO_3^-, IO_3^-, ClO_4^-, IO_4^-, and also SO_4^{2-}. The iodate ion exhibits all of the behaviours (b), (c), and (d) noted earlier. In all the crystalline salts $Ce(IO_3)_4$ (with which $Pu(IO_3)_4$ is isostructural), $Ce(IO_3)_4 . H_2O$, and $Zr(IO_3)_4$ only two of the O atoms of each IO_3^- are used to coordinate the metal ions, and in all three compounds IO_3^- behaves as a bridging ligand. In anhydrous $Ce(IO_3)_4$ columns of CeO_8 coordination groups, intermediate in shape between cubic and square antiprismatic, are formed by bridging IO_3 groups as shown diagrammatically in Fig. 7.6(a). In $Ce(IO_3)_4 . H_2O$ each of the eight O atoms around Ce^{4+} belongs to a different IO_3^- ion, and the eight IO_3^- ions connect a particular Ce^{4+} to eight others. (The H_2O molecules are not involved in the coordination groups of the metal ions, but are situated in tunnels entirely surrounded by O atoms of iodate ions.) In $Zr(IO_3)_4$ also there is antiprismatic coordination of the metal, but here Zr^{4+} is surrounded by 8 IO_3^- ions which bridge in pairs as shown diagrammatically in Fig. 7.6(b). The pairs of O atoms involved in each bridge correspond to alternate slanting edges of the antiprism coordination group, and the structure consists of layers of Zr^{4+} ions linked together in this way by IO_3^- ions.

We have seen that ions XO_3^- cannot balance the e.b.s. of an 8-coordinated ion M^{4+} without rearranging the anionic charge and behaving as either bidentate or bridging ligands. For an ion such as SO_4^{2-} the possible types of behaviour are more numerous and are rather completely illustrated by the structures of $Zr(SO_4)_2$ and its hydrates. This compound has an unexpectedly complex structural chemistry, for

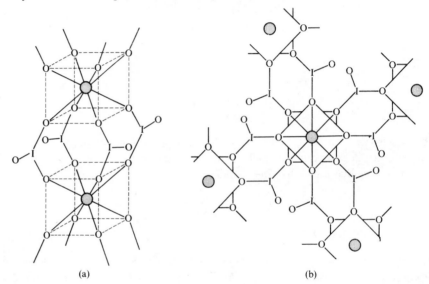

(a) (b)

FIG. 7.6. Behaviour of the IO_3^- ligand in (a) $Ce(IO_3)_4$, (b) $Zr(IO_3)_4$ (diagrammatic).

in addition to three forms of the anhydrous salt there are numerous hydrates, and these are by no means a normal series of hydrates. It is not possible to prepare the higher hydrates in succession by simply increasing the water-vapour pressure over the anhydrous salt; for example, the 5- and 7-hydrates have lower saturated solution vapour pressures than the 4-hydrate and therefore cannot be made by vapour-phase hydration of the latter. The relations between some of the phases we shall describe are set out in Table 7.17.

The normal charge distribution of the sulphate ion, (a), could be rearranged to (b) or (c):

(a) (b) (c)

For a salt $M(SO_4)_2$ charge balance can be achieved with the normal structure (a), since the e.b.s. of M^{4+}–O for 8-coordination is $\frac{1}{2}$. However, if M^{4+} is coordinated by a smaller number of O atoms of sulphate ions, either because it has a smaller c.n. (for example, 7) or because its coordination group is partly composed of H_2O or

TABLE 7.17

Relations between the anhydrous and hydrated sulphates of zirconium

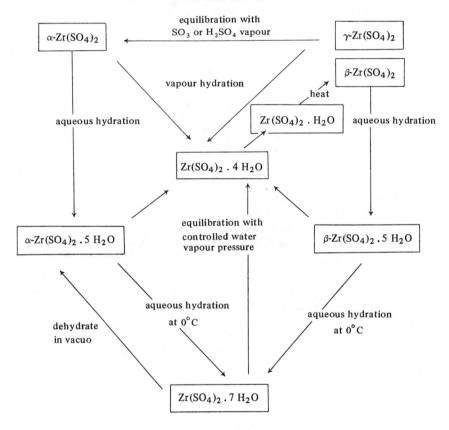

For further details see JSSC 1970 **1** 497.

other neutral molecules which do not neutralize any of the cationic charge, then rearrangement to (b) or (c) is necessary:

$$\text{c.n. of } M^{4+} \text{ by O of } SO_4^{2-}: \quad 8 \quad 6 \quad 4$$
$$\text{form of anion:} \quad\quad\quad (a) \quad (b) \quad (c)$$

An intermediate c.n. such as 7 would require a mixture of (a) and (b).

The simpler structural possibilities implied by (a)–(c) are

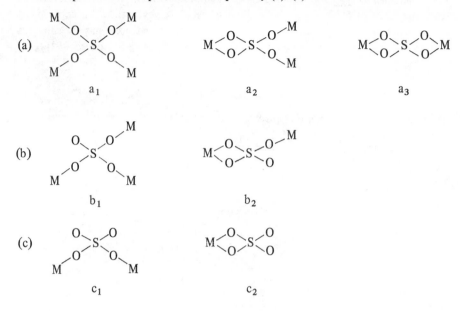

In α-Zr(SO$_4$)$_2$ the coordination group around Zr^{4+} is necessarily composed entirely of O atoms of sulphate ions. The metal is 7-coordinated, and equal numbers of SO$_4$ groups are of types a_1 and b_1, that is, bonded to 4 or 3 M^{4+} ions (Fig. 7.7). The behaviour of the SO$_4^{2-}$ ion in this salt is thus very similar to that of O^{2-} in the 7 : $\frac{3}{4}$ coordinated structure of monoclinic ZrO$_2$.

The monohydrate provides examples of (b). In both polymorphs the cation is 7-coordinated, by 6 O + 1 H$_2$O. In the γ structure all SO$_4$ groups are of type b_1; each bridges 3 cations forming a very simple layer which was illustrated in Fig. 6.15 (p. 211) to show its relation to the CdI$_2$ layer. The α form also has a layer structure, but here there are SO$_4$ groups of both types b_1 and b_2. With increasing hydration transition to (c) occurs, and in fact the coordination group of Zr^{4+} in the 4-hydrate, both forms of the 5-hydrate, and the 7-hydrate is in all cases the same, namely, 4 O + 4 H$_2$O. The tetrahydrate has a very simple layer structure based on the 4-gon net, all SO$_4$ groups being of type c_1, as shown diagrammatically in

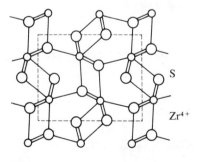

FIG. 7.7. The structure of α-Zr(SO$_4$)$_2$. The O atoms (one to each Zr–S line) are omitted.

284

Fig. 7.8. Both forms of the pentahydrate and the heptahydrate are built of bridged molecules

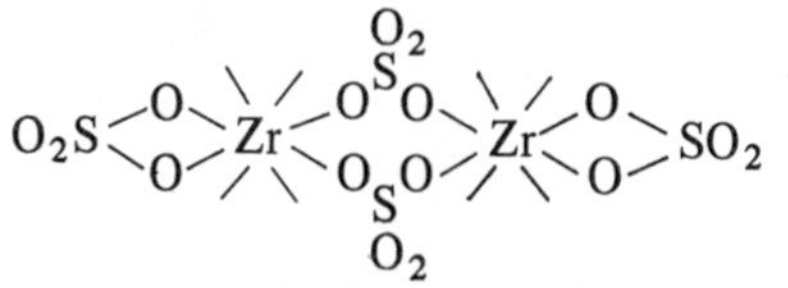

$$O_2S\diagdown{}_O^O\diagdown Zr\diagdown{}^O_O\diagdown{}^{O_2\atop S}{}^O_O\diagdown Zr\diagdown{}^O_O\diagdown SO_2$$

in which the SO_4 groups are of the two types c_1 and c_2. Dodecahedral 8-coordination groups around Zr^{4+} are completed by $4\,H_2O$ and the remaining water molecules are not associated with the metal ions. The structural formulae are therefore $[Zr_2(SO_4)_4(H_2O)_8]\,.\,2\,H_2O$ and $[Zr_2(SO_4)_4(H_2O)_8]\,.\,6\,H_2O$. (For further details see: AC 1959 **12** 719; AC 1969 **B25** 1558, 1566, 1572; AC 1970 **B26** 1125, 1131, 1140.)

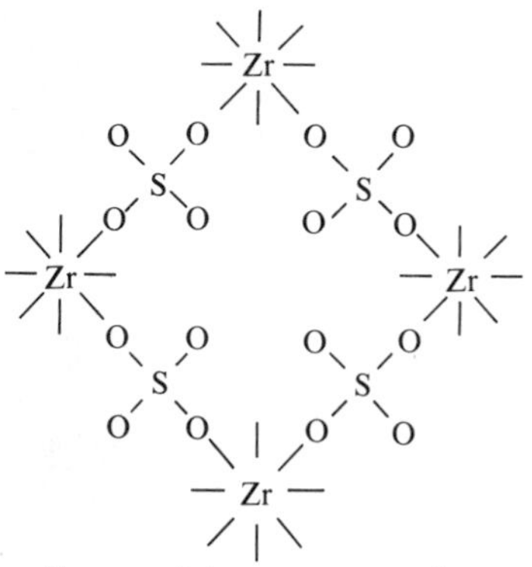

FIG. 7.8. Linking of Zr^{4+} and SO_4^{2-} ions in a layer of $Zr(SO_4)_2$. $4H_2O$ (water molecules omitted).

Summarizing, we see that problems of charge balance lead to geometrical difficulties in two distinct classes of salt:

	Charge on O of anion	Cation size	Cation charge
(a)	High	large	low
(b)	Low	small	high

We have discussed (a) as it applies to the extreme cases Cs_3BO_3, Cs_4SiO_4, and Cs_6TeO_6, but the problem exists, though in a less acute form, in pyro-salts such as $M_6Si_2O_7$ and $M_4P_2O_7$. As in the case of K_3PO_4 and KH_2PO_4 it can be overcome by the formation of salts such as $Na_2H_2P_2O_7$. At the other extreme, case (b), the difficulties arise not only with ions XO_3^- and XO_4^-, as already discussed, but also with ions such as $S_2O_7^{2-}$ if combined with cations carrying high charges. Thus the normal ionic disulphate and dichromate of Ti^{4+} or even the salts of the larger Th^{4+} ion would not be possible with the distribution of the anionic charge over all (or even six) of the O atoms. For a family of pyro-ions we encounter difficulties at both ends of the series,

$$Cs_6(Si_2O_7) \quad -----\quad Th(S_2O_7)_2$$
case (a) \qquad\qquad\qquad\qquad case (b)

just as we do for the simpler XO_3^{n-} ions

$$Cs_3BO_3 \quad -----\quad Ti(NO_3)_4.$$

285

Part II

THE PERIODIC TABLE OF THE ELEMENTS

SHORT PERIODS — Groups I, II
LONG PERIODS — A SUB-GROUPS · GROUP VIII TRIADS · B SUB-GROUPS

	I A	II A	III A	IV A	V A	VI A	VII A	VIII			I B	II B	III B	IV B	V B	VI B	VII B	(0)
	H 1																	He 2
	Li 3	Be 4											B 5	C 6	N 7	O 8	F 9	Ne 10
	Na 11	Mg 12											Al 13	Si 14	P 15	S 16	Cl 17	Ar 18
	K 19	Ca 20	Sc 21	Ti 22	V 23	Cr 24	Mn 25	Fe 26	Co 27	Ni 28	Cu 29	Zn 30	Ga 31	Ge 32	As 33	Se 34	Br 35	Kr 36
	Rb 37	Sr 38	Y 39	Zr 40	Nb 41	Mo 42	Tc 43	Ru 44	Rh 45	Pd 46	Ag 47	Cd 48	In 49	Sn 50	Sb 51	Te 52	I 53	Xe 54
	Cs 55	Ba 56	La† 57	Hf 72	Ta 73	W 74	Re 75	Os 76	Ir 77	Pt 78	Au 79	Hg 80	Tl 81	Pb 82	Bi 83	Po 84	At 85	Rn 86
	Fr 87	Ra 88	Ac‡ 89															

† And the 4f series (lanthanides)

Ce 58	Pr 59	Nd 60	Pm 61	Sm 62	Eu 63	Gd 64	Tb 65	Dy 66	Ho 67	Er 68	Tm 69	Yb 70	Lu 71

‡ And the 5f series (actinides)

Th 90	Pa 91	U 92	Np 93	Pu 94	Am 95	Cm 96	Bk 97	Cf 98	Es 99	Fm 100	Md 101	No 102	Lw 103

8

Hydrogen: The Noble Gases

Introductory

The hydrogen atom has the simplest electronic structure of all the elements, having only one valence electron and only one orbital available for bond formation. Nevertheless hydrogen combines, in one way or another, with most of the elements, other than the noble gases, and plays an extremely important role in chemistry. The three simplest ways in which H can function are the following:

(a) By loss of the electron the H^+ ion (proton) is formed. An acid may be defined as a source of protons:

$$A \rightleftharpoons H^+ + B$$
$$\text{acid} \qquad \text{base}$$

According to this definition NH_4^+ is regarded as an acid ($NH_4^+ \rightleftharpoons NH_3 + H^+$) and water as both acid and base, since

$$H_2O \rightleftharpoons H^+ + OH^-, \quad \text{and} \quad H^+ + H_2O \rightleftharpoons H_3O^+$$
$$\text{acid} \qquad \text{base} \qquad\qquad \text{base} \qquad \text{acid}$$

assuming the existence of hydrated rather than simple protons in aqueous solution.

(b) By acquiring a second electron the H^- ion is formed. This ion cannot exist in aqueous solution because it immediately combines with H^+, but it is found in a number of crystalline hydrides.

(c) Hydrogen can form one normal covalent (electron-pair) bond. This occurs in the molecular hydrides of the non-metals and metalloids and in ions such as OH^- and NH_4^+, in organic compounds, and in a limited number of hydrido compounds of transition metals.

In addition there are many structures, both molecular and crystalline, in which a H atom is bonded in some way to two (or more) atoms.

(d) If H is bonded to a very electronegative atom the dipole A^--H^+ results in an attraction to a second electronegative atom, which may be in the same or a different molecule,

$$\diagup\!\!\!\!\!\diagdown A-H\cdots A\!\!\diagdown\!\!\!\!\!\diagup$$

forming a hydrogen bond (or bridge). This type of bond is of great importance in the structural chemistry of many groups of compounds, notably acids and acid salts, hydroxy-compounds, water, and hydrates. Since a comprehensive treatment of the hydrogen bond would evidently cover a considerable part of structural chemistry we shall discuss in this chapter only acids and acid salts.

(e) In certain electron-deficient compounds (of Li, Be, B, and Al) H atoms form bridges between pairs of atoms as, for example, in B_2H_6, and metal–H–metal bridges (both linear and non-linear) occur in certain carbonyl hydride ions such as $[(CO)_5CrHCr(CO)_5]^-$. Boron hydrides are described in Chapter 24 and carbonyl hydrides in Chapter 22; see also the discussion of bonds in Chapter 7.

In many crystalline hydrides H is bonded to larger numbers of metal atoms. This is readily understandable for the ionic hydrides (for example, LiH) where H^- has an environment similar to that of F^- in LiF, but the nature of the bonding is less clear in some transition-metal hydrides, in particular the non-stoichiometric interstitial hydrides.

We shall discuss in the present chapter the following aspects of the structural chemistry of hydrogen:

Hydrides and hydrido complexes of transition metals,
The hydrogen bond, and
Acids and acid (hydrogen) salts.

Hydrides

Binary hydrides are known of most elements other than the noble gases, the notable exceptions being Sc, Y, and Mn, and the following members of the second and third long series:

Mo	Tc	Ru	Rh		
W	Re	Os	Ir	Pt	(Au)

These compounds are of three main types:
(i) The molecular hydrides,
(ii) The salt-like hydrides of the more electropositive elements,
(iii) The interstitial hydrides of the transition metals.

(i) *Molecular hydrides*

Hydrides MH_{8-N} are formed by the non-metals of the two short Periods (other than B which forms the exceptional hydrides described in Chapter 24) and by the B subgroup elements:

MH_4	MH_3	MH_2	MH
Tetrahedral	Pyramidal	Angular	
C	N	O	F
Si	P	S	Cl
Ge	As	Se	Br
Sn	Sb	Te	I
Pb	Bi	Po	

Structural studies have been made of all except PbH_4, BiH_3, and PoH_2, and details are given in other chapters; the stability of the heavy metal compounds is very low.

291

These hydrides exist in the same molecular form in all states of aggregation, and except for those of the most electronegative elements there are only weak van der Waals forces acting between the molecules in the crystals. We refer to crystalline NH_3, OH_2, and FH in our discussion of the hydrogen bond. Many short-lived hydride species are known to the spectroscopist, and the structures of some radicals have recently been studied in matrices at low temperatures (for example, CH_3 (planar), SiH_3 and GeH_3 (pyramidal)). Many of the non-metals form more complex hydrides M_xH_y in addition to the simple molecules noted above; the more important of these are included in later chapters under the chemistry of the appropriate non-metal.

(ii) Salt-like hydrides

The compounds to the left of the line in the following table are salt-like and contain H^- ions:

(1) ZPC 1935 **28B** 478
(2) PR 1948 73 842.
(3) AC 1963 **16** 352
(4) AC 1962 **15** 92
(5) IC 1969 8 18
(6) AC 1956 9 452

LiH[1]		AlH$_3$[5]	
NaH[2]	MgH$_2$[3]		
KH	CaH$_2$[4]		
RbH	SrH$_2$	EuH$_2$	YbH$_2$[6]
CsH	BaH$_2$		

They are colourless compounds which can be made by heating the metal in hydrogen (under moderate pressure in the case of MgH_2). They are more dense than the parent metals, the difference being greatest for the alkali metals (25–45 per cent) and less for the alkaline-earths (5–10 per cent). Solid LiH is as good an ionic conductor as LiCl (and 10^3 times better than LiF), and electrolysis of molten LiH, at a temperature just above its melting point with steel electrodes, yields Li at the cathode, hydrogen being evolved at the anode. These ionic hydrides have much higher melting points than the molecular hydrides of the previous section, for example, LiH, 691°C, NaH, 700–800°C with decomposition; the remaining alkali hydrides dissociate before melting. They all react readily with water, evolving hydrogen and forming a solution of the hydroxide. There are considerable structural resemblances between these compounds (in which the effective radius of H^- ranges from 1·3–1·5 Å) and fluorides (radius of F^-, 1·35 Å).

The alkali-metal hydrides have the NaCl structure, though the positions of the H atoms have been confirmed only in LiH (X-ray diffraction) and NaH (neutron diffraction); (Li–H, 2·04 Å, Na–H, 2·44 Å). The rutile structure of MgH_2 has been established by neutron diffraction of MgD_2; Mg–H, 1·95 Å, shortest H–H, 2·49 Å.

The alkaline-earth hydrides all have a $PbCl_2$ type of structure in which the metal atoms are arranged approximately in hexagonal closest packing. Of the two sets of non-equivalent H atoms one occupies tetrahedral holes while the other H atoms have (3 + 2)-coordination:

$$\text{Ca:} \quad \begin{matrix} 7 \text{ H } 2\cdot32 \text{ Å} \\ 2 \text{ H } 2\cdot85 \text{ Å'} \end{matrix} \qquad \text{H}_\text{I}: \quad 4 \text{ Ca } 2\cdot32 \text{ Å}, \qquad \text{H}_\text{II}: \quad \begin{matrix} 3 \text{ Ca } 2\cdot32 \text{ Å} \\ 2 \text{ Ca } 2\cdot85 \text{ Å} \end{matrix}$$

Note the tight packing of the H atoms; H_I has 8 H neighbours at 2·50–2·94 Å, and H_{II} has 10 H neighbours at 2·65–3·21 Å. The shortest metal–metal distances in these hydrides are *less* than in the metals (for example, Ca–Ca, 3·60 Å in CaH_2 as compared with 3·93 Å in the metal). The dihydrides of Eu and Yb, studied as the deuterides, have the same structure, and their formation from the metal is accompanied by a 13 per cent decrease in volume.

In the preparation of AlH_3 by the action of excess $AlCl_3$ on LiH in ether it has not proved possible to obtain pure AlH_3 (free from ether), but this hydride has been made by bombarding ultra-pure Al with H^+ ions. Dow Chemical Company (by undisclosed methods) have apparently prepared the unsolvated compound in a number of crystalline forms, one of which has essentially the same structure as AlF_3. The metal atoms occupy one-third of the octahedral holes in an approximately hexagonal closest packing of H atoms. The structure suggests that this is an ionic hydride. The shortest distance between Al atoms is 3·24 Å, and the distances Al–6 H, 1·72 Å, and H–H, 2·42 Å, are very similar to the corresponding distances in AlF_3, namely, Al–6 F, 1·79 Å, and F–F, 2·53 Å. On the other hand there are amine derivatives of AlH_3 which presumably contain H covalently bonded to Al, though the H atoms have not been located. Crystalline $AlH_3 . 2 N(CH_3)_3$[7] consists of molecules in which the atoms N–Al–N are collinear (Al–N, 2·18 Å) with 3 H most probably completing a trigonal bipyramidal coordination group around Al. This would also appear to be true in $AlH_3[(CH_3)_2NCH_2CH_2N(CH_3)_2]$, the diamine molecules linking the AlH_3 groups into infinite chains.[8]

By the action of $LiAlH_4$ on the halide (or in the case of Be on the dialkyls) a number of other hydrides have been prepared, including the dihydrides of Be, Zn, Cd, and Hg, trihydrides of Ga, In, and Tl, and CuH. Some of these are extremely unstable, those of Cd and Hg decomposing at temperatures below 0°C. Pure crystalline BeH_2 has not been prepared, but a crystalline amine complex has been made for which the bridged structure (a) has been suggested;[9] Be–H bridges have

(7) IC 1963 **2** 508

(8) AC 1964 **17** 1573

(9) IC 1969 **8** 976

(a)

(b)

also been postulated in the salt $Na_2[R_4Be_2H_2]$.[10] A partial description has been given of the structure of the etherate $[NaO(C_2H_5)_2]_2[(C_2H_5)_4Be_2H_2]$,[11] in which there is a linear system of H-bridged Na and Be atoms, (b). There is still some doubt about the existence of GaH_3, but in crystalline $GaH_3 . N(CH_3)_3$ the bond length Ga–N has been determined as approximately 2·0 Å in a presumably tetrahedral molecule.[12]

The red-brown CuH is amorphous when prepared from CuI and $LiAlH_4$ in organic solvents but it has been obtained as a water soluble material with the wurtzite structure by reduction of Cu^{2+} by aqueous hypophosphorous acid. (Cu–H, 1·73 Å, Cu–Cu, 2·89 Å, compare 2·56 Å in the metal).[13] The colour of

(10) JCS 1965 692
(11) CC 1965 240

(12) IC 1963 **2** 1298

(13) AC 1955 **8** 118

(14) JACS 1968 **90** 5769

this compound, which forms red solutions in organic solvents, is difficult to understand;[14] the n.d. study showed contamination with metallic Cu and Cu_2O.

Although we have included these B subgroup hydrides with the salt-like compounds it is possible that the bonding is at least partially covalent in some or all of these compounds; an obvious suggestion for BeH_2 is a hydrogen-bridged structure like that of $Be(CH_3)_2$.

We may mention here a B subgroup hydride which does not fall into any of our classes (i)–(iii), namely, $PbH_{0.19}$. This is formed by the action of atomic hydrogen at $0°C$ on an evaporated lead film, and apparently possesses considerable stability.[15]

(15) PCS 1964 173

(iii) *Transition-metal hydrides*

The known compounds are listed in Table 8.1, which shows that a considerable number of these metals have not yet been shown to form binary hydrides. The criterion is that the formation of a hydrogen-containing phase should be accompanied by a definite structural change, since many of these metals when finely divided adsorb large volumes of hydrogen; it is clearly difficult to distinguish such systems from non-stoichiometric hydrides by purely chemical means. Characteristic properties of transition-metal hydrides include metallic appearance, metallic conductivity or semi-conductivity, variable composition in many cases, and interatomic distances appreciably larger than in the parent metals. The expansion which accompanies their formation is in marked contrast to the contraction in the case of the salt-like hydrides (including EuH_2 and YbH_2). Incidentally it is interesting that the number of H atoms per cubic centimetre in a number of metal hydrides is greater than in solid H_2 or in water. This fact, combined with the high thermal stability of some of these compounds, makes them of interest as neutron-shielding materials for nuclear reactors.

Although these compounds were originally described as 'interstitial' hydrides, implying that they were formed by entry of H atoms into interstices (usually tetrahedral) in the metal structure, it is now known that the arrangement of metal atoms in the hydride is *usually* different from that in the parent metal. The hydride has a definite structure and in this respect is not different from other compounds of the metal:

$$Cr \text{ (b.c.c.)} \rightarrow CrH \text{ (h.c.p.)} \rightarrow CrH_2 \text{ (c.c.p.)}$$
$$Ti, Zr, \text{ and } Hf \text{ (h.c.p.)} \rightarrow MH_2 \text{ (c.c.p.)}.$$

The hydrides of V, Nb, and Ta probably come closest to the idea of an interstitial compound and are discussed later. In other cases where the arrangement of metal atoms in the hydride is the same as in (one form of) the metal there may be a discontinuous increase in lattice parameter when the hydride is formed (Pd) or there may be an intermediate hydride with a different metal arrangement. For example, the h.c.p. 4f metals Gd–Tm form hexagonal trihydrides, but the intermediate dihydrides have the fluorite structure with c.c.p metal atoms.

Moreover, the volume of the MH_3 phase is some 15–25 per cent greater than that of the metal. The hydride $YbH_{2.55}$ has, like Yb, a f.c.c arrangement of metal atoms but this phase is only made under pressure and has a *smaller* cell dimension

TABLE 8.1

Binary hydrides of transition metals

	TiH	$VH^{(2)}$	$CrH^{(3)}$	–	–	–	$NiH_{0.6}{}^{(4)}$
ScH_2	$TiH_2{}^{(1)}$	VH_2	CrH_2				
		$NbH^{(14)}$	–	–	–	–	$PdH^{(5)}$
YH_2	ZrH_2	NbH_2					
YH_3							
		TaH	–	–	–	–	–
LaH_2	$HfH_2{}^{(1)}$						
LaH_3							

4f *metals*

La Ce Pr Nd   Sm . Gd—Tm . Lu (Y)
   $MH_{1.9}$–MH_3 MH_2 fluorite structure
      f.c.c.$^{(6)}$ MH_3 hexagonal LaF_3 structure$^{(7)}$
                   (For EuH_2, YbH_2, and $YbH_{2.55}{}^{(8)}$ see text)

5f *metals*

	Ac	Th	Pa	U	$Np^{(13)}$	Pu	Am
MH_2	f.c.c.	f.c.t.$^{(9)}$			←——— f.c.c. ———→		
MH_3			β-UH_3	α-$UH_3{}^{(11)}$	←———— LaF_3 ———→		
				β-$UH_3{}^{(12)}$			
		$Th_4H_{15}{}^{(10)}$					

(1) AC 1956 9 607
(2) IC 1970 9 1678
(3) PSS 1963 3 K249
(4) JPP 1964 25 460
(5) JPCS 1963 24 1141
(6) JPC 1955 59 1226
(7) JPP 1964 25 454
(8) IC 1966 5 1736
(9) AC 1962 15 287
(10) AC 1953 6 393
(11) JACS 1954 76 297
(12) JACS 1951 73 4172
(13) JPC 1965 69 1641
(14) SPC 1970 14 522

(5·19 Å) than the metal (5·49 Å) and there is an intermediate hydride YbH_2 with the quite different (CaH_2) structure already noted.

The 4f *and* 5f *hydrides.*   All the 4f metals take up hydrogen at ordinary or slightly elevated temperatures to form hydrides of approximate composition $MH_{1.9}$ which, with the exception of EuH_2 and YbH_2 (CaH_2 structure), all have fluorite-type structures. In contrast to EuH_2 and YbH_2 the formation of these cubic 'dihydrides' is accompanied by expansion of the structure. Europium forms only EuH_2 and Yb forms only YbH_2 at atmospheric pressure; under a higher hydrogen pressure it forms $YbH_{2.55}$ (probably an ionic compound containing Yb^{2+} and Yb^{3+}), and a metastable cubic YbH_2 has been made by heating $YbH_{2.55}$ or YbH_2. The dihydrides of the other 4f elements react further with hydrogen at atmospheric pressure to form 'trihydrides', but the lanthanides now fall into two groups. The lighter (larger) 4f metals form continuous f.c.c. solid solutions from the approximate composition $MH_{1.9}$ to a composition approaching MH_3. Having occupied the tetrahedral holes (fluorite structure) the H atoms then occupy octahedral holes at random, as has been shown by n.d. for CeH_2 and $CeH_{2.7}$; the H atoms are slightly displaced from the ideal positions. Onwards from Sm (and Y) a new h.c.p. phase appears before the composition MH_3 is reached, and as in the dihydrides there is a range of composition over which the phase is stable. There is a gap between the

Hydrogen: The Noble Gases

stability ranges of the dihydride and trihydride which is small for Sm but larger for the other metals. Some typical figures for the H : M ratios at room temperature are:

	Composition limits (H : M) *at room temperature*	
	Cubic dihydride	*Hexagonal trihydride*
Sm	1·93–2·55	2·59–3·0
Ho	1·95–2·24	2·64–3·0
Er	1·95–2·31	2·82–3·0

The hexagonal MH_3 phase has the (revised) LaF_3 structure (p. 356) which may be regarded as derived from an expanded h.c.p. metal structure. A n.d. study of HoD_3 (with which the other phases are isostructural) shows that H (D) occupies all the tetrahedral and octahedral holes, but because of the close proximity of the pairs of tetrahedral holes there is some displacement of the H atoms from the ideal positions and this in turn necessitates some displacement of the H atoms from the octahedral holes. In the resulting structure H atoms have 3 nearest metal atom neighbours and the metal atom has 9 (+2) H neighbours:

$$Ho: \quad \begin{matrix} 9\,H & 2·10\text{–}2·29\ \text{Å} \\ (2\,H & 2·48\ \text{Å}) \end{matrix} \qquad H: \quad 3\,Ho \quad \begin{matrix} \text{at} & 2·10\ \text{Å} \\ \text{or} & 2·17\ \text{Å} \\ \text{or} & 2·24\text{–}2·29\ \text{Å} \end{matrix}$$

The typical 4f hydrides are pyrophoric and graphitic or metallic in appearance. The resistivity of the dihydrides, which may be formulated $M^{3+}(H^-)_2$ (e), is lower than that of the pure metal, but increases as more hydrogen is absorbed. For example, at 80°K there is a 10^6-fold increase in resistivity when $LaH_{1·98}$ is converted into $LaH_{2·92}$ and a 10^4-fold increase in the Ce hydrides. It seems likely that the bonding in the dihydrides (other than EuH_2 and YbH_2, which are non-conductors) is a combination of ionic and metallic bonding, and that the addition of further hydrogen leads to the formation of H^- ions by removal of electrons from the conduction band, producing an essentially ionic trihydride.

The 5f hydrides are summarized in Table 8.1. In contrast to the other elements Th forms a dihydride with a distorted (f.c. tetragonal) fluorite-like structure, resembling in this respect Ti, Zr, and Hf, but the structure becomes cubic if a little oxygen is present. Th also forms Th_4H_{15}, in which there are two types of non-equivalent H atom, and the shortest Th–Th distance (3·87 Å) is appreciably greater than in the metal (3·59 Å):

$$Th: \quad \begin{matrix} 9\,H & 2·29\ \text{Å} \\ 3\,H & 2·46\ \text{Å} \end{matrix} \qquad \begin{matrix} H_I: & 3\,Th & 2·29\ \text{Å} \\ H_{II}: & 4\,Th & 2·46\ \text{Å} \end{matrix}$$

compare Th–8 H at 2·41 Å in ThH_2.

Uranium forms only one hydride, UH_3, which is dimorphic and metallic in character. The metal atoms in β-UH_3 are in the positions of the β-W structure (p. 1017), a structure not adopted by metallic U. The U–U distances are much greater than those in α- or γ-uranium, even the shortest (U–2 U, 3·32 Å) indicating only very weak metal-metal bonds. (Compare U–8 U in γ-U, 2·97 Å, and the shortest

296

bonds in α-U, 2·76 Å.) The H atoms have been shown by neutron diffraction to occupy very large holes in which they are surrounded, approximately tetrahedrally, by four U at 2·32 Å. Since each U atom has twelve H neighbours and the compound is metallic rather than salt-like it has been suggested that the atoms are held together by some kind of delocalized covalent bonds. In α-UH_3 the metal atoms occupy the positions of the shaded circles in Fig. 29.4 (p. 1017) and the H atoms the open circles. Here again the U–U bonds are extremely weak (U–8 U, 3·59 Å), and H is surrounded tetrahedrally by 4 U at 2·32 Å as in the β form. The trihydrides of Np, Pu, and Am are isostructural with the hexagonal 4f trihydrides.

Hydrides of the 3d, 4d, *and* 5d *metals.* We now comment briefly on the remaining hydrides of Table 8.1. All the (h.c.p.) elements Ti, Zr, and Hf form dihydrides with fluorite-type structures which are cubic above their transition points and have lower (tetragonal) symmetry at ordinary temperatures. Both phases have ranges of composition, which depend on the temperature, and the following figures (for room temperature) show that for Ti the cubic phase includes the composition TiH:

Cubic phase	Tetragonal phase
TiH–TiH_2	
$ZrH_{1.50}$–$ZrH_{1.61}$	$ZrH_{1.73}$–$ZrH_{2.00}$
$HfH_{1.7}$–$HfH_{1.8}$	$HfH_{1.86}$–$HfH_{2.00}$

In addition to these phases there are solid solutions, with small hydrogen concentrations, in the h.c.p. metal, and also solid solutions in the high-temperature (b.c.c.) forms; in the latter the concentration of hydrogen may be considerable, for example, up to $ZrH_{1.5}$. The fluorite structure of TiD_2 has been confirmed by neutron diffraction. It should be noted that hydrides previously formulated Zr_2H and Zr_4H are not distinct compounds but correspond to arbitrarily selected points on the phase diagram. The fact that the arrangement of the metal atoms in the tetragonal Zr and Hf hydrides is less symmetrical (for the composition MH_2) than in the defect structures emphasizes the important part played by the metal–hydrogen bonds in these structures. The following data for the Hf hydrides show that the formation of these hydrides is by no means a question of H atoms (or ions) simply occupying interstices in a c.p. metal structure:

α-Hf (hexagonal)	$HfH_{1.7}$ (cubic)	HfH_2 (tetragonal)
Hf–6 Hf 3·13 Å	Hf–12 Hf 3·31 Å	Hf–8 Hf 3·27 Å
–6 Hf 3·20	Hf–H 2·03	–4 Hf 3·46
		Hf–8 H
		H–4 Hf 2·04

The metals V, Nb, and Ta have b.c.c. structures. The α solid solution of H in this structure extends to $VH_{0.05}$, $NbH_{0.1}$ and $TaH_{0.2}$. The next distinct phase is the β hydride, a non-stoichiometric phase which is a (tetragonal) distorted version of the b.c.c. solid solution and is stable over a wide range of composition (for example,

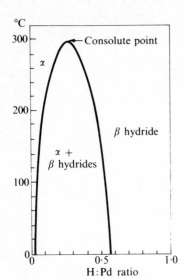

FIG. 8.1. The palladium–hydrogen
phase diagram.

$VH_{0.45}$–$VH_{0.9}$). The tetragonal b.c. β-NbH becomes cubic at about 200°C. A neutron diffraction study shows complete ordering of the H atoms in this phase when cooled below room temperature. In this region the phase diagram of these elements appears to be similar to that of Pd (Fig. 8.1), so that above a certain (consolute) temperature there is continuous absorption of hydrogen up to approximately the stage MH without radical rearrangement of the metal structure. This behaviour is presumably associated with the fact that in the b.c.c. structure there are six tetrahedral and three octahedral sites per atom as compared with two tetrahedral and one octahedral site per atom in a c.p. structure, and the coordination groups are not regular tetrahedra and octahedra. There are accordingly many possible types of disordered structure and superstructure. Special treatment (for example, high H_2 pressure) is required to make the dihydrides of V and Nb, which have the usual f.c.c. structure, and the maximum hydrogen content of a Ta hydride is reached in $TaH_{0.9}$.

Two hydrides of Cr have been made (electrolytically), CrH with an anti-NiAs type of structure (Cr–6 H, 1·91 Å, Cr–Cr, 2·71 Å) and CrH_2 with apparently the fluorite (f.c.c.) structure.

Palladium absorbs large volumes of hydrogen. The phase diagram (Fig. 8.1) shows that at room temperature there is a small solubility (α solid solution) in the metal, then a two-phase region followed by the β hydride ($PdH_{0.56}$). The cell dimensions of the two phases are: α, 3·890 Å ($PdH_{0.03}$), and β, 4·018 Å ($PdH_{0.56}$). The maximum hydrogen content of the β phase corresponds to the formula $PdH_{0.83}$, at −78°C. Above 300°C only one phase is found up to hydrogen pressures of 1000 atmospheres. Nickel hydride has been prepared only as a thin film on a Ni surface by electrolysis, with a maximum H : Ni ratio of approximately 0·6. Neutron diffraction studies of both Pd and Ni hydrides indicate a f.c.c. structure (defect NaCl structure).

Ternary hydrides

Since not very many of these compounds are at present well authenticated we group them together in Table 8.2; their structures suggest that they include both salt-like compounds and also compounds more akin to the transition-metal compounds of the last section. Methods of preparation include the action of hydrogen on a mixture of the metals (for example, $BaLiH_3$ from Ba + Li at 700°C)

TABLE 8.2
Ternary metal hydrides

$LiBeH_3$[1]	$EuLiH_3$	$LiAlH_4$[3]	$LiGaH_4$	$NiZrH_3$[5]
$NaBeH_3$	$SrLiH_3$	(also Na,	$LiInH_4$	$AlTh_2H_4$[6]
	$BaLiH_3$[2]	K, Cs)		
Li_2BeH_4		Li_3AlH_6		Mg_2NiH_4[7]
Na_2BeH_4		Na_3AlH_6[4]		Li_4RhH_4[8]
				Sr_2IrH_4[8]
		$Mg(AlH_4)_2$		
		$Ca(AlH_4)_2$		

(1) JCS A 1968 628
(2) JCP 1968 48 4660
(3) IC 1967 6 669
(4) IC 1966 5 1615
(5) JPP 1964 25 451
(6) AC 1961 14 223
(7) IC 1968 7 2254
(8) IC 1969 8 1010

or on an alloy (for example, Mg_2NiH_4 from Mg_2Ni and H_2 under pressure at $325°C$). The first product of the action of H_2 under pressure on a mixture of Na and Al in toluene at $165°C$ is Na_3AlH_6; further reaction gives $NaAlH_4$.

A n.d. study has confirmed the perovskite structure of $BaLiH_3$ (and presumably therefore the structures of the isostructural Sr and Eu compounds). The M—H distances are appreciably larger in this structure than in MH_2:

	Eu—H	Sr—H	Ba—H
12-coordination	2·68 Å	2·71 Å	2·84 Å
7-coordination	2·45	2·49	2·67

and the Li—H distance is rather larger (2·01 Å) in $BaLiH_3$ than in $SrLiH_3$ (1·92 Å) or $EuLiH_3$ (1·90 Å); compare 2·04 Å in LiH.

In $LiAlH_4$ there is nearly regular tetrahedral coordination of Al (Al—H, 1·55 Å) and rather irregular 5-coordination of Li (4 H at 1·88–2·00 Å, 1 H at 2·16 Å). The four H atoms of an AlH_4 group are of two kinds, three being 2-coordinated (to Al and Li) and the fourth 3-coordinated (to Al + 2 Li). The structure has been described as containing AlH_4^- ions (though this is not a necessary deduction from the structure); the salt $[(CH_3)_4N]AlH_4$ is much more stable to hydrolysis than $LiAlH_4$. The structure of Na_3AlH_6 is said to be of the cryolite type.

Hydrogen is readily absorbed by $AlTh_2$ ($CuAl_2$ structure, p. 1046) forming ultimately $AlTh_2H_4$, in which H atoms occupy all of the tetrahedral holes between 4 Th atoms in the alloy structure. The Th—H distance (approximately 2·4 Å) is similar to that in ThH_2. The structure of $NiZrD_3$ is more complex, for here some of the D atoms occupy tetrahedral holes and some are in positions of 5-coordination:

$$D \begin{cases} \text{Zr} & 1·96\ \text{Å} \\ 2\ \text{Zr} & 2·18 \\ \text{Ni} & 1·77 \end{cases} \quad \text{and} \quad D \begin{cases} \text{Zr} & 1·95\ \text{Å} \\ 2\ \text{Zr} & 2·38 \\ 2\ \text{Ni} & 1·78 \end{cases}$$

Hydrido complexes of transition metals

The d-type transition metals constitute the large block of elements lying between the electropositive metals which form ionic hydrides and the B subgroup and non-metallic elements which form covalent molecular hydrides. In addition to forming interstitial hydrides these transition metals also form covalent molecules MH_xL_y in which H atoms are directly bonded to the metal. Molecules of this general type are formally similar to substituted hydrides such as PHF_2, GeH_2Cl_2, etc. of non-metals and B subgroup elements, but a characteristic feature of the transition-metal compounds is that the ligands L must be of a particular kind, namely, those which cause electron-pairing in the d orbitals of the metal and are present in sufficient number to fill all the non-bonding d orbitals. They include CO, cyclopentadienyl, and tertiary phosphines and arsines; some examples are given in Table 8.3. The formation by Re and Tc of the remarkable ions $(MH_9)^{2-}$ shows that H may bond to certain transition metals in the absence of such ligands. The majority of the elements of Table 8.3 form carbonyl hydrides, but most of these

TABLE 8.3
Some hydrido compounds of transition metals

	$[Cr_2H(CO)_{10}]^-$	$MnH(CO)_5$	$FeH_2(PP)_2$	$CoH(PP)_2$	
	$CrH(C_5H_5)(CO)_3$	$Mn_2H(CO)_8(P\phi_2)^{(1)}$		$CoH(PF_3)_4{}^{(2)}$	
	$MoH_2(C_5H_5)_2{}^{(3)}$	$[TcH_9]^{2-(4)}$	$RuHCl(PP)_2$	$RhH(CO)(P\phi_3)_3{}^{(5)}$	
$TaH_3(C_5H_5)_2$	$WH_2(C_5H_5)_2$	$[ReH_9]^{2-(6)}$	$OsHBr(CO)(P\phi_3)_3{}^{(7)}$	$RhHCl_2P_3$	
	WH_6P_3	ReH_7P_2	OsH_4P_3	$IrHCl_2P_3$	$PtHBrP_2{}^{(8)}$
					$PtHClP_2{}^{(9)}$

(P stands for a tertiary phosphine, PR_3, and PP for $R_2P . CH_2 . CH_2 . PR_2$ or [cyclohexane ring with PR_2 substituents])

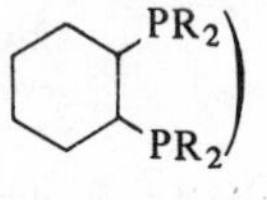

(1) JACS 1967 **89** 4323. (2) IC 1970 **9** 2403. (3) IC 1966 **5** 500. (4) IC 1964 **3** 567. (5) AC 1965 **18** 511. (6) IC 1964 **3** 558. (7) PCS 1962 333. (8) AC 1960 **13** 246. (9) IC 1965 **4** 773.

compounds are omitted since they are described in Chapter 22. We should, however, note here the following examples of molecules in which H acts as a bridge between two metal atoms:

$[(CO)_5 Cr-H-Cr(CO)_5]^-$ (linear bridge, M–H–M, 3·4 Å), (p. 771),
$HMn_3(CO)_{10}(BH_3)_2$ (non-linear bridge, Mn–H, 1·65 Å), (p. 872).
$(CO)_4 MnH(P\phi_2)Mn(CO)_4$ (non-linear bridge, Mn–H 1·87 Å).

In the remaining examples of this section H is bonded to one metal atom only, apparently as a normal covalently bound ligand, with M–H, 1·6–1·7 Å. (Earlier evidence for much shorter M–H bonds in compounds such as carbonyl hydrides and $(C_5H_5)_2MoH_2$ is now known to be unreliable.)

Molecules or ions in which the H atoms have been definitely located by n.d. or X-ray diffraction include the following:

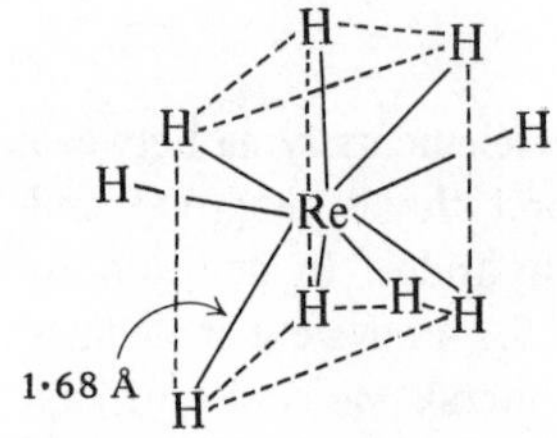

$[ReH_9]^{2-}$

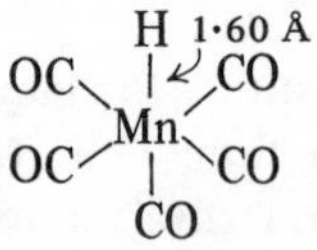

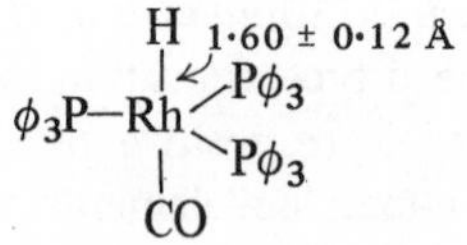

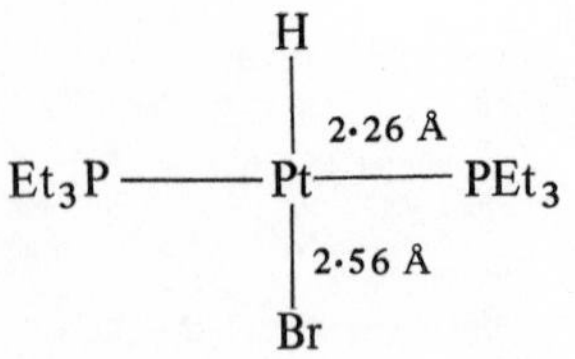

Less direct evidence for the positions of H atoms comes from X-ray studies of complexes in which the location of the other ligands strongly suggests the position of the H atom(s). In crystalline $PtHBr(PEt_3)_2$ the P and Br atoms were found to be coplanar with the Pt atom, with P–Pt–Br angles of 94°, leaving little doubt that the H is directly bonded to the Pt in the *trans* position to the Br atom. The fact that the Pt–Br bond is rather longer than expected (sum of covalent radii, 2·43 Å) may be associated with the high chemical lability of this atom.

300

In an incomplete X-ray study of the diamagnetic $Os^{II}HBr(CO)(P\phi_3)_3$ the five heavier ligands were found to occupy five of the six octahedral positions, H presumably occupying the sixth position. Still less direct evidence comes from spectroscopic[1] and dipole moment studies.[2] Configurations deduced in these ways include the following:

(1) JACS 1966 **88** 4100
(2) PCS 1962 318

(P = tertiary phosphine)

The hydrogen bond

It has been recognized for a long time that the properties of certain pure liquids and of some solutions indicate an unusually strong interaction between molecules of solvent, between solvent and solute, or between the solute molecules themselves. The first type of interaction, between the molecules in a pure liquid, led to the distinction between 'associated' and non-associated solvents. It may be illustrated by the properties of NH_3, H_2O, and HF, but many organic solvents, particularly those containing OH, COOH, or NH_2 groups, show somewhat similar abnormalities. If we compare such properties as melting point, boiling point, and heat of evaporation of a series of hydrides such as H_2Te, H_2Se, and H_2O, we find that these properties form a fairly regular sequence until we reach the last member of the series, H_2O. The hydrides of Groups VB and VIIB show similar behaviour, NH_3 and HF being 'abnormal' and HCl slightly so. The hydrides of Group IVB, including CH_4, form, on the other hand, a regular sequence; see Fig. 8.2. Instead of a melting point in the region of $-100°C$, as might be expected from extrapolation of the melting points of H_2Te, H_2Se, and H_2S, we find that H_2O melts at $0°C$. From the fact that the boiling points and heats of vaporization show the same type of abnormality as the melting points we deduce that many hydrogen bonds must exist in liquid NH_3, H_2O, and HF up to their boiling points. In the case of HF they persist in the vapour state.

Specific interaction between particular atoms of the solvent and solute molecules explains, for example, the much greater solubilities of aniline ($C_6H_5NH_2$) and phenol (C_6H_5OH) in water than in nitrobenzene in spite of the much larger dipole moment of the latter (4·19 D) as compared with water (1·85 D).

Association of solute molecules occurs when a substance like acetic acid, CH_3COOH, is dissolved in a non-associated solvent such as benzene, and is evident from determinations of molecular weight by the cryoscopic or ebullioscopic methods.

From the chemical nature of the molecules it is clear that these interactions are connected with the presence of hydrogen, usually as part of OH, COOH, NH_2, or other polar groups. If the hydrogen is replaced by, for example, alkyl groups, there is no association, showing that the H plays an essential part when it is attached to

301

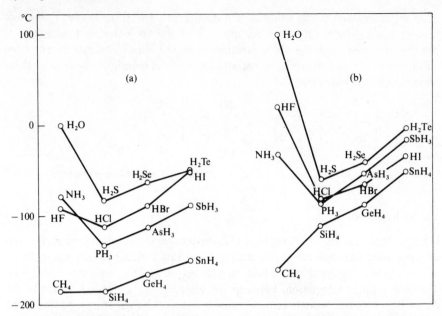

FIG. 8.2. Melting-points (a) and boiling-points (b) of series of isoelectronic hydrides.

the electronegative N, O, or F atoms. Some abnormalities are also shown by compounds containing, in addition to H, the less electronegative S or Cl, but these are much less pronounced; we include some data on these weaker interactions in Table 8.4. We shall confine our attention here largely to compounds containing N, O, and F. These intermolecular interactions, which we may write

$$N{\cdots}H{\cdots}O, \qquad O{\cdots}H{\cdots}O, \qquad \text{or} \qquad F{\cdots}H{\cdots}F,$$

are referred to as hydrogen bonds or hydrogen bridges.

In the cases we have mentioned the atoms between which the hydrogen bond is formed belong to different molecules, but there are also intramolecular hydrogen bonds. Some compounds containing both OH and COOH groups in the molecule show very little or no association, and it is found that hydrogen bond formation is possible within the molecule This had been suggested by Pfeiffer and co-workers in 1913 to account for the chemical inactivity of adjacent −OH and −CO groups in certain aromatic molecules. Examples of such intramolecular H bonds are:

With a few important exceptions the examples of H bond formation were for many years confined to organic compounds. Apart from such liquids as H_2O and HCN, the gas HF, and many liquid acids, the majority of inorganic compounds containing

302

H bonds are solids at ordinary temperatures. Early work on $NaHF_2$ (1922) suggested that the F atoms in the crystals were joined together in pairs by H bonds, and the tetrahedral arrangement of the oxygen atoms in ice could be accounted for by assuming that a H atom lay between each pair of O atoms (1923). X-ray studies of many crystalline compounds, both inorganic and organic, have emphasized the importance of hydrogen bonding in structures of many types.

In the earlier X-ray studies it was not possible to locate the H atoms, and the existence and positions of hydrogen bonds were deduced from the fact that some intermolecular contacts were closer than was to be expected for van der Waals bonds, the number of such contacts corresponding to the number of H atoms attached to O, N, etc. For example, in acid salts there is one such bond from each O atom which is involved in hydrogen bonding, and in hydroxy-compounds, two:

$$-O-H\cdots O- \qquad -O\begin{matrix} \diagup H \diagup \\ \\ {}^{\backslash}H \\ \qquad {}^{\backslash}O- \end{matrix}$$

The hydrogen atoms in many crystals have since been located using more refined X-ray techniques and particularly by neutron diffraction. Spectroscopic studies (infrared and n.m.r.) have also contributed to our knowledge of hydrogen bonding.

The properties of hydrogen bonds

The fact that strong hydrogen bonds are formed only between the most electronegative elements suggests that they are essentially electrostatic in character. The atoms concerned have lone pairs of electrons which not only influence the directional properties of bonding orbitals but also contribute substantially to the dipole moments of the molecules. Assuming various types of hybridization the orbital dipole moments of lone pairs on atoms such as N, O, and F can be calculated. These dipoles can interact with the $H^{\delta+}$ of FH, OH, NH_2, etc. to form hydrogen bonds which should be directed in accordance with the type of hybridization of the lone pair and bonding orbitals. This directional feature of hydrogen bonds is evident in ice, where each O atom has four tetrahedral neighbours, in HF, $H_2F_3^-$, and $H_4F_5^-$, and in most hydrates. There are, however, some crystals in which neither of the lone pairs of O is directed along the hydrogen bonds as, for example, in $K_2C_2O_4 \cdot H_2O$[1] where the two K^+ neighbours are nearly coplanar with the O–H bonds of the water molecule.

(1) JCP 1964 **41** 3616

Bond energies and lengths

The energies of hydrogen bonds range from $110\,kJ\,mole^{-1}$ (or possibly higher values)[2] for the strongest (in the $F-H-F^-$ ion) through values around 30 kJ $mole^{-1}$ for O–H–O to still smaller values for the weaker hydrogen bonds. There is also a considerable range of lengths for each particular type of bond X–H–X, and data for the various hydrogen bonds are summarized in Table 8.4.

(2) IC 1963 **2** 996

Position of the H *atom*

The positions of the H atoms in hydrogen bonds in many crystals have been determined by X-ray and/or neutron diffraction, and indirectly from the proton-proton separations in some hydrates by p.m.r. It should be emphasized that the positions of H atoms as determined by X-ray diffraction are different from those resulting from n.d. studies, the former showing the H atom apparently closer to the heavier atom to which it is bonded. Typically the difference for an O—H distance is about 0·15 Å (n.d., 1·0 Å, X-ray, 0·85 Å). The effect (which is smaller for heavier atoms) appears to arise from the asphericity of the electron distribution due to chemical bonding. This affects the X-ray scattering factor of the atom (which depends on the orbital electrons) but not the neutron scattering which, for a diamagnetic atom, is purely nuclear. The methods for refining crystal structures involve the use of calculated atomic scattering factors, so that if a spherical electron distribution around an atom is assumed and the position of an atomic nucleus is determined as the centre of gravity of its electron cloud, the 'X-ray position' may be different from that determined by neutron diffraction.[3]

(3) AC 1969 **B25** 2451

TABLE 8.4

Lengths of hydrogen bonds

Bond	Length	Compound	Reference
F---H---F	2·27 Å	$NaHF_2$	
F—H---F	2·45	KH_4F_5	
	2·49	HF	
O---H---O	2·40	Ni dimethyl glyoxime	See text
	2·40–2·55	Acid salts	
	2·70–2·90	Ice, hydroxy-compounds, hydrates	
O—H---F	2·65–2·72	$CuF_2 . 2 H_2O$, $Fe(H_2O)_6 SiF_6$	
O—H---Cl	2·92–2·95	$HCl . H_2O$, etc.	
	3·04	$(NH_3OH)Cl$	AC 1967 **22** 928
O—H---Br	3·04	$Cs_3 Br_3 (H_3O)(HBr_2)$	IC 1968 **7** 594
O—H---N	2·68	$N_2H_4 . 4 CH_3OH$	ACSc 1967 **21** 2669
	2·79	$N_2H_4 . H_2O$	AC 1964 **17** 1523
O—H---S	3·20 +	$MgS_2O_3 . 6 H_2O$	AC 1969 **B25** 1708
N—H---O	2·81–2·89	NH_4OOCH	AC 1968 **B24** 565
	2·99–3·01	$CO(NH_2)_2$	AC 1957 **10** 319
N—H---F	2·61–2·82	NH_4F, $N_2H_6F_2$, etc.	AC 1971 **B27** 1102
N—H---Cl	3·00, 3·11	$(CH_3)_3NHCl$, $(CH_3)_2NH_2Cl$	AC 1968 **B24** 554, 549
	3·20	$(NH_3OH)Cl$	AC 1967 **22** 928
N—H---I	3·46	$[(CH_3)_3NH] I$	AC 1970 **B26** 1334
N—H---N	2·94–3·15	NH_4N_3, dicyandiamide	See Chapter 21
	3·35	NH_3	See text
(S—H---S)?	3·94	H_2S (cryst.)	N 1969 **224** 905
(C—H---O)?	2·92 +	Organic compounds	JCS 1963 1105

It may perhaps be considered doubtful whether some of the much weaker interactions should be described as hydrogen bonds or as van der Waals or weak ionic bonds. However, if the H atom is directed towards the other atom of the 'bond' many such interactions are described in the literature as hydrogen bonds, and rather than discuss such cases in detail we give some references above. The H atoms have not been located in any of the 'C—H—O' bonds.

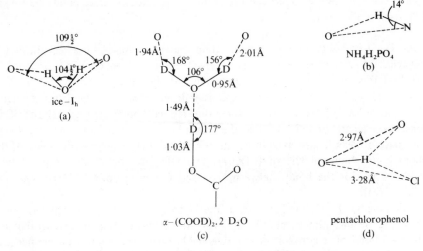

FIG. 8.3. Details of O–H–O bonds.

In many cases the H atom necessarily lies off the straight line joining the bonded atoms owing to the geometry of the molecule, as in the *o*-substituted benzene derivatives already noted. This is also true in ice, in many hydrates, acid salts, etc. (Fig. 8.3(a) and (b)). Any crystal structure represents a compromise between the bonding requirements of a number of atoms and the packing of groups of various shapes. A 'bent' hydrogen bond is energetically preferable to distortion of angles between covalent bonds. The 'bifurcated' hydrogen bond is rare (and may not exist

X–H---Y   X–H–Y   X–H⟨ Y / Y

straight   bent   bifurcated

at all in inorganic compounds) but it may occur in substituted naphthoquinones,[4] where one 'branch' would be intramolecular, and in compounds such as penta-chlorophenol[5] (Fig. 8.3(d)) but a n.d. study of $C_6Cl_4(OH)_2$[6] does not show the bifurcated bond of the type of Fig. 8.3(d) apparently shown by an earlier X-ray study.

In a limited number of hydrogen bonds the H atom lies *on* the straight line joining the bonded atoms. They include F–H–F in the bifluoride ion and O–H–O in acid salts and the short O–H–O bonds (2·52 Å) in $(COOD)_2 . 2 D_2O$. In the bent bonds the H atom is necessarily at different distances from X and Y since it is covalently bonded to X or Y as part of a group such as OH, NH_2, etc.; this is the reason why the bond is bent. The position of the H atom in straight F–H–F and O–H–O bonds has been much studied, with the following results.

F–H–F *bonds.* All physical evidence (n.d., i.r., and n.m.r.) supports the view that the short F–H–F bond (2·27 Å) in the bifluoride ion is symmetrical. In crystalline HF and KH_4F_5 the much longer hydrogen bonds (2·49 and 2·45 Å) are

(4) AC 1969 **B25** 546
(5) AC 1962 **15** 443, 1164
(6) AC 1967 **23** 107

almost certainly not symmetrical; the nature of the intermediate bond (2·33 Å) in KH_2F_3 is not known.

O–H–O *bonds*. It appeared at one time that there might be a real difference between the structures of the long O–H–O bonds, of length 2·7 Å or more, in ice, hydrates, and hydroxy-compounds, and the shorter ones of length 2·55 Å or less in acids and acid salts. In some crystals hydrogen bonds of two different lengths occur, as in $Na_3H(CO_3)_2 . 2 H_2O$, where the 'acid salt' bond between pairs of CO_3^{2-} ions has a length of 2·53 Å while the bonds between water molecules and CO_3^{2-} ions have a mean length of 2·74 Å, and $(COOH)_2 . 2 H_2O$, where the length of the short bonds between OH of $(COOH)_2$ and OH_2 is 2·53 Å as compared with 2·84–2·90 Å for the bonds between H_2O and CO groups of $(COOH)_2$ molecules (Fig. 8.3(c)).

All long O–H–O bonds in which the H atoms have been located have been shown to be unsymmetrical (e.g. H_3BO_3). Also it has been shown (n.d.) that the O–H–O bond of length 2·49 Å in tetragonal KH_2PO_4 is definitely unsymmetrical, (a), and the same is true of the short O–D–O bonds (2·52 Å) in $(COOD)_2 . 2 D_2O$, (b), and of the O–D–O bond in $DCrO_2$, (c). However, it has not been possible to

$$O \text{———} H \text{------} O \qquad\qquad \text{\Large$\diagdown$}C\text{—}C\text{\Large$\diagup$}^{O \text{———} D \text{------} OD_2}_{\quad 1\cdot03 \qquad 1\cdot49\ \text{Å}} \qquad\qquad O\text{———}D\text{------}O$$

$$\quad 1\cdot07 \quad\ \ 1\cdot42\ \text{Å} \qquad\qquad\qquad\qquad\qquad\qquad\qquad\qquad\qquad\qquad 0\cdot96 \quad 1\cdot59\ \text{Å}$$

$$(a) \qquad\qquad\qquad\qquad\qquad (b) \qquad\qquad\qquad\qquad\qquad (c)$$

decide with certainty in the case of $HCrO_2$ (O–H–O, 2·49 Å) where the i.r. data suggest a symmetrical bond, or in the case of acid salts of the type $KH(R . COO)_2$ or acids $H(PO_4R_2)$. These compounds are of particular interest because the crystals contain pairs (d), or chains (e), of anions linked by hydrogen bonds, two anions

$$(a) \qquad\qquad\qquad\qquad\qquad\qquad\qquad (e)$$

being related by a centre of symmetry or axis of 2-fold symmetry. This symmetry requires either that the H atom is centrally placed between the two O atoms or, if the bond is unsymmetrical, that there is random distribution throughout the crystal of the H atoms in positions about 1 Å from one or other of the O atoms of the hydrogen bonds. Although a n.d. study of $KH(C_6H_5CH_2COO)_2$, type (d), indicates a single H at the mid-point of the O–H–O bond (length 2·44 Å) this is not accepted as indisputable proof of the symmetry of the bond because of the limited experimental accuracy. Neutron diffraction studies have not yet been made of the shortest O–H–O bonds (2·40 Å) in $HPO_2(OC_6H_4Cl)_2$, in Ni dimethylglyoxime,[7]

(7) JACS 1959 **81** 755

(f), and the closely related amino-oxime complex, (g).[8] (References to many of these compounds will be found in Table 8.5, p. 315.)

(8) IC 1968 7 1130

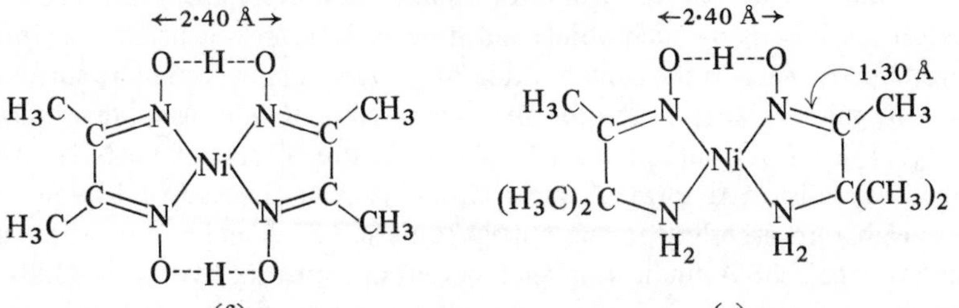

(f)   (g)

Less direct evidence for the position of the H atom also comes from proton magnetic resonance studies and from values of residual entropy. The positions of the protons in the H_2O molecules in gypsum have been determined indirectly from the fine structure of the n.m.r. lines; they are found to lie at a distance of 0.98 Å from the O atom along an O–H–O bond.[9] For n.m.r. studies of a number of hydrates see p. 564; the method has also been used to locate the H atoms in $Mg(OH)_2$ (p. 521). Measurements of residual entropy confirm the existence of two distinct locations for the proton in hydrogen bonds in ice (p. 539), salts of the type of KH_2PO_4, and $Na_2SO_4 . 10 H_2O$ (which has residual entropy about two-tenths that of ice).

(9) JCP 1948 16 327

We conclude that all O–H–O bonds, with the possible exception of some of the very shortest, are unsymmetrical.

The hydrogen bond in crystals

Hydrogen bonds play an important part in determining the structures of crystalline compounds containing N, O, or F in addition to H. We deal here with the hydrides of these elements, and with certain fluorides, oxy-acids, and acid salts. Ice and water, together with hydrates, are considered in Chapter 15, and hydroxy-acids are grouped with hydroxides in Chapter 14.

We noted in Chapter 3 that the interest, from the geometrical standpoint, of hydrogen-bonded structures lies in the variety of ways in which a group of atoms of limited extent (that is, a finite molecule or ion) can be linked by hydrogen bonds into a more extensive system, in the limit a three-dimensional network. These hydrogen-bonded systems may be:

finite groups: $[H(CO_3)_2]^{3-}$ (in $Na_3H(CO_3)_2 . 2 H_2O$), dimers of carboxylic acids,

infinite chains: HCO_3^- and HSO_4^- ions,

infinite layers: $B(OH)_3$, $B_3O_3(OH)_3$, $N_2H_6F_2$, H_2SO_4,

3-dimensional nets: ice, H_2O_2, $Te(OH)_6$, $H_2PO_4^-$ ion in KH_2PO_4.

Crystalline organic compounds containing OH, COOH, CO . NH, NH_2, and other polar groups containing H atoms provide many further examples of all the above types of structure.

Hydrides

As already mentioned, the hydrides of N, O, and F alone show evidence of strong hydrogen bonds. The simplest structures would arise if every atom was joined to all its nearest neighbours by such bonds and if every H atom was utilized in forming hydrogen bonds. Then in the solid hydride AH_n every A atom would be surrounded by $2n$ A neighbours, that is, the coordination number of A by A atoms would be 6 for NH_3, 4 for H_2O, and 2 for HF. The structure of crystalline NH_3 may be described as a distorted form of cubic closest packing. Instead of 12 equidistant nearest neighbours each N atom has 6 neighbours at 3·35 Å and 6 more at the much greater distance 3·88 Å, indicating that a N atom forms six hydrogen bonds. The precise D positions in ND_3 have been determined by n.d.[1] With this ordered structure compare the structure of ordinary ice (ice-I_h). Each O is surrounded tetrahedrally by 4 others as expected, but there is a statistical distribution of H atoms between the equivalent positions at a distance of 1 Å from O along each hydrogen bond.

The fact that there is only one H atom for each F limits the coordination numbers of both H and F in solid HF to 2, and the only possibilities are therefore infinite chains or closed rings. It is found that solid HF, studied at $-125°C$, consists of infinite planar zigzag chains:[2]

(1) JCP 1961 **35** 1730

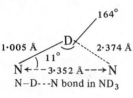

N–D---N bond in ND_3

(2) AC 1954 **7** 173

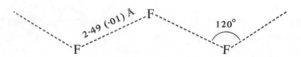

(3) JCP 1969 **50** 3611

The e.d. data for HF vapour[3] are apparently best accounted for by a mixture of monomers and puckered $(HF)_6$ rings, with mean angle F–F–F, 104°, and F–H–F, 2·53 Å. The behaviour of the other hydrogen halides in the crystalline state is complex, and careful n.d. studies have given the following results.[4]

(4) N 1967 **213** 171; N 1967 **215** 1265; N 1968 **217** 541

Solid DCl undergoes a first-order phase transition at about 105°K (98·44° for HCl). At higher temperatures the structure is f.c.c. and there is disorder of a special kind. At any given time each molecule is in one of twelve orientations, so that it is hydrogen-bonded to one of its nearest neighbours which itself is hydrogen-bonded to one of its twelve nearest neighbours, and so on. Each hydrogen bond is only temporary (being broken when the molecule reorients) so that the 'instantaneous' structure of cubic DCl (HCl) is a mixture of shortlived hydrogen-bonded polymers of various lengths and shapes. At temperatures below the transition point the structure is orthorhombic. Each D is directed towards a neighbouring Cl atom (Fig. 8.4(a)) and the structure therefore consists of chains of hydrogen-bonded molecules (D–Cl, 1·28 Å, Cl–D–Cl, 3·69 Å, Cl–Cl–Cl, $93\frac{1}{2}°$). Crystalline DBr shows two second-order (λ) phase transitions, at 93·5° and 120·3°K. At temperatures well below the lower λ region (for example, 74°K) the structure is the same ordered orthorhombic structure as for DCl (Fig. 8.4(a)), in which the bond lengths and angle are: D–Br, 1·38 Å, Br–D–Br, 3·91 Å, Br–Br–Br, 91°. As the temperature 93·5°K is approached there is a continuous change to the partially disordered

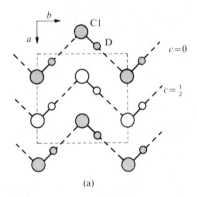

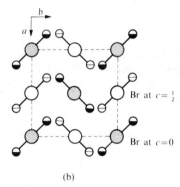

(a)   (b)

⊖ alternative positions of D atoms in

⊖ disordered molecules at $c = 0$ or $\frac{1}{2}$

FIG. 8.4. (a) The structure of crystalline DCl at 77·4°K. (b) The structure of crystalline DBr at 107°K. Small circles represent D and large circles Cl or Br.

structure of Fig. 8.4(b) in which there are two equilibrium orientations for each molecule. In contrast to the sudden loss of order in DCl this change in DBr is spread over some 20°K. At the higher λ transition there is presumably a gradual change from this disordered orthorhombic structure to a structure with the 12-fold orientational disorder of high-DCl. The behaviour of HCl is even more complex, this compound exhibiting three λ-type transitions. Details of the behaviour of HI are not yet available. (The presence of cyclic dimers in HCl condensed in solid Xe has been deduced from the i.r. spectrum, but their structure has not been determined.)[5]

(5) JCP 1967 **47** 5303

Normal fluorides

Ammonium fluoride, NH_4F, crystallizes with a structure different from those of the other ammonium (and alkali) halides. The chloride, bromide, and iodide have the CsCl structure at temperatures below 184·3°, 137·8°, and −17·6°C respectively, and the NaCl structure at temperatures above these transition points, but NH_4F crystallizes with the wurtzite structure, in which each N atom forms N–H–F bonds of length 2·71 Å[1] to its four neighbours arranged tetrahedrally around it. This is essentially the same structure as that of ordinary ice.

(1) AC 1970 **B26** 1635

Hydrazinium difluoride, $N_2H_6F_2$, provides a very interesting example of a crystal in which the structural units are linked by hydrogen bonds (N–H–F) into layers.[2] The structural units are $N_2H_6^+$ and F^- ions. In the $N_2H_6^{2+}$ ion the N–N bond length is 1·42 Å and the N–N–H bond angles 110°. Viewed along its axis the ion has the configuration shown at the right.

(2) JCP 1942 **10** 309

The position of the H atom along the hydrogen bond has been studied by p.m.r.:[3]

N————H·······F

1·08 Å   1·54 Å

(3) TFS 1954 **50** 560

309

Each hydrazinium ion forms six octahedral N–H–F bonds and every F^- ion forms three such bonds (to different $N_2H_6^{2+}$ ions) arranged pyramidally. This is exactly the same bond arrangement as in the $CdCl_2$ or CdI_2 structures. In $N_2H_6F_2$ the halogen atoms are not close-packed as in $CdCl_2$ or CdI_2. The layers have a very open structure (F–F = 4·43 Å) owing to the directed N–H–F bonds (2·62 Å, compare 2·68 Å in NH_4F and 2·76 Å in NH_4HF_2). The mode of packing of the layers on one another is geometrically similar to that in the dihalides with layer structures. In these a halogen atom of one layer rests on three of the layer below, so that throughout the structure the halogen atoms are close-packed. In $N_2H_6F_2$ the layers are superposed in a similar way, but owing to the large separation of F atoms in a layer, a F atom of one layer (shown as a large open circle in Fig. 8.5) drops between the three F atoms of the next layer and comes in contact with a N atom (and/or its three associated H atoms). This interlayer F–N distance is 2·80 Å,

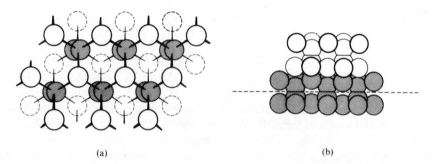

(a)   (b)

FIG. 8.5. The structure of hydrazinium fluoride $N_2H_6F_2$ (after Kronberg and Harker): (a) Plan of one layer. The lower N atoms of each $N_2H_6^{2+}$ ion (shaded) are slightly displaced. The open circles, full and dotted lines, represent F^- ions respectively above and below the plane of the paper. All lines are N–H–F bonds. (b) Elevation, showing the packing of the layers. The broken line is the trace of the plane of the paper in (a).

and the F–F, 3·38 Å. It is estimated that the change on the $N_2N_6^{2+}$ ion is about equally distributed over all the atoms, giving a charge of $+\frac{1}{4}$ on each. The interlayer bonds are therefore weak ionic bonds between F^- ions and N (and/or H) atoms of adjacent layers. The weakness of these bonds accounts for the very good cleavage parallel to the layers.

Just as NH_4F and NH_4Cl have different structures owing to the fact that F but not Cl can form strong hydrogen bonds, so the corresponding chloride, $N_2H_6Cl_2$, has a different type of structure from $N_2H_6F_2$. It forms a somewhat deformed fluorite structure in which each $N_2H_6^{2+}$ ion is surrounded by 8 Cl^- and each Cl^- by 4 $N_2H_6^{2+}$ ions.[4]

Bifluorides (acid fluorides), MHF_2
The crystal structures of the bifluorides MHF_2 of the alkali metals, NH_4, and Tl(I) are closely related to the NaCl and CsCl structures, while the structure of NH_4HF_2 differs in an interesting way from that of the ordinary form of KHF_2 owing to the formation of N–H–F bonds in the ammonium salt.

(4) JCP 1947 **15** 115

All $(F-H-F)^-$
ions parallel

$(F-H-F)^-$*ions in 4 orientations*
parallel to the body-diagonals of
the cube

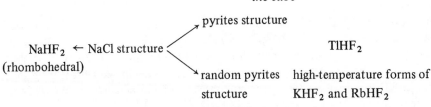

pyrites structure

$NaHF_2$ ← NaCl structure
(rhombohedral)

TlHF$_2$

random pyrites   high-temperature forms of
structure KHF_2 and $RbHF_2$

The crystal structures of bifluorides MHF_2 are as follows:

M = Li	rhombohedral $NaHF_2$ structure[a]
Na	rhombohedral (p. 198)[b]
K	low temperature, tetragonal superstructure of CsCl type[c]
	high (above 196°C), cubic (random pyrites)
Rb	low, tetragonal KHF_2 structure[d]
	high, cubic (random pyrites)
Cs	low, tetragonal[d]
	high, cubic (random CsCl, F–H–F along cube edges)
NH$_4$	Fig. 8.6 (a)[e]
Tl(I)	pyrites structure[f] (no transition to lower symmetry).

(a) AC 1962 **15** 286
(b) JCP 1963 **39** 2677; (n.d.)
(c) JCP 1964 **40** 402 (n.d.)
(d) JACS 1956 **78** 4256
(e) AC 1960 **13** 113
(f) ZaC 1930 **191** 36

The structure of the ordinary (tetragonal) form of KHF_2 is illustrated in Fig. 8.6(b). Potassium and ammonium salts are often isostructural, the K^+ and NH_4^+ ions having very similar sizes. The structures of KHF_2 and NH_4HF_2 are both superstructures of the CsCl type, but the HF_2^- ions are oriented differently in the two structures (Fig. 8.6). In KHF_2 each K^+ ion is surrounded by eight equidistant F^- neighbours, but in NH_4HF_2 the NH_4^+ ion has only four nearest F^- neighbours (at 2·80 Å) and four at greater distances (3·02 and 3·40 Å). The breakdown of the coordination group of eight into two sets of four is to be attributed to the formation of N–H–F bonds.

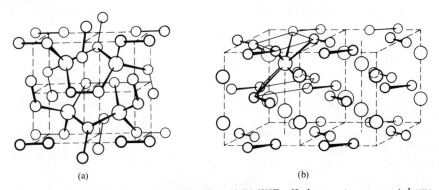

(a)   (b)

FIG. 8.6. The crystal structures of (a) NH_4HF_2 and (b) KHF_2. Hydrogen atoms are not shown. The small circles represent fluorine atoms, joined in pairs by F–H–F bonds.

Hydrogen: The Noble Gases

The F–H–F distance, 2·27 Å, may be compared with H–F, 0·92 Å, in the HF molecule.

Other fluorides MH$_n$F$_{n+1}$

In addition to KHF$_2$ at least four other 'acid' fluorides of potassium have been prepared, namely, KH$_2$F$_3$, KH$_3$F$_4$, KH$_4$F$_5$, and K$_2$H$_5$F$_7$. An X-ray study of KH$_2$F$_3$[1] shows that the anion has the structure (a) with F–H–F slightly larger than in (F–H–F)$^-$ and a bond angle of 135° as compared with 120° in solid HF. In KH$_4$F$_5$ there are (H$_4$F$_5$)$^-$ ions consisting of a tetrahedral group (b) of four F atoms surrounding a central one, with F–H–F, 2·45 Å.[2]

(1) AC 1963 **16** 58

(a)

(2) JSSC 1970 **1** 386

(b)

(1) JCS A 1966 1185

(2) IC 1963 **2** 657
(3) IC 1968 **7** 594

(4) JPC 1966 **70** 11, 20, 543

(5) BCSJ 1959 **32** 263; 1960 **33** 354

Other ions (X–H–X)$^-$ and (H$_2$O–H–OH$_2$)$^+$

We have already noted in our discussion of the hydrogen bond the formation of ions of this general type in salts KH(RCOO)$_2$, and we shall meet the ion [H(CO$_3$)$_2$]$^-$ in the later section on the structures of acid salts. With the exception of the stable bifluoride ion, which we have discussed, ions of this type, where X is Cl, Br, I, NCS, or NO$_3$, are stable only in the presence of large cations as in the salt [P(C$_6$H$_5$)$_4$](NO$_3$.H.NO$_3$).[1] Salts such as [N(CH$_3$)$_4$]HCl$_2$, [N(C$_2$H$_5$)$_4$]HBr$_2$, and [N(C$_4$H$_9$)$_4$]HI$_2$ are stable at room temperature, but all tend to lose HX rather readily and some are very unstable towards moisture and oxygen. The salt CsHCl$_2$ has been prepared but it is stable only at very low temperatures or under high HCl pressure, and salts prepared at room temperature from saturated aqueous CsCl and HCl are not CsHCl$_2$ as originally thought. One is probably CsCl . $\frac{3}{4}$(H$_3$OCl),[2] and another is CsCl . $\frac{1}{3}$(H$_3$O$^+$. HCl$_2^-$). An X-ray study of the latter,[3] and of the isostructural bromide, shows pyramidal hydronium ions and linear ions (Cl–H–Cl)$^-$ or (Br–H–Br)$^-$, of overall length 3·14 Å and 3·35 Å respectively, but a n.d. study is required to locate the H atoms. Infrared and Raman studies of salts of ions such as (Cl–H–Cl)$^-$ suggest that their structures are very sensitive to their environment, and a final statement cannot yet be made on the structures of these ions.[4]

The existence of the ion (H$_2$O . H . OH$_2$)$^+$ has been suggested in salts such as [Co(en)$_2$Br$_2$]HBr$_2$. 2 H$_2$O,[5] and it certainly seems to exist in ZnCl$_2$. $\frac{1}{2}$ HCl . H$_2$O, which is (Zn$_2$Cl$_5$)$^-$(H$_5$O$_2$)$^+$ (see p. 565); the O–O distance is reported as 2·35 Å. The (H$_5$O$_2$)$^+$ ion may also exist in ZnCl$_2$. HCl . 2 H$_2$O.

Acids and acid salts

Acids

For a discussion of their structures it is convenient to group the inorganic acids into the following classes:

(a) Acids H$_m$X$_n$. The most important of these (hydrogen halides and H$_2$S) are gases at ordinary temperatures and others are liquids (HN$_3$, H$_2$S$_n$, H$_2$O$_2$). The structures of the halogen acids HX and also H$_2$S have already been noted; the structures of other acids in this class are described under the chemistry of the appropriate element X.

312

(b) Polynuclear oxy-acids containing two or more directly bonded X atoms are few in number. They include $H_2N_2O_2$, $H_4P_2O_6$ (both crystalline solids), and the polythionic acids $H_2S_nO_6$ (n = 3–6), which are known only in aqueous solution. There are other oxy-*ions* of this type (e.g. $S_2O_3^{2-}$ and $S_2O_5^{2-}$) but we restrict our examples here to cases in which the free acid can be isolated.

(c) Per-acids. These contain the system O–O and are described after hydrogen peroxide in Chapter 11. (The description of certain ortho-acids as per-acids is regrettable but firmly established, for example, perchloric acid, $HClO_4$, and permanganic acid, $HMnO_4$.)

(d) Oxy-acids H_mXO_n. This large group contains all the simple oxy-acids. The molecule in the vapour or in the crystalline acid consists of a single X atom bonded to a number of O atoms, to one or more of which H atoms are bonded. In aqueous solution and in crystalline salts some or all of the H atoms are removed, leaving the oxy-ion. When discussing the structures of the crystalline acids it is convenient to distinguish as a subgroup the hydroxy-acids $X(OH)_n$ and to describe them under hydroxides (Chapter 14); examples include the crystalline H_3BO_3 and H_6TeO_6 — note that H_3PO_3 is not of this type. Inasmuch as all the H atoms of an oxy-acid H_mXO_n (other than certain acids of P where some H are directly bonded to P) are associated with O atoms to form OH groups (i.e. $XO_{n-m}(OH)_m$) these compounds are formally analogous to the oxyhydroxides of metals described in Chapter 14. The structural difference lies in the fact that the non-metal compounds form finite molecules in contrast to the infinite systems of metal atoms linked through M–O–M bonds in the metal oxyhydroxides.

Some oxy-acids are liquids at ordinary temperatures (HNO_3, H_2SO_4, $HClO_4$), many are solids (H_3PO_2, H_3PO_3, H_3PO_4, H_2SeO_3, H_2SeO_4, HIO_3, HIO_4, and H_5IO_6), and a further set of less stable acids are known only in aqueous solution (H_2CO_3, HNO_2, H_2SO_3, HFO, HClO, HBrO, HIO, $HClO_2$, $HClO_3$, $HBrO_3$). Replacement of OH in certain oxy-acids gives substituted acids such as HSO_3F and HSO_3NH_2.

(e) Pyro- and meta-acids. Complex oxy-ions formed from two or more XO_n groups sharing O atoms are numerous. Ions formed in this way from two ortho-ions are termed pyro-ions, and a few more complex ions of the same general type are known (for example, $P_3O_{10}^{5-}$, $S_3O_{10}^{2-}$, $P_4O_{13}^{6-}$). The sharing of two O atoms of every XO_n group leads to either cyclic or infinite linear meta-ions. Among the relatively few acids of this class which can be obtained as pure crystalline solids are metaboric acid, HBO_2 (three polymorphs) and $H_2S_2O_7$; HPO_3 and $H_4P_2O_7$ tend to form glasses. Of this same general type are the much more complex iso- and hetero-polyacids formed by metals such as Mo and W; they are described in Chapter 11.

Some of the inorganic acids, for example sulphuric acid, were among the compounds known to the earliest chemists, but comparatively little is known even today of the structures of the inorganic acids as a class. This is largely due to the facts that many are liquids at ordinary temperatures or are too unstable to be isolated. However, the extension of X-ray crystallographic studies to lower temperatures has made possible the determination of the structures of a number of acids and their hydrates which are liquid at room temperature (for example, sulphuric acid, nitric acid and its hydrates). The spectroscopy of molecules trapped in inert matrices at low temperatures offers a means of studying the structures of acids which are too unstable to isolate under ordinary conditions (for example, HNO_2, p. 657). We deal in this chapter only with crystalline anhydrous acids; the hydrates of acids are discussed with other hydrates in Chapter 15. The structures of acids such as H_2S, HN_3, HNO_3, HNCS, and HNCO, which have been studied in the vapour state, are described in other chapters.

Acid salts

These are crystalline compounds intermediate in composition between the acid and a normal salt in which there is a system of hydrogen-bonded anions, these ions being either simple (F^-) or complex (XO_n). Apart from salts containing ions $(X-H-X)^-$ and the more complex ions formed by F which have already been discussed, anhydrous acid salts are formed only by oxy-acids. In a hydrated acid salt water molecules are incorporated *between* the hydrogen-bonded anion complexes, and their presence does not, as far as is known, affect the basic principle underlying the structures of these salts, namely, that the H atoms are associated exclusively with the oxy-ions to form the complex acid-salt anion. Since a crystalline oxy-acid also consists of an array of hydrogen-bonded oxy-ions the structural possibilities are exactly the same for the acid as for an acid-salt anion $H_x(XO_n)_y$ *having the same* H : XO_n *ratio*. (In the acid salt the cations are accommodated between the hydrogen-bonded anions.)

The structures of acid salts and crystalline oxy-acids

The connectedness of the system of hydrogen-bonded anions in a crystalline oxy-acid or an acid salt is clearly *twice* the H : XO_n ratio since each hydrogen bond connects two O atoms of different anions. Therefore the same topological possibilities are presented by pairs of compounds such as a substituted phosphoric acid $H(PO_4R_2)$ and an acid salt $KHSO_4$ or by sulphamic acid, which behaves as SO_3NH_3, and $(NH_4)_2H_3IO_6$. For the first pair of compounds there is a ratio of one H to one anion, so that only cyclic systems or infinite chains are possible. For the second pair the ratio is 3 : 1, and with six hydrogen bonds from each anion to its neighbours the possible structures for the hydrogen-bonded system are 2- and 3-dimensional 6-connected nets.

The possible types of hydrogen-bonded anion complex are set out in Table 8.5, arranged in order of increasing H : XO_n ratio.

(*a*) H : XO_n ratio 1 : 2. The only possibility is the linking of the XO_n ions in pairs as in sodium sesquicarbonate, $Na_3H(CO_3)_2 . 2 H_2O$ (Fig. 8.7). Short O—H—O bonds (2·53 Å) link the carbonate ions in pairs and longer hydrogen bonds (2·75 Å) link these pairs to water molecules. The pairs of hydrogen-bonded anions in salts such as potassium hydrogen phenylacetate have already been mentioned in our discussion of the hydrogen bond.

TABLE 8.5

Hydrogen-bonded anion complexes in acid salts and crystalline oxy-acids

	$H : XO_n$ ratio	Number of O–H–O bonds from each anion	Type of $H_x(XO_n)_y$ complex	Examples	
				Acid salts	Acids
(a)	1 : 2	1	Dimer	$Na_3H(CO_3)_2 . 2 H_2O$ [1] $KH(C_6H_5CH_2COO)_2$ [2]	
(b)	2 : 3	1 and 2	Trimer	$NH_4H_2(NO_3)_3$ [3]	
(c)	1 : 1	2	Cyclic dimer	$KHSO_4$, [4a] $KHCO_3$, [5] $SnHPO_4$ [4b]	$H[PO_2(OCH_2C_6H_5)_2]$ [12]
			1-dimensional	$NaHCO_3$ [6]	$\alpha\text{-}HIO_3$ [13] $H[SeO_2(C_6H_5)]$ [14] $(CH_3)_2PO . OH$ [15] $HPO_2(OC_6H_4Cl)_2$ [16]
(d)	3 : 2	3	1-dimensional 2-dimensional	$LiH_3(SeO_3)_2$ [7] $NaH_3(SeO_3)_2$ [8]	
(e)	2 : 1	4	1-dimensional 2-dimensional	$(NH_4)_2H_2P_2O_6$ [9]	(β-oxalic acid) H_2SeO_3 [17] H_2SO_4 [18] H_2SeO_4 [19] (α-oxalic acid)
			3-dimensional	KH_2PO_4 [10]	
(f)	3 : 1	6	2-dimensional 3-dimensional	$(NH_4)_2H_3IO_6$ [11]	H_3PO_4 [20] $^-O_3S . NH_3^+$ [21]

(1) AC 1956 **9** 82 (n.d.); AC 1962 **15** 1310 (p.m.r.). (2) AC 1968 **B24** 323. (3) AC 1950 **3** 305. (4a) AC 1958 **11** 349; (4b) AC 1971 **B27** 1092. (5) AC 1968 **B24** 478. (6) AC 1965 **18** 818. (7) PR 1960 **119** 1252. (8) AC 1968 **B24** 1237. (9) AC 1964 **17** 1352. (10) PRS 1955 **A230** 359. (11) JACS 1937 **59** 2036. (12) AC 1956 **9** 327. (13) AC 1949 **2** 128. (14) AC 1954 **7** 833. (15) AC 1967 **22** 678. (16) AC 1964 **17** 1097. (17) JCS 1949 1282. (18) AC 1965 **18** 827. (19) JCS 1951 968. (20) ACSc 1955 **9** 1557. (21) AC 1960 **13** 320.

(*b*) $H : XO_n$ *ratio* 2 : 3. This leads to a hybrid system in which all the XO_n ions are not equivalent, unlike all the other cases in Table 8.5. It is included because an example is found in $NH_4H_2(NO_3)_3$, an acid salt in which the nitrate ions are joined into linear groups of three by hydrogen bonds.

(*c*) $H : XO_n$ *ratio* 1 : 1. Here the XO_n ions may be linked into cyclic systems or infinite chains. Both these possibilities are realized in $KHSO_4$, in crystals of which one-half of the sulphate ions form the simplest type of cyclic system, the dimer of type (i), below, and the remainder form chains, (ii). The length of all the O–H–O bonds is close to 2·67 Å. Dimers of type (i) also occur in $SnHPO_4$.

In the bicarbonate ion in $NaHCO_3$ there are infinite chains of composition HCO_3^- which are held together laterally by the Na^+ ions. The H atoms have been located; the O–H–O bonds are unsymmetrical, showing that the linear anion is an assembly of HCO_3^- ions as shown at (iii).

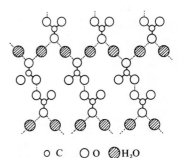

o C ◯ O ⦸ H_2O

FIG. 8.7. Layer (diagrammatic) of the structure of $Na_3H(CO_3)_2 . 2H_2O$ showing CO_3^{2-} ions joined in pairs by short O–H–O bonds and to water molecules by long O–H–O bonds. The Na^+ ions are omitted.

315

$$O\diagdown S\diagup O\text{---}H\text{---}O\diagdown S\diagup O$$
$$O\diagup {}^{S}\diagdown O\text{---}H\text{---}O\diagup {}^{S}\diagdown O$$

(i)

$$O\diagdown {}_{S}\diagup O \quad O\diagdown {}_{S}\diagup O$$
$$\text{---}O\diagup {}^{}\diagdown O\text{---}H\text{---}O\diagdown \qquad O\diagup {}^{}\diagdown O\text{---}H\text{---}$$
$$O\diagdown {}_{S}\diagup O$$
$$\diagup \quad \diagdown$$
$$O \quad O$$

(ii)

(iii)

Acids also provide examples of hydrogen bonding schemes of the same kinds. In α-HIO₃ the scheme is essentially similar to that in NaHCO₃. The distances between O atoms of different IO_3^- ions were originally interpreted in terms of 'bifurcated' hydrogen bonds. However, it appears that each I has, in addition to its three nearest O neighbours in an IO_3^- ion, three next-nearest neighbours at distances considerably less than to be expected for van der Waals bonds. These latter three I–O bonds are shown (for one iodine atom) as dotted lines in Fig. 8.8. This suggests that I is forming three weaker bonds as well as three strong bonds. If we disregard those short 'intermolecular' O–O distances which are edges of the distorted octahedral IO_6 coordination groups, then each IO_3 group is joined to its neighbours by only two O–H–O bonds forming chains, as in the bicarbonate ion. A neutron diffraction study confirms that the O–H–O bonds (length 2·68 Å) are normal straight hydrogen bonds.

Disubstituted phosphoric acids, $HPO_2(OR)_2$, provide examples of both dimers and chains exactly comparable with the two types of HSO_4^- ions, and infinite chains also occur in crystalline benzene seleninic acid. The dimer in dibenzylphosphoric

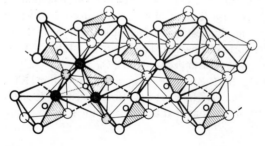

FIG. 8.8. The crystal structure of α-iodic acid shown as a system of linked IO_6 octahedra, or IO_3 groups (shaded) joined by normal hydrogen bonds (broken lines). The smaller circles represent I atoms.

acid, $HPO_2(OCH_2C_6H_5)_2$, is shown at (iv) and the chain in $HSeO_2(C_6H_5)$ at (v). A very similar chain is found in dimethylphosphinic acid, $(CH_3)_2PO(OH)$; the chain in $HPO_2(OC_6H_4Cl)_2$ has been noted earlier. This class also includes carboxylic

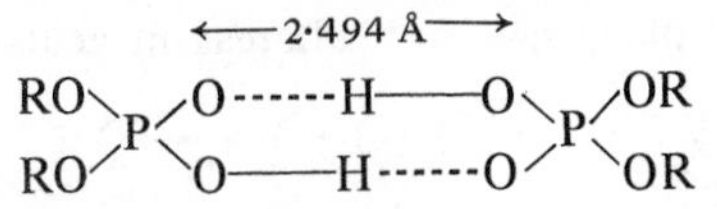

(iv)

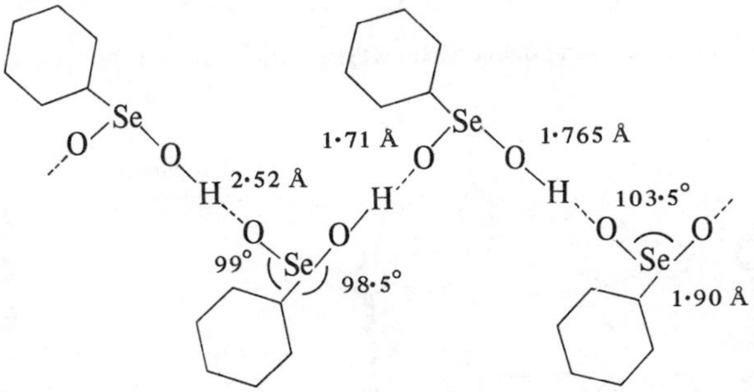

(v)

acids, in which the ratio of H atoms to R . COO is necessarily unity. The dimers present in the vapours and crystals of these compounds are of the same general type as the $(HSO_4^-)_2$ dimer shown at (i).

(*d*) H : XO_n *ratio* 3 : 2. There are now three hydrogen bonds from each oxy-ion, giving some interesting types of structure intermediate between those in classes (c) and (e). One is the chain (vi) defined by the short O–H–O bonds (2·55 Å) in

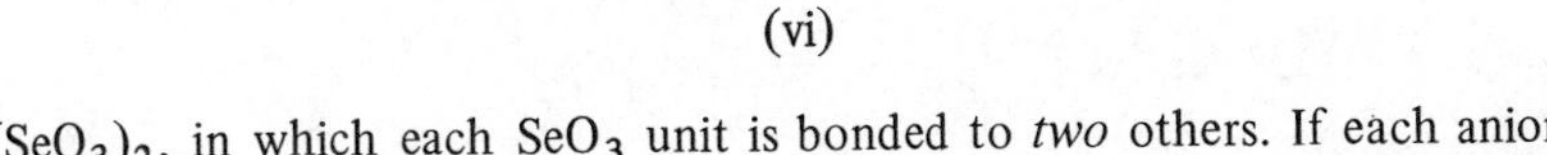

(vi)

$LiH_3(SeO_3)_2$, in which each SeO_3 unit is bonded to *two* others. If each anion is hydrogen-bonded to *three* others the topological possibilities are similar to those for a compound A_2X_3. In $NaH_3(SeO_3)_2$ each SeO_3 group is hydrogen-bonded to three others, and they form the simplest 3-connected plane (6-gon) net, as in one form of As_2O_3.

(*e*) H: XO_n *ratio* 2 : 1. This ratio implies that four hydrogen bonds link each oxy-ion to its neighbours. We therefore have the possibility of all types of hydrogen-bonded complex, from finite to 3-dimensional, as in class (d). Acid salts provide examples of 1- and 3-dimensional complexes, but layer structures do not appear to be known in this class. In $(NH_4)_2H_2P_2O_6$ short O–H–O bonds (2·53 Å) link the anions into chains (Fig. 8.9) which are then further linked into a 3D framework by N–H–O bonds. In Chapter 3, in our discussion of the diamond

317

structure, we noted that the anion in KH_2PO_4 is a 3D array of hydrogen-bonded $PO_2(OH)_2^-$ ions (Fig. 8.10).

Inorganic acids provide a number of examples of layer structures. The layers in H_2SeO_3, H_2SeO_4, and H_2SO_4 are all of the same topological type, being based on the simplest 4-connected plane net, but whereas in acids H_2XO_4 (Fig. 8.11(a))

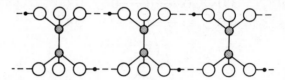

FIG. 8.9. Hydrogen-bonded chain $(H_2P_2O_6)_n^{2n-}$ in $(NH_4)_2H_2P_2O_6$.

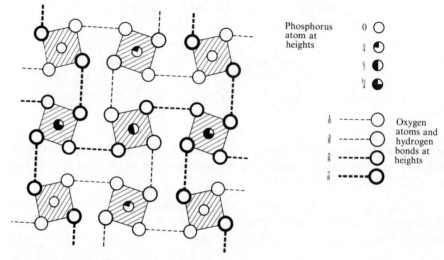

Phosphorus atom at heights

0 ○

$\frac{c}{4}$

$\frac{c}{2}$

$\frac{3c}{4}$

$\frac{1}{8}$ ----○
$\frac{3}{8}$ ----○
$\frac{5}{8}$ ----○
$\frac{7}{8}$ ----○

Oxygen atoms and hydrogen bonds at heights

FIG. 8.10. Part of the structure of KH_2PO_4 projected on to the basal plane (K^+ ions omitted). The broken lines represent hydrogen bonds which link the PO_4 tetrahedra into an infinite 3-dimensional network.

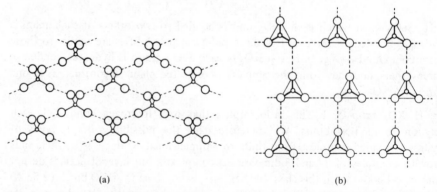

(a)

(b)

FIG. 8.11. Hydrogen bonding schemes in layers of (a) crystalline H_2SO_4, (b) orthorhombic H_2SeO_3 (diagrammatic). Broken lines indicate O—H—O bonds.

318

there is one hydrogen bond from each O atom, in H_2SeO_3 (with a smaller number of O atoms) there are two hydrogen bonds from one O and one from each of the others (Fig. 8.11(b)). This illustrates the point that while the $H : XO_n$ ratio in these crystals determines the general type of structure it is necessary to take account of the nature of the XO_n group if the structure is to be discussed in more detail.

Oxalic acid $(COOH)_2$, has two polymorphs, one of which has a layer structure and the other a chain structure:

α-oxalic acid

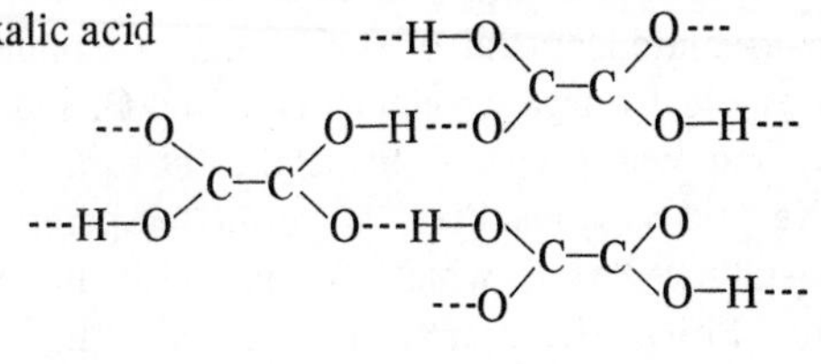

β-oxalic acid

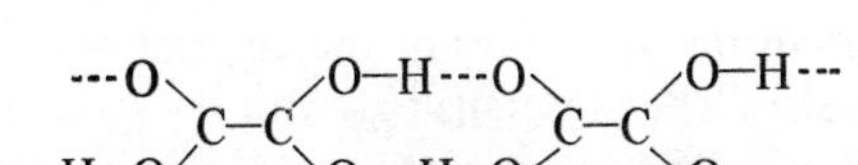

(*f*) $H : XO_n$ *ratio* 3 : 1. Crystalline H_3PO_4 has a layer structure. Within each layer each $PO(OH)_3$ molecule is linked to six others by hydrogen bonds which are of two lengths. Those between O and OH are short (2·53 Å) while those between OH groups of different molecules are longer (2·84 Å), as shown in Fig. 8.12. The salt $(NH_4)_2H_3IO_6$ is built of a 3D framework anion in which each molecule is hydrogen-bonded to six others, and the cations are accommodated in the interstices. Sulphamic acid, which apparently behaves as a zwitterion, $^-O_3S . NH_3^+$, in the crystal, is an example of an acid in which each unit is hydrogen-bonded to six others.

As already noted, the structures of hydroxy-acids, which include $B(OH)_3$ and $Te(OH)_6$, are described in Chapter 14 with hydroxides, to which they are more closely related structurally, the molecules being linked by long O–H–O bonds (two from each OH) as in many organic hydroxy-compounds. One form of metaboric acid, consisting of cyclic $B_3O_3(OH)_3$ molecules, is also of this type, for the three O atoms in the ring are not involved in hydrogen bonding.

We have confined our attention here to the structures of crystalline oxy-acids and 'acid salts', but other types of compound present similar possibilities as regards hydrogen bond systems. Two of the examples of Table 8.5 illustrate 3D systems of hydrogen-bonded ions, forming the simplest 4- and 6-connected nets. The 'hydrous oxides' of Sn and Pb extend the series with an example of a 3D system based on the 8-connected body-centred net:

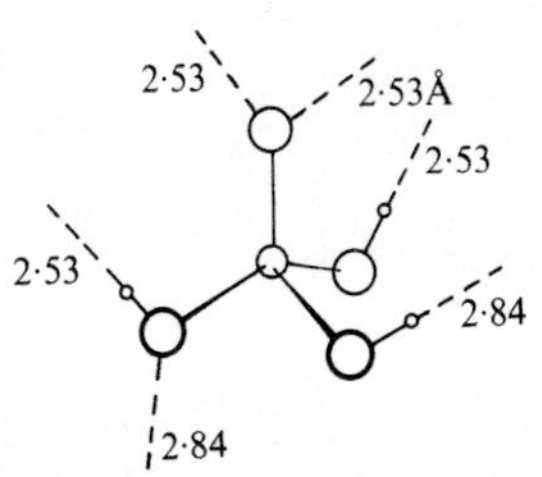

FIG. 8.12. The hydrogen bonds in crystalline H_3PO_4.

	3D *Net*
$(H_2PO_4)_n^{n-}$	Diamond
$(H_3IO_6)_n^{2n-}$	Simple cubic
$Sn_6O_4(OH)_4$	Body-centred net

319

COMPOUNDS OF THE NOBLE GASES

The existence of dimeric noble gas ions was first demonstrated in 1936 in the mass spectrometer, and there have been numerous later studies; they are formed by collisions involving an excited atom and a neutral atom ($X^* + X \rightarrow X_2^+ + e$). Other unstable species have been shown to exist in discharge tubes, for example, ArO, KrO, XeO (band spectrum), $XeCl_4^-$ in decay of ^{129}I to ^{129}Xe (Mössbauer effect), and XeF in a γ-irradiated crystal of XeF_4 (e.s.r.). Stable compounds are formed only with the most electronegative elements, F and O, and those of Xe are the most numerous. The first was prepared in 1962, the orange-yellow $XePtF_6$ formed directly from Xe and gaseous PtF_6 at room temperature. Krypton compounds include the crystalline KrF_2, which forms linear molecules, and an addition compound with $2\,SbF_5$; the preparation of KrF_4 has been described, but this compound does not seem to have been well characterized and its structure is not known.

We shall set out here the structures of the compounds of Xe without enlarging on their stereochemistry. The close analogy with the structural chemistry of iodine will be evident, for the bond arrangements are for the most part consistent with the simple view of the stereochemistry of non-transition elements set out in Chapter 7. It will be equally evident that the difficulties encountered with certain valence groups in connection with the stereochemistries of Sb, Te, and I are also encountered here.

Fluorides of xenon

These are prepared by direct combination of the elements under various conditions. They include the simple fluorides XeF_2, XeF_4, and XeF_6, which are all colourless crystalline solids with m.p. $140°$, $114°$, and $46°C$ respectively. There is no reliable evidence for an octafluoride, but there is a molecular compound $XeF_2 \cdot XeF_4$. The fluorides form numerous adducts with other halides, for example,

$$XeF_2 \cdot AsF_5, \; 2\,XeF_2 \cdot AsF_5$$
$$XeF_2 \cdot MF_5, \qquad XeF_2 \cdot 2\,MF_5, \; \text{and}\; XeF_6 \cdot MF_5 \qquad (M = Pt, Ir, etc.)$$

which form because the pentafluorides can abstract one F atom from the Xe fluoride to form cations as shown in the structural formulae:

$$(XeF)^+(M_2F_{11})^-, (Xe_2F_3)^+(AsF_6)^-, (XeF_5)^+(PtF_6)^-.$$

(On the formulation of the $(XeF)^+$ ion, see later.) Complex halides formed by XeF_6 include $RbXeF_7$, Rb_2XeF_8, and the corresponding Cs salts.

The structures of the XeF_2 and XeF_4 molecules are similar to those of the corresponding halogen ions ICl_2^- and ICl_4^-, namely, linear and square planar. In $XeF_2 \cdot XeF_4$ there are equal numbers of molecules of the two fluorides with structures the same as in the individual compounds. The same linear configuration for XeF_2 is found in $XeF_2 \cdot IF_5$, the shape of the IF_5 molecule being square pyramidal with I–F, $1 \cdot 88$ Å and bond angle to the axial I atom, $81°$.

In contrast to XeF_2 and XeF_4 the hexafluoride has presented great difficulty.

The latest e.d. data are not consistent with a regular octahedral structure or with the pentagonal bipyramidal model to be expected for a 14-electron valence group. It has been suggested that an irregular octahedral model with large amplitudes of the bending vibrations and mean length of the (non-equivalent) bonds 1·89 Å is consistent with the data.

There are four crystalline forms of XeF_6, those formed at lower temperatures having increasingly complex structures, none of which contains discrete XeF_6 molecules:

IV	—— (−180°) ——	III	—— −25° ——	II	—— +10°C ——	I
cubic		monoclinic		orthorhombic		monoclinic
Z 144		64		16		8

All the forms I, II, and III consist of tetramers. The change from I to II involves a gross rearrangement of one-half of the tetramers, while the change from II to III consists only in the ordering of the right- and left-handed forms of the tetramer. In the more complex cubic form IV there are two kinds of polymeric unit, both formed from F^- ions and square pyramidal XeF_5^- ions which are linked through the F^- ions by weaker Xe–F bonds into cyclic tetramers and hexamers, these numbering respectively 24 and 8 per unit cell. In the tetramer (Fig. 8.13(a)) each F^- forms an unsymmetrical bridge between two Xe atoms, and therefore each Xe has two additional neighbours, but in the hexamer (Fig. 8.13(b)) each F^- bridges (symmetrically) three Xe atoms. The environments of Xe are therefore as shown in Fig. 8.13(c) and (d):

	Tetramer		Hexamer
Xe	$\left\{\begin{array}{l} \underline{5\ F\ 1·84\ Å} \\ F\ 2·23 \\ F\ 2·60 \end{array}\right.$	Xe	$\left\{\begin{array}{l} \underline{\begin{array}{l}1\ F\ 1·76\ Å \\ 4\ F\ 1·92\end{array}} \\ 3\ F\ 2·56 \end{array}\right.$

Fluoro-ions

The structures of the adducts which can be formulated as salts containing the $(XeF)^+$ ion suggest that this is an over-simplification. In the compounds $XeF_2 . RuF_5$, $XeF_2 . 2\ SbF_5$, and $FXeOSO_2F$ Xe(II) forms two collinear bonds, one Xe–F being appreciably shorter than in XeF_2 (2·0 Å) and the other bond (Xe–F or Xe–O) much longer:

	a	*b*
$F\overset{a}{—}Xe\overset{b}{—}FRuF_5$	1·88 Å	2·19 Å
$F–Xe–FSb_2F_{10}$	1·84	2·35
$F–Xe–OSO_2F$	1·94	2·16 (Xe–O)

As in the case of other noble gas compounds alternative descriptions of the bonding in these compounds can be given (see Table 8.6 for references).

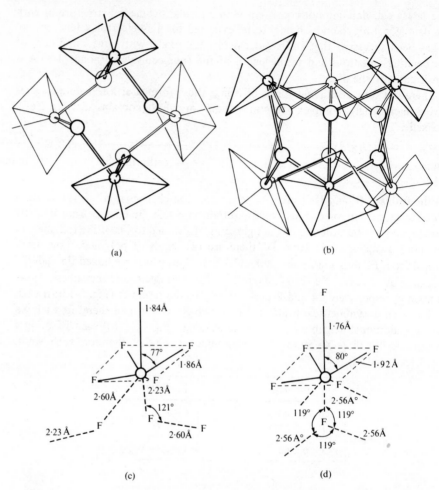

FIG. 8.13. Structure of crystalline XeF_6 (see text).

The existence of cations $Xe_2F_3^+$ and XeF_5^+ has been confirmed by structural studies. The compound with the empirical formula $2\,XeF_2 \cdot AsF_5$ contains planar ions (Fig. 8.14(a)) similar in shape to $(H_2F_3)^-$, in which the bond angle is $135°$, and I_5^-, bond angle $95°$. This is therefore an ionic compound $(Xe_2F_3)(AsF_6)$.

The salt $(XeF_5)(PtF_6)$ is made by the action of fluorine under pressure at $200°C$ on a mixture of Xe and PtF_5. It is an assembly of octahedral PtF_6^- ions (mean Pt–F, $1·89$ Å) and square pyramidal XeF_5^+ ions (Fig. 8.14(b)) in which the mean Xe–F bond length is $1·85$ Å. The Xe atom is $0·34$ Å below the base of the pyramid, and the next nearest neighbours of Xe are F atoms at $2·52$, $2·65$, and $2·95$ Å (two).

The anion XeF_8^{2-} in $(NO)_2^+(XeF_8)^{2-}$, formed from $2\,NOF$ and XeF_6, has the form of a very slightly deformed square antiprism. The mean values of the two sets of Xe–F distances to the F atoms at the vertices of the two square faces are: Xe–4 F, $1·96$ Å, Xe–4 F, $2·08$ Å.

322

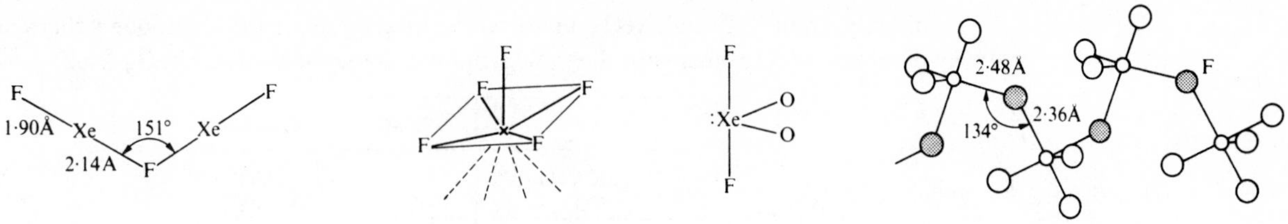

FIG. 8.14. Structures of compounds of xenon: (a) $Xe_2F_3^-$, (b) XeF_5^+, (c) XeO_2F_2, (d) XeO_3F^-.

Oxides and oxy-ions

The trioxide forms colourless crystals which are stable in dry air but deliquesce in moist air and are highly explosive. The crystal structure is similar to that of α-HIO_3 (XeO_3 is isoelectronic with IO_3^-) except, of course, that there is no hydrogen bonding:

	M—O	O—M—O (mean)
XeO_3	1·76 Å	103°
IO_3^-	1·82	97°

The tetroxide is made by the action of H_2SO_4 on the perxenate, Na_4XeO_6. In the vapour state it forms a regular tetrahedral molecule, the Xe–O bond length (1·74 Å) being intermediate between the values found in XeO_3 (1·76 Å) and $XeOF_4$ (1·70 Å).

Xenates (containing Xe^{VI}) and perxenates (containing Xe^{VIII}) have been prepared. For example, $CsHXeO_4$ has been made from XeO_3 and $CsOH$ in the presence of F^- ion, and hydrated salts containing the XeO_6^{4-} ion from XeF_6 and alkali hydroxides. The perxenate ion, of special interest because of the high oxidation state of Xe, forms a regular octahedron (Xe–O, 1·86 Å) in Na_4XeO_6. 6 H_2O and 8 H_2O and in K_4XeO_6. 9 H_2O.

Oxyfluorides and $KXeO_3F$

Two oxyfluorides are known. $XeOF_4$, a colourless liquid prepared from XeO_3 and XeF_6, shows a resemblance to halogen fluorides in forming compounds such as $CsF . XeOF_4$ and $XeOF_4. 2 SbF_5$. A microwave study indicates a square pyramidal configuration with Xe approximately in the basal plane, that is, essentially the same structure as XeF_5^-, one F being replaced by O.

The action of XeO_3 on $XeOF_4$ produces XeO_2F_2. Its molecular structure resembles that of $IO_2F_2^-$, that is, a trigonal bipyramid with a lone pair occupying one equatorial bond position (or the linear XeF_2 molecule plus two equatorial O atoms (Fig. 8.14(c)); bond angles are: O–Xe–O, 105·7°, F–Xe–F, 174·7°. Two weak additional equatorial bonds (Xe–O, 2·81 Å) from each Xe lead to a pronounced pseudo-octahedral layer of the SnF_4 type.

Closely related to these oxyfluorides is the salt $KXeO_3F$, which is prepared

323

directly from CsF and XeO_3 in solution. In crystals of this compound there are pyramidal XeO_3 units with the same structure as the molecule of XeO_3 itself:

	Xe—O	O—Xe—O
$KXeO_3F$	1·77 Å	100°
XeO_3	1·76	103°
	I—O	O—I—O
(compare IO_3^-	1·82	97°)

These XeO_3 units are joined through bridging F atoms to form infinite chains of composition XeO_3F (Fig. 8.14(d)). The Xe—F bonds are of two lengths, 2·36 and

TABLE 8.6

Structures of noble gas compounds

	Configuration	Xe—F	Xe—O	*Reference*
XeF_2	Linear	1·98 Å		JCP 1969 **51** 2355 (i.r.)
		2·00		JACS 1963 **85** 241 (cryst.)
XeF_4	Square planar	1·95		NGC p. 238 (e.d.)*
		1·93		Sc 1963 **139** 1208 (n.d.)
XeF_6	Distorted octahedral (vapour)	1·89 (mean)		JCP 1968 **48** 2460, 2466
	Crystal (see text)			{ Sc 1970 **168** 248
				{ Sc 1971 **171** 485
$XeF_2 . XeF_4$	See text: XeF_2	2·01		AC 1965 **18** 11
	XeF_4	1·96		
$XeF_2 . IF_5$		2·02		IC 1970 **9** 2264
$Xe_2F_3^+$		1·90		CC 1968 1048
		2·14		
XeF_5^+		1·85 (mean)		JCS A 1967 1190
XeF_8^{2-}	Square antiprism	1·96, 2·08		Sc 1971 **173** 1238
$FXeFSb_2F_{10}$	Linear Xe bonds	1·84	—	CC 1969 62
		2·35		
$FXeFRuF_5$	Linear Xe bonds	1·88	—	}
		2·19		} IC 1972 **11** 1124
$FXeOSO_2F$	Linear Xe bonds	1·94	2·16	}
$XeOF_4$	Square pyramid	1·90	1·70 Å	JMS 1968 **26** 410
XeO_2F_2	Trigonal bipyramidal	1·90	1·71	JCP 1973 **59** 453 (n.d.)
XeO_3	Pyramidal		1·76	JACS 1963 **85** 817
XeO_4	Tetrahedral		1·736	JCP 1970 **52** 812 (e.d.)
$KXeO_3F$		2·36	1·77	IC 1969 **8** 326
		2·48		
XeO_6^{4-}	Octahedral		1·86	IC 1964 **3** 1412, 1417
				JACS 1964 **86** 3569
$XeCl_2$	Linear			IC 1967 **6** 1758 (i.r.)
		Kr—F		
KrF_2	Linear	1·87		JACS 1968 **90** 5690 (i.r.)
		1·89		JACS 1967 **89** 6466 (e.d.)

* NGC: Noble Gas Compounds, H. H. Hyman, Ed., University of Chicago Press, 1963. For references to earlier work see also: Sc 1964 **145** 773. For discussions of the bonding in noble gas compounds see JCS 1964 1442, and IC 1972 **11** 1124.

2·48 Å, appreciably longer than in XeF_4 or XeF_6 (1·95–2·0 Å) but much shorter than would be expected for ionic or van der Waals contacts (of the order of 3·5 Å). The five bonds from Xe, three normal Xe–O bonds and two weak Xe–F bonds, are directed towards five of the vertices of an octahedron.

Structural details for the noble gas compounds and references to the literature are given in Table 8.6.

9

The Halogens–Simple Halides

Introduction

In every Periodic family the availability of only four orbitals differentiates the first member from the later ones which can make use of d in addition to s and p orbitals. We therefore find important differences between the first and later elements. In addition there are marked differences between iodine on the one hand and bromine and chlorine on the other; there is indeed little chemical similarity between the extreme members of this family, fluorine and iodine.

The first outstanding characteristic of fluorine is its extraordinary chemical reactivity. It combines directly with all non-metals except N and the noble gases (other than Kr and Xe) and with all metals. Moreover the product is (or includes) the highest fluoride of the element. For example, $AgCl \rightarrow AgF_2$, $CoCl_2 \rightarrow CoF_3$, and the platinum metals yield their highest known fluorides, RuF_5, OsF_6, IrF_6, and PtF_6. Many elements, particularly non-metals, exhibit higher oxidation states in their fluorides than in their other halides—compare AsF_5, SF_6, IF_7, and OsF_6 (earlier described as OsF_8) with the highest chlorides, $AsCl_3$, SCl_4, ICl_3, and $OsCl_4$.

Fluorine seldom forms more than one essentially covalent bond—see, for example, the interhalogen compounds and polyhalides—though two collinear bonds (with appreciable ionic character) are formed in crystalline compounds MF_3, MF_5, etc. (see later) and in the ion $(C_2H_5)_3Al-F-Al(C_2H_5)_3^-$, which has been studied in the K salt.[1] Chlorine and bromine form up to four bonds to oxygen and exceed this number only in combination with fluorine (ClF_5, BrF_5, and BrF_6^-), while iodine is 6-covalent in some oxy-compounds and also forms a heptafluoride. An ion X_3^- is not formed by fluorine, and is most stable in the case of I_3^-. Another feature of fluorine is its small affinity for oxygen, a characteristic also of bromine and incidentally of Se, the element preceding Br in the Periodic Table.

Fluorine is the most electronegative of all the elements, so that the bonds it forms with most other elements have considerable ionic character. With the exception of the alkali halides most crystalline fluorides have structures different from those of the other halides of the same metal. A number of difluorides and dioxides have the same crystal structure, whereas the corresponding dichlorides, dibromides, and diiodides have in many cases structures similar to those of disulphides, diselenides, and ditellurides. The extreme electronegativity of fluorine enables it to form much stronger hydrogen bonds than any other element, resulting in the abnormal properties of HF as compared with the other acids HX, the much

(1) AC 1963 **16** 185

greater stability of the ion $(F-H-F)^-$ than of other ions HX_2^-, and structural differences between hydrated fluorides and other halides. The structures of the ions HX_2^-, $H_2F_3^-$, and $H_4F_5^-$ are described in Chapter 8.

The anomalously low dissociation energy of the F—F bond has usually been attributed to the repulsions of the unshared electrons of the two valence shells (Table 9.1); the extrapolated value on a $1/R$ plot, R being the bond length in the X—X molecule, gives a value some 226 kJ mol^{-1} higher. It has been pointed out[2] that there are similar anomalies in quantities involving only one F atom. For example, the plot of electron affinity against ionization potential is linear for I, Br, and Cl and would extrapolate to a value 1·14 eV (108·8 kJ mol^{-1}) greater than the observed value. There are similar anomalies in the dissociation energies of covalent molecules such as HX or CH_3X or the gaseous alkali-halide molecules in which the bonds are ionic bonds. These facts suggest that an abnormally large repulsive force is experienced by an electron entering the outer shell *wherever it comes from*, being the same for an electron entering to form an ion as for one which is being shared in the formation of a covalent bond. When compared with twice the *atomic* radius (defined as the average radial distance from the nucleus of the outermost p electrons) it is seen that the bond in the F_2 molecule is abnormally long:

(2) JACS 1969 **91** 6235

	F	Cl	Br	I
2 (atomic radius):	1·14	1·94	2·24	2·64 A
X—X in X_2:	1·43	1·99	2·28	2·67 A

TABLE 9.1

Some properties of the halogens

	Ionic radius	X—X in X_2	H—X in HX	Dissociation energy X—X	Ionization potential	Electron affinity
	(Å)	(Å)	(Å)	(kJ mol^{-1})	(eV)	(eV)
F	1·36	1·43	0·92	154·8	17·42	3·448
Cl	1·81	1·99	1·28	242·7	12·96	3·613
Br	1·95	2·28	1·41	192·5	11·81	3·363
I	2·16	2·66	1·60	150·6	10·45	3·063

We note later some cations containing exclusively halogen atoms. Iodine also forms ions $(R-I-R)^+$ in which the groups R are usually aromatic radicals (Ar). The halides $(Ar-I-Ar)X$ are not very stable, but the hydroxides are strong bases, proving dissociation to $(Ar-I-Ar)^+$ ions. These iodonium compounds are the analogues of ammonium and sulphonium compounds:

$$(NR_4)X \qquad (SR_3)X \qquad (IR_2)X.$$

The expected angular shape of $(IR_2)^+$ has been demonstrated in crystalline $(C_6H_5)_2I^+(ICl_2)^-$, the angle between the iodine bonds in the cation being 94° and

(3) AC 1960 **13** 1140

I–C, 2·02 Å.[3] Other compounds peculiar to iodine include the iodoso and iodyl compounds of general formulae $Ar–I=O$ and $Ar–IO_2$ which undergo interesting reactions such as

$$C_6H_5IO + C_6H_5IO_2 + Ag^+ \rightarrow (C_6H_5)_2I^+ + AgIO_3.$$

Iodine forms a number of so-called salts, IPO_4, $I(CH_2Cl \cdot COO)_3$ and others of unknown constitution which are sometimes regarded as containing I^{3+} ions and quoted to illustrate the 'basic' properties of iodine. It is, in fact, not certain that they are 'normal' salts, for the nitrate (prepared by the action of 100 per cent

(4) CR 1954 **238** 1229

HNO_3 on iodine) is $IO(NO_3)$ and not $I(NO_3)_3$.[4] (Contrast the covalent nitrate of fluorine, p. 665.) Both Br and I form nitrato *anions* which are stable in combination with large cations, as in the $N(CH_3)_4^+$ salts of $[Br(NO_3)_2]^-$,

(5) IC 1966 **5** 2124
(6) ACSc 1964 **18** 144

$[I(NO_3)_2]^-$, and $[I(NO_3)_4]^-$.[5] On the other hand iodosyl sulphate and selenate are both anhydrous compounds with the formulae $(IO)_2SO_4$ and $(IO)_2SeO_4$.[6]

(6a) JCS 1960 3350
(7) IC 1965 **4** 257

Their structures are not known but i.r. evidence suggests that they contain some polymeric species rather than $(IO)^+$ ions.[6a] Infrared evidence suggests[7] that the compound of IO_2F and AsF_5 is $(IO_2)^+(AsF_6)^-$, and iodyl fluorosulphate has been

(8) IC 1964 **3** 1799

prepared as a hygroscopic powder stable up to 100°C,[8] but the structure of the IO_2 ion or group is not known. Magnetic, spectroscopic, and conductivity measurements appear to be consistent with the presence of I_2^+ ions (with μ_{eff}. 2 BM) in the blue solutions formed by oxidizing iodine in fluorosulphuric acid or oleum. On cooling to low temperatures these solutions become red in colour

(9) IC 1969 **8** 1751

and diamagnetic, apparently due to the conversion of 2 I_2^+ into I_4^{2+}.[9] For halogen cations see p. 333.

Nothing is yet known of the structural chemistry of astatine. The isotope At^{211} is produced by bombardment of Bi by α particles. The formation of compounds has been studied in a time-of-flight mass spectrometer, those observed being HAt, $AtCH_3$, AtCl, AtBr, and AtI. There was no evidence for the formation of At_2

(10) IC 1966 **5** 766

molecules.[10]

The stereochemistry of chlorine, bromine, and iodine

The stereochemistry of these elements is concerned with valence groups of 8, 10, 12, and 14 electrons, except for the few molecules such as ClO_2 which contain odd numbers of electrons. The shapes of all molecules and ions which are known at the present time are consistent with the following arrangements of 4, 5, 6, and 7 pairs of valence electrons: tetrahedral, trigonal bipyramidal, octahedral, and pentagonal bipyramidal, assuming that lone pairs occupy bond positions and that the gross stereochemistry is not affected by the π-bonding present in oxy-ions and molecules. Examples are given in Table 9.2. No example of the fully shared 10-electron group has yet been studied; the ion IO_5^{3-}, intermediate between IO_4^- and IO_6^{5-} may exist in Ag_3IO_5. The configuration of the IF_6^- ion, at present unknown, presents the same problem as the $SeBr_6^{2-}$ and similar ions.

TABLE 9.2

The stereochemistry of chlorine, bromine, and iodine

No. of σ-electron pairs	Type of hybrid	No. of lone pairs	Bond arrangement	Examples
4	sp^3	0	Tetrahedral	ClO_4^-, IO_4^-
		1	Trigonal pyramidal	ClO_3^-, BrO_3^-, HIO_3
		2	Angular	ClO_2^-, BrF_2^+, IR_2^+
5	$sp^3 d_{z^2}$	0	Trigonal bipyramidal	IO_5^{3-}?
		1	Distorted tetrahedral	$IO_2F_2^-$
		2	T-shape	ClF_3, BrF_3, $R . ICl_2$
		3	Collinear	I_3^-, ICl_2^-, $BrICl^-$
6	$sp^3 d_\gamma^2$	0	Octahedral	IO_6^{5-}
		1	Square pyramidal	BrF_5
		2	Square planar	BrF_4^-, ICl_4^-
7	$sp^3 d_\gamma^2 d_\epsilon$	0	Pentagonal bipyramidal	IF_7
		1	Distorted octahedral	IF_6^-?

The structures of the elements

Having seven electrons in the valence shell the halogens can acquire a noble gas structure by forming one electron-pair bond. Accordingly they form diatomic molecules in all states of aggregation. (The detailed structures of the two crystalline forms of F_2 are described on p. 665; neither structure is similar to that of crystalline Cl_2, Br_2, and I_2.) A remarkable feature is that the molecules are arranged in layers, within which the shortest intermolecular distance is considerably less than that between the layers (Table 9.3). In Fig. 9.1 (p. 330) we compare the packing of the molecules in layers of crystalline I_2, ICl, and IBr. This additional weak bonding

TABLE 9.3

Interatomic distances in crystalline halogens

	X–X (Å)	X–X (intermolecular) (Å)		Reference
		In layer	Between layers	
F	–	(See p. 1010)	–	α JSSC 1970 **2** 225
				β AC 1964 **17** 777
Cl	1·98	3·32 3·82	3·74	AC 1965 **18** 568
Br	2·27	3·31 3·79	3·99	AC 1965 **18** 568
I	2·72	3·50 3·97	4·27	AC 1967 **23** 90

is very evident also in the structures of polyiodides, which are described later in this chapter. It invites comparison with the 'charge-transfer' bonds in the numerous

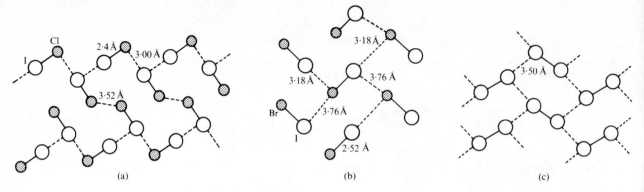

FIG. 9.1. Molecular arrangement in (a) crystalline α-ICl, (b) IBr, (c) crystalline I_2.

molecular addition compounds formed by halogens, and halogen halides, in which these bonds are collinear with the X—X molecule and much shorter than the sums of the appropriate van der Waals radii:

(1) AC 1968 **B24** 713
(2) AC 1964 **17** 712. See also: QRCS 1962 **16** 1

$$\overset{(CH_3)_2}{\underset{C}{\overset{\displaystyle O}{}}}\cdots Br\text{—}Br \cdots \overset{C(CH_3)_2}{\underset{2.82\,Å}{O}}$$

2.28 Å

$$H_3CCN\cdots Br\text{—}Br\cdots NCCH_3$$

2.84 Å   (1)

2.33 A

$$I\text{—}I \overset{2.91\,A}{\underset{2.76\,Å}{\text{--------}}} Se\overset{H_2C\text{—}CH_2}{\underset{H_2C\text{—}CH_2}{<}} \quad (2)$$

Interhalogen compounds

The compounds stable at ordinary temperatures are of the general form XX'_n where $n = 1$, 3, 5, or 7; none containing more than two different halogens is known. (Molecules with the composition Br_2Cl_2 are apparently formed when a mixture of the elements with a noble gas is passed through a microwave discharge and condensed at 20°K.[1] Their i.r. spectra suggest T-shaped molecules like ClF_3 (see below).) Interhalogen compounds are made by direct combination of the elements, except IF_7, which is prepared from IF_5. Of the six possible compounds XX' all have been prepared except IF, for the existence of which there is only spectroscopic evidence. It has not yet been possible to prepare pure BrF, which readily breaks down into $BrF_3 + Br_2$; phase-rule studies indicate that BrF may exist in the liquid state, but it does not exist in the solid state. Of the higher compounds ClF_3, BrF_3, and ICl_3, ClF_5, BrF_5 and IF_5, and IF_7 are known. As far as is known fluorine forms no compounds of the kind FX_3, FX_5, or FX_7, and iodine alone exhibits a covalency of seven. Of the penta- and hepta-compounds only fluorides have so far been prepared.

(1) IC 1968 7 1695

330

The properties of bromine trifluoride

This compound has become of special interest as a fluorinating agent by means of which a large number of new complex fluorides have been prepared.

(a) Salts of the type KVF_6 can be prepared by its action on a mixture of KCl and VCl_3, and it provides a convenient way of making salts such as KBF_4, $AgPF_6$, $KAsF_6$, $LiSbF_6$, etc.

(b) Fluorobromonium salts can be prepared in various ways, for example, $(BrF_2)NbF_6$ from Nb_2O_5, $(BrF_2)BiF_6$ from BiF_5, and $(BrF_2)AuF_4$ by dissolving Au in warm BrF_3.

If these salts react with $M(BrF_4)$ in BrF_3 solution they give salts of the type (a).

(c) By the action of NOCl in BrF_3 solution nitrosyl salts are obtained, while N_2O_4 in a similar way gives nitronium salts;

$$B_2O_3 \rightarrow NO(BF_4) \text{ and } NO_2(BF_4) \text{ respectively;}$$
$$PBr_5 \rightarrow NO(PF_6) \text{ and } NO_2(PF_6); \text{ and}$$
$$SnF_4 \rightarrow (NO)_2SnF_6 \text{ and } (NO_2)_2SnF_6, \text{ and so on.}$$

Electrical conductivity data on BrF_3 have been interpreted in terms of partial dissociation into BrF_2^+ and BrF_4^- ions, and this is supported by the preparation and reactions of the fluorobromonium salts (b):

$$2\,BrF_3 \rightleftharpoons BrF_2^+ + BrF_4^-$$

$$SbF_5 + BrF_3 \rightleftharpoons SbF_6^- + BrF_2^+$$

A solution of $AuBrF_6$ in BrF_3 immediately precipitates $Ag(AuF_4)$ from a solution of $AgBrF_4$ in the same solvent:

$$(BrF_2)(AuF_4) + Ag(BrF_4) \rightarrow Ag(AuF_4) + 2\,BrF_3$$

(The compound $(BrF_2)(AuF_4)$ offers a convenient way of preparing AuF_3, into which it decomposes on heating to $180°C$.)

The solvent action of ClF_3 also is presumably due to ionization. For example, it dissolves AsF_5 to give $(ClF_2)(AsF_6)$, and ClF_3 and BF_3 at $-78°C$, when warmed to room temperature, form $BClF_6$ as a colourless solid. It is presumably $(BF_4)(ClF_2)$.

The structures of interhalogen compounds

Bond lengths in gaseous molecules XX′ are listed in Table 9.4. In crystalline α-ICl[2] the molecules (mean I–Cl, 2·40 Å) form zigzag chains with rather strong

(2) AC 1956 **9** 274

TABLE 9.4

Interatomic distances in molecules XX′

F—Cl	1·628 Å	PR 1949 **76** 1723
F—Br	1·759	PR 1950 **77** 420
Cl—Br	2·138	PR 1950 **79** 1007
Cl—I	2·303	PR 1948 **72** 1268

interaction between I and I (3·08 Å) and between I and Cl (3·00 Å) of different molecules (Fig. 9.1(a)). The angles between the I—Cl bond and these weak additional bonds are 102° at Cl and 94° at I. Similar bond lengths and angles are found in the second (β) form of solid ICl.[3] Crystalline IBr has a structure very similar to that of I_2, and here also there are short I-Br contacts in the layers (Fig. 9.1(b)).[4] The shortest X—X distances between layers are: I—Br, 4·18 Å, I—I, Br—Br, 4·27 Å. This tendency to form additional weaker bonds is also exhibited in polyhalides (see later).

Molecules XX′$_3$. ClF_3 has been studied both by m.w. and also in the crystalline state.[5] The (planar) molecule has the T-shape shown at (a), below, the corresponding results from the X-ray study being 1·716 and 1·621 Å and 86°59′. A m.w. study[6] shows that molecules of BrF_3 have the same shape as ClF_3, with Br—F, 1·810 and 1·721 Å and F—Br—F, 86°13′. The structure of ICl_3 in the solid state is entirely different,[7] for the crystal is built of planar I_2Cl_6 molecules (b) in which the terminal I—Cl bonds are of about the same length as in ICl, ICl_2^-, and ICl_4^-, but the central bonds are considerably longer. There is no obvious theoretical explanation of the form of this molecule.

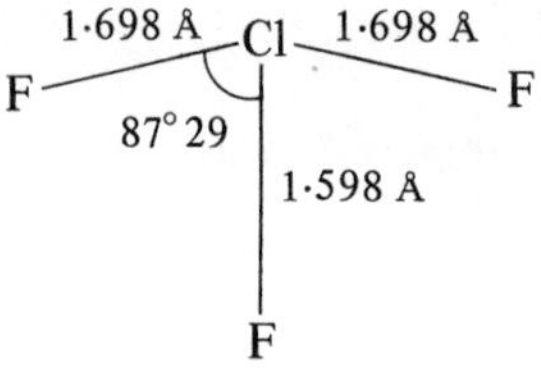

(3) AC 1962 **15** 360

(4) AC 1968 **B24** 429, 1702

(5) JCP 1953 **21** 609, 602

(6) JCP 1957 **27** 223

(7) AC 1954 **7** 417

(a) 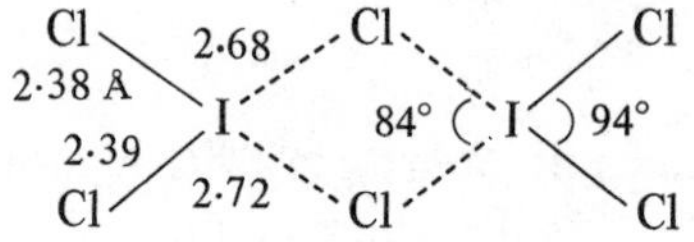 (b)

The structure of the simple ICl_3 molecule would be of great interest, but unfortunately vapour density measurements show that ICl_3 is already completely dissociated into ICl + Cl_2 at 77°C, though it melts at 101°C under its own vapour pressure (12 atm). The absorption spectrum of its solution in CCl_4 is the sum of those of ICl and Cl_2. In benzene iododichloride[8] the linear—ICl_2 system is approximately perpendicular to the plane of the benzene ring. Here the I—C bond occupies one of the equatorial bond positions. (In this molecule I—Cl is apparently rather longer (2·45 Å) than in ICl_2^-.)

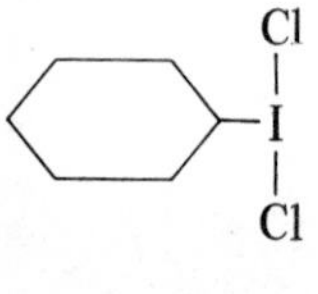

Although compounds such as $IAlCl_6$ and $ISbCl_8$, formed by combining $AlCl_3$ or $SbCl_5$ with ICl_3, may be formulated as salts containing the $(ICl_2)^+$ ions, their structures are less simple.[9] The structure of $ISbCl_8$ consists of (angular) ICl_2 and octahedral $SbCl_6$ groups, but there is weak bonding between the groups which links them into chains, each I forming two weak bonds (of length about 2·9 Å) in

(8) AC 1953 **6** 88

(9) AC 1959 **12** 859

addition to those (2·31 Å) within the ICl_2^+ ion, (a). Together with the, probably

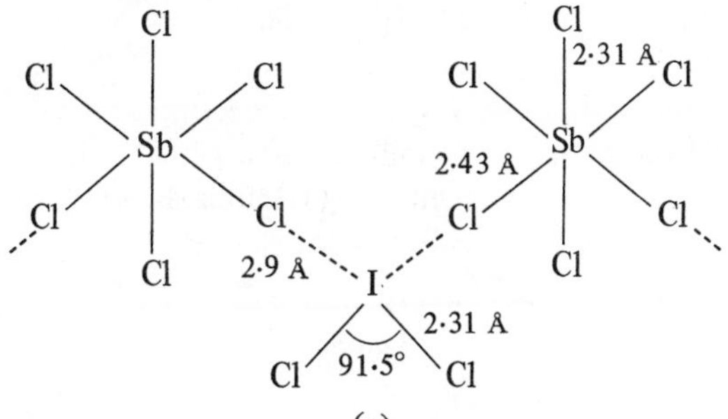

(a)

real, lengthening of two of the Sb—Cl bonds, this suggests that the state of the compound is intermediate between

$$(ICl_2)^+(SbCl_6)^- \text{ and } (ICl_4)^-(SbCl_4)^+$$

the first form predominating. Similarly the aluminium compound is in a state intermediate between $(ICl_2)^+(AlCl_4)^-$ and $(ICl_4)^-(AlCl_2)^+$. These structures are of particular interest because the bond lengths in ICl_3 are consistent with the formulation:

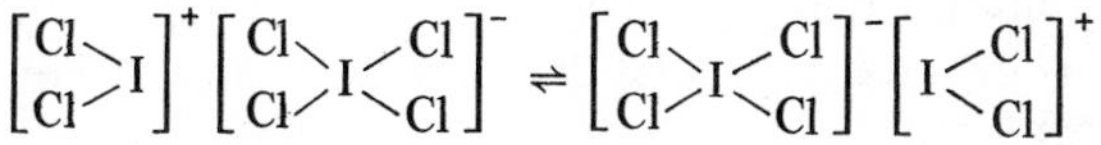

$$\begin{bmatrix} Cl \\ Cl \end{bmatrix}^+ \begin{bmatrix} Cl & Cl \\ Cl & Cl \end{bmatrix}^- \rightleftharpoons \begin{bmatrix} Cl & Cl \\ Cl & Cl \end{bmatrix}^- \begin{bmatrix} Cl \\ Cl \end{bmatrix}^+$$

Molecules XX'_5. The square pyramidal configuration expected for molecules of this type is confirmed by an X-ray study of crystalline BrF_5 (at $-120°C$) which shows that the Br atom lies just below the plane of the four F atoms.[10] One Br—F bond (1·68 Å) was found to be shorter than the other four (1·75–1·82 Å).

(10) JCP 1957 **27** 982

   The square pyramidal shape of the IF_5 molecule has been demonstrated in $XeF_2 \cdot IF_5$ (I—F, 1·88 Å), with I situated just below the base of the pyramid, giving the angle F_{apical}—I—F_{basal}, 81° (p. 320).

Molecules XX'_7. The pentagonal bipyramidal configuration suggested by i.r. and Raman data for IF_7 is consistent with the e.d. data,[11] for a molecule having all I—F bond lengths approximately equal ($1·825 \pm 0·015$ Å) but with the five equatorial F atoms not quite coplanar. The interpretation of the X-ray diffraction data, obtained from the orthorhombic form stable below $-120°C$, has presented much difficulty. The most definite statement in the latest review of the X-ray evidence is that it is not possible to demonstrate that the molecular symmetry is different from D_{5h}.[12]

(11) JCP 1960 **33** 182

(12) AC 1965 **18** 1018

Halogen and interhalogen cations
Most of our knowledge of these cations has come from work carried out during the

past decade. These ions are strongly electrophilic and therefore can exist only in the presence of molecules and anions of very low basicity, that is, in solvents such as fluorosulphuric or disulphuric acids, IF_5, or SbF_5, or in crystalline salts with anions such as AsF_6^-, SbF_6^-, or $Sb_3F_{16}^-$. The experimental techniques used for studying these ions are conductimetric, cryoscopic, spectrophotometric (visible, infrared, and Raman), and magnetic susceptibility measurements; the crystal structures of a few salts have been determined, containing the ions shown in heavy type below:

	Br_2^+	I_2^+	ClF_2^+	BrF_2^+	IF_2^+
Cl_3^+	Br_3^+	I_3^+	Cl_2F^+		ICl_2^+
		I_4^{2+}		BrF_4^+	IF_6^+
		I_5^+			

Crystalline adducts with halides have been made containing all except the three more complex iodine cations; the Cl_2^+ cation is known only as a vapour species, and is not known to be stable in solution or in the solid state.

The iodine cations were first studied, the I_2^+ ion ($\mu = 2 \cdot 0$ BM) being responsible for the blue colour of solutions of iodine in FSO_3H (previously attributed to I^+). This colour changes to the bright red due to I_4^{2+} at temperatures near the freezing point of this solvent. Blue salts $I_2(Sb_2F_{11})$ and $I_2(Ta_2F_{11})$ have been isolated, but at present the structures of the iodine cations are not known. A study of the crystal structure of the paramagnetic scarlet salt $Br_2(Sb_3F_{16})$,[1] which has $\mu = 1 \cdot 6$ BM, shows that it consists of Br_2^+ ions (Br–Br, 2·15 Å, compare 2·27 Å in Br_2) and anions consisting of three octahedral SbF_6 groups, the central one of which shares *trans* F atoms (a).

(1) JCS A 1971 2318

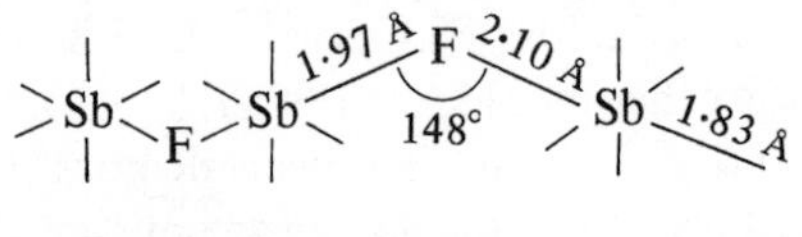

(a)

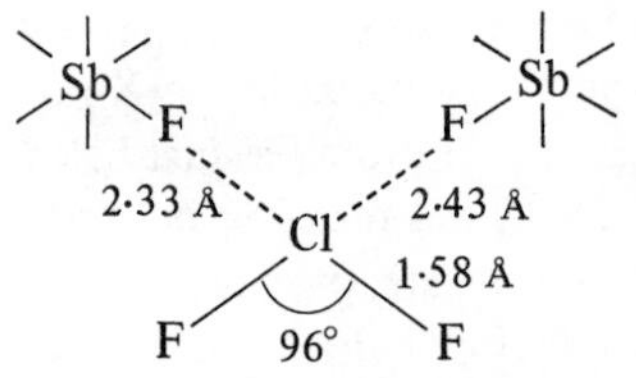

(b)

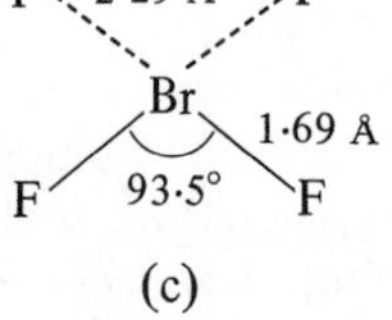

(c)

Ions $XX_2'^+$ (20 valence electrons) are expected to be angular. We have already noted (p. 332) that the structures of the adducts of ICl_3 with $SbCl_5$ and $AlCl_3$ may be formulated as containing ICl_2^+ ions but that there is considerable interaction between anions and cations in these crystals, giving an approximately square arrangement of 4 Cl around I. There is a similar type of interaction between ClF_2^+ and BrF_2^+ with the F atoms of the anions in $(ClF_2)(SbF_6)$,[2] and $(BrF_2)(SbF_6)$,[3] as shown at (b) and (c). Raman spectra of the adducts $AsF_5 . 2 ClF$ and $BF_3 . 2 ClF$ show that these compounds contain unsymmetrical $(Cl-Cl-F)^+$ cations,[4] and not the symmetrical ion as formerly supposed. The ion $(BrF_4)^+$ has been studied in $(BrF_4)(Sb_2F_{11})$;[5] it has the expected SF_4 type of structure, a trigonal bipyramid with the lone pair occupying one equatorial bond position. The mean length of the four short Br-F bonds is 1·81 Å; two weak bonds in the equatorial plane complete distorted octahedral coordination (d). The asymmetry of the

(2) JCS 1970 A 2697
(3) JCS 1969 A 1467

(4) IC 1970 **9** 811

(5) IC 1972 **11** 608

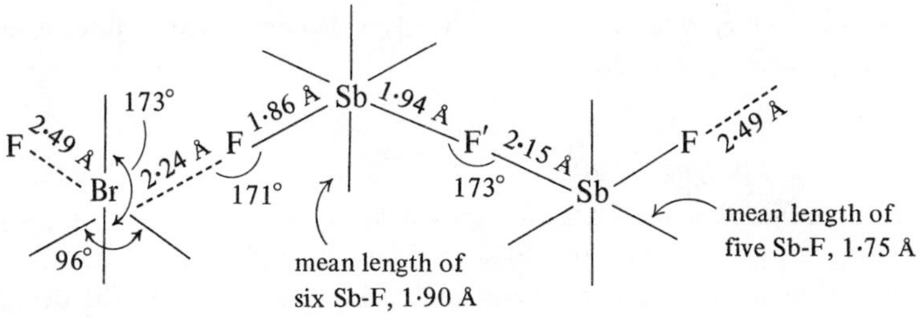

(d)

bridge in the $(Sb_2F_{11})^-$ ion suggests that this ion is not far removed from $(SbF_6)^-(SbF_5)$, the mean length of the six Sb—F bonds in SbF_6^- being 1·90 Å and of the five bonds in SbF_5 (i.e. excluding the bridge bond), 1·75 Å. The IF_6^+ ion has the expected octahedral shape (possibly slightly distorted),[6] in $(IF_6)(AsF_6)$, a salt with the NaCl structure.[7]

A review of these compounds is available.[8]

(6) JCP 1970 **52** 1960
(7) IC 1967 **6** 1783

(8) QRCS 1971 **25** 553

Polyhalides

These are salts containing *anions* of the types set out below, which with the exception of the I_8^{2-} ion in Cs_2I_8 are singly charged.

Cl_3^-	Br_3^-	I_3^-	
ClF_2^-	$BrCl_2^-$	ICl_2^-	$IBrF^-$
$ClBr_2^-$	BrI_2^-	IBr_2^-	$IClBr^-$
		I_2Br^-	
ClF_4^-	BrF_4^-	IF_4^- ICl_3F^- ICl_4^-	
	BrF_6^-	IF_6^-	
		I_8^{2-}	

They may be prepared by direct combination of metallic (or other) halide with interhalogen compound (or halogen in the case of $CsIF_4$, $CsBrF_4$, $CsClF_4$, $CsBr_3$, CsI_3, and Cs_2I_8) either in solution in a suitable solvent or by the action of the vapour of the interhalogen compound on the dry halide. $KBrF_4$ is prepared by dissolving KF in BrF_3 and distilling off the excess solvent; KIF_4 can be crystallized from a cold solution of KI in IF_5 and KIF_6 from the hot solution. Very unstable compounds must be prepared at low temperatures, for example, $NO(ClF_2)$ from NOF and ClF at $-78°C$. In general the only crystalline salts stable at ordinary temperatures are those of the larger alkali metals and large organic cations. For example, the only salts MI_3, $MICl_2$, and $MICl_4$ are those of NH_4, Rb, and Cs and of ions such as $NH(CH_3)_3^+$ and $(C_6H_5N_2)^+$. The stability increases with increasing size of cation in a series such as $KICl_2$, $RbICl_2$, and $CsICl_2$, only the last salt being stable at room temperature. No anhydrous acids are known corresponding to polyhalides, but $HICl_4 . 4 H_2O$ has been described.

For reasons which will be apparent shortly it is convenient to discuss separately the more complex polyiodide ions.

The structures of polyhalide ions

Ions $(X–X–X)^-$. Structural data for these (linear) ions (Table 9.5) emphasize two important features of these ions. First, the ions Br_3^-, ICl_2^-, and I_3^- are found to be symmetrical in some salts and unsymmetrical in others. Second, the overall length

TABLE 9.5
Structures of ions $(X^1–X^2–X^3)^-$

Ion	Salt	$X^1–X^2$ (Å)	$X^2–X^3$ (Å)	Excess over single bond length (Å)		Reference
F–Cl–F	$KClF_2$	(i.r. indicates linear)				IC 1967 6 1159
Br–Br–Br	$(C_6H_5N_2)Br_3$	2·54	2·54	0·26	0·26	ACSc 1962 **16** 1882
	$[N(CH_3)_3H]_2Br(Br_3)$	2·54	2·54	0·26	0·26	PKNAW 1958 **B61** 345
	$(C_6H_7NH)_2(SbBr_6)Br_3$	2·54	2·54	0·26	0·26	IC 1968 7 2124
	$CsBr_3$	2·44	2·70	0·16	0·42	AC 1969 **B25** 1073
	$(PBr_4)Br_3$	2·39	2·91	0·11	0·63	AC 1967 23 467
Cl–I–Cl	$(PCl_4)ICl_2$	2·55	2·55	0·23	0·23	JACS 1952 **74** 6151
	$[N(CH_3)_4]ICl_2$	2·55	2·55	0·23	0·23	AC 1964 17 1336
	$C_4H_{12}N_2(ICl_2)_2$	2·47	2·69	0·15	0·37	ACSc 1958 12 668
Cl–I–Br	$NH_4(ClIBr)$	2·91	2·51	0·59	0·04	AC 1967 **22** 812
I–I–Br	$Cs(I_2Br)$	2·78	2·91	0·12	0·43	AC 1966 20 330
I–I–I	$N(C_2H_5)_4I_3$					AC 1967 23 796
	$As(C_6H_5)_4I_3$	2·92	2·92	0·26	0·26	AC 1959 **12** 187
	$N(C_2H_5)_4I_7$	(mean)	(mean)			AC 1958 11 733
	NH_4I_3	2·79	3·11	0·12	0·44	AC 1970 **B26** 904
	CsI_3	2·83	3·03	0·16	0·36	AC 1955 8 59, 857
	Cs_2I_8	2·86	3·00	0·19	0·34	AC 1954 7 487

of each ion is approximately constant and about 0·5 Å greater than the sum of the two single bond lengths. We show in Table 9.5 by how much each bond is longer than the presumably single bond in the appropriate molecule ICl, I_2, etc. The asymmetry of these ions has been attributed to the different environments of the terminal atoms in the crystals (see for example, AC 1959 **12** 197). The effect is clearly not simply related to the size of the cation, for whereas a large asymmetry is observed in CsI_3 (and in $CsBr_3$) symmetrical ions are found in a number of salts containing very large cations. The bond lengths in the $(I–I–Br)^-$ ion, where I–I is shorter than I–Br, suggest that the ion is tending towards

$$I—I \quad \cdots \quad Br^-$$
$$2{\cdot}78 \quad 2{\cdot}91 \text{ Å}$$

and similarly the increasing asymmetry of the Br_3^- ion in a number of salts (see Table 9.5) suggests that the system is approaching the state $Br–Br \cdots Br^-$.

Ions $(XX_4')^-$. Infrared studies[1] of the ClF_4^- ion in salts of NO^+, Rb^+, and Cs^+, indicate a planar configuration. X-ray studies confirm the square planar shape of the BrF_4^- ion in $KBrF_4$[2] (Br–F, 1·89 Å) and of the ICl_4^- ion in $KICl_4 . H_2O$.[3] In this salt the I–Cl bond lengths apparently range from 2·42 Å to 2·60 Å (compare the sum of the covalent radii, 2·32 Å), the variation being attributed to the different environments of the Cl atoms. On the other hand the Br–F bond length in BrF_4^- is not appreciably greater than the sum of the covalent radii (1·86 Å).

(1) IC 1966 **5** 473

(2) JCS A 1969 1936
(3) AC 1963 **16** 243

Polyiodide ions

The polyiodides are a remarkable group of salts, most of which are of the type MI_n in which M is one of the larger alkali metals, NH_4^+, or a substituted ammonium ion, and $n = 3, 5, 7,$ or 9; the salt Cs_2I_8 is of a different kind (see later). As already noted, the tri-iodides consist of cations and I_3^- ions, but in the higher polyiodides the anion assembly is made up of a number of sub-units (I^-, I_3^-, and I_2 molecules) which are held together rather loosely to form the more complex I_n system. It is a feature of all polyiodide ions that some or all of the I–I distances are intermediate between the covalent I–I bond length (2·68 Å in I_2) and the shortest *inter*molecular separation, 3·54 Å in crystalline iodine.

In $N(CH_3)_4I_5$[1] there are approximately square nets of I atoms (Fig. 9.2(a)) and the I–I distances suggest V-shaped I_5^- ions consisting of an I^- ion very weakly bonded to two polarized I_2 molecules. The arms of the V are linear to within 6°, and the V is planar to within about 0·1 Å. This ion is obviously not related structurally to the ICl_4^- ion.

In $[N(C_2H_5)_4]I_7$[2] the anion assembly consists of I_2 molecules and apparently linear I_3^- ions (see above) in the proportion of two I_2 to one I_3^-. Between these units the shortest contacts are I–I, 3·435 Å and within them, I–I, 2·735 Å and 2·904 Å respectively.

The violet-coloured salt with the empirical formula CsI_4 is diamagnetic; the I_4^- ion would be paramagnetic, having an odd number of valence electrons. An X-ray study of the salt[2] shows that it contains rather extraordinary groupings of eight coplanar I atoms as shown in Fig. 9.2(b). In this I_8^{2-} ion the terminal I_3 groups are very similar to the I_3^- ion in tri-iodides, so that the ion may be described as an assembly of two I_3^- ions and a polarized I_2 molecule.

(1) AC 1957 **10** 596

(2) See Table 9.5

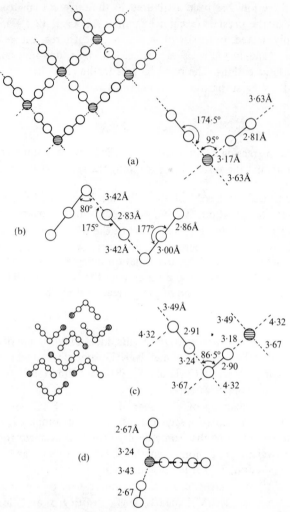

FIG. 9.2. Structures of the higher polyiodide ions: (a) I_5^- in $N(CH_3)_4I_5$, (b) I_8^{2-} in Cs_2I_8, (c) the I_5^- layer in $N(CH_3)_4I_9$, (d) environment of I^- (shaded circle) in $N(CH_3)_4I_9$.

(3) AC 1955 **8** 814

In $N(CH_3)_4I_9$ there is an even more complex arrangement.[3] Five-ninths of the I atoms form rather densely packed layers (Fig. 9.2(c)) between which there are I_2 molecules (with I—I = 2·67 Å, the same as in the ordinary I_2 molecule); the cations lie between the layers in the holes between the I_2 molecules. The other I—I distances are around 2·9, 3·2, and 3·5–3·7 Å, and the way in which the structure is described depends on the interpretation of these distances. The 2·9 and 3·2 Å bonds link I atoms into V-shaped I_5^- ions, but here it is a terminal atom which is the I^- ion, in contrast to the I_5^- ion in $N(CH_3)_4I_5$. The next shortest distances in a layer are those of length 3·49 Å, which would correspond to linking of the I_5^- ions in pairs to form closed rings. The linking of the I^- ion in a layer to I_2 molecules between layers is shown in Fig. 9.2(d).

338

We see that in iodine itself and in polyiodides there is a tendency to form dense planes with I–I distances up to about 3·5 to 3·6 Å, with much larger (van der Waals) distances of about 4·3 Å between planes. In the polyiodides there is interaction between I^- ions and the easily polarizable I_2 molecules, giving I_3^-, I_5^-, I_8^{2-}, and I_9^- systems. In the higher members there is progressively less distortion of the I_2 molecule, which has the normal I–I bond length in the ennea-iodide. The angles between the weak bonds formed by I^- ions are approximately $90°$, suggesting the use of p orbitals.

Hydrogen halides and ions HX_2^-, $H_2X_3^-$, etc.

The interatomic distances, determined spectroscopically, in the hydrogen halide molecules are:

H—F	0·917 Å	H—Br	1·410 Å
H—Cl	1·275	H—I	1·600.

The physical properties of the acids HF, HCl, HBr, and HI do not form regular sequences, HF differing from the others in much the same way as H_2O from H_2S, H_2Se, and H_2Te. The abnormal properties of HF are discussed in Chapter 8 with other aspects of hydrogen bonding, as also are the structures of the crystalline hydrogen halides.

Ions $(X–H–X)^-$ are formed by F, Cl, and Br, but in contrast to the stable and well-known bifluorides of the alkali metals the ions HCl_2^- and HBr_2^- are stable only in a few salts of very large cations. All attempts to prepare salts containing HCl_2^- ions from aqueous solution have given hydrated products (for example, $(C_6H_5)_4As . HCl_2 . 2 H_2O$) which probably do not contain these ions (see below), but anhydrous $(C_6H_5)_4As . HCl_2$ can be prepared from $(C_6H_5)_4AsCl$ and anhydrous HCl.[1] The existence of HCl_2^- ions has been disproved in a number of salts once thought to contain these ions. For example, the compound $CH_3CN . 2HCl$ is not $(CH_3CNH)^+(HCl_2)^-$ but an imine hydrochloride,[2] $[CH_3C(Cl)=NH_2]^+Cl^-$, and in $(Cl_2en_2Co)Cl . HCl . 2 H_2O$ the odd proton is associated with the two H_2O molecules in $H_5O_2^+$ units and not with the two Cl atoms in a HCl_2^- ion.[3] The structures of $Cs_3(Cl_3 . H_3O)HCl_2$ and the Br compound, which apparently do contain HX_2^- ions,[4] are described in Chapter 8.

Only F forms more complex ions $H_2X_3^-$, $H_4X_5^-$, (see Chapter 8). For the hydrates of HCl see p. 562.

(1) IC 1968 **7** 1921

(2) IC 1968 **7** 2577

(3) JINC 1961 **19** 208
(4) IC 1968 7 594

Metal hydrogen halides

The alkaline-earth salts MHX, where M = Ca, Sr, or Ba, and X = Cl, Br, or I, have been prepared by melting the hydride MH_2 with the dihalide MX_2 or by heating $M + MX_2$ in a hydrogen atmosphere at $900°C$.[1] The compound CaHCl was formerly thought to be the 'sub-halide' CaCl. All these compounds have the PbFCl structure (p. 408), but the H atoms were not directly located. Compounds described as MgHX (X = Cl, Br, I) have been shown to be physical mixtures of MgH_2 with MgX_2.[2]

(1) ZaC 1956 **283** 58; *ibid.* 1956 **288** 148, 156

(2) IC 1970 **9** 317

Oxy-compounds of the halogens

Oxides and oxyfluorides

The known oxides are listed in Table 9.6, which emphasizes the very different behaviours of the halogens towards oxygen. (The oxide ClO is apparently formed by the flash photolysis of Cl_2 and O_2 mixtures; an oxide which may be Cl_2O_3 has been produced as a very unstable dark-brown solid at $-45°C$ by u.v. photolysis of a mixture of ClO_2 and Cl_2O_6.) Structural studies have been made of those printed in heavy type in Table 9.6; the paucity of structural information being largely due to the instability of many of these compounds. For example, F_2O_2 decomposes into

TABLE 9.6
Oxides and oxyfluorides of the halogens

F_2O[1]	ClO[2] Cl_2O[3] (Cl_2O_3)[4]	Br_2O			ClO_2F ClO_3F[8]	BrO_2F BrO_3F[9]	IO_2F IO_3F[10]
F_2O_2	**ClO_2**[5]	BrO_2	I_2O_4		ClO_3OF		IOF_3
	ClO_3	BrO_3					IOF_5[11]
	Cl_2O_6 (ClO_4) **Cl_2O_7**[6]		**I_2O_5**[7]				

(1) JPC 1953 **57** 699. (2) JCS A 1968 1704. (3) JCS A 1968 658. (4) JACS 1967 **89** 2795. (5) JCS A 1970 46. (6) TFS 1965 **61** 1821. (7) RTC 1960 **79** 523. (8) JCS A 1970 872. (9) IC 1970 **9** 622. (10) JACS 1969 **91** 4561. (11) JCP 1967 **47** 1731.

the elements at temperatures above $-40°C$, and no oxides of bromine are stable at room temperature.

Electron diffraction and spectroscopic studies give the configurations (a) and (b) for the similarly constituted F_2O and Cl_2O, and a microwave study gives the structure (c) for ClO_2. This oxide is of special interest since the molecule contains

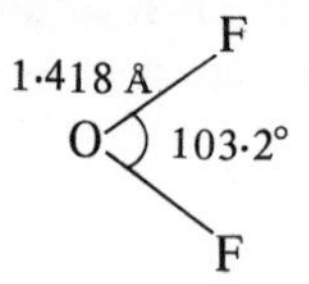

(a) 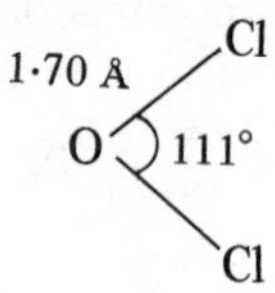   (b) 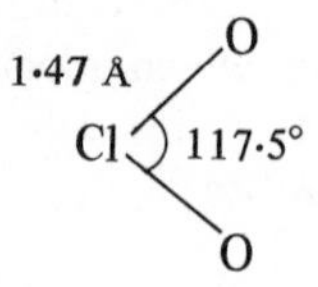   (c)

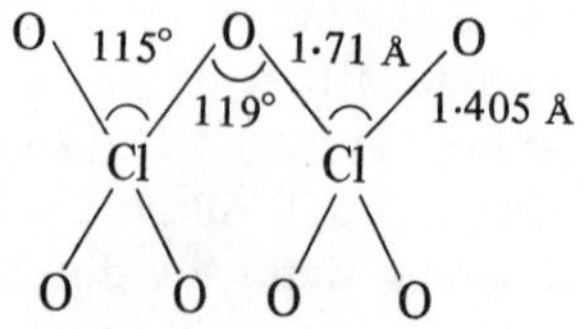

(d)

an odd number of electrons; its structure is very similar to that of SO_2. The configuration of Cl_2O_7 is approximately eclipsed, (d). Our knowledge of the structures of the oxides of iodine is slight. The pentoxide, I_2O_5, is the anhydride of iodic acid and is a white solid stable up to 300°C. From its i.r. spectrum it has been concluded that the molecule has the structure $O_2I-O-IO_2$ rather than O_2I-IO_3. Two other oxides, both yellow powders, have been assigned the formulae I_2O_4 and I_4O_9, but the existence of the latter is doubtful.[1] The diamagnetism of the dioxide rules out the existence of IO_2 molecules, but the existence of discrete I_2O_4 molecules has not yet been proved. Mass spectrometric studies of iodine oxides give no evidence for the existence of I_2O_7.

(1) ACSc 1968 **22** 3309

Little is known of the structures of the oxyfluorides listed in Table 9.6. Compounds XO_2F are formed by all three halogens (Cl, Br, and I) and possibly also XO_3F, since IO_3F has been reported. The iodine compounds include the oxide pentafluoride, a type of compound also formed by Re and Os. A m.w. study indicates C_{4v} symmetry, and presumably therefore the octahedral structure (e).

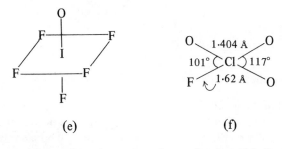

(e)   (f)

Spectroscopic or e.d. studies show that the molecules ClO_3F and BrO_3F are tetrahedral, the former having the structure (f) with multiple bonding in the Cl–O but not in the Cl–F bond.

Oxy-cations

Little is yet known of the structures of these ions. An i.r. study of $(ClO_2)(AsF_6)$ apparently indicates that this compound forms an ionic solid, presumably containing the ClO_2^+ cation.[2]

(2) IC 1969 **8** 2489

Oxy-acids and oxy-anions

In the stability of their oxy-acids there are great differences between the halogens. The existence of only one oxy-acid of fluorine has yet been firmly established, and the only types of oxy-ion formed by all three halogens Cl, Br, and I are XO^-, XO_3^-, and XO_4^- (Table 9.7). Structural studies have been made of HOCl, HOBr, α-HIO_3, $HClO_4$. H_2O, H_5IO_6, and HI_3O_8.

Acids HXO and their salts.   Until recently neither the acid HFO nor its salts were known. Following its earlier detection in low-temperature matrices HOF has now been prepared in milligram quantities by passing a stream of F_2 at low pressure over water at 0°C. The product is separated in a trap at −183°C. Trifluoromethyl

341

TABLE 9.7

Oxy-ions XO_n of the halogens

XO^-	XO_2^- angular	XO_3^- pyramidal	XO_4^- tetrahedral	XO_6^{5-} octahedral
ClO^-	ClO_2^-	ClO_3^-	ClO_4^-	–
BrO^-	(BrO_2^-)	BrO_3^-	BrO_4^-	–
IO^-		IO_3^-	IO_4^-	IO_6^{5-}

hypofluorite, F_3COF, is a very stable gas at ordinary temperatures and is prepared by fluorinating methyl alcohol or CO in the presence of AgF_2. Chlorine, bromine, and iodine all form acids HXO, the stability (in all cases small) decreasing from the chlorine to the iodine compound. The acids HOCl and HOBr have been prepared by photolysis of $Ar–HX–O_3$ mixtures at $4°K$ and studied in the Ar matrix by i.r.

(1) JACS 1967 **89** 6006

spectroscopy.[1] Assuming O–H = 0·96 Å the structures found are:

the structures of the ions XO^- have not been determined.

Acids HXO_2 and their salts. There appears to be no conclusive evidence that the

(2) CR 1965 **260** 3974; ZaC 1970 **372** 127

acids HXO_2 exist if X is F, Br, or I, and the only salts of one of these acids that have been prepared are $LiBrO_2$ and $NaBrO_2$.[2] $LiBrO_2$ is apparently formed by the dry reaction between $LiBr$ and $LiBrO_3$, and $NaBrO_2$ can be crystallized from a solution containing NaOH and Br_2. Pure chlorous acid, $HClO_2$, has not been isolated, but treatment of the barium salt with sulphuric acid gives a colourless solution which even at $0°C$ soon turns yellow owing to the formation of ClO_2 (and Cl_2). A few crystalline chlorites have been prepared, and the structure of the

(3) AC 1961 **14** 202
(4) AC 1959 **12** 867

chlorite ion has been studied in $AgClO_2$[3] and (at $-35°C$) in NH_4ClO_2.[4] The Cl–O bond length (1·57 Å) lies between the value 1·47 Å found in ClO_2 and the much larger value (1·70 Å) in the Cl_2O molecule; the bond angle is close to $111°$.

Acids HXO_3 and their salts. The acids $HClO_3$, $HBrO_3$, and HIO_3 do not form a regular sequence, as is seen from the heats of formation: 100, 52, and 234 kJ mol^{-1} respectively. The iodine acid is more stable than the chlorine and bromine compounds, for aqueous solutions of $HClO_3$ and $HBrO_3$ can be concentrated (*in vacuo*) only up to about 50 per cent molar, whereas an aqueous solution of HIO_3 may be concentrated until it crystallizes. Moreover, whereas in dilute aqueous solution iodic acid behaves as a moderately strong monobasic acid, the freezing points and conductivities of concentrated solutions suggest the presence of more

(5) QRCS 1954 **8** 123
(6) JACS 1941 **63** 278

complex ions or possibly basic dissociation to $IO_2^+ + OH^-$.[5]

The crystal structure of one polymorph (α) of HIO_3 has been determined.[6] The crystal contains pyramidal molecules (a) linked by hydrogen bonds (see

Chapter 8). The mean interbond angle O–I–O is 98°; the next nearest O neighbours, completing a very distorted octahedron around the I atom, are at 2·45, 2·70, and 2·95 Å.

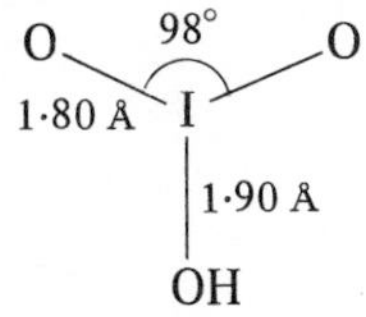

(a) 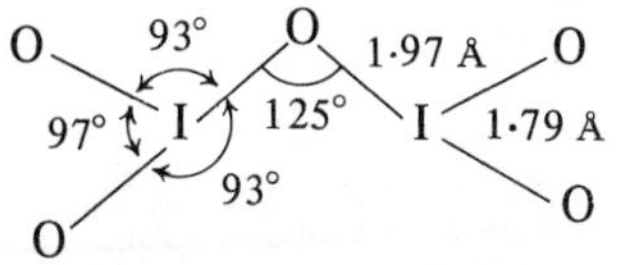 (b)

The crystal structure of the so-called anhydro-iodic acid, HI_3O_8, formed by the dehydration of HIO_3 at 110°C, shows that it is an addition compound of HIO_3 and I_2O_5 with hydrogen bonds between OH of HIO_3 and O of I_2O_5.[7] Bond lengths in the pyramidal HIO_3 molecule are similar to those in crystalline HIO_3, and those in the non-planar I_2O_5 molecule are as shown at (b). This molecule consists of two IO_3 pyramids with a common O atom.

(7) AC 1966 **20** 769

X-ray studies of crystalline salts demonstrate the pyramidal shape of the XO_3^- ions (Table 9.8). In crystalline iodates three more neighbours, at distances ranging from 2·5–3·5 Å complete very distorted octahedra of O atoms around the I atom as

TABLE 9.8

Structures of XO_3^- *ions*

	X—O (Å)	Angle O—X—O	References
ClO_3^-	1·48	106°	AC 1965 **18** 703; JCP 1968 **48** 1883
BrO_3^-	1·64	$105\frac{1}{2}°$	AC 1960 **13** 1017
IO_3^-	1·82	99°	AC 1966 **20** 758; AC 1966 **21** 841

in HIO_3. (Earlier structures assigned to $LiIO_3$ and $TlIO_3$ in which I had six equidistant neighbours were incorrect. In $LiIO_3$ I has 3 O at 1·82 Å and 3 O at 2·9 Å, the latter distance being 0·7 Å less than the van der Waals separation.)

Acids HXO_4 *and their salts.* In the formation of acids HXO_4 the halogens show still less resemblance to one another. Chlorine forms $HClO_4$ and perchlorates. Perbromates were unknown until 1969, when $KBrO_4$ was prepared (by oxidation of the bromate by F_2 in alkaline solution). Periodic acid is normally obtained (by evaporation of its solution) as H_5IO_6; it has been stated that HIO_4 results from heating H_5IO_6 at 100°C *in vacuo*, and that an intermediate product $H_4I_2O_9$ is formed at 80°C. A further peculiarity of the iodine acid is that on further heating HIO_4 does not form the anhydride (I_2O_7 apparently does not exist) but loses oxygen to form HIO_3. Whereas perchlorates are all of the type $M(ClO_4)_n$, periodates of various kinds can be obtained from a solution by altering the

343

temperature of crystallization and the hydrogen-ion concentration. For example, a solution of $Na_2H_3IO_6$ is obtained by oxidizing $NaIO_3$ with chlorine in alkaline solution, and from this solution a number of silver periodates may be prepared:

$$Na_2H_3IO_6 \xrightarrow[\substack{\text{in dilute } HNO_3 \\ \text{solution}}]{AgNO_3} Ag_2H_3IO_6 \text{ (greenish-yellow)}$$

with $Ag_2H_3IO_6$ reacting:
- $\xrightarrow{KOH} Ag_5IO_6 + K_4I_2O_9 \text{ or } K_4H_2I_2O_{10}\cdot 8H_2O$
- $\xrightarrow[HNO_3]{\text{more conc.}} AgIO_4 \text{ (orange)}$
- $\xrightarrow[H_2O]{\text{loses}} Ag_2HIO_5 \text{ (red)} \xrightarrow[\text{acid solution}]{\text{heat in}} Ag_3IO_5 \text{ (black)}$

(8a) JCS A 1970 1613
(8b) AC 1962 **15** 1201
(9) IC 1969 **8** 1190
(10) AC 1970 **B26** 1782

See, however, the later remarks on the formulae of hydrated periodates.

The structure of the $HClO_4$ molecule in the vapour state has been studied by e.d.;[8a] it is very similar to that of ClO_3F (p. 341). $Cl-O$, 1.41 Å, $Cl-OH$, 1.64 Å, $O-Cl-O$, $113°$ or $117°$.

The ions XO_4^- are tetrahedral with bond lengths similar to those in XO_3^-, namely, $Cl-O$, 1.44 Å,[8b] $Br-O$, 1.61 Å,[9] and $I-O$, 1.78 Å.[10] In $NaIO_4$ the anion is a slightly compressed tetrahedron, having two bond angles of $114°$ and four of $107°$.

Periodates containing octahedrally coordinated iodine. Of the halogens only I exhibits octahedral coordination by oxygen. Crystalline H_5IO_6 consists of nearly regular octahedral $IO(OH)_5$ molecules which are linked by $O-H-O$ bonds (ten from each molecule) into a 3D array. The bond lengths are $I-O$, 1.78 Å, $I-OH$,

(11) AC 1966 **20** 765
(12) AC 1970 **B26** 1069
(13) AC 1970 **B26** 1075

1.89 Å, and $O-H$, 0.96 Å.[11] The ion $IO_3(OH)_3^{2-}$ is present in salts such as $(NH_4)_2H_3IO_6$, $Cd[IO_3(OH)_3]\cdot 3 H_2O$,[12] and $[Mg(H_2O)_6][IO_3(OH)_3]$.[13] In this ion, (a), the bond lengths were determined as $I-O$, 1.86 Å, and $I-OH$, 1.95 Å. It seems likely that many 'hydrated periodates' have been incorrectly formulated (compare borates, Chapter 24). For example, the Cd salt just mentioned was formulated as $Cd_2H_2I_2O_{10}\cdot 8 H_2O$, apparently analogous to $K_4H_2I_2O_{10}\cdot 8 H_2O$. In fact this K salt has also been formulated as $K_2HIO_5\cdot 4 H_2O$ and $K_4I_2O_9\cdot 9 H_2O$; it is obtained from a solution of KIO_4 in concentrated KOH. It actually contains ions, (b), formed from two octahedral $IO_5(OH)$ groups sharing an edge. The H atoms were not located, but their positions were deduced from the fact that all the bonds marked *a* have the length 2.00 Å

(14a) AC 1965 **19** 629

while the length of the other (terminal) $I-O$ bonds is 1.81 Å.[14a] Dehydration of $K_4H_2I_2O_{10}\cdot 8 H_2O$, or crystallization of the solution above $78°C$ gives $K_4I_2O_9$,

(14b) ZaC 1968 **362** 301

which contains ions (c) consisting of two IO_6 octahedra sharing a face.[14b]

The simple octahedral IO_6^{5-} ion presumably occurs in anhydrous salts M_5IO_6.

(15) ACSc 1966 **20** 2886

Like TeO_6^{6-} this ion has the property of stabilizing high oxidation states of metals (for example, Ni^{IV}, Cu^{III}), as in $KNiIO_6$,[15] in which K, Ni, and I all occupy octahedral positions in an approximately hexagonal closest packing of O atoms. ($I-O$, approx. 1.85 Å.)

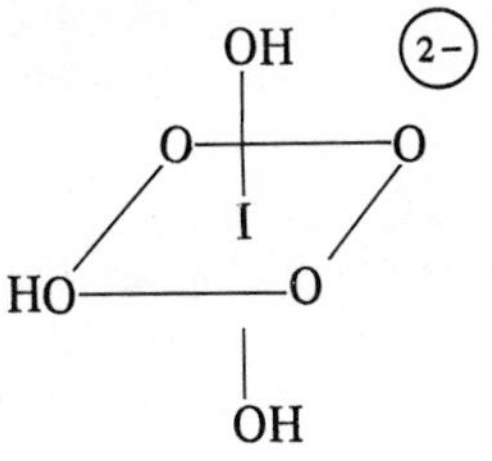

(a)

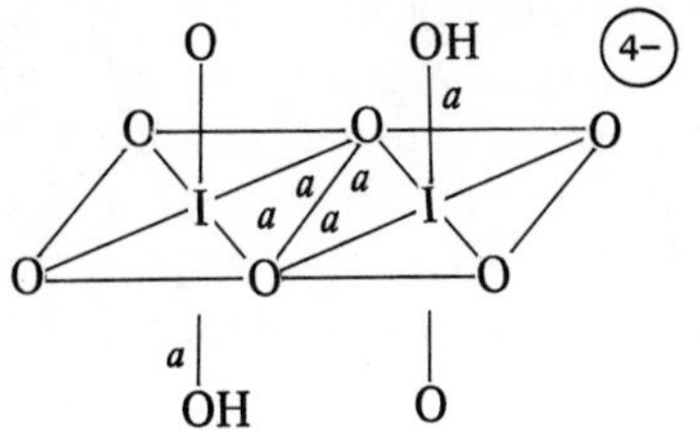

(b)

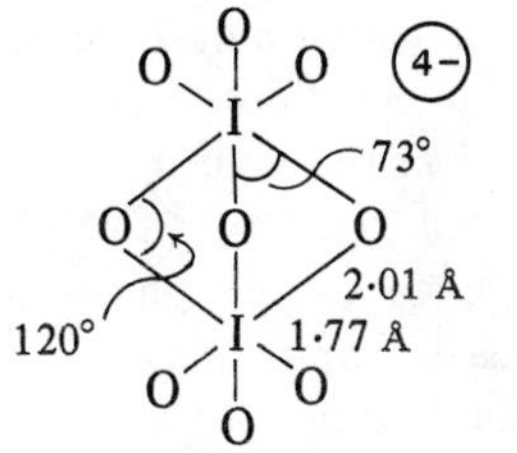

(c)

Two other ions peculiar to iodine may be mentioned here. Potassium fluoroiodate, KIO_2F_2, prepared by the action of HF on KIO_3, contains ions with the structure (d). The four bonds from I are disposed approximately towards four of the apices of a trigonal bipyramid, one bond position being occupied by the unshared pair of electrons. The plane of the I and O atoms is normal to the F—I—F

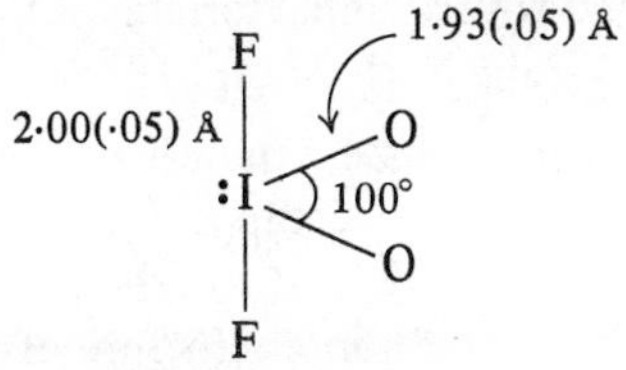

(d)

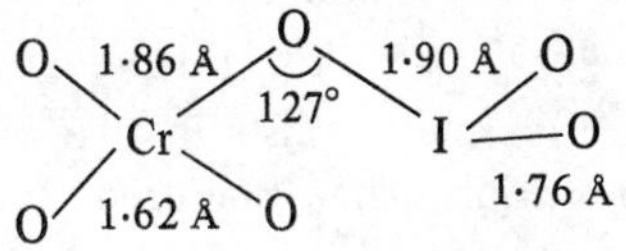

(e)

axis, and the O—I—O bond angle is contracted to $100°$.[16] (Note that in this X-ray study it was not in fact possible to distinguish between the O and F atoms.) The salt $KCrIO_6$, prepared from $K_2Cr_2O_7$ and HIO_3 in aqueous solution, contains ions (e) consisting of a tetrahedral CrO_4 group sharing one O atom with a pyramidal IO_3 group.[17]

(16) JACS 1940 **62** 1537

(17) ACSc 1967 **21** 2781

Halides of metals

The majority of metallic halides are solids at ordinary temperatures and relatively few consist of finite molecules in the crystalline state. Since these compounds melt and vaporize to finite molecules or ions a comprehensive review would call for a knowledge of their structures in the solid, liquid, and vapour states. Much of the available information relates to the solid state, but we shall indicate briefly what is known about these compounds in other states of aggregation.

Metal halides form a very large group of compounds for the following reasons.

(a) There are some 80 metals and four halogens (excluding astatine, about which very little is as yet known, see p. 328), and moreover most transition metals and many B subgroup metals form more than one compound with a given halogen. For example, Cr forms all five fluorides from CrF_2 to CrF_6 inclusive (and also Cr_2F_5, see below) and CrF has been identified as a vapour species formed from CrF_2 and Cr in a Knudsen cell. The iodides of Nb include Nb_6I_{11}, Nb_3I_8, NbI_3, NbI_4, and NbI_5. A particular halide may have more than one crystalline form; polymorphism is fairly common in halides.

(b) Apart from *solid solutions* of two or more halides and polyhalides (in which the halogen atoms are associated together in a polyhalide ion, as described earlier in this chapter) there are metal halides containing more than one halogen. Little is yet known of the structures of these compounds, but there are probably a number of types. Some are ionic (for example, SrClF, BaClF, and BaBrF),[1] others apparently molecular ($TiClF_3$, $TiCl_2F_2$, $WClF_5$, etc.). Compounds formed by Group V metals exist, like those of P, at least in some cases,[2] both as covalent liquids or low-melting molecular crystals and as ionic crystals. For example, liquid $SbCl_2F_3$ and crystalline $(SbCl_4)(SbF_6)$, molecular $NbCl_4F$ and $(NbCl_4)F$. Mixed halides such as $UClF_3$ may have structures related to those of the corresponding fluorides.

(c) In addition to normal halides MX_n there are halides in which

(i) a metal exists in two definite oxidation states with quite different environments forming a stoichiometric compound with a simple formula. Examples include $Ga^IGa^{III}Cl_4$ ($GaCl_2$), $Pd^{II}Pd^{IV}F_6$ (PdF_3), and Cr_2F_5[3] (p. 182).

(ii) a metal exists in two oxidation states in a phase exhibiting a range of composition. The coloured halides intermediate in composition between SmF_2 and SmF_3[4] (cubic $SmF_{2.00-2.14}$, tetragonal $SmF_{2.35}$, and rhombohedral $SmF_{2.41-2.46}$) result from addition of F^- ions to the fluorite structure of SmF_2 accompanied by replacement of some Sm^{2+} by Sm^{3+}.

(iii) there is metal–metal bonding, usually leading to a mean non-integral oxidation state of the metal. The metal atoms may have the same environment (Ag_2F, with the anti-CdI_2 structure, Nb_3Cl_8 and Nb_6I_{11}, p. 367) or different environments, as in Gd_2Cl_3 or in complex sub-halides such as $BiCl_{1.167}$ (p. 372). In the remarkable structure of Gd_2Cl_3[5] there are chains of Gd atoms embedded among Cl^- ions. The chains may be described either as formed from elongated octahedra sharing edges or as tightly bound Gd_2 units bridged by a second set of Gd atoms (Fig. 9.3). Note that on the one hand metal–metal bonding does not necessarily lead to non-integral oxidations states—HfCl and ZrCl have metallic properties[6]—and on the other hand there is not necessarily metal–metal bonding in a compound MX_n with n non-integral (compare the Sm fluorides mentioned above).

(1) JCP 1968 **49** 2766

(2) ZaC 1964 **329** 172

(3) AC 1964 **17** 823

(4) IC 1970 **9** 1102

(5) JACS 1970 **92** 1799

(6) IC 1970 **9** 1375

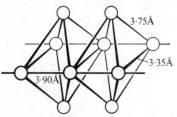

FIG. 9.3. Chain of metal atoms in crystalline Gd_2Cl_3.

346

The structures of crystalline halides MX_n

We may make two generalizations about crystalline metal halides. First, fluorides differ in structure from the other halides of a given metal except in the case of molecular halides (for example, SbF_3 and $SbCl_3$ both crystallize as discrete molecules) and those of the alkali metals, all the halides of which are essentially ionic crystals. In many cases the fluoride of a metal has a 3D structure whereas the chloride, bromide, and iodide form crystals consisting of layer, or sometimes chain, complexes. (For exceptions, particularly fluorides MF_3–MF_6, see Table 9.9.) Second, many fluorides and oxides of similar formula-type are isostructural, while chlorides, bromides, and iodides often have the same types of structure as sulphides, selenides, and tellurides. The following examples illustrate these points:

$$
\begin{array}{ll}
\left.\begin{array}{l} FeF_2 \\ PdF_2 \\ SnO_2 \end{array}\right\} \begin{array}{l} \text{rutile} \\ \text{structure} \end{array} & \begin{array}{ll} FeCl_2 & \text{layer structure} \\ PdCl_2 & \text{chain structure} \\ SnS_2 & \text{layer structure} \end{array}
\end{array}
$$

TABLE 9.9

Structures of crystalline metal halides

	C.N. of M	3D complex		Layer		Chain		Molecular
MX	4	Zn blende, wurtzite	5 + 2	TlI (yellow)	2	AuI		
	6	NaCl						
	8	CsCl						
MX₂	4	Silica-like structures	4	HgI₂ (red)	3	SnCl₂	2	HgCl₂
					3	GeF₂		
	6	Rutile	6	{ CdI₂	4	BeCl₂	4	Pt₆Cl₁₂
	6	CaCl₂		{ CdCl₂	4	PdCl₂		
	7	SrI₂, EuI₂						
	7 + 2	PbCl₂						
	8	Fluorite	8	ThI₂				
MX₃	6	ReO₃ and related structures (Table 9.16)	6	{ YCl₃	4	AuF₃	3	SbF₃
				{ BiI₃	6	ZrI₃	4	Al₂Cl₆
	7 + 2	LaF₃						
	8 + 1	YF₃	8 + 1	PuBr₃			4	Au₂Cl₆
	9	UCl₃						
MX₄	8	ZrF₄	6	PbF₄	5	TeF₄	4	SnBr₄
					6	α-NbI₄	4	SnI₄
					6	TcCl₄		
	8	UCl₄	8	ThI₄	6	ReCl₄		
MX₅	7	β-UF₅			6	BiF₅	5	SbCl₅
					6	CrF₅	6	Nb₂Cl₁₀
					7	PaCl₅	6	Mo₄F₂₀
MX₆							6	IrF₆

The Halogens—Simple Halides

Metal halides provide examples of all the four main types of crystal structure: 3D complexes, layer, chain, and molecular structures, as shown in Table 9.9. The great majority of halides MX, MX_2, and MX_3 adopt structures shown to the left of the heavy line in the table, and most monohalides and most fluorides MF_2 and MF_3 crystallize with one of the following highly symmetrical structures suited to (though not exclusive to) essentially ionic crystals:

	Structure	Environment of M	Environment of X
MX	NaCl	Octahedral	Octahedral
	CsCl	Cubic	Cubic
MX_2	Rutile	Octahedral	Triangular
	Fluorite	Cubic	Tetrahedral
MX_3	ReO_3	Octahedral	Linear

These and most of the other structures of Table 9.9 have been described and illustrated in Chapters 3–6. We shall be more concerned here with structures that have not previously been described and we shall deal in turn with the six horizontal groups of halides of Table 9.9; a note will be included on the only two known metal heptahalides.

Monohalides

Alkali halides. At ordinary temperature all the alkali halides crystallize with the NaCl structure except CsCl, CsBr, and CsI, which have the CsCl structure. The latter is more dense (for a given halide) than the NaCl structure and is adopted under pressure by the Na, K, and Rb salts, but no structural changes have been induced in Li salts under pressure.[1] The reverse change, from CsCl to NaCl structure, takes place on heating CsCl to 469°C, but neither CsBr nor CsI undergoes this transformation, at least up to temperatures within a few degrees of their melting points.[2] However, all the halides which normally crystallize with the CsCl structure can be grown with the NaCl structure from the vapour on suitable substrates (NaCl, KBr).[3] The interionic distances are listed in Table 9.10. (At 90°K LiI changes to a h.c.p. structure.)[4]

(1) JCP 1968 **48** 5123

(2) JACS 1955 **77** 2734

(3) AC 1951 **4** 487
(4) ZP 1956 **143** 591

TABLE 9.10

Interionic distances in Cs and Tl halides

	NaCl *structure*	CsCl *structure*
CsCl	3·47 Å	3·56 Å
CsBr	3·62	3·72
CsI	3·83	3·95
TlCl	3·15	3·32
TlBr	3·29	3·44
TlI	3·47	3·64

Cuprous halides.[5] The cuprous halides (other than CuF, which has not been prepared pure at ordinary temperatures) crystallize with the zinc-blende structure. At the temperatures 435°, 405°, and 390°C respectively CuCl, CuBr, and CuI transform to the wurtzite structure, and the last two exhibit a further transition at higher temperatures. *Argentous halides*[5] AgF, AgCl, and AgBr crystallize with the NaCl structure and AgI adopts this structure under pressure. The iodide crystallizes with both the zinc-blende and wurtzite structures, the former being apparently metastable. Silver iodide transforms at 145·8°C to a high-temperature form which is notable for its high ionic conductivity ($1\cdot3$ ohm^{-1} cm^{-1} at 146°C). In this form the arrangement of iodine atoms is b.c. cubic, that is, each I has only 8 I neighbours as opposed to 12 in the low-temperature polymorphs. The X-ray measurements, the high conductivity, and self-diffusion show that the silver ions move freely between positions of 2-, 3-, and 4-fold coordination between the easily deformed iodide ions. There are at least three high-pressure forms, one with the NaCl structure (Ag–I, 3·04 Å; compare 2·81 Å in the ZnS-type structures).[6]

Aurous halides. Structural information is available only for AuI, which consists of chain molecules with the structure[7]

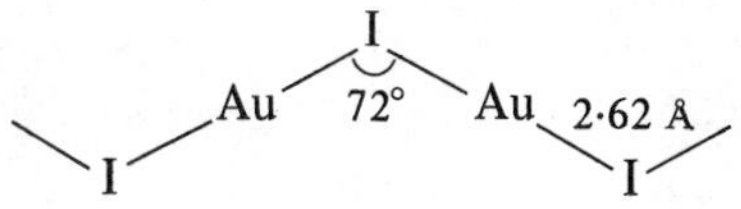

Subgroup IIIB *monohalides.* Gallium monohalides other than GaI (p. 911) are known only as vapour species, and little seems to be known about InF. The structures of other monohalides of this group are summarized in Table 9.11. TlF has a distorted NaCl structure in which the nearest neighbours of Tl are 2 F at 2·59, 2 F at 2·75, and 2 F at 3·04 Å. The low-temperature form of InCl also has a distorted NaCl structure with In–Cl ranging from 2·8 to 3·5 Å, the distortion giving In$^+$ three rather close In$^+$ neighbours (3·65 Å). InBr, InI, and the yellow form of TlI have the layer structure illustrated in Fig. 6.1 (p. 193). The metal atom has five nearest halogen neighbours at five of the vertices of an octahedron and then 2 + 2 next-nearest neighbours. In TlI the distances are: one I at 3·36, four at 3·49, and two at

TABLE 9.11

Structures of crystalline monohalides

	F	Cl	Br	I
In	–	N*[8]	T[9]	T[10]
Tl	N*	C	C	T/C

N*: distorted NaCl structure
C: CsCl structure
T: TlI structure

(5) AC 1964 **17** 1341; JPC 196 68 1111

(6) JCP 1968 **48** 2446

(7) ZK 1959 **112** 80

(8) AC 1966 **20** 905
(9) AJSR 1950 **3A** 581
(10) AC 1955 8 847

(11) JCP 1965 **43** 1381

$3 \cdot 87$ Å (Tl^+ also has two Tl^+ neighbours at $3 \cdot 83$ Å). Under high pressure TlCl, TlBr, and TlI become metallic conductors; there is a very large (33 per cent) reduction in volume of TlI under a pressure of 160 kbar.[11]

Other monohalides include HfCl and ZrCl in Group IV and the solitary OsI in Group VIII.

Dihalides

It is convenient to divide these compounds into five groups.

(i) *Tetrahedral structures.* This group includes the silica-like structures of BeF_2 (low- and high-cristobalite and quartz structures) and the chain structure of one form of $BeCl_2$, which is trimorphic. The unique chain structure is that of the form

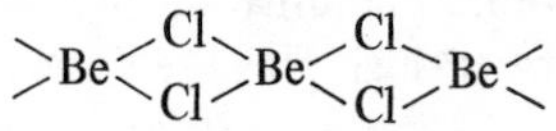

$$>Be \overset{Cl}{\underset{Cl}{<>}} Be \overset{Cl}{\underset{Cl}{<>}} Be<$$

(1) JPC 1965 **69** 3839

stable over the temperature range $403–425°$C (m.p.), but this structure is formed by quenching the melt and is also formed on sublimation.[1] The structures of $BeBr_2$ and BeI_2 are not known.

Zinc chloride also is trimorphic, and all forms consist of tetrahedral $ZnCl_4$ groups sharing all vertices. Two have 3D structures, one like cristobalite, and the

(2) ZK 1961 **115** 373

third has the layer structure of red HgI_2 (p. 162).[2] There is probably also tetrahedral coordination of Zn in the bromide and iodide, but the structures are not

(3) HCA 1960 **43** 77

known in detail.[3]

(ii) *Octahedral structures* (Mg *and the* 3d *metals*). The structures containing ions of these metals, with radii close to $0 \cdot 7$ Å, are summarized in Table 9.12. The two modified forms of the rutile structure, R* and R**, are quite different in nature, as described in Chapter 6. In CrF_2 (and the isostructural CuF_2) there is $(4 + 2)$-coordination of the metal, the four stronger bonds delineating layers based

TABLE 9.12
Crystal structures of dihalides

	Mg	Ti	V	Cr	Mn	Fe	Co	Ni	Cu	Zn
F_2	R		R	R*	R^a	R	R	R	R*	R
Cl_2	C	I	I	R**	C	C	C	C	I*	
Br_2	I	I	I	I*	I	I	C/I	I	I*	tetrahedral structures
I_2	I	I	I	I*	I	I	I	C?		

R = rutile. R* and R** = distorted rutile structures. C = $CdCl_2$. I = CdI_2 structure. I* = distorted CdI_2 structure.

^{a} For detailed studies of 3d difluorides (rutile structure) see: JACS 1954 **76** 5279; AC 1958 **11** 488.

on the simple 4-gon plane net. In $CrCl_2$, on the other hand, the four stronger bonds define chains of the same kind as in $PdCl_2$ (and $CuCl_2$). Details of the Cu compounds are given in Chapter 25; for the Cr compounds details are given in Table 9.13.

TABLE 9.13

Coordination of Cr in dihalides

	4 X *at*	2 X *at*	*Reference*
CrF_2	2·00 Å	2·43 Å	PCS 1957 232
$CrCl_2$	2·39	2·91	AC 1961 **14** 927; HCA 1961 **44** 1049
$CrBr_2$	2·54	3·00	AC 1962 **15** 672
CrI_2	2·74	3·24	AC 1962 **15** 460; AC 1973 **B29** 1463

(iii) *Dihalides of second and third series transition metals.* These elements present a very different picture from the 3d metals (Table 9.14). The following points are noteworthy: the almost complete absence of difluorides, the absence of dihalides of Tc, Rh, Hf, Ta, and Ir, the formation by Hf, Nb, and Ta of halides with non-integral oxidation numbers of the metal, and the presence of 'metal clusters' in compounds of Nb, Ta, Mo, W, Pd, and Pt. We deal with this last group of halides in a later section.

TABLE 9.14

Dihalides etc. of elements of 2nd and 3rd transition series

—	$NbF_{2.5}$	—	Tc	—	Rh	PdF_2
$ZrCl_2$	$NbCl_{2.33}$	$MoCl_2$	none known	$RuCl_2$	none known	$PdCl_2$
—	$NbBr_2$	$MoBr_2$		—		$PdBr_2$
ZrI_2	$NbI_{1.83}$	MoI_2		—		PdI_2

—	—	—	—	—	Ir	—
$HfCl_{2.5}$	$TaCl_{2.5}$	WCl_2	—	$OsCl_2$	none known	$PtCl_2$
—	$TaBr_{2.33}$	WBr_2	—	$OsBr_2$		$PtBr_2$
—	—	WI_2	ReI_2	OsI_2		PtI_2

The only difluoride in this group, PdF_2, has the normal rutile structure. $PdCl_2$ is apparently trimorphic,[1] one form (stable over the intermediate temperature range) having the simple chain structure of Fig. 9.4.[2] A second form[3] consists of hexameric molecules Pd_6Cl_{12} (isostructural with Pt_6Cl_{12})[4] with the structure described on p. 371. Crystalline $PdBr_2$ is built of chains of the same general type as in α-$PdCl_2$ but here they are puckered, and apparently two of the Pd–Br bonds are shorter than the other two (2·34 and 2·57 Å).[5] The structure of PdI_2 is not

(1) JPC 1965 **69** 3669
(2) ZK 1938 **100** 189
(3) AnC 1967 **79** 244
(4) ZaC 1965 **337** 120

(5) ZaC 1966 **348** 162

351

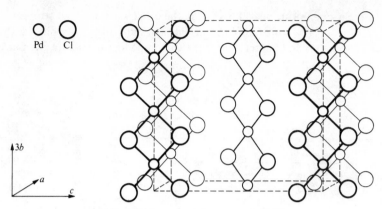

FIG. 9.4. The crystal structure of α-PdCl$_2$.

recorded. It is interesting that the dichlorides of Ni, Pd, and Pt provide examples of three different structures:

		C.N. of M
NiCl$_2$:	CdCl$_2$ layer structure	6
PdCl$_2$:	chain structure and hexamer	4
PtCl$_2$:	hexamer Pt$_6$ Cl$_{12}$	4

The chain structure of Fig. 9.4 has also been assigned to a form of PtCl$_2$, but on this point see ref. (4).

(iv) *Dihalides of the alkaline-earths,* Cd, Pb, 4f *elements, and* Th. We include here Cd, which represents a case intermediate between our classes (iii) and (iv), and Pb, which clearly belongs in this class.

The alkaline-earth halides are surprisingly complex from the structural standpoint, the twelve compounds exhibiting at least six different structures. The fluorides adopt the simple fluorite structure (and the PbCl$_2$ structure under pressure), but the complexity of the other halide structures arises because the metal ions are too large to form the polarized AX$_2$ layer structures, and instead form structures with rather irregular 7- or 8-coordination. The ions Eu^{2+}, Sm^{2+}, and Pb^{2+} are similar in size to Sr^{2+} and show a behaviour very similar to that of the alkaline-earths (Table 9.15). The di-iodides of Tm and Yb crystallize with the CdI$_2$ structure, like CaI$_2$ and PbI$_2$. The unique structure of the golden-yellow ThI$_2$, which exhibits metallic conduction, is described in Chapter 4 as a 'hybrid' intermediate between the CdI$_2$ and MoS$_2$ structures. There is both octahedral and trigonal prismatic coordination of Th atoms, the bond length Th–I (3·20, 3·22 Å) being the same as the mean value in ThI$_4$. The compound is presumably Th^{IV}(e)$_2$I$_2$, two electrons being delocalized within the layers.[1] Two black chlorides, DyCl$_2$ and DyCl$_{2·08}$, have been described, the former being isostructural

(1) IC 1968 7 2257

TABLE 9.15

Structures of some crystalline dihalides

	Ca	Sr	Ba	Eu	Sm	Pb[10]	Cd
F	F/P	F/P[1]	F/P[1]	F	F	F/P[6]	F
Cl	C[4]	C/F[3]	F/P[2]	P	P	P[3]	L
Br	C[5]	S_b[8]	P[2]	S_b[1]	S_b[1]	P	L
I	L	S_i[9]	P[2]	E[7]/S_i		L	L

L = $CdCl_2$ or CdI_2 layer structure. R = rutile structure. C = $CaCl_2$ structure. P = $PbCl_2$ structure.
E = EuI_2 (monoclinic) structure. S_b = $SrBr_2$ structure. S_i = SrI_2 structure. F = CaF_2 structure.

(1) MRB 1970 **5** 769. (2) JPC 1963 **67** 2132. (3) JPC 1963 **67** 2863. (4) AC 1965 **19** 1027.
(5) JINC 1963 **25** 1295. (6) AC 1967 **22** 744. (7) AC 1969 **B25** 1104. (8) IC 1971 **10** 1458.
(9) ZaC 1969 **368** 62. (10) AC 1969 **B25** 796 (mixed halides of Pb).

with $YbCl_2$, but the structures do not appear to be known with certainty.[2]]
Halides of Nd(II) include $NdCl_{2.3}$, $NdCl_{2.2}$, $NdCl_2$ (PbCl$_2$ structure), and $NdI_{1.95}$
($SrBr_2$ structure).[3]

(2) IC 1966 **5** 938

(3) IC 1964 **3** 993

In addition to the rutile and fluorite structures there are some 3D structures in Table 9.15 in which the environments of the ions are less symmetrical. The $CaCl_2$ structure (C) is closer to the ideal h.c.p. AX_2 structure (p. 141) than is the rutile structure; there is slightly pyramidal coordination of Cl^- (Br^-) suggesting a tendency towards the CdI_2 layer structure with its much more pronounced pyramidal coordination of the anions.

The $PbCl_2$ structure to which we refer here is that of salt-like compounds (dihalides, Pb(OH)Cl, CaH_2). There is considerable variation in axial ratios and atomic coordinates for compounds with this structure, and consequent changes in coordination number, so that in effect there are a number of different structures which should not all be described as the same '$PbCl_2$ structure'; this point is discussed in Chapter 6. The general nature of the structure is most easily visualized if the coordination group around Pb^{2+} is regarded as a tricapped trigonal prism (Fig. 6.23, p. 222), but this coordination group is very irregular. There are always two rather distant neighbours, and the structure is more accurately described as one of $7 : \frac{3}{4}$ coordination. In the accurately determined structure of the low-temperature (or high-pressure) polymorph of PbF_2 with this structure the interionic distances are:

Pb—F: 2·41, 2·45 (two), 2·53, 2·64, 2·69 (two), and 3·03 Å (two).

The mean of the first seven distances (2·55 Å) is very close to the value (2·57 Å) for 8-coordinated Pb^{2+} in the polymorph with the fluorite structure. One-half of the X^- ions have their three close cation neighbours all lying to one side. This is true for all the anions in the UCl_3 structure (Fig. 9.8, p. 359), so that although these are both 3D structures some or all of the anions have the unsymmetrical environment of anions which is a feature of layer, chain, and molecular structures.

The Eu dihalides are particularly interesting, for all four have different structures. The iodide is dimorphic; one form (orthorhombic) is isostructural with SrI_2, and the other has monoclinic symmetry. These are both structures of $7 : \frac{3}{4}$

353

coordination, the 7-coordination being similar to that of Zr in baddeleyite. The chief difference between monoclinic EuI_2 and SrI_2 lies in the bond angles at the 3-coordinated I^- ions; these are two of 100° and one of 140° and one of 101° and two of 127° respectively (compare ZrO_2, 104°, 109°, and 146°).

In the inexplicably complex $SrBr_2$ structure there are two kinds of Sr^{2+} ion with different arrangements of 8 neighbours. One type of ion has 8 Br^- at the vertices of an antiprism (3·14 Å), while the other has a very irregular arrangement of nearest neighbours: 6 at a mean distance of 3·14 Å, 1 at 3·29 Å, and 1 at 3·59 Å.

(v) B *subgroup dihalides.* The structures of these compounds, which are usually characteristic of one compound or a small number of compounds, are described in Chapters 25 and 26; they include compounds of Cu, Ag, Hg, Ga, Ge, and Sn.

Trihalides

Here it is sufficient to recognize three groups:
    (i) octahedral structures (Al, Sc, transition metals, In, Tl),
    (ii) structures of higher coordination (chiefly 4f and 5f metals),
    (iii) special structures of B subgroup compounds.

(i) *Octahedral* MX_3 *structures.* Corresponding to the rutile structure for fluorides and the h.c.p. (CdI_2) and c.c.p. ($CdCl_2$) layer structures for other dihalides we have the 3D ReO_3 structure (and variants) and the BiI_3 and YCl_3 layer structures; the last two are the structures of the low- and high-temperature forms respectively of $CrCl_3$ and $CrBr_3$. In addition, the ZrI_3 chain structure (with face-sharing octahedra) is adopted by a number of chlorides, bromides, and iodides.

Trifluorides. In Chapter 5 we surveyed the structures in which octahedral groups are joined by sharing only vertices, and we noted that the structure in which all vertices are shared can have an indefinite number of configurations. Two special cases may be recognized, one in which the X atoms are arranged in hexagonal closest packing, and the other in which these atoms occupy three-quarters of the positions of cubic closest packing. A number of trifluorides adopt the h.c.p. structure, but in most of these compounds the packing of the F atoms is of an intermediate kind (Table 9.16). Moreover it appeared that no trifluoride adopts the cubic ReO_3 structure, compounds with this structure being either non-stoichiometric (mixed valence) compounds such as $Nb(O, F)_3$ or $Mo(O, F)_3$ or compounds such as Nb^VO_2F Recently, however, it has been claimed that pure stoichiometric NbF_3 with the ReO_3 structure is obtained by heating $3 NbF_5 + 2 Nb$ at 750°C under a pressure of 3·5 kbar. The precise nature of the anion packing is of interest since it determines the M—F—M bond angle. There is an interesting connection, which is not understood, between the structures of these compounds and the position of M in the Periodic Table. Confirmation is desirable of the structure of AlF_3, in which the neighbours of Al are apparently 3 F at 1·70 and 3 F at 1·89 Å. The distorted octahedral coordination in the complex monoclinic structure of MnF_3 (Mn—F bond lengths, 1·79, 1·91, and 2·09 Å) may be contrasted with the regular octahedral coordination in VF_3, and it is interesting that in both MoF_3 and VF_3 M—F is equal to 1·95 Å yet the compounds are not isostructural. The compound PdF_3 is presumably $Pd^{II}(Pd^{IV}F_6)$.

354

TABLE 9.16

Crystal structures of trifluorides

	Angle M—F—M
H.c.p. structure	
PdF$_3$, RhF$_3$, IrF$_3$ (also Pt(O, F)$_3$)	132°
Intermediate packings	
RuF$_3$, MoF$_3$[a]	Values around
GaF$_3$, TiF$_3$, VF$_3$, CrF$_3$, FeF$_3$, CoF$_3$	150°
MnF$_3$[b]	
ScF$_3$,[c] InF$_3$[c] (AlF$_3$?)	
Cubic ReO$_3$ *structure*	
NbF$_3$[d]	180°

(a) Rhombohedral VF$_3$ type but close to h.c.p. structure.
(b) Unique distorted (monoclinic) VF$_3$ type due to Jahn–Teller effect.
(c) Close to cubic ReO$_3$ structure.
(d) CR 1971 **273** 1093.
For details and references see: Structure and Bonding, 1967 **3** 1.

Al								
Sc	Ti	V	Cr	Mn	Fe	Co		Ga
		Nb	Mo		Ru	Rh	Pd	In
					Ir			
					h.c.p.			

Other trihalides. Most of the other halides of metals of this group adopt one or more of the following structures (Table 9.17):

BiI$_3$: h.c.p. layer structure (low-CrCl$_3$)
YCl$_3$: c.c.p. layer structure (high-CrCl$_3$)
ZrI$_3$: chain structure.

TABLE 9.17

Crystal structures of trihalides

AlCl$_3$ (L)						
AlBr$_3$ (D)						
AlI$_3$ (D)						
ScCl$_3$ (L)	TiCl$_3$ (L/C)	VCl$_3$ (L)	CrCl$_3$ (L)[1]	FeCl$_3$ (L)	GaCl$_3$ (D)	
	TiBr$_3$ (L/C)		CrBr$_3$ (L)[1]	FeBr$_3$ (L)		
	TiI$_3$ (C)	VI$_3$ (L)			GaI$_3$ (D)	
YCl$_3$ (L)[2]	ZrCl$_3$ (L/C)			RuCl$_3$[3] (L/C)	InCl$_3$ (L)	
	ZrBr$_3$ (C)		MoBr$_3$ (C)		InBr$_3$ (L)	
YI$_3$ (L)	ZrI$_3$ (C)				InI$_3$ (D)	
	HfI$_3$		WCl$_3$ ReCl$_3$		TlCl$_3$ (L)	
			see pp. 366, 369			

(1) JCP 1964 **40** 1958
(2) JPC 1954 **58** 940
(3) JCS A 1967 1038.
For various compounds of this group see also IC 1964 **3** 1236; IC 1966 **5** 281; IC 1969 **8** 1994.

For the trihalides of As, Sb, and Bi see p. 706.
L = BiI$_3$ or YCl$_3$ layer structure; C = ZrI$_3$ chain structure; D = M$_2$X$_6$.

In the ZrI_3 structure the M–M distance is comparable with that in the metal because the MX_6 octahedra share opposite faces, but there is not necessarily metal–metal bonding. The magnetic moment of $ZrCl_3$ (0·4 BM) indicates considerable overlap of metal orbitals, but the moment of $TiCl_3$ (1·3 BM) is not much less than the value expected for one unpaired electron (1·75 BM). Similarly, in $CsCuCl_3$ and $CsNiCl_3$ where the anions have this chain structure, the moments are normal for one and two unpaired electrons.[1]

We have included in Table 9.17 $AlCl_3$ and some B subgroup trihalides, but we should note the following points. In contrast to $AlCl_3$ $AlBr_3$ has a molecular structure. The Br atoms are close-packed and the Al atoms occupy pairs of adjacent *tetrahedral* holes (contrast the octahedral coordination in AlF_3 and $AlCl_3$), and the crystal is built of Al_2Br_6 molecules (Fig. 9.6). The same type of molecule is found in crystalline $GaCl_3$[2] and in InI_3, with which AlI_3 and GaI_3 are probably isostructural.[3] In InI_3 the bond lengths are: $In–I_t$, 2 64 Å $In–I_b$, 2 84 Å; angles similar to those in $GaCl_3$. Dimeric molecules of the same kind exist in the vapours of both these halides.

(1) IC 1966 5 277
(2) JCS 1965 1816
(3) IC 1964 3 63

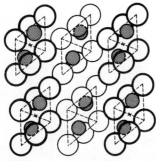

FIG. 9.5. Projection [on (010)] of the crystal structure of $AlBr_3$ showing how molecules Al_2Br_6 arise by placing Al atoms in pairs of adjacent tetrahedral holes in a close-packed array of halogen atoms. The crosses mark the corners of the monoclinic unit cell.

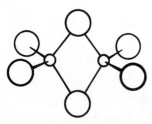

FIG. 9.6. The structure of the molecules Al_2Cl_6, Al_2Br_6, and Al_2I_6 according to Palmer and Elliot.

(ii) *Structures of higher coordination.* The octahedral structures described in (i) are not possible for larger ions, and this is nicely illustrated by the Group IIIA trihalides, which have a structural chemistry comparable as regards its complexity with that of the alkaline-earth dihalides (Table 9.18). The nine compounds studied provide examples of no fewer than seven different structures, three layer structures (below the heavy line in the table) and four 3D structures.

The LaF_3 structure is also called the tysonite structure, after the mineral of that name which is a mixed fluoride, $(Ce, La, ...)F_3$. The structure earlier assigned to LaF_3 has been revised, and the description depends on the interpretation of the various La–F distances. The neighbours of La comprise: 7 F at 2·42–2·48 Å, 2 F at 2·63 Å, and a further 2 F at 3·04 Å (means of two independent determinations). There are three kinds of non-equivalent F ion. If all the above F ions are counted as nearest neighbours the metal ions are 11-coordinated, two-thirds of F are 4-coordinated, and the remainder 3-coordinated. This 11-coordination group is a distorted trigonal prism capped on all faces. If only the 9 (or 7 + 2) F neighbours are included in the coordination group of the cation the structure is described as one of 9 : 3 coordination. There is no simple description of the cation coordination polyhedron, and the 3-coordination of the three kinds of fluoride ion ranges from nearly coplanar to definitely pyramidal. The structure is adopted by a number of 4f and 5f trifluorides (Table 9.19) and trihydrides (p. 296), some complex fluorides (e.g. $CaThF_6$, $SrUF_6$), and by $BiO_{0.1}F_{2.8}$ (p. 716). The structure of the high-temperature form of LaF_3 and the later 4f trifluorides is not known.

TABLE 9.18

Structures of Group IIIA *trihalides*

	Sc	Y	La
F	ReO_3	YF_3[1]	LaF_3[2]
Cl	YCl_3	YCl_3	UCl_3
Br	—	—	UCl_3
I	—	BiI_3	$PuBr_3$

(1) JACS 1953 **75** 2453
(2) IC 1966 **5** 1466; ZK 1965 **122** 375

In the YF_3 structure there is also a distorted 9-coordination of M^{3+} (tricapped trigonal prism), eight at approximately 2·3 Å and the ninth at 2·6 Å, so that the coordination may be described as (8 + 1). This structure is adopted by the 4f trifluorides $SmF_3 - LuF_3$ and also by TlF_3 and β-BiF_3 (for reference see p. 703).

The cubic 'yttrium trifluoride' referred to in earlier literature is apparently NaY_3F_{10}; other similar compounds include $NH_4Ho_3F_{10}$ and the Er and Tm compounds. The structure of this type of compound is illustrated in Fig. 9.7. The

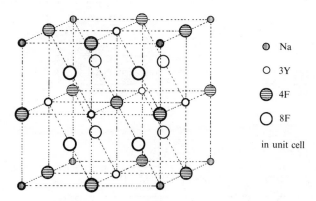

◍	Na
○	3Y
⊖	4F
◯	8F

in unit cell

FIG. 9.7. The crystal structure of NaY_3F_{10}.

content of one unit cell comprises Na, 3 Y, and 10 F arranged at random in the twelve positions shown by the larger circles.[1] A very similar cubic phase was also thought to be a cubic form of BiF_3, having a cell edge very similar in length to that of NaY_3F_{10}. It was assigned the structure of Fig. 9.7 with all metal ion positions occupied by Bi^{3+} and all twelve F^- positions occupied. (This would correspond to the fluorite structure with additional F^- ions—those shown as shaded circles at the

(1) JACS 1953 **75** 2453

357

(2) ARPC 1952 **3** 369

body-centre and at the mid-points of the edges of the cubic unit cell.) This cubic material is actually Bi_2OF_4,[2] presumably with $Bi_4(O, F)_{10}$ in the unit cell; compare $(Na + 3 Y)F_{10}$. See also p. 716 for other oxyfluorides of bismuth.

The UCl_3 structure, in which the coordination group of the metal ion is a tricapped trigonal prism, is adopted by numerous 4f and 5f trihalides (Table 9.19)

TABLE 9.19

Crystal structures of 4f *and* 5f *trihalides*

4f *trihalides*

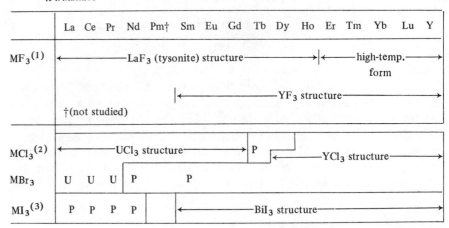

5f *trihalides*

	Ac	Th	Pa	U	Np	Pu	Am	Cm	Es	
MF_3[4]	L			L	L	L	L	L[5]		B = BiI_3 structure
MCl_3	U			U	U	U	U	U[5]	U[5a]	L = LaF_3
MBr_3	U			U / U	P	P[5]	P[5]	P[5]		P = $PuBr_3$
MI_3				P	P	P[3]	B[3]	B[5]		U = UCl_3

(1) IC 1966 **5** 1937. (2) IC 1964 **3** 185; JCP 1968 **49** 3007. (3) IC 1964 **3** 1137. (4) AC 1949 **2** 388. (5) IC 1965 **4** 985; (5a) INCL 1969 **5** 307. (6) A refinement of the (UCl_3) structure of $GdCl_3$ gives Gd–6 Cl, 2·82 A, and Gd–3 Cl, 2·91 A. (AC 1967 **23** 1112); and for $AmCl_3$, Am–6 Cl, 2·874 A, Am–3 Cl, 2·915 A (AC 1970 **B26** 1885).

and also by the trihydroxides of La, Pr, Nd, Er, Sm, Gd, and Dy. Comparison of the plans of the UCl_3 structure (Fig. 9.8) and the $PbCl_2$ structure (Fig. 6.23, p. 222) shows that these two structures are related in much the same way as are the YCl_3 and $CdCl_2$ structures, though the latter are layer structures while UCl_3 and $PbCl_2$

are 3D arrangements of ions. We have remarked in Chapter 7 that the unsymmetrical arrangement of M ions to one side of Cl in these structures is similar to that in the layer structures.

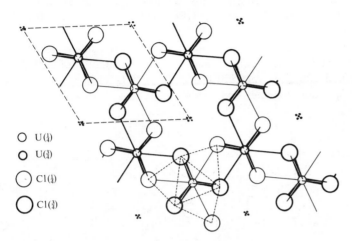

O U($\frac{1}{4}$)
o U($\frac{3}{4}$)
◯ Cl($\frac{1}{4}$)
◯ Cl($\frac{3}{4}$)

FIG. 9.8. Projection of the crystal structure of UCl$_3$ showing the tricapped trigonal prismatic coordination of U (see text).

In the UCl$_3$ structure the metal atom has nine approximately equidistant halogen neighbours (see ref. 6, Table 9.19). In the PuBr$_3$ structure, a projection of which is shown in Fig. 9.9, there is only 8-coordination of the metal atoms. This is a rather surprising structure in the sense that the halogen atoms are nearly in the positions required for 9-coordination of the metal atoms. The ninth Pu–Br bond would be that marked *c*, which would complete a coordination group of the same kind as in UCl$_3$. This distance is, however, 4·03 Å, as compared with 3·08 Å for the eight near neighbours. The layers are apparently prevented from approaching closer by the Br–Br contacts *a* and *b* which are unusually short for non-bonded Br atoms (3·81 and 3·65 Å respectively).

(iii) *B subgroup trihalides.* A number of structures peculiar to certain B subgroup halides are described in other chapters; AuX$_3$ (Chapter 25), AsX$_3$ etc. (Chapter 20).

Tetrahalides
More than thirty metals form a tetrahalide with at least one of the halogens, but in relatively few cases are all four tetrahalides of a given metal known. All four tetrahalides are known of Ge, Sn, Ti, Zr, Hf, Nb, Ta, Mo, W , Th, and U, and at the other extreme MF$_4$ is the only tetrahalide known of Cr, Mn, Pd, Ru, Rh, Ce, Pr, and Tb; no tetrahalides of Ir have been prepared. The crystal structures of more than one-half of the seventy or so known tetrahalides are known, and they include

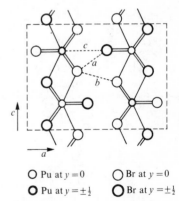

O Pu at $y = 0$ ◯ Br at $y = 0$
O Pu at $y = \pm\frac{1}{2}$ ◯ Br at $y = \pm\frac{1}{2}$

FIG. 9.9. The (layer) structure of PuBr$_3$. The planes of the layers are normal to that of the paper.

359

(1) AC 1963 **16** 446

(2) AC 1955 **8** 343

(3) AC 1962 **15** 903
(4) IC 1966 **5** 1197
(5) NW 1962 **49** 254
(6a) IC 1965 **4** 182

structures of all types, 3D, layer, chain, and molecular. Most of these structures have been described in previous chapters.

Tetrahedral molecules pack differently in $SnBr_4$[1] and SnI_4[2] with respectively h.c.p. and c.c.p. halogen atoms; compounds with the latter structure include GeI_4, $SnCl_4$, $TiBr_4$, TiI_4, $ZrCl_4$, $ZrBr_4$, and $PtCl_4$; a second form of $TiBr_4$ is isostructural with $SnBr_4$.

Octahedral structures (see Chapter 5) include two closely related chain structures, α-NbI_4[3] and $TcCl_4$[4] and the layer structure of SnF_4[5] (and the isostructural NbF_4[6a] and PbF_4). In the α-NbI_4 chain pairs of Nb atoms are alternately closer together and further apart (Fig. 9.10(a)), and the interaction

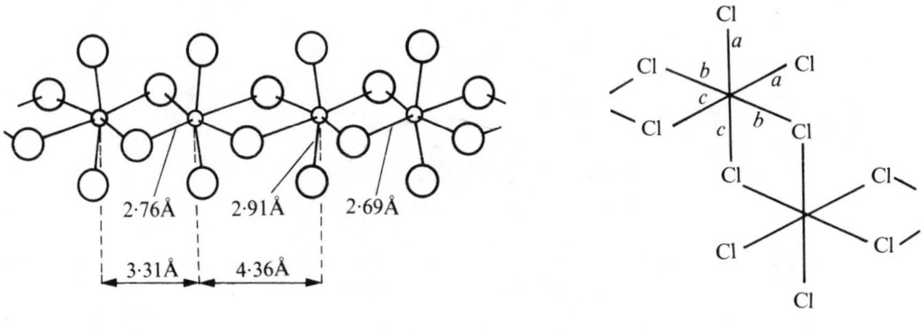

2·76Å   2·91Å   2·69Å

3·31Å   4·36Å

(a)

(b)

FIG. 9.10. Edge-sharing octahedral chains in (a) NbI_4, (b) $TcCl_4$.

(6b) ZaC 1969 **369** 154
(7) IC 1968 **7** 2602
(8) JACS 1967 **89** 2759

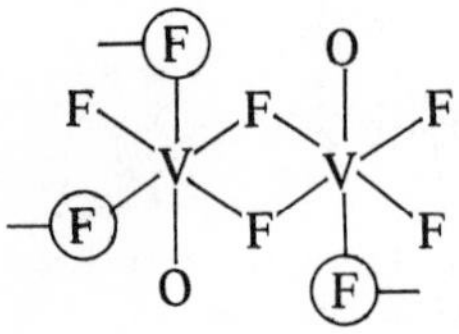

(a)

(8a) CR 1974 **278C** 1501
(8b) CC 1970 1475

between pairs of Nb atoms is sufficient to destroy the paramagnetism expected for Nb^{IV}. The bridge bonds are appreciably longer than the others, and this is also true in $TcCl_4$, where there are three different Tc—Cl bond lengths: $a = 2\cdot24$, $b = 2\cdot38$, and $c = 2\cdot49$ Å. There are no metal—metal bonds in this chain (Fig. 9.10(b)), the shortest Tc—Tc distance being 3·62 Å. The $TcCl_4$ structure has been recorded only for this compound, but the NbI_4 chain is also formed in crystalline $NbCl_4$, $TaCl_4$, TaI_4, $MoCl_4$, WCl_4, α-PtI_4[6b] and α-$ReCl_4$.[7] The second form of $ReCl_4$ (β) has a quite different chain structure[8] in which pairs of face-sharing octahedra form an infinite chain by sharing a vertex at each end, as shown at the right.

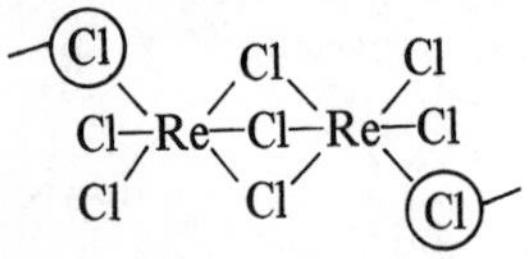

In SnF_4 octahedral SnF_6 groups form a layer by sharing four equatorial F atoms as in the K_2NiF_4 structure (p. 171).

The first 3D octahedral structure for a tetrahalide has recently been assigned to IrF_4.[8a] In this structure each octahedral IrF_6 group shares four vertices, each with one other IrF_6 group, a pair of *cis* vertices being unshared. The structure is closely related to the rutile structure, from which it is derived by removing alternate metal atoms from each edge-sharing chain. This relationship may be compared with that between the NaCl and atacamite (AX_2) structures (Fig. 4.22, p. 143).

It has been suggested, though not confirmed, that the structure of VF_4 is related to that of VOF_3.[8b] In the latter, edge-sharing pairs of octahedral coordination groups (here VOF_5) are further linked by sharing four of the remaining eight vertices to form layers as shown at (a). The structures of a number of 3d tetrahalides would be of special interest because of the presence of excess d electrons, and in this connection we may note that the VCl_4 molecule was assigned a regular tetrahedral structure as the result of an early electron diffraction study of the vapour.[8c]

Other octahedral structures include that suggested for high-$MoCl_4$[9] that is, random occupation of three-quarters of the metal positions in the BiI_3 structure. There would be isolated $MoCl_6$ octahedra and portions of edge-sharing chains.

Of three structures with 8 : 2 coordination, two are examples of antiprism coordination and one of dodecahedral coordination. The remarkable ThI_4 structure[10] consists of layers formed from square antiprisms ThI_8 which share two triangular faces and one edge; it was illustrated in Chapter 3 as an example of a layer based on the plane 6-gon net. In contrast to the molecular $ZrCl_4$ and $ZrBr_4$, ZrF_4[11] has a typically ionic structure of 8 : 2 coordination. Both types of non-equivalent Zr^{4+} ion are surrounded by 8 F forming a slightly distorted square antiprism which shares vertices with eight others. This structure is confined to tetrafluorides of the larger M^{4+} ions (Hf, Ce, Pr, Tb, and 5f elements Th–Bk). The tetrachlorides of Th, Pa, U, and Np (and also $ThBr_4$ and $PaBr_4$) adopt a different 8 : 2 structure. In the UCl_4 structure a metal atom has eight dodecahedral neighbours, and each coordination group shares one edge with each of four others in helical arrays around 4_1 axes, forming a 3D structure of 8 : 2 coordination (Fig. 9.11). In $ThCl_4$,[12] which has this structure, the M–Cl distances are four of 2·72 Å value calculated for minimal ligand repulsion (see p. 69). A very similar coordination group is found in the isostructural $PaBr_4$[13] (Pa–Br, 2·83 Å and 3·01 Å).

(8c) JACS 1945 **67** 2019

(9) ZaC 1967 **353** 281

(10) IC 1964 **3** 639

(11) AC 1964 **17** 555

(12) AC 1969 **B25** 2362

(13) N 1968 **217** 737

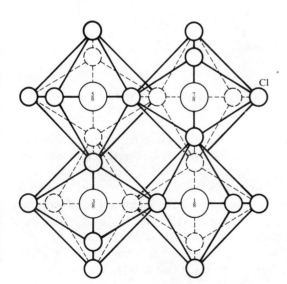

FIG. 9.11. Projection of the structure of $ThCl_4$ (UCl_4).

For the unique structure of TeF_4, in which square pyramidal TeF_5 groups form infinite chains by sharing two F atoms, we refer the reader to Chapter 16 (Fig. 16.3).

The structures of crystalline tetrahalides provide a good illustration of the ways in which the same X : M ratio can be attained by sharing of increasing numbers of atoms between MX_n coordination groups as n increases:

Coordination of M	C.N.	Elements of coordination polyhedra shared	Examples
Tetrahedral	4	None	SnI_4
Square pyramidal	5	2 vertices	TeF_4
Octahedral	6	4 vertices	SnF_4
		2 edges	$TcCl_4$, NbI_4
		1 face + 1 vertex	$ReCl_4$
Antiprismatic	8	8 vertices	ZrF_4
Dodecahedral	8	4 edges	$ThCl_4$
Antiprismatic	8	2 faces + 1 edge	ThI_4

Pentahalides

Apart from the trigonal bipyramidal $SbCl_5$ molecule, which has the same structure in the crystalline as in the vapour state, we have to deal here with 6- and 7-coordinated structures.

The octahedral structures, described in Chapter 5, include the dimeric and tetrameric molecules of types (a) and (b) (Fig. 9.12) and the '*cis*' and '*trans*' chains.

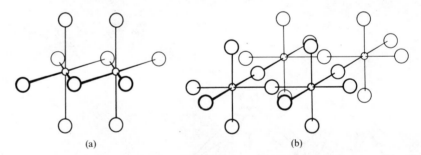

(a)   (b)

FIG. 9.12. Pentahalide molecules: (a) M_2Cl_{10}, (b) M_4F_{20}.

The dimers are chlorides and one bromide, while the cyclic tetramers and chains are fluorides. As in the case of other halides the molecules (a) or (b) can pack to give various types of closest packing of the halogen atoms. For both (a) and (b) the h.c.p. and c.c.p. structures are known, and for the dimers also the more complex *hc* packing (Table 9.20). In the dimeric molecules the metal atoms are displaced from the centres of the octahedra (by 0·2 Å in U_2Cl_{10}, and the bridge bonds are longer than the terminal ones (for example, 2·69 and 2·44 Å in U_2Cl_{10}).

The special interest of the two M_4X_{20} structures lies in the fact that the —F— bond angles are quite different, 180° (as shown at (b)) and 132° in the c.c.p. and

TABLE 9.20

Structures of crystalline pentahalides

Dimers M_2X_{10}

	Type of closest packing of X			Reference
	h	*c*	*hc*	
M_2Cl_{10}	Nb Ta Mo W			JCS A 1967 1825, 2017
		U		AC 1967 **22** 300
			Re	AC 1968 **B24** 874
M_2Br_{10}		β-Pa$_2$Br$_{10}$		AC 1969 **B25** 178

Tetrameric and linear pentafluorides

V[1] Cr		Tc Re	Ru Rh[12] Os[10] Ir Pt	*trans* chain[6]
Nb Ta	Mo W[5]	*cis* chain —F— 150°	M_4F_{20} —F— 132° h.c.p.	—F— 180° BiF$_5$ α-UF$_5$
M_4F_{20} —F— 180° c.c.p.				
	also WOF$_4$[2] NbCl$_4$F[8] TaCl$_4$F[9]	also MoOF$_4$[3] ReOF$_4$[4] TcOF$_4$[11]		pentagonal bipyramidal chain PaCl$_5$[7]

(1) JCS A 1969 1651. (2) JCS A 1968 2074. (3) JCS A 1968 2503. (4) JCS A 1968 2511. (5) JCS A 1969 909. (6) JACS 1959 81 6375. (7) AC 1967 **22** 85. (8) ZaC 1968 **362** 13. (9) ZaC 1966 **346** 272. (10) JCS A 1971 2789. (11) CC 1967 462. Also green trimeric form with —F—, 161° (JCS A 1970 2521). (12) IC 1973 **12** 2640.

h.c.p. structures respectively. (The cubic closest packing is distorted to give monoclinic symmetry in the former structures.) This is reminiscent of the trifluoride structures described earlier, as also is the choice of structure by the various pentafluorides (Table 9.20).

The two forms of octahedral MX_5 chain arise by sharing of adjacent or opposite vertices of octahedral MX_6 groups (*cis* and *trans* chains); here again it would be interesting to know what determines the choice of chain.

Of the 7-coordinated structures, that of β-UF$_5$ is a 3D structure in which U is surrounded by 7 F of which 4 are shared with other coordination groups (Fig. 28.2). In the PaCl$_5$ structure (which is unique to that compound) pentagonal bipyramidal groups share two edges to form infinite chains (Fig. 9.13). Bond lengths are: Pa—Cl (bridge), 2·73 Å, Pa—Cl (terminal), 2·44 Å.

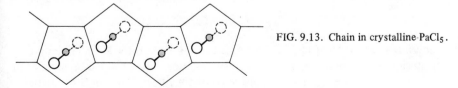

FIG. 9.13. Chain in crystalline PaCl$_5$.

The Halogens—Simple Halides

It will be observed that apart from $SbCl_5$ 5-coordination is avoided in all the pentahalide structures described, as is also the case in crystalline PCl_5 and PBr_5.

Hexahalides

Although infinite complexes (for example, chains or layers built from 7-, 8-, or 9-coordination groups) are in principle possible, only octahedral molecular structures are as yet known for crystalline hexahalides. X-ray powder photographic examination of numerous hexafluorides of second- and third-row transition metals shows that they have isostructural low-temperature (orthorhombic) and high-temperature (disordered) b.c.c. polymorphs.[1] A more complete study has been made of the structurally similar low-temperature form of $Os^{VII}OF_5$,[2] which consists of nearly regular octahedral molecules. Little recent structural work seems to have been done on crystalline hexachlorides. Like the halides M_2X_{10} the hexahalides adopt different kinds of closest packing of the halogen atoms:

(1) IC 1966 **5** 2187
(2) JCS A 1968 543
(3) AC 1974 *B* 30 1481

$$h: \quad UCl_6,^{(3)} WCl_6, MoCl_6$$
$$hc: \quad UF_6 \text{ (and } OsOF_5)$$

Heptahalides

The only known metal heptahalides are ReF_7 and OsF_7, the latter being stable only at low temperatures.[1] The only other heptafluoride known is IF_7. The evidence for the pentagonal symmetry of ReF_7 (and IF_7) has been summarized.[2]

(1) CB 1966 **99** 2652
(2) JCP 1968 **49** 1803

No higher halides are known; the compound earlier described as OsF_8 proved to be OsF_6.

Polynuclear complexes containing metal–metal bonds

Binuclear halide complexes formed by Mo, Tc, and Re

A number of elements form ions $M_2Cl_9^{n-}$ consisting of two octahedral groups sharing a face (p. 166), and similar units are joined by sharing one Cl at each end to form the infinite chain molecules in β-$ReCl_4$ (p. 87). Within the sub-units Re–Re is 2·73 Å, indicating a (single) metal–metal bond. The ion $Re_2Cl_9^-$ has a similar structure, and presumably also the $Mo_2Cl_9^{3-}$ ion. Molybdenum also forms two other binuclear ions:

	Oxidation number of Mo	M–M (Å)
$Mo_2Cl_8^{4-}$	2	2·13
$Mo_2Cl_8^{3-}$	2·5	2·38
$Mo_2Cl_9^{3-}$	3	—

The intermediate ion, $Mo_2Cl_8^{3-}$, has the same general shape as the $Mo_2Cl_9^{3-}$ ion, to which it can be oxidized electrochemically in solution, but with one bridging Cl missing and a rather strong metal–metal bond (Fig. 9.14(a)). The $Mo_2Cl_8^{4-}$ ion has

364

an entirely different structure, Fig. 9.14(b), which is also that of the $Tc_2Cl_8^{3-}$ and $Re_2Cl_8^{2-}$ ions. These ions have an extremely short M–M bond and an eclipsed configuration, the eight halogen atoms being at the vertices of an almost perfect cube. According to a m.o. treatment the metal–metal bond is regarded as quadruple ($\sigma\pi^2\delta$), the δ component accounting for the eclipsed configuration.

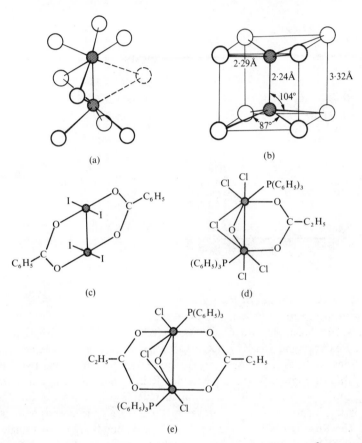

FIG. 9.14. Binuclear halide ions and related molecules: (a) $Mo_2Cl_8^{3-}$, (b) $Re_2Cl_8^{2-}$, (c) $Re_2I_4(OOC \cdot C_6H_5)_2$, (d) $Re_2OCl_5(OOC \cdot C_2H_5)(P\phi_3)_2$, (e) $Re_2OCl_3(OOC \cdot C_2H_5)_2(P\phi_3)_2$.

Substitution reactions may be performed on the Re_2X_8 nucleus, the general shape of which is retained in molecules such as $Re_2Cl_6(PEt_3)_2$, in which two (*trans*) X atoms are replaced by phosphine molecules, and Re_2I_4 (benzoate)$_2$ (Fig. 9.14(c)); in both of these molecules Re–Re is close to 2·2 Å. In molecules such as (d) and (e), where there are bridging O and Cl atoms (and six ligands attached to each metal atom), the length of the metal–metal bonds is close to 2·5 Å. Note the range of Re–Re bond lengths, 2·2, 2·5, and 2·7 Å in these binuclear complexes. (For references see Table 9.21.)

TABLE 9.21

Metal–metal bond lengths in halide complexes

Ion or molecule	M—M (Å)	Reference
$ReCl_4$ (crystalline)	2·73	JACS 1967 89 2759
$(Re_2Cl_9)^-$	2·71	P. F. Stokely, Ph.D. Thesis M.I.T. 1969
$Re_2OCl_5(O_2CC_2H_5)(P\phi_3)_2$	2·52	IC 1968 7 1784
$Re_2OCl_3(O_2CC_2H_5)_2(P\phi_3)_2$	2·51	IC 1969 8 950
$Re_2I_4(O_2CC_6H_5)_2$	2·20	IC 1969 8 1299
$Re_2Cl_6(PEt_3)_2$	2·22	IC 1968 7 2135
$(Re_2Cl_8)^{2-}$	2·24	IC 1965 4 330, 334
$(Tc_2Cl_8)(NH_4)_3 \cdot 2\,H_2O$	2·13	IC 1970 9 789
$(Mo_2Cl_8)K_4 \cdot 2\,H_2O$	2·13	IC 1969 8 7
$(Mo_2Cl_8)Cl(NH_4)_5 \cdot H_2O$	2·15	IC 1970 9 346
$(Mo_2Cl_8)Cs_3$	2·38	IC 1969 8 1060
Re_3I_9	2·44, 2·51	IC 1968 7 1563
Re_3Cl_9	2·49	IC 1964 3 1402
$Re_3Cl_9(P\phi Et_2)_3$	2·49	IC 1964 3 1094
$(Re_3Cl_{11})(As\phi_4)_2$	2·44, 2·48	IC 1966 5 1758
$(Re_3Br_{11})Cs_2$	2·43, 2·49	IC 1966 5 1763
$(Re_3Cl_{12})Cs_3$	2·48	IC 1963 2 1166
$(Re_4Br_{15})(QnH)_2$	2·47	IC 1965 4 59

Trinuclear halide complexes of Re

Rhenium forms a variety of ions and molecules based on a triangular unit of three Re atoms, for example, $Re_3Cl_9(pyr)_3$, $Re_3Cl_3(NCS)_8^{2-}$ and $Re_3Cl_3(NCS)_9^{3-}$, $Re_3X_{10}^-$, $Re_3X_{11}^{2-}$, and $Re_3X_{12}^{3-}$, (X = Cl, Br), the ions being isolated as salts of large cations such as Cs^+, $As(C_6H_5)_4^+$, etc. (Salts of the type $[N(C_2H_5)_4]_2$ Re_4Br_{15} do not contain Re_4 complexes but $ReBr_6^{2-}$ ions and Re_3Br_9 units.) The $Re_3Cl_{12}^{3-}$ ion, Fig. 9.15(a), is the anion in $CsReCl_4$. The Re—Re bond length, 2·48 Å, is much shorter than that in the metal (2·75 Å) and corresponds to bond order 2. The Re—Cl_a (2·36 Å) and Re—Cl_b (2·39 Å) bonds are normal single bonds (compare 2·37 Å in $K_4ReOCl_{10} \cdot H_2O$), and the length of Re—Cl_c (2·52 Å) has been attributed to overcrowding. The same Re_3 nucleus exists in $Re_3Cl_9(P\phi Et_2)_3$ and many other complexes, and the $Re_3Cl_{11}^{2-}$ ion has a structure very similar to $Re_3Cl_{12}^{3-}$ but devoid of one equatorial Cl atom.

The trihalides $ReCl_3$ and $ReBr_3$ have been shown by mass spectrometry to vaporize as Re_3X_9 molecules. Crystalline $ReCl_3$ contains Re_3Cl_9 molecules like the ions of Fig. 9.15(a) without the equatorial Cl_c atoms. These molecules are joined through weaker Re—Cl bridges (2·66 Å) into layers based on the plane 6-gon net (Fig. 9.15(b)). In each bridge one Cl_a atom of one molecule occupies the Cl_c position in the adjacent molecule (Fig. 9.15(c)). In ReI_3 the Re_3X_9 molecules are of the same kind, but instead of being linked into layers by the bridges they are linked into chains (Fig. 9.15(d)).

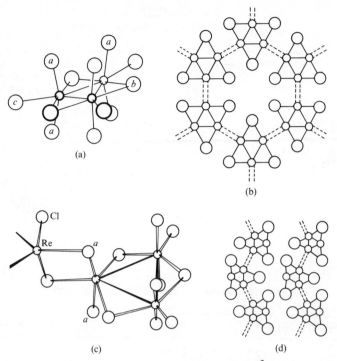

FIG. 9.15. Trinuclear halide complexes of Re: (a) $(Re_3Cl_{12})^{3-}$, (b) crystalline $ReCl_3$, (c) bridge in $ReCl_3$, (d) crystalline Re_3I_9.

Halide complexes of Nb, Ta, Mo, W, Pd, *and* Pt-*containing metal 'clusters'*
The compounds to be discussed here are formed by certain metals of the second and third transition series,

Nb Mo Pd

Ta W Pt.

A limited number of B subgroup metals also form halides in which there is metal–metal bonding, notably Hg and Bi. We shall refer briefly to the Bi compounds; for mercurous compounds see Chapter 26, and for $Cu_4I_4(AsEt_3)_4$ see Chapter 25.

A feature of the chemistry of Nb and Ta is the formation of many halogen compounds in which the metal exhibits a non-integral oxidation number (Table 9.22). The halides are made by methods such as the following: Nb_3I_8 by thermal decomposition of the higher iodides, Nb_6I_{11} by heating Nb_3I_8 with the metal, $CsNb_4Cl_{11}$ by a vapour transport reaction from a mixture of CsCl, Nb_3Cl_8, and Nb metal, and $K_4Nb_6Cl_{18}$ by heating together KCl and Nb_6Cl_{14}. A different set of compounds is obtained from solution, namely, those containing Nb_6X_{12} groups, which can be obtained in several different oxidation states. The $Nb_6Cl_{12}^{2+}$ ion may

TABLE 9.22

Some halides and halide complexes of Nb

			Oxidation number of Nb				
1·83	2	2·33	2·50	2·67	3	4	5
			Nb_6F_{15}		NbF_3	NbF_4	NbF_5
		$(Nb_6Cl_{12})Cl_2$		Nb_3Cl_8	$NbCl_3$	$NbCl_4$	$NbCl_5$
		$K_4(Nb_6Cl_{12})Cl_6$	$CsNb_4Cl_{11}$		$Cs_3Nb_2Cl_9$		
		$(Nb_6Cl_{12})^{2+}$	$(Nb_6Cl_{12})^{3+}$	$(Nb_6Cl_{12})^{4+}$			
	$NbBr_2$						
Nb_6I_{11}				Nb_3I_8	NbI_3	NbI_4	NbI_5

TABLE 9.23

Compounds containing M_6X_8 *or* M_6X_{12} *'clusters'*

Type of complex		Reference
	M_6X_8	
Finite		
$(Mo_6Cl_8)Cl_6(NH_4)_2 . H_2O$		AK 1949 **1** 353
$(Mo_6Cl_8)(OH)_4(H_2O)_2 . 12 H_2O$ }		
1-dimensional		
$(W_6Br_8)Br_4 . Br_4$	W_6Br_{16}	ZaC 1968 **357** 289
2-dimensional		
$(Mo_6Cl_8)Cl_4$	$MoCl_2, Br_2, I_2$ }	ZaC 1967 **353** 281
	WCl_2, Br_2, I_2 }	
3-dimensional		
$(Nb_6I_8)I_3$	Nb_6I_{11}	ZaC 1967 **355** 295
$(Nb_6I_8)I_3H$	$Nb_6I_{11}H$	ZaC 1967 **355** 311
	M_6X_{12}	
Finite		
Pd_6Cl_{12}	$PdCl_2$	AnC 1967 **79** 244
Pt_6Cl_{12}	$PtCl_2$	ZaC 1965 **337** 120
$(W_6Cl_{12})Cl_6$	WCl_3	AnC 1967 **79** 650
$(Ta_6Cl_{12})Cl_2(H_2O)_4 . 3 H_2O$	$Ta_6Cl_{14} . 7 H_2O$	IC 1966 **5** 1491
$(Nb_6Cl_{12}Cl_6)^{4-}$	$K_4Nb_6Cl_{18}$	ZaC 1968 **361** 235
$(Nb_6Cl_{12}Cl_6)^{2-}$	$[N(CH_3)_4]_2Nb_6Cl_{18}$	IC 1970 **9** 1347
$(Ta_6Cl_{12}Cl_6)^{2-}$	$H_2Ta_6Cl_{18} . 6 H_2O$	IC 1971 **10** 1460
2-dimensional		
$(Ta_6I_{12})I_2$	Ta_6I_{14}	JLCM 1965 **8** 388
$(Nb_6Cl_{12})Cl_2$	Nb_6Cl_{14}	ZaC 1965 **339** 155
3-dimensional		
$(Nb_6F_{12})F_3$	Nb_6F_{15}	JLCM 1965 **9** 95
$(Ta_6Cl_{12})Cl_3$	Ta_6Cl_{15}, Ta_6Br_{15}	ZaC 1968 **361** 259

be regarded as the structural unit from which $(Nb_6Cl_{12})Cl_2$ and its octahydrate are formed. Oxidation of this ion in HCl–alcohol solution by air, followed by addition of NEt_4Cl, gives $(NEt_4)_3(Nb_6Cl_{12})^{3+}Cl_6$ ($\mu = 1\cdot65$ BM), while oxidation by Cl_2 yields $(NEt_4)_2(Nb_6Cl_{12})^{4+}Cl_6$, with approximately zero moment.[1] Similarly, the $(Ta_6Cl_{12})^{n+}$ ion can be reversibly oxidized and reduced ($n = 2, 3, 4$).[2]

(1) JACS 1967 **89** 159
(2) IC 1968 **7** 631, 636

In all of the compounds of Table 9.22 except NbF_4 and the pentahalides there is some interaction between metal atoms. This ranges from binding between alternate pairs of Nb atoms in the NbX_4 chain structure, through the formation of Nb_3 groups in the halides Nb_3X_8 and groups of four Nb atoms in $CsNb_4Cl_{11}$, to the formation of octahedral M_6 groups in all the compounds of Table 9.23.

Halides Nb_3X_8 form a layer structure[3] in which Nb atoms occupy $\frac{3}{4}$ of the octahedral holes between alternate layers of halogen atoms (Fig. 5.22(d), p. 177). The interaction between the Nb atoms is such that triangular Nb_3 groups can be distinguished which may be shown in idealized form as in Fig. 9.16(a), it being understood that these groups share all the outer Cl atoms with adjacent groups as indicated. (Of the 13 Cl atoms shown in the Figure, 6 are common to two and 3 to three such groups: $4 + 6(\frac{1}{2}) + 3(\frac{1}{3}) = 8$.)

(3) JLCM 1966 **11** 31; JACS 1966 **88** 1082

In $CsNb_4Cl_{11}$ groups of 4 octahedra may be distinguished on the basis of the Nb–Nb distances, namely, $a = 2\cdot95$, $b = 2\cdot84$, $c = 3\cdot56$, and $d = 3\cdot95$ Å (Fig. 9.16(b)); the group of four octahedra shares edges and faces as shown.

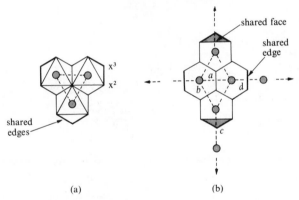

FIG. 9.16. (a) The M_3X_8 unit in Nb_3Cl_8; (b) the M_4X_{11} unit in $CsNb_4Cl_{11}$. In (a) the atoms X^2 and X^3 are common to two or three M_3X_8 groups respectively. In (b) all X atoms on shared edges or faces are common to two M_4X_{11} groups.

We now come to the compounds of Table 9.23, the structures of which are based on one of two units, both containing a central octahedral group of metal atoms. These halides include not only the compounds of Nb and Ta in which the metal has non-integral oxidation numbers but also dihalides of Mo, W, Pd, and Pt and WCl_3. The units are illustrated in idealized form in Fig. 9.17. In (a) M atoms are at the centres of the faces of a cube and X atoms at the vertices, giving a finite group of composition M_6X_8. This group was first recognized as the basic structural unit in a number of chloro complexes of Mo. The yellow 'dichloride' is soluble in alcohol and from the solution alcoholic $AgNO_3$ solution precipitates only one-third of the chlorine. A saturated solution of the dichloride in aqueous HCl produces crystals of a 'chloro acid', $H_2(Mo_6Cl_{14}) \cdot 8 H_2O$, from which salts such as $(NH_4)_2(Mo_6Cl_{14}) \cdot H_2O$ can be prepared. Controlled hydrolysis of the solution of the chloro acid gives products in which part of the Cl is replaced by OH, H_2O or both. Attachment of 6 Cl to the M_6X_8 group of Fig. 9.17(a), as shown by the X atoms, gives the ion $(Mo_6Cl_8 \cdot Cl_6)^{2-}$ in the chloro acid, while attachment of 4 Cl and 2 H_2O gives a neutral group $[Mo_6Cl_8 \cdot Cl_4(H_2O)_2]6 H_2O$ in the 'octahydrate' of the dichloride.

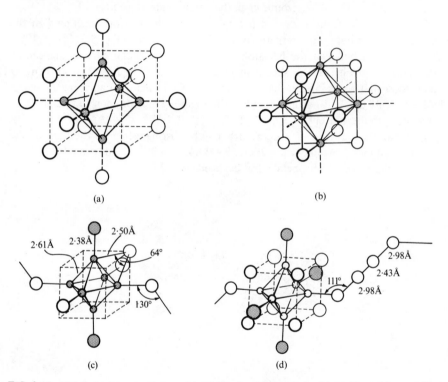

FIG. 9.17. Metal cluster complexes: (a) M_6X_8 (or $M_6X_8 \cdot X_6$), (b) M_6X_{12}, (c) crystalline $MoCl_2$ (details), (d) W_6Br_{16}.

If the four equatorial X atoms of Fig. 9.17(a) are shared with other similar groups the result is a layer of composition $(Mo_6Cl_8)Cl_4$, and this is the structure of $MoCl_2$, WCl_2, etc. (Table 9.23). Details are shown in Fig. 9.17(c). Sharing of all six X atoms with other units gives a 3D framework with the composition M_6X_{11} which represents the structure of crystalline Nb_6I_{11}. At temperatures above 300°C the paramagnetic Nb_6I_{11} absorbs hydrogen at atmospheric pressure to form a hydride with the limiting composition $Nb_6I_{11}H$, which is diamagnetic at low temperatures. A n.d. study shows the H atom at the centre of the Nb_6I_{11} group.

The action of liquid Br_2 on WBr_2 (W_6Br_{12}) gives successively W_6Br_{14}, W_6Br_{16}, and W_6Br_{18}. In W_6Br_{16} ($W_6Br_8 \cdot Br_4$) groups are linked through linear Br_4 groups (Fig. 9.17(d)) to form infinite chain molecules.

The grouping of Fig. 9.17(b) has the composition M_6X_{12}, and represents the structure of the molecules in one of the crystalline forms of $PdCl_2$ and $PtCl_2$ and in their vapours. Addition of a further X atom to each M gives M_6X_{18}, which is the structure of the molecule in tungsten trichloride. The $(Nb_6X_{12})^{n+}$ and $(Ta_6X_{12})^{n+}$ ions can be oxidized from $n = 2$ to $n = 3$ and 4, the corresponding $(M_6X_{18})^{m-}$ ions having $m = 4$, 3, and 2. Salts containing all of these ions have been made, and some of their structures have been determined (see Table 9.23):

$$(Nb_6Cl_{18})^{4-}: Nb\text{—}Nb, 2\cdot92 \text{ Å}$$
$$(Nb_6Cl_{18})^{2-}: Nb\text{—}Nb, 3\cdot02 \text{ Å} \quad \text{(see Table 9·23)}$$
$$(Ta_6Cl_{18})^{2-}: Ta\text{—}Ta, 2\cdot96 \text{ Å}.$$

As in the molybdenum complexes derived from M_6X_8 units some of the additional six ligands may be, for example, H_2O molecules, and this is so in $Ta_6Cl_{14} \cdot 7 H_2O$, in which the structural unit is $[Ta_6Cl_{12}(Cl_2)(H_2O)_4]$. The basic structure of this unit had been deduced earlier from X-ray diffraction data obtained from a concentrated alcoholic solution of the compound.

If units M_6X_{12} are joined into layers through four additional (equatorial) X atoms the composition is M_6X_{14}, as in Nb_6Cl_{14} and Ta_6I_{14}. Finally, linking of such units to six others through additional X atoms gives a 3D framework with the composition M_6X_{15}, and this is the type of structure adopted by Nb_6F_{15}, Ta_6Cl_{15}, and Ta_6Br_{15}.

It will be observed that the atoms of the M_6X_{12} group of Fig. 9.17(b) also represent a unit cell of the structure of NbO, which contains the same octahedral Nb_6 group as in the compounds we have been discussing. However, the system of Nb—Nb bonds in NbO is continuous throughout the crystal, accounting for the metallic lustre and conductivity of this oxide.

Other polyhedral metal groupings include (presumably) the trigonal bipyramidal Pt_3Sn_2 nucleus in the ion $(Pt_3Sn_8Cl_{20})^{4-}$ in $[N(CH_3)_4]_4Pt_3Sn_8Cl_{20}$, a compound produced by the interaction of $PtCl_2$ and $SnCl_2$ in acetone in the presence of $[N(CH_3)_4]^+$ ions.[4] There is as yet only i.r. evidence for the structure of the $(Pt_3Sn_8Cl_{20})^{4-}$ ion, but two other complexes containing $SnCl_3$ ligands have been

(4) IC 1966 5 109

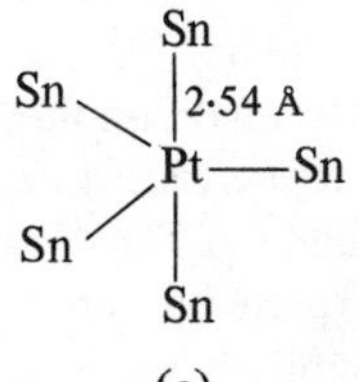

(a)

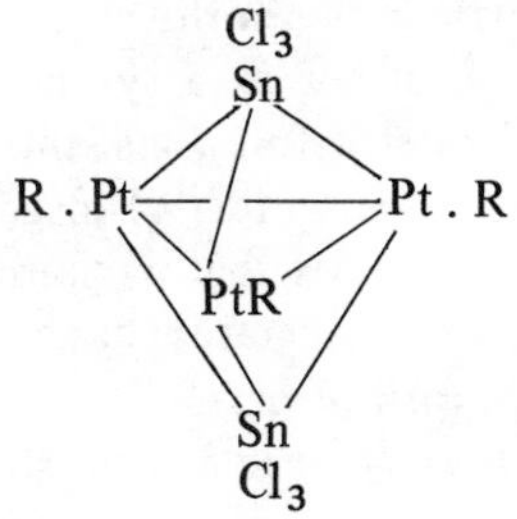

(b)

(5) JACS 1965 **87** 658

(6) CC 1968 512

(7) IC 1963 **2** 979

(8) IC 1964 **3** 1408

studied. The detailed structure of the anion in $(\phi_3PCH_3)_3[Pt(SnCl_3)_5]$, (a), could not be determined owing to disorder in the crystal;[5] this is not a polyhedral cluster compound. A trigonal bipyramidal nucleus exists in $(C_8H_{12})_3Pt_3(SnCl_3)_2$,[6] (b), in which the following bond lengths were determined: Pt–Pt, 2·58 Å, Pt–Sn, 2·80 Å, and Sn–Cl, 2·39 Å. See also the molecules $Pt_4(OH)_4(CH_3)_{12}$ etc. described in Chapter 27.

Molten $BiCl_3$ dissolves metallic Bi, and from the melt black crystals of composition $BiCl_{1·167}$ are obtained. These remarkable crystals[7] contain poly-hedral Bi_9^{5+} groups (tricapped trigonal prisms) in which Bi–Bi ranges from 3·08–3·29 Å (compare 3·10 Å in the element), distorted tetragonal pyramidal $BiCl_5^{2-}$ ions, and dimeric $Bi_2Cl_8^{2-}$ ions formed from two such pyramidal groups sharing a basal edge. A m.o. treatment of the bonding in the Bi_9^{5+} units has been given.[8]

Metal halides in the fused and vapour states

A complete picture of the structural chemistry of a compound would require a knowledge of its structure in the solid, liquid, and gaseous states. The amount of information obtainable about the structure of a liquid halide is very limited, and few X-ray studies have been made. (Examples include SnI_4,[1] $InCl_3$,[2] and CdI_2.[3]) We are therefore obliged to make direct comparisons of crystal and vapour. Strictly, these comparisons relate only to the process of sublimation, and if the compound is polymorphic the relevant crystal structure is that of the polymorph stable at the temperature of sublimation.

(1) JACS 1952 **74** 1763
(2) JACS 1952 **74** 1760
(3) JACS 1953 **75** 471

(4) JACS 1951 **73** 3151

(5) JCP 1960 **33** 366

(6) JCP 1960 **32** 1150

Physical properties are sometimes indicative of structural changes. Crystalline $AlCl_3$ has a layer structure, and the electrical conductivity of the solid increases rapidly as the melting point is approached, at which temperature it falls suddenly to nearly zero. At the melting point there is an unusually large decrease in density (about 45 per cent) as the ionic crystal, in which Al is 6-coordinated, changes to a melt consisting of Al_2Cl_6 molecules.[4] These molecules are the predominant species in the vapour at temperatures below about 400°C, and consist of two tetrahedral $AlCl_4$ groups with a common edge. The viscosity of aqueous $ZnCl_2$ solutions rises sharply at high concentrations, and the viscosity of the molten salt is much higher[5] than that of normal unassociated dihalides such as $MgCl_2$ or $CdCl_2$, being intermediate between the value for such compounds and that of BeF_2. Like $ZnCl_2$ BeF_2 crystallizes as a system of tetrahedral MX_4 groups linked through vertices (silica-like structure), and the viscosity of the molten BeF_2 at its melting point[6] is of the same order of magnitude as the viscosities of GeO_2 and B_2O_3, that is, about 10^8 times that of 'normal' dihalides. The X-ray scattering curve of glassy BeF_2 at room temperature has been interpreted in terms of a random network of the silica type formed from BeF_4 groups. Similarly the high viscosity of SbF_5 (at room temperature) suggests a chain structure formed from SbF_6 octahedra sharing two vertices.

The structures of many halide molecules have been studied in the vapour state, generally by electron diffraction. Configurations found for molecules MX_n are:

MX_2: linear; Zn, Cd, and Hg dihalides ⎱
          angular; Sn and Pb dihalides   ⎰ (Table 9·25)

MX_3: pyramidal; As, Sb, and Bi trihalides (Chapter 20)

MX_4: tetrahedral; Group IV A and B tetrahalides (Table 21.1)

MX_5: trigonal bipyramidal; Nb, Ta, As, and Sb pentahalides (Tables 9.25 and 20.3)

MX_6: octahedral; MoF_6, WCl_6, TeF_6, and Table 9.25.

The conventional methods of studying vapour species (e.d. and spectroscopic) have been developed in a number of ways. Examples are the microwave spectroscopy of molecules formed by the direct vaporization of solids (Ag–Cl, 2·28 Å, Ag–Br, 2·39 Å[7]) or of unstable species prepared in special ways. For example, the reaction of aluminium halides with metal gives monohalides (Al–F, 1·6544 Å, Al–Cl, 2·1298 Å[8]). Other methods include i.r. studies of molecules in the vapour or trapped in matrices,[9] giving estimates of bond angles in molecules such as GeF_2 (94° ± 4°) and TiF_2 (130° ± 5°), and dipole moment studies.[10] The deflection of molecular beams by inhomogeneous electric fields shows that some MX_2 molecules have permanent electric dipole moments and are therefore presumably non-linear.[11] The division of the Group II halides into two groups is somewhat unexpected and has been confirmed in a number of cases by i.r. studies of the halide in solid Kr at 20°K.

(7) JCP 1966 **44** 391

(8) JCP 1965 **42** 1013
(9) JCP 1965 **42** 902

(10) JCP 1964 **40** 3471

(11) JCP 1963 **39** 2023; JACS 1964 **86** 4544

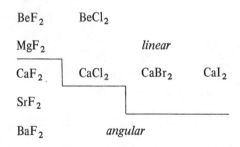

Estimates of the bond angles include CaF_2 (140°), SrF_2 (108°), and BaF_2 (100°).

The relation between structure in the crystalline and vapour states is simplest in the case of a molecular crystal which vaporizes to molecules of the same kind as those in the crystal, a process which merely involves the separation of molecules against the van der Waals forces with very little change in the internal structure of the molecule. The molecule may be mononuclear (SnI_4, WCl_6) or, rarely, polymeric (Al_2Br_6). In all other cases there is breakdown of more extensive metal–halogen systems, either directly to MX_n molecules, or in some instances to polymeric species which dissociate to monomers at higher temperatures. The vaporization is accompanied by a reduction in the coordination number of the metal, which is shown in brackets in the following examples:

Type of crystal structure	Halide	Vapour species
3D complex	LiCl (6)	Li_2Cl_2 (2) → LiCl (1)
	CuCl (4)	Cu_3Cl_3 (2) [→ CuCl (1)]
Layer	CdI_2 (6)	CdI_2 (2)
	SbI_3 (3 + 3)	SbI_3 (3)
	$AlCl_3$ (6)	Al_2Cl_6 (4) [→ $AlCl_3$ (3)]
Molecular	Nb_2Cl_{10} (6)	$NbCl_5$ (5)

The structures of the vapour molecules of some halides are closely related to the structures of the crystals (Fig. 9.18). Mass spectrometric analysis of the ions produced by electron impact shows the presence of dimers, trimers, and in some cases tetramers, in the vapours of alkali halides. In particular the lithium halide vapours contain more dimers than monomers. The structures of three of the LiX

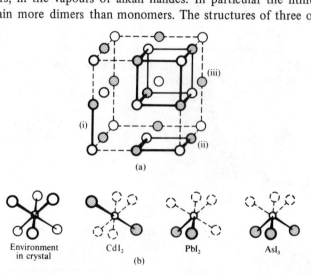

FIG. 9.18. Relation between crystal structure and configuration of molecule in the vapour state: (a) molecules AX, A_2X_2, and A_4X_4 derivable from the NaCl structure, (b) molecules AX_2 and AX_3 derived from halides with layer structures.

TABLE 9.24

Bond lengths in lithium halides

X	Li—X (Å)			Angle X—M—X in dimer	Reference
	Monomer	*Dimer*	*Crystal*		
Cl	2·02	2·23	2·57	108 (4)°	JCP 1960 **33** 685
		2·17		116°	ZPC 1960 **213** 111
Br	2·17	2·35	2·75	110 (4)°	
I	2·39	2·54	3·03	116 (4)°	

TABLE 9.25

Interatomic distances (Å) in some halide molecules

	Vapour (m.w.)[a]	crystal
LiF	1·5639	2·009
LiCl	2·021	2·566
LiBr	2·1704	2·747
LiI	2·3919	3·025
NaF	1·926	2·307
NaCl	2·3606	2·814
NaBr	2·5020	2·981
NaI	2·7115	3·231
KF	2·171	2·664
KCl	2·6666	3·139
KBr	2·8207	3·293
KI	3·0478	3·526
RbF	2·266	2·815
RbCl	2·7868	3·285
RbBr	2·9448	3·434
RbI	3·1769	3·663
CsF	2·3453	3·005
CsCl	2·9062	3·560†
CsBr	3·0720	3·713†
CsI	3·3150	3·950†

† CsCl structure.

Linear MX_2 molecules[b]

	F	Cl	Br	I
Zn	1·81	2·05	2·21	2·38
Cd	1·97	2·21	2·37	2·55
Hg	–	2·34	2·44	2·61

Non-linear molecules MX_2[c]

	Cl	Br	I
Sn	2·42	2·55	2·73
Pb	2·46	2·60	2·79

GaF[f]	1·775
GaCl	2·20
GaBr	2·35
GaI	2·57
InF	1·985
TlF	2·084
TlCl[g]	2·55
TlBr	2·68
TlI	2·87

$$M_2X_6{}^{(d)}:\quad \begin{matrix}X_2\\X_2\end{matrix}\!>\!M\!<\!\begin{matrix}X_1\\X_1\end{matrix}\!>\!M\!<\!\begin{matrix}X_2\\X_2\end{matrix}$$

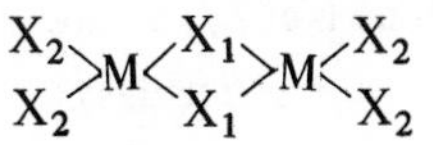

	M—X$_1$	M—X$_2$
Al$_2$Cl$_6$	2·21	2·06
Al$_2$Br$_6$	2·33	2·21
Al$_2$I$_6$	2·58	2·53
Ga$_2$Cl$_6$[e]	2·29	2·09*

Hg_2Cl_2[h]: Hg—Cl, 2·23

Regular octahedral molecules MF_6[i]

	M—F		M—F
W	1·833	U	1·996
Os	1·831	Np	1·981
Ir	1·830	Pu	1·971

Trigonal bipyramidal molecules MX_5[j]

	M—X		M—X
NbCl$_5$	2·29 Å	TaCl$_5$	2·30 Å
NbBr$_5$	2·46	TaBr$_5$	2·45

(*a*) JCP 1964 **40** 156. (*b*) ZFK 1956 **30** 155, 951; TFS 1937 **33** 852. (*c*) TFS 1941 **37** 406. (*d*) JACS 1938 **60** 1852. (*e*) K 1959 **4** 194. (*f*) JCP 1966 **45** 263. (*g*) AP 1936 **26** 1. (*h*) AP 1940 **57** 21. (*i*) JCP 1968 **48** 4001. (*j*) TFS 1940 **36** 668.
* Earlier results (JACS 1942 **64** 2514) for Ga$_2$Br$_6$ and In$_2$X$_6$ gave only mean values of bridging and terminal bond lengths.

375

dimers have been determined by electron diffraction. They form planar rhombus-shaped molecules as expected from a theoretical treatment assuming essentially Coulombic interaction between the oppositely charged ions. The bond lengths M–X have values intermediate between those in the monomer and in the crystal (Table 9.24). There are also some dimers in TlBr vapour.[12]

At temperatures above 435°C CuCl has a disordered wurtzite structure, and a mass spectrometric study shows[13] that Cu_3Cl_3 molecules are the predominant species in the saturated vapour at 450°C. The probable configuration of the trimer is an essentially planar 6-ring, but the detailed structure was difficult to establish with certainty. A later mass spectrometric study of the vapour in equilibrium with solid CuCl (280–430°C) indicated comparable concentrations of Cu_3Cl_3 and Cu_4Cl_4 molecules and a smaller concentration of Cu_5Cl_5 molecules.[14]

In contrast to the examples of Fig. 9.18 there is a more radical rearrangement of nearest neighbours in changes such as the following:

crystalline $AlCl_3$ (octahedral coordination) → Al_2Cl_6 dimer (tetrahedral coordination), and
crystalline $NbCl_5$ (octahedral Nb_2Cl_{10} dimers) → $NbCl_5$ (trigonal bipyramidal molecules).

Table 9.25 includes some interatomic distances not given in other chapters.

(12) JPC 1964 **68** 3835

(13) JACS 1957 **61** 358

(14) JCP 1971 **55** 4566

10

Complex, Oxy-, and Hydroxy-Halides

Complex halides

Complex halides are solid phases containing two or more kinds of metal ion (or other cation) and usually one kind of halogen atom. The simplest salts of this type contain only two kinds of cation and a common anion, and were formerly called 'double halides'. As generally understood the term complex halide does not include solid solutions such as (Na, K)Cl, the composition of which varies over a range dependent on the relative sizes of the two cations, but it does include stoichiometric compounds with random arrangements of two or more kinds of cation (see later). There are also compounds containing a metal and more than one halogen, for example PbFCl, p. 408, and PbClBr,[1] etc. with the $PbCl_2$ structure, p. 221, and polyhalides such as $CsICl_2$, which are included in Chapter 9. Many complex halides are anhydrous, others are hydrated to various degrees. Except in one or two special cases we shall exclude hydrated salts from the present discussion; their structures are described in Chapter 15.

(1) AC 1969 **B25** 796

The empirical formulae of complex halides are of all degrees of complexity, as may be seen from the following selection: $CsAgI_2$, $CsHgCl_3$, $TlAlF_4$, Cs_3CoCl_5, Na_3AlF_6, $Na_5Zr_2F_{13}$, $Rb_2Fe_5F_{17}$, and $Na_7Zr_6F_{31}$. As is generally true throughout the chemistry of solids, similarity in formula-type does not imply that compounds have similar structures. For example, $KMgF_3$, NH_4CdCl_3, and $CsAuCl_3$ have quite different structures. Moreover, none of these salts ABX_3 contains a finite complex ion BX_3; in fact, $CsAuCl_3$ contains ions of two kinds and is preferably written $Cs_2(AuCl_2)(AuCl_4)$.

Most of the compounds we shall describe have formulae of the type A_mBX_n, though we shall mention some with more complex formulae. In particular there are a number with binuclear complex ions, for example, B_2X_{11}, B_2X_{10}, and B_2X_9, formed from two octahedral groups with a common vertex, edge, or face; the more complex 'metal cluster' ions have been described in the last chapter. Adopting the convention that B is the more highly-charged cation (or that the oxidation state of B is higher than that of A) it follows that most complex halides must contain ions A^+ (alkali metals, Ag^+, Tl^+, NH_4^+, etc.) or A^{2+} (alkaline-earths, etc.). For values of n up to 6 the possible types of complex halide are:

$$A^IB^IX_2 \qquad A^IB^{II}X_3 \qquad A^IB^{III}X_4 \qquad A^IB^{IV}X_5 \qquad A^IB^VX_6$$
$$A_2^IB^{II}X_4 \qquad A_2^IB^{III}X_5 \qquad A_2^IB^{IV}X_6$$
$$(A^{II}B^{II}X_4) \qquad A^{II}B^{III}X_5 \qquad A^{II}B^{IV}X_6$$
$$A_3^IB^{III}X_6$$
$$(A^{III}B^{III}X_6)$$

377

The great majority of known complex halides are fluorides; in particular, those with $n > 7$ are almost exclusively fluorides, of elements of Groups IV–VI, Ce, and 5f metals.

A simple possibility for a complex halide is that it adopts a structure of a halide (or oxide) A_mX_n with A and B replacing, either statistically or regularly, the positions occupied by atoms of one kind in the binary halide (oxide); these form our class (a) in Table 10.1. Known examples are all fluorides, that is, they are ionic crystals, and the basic requirement is that the ions A and B are of similar size and carry charges appropriate to the structure, as in $Na^+Y^{3+}F_4$ or $K_2^+U^{4+}F_6$ with structures of the fluorite type. Other complex halides are conveniently classified according to the type of grouping of the B and X atoms in the crystal. These atoms may form a finite group, in the simplest case a mononuclear group BX_n, or the B

TABLE 10.1

Structures of complex halides

		A_mBX_3	A_mBX_4	A_mBX_5	A_mBX_6	A_mBX_7
(a)	*Statistical* AX_n *structures or superstructures*					
	Fluorite		α-NaYF$_4$		α-K$_2$UF$_6$	
	Trirutile				Li$_2$TiF$_6$	
	LaF$_3$				BaThF$_6$	
	ReO$_3$				CaPbF$_6$	
	VF$_3$				LiSbF$_6$	
	Ge$_3$N$_4$ (phenacite)		Li$_2$BeF$_4$			
	3D *complexes*					
	Perovskite	ABX$_3$				
(b)	*Layer structures*		K$_2$NiF$_4$			
			TlAlF$_4$			
			BaMnF$_4$			
	(layers + X$^-$ ions)				Ba(CrF$_4$)F$_2$	
(c)	*Chain structures*	K$_2$CuCl$_3$	K$_2$HgCl$_4$. H$_2$O		(NH$_4$)$_2$CeF$_6$	
		CsBeF$_3$		Tl$_2$AlF$_5$	K$_2$ZrF$_6$	
		CsNiCl$_3$		(NH$_4$)$_2$MnF$_5$	RbPaF$_6$	K$_2$PaF$_7$
		NH$_4$CdCl$_3$				
	(chains + X$^-$ ions)				Sr(PbF$_5$)F	
	Structures containing finite complex ions			For the ions		
(d)	*Mononuclear*		K$_2$PtCl$_4$	CuCl$_5^{3-}$,	ABX$_6$	K$_3$ZrF$_7$
				InCl$_5^{2-}$,	A$_2$BX$_6$	K$_2$NbF$_7$
				SnCl$_5^-$.	A$_3$BX$_6$	
				see p.	A$_4$BX$_6$	K$_2$TaF$_7$
	(+ X$^-$ ions)			Cs$_3$(CoCl$_4$)Cl		(NH$_4$)$_3$(SiF$_6$)F
(e)	*Polynuclear*					
(f)	*Complex ions of two kinds*	See text				

atoms may be linked by sharing X atoms to form 1- or 2-dimensional complex ions, giving the classes (d), (c), and (b) in Table 10.1. We have added subgroups for structures containing additional X^- ions (not coordinated to B atoms). It might seem logical to recognize also 3D system of linked B and X atoms. For example, the B and X atoms in the perovskite (ABX_3) structure form a framework of vertex-sharing octahedra (ReO_3 structure) which extends throughout the crystal, with the 12-coordinated A ions in the interstices. However, there is little advantage to be gained from distinguishing the B–X complexes in the essentially ionic fluorides of class (a), particularly those with statistical structures (see later), because this would involve allocating these structures to classes (b), (c), or (d), and it is preferable to keep them together as a group to emphasize their relation to the simpler structures. For example, we may distinguish discrete BeF_4^{2-} ions in Li_2BeF_4, but since both Li^+ and Be^{2+} are tetrahedrally coordinated the structure is equally well described as one of the type A_3X_4. Similarly, Li_2TiF_6 (superstructure of rutile, AX_2 type) could be placed in class (d) since it contains TiF_6^{2-} ions. On the other hand, if the coordination number of B (or A) is greater than the ratio of X to B (or A) atoms, then the coordination groups around B(A) must share X atoms. This is the situation in the (statistical) fluorite structure of K_2UF_6. The F : U ratio shows that there must be sharing of F atoms between UF_8 coordination groups, but the extent of the U–F complex varies in a random way throughout the structure, so that allocation to one of the classes (b)–(d) is not possible.

The formation of an ionic compound A_mBX_n implies the packing of A, B, and X ions to give the maximum lattice energy, and the sharing of X atoms between the coordination groups around B (or A) ions is presumably incidental to the attainment of suitable coordination numbers of A and B. So we find

		c.n.'s of	
		A	B
$CaPbF_6$:	ReO_3 superstructure (discrete PbF_6^{2-} ions)	6	6
$SrPbF_6$:	linear $(PbF_5)_n^{n-}$ ions and F^- ions	10	6
$BaPbF_6$:	$BaSiF_6$ structure (discrete PbF_6^{2-} ions)	12	6

The structure of $SrPbF_6$ illustrates a further point, namely, that the fact that the F : B ratio is equal to the c.n. of B does not necessarily mean that discrete BF_6 groups are present, although this is true in many ABX_6 and A_2BX_6 structures. Another example is the structure of Ba_2ZnF_6, which contains ZnF_4 layers and 2 F^- ions.

It should perhaps be emphasized that although there are undoubtedly differences between the ionic–covalent character of A–X and B–X bonds in many complex halides, particularly chlorides, bromides, and iodides of transition and B subgroup metals, the recognition of chains, layers, etc. in complex fluorides is largely for convenience in describing and classifying the structures; it does not necessarily imply any large difference in character between the A–X and B–X bonds. For example, in $NaKThF_6$,[2] there is 6-coordination of Na^+ (the smallest cation) and 9-coordination or K^+ and Th^{4+}. The coordination groups around K and Th have F atoms in common but those around Na do not, but it is not desirable to

(2) AC 1970 **B26** 1185

single out the NaF_6 groups as complex ions simply because they do not share F atoms.

The stability of a crystalline complex halide depends on the nature (size, shape, and charge) of all the ions present, and in particular the stability of a finite complex ion BX_n depends on the size of the ion A. Many new complex ions have been isolated in combination with large cations. For example, the salts $[N(C_2H_5)_4]_2CeBr_6$ and $[N(C_2H_5)_4]_2CeI_6$ can be prepared from the solid chloro salt and the anhydrous acid HX or in acetonitrile solution. Ions prepared in combination with $[(C_6H_5)_3PH]^+$ include PrI_6^{3-} and other hexaiodo 4f ions, FeI_4^-, AuI_4^-, etc. Hexachloro ions MCl_6^{3-} of the 3d metals are not known in aqueous solution but are stable in salts with large cations such as $Co(NH_3)_6^{3+}$. Many complex chlorides, bromides, and iodides which cannot be prepared from aqueous solution have been made in recent years under anhydrous conditions, i.e. using non-aqueous solvents or crystallizing from the melt.

The number of complex halides of which structural studies have been made is already very large, but many groups have yet to be studied in detail. For example, compounds formed by the alkali and alkaline–earth halides include the following:[3]

(3) IC 1965 **4** 1510; ACSc 1966 **20** 255

$KCaCl_3$	K_2SrCl_4	K_2BaCl_4
	KSr_2Cl_5	

$NaCa_2Br_5$	$LiSr_2Br_5$	K_2BaBr_4
$KCaBr_3$	K_2SrBr_4	
	KSr_2Br_5	

(4) IC 1962 **1** 220 (130 refs.)
(5) AC 1972 **B28** 1159
(6) AC 1972 **B28** 2115

There are numerous complex fluorides of alkali metals and metals of Groups I–IV.[4] The structures of $RbBe_2F_5$[5] and $CsBe_2F_5$[6] were mentioned in Chapter 3 as examples of the plane 6-gon net and the cubic (10, 3) net. Each BeF_4 tetrahedron shares 3 vertices to form respectively a layer or a 3D framework of composition Be_2F_5.

We shall deal first with the more important groups of complex halides according to formula type, that is, with the vertical columns of Table 10.1 down to class (d), including the small group of halides ABX_2, then separately with classes (e) and (f), and finally with selected groups of complex halides of particular elements.

Halides ABX_2

These are necessarily few in number since only singly-charged cations are involved. Many pairs of alkali halides form solid solutions (in some cases over the complete range of composition) and only the extreme members form compounds, $RbLiF_2$ and $CsLiF_2$. In these fluorides tetrahedral LiF_4 groups share one edge and two vertices to form layers, between which the larger alkali-metal ion occupies large holes (distorted 8-fold antiprism coordination). In $RbLiF_2$ the eight Rb–F distances range from 2·78 to 3·16 Å (mean 2·95 Å), but in $CsLiF_2$ two of the eight neighbours are appreciably more distant than the other six (Cs–6 F, 2·96–3·15, mean 3·07 Å, Cs–2 F, 3·50, 3·53 Å).[1]

(1) IC 1965 **4** 1510

Halides $A_m BX_3$

Numerous complex fluorides ABF_3 and a smaller number of chlorides (for example, $TlMnCl_3$, $KMgCl_3$[1]), bromides, and iodides (e.g. $CsPbI_3$ p. 937) crystallize with the cubic perovskite structure, with distorted variants of this structure, or with the closely related hexagonal $BaTiO_3$, $BaRuO_3$, or $BaNiO_3$ structures (Table 5.6, p. 187). Examples include $AgMF_3$ (M = Mg, Mn, Co, Ni, Zn),[2] interesting as examples of 12-coordinated Ag^+, $KNiF_3$ (cubic perovskite), $RbNiF_3$ (hexagonal $BaTiO_3$), $CsNiF_3$ ($BaNiO_3$), and $CsCoF_3$ ($BaRuO_3$).[3]

(1) JCP 1966 **45** 4652

(2) CR 1970 **270** 216
(3) ZaC 1969 **369** 117

Three simple chains of composition BX_3 are formed from (a) tetrahedral BX_4 groups sharing two vertices, (b) octahedral BX_6 groups sharing opposite faces, and

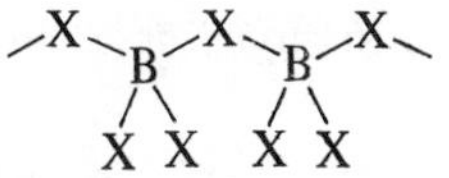

(a) 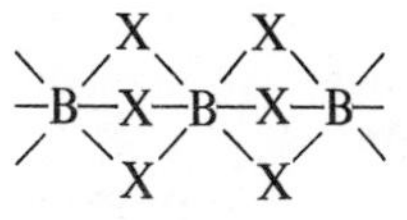   (b)

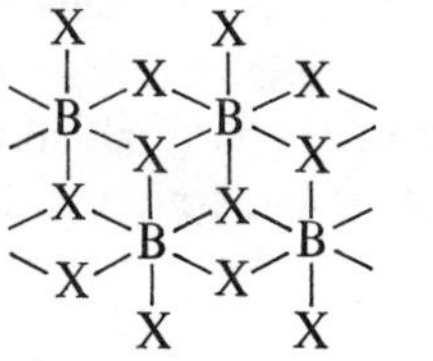

(c)

(c) octahedral BX_6 groups sharing four edges. All have been illustrated in Chapter 5; some examples are given in Table 10.2. The chains are arranged in the various

TABLE 10.2
Chain structures of halides $A_m BX_3$

Tetrahedral chains (a)	$CsBeF_3$ $KHgBr_3 . H_2O$ K_2CuCl_3, Cs_2AgCl_3, Cs_2AgI_3	AC 1968 **B24** 807 AC 1969 **B25** 647 see Chapter 26
Octahedral chains (b)	$CsNiCl_3$, $CsCuCl_3$,† $CsMgCl_3$[i] $[N(CH_3)_4]MnCl_3$,[ii] $RbCoCl_3$,[iii] $CsCoCl_3$[iv] $CsNiBr_3$[v] $[N(CH_3)_4]NiCl_3$[vi]	(i) JCP 1970 **52** 815 (ii) AC 1967 **23** 766 (iii) ACSc 1967 **21** 168 (iv) ACSc 1968 **22** 2793 (v) JACS 1966 **88** 4828 (vi) AC 1968 **B24** 330
(c)	NH_4CdCl_3, $RbCdCl_3$, $KCuCl_3$†	See Chapter 26

† distorted (4 + 2)-coordination.

structures so as to provide suitable coordination groups for the A ions, which are large, and in $KHgBr_3 . H_2O$ to accommodate also the water molecules between the chains. With the structure of this monohydrate contrast the structure of α-$KZnBr_3 . 2 H_2O$, which contains discrete tetrahedral groups $[ZnBr_3(H_2O)]$ and should therefore be formulated $K[ZnBr_3(H_2O)] . H_2O$.[4]

(4) AC 1968 **B24** 1339

The formulae of salts such as $A_2(Pt_2Br_6)$, p. 392, and $Cs_2(AuCl_2)(AuCl_4)$, p. 393, containing respectively a bridged binuclear anion and two different kinds of anion, should not be reduced to the simpler form ABX_3.

Complex, Oxy-, and Hydroxy-Halides

Halides $A_m BX_4$

The major groups of compounds are ABX_4 ($A^I B^{III} X_4$ and $A^{II} B^{II} X_4$) and $A_2^I B^{II} X_4$. The structures range from typically ionic structures of fluorides containing large ions with high coordination numbers (for example, the 9-coordination of both types of cation in one polymorph of $KCeF_4$, p. 996) through octahedral chain and layer structures (described in Chapter 5) to numerous compounds ABX_4 and $A_2 BX_4$ containing discrete BX_4 ions. For compounds ABX_4 containing the larger cations structures of the fluorite type are possible. This could be either a random fluorite structure, as adopted by the second form of $KCeF_4$, or a superstructure. An interesting distorted superstructure of fluorite is found for some fluorides ABF_4 in which A is Ca or Sr and B is Cr^{II} or Cu^{II}, where the transition-metal ions have a distorted tetrahedral arrangement of nearest neighbours instead of the normal cubic 8-coordination group. This structure is described under the fluorite structure in Chapter 6.

There are two octahedral layers (see p. 171), the planar ('*trans*') layer formed from octahedral BX_6 groups sharing all equatorial vertices and the puckered ('*cis*') layer in which the two unshared vertices are *cis* to one another. Examples include:

(1) BSFMC 1969 **92** 335
(2a) JCP 1969 **51** 4928

(2b) ZaC 1965 **336** 200

'*trans*' layer $\begin{cases} K_2 NiF_4 \text{ structure: } K_2 MF_4 \text{ (M = Mg, Zn, Co, Cu (distorted))} \\ TlAlF_4 \text{ structure: } \beta\text{-}RbFeF_4^{(1)} \end{cases}$

'*cis*' layer $\quad BaMnF_4$ structure[2a]: $SrNiF_4$, $BaMF_4$ (M = Mn, Fe, Co, Ni, Cu, Zn)

The simple octahedral BX_4 chain (rutile chain) is found with distorted $(4 + 2)$-coordination of M^{2+} in $Na_2 CrF_4$ and $Na_2 CuF_4$;[2b] see also $K_2 HgCl_4 . H_2 O$.

The majority of discrete ions BX_4 are tetrahedral and many complex halides are isostructural with oxy-salts of similar formula type:

$$NH_4 BF_4 \text{ and } RbBF_4 \text{ isostructural with } BaSO_4,$$
$$Cs_2 CoCl_4 \text{ and } Cs_2 ZnCl_4 \text{ isostructural with } K_2 SO_4.$$

(3) AC 1966 **20** 135

The phenacite structure (p. 811) of $Li_2 BeF_4$[3] may be described either as containing tetrahedral BeF_4^{2-} ions or, since both Li^+ and Be^{2+} are tetrahedrally coordinated, as a 3D ionic structure.

A number of chloro-, bromo-, and iodo- ions which cannot be prepared from aqueous solution in metallic salts can be stabilized by very large cations, for example, $(NEt_4)_2 MBr_4$ (M = Mn, Fe) and $[(C_6 H_5)_3 As(CH_3)]_2 FeI_4$,[4] and the structures of a number of MCl_4^{n-} ions have been studied in salts such as $[(C_6 H_5)_3 As(CH_3)]_2 NiCl_4$[5] and $[(C_2 H_5)_4 N] InCl_4$.[6]

(4) IC 1966 **5** 1498, 1510

(5) JACS 1962 **84** 167
(6) AC 1969 **B25** 603
(7) IC 1965 **4** 881

In $Cs_2 CuCl_4$ the anion has a flattened tetrahedral shape peculiar to this anion (Chapter 25), and in $K_2 PtCl_4$ and the isostructural Pd compound there are planar BX_4^{2-} ions (Fig. 10.1). The structure of $Pd(NH_3)_4 Cl_2 . H_2 O$ is similar, $Pd(NH_3)_4^{2+}$ replacing $PtCl_4^{2-}$ and Cl^- replacing K^+, with $H_2 O$ molecules at the points indicated by small black circles.

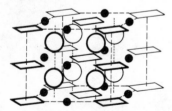

FIG. 10.1. The crystal structure of $K_2 PtCl_4$.

There is 9-coordination (t.c. t.p.) of the 4f metal ions in $NaNdF_4$ (and the isostructural Ce and La compounds), the structure being related to, but different from, the $\beta_2\text{-}Na_2 ThF_6$ structure (p. 996) originally assigned to these fluorides.[7]

The unique $Re_3 Cl_{12}^{3-}$ anion in $Cs_3 Re_3 Cl_{12}(CsReCl_4)$ is described on p. 366.

Halides $A_m BX_5$

We have seen that there are very few crystalline pentahalides which consist of molecules MX_5. Similarly, 5-coordination is avoided in many complex halides containing X : B in the ratio 5 : 1. The following types of structure are found:

	C.N. of B	Examples
(a) 3D ionic structures:	9	$LiUF_5$ (p. 996)
(b) Octahedral BX_5 chain:	6	Tl_2MF_5 (M = Al, Fe, Ga, Cr)[1], $(NH_4)_2MnF_5$[2], $CaCrF_5$[3], $BaFeF_5$[4]
(c) Finite BX_5 ion:	5	(t.b.)$(C_{28}H_{21}Cl)(SnCl_5)$, $[Cr(NH_3)_6]CuCl_5$ (t.p.) $(NEt_4)_2(InCl_5)$, K_2SbF_5, $(NH_4)_2SbCl_5$
(d) BX_4^{2-} and X^- ions:	4	Cs_3CoCl_5 and isostructural salts (p. 394)

(t.b. = trigonal bipyramidal; t.p. = tetragonal pyramidal.)

(1) JSSC 1970 **2** 269
(2) JCP 1969 **50** 1066
(3) MRB 1971 **6** 561
(4) AC 1971 **B27** 2345

We saw in Chapter 5 that there are two configurations of the simple octahedral MX_5 chain, the *trans* and *cis* chains. Examples of pentafluorides with both types of structure were given in Chapter 9, but only the *trans* chain is known as the structure of an anion. The isostructural $BaFeF_5$ and $SrAlF_5$ are of special interest since they contain equal numbers of chains of two kinds, the linear *trans* chain and a 'ramified' chain in which two additional octahedral BX_6 groups are attached to each member of the simple chain, as shown in Fig. 10.2. The composition of the chain remains MX_5. This complication is presumably due to the need to accommodate the larger Sr^{2+} and Ba^{2+} ions in positions of irregular 8- and 9-coordination between the chains—contrast the 7-coordination of Ca^{2+} in $CaCrF_5$.

Examples of the two configurations of finite BX_5 ions are at present confined to B subgroup metals, and references will be found in the chapters which deal specifically with these elements. It seems probable that ions BX_5^{n-} can be formed by many metals which form BX_4 and BX_6 ions if appropriate conditions for preparing the salts are found. A primary requirement seems to be a large cation.

Compounds in class (d) include a number of salts Cs_3MCl_5 (M = Mg, Mn, Co, Ni, Cu, and Zn) (see p. 394). There is a considerable number of monohydrated salts $A_2(MX_5 . H_2O)$ which crystallize with the K_2PtCl_6 structure (p. 387) or distorted variants of that structure. The anion is here an octahedral group $(MX_5 . H_2O)$ and A is one of the larger cations:

$(NH_4)_2(VF_5 . H_2O)$		$(NH_4)_2(FeCl_5 . H_2O)$
Rb_2	$Rb_2(CrF_5 . H_2O)$	$(NH_4)_2(InCl_5 . H_2O)$
Tl_2	Tl_2	

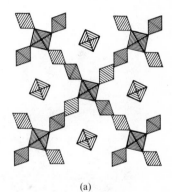

(a)

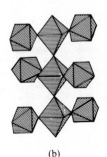

(b)

FIG. 10.2. The crystal structure of $BaFeF_5$ showing (a) the two types of $(FeF_5)_n^{2n-}$ chain, and (b) one ramified chain.

(5) AC 1963 **16** A45
(6) JCS A 1971 2653

However, all monohydrates of this type do not contain discrete $(MX_5 . H_2O)$ groups. In $K_2AlF_5 . H_2O$[5] and $K_2MnF_5 . H_2O$[6] there are infinite MX_5 chains formed from octahedral MX_6 groups sharing a pair of opposite vertices; the H_2O molecules and the cations lie between the chains.

Halides $A_m BX_6$

In some complex fluorides $A_m BF_6$ the A and B ions occupy the cation positions in simple ionic structures AX_n, either randomly or regularly (superstructure):

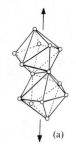

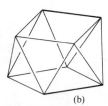

(a)

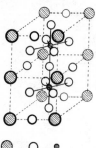

(b)

AX_n *structure*	*C.N. of* A *and* B	*Random*	*Superstructure*
Rutile	6	Mg_2FeF_6	Li_2TiF_6
Fluorite	8	$\alpha\text{-}K_2UF_6$	$\gamma\text{-}Na_2UF_6$
ReO_3	6	–	$CaPbF_6$
LaF_3	9	$BaThF_6$	–

We have already noted that there are discrete BX_6 groups in superstructures such as the trirutile structure of Li_2TiF_6 and the ReO_3 superstructure of $CaPbF_6$. In the latter structure the small open and black circles of Fig. 10.7 represent Ca^{2+} and Pb^{4+} ions respectively and the large open circles F^- ions. Such structures are therefore alternativley placed in class (d) of Table 10.1 as structures containing finite BX_6 ions. Similarly, the fluorite superstructure of $\gamma\text{-}Na_2UF_6$ contains UF_8 coordination groups which share two opposite edges to form chains, as may be seen from Fig. 28.3 (p. 995).

The sharing of F atoms between BF_n coordination groups also occurs in a number of other complex fluorides (Table 10.3); it must occur for large B ions with c.n. > 6. The two closely related K_2UF_6 structures are described in Chapter 28. Dodecahedral coordination groups share two edges in $RbPaF_6$ and in K_2ZrF_6 (Fig. 10.3(a)), while antiprism groups share the edges shown in Fig. 10.3(b) in $(NH_4)_2CeF_6$. It is convenient to include here another chain structure which illustrates a different way of attaining a F : B ratio of 6 : 1. In $SrPbF_6$ there are chains of composition PbF_5 formed from octahedral PbF_6 groups sharing opposite vertices, one-sixth of the fluorine being present as separate F^- ions (Fig. 10.3(c)). The structural formula is therefore $Sr(PbF_5)F$. Two-dimensional BX_n complexes are rare; they occur in salts Ba_2MF_6 (M = Cr, Mn, Fe, Co, Ni, Zn) which contain layers MF_4 of the same kind as in K_2NiF_4 formed from octahedral MF_6 groups sharing all equatorial vertices, that is, $Ba_2(MF_4)F_2$.

Sr F Pb (c)

FIG. 10.3. (a) Portion of the infinite chain ion $(ZrF_6)_n^{2n-}$ in K_2ZrF_6; (b) edges of CeF_8 antiprismatic coordination groups shared in $(NH_4)_2CeF_6$; (c) the structure of $Sr(PbF_5)F$.

TABLE 10.3
Structures of complex fluorides $A_m BF_6$

Compound	*C.N. of* B	*Coordination group of* B	*Reference*
1-*dimensional* BX_n *complex*			
K_2UF_6	9	t.c. trigonal prism sharing two faces	AC 1948 **1** 265
$RbPaF_6$	8 ⎫	dodecahedron sharing two edges	AC 1968 **B24** 1675
K_2ZrF_6	8 ⎭		AC 1956 **9** 929
$(NH_4)_2CeF_6$	8	antiprism sharing two edges	IC 1969 **8** 33
$\gamma\text{-}Na_2ZrF_6$	7	irregular sharing two vertices	AC 1969 **B25** 2164
$Sr(PbF_5)F$	6	octahedron sharing two vertices	ZaC 1957 **293** 251
2-*dimensional* BX_n *complex*			
$Ba_2MnF_4(F_2)$	6	octahedron sharing four vertices	ZaC 1967 **353** 13

Our group (d), comprising compounds containing discrete BX_6 groups, is an extremely large one. Not only are there compounds ABX_6, A_2BX_6, A_3BX_6, and A_4BX_6, but the first class includes $A^IB^VX_6$ and $A^{II}B^{IV}X_6$; at least one hundred fluorides $A^IB^VF_6$ have been prepared. The octahedral BX_6 ions are large, approximately spherical ions, and the crystal structures of these salts are determined in a general way by the relative sizes and charges of the A and BX_6 ions rather than by the chemical nature of the elements A and B. For example, $BaSiF_6$ has the same type of structure as $TlSbF_6$, while one form of K_2SiF_6 is isostructural with K_2PtCl_6, K_2SeBr_6, and numerous other compounds, as noted later.

The structures of many of these compounds may be described in two ways. Considering the A and BX_6 ions as the structural units we may describe the structures as derived from simple structures AX_n by replacing A or X by BX_6 ions. For example, in many compounds ABX_6 the A and BX_6 ions are arranged in the same way as the ions in the NaCl or CsCl structures. In the K_2PtCl_6 structure (Fig. 10.5) the positions of the K^+ and $PtCl_6^{2-}$ ions are those of the F^- and Ca^{2+} ions in the CaF_2 structure, hence the name 'anti-fluorite' structure. Alternatively, if the A and X atoms are close-packed, as in $KOsF_6$ and in many halides A_mBX_{3m}, we may describe the structures as close-packed assemblies of A and X atoms (ions) in which B atoms occupy certain of the octahedral holes between groups of 6 X atoms. This aspect of these structures has been emphasized in Chapter 4, and shows, for example, the relation between three A_2BX_6 structures based on close-packed (A + 3 X) layers:

Structure	Sequence of c.p. AX_3 layers
Cs_2PuCl_6 (K_2GeF_6)	*h*
K_2PtCl_6	*c*
K_2MnF_6	*hc*

ABX_6 *structures.*   Most of the compounds $A^IB^VX_6$ and $A^{II}B^{IV}X_6$ are fluorides. Compounds containing the heavier halogens are much less stable or in many cases unknown, though some ions BCl_6^{n-} form salts of this type with large cations as, for example, $[Co(NH_3)_6]TlCl_6$ and $[Ni(H_2O)_6]SnCl_6$. All fluorides ABF_6 crystallize with structures of the NaCl or CsCl types, various modifications of the ideal cubic structures being necessary to give the A ions suitable numbers of X neighbours. Five structures have been recognized for fluorides and a sixth, a tetragonal NaCl-type structure, is adopted by $Na[Sb(OH)_6]$.

NaCl type

   (a) $LiSbF_6$:   a rhombohedral structure alternatively described as a super-structure of VF_3 with 6-coordination of both A and B.

   (b) and (c):   The $NaSbF_6$ and $CsPF_6$ structures are both cubic, with 6- and 12-coordination of A respectively. There still seems to be some confusion in the literature regarding these two structures.

CsCl type

    (d) $KNbF_6$:   Tetragonal, coordination of A (8 + 4).

    (e) $KOsF_6$:   Rhombohedral—the $BaSiF_6$ structure, which is also adopted by $SrPtF_6$, $BaPtF_6$, $BaTiF_6$, and $CsUF_6$.

Table 10.4 shows that the type of structure of compounds $A^IB^VF_6$ is largely determined by the size of the A ion, the c.n. of which is 6 in (a) and (b), 8 + 4 in (d), and 12 in (c) and (e).

TABLE 10.4

Crystal structures of fluorides $A^IB^VF_6$

B	A						
	Li	Na	Ag	K	Tl	Rb	Cs
P		b	b/c	c/e	c	c	c
As	a		b/c				
V, Ru, Ir, Os				d		e	
Re, Mo, W, Sb, Nb, Ta		b					

(A few of the Ag and Tl compounds have not been studied.)

(a) $LiSbF_6$ structure: AC 1962 **15** 1098

(b) and (c) $NaSbF_6$ and $CsPF_6$ structures: AC 1956 **9** 539

(d) $KNbF_6$ structure: AC 1958 **11** 80

(e) $KOsF_6$ structure: JINC 1956 **2** 79; $BaSiF_6$: JACS 1940 **62** 3126

For a review see: JCS 1963 4408.

The $KOsF_6$ structure may alternatively be described as a slightly distorted cubic closest packing of nearly plane layers of composition KF_6 (Fig. 10.4) between which Os atoms occupy octahedral interstices.

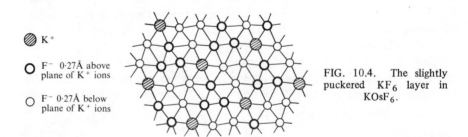

K^+

F^- 0·27Å above plane of K^+ ions

F^- 0·27Å below plane of K^+ ions

FIG. 10.4. The slightly puckered KF_6 layer in $KOsF_6$.

A_2BX_6 *structures.* We have already noted examples of halides A_2BX_6 with statistical AX_2 structures or with superstructures of AX_2 structures. Complex halides of this group necessarily contain alkali-metal ions, NH_4^+, Tl^+, or Ag^+. The majority contain cations which are comparable in size with one or more of the halide ions, and these form structures based on c.p. AX_3 layers. Such structures are not possible for the smaller Na^+ or Li^+ ions, which can be accommodated, like the B ions, in octahedral holes in a c.p. assembly of X ions. Accordingly we find a number of structures characteristic of compounds Li_2BX_6 and Na_2BX_6. These are closely related to (sometimes superstructures of) AX_2 structures, since they represent ways of filling (by $A_2 + B$) one-half of the octahedral holes in a closest packing of X ions (X_6). They are described in Chapter 4 and include three closely related structures:

Trirutile structure: high-Li_2SnF_6, low-Li_2GeF_6: JSSC 1971 **3** 525

Na_2SiF_6 structure: Na_2MF_6 (M = Si, Ge, Mn, Cr,
Ti, Pd, Rh, Ru, Pt, Ir, Os): AC 1964 **17** 1408;
JCS 1965 1559

Li_2ZrF_6 structure: Li_2NbOF_5: ACSc 1969 **23** 2949

Several of these compounds are polymorphic; for example, the low–high transitions of Li_2SnF_6 (Li_2ZrF_6 to trirutile) and Li_2GeF_6 (trirutile to Na_2SiF_6).

We now describe the three c.p. structures (K_2PtCl_6, K_2GeF_6, and K_2MnF_6) in which A + 3 X together form the c.p. assembly.

The K_2PtCl_6 *structure.* In this structure (Fig. 10.5) the K^+ and $PtCl_6^{2-}$ ions occupy respectively the F^- and Ca^{2+} positions of the fluorite structure. In contrast to the K_2GeF_6 and K_2MnF_6 structures, which are largely confined to fluorides, this structure is more generally adopted by hexafluorides, hexachlorides, hexabromides, and a few hexaiodides (e.g. Rb_2SnI_6, Cs_2TeI_6) of the larger alkali metals, NH_4, and Tl. Some hexaiodides have less symmetrical variants, for example, Rb_2PtI_6 (tetragonal), K_2ReI_6 and K_2PtI_6 (orthorhombic).

Other compounds with this structure include hydrated compounds such as $(NH_4)_2(FeCl_5 . H_2O)$, $K_2TcCl_5(OH)$,[1] $K_2(PtF_3Cl_3)$,[2] and $Cs_2Co^{IV}F_6$. The salt $(NH_4)_2SbBr_6$ is a superstructure of K_2PtCl_6 containing $SbBr_6^{3-}$ and $SbBr_6^-$ ions (p. 706). The relation of the K_2PtCl_6 to the cryolite and perovskite structures is discussed later.

The K_2GeF_6 *(Cs_2PuCl_6) structure.* This structure is confined to fluorides except for a few compounds such as Cs_2PuCl_6 and the trigonal forms of Cs_2ThCl_6 and Cs_2UCl_6. It is illustrated in Fig. 10.6. In the 'ideal' structure an A ion would have 12 equidistant X neighbours, but with the possible exception of Cs_2PuCl_6 the neighbours of A are either 9 approximately equidistant X atoms with 3 at a rather greater distance or groups of 3 + 6 + 3 neighbours. Typical figures are:

$$\text{In } K_2GeF_6: K\begin{cases} 9 \text{ F at } 2\cdot85 \text{ Å} \\ 3 \text{ F at } 3\cdot01 \text{ Å} \end{cases}; \text{ in } K_2TiF_6: K\begin{cases} 3 \text{ F at } 2\cdot75 \text{ Å} \\ 6 \text{ F } \quad 2\cdot87 \text{ Å} \\ 3 \text{ F } \quad 3\cdot08 \text{ Å} \end{cases}$$

In contrast, the neighbours of the A ions in the K_2PtCl_6 structure are always 12 equidistant X atoms, whether the close packing is perfect or whether there is some departure from perfect close packing. In the latter case the coordination group is no longer a regular cuboctahedron.

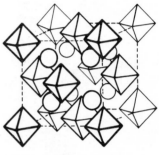

FIG. 10.5. The crystal structure of K_2PtCl_6.

(1) JCS A 1967 1423
(2) JCS A 1966 1244

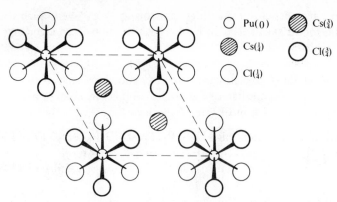

FIG. 10.6. The crystal structure of Cs_2PuCl_6 (or K_2GeF_6). Figures in brackets indicate the heights of atoms above the plane of the paper in terms of c (6·03 Å).

The K_2MnF_6 structure.   This structure is adopted by numerous hexafluorides, many of which are polymorphic (Table 10.5).[3] Of special interest are the red salts $M_2Ni^{IV}F_6$ (Ni–F, 1·70 Å) and the rose-coloured $M_2Cr^{IV}F_6$ (Cr–F, 1·72 Å). Since the feature common to all the three structures, K_2PtCl_6, K_2GeF_6, and K_2MnF_6, is the close packing of the A and X ions, there is no simple connection between the relative sizes of the A and BX_6 ions as in the ABX_6 structures of Table 10.4.

(3) JCS 1958 611

TABLE 10.5

Crystal structures of complex fluorides A_2BX_6

	K_2	Rb_2	Cs_2	$(NH_4)_2$	Tl_2
SiF_6	T C	C	C	T C	C
GeF_6	T H	T H	C	T	–
TiF_6	T H C	T H C	T C	T	T C
MnF_6	T H C	H C	C	–	–
CrF_6 (a)	H C	H C	C	–	–
ReF_6 (b)	T	T	T	–	–
NiF_6 (a)	C	C	C	–	–
PdF_6	H	H C	C	–	–
PtF_6	T	T	T	–	–

T = K_2GeF_6, H = K_2MnF_6, C = K_2PtCl_6 structure.
(a) ZaC 1956 **286** 136. (b) JCS 1956 1291.

The cryolite family of structures for compounds A_3BX_6 or $A_2(B'B'')X_6$.   The crystal structure adopted by certain salts A_3BX_6 is very simply related to that of K_2PtCl_6. In $(NH_4)_3AlF_6$ two-thirds of the NH_4^+ ions and the AlF_6^{3-} ions occupy the positions of the K^+ and $PtCl_6^{2-}$ ions in Fig. 10.5, and in addition there are NH_4^+ ions at the mid-points of the edges and the body-centre of the cubic unit cell. In this structure, as in K_2PtCl_6, the position of X along the cube edge is variable ($u00$, etc.) with $u \approx \frac{1}{4}$. It is convenient to describe this structure as the cryolite structure

388

although the mineral cryolite itself (Na_3AlF_6)[1] does not crystallize with the most symmetrical form of the structure (compare the distorted form of the 'perovskite' structure adopted by the mineral of that name).

(1) ACSc 1965 **19** 261

From the structural standpoint it is preferable to write the formulae of compounds of the cryolite family in the form $A_2(B'B'')X_6$ rather than A_3BX_6 because in this structure one-third of the A atoms in the formula A_3BX_6 are, like the B atoms, in octahedral holes in a cubic close-packed A_2X_6 (AX_3) assembly; the other two-thirds, corresponding to the K^+ ions in the K_2PtCl_6 structure, are surrounded by twelve equidistant X ions. Accordingly the cryolite structure is very closely related also to the perovskite structure. In Fig. 10.7 the A and X atoms are

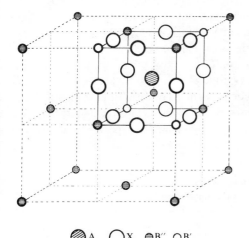

FIG. 10.7. Relation between perovskite, K_2PtCl_6, and cryolite structures.

⊘ A ◯ X ⊖ B'' ◯ B'

arranged in a c.c.p. arrangement of composition AX_3. Insertion of B^- atoms as shown gives the $A_2B''X_6$ (K_2PtCl_6) structure, which requires a unit cell (dotted lines) with eight times the volume of the small cube. Insertion of the B' atoms at the points marked gives the cryolite structure, $A_2(B'B'')X_6$, and if $B' = B''$ the structure is the perovskite structure (ABX_3), referable to the small unit cell. In other words, the very symmetrical structure of Fig. 10.7 (with $u = \frac{1}{4}$) is a superstructure of perovskite—compare the elevations of Fig. 4.31(b) and (e), p. 151.

As in the case of the perovskite structure there are variants of the cryolite structure with lower symmetry. In discussing these structures it is important to note that for full cubic symmetry (space group $Fm3m$) the X atoms must lie along the cube edges, their positions being determined by the value of the variable parameter u. Cubic symmetry may be retained, though of a lower class (space group $Pa3$) if the X atoms are not constrained to lie on the cell edges, and accordingly there is another (less symmetrical) structure, the *elpasolite* structure of compounds such as K_2NaAlF_6,[2] also with cubic symmetry. These structures are distinguished

(2) JINC 1961 **21** 253

(3) AC 1953 **6** 49

(4) IC 1969 **8** 2694

(5) AC 1968 **B24** 225

(6) AC 1951 **4** 503

(7) IC 1970 **9** 1771

by the letters C and E in Table 10.6. There are also modified cryolite structures with only tetragonal or monoclinic symmetry.[3]

Most of the compounds A_3BX_6 with cubic symmetry at room temperature appear to be ammonium salts (e.g. $(NH_4)_3AlF_6$, $(NH_4)_3FeF_6$, and $(NH_4)_3MoO_3F_3$): K_3MoF_6 is an exception.[4] Generally departures from cubic symmetry are observed, but at higher temperatures the structure becomes cubic, as shown in Table 10.6. The behaviour of Li_3AlF_6, of which five polymorphs have been described, is apparently much more complex.[5] There is very irregular coordination of three types of non-equivalent Li^+ ion in the room-temperature form.

A similar behaviour is shown by the isostructural oxides R_3WO_6 and $R_2'R''WO_6$ and the corresponding Mo compounds.[6] For example, Ba_2CaWO_6 and Ba_2CaMoO_6 are cubic at ordinary temperatures, while Ba_2SrWO_6 is non-cubic but becomes cubic at 500°C. The simpler compounds Ca_3WO_6, Sr_3WO_6, and Ba_3WO_6 also crystallize at room temperature with distorted forms of the cubic structure.

The cryolite structure is also adopted by more than twenty chlorides $Cs_2NaM^{III}Cl_6$ in which M is In, Tl, Sb, Bi, Fe, Ti, Sc, Y, La, and 4f and 5f elements,[7] and by compounds with complex anions or cations. Examples include $I_3[Co(NH_3)_6]$ and numerous hexanitrites $M_3'[M''(NO_2)_6]$, in which $M' = NH_4$, K, Rb, Cs, or Tl, and $M'' = Co$, Rh, or Ir. Salts containing two types of M' atoms are

TABLE 10.6
Compounds with cryolite-type structures

	−180°	20°	300°	550°C
K_2NaAlF_6	Cubic (E)	Cubic (E)	−	−
$(NH_4)_3AlF_6$ } $(NH_4)_3FeF_6$ }	Tetragonal	Cubic (C)	−	−
K_3AlF_6	−	Tetragonal	Cubic (C)	−
Na_3AlF_6	−	Monoclinic	−	Cubic (C)

also known, for example, $K_2Ca[Ni(NO_2)_6]$; for $K_2Pb[Cu(NO_2)_6]$ see p. 900. Oxyfluorides provide examples not only of the monoclinic and tetragonal forms of the cryolite structure, but also of the (cubic) elpasolite structure (see Table 10.11, p. 405).

(1) ZaC 1956 **284** 10

(2) Vol. 1, p. 319 (1906)

Halides A_4BX_6. These include $(NH_4)_4CdCl_6$[1] and the isostructural salts K_4MnCl_6, $K_3NaFeCl_6$, and K_4PbF_6, which contain octahedral ions $(BX_6)^{4-}$. The salt K_4ZnCl_6 is described in Groth's Chemische Kristallographie,[2] but its structure is not known. If this salt contains $ZnCl_6^{4-}$ ions it would be the only known example of 6-coordination of Zn by Cl, but it could conceivably contain tetrahedral $ZnCl_4^{2-}$ (or trigonal bipyramidal $ZnCl_5^{3-}$) ions and additional Cl^- ions. There are numerous examples of octahedral coordination of Zn by F.

Halides $A_m BX_7$

Complex fluorides of this type are formed by various metals of the IVA, VA, and VIA subgroups, by 5f elements, and by Tb in the 4f group. Cs_3TbF_7 is the first fluoro complex of Tb(IV) to be isolated.[1] These compounds provide examples of the following types of structure:

Finite BX_7 ion: pentagonal bipyramidal—K_3UF_7, K_3ZrF_7, $(NH_4)_3ZrF_7$
capped trigonal prism—K_2TaF_7, Rb_2UF_7, Rb_2PuF_7[2]
Linear $(BX_7)_n$ ion: K_2PaF_7[3] (p. 998)
Finite BX_6 ion + X^-: $(NH_4)_3(SiF_6)F$.

For further details of the IVA and VA compounds see p. 396, and for compounds of Th and U see Chapter 28. The structures of compounds such as $RbMoF_7$ and $RbWF_7$ are not yet known.

The chain ion in K_2PaF_7 (and the isostructural Rb, Cs, and NH_4 salts) is formed from tricapped trigonal prisms PaF_9 sharing two edges, and was mentioned in Chapter 1 as an interesting way of attaining an X : M ratio of 7 : 1 from a 9-coordination group.

Halides $A_m BX_8$

A considerable number of fluorides of this kind are formed by transition and 5f metals, for example, K_2MoF_8 and K_2WF_8[1] (structures not known), Na_3TaF_8, and compounds of Th, U, etc. (For $NaHTiF_8$ see p. 399.)

An ionic 3D structure with 8-coordination of both ions has been proposed for Na_2UF_8[2] (a compound earlier thought to be Na_3UF_9), and isolated UF_8^{4-} ions (distorted antiprisms) occur in $(NH_4)_4UF_8$ and the isostructural compounds of Ce, Pa, Np, Pu, and Am.[3] (There is square antiprism coordination also in Na_3TaF_8, for which see p. 398.) The corresponding Th compound, $(NH_4)_4ThF_8$, has a quite different structure.[4] There is a chain of composition ThF_7, rather similar to that in K_2PaF_7, formed from tricapped trigonal prism ThF_9 groups sharing two edges. The eighth F is a separate F^- ion, so that this compound, like $SrPbF_6$, has chain ions and separate F^- ions:

$SrPbF_6$: $(PbF_5)_n^{n-}$ chain ions + F^-,
$(NH_4)_4ThF_8$: $(ThF_7)_n^{3n-}$ chain ions + F^-;

compare the following compounds containing finite complex ions and additional X^- ions:

	Cs_3CoCl_5	$(NH_4)_3SiF_7$	$(NH_4)_5Mo_2Cl_9 . H_2O$
Ions	$\{(CoCl_4)^{2-}$	$(SiF_6)^{2-}$	$(Mo_2Cl_8)^{4-}$
	$\{Cl^-$	F^-	Cl^-

In Na_3UF_8 (and the isostructural Pa and Np compounds) there is almost perfect cubic 8-coordination of the 5f ion.[5]

We have now dealt with the vertical subdivisions of Table 10.1. It remains to comment on the classes (e) and (f) and to add short sections on the complex halides of certain metals.

(1) ZaC 1961 **312** 277

(2) JACS 1965 87 5803
(3) JCS A 1967 1429, 1979

(1) JCS 1956 1242; 1958 2170

(2) IC 1966 5 130

(3) AC 1970 **B26** 38

(4) AC 1969 **B25** 1958

(5) JCS A 1969 1161

M—X—M

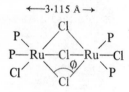

(1) See p. 904
(2) AC 1964 **17** 587

$M \overset{X}{\underset{X}{\diagup\diagdown}} M$

M—X—M

(2a) IC 1971 **10** 122
(3) AC 1957 **10** 466
(4) AC 1958 **11** 689
(5) ACSc 1968 **22** 2943
(6a) IC 1971 **10** 1453
(6b) JACS 1969 **91** 2174
(6c) IC 1963 **2** 817
(7) JCP 1935 **3** 117
(8) AC 1969 **B25** 1919

←—3·115 Å—→

$[P = P(n\text{-}C_4H_9)_3]$

(a)

(9) JCS A 1968 1981
(10) JCS A 1968 2108

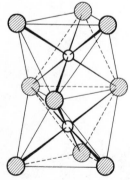

FIG. 10.8. The $Tl_2Cl_9^{3-}$ ion.

(e) *Complex halides containing finite polynuclear complex ions*

Ions in which metal-metal interactions play a prominent part are included with 'metal cluster' compounds (p. 367); they are formed notably by Nb, Ta, Mo, W, Tc, Re, and Pt. Here we mention only a few examples of simpler groupings involving halogen bridges of three main types.

Single halogen bridges occur in $NaSn_2F_5$, in which two pyramidal SnF_3 groups share one F (p. 936), in $(SeF_3)(Nb_2F_{11})$ where the anion consists of two octahedral groups sharing one vertex (p. 600), and in $Na_5Zr_2F_{13}$ (p. 398) where a single F atom links two trigonal prism ZrF_6 groups.

Double halogen bridges link the metal atoms in the anions in $K_2(Cu_2Cl_6)$[1] and $[N(C_2H_5)_4]_2\ Pt_2Br_6$[2] and in the anion in $(PCl_4)_2\ Ti_2Cl_{10}$[2a] which consists, like the Re_2Cl_{10} molecule, of two octahedral groups with a common edge.

Triple halogen bridges occur in the anions in certain enneahalides which consist of two octahedral MX_6 groups with a common face. For example, $Cs_3Tl_2Cl_9$ contains ions $Tl_2Cl_9^{3-}$ of the type shown in Fig. 10.8. The structure is alternatively described as a close-packed $CsCl_3$ array with Tl^{3+} ions in pairs of adjacent octahedral holes (p. 154). Compounds of this type include $Cs_3Ti_2Cl_9$, $Cs_3V_2Cl_9$, $Cs_3Cr_2Cl_9$,[3] $K_3W_2Cl_9$,[4] and $Cs_3Bi_2I_9$.[5] In $K_3W_2Cl_9$ there is obviously bonding between the metal atoms (W—W, 2·44 Å) and consequent reduction of magnetic moment, but in the other examples given there is no metal-metal interaction (e.g. Cr—Cr, 3·1 Å) and normal magnetic properties. In this respect the ions $Mo_2Cl_9^{3-}$ (Mo—Mo, 2·66 Å) and $Mo_2Br_9^{3-}$ (Mo—Mo, 2·82 Å) seem to occupy an intermediate position.[6a] The $W_2Cl_9^{3-}$ ion may be oxidized by Cl_2 in dichloromethane to $W_2Cl_9^{2-}$, which has been isolated in the violet salt $[(n\text{-}C_4H_9)_4N]_2(W_2Cl_9)$.[6b] The structure of this ion is not known. For $(Re_2Cl_9)^-$ see p. 364.

On the evidence of an X-ray powder photograph it is concluded that $Cs_3Fe_2Cl_9$ is isostructural with the Cr analogue and therefore contains $(Fe_2Cl_9)^{3-}$ ions.[6c] The structure of the $(Ti_2Cl_9)^-$ ion has been determined in $(PCl_4)(Ti_2Cl_9)$,[2a] where the bond lengths (terminal Ti—Cl, 2·23 Å, and bridging, 2·49 Å) are very similar to those in the $(Ti_2Cl_{10})^{2-}$ ion (above).

There are other compounds $A_mB_2X_9$ which do not contain B_2X_9 ions. In $Cs_3As_2Cl_9$[7] the nearest neighbours of As are 3 Cl at 2·25 Å and 3 more at 2·75 Å, so that this crystal would appear to consist of pyramidal $AsCl_3$ molecules embedded among Cs^+ and Cl^- ions. The hydrate of $(NH_4)_5Mo_2Cl_9$ is $(NH_4)_5(Mo_2Cl_8)Cl . H_2O$ (p. 366), while KU_2F_9 has a typical ionic structure in which U^{4+} is 9-coordinated (tricapped trigonal prism) and K^+ is 8-coordinated (distorted cube).[8]

Closely related to the enneahalide ions are molecules such as $Ru_2Cl_5[P(C_4H_9)_3]_4$,[9] (a), $Ru_2Cl_4[P(C_6H_5)(C_2H_5)_2]_5$,[10] and $Ru_2Cl_3(CO)_5$ $SnCl_3$. As in $Cr_2Cl_9^{3-}$ there is no metal-metal bonding, as is shown by the M—M distances or by the value of the angle ϕ. This is $70\frac{1}{2}°$ for regular octahedra sharing a face, $77°$ in $Cr_2Cl_9^{3-}$, $79\frac{1}{2}°$ in (a), but only $58°$ in $W_2Cl_9^{3-}$.

(f) *Complex halides containing complex anions of more than one kind*

There are numerous complex halides in which both cation and anion are complex but little is known of compounds containing two kinds of complex anion. Examples include $Na_3(HF_2)(TiF_6)$, p. 399, and the deeply-coloured salts $(NH_4)_4(Sb^VCl_6)(Sb^{III}Cl_6)$, p. 706, and $Cs_2(Au^ICl_2)(Au^{III}Cl_4)$. The structure of the last compound is closely related to the perovskite structure, the structure being distorted so that instead of six octahedral neighbours the two kinds of Au atom have two collinear or four coplanar nearest neighbours. The linear $(Cl-Au^I-Cl)^-$ and the square planar $(Au^{III}Cl_4)^-$ ions can be clearly seen in Fig. 10.9(b). For comparison a portion of the ideal cubic perovskite structure is shown in Fig. 10.9(a), the inscribed cube being the unit cell. For an alternative unit cell see Fig. 13.3(a) (p. 483).

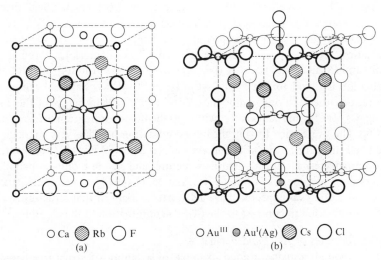

◯ Ca ⬙ Rb ◯ F   ◯ Au^{III} ⊖ Au^I(Ag) ⬙ Cs ◯ Cl
(a)   (b)

FIG. 10.9. The structures of (a) $RbCaF_3$, perovskite structure, and (b) $Cs_2Au^IAu^{III}Cl_6$ in similar orientations.

Miscellaneous complex halides

Apart from the special groups described in the present chapter other complex halides are described in other chapters: Sb, Chapter 20, B subgroup elements, Chapters 25 and 26, U and 5f elements, Chapter 28.

Complex chlorides $CsMCl_3$, Cs_2MCl_4, *and* Cs_3MCl_5 *formed by* 3d *metals*[1]

The eight elements from Ti to Cu inclusive all form ions M^{2+} and in these ions the number of d electrons increases from 2 to 9. All of these elements form one or more of the following types of complex chloride, $CsMCl_3$, Cs_2MCl_4, or Cs_3MCl_5:

$CsMCl_3$:   formed by all these elements. The structure ($CsNiCl_3$) consists of c.p. Cs^+ and Cl^- ions with M^{2+} in *octahedral* coordination, which is distorted from regular octahedral coordination in the isostructural Cr and Cu compounds.

(1) For earlier refs. see JPC 1962 **66** 65

393

Cs_2MCl_4: found in all systems investigated ($CsCl–CrCl_2$ not studied) except V and Ni. The only compound found in the $CsCl–VCl_2$ system is $CsVCl_3$, while in the $CsCl–NiCl_2$ system Cs_2NiCl_4 does not exist. The salts Cs_2MCl_4 contain *tetrahedral* MCl_4^{2-} ions.

Cs_3MCl_5: known to be formed by Mn, Co, Cu (*not* by V) and by Ni, but Cs_3NiCl_5 is stable only at high temperatures. It can be prepared from the molten salts and quenched, but on slow cooling it breaks down into $CsCl$ + $CsNiCl_3$. The salts are isostructural with Cs_3CoCl_5[2] and contain tetrahedral MCl_4^{2-} and Cl^- ions. (The ion $NiCl_4^{2-}$ can exist at ordinary temperatures in conjunction with very large cations, as in the salt $[(C_6H_5)_3CH_3As]_2NiCl_4$, in which it has a regular tetrahedral structure (Ni–Cl, 2·27 Å).)[3]

These findings agree well with the expectations from crystal field theory. A strong preference for octahedral coordination should be shown by V^{2+} (d^3) and Ni^{2+} (d^8). In the $CsMCl_3$ compounds the large distortion of the octahedral coordination group expected for the d^4 and d^9 configurations is found in the isostructural $CsCuCl_3$ and $CsCrCl_3$ ((4 + 2)-coordination). A non-regular tetrahedral coordination is found for Cu^{2+} in Cs_2CuCl_4, but Cs_2CrCl_4 has not been prepared.

As regards these chloro compounds Zn^{2+} (d^{10}) falls in line with the 3d ions, forming Cs_2ZnCl_4 and Cs_3ZnCl_5 with tetrahedral coordination of Zn but not $CsZnCl_3$. The preference for tetrahedral coordination by Cl is very marked, being found in all three forms of $ZnCl_2$ and even in $Na_2ZnCl_4 . 3 H_2O$.[4] However, Zn^{2+} is 6-coordinated by F^- in ZnF_2 (rutile structure) and also in $KZnF_3$ which, like the corresponding compounds of Mn, Fe, Co, and Ni, has the ideal perovskite structure. $KCrF_3$ and $KCuF_3$, as expected, have a distorted variant of that structure, but with (2 + 4)-coordination in contrast to the (4 + 2)-coordination in the chlorides.[5]

Complex fluorides of Al[1] and $Fe(III)$[2]

These compounds provide a good example of a family of structures based on a simple principle, namely, the sharing of vertices between octahedral AlF_6 (FeF_6) coordination groups to give various F : Al ratios. Most of the compounds discussed below are formed by K, Rb, and Tl (sometimes also NH_4), indicated by M, but certain of the structures are peculiar to the smaller Na^+ ion. An F : Al ratio of 6 : 1 implies discrete AlF_6 groups, as in the cryolite structure of compounds such as Na_3AlF_6 and in the more complex structure of cryolithionite, $Na_3Li_3Al_2F_{12}$. This mineral has the garnet structure, so that its formula may alternatively be written $Al_2Na_3(LiF_4)_3$. This illustrates the point that the description of these complex fluorides in terms of AlF_6 or FeF_6 octahedra is arbitrary in the sense that the Al–F bonds are not appreciably different in nature from the other (M–F) bonds in the structures. Sharing of two opposite vertices of each AlF_6 octahedron gives the AlF_5 chain, and of four equatorial vertices, the '*trans*' AlF_4 layer (Fig. 10.10(a)). In the intermediate layer, of composition Al_3F_{14}, one-third of the AlF_6 octahedra share four equatorial vertices and the remainder share two opposite vertices (Fig. 10.10(b)). The alternative puckered ('*cis*') layer in which two adjacent vertices are *un*shared is found in $NaFeF_4$.

(2) AC 1964 **17** 506

(3) IC 1966 **5** 1498

(4) ZK 1960 **114** 66

(5) AC 1961 **14** 583

(1) ZaC 1937 **235** 139; 1938 **238** 201; 1938 **239** 301
(2) JSSC 1970 **2** 269

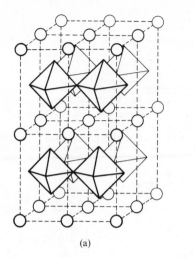
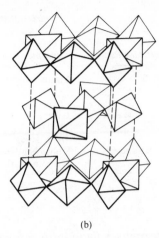

(a)   (b)

FIG. 10.10. Portions of (a) the AlF_4 layers in $TlAlF_4$ and (b) the Al_3F_{14} layers in $Na_5Al_3F_{14}$.

An interesting structure has been suggested for compounds $M_2Fe_5F_{17}$ based on the dimensional analogy with the oxide $MoW_{11}O_{36}$. In Fig. 10.11(a) each square represents a chain of vertex-sharing octahedra perpendicular to the plane of the paper (ReO$_3$-type chain), the shaded squares containing Mo. Removal of the metal atoms marked by the small black circles and of the O atoms between them would

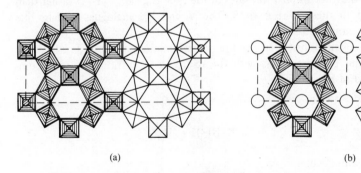

(a)   (b)

FIG. 10.11. Relation of octahedral layer M_5X_{17} (b) to 3D MX_3 structure (a).

leave layers of composition $W_{10}O_{34}$, suggesting the structure Fig. 10.11(b) for $M_2Fe_5F_{17}$ in which the complex layers are perpendicular to the plane of the paper. In this layer one-fifth of the octahedra share 6 and the remainder 5 vertices. The simplest 3D framework formed from octahedral groups sharing all their vertices represents the structures of AlF_3 or FeF_3, and a more complex framework is found in a hydrated hydroxyfluoride of Al with the pyrochlore structure (p. 209). These structures are summarized in Table 10.7.

395

TABLE 10.7

Structures of complex fluorides

Number of $Al(Fe)F_6$ vertices shared	Type of complex	Examples	
0	Finite	M_3FeF_6	Na_3AlF_6 (cryolite)
			$Na_3Li_3Al_2F_{12}$ (cryolithionite)
2	Chain	M_2FeF_5	M_2AlF_5
2 and 4	Layer	$Na_5Fe_3F_{14}$	$Na_5Al_3F_{14}$ (chiolite)
4	'cis' layer	$NaFeF_4$	
	'trans' layer	$MFeF_4$	$MAlF_4$
(5 and 6)	Multiple layer	$M_2Fe_5F_{17}$)	
6	3D framework	FeF_3	AlF_3

Some complex fluorides of group IVA *and* VA *elements*

In Group VA there is an interesting gradation in properties from V to Ta. In complex salts V prefers bonds to O rather than F—note the extensive oxygen chemistry of V and the formation of numerous oxyhalides by V and Nb; Ta forms only TaO_2F. The most stable salts obtained by the addition of metallic fluorides to a solution of Nb_2O_5 in HF are those containing the ion $NbOF_5^{2-}$. This ion is much more stable than simple NbF_n ions. For example, K_2NbF_7 is prepared from a solution of $K_2NbOF_5 . H_2O$ in HF but is easily hydrolysed back to that salt, while Rb_2NbF_6 is made by repeated crystallization of Rb_2NbOF_5 from HF. Finally, tantalum prefers to form complex fluorides of the type $R_mTaF_n . xH_2O$ rather than oxyfluorides, and in these n rises to 8. The V and Nb analogues of salts like Na_3TaF_8 are not known.

The following scheme illustrates, for the potassium-niobium fluorides, how the composition of the product varies with the relative concentrations of the component halides and of HF:

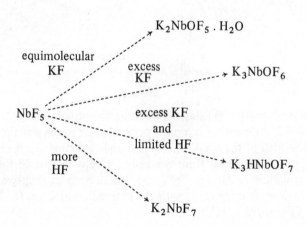

Complex fluorides and oxyfluorides provide many examples of isostructural compounds as, for example, compounds in the same horizontal row in the following table:

IV	V	VI
$K_2TiF_6 \cdot H_2O$	$K_2NbOF_5 \cdot H_2O$	$K_2WO_2F_4 \cdot H_2O$
$(NH_4)_3TiF_7$	$\begin{cases} (NH_4)_3NbOF_6 \\ (NH_4)_3TaOF_6 \end{cases}$	
K_3ZrF_7	K_3NbOF_6	
$\left.\begin{array}{l} K_3HSnF_8 \\ K_3HPbF_8 \end{array}\right\}$	K_3HNbOF_7	

The special interest of these compounds lies in the shape of the coordination group around the metal atom in the complex ion.

The dodecahedral ZrF_8 coordination group in K_2ZrF_6 has already been noted and illustrated in Fig. 10.3 (p. 384). There are also infinite chains of 8-coordination groups sharing two edges in $(N_2H_6)ZrF_6$, but here the coordination group is a bicapped trigonal prism (Fig. 10.12(a)). Well defined dodecahedral 8-coordination

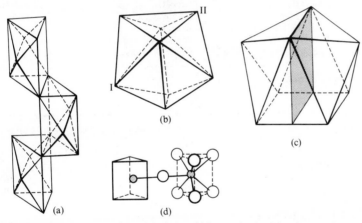

FIG. 10.12. Complex fluorides of Zr: (a) chain of bicapped trigonal prisms in $(N_2H_6)(ZrF_6)$, (b) 8-coordination group in Li_6BeZrF_{12}, (c) relation between bicapped trigonal prism and tetragonal antiprism, (d) 7-coordination of Zr in $Na_5Zr_2F_{13}$.

of Zr^{4+} is found in Li_6BeZrF_{12}, in which there is rather irregular octahedral coordination of Li^+, tetrahedral BeF_4 and discrete ZrF_8 groups. The ZrF_8 group (Fig. 10.12(b)), formed from two interpenetrating tetrahedra, one elongated (II) and the other flattened (I), may be compared with the coordination in $ThCl_4$ and related tetrachlorides (p. 361). There is 8-coordination of a different kind in $Na_7Zr_6F_{31}$ and the isostructural compounds of Pa, U, and other 5f elements. Thirty of the F atoms together with the Zr atoms form a 3D framework built from ZrF_8 antiprisms. The antiprisms share vertices in groups of six around cuboctahedral 'holes', and these subgroups consisting of six antiprisms then form the 3D framework by sharing some of their edges. The composition of this framework is $(ZrF_5)_n$; the remaining F^- ions and the Na^+ ions occupy interstices in the framework. (Some analogy may be drawn between this structure and structures such as pyrochlore (p. 209) and $La_4Re_6O_{19}$ (p. 183)). The close relation between certain pairs of polyhedra has been noted in Chapter 3 (and also in MSIC). Figure 10.12(c) illustrates this point for the bicapped trigonal prism and the antiprism. In $Na_5Zr_2F_{13}$ there is 7-coordination of Zr, two trigonal prisms being linked by F to form a Zr_2F_{13} complex (Fig. 10.12(d)).

TABLE 10.8

Coordination polyhedra in complex fluorides and oxyfluorides

F : Zr *ratio*	*Formula*	*C.N. of* Zr	*Coordination polyhedron*	*Reference*
5·17	$Na_7Zr_6F_{31}$	8	Antiprism	AC 1968 **B24** 230
5·25	$Rb_5Zr_4F_{21}$	6, 7, 8	See text	AC 1971 **B27** 1944
6	K_2ZrF_6	8	Dodecahedron	AC 1956 9 929
6	$(N_2H_6)ZrF_6$	8	Bicapped trigonal prism	AC 1971 **B27** 638
6	γ-Na_2ZrF_6	7	Irregular	AC 1969 **B25** 2164
6·5	$Na_5Zr_2F_{13}$	7	Monocapped t.p.	AC 1965 **18** 520
7	Na_3ZrF_7	7*	Irregular	AC 1948 1 265
7	$(NH_4)_3ZrF_7$	7*	Pentagonal bipyramid	AC 1970 **B26** 2136
8	$Li_6BeF_4(ZrF_8)$	8*	Dodecahedron	JCP 1964 **41** 3478

* Denotes finite complex ion BX_n

F : Ce *ratio*	*Formula*	*C.N. of* Ce	*Coordination polyhedron*	*Reference*
6	$(NH_4)_2CeF_6$	8	Antiprism	IC 1969 8 33
7	$(NH_4)_3CeF_7 . H_2O$	8	Dodecahedron	AC 1971 **B27** 1939
8	$(NH_4)_4CeF_8$	8*	Antiprism	AC 1970 **B26** 38

* Denotes finite complex ion BX_n

Empirical formula	*Type of complex ion*	*Shape*	*Reference*
$(NH_4)_3SiF_7$	SiF_6^{2-}	Octahedral	JACS 1942 64 633
K_2NbF_7, K_2TaF_7	NbF_7^{2-}, TaF_7^{2-}	Monocapped t.p.	AC 1966 20 220
K_3NbOF_6	$NbOF_6^{3-}$	(Pentagonal bipyramid)[a]	JACS 1942 64 1139
$K_2NbOF_5 . H_2O$	$NbOF_5^{2-}$	Octahedral	JACS 1941 63 11
K_3HNbOF_7	$\begin{cases} NbOF_5^{2-} \\ HF_2^- \end{cases}$	Octahedral / Linear	JACS 1941 63 11
Na_3TaF_8	TaF_8^{3-}	Antiprism	JACS 1954 76 3820
Na_3HTiF_8	$\begin{cases} HF_2^- \\ TiF_6^{2-} \end{cases}$	Linear / Octahedral	AC 1966 20 534

[a] Probably as in $(NH_4)_3ZrF_7$, but no recent study has been made.

The data of Table 10.8 show that Zr^{4+} prefers 7- or 8-coordination by F^-, the shape of the coordination polyhedron adjusting to the nature of the cations, and F : Zr ratios less than the coordination number requiring the sharing of vertices or edges of the coordination polyhedra. The inadvisability of trying to relate ionic sizes to coordination number and shape of coordination polyhedron is nicely illustrated by the structure of $Rb_5Zr_4F_{21}$. In this crystal there are four kinds of coordination of Zr^{4+} ions:

C.N. Polyhedron

8   irregular antiprism,

7   irregular antiprism less 1 vertex,

7   pentagonal bipyramid,

6   octahedron.

Complex fluorides A_mBX_7 and A_mBX_8 provide examples of a number of coordination polyhedra (Table 10.8). In contrast to $(NH_4)_3SiF_7$, which contains octahedral SiF_6^{2-} and separate F^- ions, $(NH_4)_3ZrF_7$ contains pentagonal bipyramidal ZrF_7^{3-} ions (compare K_3UF_7 and $K_3UO_2F_5$, p. 997), while the NbF_7^{2-} and TaF_7^{2-} ions are monocapped trigonal prisms. Yet another possibility is realized in $(NH_4)_3CeF_7 . H_2O$, where two dodecahedral coordination groups share an edge to form $(Ce_2F_{14})^{6-}$ ions. These compounds thus illustrate three ways of attaining a F : B ratio of 7 : 1, with c.n.'s of 6, 7, and 8 for the B atoms. Conversely, in the three ammonium ceric fluorides listed in Table 10.8 there is 8-coordination of Ce with F : Ce ratios of 6 : 1, 7 : 1, and 8 : 1, while $(NH_4)_2CeF_6$ is our third example of a salt A_2BX_6 in which the anion is an infinite chain built of 8-coordination polyhedra, all of different kinds:

Coordination polyhedron

K_2ZrF_6	Dodecahedron
$(N_2H_6)ZrF_6$	Bicapped trigonal prism
$(NH_4)_2CeF_6$	Antiprism.

In Na_3TaF_8 the anion has the shape of a square antiprism; on the other hand there is no BX_8 ion in Na_3HTiF_8 but equal numbers of HF_2^- and TiF_6^{2-} ions. Similarly, K_3HNbOF_7 contains HF_2^- and octahedral $NbOF_5^{2-}$ ions; the latter are the anions in $K_2NbOF_5 . H_2O$.

Equally interesting are the complex fluorides of Th and U(IV), which are described in Chapter 28, and in which 9-coordination also is important.

Metal nitride halides and related compounds

'Nitride halides' have been prepared in various ways, for example, Mg_2NCl (also F, Br, I) by direct combination of metal nitride and halide, Mg_3NF_3 by the action of nitrogen on a mixture of metal and halide at $900°C$, and ZrNI by heating

$ZrI_4 . n NH_3$. Little is yet known of their structures (Table 10.9). The compound $VNCl_4$ is apparently vanadium chlorimide trichloride and has been assigned a structure in which the molecules are loosely associated in pairs. The bond

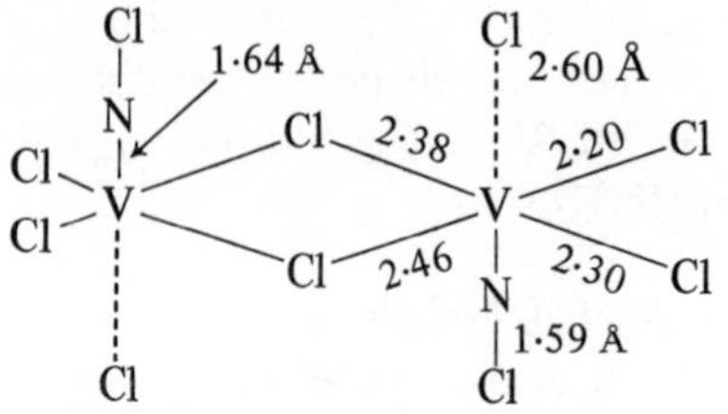

TABLE 10.9

Metal nitride halides and related compounds

	Structure	Reference
Mg_3NF_3	Defect NaCl (p. 195)	AC 1969 **B25** 1009
Mg_2NCl (F)	—	ZaC 1968 **363** 191; JSSC 1970 **1** 306
ZrNCl(Br, I)		
TiNCl(Br, I)	FeOCl (N not located)	ZaC 1964 **327** 207
ThNF	LaOF (rhombohedral)	
ThNCl(Br, I)	BiOCl	ZaC 1968 **363** 258
UNCl(Br, I)	BiOCl	ZaC 1969 **366** 43
[U(NH)Cl?]		ZaC 1966 **348** 50
LaN_xF_{3-x}	LaF_3	MRB 1971 **6** 57
VCl_3(NCl)	See text	ZaC 1968 **357** 325

arrangement around V is square pyramidal or distorted octahedral according to whether the weak bonds to Cl atoms of other molecules are included; the accuracy of the bond lengths is not clear. The V–N bond length is similar to that in other transition-metal complexes in which N and Cl are directly bonded to the metal (e.g. K_2OsNCl_5, Os–N, 1·61 Å).

The defect NaCl structure of Mg_3NF_3 is illustrated in Fig. 6.1(e) (p. 193); Mg has 4 F and 2 N (octahedral) neighbours (at 2·11 Å), N has 6 Mg (octahedral) and F has 4 Mg (coplanar) nearest neighbours. The tetragonal structure of Mg_2NF (X-ray powder data) is apparently an ordered NaCl-like structure which is extended in one direction so that the nearest neighbours of Mg are 3 N and 2 F (at a mean distance of 2·13 Å) with the sixth F at 2·86 Å. The coordination of N is distorted octahedral, and F has 4 close Mg neighbours as in Mg_3NF_3. The structure of the high-pressure form of Mg_2NF is apparently much closer to a cubic NaCl structure, as is to be expected since Mg_2NF is isoelectronic with MgO.

Thiohalides

Very little is known about the structures of metal thiohalides although a considerable number have been prepared; for the thiohalides of phosphorus see

Chapter 19. For example, all the thiohalides of In have been prepared, and from X-ray photographic evidence InSCl and InSBr have been assigned a statistical $CdCl_2$ structure, but further examination would be desirable if single crystals could be grown.[1]

Of the twelve possible compounds ABX where A is Sb or Bi, B is S or Se, and X is Cl, Br, or I, all except SbSCl and SbSeCl have been prepared, and of these all except BiSeCl are isostructural. The structure of SbSBr[2] consists of pleated chains $(SbS)_n^{n+}$ between which lie Br^- ions (Fig. 10.13). The chain is the simplest 3-connected linear system (see Fig. 3.19(a), p. 87), and it is closely related to the chain in Sb_2S_3 (p. 724). Within a chain, Sb–S = 2·49 Å (1) and 2·67 Å (2), Sb bond angles are 84° (2) and 96° (1), and S bond angles are all close to 96°. The nearest neighbours of a Br^- ion are two Sb (at 2·94 Å) and one S (at 3·46 Å). For $Hg_3S_2Cl_3$ see Chapter 26.

The compound NbS_2Cl_2 is not a simple thiohalide for it apparently contains S_2 groups, having a layer structure of the $AlCl_3$ type in which one Cl has been replaced by an S_2 group (S–S, 2·03 Å).[3] The nearest neighbours of Nb are 4 Cl at 2·61 Å and 4 S (of two S_2 groups) at 2·50 Å. There are Nb–Nb distances of 2·90 Å between pairs of metal atoms.

Molecular metal thiohalides are at present represented only by $WSCl_4$ and the similar, but not isostructural, $WSBr_4$.[4] Square pyramidal molecules are associated in pairs, the bridging W–Cl distance (3·05 Å) being much larger than the mean W–Cl (2·28 Å) within the $WSCl_4$ molecule.

Oxyhalides

Compounds containing one or more elements combined with oxygen and halogen atoms are of a number of quite different types, and as with complex halides the empirical formulae give no indication of the structures of compounds MO_xX_y. For example, calcium hypochlorite, CaO_2Cl_2, is a salt containing ClO^- ions–$Ca(OCl)_2$. Uranyl chloride, UO_2Cl_2 consists of UO_2^{2+} and Cl^- ions, that is, $(UO_2)Cl_2$, while sulphuryl chloride, SO_2Cl_2, is a molecular oxychloride. The following groups of compounds are included in other chapters:

salts of oxy-ions of the halogens (Chapter 9);
oxyhalides and oxyhalide ions of non-metals (e.g. POF_3, PO_2F^-, PO_3F^{2-}) are described under the appropriate element to which the O and X atoms are separately bonded;
complexes of $POCl_3$, $SeOCl_2$, etc. with halides are included in Chapters 19 and 16 respectively.

We include in this chapter a short section on oxyhalide ions of transition metals.

Oxyhalides of one kind or another are formed by one or more metals of most of the Periodic Groups except the alkali metals and possibly the elements of subgroup IB, though at present very few sets of compounds of the same formula-type are known containing a particular metal and all four halogens. An example is the series $NbOX_3$. We show in Table 10.10 the known compounds of some of the metals of

(1) ZaC 1962 **314** 303

(2) AC 1959 **12** 14

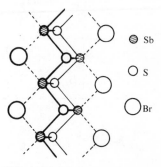

FIG. 10.13. The crystal structure of SbSBr.

(3) ZaC 1966 **347** 231

$$\begin{array}{ccc} & Cl & Cl \\ S{\diagdown}\,|\,{\diagup}Cl{\diagdown}\, |\,{\diagup}Cl \\ Cl{\diagup}\,|\,W{\diagup}{\cdots}Cl{\cdots}\,W{\diagdown}\,S \\ & Cl & Cl \end{array}$$

2·10 Å

(4) JCS A 1970 2815

TABLE 10.10
Some transition-metal oxyhalides

V			VI					VII		VIII
VOF $VOCl_2$ Br_2	VOF_3 Cl_3 Br_3				$CrOF_3$ Cl_3	CrO_2F_2 Cl_2	$CrOF_4$	MnO_3F		$FeOF$ $FeOCl$
$NbOCl_2$ Br_2 I_2	$NbOF_3$ Cl_3 Br_3 I_3	NbO_2F	$MoOCl_2$	MoO_2Cl	$MoOCl_3$ Br_3	MoO_2F_2 Cl_2 Br_2	$MoOF_4$ Cl_4	$TcOF_4$ $TcOCl_3$ Br_3		
	TaO_2F		$WOCl_2$		$WOCl_3$ Br_3	WO_2Cl_2 Br_2 I_2	WOF_4 Cl_4 Br_4	$ReOF_4$ $ReOF_5$ ReO_2F_3 ReO_3F	$ReOCl_3$ $ReOBr_3$	OsO_3F_2 $OsOF_5$ $PtOF_4$

groups IVA–VIIA and VIII, emphasizing the predominance of fluorides and chlorides and the very small number of iodides. Further study may, of course, alter this general picture. At least ten types of formula are found:

$$MOX, MOX_2, MOX_3, MOX_4, MOX_5,$$

$$MO_2X, \quad MO_2X_2, \quad MO_2X_3,$$

$$MO_3X, \quad MO_3X_2$$

and it is likely that, except in the case of molecular oxyhalides, structural differences will be found between crystalline oxyfluorides and other oxyhalides of a given metal.

Oxyhalide ions

A great variety of oxyhalide ions is formed by transition metals in their various oxidation states. The structures of some of the salts are similar to those of oxy-salts or complex halides. For example, the ion CrO_3F^- forms an almost regular tetrahedron in $KCrO_3F$, and because of the similar sizes of O and F there is random orientation of the anions in the crystal ($CaWO_4$-type structure). On the other hand, $KCrO_3Cl$ has a distorted form of that structure because of the difference between Cr–O (1·53 Å) and Cr–Cl (2·16 Å). $(NH_4)_3MoO_3F_3$ has the same structure as $(NH_4)_3AlF_6$, with octahedral $MoO_3F_3^{3-}$ ions, but whereas K_2OsCl_6 has the K_2PtCl_6 structure, the salt $K_2OsO_2Cl_4$[1] has a rather deformed version of that structure (Os–Cl, 2·38 Å and Os–O, 1·75 Å—compare 2·36 Å in K_2OsCl_6 and 1·66 Å in OsO_4). The salts $Cs_2M^VOCl_5$ (M = Nb, Cr, Mo, W) also have K_2PtCl_6-type structures,[2] and discrete octahedral ions exist in $Cs_2UO_2Cl_4$[3] (a), with the two short U–O bonds characteristic of uranyl compounds (Chapter 28).

Some oxyhalide ions are isolable only in combination with large cations. In $(C_2H_5)_4N[ReBr_4O(H_2O)]$[4] there is a square pyramidal arrangement of 4 Br and 1 O (with the metal atom 0·32 Å above the base of the pyramid), the sixth 'octahedral' bond to H_2O being very weak (b).

(1) AC 1961 **14** 1035

(2) JCS 1964 4944
(3) AC 1966 **20** 160

(4) IC 1965 **4** 1621

Examples of more complex oxyhalide ions are shown at (c) and (d). The anion in $(NH_4)_3UO_2F_5$[5] has the pentagonal bipyramidal configuration (c). The binuclear ion $(Cl_5Ru-O-RuCl_5)^{4-}$ (d), is the anion in $K_4Ru_2Cl_{10}O$. H_2O,[6] originally formulated as a compound of Ru^{IV}, $K_2RuCl_5(OH)$. (Note that the compound thought to be $K_4Tc_2Cl_{10}O$ is in fact $K_2TcCl_5(OH)$, p. 387.) Whereas $K_2Ru^{IV}Cl_6$ is paramagnetic with a moment corresponding to two unpaired electrons, this salt is diamagnetic. A m.o. treatment accounts for the diamagnetism and for the collinear arrangement of the Ru—O—Ru bonds, the length of which shows that they are nearly double bonds. An ion of the same kind exists in $K_4Re_2Cl_{10}O$. H_2O;[7] each Re atom is situated 0·12 Å out of the plane of its four Cl atoms towards the O atom.

(5) AC 1969 **B25** 67

(6) IC 1965 **4** 337

(7) AC 1962 **15** 851

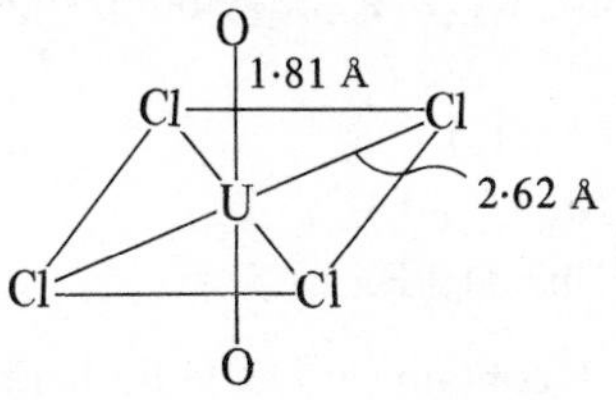

(a)

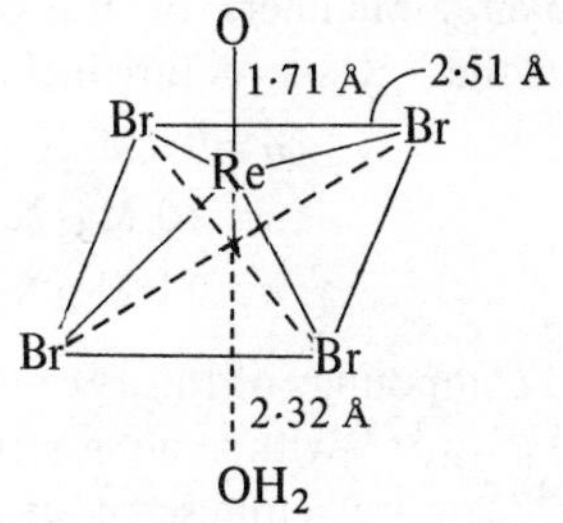

(b)

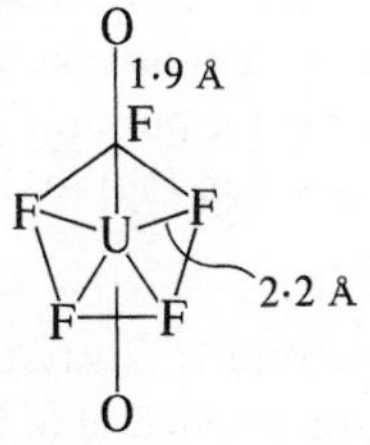

(c) 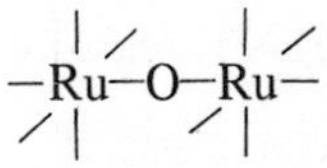   (d) 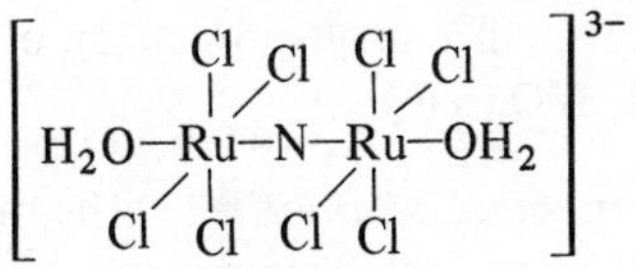   (e)

Although not an oxyhalide ion, the linear anion (e) in the salt $K_3[Ru_2NCl_8(H_2O)_2]$[8] may be mentioned here in view of its similarity to the ion (d).

(8) CC 1969 574

In addition to finite oxyhalide ions there are infinite oxyhalide ions in some complex oxyhalides. Examples include the anion in $K_2(NbO_3F)$, which has the K_2NiF_4 (layer) structure (see Table 10.11), and the *cis* octahedral AX_5 chain ion in $K_2(VO_2F_3)$,[9] F atoms forming the bridges.

(9) AC 1971 **B27** 1270

The structures of metal oxyhalides

We shall discuss these compounds in the following groups:

(a) ionic oxyfluorides,
(b) oxyhalides of transition metals in high oxidation states,
(c) oxyhalides MOCl, MOBr, and MOI,
(d) other oxyhalide structures.

Complex, Oxy-, and Hydroxy-Halides

(a) *Ionic oxyfluorides*

Examples are now known of oxyfluorides having the same structures as most of the simple oxides or fluorides MX_2 or MX_3 or structures characteristic of complex oxides or fluorides (Table 10.11). There is either random arrangement of O and F in the anion positions of such structures or, more rarely, regular arrangement (superstructure). We describe shortly the structures of the two forms of LaOF as examples of superstructures of the fluorite structure. There are also numerous oxyfluorides of transition metals which have structures similar to those of oxides of these metals and compositions very close to those of oxides. Examples noted in Chapter 13 include Nb_3O_7F and $LiNb_6O_{15}F$; more complex examples are $Nb_{17}O_{42}F$ and $Nb_{31}O_{77}F$,[1] with structures related to that of one of the forms of Nb_2O_5. Members of the family $M_{3n+1}X_{8n-2}$, built from different sized blocks of the ReO_3 structure include

$$n = 9 \quad M_{28}O_{70} \ (Nb_2O_5)$$
$$n = 10 \ M_{31}X_{78} \ (Nb_{31}O_{77}F)$$
$$n = 11 \ M_{34}X_{86} \ (Nb_{17}O_{42}F).$$

(1) ACSc 1965 **19** 1401

In contrast to compounds of this type, which contain very little F, there are others, such as $Zr_7O_9F_{10}$,[2] with unique structures (in this case with 6- and 7-coordination of Zr^{4+}) more reminiscent of complex fluorides such as $Na_7Zr_6F_{31}$ (p. 397).

(2) AC 1970 **B26** 830.

In some oxyfluorides the composition is variable within certain limits, in which case the formula given in Table 10.11 falls within the observed composition range. We noted earlier that the LaF_3 structure is apparently not stable for pure BiF_3 but becomes so if a small proportion of fluorine is replaced by oxygen (approximate formula, $BiO_{0.1}F_{2.8}$.)

Superstructures of fluorite. The cubic fluorite (CaF_2) structure is illustrated in Fig. 6.9 on p. 204. This structure may also be referred to a rhombohedral unit cell ($a = 7.02$ Å, $\alpha = 33\frac{1}{3}°$) containing two CaF_2 as shown in Fig. 10.14(a). The arrangement of the F^- and O^{2-} ions in the rhombohedral form of LaOF is such that the symmetry has dropped to rhombohedral, the ions being arranged with 3-fold symmetry around only one body-diagonal of the original cube (as at (b)). A recent reinvestigation of rhombohedral YOF shows that the positions earlier assigned to O and F should be interchanged. The distances Y–O and Y–F are different, and it was originally assumed that the latter would be the shorter ones. The later results are: Y–O, 2·24 and 2·34 Å; Y–F, 2·41 and 2·47 Å. (In ScOF the Sc–F distances are approximately 0·1 Å greater than Sc–O.) In the tetragonal superstructure there is a quite different arrangement of the anions (Fig. 10.14(c)). This form is stable over a range of composition, expressed by the formula MO_nF_{3-2n} ($0.7 < n < 1.0$). At the fluorine-rich limit the composition corresponds to the formula $MO_{0.7}F_{1.6}$, that is, there is an excess of F over the amount which can be accommodated in a fluorite-like structure. In this case it appears that the four O positions of Fig. 10.14(c) are occupied on the average by 2·8 O + 1·2 F and an excess of 1·2 F^- ions occupy some of the interstices marked by small black circles.

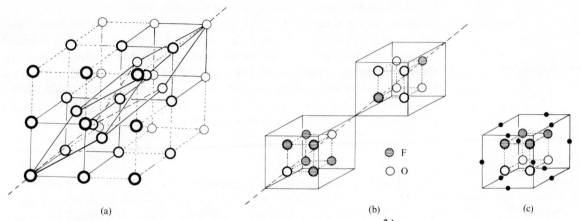

FIG. 10.14. (a) Alternative rhombohedral unit cell of the fluorite structure. Ca^{2+} ions lie at the centres of all faces of the smaller cubes, but only sufficient of these ions are shown to indicate the rhombohedral cell. (b) and (c) Structures of the two forms of LaOF (see text).

TABLE 10.11

Crystal structures of oxyfluorides

Structure	Examples	Reference
Rutile	FeOF, TiOF, VOF (high press./temp.)	JSSC 1970 **2** 49
α-PbO_2	$NaNbO_2F_2$ (superstructure)	AC 1969 **B25** 847
ZrO_2 (baddeleyite)	ScOF	ACSc 1966 **20** 1082
Fluorite	AcOF, HoOF	⎧ AC 1970 **B26** 2129
		⎨ AC 1957 **10** 788
superstructure:	rhombohedral: YOF, LaOF, SmOF, etc.	⎬ JACS 1954 **76** 4734, 5237
	tetragonal: YOF, LaOF, PuOF	⎩ IC 1969 **8** 232
	distorted: TlOF	MRB 1970 **5** 185
$PbCl_2$	LaOF and 4f metals (h.p.)	MRB 1970 **5** 769
ReO_3	$TiOF_2$	AC 1955 **8** 25
	NbO_2F, TaO_2F	AC 1956 **9** 626
	$WO_{3-x}F_x$, $MoO_{3-x}F_x$ ⎫	IC 1969 **8** 1764
MoO_3	$Mo_4O_{11}F$ ⎬	
LaF_3	$ThOF_2$	AC 1949 **2** 388
	$BiO_{0.1}F_{2.8}$	ACSc 1955 **9** 1209
Perovskite	$KNbO_2F$, $NaNbO_2F$	ZaC 1964 **329** 211
	$Tl^ITl^{III}OF_2$	MRB 1970 **5** 185
K_2NiF_4	K_2NbO_3F	JPC 1962 **66** 1318
	Sr_2FeO_3F	JPC 1963 **67** 1451
Spinel	Cu_2FeO_3F	JPCS 1963 **24** 759
	$Fe_3O_{4-x}F_x$	CR 1970 **270C** 2142
Cryolite	$Na_3VO_2F_4$ (monoclinic) ⎫	
	K_3TiOF_5 (tetragonal) ⎬	ZaC 1969 **369** 265
Elpasolite	$K_2NaVO_2F_4$ ⎭	
Tetragonal bronze	KNb_2O_5F	ACSc 1965 **19** 1510
Pyrochlore	$Cd_2Ti_2O_5F_2$	CR 1969 **269C** 228
Garnet	$Y_3Fe_5O_{12-x}F_x$	MRB 1971 **6** 63
Magnetoplumbite	$Y_3Al_4NiO_{11}F$ ⎫	JPCS 1963 **24** 759
	$BaCoFe_{11}O_{18}F$ ⎭	

(b) *Oxyhalides of transition metals in high oxidation states*

A number of the compounds of Table 10.10 have been studied in the vapour state and shown to form tetrahedral molecules (Table 10.12). Both crystalline forms of the green $OsOF_5$ consist of slightly distorted octahedral molecules; compare the regular octahedral structures of OsF_6 and the 5f compounds MF_6.[1] (It is interesting that whereas OsO_4 and OsO_3F_2 are known, the octafluoride has not been made nor have $OsOF_6$ or OsO_2F_4.) It is likely that many of the compounds of Table 10.10 form molecular crystals; for example, at ordinary temperatures

(1) JCS A 1968 536, 543

TABLE 10.12

Structures of metal oxyhalide molecules

Molecule	M—O (Å)	M—X (Å)	Angles		Method	Reference
$VOCl_3$	1·56 (·04)	2·12 (·03)	O—V—Cl	108 (2)°	e.d.	JACS 1938 **60** 2360
			Cl—V—Cl	111 (2)°		
CrO_2Cl_2	1·57 (·03)	2·12 (·02)	O—Cr—O	105 (4)°	e.d.	JACS 1938 **60** 2360
			O—Cr—Cl	$109\frac{1}{2}$ (3)°		
			Cl—Cr—Cl	113 (3)°		
MoO_2Cl_2	1·75 (·10)	2·28 (·03)	(O—Mo—O $109\frac{1}{2}$°)		e.d.	H. A. Skinner, Thesis, Oxford 1941
			O—Mo—Cl	108 (7)°		
			Cl—Mo—Cl	113 (7)°		
MnO_3F	1·586 (·005)	1·724 (·005)	O—Mn—F	108°27′ (7′)	m.w.	PR 1954 **96** 649
ReO_3F	1·692 (·003)	1·859 (·008)	O—Re—F	109° 31′ (16′)	m.w.	JCP 1959 **31** 633
ReO_3Cl	1·702 (·004)	2·229 (·004)	Cl—Re—O	109° 22′ (7′)	m.w.	

(2) AC 1959 **12** 21
(3) JCS A 1968 1061; AC 1970 **B26** 1161

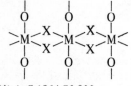

(4) AnC 1964 **76** 833

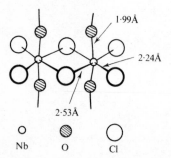

O   ⦻   ◯
Nb   O   Cl

FIG. 10.15. Portion of the infinite chain molecule in crystalline $NbOCl_3$.

CrO_2Cl_2 is a red liquid (b.p. 117°C). (For the uranyl halides see Chapter 28.) On the other hand, $NbOCl_3$ (like $VOCl_3$) is monomeric in the vapour state but has a polymeric chain structure (Fig. 10.15) in the crystal.[2] Octahedral NbO_2Cl_4 (*trans*) groups are linked into double chains by sharing an edge (2 Cl) and two opposite vertices (O atoms). There is a close relation between the structure of crystalline $NbOCl_3$ and $NbCl_5$, for if the O atoms of Fig. 10.15 are replaced by Cl the portion of chain has the composition Nb_2Cl_{10} and represents the finite molecule in the crystalline pentachloride. Other compounds with the structure of Fig. 10.15 include: $MoOCl_3$ (tetragonal), $WOCl_3$, $WOBr_3$, $TcOBr_3$, $NbOBr_3$, and $MoOBr_3$.[3]

Extension of the double chain of $NbOCl_3$ gives a layer of composition MOX_2 which is apparently the structure of $NbOCl_2$, $NbOBr_2$, and $NbOI_2$.[4]

A number of oxyhalides have structures which are similar to those of halides, some of the halogen atoms being replaced by O atoms. Since many of the ionic 3D structures of Table 10.11 are adopted by both fluorides and oxides, many of the oxyfluorides listed therein could be included here. If, however, we restrict ourselves here to structures which are peculiar to halides (that is, are not adopted by oxides) the known examples are those of Table 10.13.

TABLE 10.13

Oxyhalides related to MX_4 *and* MX_5 *structures*

Halide structure		Oxyhalides	Reference
Tetramer	Nb_4F_{20}	$(WOF_4)_4$	JCS A 1968 2074;
Octahedral chain structures	$TcCl_4$	$ReOBr_3, TcOCl_3$ $MoOCl_3$ (monoclinic)	JCS A 1969 2415
	α-UF_5	$WOCl_4, WOBr_4$	ZaC 1966 **344** 157
	CrF_5	$MoOF_4, ReOF_4, TcOF_4$	See Table 9.20
Layer	SnF_4	WO_2Cl_2	ZaC 1968 **363** 58

A number of oxychlorides and oxybromides MOX_3 have structures similar to $TcCl_4$ (octahedral '*cis*' chain structure), namely, $MoOCl_3$ (monoclinic), $TcOCl_3$, and $ReOBr_3$. In these compounds M is displaced from the centre of the octahedral coordination group towards the O atom. In $MoOCl_3$ (monoclinic), for example, the bridging atoms are Cl atoms and the bond opposite O is longer in the unsymmetrical bridge.

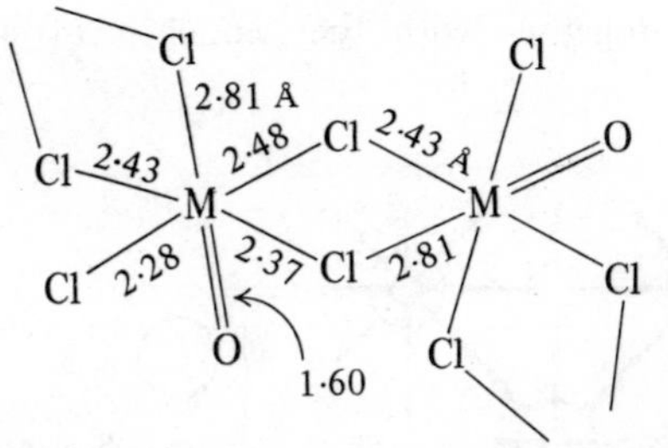

In $WOCl_4$ and $WOBr_4$ (α-UF_5 structure) the bridging atoms are O atoms, but these were not accurately located.

The structure assigned to WO_2Cl_2 consists of layers of octahedral WO_4Cl_2 groups which share four equatorial O atoms—compare the SnF_4 layer structure. The oxygen bridges are unsymmetrical, so that in both directions in the layer there are alternate short and long W–O bonds (mean lengths, 1·67 and 2·28 Å); in this connection see the structure of WO_3. The O bond angle is approximately $160°$ and W–Cl, 2·31 Å.

(c) *Oxyhalides* MOCl, MOBr, *and* MOI

All the elements Al, Ga, La, Ti, V, and the 4f and 5f metals form an oxychloride MOCl and most of them also form MOBr and MOI. Antimony and bismuth also form more complex oxyhalides, which are described in Chapter 20, and Bi forms BiOF which belongs in this group. Other oxyfluorides MOF of the above elements were included in group (a), ionic oxyfluorides.

All the compounds of this group are structurally quite different from those in (a) or (b) and crystallize with one of three layer structures, in which there is respectively 4-, 6-, and 8(9)-coordination of M.

407

GaOCl. The structure of GaOCl[1] (and the isostructural AlOCl, AlOBr, and AlOI) consists of layers of tetrahedral GaO_3Cl groups (Ga—O, 1·91 Å, Ga—Cl, 2·22 Å) which are linked as shown in Fig. 10.16. Each O atom is common to three tetrahedra, the unshared vertices of which are occupied by the Cl atoms. In a given row of tetrahedra these Cl atoms lie alternately to one side or the other of the (puckered) plane of the Al and O atoms.

FeOCl. The layers in this structure are of the same type as in γ-FeO . OH but they are packed together in a rather different way, as described on p. 528. The FeOCl structure is closely related to that of BiOCl as may be seen by comparing the elevation of FeOCl (Fig. 10.17) with that of BiOCl (Fig. 20.6(c), p. 715); the essential difference is that in FeOCl M is 6-coordinated and in BiOCl 8-coordinated. Bond lengths in FeOCl[2] are as shown. Compounds isostructural with FeOCl include InOCl and InOBr,[3] InOI,[4] VOCl and TiOCl,[5] and CrOCl.[6]

BiOCl. The remaining oxyhalides in this group, which does *not* include SbOCl (for which see p. 716) are built of layers of a different kind and have the same type of structure as PbFCl and PbFBr. Each complex layer in the BiOCl structure consists of a central sheet of coplanar O atoms with a sheet of halogen atoms on each side, and the metal atoms between the Cl—O—Cl sheets. The plan of a layer is shown in Fig. 10.18 together with the sequence of atoms in a direction

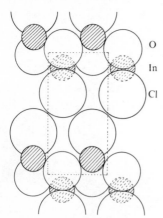

○ Cl above Ga–O plane
◌ Cl below Ga–O plane

FIG. 10.16. Layer of the structure of GaOCl (idealized).

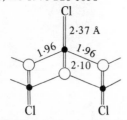

FIG. 10.17. Elevation of the crystal structure of InOCl or FeOCl.

(1) CR 1963 **256** 3477

(2) AC 1970 **B26** 1058

Cl
2·37 Å
1·96   1·96
2·10
Cl   Cl

Bond lengths in FeOCl.

(3) ACSc 1956 **10** 1287
(4) JCS A 1966 1004
(5) ZaC 1958 **295** 268
(6) ACSc 1962 **16** 777

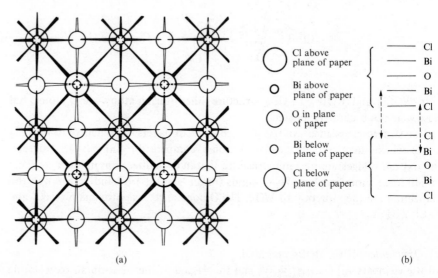

Cl above plane of paper

Bi above plane of paper

O in plane of paper

Bi below plane of paper

Cl below plane of paper

Cl
Bi
O
Bi
Cl
Cl
Bi
O
Bi
Cl

(a)   (b)

FIG. 10.18. (a) Atomic arrangement in a layer of the structure of BiOCl (or PbFCl). (b) Sequence of atoms in a direction normal to the layer (a).

perpendicular to the layers. Within one of these complex layers a Bi atom is surrounded by 4 O and 4 Cl atoms at the vertices of a square antiprism. However, the various interatomic distances in these structures (Table 10.14) suggest that, at

TABLE 10.14

Interatomic distances in oxyhalides MOX *with the BiOCl structure* (Å)

	M–4 O	M–4 X in same layer	M–X in next layer	X–X in layer	X–X between layers
LaOCl	2·39	3·18	3·14	4·11	3·46
BiOCl	2·31	3·07	3·49	3·88	3·48
LaOBr	2·40	3·28	3·47	4·15	3·54
BiOBr	2·32	3·18	4·04	3·92	3·72
LaOI	2·41	3·48	4·79	4·14	4·14
BiOI	2·33	3·36	4·88	3·99	4·17

least in the chlorides and bromides, there are appreciable interactions between the metal atoms in one layer and halogen atoms in a neighbouring layer, as indicated by the dotted lines in Fig. 10.18(b)–compare the M–4 X and M–X distances in Table 10.14. This 9-coordination is evident also in BiOF[7] (Bi–4 O, 2·27, 4 F, 2·75, and 1 F, 2·92 Å). The coordination polyhedron is a monocapped antiprism, with one X atom beyond the larger square face. The small alteration in the unit cell dimension *a* in the plane of the layer in the series LaOCl–LaOBr–LaOI compared with the large increase in the dimension *c* perpendicular to the plane of the layers,

(7) ACSc 1964 **18** 1823

	a	*c*
LaOCl	4·109 Å	6·865 Å
LaOBr	4·145	7·359
LaOI	4·144	9·126

reflects the rigidity of the M_2O_2 system consisting of layers of M atoms on each side of an oxygen layer. This type of tetragonal metal–oxygen–metal layer is a characteristic feature of the structures of a number of compounds of Bi, La, and Pb (e.g. La_2MoO_6 and PbO, p. 461), in particular the complex oxyhalides and oxides of Bi which are described on pp. 713 and 716.

Compounds which crystallize with the BiOCl (PbFCl) structure include:

MOF: Bi
MOCl: Bi, 4f compounds from La to Er (TmOCl, YbOCl, and LuOCl have a different, unknown, structure), Ac, and Pu
MOBr: Bi, all 4f compounds,[8] Ac, and Pu
MOI: Bi, La, Sm, Tm, Yb, Pu.

(8) IC 1965 **4** 1637

(d) *Other oxyhalide structures*
It is likely that many stuctures will be discovered which are peculiar to single oxyhalides or to small groups of compounds. For example there are compounds M_4OCl_6 and M_4OBr_6, made by heating together oxide and halide, suitable for the ions Ca^{2+}, Sr^{2+}, and Ba^{2+}. In Ba_4OCl_6 there is almost regular tetrahedral coordination of O by 4 Ba, and Ba is coordinated by either 7 Cl + O or by 9 Cl + O.[1]

(1) AC 1970 **B26** 16

409

(2) AC 1968 **B24** 304

The structure of $PaOCl_2$ (and the isostructural Th, U, and Np compounds) is unexpectedly complex.[2] There are tightly-knit chains of composition $Pa_3O_3Cl_4$ $(Cl_{\frac{4}{2}}^*)$ of which the Cl* atoms are shared with other similar chains, forming a 3D structure with the composition $PaOCl_2$. There is 2- and 3-coordination of Cl, 3- and 4-coordination of O, and 7-, 8-, and 9-coordination of the Pa atoms.

Hydroxyhalides

(1) ACSc 1957 **11** 676
(2) JACS 1948 **70** 105

Our knowledge of the structures of hydroxyhalides is practically confined to hydroxychlorides and hydroxybromides, little being known of the structures of hydroxyfluorides or hydroxyiodides. Anhydrous hydroxyfluorides include $Cd(OH)F$, $Zn(OH)F$ and other less well-defined phases. $In(OH)F_2$ has a distorted ReO_3 type of structure, in which In is surrounded octahedrally by 2 OH + 4 F (all at about 2·07 Å).[1] Hydrated oxyfluorides of Al with the pyrochlore structure[2] (p. 209) have been prepared by adding ammonia to solutions of $Al_2(SO_4)_3$ containing varying proportions of AlF_3.

(3) HCA 1962 **45** 479
(4) HCA 1961 **44** 2095
(5) ACSc 1959 **13** 1049

The largest groups of hydroxychlorides and hydroxybromides are the compounds $MX(OH)$ and $M_2X(OH)_3$ formed by Mg and the 3d metals Mn, Fe, Co, Ni, and Cu(II). Their structures consist of c.p. layers of composition $X(OH)$ or $X(OH)_3$ as illustrated in Fig. 4.15 (p. 135) between which the metal atoms occupy respectively one-half or one-quarter of the octahedral interstices. These c.p. structures are not suitable for larger ions such as Pb^{2+} or Y^{3+}, and we deal later with compounds containing these ions.

Hydroxyhalides MX(OH)
Partial or complete X-ray studies have been made of a number of these compounds, for example, $CoBr(OH)$,[3] $CuCl(OH)$,[4] and $\beta\text{-}ZnCl(OH)$.[5] Various layer sequences are found comparable to the differences between the *C*6, *C*19, and *C*27 structures of dihalides and dihydroxides. In CuCl(OH) there is the usual distortion of the octahedral coordination group around Cu(II)—1 Cl at 2·30, 3 OH at 2·01, and 2 Cl at 2·71 Å.

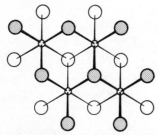

FIG. 10.19. Portion of one layer of Cd(OH)Cl. The OH groups (shaded) lie above, and the Cl atoms below, the plane of the metal atoms.

The structure of CdCl(OH)[6] differs from the above structures in that each layer contains all Cl or OH (Fig. 10.19) instead of equal numbers of Cl and OH in each c.p. layer.

(6) ZK 1934 **87** 110
(7) HCA 1964 **47** 272

Hydroxyhalides $M_2X(OH)_3$
The structures are of two general types, layers and 3D frameworks. In addition to the ordered structures described below (based on regular $X(OH)_3$ layers) the hydroxychlorides $M_2Cl(OH)_3$ of Mg, Mn, Fe, Co, and Ni form disordered phases of the *C*6 (CdI_2) type in which the Cl and OH are randomly arranged.[7]

For hydroxyhalides $M_2Cl(OH)_3$ (M_2XY_3) the structural possibilities are essentially the same as for dihalides or dioxides MX_2 based on close-packed X atoms with M occupying octahedral holes; the relevant structure types have been set out in Table 4.6 (p. 142). There are, however, two new factors to take into account: (i) There are two types of close-packed XY_3 layer, the rhombic and trigonal layers of Fig. 4.15(c) and (d), so that the number of structures possible for each kind of close-packing is doubled. (We shall suppose that all the layers in a given structure are of the same kind, as has been found up to the present time.) (ii) The number of possible structures is, on the other hand, reduced owing to the requirement that the X or Y atoms shall have their three M neighbours arranged pyramidally with X–M–X (Y–M–Y) bond angles of 90°. An asterisk in Table 10.15 indicates that structures satisfying this requirement are not possible.

The possible structures for hexagonal and cubic close packing are listed, with examples, in Table 10.15, which corresponds to the portion of the more general Table 4.6 (p. 142) for 'one-half octahedral holes occupied'. In the ideal c.p. structures the coordination group around a metal atom is an octahedral group (of 4 OH + 2 X or 5 OH + X) in which the M–O and M–X bonds have their normal lengths. As in the case of other cupric compounds there are distortions of the

TABLE 10.15
Crystal structures of hydroxyhalides $M_2X(OH)_3$

	Layer sequences	
	AB...	ABC...
Rhombic XY_3 layers (Fig. 4.15(c), p. 135)	*Layer structure* (CdI_2 C 6 type) $Co_2Br(OH)_3$ [a] $Cu_2Cl(OH)_3$ [b] (botallackite) $Cu_2Br(OH)_3$ [c] $Cu_2I(OH)_3$ [c]	*Layer structure* ($CdCl_2$ type) –
	Framework structure *	*Framework structure* $Cu_2Cl(OH)_3$ [d] (atacamite) β-$Mg_2Cl(OH)_3$ [e]
Trigonal XY_3 layers (Fig. 4.15(d), p. 135)	*Layer structure* –	*Layer structure* –
	Framework structure *	*Framework structure* β-$Co_2Cl(OH)_3$ [f]

(a) AC 1950 **3** 370. (b) MM 1957 **31** 237. (c) HCA 1961 **44** 2103. (d) AC 1949 **2** 175. (e) ASR 1954 **B3** 400. (f) AC 1953 **6** 359.

structures to give (4 + 2)-coordination (see Chapter 25). For a further note on $Cu_2Br(OH)_3$ and $Cu_2Cl(OH)_3$ and illustrations of the botallackite and atacamite structures see p. 906.

Other hydroxyhalides
The hydroxychloride PbCl(OH) occurs as the mineral laurionite and its structure is very similar to that of $PbCl_2$ (p. 221) with (7 + 2)-coordination of Pb.

Other structures with higher coordination of the metal ions include those of the two forms of $YCl(OH)_2$,[8] in both of which there is 8-coordination of Y^{3+} (bicapped trigonal prism).

(8) AC 1967 **22** 435

Hydrated hydroxychlorides

Two of these compounds present points of special interest. The action of a solution of $MgCl_2$ on MgO gives first a compound $5\,Mg(OH)_2 . MgCl_2 . 7\,H_2O$ (?) which slowly changes into $Mg_2(OH)_3Cl . 4\,H_2O$. The basis of the structure of this compound (and also apparently of a number of other hydrated hydroxy-salts of Mg) is the double chain of composition $Mg_2(OH)_3(H_2O)_3$ which has the same form as the double chain anion in $NH_4(CdCl_3)$. A portion of one of these chains is shown in Fig. 10.20(a), and (b) shows the structure viewed along the direction of

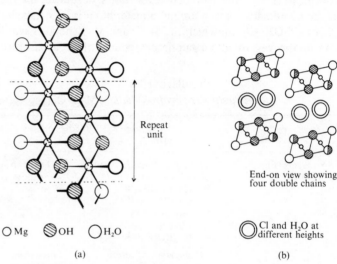

Repeat unit

End-on view showing four double chains

○ Mg ◎ OH ○ H₂O

◎ Cl and H₂O at different heights

(a)   (b)

FIG. 10.20. The crystal structure of $Mg_2(OH)_3 Cl . 4\,H_2O$.

the infinite chains, which are held together by Cl^- ions and H_2O molecules. The structural formula of this compound is accordingly $[Mg_2(OH)_3(H_2O)_3]$ $Cl . H_2O$.[9]

The relationship of the layer structure of $Zn_5(OH)_8Cl_2 . H_2O$ to the CdI_2 structure is described in Chapter 6. The structure is of interest because 3 Zn are in octahedral holes (Zn–O, 2·16 Å) and 2 Zn are in tetrahedral holes (Zn–3 O, 2·02, Zn–Cl, 2·33 Å).[10]

Amminohalides

The absorption of ammonia by many solid halides has been known for a long time. The products vary widely in composition and stability. For example, $CaCl_2 . 8\,NH_3$ and $CuCl_2 . 6\,NH_3$, prepared from the anhydrous halides and ammonia gas, readily lose ammonia, but many other amminohalides may be prepared from ammoniacal salt solutions and some possess considerable stability. The simplest compounds of this type have the general formula $MX_n . m\,NH_3$ (with sometimes water of crystallization), where m is the normal coordination number of M. They contain complex ions $M(NH_3)_m^{n+}$ and X^- ions. Typical ions of this sort are: $Ag(NH_3)_2^+$,

(9) AC 1953 6 40

(10) ZK 1962 **117** 238

$Zn(NH_3)_4^{2+}$, and $Co(NH_3)_6^{3+}$. Many salts $[M(NH_3)_6]X_2$ and $[M(NH_3)_6]X_3$ have simple structures, for example, $[Co(NH_3)_6]Cl_2$ with the antifluorite structure of K_2PtCl_6, and $[Co(NH_3)_6]I_3$ with the $(NH_4)_3AlF_6$ structure.

The coordination group around M may, however, be made up of NH_3 and X as in $[Co(NH_3)_5Cl]Cl_2$ or $[Pt(NH_3)_3Cl_3]Cl$. Some of these Werner coordination compounds are described in Chapter 27. The compounds $M(NH_3)_xCl_x$ are non-electrolytes if $2x$ is the coordination number of M. Zinc, for example, forms $Zn(NH_3)_2Cl_2$ consisting of tetrahedral molecules (a) which may be compared with the ions $Zn(NH_3)_4^{2+}$, (b), in a salt such as $Zn(NH_3)_4Cl_2 . H_2O$.

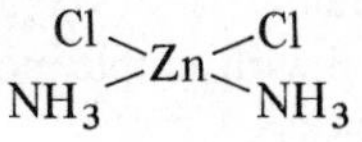

$$
\begin{array}{cc}
\begin{array}{c} Cl \\ \diagdown Zn \diagdown \\ NH_3 \quad NH_3 \end{array} Cl
&
\left[\begin{array}{c} NH_3 \\ \diagdown Zn \diagdown \\ NH_3 \quad NH_3 \end{array} NH_3 \right]^{2+}
\\[2em]
(a) & (b)
\end{array}
$$

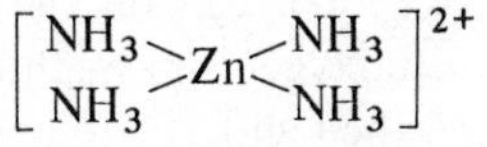

The molecule $Pd(NH_3)_2I_2$ is similar to (a) but planar instead of tetrahedral. Not all compounds with similar empirical formulae are of this type, for one form of $Pt(NH_3)_2Cl_2$ consists of planar ions $Pt(NH_3)_4^{2+}$ and $PtCl_4^{2-}$, while $Cd(NH_3)_2Cl_2$ is composed of infinite chains built from octahedral $Cd(NH_3)_2Cl_4$ groups sharing opposite edges:

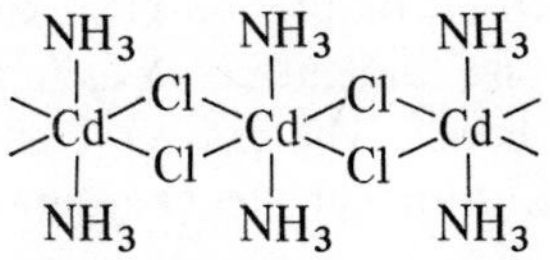

$$
\begin{array}{ccccccc}
NH_3 & & NH_3 & & NH_3 \\
| & & | & & | \\
\diagup Cd \diagdown & Cl & \diagup Cd \diagdown & Cl & \diagup Cd \diagdown \\
& Cl & & Cl & \\
| & & | & & | \\
NH_3 & & NH_3 & & NH_3
\end{array}
$$

This is one of a large family of compounds with octahedral chain structures which are listed in Table 25.4 (p. 903).

For amminohalides of Cu, Ag, and Hg see Chapters 25 and 26, and for those of Group VIII metals see Chapter 27.

Ammines are compared with hydrates in Chapter 15.

11

Oxygen

After some introductory remarks on the stereochemistry of oxygen and some of the differences between oxygen and sulphur we deal with simple molecules and ions containing oxygen and then with the following topics:

peroxo- and superoxo-salts,
oxo molecules and ions, and
oxy-ions, isopoly, and heteropoly ions.

The stereochemistry of oxygen

With six electrons in the valence shell the simplest ways in which the octet can be completed are (a) the formation of the ion O^{2-}, (b) the acquisition of one electron and the formation of one covalent bond, as in the ion OH^-, and (c) the formation of two electron-pair bonds. The ions O^{2-} and OH^- are found in the oxides and hydroxides of metals. Although the O atom could, in principle, form a maximum of four covalent bonds, since there are four orbitals available, the formation of more than two *essentially covalent* bonds is rarely observed (see below). Assuming that the bond arrangement is determined by the number of σ bonds and lone pairs we may summarize the stereochemistry of oxygen as shown in Table 11.1. We have included one case where O forms two collinear bonds, the ion $Ru_2OCl_{10}^{4-}$ (p. 403), in which the bonds from the O atoms have considerable double-bond character; other examples will be found in the section on 'Oxo-salts'. The collinear bonds in $Sc_2Si_2O_7$ are mentioned later.

TABLE 11.1

The stereochemistry of oxygen

No. of σ pairs	Type of hybrid	No. of lone pairs	Bond arrangement	Examples
2	sp	0	Collinear	$(Cl_5RuORuCl_5)^{4-}$
3	sp^2	1	Angular	O_3
4	sp^3	0	Tetrahedral	$Be_4O(CH_3COO)_6$
		1	Pyramidal	H_3O^+
		2	Angular	H_2O, F_2O, H_2O_2

Differences between oxygen and sulphur

It is not proposed to discuss here the stereochemistry of S, which is much more complex than that of O because of the availability of d orbitals, but merely to summarize some points of difference between O and S. From O to Te the atoms increase in size and we may associate the change in behaviour of the outer valency electrons with the increased screening of the nuclear charge by the intervening completed shells of electrons. This shows itself in a number of ways:

(1) The decreasing stability of negative ions. The alkali oxides, sulphides, selenides, and tellurides all crystallize with typical ionic structures, showing that all the ions O^{2-}, S^{2-}, Se^{2-}, and Te^{2-} exist in solids. These compounds are all soluble in water, but the anhydrous compounds cannot be recovered from the solutions. If Na_2O is dissolved in water the solution on evaporation yields solid NaOH. A concentrated solution of Na_2S on evaporation yields the hydrate $Na_2S . 9 H_2O$, but in dilute solution the sulphur is almost entirely in the form of SH^- ions. This hydrosulphide ion is much less stable than the hydroxyl ion and, on warming, solutions of hydrosulphides evolve H_2S. Solutions of selenides and tellurides hydrolyse still more readily, liberating the hydrides, which are easily oxidized to the elements. In the case of oxygen, therefore, we have the combination of O^{2-} with H^+ giving the very stable OH^- ion, Sulphur behaves similarly, giving the less stable SH^- ion, but with Se^{2-} and Te^{2-} the process easily goes a stage farther to H_2Se and H_2Te. The increasing tendency for the divalent ion to form the hydride ($X^{2-} \rightarrow XH^- \rightarrow XH_2$) is not due to the increasing stability of the hydrides, for this decreases rapidly from H_2O to H_2Te, the latter being strongly endothermic, but to the decreasing stability of the negative ions in aqueous solution.

These elements do not, with the exception of Te, form stable monatomic cations. The formation of the ion Te^{4+} (in TeO_2 and presumably in $Te(NO_3)_4$ and $Te(SO_4)_2$) is due to the inertness of the two s electrons (see Chapter 16 for details of Te compounds).

(2) The bonds formed by S with a particular element have less ionic character than those between O and the same element. Pauling assigns the following electronegativity coefficients:

$$\begin{array}{ll} O & 3.5 \\ S & 2.5 \end{array} \quad \text{compare} \quad \begin{array}{ll} F & 4.0 \\ Cl & 3.0. \end{array}$$

In many cases a dioxide has a simple ionic structure while the corresponding disulphide has a layer structure, and in general the same structure types are found for crystalline oxides as for fluorides of the same formula type, and similarly for sulphides and chlorides.

The dioxides of S, Se, Te, and Po, which are described later, range from molecular oxides to ionic crystals.

(3) The highest known fluoride of oxygen is OF_2, but S, Se, and Te all form hexafluorides. By the action of an electric discharge in O_2–F_2 mixtures at low temperatures O_4F_2 is produced, together with O_2F_2, but the only fluorides containing directly bonded S, Se, or Te are S_2F_2, S_2F_{10}, and Te_2F_{10}. As in many

415

other cases these three elements exhibit their highest valence in combination with fluorine; this is the only element with which S forms six bonds.

(4) Except in ozone, the alkali-metal ozonates, the unstable O_4F_2, and a number of organic peroxides and trioxides (e.g. $F_3C.OOO.CF_3$), the covalent linking of O to O does not proceed beyond O=O or –O–O–. Sulphur, on the other hand, presents a very different picture, a feature of its chemistry being the easy formation of chains of S atoms—in the element itself, in halides S_nX_2, poly-sulphides, and polythionates.

(5) Unlike sulphur oxygen seldom forms more than two covalent bonds in simple molecules or ions. There are numerous ionic crystals in which O forms three bonds (rutile, phenacite, etc.). In salt hydrates where H_2O is bonded to a metal atom, (a), or in hydroxides where OH bridges two metal atoms, (b), the bonds to the metal atoms presumably have appreciable ionic character, and this is probably also true in AlOCl (p. 408) and $[Ti(OR)_4]_4$ (p. 942). Three equivalent (covalent ?)

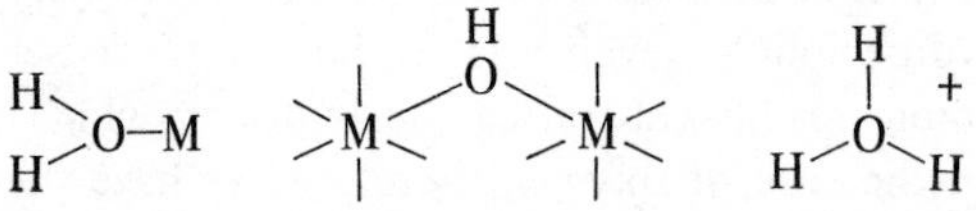

(a)   (b)   (c)

bonds are formed in the hydronium ion, (c), found in the crystalline monohydrates of HCl and other acids, and a variety of methods indicate a rather flat pyramidal structure with O–H, 1·02 Å and angle H–O–H, 115–117°. The ion $O(HgCl)_3^+$, which is nearly planar (angle Hg–O–Hg, 118°), exists in crystalline Hg_3OCl_4 (p. 922). Other examples of the formation of three nearly coplanar bonds are the borate ion in $SrB_6O_9(OH)_2.3H_2O$ (p. 859) and the cyclic molecule (d).

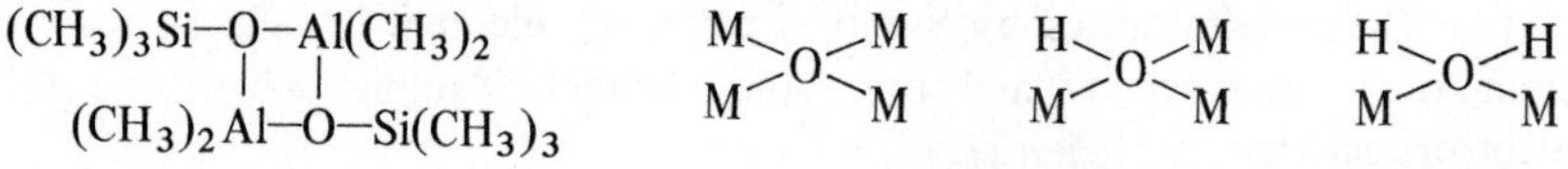

(d)   (e)   (f)   (g)

There are many molecules and crystals in which O forms four approximately regular tetrahedral bonds, but some or all of the bonds are to metal atoms and presumably are fairly polar. We find (e) in numerous monoxides such as ZnO and PdO (but note the four coplanar bonds in NbO, a structure in which there is metal-metal bonding), in tetrahedral molecules such as $OBe_4(CH_3COO)_6$ (Fig. 11.1), $OMg_4Br_6.4C_4H_{10}O$, and other similar molecules listed in Table 3.1, and also in $OZn_4B_6O_{12}$ and $OLa_4Re_6O_{18}$ (q.v.). The bond arrangement (f) represents the environment of OH⁻ in $La(OH)_3$ the environment of the central O atoms in $[Ti(OAlkyl)_4]_4$ or of O in $[TlOAlkyl]_4$, while (g) represents a bridging water molecule in a salt hydrate.

(6) The last important difference between O and S which we shall mention here

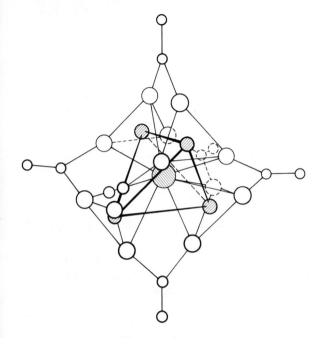

FIG. 11.1. The molecule of beryllium oxy-acetate, $OBe_4(CH_3COO)_6$, drawn (not strictly to scale) to show the central oxygen atom (large shaded circle) surrounded tetrahedrally by four Be atoms (small shaded circles). An acetate group lies beyond each edge of the tetrahedron. Each Be atom has four O atoms arranged tetrahedrally as nearest neighbours, the central one and three of different acetate groups. (Small unshaded circles represent C atoms; H atoms are omitted.)

is the formation by the former of relatively strong hydrogen bonds, a property of only the most electronegative elements. As a result many oxygen compounds have properties very different from those to be expected by analogy with the corresponding sulphur compounds.

Simple molecules and ions

We deal here with the following topics: the molecules O_2 and O_3 and the ions O_2^+ and O_3^-; simple molecules OR_2; simple molecules and ions containing the system O–O, viz.

hydrogen peroxide, H_2O_2,
oxygen fluorides,
the peroxide, O_2^{2-}, and superoxide, O_2^-, ions.

The oxygen molecule and dioxygenyl ion[1]

The magnetic moment of the O_2 molecule indicates the presence of two unpaired electrons. The valence bond treatment formulates the molecule with one electron pair and two three-electron bonds. The m.o. method, on the other hand, accounts for the paramagnetism by showing that one state of the molecule ($^3\Sigma$), with the spins of two electrons parallel, is more stable than the other possible state ($^1\Sigma$) with opposed spins, and the molecule is formulated with a double bond (O=O, 1·211 Å). The lengths of selected O–O bonds are given in Table 11.2.

The O_2 molecule can be ionized by PtF_6 to the dioxygenyl ion, O_2^+, which exists in O_2PtF_6, a red crystalline solid formed directly from equimolar quantities of

(1) IC 1969 8 828

417

TABLE 11.2

The lengths of O—O *bonds* (Å)

Molecule or ion	O—O	Reference
O_2^+	1·1227	GH 1950*
O_2	1·2107	PR 1953 **90** 537
O_3	1·278	JCP 1956 **24** 131
O_2^- (in α-KO_2)	1·28 (·02)	AC 1955 **8** 503
O_2^{2-} (in BaO_2)	1·49 (·04)	AC 1954 **7** 838
H_2O_2	$\begin{cases} 1·475 \\ 1·467 \end{cases}$	JCP 1962 **36** 1311 JCP 1965 **42** 3054

* G. Herzberg, Infra-red Spectra of Diatomic Molecules (2nd. ed.) Van Nostrand Co. N.Y.

oxygen gas and PtF_6 at 21°C. This salt exists in two crystalline forms both of which are structurally similar to salts such as $KPtF_6$ and $NO(OsF_6)$. The salts $O_2(BF_4)$ and $O_2(PF_6)$ have been prepared from O_2F_2 and BF_3 and PF_3 respectively. A careful n.d. study[2] of $O_2(PtF_6)$ has not provided a definite value for the O—O bond length; neither X-ray nor neutron diffraction data can at present distinguish between the expected values 1·12 Å for O_2^+ and 1·28 Å for O_2^-, but the formulation $O_2^-PtF_6^+$ is not consistent with known ionization energies.

A number of transition-metal complexes have been prepared which contain an O_2 group of some kind as a ligand; examples are shown below. The 'sideways' attachment of O_2 may be contrasted with the 'end-on' attachment of N_2 when that molecule functions as a ligand (p. 638). The short O—O bond length in the (diamagnetic) molecule (a), which is a reversible oxygen-carrier, may be contrasted

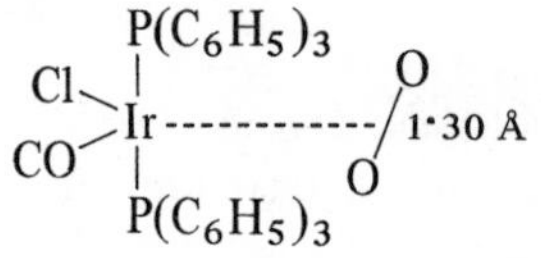

(2) JCP 1966 **44** 1748

(3) JACS 1965 **87** 2581
(4) JACS 1969 **91** 2123

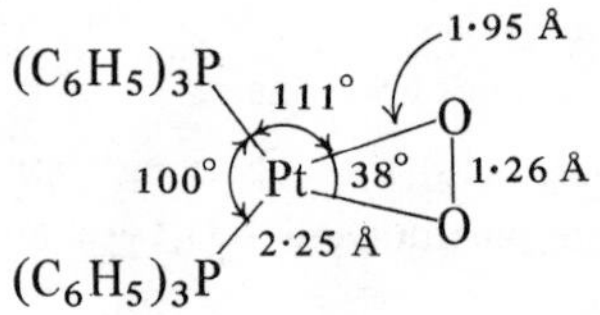

(a)[3]   (b)[4]

with the typical peroxy-bond in the iodine analogue, which does not take up oxygen reversibly (p. 423). Peroxo- and superoxo- derivatives of metals are discussed shortly.

Ozone and ozonates

Microwave studies of O_3 give an angular configuration for the molecule of ozone, which has a dipole moment of 0·49 D.

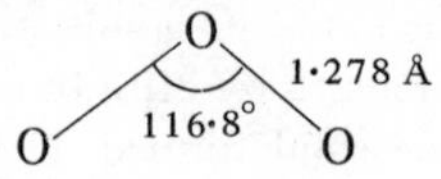

418

Red ozonates MO_3 have been prepared of Na, K, Rb, and Cs by the action of ozone on the dry powdered hydroxide at low temperatures, the ozonate being extracted by liquid NH_3. Lithium forms $Li(NH_3)_4O_3$ which apparently decomposes when the NH_3 is removed, and NH_4O_3 also has not been prepared free from excess NH_3; $N(CH_3)_4O_3$ has, however, been prepared.[1] An X-ray powder photographic study of KO_3, based on a small number of reflections, gave O–O, 1·19 Å and interbond angle, $100°$,[2] but a more detailed study of the O_3^- ion would be desirable.

(1) IC 1962 **1** 659; JACS 1962 **84** 34
(2) PNAS 1963 **49** 1

Peroxides, superoxides, and sesquioxides

All the alkali metals form peroxides M_2O_2, and peroxides MO_2 of Zn, Cd, Ca, Sr, and Ba are also known. Crystallographic studies of alkali-metal peroxides,[1] of ZnO_2 and CdO_2 (pyrites structure),[2] BaO_2, and $CaO_2 . 8 H_2O$ show that the crystals contain O_2^{2-} ions in which the O–O bond length is 1·49 Å (single bond).[3] For HgO_2 see pp. 223 and 924. Heating an alkali-metal peroxide in oxygen under pressure at $500°C$ or the action of oxygen on the metal dissolved in liquid NH_3 at $-78°C$ produces the superoxide MO_2. Superoxides are formed by all the alkali metals, but the only superoxides prepared in a pure state and known to be stable at room temperature are those of Na and the heavier alkali metals and $[(CH_3)_4N]O_2$.[4] LiO_2 has been prepared in an inert gas matrix by oxidation of an atomic Li beam by a mixture of oxygen and argon at $15°K$. The structure of the molecule is not known with certainty, but the shape is probably an isosceles triangle with the dimensions, O–O, 1·33 Å, and Li–O, 1·77 Å.[5]

Superoxides contain O_2^- ions in which the bond length (1·28 Å) corresponds to bond order 1·5, as in ozone (Table 11.2). They are paramagnetic with moments close to 2 BM, compare the theoretical value (1·73 BM) for one unpaired electron. Sodium superoxide is trimorphic. At temperatures below $-77°C$ it crystallizes with the marcasite structure (p. 203), above that temperature with the pyrites structure (p. 196), and above $-50°C$ the structure becomes disordered owing to rotation of the anions. At ordinary temperatures KO_2, RbO_2, and CsO_2 are isostructural, crystallizing with the tetragonal CaC_2 structure (p. 757), but KO_2 also has a high-temperature cubic form with the NaCl (or disordered pyrites) structure.[6]

Sesquioxides of Rb and Cs have been prepared as dark-coloured paramagnetic powders (contrast the yellow M_2O_2 and MO_2). They have been assigned the anti-Th_3P_4 structure, (p. 160), that is, they would be formulated $M_4(O_2)_3$.[7] The symmetry of this cubic structure implies equivalence of the three O_2 ions in $M_4(O_2^-)_2(O_2^{2-})$; it was not possible to determine the O–O bond length from the X-ray powder data. Further study of these compounds is desirable.

(1) ZaC 1962 **314** 12

(2) JACS 1959 **81** 3830
(3) AC 1954 **7** 838

(4) IC 1964 **3** 1798

(5) JCP 1969 **50** 4288

(6) AC 1952 **5** 851

(7) ZaC 1939 **242** 201

Molecules OR_2

Molecules of this type are angular, the bond angle usually lying in the range $100-110°$. For pure p bonds an angle of $90°$ would be expected, but partial use of the s orbital would lead to larger angles, and in the limiting cases $109\frac{1}{2}°$ for sp^3 or $120°$ for sp^2 bonds. When O is attached to two aromatic nuclei the angle is, in fact, close to 120°. In Table 11.3 we also include ethylene oxide, probably with 'bent' bonds as in cyclopropane. Much larger angles are found in $O(SiH_3)_2$, in pyro-

TABLE 11.3
Oxygen bond angles in simple molecules

Molecule	Oxygen bond angle	O—R (Å)	Method	Reference
$(CH_2)_2O$	61·6°	1·436	m.w.	JCP 1951 **19** 676
F_2O	103·3°	1·409	m.w.	JCP 1961 **35** 2240
H_2O	104·5	0·97	i.r.	JCP 1965 **42** 1147
OCl_2	110·9°	1·700	m.w.	JCS A 1966 336
$O(CH_3)_2$	111·5°	1·416	m.w.	JCP 1959 **30** 1096
$p\text{-}C_6H_4(OCH_3)_2$	121°	1·36	x	AC 1950 **3** 279
$O(SiH_3)_2$	144°	1·634	e.d.	ACSc 1963 **17** 2455
$O(GeH_3)_2$	126·5°	1·766	e.d.	JCS A 1970 315

ions and molecules X_2O_7, and also in meta-salts. Two collinear bonds are formed by O in the ReO_3 structure and in certain silicates (e.g. the $Si_2O_7^{6-}$ ion in $Sc_2Si_2O_7$) where the bonds presumably have considerable ionic character. (On the subject of apparently collinear O bonds in certain pyrophosphates see p. 689.) There are also collinear O bonds in Tc_2O_7 and in metallic oxo-compounds containing M—O—M bridges; in these cases the bond lengths indicate appreciable double bond character.

Hydrogen peroxide

The configurations of molecules R_1—O—O—R_2 (and the analogous compounds of S, Se, and Te) are of particular interest because wave-mechanical calculations indicate that the bond O—R_1 will not lie in the plane of —O—O—R_2 but that owing to the strong repulsion of the unshared (p_π) electrons of the O atoms the molecule will have the configuration shown at (a). The stereochemistry of molecules O_2R_2, S_2R_2, etc., is discussed further on p. 591 with the S compounds, of which more examples have been studied. For H_2O_2 the calculated values of θ and O are close to 100° if sp hybridization of the lone pairs is assumed, but a later treatment assuming sp^3 hybridization predicts a dihedral angle of 120°.

Many studies have been made of the structure of the H_2O_2 molecule giving values close to 1·47 Å for the O—O bond length, and O bond angle 103° (solid) or 98° (gas). However, the dihedral angle ranges from 90° in crystalline H_2O_2 to 180° in $Na_2C_2O_4 . H_2O_2$, where the H_2O_2 molecule is found to have a planar *trans* configuration ($\theta = 97°$). This large range of values for the dihedral angle indicates that the configuration of this molecule is very sensitive to its surroundings. Apparently the energy barrier corresponding to the *trans* configuration is only 3·8 kJ mol^{-1} above that of the equilibrium configuration (with $\phi = 111·5°$), as compared with 16·7 kJ mol^{-1} for the *cis* configuration. The value of ϕ would therefore easily be altered by hydrogen bonding, and also in $Na_2C_2O_4 . H_2O_2$ by the proximity of two Na^+ ions which complete a tetrahedral group around the O

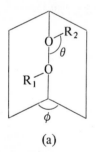

(a)

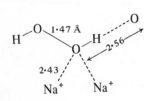

420

atom. Values of the dihedral angle found in organic peroxides also cover a wide range (Table 11.4).

TABLE 11.4
Dihedral angles ROO/OOR

Compound	Dihedral angle	Reference
H_2O_2 (solid, n.d.)	$90 \cdot 2°$	JCP 1965 **42** 3054
(gas, i.r.)	$111 \cdot 5°$	JCP 1965 **42** 1931
	$119 \cdot 8°$	JCP 1962 **36** 1311
$H_2O_2 . 2 H_2O$	$129°$	ACSc 1960 **14** 1325
$Na_2C_2O_4 . H_2O_2$	$180°$	ACSc 1964 **18** 1454
$Li_2C_2O_4 . H_2O_2$	$180°$	ACSc 1969 **23** 1871
$K_2C_2O_4 . H_2O_2$	$101 \cdot 6°$	ACSc 1967 **21** 779
$Rb_2C_2O_4 . H_2O_2$	$103 \cdot 4°$	ACSc 1967 **21** 779
O_2F_2	$87 \cdot 5°$	JCS 1962 4585
O_2R_2 (organic)	Values from 81 to 146°	AC 1967 **22** 281
		AC 1968 **B24** 277
Theoretical	$90 \text{--} 120°$	CJP 1962 **40** 765
		JCP 1962 **36** 1311
		AC 1967 **22** 281

Oxygen fluorides

The (angular) molecule OF_2 has already been mentioned. Mixtures of oxygen fluorides have been prepared by passage of $O_2 + F_2$ through an electric discharge and more recently by radiolysis of liquid mixtures of these elements at low temperatures with 3-MeV bremsstrahlung. The existence of all compounds O_nF_2 (n from 2 to 6) has been claimed but the latest data support the existence of only O_2F, O_2F_2, and O_4F_2.[1] It is thought that O_3F_2 and probably also O_5F_2 and O_6F_2 are mixtures. An infrared study[2] of O_2F trapped in solid oxygen indicates a bent molecule with bond lengths similar to those in O_2F_2 (below). The molecule O_2F_2 is of the same geometrical type as H_2O_2, with dihedral angle $87 \cdot 5°$ and bond angle $109 \cdot 5°$ but with very abnormal bond lengths: O–O, $1 \cdot 217$ Å and O–F, $1 \cdot 575$ Å.[3] The extreme length of the latter (compare $1 \cdot 40$ Å in OF_2) and the shortness of O–O (the same as in O_2) suggest an analogy with the nitrosyl halides (q.v.).

The preparation of organic compounds containing the system –O–O–O– has been described[4] (e.g. $F_3COOOCF_3$) but their structures are not known.

(1) JACS 1969 **91** 4702
(2) JCP 1966 **44** 3641

(3) JCS 1962 4585

(4) JACS 1966 **88** 3288, 4316

Per-acids of non-metals

These include compounds of the following types:

(1) Permono-acids presumably containing systems H–O–O–A (e.g. H_3PO_5, H_2SO_5). The constitution of perborates and percarbonates such as $LiBO_4 . H_2O$, $KBO_5 . H_2O$, $KHCO_4$, etc. is not known. These compounds are prepared by the action of H_2O_2 on the normal salts, and many are formulated as hydrated

(1) ACSc 1961 **15** 934

compounds. The demonstration that the sodium peroxoborate, originally written $NaBO_3 . 4 H_2O$ or $NaBO_2 . H_2O_2 . 3 H_2O$, is in fact $Na_2 [B_2(O_2)_2(OH)_4] . 6 H_2O^{(1)}$ containing the cyclic ion of Fig. 11.2(a), with O—O, 1·47 Å as in H_2O_2, suggests that speculation about the structures of these compounds is less profitable than X-ray crystallographic examination.

(2) Perdi-acids containing the grouping A—O—O—A include $H_2C_2O_6$, $H_4P_2O_8$, and $H_2S_2O_8$. The structure of the $S_2O_8^{2-}$ ion has been determined in the ammonium salt (Fig. 11.2(b)); the structures of other ions of this class have not been established.

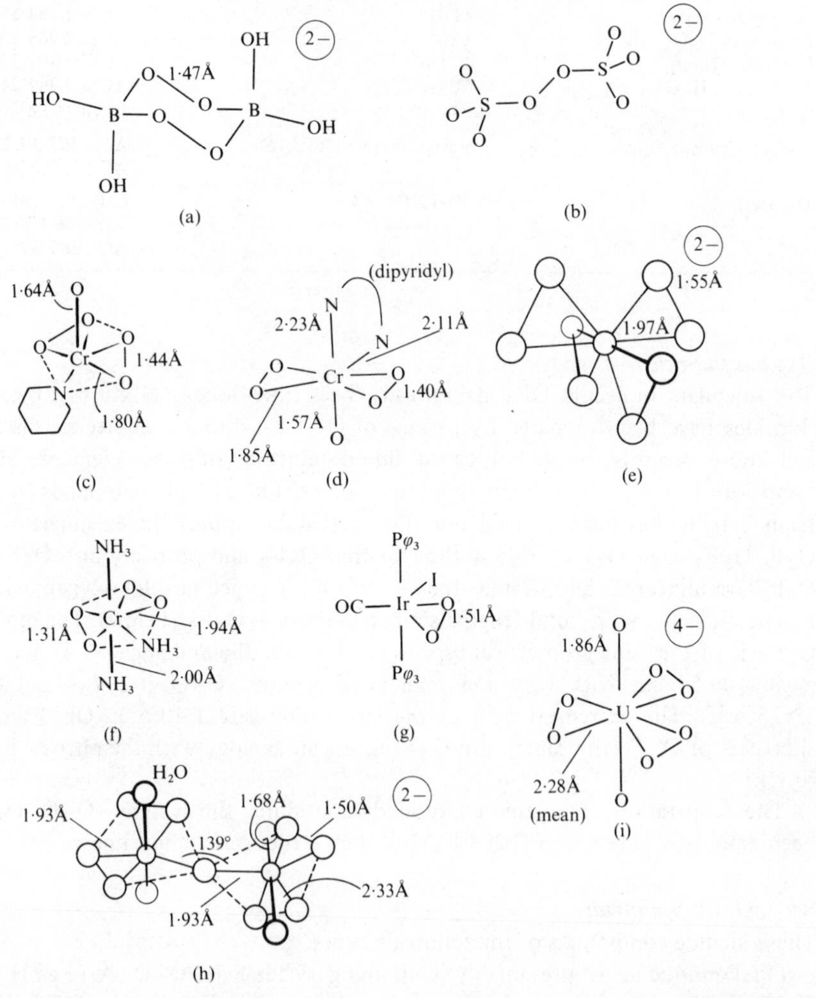

FIG. 11.2. Peroxo-ions and molecules: (a) $[B_2(O_2)_2(OH)_4]^{2-}$, (b) $S_2O_8^{2-}$, (c) $(C_5H_5N)CrO_5$, (d) (dipyridyl)$CrO(O_2)_2$, (e) $[Mo(O_2)_4]^{2-}$, (f) $(NH_3)_3Cr(O_2)_2$, (g) $Ir(P\phi_3)_2(CO)(O_2)I$, (h) $[W_2O_3(O_2)_4(H_2O)_2]^{2-}$, (i) $[UO_2(O_2)_3]^{4-}$.

Peroxo- and superoxo-derivatives of metals

Both the peroxo- and superoxo-ions can function as ligands with both oxygen atoms bonded to the same metal atom or as bridges between two metal atoms:

$$M{<}^{O}_{O} \quad \text{or} \quad M-O{^{O-M}}$$

Peroxo-ions or molecules in which O_2^{2-} ions are coordinated to one metal atom are formed by a number of transition metals, including Ti, Cr, Mo, W, Nb, Ta, Ir, and U.

The molecule $(C_5H_5N)CrO_5$ has the form of a distorted pentagonal pyramid (or trigonal pyramid if O_2 is counted as one ligand) as shown in Fig. 11.2(c), the two O atoms of each O_2 group being equidistant from the metal atom.[2] The blue form of $CrO(O_2)_2$ (dipyridyl) has a closely related structure (Fig. 11.2(d)) with ligand atoms at the vertices of a pentagonal bipyramid.[3] It has been suggested that the blue and violet perchromates are probably of this general type; for example, the explosive violet $KHCrO_6$ may be $K[Cr^{VI}O(O_2)_2OH]$. The red perchromates (and isostructural Nb and Ta compounds) such as K_3CrO_8 contain the ion $Cr^{V}(O_2)_4^{3-}$, in which four O_2^{2-} ions form a dodecahedral arrangement (as in $Mo(CN)_8^{4-}$) with apparently a slightly unsymmetrical orientation of the O_2 groups relative to the metal atom.[4] The paramagnetism of K_3CrO_8 indicates one unpaired electron, confirming its formulation as a peroxo-compound of Cr(V). The anion in the dark red $[Zn(NH_3)_4][Mo(O_2)_4]$[5] has a very similar structure (Fig. 11.2(e)) with all Mo–O, 1·97 Å, and O–O, 1·55 Å.

The action of H_2O_2 on $(NH_4)_2CrO_4$ in aqueous ammonia gives dark-reddish-brown crystals of $(NH_3)_3CrO_4$ which have a metallic lustre. They consist of pentagonal bipyramidal molecules (Fig. 11.2(f)) in which the four O atoms, one N, and Cr are coplanar (as also are Cr and the three N atoms).[6] As regards its stoichiometry this compound could be formulated as a Cr(II) superoxide or a Cr(IV) peroxide. The magnetic moment (indicating 2 unpaired electrons) and its chemistry suggest the latter formulation, for in the former Cr(II) would contribute 2 unpaired electrons and so also would the two O_2^- ions. The O–O bond is anomalously short, but the widely different lengths found for this bond in the tetraperoxo-ions (above) suggest that none of the O–O bond lengths is very accurately known. On the basis of O–O bond lengths the Ir iodo compound[7] of Fig. 11.2(g) is described as a peroxo-compound of Ir(II) and the structurally similar chloro compound[8] could be formally described as a superoxo-compound of Ir(II) (O–O, 1·30 Å), but there are then problems in accounting for its diamagnetism. For m.o. descriptions of the bonding in these complexes see references 7 and 8. The arrangement of ligands (counting O_2 as one ligand) in each of these molecules is trigonal bipyramidal.

The colourless tetraperoxoditungstate (VI), $K_2W_2O_{11}.4H_2O$, contains the binuclear ion of Fig. 11.2(h).[9] Around each W atom the configuration of ligands is pentagonal bipyramidal, the bond lengths suggesting double bonds to the single O atoms (compare W–O in $CaWO_4$, 1·78 Å). The bonds to the water molecules are obviously weak. The diamagnetic $K_2Mo_2O_{11}.4H_2O$[10] has a very similar

(2) ACSc 1963 **17** 557

(3) ACSc 1968 **22** 1439

(4) ACSc 1963 **17** 1563

(5) ACSc 1969 **23** 2755

(6) JPC 1959 **63** 1279; JCS 1962 2136

(7) IC 1967 6 2243
(8) JACS 1965 87 2581

(9) AC 1964 **17** 1127

(10) ACSc 1968 **22** 1076

1·94 Å ⋯⋯ O
Cr ⋯⋯ | 1·41 Å
1·85 Å ⋯⋯ O

(11) JCS A 1968 1588

structure. Figure 11.2(i) shows the structure of the anion in $Na_4[UO_2(O_2)_3]$. $9 H_2O^{(11)}$ in which three O_2 groups in the equatorial plane and the two uranyl O atoms form a trigonal (hexagonal) bipyramidal coordination group around U.

Niobium provides examples of peroxo-complexes in which respectively 2, 3, and 4 O_2^{2-} groups are attached to the metal atom: $(NH_4)_3[Nb(O_2)_2(C_2O_4)_2]$.

(12) AC 1971 **B27** 1572
(13) AC 1971 **B27** 1582
(14) AC 1971 **B27** 1598

$H_2O,^{(12)}$ $K[Nb(O_2)_3C_{12}H_8N_2].3 H_2O,^{(13)}$ and $KMg[Nb(O_2)_4].7 H_2O.^{(14)}$ In all of these complexes there is a dodecahedral arrangement of atoms bonded to Nb, these being 8 O atoms or 6 O and 2 N in the phenanthroline complex. The $Nb(O_2)_4^{3-}$ ion is similar in structure to the Cr and Mo ions, and in all these compounds O–O is close to 1·50 Å in the peroxo-ligands. In the dioxalato complex the two O_2 ligands occupy *cis* positions.

Cobalt provides examples of both peroxo- and superoxo-bridges between metal atoms. There are pairs of ions having the same chemical composition but different charges, one series of salts being greenish-black in colour and paramagnetic with moments corresponding to one unpaired electron, while the other salts are reddish-brown and diamagnetic:

(15) AC 1968 **B24** 246
(16) IC 1969 8 291
(17) IC 1968 7 725

(a) $[(NH_3)_5Co-O_2-Co(NH_3)_5]^{5+(15)}$ (c) $[(NH_3)_5Co-O_2-Co(NH_3)_5]^{4+(17)}$

(b) $[(NH_3)_4Co\underset{O_2}{\overset{NH_2}{<}}Co(NH_3)_4]^{4+(16)}$ (d) $[(NH_3)_4Co\underset{O_2}{\overset{NH_2}{<}}Co(NH_3)_4]^{3+}$

green—paramagnetic   red—diamagnetic

The paramagnetic compounds contain superoxo-bridges and e.s.r. data indicate that the odd electron spends equal times on both Co atoms, which are both Co(III). In (a) the O–O bond length (1·31 Å) is typical of a superoxo-compound, and the system Co–O–O–Co is coplanar, and in (b) the central 5-ring is planar. In the ion (c) the O–O bond is similar in length to that in H_2O_2 (1·47 Å), the O–O–Co angle is 113° and the dihedral angle is 146°. Note that in both these series of compounds the O_2 groups bridge by forming one bond from each O atom, as at (a), (b), and (c) below:

$$\left[(NH_3)_5Co\overset{O}{\underset{118°}{\diagdown}}\overset{1·32 Å}{\underset{O}{\diagup}}Co(NH_3)_5\right]^{5+}$$
(a)

$$\left[(NH_3)_5Co\overset{O\overset{1·47 Å}{———}O}{\diagdown\diagup}Co(NH_3)_5\right]^{4+}$$
dihedral angle 146°
(c)

$$\left[(NH_3)_4Co\overset{NH_2}{\underset{O-O}{<>}}Co(NH_3)_4\right]^{4+}$$
(b)

A different type of bridge apparently occurs in a red diamagnetic salt which was originally regarded as an isomer of the green $[(en)_2Co(NH_2)O_2Co(en)_2]$ $(NO_3)_4.H_2O$. The latter contains a normal superoxo-bridge of type (b), but the red salt is apparently$^{(18)}$

(18) JACS 1967 89 6364

$$\left[(en)_2Co \overset{\overset{\displaystyle NH_2}{79° \overset{1·95 Å}{\diagup\diagdown} }}{\underset{102° \underset{115° \overset{\displaystyle O}{\diagup\diagdown} \overset{1·42 Å}{|}}{ }}{\diagdown}} Co(en)_2 \right] (NO_3)_4 . 2 H_2O$$

with bond lengths 1·92 Å and OH

Metal oxo-ions and molecules

Finite ions and neutral molecules containing O directly bonded to a metal atom are formed by many transition metals. They include molecular oxyhalides and oxyhalide ions, which are included in Chapter 10, vanadyl and uranyl compounds, and numerous oxo-molecules and ions of metals such as Mo, Re, Os, etc. We give here a few examples of complexes of which the structures have been determined. The octahedral ion in $K_2[Os^{VI}O_2(OH)_4]$ has the same (*trans*) configuration as the anion in $K_2(OsO_2Cl_4)$. Rhenium forms many octahedral ions $(Re^V X_4OL)$ where X is halogen and L is H_2O, CH_3CN, etc.

In vanadyl compounds the characteristic coordination group of V^{IV} is a very distorted octahedron. The metal atom is displaced from the centre towards the O atom and the sixth ligand is very weakly bonded, (a), or in some cases not within bonding distance (b). The coordination is therefore more realistically described as tetragonal pyamidál; see also the structures of oxy-compounds of V^V (p. 467).

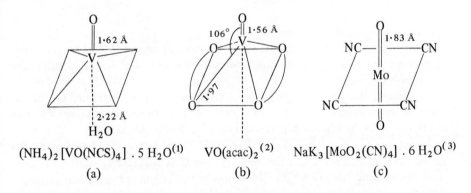

$(NH_4)_2[VO(NCS)_4] . 5 H_2O^{(1)}$ $VO(acac)_2^{(2)}$ $NaK_3[MoO_2(CN)_4] . 6 H_2O^{(3)}$

(a) (b) (c)

(1) JCS 1963 5745
(2) JCP 1961 **35** 55
(3) JACS 1968 **90** 3374

Molybdenum provides many examples of oxo-complexes. The ion $[Mo^{IV}O_2(CN)_4]^{4-}$, (c), in $NaK_3[MoO_2(CN)_4] . 6 H_2O$ is octahedral and not, as earlier supposed, an 8-coordinated Mo^{IV} complex, $[Mo(OH)_4(CN)_4]^{4-}$. Examples of Mo^{VI} complexes with two or three O atoms bonded to the metal include $Mo[(CH_3)_2NCHO]_2O_2Cl_2$, (d), and $MoO_3 . NH(CH_2CH_2NH_2)_2$, (e).

In (e) the MoO_3 portion of the molecule is very much like part of a tetrahedral MoO_4 group, having Mo–O, 1·74 Å, and O–Mo–O bond angles (106°) close to the tetrahedral value. The MoO_3 portions of the more complex ion (f) have the same bond lengths and interbond angles, as compared with Mo–O, 2·20 Å, and Mo–N, 2·40 Å, in the edta bridges.

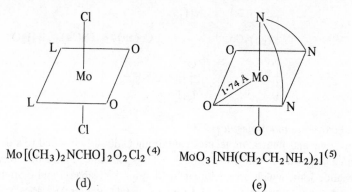

(4) IC 1968 7 722
(5) IC 1964 3 397

$Mo[(CH_3)_2NCHO]_2O_2Cl_2$ [4] $MoO_3[NH(CH_2CH_2NH_2)_2]$ [5]

(d)   (e)

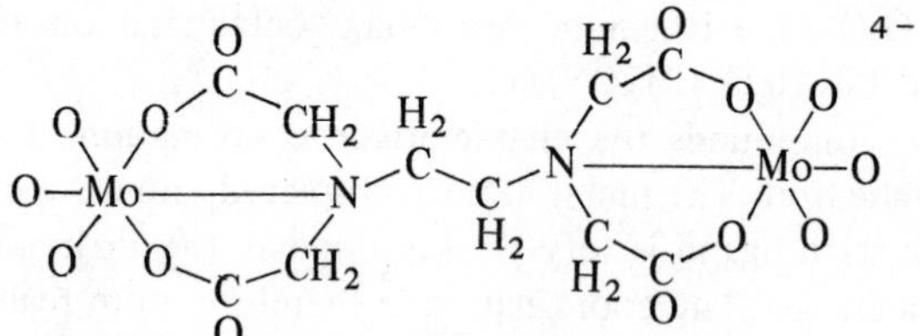

$[O_3Mo(edta)MoO_3]Na_4 \cdot 8H_2O$ [6]

(f)

(6) JACS 1969 91 301

Metal oxo-compounds containing M—O—M *bridges*

We noted earlier that two collinear bonds to metal atoms are formed by O in some finite ions and molecules. Examples include compounds of both transition and non-transition metals (Table 11.5). In all cases the short M—O bonds indicate multiple-bond character though there are still shorter bonds to terminal O atoms in some of the molybdenum complexes. Details of some of these complexes are given at (a)–(c). Note the trigonal bipyramidal coordination of Al in (a).

In the ion $[O-Re(CN)_4-O-Re(CN)_4-O]^{4-}$, which consists of two octahedra in the eclipsed configuration, the terminal Re—O is appreciably shorter (1·70 Å) than the bridging bond (1·92 Å). The cation in 'ruthenium red' is a linear system of three octahedral groups, $[(NH_3)_5Ru-O-Ru(NH_3)_4-O-Ru(NH_3)_5]^{6+}$ in which the mean oxidation state of the metal is 10/3. The mean length of the Ru—O bond has

been determined in the 'en' analogue of this ion. In double bridges $M\overset{O}{\underset{O}{<>}}M$ the O

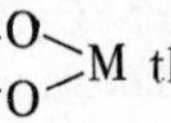

bonds are necessarily non-collinear. An example is the ion in $Ba[Mo_2O_4(C_2O_4)_2]$. $5H_2O$, (d), where the Mo—Mo bond length (2·54 Å) indicates sufficient interaction between the metal atoms to make the compound nearly diamagnetic (0·4 BM). Note the large range of Mo—O bond lengths in these compounds, ranging from the very short bonds (1·65 to 1·7 Å) to terminal O atoms, through values around 1·9 Å in the bridges to the long, presumably single, bonds of length 2·1 Å and above.

426

TABLE 11.5
Bridged metal-oxo-compounds

Compound	M—O (bridge)	Reference
Type M—O—M		
Al_2O (2-methyl 8-quinolinol)$_4$ (a)	1·68 Å	JACS 1970 **92** 91
$Cl_2(C_5H_5)Ti—O—Ti(C_5H_5)Cl_2$	1·78	JACS 1959 **81** 5510
$K_2[Ti_2O_5(C_7H_3O_4N)_2].5 H_2O$	1·83	IC 1970 **9** 2391
$Ti_2OCl_2(acac)_4.CHCl_3$	1·80	IC 1967 **6** 963
$[Fe_2O(H_2O)_2(C_{15}H_{21}N_5)_2](ClO_4)_4$	≈1·8	JACS 1967 **89** 720
(see p. 948)		
$Mo_2O_3(S_2COC_2H_5)_4$ (b)	1·86	JACS 1964 **86** 3024
$K_2[Mo_2O_5(C_2O_4)_2(H_2O)_2]$ (c)	1·88	IC 1964 **3** 1603
$Mo_2O_3[S_2P(OEt)_2]_4$	1·86	AC 1969 **B25** 2281
$Mn_2O(pyr)_2(phthalocyanin)_2$	1·71	IC 1967 **6** 1725
$(Cl_5Ru—O—RuCl_5)K_4.H_2O$	1·80	IC 1965 **4** 337
$(Cl_5Re—O—ReCl_5)K_4.H_2O$	1·86	AC 1962 **15** 851
$O_3Re—O—ReO_3(H_2O)_2$	1·80, 2·10	AnC 1968 **80** 286, 291
$[Fe_2O(HEDTA)_2]^{2-}$	1·79	IC 1967 **6** 1825
$[ORe(CN)_4—O—Re(CN)_4O]^{4-}$	1·92	IC 1971 **10** 2785
$[(NH_3)_5Ru—O—Ru(en)_2—O—Ru(NH_3)_5]^{6+}$	1·87	IC 1971 **10** 1943
Type M〈O/O〉M		
$Ba[Mo_2O_4(C_2O_4)_2].5 H_2O$ (d)	1·91	IC 1965 **4** 1377
$[Mo_2O_4(cysteine)_2]_2Na_2.5 H_2O$	1·93	AC 1969 **B25** 1857

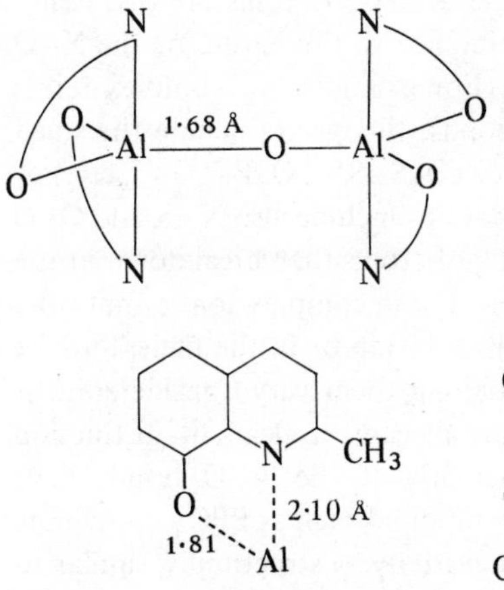

(a)

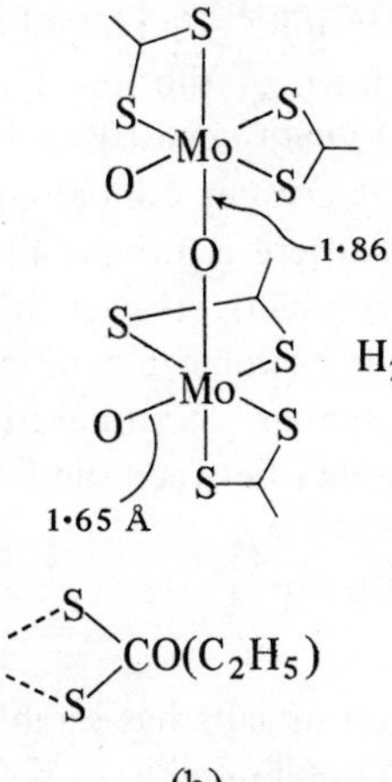

(b)

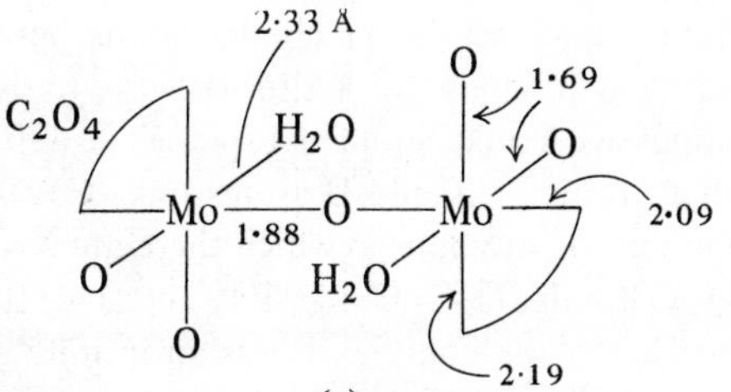

(c)

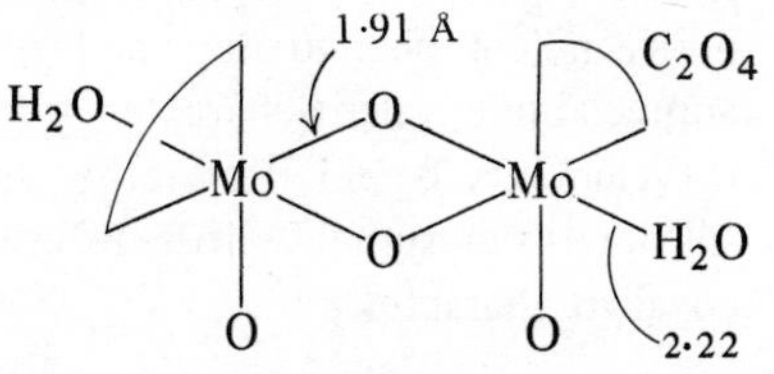

(d)

Oxygen

Oxy-ions

Oxygen accounts for some nine-tenths by volume of the earth's crust, and the greater part of inorganic chemistry is concerned with compounds which contain oxygen. In the mineral world pure oxides are rare. Compounds containing two or more elements in addition to oxygen may be roughly grouped into classes according as there is a small or large difference between the electronegativities of the elements. Since compounds $A_x X_y O_z$ containing two very electronegative elements A and X are not numerous, there are two main groups:

    (i)   A and X both comparatively electronegative,
    (ii)   A electropositive and X electronegative.

Compounds of the first group, *complex oxides*, may be regarded as assemblies of ions of two (or more) metals and O^{2-} ions. The numbers of oxygen ions surrounding the cations (their oxygen coordination numbers) are related to the sizes of the ions (Chapter 7). These coordination numbers are high (up to 12) for the largest ions, for example, Cs^+ and Ba^{2+}, and usually vary, within certain limits, in different structures. The crystal structures of complex oxides are described in Chapter 13.

Compounds of the second group are termed *oxy-salts* since they are formed by the combination of a basic oxide of the electropositive element A with the acidic oxide of the electronegative element X. The latter is usually a non-metal but may be a transition metal in a high oxidation state. Oxy-salts are assemblies of A ions and complex anions XO_n, the binding between the A ions and the oxygen atoms of the anions being essentially ionic in character. The complex ion is a charged group of atoms, which may be finite or extend indefinitely in one, two, or three dimensions, within which the bonds between the X and O atoms are essentially covalent in character. (Alternatively, to avoid reference to the nature of the X–O bonds we could adopt a topological definition. For example, the complex ion is that group of X and O atoms linked by X–O bonds, though to include the small number of oxy-ions in which there are X–X bonds ($O_3S–S^{2-}$, $O_3P–PO_3^{4-}$, etc.) or O–O bonds ($O_3S–O–O–SO_3^{2-}$, etc.) we should have to include also X–X and O–O bonds.) If the complex ion is an infinite grouping of atoms then breakdown of the crystal necessarily implies breakdown of the ion. If the complex ion is finite the same charged group of atoms XO_n *may* persist in solution or in the melt, but the stabilities of complex ions and of crystals containing them vary considerably. It should perhaps be remarked that there is not in all cases a clear-cut distinction between complex oxides and oxy-salts, particularly if the A–O bonds have appreciable covalent character. For example, a compound such as BPO_4 containing the elements B and P of rather similar electronegativity is structurally similar to silica, with presumably little difference between the B–O and P–O bonds as regards covalent character.

Types of complex oxy-ions

X-ray crystallographic examination of salts has established the structures of many mononuclear oxy-ions. There is a small number of polynuclear oxy-ions in which X or O atoms are directly bonded, as in the examples quoted above, but most polynuclear oxy-ions are formed from XO_n groups joined together by sharing O atoms. They are formed by many elements, notably by B, Si, P, Mo, and W, and to a much smaller extent by As, V, Nb, S, and Cr(VI). The possible types of complex oxy-ion may be derived by considering how XO_3 (planar or pyramidal), XO_4 (tetrahedral) or XO_6 (octahedral) groups, for example, can link up by sharing vertices (O atoms) or edges; sharing of faces of polyhedral coordination groups in complex oxy-ions is rarely observed. From the ortho-ion XO_n we derive in this way:

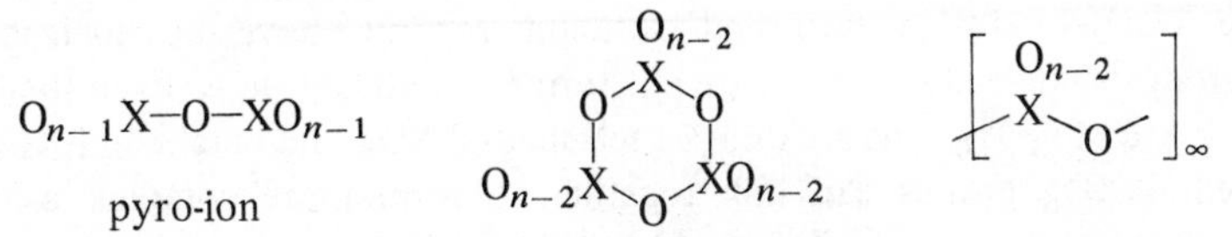

$$O_{n-1}X{-}O{-}XO_{n-1}$$

pyro-ion

and other cyclic ions

meta-ions

TABLE 11.6
Shapes of oxy-anions

Non-linear			NO_2^-	ClO_2^-
Planar	BO_3^{3-}	CO_3^{2-}	NO_3^-	
Pyramidal				SO_3^{2-} (in SO column) ClO_3^- BrO_3^- IO_3^-
Tetrahedral		SiO_4^{4-}	PO_4^{3-} AsO_4^{3-} VO_4^{3-}	SO_4^{2-} ClO_4^- IO_4^- MnO_4^-
Octahedral				TeO_6^{6-} IO_6^{5-}

Intermediate between the pyro-ion and the infinite linear meta-ion are the ions formed from three or more XO_n groups such as $S_3O_{10}^{2-}$ and $P_3O_{10}^{5-}$. The simplest ions arising from planar BO_3 groups are set out in Chapter 24 as possible complex borate ions. The oxygen chemistry of the elements Si, P, S, and Cl in their highest oxidation states is based largely on tetrahedral XO_4 groups; there are of course exceptions such as the planar SO_3 molecule in the vapour state. The types of oxy-ion that these elements can form may be summarized as follows. Simple pyro-ions are possible for Si, P, and S, the end member of the series being the neutral molecule Cl_2O_7. Phosphorus and silicon can form cyclic and linear meta-ions, and silicon also layer ions $(Si_2O_5)_n^{2n-}$. The number of complex oxy-ions containing silicon is greatly increased by the substitution of some Si by Al, making possible the formation of 3D framework ions. Accordingly the greatest variety of oxy-ions is formed by Si (see Chapter 23 for silicates and aluminosilicates).

429

Number of vertices shared		
1	Finite pyro-ions	$Si_2O_7^{6-}$, $P_2O_7^{4-}$, $S_2O_7^{2-}$
2	Closed rings or infinite chains	$(SiO_3)_n^{2n-}$, $(PO_3)_n^{n-}$
3	Infinite layer ions	$(Si_2O_5)_n^{2n-}$
4	Infinite 3D ions	Aluminosilicates

The relatively simple pyro- and meta-ions set out above, in which tetrahedral XO_4 groups share only vertices, are not formed by XO_6 groups. Even the $Mo_2O_7^{2-}$, $W_2O_7^{2-}$, and $Mo_3O_{10}^{2-}$ ions are built of tetrahedral XO_4 and octahedral XO_6 groups. Octahedral XO_6 groups can link together to form more complex oxy-ions by sharing both vertices and edges to form *isopoly* ions, or they may form a cluster around a central atom of a different element (*heteropoly* ion). These types of condensed polyoxy-ion are formed by Mo and W and to a lesser extent by V, Nb, and Ta. As regards their geometry these ions represent an extension of the relatively simple octahedral complexes described in Chapter 5 (Fig. 5.8), where they are listed in Table 5.3.

Another type of more complex finite oxy-ion is formed by the bonding of XO_n ions to a metal atom, either through X or through one or more of the O atoms. Examples are noted in other chapters and include $[Co(NO_2)_6]^{3-}$ and $[Co(NO_3)_4]^{2-}$.

Isopoly ions

Ions of V, Nb, *and* Ta

There is much experimental evidence for the formation of complex oxy-ions in solutions of vanadates, niobates, and tantalates. We describe in Chapters 12 and 13 the structures of some crystalline vanadates; here we note only certain finite complex ions which exist both in solution and in crystalline salts. The ion of Fig. 11.3(a) has been shown to exist in the salts $Na_7H(Nb_6O_{19})$. $15\,H_2O$.[1] Light scattering from an aqueous solution of $K_8(Ta_6O_{19})$. $16\,H_2O$ indicates that the anion species contains 6 Ta atoms and is presumably similar to the ion in the crystal.[2]

The existence of a decavanadate ion, previously proved to exist in solution, has been demonstrated in crystalline salts, at least two of which occur as minerals, $Ca_3V_{10}O_{28}$. $27\,H_2O$ and $K_2Mg_2V_{10}O_{28}$. $16\,H_2O$. Salts of this type are formed by evaporation of vanadate solutions having pH 2–6. The $V_{10}O_{28}^{6-}$ ion (Fig. 11.3(b)), studied in the Ca[3] and Zn[4] salts, contains ten V atoms situated at the vertices of a pair of octahedra which share an edge. As regards their geometry both of the ions of Fig. 11.3 may be described as portions of NaCl structure, substituting Nb or Ta for Na and O for Cl.

(1) AK 1953 **5** 247; ibid 1954 **7** 49

(2) IC 1963 **2** 985

(3) AC 1966 **21** 397
(4) IC 1966 **5** 967

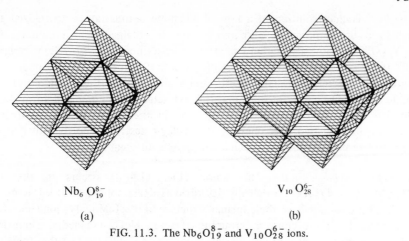

$Nb_6O_{19}^{8-}$ $V_{10}O_{28}^{6-}$

(a)   (b)

FIG. 11.3. The $Nb_6O_{19}^{8-}$ and $V_{10}O_{28}^{6-}$ ions.

Ions of Mo *and* W

Only the simplest molybdates and tungstates, namely the ortho salts A_nMoO_4 and A_nWO_4, are structurally similar to sulphates and chromates in that they contain tetrahedral XO_4 ions (e.g. $PbMoO_4$, $PbWO_4$, $CaMoO_4$, and $CaWO_4$ with the scheelite structure, Li_2MoO_4 and Li_2WO_4 with the phenacite structure). Unlike pyrosulphates and dichromates, which contain X_2O_7 ions consisting of two tetrahedral groups with a common vertex, the isostructural $Na_2Mo_2O_7$[1] and $Na_2W_2O_7$ contain infinite chain ions built from tetrahedral and octahedral groups—see Fig. 5.43 (p. 191). Again, unlike the trisulphate ion $S_3O_{10}^{2-}$, which is a linear system of three tetrahedral SO_4 groups, the anion in $K_2Mo_3O_{10}$[2] is also an infinite chain built from MoO_4 and MoO_6 groups—also illustrated in Fig. 5.43. In the tetrahedral MoO_4 groups in $Na_2Mo_2O_7$ Mo–O ranges only from 1·71 to 1·79 Å, but in $K_2Mo_3O_{10}$ these groups are apparently much less regular, with Mo–O from 1·64 to 1·95Å and a fifth O at only 2·08 Å; the description in terms of 5-coordinated Mo is probably more realistic (distorted tetragonal-pyramidal or trigonal bipyramidal). In the MoO_6 groups in both these molybdates Mo–O ranges from around 1·7 to values in the range 2·2–2·3 Å, and this wide variation is found in MoO_3 itself and in other polymolybdates.

In more complex polymolybdates Mo is found exclusively in octahedral coordination, and the same is true of W in polytungstates. The alkali di- and tri-molybdates and tungstates can be made by fusing together the oxides (or normal salt and trioxide) and crystallizing from the melt. The more complex isopoly acids and their salts are prepared from aqueous solution; they are invariably hydrated and contain complex ions containing at least six Mo or W atoms. The old nomenclature of some of these salts (meta- and para-salts), though still used, is unsatisfactory because paramolybdates and paratungstates contain different types of complex ion (a 7-ion and 12-ion respectively) and the same is probably true of the meta-salts. We note elsewhere that relatively few oxy-compounds of Mo^{VI} and W^{VI} are isostructural, and it would seem that this is true also of the polyacids and their salts.

(1) ACSc 1967 **21** 499

(2) ACSc 1967 **21** 499; JCS A 1968 1398

Oxygen

If the (alkaline) solution of a normal alkali-metal tungstate is neutralized and crystallized by evaporation, a 'paratungstate' is obtained, which contains the $(H_2W_{12}O_{42})^{10-}$ ion. By boiling a solution of a paratungstate with yellow 'tungstic acid' (which results from acidification of a hot solution of a normal or a paratungstate) a 'metatungstate' results, and acidification of the solution gives the soluble metatungstic acid, which may be extracted with ether and forms large colourless crystals. The metatungstates contain the $(H_2W_{12}O_{40})^{6-}$ ion. The structures of these two polytungstate ions will be described shortly, but first we deal with two polymolybdates which are of quite a different kind.

(3) JACS 1968 **90** 3275

The heptamolybdate ion. This anion (Fig. 11.4(a)) occurs in the salt $(NH_4)_6Mo_7O_{24} . 4H_2O$,[3] originally described as a paramolybdate. It is interesting that this ion does not have the coplanar structure of the $TeMo_6O_{24}^{6-}$ ion (see later) but is instead a more compact grouping of seven MoO_6 octahedra. There is a considerable range of Mo—O bond lengths, corresponding to the fact that there are metal atoms bonded to 1, 2, 3, and 4 oxygen atoms.

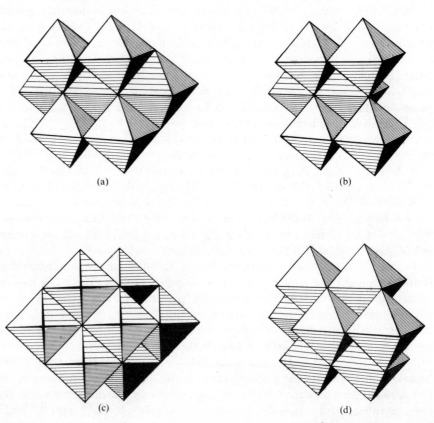

(a)　　　　　　　　　　(b)

(c)　　　　　　　　　　(d)

FIG. 11.4. (a) The $Mo_7O_{24}^{6-}$ ion. (c) and (d) Views of the $Mo_8O_{26}^{4-}$ ion. (b) The portion common to the two ions.

The octamolybdate ion. Salts with the ratio $MoO_3 : M_2O$ equal to 4 : 1 have been called tetramolybdates. An X-ray study of the ammonium salt[4] showed that it contains ions $Mo_8O_{24}^{4-}$ illustrated in Fig. 11.4(c) and (d); it should therefore be formulated $(NH_4)_4Mo_8O_{26} . 5 H_2O$. In this ion six of the MoO_6 octahedra are arranged in the same way as in the $Mo_7O_{24}^{6-}$ ion, as shown in Fig. 11.4(b).

(4) AK 1950 **2** 349

The paratungstate and metatungstate ions. The paratungstate ion has been studied in the salt $(NH_4)_{10}(H_2W_{12}O_{42}) . 10 H_2O$.[5] It is built from 12 WO_6 octahedra which are first arranged in four edge-sharing groups of three, two of type (a) and two of type (b), Fig. 11.5. These groups of three octahedra are then joined

(5) AC 1971 **B27** 1393

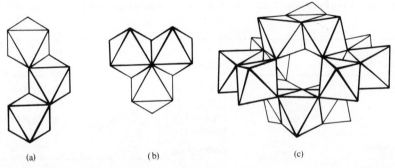

(a)　　　　(b)　　　　(c)

FIG. 11.5. (a) and (b) 3-octahedron sub-units from which the paratungstate ion (c), is built.

together by sharing vertices as indicated in the figure. In the resulting aggregate the 42 oxygen atoms are arranged approximately in positions of hexagonal closest packing. As in the polymolybdates there is a considerable range of metal–oxygen bond lengths, here approximately 1·8, 2·0, and 2·2 Å to O atoms bonded to 1, 2, and 3 W atoms respectively. It is interesting to compare with this ion the metatungstate ion, which is also built of 12 WO_6 octahedra but in this case arranged in four groups of type (b) to give an ion of composition $(H_2W_{12}O_{40})^{6-}$. This ion has the same basic structure as the heteropolyacid ion $(PW_{12}O_{40})^{3-}$ which is described in more detail later. The evidence for the structure of the metatungstate ion is indirect, for example, the fact that $H_8W_{12}O_{40} . 5 H_2O$ is isostructural with $H_3PW_{12}O_{40} . 5 H_2O$ and $Cs_3H_5W_{12}O_{40} . 2 H_2O$ with $Cs_3PW_{12}O_{40} . 2 H_2O$. There is still discussion of the number (probably 2) and mode of association of the protons with the $W_{12}O_{40}$ nucleus in metatungstates.

We noted in Chapter 5 that the very simple A_4X_{16} unit illustrated in Fig. 5.8(j) (p. 165) is not known as an ion in solution. However, a discrete W_4O_{16} unit with this configuration has been recognized in tungstates of the type $Li_{14}(WO_4)_3W_4O_{16} . 4 H_2O$[6] and the isostructural $(Li_{11}Fe)(WO_4)_3W_4O_{16}$,[7] in which it occurs together with tetrahedral WO_4^{2-} ions. There is, however, some disagreement as to whether one Li occupies the central tetrahedral cavity in the W_4O_{16} group or whether all the cations are situated between the $W_4O_{16}^{8-}$ and WO_4^{2-} ions in positions of octahedral coordination.

(6) BB 1966 **70** 598
(7) K 1968 **13** 980

Oxygen

Heteropoly ions

These ions consist of compact groupings of (usually) MoO_6 or WO_6 octahedra incorporating one or more atoms of a different element (hetero-atom), these atoms also being completely surrounded by O atoms of the octahedra. In some heteropoly ions the $MoO_6(WO_6)$ octahedra surround the hetero-atom(s) on all sides forming a globular cluster within which there are positions for the hetero-atom(s) surrounded by 4 (tetrahedral), 6 (octahedral), or 12 (icosahedral) O neighbours. These ions are distinguished by heavy type in the accompanying list. Other heteropoly ions do not have this globular shape, either because there are insufficient octahedra to surround the hetero-atom completely (as in $TeMo_6O_{24}^{6-}$) or because the ion is obviously composed of two portions linked by the hetero-atom (for example, $MnNb_{12}O_{38}^{12-}$).

Coordination of hetero-atom	Examples	
Tetrahedral	$PW_{12}O_{40}^{3-}$ etc.	$P_2W_{18}O_{62}^{6-}$
Octahedral	$TeMo_6O_{24}^{6-}$	
	$\mathbf{MnMo_9O_{32}^{6-}}$	$H_4Co_2Mo_{10}O_{38}^{6-}$
	$MnNb_{12}O_{38}^{12-}$	
Icosahedral	$\mathbf{CeMo_{12}O_{42}^{8-}}$	

Tetrahedral coordination of hetero-atom

This class includes the best known of the heteropolyacids, the large group of 12-acids and their salts which were the first to be studied by the X-ray method. These compounds contain ions $MW_{12}O_{40}^{n-}$ where M is P in the well known phosphotungstate ion but may be one of a variety of elements such as B, Al, Si, As, Fe, Cu, or Co(III). On long standing the 12-ions transform into more complex ions such as $P_2W_{18}O_{62}^{6-}$, and possibly others, though at present only the structure of the 2 : 18 ion is known. The following are typical of the numerous methods of preparing heteropoly acids. Ammonium phosphomolybdate is precipitated from a solution containing a phosphate, ammonium molybdate, and nitric acid. Alkali phosphotungstates result from boiling a solution of the alkali phosphate with WO_3. The heteropoly acids are very soluble in water and ether and crystallize very well, often with large numbers of molecules of water of crystallization. For example, hydrates of $H_3PW_{12}O_{40}$ are described with 5, 14, 21, 24, and 29 H_2O. These acids have the remarkable property of forming insoluble precipitates with many complex organic compounds such as dyes, albumen, and alkaloids. Both the acids and salts possess considerable stability towards acids, but are readily reduced by sulphur dioxide and other reducing agents. Like the acids many salts are highly hydrated as, for example, $K_4SiW_{12}O_{40} . 18 H_2O$. Caesium, however, forms only the salt $Cs_3HSiW_{12}O_{40} . 0$ or $2 H_2O$, no matter what proportions of caesium salt and acid are mixed. This salt has a structure similar to that of $H_4SiW_{12}O_{40} . 5 H_2O$, the positions of three of the water molecules being occupied by the large Cs^+ ions. The

434

acids $H_5BW_{12}O_{40}$, $H_6(H_2W_{12}O_{40})$, and $H_3PW_{12}O_{40}$ all give similar tri-caesium salts in spite of their different apparent 'basicities'. It would therefore appear that the primary factor determining the formula of a salt is the packing of the large complex ions rather than the number of H^+ ions in the acid.

The $PW_{12}O_{40}^{3-}$ *ion.* An X-ray study[1] of the pentahydrate of 12-phosphotungstic acid showed that the complex ion has the structure shown in Fig. 11.6(b). The five water molecules are arranged between these large negative ions so that the

(1) PRS 1936A **157** 113

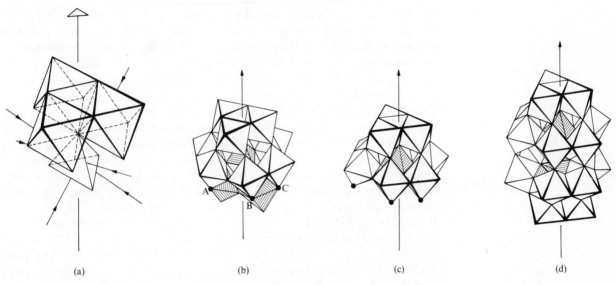

FIG. 11.6   (a) Arrangement of one group of three WO_6 octahedra relative to the central PO_4 tetrahedron in the $PW_{12}O_{40}^{3-}$ anion (b). (c) The 'half-unit'. (d) The $P_2W_{18}O_{62}^{6-}$ anion.

water molecules, and the three H^+ ions associated with them, are responsible for the cohesion of those ions. The pentahydrates of $H_5BW_{12}O_{40}$, $H_4SiW_{12}O_{40}$, and of metatungstic acid, $H_6(H_2W_{12}O_{40})$, are all isostructural with $H_3PW_{12}O_{40} \cdot 5\,H_2O$.

The $PW_{12}O_{40}^{3-}$ ion consists of a group of twelve WO_6 octahedra arranged around, and sharing O atoms with, a central PO_4 tetrahedron. The twelve WO_6 octahedra are arranged in four groups of three (of type (b) in Fig. 11.5) and within each group each octahedron shares two edges. These groups of three octahedra are then joined by sharing *vertices* to form the complete anion (Fig. 11.6(b)). Figure 11.6(a) shows the disposition of one group of three WO_6 octahedra relative to the central PO_4 tetrahedron. Each WO_6 octahedron shares two edges, one with each of the two neighbouring octahedra, and each O of the PO_4 tetrahedron belongs to three WO_6 octahedra. The formation of this ion from four similar sub-units, the groups of three edge-sharing octahedra, may be contrasted with the two types of sub-unit in the paratungstate ion described earlier.

435

Oxygen

It is noted in MSIC (pp. 90, 144) that there are three ways of joining together four edge-sharing groups of 3 octahedra (as in Fig. 11.5(b)) to form complexes with cubic symmetry:

	(a) $M_{12}X_{38}$	(b) $M_{12}X_{40}$	(c) $M_{12}X_{40}$
Shape of central cavity	Octahedral	Tetrahedral	Tetrahedral
Type of packing of X atoms	c.c.p.	c.c.p.	—

(2) AC 1960 **13** 1139

In (c) the 16 X atoms in the central nucleus are c.c.p., but on the four c.p. faces of this truncated tetrahedral group the next layers (of six atoms) are placed in positions of h.c.p. This is the structure of the $PW_{12}O_{40}^{3-}$ ion and also apparently of the metatungstate ion (p. 433). The $CoW_{12}O_{40}^{5-}$ ion has been shown to have the same structure.[2] In the $M_{12}X_{40}$ group (c) the M atoms are arranged at the vertices of a cuboctahedron. The group does not, however, have the full cubic symmetry of a regular cuboctahedron, for four of the triangular faces correspond to edge-sharing groups of three octahedra and the other four to vertex-sharing groups, with the result that the whole complex has only tetrahedral symmetry. The structures (a) and (b) are not known for heteropolyacid ions, but (b), which corresponds to a portion of the spinel structure, represents the structure of a hydroxy-complex formed in partially hydrolysed solutions of aluminium salts. It has been studied in $Na[Al_{13}O_4(OH)_{24}(H_2O)_{12}](SO_4)_4 . 13 H_2O$[2a] and $[Al_{13}O_4(OH)_{25}(H_2O)_{11}]$ $(SO_4)_3 . 16 H_2O$[2b] and is illustrated in Fig. 11.7(a). The central tetrahedral cavity and the centres of the twelve octahedra are all occupied by Al atoms. (For another view of this group of twelve octahedra see MSIC Fig. 107(b).) Figure 11.7(b) shows the $PW_{12}O_{40}^{3-}$ ion built from groups of three octahedra sharing additional vertices instead of edges, as in Fig. 11.7(a).

(2a) ACSc 1960 **14** 771
(2b) AK 1963 **20** 321

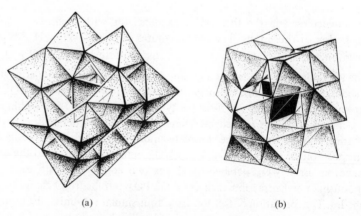

(a)   (b)

FIG. 11.7. (a) The c.c.p. $M_{12}X_{40}$ grouping of $[Al_{13}O_4(OH)_{24}(H_2O)_{12}]^{7+}$ and related ions. (b) The $PW_{12}O_{40}^{3-}$ ion.

436

The $P_2W_{18}O_{62}^{6-}$ *ion.*[3]   This ion is related in a very simple way to the 12-ion. Removal of three WO_6 octahedra from the base of the 12-ion leaves the 'half-unit' shown in Fig. 11.6(c). If two half-units are joined by sharing the exposed O atoms (three of which are shown as black dots in (c)) we obtain the $P_2W_{18}O_{62}^{6-}$ ion of Fig. 11.6(d). The upper half of the ion is related to the lower half by the plane of symmetry *ABC* in (b).

Octahedral coordination of hetero-atom

The $TeMo_6O_{24}^{6-}$ *ion.*   An example of the simplest ion of this type is the ion $TeMo_6O_{24}^{6-}$ (Fig. 11.8) which exists in salts such as $(NH_4)_6(TeMo_6O_{24}) . 7 H_2O$ and has been studied in detail in $(NH_4)_6TeMo_6O_{24} . Te(OH_6 . 7 H_2O$.[4] Crystals of this compound also contain octahedral $Te(OH)_6$ molecules (almost regular, with Te–OH, 1·91 Å). The atomic arrangement in this ion is clearly the same as in the CdI_2 layer, with some distortion from the idealized arrangement of Fig. 11.8 due to the different Mo–O and Te–O bond lengths. The heteropoly ion in $Na_3H_6Cr^{III}Mo_6O_{24} . 8 H_2O$ has the same structure (Cr–O, 1·975 Å)[4].

The $MnMo_9O_{32}^{6-}$ *ion.*   The structure of a 9-ion has been determined in $(NH_4)_6MnMo_9O_{32} . 8 H_2O$.[5] In this ion, illustrated in Fig. 11.9, nine MoO_6 octahedra are grouped around, and completely enclose, the Mn^{4+} ion.

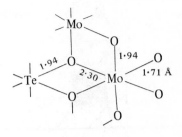

(3) AC 1953 **6** 113
(4) JACS 1968 **90** 3275
(5) AC 1954 **7** 438

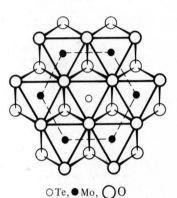

○ Te, ● Mo, ◯ O

FIG. 11.8. The $TeMo_6O_{24}^{6-}$ ion in $(NH_4)_6(TeMo_6O_{24}) . 7H_2O$, showing the mode of linking of the MoO_6 octahedra.

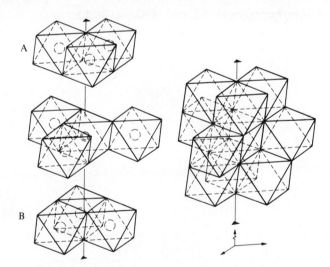

FIG. 11.9. The structure of the $MnMo_9O_{32}^{6-}$ anion with 'exploded' view at left. The single circles represent Mo and the double circle Mn.

The $MnNb_{12}O_{38}^{12-}$ *ion.*[6]   This ion may be constructed from the same edge-sharing groups of three octahedra from which $PW_{12}O_{40}^{3-}$ is built and which are also sub-units in $MnMo_9O_{32}^{6-}$. We may refer to the unit as A or as B if it is upside-down (Fig. 11.9). If B is placed on A there is formed an octahedral group of six octahedra which is the structure of the $Nb_6O_{19}^{8-}$ ion of Fig. 11.3(a).

(6) IC 1969 **8** 335

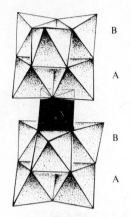

FIG. 11.10. The $[MnNb_{12}O_{38}]^{12-}$ anion.

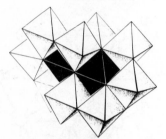

FIG. 11.11. The $[H_4Co_2Mo_{10}O_{38}]^{6-}$ anion. The black octahedra contain the Co atoms.

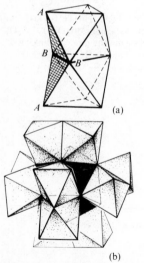

(a)

(b)

FIG. 11.12. The $[H_6CeMo_{12}O_{42}]^{2-}$ anion.

(7) JACS 1969 **91** 6881
(8) JACS 1968 **90** 3589

Oxygen

Two of these units joined together by a Mn atom form the anion in $Na_{12}Mn^{IV}Nb_{12}O_{38}$. H_2O, as shown in Fig. 11.10.

The $H_4Co_2Mo_{10}O_{38}^{6-}$ *ion.*[7] A model of this ion, studied in $(NH_4)_6(H_4Co_2Mo_{10}O_{38})$. $7H_2O$, is readily made from a chain of four edge-sharing octahedra by adding eight more octahedra, four in front and four behind, as shown in Fig. 11.11. Alternatively the ion may be constructed from two planar $CoMo_6$ units (similar to the $TeMo_6O_{24}^{6-}$ ion), from which one Mo is removed, and one unit is rotated relative to the other so that the Co atom of one unit occupies the position of a Mo atom in the other. The four H atoms were not located.

Icosahedral coordination of hetero-atom

The $CeMo_{12}O_{42}^{8-}$ *ion.*[8] Six pairs of face-sharing octahedra may be joined together by coalescing vertices of types *A* and *B* (Fig. 11.12(a)) of different pairs. The faces shaded in Fig. 11.12(a) then become twelve of the faces of an icosahedron and the hetero-atom occupies the nearly regular icosahedral hole at the centre of the group. The Ce–O distance (2·50 Å) is the same as in ceric ammonium nitrate, and in fact the 12 O atoms of the six bidentate nitrate groups in the ion $Ce(NO_3)_6^{2-}$ are arranged in the same way as the inner 12 O atoms of Fig. 11.12(b). There is a considerable range of Mo–O bond lengths. 1·68, 1·98, and 2·28 Å for bonds to O atoms belonging to 1, 2, or 3 MoO_6 octahedra.

438

12

Binary Metal Oxides

Introduction

We shall be largely concerned here with the structures of metal oxides in the crystalline state since nearly all these compounds are solids at ordinary temperatures. We shall mention a number of suboxides, but we shall exclude peroxides and superoxides (and ozonates), for these compounds, in which there are O–O bonds, are included in Chapter 11. Little is known of the structures of metal oxides in the liquid or vapour states, though several have been studied as vapours (Table 12.1). The structures of the oxides of the semi-metals and of the B subgroup elements are described in other chapters.

TABLE 12.1

Metal oxides in the vapour state

Oxide	M–O (Å)	μ(D)	M–O–M	*Method*	*Reference*
Li_2O	1·59		180°	m.sp., i.r.	JCP 1963 **39** 2463
	1·55			m.sp., i.r.	JCP 1963 **39** 2299
LiO	1·62			m.sp., i.r.	JCP 1963 **39** 2463
Li_2O_2	1·90		64°		JCP 1967 **46** 605
Cs_2O			$\neq 180°$	m.sp.	JCP 1965 **43** 943
SrO	1·92	8·91		m.sp.	
BaO	1·94	7·93		m.w.	JCP 1963 **38** 2705
MoO_3	See p.474				
OsO_4	1·71			e.d.	ACSc 1966 **20** 385
RuO_4	1·705			e.d.	ACSc 1967 **21** 737

Metal oxides range from the 'suboxides' of Cs and of transition metals such as Ti and Cr, in which there is direct contact between metal atoms, and the ionic compounds of the earlier A subgroups and of transition metals in their lower oxidation states, to the covalent oxides of Cr(vi), Mn(vii), and Ru(viii). In contrast to the basic oxides of the most electropositive metals there are the acidic CrO_3 and Mn_2O_7 which are the anhydrides of H_2CrO_4 and $HMnO_4$. As regards physical properties metal oxides range from high melting ionic compounds (HfO_2 melts at 2800°C) to the volatile molecular RuO_4 (m.p. 25°, b.p. 100°C) and OsO_4 (m.p. 40°, b.p. 101°C), both of which are soluble in CCl_4 and H_2O, and Mn_2O_7 (m.p. 6°C). The colours of metal oxides range over the whole spectrum, and their electrical properties show an equally wide variation, from insulators

439

(Na_2O, CaO), through semiconductors (VO_2), to 'metallic' conductors (CrO_2, RuO_2). Binary oxides also include two of the rare examples of ferromagnetic compounds, EuO and CrO_2.

The crystal structures of metallic oxides include examples of all four main types, molecular, chain, layer, and 3D structures, though numerically the first three classes form a negligible fraction of the total number of oxides. The metals forming oxides with molecular, chain, or layer structures are distributed in an interesting way over the Periodic Table.

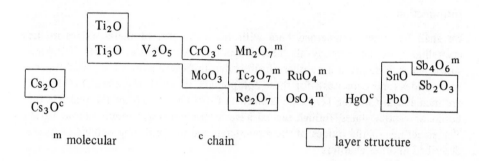

^{m} molecular ^{c} chain   [ ] layer structure

The following generalizations may be found useful, but it should be remembered that like all generalizations they are subject to many exceptions, some of which will be noted later.

(1) The majority of metal oxides have essentially ionic structures with high coordination number of the metal atom (often 6 or 8), the structures being in many cases similar to those of fluorides of the same formula type. However, the adoption of a simple structure characteristic of ionic compounds does not necessarily preclude some degree of covalent or metallic bonding—witness the NaCl structure not only of UO but also of UC and UN.

(2) The ionic radii of metals other than K^+, Rb^+, Cs^+, Ba^{2+}, and Tl^+ are smaller than that of O^{2-}. For example, the ionic radii of Al, Mg, and all the 3d metals lie in the range 0.5–0.8 Å. For this reason the oxygen ions in many oxides, both simple and complex, are close-packed or approximately so, with the smaller metal ions usually in octahedral holes.

(3) Comparison of Tables 12.2 and 17.1 shows that there is little resemblance between the structures of the corresponding oxides and sulphides of a particular metal except in the case of the ionic compounds of the most electropositive metals and the compounds of Be and Zn. In some cases oxides and sulphides of the same formula type do not exist or have very different stabilities. For example, PbO_2 is stable at atmospheric pressure but PbS_2 can only be made under higher pressures. In the case of iron the disulphide FeS_2 has the pyrites and marcasite structures, both containing S_2 groups, but FeO_2 is not known. A comparison of the highest known oxides of the Group VIII metals emphasizes the individuality of the

TABLE 12.2
The crystal structures of metal oxides

Type of structure	Formula type and coordination numbers of M and O	Name of structure	Examples
Infinite 3-dimensional complex	MO_3 6 : 2	ReO_3	WO_3
	MO_2 8 : 4	Fluorite	ZrO_2, HfO_2, CeO_2, ThO_2, UO_2, NpO_2, PuO_2, AmO_2, CmO_2, PoO_2, PrO_2, TbO_2
	6 : 3	Rutile	TiO_2, VO_2, NbO_2, TaO_2, CrO_2, MoO_2, WO_2, MnO_2, TcO_2, ReO_2, RuO_2, OsO_2, RhO_2, IrO_2, PtO_2, GeO_2, SnO_2, PbO_2, TeO_2
	M_2O_3 6 : 4	Corundum	Al_2O_3, Ti_2O_3, V_2O_3, Cr_2O_3, Fe_2O_3, Rh_2O_3, Ga_2O_3
	See text	$\begin{cases} A\text{-}M_2O_3 \\ B\text{-}M_2O_3 \end{cases}$	4f oxides
	6 : 4	$C\text{-}M_2O_3$	Mn_2O_3, Sc_2O_3, Y_2O_3, In_2O_3, Tl_2O_3
	MO 6 : 6	Sodium chloride	MgO, CaO, SrO, BaO, TiO, VO, MnO, FeO, CoO, NiO, CdO, EuO,
	4 : 4	Wurtzite	BeO, ZnO
	MO_2 4 : 2	Silica structures	GeO_2
	M_2O 2 : 4	Cuprite	Cu_2O, Ag_2O
	4 : 8	Antifluorite	Li_2O, Na_2O, K_2O, Rb_2O
Layer structures			MoO_3, As_2O_3, PbO, SnO, Re_2O_7
Chain structures			HgO, SeO_2, CrO_3, Sb_2O_3
Molecular structures			RuO_4, OsO_4, Tc_2O_7, Sb_4O_6

Note: this table does not distinguish between the most symmetrical form of a structure and distorted variants, superstructures, or defect structures; for more details the text should be consulted.

elements and the difficulty of making generalizations:

Fe_2O_3	Co_3O_4	NiO
RuO_4	Rh_2O_3	PdO
OsO_4	IrO_2	PtO_2

(4) In contrast to earlier views on metal oxides, detailed diffraction studies and measurements of density and other properties have shown that many metal-oxygen systems are complex. In addition to oxides with simple formulae such as M_2O, MO, M_2O_3, MO_2, etc. containing an element in one oxidation state and the well-known oxides M_3O_4 containing M((II) and M(III) there are numerous examples of more

441

complex stoichiometric oxides containing a metal in two oxidation states, for example,

$$U_3O_8 \qquad U_4O_9 \qquad Mn_5O_8 \qquad V_6O_{13} \qquad Tb_7O_{12}$$
$$Cr_5O_{12} \qquad Pr_6O_{11}.$$

Some metals form extensive series of oxides such as Ti_nO_{2n-1} or Mo_nO_{3n-1} with structures related to simple oxides MO_2 or MO_3. Many transition-metal oxides show departures from stoichiometry leading to semiconductivity and others have interesting magnetic and electrical properties which have been much studied in recent years. We shall illustrate some of these features of oxides by dealing in some detail with selected metal–oxygen systems and by noting peculiarities of certain oxides.

The structures of binary oxides

The more important structure types are set out in Table 12.2; others are mentioned later. Most of the simpler structures have already been described and illustrated in earlier chapters. In Table 12.2 the structures are arranged according to the type of complex in the crystal (3-, 2-, or 1-dimensional or finite) and for each structure the coordination number of the metal atom is given before that of oxygen. In a particular family of oxides the changes in structure type may be related in a general way to the change in bond type from the essentially ionic structures through the layer and chain structures to the essentially covalent molecular oxides. These changes may be illustrated by the structures of the dioxides of the elements of the fourth Periodic Group. After each compound we give the c.n.'s of M and O and the structure type. Starting with the molecular CO_2 we pass through the silica structures with ionic–covalent bonds to the ionic structures of the later metal dioxides.

$$CO_2 \ (2:1) \ \text{Molecular}$$
$$SiO_2 \ (4:2) \ \text{various 3D networks}$$

TiO_2	$(6:3)$ rutile, etc.	GeO_2 $\begin{cases}(4:2)\\(6:3)\end{cases}$	quartz rutile
ZrO_2 HfO_2 ThO_2 $\Big\}$	$(8:4)$ fluorite	$SnO_2 \quad (6:3)$	rutile
		$PbO_2 \begin{cases}(6:3)\\(6:3)\end{cases}$	rutile columbite

The structures of these compounds may be compared with those of the corresponding disulphides, which are set out in a similar way at the beginning of Chapter 17.

After noting some suboxides which are peculiar to Rb and Cs we shall describe the crystal structures of metal oxides in groups in the following order: M_3O, M_2O, MO, MO_2, MO_3, MO_4; M_2O_3, M_2O_5, M_2O_7, M_3O_4, and miscellaneous oxides M_xO_y. The remainder of the chapter is devoted to brief surveys of the oxides of certain metals which present points of special interest.

Suboxides of Rb *and* Cs

The elements Rb and Cs are notable for forming a number of suboxides including Rb_6O and Rb_9O_2,[1] Cs_7O,[2] Cs_4O, Cs_7O_2, and Cs_3O. Both of the rubidium oxides contain Rb_9O_2 groups consisting of two ORb_6 octahedra sharing one face (that is, an anti-M_2X_9 group). In Rb_6O layers of these groups alternate with layers of c.p. metal atoms, so that the structural formula is $(Rb_9O_2)Rb_3$. The oxide Rb_9O_2, which is copper-red with metallic lustre, consists of these Rb_9O_2 groups held together by Rb—Rb bonds of lengths 5·11 Å, rather longer than those in the element (4·85 Å) and much longer than the shortest Rb—Rb bonds within the groups (3·52 Å).

The bronze coloured Cs_7O contains units of a different kind, consisting of three octahedral OCs_6 groups each sharing two adjacent faces to form a trigonal group (Fig. 12.1) similar to the sub-unit in Nb_3S_4 (p. 623). These units are surrounded by and bonded to other similar units through Cs atoms (10 per unit) giving the formula $(Cs_{11}O_3)Cs_{10}$, or Cs_7O. The Cs—Cs bonds within the units are much shorter (3·76 Å) than those between the units (5·27 Å), the latter having the same length as in metallic Cs.

Oxides M_3O

Two structures established for suboxides M_3O are 'anti' structures of halides MX_3. In Cs_3O,[3] which forms dark-green crystals with metallic lustre, there are columns of composition Cs_3O consisting of OCs_6 octahedra each sharing a pair of opposite faces (Fig. 12.2); this is the anti-ZrI_3 structure. The Cs—O bond length (2·89 Å) is similar to that in Cs_2O. The Cs—Cs bonds between the chains (around 5·8 Å) are

(1) NW 1971 **58** 623
(2) NW 1971 **58** 622

(3) JPC 1956 **60** 345

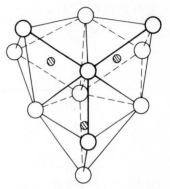

FIG. 12.1. The O_3Cs_{11} unit in Cs_7O.

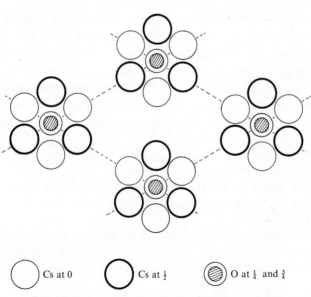

$\bigcirc$ Cs at 0	$\bigcirc$ Cs at $\frac{1}{2}$	$\oslash$ O at $\frac{1}{4}$ and $\frac{3}{4}$

FIG. 12.2. Projection of the crystal structure of Cs_3O.

somewhat longer than in the element (5·26 Å), but in the distorted octahedra in the chains the lengths of the edges are 3·80 Å (six) and 4·34 Å (six). The former Cs—Cs distance (omitted from ref. 3) is similar to that in Cs_7O.

(4) AC 1968 **B24** 211

The oxide $Ti_3O^{(4)}$ has a structure closely related to the anti-BiI_3 layer structure which has been described in Chapter 4.

For Cr_3O, Mo_3O, and W_3O see p. 473.

Oxides M_2O

The structure adopted by the alkali-metal oxides M_2O and sulphides M_2S *other than those of Cs* is the anti-fluorite structure, with 4 : 8 coordination. In this structure the alkali-metal ions occupy the F^- positions and the anions the Ca^{2+} positions of the CaF_2 structure. We have commented in Chapter 7 on the structures of Cs compounds. The structure of Cs_2S is not known, and the structure of the orange-yellow Cs_2O is quite different from that of the other alkali-metal oxides and

(1) JPC 1956 **60** 338

more closely related to Cs_3O. This oxide has the anti-$CdCl_2$ (layer) structure,[1] with Cs—Cs distances of 4·19 Å between layers and Cs—O distances (2·86 Å) within the layers, similar to those in Cs_3O. The corresponding sums of Pauling ionic radii are 3·38 and 3·09 Å. There is considerable polarization of the large Cs^+ ions.

The crystal structure of Cu_2O (and the isostructural Ag_2O) has been described in Chapter 3 together with other structures related to the diamond net.

(2) ACSc 1966 **20** 1996
(3) IC 1968 7 660

Evidence for the existence of In_2O in the solid state is not satisfactory,[2] and the compound described as Sm_2O is almost certainly SmH_2.[3]

The anti-CdI_2 structure of Ti_2O is mentioned in the later section on 'Oxides of titanium'.

Oxides MO

The structures of metallic monoxides are set out in the self-explanatory Table 12.3. The symbols mean that the structure of a particular compound is of the general type indicated and is not necessarily the ideal structure with maximum symmetry. An interesting point about the wurtzite structure is that unless $4a^2 = (12u - 3)c^2$, where u is the z coordinate of O referred to Zn at (000), there is a small but real difference between one M—O distance and the other three, for example, in ZnO,

(1) AC 1969 **B25** 1233, 2254

1·973 Å (3) and 1·992 Å (1).[1] For the ideal structure with regular tetrahedral bonds $u = 3/8$, and $c/a = 2\sqrt{2}/\sqrt{3} = 1·633$.

(2) IC 1968 7 660
(3) IC 1970 9 851

Note that Eu is probably the only lanthanide which forms a monoxide; 'SmO' is Sm_2ON,[2] and 'YbO' is Yb_2OC.[3]

Whereas the oxides of Mg, the alkaline-earths, and Cd have the normal NaCl structure those of certain transition metals depart from the ideal structure in the following ways.

(a) The structure has full (cubic) symmetry only at higher temperatures, transition to a less symmetrical variant taking place at lower temperatures, namely, FeO, NiO, MnO (rhombohedral), and CoO (tetragonal). For references see PR 1951 **82** 113.

(b) The oxide has a defect structure, that is, there is not 100 per cent

TABLE 12.3

Crystal structures of monoxides

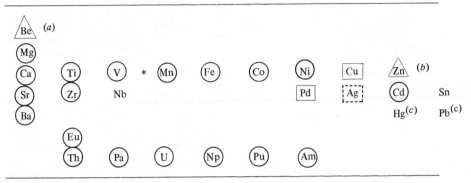

$\triangle$ Wurtzite structure. $\bigcirc$ NaCl structure. $\square$ PtS structure.
(*a*) Also a high-temperature polymorph.
(*b*) NaCl structure at high pressure [4].
(*c*) Polymorphic.
* There is no reliable evidence for the existence of CrO.

(4) Sc 1962 **137** 993

occupancy of both anion and cation sites. At atmospheric pressure FeO is non-stoichiometric (see p. 456). Stoichiometric NiO is pale-green (like Ni^{2+} in aqueous solution) and is an insulator. By heating in O_2 or incorporating a little Li_2O it becomes grey-black and is then a semiconductor (due to the presence of Ni^{2+} and Ni^{3+} in the same crystal).[5]

(5) JPC 1961 **65** 2154

The oxides TiO, NbO, and VO show appreciable ranges of composition, having either a superstructure or a defect structure at the composition MO. The oxides of Ti and V are discussed later in this chapter. The structure of NbO is described shortly.

The larger unit cells necessary to account for the neutron diffraction data from antiferromagnetic compounds such as MnO are due to the antiparallel alignment of the magnetic moments of the metal ions. They are not to be confused with the larger unit cells of many superstructures which are due to ordered as opposed to random arrangement of atoms of two or more kinds.

Table 12.3 shows that the majority of monoxides adopt the NaCl structure. Other structures are peculiar to one oxide or to a small number of oxides; they include the following.

BeO. In the high-temperature (β) form of BeO[6] pairs of BeO_4 tetrahedra share an edge and these pairs are then joined through vertices (Fig. 12.3). The arrangement of O atoms is essentially the same as in rutile (see Fig. 6.5) and since β-BeO is approximately h.c.p. it is closely related to the wurtzite structure of low-temperature BeO. One structure is converted into the other by moving one-half of the Be atoms into adjacent tetrahedral holes. Owing to the edge-sharing by BeO_4 tetrahedra in β-BeO there are necessarily short Be—Be distances (2·24 Å), the same as in metallic beryllium, but this does not imply metal–metal bonding. γ-$LiAlO_2$ has the same structure.

(6) AC 1965 **18** 393

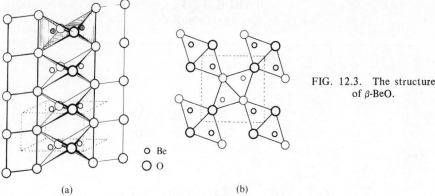

FIG. 12.3. The structure of β-BeO.

O Be
O O

(a)   (b)

7) AC 1966 **21** 843

NbO.   This oxide is unique in having the structure of Fig. 6.1(d) (p. 193)[7] in which both Nb and O form four coplanar bonds, occupying alternate positions in one of the simplest 3D 4-connected nets. Alternatively the structure is described as a defect NaCl structure, having 3 NbO in the unit cell with vacancies at (000) and $(\frac{1}{2}\frac{1}{2}\frac{1}{2})$. Note the existence of a 3D framework built from octahedral Nb_6 units (Nb–Nb, 2·98 Å) reminiscent of the halide complexes of Nb and Ta (p. 367). This structure has been confirmed by neutron diffraction.

PdO.   In this structure also (which is that of PtS) the metal atoms form four coplanar bonds; the coordination of O is tetrahedral (Fig. 12.4). As noted in Chapter 1 the bond angles represent a compromise between the ideal values of 90° and $109\frac{1}{2}$° due to purely geometrical factors. CuO has a less symmetrical variant of the PtS structure, and it is interesting that whereas CuO has the less symmetrical

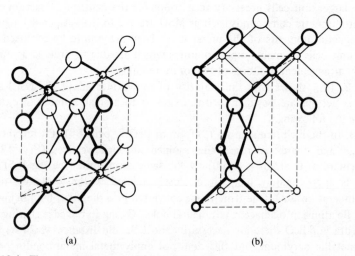

(a)   (b)

FIG. 12.4. The crystal structures of (a) CuO (tenorite) and AgO, and (b) PdO and PtO.

(monoclinic) structure and PtO apparently does not exist yet solid solutions $Cu_xPt_{1-x}O$ may be prepared (under high pressure) which have the tetragonal PtS structure $(0·865 > x > 0·645)$.[8] The structures of CuO and AgO are described in more detail in Chapter 25.

For the structures of the other B subgroup monoxides see Chapters 25 and 26.

Oxides MO_2

Many dioxides crystallize with one of two simple structures, the larger M^{4+} ions being 8-coordinated in the fluorite structure and the smaller ions 6-coordinated in the rutile structure. The term 'rutile-type' structure in Table 12.4 includes the most symmetrical (tetragonal) form of this structure and the less symmetrical variants referred to later—as indicated by the broken lines in the Table. Polymorphism is common; for example, PbO_2 and ReO_2 also crystallize with the columbite structure (p. 147), and GeO_2 with the α-quartz structure, while TiO_2 and ZrO_2 are trimorphic at atmospheric pressure. TiO_2 also adopts the columbite structure under pressure, and under shock-wave pressures greater than 150 kbar it converts to a still more dense form which is not, however, like the columbite structure, retained after release of pressure.[1] PtO_2 is a high-pressure compound with two forms, α (hexagonal, structure not known) and β with the $CaCl_2$ structure.[2] This is the only dioxide known to have this structure, which is closer to the ideal h.c.p. structure than is the tetragonal rutile structure (see Chapters 4 and 6).

Although the coordination group around the metal ion in the tetragonal rutile structure approximates closely to a regular octahedron accurate determinations of the M—O distances show small differences. For example, in TiO_2,[3] Ti—4 O, 1·944 (0·004) and Ti—2 O, 1·988 (0·006) Å, and in RuO_2,[4] Ru—4 O, 1·917 (0·008) and Ru—2 O, 1·999 (0·008) Å. This effect is not confined to the rutile structure for in anatase[5] there are comparable differences: Ti—4 O, 1·937 (0·003), and Ti—2 O, 1·964 (0·009) Å, while in brookite[6] there is a much less symmetrical environment of Ti^{4+} with Ti—O ranging from 1·87 to 2·04 Å. It follows from our discussion of the geometry of octahedral structures (Chapter 5) that the significance of these differences in bond lengths is by no means obvious.

(8) JLCM 1969 **19** 209

(1) JCP 1969 **50** 319
(2) JINC 1969 **31** 3803

(3) AC 1956 $\overline{9}$ 515
(4) IC 1966 **5** 317

(5) JACS 1955 $\overline{77}$ 4708
(6) AC 1961 **14** 209

TABLE 12.4

Crystal structures of dioxides

Ti*	V*	Cr	Mn*			rutile-type structure	Ge*	
Zr*	Nb	Mo	Tc	Ru	Rh†		Sn	
Hf*	Ta	W	Re*	Os	Ir	Pt†	Pb*	Po
	Ce	Pr			Tb		fluorite structure	
	Th	Pa	U	Np	Pu	Am	Cm	

*Polymorphic. † Amorphous when prepared under atm. press. but crystallizes under pressure. For a survey (with many references) see: IC 1969 8 841.

Binary Metal Oxides

In the normal rutile structure each M is equidistant from two others in each chain of octahedra. Some dioxides (of elements within the broken lines of Table 12.4) crystallize with less symmetrical variants of this structure, the distortion being of such a nature as to bring successive pairs of metal atoms alternately closer together and further apart. The same effect is observed in NbO_2, which crystallizes with a complex superstructure of the rutile type. The close approach of M atoms can give rise to metallic conductivity (resistivity decreasing as temperature is lowered), but the relations between structure, number of d electrons, and physical properties are complex. For example, metallic conductivity can occur in compounds with the undistorted rutile structure (CrO_2, RuO_2, OsO_2, and IrO_2, and the high-temperature form of VO_2), and on the other hand the low-temperature ($<340°K$) form of VO_2 is a semiconductor with the distorted rutile structure. If the very short M–M bonds of length close to 2·5 Å found in certain of these dioxides (Table 12.5) are multiple bonds it is necessary to recognize at least four types of structure in this family, namely, the undistorted structure (with or without M–M interactions) and distorted structures (with single or multiple M–M bonds). On this subject see also p. 202.

TABLE 12.5

Metal–metal distances in some dioxides

Oxide	Number of d electrons in metal	M–M in chain (Å)		Reference
TiO_2	0	2·96		
VO_2 {tetrag.	1	2·85		JPSJ 1967 **23** 1380
{monocl.		2·62	3·17	ACSc 1970 **24** 420
NbO_2	1	2·80	3·20	AK 1962 **19** 435
MoO_2	2	2·51	3·10	ACSc 1967 **21** 661
WO_2	2 ⎫			
TcO_2	3 ⎬	2·5	3·1	ACSc 1955 **9** 1378
ReO_2	3 ⎭			
RuO_2		3·107		ACSc 1970 **24** 116
OsO_2		3·184		ACSc 1970 **24** 123

The polymorphism of ZrO_2 presents several points of interest. The normal (monoclinic) form, the mineral baddeleyite, changes at around 1100°C to a tetragonal form and at around 2300°C to a cubic form (fluorite structure). The transition at 1100°C is of practical importance since it restricts the use of the pure oxide as a refractory material, thermal cycling through this temperature region causing cracking and disintegration. Solid solutions containing CaO or MgO with the fluorite structure are not subject to this change.[7] The tetragonal form cannot be quenched to room temperature, but tetragonal ZrO_2 can exist at room temperature if prepared by precipitation from aqueous solution or by calcining salts at a low temperature. It is apparently stabilized by the higher surface energy arising from the smaller particle size, which must not exceed 300 Å.[8] This tetragonal

(7) PRS 1964 **279A** 395

(8) JPC 1965 **69** 1238

form[9] has a distorted fluorite structure. Instead of 8 O at 2·20 Å (the distance from Zr^{4+} to eight equidistant O^{2-} neighbours in the cubic high-temperature form) there are two sets of 4 (at 2·065 and 2·455 Å) forming flattened and elongated tetrahedral groups—a distorted cubic arrangement similar to that in $ZrSiO_4$. In monoclinic ZrO_2[10] Zr^{4+} is 7-coordinated and equal numbers of O^{2-} are 3- and 4-co-ordinated. The 3-coordinated ions (O_I) have a practically coplanar arrangement of 3 Zr neighbours (Zr–O, 2·07 Å) with interbond angles of 104°, 109°, and 143°, while the remainder (O_{II}) have 4 tetrahedral neighbours at a mean distance of 2·21 Å. All the interbond angles lie within the range 100–108° except one (134°). The 7-coordination group of Zr^{4+} is shown in idealized form in Fig. 12.5; the next nearest O^{2-} neighbour is at a distance of 3·58 Å, and accordingly the 7-coordination group is well defined.

The relation of the $C-M_2O_3$ structure to fluorite is described on p. 451. The structures of certain oxides of 4f and 5f elements are related in a more complex way to the structures of their dioxides (fluorite structure). For example, PrO_2 loses oxygen on heating to form Pr_6O_{11} with a fluorite-like structure from which one-twelfth of the anions have been removed at random.[11] Terbium forms, in addition to TbO_2, $TbO_{1·715}$, $TbO_{1·809}$, and $TbO_{1·823}$,[12] of which the first (Tb_7O_{12}) is noted on p. 501 in connection with the isostructural complex oxides $M'M''_6O_{12}$. PuO_2 loses oxygen to form substoichiometric oxides,[13] but in contrast UO_2 adds oxygen at high temperatures. A n.d. study shows that as interstitial oxygen is added to UO_2 to form UO_{2+x} the original O atoms move to more general positions and finally at the composition U_4O_9 the interstitial O atoms adopt a new (body-centred) ordered arrangement.[14][15]

(9) AC 1962 **15** 1187

(10) AC 1965 **18** 983

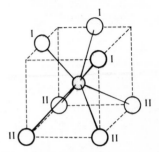

FIG. 12.5. The coordination of Zr^{4+} in monoclinic ZrO_2 (idealized).

(11) JACS 1950 **72** 1386
(12) JACS 1961 **83** 2219

(13) JINC 1965 **27** 541

(14) PRS 1963 **274A** 122, 134
(15) AC 1968 **A24** 657

Oxides MO_3

These are few in number and include CrO_3, MoO_3, WO_3, ReO_3, TeO_3, and UO_3. The (tetrahedral) chain structure of CrO_3 is described under the structural chemistry of Cr(VI) in Chapter 27. One of the forms of TeO_3 is isostructural with FeF_3 (p. 355). The very simple ReO_3 structure is described on p. 173, and the structures of MoO_3 and WO_3 are described later in this chapter. For UO_3 see p. 999.

Oxides MO_4

There are only two metal tetroxides, the volatile and low-melting RuO_4 and OsO_4. Both exist as regular tetrahedral molecules in the crystalline and vapour states:

	M—O (Å)	Reference
RuO_4	1·705 (vapour)	ACSc 1967 **21** 737
OsO_4	1·71 (vapour)	ACSc 1966 **20** 385
	1·74 (crystal)	AC 1965 **19** 157

449

Binary Metal Oxides

Oxides M_2O_3

Of the structures listed in Table 12.2 only two are of importance for elements other than the 4f and 5f metals; these are the corundum (α-Al_2O_3) and the C-M_2O_3 structures. Many oxides M_2O_3 are polymorphic, including most (if not all) of the 4f and 5f oxides. References to recent work, particularly refinements of earlier structures, are included in Table 12.6.

TABLE 12.6
Crystal structures of sesquioxides

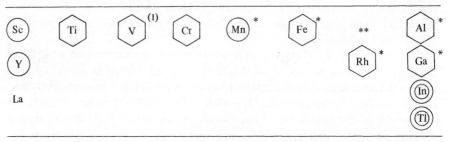

○ C structure. ◎ C structure under atm. pressure; corundum structure under high pressure.

◯ Corundum structure.

* Also other polymorphs.
(1) Changes to a semiconducting form at low temperatures.
** There is no evidence for the existence of Co_2O_3 or Ni_2O_3.

Crystal structures of 4f *and* 5f *sesquioxides*

La	Ce	Pr	Nd	Sm	Eu	Gd	Tb	Dy	Ho----Lu	U	Np	Pu	Am	Cm
		A										A	A	A
		?	?			B			under pressure					
?	?	?				C						C	C	C

Corundum structure: IC 1969 **8** 1985. Rh_2O_3, AC 1970 **B26** 1876.

In_2O_3, ACSc 1967 **21** 1046. Fe_2O_3, AM 1966 **51** 123, α-Ga_2O_3, JCP 1967 **46** 1862.

A-M_2O_3 structure: AC 1953 **6** 741. Pr_2O_3, IC 1963 **2** 791.

B-M_2O_3 structure: Gd_2O_3, AC 1958 **11** 746. Sm_2O_3, JPC 1957 **61** 753. Tb_2O_3, ZaC 1968 **363** 145.

C-M_2O_3 structure: In_2O_3, AC 1966 **20** 723. Mn_2O_3, ACSc 1967 **21** 2871. Y_2O_3, AC 1969 **B25** 2140. Sc_2O_3, ZK 1967 **124** 136.

4f sesquioxides: IC 1965 **4** 426; IC 1966 **5** 754; IC 1969 **8** 165.

Other structures: β-Ga_2O_3, JCP 1960 **33** 676. Rh_2O_3 (high temp.), IC 1963 **2** 972; (high press.), JSSC 1970 **2** 134. Au_2O_3, JINC 1969 **31** 2966 (preparation only).

The corundum structure is an approximately h.c.p. array of O atoms in which Al^{3+} ions occupy two-thirds of the octahedral holes. In one sense the structure is surprisingly complex, for there is sharing of vertices, edges, and faces of AlO_6

450

coordination groups. Models of this and other c.p. structures suggest that this structure is adopted in preference to other geometrically possible ones because the arrangement of 4 Al^{3+} ions around O^{2-} approximates most closely to a regular tetrahedral one (see MSIC, p. 63). We have noted elsewhere that for purely geometrical reasons a 3D M_2X_3 structure with regular octahedral coordination of M *and* regular tetrahedral coordination of X is not possible. Because faces are shared between pairs of AlO_6 octahedra in this structure there are two sets of M–X distances; a refinement of the structure of α-Fe_2O_3 gives Fe–3 O, 1·945 Å, and Fe–3 O, 2·116 Å. (For γ-Fe_2O_3 see p. 456.)

The structures of three forms of Rh_2O_3 have been determined, the normal (corundum) form, a high-temperature orthorhombic form, and a high-pressure polymorph made at 1200°C under 65 kbar. In this high-pressure form there are pairs of face-sharing octahedra (as in corundum) but a different selection of shared edges; the structure may be regarded as built from slices of the corundum structure.

The C-M_2O_3 structure is related to that of CaF_2, from which it may be derived by removing one-quarter of the anions (shown as dotted circles in Fig. 12.6) and then rearranging the atoms somewhat. The 6-coordinated M atoms are of two types.

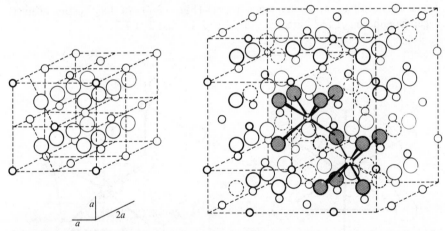

FIG. 12.6. The C-M_2O_3 structure of α-Mn_2O_3 and other sesquioxides, showing its relationship to the fluorite structure (left). The two kinds of coordination group of the metal ions are shown in the right-hand diagram.

Instead of 8 neighbours at the vertices of a cube, two are missing. For one-quarter of the M atoms these two are at the ends of a body-diagonal, and for the remainder at the ends of a face-diagonal. Both coordination groups may be described as distorted octahedra. The O atoms are 4-coordinated, and in this structure also it is probably the approximately regular tetrahedral coordination of O^{2-} which is responsible for the less regular coordination of the metal ions. In contrast to the corundum structure the C-M_2O_3 structure is an assembly of *edge*-sharing (distorted) octahedral coordination groups. In fluorite each cubic MX_8 coordination group shares an edge with each of 12 neighbouring MX_8 groups. Removal of

451

2 X at the ends of either a face-diagonal or body-diagonal leaves intact six of the original cube edges, so that in the $C\text{-}M_2O_3$ structure each octahedral coordination group shares six edges.

Not all compounds with the $C\text{-}M_2O_3$ structure have the most symmetrical (cubic) form of the structure, which is that of the mineral bixbyite, $(Fe, Mn)_2O_3$. In (cubic) In_2O_3 the first type of metal ion has 6 equidistant O neighbours (at 2·18 Å) and the other type has 2 O at each of the distances 2·13, 2·19, and 2·23 Å (mean 2·18 Å), and there is some distortion of the tetrahedral arrangement of neighbours around O^{2-} (four angles of 100° and two of 126°). There is much less regular coordination of M^{3+} in (synthetic) Mn_2O_3, which is not cubic. Each metal ion has 4 closer neighbours (at approximately 1·96 Å) and 2 more distant (at 2·06 and 2·25 Å for the two types of metal ion); compare the similar environment of Mn^{3+} (d^4) in $ZnMn_2O_4$, 4 O at 1·93 and 2 O at 2·28 Å.

The $A\text{-}M_2O_3$ and $B\text{-}M_2O_3$ structures are associated particularly with the 4f and 5f sesquioxides. When there is polymorphism the A, B, and C structures are characteristically those of high-, medium-, and low-temperature forms respectively. The structures of some 4f and 5f oxides are set out in Table 12.6. A striking feature of the $A\text{-}M_2O_3$ structure, which has been confirmed by neutron diffraction, is the unusual 7-coordination of the metal atoms (Fig. 12.7). In La_2O_3 the nearest oxygen neighbours of La^{3+} are 3 at 2·38, 1 at 2·45, and 3 at 2·72 Å.

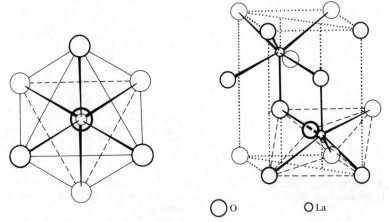

$\bigcirc$ O $\bigcirc$ La

FIG. 12.7. The $A\text{-}M_2O_3$ structure of La_2O_3 and rare-earth sesquioxides. At the left is shown in projection the 7-coordination group around a M^{3+} ion, consisting of an octahedral group of O^{2-} ions with an additional O^{2-} ion (heaviest circle) above one of the octahedron faces.

In the rather complex $B\text{-}M_2O_3$ structure there are three kinds of non-equivalent metal ion, some with 6-coordination (octahedral) and the remainder with 7-coordination (monocapped trigonal prism). The seventh neighbour is appreciably more distant than the six at the vertices of the prism—at 2·73 Å (mean) as compared with 2·39 Å (in Sm_2O_3). On the basis of density differences the expected order of stability with increasing pressure would be expected to be

$$C \rightarrow \text{corundum} \rightarrow B$$
$$\text{c.n. of M} \quad 6 \qquad\qquad 6 \qquad\qquad 6 \text{ and } 7$$

In fact, In_2O_3 and Tl_2O_3 (C structure) do change to the corundum structure under pressure, though the isostructural oxides of Mn, Sc, and Y, containing ions of comparable size, do not form the corundum structure but go straight to the B structure. The corundum structure is the high-pressure structure of Ga_2O_3 and also the structure at ordinary temperatures, but at high temperatures this oxide transforms to yet another sesquioxide structure. In this β-Ga_2O_3 structure one-half of the metal ions occupy tetrahedral and the remainder octahedral holes in an approximately cubic closest packing of O^{2-} ions. The same structure is adopted by θ-Al_2O_3; other polymorphs of this oxide are noted in a later section on the oxides of Al.

Two gold oxides have been prepared under high oxygen pressure, one of which is presumably Au_2O_3 since it is apparently structurally similar to a phase $Na_xAu_2O_3$ obtained when sodium is present (compare Pt_3O_4 and $Na_xPt_3O_4$). . All these phases gradually decompose when kept in a desiccator, evolving oxygen and leaving metallic gold. Their structures are not yet established.

The structures of Sb_2O_3, Bi_2O_3, and Pb_2O_3 are described elsewhere.

Oxides M_2O_5

The pentoxides of As, Sb, and Bi are included in Chapter 20; nothing is known of the structures of these oxides. We have to deal here with the Group VA pentoxides; for Pa_2O_5 and $UO_{2.6}$ see Chapter 28.

The structure of V_2O_5 is described under the oxygen chemistry of V (p. 467); there is a characteristic 5-coordination with one very short V–O bond. In marked contrast to V_2O_5 is the complexity of the crystal chemistry of Nb_2O_5 and Ta_2O_5. Many Roman and Greek letters have been used to distinguish the numerous polymorphs of Nb_2O_5,[1] in all of which there is sharing of vertices and edges of octahedral NbO_6 groups in various combinations. We refer the reader first to Chapter 5 (p. 184) and then to the discussion of complex oxides of Ti, V, Nb, etc. in Chapter 13 (p. 502) where we describe the formation of more complex structures from slices or blocks of ReO_3-like structure. The reason for the complexity of the crystal chemistry of Nb_2O_5 is that these sub-units can be joined together in an indefinitely large number of ways. Perhaps these 'forms' should be regarded as polytypes rather than polymorphs, since they result from a particular kind of growth mechanism and in some cases may be stabilized by foreign atoms. In this respect they have more in common with the polytypes of, for example, SiC or ZnS rather than with normal polymorphs. The sharing of edges of octahedral coordination groups reduces the O : M ratio below 3 : 1, and the same processes that give the exact ratio 5 : 2 also give oxides such as $Nb_{12}O_{29}$, $Nb_{22}O_{54}$, and $Nb_{25}O_{62}$ and numerous oxides of Mo and W intermediate between MO_2 and MO_3. We give here only a few examples. The ReO_3 blocks can be joined as in Fig. 5.38 or as in Fig. 5.39; in the latter case tetrahedral holes are formed along the junction

(1) AnC 1966 **5** 40

453

(2) NW 1965 **52** 617
(2a) JSSC 1970 **1** 419
(3) ZaC 1967 **351** 106

lines of the ReO$_3$ blocks. Examples of the first type include P-Nb$_2$O$_5$,[2] M-Nb$_2$O$_5$,[2a] and N-Nb$_2$O$_5$;[3] H-Nb$_2$O$_5$ (Nb$_{27}$NbO$_{70}$) is described on p. 185. Figure 12.8(a) shows a projection of the relatively simple structure of P-Nb$_2$O$_5$.

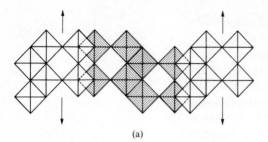

(a)

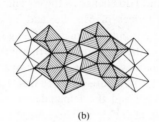

(b)

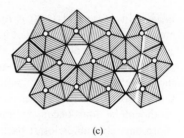

(c)

FIG. 12.8. The structures of crystalline oxides: (a) P-Nb$_2$O$_5$, (b) H-Ta$_2$O$_5$, (c) α-U$_3$O$_8$. The chains of vertex-sharing octahedra or pentagonal bipyramids are normal to the plane of the paper.

There are at least two structurally distinct forms of Ta$_2$O$_5$, with a transition point at 1360°C. As first prepared, by heating Ta in air or oxygen at 600°C, Ta$_2$O$_5$ is poorly crystalline, but the crystallinity is increased by heating at 1350°C and the exact structure (as indicated by the superstructure lines) depends on the heat treatment. After extensive heat treatment the structure reaches an equilibrium state. (This is also true of a number of phases containing W up to the limiting composition 11 Ta$_2$O$_5$. 4 WO$_3$.) The structures of the low-temperature form, L-Ta$_2$O$_5$,[4] and of all these phases are of the same general type, consisting of chains built from octahedral or pentagonal bipyramidal groups sharing two opposite vertices, these chains being joined by vertex- or edge-sharing to give the 3D structure. The high-temperature form, H-Ta$_2$O$_5$, apparently undergoes various phase transitions on cooling, but the tetragonal structure can be stabilized by the addition of 2–4 mole per cent of Sc$_2$O$_3$. The structure[5] Fig. 12.8(b)) is of the same general type as L-Ta$_2$O$_5$ and consists of blocks of octahedral and pentagonal bipyramidal coordination groups at two different 'levels' (see p. 185) so that there is both 6- and 7-coordination of the metal atoms. These Ta$_2$O$_5$ structures may be compared with the structure of α-U$_3$O$_8$ (Fig. 12.8(c)), in which all metal atoms are in chains of vertex-sharing pentagonal bipyramids, and with WNb$_2$O$_8$ and NaNb$_6$O$_{15}$F (Fig. 13.14, p. 510) in which there is both distorted octahedral and pentagonal bipyramidal coordination of the transition-metal atoms. A third 'form' of Ta$_2$O$_5$, isostructural with one form of Nb$_2$O$_5$, may contain K as an essential constituent of the structure.[6]

(4) AC 1971 **B27** 1037

(5) JSSC 1971 **3** 145

(6) JSSC 1970 **2** 24

Oxides M$_2$O$_7$

This small group consists of Mn$_2$O$_7$, which at ordinary temperatures is a deeply coloured oil, Tc$_2$O$_7$, and Re$_2$O$_7$. The last two are yellow, volatile, water-soluble solids. Mn$_2$O$_7$ is presumably molecular in all states of aggregation, and Tc$_2$O$_7$ consists of molecules in the crystalline state. In the centrosymmetrical molecule[1]

(1) AnC 1969 **81** 328

the Tc–O–Tc system is linear, the bridge bonds are similar in length to those in the $Re_2OCl_{10}^{4-}$ and $Ru_2OCl_{10}^{4-}$ ions, and the terminal bonds have approximately the same length as Re–O in ReO_3Cl and Re_2O_7 (1·70 Å).

Re_2O_7 vaporizes as molecules of the same type, but the crystal has an interesting layer structure[2] (Fig. 12.9(a)) in which equal numbers of metal atoms are in tetrahedral and octahedral coordination. (Contrast the chain ion in

(2) IC 1969 **8** 436

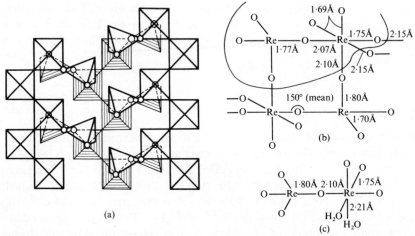

FIG. 12.9. (a) One layer of the crystal structure of Re_2O_7; the small circles represent O atoms shared between octahedral and tetrahedral groups, that is, the O atoms of the ring in (b). (b) Details of one ring of two tetrahedra and two octahedra. (c) The $Re_2O_7(H_2O)_2$ molecule.

$Na_2Mo_2O_7$ which illustrates a third way of realizing the required O : M ratio of 7 : 2, also with equal numbers of MO_4 and MO_6 groups.) If chains of vertex-sharing octahedra at two different levels are linked by tetrahedra, rings consisting of two tetrahedra and two octahedra are formed which are perpendicular to the plane of the paper in Fig. 12.9(a). In each octahedral ReO_6 group three of the Re–O bonds are appreciably longer than the others. If two of these bonds are broken, as in Fig. 12.9(b), which shows one of the rings in crystalline Re_2O_7, molecules are produced, accounting for the volatility of the solid oxide. The structure of Re_2O_7 is also very closely related to that of the molecule $Re_2O_7(OH_2)_2$,[3] the first hydration product of the heptoxide (Fig. 12.9(c).)

(3) AnC 1968 **80** 286

Oxides M_3O_4

Relatively few elements form an oxide M_3O_4, which necessarily contains the element in two different oxidation states. Of the three possibilities $M_2^I M^{VI} O_4$, $M_2^{II} M^{IV} O_4$, and $M_2^{III} M^{II} O_4$, the first corresponds to an oxy-salt such as K_2WO_4, the second is realized in Pb_3O_4, and the third in Fe_3O_4, Co_3O_4, Mn_3O_4, and Eu_3O_4. The structure of Pb_3O_4 is described later in this chapter in the section on the oxides of lead and that of Eu_3O_4 on p. 496. The 3d oxides M_3O_4 have the spinel structure, which is described in Chapter 13 since it is adopted by many complex oxides XY_2O_4.

455

(1) JCP 1952 **20** 199

(2) JLCM 1968 **16** 129

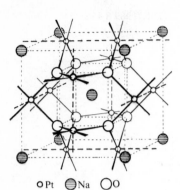

O Pt ⊜Na ◯O

FIG. 12.10. Structure proposed for $Na_xPt_3O_4$.

The structure of Fig. 12.10 was proposed many years ago for a compound with the limiting composition $NaPt_3O_4$.[1] Although it is doubtful if the full complement of Na atoms is ever present there are certainly phases $Na_xPt_3O_4$ ($0 < x < 1$), and Pd forms phases of the same type. Also it appears that the basic framework of Pt and O atoms can exist in the absence of Na atoms, when it is the oxide Pt_3O_4, though the phase $Na_xPd_3O_4$ cannot be prepared free from sodium.[2] In the Pt_3O_4 framework each O is bonded to 3 Pt (Pt–O, 1·97 Å) and each Pt forms 4 coplanar bonds to O atoms. Each metal atom also has 2 Pt atoms (at 2·79 Å) completing an elongated octahedral coordination group, as shown by the heavy broken lines in Fig. 12.10. The Na atoms occupy holes surrounded by 8 O at the vertices of a cube (Na–O, 2·46 Å).

The oxides of iron

(1) AC 1953 **6** 565; AC 1955 **8** 257

The structures of FeO (NaCl structure) and of α-Fe_2O_3 (corundum structure) have already been noted. The structure of the second (γ) form of Fe_2O_3 is related in an interesting way to the structures of FeO and Fe_3O_4. There are 32 oxygen atoms in the unit cell of the spinel structure (p. 490) and a total of 24 cations. Magnetite, Fe_3O_4, has the spinel structure with the full complement of cations, of which one-third are Fe^{2+} and the remainder Fe^{3+}–$Fe^{3+}(Fe^{2+}Fe^{3+})O_4$. It appears that the inverse spinel structure of magnetite is the reason for its high electronic conductivity compared with that of Mn_3O_4 (tetragonally distorted normal spinel structure), continuous interchange of electrons between Fe^{2+} and Fe^{3+} ions in the 16-fold position being possible. (At 119°K magnetite becomes antiferromagnetic, and the symmetry drops to rhombohedral or lower.[1])

In γ-Fe_2O_3 there are on the average only $21\frac{1}{3}$ Fe atoms per unit cell distributed at random among the eight tetrahedral and sixteen octahedral sites. Accordingly γ-Fe_2O_3 and Fe_3O_4 are easily interconvertible. Careful oxidation of Fe_3O_4 yields γ-Fe_2O_3, which is converted back into Fe_3O_4 by heating *in vacuo* at 250°C. The relation of FeO to γ-Fe_2O_3 and Fe_3O_4 is of interest in this connection. FeO has, ideally, the NaCl structure with four Fe^{2+} and four O^{2-} ions per unit cell, though when prepared at atmospheric pressure it is deficient in Fe. (Stoichiometric FeO has been prepared at pressures greater than 36 kbar at 770°C from $Fe_{0.95}O$ and metallic iron.[2])

(2) JCP 1967 **47** 4559

(3) CR 1956 **242** 776

(4) AC 1953 **6** 827

At about 1000°C the composition range is $Fe_{0.946}O$–$F_{0.875}O$, but at lower temperatures the composition limits converge. At 570°C the composition is $Fe_{0.93}O$, and at still lower temperatures disproportionation to α-Fe + Fe_3O_4 occurs,[3] so that this oxide must be prepared at temperatures above 570°C and quenched. At about 200°K a second-order change takes place over a range of temperature, accompanied by the development of antiferromagnetism and rhombohedral symmetry, but involving only a very minor structural change.[4] The length of the side of the unit cell varies, approximately linearly, with the composition, and typical values are: a = 4·3010 Å for 48·56 atomic per cent. Fe and 4·2816 Å for 47·68 per cent Fe. Fe_3O_4 has the spinel structure (a = 8·37 Å) with eight Fe^{2+}, sixteen Fe^{3+}, and 32 O^{2-} ions in the unit cell. The arrangement of oxygen ions is therefore the same (cubic close-packed) in FeO, γ-Fe_2O_3, and Fe_3O_4. If we double the above spacings for iron-deficient specimens of FeO (viz. 8·6020 for 48·56 per cent and 8·5632 Å for 47·68 per cent) and plot them against composition we find that the line passes through the points for Fe_3O_4 and γ-Fe_2O_3 (viz. *c*. 8·37 Å, 42·86 per cent Fe, and *c*. 8·30 Å, 40 per cent Fe). The relation between these structures is therefore as follows. A cube with edge about 8·6 Å containing 32 oxygen ions can accommodate 32 Fe^{2+} ions in the octahedral holes between the cubic close-packed oxygen ions. This would give the ideal structure of stoichiometric FeO. If we remove a small proportion of these Fe^{2+} ions and replace them by two-thirds their number of Fe^{3+} ions, so maintaining electrical neutrality, we have FeO deficient in iron, alternatively (but incorrectly) described as FeO containing excess oxygen (see also FeS, p. 610). (It has been suggested that in quenched specimens of $Fe_{0.9}O$ there are periodically spaced clusters of vacancies (probably about 13) and 4 cations in tetrahedral sites.)[5]

(5) AC 1969 **B25** 275

If the removal of Fe atoms is continued until there are only 24 Fe atoms in the volume containing 32 O atoms, the composition is Fe_3O_4. Further removal of Fe gives γ-Fe_2O_3, with an average of $21\frac{1}{3}$ metal atoms in the same volume. We have now a physical picture of the oxidation of FeO $\rightarrow$ Fe_3O_4 $\rightarrow$ γ-Fe_2O_3. The oxygen lattice of FeO is extended by the addition of more oxygen, added as new layers of close-packed oxygen atoms. Into these Fe atoms (ions) migrate, resulting in a continuous decrease in the concentration of Fe atoms in a given volume of the oxygen lattice.

The oxides of aluminium

The hydroxide, $M(OH)_3$, oxyhydroxide, MO(OH), and oxide, M_2O_3, of Al (and certain other trivalent elements) exist in α and γ forms. We give the mineral names of these Al compounds because they are often referred to by name and also because the American nomenclature is sometimes different from that used in England.

$Al(OH)_3$	α	bayerite	γ	gibbsite, hydrargillite
AlO(OH)	α	diaspore	γ	boehmite
Al_2O_3	α	corundum	γ	—

On heating diaspore dehydrates directly to α-Al_2O_3 but boehmite, both forms of $Al(OH)_3$, and compounds such as ammonium alum produce α-Al_2O_3 only after prolonged heating at high temperatures. At lower temperatures a variety of phases are formed which are collectively known as γ-Al_2O_3, and in various studies most of the Greek alphabet has been used in naming them. These phases represent various degrees of ordering of Al atoms in an essentially cubic closest packing of O atoms, described as a defect spinel structure (q.v.), since there are only $21\frac{1}{3}$ metal atoms arranged at random in the 16 octahedral and 8 tetrahedral positions of that structure. This γ-Al_2O_3 is important technically because of its great adsorptive power ('activated alumina') and its catalytic properties, and it is also used in the manufacture of synthetic sapphire in the Verneuil furnace. The degree of order and the pore structure depend on the starting material. For example, the arrangement of the O atoms in bayerite is h.c.p., and conversion to the c.c.p. spinel structure involves rebuilding the whole structure. In the case of boehmite, although the structure as a whole is not close-packed the O atoms within a layer are c.c.p., so that less rearrangement is required to form the γ-Al_2O_3 structure. The following[1] are typical of the schemes postulated as the result of X-ray or electron diffraction studies:

(1) IEC 1950 **42** 1398; AC 1964 **17** 1312

$$\text{boehmite} \xrightarrow{450°} \gamma \xrightarrow{750°} \delta \xrightarrow{1000°} \theta + \alpha \xrightarrow{1200°C} \alpha$$
$$\text{bayerite} \xrightarrow{230°} \eta \xrightarrow{850°} \theta \xrightarrow{1200°C} \alpha$$

The δ and θ phases formed at intermediate temperatures are both highly crystalline phases, the former with a superstructure of the spinel type and the latter with the β-Ga_2O_3 structure mentioned in the section on 'Oxides M_2O_3'.

For β-alumina and related compounds see p. 494.

The oxides of manganese

These comprise MnO, Mn_3O_4, Mn_2O_3, Mn_5O_8, MnO_2, and Mn_2O_7. The last is the anhydride of permanganic acid (a compound which is a volatile solid stable below $1°C$)[1] and forms an oil or dark green crystals. It breaks down irreversibly at temperatures above about $0°C$ with evolution of oxygen and formation of MnO_2; it is a powerful oxidizing agent, and obviously resembles Cl_2O_7 rather than metallic oxides.

(1) JACS 1969 **91** 6200

The first three oxides, on the other hand, are closely similar to those of iron. The stability of the green MnO to oxygen depends on the temperature of preparation, low-temperature ignition of the carbonate giving very reactive material. It apparently oxidizes to the stage $MnO_{1.13}$ without forming new phases, and it has the $NaCl$ structure, like FeO. There are two forms of the black Mn_2O_3, α having the C sesquioxide structure and γ being related to Mn_3O_4 in the same way as γ-Fe_2O_3 to Fe_3O_4 (p. 456). All oxides and oxyhydroxides of Mn on heating in air to about $1000°C$ form the purplish-red Mn_3O_4, which has a distorted spinel structure at ordinary temperatures but apparently becomes cubic at temperatures above $1170°C$.[2] Both γ-Mn_2O_3 and Mn_3O_4 show the same type of tetragonal distortion ($c : a = 1.16$).

(2) JRNBS 1950 **45** 35

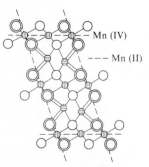

— Mn (IV)

--- Mn (II)

FIG. 12.11. Elevation of the structure of Mn_5O_8; the Cdl_2-like layers are perpendicular to the plane of the paper.

The oxide Mn_5O_8 is $Mn_2^{II}Mn_3^{IV}O_8$ and is isostructural with $Cd_2Mn_3O_8$. The structure may be visualized as built from CdI_2-like layers from which one-quarter of the metal atoms are removed ($Mn_3^{IV}O_8$), Mn^{II} atoms then being added above and below the empty sites to give the composition M_5X_8 as in $Zn_5(OH)_8Cl_2 . \dot{H}_2O$ (p. 213) or $Zn_5(OH)_8(NO_3)_2 . 2 H_2O$ (p. 536). These Mn^{II} atoms join the 'layers' into a 3D structure. The coordination of Mn^{IV} is distorted octahedral (mean Mn^{IV}–O, 1.87 Å) and that of Mn^{II} is trigonal prismatic (mean Mn^{II}–O, 2.22 Å), (Fig. 12.11). There is considerable distortion of the latter coordination group (2.05, 2.17 (two), 2.28, and 2.31 Å (two)).[3]

(3) HCA 1967 **50** 2023

The dioxide has long been known as the mineral pyrolusite with the simple tetragonal rutile structure, but many materials approximating in composition to MnO_2 have been prepared in the laboratory and recognized in the mineral world. They have been described as polymorphs of MnO_2, though their compositions are often appreciably different from pure MnO_2, for example, γ-MnO_2 (analysis corresponding to $MnO_{1.93}$) and α-MnO_2 (91·5 per cent MnO_2). The β form (pyrolusite) can be prepared pure, but not by hydrothermal methods. There is a large literature on the oxides of Mn, particularly the dioxide, which is important as a catalyst and as a component of dry batteries.[4] X-ray studies have now shown why the MnO_2 system is so complex and why different structures appear when the oxide is prepared from solutions containing ions of various kinds. Two main types of structure have been identified, 3D frameworks and layer structures.

Framework structures

These are built of single or multiple octahedral chains joined up along their lengths by sharing corners, and they can conveniently be illustrated as projections along the length of the chain. In this way Fig. 12.12(a) represents the rutile structure of β-MnO_2. The mineral ramsdellite and Glemser's γ-MnO_2 have the structure shown at (b), built of double octahedral chains.[5] This is essentially the diaspore structure (Fig. 14.12) except that the short O—O contacts of 2·65 Å due to hydrogen bonding in AlO . OH (and MnO . OH) have lengthened to the much larger value 3·34 Å in ramsdellite. The collective name γ-MnO_2 is given to a range of materials. The X-ray diffraction patterns of the more poorly crystalline materials resemble that of pyrolusite, but the patterns of more crystalline preparations show a stronger resemblance to that of ramsdellite. It has been suggested[6] that these γ phases may have structures intermediate between those of Fig. 12.12(a) and (b), being built of slices of the β and γ structures, as shown at (e).

(4) JACS 1950 **72** 856; RPAC 1951 **1** 203; ZaC 1961 **309** 1, 20, 121

(5) ACSc 1949 **3** 163

(6) AC 1959 **12** 341

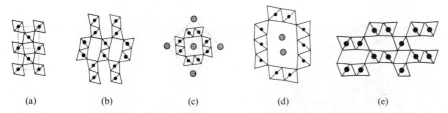

(a)   (b)   (c)   (d)   (e)

FIG. 12.12. The structures of (a) β-MnO_2, (b) ramsdellite, (c) α-MnO_2 and hollandite, (d) psilomelane, (e) a 'γ-MnO_2' intermediate between (a) and (b).

If some of the Mn^{4+} ions in a MnO_2 framework are replaced by Mn^{2+} or other ions carrying smaller positive charges, then the framework is negatively charged and can accommodate other positive ions in its interstices. The structure of Fig. 12.12(c) corresponds to a group of minerals, hollandite, cryptomelane, and coronadite, which with α-MnO_2 form an isostructural series of the general formula $A_{2-y}B_{8-z}X_{16}$.[7] (A represents large ions such as Ba^{2+}, Pb^{2+}, or K^+, B is

(7) AC 1950 **3** 146; AC 1951 **4** 469

Binary Metal Oxides

Mn^{4+}, Fe^{3+}, or Mn^{2+}, and X is O^{2-} or OH^{-}.) In the compounds studied $0.8 < y < 1.3$ and $0.1 < z < 0.5$. It is evident that in the more open structures such as (c) and (d) some large ions are necessary to prevent collapse of the framework, and accordingly the so-called α-MnO_2 can be prepared only in the presence of a large ion such as K^{+}. Some of these compounds show cation exchange like the zeolites. An even more open structure is found in psilomelane,[8] $(Ba, H_2O)_2Mn_5O_{10}$, in which there are double and triple chains of octahedra and larger tunnels accommodating Ba^{2+} ions and water molecules (Fig. 12.12(d)). Dehydration is accompanied by structural change to hollandite, into which psilomelane is converted on heating to $550°C$.

(For a complex oxide containing Mn with a quite different type of structure see Mg_6MnO_8 (p. 501).)

Layer structures

There are many ill-defined hydrous manganese oxide minerals somewhat reminiscent of clay minerals, and probably like them having layer structures. A well-crystallized lithiophorite, $(Al_{0.68}Li_{0.32})Mn^{2+}_{0.17}Mn^{4+}_{0.82}O_2(OH)_2$, has a 2-layer structure in which both layers are of the $Mg(OH)_2$ type (p. 209) with ideal compositions MnO_2 and $(Li, Al)(OH)_2$.[9] The layers are held together by O—H—O bonds as in $Al(OH)_3$, though it may be deduced from the above formula that if all anions in the MnO_2 layer are O^{2-} and all are OH^{-} in the $(Li, Al)(OH)_2$ layer then the former carries a charge of -0.38 and the latter $+0.38$ per unit of formula. The detailed arrangement of the cations and the positions of the H atoms are not known with certainty.

A Mn-containing layer of essentially the same kind occurs in chalcophanite, $ZnMn_3O_7 . 3H_2O$,[10] but with one-seventh of the metal sites unoccupied so that the composition is Mn_3O_7 instead of MnO_2. Associated with each Mn layer and directly above and below the unoccupied Mn positions are the Zn^{2+} ions

(8) AC 1953 6 433

(9) AC 1952 5 676

(10) AC 1955 8 165

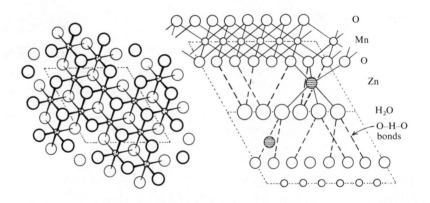

Plan of one O—Mn—O layer   Elevation

FIG. 12.13. The crystal structure of chalcophanite, $ZnMn_3O_7 . 3H_2O$.

460

coordinated by three O atoms of the layer and on the other side by three H_2O, these six neighbours forming a somewhat distorted octahedron (Zn–3 O 1·95 (·05) Å, Zn–3 H_2O 2·15 (·05) Å). The layers are held together by O–H–O bonds between water molecules as shown in Fig. 12.13. The nature of the Zn–O bonds in this and other crystals is discussed on p. 914.

The oxides of lead

The existence of three oxides of lead, PbO, Pb_3O_4, and PbO_2, has been recognized for a long time, but there has been far less general agreement concerning other oxides intermediate between PbO and PbO_2.[1] The black Pb_2O_3 is prepared either hydrothermally or by heating PbO at 600°C under 1 kbar pressure of O_2. A brown-black $Pb_{12}O_{19}$ is described as formed by heating PbO or PbO_2 at 500°C under 175 bar pressure of O_2; the existence of a further oxide ($Pb_{12}O_{17}$?) is not fully confirmed.[2] Since the oxides PbO, Pb_3O_4, and PbO_2 are to some extent interconvertible (by heating in air or in other ways) mixtures are often obtained. For example, it is stated that Pb_3O_4 results from heating either PbO or PbO_2 in air to 500°C, whereas above 550°C only PbO is stable. In practice, therefore, Pb_3O_4 and PbO_2 are not prepared thermally without further treatment if wanted pure in small quantities. Careful decomposition of $PbCO_3$ or $Pb(OH)_2$ gives PbO, oxidation of PbO followed by removal of unchanged PbO by acetic acid gives Pb_3O_4, while PbO_2 may be prepared by the action of nitric acid on the latter or by oxidation of an alkaline suspension of PbO by a hypochlorite. These oxides have striking colours—PbO yellow and red, Pb_3O_4 red, Pb_2O_3 black, PbO_2 maroon and black, showing, as in the case of PbS with its brilliant metallic lustre and semiconductivity, that the bonds are not of simple types.

Lead monoxide, PbO
This oxide has two polymorphs, a red (tetragonal) form prepared by heating PbO_2 in air to 550°C, and a yellow (orthorhombic) form made by heating PbO_2 to 650°C or by rapidly cooling molten PbO. Tetragonal PbO has the layer structure of Fig. 12.14 in which the metal atom is bonded to 4 O atoms which are arranged in a square to one side of it, with the lone pair of electrons presumably occupying the apex of the tetragonal pyramid;[3] for this structure see also pp. 100 and 137. The structural chemistry of lead is summarized in Chapter 26. It has not proved possible to locate the O atoms in the yellow form of PbO by X-ray diffraction, but their positions have been determined by neutron diffraction.[4] The structure is built of layers which are recognizable as very distorted versions of that of Fig. 12.14, but instead of four equal Pb–O bonds of length 2·30 Å in the red form there are two of 2·20 Å and two of 2·49 Å. The shorter bonds delineate zigzag chains (O–Pb–O, 148°) which are bonded into layers by the longer Pb–O bonds. The shortest Pb–Pb distances in and between the layers are respectively 3·47 and 3·63 Å.

(1) JCS 1956 725

(2) JACeS 1964 **47** 242

(3) AC 1961 **14** 1304 (n.d.)

(4) AC 1961 **14** 66, 80 (n.d.)

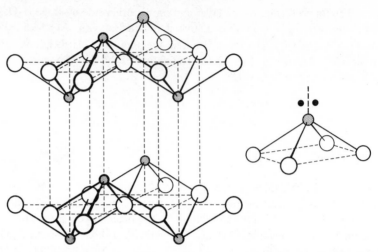

FIG. 12.14. The crystal structure of tetragonal PbO (and SnO). The small shaded circles represent metal atoms. The arrangement of bonds from a metal atom is shown at the right, where the two dots represent the 'inert pair' of electrons (see p. 937).

Red lead, Pb_3O_4

The structure of Pb_3O_4 is illustrated in Fig. 12.15. It consists of chains of $Pb^{IV}O_6$ octahedra sharing opposite edges, these chains being linked by the Pb^{II} atoms, each with a pyramidal arrangement of three O neighbours. The bond lengths and interbond angles are: Pb^{IV}–O, 6 at 2·14 Å, Pb^{II}–O, 2 at 2·18 Å and 1 at 2·13 Å; mean O–Pb–O bond angle 76° (as in PbO).[5]

(5) ZaC 1965 **336** 104 (n.d.)

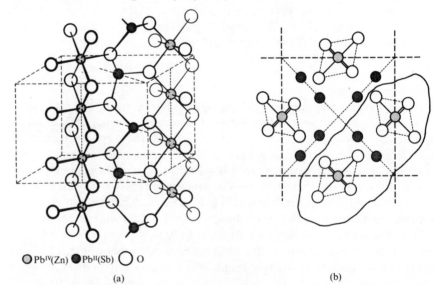

○ Pb^{IV}(Zn) ● Pb^{II}(Sb) ○ O

(a)   (b)

FIG. 12.15. The crystal structure of Pb_3O_4 (and $ZnSb_2O_4$). (a) Portion of the structure outlined in the projection (b) showing the chains of $Pb^{IV}O_6$ octahedra joined by pyramidally coordinated Pb^{II} atoms.

Lead sesquioxide, Pb_2O_3

In this (monoclinic) oxide the Pb^{II} atoms are situated between layers of distorted $Pb^{IV}O_6$ octahedra (Pb^{IV}–O, 2·08–2·28 Å, mean 2·18 Å). The Pb^{II} atoms are in positions of very irregular 6-coordination, these neighbours being arranged at six of the vertices of a distorted cube (those at the ends of one face-diagonal being absent). The six Pb^{II}–O distances fall into two sets, three short (2·31, 2·43, and 2·44 Å) and three longer (2·64, 2·91, and 3·00 Å), and there are Pb^{II}–Pb^{II} distances around $3\frac{1}{2}$ Å, as in the metal.[6] These bond lengths are clearly difficult to reconcile with those in the other oxides and with the sum of the ionic radii (for 6-coordination), about 2·6 Å.

(6) AC 1970 **A26** 501 (X-ray and n.d.)

Lead dioxide, PbO_2

The usual maroon form of this oxide has the rutile structure, the mean distance of Pb to the six octahedral neighbours being 2·18 Å. At 300° under 40 kbar pressure this converts to a black orthorhombic polymorph, the structure of which has been described in the discussion of close-packed structures in Chapter 4.

Some complex oxides of Pb(IV)

Closely related to PbO_2 are some complex oxides in which Pb(IV) is octahedrally coordinated. In Sr_2PbO_4[7] there are rutile-like chains which are joined through 7-coordinated Sr^{2+} ions (instead of Pb^{II} in Pb_3O_4), while Ba_2PbO_4 has the K_2NiF_4 (layer) structure (p. 171) which accommodates the larger Ba^{2+} ions in positions of 9-coordination:

(7) ZaC 1969 **371** 225

	Isostructural compounds
Ca_2PbO_4	Ca_2SnO_4
Sr_2PbO_4	Cd_2SnO_4
Ba_2PbO_4	Sr_2SnO_4
	Ba_2SnO_4

In contrast to the octahedral coordination of M(IV) in these compounds there is 5-coordination in the hygroscopic oxides $K_2M^{IV}O_3$ (M = Zr, Sn, Pb).[8] These contain the unusual MX_3 chain of Fig. 12.16 in which M has a tetragonal pyramidal arrangement of 5 O neighbours (one closer than the other four), the M atom lying above the base of the pyramid. The detailed structure of the Pb compound was not determined; in K_2SnO_3 bond lengths are: Sn–O, 1·93 Å, Sn–2 O, 2·03 Å, and Sn–2 O, 2·21 Å. Note the numerous examples of isostructural compounds of Sn(IV) and Pb(IV).

(8) JSSC 1970 **2** 410

The oxygen chemistry of some transition elements

For a discussion of certain aspects of their oxygen chemistry it is convenient to group together the six elements Ti, V, Nb, Mo, W, and Re. The following

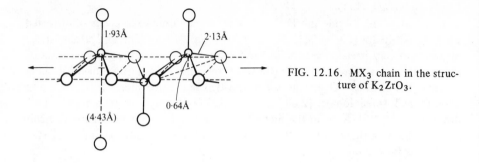

FIG. 12.16. MX_3 chain in the structure of K_2ZrO_3.

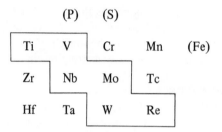

generalizations may be made. The diagonal relationship emphasized above is associated on the one hand with the larger size, more electropositive character, and reduced stability of lower oxidation states of Zr, Hf, and Ta as compared with the elements to their right, and on the other with the increasing resemblance in their highest oxidation states to the typical elements Si, P, S, and Cl as we go along the series T, V, Cr, and Mn. Thus Ti is typically ionic in its oxy-compounds, and while it can exist in lower oxidation states the ionic form Ti^{4+} in octahedral coordination is the preferred state of Ti in its oxy-compounds. In some of its oxy-compounds V^V shows a resemblance to P and the other 'tetrahedral' elements Si, Ge, and As (as in forming the tetrahedral VO_4^{3-} ion), while in others there is octahedral coordination by oxygen. However, in many crystals the coordination is more realistically described as trigonal bipyramidal or square pyramidal, the distinction between these two descriptions being somewhat arbitrary when there is appreciable distortion from the ideal arrangement.

As far as is known the oxygen chemistry of Cr^{VI} is based exclusively on tetrahedral coordination. Cr^{VI} is present in some, if not all, of the black oxides intermediate in composition between CrO_2 and CrO_3 (p. 947) which are obviously metallic oxides more akin to those of Mo and W. Not much is known of the structural chemistry of oxy-compounds of Mn^{VII} apart from the fact that the permanganate ion MnO_4^- is tetrahedral. We have already referred to Mn_2O_7. Both Mo^{VI} and W^{VI} form tetrahedral oxy-ions but their complex oxy-chemistry is largely based on octahedral coordination, which appears even in the 'pyro'-salts $Na_2Mo_2O_7$ and $Na_2W_2O_7$ (p. 431). Some finite oxy-ions of V, Nb, Mo, and W are described in

Chapter 11. Comparatively little is yet known of the complex oxy-chemistry of Re and still less in the case of Tc, while that of Ta has yet to be developed, but from what we know of these elements it would seem justifiable to group together Ti, V, Nb, Mo, W, and Re for a discussion of their complex oxy-chemistry.

The key to the structures of many of the complex oxy-compounds of these elements is the relation to the structures of TiO_2 (rutile) and ReO_3. VO_2 has normal and distorted forms of the rutile structure and NbO_2 has a complex superstructure of the rutile type. WO_3 crystallizes with distorted forms of the ReO_3 structure, but MoO_3 has a unique layer structure (p. 473). Octahedral MO_6 groups share only vertices in ReO_3, but by the sharing of various numbers and combinations of vertices and edges a great variety of more complex structures can be derived from portions of ReO_3 structure with formulae in the range MO_2–MO_3. Similarly, more complex structures may be built from portions of the rutile structure by the sharing of octahedral faces. In this chapter we shall devote sections to the following topics:

the oxides of Ti,
the oxygen chemistry of V, and
the oxides of Mo and W.

In Chapter 13 we discuss complex oxides of these elements, including the 'bronzes'.

The oxides of titanium

The elements Ti and V illustrate very well the extraordinary complexity of some metal–oxygen systems.

The h.c.p. metal Ti takes up oxygen to the stage $TiO_{0.50}$, and in this range three distinct phases have been recognized. First there is a random solid solution of O in Ti, and this is followed by preferential occupancy of certain positions leading to essentially ordered structures at compositions close to $TiO_{0.33}$ (anti-AX_3 layer structure) and $TiO_{0.50}$ (anti-CdI_2 structure). At higher oxygen content the hexagonal closest packing of the metal atoms breaks down. An oxide within the composition range $TiO_{0.68}$–$TiO_{0.75}$ results from oxidizing the metal under a low pressure of oxygen at temperatures below 900°C. This oxide has the ϵ-TaN structure (Fig. 29.22, p. 1056) but deficient in oxygen. The metal atoms are no longer close-packed but each Ti has two nearest Ti neighbours at 2·88 Å and twelve more at 3·22 Å forming a hexagonal prism. Each O has six Ti neighbours at the vertices of an elongated octahedron.

At the composition $TiO_{1.00}$ the phase has a defective NaCl structure. At high temperatures the structure has cubic symmetry with equal numbers of random vacancies in the cation and anion positions. At ordinary temperatures it has an ordered defective NaCl-type structure (Fig. 12.17) in which one-sixth of the sites for each type of ion are unoccupied. The symmetry is monoclinic with the angle β very close to 90° (89·9°). At 800°C the cubic NaCl-type structure has a narrow homogeneity range around $TiO_{1.18}$; at higher temperatures the range of composition increases, with a corresponding variation in the percentages of vacant sites.

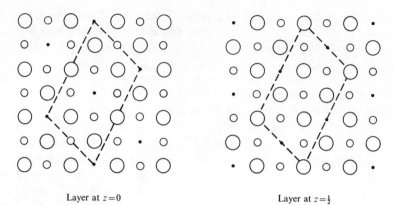

Layer at $z=0$ Layer at $z=\frac{1}{2}$

FIG. 12.17. Successive layers in the structure of TiO (*defect* NaCl superstructure). The small black dots indicate vacant cation or anion sites (one-sixth of the total of each kind).

For example:

O : Ti *ratio*	Percentage of sites occupied	
	Ti	O
1·33	74	98
1·12	81	91
0·69	96	66

Interpolation gives 85 per cent of each type of site occupied at the composition TiO.

These phases are followed by Ti_2O_3, which has the corundum structure at ordinary temperatures and undergoes a structural change at 200°C, Ti_3O_5, and no fewer than seven distinct phases in the range $TiO_{1.75}-TiO_{1.90}$. These form a series Ti_nO_{2n-1} (*n* from 4 to 10) comparable with the series of oxides formed by V, Mo, and W (see later). At temperatures below 100°C Ti_3O_5 consists of a 3D array of TiO_6 octahedra sharing edges and vertices (Fig. 12.18). At 100°C this oxide undergoes a rapid and reversible transformation to a form with a slightly distorted pseudobrookite structure (p. 498), a structure which is apparently stabilized at lower temperatures by the presence of a little Fe. In the low-temperature Ti_3O_5 there are metal–metal distances of three kinds across shared octahedron edges, approximately 3·1, 2·8, and 2·6 Å (compare 2·93 Å in α-Ti). The oxides Ti_4O_7 and Ti_5O_9, of the Ti_nO_{2n-1} series, are built of slabs of rutile structure, infinite in two dimensions and respectively four and five octahedra thick, joined by sharing octahedron *faces* (compare corundum).

The highest oxide is the familiar TiO_2. The three forms stable at atmospheric pressure, rutile, anatase, and brookite, all occur as minerals. These three polymorphs, together with the high-pressure form with the α-PbO_2 structure, are all

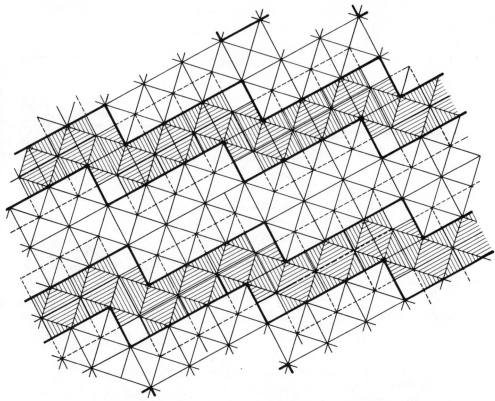

FIG. 12.18. Crystal structure of the low-temperature form of Ti_3O_5.

built of octahedral coordination groups, the Ti–O bond lengths in which have been noted (p. 447). A fifth polymorph, produced at pressures in excess of 150 kbar, is not stable at atmospheric pressure; it may contain Ti in a higher coordination state. Numerous complex oxides containing Ti are described in Chapter 13. The binary oxides are summarized in Table 12.7.

The oxygen chemistry of vanadium

Lower oxides.   Thermal and X-ray studies have shown that the vanadium–oxygen system shows marked similarities to the titanium–oxygen system up to the composition MO_2. A notable difference is that the structure of V_3O_5 is quite different from the structures of the Ti_3O_5 polymorphs. In V_3O_5 there is sharing not only of vertices and edges of octahedral coordination groups (as in Ti_3O_5) but also of faces. Across shared faces V–V is 2·74 Å and across shared edges 2·96–3·10 Å.

In addition to the α phase, a solid solution of O in b.c.c. V (up to $VO_{0·03}$ at 900°C), and the β phase ('V_4O'), $VO_{0·15}$–$VO_{0·25}$, consisting of a tetragonally

467

TABLE 12.7

The oxides of titanium

Oxide	Structure	Reference
Ti_3O	Anti-AX_3 (layer)*	AC 1968 **B24** 211
Ti_2O	Anti-CdI_2 (layer)	ACSc 1957 **11** 1641
δ-TiO_x	O-defective ϵ-TaN	ACSc 1959 **13** 415
TiO	NaCl superstructure	AC 1967 **23** 307
Ti_2O_3	Corundum	—
Ti_3O_5		AC 1959 **12** 575
Ti_4O_7	Members of homologous series Ti_nO_{2n-1} ($4 \leqslant n \leqslant 9$)	JSSC 1971 **3** 340
Ti_5O_9		ACSc 1960 **14** 1161
TiO_2–I	Rutile, anatase, brookite	—
—II	α-PbO_2	JCP 1969 **50** 519

Ti—O system: ACSc 1962 **16** 1245; AK 1963 **21** 413; JCP 1967 **46** 2461, 2465.

* There is an indefinitely large number of c.p. layer structures AX_3 in which a particular set of octahedral sites is occupied between alternate pairs of c.p. layers. For the pattern of sites of Fig. 4.24(e) the two simplest layer sequences give the h.c.p. BiI_3 (low-$CrCl_3$) and the c.c.p. YCl_3 structures. In Ti_3O the arrangement of the (c.p.) Ti atoms is h.c.p. and the pattern of sites of O atoms is that of Fig. 4.24(e), but the relative translations (parallel to the c.p. layers) of the sets of O atoms are different from those of the metal atoms in the BiI_3 structure.

(1) JM 1953 **5** 292
(2) ACSc 1954 **8** 221, 1599;
  ACSc 1960 **14** 465

deformed V lattice containing a small concentration of oxygen,[1] the oxides listed in Table 12.8 have been recognized.[2]

In V_2O_3 and the oxides intermediate between V_2O_3 and VO_2 there is octahedral coordination of V, appreciably distorted in some cases. For example, in V_4O_7 V—O ranges from 1·78 to 2·12 Å, and in monoclinic VO_2 from 1·76 to 2·06 Å, but in the

TABLE 12.8

The oxides of vanadium

Oxide	Structure		Reference
$VO_{0.53}$	—		ACSc 1960 **14** 465
VO	NaCl structure (homogeneous at 900°C over range $VO_{0.80}$–$VO_{1.20}$)		Refs. 1 and 2 above
$VO_{1.27}$	B.c. tetragonal superstucture of NaCl type (V-deficient)		ACSc 1960 **14** 465
$VO_{1.50}$	Corundum structure		
$VO_{1.67}$ to	A series of six oxides forming a homologous series	V_3O_5	MRB 1971 **6** 833
$VO_{1.87}$	V_nO_{2n-1} (V_3O_5 to V_8O_{15})	V_4O_7	AC 1972 **B28** 1404
VO_2	Monoclinic (MoO_2-type distorted rutile) (>68°C) Tetragonal rutile structure		ACSc 1970 **24** 420 JPSJ 1967 **23** 1380
$VO_{2.17}$	V_6O_{13}		ACSc 1971 **25** 2675
$VO_{2.25}$	V_4O_9	see text	ACSc 1970 **24** 3409
$VO_{2.33}$	V_3O_7		ACSc 1970 **24** 1473
$VO_{2.5}$	V_2O_5		ZK 1961 **115** 110

oxides between VO_2 and V_2O_5 (i.e. with oxidation states IV and V) the distortion of some of the octahedra is so great that the coordination is better described as 5-coordination. In V_6O_{13} there are equal numbers of three kinds of non-equivalent V atoms with nearest neighbours:

V^1	V^2	V^3
1·77 Å	1·66 Å	1·64 Å
1·88 (two)	1·76	1·92 (two)
1·96	1·90 (two)	1·93
1·99	2·08	1·98
2·06	2·28	2·26

These environments would appear to distinguish V^1 from V^2 and V^3, but the sums of the bonds strengths favour (though not in a clear-cut way) describing V^1 and V^3 as V(IV) and V^2 as V(V). In V_3O_7 the coordination of one-third of the metal atoms is (distorted) octahedral, while the remainder have only five nearest neighbours ($V^{IV}V_2^V O_7$?). However, in the compound described as V_4O_9 all V atoms form one short bond (1·60–1·65 Å), four bonds in the range 1·87–2·02 Å and a sixth long bond. The lengths of the latter for the four non-equivalent V atoms are 2·23, 2·40, 2·50, and 3·00 Å, so that one-quarter of the V atoms are described as 5-coordinated. It does not seem possible to correlate the type of coordination with the oxidation state of individual atoms (unless the compound were $V_3^{IV}V^V O_8(OH)$).

The oxide V_6O_{13} containing V^{IV} and V^V is thought to play an important part in the action of 'vanadium pentoxide' catalysts used in the oxidation of SO_2 to SO_3 and of naphthalene to phthalic anhydride. Its structure has been described in Chapter 5.

A number of complex oxides MVO_3 of V^{III} with La and the rare-earths have the perovskite structure (p. 153), and $CaV^{IV}O_3$ also has this structure.

(In the Ta–O system[3] the following phases have been recognized: α, Ta metal with up to about 5 atomic per cent O; β, a deformed version of the body-centred cubic α structure with a maximum O content corresponding to Ta_4O; γ, TaO with the NaCl structure; δ, TaO_2 with the rutile structure; and Ta_2O_5.)

(3) ACSc 1954 **8** 240

Vanadium pentoxide and vanadates. The structural chemistry of pentavalent V in V_2O_5 and vanadates is quite different from that of the lower oxides. Ortho-vanadates, containing tetrahedrally coordinated V^V are often isostructural with orthophosphates and orthoarsenates. For example, the dodecahydrates of Na_3PO_4, Na_3AsO_4, and Na_3VO_4 are isostructural, as also are the complex salts $Pb_5(PO_4)_3Cl$, $Pb_5(AsO_4)_3Cl$, and $Pb_5(VO_4)_3Cl$, while the rare-earth compounds $NdVO_4$, $SmVO_4$, $EuVO_4$, and YVO_4 all crystallize with the zircon ($ZrSiO_4$) structure like YPO_4 and $YAsO_4$. The pyrovanadate ZrV_2O_7 is isostructural with ZrP_2O_7.

Although V^V thus behaves like P^V in forming the ions VO_4^{3-} and $V_2O_7^{4-}$ it shows a much greater tendency than P to form condensed oxy-ions. Soluble ortho-, meta, and pyro-vanadates are known, but the order of stability of these salts in aqueous solution is the reverse of that of the corresponding phosphates. Whereas meta- and pyro-phosphates are converted into orthophosphates on boiling in solution, that is, the complex ions break down into simple PO_4^{3-} ions, the orthovanadate ion VO_4^{3-} is rapidly converted into the pyro-ion $V_2O_7^{4-}$ in the cold, and on boiling further condensation into metavanadate ions $(VO_3)_n^{n-}$ takes place. The particular type of vanadate which is obtained from a solution depends on the temperature and also on the acidity of the solution. Acidification of a vanadate solution leads to the formation of deeply-coloured polyvanadates. A weakly acid solution of potassium vanadate contains the $V_{10}O_{28}^{6-}$ ion (p. 430) but on long standing or heating less soluble polyvanadates are precipitated, of which KV_3O_8[4] and $K_3V_5O_{14}$[5] are examples. Both these salts have layer structures, and there is apparently somewhat irregular coordination of V by 5 or 6 O atoms. In KV_3O_8 the metal atom has one close O neighbour (at 1·6 Å), four more at about 1·9 Å, these five forming a square pyramid, and a sixth O at a much greater distance (2·5 Å) completing a very distorted octahedron. This irregular coordination, which is characteristic of many complex oxy-compounds of vanadium, is in marked contrast to that of P, which retains tetrahedral coordination in all phosphates. The detailed environment of V in the colourless, hygroscopic $NaVO_3$ (diopside structure) has not been determined, but in both $KVO_3 . H_2O$[6] and V_2O_5[7] there is 5-coordination of V, though of rather different kinds.

In Fig. 12.19(a) we show a 'meta' chain formed of tetrahedral groups linked through two corners. If now we place together two chains of type (a) each V has a fifth O neighbour in the other chain, as at (b). This double chain occurs in $KVO_3 . H_2O$ with the bond lengths shown, the five O atoms around V being arranged at the apices of a distorted trigonal bipyramid with V displaced towards two of the equatorial O atoms. (The K^+ ions and H_2O molecules are accommodated between the chains.) In $Sr(VO_3)_2 . 4 H_2O$[8] there are chains similar to those in $KVO_3 . H_2O$.

If these double chains are now joined through the atoms marked *a* to form layers we have the structure of V_2O_5, illustrated diagrammatically at (c). We may distinguish as O_1, O_2, and O_3 oxygen atoms attached to one two, and three V atoms respectively. The bond lengths in $KVO_3 . H_2O$ and V_2O_5 are as follows:

(4) IC 1966 **5** 1808
(5) ACSc 1959 **13** 377

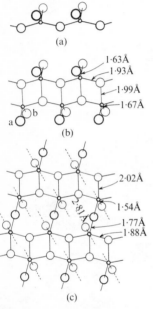

(a)

1·63Å
1·93Å
1·99Å
1·67Å

(b)

2·02Å
2·81Å
1·54Å
1·77Å
1·88Å

(c)

FIG. 12.19. (a) Metavanadate ion $(VO_3)_n^{n-}$, (b) double chain ion in $KVO_3 . H_2O$, (c) layer of the structure of V_2O_5.

(6) AC 1954 **7** 801
(7) ZK 1961 **115** 110

	$KVO_3 . H_2O$			V_2O_5	
V	$2 O_1$	1·63, 1·67 Å	V	$1 O_1$	1·59 Å
	$2 O_3$	1·93		$1 O_2$	1·78
	$1 O_3$	1·99		$2 O_3$	1·88
				$1 O_3$	2·02

In $KVO_3 . H_2O$ there are no further O neighbours which could be included in the coordination group, but in V_2O_5 V has a sixth neighbour in the adjacent layer at a distance of 2·8 Å, as indicated by the dotted lines in Fig. 12.19(c). The idealized structure of V_2O_5 is shown in Fig. 12.22(g) as built of octahedra to indicate its relation to the structure of MoO_3. In view of the V–O distances it is more realistic to describe the coordination as square pyramidal. If the structure of Fig. 12.22(g) is viewed in the direction of the arrow (top left) it appears as in Fig. 12.20(a), each square representing, as before, a ReO_3-type chain perpendicular to the paper, these chains now being those along the direction of the arrow. The longer (sixth) V–O bond in each 'octahedron' is shown as a broken line. Representing the coordination group of V as a square pyramid, that is, removing the sixth O from each octahedron, the structure is then represented as in Fig. 12.20(b). This characteristic coordination group is also found for all the V atoms in two vanadium bronzes, one of which has essentially the same structure as V_2O_5, while in some other bronzes there is both 5- and 6-coordination (for example, $Li_{1+x}V_3O_8$) or only 6-coordination ($Ag_{0.68}V_2O_5$)–see Chapter 13.

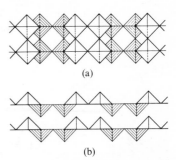

(a)

(b)

FIG. 12.20. The structure of V_2O_5 (see text).

Vanadium oxyhydroxides. As noted in Chapter 5 some interesting layer structures arise by lateral vertex–vertex linking of rutile-type chains. A number of these are found as the structures of minerals, some of which are dark-coloured compounds containing V in more than one oxidation state. They are listed in Table 12.9 which shows the apparent (mean) oxidation state of the metal and also includes the end-member of the series, VO(OH), which is isostructural with goethite and diaspore. The layers in these structures are held together by O–H · · · O bonds as shown in Fig. 12.21. Comparison of Fig. 12.21(c) with the idealized structure of V_2O_5, Fig. 12.22(g), shows that häggite is closely related to V_2O_5, into which it is converted on heating. There is considerable distortion of the VO_6 octahedra in

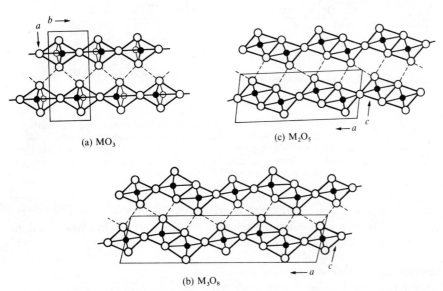

(a) MO_3

(c) M_2O_5

(b) M_3O_8

FIG. 12.21. Structures of vanadium oxyhydroxides: (a) $VO(OH)_2$, (b) $V_3O_4(OH)_4$, (c) $V_4O_4(OH)_6$.

471

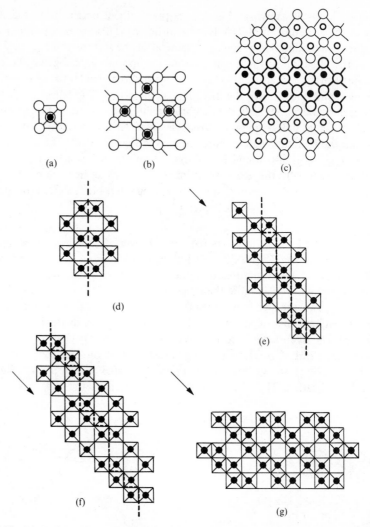

FIG. 12.22. (a) End-on view of infinite chain of MO_6 octahedra sharing opposite vertices, (b) ReO_3 structure viewed along direction of chains, (c)–(g) oxide structures in which chains of octahedra also share equatorial edges: (c) MoO_3 (layer) structure, (d) hypothetical structure for oxide M_nO_{3n}, (e) and (f) modes of linking slices of ReO_3 structure in oxides M_nO_{2n-1} and M_nO_{3n-2}, (g) idealized structure of V_2O_5.

these minerals; for example in häggite, V has 1 O at 1·82 Å and 5 more at a mean distance of 2·00 Å, and in doloresite the short bond is 1·70 Å with again 5 more at 2·00 Å (mean).

The oxides of molybdenum and tungsten

For an introduction to these and other oxides based on octahedral MO_6 coordination groups we refer the reader to the survey of such structures in Chapter 5.

472

TABLE 12.9
Structures of vanadium oxyhydroxides

Figure 12.21	Formula	Mineral name	Oxidation state of V	Reference
(a)	$VO_2(OH)$	Paraduttonite	5 ⎫	AC 1958 **11** 56
	$VO(OH)_2$	Duttonite	4 ⎭	
(b)	$V_3O_4(OH)_4$	Doloresite	4	AM 1960 **45** 1144
	$V_3O_3(OH)_5$	Protodoloresite	$3\frac{2}{3}$	AM 1960 **45** 1144
(c)	$V_4O_4(OH)_6$	Häggite	$3\frac{1}{2}$	AM 1960 **45** 1144
	$VO(OH)$	Montroseite	3	AM 1953 **38** 1242

We shall discuss the oxides of Mo and W together, but it should be noted that there is only a general structural similarity between these compounds. For example, both trioxides are built of octahedral MO_6 groups and there are close relationships between certain of the intermediate complex oxides, but of the oxides of these two metals only the dioxides are isostructural and Mo does not, for example, form compounds analogous to the W bronzes at atmospheric pressure, though at higher pressures Mo does form bronzes with structures similar to those of the tungsten bronzes (p. 510).

In general the structural chemistry of Mo oxy-compounds is more complex than that of the analogous W compounds; the reason for this difference is not known.

Before proceeding to the higher oxides of these metals we should mention that both are said to form an oxide M_3O. The second form of tungsten metal described as 'β-tungsten' is produced by such methods as electrolysis of fused mixtures of WO_3 and alkali-metal phosphates or of fused alkali-metal tungstates at temperatures below 700°C; above this temperature it turns irreversibly into α-W. It has been suggested[1] that β-W is in fact an oxide W_3O, and that the six W and two O atoms in the unit cell are arranged statistically in the eight positions (6-fold, open circles, and 2-fold, shaded circles) of Fig. 29.4 (p. 1017). (In Cr_3O[1a] the metal atoms are supposed to occupy the 6-fold and the O atoms the 2-fold positions.) On the other hand, it has been shown[2] that β-W can be prepared with less than 0·01 atoms of O per atom of W (by reducing $WO_{2.9}$ by hydrogen), though the presence of a small number of impurity atoms seems essential to the stability of the β-W structure. The oxide Mo_3O apparently has a defective anti-BiF_3 structure (p. 357) with nine Mo arranged at random in nine of the twelve positions (000), etc., (4-fold), $(\frac{1}{4}\frac{1}{4}\frac{1}{4})$, etc., (8-fold), and three O in the 4-fold position $(\frac{1}{2}\frac{1}{2}\frac{1}{2})$, etc.[3] More recently doubt has been expressed about the existence of Mo_3O (or Cr_3O).[4]

Dioxides and trioxides. The (isostructural) dioxides have distorted rutile structures (p. 448), but the trioxides are not isostructural. MoO_3 has a unique layer structure (Fig. 12.22(c)).[5] Each octahedral MoO_6 group shares two adjacent edges with similar groups and, in a direction perpendicular to the plane of the paper in Fig. 12.22, the octahedra are linked through vertices. Three O atoms of each MoO_6 octahedron are therefore common to three octahedra, two are shared between two octahedra, and the sixth is unshared. (See also p. 181.)

(1) AC, 1954 **7** 351
(1a) ACSc 1954 **8** 221

(2) ZaC, 1957 **293** 241

(3) ACSc 1954 **8** 617
(4) ACSc 1962 **16** 2458

(5) AK 1963 **21** 357

(6) JCP 1957 **26** 842

(7) AC 1956 **9** 475

(8) AC 1966 **21** 158, ibid., 1969 **B25** 1420

(1) ACSc 1959 **13** 954; IC 1966 **5** 136

(2) ACSc 1964 **18** 1571
(3) AK 1963 **21** 427
(4) ACSc 1948 **2** 501
(5) ACSc 1963 **17** 1485
(6) AK 1963 **21** 471
(7) AK 1950 **1** 513
(8) ACSc 1965 **19** 1514

(9) AC 1953 **6** 495

A mass-spectrometric study has been made of MoO_3 vapour (at 850°C),[6] in which the most abundant species are Mo_3O_9, Mo_4O_{12}, and Mo_5O_{15}.

WO_3 is monoclinic at ordinary temperatures and tetragonal at temperatures above 710°C; a third polymorph has been found in the temperature range 200–300°C.[7] All forms have structures of the ReO_3-type. X-ray and n.d. studies of the monoclinic form show that there is a range of W–O bond lengths (1·72–2·16 Å), four short and two long bonds in each WO_6 octahedron; long and short bonds alternate along two of the axial directions.[8]

Intermediate oxides. By heating together the metal and trioxide *in vacuo* to temperatures up to 700°C for varying times a number of intermediate oxides of these metals have been prepared.[1] They have been characterized by means of X-ray powder photographs and most of their structures have been determined from single-crystal data. These oxides, which are blue to violet in colour. include Mo_4O_{11},[2] Mo_5O_{14},[3] Mo_8O_{23},[4] Mo_9O_{26},[4] and $Mo_{17}O_{47}$,[5] $W_{18}O_{49}$[6] (in earlier literature W_2O_5 or W_4O_{11}), $W_{20}O_{58}$[7] and $W_{40}O_{118}$.[8]

In the orthorhombic form of Mo_4O_{11} three-quarters of the Mo atoms are 6-coordinated and the remainder tetrahedrally coordinated, and the structure may be regarded as consisting of slices of a ReO_3-like structure connected by 4-coordinated Mo atoms. The oxides Mo_5O_{14}, $Mo_{17}O_{47}$, and $W_{18}O_{49}$ are of a different type, having both octahedral and pentagonal bipyramidal coordination of the metal atoms (see 'Bronzes', Chapter 13).

The oxides Mo_8O_{23} and Mo_9O_{26}, a number of mixed oxides, and $W_{20}O_{58}$ form a group based on a common structural principle. In the ReO_3 structure octahedra are joined through vertices only (Fig. 12.22(b)). If there is sharing of edges of octahedra at regular intervals there arise series of related structures which may be regarded as built of slices of ReO_3-like structure. The simplest possibility would be that shown in Fig. 12.22(d), with octahedra of pairs of chains sharing edges. The dotted line marks the junction of the two portions of ReO_3 structure. Since no O atom is common to more than two octahedra the formula is still M_nO_{3n}. No example of this structure is known. More complex arrangements would involve octahedra belonging to groups of 4, 6, etc. ReO_3 chains sharing equatorial edges, as at (e) and (f), with formulae M_nO_{3n-1} and M_nO_{3n-2} respectively, the value of n depending on the repeat distance in the direction of the arrows in Fig. 12.22(e) and (f).[9]

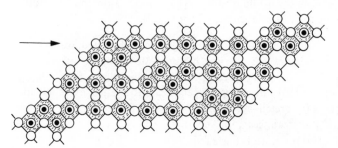

FIG. 12.23. The crystal structure of Mo_8O_{23}.

Examples of binary oxides with these structures are Mo_8O_{23} and Mo_9O_{26} in the M_nO_{3n-1} series and $W_{20}O_{58}$ and $W_{40}O_{118}$ in the M_nO_{3n-2} series. A larger portion of the structure of Mo_8O_{23} is shown in Fig.12.23. Higher members of the M_nO_{3n-1} series are represented by mixed (Mo, W) oxides, in which it appears that the proportion of W is of fundamental importance in determining the width of the ReO_3-like slices. The increasing value of n with increasing W content may be compared with the sharing of only vertices of WO_6 octahedra in WO_3 — contrast the complex structure of MoO_3.

Per cent W	Structure found [10]
0	Mo_8O_{23}, Mo_9O_{26}
14–24	$(Mo, W)_{10}O_{29}$, $(Mo, W)_{11}O_{32}$
32–44	$(Mo, W)_{12}O_{35}$
48	$(Mo, W)_{14}O_{41}$

[10] ACSc 1955 9 1382

In Fig. 12.22(d), (e), and (f) there are respectively 1, 2, and 3 *pairs* of adjacent MO_6 octahedra sharing edges; the limiting case would be the structure shown in (g), corresponding to the composition M_2O_5. This represents the idealized structure of V_2O_5 (p. 469).

The rather irregular coordination of V by five O may be described as square pyramidal or, since there is a sixth O at a much greater distance, as highly distorted octahedral. Although the sixth O is at a distance nearly twice as great as that to the nearest O, this interesting relation between the idealized structure of Fig. 12.22(g) and that of MoO_3 in Fig. 12.22(c) is apparently the reason for the considerable solubility of MoO_3 in V_2O_5. At 650°C about 30 per cent of the V atoms can be substituted by Mo, and the structure of $(Mo_{0.3}V_{0.7})_2O_5$[11] is of the same general type as V_2O_5. MoO_3 does not, however, take up V_2O_5 in solid solution; this would require an excess of metal atoms in the structure.

[11] ACSc 1967 21 2495

Another series of molybdenum oxides, Mo_nO_{3n-m+1} ($n = 13$, $m = 2$; $n = 18$, $m = 3$; $n = 26$, $m = 4$), based on the MoO_3 structure, has been described.[12]

[12] AK 1963 21 443

13

Complex Oxides

Introduction

In Chapter 11 we distinguished between oxy-salts and complex oxides, but observed that there is no hard and fast dividing line between the two groups of compounds. As regards their crystal structures we may distinguish two main classes of complex oxide.

I. The positions of the atoms are the same (or essentially the same) as in a binary oxide.

(a) In most binary oxides in which all metal atoms are in the same oxidation state the environment of all the metal atoms is the same or approximately the same. (Exceptions include the minor difference between the two types of distorted octahedral coordination group in the C-M_2O_3 structure and the much larger difference in β-Ga_2O_3, with tetrahedral and octahedral coordination of the metal, and the B-M_2O_3 structure, with 6- and 7-coordination of the metal.) In the complex oxide with such a structure there may be random arrangement of atoms of two or more metals (*statistical structure*) or a regular arrangement (*superstructure*).

(b) If the binary oxide contains the metal in two oxidation states there may be appreciably different environments of the two kinds of metal ion as, for example, in Pb_3O_4 (3- and 6-coordination of Pb(II) and Pb(IV)) or Eu_3O_4 (6- or 8-coordination of Eu(III) and Eu(II)). Such structures are also possible for complex oxides, the structure usually being a regular one (like that of the binary oxide) rather than a statistical one. Some structures common to simple and complex oxides are listed in Table 13.1.

II. In some complex oxide structures the environments of the different kinds of metal ion are so different that the structure is not possible for a binary oxide. The size difference between the ions necessary for the stability of the structure may be too large (as in the perovskite and related structures) or the two (or more) oxidation states required for charge balance in the structure may not be possible for one metal. (The scheelite structure, for example, calls for equal numbers of 8- and 4-coordinated atoms in oxidation states totalling 8.) It should be noted that positions of different coordination number in complex oxide structures are not necessarily occupied by atoms of different metals. Just as one kind of ion occupies coordination groups of two kinds in certain exceptional binary oxides, as noted above, so we find the same phenomenon in some complex oxides. In the garnet structure (p. 500) there are positions of 4-, 6-, and 8-coordination for metal ions. In

TABLE 13.1

Structures common to simple and complex oxides

Structure	Simple oxide	Complex oxide	
		statistical	*superstructure*
NaCl	MgO etc.	Li_2TiO_3, $LiFeO_2$	$LiNiO_2$, $LiInO_2$
Wurtzite	ZnO		$LiGaO_2$
β-BeO	β-BeO		$LiAlO_2$
Rutile	TiO_2 etc.	$CrTaO_4$, $CrNbO_4$	$ZnSb_2O_6$
Corundum	α-Al_2O_3		$FeTiO_3$
			$LiNbO_3$
C-M_2O_3	Mn_2O_3, etc.	$CaUO_3$	
Columbite ⎫	ReO_2 (high)		$FeNb_2O_6$
Wolframite ⎭	α-PbO_2		$NiWO_4$

	Oxide with same metal in two oxidation states	Statistical	Regular structure
	α-Sb_2O_4		$SbNbO_4$, $SbTaO_4$
	β-Sb_2O_4		$BiSbO_4$
Spinel	Fe_3O_4 etc.	Inverse spinels	$MgAl_2O_4$
			$LiAl_5O_8$ (low)
	Pb_3O_4		$ZnSb_2O_4$
$CaFe_2O_4$	Eu_3O_4		$SrEu_2O_4$
Pseudobrookite	Ti_3O_5		Fe_2TiO_5
	Mn_5O_8		$Cd_2Mn_3O_8$
	Tb_7O_{12}		$U^{VI}M^{III}_6O_{12}$
	$Nb_{12}O_{29}$	$Ti_2Nb_{10}O_{29}$	

some garnets ions of the same kind occupy sites of 4- and 6-coordination and in others sites of 6- and 8-coordination. It is, however, unlikely that all three types of site would be occupied by ions of the same kind in a particular crystal. Structures of this second main group could be described as characteristically or exclusively *complex oxide structures*.

Many complex oxide structures can be described in such a way as to emphasize a particular feature of the structure and have been so described in previous chapters, in particular the following groups:

structures containing chains, layers based on simple plane nets, or frameworks based on simple 3D nets (Chapter 3);

structures in which the O atoms are close-packed and M atoms occupy tetrahedral and/or octahedral interstices. In a special group $A_nB_mO_{3n}$ the A and 3 O atoms together form the c.p. array (Chapter 4);

structures, not necessarily close packed, built from tetrahedral and/or octahedral coordination groups (Chapter 5).

The grouping of structures in this way stresses the geometry or topology of the structure; we adopt here a more 'chemical' classification. We first deal with the simpler structures arranged in order of increasing complexity of chemical formula and then devote separate sections to certain selected groups of compounds.

477

Complex Oxides

Oxides ABO_2

A number of oxides $M^IM^{III}O_2$ have structures closely related to oxides $M^{II}O$ (Table 13.2). The tetragonal NaCl superstructure is illustrated in Fig. 6.2, p. 196, and the

TABLE 13.2

Structures of oxides $M^IM^{III}O_2$

Coordination of metal atoms	Structure	Examples
Both tetrahedral	Zinc-blende superstructure	h.p. $LiBO_2$[1]
	Wurtzite superstructure	$LiGaO_2$[2]
	β-BeO	γ-$LiAlO_2$[3]
Both octahedral	NaCl superstructure:	
	Tetragonal	$LiFeO_2$, $LiInO_2$[4]
		$LiScO_2$, $LiEuO_2$
	Rhombohedral	$LiNiO_2$, $LiVO_2$
		$NaFeO_2$, $NaInO_2$[5]
M^I 2 collinear $\}$ M^{III} 6 octahedral	$HNaF_2$	$CuCrO_2$[6], $CuFeO_2$ (see also p. 219).

(1) JCP 1966 **44** 3348
(2) AC 1965 **18** 481
(3) AC 1965 **19** 396
(4) JCP 1966 **44** 3348 (review of oxides $LiMO_2$)
(5) ZaC 1958 **295** 233
(6) JACS 1955 **77** 896

relation between the rhombohedral NaCl superstructure and the $NaHF_2$ structure (which should be written $HNaF_2$ in the present context) has been described on p. 219. The latter structure is adopted by numerous Ag and Cu compounds $M^IM^{III}O_2$ (e.g. M^{III} = Al, Cr, Co, Fe, Ga, Rh), and also by $PdCoO_2$, $PdCrO_2$, $PdRhO_2$, and $PtCoO_2$.[1] It has been confirmed (Mössbauer) that $CuFeO_2$ contains Fe^{3+} (and therefore Cu^+), and apparently the Pd and Pt compounds contain the noble metal in the formal oxidation state +1. (There is some doubt about the stoichiometry of $PtCoO_2$; if it is metal-deficient it could be $Pt_{0.8}^{2+}Co_{0.8}^{3+}O_2$.) However, there is apparently metal–metal bonding in the Pd and Pt compounds. In the $HNaF_2$ structure for compounds ABO_2 there is linear 2-coordination of A by O, but A also has 6 A neighbours, hence the coordination of Pt and Pd in these compounds should probably be described as 2 + 6 (hexagonal bipyramidal), the six equatorial neighbours being metal atoms, (Table 13.3). Delocalization of electrons in the hexagonal layers of metal atoms is responsible for the metallic conduction, suggesting the formulation $Pt^{II}Co^{III}O_2(e)$.

(1) JSSC 1971 **10** 713, 719, 723

TABLE 13.3

Metal–metal distances in oxides ABO_2

Compound	A-6 A	Distance in metal	
$CuFeO_2$	3·04 Å	2·56 Å (Cu)	semiconductors
$AgFeO_2$		2·89 Å (Ag)	
$PtCoO_2$	2·83 Å	2·77 Å (Pt)	metallic conductors
$PdCoO_2$		2·75 Å (Pd)	

478

Metaborates MBO_2 generally contain metaborate ions, but under high pressures $LiBO_2$ crystallizes as a zinc-blende superstructure.

Several structures of oxides $M^{II}M^{II}O_2$ have been determined in which ions of appreciably different size are accommodated. In $SrZnO_2$[2] layers of ZnO_4 tetrahedra (sharing all vertices) provide sites for Sr^{2+} ions in 7-coordination between the layers (see Fig. 5.6, p. 162 for the layer), while in $BaZnO_2$[3] the ZnO_4 tetrahedra form a quartz-like framework containing Ba^{2+} ions in positions of irregular 8-coordination. In contrast to $BaZnO_2$ (and the isostructural Co and Mn compounds) $BaCdO_2$[4] has a structure in which there is rather irregular $(4+1)$-coordination of Cd and 7-coordination of Ba. In $BaNiO_2$[5] there are perovskite-like layers from which one-third of the O atoms are missing. The layers are stacked in approximate h.c.p. with Ni atoms between the layers in what would have been positions of octahedral coordination, but because of the absence of two of the O neighbours Ni has only 4 (coplanar) neighbours.

(2) ZaC 1961 **312** 87

(3) AC 1960 **13** 197; ZaC 1960 **305** 241

(4) ZaC 1962 **314** 145

(5) AC 1951 **4** 148

Oxides ABO_3

The two largest groups of oxides ABO_3 are (a) those containing ions A and B of approximately the same size and of a size suitable for octahedral coordination by oxygen (e.g. 3d ions M^{2+} and M^{3+}, Zn^{2+}, Mg^{2+}, In^{3+}, etc.), and (b) those containing a much larger ion which together with O^{2-} can form c.p. layers AO_3. An intermediate class contains the defect pyrochlore structure (Table 13.4). Oxides of class (a) adopt a sesquioxide structure with either random or ordered arrangement of A and B ions, or in some cases a structure peculiar to a small number of compounds (e.g. $LiSbO_3$). The ilmenite ($FeTiO_3$) and $LiNbO_3$ structures differ in the way shown in Fig. 6.19, p. 216, that is, in the distribution of the two kinds of ion among a given set of octahedral sites. On the other hand a different selection of metal-ion sites is occupied in $LiSbO_3$, as shown in Fig. 4.24(f); $NaSbO_3$ and $NaBiO_3$ have the ilmenite structure. Typically, transition-metal oxides $M^{3+}M^{3+}O_3$ adopt a random corundum structure, and oxides $M^{2+}M^{4+}O_3$ the ilmenite superstructure, for example:

Corundum	FeV, TiV	MnFe,	FeCr, VCr NiCr	InFe GaFe	(under pressure)
Ilmenite	FeTi CoTi NiTi		$CoMn^{4+}$ NiMn MgMn	CoV^{4+} NiV MgV	(distorted)

In some systems the oxygen-deficient pyrochlore structure $A_2B_2O_{7-x}$ is stable at $x = 1$ (e.g. $PbReO_3$, $BiYO_3$, $AgSbO_3$, $BiScO_3$, $PbTcO_3$). Pressure may convert to the perovskite structure, as in the case of $BiYO_3$ and $BiScO_3$. This change involves a rearrangement of the vertex-sharing BO_6 octahedra to form a different 3D framework and an increase in the c.n. of A to 12. In other systems the phase stable

TABLE 13.4

Structures of oxides ABO_3

C.N.'s of A and B	Structure	Examples	Reference
6 : 6	C-M_2O_3 Corundum: random superstructure ilmenite	$ScTiO_3$, $ScVO_3$ $A^{3+}B^{3+}O_3$ $A^{2+}B^{4+}O_3$	IC 1967 **6** 521 AC 1964 **17** 240; JSSC 1970 **1** 512
	$LiNbO_3$ $LiSbO_3$	$A^+B^{5+}O_3$ $BiSbO_3$	AC 1968 **A24** 583 ACSc 1954 **8** 1021 ACSc 1955 **9** 1219
6 : 6 12 : 6	Defect pyrochlore C.P. structures	$PbReO_3$ (see Table 13.5)	MRB 1969 **4** 191

at atmospheric pressure has a defect pyrochlore structure with $x < 1$. For example, heating the components in stoichiometric proportions gives $Pb_2Ru_2O_{7-x}$ which at $1400°C$ under a pressure exceeding 90 kbar converts to a perovskite-type $PbRuO_3$ with loss of oxygen.

The structures based on c.p. AO_3 layers have the composition ABO_3 if all the octahedral holes are occupied ·by B atoms. Together with some closely related structures, in which some of these holes are unoccupied, they form the subject-matter of the next section.

Structures based on close-packed AO_3 layers

Certain cations comparable in size with O^{2-} form c.p. layers AO_3 which can be stacked in various c.p. sequences. Smaller cations can then occupy the octahedral holes between groups of six O^{2-} ions to form structures of the type $A_nB_mO_{3n}$. Some of these structures have been described in Chapter 4, and it was noted in Chapter 5 that these structures may alternatively be described in terms of the way in which the BO_6 octahedra are linked together. Only vertices and/or faces are shared, and the extent of face-sharing is indicated in Table 13.5. We shall deal in detail only with the simplest of the c.p. ABO_3 structures, the perovskite structure. The structure of hexagonal $BaTiO_3$ is compared with perovskite in Fig. 13.1, and in Fig. 13.2 we show sections through the 5-, 9-, and 12- layer structures to illustrate the relations between the BO_6 octahedra.

As regards magnetic properties it is important to distinguish between those structures in which octahedra share only vertices and those in which face-sharing occurs. The sharing of faces is associated with short metal–metal separations (e.g. Ru–Ru, 2.55 Å in $BaRuO_3$) and often with abnormal magnetic and electrical properties. Face-sharing occurs in those structures which are marked with an asterisk in Table 13.5; it ranges from the formation of pairs, in hexagonal $BaTiO_3$ and high-$BaMnO_3$, through groups of 3 in $BaRuO_3$, to infinite chains of octahedra in $BaNiO_3$ in which every NiO_6 octahedron shares a pair of opposite faces. Face-sharing does not occur in $Sr_5Ta_4O_{15}$ (and the isostructural $Ba_5Ta_4O_{15}$ and

TABLE 13.5

Structures with close-packed AO_3 *layers*

Number of layers	Type of closest packing	Face-sharing groups	Example	Reference
2*	h	Chains	$BaNiO_3$	AC 1951 **4** 148
3	c	–	Perovskites	
4*	hc	All pairs	High-$BaMnO_3$	AC 1962 **15** 179
5	hhccc	–	$Sr_5Ta_4O_{15}$	AC 1970 **B26** 102
6*	hcc	Single + pairs	hex. $BaTiO_3$	AC 1948 **1** 330
8	ccch	–	$Sr_4Re_2SrO_{12}$	IC 1965 **4** 235
9*	chh	Groups of 3	$BaRuO_3$	IC 1965 **4** 306
12	hhcc	–	$A_4Re_2BO_{12}$	IC 1965 **4** 235
			(A = Sr, Ba)	
			(B = Mg, Zn, Co)	

Structures containing other groups of face-sharing octahedra (which are further linked by sharing terminal vertices) apparently occur in the system $BaMnO_{3-x}$ (JSSC 1971 **3** 323) as follows: 4 + 2 (6-layer). 4 (8-layer), 3 + 2 (10-layer), and 5 (15-layer). Complete single crystal structure determinations have not yet been published.

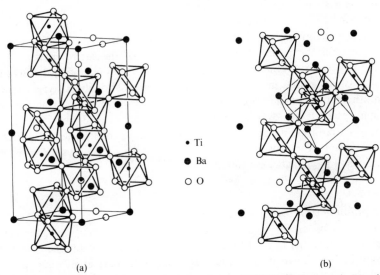

- • Ti
- ● Ba
- ○ O

(a)   (b)

FIG. 13.1. The crystal structures of barium titanate, $BaTiO_3$: (a) hexagonal, (b) cubic.

$Ba_5Nb_4O_{15}$) or in $A_4Re_2BO_{12}$ (Fig. 13.2). We were careful to say above that face-sharing is *associated with* short metal–metal separations, which are often an indication of bonding between the metal atoms. It is important to remember that this bonding (and any abnormal physical properties to which it may give rise) is not a *result* of the geometry of the structure but rather the reverse; the structure results from the various interactions between the component atoms. Structural studies, not

481

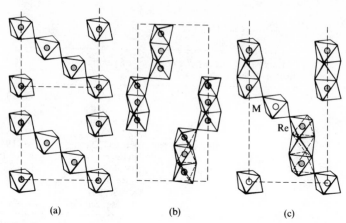

FIG. 13.2. Sections through 5-, 9-, and 12-layer structures built of AO_3 layers, showing how the BO_6 octahedra are linked through vertex- or face-sharing: (a) $Ba_5Ta_4O_{15}$, (b) $BaRuO_3$, (c) $A_4Re_2MO_{12}$. In all cases the A atoms are omitted, and in (c) only eight layers are shown.

(1) JPC 1964 **68** 3786

(2) IC 1962 **1** 245

only of oxides but also of some halides, sulphides, etc., show that certain structures—and hence certain compounds—arise because of the contribution made to the lattice energy by metal–metal bonding. In compounds such as Ta_6S and Gd_2Cl_3 this must be responsible for a major part of the lattice energy; in other cases it is less important but still a determining factor. For compounds ABO_3 there is a choice of structures, in some of which metal–metal bonding plays a larger part than in others. It appears, for example, that the hexagonal $BaTiO_3$ structure, which is in some cases stable over a range of composition (e.g. $BaFeO_{2.47-2.92}$),[1] requires the transition-metal atoms to stabilize it by metal–metal interactions, though it is not necessary for all the B atoms to be of this type; for example, Ba_2IrScO_6 and Ba_2IrInO_6 have this structure.[2]

Polymorphism of c.p. oxides ABO_3

The extension of earlier studies of polymorphism, involving only temperature changes, to include the effect of pressure (which also unavoidably includes a rise in temperature) has shown many structural changes in these as in other groups of compounds. Five structures are commonly encountered for these c.p. oxides (Table 13.6). It appears that the amount of 'hexagonal character' in a particular compound (or series of compounds) decreases with pressure and also with decreasing size of A in a series such as $BaMnO_3$, $SrMnO_3$, $CaMnO_3$. It is by no means certain that all the earlier literature refers to stoichiometric compounds. For example, $BaMnO_3$–4 H prepared at atmospheric pressure is oxygen-deficient with *larger* cell dimensions than those of stoichiometric $BaMnO_3$ prepared under pressure. The preparation of a series of non-stoichiometric polymorphs (polytypes ?) of $BaMnO_3$ with 4-, 6-, 8-, 10-, and 15-layer sequences has been described.

TABLE 13.6

Effect of pressure on c.p. ABO_3 and ABX_3 structures

C.P. type	h	hhc	hc	cch	c
Symbol †	2 H	9 R	4 H	6 H	3 C
Groups of face-sharing octahedra	infinite chains	(3)	(2)	(2) + (1)	–

$CsMnF_3$-I $\longrightarrow CsMnF_3$-II

$CsNiF_3$-I $\longrightarrow CsNiF_3$-II $\longrightarrow CsNiF_3$-III
     5 kbar   50 kbar

$BaRuO_3$-I→$BaRuO_3$-II→$BaRuO_3$-III
      15 kbar   30 kbar

$BaMnO_3$-I $\longrightarrow BaMnO_3$-II→$BaMnO_3$-III
  30 kbar   90 kbar

$SrMnO_3$-I $\longrightarrow SrMnO_3$-II
          50 kbar

$CaMnO_3$

(3) IC 1965 **4** 71

(4) IC 1967 **6** 1474

† H = hexagonal, R = rhombohedral, C = cubic.

For $SrMnO_3$ and $BaMnO_3$ see. JSSC 1969 **1** 103, 506; JSSC 1971 **3** 323.

The perovskite structure

This structure is adopted by a few complex halides and by many complex oxides. The latter are very numerous because the sum of the charges on A and B (+6) may be made of 1 + 5, 2 + 4, or 3 + 3, and also in more complex ways as in $Pb(B'_{\frac{1}{2}}B''_{\frac{1}{2}})O_3$, where B' = Sc or Fe and B'' = Nb or Ta, or $La(B'_{\frac{1}{2}}B''_{\frac{1}{2}})O_3$, where B' = Ni, Mg, etc. and B'' = Ru(IV) or Ir(IV).[3] Also, many compounds ABO_3 are polymorphic with as many as four or five forms, some of which represent only small distortions of the most symmetrical form of the perovskite structure. The study of these compounds under higher pressures has produced many more examples of polymorphism (e.g. $InCrO_3$ and $TlFeO_3$).[4] The perovskite structure of compounds such as $CaFeO_3$ (high pressure), $SrFeO_3$, and $BaFeO_3$, is of interest as an example of the stabilization of a high oxidation state (Fe^{IV}) in an oxide structure.

The 'ideal' perovskite structure, illustrated in Fig. 13.3(a), is cubic, with A surrounded by 12 O and B by 6 O. Comparatively few compounds have this ideal cubic structure, many (including the mineral perovskite, $CaTiO_3$; itself) having slightly distorted variants with lower symmetry. Some examples are listed in Table 13.7. These departures from the most symmetrical structure are of great interest because of the dielectric and magnetic properties of these compounds. For example, many are ferroelectric, notably $BaTiO_3$, some are antiferroelectric, for example, $PbZrO_3$ and $NaNbO_3$ (see p. 486), and ferromagnetic ($LaCo_{0.2}Mn_{0.8}O_3$) and antiferromagnetic ($GdFeO_3$, $LaFeO_3$, etc.) compounds are known.

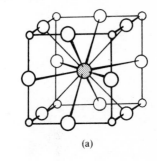

(a)

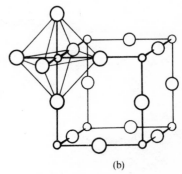

(b)

FIG. 13.3. (a) The perovskite structure for compounds ABO_3 (or ABX_3). Large open circles represent O (or F) ions, the shaded circle A, and the small circles B ions. (b) The crystal structure of ReO_3.

TABLE 13.7

Compounds with the perovskite type of structure

Ideal cubic structure	$SrTiO_3$, $SrZrO_3$, $SrHfO_3$ $SrSnO_3$, $SrFeO_3$ $BaZrO_3$, $BaHfO_3$, $BaSnO_3$ $BaCeO_3$ $EuTiO_3$, $LaMnO_3$

At least one form with distorted small cell ($a \approx 4$ Å):
    Cubic (C)
    Tetragonal (T)
    Orthorhombic (O)
    Rhombohedral (R)
	$BaTiO_3$ (C, T, O, R)
	 $KNbO_3$ (C, T, O, R)
	 $KTaO_3$ (C, ?)
	 $RbTaO_3$ (C, T)
	 $PbTiO_3$ (C, T)

Distorted multiple cells	$CaTiO_3$, $NaNbO_3$, $PbZrO_3$ $PbHfO_3$, $LaCrO_3$ low-$PbTiO_3$ low-$NaNbO_3$, high-$NaNbO_3$

The complex oxide $BaTiO_3$ ('barium titanate') is remarkable in having five crystalline forms, of which three are ferroelectric. The structure of the high-temperature hexagonal form, stable from 1460°C to the melting point (1612°C), has already been described. The other forms are:

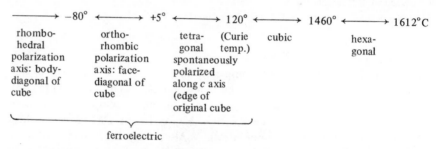

KNbO₃ shows the closest resemblance to $BaTiO_3$, having three transitions (to tetragonal, orthorhombic, and rhombohedral forms), while $NaNbO_3$ shows quite different dielectric behaviour, its transitions (at 350°, 520°, and 640°C) being of a different kind from those in $KNbO_3$ and $BaTiO_3$.[5]

In attempts to understand the distortions from the cubic structure these oxides ABO_3 were first regarded as purely ionic crystals. From the geometry of the structure it follows that for the 'ideal' structure there is the following relation between the radii of the A, B, and O^{2-} ions:

$$r_A + r_O = \sqrt{2}(r_B + r_O).$$

Actually the cubic perovskite structure or slightly deformed variants of it are found for ions which do not obey this relation exactly, and this was expressed by introducing a 'tolerance factor':

$$r_A + r_O = t\sqrt{2}(r_B + r_O).$$

(5) AC 1956 9 256

Provided that the ionic sizes are approximately right, the only other condition to be fulfilled is that the structure is electrically neutral, that is, that the sum of the charges on A and B is 6.

It then appeared that for all the compounds with the perovskite-type of structure the value of t lies between approximately 0·80 and 1·00—for lower values of t the ilmenite structure is found—and that for the ideal cubic structure t must be greater than 0·89. However, it is now evident that although in a given ABO_3 series (with either A or B constant) the use of a self-consistent set of radii does enable one to use the tolerance factor t to predict the approach to the ideal structure, it is not possible to compare two series of perovskites in which both A and B have been changed because the effective ionic radii are not constant in all the crystals. For example, in yttrium compounds B—O varies by as much as 0·09 Å, and in Fe compounds Fe—O varies by 0·05 Å. Moreover, it has been assumed in the above discussion that the ionic radii of Goldschmidt were used. If other 'ionic radii' (e.g. those of Pauling) are used, no such simple relation is found between t and the departures from cubic symmetry. Moreover, the properties of solid solutions suggest that the above picture is much over-simplified. For example, replacement of Ba^{2+} by a smaller ion such as Sr^{2+} or Pb^{2+} might be expected to have the same effect as replacing Ti^{4+} by the larger Zr^{4+} or Sn^{4+}. In fact, replacement of Ti by Zr or Sn lowers the Curie temperature of $BaTiO_3$ and replacement of Ba by Sr has the same effect, but the smaller Pb^{2+} has the opposite effect. Indeed, the situation is even more complex, for in the $(Ba, Pb)TiO_3$ system the upper transition temperature rises but the two lower transition temperatures fall as Pb replaces Ba. In pure $PbTiO_3$ the transition (at 490°C) is associated with structural changes so extensive as to crack large crystals as they cool through the Curie temperature.

The determinations of the structures of the tetragonal forms of $BaTiO_3$[6] and $PbTiO_3$[7] provide good examples of the power of combined X-ray and neutron diffraction studies in cases where the X-ray method alone fails to give an unambiguous answer. The shifts of Ti and Ba (Pb) are conventionally shown relative to the O_6 octahedra around the original Ti position. In $BaTiO_3$ Ti shifts by 0·12 Å and Ba in the same direction by 0·06 Å (Fig. 13.4(a)). In tetragonal $PbTiO_3$ Ti is shifted by 0·30 Å with respect to the oxygen octahedra and Pb moves in the same direction by 0·47 Å (Fig. 13.4(b)). The Ti environments in these crystals are shown at (c) and (d). In both cases Ti is displaced from the centre of its octahedron giving one short Ti—O distance of 1·86 Å in $BaTiO_3$ and 1·78 Å in $PbTiO_3$ (compare the mean value 1·97 Å in rutile). As regards the Ba, the shift has a negligible effect on the twelve Ba—O distances, but in the Pb compound the large shift of Pb does have an appreciable effect on the Pb—O distances as shown at (e). These distances may be compared with 2·78 Å, the sum of the Goldschmidt ionic radii for 12-coordination. This difference is the most marked one between ferroelectric $BaTiO_3$ and $PbTiO_3$, and the environment of Pb, with four short Pb—O distances to one side, is reminiscent of that in tetragonal PbO. Whether the important factor in the transition to ferroelectric forms is the long-range forces of dipole interaction or an increase in the homopolar character of the Pb—O bonds is still a matter for discussion and is outside the scope of this book. Detailed studies have also been

(6) PR 1955 **100** 745
(7) AC 1956 **9** 131

485

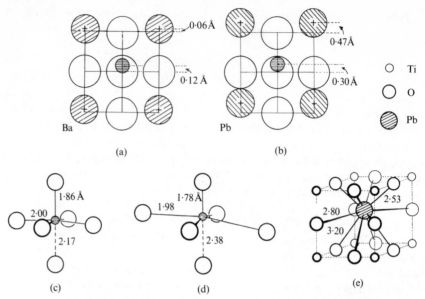

FIG. 13.4. Environments of ions in perovskite-type structures (see text).

(8) AC 1967 **22** 639
(9) AC 1969 **B25** 851

made of the ionic displacements in other perovskites, including $CaTiO_3$, $CdTiO_3$, $NaTaO_3$, $KNbO_3$,[8] and $NaNbO_3$.[9]

The fact that the structures of compounds ABO_3 are dependent not only on size factors but also on the nature of B has been demonstrated in many comparative studies. For example, while $AFeO_3$ (A = lanthanide) all have perovskite-type structures this is true for $AMn^{3+}O_3$ only if A is La or Ce–Yb. The compounds in which A = Ho–Lu adopt a new hexagonal structure with 5- and 7-coordination of

(10) AC 1963 **16** 957

Mn and A respectively.[10] In the series $Ba_{1-x}Sr_xRuO_3$ the structure changes from the 9-layer to the 4-layer and then to the perovskite structure for $x = 0$, $\frac{1}{6}$, and $\frac{1}{3}$, but

(11) IC 1966 **5** 335, 339

$Ba_{1-x}Sr_xIrO_3$ shows a more complex behaviour.[11]

Superstructures of perovskite. If B is progressively replaced by a second metal a large size difference tends to lead to superstructures rather than random arrangements of the two kinds of ion. The relation of the cryolite structure to perovskite has already been described in Chapter 10, where we noted oxides such as Ba_2CaWO_6 as having this type of superstructures. (A distorted form of the cryolite

(12) AC 1966 **20** 508

structure is adopted by $Ca_2'(Ca''U)O_6$,[12] possibly due to the tendency of U(VI) to form two stronger bonds; the Ca″ ions have a similar environment (compare Mg in $MgUO_4$).) Similarly in compounds $Ba_3M^{II}Ta_2O_9$ there is random arrangement of M^{II} and Ta in the octahedral positions when M^{II} is Fe, Co, Ni, Zn, or Ca, but

(13) JACS 1961 **83** 2830

$Ba_3SrTa_2O_9$ has a hexagonal (ordered) superstructure.[13]

In a number of oxides

MNb_3O_9 (M = La, Ce, Pr, Nd); and
MTa_3O_9 (M = La, Ce, Pr, Nd, Sm, Gd, Dy, Ho, Y, Er)

486

there is an octahedral framework of the ReO_3-type but incomplete occupancy of the 12-coordinated sites (i.e. a *defect* perovskite structure).[14] The B sites in Fig. 13.5 are all vacant and there is two-thirds occupancy of the A sites.

(14) AC 1967 **23** 740

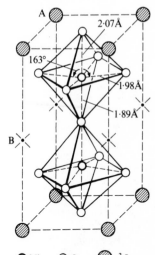

O Nb   O O   ▨ $\frac{2}{3}$ La

FIG. 13.5. Structure of $LaNb_3O_9$.

Oxides ABO_4

Numerous structures are found for compounds ABO_4, many of which are polymorphic at atmospheric pressure and also adopt different structures at higher pressures. These compounds range from those with silica-like structures, with tetrahedral coordination of A and B to oxy-salts containing well-defined oxy-ions, BO_4, and high coordination number of A. Some of these structures are listed in Table 13.8, and we comment here only on a number of general points. (For uranates $CaUO_4$ etc. see Chapter 28.)

Some series of compounds illustrate the change in environment of A with its size, as in chromates of divalent metals:

Structure	C.N. of A	
$BaSO_4$	12	$PbCrO_4$, $BaCrO_4$, $SrCrO_4$
$ZrSiO_4$	8	$CaCrO_4$
$CrVO_4$	6	$ZnCrO_4$, $CdCrO_4$, $CuCrO_4$

TABLE 13.8
Crystal structures of compounds ABO_4

C.N. B	C.N. A	Structure	Examples	Reference
4	4	Silica-like structures	BPO_4, $BeSO_4$, $AlAsO_4$	
	6	$CrVO_4$		AC 1967 **22** 321
		α-$MnMoO_4$		JCP 1965 **43** 2533
	8	Scheelite ($CaWO_4$)	$CaWO_4$	JCP 1964 **40** 501, 504
		Fergusonite	$M^{III}NbO_4$	AC 1963 **16** 888
				IC 1964 **3** 600
		Zircon ($ZrSiO_4$)	YVO_4	AC 1968 **B24** 292
		Anhydrite ($CaSO_4$)		
	12	Barytes ($BaSO_4$)		
6	6	Rutile (statistical)	h.p. $CrVO_4$	AC 1962 **15** 1305
			$AlAsO_4$	AC 1964 **17** 1476
		Wolframite (statistical)	h.p. $FeVO_4$	AC 1964 **17** 1476
			$\{FeWO_4$	ZK 1967 **124** 192
		Wolframite	$MnWO_4$	ZK 1967 **125** 120
			$NiWO_4\}$	AC 1957 **10** 209
		Wolframite (distorted)	$CuWO_4$	AC 1970 **B26** 1020
		$CoMoO_4$		AC 1965 **19** 269
[4	8	Defect scheelite structure	$Eu_2(WO_4)_3$	JCP 1969 **50** 86]

The tetrahedral coordination of Mo(VI) in α-MnMoO$_4$ may be compared with its octahedral coordination in CoMoO$_4$, and the 6-coordination of Cr(III) in CrVO$_4$ with the 4-coordination of Cr(VI) in the isostructural CuCrO$_4$. (The tetrahedral coordination of V in CrVO$_4$ has been confirmed by examining the fine structure of the K absorption edges of Cr in NiCrO$_4$ (tetrahedral Cr(VI)) and CrVO$_4$ and of V in CrVO$_4$ and Na$_3$VO$_4$ (tetrahedral V(v)), the characteristic shape of the absorption curve showing that it is V that is tetrahedrally coordinated in CrVO$_4$.) Other compounds with the CrVO$_4$ structure include InPO$_4$, TlPO$_4$, and an unstable form of CrPO$_4$.

The high-pressure form of CrVO$_4$ has the (statistical) rutile structure, with 6-coordination of both Cr and V, and a change from 4- to 6-coordination of As occurs in the high-pressure form of AlAsO$_4$. There are many examples of high-pressure oxides having the same type of structure as ambient-pressure oxides containing the next element in the same vertical Periodic family (or subgroup), as for example h.p. ZnO and normal CdO (NaCl structure), or h.p. SiO$_2$ and normal GeO$_2$ (rutile structure). Examples among oxides ABO$_4$ include:

Wolframite structure	Statistical wolframite	Rutile
h.p. MgMoO$_4$, etc.	h.p. FeVO$_4$	h.p. CrVO$_4$
MgWO$_4$, etc.	FeNbO$_4$	CrNbO$_4$

The statistical rutile structure is the normal form of many M^{III}M^{V}O$_4$ compounds, for example:

CrTaO$_4$	CrNbO$_4$	AlSbO$_4$	RhVO$_4$
Fe	Fe	Cr	
Rh	Rh	Fe	
V		Rh	
		Ga	

The structure of AlNbO$_4$ (and the isostructural GaNbO$_4$) is mentioned later (p. 504).

The wolframite structure

This structure takes its name from the mineral with composition (Fe, Mn)WO$_4$. The metal atoms each occupy one-quarter of the octahedral holes in a somewhat distorted hexagonal close packing of oxygen atoms. The pattern of sites occupied between the close-packed O layers is that of Fig. 4.24 (b) (p. 145), in which the small black and open circles represent Ni and W atoms in octahedral sites between alternate pairs of c.p. layers. In NiWO$_4$ (isostructural with the Mg, Mn, Fe, Co, and Zn compounds) Ni has six equidistant O neighbours (at 2·08 (0·05) Å), but the octahedral group around W is considerably distorted, there being four O at 1·79 Å and two at 2·19 Å. As noted above this structure is also adopted by the high-pressure forms of the corresponding molybdates, and there are also oxides with the statistical variant of the structure. In the distorted wolframite structure of CuWO$_4$ Cu has 4 O at 1·98 Å and 2 O at 2·40 Å. As in all W^{VI}O$_6$ octahedra there is a range of W—O distances from 1·76–2·2 Å.

The scheelite and fergusonite structures

The scheelite structure is named after the mineral with the composition $CaWO_4$. Examples of compounds with this structure include:

$$M^I M^{VII} O_4: \quad NaIO_4, KIO_4, KRuO_4,$$
$$M^{II} M^{VI} O_4: \quad MMoO_4 \text{ and } MWO_4 \ (M = Ca, Sr, Ba, Pb)$$
$$M^{III} M^V O_4: \quad MNbO_4 \text{ and } MTaO_4 \ (M = Y \text{ or } 4f \text{ metal}).$$

Detailed X-ray and n.d. studies of the (tetragonal) scheelite structure show very slight distortion of the WO_4 tetrahedra (four angles of $107.5°$ and two of $113.4°$, $W{-}O$, 1.784 Å). Many compounds of the third group listed above have a distorted (monoclinic) variant of the structure (*fergusonite* structure). The oxides $YTi_{0.5}^{IV} Mo_{0.5}^{VI} O_4$ and $YTi_{0.5} W_{0.5} O_4$ show an interesting difference, the Mo compound having the tetragonal scheelite structure and the W compound the fergusonite structure. Such differences between compounds of Mo and W are numerous, and other examples are noted elsewhere.

A different deformed version of the scheelite structure is found for $KCrO_3Cl$.

One of several structures found for molybdates and tungstates of trivalent metals is the 'defect scheelite' structure, adopted by $Eu_2(WO_4)_3$ and one polymorph of $Nd_2(MoO_4)_3$. In this structure M^{3+} ions occupy two-thirds of the Ca^{2+} sites in the scheelite structure.

Oxides AB_2O_4

Three structures of this type were described in our discussion of the closest packing of equal spheres, namely, those in which certain proportions of the tetrahedral and/or octahedral holes are occupied:

$\frac{3}{8}$ of tetrahedral holes: phenacite (Be_2SiO_4),
$\frac{1}{8}$ tetrahedral $\Big\}$. h.c.p. olivine (Mg_2SiO_4), chrysoberyl (Al_2BeO_4),
$\frac{1}{2}$ octahedral $\Big\}$. c.c.p. spinel (Al_2MgO_4).

Phenacite and olivine are orthosilicates, and since the present group includes all compounds $B_2(AO_4)$ containing tetrahedral AO_4 ions and ions B of various sizes it is a very large one and there are numerous structures, of which a selection is included in Table 13.9. For example, the alkali-metal tungstates have four different crystal structures:

Li_2WO_4 (phenacite), Na_2WO_4 (modified spinel), K_2WO_4 (and the isostructural Rb_2WO_4), and Cs_2WO_4 (β-K_2SO_4 structure), with respectively 4-, 6-, 8-, and (9 and 10)-coordination of M^+ ions.

However, we are concerned here with complex oxides rather than oxy-salts containing well-defined oxy-ions, though the dividing line is somewhat arbitrary, as is illustrated by compounds such as Li_2BeF_4, Zn_2GeO_4 and $LiAlGeO_4$ with the phenacite structure and $LiNaBeF_4$, Cs_2BeF_4, and Ba_2GeO_4 with the olivine structure.

By far the most important of the structures of Table 13.9 is the spinel structure. A recent survey shows over 130 compounds with the cubic spinel or closely related

TABLE 13.9

Crystal structures of compounds AB_2O_4

C.N. of B	4	6	8	9 and/or 10
C.N. of A				
4	Phenacite	Olivine Spinel	K_2WO_4	β-K_2SO_4
6			*	K_2NiF_4
8		$CaFe_2O_4$ $CaTi_2O_4$		

For a discussion of the crystal chemistry of compounds AB_2O_4 see, for example: IC 1968 **7** 1762; JSSC 1970 **1** 557.

* We could include here a number of structures in which rutile-like chains containing A are held together by B in positions of 7-coordination (Sr_2PbO_4, p. 176) or 6-, 7-, and 9-coordination (Ca_2IrO_4, ZaC 1966 **347** 282).

structures; some 30 of these are sulphides but most of the remainder are oxides. After the spinel structure we shall deal briefly with two other structures in which there is 6-coordination of B, namely, those of $CaFe_2O_4$ and $CaTi_2O_4$, and then with the K_2NiF_4 structure.

There is considerable interest in the high-pressure forms of oxides AB_2O_4, for high-pressure forms of olivine, $(Fe, Mg)_2SiO_4$, are believed to be important constituents of the earth's mantle. High pressure tends to convert a structure into one with higher c.n.s, that is, from top left to bottom right of Table 13.9. Examples include:

Ca_2GeO_4: olivine $\rightarrow$ K_2NiF_4 structure (density increase 25 per cent)

Mn_2GeO_4: olivine $\rightarrow$ Sr_2PbO_4 structure (density increase 18 per cent).

The normal and 'inverse' spinel structures

The spinel structure is illustrated in Fig. 7.4(a). It is most easily visualized as an octahedral framework of composition AX_2 (atacamite) which is derived from the NaCl structure by removing alternate rows of metal ions (Fig. 4.22(b), p. 143). Additional metal ions may then be added in positions of tetrahedral coordination (Fig. 7.4(a), *facing* p. 268). In the resulting structure each O^{2-} ion also has tetrahedral coordination, its nearest neighbours being three metal atoms of the octahedral framework and one tetrahedrally coordinated metal atom.

The crystallographic unit cell of the spinel structure contains thirty-two approximately cubic close-packed oxygen atoms, and there are equivalent positions in this cell for eight atoms surrounded tetrahedrally by four O atoms and for sixteen atoms surrounded octahedrally by six O atoms. (These are not, of course, the total numbers of tetrahedral and octahedral holes, which are respectively 64 and 32 (see p. 127).) Since there are eight A atoms and sixteen B atoms to place in the unit cell (which contains 8 AB_2O_4) it was natural to place the A atoms in the positions of tetrahedral and the B atoms in those of octahedral coordination. The spinels MAl_2O_4 (where M is Mg, Fe, Co, Ni, Mn, or Zn) have this structure, but in certain other spinels the A and B atoms are arranged differently. In these the eight tetrahedral positions are occupied, not by the eight A atoms, but by one-half of the B atoms, the rest of which together with the A atoms are arranged at random in the 16 octahedral positions. These 'inverse' spinels are therefore conveniently formulated $B(AB)O_4$ to distinguish them from those of the first type, AB_2O_4. Examples of inverse spinels include $Fe(MgFe)O_4$ and $Zn(SnZn)O_4$.

The nature of a spinel is described by a parameter λ, the fraction of B atoms in tetrahedral holes; some authors refer to the degree of inversion y (= 2λ). For a normal spinel $\lambda = 0$, and for an inverse spinel $\lambda = \frac{1}{2}$. Intermediate values are found (e.g. $\frac{1}{3}$ in a random spinel), and λ is not necessarily constant for a given spinel but can in some cases be altered by appropriate heat treatment. For $NiMn_2O_4$ λ varies from 0·37 (quenched) to 0·47 (slow-cooled).[1] Values of λ have been determined by X-ray and neutron diffraction, by measurements of saturation magnetization, and also by i.r. measurements. In favourable cases i.r. bands due to tetrahedral AO_4 groups can be identified showing, for example, that in $Li(CrGe)O_4$ Li occupies tetrahedral positions.[2]

If there is sufficient difference between the X-ray scattering powers of the atoms A and B it is possible to determine the distribution of these atoms by the usual methods of X-ray crystallography, but in spinel itself ($MgAl_2O_4$), for example, this is not possible. However, the scattering cross-section for neutrons of Mg is appreciably greater than that of Al, and this makes it possible to show that $MgAl_2O_4$ has the normal spinel structure.[3] Many 2 : 3 spinels have the normal structure, though some (including most 'ferrites') have the inverse structure, as for example $Ga(MgGa)O_4$ ($\lambda = 0·42$).[4] All 4 : 2 spinels so far examined have the inverse structure, for example, $Zn(ZnTi)O_4$ and $Fe(FeTi)O_4$ ($\lambda = 0·46$).[5]

The spinels of composition between $MgFe_2O_4$ and $MgAl_2O_4$, which have been studied magnetically and also by neutron diffraction,[6] are of interest in this connection. $MgFe_2O_4$ has an essentially inverse structure ($\lambda \approx 0·45$), that is, nine-tenths of the Mg^{2+} ions are in octahedral (B) sites. As Fe is replaced by Al the latter goes into B sites and forces Mg into tetrahedral (A) sites, so that there is a continuous transformation from the inverse structure of $MgFe_2O_4$ to the normal structure of $MgAl_2O_4$.

The structures of the spinels present two interesting problems. First, why do some compounds adopt the normal and others the inverse spinel structure? Second, there are some spinels which show distortions from cubic symmetry. This is part of the more general problem of minor distortions from more symmetrical structures which was mentioned in connection with ligand field theory in Chapter 7. We deal with these points in turn.

Calculations of the lattice energy on the simple electrostatic theory, without allowance for crystal field effects, indicate that while the inverse structure should be more stable for 4 : 2 spinels, the preferrred structure for 2 : 3 spinels should be the normal structure. In fact a number of the latter have the inverse structure, as

(1) AC 1969 **B25** 2326

(2) AC 1963 **16** 228

(3) AC 1952 **5** 684

(4) AC 1966 **20** 761
(5) AC 1965 **18** 859

(6) AC 1953 **6** 57

TABLE 13.10

Cation distribution in 2:3 spinels (values of λ)

B^{3+} \ A^{2+}	Mg^{2+}	Mn^{2+}	Fe^{2+}	Co^{2+}	Ni^{2+}	Cu^{2+}	Zn^{2+}
Al^{3+}	0	0	0	0	0·38	–	0
Cr^{3+}	0	0	0	0	0	0	0
Fe^{3+}	0·45	0·1	0·5	0·5	0·5	0·5	0
Mn^{3+}	–	0	–	–	–	–	0
Co^{3+}	–	–	–	0	–	–	0

(7) JPCS 1957 **3** 318; PR 1955 **98** 391

shown by the (approximate) values of λ in Table 13.10. The cation distribution in spinels has been discussed in terms of crystal field theory.[7] Although values of Δ (see p. 272) appropriate to spinels are not known, estimates of these quantities may be made and from these the stabilization energies for octahedral and tetrahedral coordination obtained. The differences between these quantities, of which the former is the larger, give an indication of the preference of the ion for octahedral as opposed to tetrahedral coordination.

Excess octahedral stabilization energy (kJ mol^{-1})

Mn^{2+}	Fe^{2+}	Co^{2+}	Ni^{2+}	Cu^{2+}	Ti^{3+}	V^{3+}	Cr^{3+}	Mn^{3+}	Fe^{3+}
0	17	31	86	64	29	54	158	95	0

Thus while Cr^{3+} and Mn^{3+} occupy octahedral sites, most 'ferrites' are inverse spinels, Fe^{3+} having no stabilization energy for octahedral sites. The only normal, or approximately normal, ferrites are those of Zn^{2+} and Mn^{2+}, divalent ions which have no octahedral stabilization energy. The only inverse 'aluminate' is the Ni^{2+} spinel, and of the above ions Ni^{2+} has the greatest preference for octahedral coordination. Note that we are disregarding all other factors (covalent bonding and differences in normal lattice energies) when we take account only of the crystal field stabilization energies. For example, for NiAl$_2$O$_4$ a classical calculation of the lattice energy, as the sum of the electrostatic (Madelung) potential, the polarization energy, and the Born repulsion energy, shows that the normal structure would be some 105 kJ mol^{-1} more stable than the inverse one, so that the crystal field stabilization of octahedral Ni^{2+} almost compensates for this. The observed structure is in fact very close to the random one, for which λ = 0·33. The fact that Fe$_3$O$_4$ is an inverse spinel with cubic symmetry while Mn$_3$O$_4$ is a normal spinel with some tetragonal distortion (for which see p. 458) is explained by the much greater preference of Mn^{3+} than Fe^{3+} for octahedral coordination by oxygen, as shown by the stabilization energies quoted above. The simple crystal field theory is seen to be very helpful in accounting for the cation distributions in these 2 : 3 spinels.

The second problem concerns the departures from cubic symmetry. For example, $CuFe_2O_4$ is cubic at high temperatures (and at room temperature if quenched from temperatures above $760°C$),[8] but if cooled slowly it has tetragonal symmetry ($c : a = 1·06$ at room temperature). It is an inverse spinel with Fe^{3+} in the tetrahedral sites, and it is interesting that the FeO_4 tetrahedra are not distorted. The tetragonal symmetry appears to be due to the tetragonal distortion of the $Cu^{II}O_6$ octahedra (p. 273). $CuCr_2O_4$ is also tetragonal, but with $c : a = 0·91$.[9] Here the Cu atoms are in tetrahedral sites, but the bond angles of $103°$ and $123°$ show a tendency towards square coordination, or alternatively, the tetragonal distortion (type (c) of Table 7.14, p. 272) predicted by ligand field theory. The spinels of intermediate composition, $CuFe_{2-x}Cr_xO_4$, are even more interesting.[10] For $x = 0$ ($CuFe_2O_4$) there is a tetragonal distortion with $c : a = 1·06$, and for $x = 2$ ($CuCr_2O_4$) the structure is tetragonal with $c : a = 0·91$. Over the range $0·4 < x < 1·4$ the structure is cubic. Since for $x = 0$ the structure is almost completely inversed, and at $x = 2$ nearly normal, there is a continuous displacement of Fe by Cu in the tetrahedral sites. This is similar to the $MgFe_2O_4$–$MgAl_2O_4$ series but with the added complication that at one end of the series there is distortion due to Cu in octahedral sites and at the other due to Cu in tetrahedral sites, the elongation of the CuO_6 octahedra leading to $c : a > 1$ and the flattening of the tetrahedra to $c : a < 1$.

(Some sulphides with structures closely related to the spinel structure are noted on p. 618.)

(8) AC 1956 **9** 1025

(9) AC 1957 **10** 554

(10) JPSJ 1956 **12** 1296

Spinel superstructures

There are many spinel-like structures containing metal atoms of more than one kind. If these are present in suitable numbers they can arrange themselves in an orderly way; for example, the 16-fold position may be occupied by 8 A + 8 B or by 4 A + 12 B. Large differences in ionic charge favour ordering of cations, and since in 'defect' structures holes also tend to order, superstructures with ordered vacancies are possible. Some examples of spinel superstructures are summarized in Table 13.11. A compound may be ordered at temperatures below a transition point and disordered at higher temperatures. For example, below $120°K$ there is ordering of Fe^{2+} and Fe^{3+} in octahedral sites in Fe_3O_4 but random arrangement at higher temperatures, leading to free exchange of electrons. The compounds $LiAl_5O_8$ and $LiFe_5O_8$ have ordered structures at temperatures below $1290°$ and $755°C$ respectively and disordered structures at higher temperatures. In the low-temperature forms the $4 Li^+$ and $12 M^{3+}$ ions are arranged in an ordered way in the 16 octahedral sites but in a random way at temperatures above the transition points.

'Ferrites' etc. with structures related to spinel

Certain oxides, notably those of the alkali metals, the alkaline-earths, and lead, form well-defined 'aluminates' and 'ferrites' with general formulae $M_2O . m(Al, Fe)_2O_3$ or $MO . n(Al, Fe)_2O_3$. Some of the simpler 'ferrites', for example, $LiFeO_2$, $NaFeO_2$,

TABLE 13.11
Spinel superstructures

Compound	Occupancy of positions				Reference
	tetrahedral	octahedral			
$Zn(LiNb)O_4$	Zn	Nb	Li	O_4	
$V^V(LiCu)O_4$	V	Li	Cu	O_4	
Fe_3O_4 (<120°K)	$\underline{Fe^{3+}}$	Fe^{2+}	Fe^{3+}	O_4	AC 1955 **8** 257
	8	16		O_{32}	
Fe_5LiO_8 (Al_5LiO_8)	8 Fe	4 Li	12 Fe	O_{32}	JCP 1964 **40** 1988
γ-Fe_2O_3	8 Fe	$\frac{4}{3}Fe^{3+}$ $\frac{8}{3}$ □	12 Fe	O_{32}	N 1958 **181** 44
$(LiFe)Cr_4O_8$	4 Li 4 Fe	16 Cr		O_{32}	
In_2S_3	$\frac{16}{3}$ In $\frac{8}{3}$ □	16 In		S_{32}	
$Zn_2Ge_3O_8$	2 Zn	3 Ge □		O_8	
$LiZn(LiGe_3)O_8$	Li, Zn	3 Ge Li		O_8	
$Co_3(VO_4)_2$ (low)	2 V	3 Co □		O_8	J.- C. Joubert Thesis, Grenoble, 1965
$LiGaTiO_4$	2 Ga, Li	3 Ti, Ga, 2 Li		O_{12}	

□ represents vacancy

and $CuFeO_2$ (all with different structures) and $MgFe_2O_4$ and $CaFe_2O_4$ are included in other sections of this chapter. Examples of more complex compounds include:

$Na_2O \cdot 11\ Al_2O_3$, also formed by K
$K_2O \cdot 11\ Fe_2O_3$, also formed by Rb
$3\ CaO \cdot 16\ Al_2O_3$, also formed by Sr and Ba
$SrO \cdot 6\ Fe_2O_3$, also formed by Ba and Pb.

The first of these, $Na_2O \cdot 11\ Al_2O_3$, was originally thought to be a form of Al_2O_3, hence its name—β-alumina. X-ray investigation, however, showed it to have a definite structure closely related to that of spinel, no fewer than 50 of the 58 atoms in the unit cell being arranged in exactly the same way as in the spinel structure. The large Na or K ions are situated between slices of spinel structure and their presence is essential to the stability of the structure (Fig. 13.6(a)).

Later work indicates[11] that although the ideal composition of β-alumina is $Na_2O \cdot 11\ Al_2O_3$ the composition of clear single crystals varies between 9–$10\frac{1}{2}$ $Al_2O_3 : Na_2O$. The Na^+ ions can be replaced by Ag^+, K^+, Rb^+, etc. by heating with the appropriate molten salts. Furthermore, the Ag^+ may then be replaced by NO^+ (by heating in a $NOCl$–$AlCl_3$ melt) or by Ga^+. Because of the very high mobility of the Na^+ ions in this structure the compound has become of interest in connection with 'solid' batteries. The composition of a crystal used in a refinement of the crystal structure[12] corresponded to the formula $Na_{2 \cdot 58}Al_{21 \cdot 8}O_{34}$. It appears

(11) IC 1969 **8** 994

(12) AC 1971 **B27** 1826

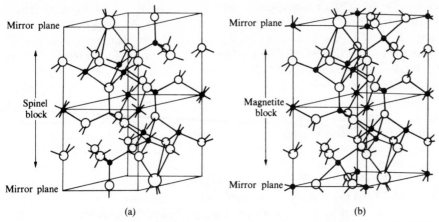

FIG. 13.6. (a) The relation of the structure of β-alumina, $NaAl_{11}O_{17}$, to that of spinel (after Beevers and Ross). Large circles represent Na, small ones O, and black circles Al. (b) The relation of the structure of magneto-plumbite, $PbFe_{12}O_{19}$, to that of magnetite Fe_3O_4 (after Adelsköld). Large circles represent Pb, small ones O, and black circles Fe.

that the distribution of the Na^+ ions is much more complex than in the original structure, and for details the reader is referred to the literature reference.

The structure of magnetoplumbite, $PbFe_{12}O_{19}$,[13] is similarly built of slices of magnetite (Fe_3O_4) structure alternating with layers containing Pb^{2+} ions. In this structure (Fig. 13.6(b)) there is a sequence of 5 c.p. layers, four of O atoms (O_4 per cell) and one in which Pb^{2+} ions occupy one-quarter of the c.p. positions (PbO_3). The composition is therefore PbO_{19} plus the associated Fe atoms in tetrahedral and octahedral holes. The compounds containing Ba instead of Pb have been extensively studied in recent years, for they form a large family of ferrimagnetic compounds valuable in electronic devices. Early studies produced $BaFe_{12}O_{19}$, isostructural with magnetoplumbite, $BaFe_{15}O_{23}$[14] and $BaFe_{18}O_{27}$ with respectively five and six O_4 layers in the repeat unit. Already the hexagonal ferrites are more numerous than the polytypes of SiC, of which at least fifty are known, with c cell dimensions up to 990 Å.

These compounds fall into two distinct series, each formed by different stacking sequences based on three kinds of sub-unit formed from O_4 or BaO_3 layers:

(13) AKMG 1938 **12A** No. 29

(14) AC 1954 **7** 640

	O_4		O_4		O_4
O_4		O_3Ba		O_3Ba
2 layers		O_4		O_3Ba
S unit		O_4		O_4
$Me_2Fe_4O_8$		5 layers		O_4
		M unit		6 layers
		$BaFe_{12}O_{19}$		Y unit
(Me is usually Zn, Ni, Co, or Fe)				$Ba_2Me_2Fe_{12}O_{22}$

The blocks S, M, and Y stack in various ratios and permutations but as regards sequences the units show decided preferences: M is usually associated with Y (less

495

commonly with S), Y not with S, and M, Y, and S not all together. There are therefore two main series, (i) M_nS (one S and n M blocks), giving in the limit, M ($BaFe_{12}O_{19}$), and (ii) M_pY_n, of which the limiting member is Y ($Ba_2Me_2Fe_{12}O_{22}$). Five of type (i) have been prepared, for example, M_6S is $Ba_6Me_2Fe_{70}O_{122}$ with $c = 223$ Å, but the much more extensive family (ii) includes nearly sixty members. For example, M_4Y_{33} has the sequence: MY_6MY_{10} MY_7MY_{10}, with composition $Ba_{70}Me_{66}Fe_{444}O_{802}$ and $c = 1577$ Å. It would not be feasible to elucidate these complex structures by X-ray diffraction alone, but fortunately the stacking sequence can be determined by electron microscopy of HNO_3-etched samples, using Pt-shadowing of carbon replicas. The etch patterns take the form of fine terracing along the sides of gently sloping etch pits, and knowing the absolute magnification the heights of the two types of steps can be determined from the replicas. For example, a M_2Y_7 specimen shows large and small steps corresponding to the units MY and MY_6 stacked in the sequence MY MY_6 MY MY_6 .... The more complex four-step etch pattern of a M_4Y_{15} ferrite is interpreted as indicating the sequence MY MY_2 MY_3 MY_9 ... ($c \approx 793$ Å). (For references to the literature see ref. 15.)

(15) Sc 1971 **172** 519
(16) ZK 1967 **125** 437

The structure of $BaFe_{12}O_{19}$[16] is of interest as an example of a 10-layer c.p. sequence (*cchh*):

$$B\,A\,B'\,A\,B\,C\,A\,C'\,A\,C\ldots$$

where B' and C' indicate BaO_3 layers. It is also remarkable for the three types of coordination of the Fe atoms:

$$\left.\begin{array}{rr} \text{octahedral} & 18\ Fe \\ \text{tetrahedral} & 4\ Fe \\ \text{trigonal bipyramidal} & 2\ Fe \end{array}\right\} \quad \text{for } 2(BaFe_{12}O_{19}).$$

The $CaFe_2O_4$, $CaTi_2O_4$, *and related structures*

For larger A ions the spinel structure is not stable, and we find a number of structures built from 'double rutile' chains which form 3D frameworks enclosing the larger A ions in positions of 8-coordination. These structures have been described in Chapter 5.

No other compounds are yet known to have the $CaTi_2O_4$ structure, but there are numerous compounds with the $CaFe_2O_4$ structure, not only oxides $M^{II}M_2^{III}O_4$ (and some sulphides) but also more complex compounds in which ions M^{3+} and M^{4+} occupy at random the Fe^{3+} positions and Na^+ the Ca^{2+} positions:

$Eu^{II}Eu_2^{III}O_4$	$CaFe_2O_4$	$NaScTiO_4$	Eu	Y_2	
	V_2	$NaScSnO_4$	Sm	Ho_2	S_4
	Cr_2	$NaAlGeO_4$	Pb	etc.	
	In_2	etc.	Sr		
	etc.		Ba		

(For references see: IC 1967 **6** 631; IC 1968 **7** 113, 1762; AC 1967 **23** 736).

In all cases the environment of Ca^{2+} (or the equivalent ion) consists of 8 O (S) neighbours, the ninth vertex of the tricapped trigonal prism coordination group being at a much greater distance.

In Eu_3O_4 $\qquad$ Eu^{2+}—6 O (octahedral) at 2·34 Å

$$Eu^{3+}\begin{cases} 6 \text{ O (prism) } 2\cdot67 \text{ Å} \\ 2 \text{ O } 2\cdot72 \text{ and } 2\cdot96 \text{ Å} \\ 1 \text{ O } 3\cdot99 \text{ Å} \end{cases}$$

Similarly in $NaScTiO_4$ Na^+ has 8 O neighbours at 2·43–2·65 Å and the ninth at 3·34 Å.

In the $CaFe_2O_4$ and $CaTi_2O_4$ structures a *larger* ion (Ca^{2+}) is accommodated between double octahedral chains which contains the *smaller* Fe^{3+} or Ti^{3+} ions. A

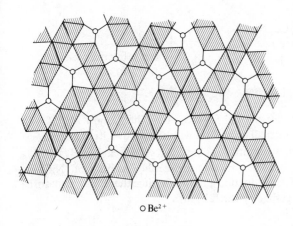

○ Be^{2+}

FIG. 13.7. The structure of BeY_2O_4 projected along the direction of the quadruple rutile chains.

very interesting structure is adopted by BeY_2O_4[1] in which the *smaller* Be^{2+} ion (Be–O, 1·55 Å, Y–O, 2·29 Å) is accommodated in positions of 3-coordination (coplanar) between quadruple chains. These chains are in contact not only at two terminal points but also at the mid-points of both sides, as may be seen from the projection of the structure (Fig. 13.7). (This unusual 3-coordination is also found for four of the seventeen Be^{2+} ions in $Ca_{12}Be_{17}O_{29}$.[2]) The 3D framework of Fig. 13.7 has the composition MO_2 (M_2O_4) and is found also in a number of oxyborates. In the synthetic $FeCoBO_4$ there is ordered arrangement of Fe and Co in the framework, and B replaces Be in Fig. 13.7. Since there are planar BO_3 groups in this structure the formula may be written $FeCo(BO_3)O$.[3] In the mineral warwickite, $Mg_3TiB_2O_8$, Ti atoms occupy one-quarter of the metal positions in the framework, and Mg may be replaced by Fe.[4] We may therefore write the formulae of these compounds

$$Be(Y_2O_4), \qquad B(FeCoO_4), \qquad \text{and} \qquad B_2(Mg_3TiO_8)$$

and it would be interesting to know if the series can be completed by making compounds such as $Li(M^{III}M^{IV}O_4)$. (See also p. 228)

(1) AC 1967 **22** 354

(2) AC 1966 **20** 295

(3) ZK 1972 **135** 321

(4) AC 1950 3 98, 473

497

Complex Oxides

The K_2NiF_4 structure

The structure of a number of complex oxides X_2YO_4 is closely related to the perovskite structure, which is that of the corresponding oxide XYO_3. A slice of the perovskite structure one unit cell thick has the composition X_2YO_4. If such slices are displaced relative to one another as shown in Fig. 13.8, there is formed a tetragonal structure with $c = 3a$, a being the length of the cell edge of the perovskite structure. In this structure the Y atoms have the same environments as in perovskite, namely six O arranged octahedrally, but the X atoms have an unusual arrangement of nine O instead of the original twelve neighbours. If the slices of the perovskite structure are n unit cells thick the composition of the crystal is $X_{n+1}Y_nO_{3n+1}$. In the complex Sr–Ti oxides the first three members of this series have been characterized, namely, Sr_2TiO_4, $Sr_3Ti_2O_7$, and $Sr_4Ti_3O_{10}$. The X_2YO_4 structure was first assigned to K_2NiF_4. Examples of compounds with this structure are:

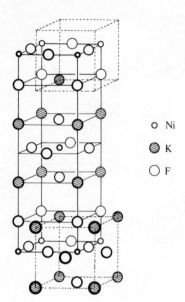

FIG. 13.8. The K_2NiF_4 structure.

O Ni

◉ K

O F

$$K_2MF_4: \quad M = Mg, Zn, Co, Ni$$
$$M_2UO_4: \quad M = K, Rb, Cs$$
$$Sr_2MO_4: \quad (M = Ti, Sn, Mn, Mo, Ru, Ir, Rh)$$
$$Ba_2MO_4: \quad (M = Sn, Pb)$$
$$Sr_2FeO_3F, \; K_2NbO_3F, \; La_2NiO_4.$$

For references see JPC 1963 **67** 1451.

It is interesting to compare the isostructural Sr_2TiO_4, Ca_2MnO_4, and K_2NiF_4 with $SrTiO_3$, $CaMnO_3$, and $KNiF_3$ which all crystallize with the perovskite structure.

The distorted variants of this structure adopted by K_2CuF_4 and $(NH_4)_2CuCl_4$ are described in Chapter 5, where the structure is described as built of $(NiF_4^{2-})_n$ layers formed from octahedral coordination groups sharing their four equatorial vertices; the 2D ions are held together by the K^+ ions.

Further complex oxide structures

The pseudobrookite structure, A_2BO_5

In this structure[1] there is octahedral coordination of the B ions, but the coordination of A is unusual. It can be described either as a highly deformed octahedron or as a distorted tetrahedron, for in Fe_2TiO_5 Fe has four neighbours at a mean Fe–O distance of 1·92 Å and two more at 2·30 Å. Ti has six octahedral neighbours as in the forms of TiO_2, $FeTiO_3$, etc. Al_2TiO_5 also has the pseudobrookite structure for Ti_3O_5 see p. 466.

Oxides AB_2O_6

We noted in Chapter 4 a number of structures in which metal ions occupy one-half of the octahedral interstices in c.p. structures. Three h.c.p. structures of this kind are:

trirutile structure: for which see pp. 147 and 203,
columbite (niobite) structure (p. 147),[2]
Na_2SiF_6 structure (pp. 147 and 387).

(1) AC 1953 **6** 812

(2) AK 1963 **21** 407

498

Since both A and B ions occupy octahedral interstices in all of these structures they are suitable only for the smaller metal ions, as may be seen from the following examples: $ZnSb_2O_6$ (trirutile), $(Fe, Mn)Nb_2O_6$ (niobite), and NiU_2O_6[3] (Na_2SiF_6 structure).

(3) ZaC 1968 358 226

The $CaTa_2O_6$ structure[4] has been illustrated with other octahedral structures in Chapter 5. As in the $CaFe_2O_4$ structure Ca^{2+} is 8-coordinated: the ninth, more distant, O would complete a tricapped trigonal prismatic coordination group.

(4) ACSc 1963 17 2548

The pyrochlore structure, $A_2B_2O_7$

We have already mentioned this structure in connection with 3D nets in Chapter 3, and also as an example of a 3D framework of octahedra in Chapter 5. The geometry of the structure is further discussed in Chapter 6 (p. 209). We noted that this framework has the composition B_2O_6 (BO_3) and that it can exist without the A atoms or without the seventh O atom. In the latter case it serves as the structure of a number of oxides ABO_3 ('defect pyrochlore structure'). The structure is named after the mineral pyrochlore which has the approximate composition $(CaNa)Nb_2O_6F$, and since the seventh anion position may be occupied by OH^-, F^-, or even H_2O, we write the general formula $A_2B_2X_6'X''$ in Table 13.12. The largest group of compounds with this structure are oxides $A_2B_2O_7$ which include the following:

$M_2^{II}M_2^VO_7$: $Cd_2Nb_2O_7, Cd_2Re_2O_7, Ca_2Ta_2O_7, Hg_2Nb_2O_7,$
$M_2^{III}M_2^{IV}O_7$: $M_2Ti_2O_7$ (M = Sc, Y, La, and all lanthanides except Pm).
$\qquad\qquad$ $M_2Pt_2O_7$ (M = Sc, Y, In, Tl; made under high O_2 pressure).

The coordination group around A in $A_2B_2X_7$ consists of six X atoms of the B_2X_6 framework and two of the additional X atoms. In some of the oxide structures the 8 O atoms are approximately equidistant from A, forming an approximately cubic coordination group, but in other cases there is an appreciable difference between A–6 O and A–2 O. It is then justifiable to describe the structure as an A_2O framework interpenetrating the B_2O_6 octahedral framework. $Hg_2Nb_2O_7$ has been described in this way in Chapter 3 as related to the diamond structure, Hg forming two collinear bonds to O (Hg–2 O, 2·26 Å) in a 3D framework of the same kind as

TABLE 13.12
Compounds with the pyrochlore structure $A_2B_2X_7$

Compound	A_2	B_2	X_6'	X''
$AgSbO_3$	Ag	Sb_2	O_6	–
Sb_3O_6OH	Sb^{III}	Sb_2^V	O_6	OH
$BiTa_2O_6F$	Bi^{III}	Ta_2	O_6	F
$Al(OH, F)_3 \cdot \frac{3}{8} H_2O$	–	Al_2	$(OH, F)_6$	$(H_2O)_{\frac{3}{4}}$
$Hg_2Nb_2O_7$	Hg_2	Nb_2	O_6	O

Recent references to complex oxides with this structure include: IC 1965 4 1152; IC 1968 7 1649, 1704, 2553; IC 1969 8 1807.

one of the two equivalent nets in Cu_2O. (The other six O atoms are at a distance of 2·61 Å.) This aspect of the pyrochlore structure is emphasized in Fig. 7.4(b).

The garnet structure

The garnets have been known for a long time as a group of orthosilicate minerals $M_3^{II}M_2^{III}(SiO_4)_3$ in which M^{II} is Ca, Mg, or Fe, and M^{III} is Al, Cr, or Fe^{3+}. Since the total negative charge of -24 can be made up in a variety of ways the garnet structure can be utilized by a great number of complex oxides, of which some examples are given in Table 13.13. Interest in the optical properties of these compounds has led in recent years to many structural studies.

TABLE 13.13
Compounds with the garnet structure

C.N.	M' 8	M'' 6	M''' 4		Reference
	Mg_3	Al_2	Si_3,	O_{12}	AC 1961 **14** 835; AM 1965 **50** 2023
	Ca_3	Al_2	Si_3		ZK 1966 **123** 81
	Y_3	Al_2	Al_3		
	$NaCa_2$	Mg_2	As_3		
	$NaCa_2$	Zn_2	V_3		
	Y_3	YAl	Al_3		
	$CaNa_2$	Ti_2	Ge_3		
	Ca_3	Te_2	Zn_3		IC 1969 **8** 1000
	Na_3	Te_2	Ga_3		IC 1969 **8** 1000
	Ca_3	$CaZr$	Ge_3		IC 1969 **8** 183
	Cd_3	$CdGe$	Ge_3		Sc 1969 **163** 386
	Ca_3	Al_2		$(OH)_{12}$	AC 1964 **17** 1329; JCP 1968 **48** 3037 (n.d.)

The backbone of the garnet structure is a 3D framework built of $M''O_6$ octahedra and $M'''O_4$ tetrahedra in which each octahedron is joined to six others through vertex-sharing tetrahedra. Each tetrahedron shares its vertices with four octahedra, so that the composition of the framework is $o_2t_3 = (M''O_3)_2(M'''O_2)_3 = M_2''M_3'''O_{12}$. Larger ions ($M'$) occupy positions of 8-coordination (dodecahedral) in the interstices of the framework, giving the final composition $M_3'M_2''M_3'''O_{12}$ or $M_3'M_2''(M'''O_4)_3$ if it is wished to emphasize the tetrahedral groups as in an orthosilicate. In some compounds the M'' and M''' positions are all occupied by atoms of the same element, when the formula becomes, for example, $Y_3Al_5O_{12}$. If, in addition, one-half of the M'' positions are occupied by M' atoms the formula becomes, for example, $Y_4Al_4O_{12}$ (or $YAlO_3$).

The ordered arrangement of two kinds of ion in the octahedral sites is also found in the tetragonal (pseudocubic) high-pressure forms of $CaGeO_3$ and $CdGeO_3$. The structures of the polymorphs of $CaGeO_3$ provide an excellent illustration of the increase in cation coordination number with increasing pressure:

c.n.'s of cations:	6, 4	(8 and 6)	(6 and 4)	12, 6
	$CaGeO_3$-II	$CaGeO_3$-II		$CaGeO_3$-III
structure:	wollastonite	garnet-type		perovskite

The garnet framework can exist without the M' ions, as in $Al_2(WO_4)_3$, or $M_2'' M_3''' O_{12}$, when there is considerable distortion, with larger angles at the O atoms shared between M'' and M''' (143–175°). The same framework is also found in the so-called 'hydrogarnets'. A n.d. study of hydrogrossular, $Ca_3 Al_2 (OH)_{12}$, shows that tetrahedral $O_4 H_4^{4-}$ groups take the place of SiO_4^{4-}, the centre of the group being the position of Si in the garnet structure. A slightly distorted tetrahedron of O atoms is surrounded by a tetrahedral group of H atoms. The coordination group of Ca is intermediate between cube and antiprism.

Miscellaneous complex oxides

(1) JINC 1964 **20** 693

The structure of $Li_4 UO_5$[1] (with which $Na_4 UO_5$ is isostructural) may be described as a NaCl structure in which U occupies $\frac{1}{5}$ and Li $\frac{4}{5}$ of the cation positions, or alternatively as containing chains of UO_6 octahedra sharing opposite vertices. The UO_6 groups are tetragonally distorted, having four shorter equatorial bonds (1·99 Å) and two longer bonds (2·32 Å); contrast the oxides UMO_{12} (below) with six equal U–O bonds and $BaUO_4$ with two shorter collinear bonds.

(2) AC 1954 **7** 246

$Mg_6 MnO_8$ and $Cu_6 PbO_8$ have a slightly distorted NaCl structure[2] in which one-quarter of the cations M^{2+} have been replaced by half their number of M^{4+} ions. Since the Mn^{4+} ion is rather smaller than Mg^{2+} (Mg–6 O, 2·10 Å, Mn–6 O, 1·93 Å) the cubic close packing of the oxygen ions is slightly distorted. From calculations of the electrostatic energies of defect structures it has been concluded that in general ionic crystals with vacancies will have ordered rather than random structures.

(3) AC 1966 **21** 482

The structure of $Zn_2 Mo_3^{IV} O_8$ presents several points of interest.[3] There is *ABCB* . . . type of c.p. oxygen atoms. The Mo atoms occupy octahedral holes in groups of three, as shown in Fig. 5.22(d), p. 177. Owing to the metal–metal interactions (Mo–Mo, 2·52 Å) the compound is diamagnetic. One-half of the Zn atoms occupy tetrahedral and the remainder octahedral holes, the mean bond lengths being: Zn–4 O, 1·98 Å, Zn–6 O, 2·10 Å.

(4) IC 1964 **3** 949; IC 1966 **5** 749

Solid solutions UO_{2+x} : $Y_2 O_3$[4] adopt a defect fluorite structure with a degree of anion deficiency depending on composition and partial pressure of oxygen. When the oxygen deficiency becomes too large rearrangement takes place to a rhombohedral structure $M' M_6'' O_{12}$. This structure is derivable from fluorite by removing one-seventh of the O atoms and rearranging to provide both M^{VI} and M^{III} with octahedral coordination. It is formed by

$$UM_6 O_{12} \quad M = La \text{ and all 4f metals from Pr to Lu,}$$
$$WM_6 O_{12} \quad M = Y \text{ and all 4f metals from Ho to Lu, and}$$
$$MoM_6 O_{12} \quad M = Y \text{ and all 4f metals from Er to Lu,}$$

and also by the oxides $Tb_7 O_{12}$, $Pr_7 O_{12}$, and $Ce_7 O_{12}$. The structure is restricted to binary oxides of these particular elements because it requires M^{3+} and M^{4+} ions. The oxides UM_6O_{12} are of interest as rare examples of U(VI) with essentially regular octahedral coordination by O (6 O at 2·07 Å); see also δ-UO_3.

Complex oxides tend to be isostructural with complex fluorides rather than with complex chlorides, but note the isostructural oxides and chlorides with c.p. $A_m B_n X_{3m}$ structures. Other exceptions include Ba_3VO_5,[5] isostructural with Cs_3CoCl_5 (and the Mn and Fe salts), and Sr_4MO_6 (M = Pt, Ir, Rh),[6] isostructural with K_4CdCl_6.

In $Ca_2Fe_2O_5$ layers are formed from FeO_6 octahedra sharing 4 equatorial vertices, and these layers are joined through vertex-sharing tetrahedra FeO_4 to form 3D framework of composition Fe_2O_5. This framework consists of equal numbers of tetrahedra and octahedra which share vertices:

$$ t \begin{cases} 2\,o \\ 2\,t \end{cases} \quad \text{and} \quad o \begin{cases} 4\,o \\ 2\,t \end{cases} \text{(see p. 188)} $$

The Ca^{2+} ions occupy irregular interstices in the framework.[7]

(5) RTC 1965 **84** 821

(6) AC 1959 **12** 519

(7) AC 1970 **B26** 1469

Complex oxides containing Ti, V, Nb, Mo, or W

In Chapter 5 we derived some of the simpler structures that may be constructed by linking together octahedral AX_6 coordination groups by sharing vertices and edges and/or faces to give arrangements extending indefinitely in 1, 2, or 3 dimensions. Here we extend the treatment to include some rather more complicated structures. In Chapter 5 we used as building units first the AX_5 (ReO_3) chain formed from octahedra sharing two opposite vertices and the AX_4 (rutile) chain formed from octahedra sharing two opposite edges and then the corresponding double chains formed from two simple chains by edge-sharing. We also mentioned the 'shear' structures of certain Mo and W oxides which may be built from slices of the ReO_3 structure and some more complex structures built from blocks of that structure (V_6O_{13}, high-Nb_2O_5).

We saw that there are two ways in which blocks of ReO_3 structure can share edges to form more extended structures (Fig. 5.35(a) and (b)), that edge-sharing only of type (a) occurs in some Mo and W oxides, and only of type (b) in $WNb_{12}O_{33}$, but of both types (a) and (b) in one form of Nb_2O_5 and in the simple 'brannerite' structure of $NaVMoO_6$, and the metamict minerals $ThTi_2O_6$ and UTi_2O_6. The general types of structure that result are the following.

(a) Structures of binary or more complex oxides in which all the metal atoms are in octahedral coordination groups,

(b) 3-dimensional frameworks built around ions of Na, K, etc. in positions of higher coordination ($>$6), these ions being essential to the stability of the framework, which does not exist as the structure of an oxide of the transition element. Normal stoichiometric compounds of this kind include $Na_2Ti_6O_{13}$ and KTi_3NbO_9, with respectively 8- and 10-coordination of the alkali-metal ions. In certain frameworks with compositions MO_2 or MO_3, corresponding to a normal oxide, a valence change in a proportion of the M atoms leads to the formation of non-stoichiometric phases such as Na_xMO_2 or Na_xMO_3 which have very unusual physical and chemical properties. These *bronzes* are described in a later section.

(c) Layer structures of compounds such as $K_2Ti_2O_5$, $Na_2Ti_3O_7$, and $KTiNbO_5$, in which the c.n.'s of the alkali-metal ions are respectively 8, 7 and 9, and 8, represent an alternative way of providing positions for larger cations.

(d) We shall also mention a further group of structures which are closely related to the bronzes in which there is pyramidal coordination of certain metal atoms. This group includes the binary oxides Mo_5O_{14}, $Mo_{17}O_{47}$, and $W_{18}O_{49}$.

A very simple way of utilizing both types of edge-sharing by ReO_3 chains is to form first a double (or other multiple) chain and to put two of these together to form the more complex chain of Fig. 13.9(a) and (b). These chains may then join

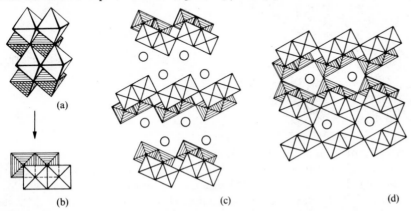

FIG. 13.9. (a) and (b) Multiple chain formed from four 'ReO$_3$' chains, (c) layer in KTiNbO$_5$, (d) 3D framework in KTi$_3$NbO$_9$.

by vertex-sharing into a layer, (c), or by further vertex-sharing to form a 3D structure, (d). Both types of structure can accommodate alkali or alkaline-earth ions, as in the examples shown. Starting with a triple chain we have the same possibilities (Table 13.14). More complex structures of the same general type can be visualized with $(2+3)$ or $(3+4)$ octahedra in the primary 'chain', giving 3D structures with intermediate formulae M_5O_{11}, M_7O_{15}, etc. An example of the latter is $Na_2Ti_7O_{15}$.[1]

(1) AC 1968 **B24** 392

TABLE 13.14

Structures built from multiple octahedral chains

Type of chain	Layer structure	3D structure	Reference
4-octahedra (Fig. 13.9)	M_2O_5	M_4O_9	
	KTiNbO$_5$	KTi$_3$NbO$_9$ BaTi$_4$O$_9$	AC 1964 17 623 JCP 1960 **32** 1515
6-octahedra	M_3O_7	M_6O_{13}	
	Na$_2$Ti$_3$O$_7$	Na$_2$Ti$_6$O$_{13}$	AC 1961 14 1245 AC 1962 **15** 194

Complex Oxides

A second building principle is to take an infinite block of ReO_3 structure, join it to others to form either (i) isolated multiple blocks or (ii) an infinite layer, and then to join such units into 3D structures by edge-sharing of the second type to similar units at a different level. The structures are rather simpler in (ii) than in (i) and are therefore illustrated first. The smallest possible block of ReO_3 structure contains 4 ReO_3 chains and the 3D framework of $AlNbO_4$[2] is built of such sub-units. This framework is not known as the structure of a simple dioxide but it is found in Na_xTiO_2,[3] a bronze containing Ti^{3+} and Ti^{4+} in which there is partial occupancy by Na^+ of the positions indicated in Fig. 13.10. A number of oxide

(2) ACSc 16 421; AC 1965 18 874 874

(3) AC 1962 15 201

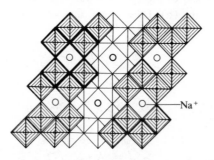

FIG. 13.10. The structure of Na_xTiO_2.

phases have structures of this type utilizing larger ReO_3 blocks, for example V_6O_{13} ((3 x 2) blocks), illustrated in Fig. 5.38 (p. 185), $TiNb_2O_7$ ((3 x 3) blocks), $Nb_{12}O_{29}$ and $Ti_2Nb_{10}O_{29}$ ((3 x 4) blocks). All members of this family, in which the blocks of ReO_3 structure contain (3 x n) chains, can be represented by the general formula $M_{3n}O_{8n-3}$. The end-member is represented by Nb_3O_7F and V_2MoO_8 (Table 13.15).

TABLE 13.15

The $M_{3n}O_{8n-3}$ family of oxide structures

n	Formula	Example	Reference
2	M_6O_{13}	V_6O_{13}	HCA 1948 31 8
3	M_9O_{21} (= M_3O_7)	$TiNb_2O_7$	AC 1961 14 660
4	$M_{12}O_{29}$	$Nb_{12}O_{29}$	ACSc 1966 20 871
		$Ti_2Nb_{10}O_{29}$	AC 1961 14 664
∞	M_3O_8	V_2MoO_8	ACSc 1966 20 1658
		Nb_3O_7F	ACSc 1964 18 2233

(4) AC 1965 18 724

In the structures we have just described the basic ReO_3 blocks are linked into infinite 'layers' by the first type of edge-sharing. There are also structures in which the blocks at both levels are joined in pairs, and here also there is the possibility of blocks of various sizes. Figure 13.11(a) shows the structure of $Nb_{25}O_{62}$[4] (and $TiNb_{24}O_{62}$) where the block size is (3 x 4). In an intermediate class of structure there is linking of the ReO_3 blocks of one family into infinite layers (those lightly shaded in Fig. 13.11(b)) while the blocks at the other level are discrete. This group

504

of structures includes $Nb_{22}O_{54}$[5] and the high-temperature form of Nb_2O_5,[6] which differ in the sizes of the ReO_3 blocks at the two levels. In high-Nb_2O_5 (Fig. 13.11(b)) blocks of (3 x 5) octahedra are joined at one level to form infinite planar slabs, and these slabs are further linked by (3 x 4) blocks (heavy outlines). In both the structures of Fig. 13.11 there are tetrahedral holes, indicated by the black circles, some of which are occupied—in a regular way—by metal atoms. In high-Nb_2O_5 27 Nb atoms in the unit cell are in positions of octahedral co-ordination and 1 Nb occupies a tetrahedral hole.

(5) ACSc 1965 **19** 1401
(6) AC 1964 **17** 1545
For a survey of some of these compounds see HCA (Fasc. extra Alfred Werner) 1967 207.

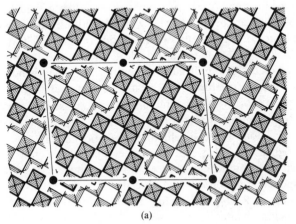
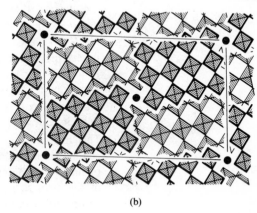

(a)   (b)

FIG. 13.11. Structures built from ReO_3 blocks: (a) $TiNb_{24}O_{62}$, (b) high-Nb_2O_5.

Bronzes and related compounds

The name *bronze*, originally given to the W compounds by Wohler in 1824, is now applied to solid oxide phases with the following characteristic properties; intense colour (or black) and metallic lustre, metallic conductivity or semiconductivity, a range of composition, and resistance to attack by non-oxidizing acids. The bronzes Na_xWO_3, for example, have colours ranging from golden yellow ($x \approx 0 \cdot 9$) through red ($x \approx 0 \cdot 6$) to deep violet ($x \approx 0 \cdot 3$). Most, but not all, bronzes consist of a host structure of composition MO_2 or MO_3 in which M is a (transition) metal capable of exhibiting a valence less than 4 or 6 respectively. If a proportion of the M(VI) atoms in MO_3 are converted to M(V) the requisite cations (alkali, alkaline-earth, La^{3+}, etc.) are incorporated into the structure to maintain electrical neutrality. The electrons liberated in this process are not, however, captured by individual metal ions of the host structure, but are distributed over the whole structure giving rise to the metallic or semi-metallic properties. From the geometry of the structures it would appear that the metallic conductivity of bronzes is not due to direct overlap of metal orbitals but to interactions through the oxygen atoms.

Bronzes have been prepared in which M is Ti, V, Nb, Ta, Mo, W or Re, (see also $Na_xPt_3O_4$, p. 456). Typical methods of preparation include heating a mixture of Na_2WO_4, WO_3, and WO_2 *in vacuo* or reducing Na_2WO_4 by hydrogen or molten zinc (for Na–W bronzes), fusing V_2O_5 with alkali oxide, when oxygen is lost and V bronzes are formed, or reducing $Na_2Ti_3O_7$ by hydrogen at 950°C, when blue-black crystals of Na_xTiO_2 are formed. The properties of bronzes and the reasons for their formation are by no means fully understood; in particular there is no clear connection between the appearance of the 'bronze' properties and the structure of the phase. For example, there are phases with both the tetragonal and hexagonal W bronze structures which are not bronzes (see later). Also, $Li_xTi_{4-x/4}O_8$ is a non-stoichiometric phase with the ramsdellite structure (p. 459) but it is colourless (i.e. it is not a bronze), whereas $K_{0.13}TiO_2$ has the closely related hollandite structure and is a bronze. Some bronzes have layer structures, examples of which are described later, but most bronzes have 3D framework structures, and the composition of the host structure corresponds to a normal oxide (as in Na_xTiO_2, $Na_xV_2O_5$, and Na_xWO_3) though the framework found in the bronze is not necessarily stable for the pure oxide. The hollandite structure of K_xTiO_2 is not stable for $x = 0$, and for all the isostructural K, Rb, and Cs compounds x is approximately equal to 0.13. The cubic W bronze structure is, however, known not only as an oxide structure (ReO_3) and in distorted forms for WO_3 itself, but also with its full complement of Na^+ ions as the stoichiometric compound NaW^VO_3. This high-pressure phase (reference in Table 13.16) has the appearance (bronze colour) and metallic conductivity characteristic of bronzes. However, the description of such a stoichiometric compound as a bronze would logically imply the extension of the term to compounds such as ReO_3, CoS_2, ThI_2, and other electron-excess compounds to which we referred in Chapter 7.

Numerous analogues of the alkali-metal bronzes which have been prepared include La_xTiO_3 and Sr_xNbO_3 (cubic),[1] In_xWO_3 (hexagonal),[2] Sn_xWO_3,[3] $Ba_{0.12}WO_3$ and $Pb_{0.35}WO_3$ (tetragonal),[4] Cu_xWO_3,[5] and 4f compounds $M_{0.1}WO_3$.[6] It is possible to replace some W by Ta in both the tetragonal and hexagonal bronze structure, as in $K_{0.5}(Ta_{0.5}W_{0.5})O_3$ (tetragonal) and $Rb_{0.3}(Ta_{0.3}W_{0.7})O_3$ (hexagonal).[7] These cream-white compounds are not bronzes but normal oxides. Bronzes containing Re have been made under high pressures.[8]

(1) JACS 1955 77 6132, 6199
(2) IC 1968 7 969
(3) IC 1968 7 1646
(4) IC 1965 4 994
(5) JACS 1957 79 4048
(6) IC 1966 5 758
(7) JPC 1964 68 1253
(8) SSC 1969 7 299

Tungsten bronzes

The approximate limits of stability of various bronzes are shown in Table 13.16, but somewhat different ranges have been found by different workers, and it is probable that the limits depend to some extent on the temperature of preparation.

The structures of these compounds provide a very elegant illustration of the formation of 3-dimensional networks MO_3 by the joining of MO_6 octahedra through all their vertices. The simplest and most symmetrical structure of this kind is the ReO_3 structure; addition of alkali-metal atoms at the centres of the unit cells gives the perovskite type of structure which is that of the cubic phases in Table 13.16 (Fig. 13.12(a) and (b)). Essentially the same basic framework occurs in the tetragonal-II Li and Na compounds, which have a slightly distorted perovskite-type structure.[1] Two quite different structures are found for the tetragonal K bronze ($x \approx 0.48$–0.57) and the hexagonal bronze formed by K, Rb, and Cs with $x \approx 0.3$.[2]

(1) ACSc 1951 5 372 670
(2) ACSc 1953 7 315

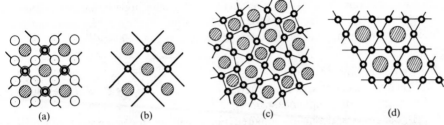

FIG. 13.12. Projections of the structures of tungsten bronzes: (a) perovskite structure of cubic bronzes, (b) the same showing only alkali-metal ions (shaded circles) and W atoms (small circles), (c) and (d) the W frameworks and alkali-metal ions in the tetragonal and hexagonal bronzes.

(The tetragonal-I Na compound[3] has a structure closely related to the tetragonal K bronze.) Projections of these two structures are shown in Fig. 13.12(c) and (d), where it will be seen that the larger alkali-metal ions are situated in tunnels bounded by rings of five and six, instead of four, WO_6 octahedra. Note that these ions do not lie at the same level as the W atoms but $c/2$ above or below them. It will

(3) AK 1949 **1** 269

TABLE 13.16
Stability limits of tungsten bronzes

Li	Na	K	Rb	Cs	$\dfrac{x}{1\cdot00}$
		cubic†			
	cubic†				
		tetragonal			— 0·50
cubic					
	tetragonal-I				
		hexagonal	hexagonal	hexagonal	
tetragonal	tetragonal				
II	II				— 0·00

† The upper limits shown by the full heavy lines refer to compounds made at atmospheric pressure. Under high pressures the stability range of the cubic Na bronze includes stoichiometric $NaWO_3$, and K bronzes have been prepared with metal content up to $K_{0\cdot9}$ (IC 1969 **8** 1183).

be appreciated that the upper limit of x in M_xWO_3 is determined by purely geometrical factors. The ratio of the number of holes for alkali-metal atoms to the number of W atoms falls from 1 in (b) to 0·6 in (c) and 0·33 in (d), so that from

507

Complex Oxides

the compositions determined experimentally it would appear that nearly the full complements of these atoms can be introduced into the structures, and indeed are necessary for the stability of the tetragonal and hexagonal phases.

The tetragonal bronze structure

We deal in more detail with this structure because it forms the basis of the structures of three groups of compounds, namely:

(i) the bronzes, the highly coloured, non-stoichiometric compounds to which we have just referred,

(ii) ferroelectric, usually colourless, compounds mostly with formulae of the type $M^{II}Nb_2O_6$ or $M^{II}Ta_2O_6$ but in some cases containing additional atoms, as in $K_6Li_4Nb_{10}O_{30}$, and

(iii) a large family of (stoichiometric) compounds with formulae such as $mNb_2O_5 . nWO_3$ in which metal atoms and additional O atoms occupy some of the pentagonal tunnels in the bronze structure. We shall also include here compounds such as $LiNb_6O_{15}F$ with structures of the same general type based on closely related octahedral frameworks.

In the tetragonal bronze structure there are tunnels of three kinds (Fig. 13.13), of which only two (S and P) are occupied in the bronzes (class (i)). The environment of an atom in a tunnel depends on its height relative to the atoms in the framework. In compounds of classes (i) and (ii) the 'tunnel' atom is at height $\frac{1}{2}$, when the coordination groups of such atoms are:

T: tricapped trigonal prism (9-coordination), site suitable only for very small ion (Li^+);

S: distorted cuboctahedral (12-coordination);

P: in a regular pentagonal tunnel the coordination group would be a pentagonal prism with atoms beyond each vertical (rectangular) face. In fact the pentagonal tunnels have an elongated cross-section, and the coordination group is closer to a tricapped trigonal prism (9-coordination).

(F is the (distorted) octahedral site in the MO_3 framework.)

The examples in Table 13.17 show how certain of these positions are occupied in some of the ferroelectric compounds of class (ii), the numbers at the heads of the columns indicating the numbers of sites of each type in a unit cell containing 30 O atoms. In $Li_4K_6Nb_{10}O_{30}$ all the metal sites are occupied, including the T sites for very small ions. If the oxidation number of all the atoms in the framework sites is 5, only 5 M^{2+} ions are required for charge balance and these occupy statistically the $2S + 4P$ sites. The chemical formula reduces to a simple form if all these atoms are of the same element, as in $PbNb_2O_6$. If, however, some M^V is replaced (again statistically) by M^{IV} in the $10F$ sites, the full complement of M^{II} atoms can be taken up, as in $Ba_6Ti_2Nb_8O_{30}$.

In class (iii) a new principle is introduced. A metal atom is placed in a P tunnel, assumed now to have approximately regular pentagonal cross-section, at the height 0 (see Fig. 13.13) so that it is surrounded by a ring of five O atoms at the same height. An additional O atom is introduced with each metal atom in the tunnel, the

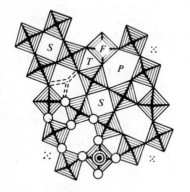

○   O at height 0

◎ $\left\{ \begin{array}{l} M \text{ at height 0} \\ O \text{ at heights } \pm \frac{1}{2} \end{array} \right.$

FIG. 13.13. The tetragonal bronze structure, showing the three kinds of tunnel.

TABLE 13.17

Compounds with structures related to the tetragonal bronze structure

Ferroelectrics (class (ii))

Cell content	4T	2S	4P	10F	30 O	*Reference*
	Li_4	K_6	Nb_{10}		O_{30}	JCG 1967 **1** 315, 318
		Ba_6	Ti_2Nb_8		O_{30}	AC 1968 **B24** 984
		$(Ba, Sr)_5$	Nb_{10}		O_{30}	JCP 1968 **48** 5048
		Pb_5	Nb_{10}		O_{30}	AC 1958 **11** 696

Superstructures (class (iii)) and structures based on other MO_3 frameworks of the same general type (with T, S, and P tunnels).

Formula type	Composition	*Reference*
M_3O_8	WNb_2O_8	NBS 1966 **70A** 281
	$NaNb_6O_{15}F$ (or OH)	ACSc 1965 **19** 2285
	$LiNb_6O_{15}F$	ACSc 1965 **19** 2274
	Ta_3O_7F	ACSc 1967 **21** 615
$M_{17}O_{47}$	$Mo_{17}O_{47}$	ACSc 1963 **17** 1485
	$Nb_8W_9O_{47}$	AC 1969 **B25** 2071
$M_{23}O_{63}$	$Nb_{12}W_{11}O_{63}$	AC 1968 **B24** 637
Miscellaneous	Mo_5O_{14}	AK 1963 **21** 427
	$W_{18}O_{49}$	AK 1963 **21** 471

On the subject of the nets in these compounds see: AC 1968 **B24** 50.

two kinds of atom alternating, giving M seven O neighbours at the vertices of a pentagonal bipyramid. The formula of the compound obviously depends on the proportion of P tunnels occupied in this way. For example, if one-half of the P tunnels in the bronze structure are occupied (by 2 M + 2 O) the composition becomes $M_{10+2}O_{30+2}$ or M_3O_8. Occupation of one-third of the P tunnels (which implies a larger unit cell) would give the composition $M_{17}O_{47}$ ($M_{15+2}O_{45+2}$). There are numerous oxide phases with structures of this kind. Some (class (iii)) are superstructures of the tetragonal bronze structure (with multiple cells), while others are based on closely related but topologically different 3-dimensional MO_3 frameworks. The latter, of which there is an indefinitely large number, have different relative numbers of T, S, and P tunnels and/or different spatial arrangements of these tunnels (see reference in Table 13.17). It is interesting that compounds so closely related chemically as $LiNb_6O_{15}F$ and $NaNb_6O_{15}F$ have structures based on different 3D octahedral nets; both, like WNb_2O_8, are of the M_3O_8 type, but with additional alkali-metal atoms (not shown in Fig. 13.13).

These structures may alternatively be described as assemblies of composite units of the kind indicated in Fig. 13.14. Such a unit consists of a column of pentagonal bipyramidal coordination groups sharing edges with the surrounding columns of octahedra; these multiple columns then form the 3D framework by vertex-sharing. In the series of stoichiometric oxides formed between Nb_2O_5 and WO_3 the structures are based on blocks of ReO_3 structure up to the composition $W_8Nb_{18}O_{69}$ but phases richer in W, starting at WNb_2O_8, have structures of the kind we have just described.

Complex Oxides

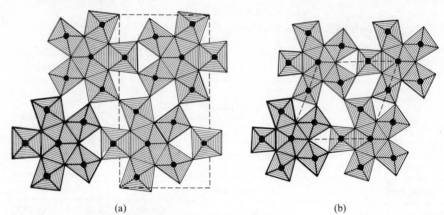

(a)   (b)

FIG. 13.14. Unit cells of the structures of (a) $LiNb_6O_{15}F$, (b) $NaNb_6O_{15}F$. The alkali-metal ions are omitted.

Molybdenum bronzes

The preparation under high pressure (65 kbar) of Mo bronzes with tungsten bronze structures has been described; for example, cubic Na_xMoO_3 and K_xMoO_3 ($x \approx 0.9$) and tetragonal K_xMoO_3 ($x \approx 0.5$).[1] The Mo bronzes prepared by the electrolysis of fused mixtures of alkali molybdates and MoO_3[2] have more complex structures. A bronze with the approximate composition $Na_{0.9}Mo_6O_{17}$ has a hexagonally distorted perovskite structure with an ordered arrangement of Na in one-sixth of the available sites and apparently a statistical distribution of 17 O atoms over 18 sites.[3] (The observed average structure apparently results from twinning of a monoclinic structure in which the O atoms are regularly arranged.)

Two potassium–molybdenum bronzes, a red $K_{0.33}MoO_3$[4] and a blue-black $K_{0.30}MoO_3$[5] have closely related layer structures. Edge-sharing groups of 6 and 10 octahedra respectively are further linked into layers by sharing vertices as indicated in Fig. 5.34 (p. 183), the composition of the layer being in each case MoO_3. The layers are held together by the K^+ ions which occupy positions of 8-coordination in the first compound and of 7- and (6 + 4)-coordination in the second. There is a closely related layer in $Cs_{0.25}MoO_3$,[6] in which the basic repeat unit is a different grouping of six octahedra. These units share the edges and vertices marked in Fig. 13.15 to form layers between which the larger Cs^+ ions can be accommodated.

Vanadium bronzes

These are made by methods such as fusing V_2O_5 with alkali-metal oxide or vanadate. Oxygen is lost and black semiconducting phases are formed which contain some alkali metal and V^{IV} atoms. Bronzes have been made containing Li, Na, K, Cu, Ag, Pb, etc., and in some systems (e.g. $Li_2O–V_2O_5$ and $Na_2O–V_2O_5$) there are several bronze phases with different composition ranges and structures.

510

Sidebar references:

(1) IC 1966 **5** 1559
(2) IC 1964 **3** 545

(3) AC 1966 **20** 59

(4) AC 1965 **19** 241; IC 1967 **6** 1682
(5) AC 1966 **20** 93

(6) JSSC 1970 **2** 16
For later work on lithium Mo bronzes see: JSSC 1970 **1** 327, and for Mo fluoro bronzes; JSSC 1970 **1** 332.

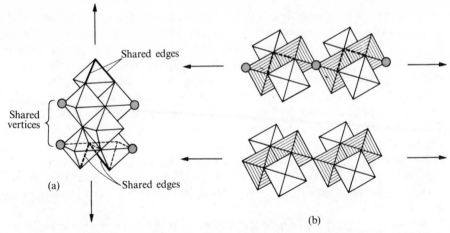

FIG. 13.15. The structure of the bronze $Cs_{0.25}MoO_3$: (a) the 6-octahedron unit which shares the edges and vertices indicated to form the layers (b), which are perpendicular to the plane of the paper.

The structure of $Li_{0.04}V_2O_5$ (Fig. 13.16(a)) is essentially that of V_2O_5 with a small number of Li^+ ions in positions of trigonal prism coordination. In LiV_2O_5 (Fig. 13.16(b)) also there is 5-coordination of the V atoms; both these compounds have layer structures.

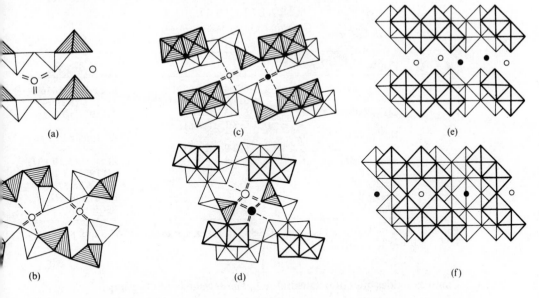

FIG. 13.16. The structures of vanadium and titanium bronzes: (a) $Li_{0.04}V_2O_5$, similar to V_2O_5 but with Li^+ ions (small circles) in trigonal prismatic coordination, (b) LiV_2O_5 (Li^+ in octahedral coordination), (c) $Li_{1+x}V_3O_8$ (octahedral coordination of Li^+), (d) $Na_{0.15}V_2O_5$ (7-coordinated Na^+), (e) $Ag_{0.68}V_2O_5$ (Ag^+ with 5 near neighbours), (f) $Na_{0.2}TiO_2$ (8-coordinated Na^+).

In the next two compounds of Fig. 13.16 there is both 5- and 6-coordination of V. $Li_{1+x}V_3O_8$, (c), has a layer structure; in (d), which represents the structure of $\beta\text{-}Li_{0.30}V_2O_5$ and $Na_{0.15}V_2O_5$, the VO_5 and VO_6 coordination groups form a 3D framework. The sodium bronze $Na_xV_2O_5$ is stable over the range $0.15 < x < 0.33$, the upper limit corresponding to the formula NaV_6O_{15}. This corresponds to occupation of only one-half of the available Na^+ sites; if they were all occupied in each tunnel Na^+ would have a close Na^+ neighbour in addition to 7 O atoms. The formula may therefore be written $Na_{2-y}V_6O_{15}$. A steel-blue $Pb_{0.20}V_2O_5$ with the structure of Fig. 13.16(d) has also been prepared.

In the $Ag_2O\text{-}V_2O_5$ system there is a bronze $Ag_{2-x}V_6O_{15}$ isostructural with the Na bronze of Fig. 13.16(d) and also $Ag_{0.68}V_2O_5$, (e), in which the V atoms may be described as octahedrally coordinated, though the V—O bond lengths range from 1.5–2.4 Å.

The structure of $Na_{0.2}TiO_2$ is included at (f) to show the close relation of its structure to that of the bronze (e). Further sharing of vertices of octahedra converts the layer structure (e) to the 3D framework (f) of the titanium bronze.

In the systems $M_2O_3\text{-}VO_2\text{-}V_2O_5$ (M = Al, Cr, Fe) phases include one $(Al_{0.33}V_2O_5)$ closely related to V_6O_{13}, from which it is derived by addition of Al and O atoms in the tunnels of that structure. There are also phases $M_zV_{2-z}O_4$ with a superstructure of the rutile type, having equal numbers of M^{III} and V^V replacing part of the V^{IV}, as in $Fe^{III}_{0.07}V^V_{0.07}V^{IV}_{1.86}O_4$). For references see Table 13.18.

TABLE 13.18
Coordination of V in bronzes

Fig. 13.16	Bronze	C.N. of V	V—O (Å)	Reference
a	$Li_{0.04}V_2O_5$	5	1.6–2.45 (then 2.82)	BSCF 1965 1056
b	LiV_2O_5	5	1.61–1.98 (3.09)	AC 1971 **B27** 1476
c	$Li_{1+x}V_3O_8$	6	1.6–2.3 —	AC 1957 **10** 261
		5	1.6–2.1 (2.86)	
d	$Na_{0.15}V_2O_5$	6	1.6–2.3 —	AC 1955 **8** 695
		5	1.6–2.0 (2.68)	
e	$Ag_{0.68}V_2O_5$	6	1.5–2.4 —	ACSc 1965 **19** 1371
f	$Na_{0.2}TiO_2$			AC 1962 **15** 201; IC 1967 **6** 321
	$Pb_{0.20}V_2O_5$			BSMC 1969 **92** 17
	$Al_{0.33}V_2O_5$			BSCF 1967 227

For later work on $M_xV_2O_5$ phases see: JSSC 1970 **1** 339, and for a new Ti bronze, $K_3Ti_8O_{17}$; JSSC 1970 **1** 319.

Complex oxides built of octahedral AO_6 and tetrahedral BO_4 groups

Some of the simpler structures of this type were noted in the discussion in Chapter 5 of structures built from tetrahedra and octahedra. Here we describe some of the structures listed in Table 5.7 (p. 190) and also some more complex structures of the same general type. The composition depends on the relative numbers of octahedral and tetrahedral groups and on the way in which the vertices are shared. In the simplest case each tetrahedron shares all its vertices with octahedra and each octahedron shares four vertices with tetrahedra and two with other octahedra. The composition is then ABO_5.

Two simple structures of this kind may be regarded as built from ReO_3-type chains (in which each octahedron shares two opposite vertices) which are then cross-linked by BO_4 tetrahedra to form 3D structures. In $NbOPO_4$ (and the isostructural $MoOPO_4$, $VOMoO_4$ and tetragonal $VOSO_4$) each BO_4 tetrahedron links four such chains (Fig. 13.17(a)) forming a structure which approximates to a c.c.p. assembly of O atoms in which Nb atoms occupy $\frac{1}{5}$ of the octahedral holes and P atoms $\frac{1}{10}$ of the tetrahedral holes (Fig. 13.17(b)). In a second form of $VOSO_4$ also each tetrahedral group (SO_4) shares its vertices with four octahedral groups (VO_6) but two of these belong to the same chain, so that the tetrahedral group links three ReO_3-type chains as compared with four in $NbOPO_4$.

Although it is convenient to describe these structures in terms of octahedral AO_6 and tetrahedral BO_4 groups it should be emphasized that while the BO_4 groups are essentially regular tetrahedra the AO_6 groups are far from regular. The metal atom is displaced from the centre towards one vertex of the octahedron, and the arrangement of the five bonds may be described as tetragonal pyramidal. The distortion of the octahedral coordination group is of a kind found in many oxides of V^{IV}, Nb^V, and Mo^V, with one very short and one very long bond and four bonds of intermediate length. Some examples are given in Table 13.19.

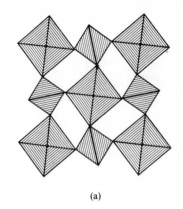

(a)

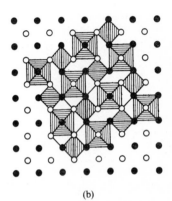

(b)

FIG. 13.17. Projection of the crystal structure of $VOMoO_4$.

TABLE 13.19

Metal coordination groups in some oxy-compounds

Compound	Bond lengths in octahedron (Å)			Reference
	short (one)	(four)	long (one)	
$NbOPO_4$	1·78	1·97	2·32	ACSc 1966 **20** 72
Nb_3O_7F	1·86	1·98	2·20	ACSc 1964 **18** 2233
$MoOPO_4$	1·66	1·97	2·63	ACSc 1964 **18** 2217
$VOMoO_4$	1·68	1·97	2·59	ACSc 1966 **20** 722
$VOSO_4$ (β) (rh.)	1·59	2·03	2·28	ACSc 1965 **19** 1906
(α) (tetrag.)	1·63	2·04	2·47	JSSC 1970 **1** 394

A number of oxy-compounds containing P and Mo (or W) with low Mo(W) : P ratios have been prepared by methods such as the following: $Mo(OH)_3PO_4$ by heating a solution of MoO_3 in concentrated H_3PO_4 at 180°C and then diluting with conc. HNO_3, and WOP_2O_7 from WO_3 and P_2O_5 on prolonged heating in an autoclave at 550°C. These compounds are built from PO_4 tetrahedra and MoO_6 (WO_6) octahedra sharing vertices only, and a structural feature common to several of them is a chain consisting of alternate PO_4 and MoO_6 groups. In some compounds PO_4 shares vertices entirely with MoO_6 while in others there is linking of PO_4 groups in pairs or infinite chains, the further linking of these units through MoO_6 octahedra leading to infinite chains, layers, or 3D networks.

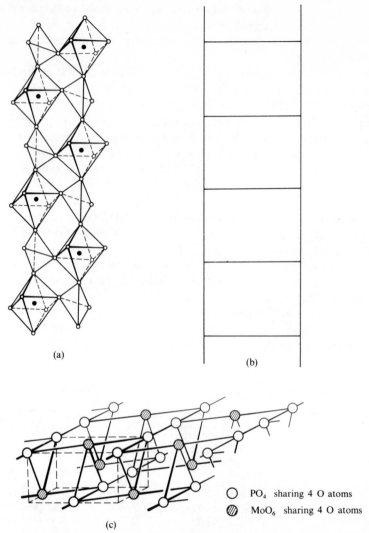

FIG. 13.18. (a) and (b) The double chain $MoPO_4(OH)_3$. (c) The double layer P_2MoO_8.

The formulae assigned to these compounds in column 2 of Table 13.20 indicate the extent of sharing of O atoms between PO_4 groups, and the more elaborate structural formulae of column 3 show how many vertices of each PO_4 or MoO_6 group are shared and hence the nature of the fundamental net (column 4). The symbol ϕ represents an O atom shared between two coordination groups (PO_4 or MoO_6). We comment elsewhere on the surprisingly small number of pairs of isostructural Mo and W compounds. Here $Na(PMoO_6)$ and $Na(PWO_6)$ are isostructural, but the pairs P_2MoO_8 and P_2WO_8, $P_2Mo_2O_{11}$ and $P_2W_2O_{11}$ are not. We note now the structural principles in three of the compounds of Table 13.20.

TABLE 13.20

Some oxy-compounds built of PO_4 *and* MoO_6 (WO_6) *groups*

Compound		Structural formula	Nature of structure
$Mo(OH)_3PO_4$		$(PO\phi_3)[Mo\phi_3(OH)_3]$	3-connected double chain
P_2MoO_8	$MoO_2(PO_3)_2$	$(P\phi_4)_{2n}(MoO_2\phi_4)_n$	4-connected double layer
P_2WO_8	WOP_2O_7	$(P\phi_4 . PO\phi_3)(WO\phi_5)$	3,4,5-connected layer
$NaPWO_6$	$NaWO_2PO_4$	$Na[(P\phi_4)(WO_2\phi_4)]$	4-connected
$P_2W_2O_{11}$	$W_2O_3(PO_4)_2$	$(P\phi_4)(WO\phi_5)$	4,5-connected } 3D net
$P_2Mo_2O_{11}$	$(MoO_2)_2P_2O_7$	$(P\phi_4 . P\phi_4)(MoO\phi_5)_2$	4,5-connected

References: AK 1962 **19** 51; ACSc 1964 **18** 2329

The structural unit in $MoPO_4(OH)_3$ is the double chain of Fig. 13.18(a). The H atoms were not located but are presumably attached to the three unshared O atoms of each MoO_6 group and involved in hydrogen bonding between the chains. Since each PO_4 and MoO_6 shares 3 vertices the topology of the structure may be illustrated by the double chain of Fig. 13.18(b). In P_2MoO_8 chains of PO_4 tetrahedra sharing two O atoms are further linked by MoO_6 octahedra to form the double layer illustrated diagrammatically in Fig. 13.18(c). The simplest type of 3D network is found in $Na(PWO_6)$, in which every PO_4 shares all four vertices with MoO_6 groups and each of the latter shares four vertices with PO_4 groups. If we separate the two unshared O atoms of each MoO_6 the formula becomes $Na(PWO_4 . O_2)$–compare $NaAlSiO_4$. Basically the anion is a 4-connected net analogous to those found in aluminosilicates and in fact is very similar to those of the felspars (Chapter 23).

515

14

Metal Hydroxides, Oxyhydroxides, and Hydroxy-Salts

We shall be concerned in this chapter with the following groups of compounds:

Hydroxides $M(OH)_n$ (with which we shall include the alkali-metal hydrosulphides MSH)

Complex hydroxides $M'_x M''_y (OH)_z$

Oxy-hydroxides MO(OH)

Hydroxy-salts (basic salts)

Metal hydroxides

Compounds $M(OH)_n$ range from the strongly basic compounds of the alkali and alkaline-earth metals through the so-called amphoteric hydroxides of Be, Zn, Al, etc. and the hydroxides of transition metals to the hydroxy-acids formed by non-metals $(B(OH)_3)$ or semi-metals $(Te(OH)_6)$. The latter are few in number and are included in other chapters.

Apart from the rather soluble alkali-metal (and Tl^I) hydroxides and the much less soluble alkaline-earth compounds, most metal hydroxides are more or less insoluble in water. Some of them are precipitated in a gelatinous form by NaOH and redissolve in excess alkali, for example, $Be(OH)_2$, $Al(OH)_3$, and $Zn(OH)_2$, while others such as $Cu(OH)_2$ and $Cr(OH)_3$ show the same initial behaviour but redeposit on standing. Such hydroxides, which 'dissolve' not only in acids but also in alkali have been termed amphoteric, the solubility in alkali being taken to indicate acidic properties. In fact the solubility is due in some cases to the formation of hydroxy-ions, which may be mononuclear, for example, $Zn(OH)_4^{2-}$,[1] or polynuclear. From a solution of $Al(OH)_3$ in KOH the potassium salt of the bridged ion (a) can be crystallized. (No difference was found between the lengths of the Al–O and Al–OH bonds (1·76 Å, mean).[2]) It is likely that hydroxy-ions of some kind exist in the solution. Hydroxy- or hydroxy-aquo complexes also exist in the 'basic' salts formed by the incomplete hydrolysis of normal salts (see later section), and some interesting polynuclear ions have been shown to exist in such solutions. The Raman and i.r. spectra[3] of, and also the X-ray scattering[4] from, hydrolysed Bi perchlorate solutions show that the predominant species over a wide range of Bi and H ion concentration is $Bi_6(OH)_{12}^{6+}$ (Fig. 14.1(a)). Similarly, X-ray scattering from hydrolysed Pb perchlorate solutions

(1) JCP 1965 **43** 2744

(2) ACSc 1966 **20** 505

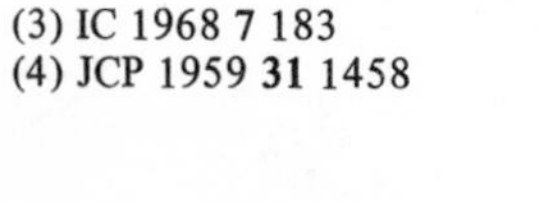

$$\begin{array}{ccc} HO & & OH^{2-} \\ \diagdown & & \diagup \\ HO-Al & \underset{O}{\overset{132°}{\diagup\diagdown}} & Al-OH \\ \diagup & & \diagdown \\ HO & & OH \end{array}$$

(a)

(3) IC 1968 **7** 183
(4) JCP 1959 **31** 1458

indicates at a OH : Pb ratio of 1·00 a complex based on a tetrahedron of Pb atoms, and at a OH : Pb ratio of 1·33 (the maximum attainable before precipitation occurs) a Pb_6 complex similar to that in the crystalline basic salt $Pb_6O(OH)_6(ClO)_4)_4 . H_2O$—see Fig. 14.1(b).[5]

In other cases the conductivities of 'redissolved' hydroxide solutions are equal to those of alkali at the same concentration, suggesting that the 'solubility' of the hydroxide is more probably due to the formation of a sol. For example, the freshly precipitated cream-coloured $Ge(OH)_2$ 'dissolves' in strongly basic solutions to form reddish-brown colloidal solutions.[6] Hydrolysis of metal-salt solutions often yields colloidal solutions of the hydroxides, as in the case of the trihydroxides of Fe and Cr and the tetrahydroxides of elements of the fourth Periodic Group such as Sn, Ti, Zr, and Th. It is unlikely that the hydroxides $M(OH)_4$ of the latter elements are present in such solutions, for the water content of gels $MO_2 . xH_2O$ is very variable.

We must emphasize that our picture of the structural chemistry of hydroxides is very incomplete. Crystalline material is required for structural studies, preferably single crystals, and some hydroxides readily break down into the oxide, even on boiling in water (e.g. $Cu(OH)_2$, $Sn(OH)_2$, $Tl(OH)_3$). Many simple hydroxides are not known, for example, CuOH, AgOH, $Hg(OH)_2$, $Pb(OH)_4$. Though many of the higher hydroxides do not appear to be stable under ordinary conditions, nevertheless compounds such as $Ni(OH)_3$, $Mn(OH)_4$, $U(OH)_4$, $Ru(OH)_4$ etc. are described in the literature; confirmation of their existence and knowledge of their structures would be of great interest. Some may not be simple hydroxides; for example, $Pt(OH)_4 . 2 H_2O$ may be $H_2[Pt(OH)_6]$. Conductometric titrations and molecular weight measurements show that 'aurates' and the gelatinous 'auric acid' prepared from them contain the $Au(OH)_4^-$ ion.[7] The growth of crystals of $Pb(OH)_2$ in aged gels is described in the literature, but the most recent evidence suggests that this hydroxide does not exist.[8] The crystalline solid obtained by the slow hydrolysis of solutions of plumbous salts is $Pb_6O_4(OH)_4$ (p. 936).

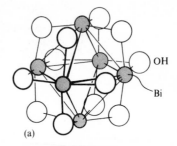

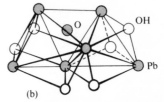

FIG. 14.1. Hydroxy-complexes: (a) $Bi_6(OH)_{12}^{6+}$, (b) $Pb_6O(OH)_6^{4+}$. In (b) there is an O atom at the centre of the central tetrahedron and OH groups above each face of the terminal tetrahedra.

(5) IC 1969 **8** 856
(6) JINC 1964 **26** 2123
(7) ZaC 1960 **304** 154, 164
(8) IC 1970 **9** 401

The structures of hydroxides $M(OH)_n$

From the structural standpoint we are concerned almost exclusively with the solid compounds. Only the hydroxides of the most electropositive metals can even be melted without decomposition, and only the alkali hydroxides can be vaporized. Microwave studies of the vapours of KOH, RbOH, and CsOH and i.r. studies of the molecules trapped in argon matrices show that the (ionic) molecules are linear:

	M—O (Å)	O—H (Å)	μ(D)	Reference
KOH	2·18	—	—	JCP 1966 **44** 3131
RbOH	2·31	0·96	—	JCP 1969 **51** 2911;
CsOH	2·40	0·97	7·1	JCP 1967 **46** 4768
(cf. CsF	2·35 (Cs-F)	—	7·87	PR 1965 **138A** 1303)

The structures of crystalline metal hydroxides and oxy-hydroxides MO(OH) are determined by the behaviour of the OH^- ion in close proximity to cations. We may envisage three types of behaviour: (a) as an ion with effective spherical symmetry,

(b) with cylindrical symmetry, or (c) polarized so that tetrahedral charge distribution is developed. Attractions between the positive and negative regions of

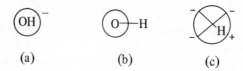

| (a) | (b) | (c) |

different OH^- ions in (c) leads to the formation of hydrogen bonds. The transition from (b) to (c) is to be expected with increasing charge and decreasing size of the cation. If the polarizing power of a cation M^{n+} is proportional to ne/r^2 it can be readily understood why there is hydrogen bonding in, for example, $Al(OH)_3$ but not in $Ca(OH)_2$ or $La(OH)_3$, the relative values of the polarizing powers being

Ca^{2+}	La^{3+}	Al^{3+}
1·8	2·0	9·2

Case (a), implying random orientation or free rotation, is apparently realized in the high-temperature form of KOH (and RbOH ?). The high-temperature form of NaOH was originally supposed to have the same (NaCl) structure but this is not confirmed by later work (see later). Apparently the SH^- ion behaves in this way in CsSH and in the high-temperature forms of NaSH, KSH, and RbSH, which is consistent with the fact that S forms much weaker hydrogen bonds than oxygen. Behaviour of type (a) is confined to (at most) one or two compounds, and on the basis of their structures we may recognize two main groups of crystalline hydroxides. The feature which distinguishes hydroxides of the first class is the absence of hydrogen bonding between the OH^- ions.

I. *Hydroxides MOH, $M(OH)_2$, and $M(OH)_3$ with no hydrogen bonding*

(a) *Random orientation or rotation of* OH^- *ions*.   In the high-temperature form of KOH the OH^- ion is behaving as a spherical ion of effective radius 1·53 Å, intermediate between those of F^- and Cl^-, and this compound has the NaCl structure. It is not known whether the cubic symmetry arises from random orientation of the anions or from free rotation.

Only the most electropositive metals, the alkali metals and alkaline-earths, form hydrosulphides. LiSH is particularly unstable, being very sensitive to hydrolysis and oxidation, and decomposes at temperatures above about 50°C. It may be prepared in a pure state by the action of H_2S on lithium n-pentyl oxide in ether solution. Its structure is of the zinc-blende type and similar to that of $LiNH_2$, but there is considerable distortion of the tetrahedral coordination groups:

$$SH \begin{cases} 2\ Li\ \ 2\cdot43\ Å \\ \ Li\ \ 2\cdot46 \\ \ Li\ \ 2\cdot67 \end{cases} \qquad Li_I - 4\ SH\ 2\cdot43\ Å \qquad Li_{II} \begin{cases} 2\ SH\ 2\cdot46\ Å \\ 2\ SH\ 2\cdot67 \end{cases}$$

The high-temperature forms of NaSH, KSH, and RbSH have the NaCl structure; CsSH crystallizes with the CsCl structure. The fact that the unit cell dimension of

CsSH is the same as that of CsBr shows that SH^- is behaving as a spherically symmetrical ion with the same radius as Br^-. In contrast to the low-temperature forms of the hydroxides of Na, K, and Rb, which are described in (b), the hydrosulphides have a rhombohedral structure at room temperature. In this calcite-like structure the SH^- ion has lower symmetry than in the high-temperature forms and appears to behave like a planar group. It is probable that the proton rotates around the S atom in a plane forming a disc-shaped ion like a rotating CO_3^{2-} ion.

(b) *Oriented* OH^- *ions.* The closest resemblance to simple ionic AX_n structures is shown by the hydroxides of the larger alkali metals, alkaline-earths, and La^{3+}. The hydroxides of Li, Na, and the smaller M^{2+} ions form layer structures indicative of greater polarization of the OH^- ion.

The forms of KOH and the isostructural RbOH stable at ordinary temperatures have a distorted NaCl structure in which there are irregular coordination groups (K—OH ranging from 2·69 to 3·15 Å). The shortest O—O distances (3·35 Å) are between OH groups coordinated to the same K^+ ion and delineate zigzag chains along which the H atoms lie. The weakness of these OH—O interactions is shown by the low value of the heat of transformation from the low- to the high-temperature form (6·3 kJ mol^{-1} compare 21-25 kJ mol^{-1} for the O—H—O bond in ice). The structures of CsOH and TlOH are not known.

The larger Sr^{2+} and Ba^{2+} ions are too large for the CdI_2 structure which is adopted by $Ca(OH)_2$. $Sr(OH)_2$ has a structure of 7 : (3, 4)-coordination very similar to that of YO . OH which is a 3D system of edge-sharing monocapped trigonal prisms.[1] There are no hydrogen bonds (shortest O—H—O, 2·94 Å). A preliminary examination has been made of crystals of one form of $Ba(OH)_2$ but the complex structure has not been determined.[2] (For $Sr(OH)_2$. H_2O see p. 562.)

The trihydroxides of La, Y,[3] the rare-earths Pr, Nd, Sm, Gd, Dy, Er, and Yb, and also $Am(OH)_3$[4] crystallize with a typical ionic structure in which each metal ion is surrounded by 9 OH^- ions and each OH^- by 3 M^{3+} ions. This is the UCl_3 structure illustrated in Fig. 9.8 (p. 359) in which the coordination group around the metal ion is the tricapped trigonal prism. The positions of the D atoms in $La(OD)_3$ have been determined by n.d.[5] In Fig. 14.2 all the atoms (including D) lie at heights $c/4$ or $3c/4$ above and below the plane of the paper ($c = 3·86$ Å). It follows that the D atom marked with an asterisk (at height $c/4$) lies approximately at the centre of a square group of four O atoms, two at $c/4$ below and two at $3c/4$ above the plane of the paper, with D—O (mean), 2·74 Å. In the OD^- ion O—D is equal to 0·94 Å.

A layer in the structure of LiOH is illustrated in Fig. 3.33 as an example of the simplest 4-connected plane layer. In a layer each Li^+ is surrounded tetrahedrally by 4 OH^- and each OH^- has 4 Li^+ neighbours all lying to one side. A n.d. study shows that the O—H bonds are normal to the plane of the layer (O—H, 0·98 Å), and from the elevation of the structure in Fig. 14.3(b) it is evident that there is no hydrogen bonding between the layers.

(1) ZaC 1969 **368** 53

(2) AC 1968 **B24** 1705
(3) ACSc 1967 **21** 481
(4) AC 1968 **B24** 979

(5) JCP 1959 **31** 329

519

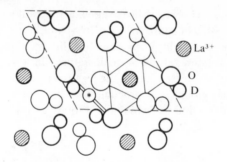

FIG. 14.2. The crystal structure of La(OD)$_3$. All atoms are at a height $c/4$ (light) or $3c/4$ (heavy), i.e. 0·96 Å above or below the plane of the paper.

The form of NaOH stable at ordinary temperatures has the TlI (yellow) structure. This is built of slices of NaCl structure in which Na$^+$ has five nearest neighbours at five of the vertices of an octahedron (Na—OH, 2·35 Å). The nearest neighbours of Na$^+$ in the direction of the 'missing' octahedral neighbour are 2 OH$^-$ at 3·70 Å, and the shortest interlayer OH—OH contacts are 3·49 Å. The H atoms

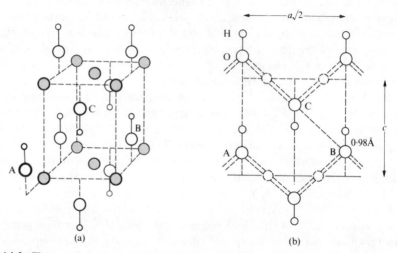

FIG. 14.3. The crystal structure of LiOH showing positions of protons: (a) unit cell, (b) section of structure parallel to (110) showing contacts between OH groups of adjacent layers.

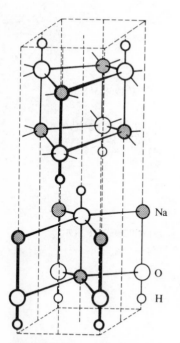

FIG. 14.4. The (yellow) TlI structure of low-temperature NaOH. Since the coordinates of the H atoms have not been determined O—H has been made equal to 1 Å.

have been located by n.d. at the positions indicated in Fig. 14.4, the system Na—O—H being collinear. The high-temperature form, which is stable over a very small temperature range (299·6–318·4°C (m.p.)), is built of layers of the same kind. These apparently move relative to one another as the temperature rises (the monoclinic β angle varying) so that the structure tends towards the NaCl structure.

The third type of layer found in this group of compounds is the CdI$_2$ layer, this structure being adopted by the dihydroxides of Mg, Ca, Mn, Fe, Co, Ni, and Cd. (In addition to the simple CdI$_2$ structure some of these compounds have other structures with different layer sequences, as in the case of some dihalides.) Each OH$^-$ forms three bonds to M atoms in its own layer and is in contact with three

OH$^-$ of the adjacent layer. The environments of OH$^-$ in LiOH and Mg(OH)$_2$ show that there are no hydrogen bonds in these structures:

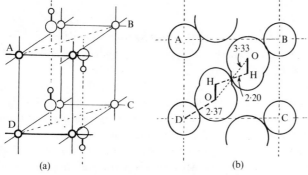

compare

As in LiOH the OH$^-$ ions in these hydroxides are oriented with their H atoms on the outer surfaces of the layers, as shown by an n.m.r. study of Mg(OH)$_2$[6] and a n.d. study of Ca(OH)$_2$.[7] The elevation of the structure of Ca(OH)$_2$ (Fig. 14.5(b)) may be compared with that of LiOH in Fig. 14.3(b). It is interesting that a redetermination of the structure of Mn(OH)$_2$[8] gives Mn–OH equal to 2·21 Å, as compared with 2·22 Å for Mn–O in MnO, indicating that the OH$^-$ radius of 1·53 Å is relevant only to structures such as high-KOH.

(6) JCP 1956 **25** 742
(7) JCP 1957 **26** 563 (n.d.);
AM 1962 **47** 1231 (n.m.r.)

(8) ACSc 1965 **19** 1765

FIG. 14.5. The crystal structure of Ca(OH)$_2$ showing the positions of the protons: (a) unit cell, (b) section parallel to (110) plane. Ca^{2+} ions are at the corners of the cell and OH$^-$ ions on the lines $\frac{1}{3}\frac{2}{3}z$ and $\frac{2}{3}\frac{1}{3}z$.

(9) AC 1961 **14** 1041
(10) AC 1967 **22** 441

From X-ray powder photographic data Cu(OH)$_2$ has been assigned a structure similar to that of γ-FeO . OH (which is described later), modified to give Cu^{II} a distorted octahedral arrangement of neighbours (4 OH at 1·94 and 2 OH at 2·63 Å).[9] This is a surprising structure for this compound since it has two types of non-equivalent O atom and is obviously more suited to an oxyhydroxide. Since the shortest distance between OH$^-$ ions attached to different Cu atoms is 2·97 Å there is presumably no hydrogen bonding. The H positions in this crystal would be of great interest, but a detailed study would require larger single crystals than it has yet been possible to grow.

A number of dihydroxides are polymorphic. A new AX$_2$ structure has been proposed for γ-Cd(OH)$_2$,[10] consisting of octahedral coordination groups linked into double chains by *face*-sharing, these double chains sharing vertices to form a 3D framework of a new type (Fig. 14.6).

References to the alkali-metal hydroxides and hydrosulphides are given in Table 14.1.

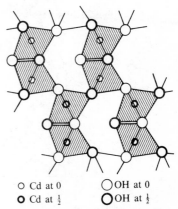

○ Cd at 0   ◯ OH at 0
● Cd at $\frac{1}{2}$ ◯ OH at $\frac{1}{2}$

FIG. 14.6. Projection along direction of double chains of the structure of γ-Cd(OH)$_2$.

TABLE 14.1

Crystal structures of MOH *and* MSH

	LiOH	NaOH	KOH	RbOH	CsOH
Low-temperature form		Orthorhombic[b], [e] (TlI structure) 299·6°C	Monoclinic[d]		?
Transition temp. High-temperature form	Fig. 14.3 (a)	Monoclinic[e]	248° NaCl[c]	?	?

	LiSH	NaSH	KSH	RbSH	CsSH
Low-temperature form Transition temperature High-temperature form	See text[f]	90° Rhombohedral structure NaCl structure[g]	160–170°	?	CsCl structure

(a) ZK 1959 **112** 60 (n.d.); JCP 1962 **36** 2665 (i.r.). (b) JCP 1955 **23** 933. (c) ZaC 1939 **243** 138. (d) JCP 1960 **33** 1164. (e) ZK 1967 **125** 332. (f) ZaC 1954 **275** 79. (g) ZK 1934 **88** 97; ZaC 1939 **243** 86.

II. *Hydroxides* $M(OH)_2$ *and* $M(OH)_3$ *with hydrogen bonds*

Zinc hydroxide. A number of forms of this hydroxide have been described, one with the C 6 (CdI_2) structure and others with more complex layer sequences, though some of these phases may be stable only in the presence of foreign ions.[1] This hydroxide may be crystallized by evaporating the solution of the ordinary precipitated gelatinous form in ammonia. The crystals of the ϵ form are built of $Zn(OH)_4$ tetrahedra sharing all vertices with other similar groups.[2] Representing the structure as basically a 3-dimensional network of Zn atoms each linked to four others through OH groups, the net is simply a distorted version of the diamond (or cristobalite) net (see p. 104). The distortion is of such a kind as to bring each OH near to two others attached to different Zn atoms, so that every OH is surrounded, tetrahedrally, by two Zn and two OH. This arrangement and the short OH–OH distances (2·77 and 2·86 Å) show that hydrogen bonds are formed in this structure. The β form of $Be(OH)_2$ has the same structure;[3] for α-$Be(OH)_2$ see reference (4).

One of the less stable forms of $Zn(OH)_2$, the so-called γ form, has an extraordinary structure[5] consisting of rings of three tetrahedral $Zn(OH)_4$ groups which are linked through their remaining vertices into infinite columns (Fig. 14.7). The upper vertices (1) of the tetrahedra of one ring are the lower vertices (2) of the ring above, so that all the vertices of the tetrahedra are shared within the column, which therefore has the composition $Zn(OH)_2$. Each column is linked to other similar ones only by hydrogen bonds (O–H–O, 2·80 Å, Zn–O, 1·96 Å).

Scandium and indium hydroxides. Just as $Zn(OH)_2$ and $Be(OH)_2$ have the simplest 3-dimensional framework structure possible for a compound AX_2 with 4 : 2 coordination, distorted so as to bring together OH groups of different coordination groups, so $Sc(OH)_3$ and $In(OH)_3$ have the simplest 3-dimensional framework structure of the AX_3 type, namely the ReO_3 structure, again distorted so as to permit hydrogen bonding between OH groups of different $M(OH)_6$

(1) HCA 1966 **49** 1971

(2) ZaC 1964 **330** 170

(3) ZaC 1950 **261** 94
(4) JNM 1964 **11** 310

(5) ACSc 1969 **23** 2016

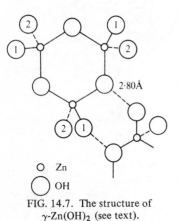

2·80Å

○ Zn

◯ OH

FIG. 14.7. The structure of γ-$Zn(OH)_2$ (see text).

octahedra.[6] The nature of the distortion can be seen from Fig. 14.8. Instead of the OH groups lying on the straight lines joining metal atoms they lie off these lines, so that each is hydrogen-bonded to two others. The OH group A is bonded to the metal atom M and to a similar atom vertically above M and hydrogen-bonded to the OH groups B and C. A neutron diffraction study of $In(OH)_3$[7] indicates statistical distribution of the H atoms, with O–H–O, 2·744 and 2·798 Å.

(6) ZaC 1948 **256** 226

(7) ACSc 1967 **21** 1046

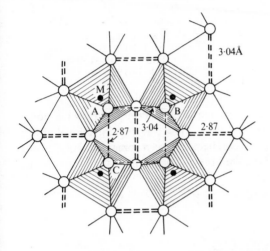

FIG. 14.8. The crystal structure of $Sc(OH)_3$.

Aluminium hydroxide. A number of forms of $Al(OH)_3$ are recognized which are of two main types, α (bayerite), and γ (gibbsite), but there are minor differences within each type arising from different ways of superposing the layers. All forms are built of the same layer, the AX_3 layer shown in idealized form in Fig. 14.9. This may be described as a system of octahedral AX_6 coordination

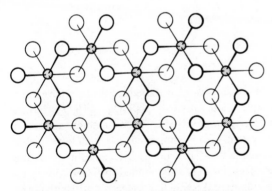

FIG. 14.9. Part of a layer of $Al(OH)_3$ (idealized). The heavy and light open circles represent OH groups above and below the plane of the Al atoms (shaded)

groups each sharing three edges, or as a pair of approximately close-packed X layers with metal atoms in two-thirds of the octahedral interstices. The structures differ in the way in which the layers are superposed. In bayerite there is approximate hexagonal closest packing of the O atoms throughout the structure (as in $Mg(OH)_2$), but in hydrargillite (monoclinic gibbsite) the OH groups on the underside of one layer rest directly above the OH groups of the layer below, as shown in the elevation of Fig. 14.10. Nordstrandite appears to represent an intermediate case, with the layers superposed in nearly the same way as in hydrargillite; compare the interlayer spacings (d):

	d (Å)	Reference
Bayerite	4·72	ZK 1967 **125** 317
Nordstrandite	4·79	AC 1970 **B26** 649
Hydrargillite (gibbsite)	4·85	PRS 1935 A **151** 384; N 1959 **183** 944

The mode of superposition of the layers in hydrargillite suggested directed bonds between OH groups of adjacent layers rather than the non-directional forces operating in $Mg(OH)_2$ and similar crystals. There has been some discussion of the proton positions in gibbsite and bayerite, and these may still be in doubt; the reader is referred to the references given above.

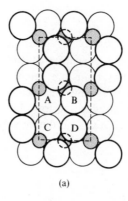

(a)

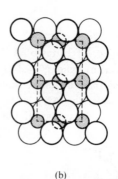

(b)

FIG. 14.10. The structures of (a) $Al(OH)_3$ and (b) $Mg(OH)_2$, viewed in a direction parallel to the layers, to illustrate the difference in packing of OH groups of different layers.

Complex hydroxides

These compounds are more numerous than was once thought, for many have been formulated as hydrates of, for example, meta-salts (e.g. $NaSbO_3 . 3 H_2O$ is in fact $NaSb(OH)_6$). Compounds $M'_x M''_y (OH)_z$ present the same general structural possibilities as complex halides. They range from compounds in which both metals are electropositive, when the crystal consists of an array of OH^- ions incorporating metal ions in holes of suitable sizes entirely surrounded by OH^- ions, to compounds in which one metal (or non-metal) is much less electropositive than the other and forms an essentially covalent hydroxy-ion. As examples we may quote $Ca_3 Al_2 (OH)_{12}$[1] (formerly 3 CaO . Al_2O_3 . 6 H_2O), in which Ca^{2+} and Al^{3+} ions are surrounded by 8 OH^- and 6 OH^- respectively (Ca–OH, 2·50 Å, Al–OH, 1·92 Å), and the group of isostructural compounds $Na[Sb(OH)_6]$, $Fe[Ge(OH)_6]$, and $Fe[Sn(OH)_6]$, in which the bonds from the B subgroup metals probably have

(1) AC 1964 **17** 1329

524

appreciable covalent character. The structure of these compounds[2] may be described as a NaCl-like packing of, for example, Fe^{2+} and $[Ge(OH)_6]^{2-}$ ions (Fe–OH, 2·14 Å, Ge–OH, 1·96 Å) or alternatively as a superstructure of the ReO_3 ($Sc(OH)_3$) type, There is presumably a larger difference in bond type in salts such as $Na[B(OH)_4]$.

In view of the behaviour of the OH^- in simple hydroxides we may expect layer structures to occur frequently. Orderly replacement of two-thirds of the Mg atoms in the $Mg(OH)_2$ layer by K and of the remainder by M^{IV} gives the structure of $K_2Sn(OH)_6$ and the isostructural Pb and Pt compounds. Like $NaSb(OH)_6$ these crystals contain discrete $M(OH)_6$ groups.

Binuclear hydroxy-ions (a) occur in $Ba_2Al_2(OH)_{10}$,[3] consisting of a pair of edge-sharing octahedra. There is a small difference between the lengths of the bridging (1·98 Å) and terminal (1·89 Å) Al–OH bonds.

(2) AC 1961 **14** 205; ACSc 1969 **23** 1219

(3) AC 1970 **B26** 867

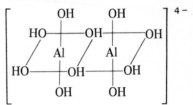

Oxyhydroxides

The largest class of compounds of which the structures are known are of the type MO . OH, formed by Al, Sc, Y, V, Cr, Mn, Fe, Co, Ga, and In. A number of these compounds, like the trihydroxides and sesquioxides of Al and Fe, exist in α and γ forms. (The so-called β-FeO . OH is not a pure oxyhydroxide; it has the α-MnO_2 structure, and is stable only if certain interstitial ions such as Cl^- are enclosed within the framework.[1]) We refer later to types of oxyhydroxide other than MO . OH. As in the case of hydroxides we may distinguish between structures in which there is hydrogen bonding and those in which this does not occur; it is convenient to deal first with a structure of the latter type.

(1) MM 1960 **32** 545
(2) ACSc 1966 **20** 896
(3) ACSc 1966 **20** 2658

The YO . OH *structure*

All the 4f compounds MO . OH and YO . OH have a monoclinic structure in which M has 7 O (OH) neighbours arranged at the vertices of a monocapped trigonal prism (as in monoclinic Sm_2O_3).[2] The interatomic distances suggest that the 3-coordinated O atoms belong to OH^- ions and the 4-coordinated ions are O^{2-}:

$$\begin{matrix} M & M \\ \diagdown & \diagup 2\cdot41\ \text{Å} \\ & OH \\ & | \\ & M \end{matrix} \qquad \text{and} \qquad \begin{matrix} M & M \\ \diagdown & \diagup 2\cdot27\ \text{Å} \\ & O \\ \diagup & \diagdown \\ M & M \end{matrix}$$

The mean O–OH distance, 3·01 Å, indicates that there is no hydrogen bonding, a conclusion confirmed by a n.d. study of YO . OD.[3]

The InO . OH *structure*

This is a rutile-like structure modified by the formation of hydrogen bonds (2·58 Å) between certain pairs of atoms (Fig. 14.11).[4] One form of CrO . OH also has this structure,[5] an interesting feature of which is the crystallographic equivalence of the O atoms, one-half of which belong to OH groups.

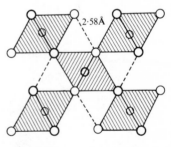

○ In at 0
○ In at ½
○ O at 0
○ O at ½

FIG. 14.11. The distorted rutile-like structure of InO(OH) showing hydrogen bonds.

(4) ACSc 1967 **21** 1046
(5) IC 1966 **5** 1452

The α-MO . OH structure

(6) ACSc 1967 **21** 121

Compounds with this structure include: α-AlO . OH (diaspore), α-FeO . OH (goethite), α-MnO . OH (groutite), α-ScO . OH,[6] and VO . OH (montroseite).

In our survey of octahedral structures in Chapter 5 we saw that double chains of the rutile type could be further linked by sharing vertices to form either 3-dimensional framework structures, of which the simplest is the diaspore structure, or a puckered layer as in lepidocrocite. These two structures are shown in perspective in Fig. 14.12. In Fig. 14.13(a) the double chains are seen end-on, as also are the c.p.

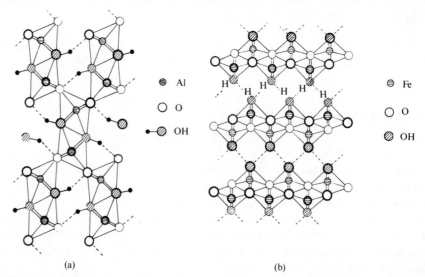

(a)   (b)

FIG. 14.13. Elevations of (a) the α-AlO(OH), diaspore, (b) the γ-FeO(OH), lepidocrocite, structures.

layers of O atoms. The H atoms in diaspore have been located by n.d. and are indicated as small black circles attached to the shaded O atoms. The O–H–O bonds, of length 2·65 Å, are shown as broken lines, and the H atom lies slightly off the straight line joining a pair of O atoms:

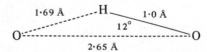

It will be seen that OH has 3 Al neighbours arranged pyramidally to one side, opposite to that of the H atom (Al–3 OH, 1·98 Å), while O is bonded to three more nearly coplanar Al neighbours (Al–3 O, 1·86 Å), the O–H–O bond being directed along the fourth tetrahedral bond direction. In the α-MO . OH structure each OH is hydrogen-bonded to an O atom; contrast the γ-MO . OH structure.

Bond lengths in the α structures are given in Table 14.2, which also includes γ-MnO . OH for comparison with α-MnO . OH. In groutite the length of the O–H–O bond is 2·63 Å. These Mn[III] compounds are of special interest because of the expected Jahn–Teller distortion of the octahedral coordination group. This

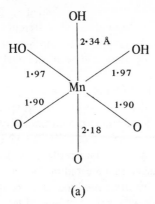

(a)

major distortion is superposed on the smaller difference between M—O and M—OH, as is seen by comparing MnO . OH with AlO . OH and VO . OH (Table 14.2). Details of the coordination group in groutite are shown at (a).

TABLE 14.2

Bond lengths in MO . OH *structures*

	α-AlO . OH[a] (diaspore)	α-VO . OH[b] (montroseite)	α-FeO . OH[c] (goethite)	α-MnO . OH[d] (groutite)	γ-MnO . OH[e] (manganite)
M—O (one)	1·858 Å	1·94 Å	1·89 Å	2·18 Å	2·20 Å
(two)	1·851	1·96	2·02	1·90	1·87
M—OH (one)	1·980	2·10	2·05	2·34	2·33
(two)	1·975	2·10	2·12	1·97	1·97

(a) AC 1958 **11** 798. (b) AM 1955 **40** 861. (c) ZK 1941 **103** 73. (d) AC 1968 **B24** 1233. (e) ZK 1963 **118** 303.

The γ-MO . OH *structure*

Compounds with this structure include: γ-AlO . OH (boehmite), γ-FeO . OH (lepidocrocite), γ-MnO . OH (manganite), and γ-ScO . OH.[6]

Two X-ray studies of boehmite were based on the centrosymmetrical space group *Cmcm*, and gave 2·47 Å and 2·69 Å for the length of the O—H—O bond. A p.m.r. study[7] shows that this bond is not symmetrical; further study of this structure appears to be required.   (7) JPC 1958 **62** 992

The structure of lepidocrocite, γ-FeO . OH, is shown as a perspective drawing in Fig. 14.12(b). In the elevation of this structure (Fig. 14.13(b)) the broken circles indicate atoms *c*/2 (1·53 Å) above and below the plane of the full circles, which themselves repeat at *c* (3·06 Å) above and below their own plane. Each Fe atom is therefore surrounded by a distorted octahedral group of O atoms, and these groups are linked together to form corrugated layers. The H atoms have not been located in this structure, but from the environments of the O atoms it is possible to distinguish between O and OH. The oxygen atoms within the layers are nearly equidistant from 4 Fe atoms (2 Fe at 1·93 Å and 2 Fe at 2·13 Å), whereas the atoms on the surfaces of the layers are bonded to 2 Fe atoms only (at 2·05 Å). Also, the distance between the O atoms of different layers is only 2·72 Å, so that there are clearly OH groups on the outer surface of each layer. In this structure there is hydrogen bonding only between the OH groups, each forming two hydrogen bonds.

The formation of O–H–O bonds between the OH groups of different layers accounts for the fact that the O atoms of adjacent layers are not packed together in the most compact way, though there is approximately cubic closest packing within a particular layer. As may be seen from Figs. 14.12 and 14.13, a hydroxyl group lies in the same plane as those of the next layer with which it is in contact, so that the H atoms must be arranged as indicated in Fig. 14.13(b). The oxychloride FeOCl is built of layers of exactly the same kind as in lepidocrocite but the layers pack together so that the Cl atoms on the outsides of adjacent layers are close-packed. This difference between the structures of FeOCl and γ-FeO.OH is quite comparable to that between $Mg(OH)_2$ and $Al(OH)_3$, being due to the formation of O–H–O bonds between the layers in the second of each pair of compounds.

The structures of the two forms of AlO.OH are of interest in connection with the dehydration of $Al(OH)_3$ and AlO.OH to give catalysts and adsorbents. In diaspore, α-AlO.OH, the O atoms are arranged in hexagonal closest packing, and this compound dehydrates directly to α-Al_2O_3 (corundum) in which the O atoms are arranged in the same way. In boehmite, γ-AlO.OH, the structure as a whole is not close-packed but within a layer the O atoms are arranged in cubic closest packing. On dehydration boehmite does not go directly to γ-Al_2O_3 with the spinel type of structure, but instead there are a number of intermediate phases, and there is still not complete agreement as to the number and structures of these phases.[8] Among schemes suggested are the following:

$$Al(OH)_3 \text{ (gibbsite)} \rightarrow \text{boehmite} \rightarrow \chi \rightarrow \delta \rightarrow \kappa \rightarrow \theta \rightarrow \alpha$$

$$\text{or gibbsite} \rightarrow \text{boehmite} \longrightarrow \kappa' \longrightarrow \kappa$$

$$\chi \rightarrow \eta \rightarrow \theta \longrightarrow \alpha$$

It seems likely that these intermediate phases, collectively known as forms of γ-alumina, represent different degrees of ordering of the Al atoms in a more or less perfect closest packing of O atoms (see also p. 457).

The CrO.OH (HCrO$_2$) *structure*

CrO.OH is normally obtained in a red (rhombohedral) form. This is built of CdI$_2$-type layers which are superposed so that O atoms of one layer fall directly above those of the layer below (Fig. 14.14).[9] The structure as a whole is therefore not close-packed. This packing of the layers is due to the formation of O–H–O bonds which are notable for their shortness, OH–O (2·49 Å), OD–O (2·55 Å). The n.d. and i.r. data show that the latter bonds are certainly asymmetric:

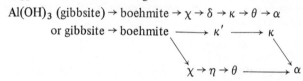

(8) JCS A 1967 2106
(9) AC 1963 **16** 1209

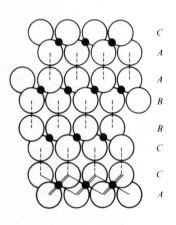

FIG. 14.14. Elevation of the crystal structure of HCrO$_2$ showing the way in which the layers are superposed. The broken lines represent O–H–O bonds.

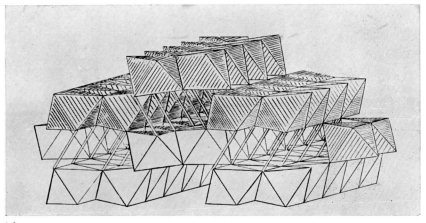

(a)

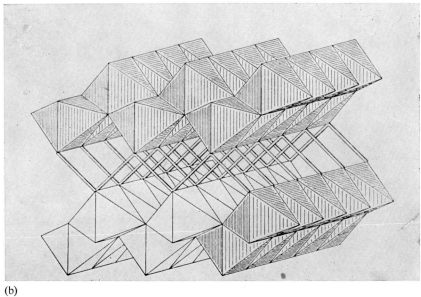

(b)

FIG. 14.12 The structures of (a) diaspore, AlO(OH), and (b) lepidocrocite, FeO(OH) (after Ewing). Oxygen atoms are to be imagined at the vertices of each octahedron and an Al (or Fe) atom at its centre. The double lines indicate O—H—O bonds.

but it has not been possible to decide whether the O—H—O bonds are asymmetric or symmetrical. This is perhaps a somewhat academic point since the barrier between the two minima is apparently very low. The same structure is adopted by CoO . OH, obtained by oxidizing β-Co(OH)$_2$.[10] A second (green) form of CrO . OH is related to the black CrO_2 in the following way:

(10) JCP 1969 **50** 1920

$$2\,CrO_2 + H_2O \underset{350°C\ in\ air}{\overset{450°C\ under\ pressure}{\rightleftharpoons}} 2\,CrO\,.\,OH + \tfrac{1}{2}\,O_2$$

This form is isostructural with InO . OH (distorted rutile structure), the structure of which has already been described.

Other oxyhydroxides

For other oxyhydroxides of V see p. 471. Uranyl hydroxide, $UO_2(OH)_2$, is not a compound of the type we have been considering but is the dihydroxide of the UO_2^{2+} ion. The structures of two of its forms are described in Chapter 28.

The reduction of MoO_3, by a variety of methods, gives compounds $MoO_{3-x}(OH)_x$ ($0.5 \leqslant x \leqslant 2$) of which the first member is $Mo_2O_5(OH)$. This compound has essentially the same (layer) structure as MoO_3. The H atoms have not been located but some O—O separations of 2·80 Å between the layers presumably indicate hydrogen bonds.[11] In MoO_3 the metal atoms are displaced

(11) ACSc 1969 **23** 419

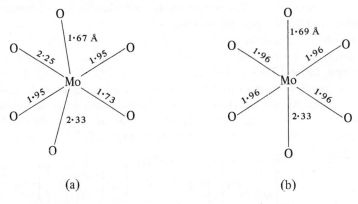

(a) (b)

from the centres of the octahedral O_6 groups to give the rather unsymmetrical coordination group shown at (a), which might be described as (2 + 2 + 2)-coordination. In $Mo_2O_5(OH)$, where there is 1 OH to 5 O, the coordination is that shown at (b), with one very short and one very long bond, (1 + 4 + 1)-coordination.

Hydroxy (basic) salts

The term 'basic salt' is applied to a variety of compounds intermediate between the normal salt and the hydroxide or oxide. This broad definition includes compounds such as Be and Zn oxyacetates (p. 416), oxy- and hydroxy-halides (Chapter 10), and hydroxy-oxysalts. We shall confine our attention here almost entirely to the last group of compounds. Coordination compounds with bridging OH groups, such as $[(NH_3)_4Co(OH)_2Co(NH_3)_4]Cl_4$, are mentioned under the element concerned.

Basic salts are numerous and of considerable interest and importance. They are formed by Be, Mg, Al, many of the A subgroup transition elements (e.g. Ti, Zr), 3d elements such as Fe, Co, and Ni, 4f and 5f elements (Ce, Th, U), and most of the B subgroup elements, particularly Cu(II), Zn, In, Sn, Pb, and Bi. Being formed by the action of oxygen and moisture on sulphide and other ores they form a large class of secondary minerals, some of which are also important as corrosion products of metals. The minerals brochantite, $Cu_4(OH)_6SO_4$, and atacamite, $Cu_2(OH)_3Cl$, form as patinas on copper exposed to town or seaside atmospheres, lepidocrocite, γ-FeO . OH, is formed during the rusting of iron, and hydrozincite, $Zn_5(OH)_6(CO_3)_2$, is the usual corrosion product of zinc in moist air. White lead, $Pb_3(OH)_2(CO_3)_2$, is one of a considerable number of basic salts which have been used as pigments, while $Mg_2(OH)_3Cl . 4 H_2O$ is formed during the setting of Sorel's cement.

Basic salts may be prepared in various ways, which usually involve—directly or indirectly—hydrolysis of a normal salt. The hydrolysis may be carried out under conditions of controlled temperature, acidity, and metal-ion concentration, or indirectly by heating a hydrated salt. Many hydroxy-salts are formed by the latter process instead of the anhydrous normal salts, for example, $Cu_2(OH)_3NO_3$ by heating hydrated cupric nitrate. The precipitates formed when sodium carbonate solution is added to metallic salt solutions are often hydroxy-carbonates. Some metals do not form normal carbonates (e.g. Cu, see p. 887); others such as Pb, Zn, Co, and Mg form normal or hydroxy-salts according to the conditions of precipitation. Lead, for example, forms $2 PbCO_3 . Pb(OH)_2$ and $PbCO_3 . Pb(OH)_2$, both of which occur as minerals, while Co forms $Co_4(OH)_6CO_3$ in addition to the normal carbonate.

Before describing the structures of a few hydroxy-salts we should mention that 'structural formulae' have been assigned to some of these compounds on the supposition that they are Werner coordination compounds. We may instance:

$$\left[Cu\left\{{OH \atop OH}{>}Cu\right\}_3\right]Cl_2 \quad \text{and} \quad \left[Cu\left\{{OH \atop OH}{>}Cu\right\}\right]CO_3$$

for atacamite, $Cu_2(OH)_3Cl$, and malachite, $Cu_2(OH)_2CO_3$, respectively. Such formulae are quite erroneous, for these compounds do not contain complex ions with a central copper atom. Malachite, for example, consists of an infinite array in three dimensions of Cu^{2+}, CO_3^{2-}, and OH^- ions. (For the structure of atacamite see p. 906.) On the other hand, finite complexes containing M atoms bridged by OH groups do occur in some basic salts; examples will be given shortly.

The crystal structures of basic salts

The first point to note that is relevant to the structures of hydroxy-salts is that there is a wide range of OH : M ratios:

$$Ca_5(OH)(PO_4)_3 \quad Cu_2(OH)PO_4 \quad Cu(OH)IO_3 \quad Zn_5(OH)_6CO_3$$

OH : M ratio $\quad \frac{1}{5} \qquad\qquad \frac{1}{2} \qquad\qquad 1 \qquad\qquad \frac{6}{5}$

$$Cu_2(OH)_3NO_3 \quad Th(OH)_2SO_4$$

OH : M ratio $\quad \frac{3}{2} \qquad\qquad 2$

The OH$^-$ ions are associated with the metal ions, so that for small OH : M ratios the coordination group around M consists largely of O atoms of oxy-ions; with increasing OH : M ratio OH$^-$ ions form a more important part of the coordination group of M. This is rather nicely illustrated by the hydrolysis of $Zr(SO_4)_2 . 4 H_2O$ at $100°C$ to $Zr_2(OH)_2(SO_4)_3 . 4 H_2O$ and at $200°C$ to $Zr(OH)_2SO_4$, the environment of Zr^{4+} being compared with that in (cubic) ZrO_2, the final product of heating these salts in air. (There is 7-coordination of the cations in the monoclinic form of ZrO_2.) The O atoms in the coordination groups around the metal ions in the following table are, of course, oxygen atoms of SO_4^{2-} ions.

Coordination groups around Zr^{4+}				
$Zr(SO_4)_2 . 4 H_2O^{(a)}$	$Zr_2(OH)_2(SO_4)_3 . 4 H_2O^{(b)}$	$Zr(OH)_2SO_4^{(b)}$	ZrO_2	(a) AC 1959 **12** 719
4 H$_2$O	2 H$_2$O	4 OH	8 O	(b) IC 1966 **5** 284
4 O	2 OH	4 O		
	4 O			
Antiprism	Dodecahedron	Antiprism	Cube	

In crystals such as those of hydroxyapatite, $Ca_5(OH)(PO_4)_3$, which is isostructural with $Ca_5F(PO_4)_3$, no hydroxy-metal complex can be distinguished, but with higher OH : M ratios there is generally sharing of OH between coordination groups of different M atoms so that a connected system of OH and M can be seen. Such a hydroxy-metal complex may be finite or infinite in 1, 2, or 3 dimensions, and in describing the structures of hydroxy-oxysalts it is convenient first to single out the M–OH complex and then to show how these units are assembled in the crystalline basic salt. The description of a structure in terms of the M–OH complex is largely a matter of convenience. For example, the M–OH system may extend only in one or two dimensions but the oxy-ions may then link these sub-units into a normal 3-D structure, that is, one which is not a chain or layer structure. Thus, while $Cu_2(OH)_3NO_3$ and $Zn_5(OH)_8Cl_2 . H_2O$ are layer structures. $Cu(OH)IO_3$, $Fe(OH)SO_4$, $Zn_2(OH)_2SO_4$, and $Zn_5(OH)_6(CO_3)_2$ are all 3D structures, though we shall see how they are constructed from 1- or 2-dimensional M–OH sub-units. It should be emphasized that there is no simple connection between the OH : M ratio and the possible types of M : OH complex. Assuming that every OH is shared between two M atoms, cyclic and linear M : OH complexes become possible for OH : M $\geqslant 1$. An additional reason for distinguishing the M : OH complex is that some of the M : OH complexes in basic salts formed from dilute solution, with high OH : M ratios, are related in simple ways to the structures of the final hydrolysis products, namely, the hydroxide or oxide of the metal.

Finite hydroxy–metal complexes

(1) ACSc 1962 **16** 403
(2) ACSc 1968 **22** 389

One product of the hydrolysis of $Al_2(SO_4)_3$ by alkali is a salt with the empirical formula $Al_2O_3 . 2 SO_3 . 11 H_2O$ which contains bridged hydroxy-aquo complexes (a) and should therefore be formulated as $[Al_2(OH)_2(H_2O)_8](SO_4)_2 . 2 H_2O$.[1] A more complex example is the dimeric $Th_2(OH)_2(NO_3)_6(H_2O)_6$ complex, (b),[2] formed by hydrolysis of the nitrate in solution. Three bidentate NO_3^- ions and 3 H_2O complete an 11-coordination group about each metal atom.

(3) AC 1956 **9** 555

The compounds originally formulated $ZrOCl_2 . 8 H_2O$ and $ZrOBr_2 . 8 H_2O$ also contain hydroxy-complexes. The complicated behaviour of these compounds in aqueous solution suggests polymerization, and in fact the crystals contain square complexes (c) in which there is a distorted square antiprismatic arrangement of 8 O atoms around each Zr atom. The halide ions and remaining H_2O molecules lie between the complexes, and the structural formula of these compounds is therefore $[Zr_4(OH)_8(H_2O)_{16}]X_8 . 12 H_2O$.[3] We refer again to this type of complex later.

$$\left[\begin{array}{c}
H_2O \qquad H_2O \\
H_2O\diagdown \;\;\Big|\; \diagup OH \diagdown\; \Big|\; \diagup H_2O \\
\qquad Al \qquad\qquad Al \\
H_2O\diagup \;\;\Big|\; \diagdown OH \diagup\; \Big|\; \diagdown H_2O \\
H_2O \qquad H_2O
\end{array}\right]^{4+}$$

(a)

$$\left(\; O{-}N\diagup\!\!\!\diagdown\!\!\begin{array}{c}O\\O\end{array}\!\!\!\Big\}Th\!\begin{array}{c}(H_2O)_3\\ \diagup OH\\ \diagdown OH \end{array}\!Th\!\begin{array}{c}\diagup O\\ \diagdown O\end{array}\!\!N{-}O \;\right)_3$$

$(H_2O)_3$

(b)

$$\left[\begin{array}{c}
OH \\
(H_2O)_4 Zr \diagup OH \diagdown Zr(H_2O)_4 \\
OH \quad OH \quad OH \quad OH \\
(H_2O)_4 Zr \diagdown OH \diagup Zr(H_2O)_4 \\
OH
\end{array}\right]^{8+}$$

(c)

A polyhedral complex is found in $U_6O_4(OH)_4(SO_4)_6$[4] and the isostructural Ce compound. This complex (Fig. 14.15) consists of an octahedron of U atoms and an associated cubic group of 8 O atoms (the positions of the H atoms are not known) of the same general type as the $Mo_6Cl_8^{4+}$ ion (p. 370), but whereas in the Mo complex the metal–metal distances are rather shorter than in the metal, here the U–U distances (3·85 Å) are as large as in UO_3, showing that the complex is held together by U–O bonds. The oxyhydroxides $M_6O_4(OH)_4$ of Sn and Pb form molecules of this type (p. 936).

(4) AK 1953 **5** 349

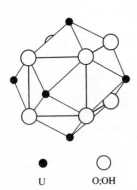

●	○
U	O;OH

FIG. 14.15. The finite $U_6O_4(OH)_4$ complex in $U_6O_4(OH)_4(SO_4)_6$.

1-dimensional hydroxy–metal complexes

A type of chain found in a number of hydroxy-salts with OH : M = 1 consists of octahedral coordination groups sharing opposite edges which include the (OH) groups. The simplest possibility is that these chains are then joined through the oxy-ions to form a 3-dimensional structure, as illustrated by the plan and elevation of the structure of $Cu(OH)IO_3$[5] (Fig. 14.16). The other four atoms of the octahedral coordination group of M in the chain are necessarily O atoms of oxy-ions. The structures of $Fe(OH)SO_4$[6] and $In(OH)SO_4$ are of the same general type. A further interesting possibility is realized in $Zn_2(OH)_2SO_4$.[7] Instead of the

(5) AC 1962 **15** 1105

(6) ACSc 1962 **16** 1234
(7) AC 1962 **15** 559

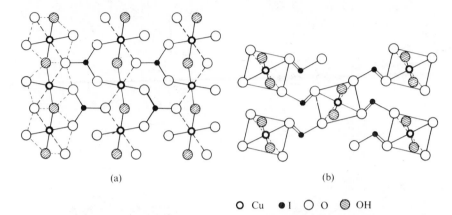

O Cu ● I ○ O ⊘ OH

FIG. 14.16. The structure of $Cu(OH)IO_3$: (a) plan, (b) elevation, in which the octahedral chains are perpendicular to the plane of the paper.

octahedral chains being cross-linked by finite oxy-ions as in $Cu(OH)IO_3$ they are linked by chains of a second type running in a direction perpendicular to that of the first set, as shown in Fig. 14.17. This structure is one of a number in which there is both tetrahedral and octahedral coordination of Zn^{2+}, a point referred to in Chapters 4 and 5.

A second kind of M–OH chain is related to the cyclic Zr complex already mentioned; it is a zigzag chain of M atoms joined by double hydroxy-bridges. The 4 OH all lie to one side of a particular M atom, and this chain is characteristic of larger metals such as Zr, Th, and U forming M^{4+} ions which are 8- (sometimes 7-) coordinated by oxygen. The chain can be seen in the plan (Fig. 14.18) of the structure of $Th(OH)_2SO_4$,[8] with which the Zr and U compounds are isostructural. The chains are bound together by the SO_4^{2-} ions, O atoms of which complete the antiprismatic coordination groups around the M atoms in the chains. In $Hf(OH)_2SO_4 \cdot H_2O$[9] 4 OH and H_2O form a planar pentagonal group around Hf, and 2 O atoms of sulphate ions complete the pentagonal bipyramidal coordination group.

(8) AK 1952 **4** 421

(9) ACSc 1969 **23** 3541

$$\begin{array}{ccccccc} & & H_2O & & H_2O & & \\ \diagdown & OH{-}OH & & OH{-}OH & & OH{-} & \\ & | & Hf & | & Hf & | & \\ {-}OH & & OH{-}OH & & OH{-}OH & & \\ & H_2O & & H_2O & & & \end{array}$$

The dioxides of Ce, Zr, Th, and U all crystallize with structures of the fluorite type, and it is interesting to note that three of the complexes we have described

533

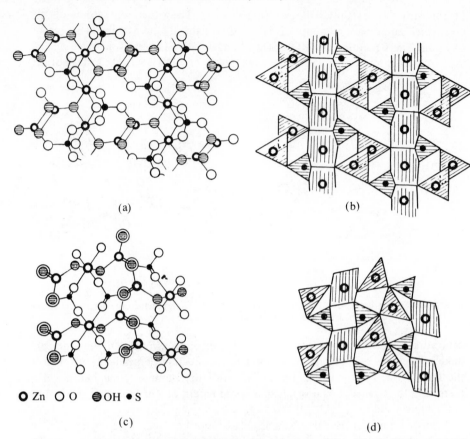

(a)

(b)

O Zn O O ⊖ OH • S

(c)

(d)

FIG. 14.17. The structure of $Zn_2(OH)_2SO_4$: (a) plan, (c) elevation, (b) and (d) diagrammatic representations of (a) and (c). In (a) and (b) the octahedral chains run parallel to the plane of the paper and the tetrahedral chains perpendicular to the paper; in (c) and (d) these directions are interchanged.

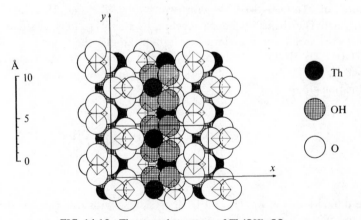

FIG. 14.18. The crystal structure of $Th(OH)_2SO_4$.

may be regarded as portions of this structure. In the $[Zr_4(OH)_8(H_2O)_{16}]^{8+}$ ion the hydroxy-bridges are perpendicular to the plane of the Zr atoms, so that the arrangement of the 4 Zr and 8 OH is approximately that of Fig. 14.19(a), which is simply a portion of the fluorite structure. Similarly the $M_6O_4(OH)_4^{12+}$ ion is the portion (b), and the $[M(OH)_2]_n$ chain in $Th(OH)_2SO_4$ is the system of atoms (c) of the same structure.

2-dimensional hydroxy–metal complexes

The hydroxides of Ca, Mg, Mn, Fe, Co, and Ni crystallize with the CdI_2 (layer) structure. Although $Cu(OH)_2$ does not adopt this structure a number of basic cupric salts have structures closely related to it, for example, $Cu_2(OH)_3Br$ and the hydroxy-nitrate to be described shortly. Also, although Zn prefers tetrahedral coordination in $Zn(OH)_2$ itself, various hydroxy compounds with CdI_2-like structures appear to exist. In a number of basic salts Zn adopts a compromise, as in $Zn_2(OH)_2SO_4$, between the tetrahedral coordination in $Zn(OH)_2$ and the octahedral coordination in $ZnCO_3$ and other oxy-salts, by exhibiting both kinds of coordination in the same crystal. This results in more complicated structures, and for this reason we shall describe first the very simple relation between $Cu_2(OH)_3NO_3$ and the CdI_2-type structure of $CuCl_2$ and $Cu_2(OH)_3Br$.

Replacement by Br of one-quarter of the OH groups in a (hypothetical) $Cu(OH)_2$ layer of the CdI_2 type gives the layer of $Cu_2(OH)_3Br$, with suitable distortion to give Cu^{II} (4 + 2) instead of regular octahedral coordination. If instead of Br we insert one O atom of a NO_3^- ion, the plane of which is perpendicular to the layer, we have the (layer) structure of $Cu_2(OH)_3NO_3$,[10] illustrated in elevation in Fig. 14.20(a). The other O atoms of the NO_3^- ions are hydrogen-bonded to OH^- ions of the adjacent layer.

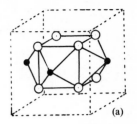

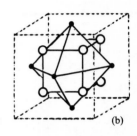

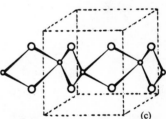

FIG. 14.19. The complexes (a) Zr_4O_8, (b) M_6O_8, and (c) the $(MO_2)_n$ chain shown as portions of the fluorite structure.

(10) HCA 1952 **35** 375

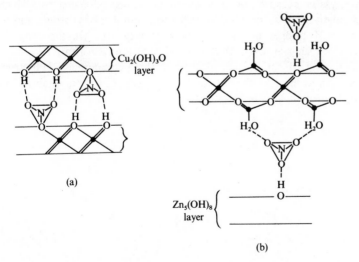

(a)

(b)

FIG. 14.20. Diagrammatic elevations of the structures of (a) $Cu_2(OH)_3NO_3$. (b) $Zn_5(OH)_8(NO_3)_2 \cdot 2H_2O$.

535

We showed in Chapter 6 how the CdI_2 layer is used as the basic structural unit in a number of more complex structures, including

$Zn_5(OH)_8(NO_3)_2 . 2 H_2O$	AC 1970 **B26** 860
$Zn_5(OH)_8Cl_2 . H_2O$	ZK 1968 **126** 417
$Zn_5(OH)_6(CO_3)_2$	AC 1964 **17** 1051

The first two structures are layer structures, the layer being derived from a hypothetical $Zn(OH)_2$ layer of the CdI_2 type. One-quarter of the (octahedrally coordinated) Zn atoms are removed and replaced by pairs of tetrahedrally coordinated Zn atoms, one on each side of the layer, giving a layer of composition $Zn_5(OH)_8$. In the hydroxynitrate a water molecule completes the tetrahedral coordination group of the added Zn atoms, and the NO_3^- ions are situated between the layers, being hydrogen-bonded to 2 H_2O of one layer and 1 OH of the adjacent layer (Fig. 14.20 (b)). In $Zn_5(OH)_8Cl_2 . H_2O$ the fourth bond from the tetrahedrally coordinated Zn is to Cl, and the water molecules are situated between the layers. In the carbonate there is also replacement of one-quarter of the OH groups in the original $Zn(OH)_2$ layer by O atoms of CO_3^{2-} ions, so that the OH : Zn ratio becomes 6 : 5 instead of 8 : 5 as in the other salts. There is direct bonding between the tetrahedrally coordinated Zn atoms of one layer and the CO_3 groups of the adjacent layers, with the result that the structure is no longer a layer structure (see p. 213).

Some basic salts with layer structures are always obtained in a poorly crystalline state. For example, crystals of 'white lead', $Pb_3(OH)_2(CO_3)_2$, are too small and too disordered to give useful X-ray photographs, but from electron diffraction data it is concluded that there are probably $Pb(OH)_2$ layers interleaved with Pb^{2+} and CO_3^{2-} ions in a rather disordered structure.[11]

Three-dimensional hydroxy-metal frameworks are unlikely to occur in hydroxy-oxysalts because the number of O atoms belonging to OH groups is usually a relatively small proportion of the total number of O atoms, and therefore many M–O–M bridges involve O atoms of oxy-ions. On the other hand, in hydroxyhalides such as $M_2(OH)_3X$ the hydroxyl ions constitute three-quarters of the total number of anions, and in some of these compounds 3D hydroxy-metal frameworks do occur. (See the note on hydroxy-salts of Cu^{II} in Chapter 25 and the section on hydroxyhalides in Chapter 10.)

(11) AC 1956 9 391

15

Water and Hydrates

The structures of ice and water

The structure of the isolated water molecule in the vapour is accurately known from spectroscopic studies, but we have seen in Chapter 8 that compounds containing hydrogen are often abnormal owing to the formation of hydrogen bonds. Therefore it may not be assumed that the structural unit in the condensed phases water and ice has precisely this structure. It is convenient to discuss the structure of ice before that of water because we can obtain by diffraction methods much more information about the structure of a solid than about that of a liquid. In a liquid there is continual rearrangement of neighbours, and we can determine only the mean environment, that is, the number and spatial arrangement of nearest neighbours of a molecule averaged over both space and time. The only information obtainable by X-ray diffraction from a liquid at a given temperature is the scattering curve, from which is derived the radial distribution curve, and the interpretation of the various maxima in terms of nearest, next nearest neighbours, and so on, is a matter of great difficulty.

Ice

At atmospheric pressure water normally crystallizes as ice-I_h, which has a hexagonal structure like that of tridymite. It may also be crystallized directly from the vapour, preferably *in vacuo*, as the cubic form ice-I_c with a cristobalite-like structure provided the temperature is carefully controlled ($-120°$ to $-140°C$). (This cubic form is more conveniently prepared in quantity by warming the high-pressure forms from liquid nitrogen temperature.) Ice-I_c is metastable relative to ordinary ice-I_h at temperatures above $153°K$. The existence of a second metastable crystalline form, ice-IV, has been firmly established for D_2O but less certainly for H_2O. A vitreous form of ice is formed by condensing the vapour at temperatures of $-160°C$ or below.

In addition to I_h and I_c there are a number of crystalline polymorphs stable only under pressure, though the complete phase diagram is not yet established (Fig. 15.1). There are five distinct structures (differing in arrangement of O atoms) and there are low-temperature forms of two of them; the numbering is now unfortunately unsystematic:

II	III	V	VI	VII
	↓			↓
	IX			VIII	(low-temperature forms).

The forms II–VII are produced by cooling liquid water under increasingly high pressures; cooling to −195°C is necessary for II, which also results from decompressing V at low temperature and 2 kbar pressure. Ice-III converts to IX at temperatures below about −100°C and VII to VIII below 0°C. All the high-pressure forms II–VII can be kept and studied at atmospheric pressure if quenched to the temperature of liquid nitrogen.

The main points of interest of the structures of these polymorphs are (i) the analogies with silica and silicate structures, (ii) the presence of two interpenetrating frameworks in the most dense forms VI and VII (VIII), and (iii) the ordering of the protons. Analogies with silica and silicate structures are noted in Table 15.1, namely, ice-III with a keatite-like structure, ice-VI with two interpenetrating frameworks of the edingtonite type (p. 828), and ice-VII (and VIII) with two interpenetrating cristobalite-like frameworks. In these structures, related to those of

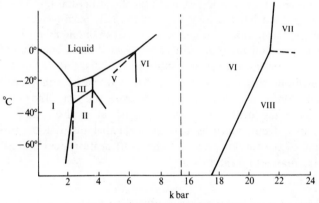

FIG. 15.1. Partial phase diagram for water.

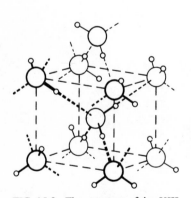

FIG. 15.2. The structure of ice-VIII.

forms of silica or silicates, O atoms of H_2O molecules occupy the positions occupied by Si atoms in the silica structure or by Si or Al in the case of edingtonite. A projection of the Si atoms in keatite is similar to that of the Ge atoms in a high-pressure form of that element and has been illustrated in Fig. 3.41 (p. 110). The arrangement of the O atoms in ice-VII (or VIII) is body-centred cubic (Fig. 15.2). Each H_2O molecule has 8 equidistant neighbours but is hydrogen-bonded only to the 4 of its own framework. At atmospheric pressure (quenched in liquid N_2) the length of an O–H–O bond is 2·95 Å, and at −50°C under a pressure of 25 kbar it is 2·86 Å.

The hydrogen-bonded frameworks in two of the ice polymorphs (II and V) are different from any of the 4-connected Si(Al)–O frameworks. The (rhombohedral) structure of ice-II has some similarity to the tridymite-like structure of ice-I, the greater density being achieved by the proximity of a fifth neighbour at 3·24 Å in addition to the four nearest neighbours at 2·80 Å—compare the distance of the next nearest neighbours (4·5 Å) in ice-I_h. The framework of ice-V also has no obvious silica or silicate analogue, in spite of the fact that the ratio of the densities of ice-V

538

to ice-I is the same as that of coesite (SiO_2) to tridymite. In ice-V H_2O has four nearest neighbours at a mean distance of 2·80 Å and then next nearest neighbours at 3·28 Å and 3·45 Å, but more striking is the very distorted tetrahedral arrangement of the nearest neighbours, the angles O–O–O ranging from 84°–128°.

Proton positions in ice polymorphs. In the early X-ray study of ice-I_h only the O atoms were located. Each is surrounded by a nearly regular tetrahedral arrangement of 4 O atoms, O–O being 2·752 Å for the bond parallel to the *c* axis and 2·765 Å for the other three, and the O–O–O angles are very close to $109\frac{1}{2}°$. One H is to be placed somewhere between each pair of O atoms. The i.r. absorption frequencies indicate that the O–H distance in a H_2O molecule cannot have changed from 0·96 Å to 1·38 Å (one-half O–O in ice), therefore a H atom must be placed unsymmetrically along each O–O line. There are two possibilities: either H atoms occupy fixed positions (ordered protons) or there is random arrangement of H atoms (disordered protons) with the restriction that at any given time only two are close to a particular O atom, that is, there are normal H_2O molecules. It was suggested by Pauling that an 'average' structure, with $\frac{1}{2}$ H at each of 4 *N* sites (each 0·96 Å from an O atom along an O–O line) within an array of *N* oxygen atoms would account for the observed value (0·82 e.u.) of the residual entropy of ice.

It has since been shown by a variety of physical techniques that in some of the high-pressure forms of ice the protons are ordered and in others disordered, in particular, the low-temperature forms IX and VIII are the ordered forms of ice-III and ice-VII respectively. The nature of the far infrared absorption bands indicates that the protons are ordered in II and IX (sharp bands) but disordered in V (diffuse bands).[1] The dielectric properties of all the following forms of ice have been studied: I_c,[2] II, III, V, and VI,[3] VII and VIII,[4] and the conclusions as to proton order/disorder are included in Table 15.1.

(1) JCP 1968 **49** 775
(2) JCP 1965 **43** 2376
(3) JCP 1965 **43** 2384
(4) JCP 1966 **45** 3976

TABLE 15.1

The polymorphs of ice

Polymorph	Density (g/cc)	Ordered or disordered protons	Analogous silica or silicate structure	Reference
I_h	0·92	D	Tridymite	AC 1957 **10** 70 (n.d.)
I_c	0·92	D	Cristobalite	JCP 1968 **49** 4365 (n.d.)
II	1·17	O	–	AC 1964 **17** 1437
				JCP 1968 **49** 4361 (n.d.)
III⎫	1·16	D	Keatite	AC 1968 **B24** 1317
IX⎭		O	Keatite	JCP 1968 **49** 2514 (n.d.)
[IV	–	–	–	JCP 1937 **5** 964]
V	1·23	D	–	AC 1967 **22** 706
VI	1·31	D	*Edingtonite	PNAS 1964 **52** 1433
VII⎫	1·50	D	*Cristobalite	JCP 1965 **43** 3917
				NBS 1965 **69C** 275
VIII⎭		O	*Cristobalite	JCP 1966 **45** 4360

* Structure consists of 2 interpenetrating frameworks.

Water and Hydrates

The second problem is the relation of the protons to the O—O lines. In ice-I_h the angles O—O—O are close to $109\frac{1}{2}°$, and n.d. shows that the H atoms lie slightly off these lines in positions consistent with an H—O—H angle close to $105°$. The regular tetrahedral arrangement of the four O neighbours is due to the randomness of orientation of the molecules in the structure as a whole. The very large range of angles in the (proton-disordered) ice-v has already been noted. In the ordered II and IX phases also n.d. and p.m.r. studies[5] confirm that the H—O—H angle is close to $105°$. However, the O—O—O angles are respectively $88°$ and $99°$, and $87°$ and $99°$; the protons therefore lie appreciably off the O—O lines.

(5) JCP 1968 49 4660

Water

The fact that water is liquid at ordinary temperatures whereas all the hydrides CH_4, NH_3, HF, PH_3, SH_2, and HCl of elements near to oxygen in the Periodic Table are gases, indicates interaction of an exceptional kind between neighbouring molecules. That these interactions are definitely directed towards a small number of neighbours is shown by the low density of the liquid compared with the value ($1·84$ g/cc) calculated for a close-packed liquid consisting of molecules of similar size, assuming a radius of $1·38$ Å as in ice. When ice melts there are two opposing effects, the breakdown of the fully hydrogen-bonded system in the tridymite-like structure of ice to give a denser liquid, and thermal expansion operating in the opposite sense. The first process must take place over a range of temperature in order to explain the minimum in the volume/temperature curve. There has been considerable discussion of the concentrations and energies of hydrogen bonds in water at various temperatures. Raman spectroscopic studies and measurements of the viscosity indicate that the number of hydrogen bonds in water at $20°C$ is about half that in ice. The idea that large numbers of hydrogen bonds are broken when ice melts or when water is heated is not generally accepted, and certainly does not appear probable if we accept Pauling's value of 19 kJ mol^{-1} for the hydrogen-bond energy. Alternative suggestions are the bending rather than breaking of these bonds or the breaking of weaker hydrogen bonds. Before describing actual structural models suggested for water it is relevant to mention two other properties of water, namely, the abnormal mobilities of H^+ and OH^- ions and the fact that the dielectric constant of water is practically the same as that of ice for low frequencies.

The mobilities of the H^+ and OH^- ions are respectively $32·5$ and $17·8 \times 10^{-4}$ cm/sec for an applied field of 1 volt/cm, whereas the values for other ions are of the order of 6×10^{-4} cm/sec. Calculations show that very little energy is required to remove a proton from one water molecule, to which it is attached as $(H_3O)^+$, to another, for the states $H_3O, H_2O \rightleftharpoons H_2O, H_3O$ have the same energy. If we assume that the H^+ ion acts as a bond between two molecules, then the process shown diagrammatically in Fig. 15.3(a) results in the movement of H^+ from A to B. The analogous mechanism for the effective movement of OH^- ions through water is illustrated in Fig. 15.3(b). Whereas other ions have to move bodily through the water, the H^+ and OH^- ions move by what Bernal termed a kind of relay race, small shifts of protons only being necessary.

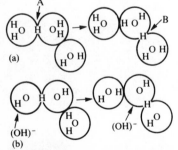

FIG. 15.3. The effective movement of (a) H^+ ion, and (b) OH^- ion resulting from small shifts of protons (after Bernal and Fowler).

540

In order to account for the high dielectric constant of water it is necessary to suppose that there exist groups of molecules with a pseudocrystalline structure, that is, with sufficient orientation of the O–H–O bonds to give an appreciable electric moment. The upper limit of size of these 'clusters' has been estimated from studies of the infrared absorption bands in the $1\cdot1$–$1\cdot3\ \mu$ region as approximately 130 molecules at $0°$, 90 at $20°$, and 60 at $72°C$. In a liquid there is constant rearrangement of the molecules, and it is postulated that a given cluster persists only for a very short time, possibly of the order of 10^{-11} to 10^{-10} seconds.

The idea that water is in a general sense structurally similar to ice allowing, of course, for greater disorder in the liquid than the solid, is confirmed by the X-ray diffraction effects. Radial distribution curves have been derived from X-ray photographs taken at a number of temperatures from $1\cdot5°$ to $83°C$ and from the areas under the peaks the average numbers of neighbours at various distances can be deduced, at least in principle. The first peak on the curve suggests that at $1\cdot5°C$ there are on the average $4\cdot4$ neighbours at a mean distance of $2\cdot90$ Å; at $83°$ the corresponding figures are $4\cdot9$ at $3\cdot05$ Å. This first peak is succeeded by a curve which rises gradually to a poorly resolved maximum in the region $4\cdot5$–$4\cdot9$ Å, indicating the presence of molecules at distances between those of the nearest and next nearest neighbours in ice ($2\cdot8$ and $4\cdot5$ Å). Because of the small number of well-defined peaks the radial distribution curve has been interpreted in many ways.

Various arrangements of water molecules are possible if the only condition is that each is surrounded tetrahedrally by four neighbours, but the structure must be more dense than the tridymite-like structure of ice-I. Three suggestions have been made for the structure of the pseudocrystalline regions. The first is a quartz-like packing tending at high temperatures towards a more close-packed liquid, the density of quartz being $2\cdot66$ g/cc compared with $2\cdot30$ for tridymite. The second is a structure resembling the water framework in the chlorine hydrate structure (p. 545) with non-hydrogen-bonded water molecules in all the polyhedral interstices, of which there are 8 in a unit cell containing 46 framework molecules. The third is a slightly expanded tridymite-like structure with interstitial molecules. It seems most likely that the higher density of water as compared with ice is due to the insertion of neighbours between the nearest and next nearest neighbours of an ice-like structure. In one model which has been refined by least squares to give a good fit with the observed radial distribution curve each framework H_2O has 4 framework molecules as nearest neighbours (1 at $2\cdot77$, 3 at $2\cdot94$ Å) and each interstitial molecule has 12 framework H_2O neighbours (3 at each of the distances $2\cdot94$, $3\cdot30$, $3\cdot40$, and $3\cdot92$ Å). In this model the ratio of framework to interstitial molecules is $4:1$.

In view of the very extensive literature on the structure of water we give here only a few references to the more recent papers which include many references to earlier work.[1]

Aqueous solutions

X-ray diffraction studies have been made of a number of aqueous solutions. It is possible to obtain some information about the nearest neighbours of a particular

(1) RTC 1962 **81** 904; JACS 1962 **84** 3965; JPC 1965 **69** 2145; JCP 1965 **42** 2563; JCP 1966 **45** 4719.

ion from the $\sigma(r)$ curve derived from concentrated solutions, though it is doubtful if the more detailed deductions made about the structures of solutions of HNO_3, H_2SO_4, and $NaOH$ represent unique interpretations of the experimental data.[1] In solutions containing complex ions interatomic distances corresponding to bonds within the ions can be identified, for example, peaks at $1\cdot33$ Å and $1\cdot49$ Å in the radial distribution curves of aqueous solutions of NH_4NO_3 and $HClO_4$ are attributable to the N–O bond and Cl–O bonds.[2] Deductions have been made about the environment of the cations in solutions of $EuCl_3$,[3] KOH,[4] $FeCl_3$,[5] $ZnCl_2$,[6] $ZnBr_2$,[7] and $CoCl_2$.[8] (For solutions of hydrolysed Pb and Bi salts containing hydroxy-complexes see p. 516.)

There is inevitably some doubt about the existence of octahedral as opposed to tetrahedral complexes in neutral solutions of $FeCl_3$ since the peak at $2\cdot25$ Å can be interpreted either in terms of tetrahedral coordination of Fe^{3+} by Cl^- (Fe–Cl, $2\cdot25$ Å) or as the mean of 4 Fe–O ($2\cdot07$ Å) and 2 Fe–Cl ($2\cdot30$ Å) in octahedral complexes of the type that exist in the crystalline hexahydrate (q.v.). On the other hand there seems to be no doubt about the existence of $FeCl_4^-$ in dilute solutions containing excess Cl^- ions or of Fe_2Cl_6 molecules in non-aqueous solvents of low dielectric constant such as methyl alcohol.

Zinc chloride is extremely soluble in water even at room temperature, and the very viscous $27\cdot5$ molar solution approaches the molten salt in composition $(Zn_{0\cdot20}Cl_{0\cdot40}(H_2O)_{0\cdot40})$. Its X-ray radial distribution curve is consistent with tetrahedral groups $ZnCl_3(H_2O)$, which must share, on the average, two Cl atoms with one another. In more dilute solutions ($5M$) groups $ZnCl_2(H_2O)_2$ predominate, and similar coordination groups are found in concentrated aqueous $ZnBr_2$. These change to $ZnBr_4^{2-}$ in the presence of sufficient excess Br^- ions. X-ray diffraction data from concentrated solutions of $CoCl_2$, on the other hand, have been interpreted in terms of octahedral $Co(H_2O)_6^{2+}$ groups (Co–O around $2\cdot1$ Å), but in methyl and ethyl alcohols there are apparently highly associated tetrahedral $CoCl_4$ groups (Co–Cl, approx. $2\cdot3$ Å).

Hydrates

One of the simplest ways of purifying a compound is to recrystallize it from a suitable solvent. The crystals separating from the solution may consist of the pure compound or they may contain 'solvent of crystallization'. For most salts water is a convenient solvent, and accordingly crystals containing water–hydrates–have been known from the earliest days of chemistry, and many inorganic compounds are normally obtained as hydrates. We have already noted in our survey of hydroxides that many compounds containing OH groups were originally formulated as hydrates, for example, $NaSb(OH)_6$ as $NaSbO_3 . 3 H_2O$, $NaB(OH)_4$ as $NaBO_2 . 2 H_2O$, and many others. The term hydrate should be used only for crystalline compounds containing H_2O molecules or in the case of hydrated acids, ions such as H_3O^+, $H_5O_2^+$, etc. (see the discussion of these compounds later). Apart from inorganic and organic acids, bases, and salts, certain essentially non-polar compounds form hydrates, among them being chlorine and bromine, the noble

(1) TKBM 1943 **3** 83; 1944 **4** 26, 5
(2) ACSc 1952 **6** 801
(3) JACS 1958 **80** 3576
(4) JCP 1958 **28** 464
(5) JCP 1969 **50** 4013
(6) IC 1962 **1** 941
(7) JCP 1965 **43** 2163
(8) JCP 1969 **50** 4313

gases, methane, and alkyl halides. Many of these hydrates contain approximately 6–8 H_2O or 17 H_2O for every molecule (or atom) of 'solute'. The hydrates of the gaseous substances are formed only under pressure and are extremely unstable, and the melting points are usually close to $0°C$. In these 'ice-like' hydrates there is a 3D framework built of H_2O molecules which encloses the solute molecules in tunnels or polyhedral cavities ('clathrate' hydrates). At the other extreme there are hydrates in which water molecules (and cations) are accommodated in tunnels or cavities within a rigid framework such as the aluminosilicate frameworks of the zeolites. Such a hydrate can be reversibly hydrated and dehydrated without collapse of the framework, in contrast to the clathrate hydrates in which the framework itself is composed of water molecules. Between these two extremes lie the hydrates of salts, acids, and hydroxides, in which H_2O molecules, anions, and cations are packed together to form a structure characteristic of the hydrate. Removal of some or all of the water normally results in collapse of the structure, which is usually unrelated to that of the anhydrous compound. The monohydrate of $Na_3P_3O_9$ is exceptional in this respect. In the structure of the anhydrous salt there is almost sufficient room for a water molecule, so that very little rearrangement of the bulky $P_3O_9^{3-}$ ions is necessary to accommodate it. For this reason the structures of $Na_3P_3O_9$ and $Na_3P_3O_9 . H_2O$ are very similar.

We shall be concerned in this chapter with the clathrate hydrates and the hydrates of salts, acids, and hydroxides. The structures of zeolites are described in Chapter 23, as also are the clay minerals, which can take up water between the layers. Examples of hydrated 'basic salts' and of hydrates of heteropoly acids and their salts are also discussed in other chapters.

Clathrate hydrates

In this type of hydrate the water molecules form a connected 3D system enclosing the solute atoms or molecules. We may list the latter roughly in order of increasing interaction with the water framework: atoms of noble gases (Ar, Xe) and molecules such as Cl_2, CO_2, C_3H_8 and alkyl halides, compounds such as $N_4(CH_2)_6$ which are weakly hydrogen-bonded to the framework, and finally ionic compounds in which ions of one or more kinds are associated with or incorporated in the water framework. As regards their structures, the frameworks range from those in which there are well defined polyhedral cavities to 3- or (3 + 4)-connected nets in which there are no obvious voids of this kind. In the first group a H_2O molecule is situated at each point of a 4-connected net, the links of which (O–H–O bonds) form the edges of a space-filling arrangement of polyhedra. These 'ice-like' structures are not stable unless all (or most) of the larger voids are occupied by solute molecules, and the melting points are usually not far removed from $0°C$. In Table 15.2 we include some less regular structures to which we refer later. The polyhedral frameworks are divided into two classes, (a) and (b), the latter involving the pentagonal dodecahedron and related polyhedra with hexagonal faces to which reference was made in Chapter 3.

TABLE 15.2
Clathrate hydrates

	H_2O molecules only in framework	H_2O and F, N, S in framework	Reference
	Polyhedral frameworks		
Class (a)			
(i)	$HPF_6 . 6 H_2O$		AC 1955 **8** 611
		$(CH_3)_4NOH . 5 H_2O$	JCP 1966 **44** 2338
(ii)	$(CH_3)_3CNH_2 . 9\frac{3}{4} H_2O$		JCP 1967 **47** 1229
Class (b)			
(i)	$Cl_2 . 7\frac{1}{4} H_2O$		PNAS 1952 **36** 112
			JCS 1959 4131
	$6\cdot4 C_2H_4O . 46 H_2O$		JCP 1965 **42** 2725
		$(n\text{-}C_4H_9)_3SF . 20 H_2O$	JCP 1962 **37** 2231
(ii)	$CHCl_3 . 17 H_2O$		JCP 1951 **19** 1425
	$7\cdot33 H_2S . 8 C_4H_8O . 136 H_2O$		JCP 1965 **42** 2732
(iii)		$(i\text{-}C_5H_{11})_4NF . 38 H_2O$	JCP 1961 **35** 1863
(iv)	$Br_2 . 8\cdot6 H_2O$		JCP 1963 **38** 2304
	Intermediate types		
Class (c)			
(i)		$(n\text{-}C_4H_9)_3SF . 23 H_2O$	JCP 1964 **40** 2800
(ii)	$(C_2H_5)_2NH . 8\frac{2}{3} H_2O$		JCP 1967 **47** 1222
(iii)	$4 (CH_3)_3N . 41 H_2O$		JCP 1968 **48** 2990
	Non-polyhedral frameworks		
Class (d)			
(i)	$N_4(CH_2)_6 . 6 H_2O$		JCP 1965 **43** 2799
(ii)		$(CH_3)_4NF . 4 H_2O$	JCP 1967 **47** 414

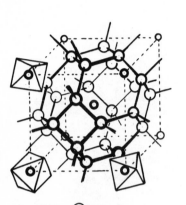

○ P   ◯ H_2O

FIG. 15.4. The crystal structure of $HPF_6 . 6 H_2O$.

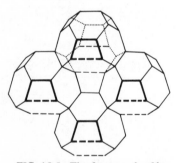

FIG. 15.5. The framework of hydrogen-bonded OH^- ions and H_2O molecules in $(CH_3)_4N.OH.5 H_2O$ showing distortion of the truncated octahedron in the 'Fedorov' net. The broken lines indicate O–O edges of length 4·36 Å.

Polyhedral frameworks

Class (a). The analogy between the structural chemistry of H_2O and SiO_2 is illustrated by the structure of $HPF_6 . 6 H_2O$. In this crystal the H_2O molecules are situated at the apices of Fedorov's space-filling by truncated octahedra, that is, at the same positions as the Si(Al) atoms in the framework of ultramarine (p. 832). Each H_2O molecule in the framework (Fig. 15.4) is hydrogen-bonded to its four neighbours at a distance of 2·72 Å, and the PF_6^- ions occupy the interstices (H_2O–F, 2·74 Å; P–F, 1·73 Å).

If each link in the 'Fedorov' net is to be a hydrogen bond there are sufficient H atoms in M . 6 H_2O. In $HPF_6 . 6 H_2O$ there are 13 H, and H^+ is presumably associated in some way with the framework. An interesting distortion of this net occurs in $(CH_3)_4NOH . 5 H_2O$, where $OH^- + 5 H_2O$ are situated at the vertices of the truncated octahedra. Since there are only 11 H all the edges of the polyhedra cannot be hydrogen bonds. The inclusion of the cations in the truncated octahedra expands them by extending certain of the edges to a length of 4·36 Å (Fig. 15.5).

Much less symmetrical polyhedral cavities are found in $(CH_3)_3C . NH_2 . 9\frac{3}{4} H_2O$. The framework represents a space-filling by 8-hedra ($f_4 = 4$, $f_5 = 4$) and 17-hedra ($f_3 = 3$, $f_5 = 9$, $f_6 = 2$, $f_7 = 3$). The unit cell contains 156 H_2O (16 formula units), and the amine molecules occupy the 16 large voids (17-hedra); the 8-hedra are not occupied.

Class (b). In this class the (4-connected) networks are the edges of space-filling arrangements of pentagonal dodecahedra and one or more of the related polyhedra: $f_5 = 12, f_6 = 2, 3, 4$ (Table 15.3).

TABLE 15.3

The dodecahedral family of hydrate structures

Hydrate	Z	Vertices	Voids (n-hedra)				Total of large voids
			12	14	15	16	
$Cl_2 . 7\frac{1}{4} H_2O$	6	46	2	6	–	–	6
$CHCl_3 . 17 H_2O$	8	136	16	–	–	8	8
$(i\text{-}C_5H_{11})_4N^+F^- . 38 H_2O$	2	80	6	4	4	–	8
$Br_2 . 8\cdot6 H_2O$	20	172	10	16	4	–	20

The hydrate of chlorine is of special interest as the solid phase originally thought to be solid chlorine but shown (in 1811) by Humphry Davy to contain water. It was later given the formula $Cl_2 . 10 H_2O$ by Michael Faraday. The unit cell of this (cubic) structure ($a \approx 12$ Å) contains $46 H_2O$ which form a framework (Fig. 15.6) in which there are 2 dodecahedral voids and 6 rather larger ones (14-hedra). If all

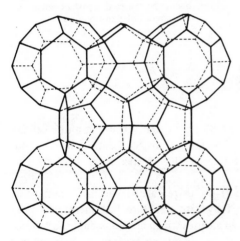

FIG. 15.6. The oxygen framework of the Type I gas hydrate structure. At the centre of the diagram are two of the six 14-hedra voids in the unit cell.

the voids are filled, as is probable for Ar, Xe, CH_4, and H_2S, this corresponds to $46/8 = 5\frac{3}{4} H_2O$ per atom (molecule) of solute. If only the larger holes are filled the formula of chlorine hydrate would be $Cl_2 . 7\frac{2}{3} H_2O$. Earlier analyses suggested $8 H_2O$, which would correspond to $2 H_2O$ in the smaller voids. However, more recent chemical analyses and density measurements indicate a formula close to $Cl_2 . 7\frac{1}{4} H_2O$, suggesting partial ($\not> 20$ per cent) occupancy of the smaller voids by Cl_2 molecules.

The sulphonium fluoride $(n\text{-}C_4H_9)_3SF$ forms three hydrates, one of which ($20 H_2O$) has this structure, apparently with $2 S^+$ statistically occupying framework sites, leaving 4 vacant sites, two of which may be occupied by 2 F^-. In hydrates of substituted sulphonium and ammonium salts the bulky alkyl groups occupy voids adjacent to the framework site occupied by S^+ or N^+. The host structure invariably has higher symmetry than is compatible with the arrangement of the guest ions in any one unit cell, and there is accordingly disorder of various kinds in these structures.

The second structure of Table 15.3, which is also cubic ($a \approx 17\cdot2$ Å), is adopted by hydrates of liquids such as $CHCl_3$; a portion of the structure is illustrated in Fig. 15.7. The unit cell contains 136 H_2O and there are voids of two sizes, 16 smaller and 8 larger. Filling of only the larger voids gives a ratio of $136/8 = 17\,H_2O$ per molecule of $CHCl_3$, CH_3I, etc., but molecules of two quite different sizes can be accommodated, each in the holes of appropriate size, as in $CHCl_3 \cdot 2\,H_2S \cdot 17\,H_2O$. In the H_2S–tetrahydrofuran hydrate listed in Table 15.2 there is rather less than half-occupancy of the smaller holes by molecules of H_2S.

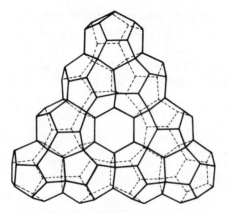

FIG. 15.7. Part of the framework of water molecules in a hydrate $M \cdot 17\,H_2O$ shown as a packing of pentagonal dodecahedra and hexakaidecahedra. At the centre part of one of the larger holes can be seen.

In the third structure of Table 15.3 the 80 polyhedral vertices in a unit cell are occupied by 76 H_2O, 2 N^+, and 2 F^-. Alkyl groups project from N^+ into the four (tetrahedrally disposed) polyhedra meeting at that vertex (two 14- and two 15-hedra).

Class (c). In $(n\text{-}C_4H_9)_3SF \cdot 23\,H_2O$ there are layers of pentagonal dodecahedra similar to those in the 38-hydrate structure just mentioned, and between them are large irregularly shaped cavities which accommodate the cations. The formation of this structure apparently represents an attempt to accommodate the $(n\text{-}C_4H_9)_3S^+$ cation in a hydrate with composition close to that of the 20-hydrate which forms the cubic structure already described.

Amines form numerous hydrates, with melting points ranging from $-35°$ to $+5°C$ and containing from $3\frac{1}{2}$ to 34 H_2O per molecule of amine. The structures include the cubic gas hydrate structure and some less regular ones. For example, in $(C_2H_5)_2NH \cdot 8\frac{2}{3}\,H_2O$ layers of 18-hedra ($f_5 = 12$, $f_6 = 6$) are linked by additional water molecules to form less regular cages (12 in a cell containing 104 H_2O). The N atoms are not incorporated in the framework but are hydrogen-bonded to the H_2O molecules as in $N_4(CH_2)_6 \cdot 6\,H_2O$ (see later).

Class (d). These hydrates are distinguished on the grounds that there are no well defined polyhedral cavities, the nets being 3- or (3 + 4)-connected. The 3-connected framework in $N_4(CH_2)_6 . 6 H_2O$ (m.p. 13·5°C) is that of Fig. 3.31 (p. 97), the same as one of the two identical interpenetrating frameworks in β-quinol clathrates. Since a framework of this kind built of H_2O molecules has only 9 links (O–H–O bonds) for every 6 H_2O there are 3 H atoms available to form hydrogen bonds to the guest molecules. The latter are suspended 'bat-like' in the cavities halfway between the 6-rings (Fig. 15.8), that is, they occupy the

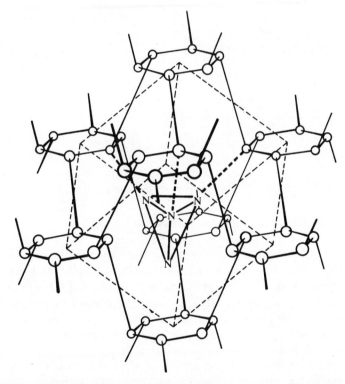

FIG. 15.8. The structure of $(CH_2)_6N_4 . 6 H_2O$ showing one molecule of $(CH_2)_6N_4$ 'suspended' in one of the interstices. The $(CH_2)_6N_4$ molecule is represented diagrammatically as a tetrahedron of N atoms attached to the hydrogen-bonded water framework by N . . . H–O bonds (heavy broken lines).

positions of the rings of the second network in the β-quinol structure. One-half of the H_2O molecules form three pyramidal O–H–O bonds and the remainder four tetrahedral hydrogen bonds (one to N). One-half of the protons are disordered (those in the 6-rings), the remainder are ordered—compare ice-I, in which all the protons are disordered, and ice-II, in which all protons are ordered.

In $(CH_3)_4NF . 4 H_2O$ the F^- ions and H_2O molecules form a hydrogen-bonded framework (Fig. 15.9) in which F^- has a 'flattened tetrahedral' arrangement of

547

4 H$_2$O neighbours (O–F–O, 156° (two) and 92$\frac{1}{2}$° (four)) and H$_2$O has three nearly coplanar neighbours (O–O, 2·73 Å, O–F, 2·63 Å). The cations occupy cavities between pairs of F$^-$ ions.

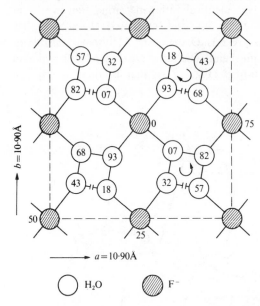

FIG. 15.9. The framework of F$^-$ ions (4-connected) and H$_2$O molecules (3-connected) in (CH$_3$)$_4$NF.4 H$_2$O. The heights of atoms above the plane of the paper are in units of *c*/100 (*c* = 8·10 Å).

Hydrates of oxy-salts, hydroxides, and halides

We exclude from the major part of our discussion the structures of hydrated complex salts, since the principles determining their structure are much less simple. For example, in hydrated complex halides A$_m$(BX$_n$).pH$_2$O, which we consider briefly later, the water is in some cases attached to B (e.g. in (NH$_4$)$_2$(VF$_5$. H$_2$O)) while in others (e.g. K$_2$(MnF$_5$) . H$_2$O, p. 383) it is situated together with the A ions between BX$_n$ complexes (here infinite octahedral chain ions). Similarly there are octahedral (HgCl$_4$)$_n^{2n-}$ chains in K$_2$HgCl$_4$. H$_2$O between which lie the K$^+$ ions and the H$_2$O molecules.

Before reviewing the crystal structures of these compounds we note some general points. They form an extremely large group of compounds, ranging from highly hydrated salts such as MgCl$_2$. 12 H$_2$O and FeBr$_2$. 9 H$_2$O to monohydrates and even hemihydrates, for example, Li$_2$SO$_4$. H$_2$O and CaSO$_4$. $\frac{1}{2}$ H$_2$O. Moreover, a particular compound may form a series of stoichiometric hydrates. Many simple halides form three, four, or five different hydrates, FeSO$_4$ crystallizes with 1, 4, 5, 6, and 7 H$_2$O, and NaOH is notable for forming hydrates with 1, 2, 3$\frac{1}{2}$, 4, 5, and 7 H$_2$O. The degree of hydration depends on the nature of both anion and cation. In some series of alkali-metal salts containing large anions such as SO$_4^{2-}$ or SnBr$_6^{2-}$ the Li and Na salts are hydrated while those containing the larger ions of K, Rb, and Cs are anhydrous. The alkali-metal chlorides behave similarly, but the fluorides show the reverse effect, and the figures in Table 15.4 illustrate the difficulty of generalizing about the degree of hydration of series of salts.

548

TABLE 15.4

Hydrates of some alkali-metal salts

	Li	Na	K	Rb	Cs
MF	—	—	2, 4	$1\frac{1}{2}$	$\frac{2}{3}, 1\frac{1}{2}$
MCl	1, 2, 3, 5	2	—	—	—
M_2CO_3	—	1, 7, 10	2, 6	$1, 1\frac{1}{2}$	$3\frac{1}{3}$
M_2SO_4	1	1, 7, 10	—	—	—

We have seen that the structures of the ice polymorphs and of the ice-like hydrates indicate that the H_2O molecule behaves as if there is a tetrahedral distribution of two positive and two negative regions of charge. The arrangement of nearest neighbours of water molecules in many crystalline hydrates is consistent with this tetrahedral character of the water molecule. In hydrated oxy-salts we commonly find a water molecule attached on the one side to two O atoms of oxy-ions and on the other to two ions M^+ or to one ion M^{2+} thus:

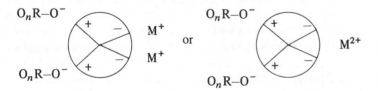

The neighbours of a water molecule may equally well be other water molecules suitably oriented so that oppositely charged regions are adjacent. In this way groups of water molecules may be held together as in $H_3PW_{12}O_{40} \cdot 29\,H_2O$, and water molecules in excess of those immediately surrounding the metal ions may be present, as in $NiSO_4 \cdot 7\,H_2O$, which may be written $[Ni(H_2O)_6]\,H_2O \cdot SO_4$. In our discussion of the structures of hydroxides we saw that in suitable environments, the OH group is polarized to the stage where an O–H⋯O bond is formed, and in hydrates we find an analogous effect. In hydrated oxy-salts and hydroxides the short distances (2·7–2·9 Å) between O atoms of water molecules and those of the oxy-ions are similar to those found in certain hydroxides and oxyhydroxides, and indicate the formation of hydrogen bonds. In hydrated fluorides there are O–H⋯F bonds of considerable strength, and we give examples later of O–H⋯Cl bonds. We deal separately with hydrated acids and acid salts in which protons are associated with some or all of the water molecules to form H_3O^+ or more complex groupings.

Some of the numerous possible environments of a water molecule in hydrates are shown in Fig. 15.10. It might seem logical to classify hydrates according to the way in which the water molecules are bonded together. If all the H_2O molecules in a particular hydrate have environments of one of the types shown in Fig. 15.10, with 4, 3, 2, 1, or zero H_2O molecules as nearest neighbours then the systems of linked H_2O molecules (aquo-complex) would be as follows: (a) and (b), all possible

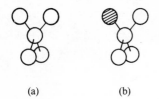 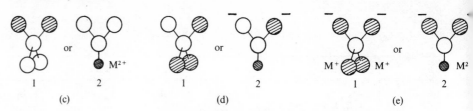

FIG. 15.10. Environment of water molecules in crystals. The larger shaded circles represent M^+, OH^-, F^-, or oxygen of oxy-ion.

types up to and including 3D frameworks, (c), rings or chains, (d), pairs of H_2O molecules, and (e), no aquo-complex. For example, each H_2O in $Li_2SO_4 \cdot H_2O$ is of type (c), hydrogen-bonded to two others, so that chains of water molecules can be distinguished in the crystal, whereas in $KF \cdot 2H_2O$ each H_2O is surrounded tetrahedrally by $2 F^-$ and $2 K^+$ ions, as at (e), that is, there is no aquo-complex. However, any classification of this kind would be impracticable because the water molecules in many hydrates do not all have the same kind of environment. There may be a major difference in environment, as when some of the H_2O molecules are bonded to the metal atoms and others are accommodated between the complexes as, for example, in $[CoCl_2(H_2O)_4] \cdot 2H_2O$, or there may be differences between the environments of the various water molecules of one $M(H_2O)_n$ complex. Thus in $NiSO_4 \cdot 7H_2O$ the seventh H_2O molecule is not in contact with a metal ion but in addition there is the further complication that the environments of the six H_2O molecules in a $Ni(H_2O)_6^{2+}$ group are not the same and are in fact of no fewer than four different types. Some have three approximately coplanar neighbours (Ni^{2+} and 2 O of SO_4^{2-} or $2 H_2O$) and others four tetrahedral neighbours (Ni^{2+}, 2 O and H_2O, or Ni^{2+}, O, and $2 H_2O$). Much of the structural complexity of hydrates is due to this non-equivalence of water molecules which in turn is associated with the fact that so many arrangements of nearest neighbours are compatible with the tetrahedral charge distribution of the H_2O molecule as indicated in Fig. 15.10.

A detailed description of the bonding in hydrates evidently requires a knowledge of the positions of the H atoms. In the earlier X-ray studies it was not possible to locate these atoms, and it was assumed that they were responsible for certain unusually short O–O or O–X distances in the crystals. Later studies, particularly n.d. and n.m.r., have confirmed this and led to the precise location of the H atoms. It is now becoming possible to discuss not only the gross structures of the compounds, that is, the spatial arrangement of the heavier atoms, but also two aspects of the finer structure, namely, the positions of the H atoms in hydrogen bonds and the ordering of the protons. Reference to these topics will be made later.

Since in the hydrates under discussion the H_2O molecules tend to associate, albeit not exclusively, with the cations, it is convenient to show in any classification the nature of the coordination group around the cations. In a hydrated oxy-salt or halide this will be made up of O atoms of oxy-ions, halide ions, or O atoms of H_2O molecules (the H atoms of which will be directed away from M). If n is the coordination number of M in $M_xX_y \cdot zH_2O$, where X represents the anion (Cl^-,

550

SO_4^{2-}, O_2^{2-}, OH^-, etc.) we may list hydrates according to the value of z/xn (Table 15.5). Evidently, very simple structures may be expected if $z/xn = 1$ ($z = xn$), since there is in this case exactly the number of H_2O molecules necessary to form complete coordination groups $M(H_2O)_n$ around every M ion. However, the situation is more complicated than this because if H_2O molecules are common to two $M(H_2O)_n$ coordination groups there can be complete hydration of M with smaller values of z/xn. For example, for octahedral coordination of M ($n = 6$):

z/xn		Example
1	Discrete $M(H_2O)_6$ groups	$MgCl_2 . 6 H_2O$
$\frac{2}{3}$	$M(H_2O)_6$ sharing two edges	$KF . 4 H_2O$
$\frac{1}{2}$	$M(H_2O)_6$ sharing two faces	$LiClO_4 . 3 H_2O$

There is therefore no simple relation between the value of z/xn and the composition of the coordination group around M, as is clearly seen in Table 15.5.

For descriptive purposes it is convenient to make three horizontal subdivisions of the Table.

A. $(z/xn) \geqslant 1$

In these hydrates there is sufficient (or more than sufficient) water for complete hydration of the cations without sharing of H_2O molecules between $M(H_2O)_n$ coordination groups. There are no known exceptions to the rule that if $z/xn > 1$ (class AI) M is fully hydrated and the excess water is accommodated between the $M(H_2O)_n$ complexes or, alternatively, associated with the anions. A set of very simple structures is found if $z/xn = 1$ (class AII), but the structures of greatest interest are those of class AIII, at present represented by only two structures, both of halides of 3d metals. Although $z/xn = 1$ some of the coordination positions around M are occupied by Cl in preference to H_2O. There are no entries in Class AIV, for if $z/xn \geqslant 1$ either the cation is fully hydrated or there is excess water of crystallization.

B. $1 > (z/xn) \geqslant \frac{1}{2}$

Here there is sufficient water for complete hydration of M assuming that H_2O molecules can be shared between two (and only two) $M(H_2O)_n$ coordination groups. Structures of all four types I–IV are known, and those of greatest interest are perhaps those of types I and III where, as in AIII, structures could be envisaged in which all the water would be associated with M ions; instead, only part of the water hydrates the cations.

C. $(z/xn) < \frac{1}{2}$

There is insufficient water for complete hydration of M even allowing sharing of H_2O molecules between two $M(H_2O)_n$ coordination groups. Apart from one case the examples of Table 15.5 are all mono- or di-hydrates, and as might be expected are of type IV. Because of its charge distribution a water molecule is unlikely to have more than two cation neighbours, and accordingly there are no hydrates in classes CI or CII.

551

TABLE 15.5

A classification of salt hydrates $M_x X_y . z H_2O$

	z/xn	M *fully hydrated*		M *incompletely hydrated*	
		excess H_2O I	II	*excess* H_2O III	IV
A	2 $\frac{9}{6}$ $\frac{8}{6}$ $\frac{7}{6}$ 1	$MgCl_2 . 12 H_2O^{(1)}$ $FeBr_2 . 9 H_2O$ see text $NiSO_4 . 7 H_2O^{(2)}$ $NaOH . 7 H_2O^{(3)}$	$Sm(BrO_3)_3 . 9 H_2O^{(4)}$ $Nd(BrO_3)_3 . 9 H_2O^{(5)}$ $CaO_2 . 8 H_2O^{(6)}$ $Sr(OH)_2 . 8 H_2O^{(7)}$ $Zn(BrO_3)_2 . 6 H_2O^{(8)}$ $CoI_2 . 6 H_2O^{(9)}$ $MgCl_2 . 6 H_2O^{(10)}$ $AlCl_3 . 6 H_2O^{(11)}$ $CrCl_3 . 6 H_2O^{(12)}$ $Mg(ClO_4)_2 . 6 H_2O^{(13)}$ $BeSO_4 . 4 H_2O^{(14)}$	$CoCl_2 . 6 H_2O^{(15)}$ $NiCl_2 . 6 H_2O^{(16)}$ $FeCl_3 . 6 H_2O^{(17)}$ $CrCl_3 . 6 H_2O^{(18)}$	
B	$\frac{5}{6}$ $\frac{3}{4}$ $\frac{2}{3}$ $\frac{7}{12}$ $\frac{1}{2}$	$Na_2SO_4 . 10 H_2O^{(19)}$	$Na_2CO_3 . 10 H_2O^{(20)}$ $SrCl_2 . 6 H_2O^{(21)}$ $KF . 4 H_2O^{(22)}$ $Na_2HAsO_4 . 7 H_2O^{(23)}$ $NaOH . 3\frac{1}{2} H_2O^{(24)}$ $LiClO_4 . 3 H_2O^{(25)}$	$CuSO_4 . 5 H_2O^{(26)}$ $SnCl_2 . 2 H_2O^{(27)}$ $FeF_3 . 3 H_2O^{(28)}$	$CaCO_3 . 6 H_2O^{(29)}$ $GdCl_3 . 6 H_2O^{(30)}$ $FeF_2 . 4 H_2O^{(31)}$ $FeCl_2 . 4 H_2O^{(32)}$ $MnCl_2 . 4 H_2O^{(32a)}$ $CuSO_4 . 3 H_2O^{(33)}$
C	$\frac{4}{9}$ $\frac{1}{3}$ $\frac{1}{4}$ $\frac{1}{6}$ $\frac{1}{8}$ $\frac{1}{9}$ $\frac{1}{12}$			$CdSO_4 . \frac{8}{3} H_2O^{(34)}$	$KF . 2 H_2O^{(35)}, NaBr . 2 H_2O^{(36)}$ $CoCl_2 . 2 H_2O^{(37)}$ $NiCl_2 . 2 H_2O^{(37a)}$ $BaCl_2 . 2 H_2O^{(38)}$ $SrCl_2 . 2 H_2O^{(39)}$ $CaSO_4 . 2 H_2O^{(40)}$ $LiOH . H_2O^{(41)}$ $CuSO_4 . H_2O^{(42)}$ $Li_2SO_4 . H_2O^{(43)}$ $Sr(OH)_2 . H_2O^{(44)}$ $SrBr_2 . H_2O^{(45)}$ $Na_2CO_3 . H_2O^{(46)}$

(1) AC 1966 **20** 875. (2) AC 1964 **17** 1167, 1361; AC 1969 **B25** 1784. (3) CR 1953 **236** 1579. (4) AC 1969 **A25** 621. (5) JACS 1939 **61** 1544. (6) AC 1951 **4** 67. (7) AC 1953 **6** 604. (8) ZK 1936 **95** 426. (9) ZSK 1963 **4** 63. (10) ZK 1934 **87** 345. (11) AC 1968 **B24** 954. (12) ZK 1934 **87** 446. (13) ZK 1935 **91** 480. (14) AC 1969 **B25** 304, 310. (15) JPSJ 1961 **16** 1574. (16) JCP 1969 **50** 4690. (17) JCP 1967 **47** 990. (18) AC 1966 **21** 280. (19) JACS 1961 **83** 820. (20) AC 1969 **B25** 2656. (21) KDV 1940 **17** Nr.9 (22) JCP 1964 **41** 917. (23) AC 1970 **B26** 1574, 1584. (24) BSFMC 1958 **81** 287. (25) AC 1952 **5** 571. (26) PRS A 1962 **266** 95. (27) JCS 1961 3954. (28) AC 1964 **17** 1480. (29) IC 1970 **9** 480. (30) AC 1961 **14** 234. (31) AC 1960 **13** 953. (32) AC 1971 **B27** 2329; (32a) IC 1964 **3** 529; IC 1965 **4** 1840. (33) AC 1968 **B24**

We have already remarked that in many hydrates the H_2O molecules are not all equivalent, often having very different environments. It is also found that in some hydrates there are two or more kinds of non-equivalent cation. This complication is less frequently encountered; an example is $Na_4P_2O_7 . 10 H_2O$,[1] in which one-half of the Na^+ ions have 6 H_2O while the remainder have 4 H_2O and O atoms of anions as nearest neighbours. We confine our examples in Table 15.5 and the following account to hydrates in which all the cations have similar arrangements of nearest neighbours, as is the case in most hydrates.

(1) AC 1957 **10** 428
(2) AC 1964 **18** 698
(3) JCP 1963 **39** 2881

Hydrates of Class A: $z/xn \geqslant 1$.

Type AI.   This is apparently a very small group of compounds, of which very few structures have been determined.

$MgCl_2 . 12 H_2O$.   This compound is of special interest as the most highly hydrated simple salt of which we know the structure. It is stable only at low temperatures, as shown by the transition points:

$$MgCl_2 . 12 H_2O \xrightarrow{-16\cdot4^\circ} 8 H_2O \xrightarrow{-3\cdot4^\circ} 6 H_2O \xrightarrow{116\cdot7^\circ} 4 H_2O \xrightarrow{181^\circ C} 2 H_2O$$

(The known hydrates of $MgBr_2$ and MgI_2 contain respectively 6 and 10, and 8 and 10 H_2O.) The structure consists of very regular $Mg(H_2O)_6$ octahedra and very distorted $Cl(H_2O)_6$ octahedra (Cl–O, $3\cdot11$–$3\cdot26$ Å, but edges, $3\cdot86$–$5\cdot55$ Å). Each octahedral coordination group shares four vertices, $Mg(H_2O)_6$ with 4 $Cl(H_2O)_6$ and $Cl(H_2O)_6$ with 2 $Cl(H_2O)_6$ and 2 $Mg(H_2O)_6$, as shown diagrammatically in Fig. 15.11. The layers, of composition $MgCl_2(H_2O)_{12}$, are held together by O–H–O bonds between H_2O molecules. The H_2O molecules attached to Mg^{2+} have two other neighbours (H_2O or Cl^-); the others, one-half of the total, have four tetrahedral neighbours (2 Cl^- + 2 H_2O or 1 Cl^- + 3 H_2O). The structure provides a beautiful illustration of the behaviour of the water molecule in this type of hydrate.

($z/xn = 8/6$. We may include here two salts containing hexanitrato-ions, namely, $Mg[Th(NO_3)_6] . 8 H_2O$[2] and $Mg_3[Ce(NO_3)_6]_2 . 24 H_2O$,[3] in both of which Mg^{2+} is completely hydrated and one-quarter of the water of hydration is not attached to cations.)

$NiSO_4 . 7 H_2O$.   Reference has already been made to the fact that as regards their environment in the crystal there are five different kinds of H_2O molecule in this hydrate. From the chemical standpoint, however, we need only distinguish between those forming octahedral groups around Ni^{2+} ions and the seventh H_2O which is situated between three water molecules of $Ni(H_2O)_6$ groups and one O atom of a sulphate ion.

Type AII.   In this, the simplest type of salt hydrate, all the water molecules are associated with the cations. The number of O atoms which can be

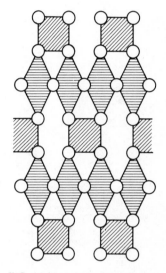

FIG. 15.11. Layer formed from vertex-sharing $[Mg(H_2O)_6]^{2+}$ and $[Cl(H_2O)_6]^-$ groups in $MgCl_2 . 12$ H_2O (diagrammatic). The squares represent $[Mg(H_2O)_6]^{2+}$ and the rhombuses $[Cl(H_2O)_6]^-$ groups. Two (unshared) vertices of each octahedron are not shown.

508. (34) PRS A 1936 **156** 462. (35) AC 1951 **4** 181. (36) AC 1964 **17** 730. (37) AC 1963 **16** 1176; (37a) AC 1967 **23** 630. (38) AC 1966 **21** 450. (39) KDV 1943 **20** Nr.5. (40) JCP 1958 **29** 1306. (41) AC 1971 **B27** 1682. (42) CR 1966 **B262** 722. (43) JCP 1968 **48** 5561. (44) AC 1967 **22** 252. (45) JPC 1964 **68** 3259. (46) ZK 1936 **95** 266.
Miscellaneous: NaOH . 4 H_2O JCP 1964 **41** 924.

accommodated around the ion M is determined by the radius ratio $r_M : r_O$, and since the bonds M–O are essentially electrostatic in nature the coordination polyhedra are those characteristic of ionic crystals.

Shape of $M(H_2O)_n$ *complex*

Tetrahedron:	$BeSO_4 . 4 H_2O$
Octahedron:	$MgCl_2 . 6 H_2O$; $CoI_2 . 6 H_2O$; $AlCl_3 . 6 H_2O$; $CrCl_3 . 6 H_2O$; $Mg(ClO_4)_2 . 6 H_2O$, etc.
Square antiprism:	$CaO_2 . 8 H_2O$; $Sr(OH)_2 . 8 H_2O$
Tricapped trigonal prism:	$Nd(BrO_3)_3 . 9 H_2O$; $Sm(BrO_3)_3 . 9 H_2O$

The environment of the water molecules in, for example, $BeSO_4 . 4 H_2O$ is consistent with our description of the H_2O molecule; it corresponds to (e) 2 of Fig. 15.10.

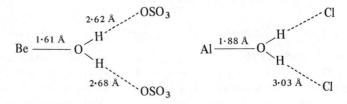

If the only factor determining the structure of $Mg(H_2O)_6Cl_2$ were the relative sizes of $Mg(H_2O)_6^{2+}$ and Cl^- this hydrate could have the fluorite structure, but this would mean that each H_2O would be in contact with 4 Cl^- ions. In fact, the corresponding ammine, $Mg(NH_3)_6Cl_2$, does crystallize with this structure, but the hexahydrate has a less symmetrical (monoclinic) structure in which each water molecule is in contact with only 2 Cl^- ions. Similarly, the ammine $Al(NH_3)_6Cl_3$ has the YF_3 structure, but $Al(H_2O)_6Cl_3$ has a much more complex rhombohedral structure in which every H_2O molecule is adjacent to only 2 Cl^- ions, giving it an environment very similar to that in $BeSO_4 . 4 H_2O$ as shown above.

Type AIII. The two structures in this class, both of hexahydrates of 3d metal halides, form a striking contrast to two structures of AII. Both contain octahedral coordination groups $M(H_2O)_4Cl_2$ with the *trans* configuration. ($NiCl_2 . 6 H_2O$ is similar to the cobalt compound.)

	Coordination group of M		Coordination group of M	Structural formula
$MgCl_2 . 6 H_2O$	$6 H_2O$	$CoCl_2 . 6 H_2O$	$4 H_2O$ $2 Cl$	$[CoCl_2(H_2O)_4] . 2 H_2O$
$AlCl_3 . 6 H_2O$	$6 H_2O$	$FeCl_3 . 6 H_2O$	$4 H_2O$ $2 Cl$	$[FeCl_2(H_2O)_4] Cl . 2 H_2O$

In contrast to $CoCl_2 . 6 H_2O$ the iodide is $Co(H_2O)_6I_2$ (type AII). For the trichlorides the stabilities of the two hydrate structures are presumably not very different, for the hexahydrate of $CrCl_3$ forms structures of both types. Early chemical evidence (for example, the proportion of the total chlorine precipitated by $AgNO_3$) indicated the following structures for the hydrates of chromic chloride:

blue hydrate: $[Cr(H_2O)_6]Cl_3$

green hydrates: $[CrCl(H_2O)_5]Cl_2 . H_2O$ and $[CrCl_2(H_2O)_4]Cl . 2 H_2O$

X-ray studies have confirmed the first and third structures. With these hydrates compare $GdCl_3 . 6 H_2O$ of type BIV (later).

Hydrates of Class B: $1 > z/xn \geqslant \frac{1}{2}$

Type BI.

$Na_2SO_4 . 10 H_2O$. In two well-known decahydrates, those of Na_2SO_4 and Na_2CO_3, the $5:1$ ratio of $H_2O:M$ is achieved in different ways, Na^+ being completely hydrated in both hydrates. In the sulphate there are infinite chains of octahedral $Na(H_2O)_6$ groups sharing two edges so that there is an excess of $2 H_2O$ to be accommodated between the chains: $[Na(H_2O)_4]_2SO_4 . 2 H_2O$. Note that octahedral chains of composition $Na(H_2O)_5$ in which the octahedral groups share vertices have not been found in a hydrate, possibly because the minimum value of the angle $Na-O-Na$ would be approximately $130°$ (see p. 157).

Type BII.   This is an interesting group of hydrates, in all of which the cations are completely hydrated, in which different $H_2O:M$ ratios arise as the result of sharing different numbers of H_2O molecules of each coordination group, as illustrated by the following examples in which there is octahedral coordination of the cations:

Number of H_2O of each $M(H_2O)_n$ group shared	$H_2O:M$ ratio in hydrate	Examples
2	5	$Na_2CO_3 . 10 H_2O$
4	4	$KF . 4 H_2O$
5	$3\frac{1}{2}$	$Na_2HAsO_4 . 7 H_2O$, $NaOH . 3\frac{1}{2} H_2O$
6	3	$LiClO_4 . 3 H_2O$

$Na_2CO_3 . 10 H_2O$.   Here the $Na(H_2O)_6$ groups are associated in pairs to form units $Na_2(H_2O)_{10}$ of the same general type as the dimers of certain pentahalides. Contrast the structure of $Na_2SO_4 . 10 H_2O$ above.

$KF . 4 H_2O$.   The $K(H_2O)_6$ coordination groups share two edges to give a $H_2O:M$ ratio of $4:1$.

$Na_2HAsO_4 . 7 H_2O$.   This hydrate contains a unique chain of composition $Na_2(H_2O)_7$ (Fig. 15.12) formed from face-sharing pairs of $Na(H_2O)_6$ octahedra which also share two terminal edges. The chains are cross-linked at intervals by $O-H-O$ bonds, and the $AsO_3(OH)^{2-}$ ions are accommodated between the chains. There is the same $H_2O:M$ ratio in $NaOH . 3\frac{1}{2} H_2O$, in which there is complete hydration of the cations as in the arsenate.

$LiClO_4 . 3 H_2O$.   The value $\frac{1}{2}$ for z/xn in this hydrate results from the formation of infinite chains by the sharing of opposite faces of octahedral

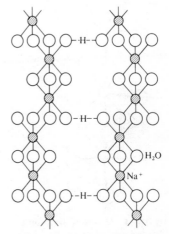

FIG. 15.12. Chains of composition $Na_2(H_2O)_7$ in $Na_2HAsO_4 . 7 H_2O$.

Li(H$_2$O)$_6$ groups (Fig. 15.13). The four nearest neighbours of a water molecule (other than water molecules of the same chain, contacts with which are clearly not contacts between oppositely charged regions of H$_2$O molecules) are 2 Li$^+$ of the chain and 2 O atoms of different ClO$_4^-$ ions. These Li–H$_2$O–Li and O–H$_2$O–O bonds lie in perpendicular planes, so that the four neighbours of H$_2$O are arranged tetrahedrally. The structure of LiClO$_4$. 3 H$_2$O is closely related to that of a large group of isostructural hexahydrates M(BF$_4$)$_2$. 6 H$_2$O and M(ClO$_4$)$_2$. 6 H$_2$O in

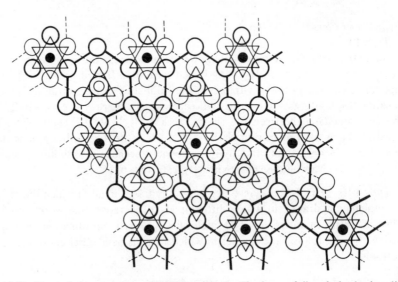

FIG. 15.13. Plan of the structure of LiClO$_4$. 3 H$_2$O. The heavy full and the broken lines indicate O–H–O bonds between H$_2$O molecules and oxygen atoms (of ClO$_4^-$ ions) which are approximately coplanar. These bonds thus link together the columns of Li(H$_2$O)$_6$ octahedra, which share opposite faces, and the ClO$_4^-$ ions. The small black circles represent Li$^+$ ions in planes midway between the successive groups of 3 H$_2$O.

which M is Mg, Mn, Fe, Co, Ni, or Zn. In LiClO$_4$. 3 H$_2$O there is a Li$^+$ ion at the centre of each octahedral group of water molecules in the infinite chain of Fig. 15.13. If we remove these ions and place a Mg^{2+} ion in each alternate octahedron the crystal has the composition Mg(ClO$_4$)$_2$. 6 H$_2$O, and there are discrete Mg(H$_2$O)$_6^{2+}$ groups (Type AII). Instead of 2 Li$^+$ neighbours, each H$_2$O has now only 1 Mg^{2+} neighbour; compare

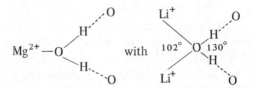

SrCl$_2$. 6 H$_2$O. Here there are columns of tricapped trigonal prisms, each sharing a pair of opposite faces (Fig. 15.14). This structure is adopted by the

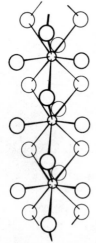

FIG. 15.14. Portion of the infinite 1-dimensional cation–water complex in crystalline SrCl$_2$.6 H$_2$O, in which each Sr^{2+} ion (small circle) is surrounded by nine water molecules.

556

hexahydrates of all the following halides:

$$CaCl_2 \qquad SrCl_2$$
$$CaBr_2 \qquad SrBr_2$$
$$CaI_2 \qquad SrI_2 \qquad BaI_2$$

Type BIII.   Although there is sufficient water to hydrate M completely, not only is there incomplete hydration of the cations but there are H_2O molecules not attached to cations.

$CuSO_4 . 5 H_2O$.   In this hydrate the metal ion is 6-coordinated, but although there are only 5 H_2O molecules for every Cu^{2+} ion the fifth is not attached to a cation. Instead, the coordination group around the cupric ion is composed of 4 H_2O and 2 O atoms of sulphate ions, and the fifth H_2O is held between water molecules attached to cations and O atoms of sulphate ions, as shown in Fig. 15.15.

$SnCl_2 . 2 H_2O$.   This hydrate is of more interest in connection with the structural chemistry of divalent tin. Discrete pyramidal complexes $SnCl_2(H_2O)$, in which the mean interbond angle is 83°, are arranged in double layers which alternate with layers of water molecules.

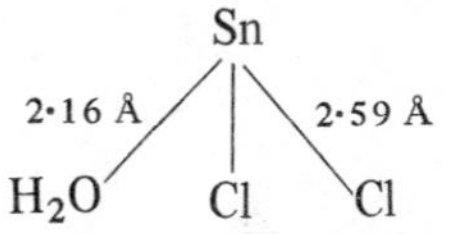

$FeF_3 . 3 H_2O$.   Instead of other simpler structural possibilities, such as finite $FeF_3(H_2O)_3$ groups, one form of this hydrate consists of infinite chains (Fig. 15.16(f)) of composition $FeF_3(H_2O)_2$ between which the remaining water molecule is situated. In the chain there is random distribution of 2 F and 2 H_2O among the four equatorial positions in each octahedron. In the variant of this chain in $RbMnCl_3 . 2 H_2O$ there is, however, a regular arrangement of 2 Cl and 2 H_2O in these positions (Fig. 15.16(g)).

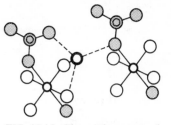

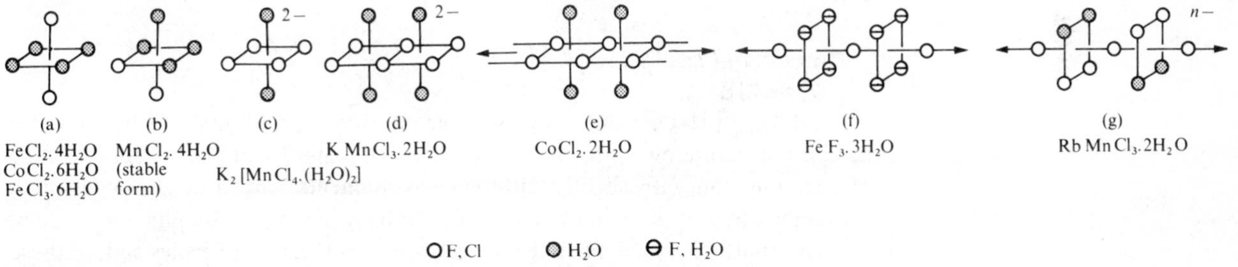

FIG. 15.15. The environment of the fifth H_2O molecule (centre) in $CuSO_4.5 H_2O$. The octahedral co-ordination group around a copper atom is composed of four H_2O molecules and two oxygen atoms of SO_4^{2-} ions (shaded circles).

$$\overset{\displaystyle Sn}{\underset{\displaystyle H_2O \quad Cl \quad Cl}{\diagup \; \big| \; \diagdown}}$$

2·16 Å   2·59 Å

FIG. 15.16. Coordination groups of 3d ions in some hydrated halides and complex halides (see text).

(a) $FeCl_2.4H_2O$, $CoCl_2.6H_2O$, $FeCl_3.6H_2O$

(b) $MnCl_2.4H_2O$ (stable form)

(c) $K_2[MnCl_4.(H_2O)_2]$

(d) $K MnCl_3.2H_2O$

(e) $CoCl_2.2H_2O$

(f) $Fe F_3.3H_2O$

(g) $Rb MnCl_3.2H_2O$

○ F, Cl   ◉ H_2O ⊖ F, H_2O

Type BIV.

$CaCO_3 . 6 H_2O$.   In addition to the well-known anhydrous forms $CaCO_3$ also crystallizes with 1 and 6 H_2O. In the remarkable structure of the hexahydrate there are isolated ion-pairs surrounded by an envelope of 18 water molecules. Of these, six complete the 8-coordination group around Ca^{2+} and the remainder are bonded to other cations. (Each water molecule is adjacent to (only) one Ca^{2+} ion.) This type of hydrate, in which ion pairs (resembling the classical picture of a $CaCO_3$ 'molecule') are embedded in a mass of H_2O molecules, may be contrasted with

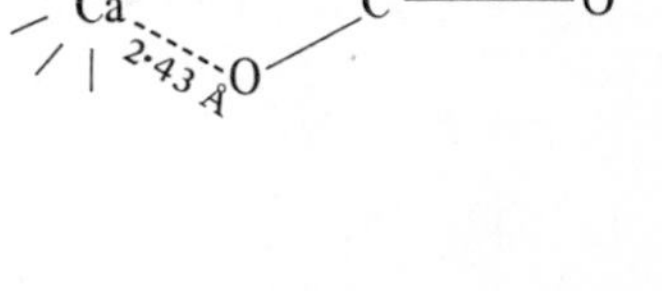

557

$MgCl_2 . 12 H_2O$ in which separate Mg^{2+} and Cl^- ions are completely surrounded by water molecules (some of which are shared between the coordination groups) and with the clathrate hydrates described earlier.

$GdCl_3 . 6 H_2O$. Here also there is 8-coordination of the cations (by 6 H_2O and 2 Cl^-)–contrast $[Al(H_2O)_6]Cl_3$ and $[Fe(H_2O)_4Cl_2]Cl . 2 H_2O$. This type of structure is adopted by the hexahydrates of the trichlorides and tribromides of the smaller 4f metals and by $AmCl_3 . 6 H_2O$ and $BkCl_3 . 6 H_2O$.[1] The larger M^{3+} ions of La, Ce, and Pr form $MCl_3 . 7 H_2O$, the structure of which is not yet known.

$FeF_2 . 4 H_2O$ and $FeCl_2 . 4 H_2O$. In both of these hydrates there are discrete octahedral molecules $FeX_2(H_2O)_4$. In the fluoride it was not possible to distinguish between F and H_2O; the chloride has the *trans* configuration. Although the water content of these hydrates is sufficient to form an infinite chain of octahedral $Fe(H_2O)_6$ groups sharing opposite edges, finite groups with 2 Cl attached to the cation are preferred.

$MnCl_2 . 4 H_2O$. Two polymorphs of this hydrate have been known for a long time (both monoclinic), one of which is described as metastable at room temperature. This form has the same structure as $FeCl_2 . 4 H_2O$, i.e. it consists of *trans* $MnCl_2(H_2O)_4$ molecules (Fig. 15.16(a)). Somewhat unexpectedly the stable form is also built of molecules with the same composition but with the *cis* configuration (Fig. 15.16(b)).

$CuSO_4 . 3 H_2O$. Cupric sulphate forms hydrates with 1, 3, and 5 H_2O. In contrast to the pentahydrate, in which only 4 H_2O are associated with the cations, all the water is coordinated to Cu^{2+} in the trihydrate. The coordination group around the metal ion consists of 3 H_2O + 1 O (mean Cu–O, 1·94 Å) with two more distant O atoms of SO_4^{2-} ions (at 2·42 Å) completing a distorted octahedral group.

Hydrates of Class C: $z/xn < \frac{1}{2}$.
   Type CIII.
   $CdSO_4 . \frac{8}{3} H_2O$. Our only example of this type of hydrate has a rather complex structure owing to the unusual ratio of water to salt. There are two kinds of cadmium ion with slightly different environments, but both are octahedrally surrounded by two water molecules and four oxygen atoms of sulphate ions. There are four kinds of crystallographically non-equivalent water molecules and, of these, three-quarters are attached to a cation. The remaining water molecules have no contact with a metal ion, but have four neighbours, two other water molecules and two oxygen atoms of oxy-ions. It appears that the important point is the provision of three or four neighbours, which can be

$$Cd^{2+} \quad \begin{matrix} ⁻H_2O^+_+ \\ ⁻H_2O^+_+ \\ ⁻H_2O^+_+ \end{matrix} \quad \text{or} \quad Cd^{2+} \quad ⁻H_2O^+_+ \quad \text{or} \quad \begin{matrix} O^- \quad H_2O_+ \quad O^- \\ ⁻H_2O^+_+ \\ ⁻H_2O^+_+ \quad H_2O^+ \quad O^- \end{matrix}$$

It is well to remember that in the present state of our knowledge we are far from understanding why a hydrate with a formula so extraordinary as $CdSO_4 . \frac{8}{3} H_2O$

(1) IC 1971 **10** 147

should form at all; its structure and therefore its chemical formula represent a compromise between the requirements of Cd^{2+}, SO_4^{2-}, and H_2O.

Type CIV.   Our examples here are of mono- and dihydrates. We describe first the structures of some dihydrated halides.

KF . $2 H_2O$.   In this hydrate each $K^+(F^-)$ ion is surrounded by $4 H_2O$ and $2 F^- (K^+)$ at the vertices of a slightly distorted octahedron. Each H_2O has $2 K^+$ and $2 F^-$ ions as nearest neighbours arranged tetrahedrally (Fig. 15.17).

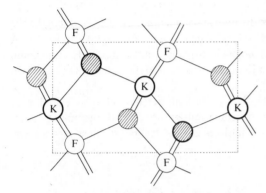

FIG. 15.17. Projection of the crystal structure of KF . $2 H_2O$. Heavy circles represent atoms lying in the plane of the paper and light circles atoms lying in planes 2 Å above and below that of the paper. Shaded circles represent water molecules.

NaBr . $2 H_2O$.   There are similar coordination groups (i.e. *cis* $NaBr_2(H_2O)_4$) in this crystal. The (layer) structure is related to that of $MnCl_2 . 4 H_2O$ in the following way. If an octahedral MX_3 layer of the $AlCl_3$ ($Al(OH)_3$) type is built of *cis* $MX_2(H_2O)_4$ groups its composition is $MX . 2 H_2O$, and this is the layer in NaBr . $2 H_2O$, shown diagrammatically in Fig. 15.18. Removal of one-half of the cations (those shown as small open circles in Fig. 15.18) leaves a layer of composition $MX_2 . 4 H_2O$, consisting of isolated $MX_2(H_2O)_4$ molecules. This represents the structure of the stable form of $MnCl_2 . 4 H_2O$ to which we have already referred. The relation between these two hydrates should be compared with that between the structures of $LiClO_4 . 3 H_2O$ and $Mg(ClO_4)_2 . 6 H_2O$ noted above.

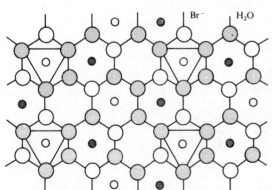

FIG. 15.18. Layer in NaBr . $2 H_2O$.

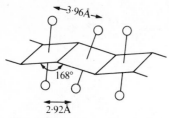

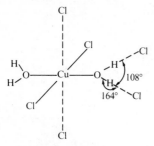

FIG. 15.19. Distorted octahedral chain in $NiCl_2 . 2 H_2O$.

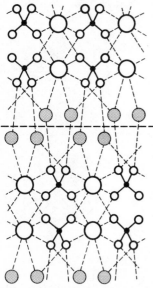

FIG. 15.20. The structure of $CuCl_2(H_2O)_2$.

FIG. 15.22. Section through the crystal structure of gypsum perpendicular to the layers (diagrammatic). The layers are composed of SO_4^{2-} and Ca^{2+} ions (large open circles) with H_2O molecules on their outer surfaces (shaded). The heavy broken line indicates the cleavage, which breaks only O—H—O bonds.

$CoCl_2 . 2 H_2O$. The dihydrates of a number of 3d dihalides are built of the edge-sharing octahedral MX_4 chains of Fig. 15.16(e); they include the dichlorides and dibromides of Mn, Fe, Co, and Ni. There is an interesting distortion of the chain in $NiCl_2 . 2 H_2O$, which is not isostructural with the Co compound (Fig. 15.19). The planes of successive equatorial $NiCl_4$ groups are inclined to one another at a small angle and the distances between successive H_2O molecules along each side of the chain are alternately 2·92 Å and 3·96 Å.

A quite different type of distortion of the octahedron occurs in $CuCl_2 . 2 H_2O$ where the nearest neighbours of Cu are: 2 H_2O at 1·93 Å, 2 Cl at 2·28 Å, and 2 Cl at 2·91 Å. In view of the very large difference between the Cu—Cl bond lengths the structure is preferably described as built of planar *trans* $CuCl_2(H_2O)_2$ molecules. A n.d. study showed that the whole molecule is planar, including the H atoms, and that the latter lie close to the lines joining O atoms of H_2O molecules to their two nearest Cl neighbours. (Fig. 15.20.)

$SrCl_2 . 2 H_2O$ and $BaCl_2 . 2 H_2O$. These hydrates have closely related layer structures in both of which the coordination group around the metal ion is composed of 4 Cl^- ions and 4 H_2O molecules. The layers are illustrated in Fig. 15.21.

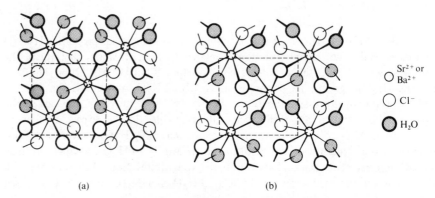

(a)   (b)

Sr²⁺ or Ba²⁺
Cl⁻
H₂O

FIG. 15.21. Plans of layers in the crystal structures of (a) $SrCl_2 . 2 H_2O$, and (b) $BaCl_2 . 2 H_2O$. Atoms above or below the plane of the metal ions are shown as heavy or light circles respectively.

As an example of a dihydrate of an oxy-salt we describe the structure of $CaSO_4 . 2 H_2O$.

$CaSO_4 . 2 H_2O$. This hydrate, the mineral gypsum, has a rather complex layer structure (Fig. 15.22) in which the layers are bound together by hydrogen

560

bonds between water molecules and O atoms of sulphate ions. Each H_2O molecule is bonded to a Ca^{2+} ion and to a sulphate O atom of one layer and to an O atom (of SO_4^{2-}) of an adjacent layer, so that its environment is similar to that of a water molecule in $BeSO_4 . 4 H_2O$ or $Zn(BrO_3)_2 . 6 H_2O$. These O–H–O bonds are the weakest in the structure, accounting for the excellent cleavage and the marked anisotropy of thermal expansion, which is far greater normal to the layers than in any direction in the plane of the layers. The location of the H atoms on the lines joining water O atoms to sulphate O atoms has been confirmed by n.m.r. and n.d. studies.

The structures of $CaHPO_4 . 2 H_2O$ (brushite) and $CaHAsO_4 . 2 H_2O$ (pharmacolite) are very similar to that of gypsum but with additional hydrogen bonding involving the OH groups of the anions.[1]

(1) AC 1969 **B25** 1544

In monohydrates H_2O molecules form a decreasingly important part of the coordination group of the cations as the coordination number of cation increases:

	C.N. of M	Coordination group
$LiOH . H_2O$	4	2 H_2O 2 OH
$Li_2SO_4 . H_2O$	4	1 H_2O 3 O 4 O
$Na_2CO_3 . H_2O$	6	2 H_2O 4 O 1 H_2O 5 O
$SrBr_2 . H_2O$	9	2 H_2O 7 Br

(In two of these monohydrates there are two sets of cations with different environments.)

$LiOH . H_2O$. Each structural unit, Li^+, OH^-, and H_2O, has four nearest neighbours. The Li^+ ions are surrounded by 2 OH^- and 2 H_2O and these tetrahedral groups share an edge and two vertices to form double chains (Fig. 15.23) which are held together laterally by hydrogen bonds between OH^- ions and H_2O molecules. Each water molecule also has four tetrahedral neighbours, 2 Li^+ and 2 OH^- ions of different chains, a similar type of environment to that in $LiClO_4 . 3 H_2O$.

$Li_2SO_4 . H_2O$. This structure provides a good example of the tetrahedral arrangement of four nearest neighbours (2 H_2O, Li^+, and O of SO_4^{2-}) around a water molecule and of water molecules hydrogen-bonded into infinite chains. The orientation of the H–H vector in the unit cell agrees closely with the value derived from p.m.r. measurements. The O–H–O bonds from water molecules are of two kinds, those to sulphate O atoms (2·87 Å) and to H_2O molecules (2·94 Å).

$Na_2CO_3 . H_2O$. The larger Na^+ ion is usually surrounded octahedrally by six neighbours in oxy-salts. In this hydrate the cations are of two types, one-half being surrounded by one water molecule and five O atoms and the remainder by two water molecules and four O atoms of carbonate ions. Here also the H_2O molecule has four tetrahedral neighbours, 2 Na^+ ions and 2 O atoms of CO_3^{2-} ions.

$SrBr_2 . H_2O$. It is interesting to note that Sr^{2+} is not 8-coordinated in this hydrate as in $SrBr_2$ but has an environment more resembling that in $BaBr_2$, two of the nine nearest neighbours being H_2O molecules.

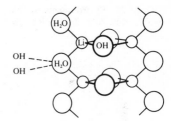

FIG. 15.23. Portion of one double chain in $LiOH . H_2O$ (diagrammatic).

Water and Hydrates

$Sr(OH)_2 . H_2O$. The structure of this hydrate (and the isostructural $Eu(OH)_2 . H_2O$) is of interest as an example of bicapped trigonal prismatic coordination (Sr–6 OH and 2 H_2O).

The classification adopted here is not, without undue elaboration, applicable to hydrates in which there are different degrees of hydration of some of the cations. For example, in $ZnCl_2 . 1\frac{1}{3} H_2O$[2] two-thirds of the Zn atoms are tetrahedrally coordinated by Cl in infinite chains of composition $(ZnCl_3)_n^{n-}$, and the remaining Zn atoms lie between these chains surrounded (octahedrally) by 4 H_2O and 2 Cl (of the chains), as shown in Fig. 15.24. Bond lengths found are: Zn–4 Cl, 2·28 Å, Zn–2 Cl, 2·60 Å and 4 H_2O, 2·03 Å.

Hydrates of 3d *halides and complex halides.* The structures described above suggest the following generalizations.

(i) $M(H_2O)_n$ groups share edges or faces but not vertices; this is true generally in hydrates.

(ii) There is complete hydration of M in $CoI_2 . 6 H_2O$ and in one form of $CrCl_3 . 6 H_2O$, but in a number of cases where there is sufficient water for complete hydration of M this does not occur (Class AII). Usually H_2O is displaced by X from the coordination group around M; sometimes both H_2O and X are partly coordinated to M and partly outside the cation coordination group, as in $[CrCl_2(H_2O)_4]Cl . 2 H_2O$. In a 4f trihalide, Cl is displaced (from a larger coordination group) in preference to H_2O, namely, in $[GdCl_2(H_2O)_6]Cl$.

(iii) If sharing of X and/or H_2O is necessary X is shared in preference to H_2O, as in numerous $MX_2 . 2 H_2O$.

(iv) Many of these compounds are polymorphic, and there is probably not very much difference between the stabilities of structures containing different types of cation coordination group, witness the three forms of $CrCl_3 . 6 H_2O$, the two forms of $MnCl_2 . 4 H_2O$, and the two forms of $RbMnCl_3 . 2 H_2O$, one containing the dimeric units of Fig. 15.16(d)[1] and the other the chains of Fig. 15.6(g).[2] Contrast the finite $[MnCl_4(H_2O)_2]^{2-}$ ions in $K_2MnCl_4 . 2 H_2O$[3] (Fig. 15.16(c)).

(v) We noted earlier that hydrated complex halides of the same formula type may have quite different structures, and that in $A_m(BX_n) . pH_2O$ there is not necessarily any H_2O attached to B; compare the coordination of Mn by 6 F in $K_2MnF_5 . H_2O$ with that of V by 5 F + H_2O in $(NH_4)_2(VF_5 . H_2O)$.

Hydrated acids and acid salts

Special interest attaches to the hydrates of acids and acid salts because of the possibility that some or all of the protons may be associated with the H_2O molecules to form ions H_3O^+ or more complex species. The acids HX form hydrates with the following numbers of molecules of water of crystallization:

HF	$\frac{1}{4}$	$\frac{1}{2}$	1		
HCl			1	2	3
HBr			1	2	3	4
HI				2	3	4

(2) AC 1970 B26 1679

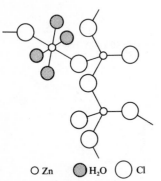

FIG. 15.24. Environments of the two kinds of Zn^{2+} ion in $ZnCl_2 . 1\frac{1}{3} H_2O$.

○ Zn   ● H_2O ○ Cl

(1) ACSc 1968 22 641
(2) ACSc 1967 21 889
(3) ACSc 1968 22 647

For references to structural studies see Table 15.6.

562

HF behaves quite differently from the other halogen acids, and nothing is yet known of the structures of its hydrates. The hydrates of HCl melt at progressively lower temperatures ($-15.4°$, $-17.4°$, and $-24.9°C$) and the crystal structures of all three are known. The i.r. spectrum of the monohydrate (at $-195°C$) indicates the presence of pyramidal H_3O^+ ions structurally similar to the isoelectronic NH_3 molecule.[1] This has been confirmed by the determination of the crystal structure. In corrugated 3-connected layers of the simplest possible type (Fig. 15.25(a)) each O or Cl atom has three pyramidal neighbours, so that the units are Cl^- and H_3O^+ ions, (i). The distance $O-H\cdots Cl$ is 2.95 Å and the interbond angle $O-H-O$ is $117°$.

(1) JCP 1955 **23** 1464

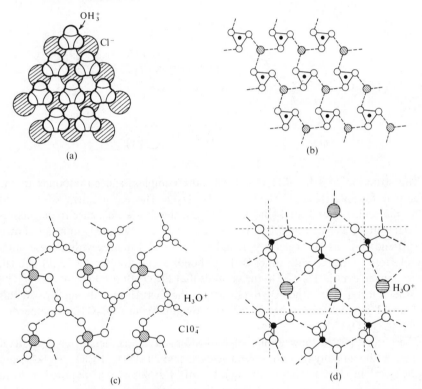

FIG. 15.25. Atomic arrangement in layers of the structures of (a) HCl . H_2O, (b) HNO_3 . H_2O, (c) $HClO_4$. H_2O, (d) H_2SO_4 . H_2O.

The angle subtended at O by two Cl^- is $110°$, the H atoms lying about 0.1 Å off the $O-H\cdots Cl$ lines. (The shortest distance between O and Cl of adjacent layers is 3.43 Å.)

The dihydrate of HCl contains $H_5O_2^+$ units, (ii), in which the central $O-H-O$ bond is extremely short, and the $O-H\cdots Cl$ bonds have a mean length of 3.07 Å (range 3.04–3.10 Å), appreciably longer than in $(H_3O)^+Cl^-$. The same units are found in the trihydrate, here hydrogen-bonded to $2 H_2O$ ($O-H-O$, 2.70 Å) and to

563

2Cl^- (O–H–Cl, 3·03 Å) instead of to 4Cl^- as in $(\text{H}_3\text{O})^+\text{Cl}^-$. The third water molecule has a normal tetrahedral environment, being surrounded by 2Cl^- (mean O–H–Cl, 3·10 Å) and $2 \text{H}_2\text{O}$ (mean O–H–O, 2·70 Å). These three hydrates are therefore

$$(\text{H}_3\text{O})^+\text{Cl}^-, \quad (\text{H}_5\text{O}_2)^+\text{Cl}^-, \quad \text{and} \quad (\text{H}_5\text{O}_2)^+\text{Cl}^- . \text{H}_2\text{O}$$

(i)

(ii)

(iii)

(iv)

The structure of $\text{HBr} . 4\,\text{H}_2\text{O}$ is even more complex, and corresponds to the structural formula $(\text{H}_9\text{O}_4)^+(\text{H}_7\text{O}_3)^+(\text{Br}^-)_2 . \text{H}_2\text{O}$. The environment of the odd water molecule is tetrahedral (Br^- and $3\,\text{H}_2\text{O}$), the shortest distance to neighbouring water molecules being 2·75 Å. The structures of the O_3 and O_4 units are shown at (iii) and (iv), the latter being pyramidal. The O–O distances within these units are all short compared with other O–H–O bonds in the structure (2·75 Å), though rather longer than in H_5O_2^+. On the grounds that the O–Br distance in the unit (iii) is rather shorter than other O–Br distances, the next shortest being 3·28 Å, this unit could alternatively be described as a pyramidal unit $(\text{H}_7\text{O}_3^+\text{Br}^-)$ somewhat similar to $(\text{H}_9\text{O}_4)^+$.

(2) TFS 1951 **47** 1261; JCP 1953 **21** 1421

The p.m.r. spectra show groups of 3 H atoms at the corners of an equilateral triangle in the monohydrates of several acids, including HNO_3, HClO_4, H_2SO_4, and H_2PtCl_6.[2] In $(\text{H}_3\text{O})^+(\text{NO}_3)^-$ the number of H atoms is that required to form three hydrogen bonds from each ion to three neighbours, and the structure (Fig. 15.25(b)) is a further example of the simple hexagonal net. In the form of $\text{HClO}_4 . \text{H}_2\text{O}$ stable at temperatures above $-30°\text{C}$ there is some rotational disorder of the H_3O^+ ions, but the low-temperature form has a structure of exactly the same topological type as that of $\text{HNO}_3 . \text{H}_2\text{O}$. As in $\text{HNO}_3 . \text{H}_2\text{O}$ there are three hydrogen bonds connecting each ion to its neighbours, so that one O of each ClO_4^- ion is not involved in hydrogen bonding (Fig. 15.25(c)). In the pyramidal H_3O^+ ion the angle H–O–H is 112° and the mean O–H–O bond length is 2·66 Å.

The system $\text{H}_2\text{SO}_4\text{–H}_2\text{O}$ is complex, freezing-point curves indicating the existence of six hydrates, with 1, 2, 3, 4, 6·5, and 8 H_2O. In $\text{H}_2\text{SO}_4 . \text{H}_2\text{O}$ the number of H atoms is sufficient for four hydrogen bonds per unit of formula. The crystals consist of SO_4H^- and H_3O^+ ions arranged in layers (Fig. 15.25(d)) and linked by hydrogen bonds, three from H_3O^+ and five from SO_4H^-. Since there are equal numbers of the two kinds of ion SO_4H^- is necessarily hydrogen-bonded to $3 H_3O^+$ and $2 SO_4H^-$; there are no hydrogen bonds between the layers. The three hydrogen bonds from H_3O^+ are disposed pyramidally (interbond angles, 101°, 106°, and 126°) and have lengths 2·54, 2·57, and 2·65 Å. Note that it is O_2 and not O_3 that forms two hydrogen bonds. In contrast to the monohydrate, which is $(HSO_4)^-(H_3O)^+$, the dihydrate is $(SO_4)^{2-}(H_3O)_2^+$. Each H_3O^+ is hydrogen-bonded to three different sulphate ions forming a 3D framework in which SO_4^{2-} forms six hydrogen bonds, two from two of the O atoms and one from each of the others (O—H—O, 2·52–2·59 Å, S—O, 1·474 Å).

In $HClO_4 . 2 H_2O$, with a ratio of one proton to two water molecules, we find $H_5O_2^+$ units as in $HCl . 2 H_2O$. The O—H—O bond length in this unit is very similar to that in $HCl . 2 H_2O$ (2·424 Å) but the geometry of this ion is somewhat variable. It has a staggered configuration in $HClO_4 . 2 H_2O$, nearly eclipsed in $HCl . 3 H_2O$, and an intermediate shape in $HCl . 2 H_2O$.

Very few structural studies have been made of hydrated acid salts. In $Na_3H(CO_3)_2 . 2 H_2O$ the proton is associated with a pair of anions, $(CO_3—H—CO_3)$, and the water is present as normal H_2O molecules. On the other hand, in salts such as $(Coen_2Br_2)Br . HBr . 2 H_2O$ (p. 312) and $ZnCl_2 . \frac{1}{2} HCl . H_2O$ in which there is the same ratio of $H^+ : H_2O$, namely, 1 : 2, there are $H_5O_2^+$ ions similar to those in certain acid hydrates. As noted in Chapter 3, the addition of one Cl^- to two $ZnCl_2$ permits the formation of a 3D network of composition Zn_2Cl_5 built of $ZnCl_4$ tetrahedra sharing three vertices (compare one form of P_2O_5). This framework forms around, and encloses, the $H_5O_2^+$ ions, so that the structural formula is $(Zn_2Cl_5)^-(H_5O_2)^+$. The structure approximates to a hexagonal closest packing of Cl in which one-sixth of these atoms have been replaced by the rather more bulky $H_5O_2^+$ ions.

TABLE 15.6

Proton–water complexes in hydrates

Complex	$H^+ : H_2O$ ratio	Hydrate	Reference
H_3O^+	1 : 1	$HCl . H_2O$	AC 1959 **12** 17
		$HNO_3 . H_2O$	AC 1951 **4** 239
		$HClO_4 . H_2O$ (low)	AC 1962 **15** 18
		$HClO_4 . H_2O$ (high)	AC 1961 **14** 318
	But also in	$H_2SO_4 . H_2O$	AC 1968 **B24** 299
		$H_2SO_4 . 2 H_2O$	JCP 1969 **51** 4213
$H_5O_2^+$	1 : 2	$HCl . 2 H_2O$	AC 1967 **23** 966
		$HClO_4 . 2 H_2O$	JCP 1968 **49** 1063
		$(ZnCl_2)_2 . HCl . 2 H_2O$	AC 1970 **B26** 1544
	But also in	$HCl . 3 H_2O$	AC 1967 **23** 971
$H_7O_3^+$	1 : 3	$HBr . 4 H_2O$	JCP 1968 **49** 1068
$H_9O_4^+$	1 : 4		

Water and Hydrates

We summarize these hydrate structures in Table 15.6, which shows that when the ratio $H^+ : H_2O$ is $1 : 1$ H_3O^+ is formed; in the mono- and dihydrates of H_2SO_4 this ratio is achieved by ionizing to $(HSO_4)^-(H_3O)^+$ and $(SO_4)^{2-}(H_3O)_2^+$ respectively. If the ratio $H^+ : H_2O$ is $1 : 2$ $H_5O_2^+$ is formed in a number of dihydrates (both of acids and acid salts), but with higher water contents the systems become more complex.

(3) AC 1953 6 385

In contrast to the hydrates we have been describing, the dihydrate of oxalic acid contains H_2O molecules rather than H_3O^+ ions.[3] Each H_2O molecule is hydrogen-bonded to 3 O atoms of $(COOH)_2$ molecules, but one of the three hydrogen bonds is a short one (2·49 Å) to a CO group while the other two are longer bonds (2·88 Å) to OH groups, which are linked in this way to two water molecules.

The location of H atoms of hydrogen bonds; residual entropy

We noted in Chapter 8 that in many hydrogen bonds the H atom lies to one side of the line joining the hydrogen-bonded atoms, and we have seen that this is also true in ice-I and hydrates such as $CuCl_2 . 2 H_2O$; in the latter the angle $O–H\cdots Cl$ is

(1) AC 1962 15 353

$164°$. Very similar values are found for $O–H\cdots F$ in $FeSiF_6 . 6 H_2O$[1] (which is

(2) JCP 1962 36 50

isostructural with $Ni(H_2O)_6SnCl_6$ and consists of a CsCl-like packing of $Fe(H_2O)_6^{2+}$ and SiF_6^{2-} ions) and in $CuF_2 . 2 H_2O$.[2] The angle subtended at the O atom of H_2O by the two O atoms to which it is hydrogen-bonded in oxy-salts and acids ranges from the value of $84°$ for $(COOH)_2 . 2 H_2O$ to around $130°$ (possibly larger).

(3) PRS A 1958 246 78

In chrome alum, $[K(H_2O)_6][Cr(H_2O)_6](SO_4)_2$[3] this angle is less than $104\frac{1}{2}°$ for both types of H_2O molecule, while in $CuSO_4 . 5 H_2O$[4] all but one of these angles

(4) PRS A 1962 266 95

are greater than H–O–H, the largest being $130°$ and the mean value $119°$. Probably the only conclusion to be drawn from these angles is that they show that the H_2O molecule retains its normal shape in hydrates and that hydrogen bonds are formed in directions as close to the O–H bond directions as are compatible with the packing requirements of the other atoms in the crystal.

Like ice-I certain hydrates possess residual entropy owing to the possibility of alternative arrangements of H atoms. The crystal structure of $Na_2SO_4 . 10 H_2O$ has already been described as having 8 H_2O in octahedral chains $[Na(H_2O)_4]_n$ and 2 H_2O not attached to cations. For the H_2O molecules in the chains there is no alternative orientation, but the hydrogen bonds involving the remaining 2 H_2O are arranged in circuits such that there are two ways of arranging the H atoms in any given ring so as to place one on each O–H–O bond. If the choice of arrangement of the H atoms in different rings throughout the crystal is a random one this would account for the observed entropy of 2 ln 2 per mole.

It is not feasible to refer here to all the studies of the positions of H atoms in hydrates. A comparison of n.m.r. with n.d. results on twelve hydrates[5] showed

(5) JCP 1966 45 4643

that there is generally very close agreement between the two methods. Many references are given in a paper[6] summarizing n.m.r. determinations of proton

(6) AC 1968 B24 1131

positions in hydrates.

We have now considered a number of hydrates of different kinds and have seen that the behaviour of the water molecules in these structures can be accounted for quite satisfactorily. The surrounding of small positive ions with a layer of H_2O molecules does not lead to structures as simple as might be expected, whereas the corresponding ammines usually have the simple structures predicted from the radius ratio. The structural differences between hydrates and ammines are due to the fact that the neighbours of NH_3 do not, as do those of H_2O, have to conform with any tetrahedral nature of the molecule. We shall conclude this chapter with a few remarks on ammines and their relation to hydrates.

Ammines and hydrates

Ammines are prepared by the action of ammonia on the salt, using either the gas or liquid and the anhydrous salt, or crystallizing the salt from ammoniacal solution. In preparing ammines of trivalent cobalt from cobaltous salts, oxidation of the cobalt to the trivalent state must accompany the action of ammonia, and this may be accomplished by passing air through the ammoniacal solution. No general statement can be made about the stability of ammines as compared with hydrates, for, as we shall see, the term ammine covers a very large number of compounds which differ greatly in constitution and stability. As extremes we may cite the very unstable ammines of lithium halides and the stable cobalt- and chrom-ammines.

A comparison of the empirical formulae of hydrates and ammines leads to the following conclusions:

(i) *Oxy-salts*.   Many sulphates crystallize with odd numbers of molecules of water of crystallization, e.g. the vitriols $MSO_4 . 7 H_2O$ (M = Fe, Co, Ni, Mn, Mg, etc.), $MnSO_4 . 5 H_2O$, and $CuSO_4 . 5 H_2O$, $MgSO_4 . H_2O$, and many others. Many of the salts which form heptahydrates also form hexahydrates, but the higher hydrate is usually obtained if the aqueous solution is evaporated at room temperature. However, ammines of these salts contain even numbers of NH_3 molecules, usually four or six, for example, $NiSO_4 . 6 NH_3$; note particularly the ammine of cupric sulphate, $Cu(NH_3)_4SO_4 . H_2O$.

The nitrates of many divalent metals crystallize with $6 H_2O$ (Fe, Mn, Mg, etc.), but a few nitrates have odd numbers of water molecules, e.g. $Fe(NO_3)_3 . 9$ (and 6) H_2O, $Cu(NO_3)_2 . 3 H_2O$. Where the corresponding ammine is known it has an even number of NH_3, as in $Cu(NH_3)_4(NO_3)_2$. There are many other series of hexahydrated salts—perchlorates, sulphites, bromates, etc.—and although in such cases the ammines often contain $6 NH_3$, the structures of ammine and hydrate are quite different—compare, for example, the structure of $[Co(H_2O)_6](ClO_4)_2$, described on p. 556, with the simple fluorite structure of $[Co(NH_3)_6](ClO_4)_2$.

The reason for the non-existence of ammines containing *large odd* numbers of NH_3 molecules is presumably that some of these molecules would have to be accommodated between the cation coordination groups and the anions (compare the environment of the fifth H_2O in $CuSO_4 . 5 H_2O$ or the seventh H_2O in $NiSO_4 . 7 H_2O$). The NH_3 molecule does not have the same hydrogen-bond-forming capability as H_2O, particularly to other NH_3 molecules.

567

(ii) *Halides.* In contrast to the oxy-salts, halides rarely crystallize with an odd number of water molecules. When they do, the number is usually not greater than the coordination number of the metal (see Table 15.5 and note, for example, $FeBr_2 \cdot 9\,H_2O$). In a hydrated oxy-salt containing water in excess of that required to complete the coordination groups around the metal ions additional water molecules can be held between the $M(H_2O)_n$ groups and the oxy-ions by means of O–H–O bonds. It seems, however, that additional water molecules are very weakly bonded between $M(H_2O)_n$ groups and Cl^- ions. The electrostatic bonding is sufficiently strong in the system (a) in a crystal such as $[Mg(H_2O)_6]Cl_2$ but much weaker in (b), where H_2O is bonded between a hydrated metal ion and Cl^-. For example, the

$$\overset{\diagdown}{\underset{\diagup}{>}}M{-}OH_2{-}Cl^-$$

$$\overset{\diagdown}{\underset{\diagup}{>}}M{-}OH_2{-}H_2O{-}Cl^-$$

(a)   (b)

hydrates $MgCl_2 \cdot 8\,H_2O$ and $MgCl_2 \cdot 12\,H_2O$ are stable only at low temperatures (p. 553). The ammines of those halides which contain *odd* numbers of H_2O in their hydrates contain *even* numbers of NH_3 molecules, e.g. $ZnCl_2 \cdot H_2O$ but $ZnCl_2 \cdot 2\,NH_3$. Although the ammines and hydrates of halides usually contain the same numbers of molecules of NH_3 and H_2O respectively:

$NiCl_2 \cdot 6\,H_2O,$ $NiCl_2 \cdot 6\,NH_3$ $NiSO_4 \begin{cases} 7\,H_2O \text{ but } 6\,NH_3 \\ 6\,H_2O \end{cases}$

                            contrast

$CuBr_2 \cdot 4\,H_2O,$ $CuBr_2 \cdot 4\,NH_3$ $CuSO_4 \cdot 5\,H_2O$ $4\,NH_3 \cdot H_2O$

$CuCl_2 \cdot 2\,H_2O,$ $CuCl_2 \cdot 2, 4,$ or $6\,NH_3,$ $Cu(NO_3)_2 \cdot 3\,H_2O$ $4\,NH_3$

             $(CuCl_2 \cdot 4\,NH_3 \cdot 2\,H_2O$

             from solution)

the crystal structures of the two sets of compounds are different, as has already been pointed out for the pairs $Al(NH_3)_6Cl_3$ and $Al(H_2O)_6Cl_3$, and $Mg(NH_3)_6Cl_2$ and $Mg(H_2O)_6Cl_2$.

From this brief survey we see that the ammines $MX_x \cdot n\,NH_3$ with n greater than the coordination number of M are likely to be rare and that n is commonly 4 or 6, depending on the nature of the metal M. In these ammines the NH_3 molecules form a group around M, and these groups $M(NH_3)_n$ often pack with the anions into a simple crystal structure. The linking of the NH_3 to M in the ammines of the more electropositive elements such as Al and Mg is of the ion-dipole type. The stability of these compounds is of quite a different order from that of the very stable ammines of trivalent cobalt and chromium (see Chapter 27). Magnetic measurements show that the (much stronger) M–NH_3 bonds in the ammine complexes of Co^{III} (and also of Ir^{III}, Pd^{IV}, Pt^{IV}) would be described as 'covalent' or, in ligand-field language, 'strong-field' bonds. In the earlier Periodic Groups the more covalent ammines of the B subgroup metals are more stable than, and quite different in structure from, the ammines of the A subgroup metals. We may conclude, therefore, that the formal similarity between hydrates and ammines is limited to those ammines in which the

M–NH$_3$ bonds are essentially electrostatic, and that even in these cases there are significant differences between the crystal structures of the two compounds MX$_x$. n NH$_3$ and MX$_x$. n H$_2$O. This latter point has already been dealt with; the former is best illustrated by a comparison of the ammines and hydrates of the salts of the metals of Groups I and II of the Periodic Table.

Of the alkali-metal chlorides only those of Li and Na form ammines, MX . y NH$_3$ (y_{max} = 6), and these are very unstable. Similarly LiCl forms a pentahydrate, NaCl a dihydrate (stable only below 0°C), while KCl, RbCl, and CsCl crystallize without water of crystallization. The formation of ammines thus runs parallel with the formation of hydrates. As the size of the ion increases, from Li to Cs, the charge remaining constant, the ability to polarize H$_2$O or NH$_3$ molecules and so attach them by polar bonds decreases. In the B subgroup we find an altogether different relation between hydrates and ammines. Ammino-compounds are commonly formed by salts which form no hydrates and they resemble in many ways the cyanido- and other covalent complexes. The only halide of silver which forms a hydrate is the fluoride, but AgCl and the other halides form ammines. Again, the nitrate and sulphate are anhydrous but form ammines, AgNO$_3$. 2 NH$_3$ and Ag$_2$SO$_4$. 4 NH$_3$ respectively. In the latter, the structure of which has been determined, and presumably also in the former, there are linear ions (NH$_3$–Ag–NH$_3$)$^+$ exactly analogous to the ion (NC–Ag–CN)$^-$ in KAg(CN)$_2$. Gold also forms ammines, e.g. [Au(NH$_3$)$_4$](NO$_3$)$_3$, and several of the ammines of copper salts have already been mentioned.

In the second Periodic Group we find a similar difference between the stability and constitution of ammines in the A and B subgroups. Magnesium forms a number of ammines, but the tendency to form ammines falls off rapidly in the alkaline-earths. The compound CaCl$_2$. 8 NH$_3$, for example, readily loses ammonia. The degree of hydration of salts also decreases from Ca to Ba except for the octahydrates of the peroxides and hydroxides (SrO$_2$. 8 H$_2$O, Ba(OH)$_2$. 8 H$_2$O), which presumably owe their stability to hydrogen bond formation between the O$_2^{2-}$ or OH$^-$ ions and the water molecules. In the B subgroup, however, as in IB, the ammines and hydrates are structurally unrelated. We find

$$\text{ZnCl}_2 . \text{H}_2\text{O} \quad \text{but} \quad \text{ZnCl}_2 . 2\,\text{NH}_3$$
$$\text{and}$$
$$\text{CdCl}_2 . 2\,\text{H}_2\text{O} \quad \text{and} \quad \text{CdCl}_2 . 2\,\text{NH}_3$$

but these two diamminodichlorides have quite different structures, in which the bonds M–NH$_3$ must possess appreciable covalent character. These structures have already been noted in the section on amminohalides in Chapter 10.

16

Sulphur, Selenium, and Tellurium

We describe in this chapter the structural chemistry of sulphur, selenium, and tellurium, excluding certain groups of compounds which are discussed elsewhere. Metal oxysulphides and sulphides (simple and complex) are described in Chapter 17, sulphides of non-metals are included with the structural chemistry of the appropriate element, and alkali-metal hydrosulphides are grouped with hydroxides in Chapter 14.

The stereochemistry of sulphur

The principal bond arrangements are set out in Table 16.1 which has the same general form as the corresponding Table 11.1 for oxygen. For the sake of simplicity we have assumed sp^3 hybridization in H_2S though the interbond angle ($92°$) is nearer to that expected for p bonds. For other molecules SR_2 see Table 16.2

TABLE 16.1

The stereochemistry of sulphur

No. of σ pairs	Type of hybrid	No. of lone pairs	Bond arrangement	Examples
3	sp^2	0	Plane triangular	SO_3
		1	Angular	SO_2
4	sp^3	0	Tetrahedral	SO_2Cl_2, $O_2S(OH)_2$
		1	Trigonal pyramidal	$SOCl_2$, $S(CH_3)_3^+$
		2	Angular	SH_2, SCl_2, S_8
5	$sp^3d_{z^2}$	0	Trigonal bipyramidal	SOF_4
		1	'Tetrahedral'	SF_4
6	$sp^3d_\gamma^2$	0	Octahedral	SF_6, S_2F_{10}

(p. 575). The stereochemistry of S is more complex than that of O because S utilizes d in addition to s and p orbitals. Although covalencies of two and four are more usual than three the latter is found in SO_3, sulphoxides, $R_2S=O$, sulphonium salts, $(R_3S)X$, and in a number of oxy-ions. The multiple character of most S–O bonds does not complicate the stereochemistry because the interbond angles are determined by the number of σ bonds and lone pairs, as shown in Table 16.1.

Elementary sulphur, selenium, and tellurium

At high temperatures the vapours of all three elements consist of diatomic molecules (in which the bond lengths are: S=S, 1.89 Å, Se=Se, 2.19 Å, and Te=Te, 2.61 Å), but at lower temperatures and in solution (e.g. in CS_2) S and Se form S_8 and Se_8 molecules respectively. Owing to the insolubility of Te there is no evidence for the formation of a similar molecule by this element.

Sulphur

Sulphur is remarkable for the number of solid forms in which it can be obtained. These include at least four well-known 'normal' polymorphs stable under atmospheric pressure, numerous high-pressure forms[1] of which one is the fibrous form made from 'plastic' sulphur, so-called amorphous forms characterized by small solubility in CS_2, and coloured forms produced by condensing the vapour on surfaces cooled to the temperature of liquid nitrogen. Some at least of the 'amorphous' sulphur preparations (e.g. milk of sulphur) give X-ray diffraction lines indicative of some degree of crystallinity, and it is perhaps preferable to use the term μ-sulphur rather than amorphous sulphur.[2] The term 'normal' polymorph used above refers to the forms now known to consist of S_6 or S_8 molecules. In recent years cyclic molecules S_n ($n = 7, 9, 10, 12, 18, 20$) have been prepared by special methods (see below).

In the orthorhombic form of sulphur stable at ordinary temperatures the unit of structure is the cyclic S_8 molecule (crown configuration)† in which the bond length is 2.06 Å, interbond angle $108°$, and dihedral angle $99°$.[3] At $95.4°C$ orthorhombic S_α changes into a monoclinic form which normally reverts fairly rapidly to rhombic S but can be kept at room temperature for as long as a month if it is pure and has been annealed at $100°C$. This form, S_β, consists of S_8 molecules with the same configuration as in rhombic S and is remarkable for the fact that two-thirds of the molecules are in fixed orientations but the remainder are randomly oriented.[4] This randomness leads to a residual entropy of $\frac{1}{3} R \ln 2$ per mole of S_8 molecules (0.057 e.u./g atom), which agrees reasonably well with the observed value (0.045 e.u./g atom). A second monoclinic form, Muthmann's S_γ, also consists of crown-shaped S_8 molecules.[5] This form is isostructural with $S_6(NH)_2$; for other molecules $S_{8-n}(NH)_n$ see 'sulphides of nitrogen', Chapter 16.

The rhombohedral S prepared by Engel in 1891 by crystallizing from toluene rapidly breaks down into a mixture of amorphous and orthorhombic S. The crystals consist of S_6 molecules (chair configuration) with S–S, 2.06 Å, interbond angle $102°$, and dihedral angle $75°$.[6] A mass spectrometric study shows that Engel's sulphur vaporizes as S_6 molecules,[7] whereas the vapour of rhombic S at temperatures just above the boiling point consists predominantly of S_8 molecules with the same configuration as in the crystal.[8] (The vapour from FeS_2 at $850°C$ consists of S_2 molecules, the same units as those in the crystal,[9] and mass spectrometric studies of solid solutions S–Se show peaks due to S_7Se, S_6Se_2, S_5Se_3, and possibly other molecules.[10])

By the interaction of H_2S_8 and S_4Cl_2 in CS_2 a further crystalline form of

† For models of the most symmetrical non-planar forms of 8-membered rings see MSIC p. 34.

(1) IC 1970 **9** 1973, 2478

(2) JACS 1957 **79** 4566

(3) AC 1965 **18** 562, 566

(4) JACS 1965 **87** 1395

(5) AC 1974 **B30** 1396

(6) JPC 1964 **68** 2363
(7) JCP 1964 **40** 287

(8) JACS 1944 **66** 818
(9) JCP 1963 **39** 275

(10) JINC 1965 **27** 755

(11) AnC 1966 **78** 1020, 1021
(11a) AnC 1970 **82** 390; NW 1973 **60** 49, 300

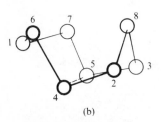

(a)

(b)

FIG. 16.1. (a) The cyclic S_{12} molecule, (b) the Se_8^{2+} cation.

(12) JCP 1963 **39** 3158
(13) JCP 1960 **33** 774
(14) JCP 1969 **51** 348
(15) JPC 1966 **70** 3528, 3531, 3534
(16) TFS 1963 **59** 559
(17) JCP 1959 **31** 1598

Sulphur, Selenium, and Tellurium

sulphur has been prepared[11] which consists of cyclic S_{12} molecules (S–S, 2·06 Å, interbond angle, $106\frac{1}{2}°$, and dihedral angle 107°). Six of the atoms of the ring, those marked a in Fig. 16.1(a), are coplanar, three b lie above and three c below the plane of the hexagon of a atoms. By reactions such as those between sulphanes and chlorophanes or between $(C_5H_5)_2TiS_5$ and SCl_2, S_2Cl_2, or SO_2Cl_2 other new cyclic S_n molecules have been made in which $n = 7, 9, 10, 18,$ or 20.[11a] For cyclic S_n^{2+} cations see p. 573.

In sulphur vapour dissociation to S atoms is appreciable only at temperatures above 1200°C, even at a pressure of 0·1 mm. If the vapour at 500°C and 0·1–1·0 mm, consisting essentially of S_2 molecules, is condensed on a surface cooled to the temperature of liquid nitrogen, sulphur condenses as a purple solid which on warming reverts to a mixture of crystalline and amorphous sulphur. The paramagnetism of this form supports the view that it consists of S_2 molecules. (The spectra of the blue solutions of S in various molten salts (LiCl–KCl, KSCN) have been interpreted in terms of S_4 molecules.[12]) The vapour from *liquid* sulphur at lower temperatures (200–400°C) can be condensed to a green solid which is not a physical mixture of the purple and yellow forms; the vapour from *solid* sulphur condenses to yellow S. An e.s.r. study of these coloured forms of sulphur suggests the presence of at least two types of trapped sulphur radicals.[13]

Fibrous or plastic sulphur results from quenching the liquid from temperatures above 300°C and stretching. It contains some S_γ, the monoclinic S_8 form noted above, which is extracted by dissolving in CS_2. The crystallinity of the residual S_ψ can be improved by heating the stretched fibres at 80°C for 40 hours. An X-ray study of such fibres shows them to consist of an approximately hexagonal close-packed assembly of equal numbers of left- and right-handed helical molecules in which the S–S bond length is 2·07 Å, the interbond angle 106°, and dihedral angle 95°.[14] The helices are remarkable for containing 10 atoms in the repeat unit of 3 turns; contrast Se and Te (below). This fibrous sulphur is apparently identical with one of the high-pressure forms.

The properties of liquid S are complex, and although there has been much theoretical and practical work on this subject it seems that no existing theory accounts satisfactorily for all the properties.[15] On melting sulphur forms a highly mobile liquid (S_λ) consisting of cyclic S_8 molecules, but at 159°C an extremely rapid and large increase in viscosity begins, which reaches a maximum at around 195°C (S_μ), above which temperature the viscosity falls off. The specific heat also shows a sudden rise at 159°C. The typical λ-shaped curve is due to sudden polymerization, and estimates have been made from e.s.r. and static magnetic susceptibility of the average chain length, which ranges from 10^6 atoms at about 200°C to 10^3 at 550°C.[16] The X-ray diffraction effects from liquid sulphur show that a S atom has an average of two nearest neighbours at approximately the same distance as in crystalline S.[17]

Selenium

The ordinary form of commerce is vitreous Se which, if heated and cooled slowly, is converted into the grey 'metallic' form, the stable polymorph. From CS_2 solution two red monoclinic polymorphs can be crystallized; both revert spontaneously to metallic Se, the β form much more rapidly than α-Se. This is an unusual case of polymorphism, for both α-[1] and β-Se[2] consist of cyclic Se_8 molecules (Se–Se, 2·34 Å, interbond angle 105° in both) and the intermolecular distances also are very similar. The metallic form consists of infinite helical chains with Se–2 Se, 2·37 Å and interbond angle 103°.[3] The atomic radial distribution curve of vitreous Se has been determined from both X-rays and neutrons, and shows that this form contains the same helical chains as the metallic form (peaks at 2·33, 3·7, and 5·0 Å).[4]

(1) AC 1972 **B28** 313
(2) AC 1953 **6** 71

(3) IC 1967 **6** 1589

(4) JCP 1967 **46** 586

Tellurium

The normal form stable at atmospheric pressure is the hexagonal (metallic) form isostructural with metallic selenium, in which Te–Te is 2·835 Å and interbond angle, 103·2°.[1] Two high-pressure forms have been recognized, one stable at 40–70 kbar (structure not known) and one above 70 kbar. The latter is apparently isostructural with β-Po, having a structure resulting from the compression of a simple cubic structure along the [111] axis.[2] The nearest neighbours are

(1) AC 1967 **23** 670

(2) JCP 1965 **43** 1149

$$\text{Te} \begin{cases} 6\ \text{Te at } 3\cdot00\ \text{Å} \\ 6\ \text{Te} \quad\ 3\cdot72 \\ 2\ \text{Te} \quad\ 3\cdot82 \end{cases} \text{compare Te} \begin{cases} 2\ \text{Te at } 2\cdot84\ \text{Å} \\ 4\ \text{Te} \quad\ 3\cdot47 \end{cases} \text{in hexagonal Te}$$

Cyclic S, Se, and Te cations

A recent development in the chemistry of these elements has been the demonstration of the existence and in some cases the structures of ions S_n^{2+} etc. These elements dissolve in solvents such as concentrated H_2SO_4, oleum, and HSO_3F, in which they are oxidized to polyatomic ions which can be isolated in salts such as the pale-yellow $S_4(SO_3F)_2$ and the dark-blue $S_8(AsF_6)_2$. (The formulae were originally doubled to be consistent with the diamagnetism of the compounds.) Elementary Se can be oxidized in HSO_3F or oleum to give intensely yellow or green solutions from which salts such as $Se_4(SO_3F)_2$, $Se_4(HS_2O_7)_2$, and $Se_4(S_4O_{13})$ may be isolated, and similar compounds of Te have been made. Cyclic cations are also formed in compounds such as $Se_8(AlCl_4)_2$, grown by vapour-phase transport, $Te_4(AlCl_4)_2$, and $Te_4(Al_2Cl_7)_2$.

At present the structures of two kinds of cyclic cation have been established by X-ray studies. Ions Se_4^{2+} and Te_4^{2+} are planar and almost exactly square. The first has been studied in $Se_4(HS_2O_7)_2$.[1] The Se–Se bond length (2·28 Å) is slightly shorter than in the element (2·34 Å) and the anions are hydrogen-bonded chains of $S_2O_7^{2-}$ ions, as in $NO_2(HS_2O_7)$, p. 656. The Te_4^{2+} ion is the cation in $Te_4(AlCl_4)_2$ and $Te_4(Al_2Cl_7)_2$,[2] the latter salt being of interest also as containing $Al_2Cl_7^-$ ions with a staggered configuration. The Te–Te bond length (2·67 Å) is appreciably smaller than in elementary Te (2·84 Å); the bond angle (90°) is, of course, smaller than in Te (103·2°). The corresponding S–N ring is the planar S_2N_2 ring in $S_2N_2(SbCl_5)_2$.

(1) IC 1971 **10** 2319

(2) IC 1972 **11** 357

(3) IC 1971 **10** 2781
(4) IC 1971 **10** 1749

The 8-ring cations in $S_8(AsF_6)_2$[3] and $Se_8(AlCl_4)_2$[4] have very similar shapes (Fig. 16.1(b)), intermediate between the crown shape of S_8 and that of S_4N_4:

$$S_8 \ (exo\text{-}exo) : S_8^{2+} \ (exo\text{-}endo) : S_4N_4 \ (endo\text{-}endo).$$

These rings differ from the crown ring of S_8 in having one very short distance across the centre of the ring and two others which are also much shorter than would correspond to van der Waals contacts:

Distance between atoms in Fig. 16.1 (b)	S_8^{2+}	Se_8^{2+}
4–5	2·86 Å	2·85 Å
2–3 ⎱ 6–7 ⎰	2·97	3·33

It would appear that as electrons are removed from S_8 (48 valence electrons) bonds are formed across the ring; the 44-electron isoelectronic S_4N_4 ring is shown since S_4^{2+} is formed in preference to S_8^{4+}. In S_8^{2+} the S—S bond length is apparently similar (2·04 Å) to that in S_8; the bond angles range from 92 to 104° (compare

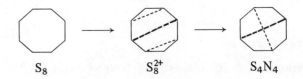

$$S_8 \qquad\qquad S_8^{2+} \qquad\qquad S_4N_4$$

108° in S_8). In Se_8^{2+} also the bonds are normal single bonds (2·32 Å); the mean interbond angle is 98° except at Se^1 and Se^8 (90°).

Molecules SR_2, SeR_2, and TeR_2

Molecules in which R is an atom or group forming a single bond to S, Se, or Te are non-linear. Bond lengths (rounded off to two decimal places) and interbond angles are summarized in Table 16.2. For S the approximate range 100–110° includes most examples except those in certain cyclic molecules such as C_2H_4S (a) and C_4H_4S (b), and in H_2S. In compounds SeR_2 the Se bond angles lie within the range 97–104° except in H_2Se (91°), and in Te compounds the bond angles are close to 100°.

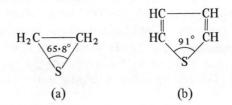

(a)   (b)

TABLE 16.2

Structures of molecules SR_2, SeR_2, *and* TeR_2

Molecule	R—S—R	S—R	Method	Reference
C_2H_4S (a)	66°	1·82 Å	m.w.	JCP 1951 **19** 676
C_4H_4S (b)	91°	1·74	e.d.	JACS 1939 **61** 1769
CH_3SH	99°	1·35 (S—H)	m.w.	PR 1953 **91** 464
		1·81 (S—C)		
C_2H_5SH	113°	1·80	e.d.	PR 1949 **75** 1319
SH_2	92°	1·33	m.w.	JCP 1967 **46** 2139
SF_2	98°	1·59	m.w.	Sc 1969 **164** 950
SCl_2	100°	2·00	R	JCP 1955 **23** 972
$S(CH_3)_2$	99°	1·80	m.w.	JCP 1961 **35** 479
$S(CF_3)_2$	106°	1·83	e.d.	TFS 1954 **50** 452
$S(CN)_2$	98°	1·70	m.w., i.r.	JCP 1965 **43** 3423
$S(SiH_3)_2$	97°	2·14	e.d.	ACSc 1963 **17** 2264
$S(GeH_3)_2$	99°	2·21	e.d.	JCS A 1970 315
	R—Se—R	Se—R		
SeH_2	91°	1·46	m.w.	JMS 1962 **8** 300
$Se(CH_3)_2$	98°	1·98	e.d.	JACS 1955 **77** 2948
$Se(SiH_3)_2$	97°	2·27	e.d.	ACSc 1968 **22** 51
$Se(SCN)_2$	101°	2·21	X	JACS 1954 **76** 2649
$Se(CF_3)_2$	104°	1·96	e.d.	TFS 1954 **50** 452
$Se[SeP(C_2H_5)_2Se]_2$	104°	2·35	X	ACSc 1969 **23** 1398
	R—Te—R	Te—R		
$TeBr_2$	98°	2·51	e.d.	JACS 1947 **69** 2102
$Te[S_2P(OCH_3)_2]_2$	98°	2·44	X	ACSc 1966 **20** 24
$Te[SeP(C_2H_5)_2S]_2$	101°	2·50	X	ACSc 1969 **23** 1389

Cyclic molecules

Bond angles C—S—C close to 100° are found in cyclic molecules based on strain-free 6-rings. All those studied have the chair conformation; examples include 1,4-dithiane,[1] 1,3,5-trithiane,[2] and hexathia-adamantane.[3] The last compound, $S_6(CH)_4$, is structurally similar to $(CH_2)_6N_4$ and $(CH_2)_6(CH)_4$. Trithiane forms many metal complexes; the structure of $S_3(CH_2)_3 . AgClO_4 . H_2O$ is mentioned in Chapter 3 (p. 90).

(1) AC 1955 **8** 91
(2) AC 1969 **B25** 1432
(3) AK 1956 **9** 169

The halides of sulphur, selenium, and tellurium

These compounds form a very interesting group, not only as regards their chemical properties but also from the stereochemical standpoint. There has in the past been considerable doubt as to which compounds of a particular formula-type actually exist, and further work is still required on, for example, the lower fluorides of tellurium. The existence of the following compounds can be regarded as reasonably certain; some have been well known for many years:

(1) AC 1956 9 295

S_2F_2	S_2Cl_2	S_2Br_2		Se_2Cl_2	Se_2Br_2				
SF_2	SCl_2			$SeCl_2$	$SeBr_2$				
SF_4	SCl_4		SeF_4	$SeCl_4$	$SeBr_4$	TeF_4	$TeCl_2$	$TeBr_2$	
SF_6			SeF_6			TeF_4	$TeCl_4$	$TeBr_4$	TeI_4[1]
S_2F_{10}	S_nCl_2	S_nBr_2				Te_2F_{10}			Te_nI_n[1]

SF_2 is an unstable species which was studied in a microwave cell but not isolated; its structure (Table 16.2) was determined from the moments of inertia of $^{32}SF_2$ and $^{34}SF_2$. Chlorides S_nCl_2 up to S_5Cl_2, consisting of S_n chains terminated by Cl atoms, can be prepared from SCl_2 and H_2S_n ($n = 1, 2, 3$)[2] and those up to S_8Cl_2 have been made by the reaction

(2) ZaC 1969 **364** 241

$$ClSSCl + HS_nH + ClSSCl \rightarrow ClS_{n+4}Cl + 2\ HCl.$$

By the action of HBr the chlorides can be converted into S_nBr_2 (n from 2 to 8).

In contrast to the numerous compounds formed with the other halogens, TeI_4 and Te_nI_n are the only known binary compounds of iodine with these elements. Whereas Te forms all four tetrahalides, S forms only SF_4 and SCl_4. The latter compound forms a yellow solid melting at $-30°C$ to a red liquid which is a mixture of SCl_2 and Cl_2 and evolves the latter gas, and it is completely dissociated into SCl_2 + Cl_2 at ordinary temperatures. The dielectric constant of the liquid of composition SCl_4 rises from 3 to 6 on freezing, and it has been suggested that the solid is a salt, $(SCl_3)Cl$. $SeCl_4$ forms a colourless solid which sublimes at $196°C$ to a vapour which is completely dissociated to $SeCl_2 + Cl_2$. Like SCl_4 it apparently exists only in the solid state, and here also the Raman spectrum suggested the ionic structure $(SeCl_3)Cl$.[3] However, the Raman data now require reinterpretation in terms of the tetrameric species found in crystalline $SeCl_4$, $TeCl_4$, and $TeBr_4$ (see later). $SeBr_4$ also dissociates on sublimation (to $SeBr_2 + Br_2$) but in CCl_4 solution to an equilibrium mixture of $SeBr_2$, Se_2Br_2, and Br_2. The conductivity of solutions in polar solvents is apparently not due to species containing Se^{IV} (e.g. $SeBr_3^+$) but to ionization of the lower bromides.[4] The vapour density of $TeCl_4$ corresponds to the formula $TeCl_4$ up to $500°C$, and this is the only one of these tetrachlorides which can be studied in the vapour state.

(3) IC 1967 6 903

(4) IC 1969 8 759

Halides SX_2 etc.
Structural details of the angular molecules SCl_2 and $TeBr_2$ are included in Table 16.2.

Halides S_2X_2 etc.
In addition to the isomer with the structure XS–SX (which is enantiomorphic) a topological isomer S–SX$_2$ is possible. S_2F_2 exhibits this type of isomerism. Both isomers are formed when a mixture of dry AgF and S is heated in a glass vessel *in vacuo* to $120°C$. The isomer $S=SF_2$ is pyramidal[5] and very similar geometrically

(5) JACS 1963 85 2028

to OSF_2 (Fig. 16.2(a) and (b)); for OSF_2 see Table 16.3. The isomer FS–SF has the H_2O_2 type of structure with an unexpectedly short S–S bond (1·89 Å, compare 1·97 Å in S_2Cl_2) and S–F abnormally long (1·635 Å). For details of the structures of S_2F_2, S_2Cl_2, and S_2Br_2 see Table 16.6 (p. 593).

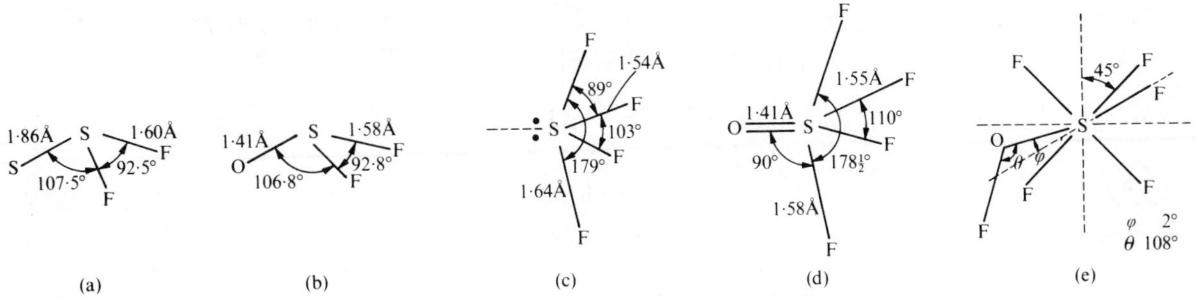

FIG. 16.2. Structures of molecules: (a) S_2F_2, (b) SOF_2, (c) SF_4, (d) SOF_4, (e) SF_5OF.

Halides SX_4 etc.; molecules R_2SeX_2 and R_2TeX_2

As already noted, few of the tetrahalides can be studied in the vapour state; SF_4 and molecules R_2SeX_2 and R_2TeX_2 have a trigonal bipyramidal configuration in which one of the equatorial bond positions is occupied by the lone pair of electrons. The stereochemistry of these molecules has been discussed in Chapter 6. A feature of these molecules is that the axial bonds are collinear to within a few degrees and longer than the equatorial ones in a molecule SX_4 (or longer than the covalent radius sum in a molecule R_2SeX_2). The lone pair of electrons in SF_4 is used to form the fifth bond in OSF_4 and these molecules have rather similar configurations (Fig. 16.2):

(6) JCP 1963 **39** 3172
(7) JCP 1969 **51** 2500

		SF_4[6]	OSF_4[7]
axial	S–F	1·643 Å	1·583 Å
equatorial	S–F	1·542	1·550

(8) JMS 1968 **28** 454

A m.w. study of SeF_4[8] shows that this molecule has the same general shape as SF_4 with a similar difference between the equatorial and axial bond lengths: Se–F, 1·68 Å (equatorial), 1·77 Å (axial); F–Se–F (equatorial) 100·55°, 169·20° (axial). For compounds with NbF_5 containing the ion SeF_3^+ see p. 600. It is not profitable to discuss the early e.d. results for $TeCl_4$ vapour[8a] since it was assumed that all four bonds were of equal length (2·33 Å); a molecular configuration of the SF_4 type would be consistent with the fact that $TeCl_4$ has a dipole moment (2·54D) in benzene solution.

(8a) JACS 1940 **62** 1267

Other molecules with the same general shape as SF_4 include $Se(C_6H_5)_2Cl_2$ and $Se(C_6H_5)_2Br_2$, $Te(CH_3)_2Cl_2$ and $Te(C_6H_5)_2Br_2$. In all these molecules the halogen atoms occupy the axial positions; for details see p. 601.

Crystalline $SeCl_4$, $TeCl_4$, and $TeBr_4$ are isostructural, and the structural unit is a tetramer of the same general type as $[Pt(CH_3)_3Cl]_4$ which has been studied in

(9) IC 1971 **10** 2795

detail for $TeCl_4$.[9] Four Te and four Cl atoms are arranged at alternate vertices of a cube and three additional Cl are attached to each Te. The Te atoms are displaced outwards along 3-fold axes so that the coordination of Te is distorted octahedral:

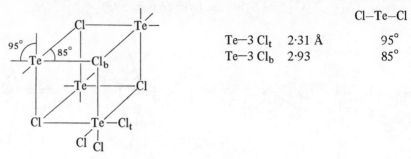

		Cl–Te–Cl
Te–3 Cl_t	2·31 Å	95°
Te–3 Cl_b	2·93	85°

The limiting case would be separation into $(TeCl_3)^+$ and Cl^- ions. Comparing the Te–Cl distances with those in $(TeCl_3)(AlCl_4)$ and $(TeCl_6)^{2-}$, namely: 3 Cl at 2·28 Å and 3 Cl at 3·06 Å, and 6 Cl at 2·54 Å, it would appear that the tetramer Te_4Cl_{16} is close to $(TeCl_3)^+Cl^-$. (An alternative description of the crystal is a distorted cubic closest packing of Cl atoms with Te atoms in one-quarter of the octahedral holes, but displaced from their centres.)

Crystalline TeF_4 is not a molecular crystal but consists of chains of square pyramidal TeF_5 groups sharing a pair of *cis* F atoms (Fig. 16.3(a)) similar to the isoelectronic $(SbF_4)_n^{n-}$ ion in $NaSbF_4$. The Te atom is 0·3 Å below the basal plane

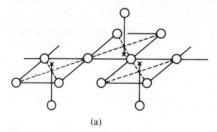

(a)

FIG. 16.3. The structure of crystalline TeF_4: (a) linear molecule, (b) bond lengths.

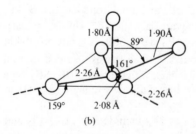

(b)

(10) JCS A 1968 2977

and the bond lengths are as shown in Fig. 16.3(b).[10] The simplest view of this structure would be to regard the coordination of Te as octahedral (valence group 2, *10*), the lone pair occupying the sixth bond position. However, the bridge is unsymmetrical (Fig. 16.3(b)) and if only the four nearest F are counted as neighbours (shaded circles) the configuration around Te would be somewhat similar to the GeF_4 sub-unit in GeF_2, that is, trigonal bipyramidal with an equatorial lone pair. (The three nearest F atoms form a pyramidal group with average interbond angle $86\frac{1}{2}°$.)

Hexahalides

Electron diffraction or microwave studies have shown that the molecules SF_6,[11] SeF_6,[11] TeF_6,[12] and also SF_5Cl[13] and SF_5Br[14] have octahedral configurations—regular in the case of the hexafluorides:

(11) TFS 1963 **59** 1241
(12) ACSc 1966 **20** 1535
(13) TFS 1960 **56** 1732
(14) JCP 1963 **39** 596

S—F	1·56 Å	S—Cl	2·03 Å	S—Br	2·19 Å
Se—F	1·69				
Te—F	1·82				

Disulphur decafluoride, S_2F_{10}

The e.d. data for this halide are consistent with a model consisting of two $-SF_5$ groups joined in the staggered configuration so that the central S—S bond is collinear with two S—F bonds, with all interbond angles 90° (or 180°), S—F, 1·56 Å, and S—S, 2·21 Å.[15] The length of the S—S bond may be due to repulsion of the F atoms.

(15) TFS 1957 **53** 1545, 1557

Oxyhalides of S, Se, and Te

The known oxyhalides are set out below, the salient points being the absence (at present) of iodides and of any well-characterized compounds of Te, and the existence of a number of more complex compounds of S peculiar to that element.

	SOF_2		$SeOF_2$	SO_2F_2	SeO_2F_2
	Cl_2		Cl_2	Cl_2	
	Br_2		Br_2	FCl	
	SOFCl			FBr	
SOF_4	SO_3F_2	$S_2O_5F_2$	$S_2O_6F_2$	$S_3O_8F_2$	$Te_3O_2F_{14}$
SOF_6	$F_5S \cdot OCl$		$S_2O_2F_{10}$	$S_3O_8Cl_2$	

Details of the structures of the pyramidal molecules of thionyl halides (SOX_2) and $SeOF_2$ and of the tetrahedral molecules of sulphuryl halides (SO_2X_2), which have been studied as gases, are included in Tables 16.3 and 16.4. The structures of $SeOCl_2$ (and also $SeOF_2$) have been determined in a number of adducts (p. 600).

The structure of OSF_4 is compared with that of SF_4 in Fig. 16.2. Apart from the expansion of the equatorial F—S—F bond angle from 103° (mean of e.d. and m.w. values) to 110° the molecules are very similar in shape (S—O, 1·41 Å). The S—F bond lengths were compared earlier.

The constitution of $SF_5 \cdot OF$ as *pentafluorosulphur hypofluorite* has been confirmed by an e.d. study (Fig. 16.2(e)).[1] Bond lengths are: S—F, 1·53, S—O, 1·64, and O—F, 1·43 Å in the essentially octahedral molecule. The values of the angles are: $\phi = 2°$, $\theta = 108°$.

(1) JACS 1959 **81** 5287

Disulphur decafluorodioxide, $S_2F_{10}O_2$, is a derivative of hydrogen peroxide, to which it is related by replacing H by $-SF_5$. The dihedral angle is similar to that in H_2O_2 ($\approx 107°$) and the angle S–O–O is 105°. Approximate bond lengths found by e.d. are: O–O, 1·46, S–O, 1·66, and S–F, 1·56 Å.[2]

Peroxydisulphuryl difluoride, $S_2O_6F_2$, is also apparently of this type. It is a colourless liquid prepared by the reaction of F_2 with SO_3 at a temperature above 100°C, and its chemical reactions together with the i.r. and n.m.r. spectra suggest the formula $O_2FS–O–O–SO_2F$,[3] (a).

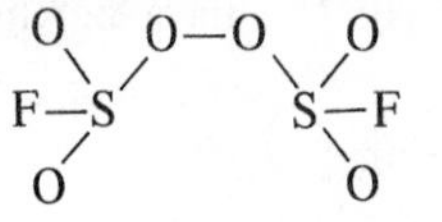

(a) 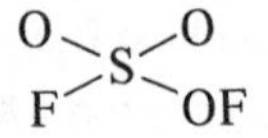 (b) 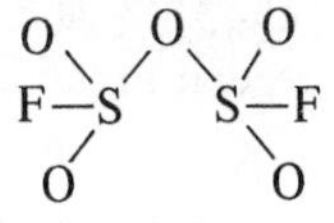 (c)

The structures suggested for SO_3F_2, (b), and $S_2O_5F_2$, (c), (and for $Te_3O_2F_{14}$[4]) have not been confirmed by structural studies.

The oxides of S, Se, **and** Te

Disulphur monoxide, S_2O

This very unstable oxide is formed when a glow discharge is passed through a mixture of sulphur vapour and SO_2 and in other ways. A microwave study[1] shows that the S–O bond is essentially a double bond as in SO_2, and the bond angle, (a), is very similar to that in SO_2. The molecule can therefore be formulated as at (b).

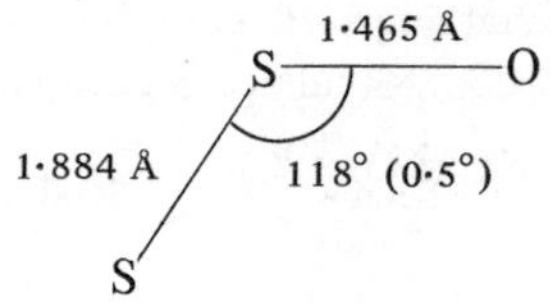

(a) 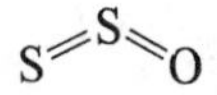 (b)

Sulphur monoxide, SO

The term 'sulphur monoxide' was originally applied (erroneously) to an equimolecular mixture of S_2O and SO_2. The oxide SO is a short-lived radical produced, for example, by the combination of O atoms produced in an electric discharge with solid sulphur. By pumping the products rapidly through a wave-guide cell the microwave spectrum can be studied: S–O, 1·48 Å, $\mu = 1·55$D (compare SO_2, 1·59D).[2]

Sulphur dioxide, SO_2

The structure of this molecule has been studied in the crystalline and vapour states by spectroscopic and diffraction methods, most recently by i.r. in a solid Kr matrix.[3] All these studies give a bond angle close to 119·5° and the bond length 1·43 Å. The dipole moment in the gas phase is noted above.

(2) JACS 1954 **76** 859

(3) JACS 1957 **79** 513

(4) JCS 1956 3454

(1) JMS 1959 **3** 405

(2) JCP 1964 **41** 1413

(3) JCP 1969 **50** 3399

Sulphur dioxide can behave as a ligand in transition-metal complexes in two ways: (i) the S atom forms a single bond M–S to the metal atom, the three bonds from S being arranged pyramidally as in $Ir(P\phi_3)_2Cl(CO)SO_2$,[4] or (ii) the S atom uses its lone pair to form the σ bond to the metal which then forms a π-bond

(4) IC 1966 **5** 405

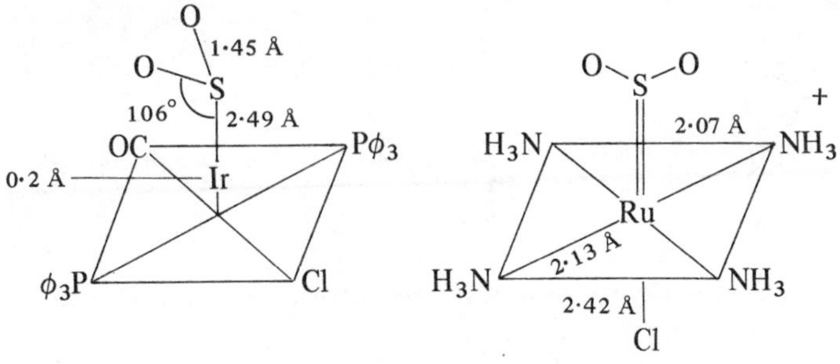

making M=S a multiple bond, as in $[Ru^{II}(NH_3)_4SO_2Cl]Cl$.[5] In (i) the metal atom is situated 0·2 Å above the base of the pyramid. In (ii) the atoms O_2S–Ru–Cl are coplanar and the SO_2 ligand has the same structure as the free molecule. Compare the similar behaviour of NO (p. 654).

(5) IC 1965 **4** 1157

Selenium dioxide, SeO_2

Whereas crystalline sulphur dioxide consists of discrete SO_2 molecules, crystalline SeO_2[6] is built of infinite chains of the type

(6) JACS 1937 **59** 789

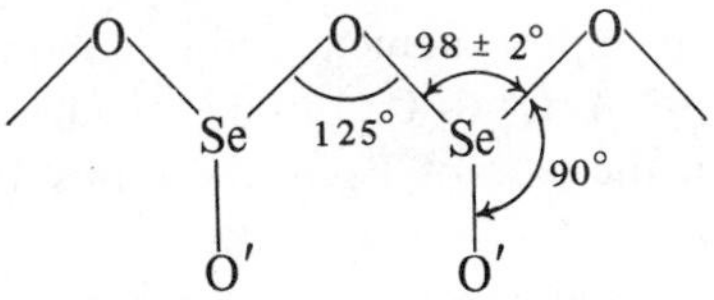

in which Se–O, 1·78 (0·03) Å and Se–O′, 1·73 (0·08) Å. In SeO_2 molecules in the vapour, studied by electron diffraction, Se–O, 1·61 Å, (angle not determined).[7]

(7) JACS 1938 **60** 1309

Tellurium dioxide, TeO_2

There are two crystalline forms, a yellow orthorhombic form (the mineral tellurite) and a colourless tetragonal form (paratellurite). There is 4-coordination of Te in both forms, the nearest neighbours being arranged at four of the vertices of a trigonal bipyramid, suggesting considerable covalent character of the Te–O bonds. Tellurite has a layer structure[8] in which TeO_4 groups form edge-sharing pairs (Fig. 16.4(a)) which form a layer, (b), by sharing the remaining vertices. The short Te–Te distance, 3·17 Å (compare the shortest, 3·74 Å, in paratellurite) may be connected with its colour. In paratellurite[9] very similar TeO_4 groups share all vertices to form a 3D structure of 4 : 2 coordination in which the O bond angle is 140°. (Figure 16.4(c).) (There are Te_2O_6 groups very similar to those in tellurite in the mineral denningite, $(Mn, Ca, Zn)Te_2O_6$.)

(8) ZK 1967 **124** 228

(9) ACSc 1968 **22** 977

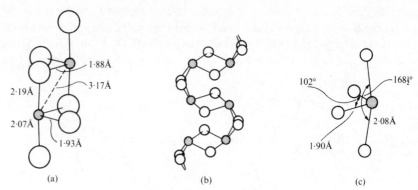

FIG. 16.4. Crystalline TeO_2: (a) and (b) tellurite, (c) paratellurite.

These group VIB dioxides form an interesting series:

	C.N.	Type of structure
SO_2	2	molecular
SeO_2	3	chain
TeO_2	4	layer and 3D
PoO_2	8	3D (fluorite)

(10) JACS 1938 **60** 2360
(11) AC 1967 **22** 48
(12) AC 1954 **7** 764
(13) JCS 1957 2440

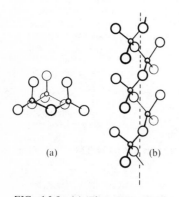

FIG. 16.5. (a) The cyclic S_3O_9 molecule in the orthorhombic form of sulphur trioxide, (b) the infinite chain molecule in the asbestos-like form.

Sulphur trioxide, SO_3

The molecule SO_3 has zero dipole moment and e.d. shows it to be a symmetrical planar molecule (S–O, 1·43 Å, O–S–O, 120°).[10] Like P_2O_5 SO_3 has a number of polymorphs, and as in the case of P_2O_5 the forms differ appreciably in their stability towards moisture.

The orthorhombic modification consists of trimers (Fig. 16.5(a)). The bond length in the ring is 1·62 Å, and there is an unexpected (and unexplained) difference between the lengths of the axial and equatorial bonds, 1·37 Å and 1·43 Å respectively.[11] The O bond angle is 121·5°.

In the asbestos-like form[12] the tetrahedral SO_4 groups are joined to form infinite chain molecules (Fig. 16.5(b)), with S–O in the chain 1·61 Å and to the unshared O, 1·41 Å. In addition to these two well-defined forms of known structure the existence of others has been postulated to account for the unusual physical properties of solid and liquid sulphur trioxide. The properties of the liquid are still incompletely understood and the system is obviously complex—compare sulphur itself, to which SO_3 is topologically similar in that both can form chain and ring polymers, and units being –S– and –O(SO_2)O–.

Other oxides of sulphur which have been described include the blue-green S_2O_3[13] (formed by the reaction of S with liquid SO_3), and SO_4, produced by the action of an electric discharge on a mixture of O_2 or O_3 with SO_2 or SO_3. The existence of S_2O_7 seems less certain.

Selenium trioxide, SeO_3

Pure SeO_3 may be prepared by heating K_2SeO_4 with SO_3, which gives a liquid mixture of SeO_3 and SO_3 from which the more volatile SO_3 may be distilled. Like SO_3 it is at least dimorphic. Although the cyclic modification forms mixed

crystals with S_3O_9 containing 25–100 molar per cent SeO_3 it consists[14] of tetrameric molecules with a configuration very similar to that of $(PNCl_2)_4$.

A compound Se_2O_5 formed by the thermal decomposition of SeO_3 is presumably $Se^{IV}Se^{VI}O_5$ since it is decomposed by water to an equimolecular mixture of selenious and selenic acids.

(14) AC 1965 **18** 795

Tellurium trioxide, TeO_3

The trioxide formed by dehydrating H_6TeO_6 with concentrated H_2SO_4 in an oxygen atmosphere is a relatively inert substance, being unattacked by cold water or dilute alkalis, though it can act as a strong oxidizing agent towards, for example, hot HCl. There is a second, even more inert, form which has been crystallized under pressure.[15a] The first form is isostructural with FeF_3, that is, it is a 3D structure formed from TeO_6 octahedra sharing all vertices.[15b]

(15a) ZaC 1968 **362** 98
(15b) CR 1968 **266** 277

Other oxides of tellurium include Te_2O_5 and Te_4O_9; there is also a compound $TeO_2(OH)$. In Te_2O_5[16] $Te^{VI}O_6$ octahedra form layers by sharing four vertices (Te–O, 1·92 Å) and the Te^{IV} atoms are situated between the layers bonded to two O of the layers and to two additional O atoms (Te–2 O 1·90 Å, Te–2 O, 2·08 Å). The coordination of Te^{IV} is similar to that in α- and β-TeO_2 (p. 581). The structure $(Te^{VI}O_4)Te^{IV}O$, has therefore considerable similarity to that of Sb_2O_4 (p. 720), $(Sb^VO_4)Sb^{III}$, there being an additional O atom for each Te^{IV} atom between the layers.

(16) AC 1973 **B29** 643

Oxy-ions and molecules formed by S, Se, **and** Te

Our chief concern will be with the compounds of sulphur since these are the most numerous and most important. We shall, however, include those species formed by Se and Te which are structurally similar to S compounds, for example, XO_3^{2-} and XO_4^{2-} ions. We deal in a later section with ions and molecules with stereochemistries peculiar to Se and Te such as complex oxy-ions of Te. First we consider pyramidal and tetrahedral molecules and ions, then ions formed from tetrahedral SO_4 groups by sharing O atoms, and finally the ions $S_2O_4^{2-}$ and $S_2O_5^{2-}$. Thionates are included in the next main section with compounds containing S_n chains. Per-ions are included in Chapter 11, the crystal structures of acids (H_2SeO_3, $C_6H_5SeO . OH$, H_2SO_4, SO_3NH_3) in Chapter 8, and $Te(OH)_6$ in Chapter 14.

Pyramidal ions and molecules
These include the following types:

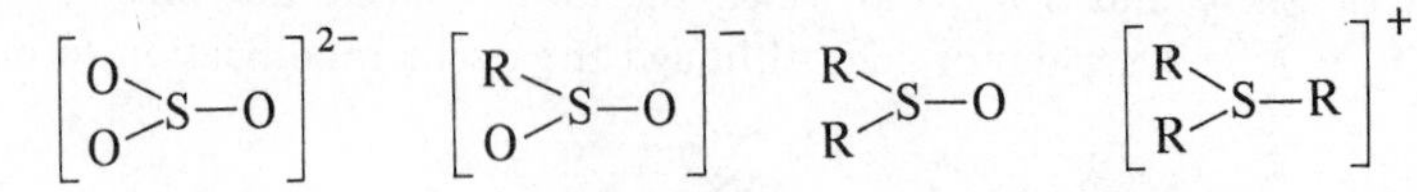

$$\left[\begin{array}{c}O\\O\end{array}\!\!\!>\!\!S\!-\!O\right]^{2-} \quad \left[\begin{array}{c}R\\O\end{array}\!\!\!>\!\!S\!-\!O\right]^{-} \quad \begin{array}{c}R\\R\end{array}\!\!\!>\!\!S\!-\!O \quad \left[\begin{array}{c}R\\R\end{array}\!\!\!>\!\!S\!-\!R\right]^{+}$$

$$\qquad\quad (a) \qquad\qquad\quad (b) \qquad\qquad (c) \qquad\qquad (d)$$

(a) The sulphite, selenite, and tellurite ions have all been shown to have the shape of flat pyramids with the structures given in Table 16.3.

As a ligand SO_3^{2-} bonds to transition metals through S in compounds such as $Coen_2(NCS)SO_3$,[1] $Pd(SO_3)(NH_3)_3$,[2] and $Na_2[Pd(SO_3)_2(NH_3)_2] \cdot 6\,H_2O$.[3]

(b) Substituted sulphite ions include SO_2F^- (salts of the alkali metals being prepared by the direct action of SO_2 on MF), and $O_2S \cdot CH_2OH^-$. Sodium hydroxy-methanesulphinate (dihydrate) is used as a reducing agent in vat dyeing. It is made by reducing a mixture of $NaHSO_3$ and formaldehyde by zinc dust in alkaline solution. The ion has the structure:

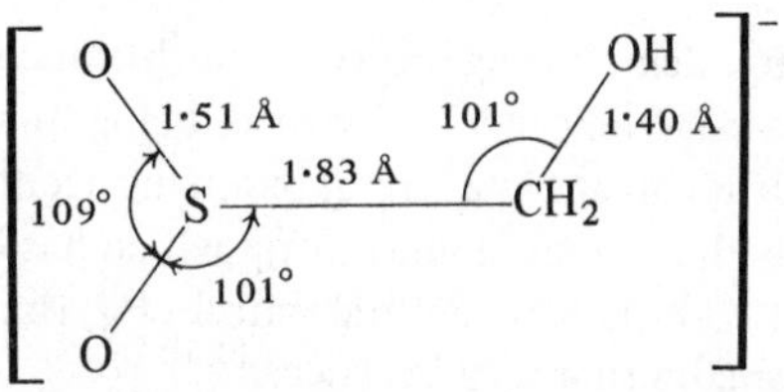

$$\left[\begin{array}{c} O \xrightarrow{1 \cdot 51\ \text{Å}} \\ 109° \quad S \xrightarrow{1 \cdot 83\ \text{Å}} CH_2 \xrightarrow[101°]{OH}_{1 \cdot 40\ \text{Å}} \\ O \quad 101° \end{array}\right]^{-}$$

(c) The simplest members of this class are the thionyl halides SOX_2 and analogous Se compounds (Table 16.3): for adducts of $SeOF_2$ and $SeOCl_2$ see p. 600. The optical activity of unsymmetrical sulphoxides such as $OS(CH_3)(C_6H_4COOH)$ and structural studies of $(CH_3)_2SO$ and $(C_6H_5)_2SO$ have confirmed the pyramidal shape of these molecules.

(d) The non-planar configuration of sulphonium ions was deduced from the optical activity of unsymmetrical ions such as $[S(CH_3)(C_2H_5)(CH_2COOH)]^+$ and has been confirmed by X-ray studies of the salts $[S(CH_3)_3]I$ and $[S(CH_3)_2C_6H_5]ClO_4$ (Table 16.3).

Tetrahedral ions and molecules
By the addition of one O atom to the four kinds of pyramidal ions and molecules of the previous section we have:

$$\left[\begin{array}{cc}O\!\!&\!\!O\\O\!\!&\!\!O\end{array}\!\!S\right]^{2-} \quad \left[\begin{array}{cc}R\!\!&\!\!O\\O\!\!&\!\!O\end{array}\!\!S\right]^{-} \quad \begin{array}{cc}R\!\!&\!\!O\\R\!\!&\!\!O\end{array}\!\!S \quad \left[\begin{array}{cc}R\!\!&\!\!R\\R\!\!&\!\!O\end{array}\!\!S\right]^{+}$$

$$\qquad\quad (e) \qquad\qquad\quad (f) \qquad\qquad (g) \qquad\qquad (h)$$

(1) AC 1969 **B25** 946
(2) JCS A 1967 1194
(3) JCS A 1969 260

TABLE 16.3

Pyramidal molecules and ions

(a)

Salt	M—O	O—M—O	Reference
$(SO_3)^{2-}NH_4Cu, Na_2$	1·51 Å	106°	ACSc 1968 **22** 581
			ACSc 1969 **23** 2253
$(SeO_3)^{2-}Mg . 6 H_2O$	1·69	101°	AC 1966 **20** 563
$(TeO_3)^{2-}Cu . 2 H_2O$	1·88	≈100°	AC 1962 **15** 698
$(TeO_3)_4^{2-}Cl . Fe_2H_3$	1·89	96°	AC 1969 **B25** 1551
$(TeO_3)^{2-}Ba . H_2O$	1·86	99° (mean)	ACSc 1971 **25** 3037

(b)

Molecule or ion	S—O	S—C	O—S—O	Reference
$O_2S(CH_2OH)^-$	1·51 Å	1·83 Å	109°	JCS 1962 3400
$O_2S . C(NH_2)_2$	1·49	1·85	112°	AC 1962 **15** 675
$HO . OS(CH_3)$	1·50 (S—O)	1·79	108°	AC 1969 **B25** 350
	1·60 (S—OH)			

(c)

Molecule	R—S—O	R—S—R	S—O	S—R	Method	Reference
F_2SO	107°	93°	1·41 Å	1·59 Å	m.w.	JACS 1954 **76** 850
Cl_2SO	106°	114°(a)	1·45	2·07	e.d.	JACS 1938 **60** 2360
Br_2SO	108°	96°	1·45(b)	2·27	e.d.	JACS 1940 **62** 2477
$(CH_3)_2SO$	107°	97°	1·53	1·80	X	AC 1966 **21** 12
$(C_6H_5)_2SO$	106°	97°	1·47	1·76	X	AC 1957 **10** 417
	R—Se—O	R—Se—R	Se—O	Se—R		
$SeOF_2$	105°	92°	1·58	1·73	m.w.	JMS 1968 **28** 461

(d)

Salt	S—C	C—S—C	Reference
$S(CH_3)_3I$	1·83 Å	103°	ZK 1959 **112** 401
$[S(CH_3)_2(C_6H_5)]ClO_4$	1·82	103°	AC 1964 **17** 465

(*a*) Angles less than 106° not tested. (*b*) Assumed.

Replacement of one O by S in (e) and (f) gives respectively the thiosulphate ion, (i), and ions such as the methane thiosulphonate ion, (j):

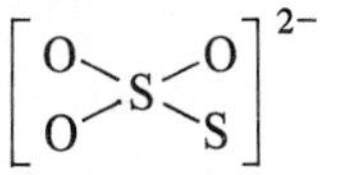

(i) 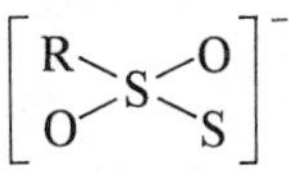   (j)

(e) The regular tetrahedral shape of the sulphate ion has been demonstrated in many salts, and numerous selenates are isostructural with the sulphates. The existence of the tetrahedral TeO_4^{2-} ion has not yet been proved by a

(1) IC 1972 **11** 1157
(2) CR 1971 **272C** 212

(3) JCS A 1970 1665

(4) CR 1971 **273C** 852

(5) MH 1969 **100(6)** 1809

(6) JCS A 1971 3074

structural study. In the wolframite structure of $MgTeO_4$[1] and in the inverse spinel structure assigned to Li_2TeO_4[2] (on the basis of the similarity of its X-ray powder photograph to that of $Zn(LiNb)O_4$) there is octahedral coordination of Te(VI) as in numerous other complex oxides. The i.r. spectrum of Na_2TeO_4[3] (made by dehydrating $Na_2H_4TeO_6$) shows that this salt is structurally different from the K, Rb, and Cs salts, but can hardly be regarded as proving the existence of TeO_4^{2-} ions in the latter,[4] for there are numerous edge-sharing octahedral structures for compounds A_2BX_4. Salts $M_2H_4TeO_6$, which could be formulated $M_2TeO_4 . 2H_2O$, contain octahedral $[TeO_2(OH)_4]^{2-}$ ions, as in the crystalline Ag salt.[5]

Examples of SO_4^{2-} behaving as a bridging ligand include the finite ion $[(en)_2Co(NH_2)(SO_4)Co(en)_2]^{3+}$ (p. 963) and the infinite chain ion in $K_2MnF_3SO_4$[6]:

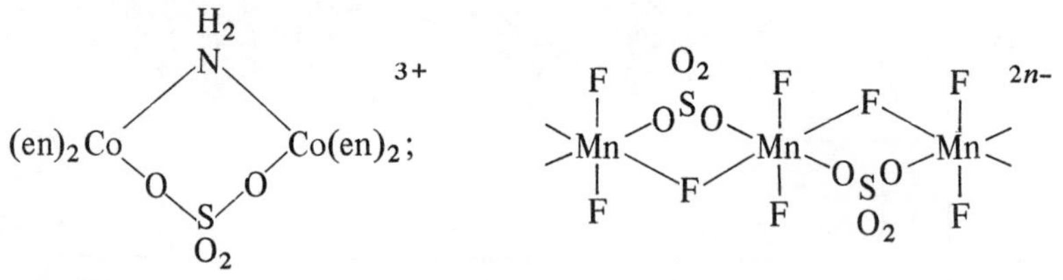

(f) Sulphonate ions RSO_3^- have been studied in a number of salts. The very stable alkali fluorosulphonates, KSO_3F etc., are isostructural with the perchlorates. In the ethyl sulphate ion in $K(C_2H_5O . SO_3)$ and in the bisulphate ion in salts such as $NaHSO_4 . H_2O$ and $KHSO_4$ the bond S—OR or S—OH is appreciably longer than the other three S—O bonds (Table 16.4). There are minor variations in these bond lengths depending on the environment of the O atoms:

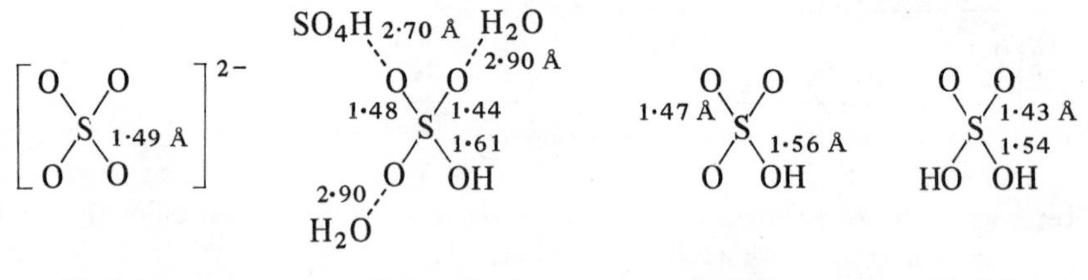

The structure of the aminosulphonate (sulphamate) ion, $SO_3NH_2^-$, has been accurately determined in the K salt, (I). Since the angles S—N—H and H—N—H are both $110°$ the N bonds are not coplanar, as suggested by an earlier X-ray study. We may include here two zwitterions. A n.d. study of sulphamic acid, $^-O_3S . NH_3^+$, shows that the molecule has approximately the staggered configuration, (II). Of special interest are the N—H···O bonds (2·93–2·98 Å) in which the N—H bond

586

TABLE 16.4

Tetrahedral molecules and ions

(e)

Ion	M—O	Reference
$(SO_4)^{2-}$	1·49 Å	AC 1965 **18** 717
$(SeO_4)^{2-}$	1·65	AC 1970 **B26** 436, 1451

(f)

Salt or zwitterion	S—O	S—X	O—S—O	O—S—X	
$(SO_3F)^-K$	1·43 Å	1·58 Å	113°	106°	JCS A 1967 2024
$(SO_3OH)^-Na \cdot H_2O$	1·44, 1·48	1·61	113°	106°	AC 1965 **19** 426
$(SO_3OH)^-K$	1·47	1·56	113°	106°	AC 1964 **17** 682
$(SO_3OC_2H_5)^-K$		see text			AC 1958 **11** 680
$(SO_3NH_2)^-K$	1·46	1·67	113°	110°	AC 1967 **23** 578
$(SO_3NHOH)^-K$	1·46	1·69	113°	$108\frac{1}{2}°$	AC 1966 **21** 819
$^-(O_3SONH_3)^+$	1·45	1·68	116°	102°	IC 1967 **6** 511
		$(S—ONH_3)$			
$^-(O_3SNH_3)^+$	1·44	1·76	115°	103°	AC 1960 **13** 320

(g)

Molecule	R—S—R	O—S—O	S—O	S—R	
F_2SO_2	96°	124°	1·41 Å	1·53 Å	JCP 1957 **26** 734
Cl_2SO_2	111°	120°	1·43	1·99	JACS 1938 **60** 2360
$(NH_2)_2SO_2$	112°	$119\frac{1}{2}°$	1·39	1·60	AC 1956 **9** 628
$(OH)_2SO_2$	104°	119°	1·43	1·54	AC 1965 **18** 827

(h)–(j)

Salt	S—O	S—S	S—C	
$[(CH_3)_3SO]^+ClO_4$	1·45 Å	—	1·78 Å	AC 1963 **16** 676, 883
$(S_2O_3)^{2-}Mg \cdot 6 H_2O$	1·47	2·01 Å	—	AC 1969 **B25** 2650
$(CH_3S_2O_2)^-Na \cdot H_2O$	1·45	1·98	1·77	ACSc 1964 **18** 619

makes angles of 7°, 13°, and 14° with the N—O vectors; there are three hydrogen bonds from each NH_3^+. Hydroxylamine-*O*-sulphonic acid has the structure (III).

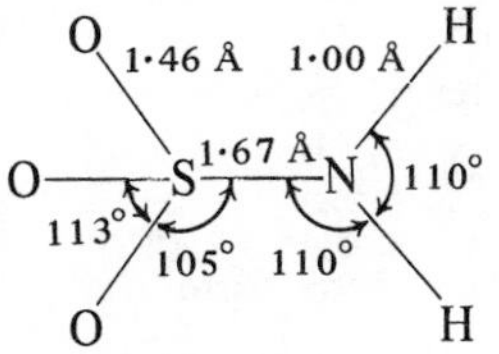

(I) 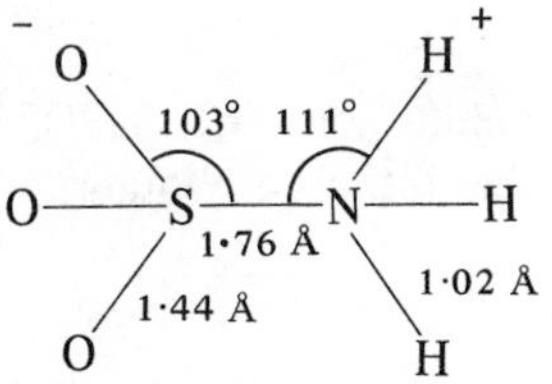   (II) 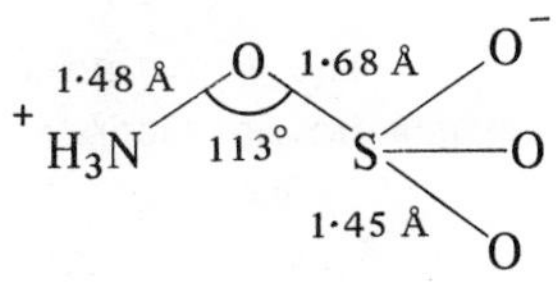   (III)

Progressive substitution of SO_3^- for H in NH_3 gives the series of ions

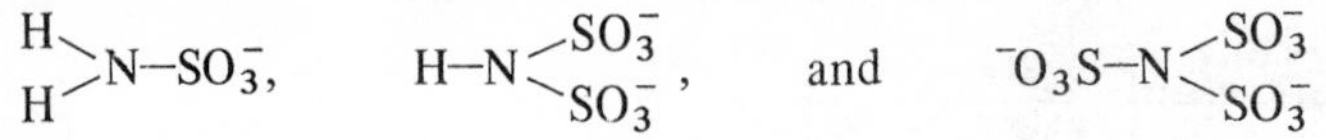

$$H{\scriptstyle\diagdown}_{H}{\scriptstyle\diagup}N{-}SO_3^-, \qquad H{-}N{\scriptstyle\diagup}^{SO_3^-}_{\diagdown SO_3^-}, \qquad \text{and} \qquad {}^-O_3S{-}N{\scriptstyle\diagup}^{SO_3^-}_{\diagdown SO_3^-}$$

The action of $KHSO_3$ on KNO_2 gives the trisulphonate, $K_3N(SO_3)_3$, which may be hydrolysed successively to the disulphonate, $K_2NH(SO_3)_2$, then to the amino-sulphonate (sulphamate), KNH_2SO_3, or to the hydroxylamine-*N*-sulphonate, $K[NH(OH)SO_3]$. The imidodisulphonate ion is the NH analogue of the pyro-sulphate ion, with which it is considered later. In $K[NH(OH)SO_3]$ the ion has the configuration (IV). The closely related nitrosohydroxylaminesulphonate ion, (V),

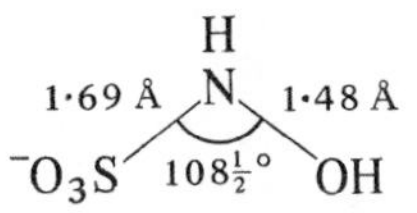

(IV) 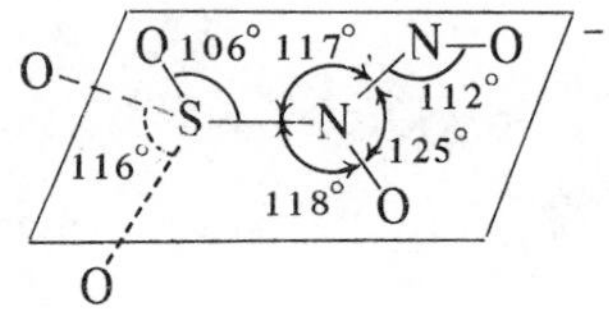　　　　　　　　　　(V) 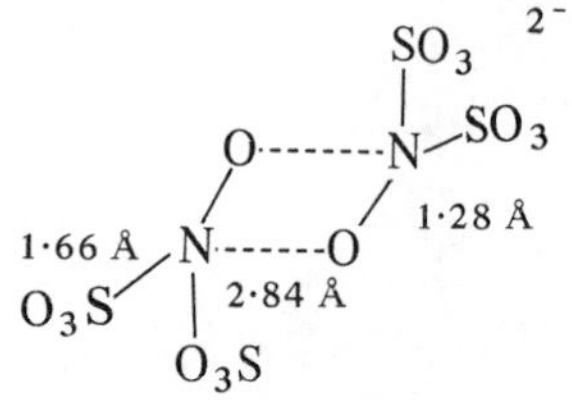　　　　　　　　　　(VI)

(1) JCS 1951 1467

(2) JCS A 1968 3043

has been studied in the K salt, made by absorbing NO in alkaline K_2SO_3 solution, and later in the ammonium salt.[1] The bonds drawn as full lines all lie in one plane, and the bond lengths indicate a single S–N bond (1·79 Å) but considerable π-bonding in the N–N bond (1·33 Å); N–O and S–O are respectively 1·28 and 1·44 Å. In Frémy's salt, $K_2[ON(SO_3)_2]$,[2] the bonds from N are coplanar, as in (V), but N–S is appreciably shorter (1·66 Å). The anions are very weakly associated in pairs, (VI), related by a centre of symmetry, reminiscent of the dimers in crystalline NO (p. 651).

(g) Structural studies have been made of the sulphuric acid molecule, of several sulphones, R_2SO_2, of sulphamide, $(NH_2)_2SO_2$, and of the sulphuryl halides SO_2F_2 and SO_2Cl_2. Microwave results for SO_2F_2 may be compared with those for SOF_2:

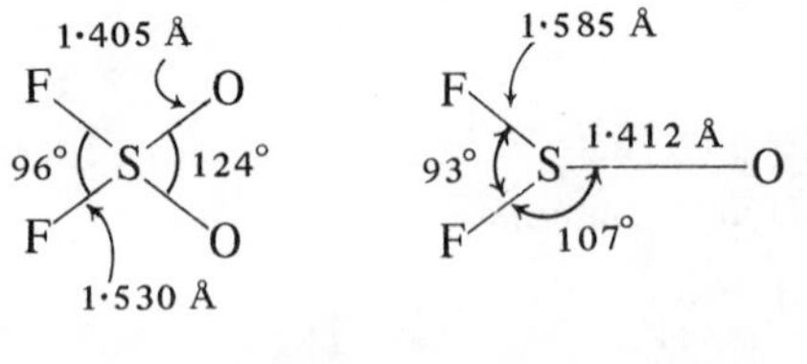

(h) The trimethyl oxosulphonium ion, $(CH_3)_3SO^+$, has been shown to be tetrahedral in the perchlorate and tetrafluoroborate, with C–S–O, 112·5°, and C–S–C, 106°.

(i) A warm solution of an alkali sulphite M_2SO_3 dissolves S to form the thiosulphate. In this reaction there is simple addition of S to the pyramidal sulphite ion forming the tetrahedral $S_2O_3^{2-}$ ion which has been studied in a number of salts (S–S, 2·01 Å). Thiosulphuric acid has been prepared by the direct combination of

H_2S with SO_3 in ether at $-78°C$ and isolated as the ether complex, $H_2S_2O_3 . 2 (C_2H_5)_2O$.

As a ligand the $S_2O_3^{2-}$ ion appears to behave in several ways, as a bridging ligand utilizing only one S atom in $Na_4[Cu(NH_3)_4][Cu(S_2O_3)_2]_2$,[3] as a monodentate ligand bonded through a S atom, in $[Pd(en)_2][Pd(en)(S_2O_3)_2]$,[4] and as a bidentate ligand forming a weak bond from S, in $Ni(tu)_4S_2O_3 . H_2O$.[5] The Ni—O

(3) AC 1966 **21** 605
(4) AC 1970 **B26** 1698
(5) AC 1969 **B25** 203.

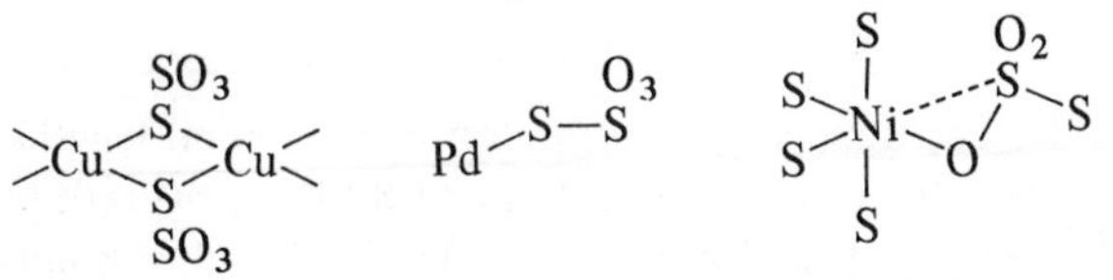

bond has the normal length (*c.* $2·1$ Å) but Ni—S ($2·70$ Å) is longer than normal ($2·4$–$2·6$ Å).

(j) An example of an ion of this type is the methane thiosulphonate ion, which has the structure shown in the monohydrated Na salt.

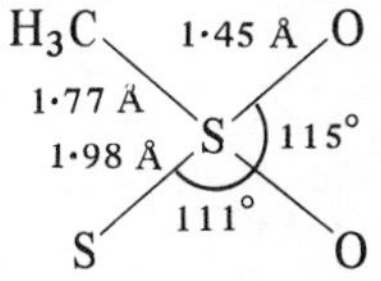

Details of the structures of tetrahedral ions and molecules (e)–(j) are given in Table 16.4

The pyrosulphate and related ions
The three related ions

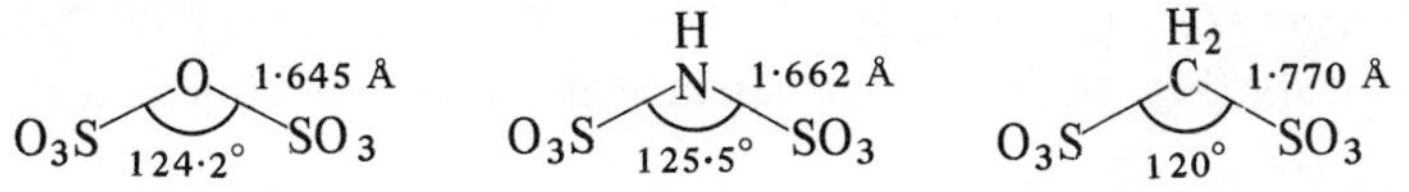

pyrosulphate   imidodisulphonate   methylene disulphonate

have very similar structures (Table 16.5).

TABLE 16.5
The pyrosulphate and related ions (in K salts)

	S—O terminal	S—O (N, C) bridging	S—M—S	Reference
$(O_3S-O-SO_3)^{2-}$	$1·437$ Å	$1·645$ Å	$124·2°$	JCS 1960 5112
$(O_3S-NH-SO_3)^{2-}$	$1·453$	$1·662$	$125·5°$	AC 1963 **16** 877
$(O_3S-CH_2-SO_3)^{2-}$	$1·461$	$1·770$	$120°$	JCS 1962 3393

In $(NO_2)(HS_2O_7)$ and $Se_4(HS_2O_7)_2$ (q.v.) the anion consists of a chain of hydrogen-bonded $S_2O_7H^-$ ions:

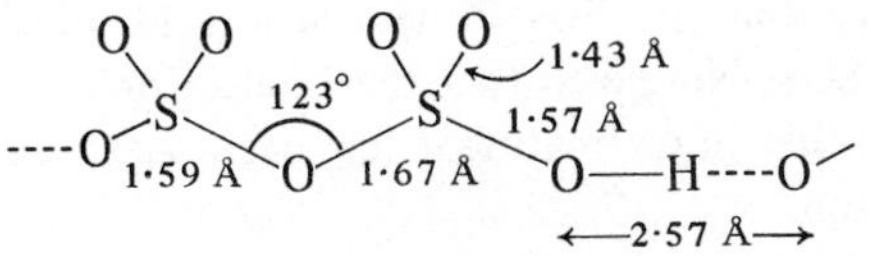

The trisulphate and pentasulphate ions, $S_3O_{10}^{2-}$ and $S_5O_{16}^{2-}$, have been studied in the K salts but accurate bond lengths were not determined in the former ion.[1][2]

(1) AC 1964 **17** 684
(2) AC 1969 **B25** 1696

Sulphur, Selenium, and Tellurium

They are linear chains of three and five SO_4 groups respectively. The configuration found for $S_5O_{16}^{2-}$ has nearly planar terminal SO_3 groups:

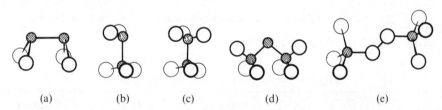

In addition to the apparent slight alternation in S–O bond lengths in the central portion there is an extremely long S–O bond (1·83 Å) to the terminal SO_3 groups, suggesting that the system is closer to a $S_3O_{10}^{2-}$ ion weakly bonded to two SO_3 molecules. It would be desirable to have an independent confirmation of the long S–O bond in another salt of this type. A long terminal bridging bond (1·72 Å) was also found in $S_3O_{10}^{2-}$, but high accuracy was not claimed. The ions $HS_2O_7^-$ and $S_3O_{10}^{2-}$ exist, together with nitronium ions, in the salts formed from SO_3 and N_2O_5 or HNO_3, namely, $(NO_2)HS_2O_7$[3] and $(NO_2)_2S_3O_{10}$.[4]

(3) AC 1954 **7** 402
(4) AC 1964 **17** 680

Dithionites

The $S_2O_4^{2-}$ ion exists in sodium dithionite, the reducing agent, 'hydrosulphite'. This salt is prepared from the Zn salt, which results from the reduction of SO_2 by Zn dust in aqueous suspension. In solution it rapidly decomposes to disulphite and thiosulphate: $2 S_2O_4^{2-} \rightarrow S_2O_5^{2-} + S_2O_3^{2-}$. The structure of this ion is rather extraordinary (Fig. 16.6(a)).[5] It has an eclipsed configuration with the planes of

(5) AC 1956 **9** 579

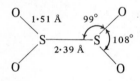

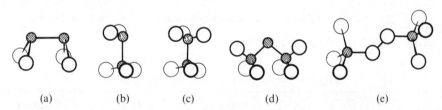

FIG. 16.6. The structures of some complex sulphur oxy-ions: (a) the dithionite ion, $S_2O_4^{2-}$, (b) the metabisulphite ion, $S_2O_5^{2-}$, (c) the dithionate ion, $S_2O_6^{2-}$, (d) the trithionate ion, $S_3O_6^{2-}$; (e) the perdisulphate ion, $S_2O_8^{2-}$.

the two SO_2 nearly parallel (angle between normals 30°), and the S–S bond is very long, corresponding to a (Pauling) bond-order of only about 0·3. This very weak bond is consistent with the fact that almost instantaneous exchange of ^{35}S occurs between the dithionite ion and SO_2 in neutral or acid solution; none takes place between $S_3O_6^{2-}$ and SO_2. Note also that the isoelectronic ClO_2 does not dimerize appreciably. A simplified m.o. treatment has been suggested to account for the configuration of this ion.

Disulphites ('metabisulphites')

If a solution of K_2SO_3 is saturated with SO_2 the salt $K_2S_2O_5$ may be crystallized

590

from the solution. The disulphite ion has the configuration shown in Fig. 16.6(b);[6] addition of SO_2 to the pyramidal SO_3^{2-} ion has produced the unsymmetrical ion $O_2S-SO_3^{2-}$:

(6) AC 1971 **B27** 517

$$\underset{1\cdot45\ \text{Å}}{\overset{\displaystyle O}{\underset{\displaystyle O}{O-S}}}\ \overset{2\cdot17\ \text{Å}}{\rule{3cm}{0.4pt}}\ \underset{1\cdot50\ \text{Å}}{\overset{\displaystyle O}{S{-}O}}$$

Note the long S–S bond.

The stereochemistry of molecules and ions containing S_n chains

The atoms of a chain S_3 are necessarily coplanar, but chains of four or more S atoms present interesting possibilities of isomerism because in a system $S_1S_2S_3S_4\cdots$ the dihedral angles $S_1S_2S_3/S_2S_3S_4$, etc., are in the region of $90°$. We have already noted that in O_2H_2 the O–H bonds are not coplanar, and this applies also to S_2R_2 molecules and the similar compounds of Se and Te. To describe the structure of such a molecule it is necessary to specify not only the interbond angles and bond lengths but also the dihedral angle ϕ (see (a), p. 420). For S these dihedral angles range from $74°$ in the S_6^{2-} ion to $110°$ in the $S_5O_6^{2-}$ ion; for Se recorded values range from $74\cdot5°$ to $102°$ and for Te the only values determined are $72°$ and $101°$. Some of these dihedral angles are listed in Table 16.6 and further values, in polythionate ions and related molecules, in Table 16.8.

Since the stereochemistry of a molecule $S_{n-2}R_2$ is similar to that of a sulphur chain S_n the following remarks apply equally to three main groups of compounds: molecules $S_{n-2}R_2$, polysulphide ions S_n^{2-}, polythionate ions $S_nO_6^{2-}$ and closely related molecules such as $TeS_{n-1}O_4R_2$.

The stereochemistry of azo compounds, RN=NR, is similar to that of, for example, dihalogenoethylenes, the compounds existing in planar *cis* and *trans* forms:

$$\overset{\displaystyle H}{\underset{\displaystyle H}{\overset{\textstyle \diagdown}{C}\overset{\textstyle \diagup}{\underset{\textstyle \diagup}{\overset{\|}{C}}}}}\overset{\displaystyle I}{\underset{\displaystyle I}{}}\quad\text{and}\quad\overset{\displaystyle H}{\underset{\displaystyle I}{\overset{\textstyle \diagdown}{C}\overset{\textstyle \diagup}{\underset{\textstyle \diagup}{\overset{\|}{C}}}}}\overset{\displaystyle I}{\underset{\displaystyle H}{}}\ ;\quad\overset{\displaystyle N}{\underset{\displaystyle N}{\overset{\|}{}}}\overset{\displaystyle R}{\underset{\displaystyle R}{}}\quad\text{and}\quad\overset{\displaystyle N}{\underset{\displaystyle N}{\overset{\|}{}}}\overset{\displaystyle R}{\underset{\displaystyle R}{}}$$

The two forms of a molecule R_1SSR_2, on the other hand, differ in having R_2 on different sides of the plane SSR_1:

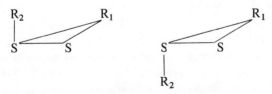

Although they appear to correspond to the *cis* and *trans* isomers of an azo compound they are in fact *d* and *l* enantiomorphs. Examples include S_2H_2, S_2F_2, and S_2Cl_2, in all of which the dihedral angle is close to $90°$.

591

In order to illustrate the possible types of isomers of longer S_n chains it is convenient to idealize the dihedral angle to 90°. In Fig. 16.7(a) and (b) we show the *d* and *l* forms of RSSR or $-S_4-$. The possible structures of the S_5 chain are found by joining a fifth S atom to S_4 in (a) so that the dihedral angle of 90° is maintained. Three of the five positions for S_5 are coplanar with $S_2 S_3 S_4$ leaving the two possibilities shown in (c) and (d), which are described as the *cis* and *trans* forms of the $-S_5-$ chain (or RS_3R molecule). Of these the *trans* form is enantiomorphic. A similar procedure shows that there are three forms of the $-S_6-$ chain, all of which are enantiomorphic (Table 16.7). The *cis* form of S_5 and the *cis-cis* form of S_6 correspond to portions of the S_8 ring, while the *trans* form of S_5 and the *trans-trans* form of S_6 correspond to portions of the infinite helical chain of 'metallic' Se or Te (or fibrous S)–Fig. 16.7(e).

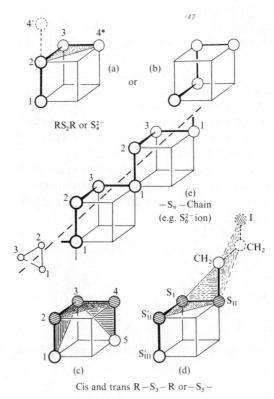

FIG. 16.7. Stereochemistry of molecules $S_2 R_2$ and $S_3 R_2$ and ions S_n^{2-}, idealized with interbond and dihedral angles equal to 90°. (a) is given for comparison with (c), (d), and (e), and (b) for comparison with the sketch of $H_2 O_2$ on p. 420.

Molecules $S_2 R_2$ and $S_3 R_2$

Structural data and references for these compounds are summarized in Table 16.6 (p. 593). The halides $S_2 X_2$ have already been mentioned (p. 576).

Molecules of the type RS_3R that have been studied include 2:2′-diiododiethyl trisulphide, $S_3 (C_2 H_4 I)_2$, dimethyl trisulphide, and cyanogen triselenide (selenium diselenocyanate). The shape of the first molecule can be appreciated from Fig. 16.7(d). The three shaded circles represent S atoms, and one of the $-CH_2-CH_2-I$

592

groups is shown at the right. The terminal portion $-S-CH_2-CH_2-I$ is coplanar and has the *trans* configuration, so that $S_{II}S_IS_{II}$, $S_IS_{II}C$, and $S_{II}CCI$ lie in three mutually perpendicular planes. The corresponding dihedral angles are close to $90°$ ($82°$ and $85°$ respectively), and the S–S–S bond angle is $113°$. An electron diffraction study of $S_3(CH_3)_2$ gave the data listed in Table 16.6 where the reference for crystalline $Se_3(CN)_2$ is given. In the crystal the molecules have the *cis*

TABLE 16.6

Structures of molecules S_2R_2, Se_2R_2, Te_2R_2, *and molecules containing chains of* S (Se) *atoms*

Compound	Dihedral angle	S—S—R	S—S	S—R	Method	Reference
S_2H_2	$\approx 90°$	$(95°)$	2·05 Å	(1·33 Å)	e.d.	JACS 1938 **60** 2872
S_2F_2	$88°$	$108°$	1·89	1·64	m.w.	JACS 1964 **86** 3617
S_2Cl_2	$83°$	$107°$	1·97	2·07	e.d.	BCSJ 1958 **31** 130
S_2Br_2	$84°$	$105°$	1·98	2·24	e.d.	
$S_2(CH_3)_2$	–	$107°$	2·04	1·78	e.d.	JACS 1938 **60** 2872
$S_2(CF_3)_2$	–	$105°$	2·05	1·83	e.d.	TFS 1954 **50** 452
$S_2(C_6H_5)_2$	$96°$	$106°$	2·03	1·80	X	AC 1969 **B25** 2094
$S_2(C_6H_5CH_2)_2$	$92°$	$103°$	2·02	1·85	X	AC 1969 **B25** 2497
$(S_3)^{2-}Ba$	–	$103°$	[2·2]	–	X	ZK 1936 **94** 439
$S_3(CH_3)_2$	$93°$	$104°$	2·04	1·78	e.d.	JCP 1948 **16** 92
$S_3(CF_3)_2$	–	$104°$	2·06	1·85	e.d.	TFS 1954 **50** 452
$S_3(CCl_3)_2$	$94°$	$106°$ (SSS) $102°$ (SSC)	2·03	1·90	X	ZK 1961 **116** 290
$S_3(C_2H_4I)_2$	$84°$	$113°$ (SSS) $98°$ (SSC)	2·05	1·86	X	JACS 1950 **72** 2701
$(S_4)^{2-}Ba.H_2O$					X	AC 1969 **B25** 2365
$[(S_5)_3Pt](NH_4)_2.2H_2O$	see text				X	AC 1969 **B25** 745
$(S_6)^{2-}Cs_2$					X	AC 1953 **6** 206
		R—Se—R	Se—Se	Se—R		
$Se_2(CF_3)_2$	–	–	2·33	1·93	e.d.	TFS 1954 **50** 452
$Se_2(C_2H_5)_2$	$83°$	–	–	–	μ	JACS 1952 **74** 4742
$Se_2(C_6H_5)_2$	$82°$	$106°$	2·29	1·93	X	AC 1952 **5** 458
$Se_2[(C_6H_5)_2HC]_2$	$82°$	$100°$	2·39	1·97	X	AC 1969 **B25** 1090
$Se_2[S(C_2H_5)_2P]_2$	$104\frac{1}{2}°$	$106°$	2·33	2·28	X	ACSc 1966 **20** 51
$Se_2(p\text{-}Cl.C_6H_4)_2$	$74\frac{1}{2}°$	$101°$	2·33	1·93	X	AC 1957 **10** 201
		R—Te—R	Te—Te	Te—R		
$Te_2(p\text{-}Cl.C_6H_4)_2$	$72°$	$94°$	2·70	2·13	X	
$Se_3(CN)_2$	see text				X	ACSc 1954 **8** 1787

configuration (Fig. 16.7(c)), with dihedral angle SeSeSe/SeSeC equal to $94°$. Examples of S_n chains with the various configurations are listed in Table 16.7.

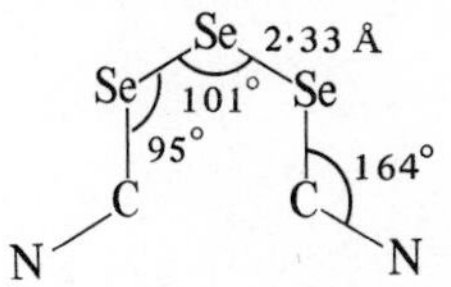

Polysulphides

These are formed, with the normal sulphide, when an alkali or alkaline-earth carbonate is fused with S or, preferably, when S is digested with a solution of the sulphide or hydrosulphide. From the resulting deeply-coloured solutions various

TABLE 16.7

Configurations of S_n chains

	Forms	Figure	Examples
S_4 or RSSR	*d or l*	(a), (b)	S_4^{2-}, $S_4O_6^{2-}$, $S_4O_6(CH_3)_2$
S_5 or RS_3R	*cis*	(c)	$S_5O_6^{2-}$
	trans (d and l)	(d)	$S_3(C_2H_4I)_2$
S_6 or RS_4R	*cis-cis (d and l)*		$S_6O_6^{2-}$ in $K_2Ba(S_6O_6)_2$
	cis-trans (d and l)		
	trans-trans (d and l)	(e)	S_6^{2-}, $S_6O_6^{2-}$ in
			$[Co(en)_2Cl_2]S_6O_6 . H_2O$

salts can be obtained containing ions from S_2^{2-} to S_6^{2-}. The alkali-metal compounds may also be prepared directly from the elements in liquid ammonia. The relative stabilities of the crystalline alkaline polysulphides vary from metal to metal. The following well-defined compounds have been prepared: for M_2S_n, $n = 2$, 4, and 5 for Na, 2, 3, 4, 5, and 6 for K, 2, 3, and 5 for Rb, and 2, 3, 5, and 6 for Cs.

By treating a solution of Cs_2S with S the salt $Cs_2S_5 . H_2O$ is obtained. This can be dehydrated *in vacuo* to Cs_2S_5, but evaporation of an alcoholic solution yields Cs_2S_6, suggesting that a solution of Cs_2S_5 contains S_6^{2-} and also lower S_n^{2-} ions. In fact, it seems probable that all polysulphide solutions are equilibrium mixtures of various S_n^{2-} ions (and M^+ or M^{2+} ions). Such solutions are immediately decomposed by acids, but if a polysulphide solution is poured into aqueous HCl a yellow oil separates from which pure acids H_2S_n (n, 2–5) may be separated by fractional distillation as yellow liquids. H_2S_6, H_2S_7, and H_2S_8 have been prepared by the reaction: $2 H_2S_2 + S_nCl_2 \rightarrow 2 HCl + H_2S_{4+n}$.

K_2S_2 and one form of Na_2S_2 are isostructural with Na_2O_2 and contain S_2^{2-} ions (S–S approx. 2·13 Å). An early X-ray study of BaS_3 gave a bond angle of 103° in the angular S_3^{2-} ion and a rather long S–S bond length (2·2 Å). The ions S_4^{2-} and S_6^{2-} have been studied in $BaS_4 . H_2O$ and Cs_2S_6; they are shown, in idealized form, in Fig. 16.7. In the former the S–S bond length is 2·07 Å, S–S–S bond angle, 104°, and mean dihedral angle, 76·4°. In Cs_2S_6 the ion has the form of an unbranched chain (mean dihedral angle 106°) with relatively short van der Waals distances of 3·4 Å between the ends of chains. These weak attractions between successive S_6^{2-} ions link them into infinite helices of S atoms very similar to the helical chains in metallic Se and Te, though with a larger bond angle (mean S–S–S, 100°). [There is apparently an alternation in bond lengths in the S_6^{2-} chain (2·02 and 2·11 Å).] By bonding to a metal atom through the two terminal S atoms the S_5^{2-} ion forms chelate complexes $M(S_5)_2$ and $M(S_5)_3$. In $(NH_4)_2Pt(S_5)_3 . 2 H_2O$ the 6-rings so formed have the chair configuration, with Pt–S, 2·39 Å, S–S, 2·05 Å, S–S–S, 111°, and dihedral angles 72–83°.

Thionates and molecules of the type $S_n(SO_2R)_2$

The lower thionate ions have the general formula $S_nO_6^{2-}$, where $n = 2$, 3, 4, 5, or 6.

The dithionate ion, $S_2O_6^{2-}$, is formed by the direct union of two SO_3^{2-} ions, for example by the electrolysis of neutral or alkaline solutions of a soluble sulphite or by passing SO_2 into a suspension of MnO_2, in water. This ion is not formed by the methods which produce the higher thionates, for these require S in addition to SO_3 groups, as may be seen from their formulae, $(O_3S-S_n-SO_3)^{2-}$, and they are formed whenever S and SO_3^{2-} ions are present together in aqueous solution. A mixture of all the ions from $S_3O_6^{2-}$ to $S_6O_6^{2-}$ is usually produced in such circumstances (compare the polysulphides), and, moreover, a solution of any one of the higher thionates is converted, though very slowly, into a mixture of them all. The alkali and alkaline-earth salts, however, are very stable in solution and interconversion of ions takes place only in the presence of hydrogen ions. Certain reactions tend to produce more of one particular $S_nO_6^{2-}$ ion, a fact which is utilized in their preparation. The controlled decomposition of a thiosulphate by acid gives largely $S_5O_6^{2-}$, while the oxidation of $S_2O_3^{2-}$ by H_2O_2 gives $S_3O_6^{2-}$ and by iodine (or electrolytically) yields the tetrathionate ion, $S_4O_6^{2-}$, e.g.

$$2\,S_2O_3^{2-} + I_2 \rightarrow S_4O_6^{2-} + 2\,I^-$$

Some of the relations between these ions are indicated in the following scheme:

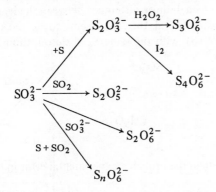

It was noted earlier that $H_2S_2O_3$ may be prepared directly from H_2S and SO_3 in ether solution at $-78°C$. If H_2S_n is used instead of H_2S, acids $H(O_3S . S_nH)$ are produced which by the action of further SO_3 give polythionic acids $H_2(O_3S . S_n . SO_3)$. Oxidation by iodine gives acids $H_2(O_3S . S_{2n} . SO_3)$, by which means $H_2S_8O_6$, $H_2S_{10}O_6$, and $H_2S_{12}O_6$ have been prepared.

The structure of the dithionate ion has been studied in a number of salts including $Na_2S_2O_6 . 2\,H_2O$,[1] $K_2S_2O_6$,[2] $NaK_5Cl_2(S_2O_6)_2$ and NaK_2Cl (S_2O_6),[3] and $BaS_2O_6 . 2\,H_2O$.[4] The normal, centrosymmetrical, form of the ion is shown in Fig. 16.6(c), but in the potassium salt the ions are of two kinds, some having the centrosymmetrical form and others almost the 'eclipsed' configuration. The length of the S—O bonds is around 1·43 Å and S—S close to 2·15 Å.

The structure of the trithionate ion is shown in Fig. 16.6(d), derived from the crystal structure of $K_2S_3O_6$[5] (S—S, 2·15 Å).

(1) AC 1956 9 145
(2) AC 1956 9 897
(3) AC 1953 6 187
(4) AC 1966 21 672

(5) ZK 1934 89 529

(6) JCS 1949 724
(7) ACSc 1954 8 42

Closely related to the trithionate ion are molecules $S(SO_2R)_2$[6] and $Se(SO_2R)_2$[7] of which the following have been studied in the crystalline state:

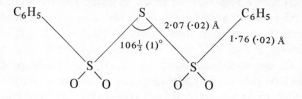

and

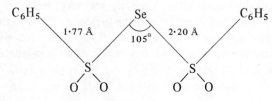

The S—O bonds are all of length 1·41 (·04) Å.

(8) ACSc 8 459
(9) ACSc 1964 18 662

In the tetrathionate ion, studied in $BaS_4O_6 . 2 H_2O$[8] and $Na_2S_4O_6 . 2 H_2O$,[9] the S_4 chain has the configuration of Fig. 16.7(b), with dihedral angles close to 90°. There appears to be a definite difference between the lengths of the central and terminal S—S bonds, (a), and this also seems to be the case in dimethane

(10) ACSc 1953 7 1

sulphonyl disulphide,[10] (b), with the same S_4 skeleton, though it is doubtful if

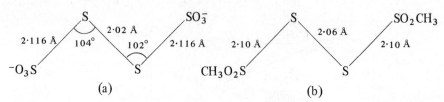

the differences are larger than the probable experimental error in the latter case (see also the S_6^{2-} ion, p. 594).

Structural studies have also been made of the pentathionate ion and the analogous $SeS_4O_6^{2-}$ and $TeS_4O_6^{2-}$ ions. In $BaS_5O_6 . 2 H_2O$ (and the isostructural $BaSeS_4O_6 . 2 H_2O$) the ion has the *cis* configuration (Fig. 16.7(c)), in which the S_5 skeleton may be regarded as a portion of an S_8 ring from which three S atoms have been removed.

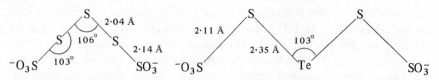

The dihedral angles of around 110° in the pentathionate and the related Se and Te ions may be compared with the much smaller values in polysulphide ions (p. 594). There appears to be a difference in length between the central and terminal S—S bonds comparable with that in the S_6^{2-} ion.

596

In contrast to $S_5O_6^{2-}$ and $SeS_4O_6^{2-}$, the $TeS_4O_6^{2-}$ ion in the ammonium salt has the *trans* configuration of Fig. 16.7(d). This difference is also found in closely related molecules of the type

$$RO_2S\diagdown S\diagdown \overset{S(Se)}{\diagup}\diagdown S\diagup SO_2R$$

cis

$$R\diagdown \underset{O\diagup\diagdown O}{S}\diagdown \overset{S}{\diagup}\diagdown Te\diagup \overset{S}{\diagdown}\underset{O\diagup\diagdown O}{S}\diagup R$$

trans

where R is CH_3, C_6H_5 etc.

Structural data and references for this last group of compounds are summarized in Table 16.8.

(11) ACSc 1965 **19** 2207

(12) ACSc 1965 **19** 2219

TABLE 16.8
Structural data for pentathionates and related compounds

Compound	Configuration	Dihedral angle	Reference
$BaS_5O_6 . 2 H_2O$			
(orthorhombic)	*cis*	110°	ACSc 1954 **8** 473
(triclinic)	*cis*	$107\frac{1}{2}°$	ACSc 1956 **10** 288
$BaSeS_4O_6 . 2 H_2O$	*cis*	109°	ACSc 1954 **8** 1701
$BaTeS_4O_6 . 2 H_2O$	*cis*	103°	ACSc 1958 **12** 52
$(NH_4)_2TeS_4O_6$	*trans*	90° (mean)	ACSc 1954 **8** 1042
$TeS_4O_4(CH_3)_2$	*trans*	81°	ACSc 1954 **8** 1032
$TeS_4O_4(C_6H_5)_2$	*trans*	79°	ACSc 1956 **10** 279

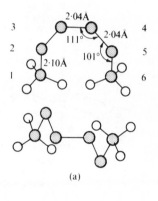

(a)

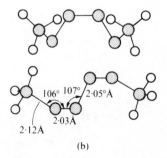

(b)

FIG. 16.8. Configurations of the $S_6O_6^{2-}$ ion in (a) $K_2Ba(S_6O_6)_2$, (b) $[Co(en)_2 Cl_2]_2 S_6O_6 . H_2O$.

The hexathionates are the most complex salts of this family which have been isolated in the pure state. As already mentioned, there are three possible isomers of a $-S_6-$ chain (each capable of existing in d and l configurations), assuming that free rotation around the S–S bonds is not possible. In $K_2Ba(S_6O_6)_2$ the anion has the *cis–cis* configuration[11] which, like the *cis* configuration of the pentathionate ion, corresponds to a portion of the S_8 ring. In $[Co(en)_2Cl_2]_2S_6O_6 . H_2O$, on the other hand, it has the extended *trans–trans* configuration.[12]

Two views of the $S_6O_6^{2-}$ ion in these two salts are shown in Fig. 16.8. There are two different dihedral angles in each configuration, the values being:

	$S_1S_2S_3/S_2S_3S_4$	$S_2S_3S_4/S_3S_4S_5$
$K_2Ba(S_6O_6)_2$	107°	89°
$[Co(en)_2Cl_2]_2S_6O_6 . H_2O$	85°	73°

The structural chemistry of selenium and tellurium

This is summarized in Table 16.9 where the horizontal divisions correspond to the formal oxidation states. The examples given have been established by structural studies unless marked by a query (?). (For example, the structures of the gaseous molecules of the dioxides and trioxides are not known.) Configurations to the left of the heavy line are also known for sulphur compounds. Simple examples of some of these valence groups have already been given, for example, elementary Se and Te and molecules SeR_2 and Se_2R_2 (4,4), SeO_3^{2-} (2,6), SeO_4^{2-} (8), molecules SeR_2X_2 (2,8) and SeX_6 (12), and the Te analogues. We are concerned here particularly with stereochemistries exhibited by Se and/or Te and not by S, that is, those to the right of the heavy line, but we shall also note some additional examples of the above-mentioned valence groups found in compounds peculiar to Se or Te.

Se(VI) *and* Te(VI)

Valence group (12). Reference has already been made to the regular octahedral shape of the SeF_6 and TeF_6 molecules. Of the elements S, Se, and Te, only Te is found octahedrally bonded to 6 O atoms, in TeO_3, $Te(OH)_6$, and various tellurates. The resemblance to iodine, the only halogen exhibiting 6-coordination by oxygen, is marked. Corresponding to $IO(OH)_5$ there is the octahedral $[TeO(OH)_5]^-$ ion in $K[TeO(OH)_5] \cdot H_2O$. There are also bridged ions, (a) in $K_4[Te_2O_6(OH)_4] \cdot 7.3 H_2O$, isoelectronic with $I_2O_8(OH)_2^{4-}$, and (b) in $Na_2K_4[Te_2O_8(OH)_2] \cdot 14 H_2O$ (Fig. 16.9). In addition there are salts containing infinite chain ions built from octahedral groups sharing vertices, (c) in $K[TeO_2(OH)_3]$, or edges, (d) in $K[TeO_3(OH)]$. Note the absence of the TeO_4^{2-} ion (p. 585), in which the valence group would be (8). An n.d. study of $Te(OH)_6$ gives Te—O, 1.91 Å.

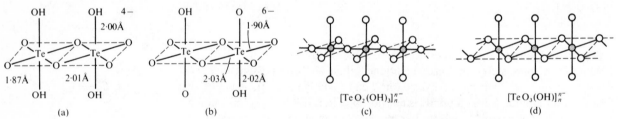

FIG. 16.9. Structures of Te(VI) compounds: (a) and (b) bridged binuclear ions, (c) and (d) infinite anions.

Se(IV) *and* Te(IV)

Valence group (2,6). Simple examples of this valence group include ions such as $Se(CH_3)_3^+$ and $Te(CH_3)_3^+$ (see later) and SeF_3^+. In crystalline $Se(CH_3)_3I$ there are pyramidal cations but each is associated rather closely with only one I^-. Although Se···I is only a weak bond (3.78 Å) this is less than the estimated van der Waals separation (4.15 Å) and in fact shorter than S···I (3.89 Å) in the sulphur analogue. Moreover this weak Se···I bond is almost collinear with one of the Se—C bonds, and the ion pair has been described as a charge-transfer complex. Similarly, the Se···I distance in the complex of I_2 with 1,4-diselenane is slightly less than the S···I distance in the S analogue. At the same time there is an increase in I—I from the value 2.66 Å in the free I_2 molecule to 2.79 Å in the S compound and to 2.87 Å in

598

TABLE 16.9

The stereochemistry of Se and Te

Total number of σ pairs	3	4	5	6	7
Arrangement	*Triangular*	*Tetrahedral*	*Trigonal bipyramidal*	*Octahedral*	*[Octahedral]*
Valence group	(6)	(8)	(10)	(12)	(14)
Te(VI)	TeO_3 molecule?			TeO_3 (cryst.), TeF_6, $Te(OH)_6$[32]. TeO_6^{6-}[1a] $[TeO(OH)_5]^{-}$[1b] $[Te_2O_6(OH)_4]^{4-}$[2] $[Te_2O_8(OH)_2]^{6-}$[3] $[TeO_2(OH)_3]^{-}$[4] $[TeO_3(OH)]^{-}$[5]	
Se(VI)	SeO_3?	SeO_4^{2-}		SeF_6	
Valence group	(2,4)	(2,6)	(2,8)	(2,10)	(2,12)
Te(IV)	TeO_2 molecule?	TeO_3^{2-} $Te(CH_3)_3^{+}$[6]	TeX_4(?) TeR_2X_2[14], TeO_2 $Te_2O_4 \cdot HNO_3$[15] $Te(C_6H_4O_2)_2$[16]	TeF_4 (cryst.) $(TeF_5)K$[20] $Te(CH_3)I_4^{-}$[21]	$TeCl_6^{2-}$[23], $TeBr_6^{2-}$[24],[24a] $Te(tmtu)_2Cl_4$[25]
Se(IV)	SeO_2 molecule	SeO_2 (cryst.), SeO_3^{2-} $SeOF_2$, $SeOCl_2$ $Se(CH_3)_3^{+}$[7] $(SeF_3)(NbF_6)$[8a] $(SeF_3)(Nb_2F_{11})$[8b] $SeOF_2 \cdot NbF_5$[9] $SeOCl_2 \cdot SbCl_5$[10] $(SeOCl_2)_2 \cdot SnCl_4$[11] $(C_9H_8NO)SeOCl_3$[12] $NMe_4Cl \cdot 5 SeOCl_2$[13]	SeR_2X_2[17] $Br_2 \cdot SeC_4H_8S$[18] $Cl_2SeC_4H_8SeCl_2$[19]	$SeOCl_2$ (pyr)$_2$[22]	$SeCl_6^{2-}$?
Valence group		(4,4)	(4,6)	(4,8)	
Te(II)		Te (element) $TeBr_2$	$Te\phi Cl(tu)$[26] $[Te\phi(tu)_2]^{+}Cl^{-}$[27]	$Te(tu)_4^{2+}$[28], $Te(tu)_2X_2$[29] $Te_2(tu)_6^{4+}$[30], $Te(etu)_2X_2$[31]	
Se(II)		Se (element) SeR_2, Se_2R_2			

(1a) IC 1964 **3** 1417. (1b) IC 1964 **3** 634. (2) ACSc 1966 **20** 2138. (3) ACSc 1969 **23** 3062. (4) ZaC 1965 **334** 225. (5) NW 1964 **51** 634. (6) JCS A 1967 2018. (7) AC 1966 **20** 610. (8a) JCS A 1970 1891. (8b) JCS A 1970 1491. (9) JCS A 1969 2858. (10) ACSc 1967 **21** 1313. (11) AC 1960 **13** 656. (12) IC 1967 **6** 1204. (13) ACSc 1967 **21** 1328. (14) AC 1962 **15** 887; IC 1970 **9** 797. (15) AC 1966 **21** 578. (16) ACSc 1967 **21** 1473. (17) AC 1953 **6** 746. (18) IC 1967 **6** 958. (19) AC 1961 **14** 940. (20) IC 1970 **9** 2100. (21) JCS A 1967 2018. (22) AC 1959 **12** 638. (23) ACSc 1966 **20** 165. (24) CJC 1964 **42** 2758. (24a) JACS 1970 **92** 307. (25) IC 1969 **8** 313. (26) ACSc 1966 **20** 132. (27) ACSc 1966 **20** 123. (28) ACSc 1965 **19** 2336. (29) ACSc 1966 **20** 113. (30) ACSc 1965 **19** 2395. (31) ACSc 1965 **19** 2349. (32) ACSc 1973 **27** 85.

the Se compound. It appears that the stronger the bond from S or Se to I the weaker is the I—I bond:

	S (Se)—I	I—I	Reference
$C_4H_8S_2 . 2 I_2$	2·87 Å	2·79 Å	AC 1960 **13** 727
$C_4H_8Se_2 . 2 I_2$	2·83	2·87	AC 1961 **14** 940
$C_4H_8OSe . I_2$	2·76	2·96	IC 1966 **5** 522
$C_4H_8OSe . ICl$	2·63	2·73 (I—Cl)	IC 1968 **7** 365

We have noted in our description of the crystalline elements and of polyiodides that the formation of weak additional bonds intermediate in length between normal covalent bonds and van der Waals bonds is a feature of Se, Te, and I. Note the different molecular structures of $I_2Se(C_4H_8)SeI_2$ and $Cl_2Se(C_4H_8)SeCl_2$ (Fig. 16.10).

The reaction of SeF_4 with NbF_5 produces $(SeF_3)(NbF_6)$ and $(SeF_3)(Nb_2F_{11})$.

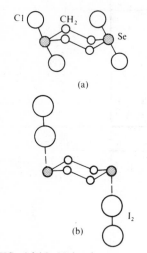

(a)

(b)

FIG. 16.10. Molecular structures of (a) $Cl_2SeC_4H_8SeCl_2$, (b) $I_2SeC_4H_8SeI_2$.

Both salts contain pyramidal SeF_3^+ ions, (a), and Se also forms three weaker bonds to F atoms of the $(NbF_6)^-$ or $(Nb_2F_{11})^-$ ions, (b), so that in addition to its three nearest (pyramidal) neighbours Se has three more distant F neighbours completing a very distorted octahedral coordination group. The large F—F distances between these three more distant neighbours (3·4–4·0 Å) may be due to the lone pair directed tetrahedrally as shown at (c)—compare the 7-coordination group in the A-La_2O_3 structure. In $(SeF_3)(NbF_6)$ Se—F was found to be 1·73 Å (and 2·35 Å) and the interbond angle 95°.

Details of the $SeOF_2$ molecule are included in Table 16.3 (p. 585); the structure of the gaseous $SeOCl_2$ molecule is not known—this compound is a colourless liquid, b.p. 177°C. The term adduct is used for compounds in which the $SeOX_2$ (or other molecule) retains essentially the same structure as that of the free molecule, in contrast to compounds such as $SeOCl_2$, $(pyridine)_2$—see later—in which there is rearrangement of the Se, O, and Cl atoms and formation of a (2,10) valence group. For example, in $SeOF_2 . NbF_5$ the O atom forms part of an octahedral group around Nb (Nb—O, 2·13 Å, Nb—F, 1·85 Å) and the structure of the $SeOF_2$ portion of the molecule is very similar to that of the free molecule (which has a pyramidal shape). Similarly, in $SbCl_5 . SeOCl_2$ and $SnCl_4 . 2 SeOCl_2$ the O atoms of $SeOCl_2$ complete the octahedral group around the metal atom. Weaker Se···Cl bonds usually complete a distorted octahedral coordination group around Se. For example, in $SnCl_4$. 2 $SeOCl_2$ these are Se—Cl, 3·01 Å (intramolecular) and Se—2 Cl, 3·34 and 3·38 Å to Cl in adjacent molecules. In 8-hydroxyquinolinium trichloro-oxyselenate, $C_9H_8NO^+(SeOCl_3)^-$, there are two additional bonds from each Se (2·98 Å) to Cl^- ions and together the $SeOCl_2$ molecules and Cl^- ions form infinite

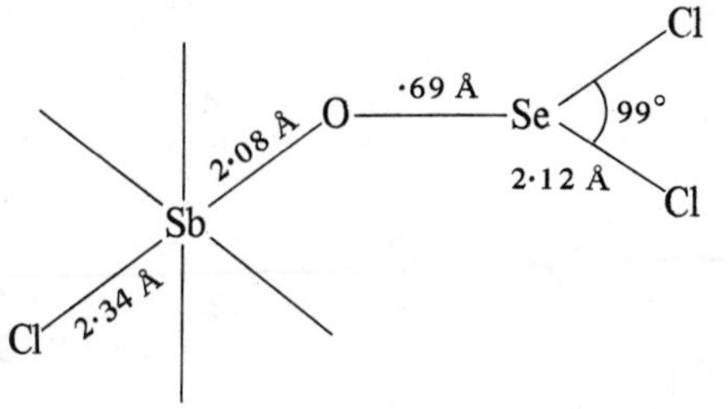

3-dimensional $SeOCl_3^-$ complexes. The sixth Se···Cl contact is at 3·3 Å, still shorter than the van der Waals radius sum (3·8 Å). The compound $[N(CH_3)_4]Cl$. 5 $SeOCl_2$ is an assembly of $N(CH_3)_4^+$ and Cl^- ions and $SeOCl_2$ molecules.

Valence group (2,8). This valence group occurs in molecules of the type SeX_2R_2 and TeX_2R_2 which have already been mentioned (p. 577), and it was noted that in these molecules the halogen atoms occupy the axial positions and the lone pair one equatorial bond position. Molecules studied include $(C_6H_5)_2SeCl_2$ and dibromide, $(C_6H_5)_2TeBr_2$ and substituted derivatives, and molecules such as Br_2 . $Se(CH_2)_4S$ and $Cl_2.Se(CH_2)_4Se$. Cl_2 (Fig. 16.10(a)). In molecules $R_2Se(Te)X_2$ the equatorial angle C—Se—C is 106–110° and in the Te compounds 96–101°, and the X—Se(Te)—X angle is close to 180°. The (axial) Se—X and Te—X bonds are much longer than the sums of the Pauling covalent radii, while the Se—C and Te—C bond lengths are close to the radius sums (Se—C, 1·94 Å, Te—C, 2·14 Å):

	Observed	Radius sum
Te—Cl	2·51 Å	2·36 Å
Te—Br	2·68	2·51
Te—I	2·93	2·70

We noted earlier the structures of the two crystalline forms of TeO_2. The similarity of the bond lengths in Te(IV) catecholate and of the four shorter Te—O bonds in both forms of TeO_2 and in Te_2O_4. HNO_3 suggests that the same bonding orbitals are used in all these compounds:

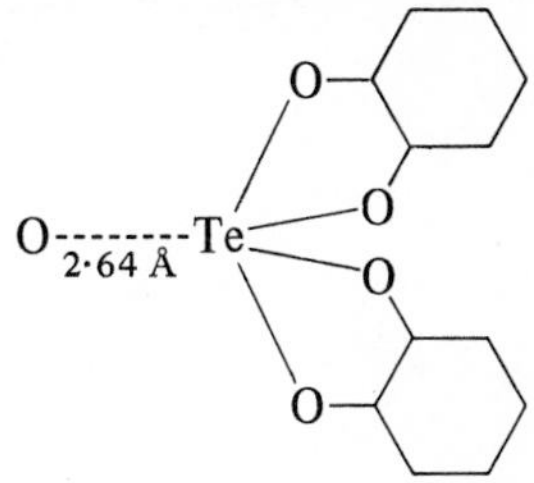

	α-TeO_2	β-TeO_2	Te_2O_4 . HNO_3	$Te(C_6H_4O_2)_2$
Te—O (axial)	(2) 2·08 Å	2·19 Å	2·16 Å	2·11 Å
		2·07	2·02	2·01
Te—O (equatorial)	(2) 1·90	1·93, 1·88	1·95, 1·88	(2) 1·98
O_{ax}—Te—O_{ax}	169°	169°	148°	154°
O_{eq}—Te—O_{eq}	102°	101°	100°	98°

The 'basic nitrate', $Te_2O_4 . HNO_3$, is made by dissolving Te in HNO_3 and crystallizing the solution. The structure is built of puckered layers (Fig. 16.11) in which there are single and double oxygen bridges between pairs of Te atoms, which form four 'trigonal bipyramidal' bonds.

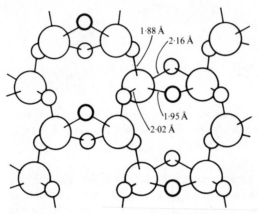

FIG. 16.11. Tellurium–oxygen layer in $Te_2O_4 . HNO_3$ (HNO_3 omitted).

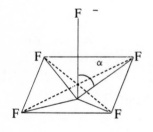

Valence group (2,10).   The unique structure of TeF_4 has been described with the other halides (p. 578). Te(IV) apparently does not form TeF_6^{2-} but only TeF_5^-. This ion exists as discrete units with the same pyramidal configuration as in TeF_4 and a structure very similar to that of the isoelectronic XeF_5^+ ion in $(XeF_5)(PtF_6)$. The Te atom lies about 0·4 Å below the basal plane of the square pyramid.

Ion or molecule	M—X$_{apical}$	M—X$_{basal}$	α
TeF_5^-	1·85 Å	1·96 Å	79°
XeF_5^+	1·81	1·88	79°
SbF_5^{2-}	2·00	2·04	83°
BrF_5	1·68	1·81	84°

The (2,10) valence group is also found in the $Te(CH_3)I_4^-$ ion in $[Te(CH_3)_3]^+$ $[Te(CH_3)I_4]^-$, a compound originally thought to be a geometrical isomer of $Te(CH_3)_2I_2$. This salt, which reacts with KI to give $Te(CH_3)_3I$ and $K[Te(CH_3)I_4]$, consists of pyramidal $[Te(CH_3)_3]^+$ and square pyramidal $[Te(CH_3)I_4]^-$ ions. Te $\cdots$ I contacts (3·84–4·00 Å) complete the octahedral environment of both types of Te(II) atom, (a) and (b).

The molecule $SeOCl_2 . 2 C_5H_5N$, (c) provides a further example of this valence group. The five ligands form a slightly distorted square pyramidal coordination group, the nearest neighbour in the direction of the sixth octahedral bond being a Cl atom of another molecule.

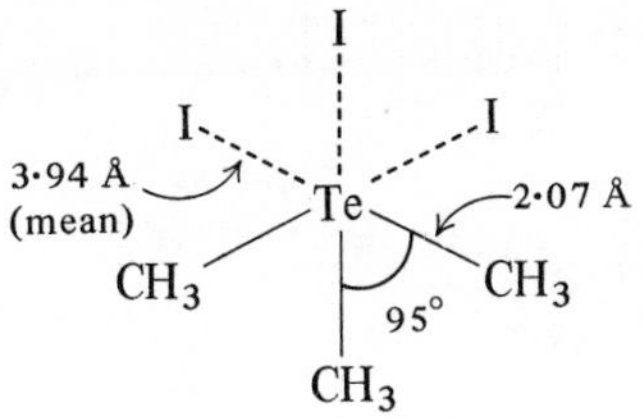

(a) 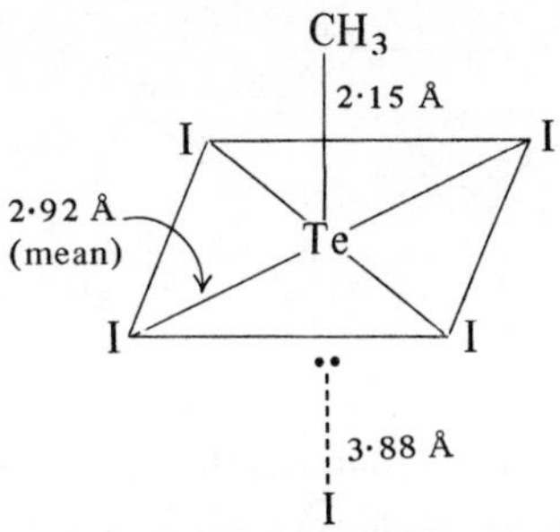   (b) 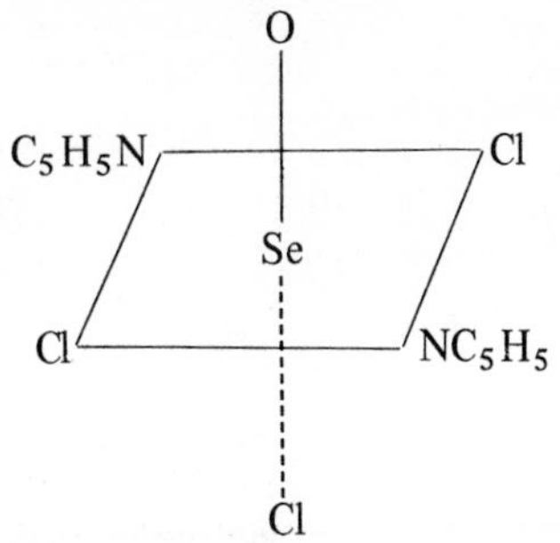   (c)

Valence group (2,12). Octahedral Te(IV) complexes imply a valence group of 14 electrons. Careful X-ray studies of $(NH_4)_2TeCl_6$ and K_2TeBr_6 show that the anions have a *regular* octahedral shape, and are therefore exceptions to the generalization that the arrangement of bonds formed by non-transition elements corresponds to the most symmetrical disposition of the total number of bonding and lone pairs of electrons. The electronic spectra suggest that the $5s^2$ electrons are

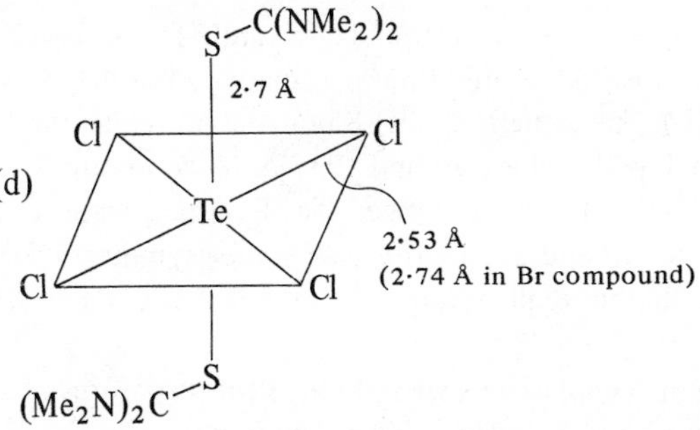

partially delocalized to the halide ligands (ref. 24a, Table 16.9). The same problem is presented by SbX_6^{3-} and IF_6^-. The Te–Cl bond length in $TeCl_6^{2-}$ is 2·54 Å, similar to that in molecules TeX_2R_2. In the octahedral molecule, $TeCl_4(tmtu)_2$ (where tmtu is tetramethyl thiourea), there is negligible angular distortion of the bonds (d).

Se(II) *and* Te(II)

Valence groups (4,6) *and* (4,8). For Se(II) and Te(II) the valence groups of 8, 10, and 12 electrons include in each case two lone pairs. The non-linear bond arrangement arising from (4,4) has been encountered in elementary Se and Te, molecules SeR_2 and Se_2R_2 and their Te analogues. The 10-electron group (4,6) leads to a T-shaped molecule, as in ClF_3. Examples include the molecule (e) and the ion (f), in which S represents the sulphur atom of thiourea in $C_6H_5 . Te(tu)Cl$ or $[C_6H_5 . Te(tu)_2]Cl$. In (e) and its Br analogue the Te–X bonds are remarkably long (but Te–C is equal to the radius sum) and Te–S is 2·50 Å, while in (f) Te–C is again normal but Te–S is abnormally long (2·68 Å).

Four coplanar bonds from Te(II) are expected for the valence group (4,8), as in

603

(e)   (f)

ICl_4^-, that is, the four equatorial bonds of an octahedral group. This bond arrangement occurs in molecules such as $Te(tu)_2X_2$, in the cation in salts $[Te(tu)_4]X_2$, and in the bridged ion in $[Te_2(tu)_6](ClO_4)_4$. There appear to be

(g)   (h)   (i)

interesting differences between the Te–S and Te–X bond lengths in the *cis* molecule (g) and those in the ethyl thiourea analogue $Te(etu)_2X_2$ which have the *trans* configuration, (h). The same long Te–S bond (2·68 Å) is found in the $[Te(tu)_4]^{2+}$ ion, as compared with values around 2·37 Å in compounds of 2-covalent Te(II) such as $Te(S_2O_2CH_3)_2$. Even longer Te–S bonds were found in the bridged $[Te_2(tu)_6]^{4+}$ ion, (i), and apparently also some asymmetry for which there would appear to be no obvious explanation.

For the sake of completeness we note here the structures of simple selenides and tellurides which are not included in other chapters:

antifluorite structure: Li_2Se, Na_2Se, K_2Se,
Li_2Te, Na_2Te, K_2Te

	Be	Mg	Ca	Sr	Ba	Zn	Cd	Hg
Se	Z	N	N	N	N	WZ	W	Z
Te	Z	W	N	N	N	WZ	Z	Z

N = NaCl structure: W = wurtzite structure: Z = zinc-blende (sphalerite) structure

17

Metal Sulphides and Oxysulphides

The structures of binary metal sulphides

Introduction

All metal sulphides are solid at ordinary temperatures, and we are therefore concerned here only with their crystal structures. These compounds may be divided into three groups, the sulphides of

(a)   the more electropositive elements of the I, II, and IIIA subgroups,
(b)   transition metals (including Cu^{II}),
(c)   elements of the II, III, IV, and VB subgroups.

The crystal structures of some sulphides are set out in Table 17.1, and comparison with the corresponding Table 12.2 for oxides shows that with few exceptions (e.g. MnO and MnS) structural resemblances between sulphides and oxides are confined to the ionic compounds of group (a) and a few compounds of group (c) such as ZnO and ZnS, HgO and HgS. The B subgroup sulphides are for the most part covalent compounds in which the metal atom forms a small number of directed bonds, and while the oxide and sulphide of some elements may be of the same topological type (e.g. with $3:2$ or $4:2$ coordination) the compounds are not usually isostructural. For example, GeS_2 consists of a 3D framework of GeS_4 tetrahedra linked through all vertices (compare the silica-like structures of GeO_2), but the actual framework is peculiar to GeS_2 and is not found in any form of GeO_2 (or SiO_2).

A bond M—S is more covalent in character than the bond M—O, and accordingly while there is often a structural resemblance between oxides and fluorides, the sulphides tend to crystallize with the same type of structure as chlorides, bromides, or iodides of the same formula type—compare MgF_2, MnF_2, TiO_2, and SnO_2 with the ionic rutile structure with $MgBr_2$, MnI_2, TiS_2, and SnS_2 (CdI_2 or $CdCl_2$ layer structures). Layer structures are rare in oxides and fluorides but are commonly found in sulphides and the other halides. When discussing the change in type of structure going down a Periodic Group towards the more electropositive elements the dioxides of the elements of Group IV were taken as examples. The structures of the corresponding disulphides are set out below, and they may be compared with those of the dioxides shown on p. 442.

CS_2 (finite molecules)

SiS_2 (infinite chains in normal form)

$\left.\begin{array}{l} TiS_2 \\ ZrS_2 \\ HfS_2 \end{array}\right\}$ (CdI_2 layer structure)

GeS_2 (3D framework of GeS_4 groups sharing all vertices)

SnS_2 (CdI_2 structure)

PbS_2 (stable only under pressure; see p. 934)

TABLE 17.1

The crystal structures of metal sulphides

Type of structure	Coordination numbers of M and S	Name of structure	Examples
Infinite 3-dimensional complexes	4 : 8	Antifluorite	Li_2S, Na_2S, K_2S, Rb_2S
	6 : 6	Sodium chloride	MgS, CaS, SrS, BaS, MnS, PbS, LaS, CeS, PrS, NdS, SmS, EuS, TbS, HoS, ThS, US, PuS
	6 : 6	Nickel arsenide	FeS, CoS, NiS,[a] VS, TiS
	6 : 6[b]	Pyrites or marcasite	FeS_2, CoS_2, NiS_2, MnS_2, OsS_2, RuS_2
	4 : 4	Zinc-blende	BeS, ZnS, CdS, HgS
		Wurtzite	ZnS, CdS, MnS
	4 : 4	Cooperite	PtS
Layer structures	6 : 3	Cadmium iodide (C 6)	TiS_2, ZrS_2, SnS_2, PtS_2, TaS_2, HfS_2
		Cadmium chloride (C 19)	TaS_2
	6 : 3	Molybdenum sulphide	MoS_2, WS_2
Chain structures			Sb_2S_3 Bi_2S_3, HgS

[a] Also the millerite structure (5 : 5 coordination)

[b] The coordination numbers here are those of Fe by S and of S_2 groups by Fe or other metal.

A number of important structure types are found in transition-metal sulphides which have no counterparts among oxide structures, notably the various layer structures and the pyrites, marcasite, and NiAs structures. Further, many sulphides, particularly of the transition metals, behave like alloys, the resemblance being shown by their formulae (in which the elements do not exhibit their normal chemical valences, as in Co_9S_8, Pd_4S, TiS_3), their variable composition, and their physical properties—metallic lustre, reflectivity, and conductivity. The crystal structures of many transition-metal sulphides show that in addition to M—S bonds there are metal–metal bonds as, for example, in monosulphides with the NiAs structure (see later), in chromium sulphides, and in many 'sub-sulphides' such as Hf_2S,

Ta_2S, Pd_4S, and Ta_4S, some of which are clearly to be regarded as closer to inter-metallic compounds than to normal sulphides. The resemblance of alloys is even more marked in some selenides and tellurides, and the compounds CoTe and $CoTe_2$ are described later to illustrate this point.

The sulphides and oxides of many metals of groups (b) and (c) do not have similar formulae; for example, there is no sulphur analogue of Fe_2O_3 or of Pb_3O_4 or oxygen analogue of FeS_2. In cases where the oxide and sulphide of the same formula type do exist they generally have quite different structures, as in the following pairs: CuO and CuS, Cu_2O and Cu_2S, NiO and NiS, PbO and PbS. The structures of all these oxides have been noted in Chapter 12; each is different from that of the sulphide.

We saw in Chapter 12 that from the structural standpoint many transition metal–oxygen systems are surprisingly complex. This is also true of many metal–sulphur systems, as we shall show later for the sulphides of Cr, Ti, V, Nb, and Ta. Before doing this we shall note some of the simpler binary sulphide structures, taking them in the order: M_2S, MS, MS_2, M_2S_3 and M_3S_4. The chapter concludes with a short account of thio-salts and complex sulphides.

Sulphides M_2S

The alkali-metal compounds have the antifluorite structure; the structure of Cs_2S is not known. The structure of Tl_2S (a layer structure of the anti-CdI_2 type) is noted in Chapter 26. In Group IB there are Cu_2S and Ag_2S, both known as minerals; for Cu_2S see Chapter 25. There are three polymorphs of Ag_2S:

$$\text{monoclinic} \xrightarrow{176°C} \text{b.c. cubic} \xrightarrow{586°C} \text{f.c. cubic}$$

In the monoclinic form (acanthite) there are two kinds of non-equivalent Ag atoms with respectively 2 and 3 close S neighbours, but in the high-temperature forms, which are notable for their electrical conductivity, there is movement of Ag atoms between the interstices of the sulphur framework.[1],[2] A structure of the cuprite type has been assigned to Au_2S.[3]

Transition-metal sulphides M_2S include those of Ti, Zr, and Hf. The physical properties of these compounds and the fact that Ti_2S[4] and Zr_2S[5] (and also the selenides) are isostructural with Ta_2P[6] show that these are not normal valence compounds but essentially metallic phases. In the complex structure of Ti_2S there are six kinds of non-equivalent Ti atom with from 3 to 5 close S neighbours and also various numbers of Ti neighbours at distances from 2·8 Å upwards. Each S has at least 7 close Ti neighbours arranged at the vertices of a trigonal prism with additional atoms capping some rectangular faces. Metal–metal bonding clearly plays an important part in this structure, which may be compared with the anti-CdI_2 structure of Ti_2O. In contrast to Ti_2S and Zr_2S, Hf_2S[7] has the anti-$2H_2$ (hexagonal NbS_2) structure which has been described in Chapter 4. Within a layer S has 6 (trigonal prism) Hf neighbours, while Hf has an octahedral arrangement of nearest neighbours, 3 S (at 2·63 Å) in the layer and 3 Hf (at 3·06 Å) of the adjoining layer. This compound is diamagnetic and a metallic conductor, and if

(1) ZaP 1955 **7** 478
(2) JSSC 1970 **2** 309
(3) CR 1966 **B263** 1327

(4) AC 1967 **23** 77
(5) MRB 1967 **2** 1087
(6) ACSc 1966 **20** 2393

(7) ZK 1966 **123** 133

bond strengths are calculated from Pauling's equation (p. 1025), including the six weaker Hf–Hf contacts at 3·37 Å, the total is close to a valence of 4 for Hf:

Number of bonds	Bond	Length	Bond order	
3	Hf–S	2·63Å	0·56	Total
3	Hf–Hf	3·06	0·50	4·08
6	Hf–Hf	3·37	0·15	

suggesting that all the $5d^2s^2$ electrons are used for Hf–S and Hf–Hf bonds.

Monosulphides

These compounds provide examples of all our classes (a), (b), and (c). MgS and the alkaline-earth compounds form ionic crystals, but the same (NaCl) structure is adopted by MnS (but no other 3d monosulphide) and by the 4f and 5f compounds. This illustrates the point that similarity in geometrical structure type does not imply similar bond character; witness the silver coloured PbS and the metallic gold colour of LaS. The d transition-metal monosulphides (class (b)) are considered shortly.

Monosulphides are formed by the following B subgroup metals (class (c)): Cu, Zn, Cd, and Hg, Ga, In, and Tl, Ge, Sn, and Pb. Details of structures peculiar to one sulphide are given in Chapters 25 and 26 (CuS, hexagonal HgS, GaS, InS, and TlS). Compounds with the zinc-blende and wurtzite structures are listed in Table 17.1.

The structures of the monochalconides of the Group IVB metals are set out in Table 17.2. Both the black P and As structures are layer structures in which M forms only three strong bonds, forming corrugated versions of the simple 6-gon net as explained in Chapter 3. Further details of the sulphides are given in Chapter 26.

TABLE 17.2
Structures of Group IVB chalconides[1]

	S	Se	Te
Ge	P	P	As[a]
Sn	P	P	N
Pb	N	N	N

[a] Below 400°C. (1) IC 1965 4 1363

P = black P structure.
As = As structure.
N = NaCl structure.

Monosulphides of transition metals. Our present knowledge of the crystal structures of these compounds is summarized in Table 17.3, concerning which we may note the following points.

(i) In contrast to the monosulphides with the NiAs structure, the monoxides of Ti, V, Fe, Co, and Ni crystallize with the NaCl structure.

(ii) Note the special behaviour of Mn, with a half-filled 3d shell (d^5).

608

(iii) The structures of the d^4 (Cr^{II}) and d^9 (Cu^{II}) compounds are peculiar to these monosulphides. That of Cr is described shortly; that of CuS (covellite) is quite inexplicable as a compound of Cu^{II}, and in fact the Cu–S system is extremely complex. There are four distinct compounds, Cu_2S, $Cu_{1.96}S$, $Cu_{1.8}S$, and CuS, with a total of probably as many as nine different structures, though there is still some doubt about the structures of some of these phases. The complicated CuS structure is described in Chapter 25.

(iv) PdS and PtS have structures with planar 4-coordination of the metal atoms, that of PdS being rather less symmetrical than that of PtS, which is described later.

The nickel arsenide structure. The structure most frequently encountered is the NiAs structure (Fig. 17.1), which is also that of many phases MX in which M is a transition metal and X comes from one of the later B subgroups (Sn, As, Sb, Bi, S,

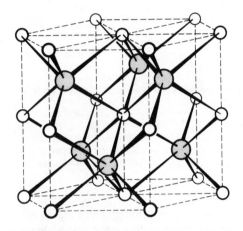

FIG. 17.1. The structure of NiAs (As atoms shaded). The Ni atom in the centre of the diagram is surrounded octahedrally by six As atoms and has also two near Ni neighbours situated vertically above and below.

Se, Te). We noted this structure in Chapter 4 as having h.c.p. X atoms with M in all the octahedral interstices, X therefore having 6 M neighbours at the apices of a trigonal prism. The immediate neighbours of a Ni atom are 6 As arranged octahedrally (at 2.43 Å), but since the $NiAs_6$ octahedra are stacked in columns in which each octahedron shares a pair of opposite faces with adjacent octahedra there are also 2 Ni atoms (at 2.52 Å) sufficiently close to be considered bonded to the first Ni atom. In the more metallic phases with this structure (e.g. CoTe or CrSb) these 8 neighbours are in fact equidistant from the transition-metal atom. It seems likely that the metal–metal bonds are essential to the stability of the structure. Compounds with this structure have many of the properties characteristic of intermetallic phases, opacity and metallic lustre and conductivity, and from the chemical standpoint their most interesting feature is their variable composition. In some systems the phase MX with the ideal NiAs structure is stable only at high temperatures. There are several possibilities for the structure of the phase MX at lower temperatures. A less symmetrical variant of the NiAs structure may be formed, an entirely new structure may be more stable, or there may be disproportionation to $M + M_{1-x}X$.

TABLE 17.3
Structures of transition-metal monosulphides

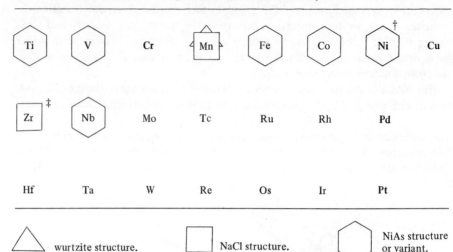

(1) AK 1954 **7** 371
(2) ZaC 1957 **292** 82
(3) AJC 1958 **11** 445

△ wurtzite structure. □ NaCl structure. ⬡ NiAs structure or variant.

Heavy type indicates structures peculiar to single compounds MS to which reference is made in the text.

† Also the millerite structure (5 : 5 coordination)

‡ An earlier study[1] indicated a tetragonal structure which was a distorted NaCl structure. This phase was not found in later studies[2],[3] which indicate a cubic structure, apparently a defect NaCl structure with ordered vacancies. The same cell ($a = 10.25$ Å) is found over the entire composition range Zr_5S_8–Zr_9S_8.

Some phases $M_{1-x}S$ have a sequence of c.p. S layers different from that in MS and are therefore recognized as definite compounds, as in the case of the sulphides of Ti described later. Alternatively the packing of the S atoms remains the same as in MS and there are vacant metal sites. For the removal of M atoms from the NiAs structure there are two simple possibilities:

(a) random vacancies in (000) and ($00\frac{1}{2}$)—see Fig. 17.2;
(b) (000) fully occupied but ($00\frac{1}{2}$) only partly filled.

The further alternatives are then (i) random, or (ii) ordered vacancies in alternate metal layers.

All examples of (b) represent structures intermediate between the NiAs structure and the $C\ 6$ (CdI_2) structure. The sulphides of Cr, which are described in a later section, provide examples of (b) (i) and (ii), and the Fe–B system illustrates (a) and (b) (ii).

The sulphides of iron include Fe_3S_4 (spinel structure), Fe_7S_8 (pyrrhotite), FeS (troilite), and FeS_2 (pyrites and marcasite). 'Ferrous sulphide' rarely has the Fe : S ratio precisely equal to unity, though stoichiometric FeS can be prepared. A microcrystalline form prepared by precipitation has been examined by X-ray powder photography and also by electron diffraction. This form (mackinawite) has

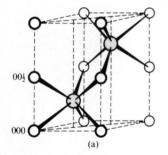

$00\frac{1}{2}$

000

(a)

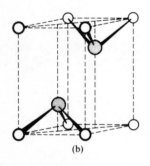

(b)

FIG. 17.2. Unit cells of the structures of (a) CoTe (NiAs structure), and (b) $CoTe_2$ (CdI_2 structure). Shaded circles represent Te atoms.

Metal Sulphides and Oxysulphides

a structure similar to that of LiOH.[1] The composition of 'FeS' ranges from 50 to approximately 46·6 atomic per cent Fe, and density measurements show that the departures from the composition FeS are due to iron deficiency, that is, $Fe_{1-x}S$. At temperatures above 138°C stoichiometric FeS has the NiAs structure, but at lower temperatures there are small displacements of the Fe atoms necessitating a larger (hexagonal) unit cell.[2] In the low-temperature forms of Fe_7S_8 one-eighth of the metal positions are unoccupied, and these vacancies are distributed in an orderly way in alternate metal layers (type (b) (ii) structure).[3] There are both monoclinic and 'hexagonal' forms of Fe_7S_8 which are superstructures of the defect NiAs type; a detailed study of one with trigonal symmetry has been made.[4] At temperatures above 340°C Fe_7S_8 is an example of (a), having random vacancies in all the metal positions.

In the Fe–S system the simple NiAs structure is stable over only a small range of composition, and if we include the defect structures the range is still only a few atomic per cent S. A number of metal–selenium and metal–tellurium systems show a different behaviour. In the Co–Te system the phase with the NiAs structure is homogeneous over the range 50–66·7 atomic per cent Te (using quenched samples), and at the latter limit the composition corresponds to the formula $CoTe_2$. Over this whole range the cell dimensions vary continuously, but change only by a very small amount:

(1) ZK 1968 **361** 94

(2) ACSc 1960 **14** 919

(3) AC 1953 **6** 557

(4) AC 1971 **B27** 1864

	a	c
50 per cent Te	3·882 Å	5·367 Å
66·7 per cent Te	3·784	5·403

The explanation of this unusual phenomenon, a continuous change from CoTe to $CoTe_2$, is as follows. The compound $CoTe_2$ crystallizes at high temperatures with the CdI_2 structure which is retained in the quenched specimen. (On prolonged annealing it changes over to the marcasite structure.) This CdI_2 structure is very simply related to the NiAs structure of CoTe, as shown in Fig. 17.2. In both the extreme structures the Te atoms are arranged in hexagonal close packing. At the composition CoTe all the octahedral holes are occupied by Co. As the proportion of Co decreases some of these positions become vacant, and finally, at the composition $CoTe_2$, only one-half are occupied, and in a regular manner, forming the CdI_2 structure. It should be emphasized that the homogeneity range and structures of samples depend on the temperature of preparation and subsequent heat treatment. For samples annealed at 600°C it is found that at the composition CoTe the product is a mixture of Co + $Co_{1-x}Te$ (defect NiAs structure). As the Te content increases the structure changes towards the $C6$ structure, but before the composition $CoTe_2$ is reached the structure changes to the marcasite structure.

The PtS *(cooperite) structure.* We have seen that FeS, CoS, and NiS crystallize with the NiAs structure, in which the metal atoms form six (or eight) bonds. Palladium and platinum however, form four coplanar bonds in their monosulphides. The structure of PtS[5] is illustrated in Fig. 17.3. Each Pt atom forms

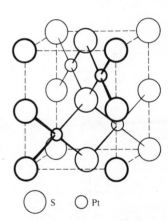

◯ S ◯ Pt

FIG. 17.3. The structure of PtS, showing the planar coordination of Pt by four S and the tetrahedral coordination of S by four Pt.

(5) MM 1932 **23** 188

four coplanar bonds and each S atom four tetrahedral bonds. The bond angles in this structure (and in the isostructural PtO and PdO) are not exactly 90° for Pt and $109\frac{1}{2}°$ for S because they represent a compromise. The angle α in the planar chain has a value ($97\frac{1}{2}°$) intermediate between that required for square Pt bonds (90°) and the tetrahedral value ($109\frac{1}{2}°$). The Pt bond angles are accordingly two of $82\frac{1}{2}°$ and two of $97\frac{1}{2}°$, while those of S are two of $97\frac{1}{2}°$ and four of 115°. In PdS[6] the coordination of the two kinds of atom is similar to that in PtS, but the structure is not quite so regular.

Disulphides

Disulphides are formed by the elements of Group IV and also by many transition elements. At ordinary temperature and pressure GeS_2 has a 3D framework structure in which GeS_4 tetrahedra share all vertices. At higher temperature and pressure GeS_2 and also SiS_2 (which normally crystallizes with a chain structure) transform into cristobalite-like structures[1] in which, however, the M—S—M bond angles are close to the tetrahedral value in contrast to the values 140–150° in silica structures. The GeS_4 and SiS_4 coordination groups are not regular tetrahedra but bisphenoids (foreshortened along the $\bar{4}$ axis) giving two S—M—S angles of 118° and four of 105°. (For SiS_2 and GeS_2 see also pp. 785 and 929.) La and the 4f metals form sulphides MS_x with x 1·7–2·0 for La and the lighter 4f elements and 1·7–1·8 for the heavier lanthanides. The detailed structures of these 'disulphides' are not known.[2] The disulphide BiS_2 has been made under pressure.[3] Most of the disulphides of the transition metals have either a layer structure or the pyrites or marcasite structures (Table 17.4), in contrast to the essentially ionic rutile-type structures of the dioxides.

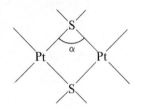

(6) ZK 1937 **96** 203

(1) Sc 1965 **149** 535

(2) IC 1970 **9** 1084
(3) IC 1964 **3** 1041

TABLE 17.4

Crystal structures of disulphides

Ti	V	Cr	Mn	Fe	Co	Ni	Cu	Zn
C	–	–	P	P/M	P	P	P[a]	P[a]
Zr	Nb	Mo	Tc	Ru	Rh	Pd	Ag	Cd
C	W[b]	W[b]	–	P	P	PdS_2	–	P[a]
Hf	Ta	W	Re	Os	Ir	Pt		
C	C[b]	W	W	P	P[c]	C		

C = CdI_2 (C 6) structure; W = MoS_2 structure (or polytype); P = pyrites structure; M = marcasite structure.

[a] Synthesized under pressure (IC 1968 7 2208).

[b] Also more complex layer sequences.

[c] Cation vacancies at atmospheric pressure; two different pyrites-like structures under 60 kbar. (IC 1968 7 389).

The layer structures are of two main types, with octahedral or trigonal prismatic coordination of the metal atoms The simplest structure of the first kind is the CdI_2 (C 6) structure of TiS_2, ZrS_2, HfS_2, TaS_2, and PtS_2. (It is stated that TaS_2 has been prepared with all the four layer structures adopted by the cadmium halides, namely, $C\,6$, $C\,27$, $C\,19$, and the 12-layer rhombohedral structure of CdBrI (p. 209)). Two distorted forms of the $C\,6$ structure have been described in which there are short metal–metal distances indicating metal–metal bonds. In one, the structure of WTe_2 and the high-temperature form of $MoTe_2$,[4] the metal atoms are situated off-centre in the octahedra leading to corrugated layers and two short M–M distances (compare the 8-coordination in NiAs), for example:

(4) AC 1966 **20** 268

$$WTe_2 \quad \begin{matrix} W-2\,W \quad 2 \cdot 86 \text{ Å} \\ \\ W-6\,Te \quad 2 \cdot 71\text{-}2 \cdot 82 \end{matrix}$$

In $ReSe_2$[5] the shifting of Re from the octahedron centres leads to three nearest Re neighbours at distances from $2 \cdot 65$–$3 \cdot 07$ Å which may be compared with $2 \cdot 75$ Å in the metal.

The structures in which there is trigonal prism coordination of the metal atom contain pairs of adjacent S layers which are directly superposed (and therefore not close-packed), but the multiple S–M–S layers are then packed in the same way as simple layers in normal c.p. sequences. The simplest structures of this kind are illustrated in Fig. 4.11 (p. 130), in which the nomenclature is similar to that used for mica polytypes. No example of the (2T) structure is yet known, but examples of three structures are:

2 H_1 ($C\,7$): MoS_2 (molybdenite), WS_2, $MoSe_2$, WSe_2, low-$MoTe_2$[4]

2 H_2:   hexagonal NbS_2[6].

3 R: rhombohedral MoS_2 [7]NbS_2, TaS_2, WS_2, ReS_2

(5) ACSc 1965 **19** 79
(6) N 1960 **185** 376
(7) SMPM 1964 **44** 105

The plan of a layer of the structures of Fig 4.11 has been illustrated in Fig. 4.9(b) (p. 128). As noted in Chapter 4 polytypes of some chalconides have been characterized in which the sequence of c.p. layers leads to both octahedral and trigonal prism coordination of M (e.g. TaS_2 and $TaSe_2$[8]); the classification of such structures has been discussed.[9]

(8) ACSc 1967 **21** 513
(9) AC 1965 **18** 31

The pyrites and marcasite structures.   The pyrites and marcasite structures, named after the two forms of FeS_2, are of quite a different kind. They contain discrete S_2 groups in which the binding is homopolar, the S–S distances being $2 \cdot 17$ Å and $2 \cdot 21$ Å respectively. Geometrically the pyrites structure is closely related to that of NaCl, the centres of the S_2 groups and the Fe atoms occupying the positions of Na^+ and Cl^- in the NaCl structure. Every Fe atom lies at the centre of an octahedral group of six S atoms, and the coordination of S is tetrahedral (S + 3 Fe). The c.n.'s are the same in the marcasite structure, which is derived from the rutile structure by rotating the chains of edge-sharing octahedra so that there are short S–S distances ($2 \cdot 21$ Å) between S atoms of different chains, as shown in Fig. 6.5(d) (p. 200). The two structures are illustrated in Fig. 17.4 and are further described in Chapter 6, the pyrites structure on p. 196, and the marcasite structure on p. 203.

There are interesting changes in the S—S and M—S distances in the 3d disulphides with the pyrites structure, differences which have been correlated with the numbers of dγ electrons.[10]

(10) JCP 1960 **33** 903

	S—S	M—S	Number of dγ electrons
MnS_2	2·086 Å	2·59 Å	2
FeS_2	2·171	2·259	0
CoS_2	2·124	2·315	1
NiS_2	2·065	2·396	2

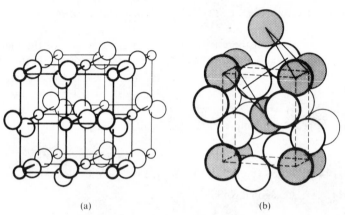

(a)   (b)

FIG. 17.4. The structures of the two forms of FeS_2: (a) pyrites, and (b) marcasite. In (a) the S—S distance has been reduced to accentuate the resemblance of this structure to that of NaCl. In (b) the shaded circles represent Fe atoms, six of which surround each S_2 group as is also the case in the pyrites structure.

Note the abnormal distances in MnS_2, attributed to the stability of the half-filled 3d shell. A cupric sulphide with the composition $CuS_{1.9}$ having the pyrites structure has been prepared from covellite (CuS) and S under high pressure at a temperature of 350°C or above.[11]

(11) IC 1966 **5** 1296

Just as the NiAs structure is adopted by a number of arsenides, stibides, etc. in addition to sulphides, selenides, and tellurides, so the pyrites structure is adopted by compounds such as $PdAs_2$, $PdSb_2$ (but not PdP_2), PtP_2, $PtAs_2$, and $PtSb_2$. Also, instead of S_2 or As_2 groups we may have mixed groups such as AsS or SbS, and we find a number of compounds structurally related to pyrites and marcasite but with lower symmetry due to the replacement of the symmetrical S_2 or As_2 group by AsS, etc. So FeS_2 and $PtAs_2$ have the pyrites structure but NiSbS a less symmetrical structure related to it. Similarly, FeS_2 and $FeAs_2$ have the marcasite structure, but FeAsS and FeSbS have related structures of lower symmetry (arsenopyrite structure).

All three minerals, CoAsS, cobaltite, NiAsS, gersdorffite, and NiSbS, ullmannite, have structures which are obviously closely related to pyrites. There are three simple structural possibilities:

(i)   Each S_2 group in pyrites has become As–S (Sb–S).
(ii) One-half of the S_2 groups have become As_2 (Sb_2).
(iii) There is random arrangement of S and As (Sb) in the S positions of pyrites.

Structure (i) was originally assigned to all three compounds, but as the result of later studies[12] the following structures have been proposed: CoAsS, (ii), NiAsS, (iii), and NiSbS, (i). It would seem that the greatest reliance may be placed on the later work on CoAsS, which shows that the apparently cubic structure is a polysynthetic twin of a monoclinic structure, and it is not impossible that other structures in this family are in fact superstructures of lower symmetry. As a result of the difference between Ni–Sb, 2·57 Å, and Ni–S, 2·34 Å, the symmetry of NiSbS has dropped to the enantiomorphic crystal class 23; the *absolute* structure has been determined.[13]

(12) AC 1957 **10** 764

The marcasite structure is adopted by only one disulphide (FeS_2) but also by a number of other chalconides and pnictides. Their structures fall into two groups, with quite differently proportioned (orthorhombic) unit cells, and there are apparently no intermediate cases: the differences in bonding leading to the two types of marcasite structure are not yet understood.

(13) MJ 1957 **2** 90

	c/a	c/b			
Normal marcasite structure	0·74	0·62	FeS_2		
			$FeSe_2$	$CoSe_2$	$\beta\text{-}NiAs_2$
			$FeTe_2$	$CoTe_2$	$NiSb_2$
Compressed marcasite structure (löllingite structure)[14]	0·55	0·48	$CrSb_2$	FeP_2	RuP_2
				$FeAs_2$	$RuAs_2$
				$FeSb_2$	$RuSb_2$
			OsP_2	$OsAs_2$	$OsSb_2$

(14) ACSc 1969 **23** 3043

In marked contrast to Pt, which forms only PtS and PtS_2 (CdI_2 structure), Pd forms a variety of sulphides, selenides, and tellurides,[15] the sulphides including Pd_4S, Pd_3S, $Pd_{2\cdot2}S$, PdS, and PdS_2. Some are high-temperature phases, for example, Pd_3S, which can be quenched but on slow cooling converts to $Pd_{2\cdot2}S + Pd_4S$. Both Pd_3S and Pd_4S are alloy-like phases, with high c.n.'s of Pd:

(15) ACSc 1968 **22** 819

$$Pd_3S: \quad \begin{matrix} Pd_I: 2\,S + 10\,Pd \\ Pd_{II}: 2\,S + 7\,Pd \end{matrix} \qquad Pd_4S: \quad Pd \left\{ \begin{matrix} 2\,S \\ 10\,Pd \end{matrix} \right.$$

Whereas RhS_2 and $RhSe_2$ have the normal pyrites structure, PdS_2 and $PdSe_2$ have a very interesting variant of this structure[16] which results from elongating that structure in one direction so that Pd has four nearest and two more distant S (Se) neighbours instead of the octahedral group of six equidistant neighbours. Alternatively the structure can be described as a layer structure, the layer consisting

(16) AC 1957 **10** 329

of Pd atoms forming four coplanar bonds to S_2 (Se_2) groups, as shown in Fig. 17.5. With this structure compare that of CuF_2, with (4 + 2)-coordination, derived in a somewhat similar way from the 6-coordinated rutile structure (p. 202). The following interatomic distances were found:

In PdS_2	Pd—4 S, 2·30 Å	
	Pd—2 S, 3·28	S—S, 2·13 Å
In $PdSe_2$	Pd—4 Se, 2·44 Å	
	Pd—2 Se, 3·25	Se—Se, 2·36 Å

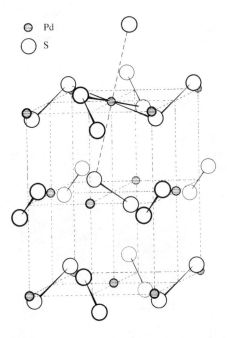

FIG. 17.5. The crystal structure of PdS_2. The large open circles represent S atoms.

There is also a high-pressure form of PdS_2, apparently with a less elongated pyrites-like structure.[17]

Sulphides M_2S_3 and M_3S_4

Most of the known sesquisulphide structures may be placed in one of three groups, corresponding to metal coordination numbers of 3, 4 and/or 6, or greater than 6 (Table 17.5).

Structures which do not fit into this simple classification include Sn_2S_3 and Rh_2S_3. In Sn_2S_3[1] double rutile chains (as in NH_4CdCl_3) of composition $Sn^{IV}S_3$ are connected through Sn^{II} atoms. In the octahedra Sn—S ranges from 2·50–2·61 Å (mean 2·56 Å) and Sn^{II} has 2 S at 2·64 Å and 1 S at 2·74 Å (mean 2·67 Å). The structure of Rh_2S_3 (and the isostructural Ir_2S_3)[2] consists of pairs of *face*-sharing octahedra which are linked into a 3D structure by further sharing of S atoms, so that a structure of 6 : 4 coordination (distorted octahedral and tetrahedral) results.

(17) IC 1969 **8** 1198

(1) AC 1967 **23** 471

(2) AC 1967 **23** 832

TABLE 17.5

Crystal structures of sesquisulphides M_2S_3 and related compounds

Class (i)

Characteristic M_2X_3 structures with 3-coordination of M (but see text):

As_2S_3 (layer) structure

Sb_2S_3 (chain) structure: Bi_2S_3, Th_2S_3, U_2S_3, Np_2S_3

Class (ii)

Structures with close-packed S and 4- or 6-coordination of M

Coordination of M	h.c.p.	c.c.p.	More complex sequences
Tetrahedral	Random wurtzite (Al_2S_3, β-Ga_2S_3) Ordered defect wurtzite (α-Ga_2S_3)	Random zinc-blende (γ-Ga_2S_3)	
Tetrahedral and octahedral		β-In_2S_3 (see Table 17.6)	
Octahedral	NiAs superstructure (Cr_2S_3) Corundum structure (Al_2S_3)	Ordered defect NaCl (Sc_2S_3)	Mo_2S_3, Bi_2Se_3 Bi_2Te_3 Sc_2Te_3

Other octahedral structures: Rh_2S_3

Class (iii)

Structures with higher coordination of M:

6 and 7: Ho_2S_3

7 and 8: Gd_2S_3

8: Ce_2S_3 (La_2S_3, Ac_2S_3, Pu_2S_3, Am_2S_3)

This structure has obvious resemblances to the corundum structure. The shortest Rh–Rh distance (3·2 Å) shows that there are no metal–metal bonds.

Class (i): M_2S_3 *structures with 3-coordination of* M. It might be expected that some of the simplest structures for sulphides M_2S_3 would be found among the compounds of Group V elements. Strangely enough, phosphorus forms no sulphide P_2S_3 (or P_4S_6), though it forms four other sulphides (p. 694). As_2S_3 has a simple layer structure[3] (p. 723), but that of Sb_2S_3 is much more complex (p. 724). The structure of Sb_2S_3 is illustrated in Fig. 20.13; it has been confirmed by a later study of the isostructural Sb_2Se_3.[4]

Class (ii): *structures with close-packed* S. In these structures metal atoms occupy tetrahedral and/or octahedral holes in a c.p. assembly of S atoms. With the exception of one form of Al_2S_3 which crystallizes with the corundum structure these are not typically M_2X_3 structures but are defect structures, that is, MX structures (zinc-blende, wurtzite NiAs, or NaCl) from which one-third of the M atoms are missing. In some structures the arrangement of the vacancies is random and in others regular.

The structures of a number of sulphides are related in the following way. A

(3) MJ 1954 **1** 160

(4) AC 1957 **10** 99

(5) ZaC 1955 **279** 241; AC 1963 **16** 946

(6) AC 1962 **15** 1195

(7) AC 1962 **15** 1198

(8) PNAS 1955 **41** 199

(9) AC 1965 **19** 967

(10) AC 1966 **20** 566; AC 1967 **23** 111

(11) JSSC 1970 **2** 6

(12) IC 1964 **3** 1220

cubic block of the zinc-blende structure with edges equal to twice those of the unit cell (Fig. 3.35(b), p. 102) contains 32 c.c.p. S atoms. Removal of one-third of the metal atoms at random gives the structure of γ-Ga_2S_3.[5] (At temperatures above 550°C β-Ga_2S_3 has a random wurtzite structure.) In this structure an average of $21\frac{1}{3}$ tetrahedral holes are occupied in each block of 32 S atoms In Co_9S_8[6] the metal atoms occupy the same number of tetrahedral holes (32) as in ZnS—but a different selection of 32 holes—and in addition 4 octahedral holes. The neighbours of a Co atom in a tetrahedral hole are 1 S at 2·13 Å and 3 S at 2·21 Å but also 3 Co at 2·50 Å (the same as Co–Co in the metal). In $Rh_{17}S_{15}$,[7] the corresponding phase in the Rh–S system, the metal–metal bonds (2·59 Å) are even shorter than in the metal (2·69 Å).

If the appropriate sets of 8 tetrahedral and 16 octahedral holes are occupied in an assembly of 32 c.c.p. S atoms we have the spinel structure, and this is the structure of Co_3S_4, though in this case the cubic closest packing is somewhat distorted. (Zr_3S_4 is another example of a sulphide of this type.) The spinel-type structure of Co_3S_4 extends over the composition range $Co_{3.4}S_4$ to $Co_{2.06}S_4$ for solid phases prepared from melts, i.e. it includes the composition Co_2S_3, but Co_2S_3 prepared by heating together Co, S, and a flux has a statistical spinel structure, so that Co_2S_3 is related to Co_3S_4 in the same way as Fe_2O_3 (cubic) is to Fe_3O_4.[8]

In γ-Al_2O_3 $21\frac{1}{3}$ metal atoms are distributed at random over the 8 tetrahedral and 16 octahedral sites of the spinel structure. In the low-temperature (α) form of In_2S_3 there is believed to be preferential occupation of the octahedral sites, as has been suggested for γ'-Al_2O_3. The structure of the high-temperature (β) form of In_2S_3 may be described as an ordered defect spinel superstructure. The dimensions of the very elongated tetragonal unit cell are: $a_{tetr.} = a_{cubic}/\sqrt{2}$, $c_{tetr.} = 3a_{cubic}$, a_{cubic} being the edge of the cubic spinel cell. This cell therefore contains 48 O atoms, and In atoms occupy 8 tetrahedral and 24 octahedral sites, as compared with 12 tetrahedral and 24 octahedral sites in the spinel structure.[9] (In addition to the two forms of In_2S_3 indium forms InS, In_6S_7, and In_3S_4 (stable above 370°C).)[10] Four forms of Al_2S_3 have been described, α and β with defect wurtzite-like structures, γ (corundum), and a high-pressure tetragonal form with a defect spinel structure like β-In_2S_3.[11]

The (unique) structure of Sc_2S_3[12] is, like that of β-In_2S_3, referable to a cell containing 48 c.c.p. S atoms, but this cell has dimensions $2a$, $\sqrt{2}a$, and $3\sqrt{2}a$, where a would be the cell dimension of a simple NaCl structure. All Sc atoms occupy octahedral holes, so that the structure is a defect NaCl structure with ordered vacancies. Each S atom has 4 Nb neighbours at four of the vertices of an octahedron (*cis* vertices vacant) and there is very little disturbance of the original NaCl structure, for all the bond angles are 90° to within 1·2°. This is the structure of the yellow stoichiometric Sc_2S_3, which is a semiconductor. There is also a black non-stoichiometric Sc_2S_3 which is a metallic conductor ($Sc_{2+x}^{3+}(e)_{3x}S_3^{2-}$), also with a NaCl-type structure referable to the simple rhombohedral cell of Fig. 6.3(b). The cell content is presumably 1 Sc at (000), 0·37 Sc at $(\frac{1}{2}\frac{1}{2}\frac{1}{2})$, and 2 S at $(\frac{1}{4}\frac{1}{4}\frac{1}{4})$. This group of c.p. structures is summarized in Table 17.6.

TABLE 17.6

Sulphides with cubic close-packed S atoms

	Structure	Interstices occupied	
		Octahedral	Tetrahedral
CeS	NaCl	All	—
Sc_2S_3	Defect NaCl	2/3	—
β-In_2S_3	Defect spinel (superstructure)	1/2	1/12
Co_3S_4	Spinel	1/2	1/8
Co_9S_8		1/8	1/2
γ-Ga_2S_3	Defect zinc-blende (random)	—	1/3
ZnS	{ Zinc-blende { Wurtzite	— —	} 1/2

It is interesting that although ScTe has, like ScS, the NaCl structure, the close packing of Te in Sc_2Te_3 is of the *cchh* (12-layer) type, with alternate layers of octahedral metal sites one-third occupied (statistically).[13] The same layer sequence is found in Fe_3S_4, with every fourth layer of octahedral sites unoccupied. For Cr_3S_4 and Ti_3S_4 see pp. 622 and 625.

In contrast to Bi_2S_3, the corresponding selenide and telluride have structures in which Bi occupies octahedral holes in close-packed assemblies of Se or Te atoms. Interesting examples of some of the more complex types of close packing are found in Bi_2Se_3, Bi_2Te_2S, Bi_2Te_3 (the last two being the minerals tetradymite and tellurobismuthite respectively), and in Bi_3Se_4.

Representing the S, Se or Te layers by *A, B,* or *C* (p. 127), and the Bi atoms as *a, b,* or *c* in the octahedral positions between the layers we find the 9-layer sequence

$$chh: \quad A \quad BA \quad B \quad CB \quad C \quad AC \quad A \dots$$
$$ c \quad\; c \quad\;\; a \quad\;\; b \quad\;\; b$$

in Bi_2Se_3, Bi_2Te_2S, and Bi_2Te_3, and the 12-layer sequence

$$cchh: \quad C \quad B \quad AB \quad A \quad C \quad BC \quad B \quad A \quad CA \quad C \dots$$
$$ a \quad c \quad\;\; c \quad b \quad a \quad\;\; a \quad\;\; c \quad b \quad\;\; b \quad a$$

in Bi_3Se_4.

The *chh* sequence also occurs in Mo_2S_3,[14] but this is not a layer structure since the following fractions of octahedral sites are occupied between successive pairs of (approximately) c.p. layers:

$$c \quad h \quad h \quad c \quad h \quad h$$
$$1 \quad \tfrac{1}{2} \quad \tfrac{1}{2} \quad \text{etc.}$$

$$M_I \quad M_{II} \quad M_{II}$$

(13) IC 1965 **4** 1760

(14) JSSC 1970 **2** 188

Metal Sulphides and Oxysulphides

There is appreciable distortion from the ideal c.p. structure (in which the metal atoms would be at the centres of the octahedral holes) owing to the formation of zigzag chains of metal–metal bonds in both the fully occupied and the half-occupied metal layers. These bonds are not much longer than in the metal. In the isostructural Nb_2Se_3 (and Ta_2Se_3)[15] there is metal–metal bonding only in the fully occupied layers, possibly because Nb and Ta each has one fewer d electrons than Mo:

(15) AC 1968 **B24** 1102

| | M—M in chains within metal layers | |
	Mo_2S_3	Nb_2Se_3
M_I layer	2·85 Å	2·97 Å
M_{II} layer	2·87	3·13
Compare b.c. metal	2·73	2·86

The compounds we have been discussing illustrate three ways of attaining the composition M_2X_3 by occupying two-thirds of the octahedral holes in c.p. assemblies. The fractions of holes occupied between successive pairs of c.p. layers are:

											c.p. sequence
Bi_2Se_3	0	1	1	.	0	1	1	.			*chh*
Mo_2S_3	1	$\frac{1}{2}$	$\frac{1}{2}$	.	1	$\frac{1}{2}$	$\frac{1}{2}$	.			*chh*
Sc_2Te_3	1	$\frac{1}{3}$	.	1	$\frac{1}{3}$	.	1	$\frac{1}{3}$	.		*cchh*

Class (iii): *structures with higher coordination of* M.   We now come to a group of structures adopted by sesquisulphides of Y, La, and the 4f and 5f elements in which the c.n. of some or all of the metal atoms exceeds 6.

There has been some confusion about the structures of certain 4f metal sesquisulphides, probably because they lose S if not made in a closed apparatus. For example, the Ce_2S_3 (defect Th_3P_4) structure has been assigned to all the compounds from La_2S_3 to Dy_2S_3, but a later study showed only Eu_3S_4 to have this structure. It is still not certain that the Ce_2S_3 structure can exist for the exact composition M_2S_3 for any Ln_2S_3. Another structure (β or B) as yet undetermined, has been assigned to 'sesquisulphides' of some of these elements, but this also may be characteristic only of S-deficient compounds. Table 17.7 shows the structures of the Ln_2S_3 compounds; La, Y, and Sc are added at places appropriate to their ionic radii.

620

TABLE 17.7

Crystal structures of 4f sesquisulphides

(La)	Ce	Pr	Nd	Pm	Sm	Eu	Gd	Tb	Dy	(Y)	Ho	Er	Tm	Yb	Lu	(Sc)
				?		*										

←——————— Gd_2S_3 structure ————————→

←—— Ho_2S_3 structure →   ←——→ Sc_2S_3

Corundum
structure

* Only Eu_3S_4 prepared.

With increasing ionic size the c.n. increases from 6 (octahedral) in Sc_2S_3 and in Lu_2S_3 and Yb_2S_3, to 8 (dodecahedral) in phases M_2S_3–M_3S_4 with the defect Th_3P_4 structure:

	C.N. of M	Mean c.n.	Reference
Ho_2S_3	6 and 7	$6\frac{1}{2}$	IC 1967 6 1872
α-Gd_2S_3	7 and 8	$7\frac{1}{2}$	IC 1968 7 1090
Ce_2S_3	8	8	IC 1968 7 2282; IC 1969 8 2069

The Ho_2S_3 structure is not a simple c.p. structure but has one-half of the metal atoms 6- and the remainder 7-coordinated, and two-thirds of the S 4-coordinated, and one-third 5-coordinated. The Gd_2S_3 structure also is complex, with equal numbers of metal atoms 7- and 8-coordinated (mono- and bi-capped trigonal prism), and all S atoms 5-coordinated (two-thirds square pyramidal and one-third trigonal bipyramidal).

In the Ce_2S_3 structure metal atoms occupy $\frac{8}{9}$ of the metal positions in the Th_3P_4 structure, that is, $10\frac{2}{3}$ Ce are distributed over 12 positions in a cell containing 16 S atoms. The formula is therefore preferably written $Ce_{2.68}S_4$. The coordination polyhedron CeS_8 is a triangulated dodecahedron. The cell dimensions of the La and Ce phases with this structure remain nearly constant over the composition range $M_{2.68}S_4$ (M_2S_3) to M_3S_4. This would not be expected if some M^{3+} are changing to the larger M^{2+} ions; possibly the metal remains as M^{3+}, the extra electrons being delocalized. On the other hand, the cell dimension *does* increase on going from Sm_2S_3 to Sm_3S_4, Sm^{2+} being more stable than Ce^{2+}; compare the difference between CeS and SmS. The former is metallic with a magnetic moment corresponding to Ce^{3+}, that is, it is $Ce^{3+}(e)S^{2-}$, whereas SmS is a semi-conductor with magnetic moment corresponding to $Sm^{2+}S^{2-}$. Other compounds with the Ce_2S_3 structure include Ac_2S_3, Pu_2S_3, and Am_2S_3.

The sulphides of chromium

The Cr–S system is much more complex than it was originally thought to be.

Metal Sulphides and Oxysulphides

Between the compositions CrS and Cr_2S_3 there are three definite solid phases:

CrS	monoclinic	$\approx CrS_{0.97}$
Cr_7S_8	trigonal	$Cr_{0.88}S - Cr_{0.87}S$
Cr_5S_6	trigonal	$Cr_{0.85}S$
Cr_3S_4	monoclinic	$Cr_{0.79}S - Cr_{0.76}S$
Cr_2S_3	trigonal	$Cr_{0.69}S$
Cr_2S_3	rhombohedral	$Cr_{0.67}S$

In all the trigonal Cr sulphides Cr has six S neighbours at 2·42–2·46 Å but there are also Cr–Cr bonds of length approximately 2·80 Å, that is, there are ionic Cr–S bonds but also metal–metal bonds.

(1) AC 1957 **10** 620

All these sulphides except CrS[1] have structures intermediate between the NiAs and CdI_2 (*C* 6) structures. In Cr_7S_8 there are random vacancies in alternate metal layers of the NiAs structure, while the others have ordered vacancies in alternate metal layers. The proportions of *occupied* metal sites between c.p. layers are:

$$1\,\tfrac{2}{3} \qquad 1\,\tfrac{1}{2} \qquad 1\,\tfrac{1}{3} \qquad \text{and} \quad 1\ 0$$
$$M_5S_6 \qquad M_3S_4 \qquad M_2S_3 \qquad\qquad MS_2$$

The patterns of *vacant* metal sites in Cr_2S_3 (trigonal and rhombohedral forms) and Cr_5S_6 are shown in Fig. 17.6; Cr_3S_4 is also of this type. A further sulphide, Cr_5S_8, has been produced under pressure;[2] it also has a structure of the same general type, as also does the isostructural V_5S_8.[3]

(2) IC 1969 8 566
(3) CR 1964 **258** 5847

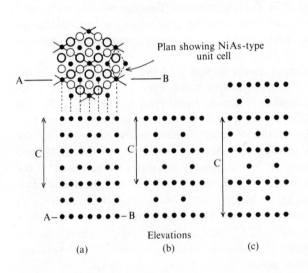

Plan showing NiAs-type unit cell

Elevations

(a)   (b)   (c)

FIG. 17.6. Ordered vacancies in the NiAs structure: (a) Cr_5S_6, (b) trigonal Cr_2S_3, (c) rhombohedral Cr_2S_3.

The structure of CrS is unique and intermediate between that of NiAs and PtS. The neighbours of Cr are four S at 2·45 Å (mean) and two much more distant (2·88 Å)—compare CrF_2 with a deformed rutile structure and also (4 + 2)-

coordination. Although CrS is formally isotypic with PtS the cells are of very different shapes and CrS is best regarded as a new structure type. It is illustrated in Fig. 17.7.

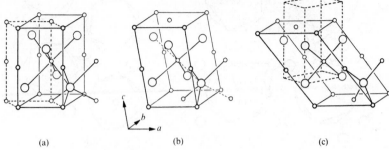

FIG. 17.7. The relation between the structures of (a) NiAs, (b) CrS, and (c) PtS. In (a) and (c) the broken lines indicate the conventional unit cells.

The sulphides of vanadium, niobium, and tantalum

The formulae and structures of the sulphides and a comparison with the oxides formed by these metals illustrate the general points noted at the beginning of this chapter. For example, at least nine crystalline Nb–S phases have been characterized, and none has its counterpart among niobium oxides. There are two forms of NbS_{1-x} (low temperature, NiAs superstructure, high temperature, MnP structure), two forms of $Nb_{1+x}S_2$, and two of NbS_2 (hexagonal and rhombohedral MoS_2 structures), in addition to the other sulphides noted:

$V_3S^{(1)}$	$Nb_{21}S_8^{(3)}$	$Ta_6S^{(5)}$
		$Ta_2S^{(6)}$
VS	NbS_{1-x} (2 forms)	
V_3S_4	$Nb_3S_4^{(4)}$	
$V_5S_8^{(2)}$	$Nb_{1+x}S_2$ (2 forms)	
	NbS_2 (2 forms)	TaS_2 (CdI$_2$, Cd(OH)Cl, CdCl$_2$ structures)
	NbS_3	TaS_3
VS_4		

(1) AC 1959 **12** 1022
(2) CR 1964 **258** 5847
(3) AC 1968 **B24** 412
(4) AC 1968 **B24** 1614
(5) AC 1970 **B26** 125
(6) AC 1969 **B25** 1736

Some of these compounds have complex structures which are not easily described, such as the two forms of V_3S and $Nb_{21}S_8$. The latter is a compound with metallic properties in which there are six kinds of Nb atoms with from 1 to 4 S neighbours.

The structure of Nb_3S_4 was illustrated in Fig. 5.41 as built of triple columns of face-sharing octahedral NbS_6 groups, each of which also shares four edges to form a 3D structure with a general geometrical similarity to the UCl$_3$ structure. The simplified projection of Fig. 17.8(a) shows that there are empty tunnels through the structure and that the S atoms are of two kinds Those (S_1) on the surfaces of the tunnels have a very unsymmetrical arrangement of 4 Nb neighbours, while those (S_2) on the central axes of the columns have 6 (trigonal prism) Nb neighbours. The

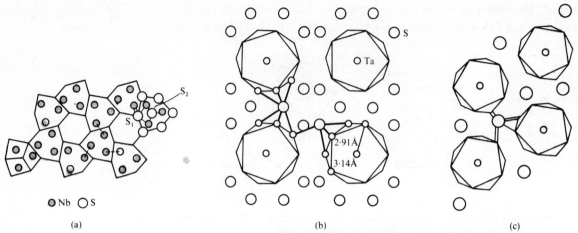

FIG. 17.8. Projections of the structures of (a) Nb_3S_4, (b) Ta_2S, (c) Ta_6S.

Nb atoms are displaced from the centres of their octahedral coordination groups so that Nb–Nb bonds are formed as zigzag chains perpendicular to the paper as indicated in projection by the broken lines in Fig. 17.8(a). The compound is a metallic conductor; Nb–Nb, 2·88 Å, compare 2·86 Å in the metal.

The structures of Ta_2S and Ta_6S are closely related and form a link between the metal-rich chalconides (and phosphides), with extensive metal–metal bonding, and the 'metal cluster' halides of Nb, Ta, etc. discussed in Chapter 9. In both these sulphide structures (Fig. 17.8(b) and (c)) the metal atoms form columns consisting of body-centred pentagonal antiprisms sharing their basal (pentagonal) faces. (Since the Ta atoms at the centres of the antiprisms have an icosahedral arrangement of 12 nearest neighbours—10 forming the antiprism and 2 at the body-centres of adjacent antiprisms—the columns could also be described as built of interpenetrating icosahedra.) These columns of metal atoms are held together by the S atoms, which in Ta_2S are of two kinds (with 4 or 6 neighbours) and in Ta_6S are all similar and have 7 Ta neighbours (monocapped trigonal prism). The Ta atoms at the centres of the columns are entirely surrounded by (12) Ta atoms; those on the periphery have 2 or 3 S neighbours (in Ta_2S) or 1 or 2 S neighbours (in Ta_6S), the remaining close neighbours being Ta atoms of the same column. There are no very short Ta–Ta contacts between the columns, but although the main metal–metal interactions are within the columns the interatomic distances indicate that there may be some interaction of this kind between the columns. Thus in Ta_2S the shortest Ta–Ta distances are those between atoms at the centres of the antiprisms (2·80 Å), as compared with 3·14 Å between those on the surfaces of the columns, but there are some comparable contacts (3·10 Å) between the columns.

It is noteworthy that although S and P are similar in size, and Ti_2S and Zr_2S are isostructural with Ta_2P, yet Ta_2S has a unique structure with largely Ta–Ta interactions, that is very low Ta–S coordination numbers; compare the 3–5 close S neighbours of a metal atom in Ti_2S.

624

The sulphides of titanium

Apart from the extreme compounds TiS_3 and Ti_2S the sulphides of Ti include TiS_2 and TiS, with the simple (h.c.p.) CdI_2 and NiAs structures, and a series of phases with compositions intermediate between those of the disulphide and monosulphide (Table 17.8). These phases have structures based on more complex c.p. sequences in which Ti atoms occupy octahedral interstices. Certain layers of metal atom sites are fully (or almost fully) occupied while others are only partially occupied (at random)—contrast the Cr sulphides Cr_7S_8, Cr_5S_6, and Cr_3S_4, in which the packing of the S atoms remains the same (h.c.p.).

TABLE 17.8

The sulphides of titanium: TiS_2-TiS; $Li_xTi_{1.1}S_2$

Sulphide	S Layer sequence	Fractional site occupancy between successive S layers	Reference
TiS_2	h	$h_{1.0}h_0$	—
$Ti_5S_8(Ti_3S_5)$	$cchh$	$c_{1.0}c_{0.2}h_{1.0}h_{0.2}$	RTC 1966 85 869
"Ti_2S_3"	ch	$c_{1.0}h_{0.4}$	AC 1957 10 715
$Ti_{2.45}S_4$		$c_{1.0}h_{0.23}$	JSSC 1970 2 36
Ti_3S_4	$chhchch$	$c_{0.5}h_{1.0}h_{0.5}c_{0.95}h_{0.7}c_{0.7}h_{0.95}$	JSSC 1970 1 519
Ti_4S_5	$chchh$	$c_{0.9}h_{0.9}c_{0.6}h_{1.0}h_{0.6}$	JSSC 1970 1 519
Ti_8S_9	chh	$c_{0.83}h_{1.0}h_{0.83}$	JSSC 1970 1 519
TiS	h	$h_{1.0}h_{1.0}$	—
$Li_xTi_{1.1}S_2$			Sc 1972 175 884

Compounds $Li_xTi_{1.1}S_2$ prepared by melting together metallic Li and $Ti_{1.1}S_2$ (CdI_2 structure) have the *ch* packing for *x* between 0·1 and 0·3, with apparently random occupancy by Li and Ti atoms of octahedral sites in the partially filled layers. At higher Li concentrations ($0.5 < x < 1.0$) a completely different (tetragonal) structure is adopted. These compounds, and also compounds Na_xMoS_2 and M_xZrS_2 and M_xHfS_2 (M = Na, K, Rb, or Cs) are notable for developing superconductivity at very low temperatures.

Complex sulphides and thio-salts

An extraordinary variety of solid phases consisting of sulphur combined with more than one kind of metal is found in the mineral world. As in oxides isomorphous replacement is widespread, leading to random non-stoichiometric compounds, but we may recognize three main types of compound. If for simplicity we describe the bonds A–S and B–S in a compound $A_xB_yS_z$ as essentially ionic or essentially covalent we might expect to find three combinations:

	A–S	B–S
(a)	Ionic	Ionic
(b)	Ionic	Covalent
(c)	Covalent	Covalent

In (a) and (c) there would be no great difference between the characters of the A–S and B–S bonds in a particular compound, while in (b) the B and S atoms form a covalent complex which may be finite or infinite in one, two, or three dimensions. By analogy with oxides we should describe (a) and (c) as complex sulphides and (b) as thio-salts. Compounds of type (c) are not found in oxy-compounds, and moreover the criterion for isomorphous replacement is different from that applicable to complex oxides because of the more ionic character of the bonding in the latter. In ionic compounds the possibility of isomorphous replacement depends largely on ionic radius, and the chemical properties of a particular ion are of minor importance. So we find the following ions replacing one another in oxide structures: Fe^{2+}, Mg^{2+}, Mn^{2+}, Zn^{2+}, in positions of octahedral coordination, while Na^+ more often replaces Ca^{2+} (which has approximately the same size) than K^+, to which it is more closely related chemically. In sulphides, on the other hand, the criterion is the formation of the same number of directed bonds, and we find atoms such as Cu, Fe, Mo, Sn, Ag, and Hg replacing Zn in zinc-blende and closely related structures.

Obviously this naïve classification is too simple to accommodate all known compounds, and because of its basis it has the disadvantage of prejudging the bond type. An essentially geometrical classification based on known crystal structures would, however, be of the same general type. Class (a) includes structures like those of complex oxides (see Table 17.9) but will tend to merge into class (c) as bond character changes from ionic to covalent or covalent–metallic. In class (a) the ions

TABLE 17.9
Crystal structures of some complex sulphides and thio-salts

$A_xB_yS_z$	C.N.'s of A and B	Type of thio-ion	Reference
$(NH_4)_2WS_4$	(9,10) : 4	Finite	AC 1963 **16** 719
Tl_3VS_4	(4 + 4) : 4	Finite	AC 1964 **17** 757
$Na_3SbS_4 . 9 H_2O$	6 : 4	Finite	AC 1950 **3** 363
$KFeS_2$	8 : 4	Chain (edge-sharing)	RTC 1942 **61** 910
$NH_4(CuMoS_4)$	12 : 4	Chain (edge-sharing)	IC 1970 **9** 1449
Ba_2MnS_3	7 : 4	Chain (vertex-sharing)	IC 1971 **10** 691
Ba_2ZnS_3	7 : 4	Double chain	ZaC 1961 **312** 99
KCu_4S_3	8 : 4	Double layer	ZaC 1952 **269** 141
$(NH_4)Cu_7S_4$	8 : 4	3D framework	AC 1957 **10** 549
		Structure	
$BaZrS_3$	12 : 6	Perovskite	} AC 1963 **16** 134
$BaTiS_3$			
$BaVS_3$	12 : 6	$CsNiCl_3$	AC 1969 **B25** 781;
$BaTaS_3$			IC 1969 **8** 2784
$BaSnS_3$	9 : 6	NH_4CdCl_3	MRB 1970 **5** 789
$NaCrS_2$, $NaInS_2$	6 : 6	NaCl superstructure	JPCS 1968 **29** 977
$LiCrS_2$	6 : 6	NiAs superstructure	IC 1970 **9** 2581
$NiCr_2S_4$	6 : 6	NiAs superstructure	IC 1966 **5** 977
$NaBiS_2$	6 : 6	NaCl (statistical)	RTC 1944 **63** 32
$FeCr_2S_4$, $CuCr_2S_4$ } $ZnAl_2S_4$	4 : 6	Spinel	AKMG 1943 **17B** No. 12 ZaC 1967 **23** 142

A and B are usually those of the more electropositive elements of the earlier A subgroups or of certain B subgroup elements (e.g. In^{3+}, Bi^{3+}). In thio-salts A may be an alkali metal, Ag, Cu(I), NH_4, or Tl(I), and B a non-metal or metalloid (Si, As, Sb) or a transition metal in a high oxidation state (V^V, Mo^{VI}). In compounds of class (c) both metals are typically from the B subgroups (Cu, Ag, Hg, Sn, Pb, As, Sb, Bi) but include some transition elements such as Fe.

We have noted one difference between complex oxides and sulphides, namely, the compounds of class (c) have no counterpart among oxy-compounds. A second difference is that sulphides other than those of the most electropositive elements show more resemblance to metals than do oxides. Metal–metal bonding occurs only rarely in simple oxides whereas it is more evident in many transition-metal sulphides. In many complex sulphides of class (c), as indeed in simple sulphides such as those of Cu, it is not possible to interpret the atomic arrangements and bond lengths in terms of normal valence states of the metals, suggesting a partial transition to metallic bonding, as is also indicated by the physical properties of many of these compounds.

Thio-salts

A considerable number of thio-salts containing alkali metals have been prepared in one of two ways:

(i) The sulphides of some non-metals and of certain of the more electronegative metals dissolve in alkali sulphide solutions, and from the resulting solutions compounds may be crystallized or precipitated by the addition of alcohol. These compounds are usually very soluble, often highly hydrated, and often easily oxidized and hydrolysed. Thiosilicates and thiophosphates have been prepared, and other compounds of this kind include $(NH_4)_3VS_4$, $Na_6Ge_2S_7 . 9 H_2O$, K_3SbS_3, $Na_3SbS_4 . 9 H_2O$, thiomolybdates, M_2MoS_4, and thiotungstates, M_2WS_4. The existence of tetrahedral thio-ions has been established in $Na_3SbS_4 . 9 H_2O$, and in $(NH_4)_2MoS_4$ and $(NH_4)_2WS_4$, the last two being isostructural with one form of K_2SO_4. Many other soluble thio-salts presumably also contain finite thio-ions analogous to the more familiar oxy-ions.

(ii) Prolonged fusion of a transition metal or its sulphide with sulphur and an alkali-metal carbonate, followed by extraction with water, yields compounds such as $KFeS_2$, $NaCrS_2$, and KCu_4S_3. (This method can also be used to give complex sulphides such as $KBiS_2$.) In all these compounds, which are insoluble and highly coloured, the alkali metal is apparently present as ions M^+, but the thio-ions are of various types, chain, layer, or 3D frameworks.

The steel-blue fibrous crystals of $KFeS_2$ are built of infinite chain ions formed of FeS_4 tetrahedra sharing opposite edges, and between these chains lie the K^+ ions surrounded by 8 S atoms (Fig. 17.9). An interesting elaboration of this chain occurs in $NH_4(Cu^I Mo^{VI} S_4)$, where Cu^I and Mo^{VI} alternate along the chain. In ammoniacal cuprous chloride solution crystals of $KFeS_2$ change into the brassy, metallic $CuFeS_2$ with the zinc-blende type of structure which is described shortly, in which no thio-ions can be distinguished. The action of H_2S on a mixture of BaO and ZnO at $800°C$ gives Ba_2ZnS_3, also containing an infinite one-dimensional

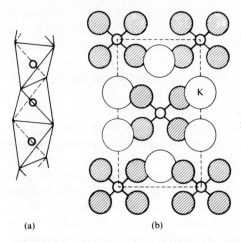

FIG. 17.9. (a) Arrangement of Fe atoms (small circles) in chains of FeS_4 tetrahedra in $KFeS_2$. (b) Projection of the structure of $KFeS_2$ along the direction of the chains.

(a)   (b)

thio-ion, in this case the double chain of tetrahedra found in the isostructural K_2CuCl_3 (q.v.). Ba_2MnS_3, with single vertex-sharing chains is isostructural with K_2AgI_3.

Copper forms a number of complex thio-salts with alkali metals. Dark-blue crystals of KCu_4S_3 can be prepared by fusing the metal with alkali carbonate and sulphur and extracting with water. The crystals are good conductors of electricity. Their structure consists of double layers built of CuS_4 tetrahedra, the layers being interleaved with K^+ ions surrounded by 8 S at the vertices of a cube. A 3D thio-ion is found in $NH_4Cu_7S_4$. Finely divided copper reacts slowly with ammonium sulphide solution in the absence of air to give, among other products, black, lustrous (tetragonal) crystals of $NH_4Cu_7S_4$. We are interested here only in the general nature of the structure (Fig. 17.10) which is a charged 3D framework of composition $Cu_7S_4^-$ built of columns cross-linked at intervals by Cu atoms arranged statistically in three-quarters of the positions indicated by the smallest circles. The framework encloses cubical holes between 8 S atoms which are occupied by NH_4^+ ions.

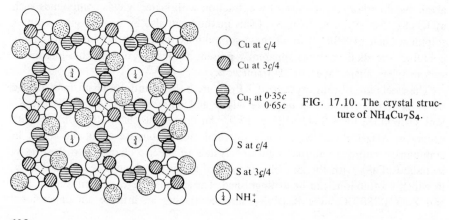

Cu at $c/4$

Cu at $3c/4$

$Cu_{\frac{1}{2}}$ at $\frac{0\cdot35c}{0\cdot65c}$

S at $c/4$

S at $3c/4$

NH$_4^+$

FIG. 17.10. The crystal structure of $NH_4Cu_7S_4$.

628

In all the above compounds there is tetrahedral coordination of the metal forming the thio-ion. We now give examples of thio-ions in which there is octahedral coordination of the metal. The (distorted) perovskite structure of $BaZrS_3$ is probably to be regarded as an ionic structure in which Ba^{2+} is 12- and Zr^{4+} 6-coordinated. (Inasmuch as the ZrS_6 octahedra are linked by vertex-sharing into a 3D framework the ZrS_3 complex could be distinguished as a 3D 'thio-ion'.) A number of compounds ABS_3 have h.c.p. structures built of AS_3 layers between which B atoms occupy columns of face-sharing octahedral holes (the $BaNiO_3$ or $CsNiCl_3$ structure), so forming infinite linear thio-ions. In $BaVS_3$ the V–V distance (2·81 Å) indicates metal–metal bonding consistent with the metallic conductivity, though $BaTaS_3$ (Ta–Ta, 2·87 Å) is only a semi-conductor. Double octahedral chain ions are found in $BaSnS_3$, $SrSnS_3$, and $PbSnS_3$ (NH_4CdCl_3 structure). Infinite 2D octahedral thio-ions occur in $NaCrS_2$ and other similar compounds. Crystals of $NaCrS_2$ are thin flakes which appear dark-red in transmitted light and greyish-green with metallic lustre in reflected light. In these crystals there are CrS_2^- layers of the same kind as in CdI_2 held together by the Na^+ ions (Fig. 17.11); both Cr and Na have 6 octahedral neighbours. Alternatively this structure may be described as a superstructure of NaCl; $NaInO_2$ and $NaInS_2$ are isostructural with $NaCrS_2$, as also are $KCrS_2$ and $RbCrS_2$. On the other hand, $LiCrS_2$ is a superstructure of NiAs, that is, Li and Cr alternate in columns of face-sharing octahedral coordination groups in a hexagonal closest packing of S atoms. Alternatively the structure may be described as CdI_2-type layers of composition CrS_2 interleaved by Li^+ ions. There are apparently no very strong Li–Cr bonds across the shared octahedron faces (Li–Cr, 3·01 Å), for the compound has a high resistivity. Just as $NaBiS_2$ (statistical NaCl structure) is related, in a structural sense, to PbS and other simple sulphides with the NaCl structure, so $FeCr_2S_4$ is isostructural with Co_3S_4 and $NiCr_2S_4$ and NiV_2S_4 with Cr_3S_4, and many of the sulphides of class (c) are related to the simplest covalent sulphide ZnS.

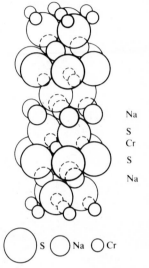

Na
S
Cr
S
Na

○ S ◯ Na ◯ Cr

FIG. 17.11. The crystal structure of $NaCrS_2$.

Sulphides structurally related to zinc-blende or wurtzite

This relationship is most direct for compounds such as α-$AgInS_2$ in which Ag and In atoms occupy at random the Zn positions in the wurtzite structure. Alternatively, there may be regular replacement of the Zn atoms in one of the two ZnS structures by atoms of two or more kinds. In both the random and regular structures the metal : sulphur ratio remains 1 : 1. We have seen that in some sulphides M_2S_3 and M_3S_4 only a proportion of the metal sites (tetrahedral or octahedral) in c.p. S assemblies are occupied. In the case of ZnS structures this implies occupancy of $\frac{2}{3}$ or $\frac{3}{4}$ of the Zn sites (that is, $\frac{1}{3}$ or $\frac{3}{8}$ of all the tetrahedral holes in the c.p. assembly). If instead of an incomplete set of metal atoms some of the S atoms are omitted, we have structures in which some of the metal atoms have 4 tetrahedral S neighbours and others 3 pyramidal S neighbours. Since the latter is a suitable bond arrangement for As or Sb we find compounds such as Cu_3AsS_3 with this kind of structure. A further possibility, the substitution of both Zn and S by other atoms, occurs in lautite, $CuAsS$,[1] in which the Zn and S positions are occupied (in a regular way) by equal numbers of Cu, As, and S. This structure is,

(1) AC 1965 **19** 543

629

however, preferably described as a substituted diamond structure, and it has been described in this way in Chapter 3. We may therefore recognize the following types of structure:

M : S ratio

(i)   1:   statistical or regular replacement of Zn in zinc-blende or wurtzite structures

(ii)   <1: $[Al_2Cd]S_4$

(iii)   >1: $[Cu_3As]S_3$

Examples of classes (i) and (ii) are set out in Table 17.10.

TABLE 17.10
Structures related to the zinc-blende and wurtzite structures

Fraction of metal positions occupied	Zinc-blende				Wurtzite	
		Regular		*Random*	*Regular*	*Random*
All	Zinc-blende ZnS	BN, BP XY (X=Al, Ga, In Y=P, As, Sb)			AlN, GaN, InN	
	Chalcopyrite $CuFeS_2$	$AgMX_2$, $CuMX_2$ (M=Al, Ga, In X=S, Se, Te) $ZnSnP_2$ $CdGeAs_2$		$CdZnSe_2$ $ZnSnAs_2$ $GaInSb_2$	$BeSiN_2$ $CuFe_2S_3$	α-$AgInS_2$
	Stannite Cu_2FeSnS_4	–		–	–	–
		Cu_3AsS_4 (luzonite), Cu_3PS_4		–	Cu_3AsS_4 (enargite)	–
$\frac{3}{4}(a)$	*Regular*		*Random*		*Random*	
	Al_2CdS_4, β-Cu_2HgI_4 β-Ag_2HgI_4 In_2CdSe_4		α-Ag_2HgI_4 α-Cu_2HgI_4 Ga_2HgTe_4		β-Al_2ZnS_4	
$\frac{2}{3}(a)$	*Random*				*Random*	
	γ-Ga_2Se_3 γ-Ga_2S_3 γ-Ga_2Te_3 γ-In_2Te_3				Al_2Se_3 β-Ga_2S_3	

(a) These fractions correspond to three-eighths and one-third of the *total* number of tetrahedral holes in close-packed assemblies as listed in Table 4.5 (p. 137).

(i) The simplest example of a superstructure of zinc-blende is the structure of chalcopyrite, or copper pyrites, $CuFeS_2$, which arises by replacing Zn by equal numbers of Cu and Fe atoms in a regular way. As a result of this substitution the atoms at the corners of the original ZnS unit cell are not all of the same kind so that the repeat unit is doubled in one direction. The unit cell in $CuFeS_2$ is therefore twice as large as that of ZnS. We may proceed a stage further by replacing one-half of the Fe atoms in $CuFeS_2$ by Sn, and so arrive at the structure of stannite, Cu_2FeSnS_4. The structures of ZnS, $CuFeS_2$, and Cu_2FeSnS_4 are shown in Fig. 17.12. The nearest neighbours of an S atom in the three structures are:

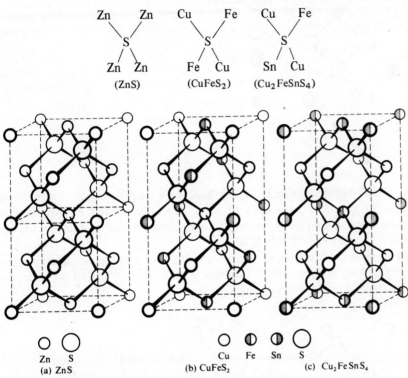

FIG. 17.12. The crystal structures of (a) ZnS, showing two unit cells, (b) $CuFeS_2$, and (c) Cu_2FeSnS_4.

It is worth noting that neither FeS nor SnS has the 4-coordinated zinc-blende structure, and also that CuS is apparently not a true sulphide of Cu^{II}. The valence of Cu (and other elements) in CuS, $CuFeS_2$, etc., is discussed on p. 908. Examples of compounds[2] with the $CuFeS_2$ structure are given in Table 17.10.

Some compounds have the ordered $CuFeS_2$ structure at ordinary temperatures and a statistical zinc-blende structure at higher temperatures, for example, $ZnSnAs_2$.[3]

Another example of a compound of this type is Cu_3AsS_4 (luzonite),[4a] the structure of which is of the zinc-blende type with three-quarters of the Zn atoms

(2) AC 1958 **11** 221

(3) AC 1963 **16** 153

(4a) ZK 1967 **124** 1

(4b) AC 1970 **B26** 1878

(5) ZK 1964 **119** 437

replaced by Cu and the remainder by As. This regular replacement of 4 Zn by 3 Cu + As leads to the larger unit cell of Fig. 17.13. Another form of Cu_3AsS_4 (enargite) is a superstructure of wurtzite.[4b]

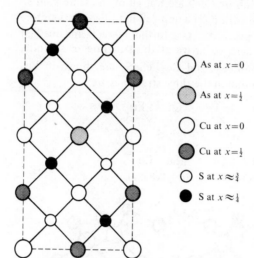

○ As at $x=0$

◐ As at $x=\frac{1}{2}$

○ Cu at $x=0$

◉ Cu at $x=\frac{1}{2}$

○ S at $x \approx \frac{3}{4}$

● S at $x \approx \frac{1}{4}$

FIG. 17.13. Projection of the structure of luzonite, Cu_3AsS_4, along the a axis.

(ii) Three regular structures of this kind have been described, and although examples of *sulphides* with each of these structures are not known they are illustrated here (Fig. 17.14) because of their close relation to chalcopyrite and stannite; for examples see Table 17.10.

(iii) Examples of structures related to ZnS by omission of some S atoms include Cu_3AsS_3 and Cu_3SbS_3,[5] though minerals of this family usually contain iron and their formulae are more complex.

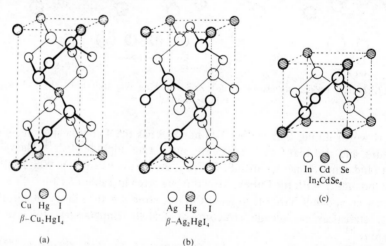

○ ◐ ○
Cu Hg I
β-Cu_2HgI_4

(a)

○ ◐ ○
Ag Hg I
β-Ag_2HgI_4

(b)

○ ◐ ○
In Cd Se
In_2CdSe_4

(c)

FIG. 17.14. Structures related to the zinc-blende structure: (a) β-Cu_2HgI_4, (b) β-Ag_2HgI_4, (c) In_2CdSe_4.

Other complex sulphides

Cubanite, $CuFe_2S_3$,[6] is of some interest since it is ferromagnetic. Although the metal–sulphur ratio is unity this is not a compound of type (i), for its structure is related to wurtzite in a more complex way. It is built of slices of the wurtzite structure joined together in such a way that pairs of FeS_4 tetrahedra share edges (Fig. 17.15). The resulting Fe–Fe distances (2·81 Å) are rather long for metal–metal bonds, but presumably indicate appreciable interaction between the Fe atoms.

(6) AM 1955 **40** 213

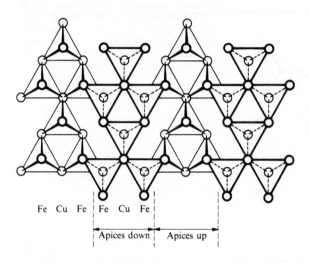

Fe Cu Fe ¦ Fe Cu Fe ¦

¦ Apices down ¦ Apices up ¦

FIG. 17.15. Portion of the structure of cubanite ($CuFe_2S_3$) showing slabs of the wurtzite structure with the tetrahedra pointing alternately up and down.

The structural relation of wolfsbergite, $CuSbS_2$[7] to wurtzite is less simple. This compound has a layer structure in which each Sb atom forms the usual three pyramidal bonds and Cu its four tetrahedral bonds to S atoms. In Fig. 17.16(a) is shown a plan of part of the wurtzite structure, in which plan Zn and S are superposed since they lie (at different heights) on the same lines perpendicular to the plane of the paper. If we take a vertical section through this structure between the dotted lines we see that a metal atom lying on these dotted lines loses one of its S neighbours, shown as a dotted circle, whereas a metal atom lying between the dotted lines retains all its four S neighbours. (It must be remembered that each metal atom is joined to one S atom lying vertically above or below it, so that the number of M–S bonds is one more than the number seen in the plan.) These latter atoms are Cu and the former Sb in the layers of $CuSbS_2$ which are viewed end-on in Fig. 17.17. Comparison of Figs. 17.16 and 17.17 shows the similarity in structure of a $CuSbS_2$ layer and the section of the wurtzite structure between the dotted lines (parallel to the crystallographic 1120 plane).

(7) ZK 1933 **84** 177

The number of complex sulphide minerals is very large, and although it is possible to regard many of them as derived from simple sulphide structures (e.g. ZnS, PbS) the relationship is not always very close. For example, the Ag and Sb positions in miargyrite, $AgSbS_2$,[8] are close to those of alternate Pb atoms in galena (PbS), but the S atoms are so far removed from the positions of the ideal

(8) AC 1964 **17** 847

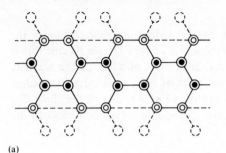

(a)

FIG. 17.16. (a) Plan of a portion of the wurtzite structure. The metal atoms, which are all Zn atoms in wurtzite, are shown as small circles of two types to facilitate comparison with Fig. 17.17. (b) The same, with certain atoms removed, to show the relation to the layers in $CuSbS_2$.

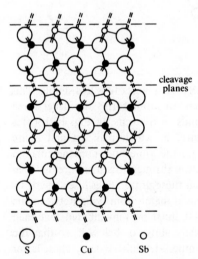

(b)

O — S   ● — Cu   o — Sb

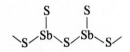

cleavage planes

FIG. 17.17. Elevation of the structure of $CuSbS_2$, viewed parallel to the plane of the layers. The neighbours of an Sb atom in a layer are one S at 2·44 A and two S at 2·57 A. The broken lines between the layers indicate much weaker Sb—S bonds (Sb—S = 3·11 A) which account for the very good cleavage parallel to the layers.

O — S   ● — Cu   o — Sb

(9) AM 1955 **40** 226
(10) ZK 1957 **109** 161
(11) SMPM 1969 **49** 109 (180 references).

PbS structure that there is no very close resemblance between the structures. A more important feature of $AgSbS_2$ is the pyramidal bonding of Sb to 3 S. In many complex sulphides Sb forms three such pyramidal bonds, and sometimes also two weaker ones, and these structures are best classified according to the nature of the Sb—S complex. This is the SbS_2 chain in berthierite, $FeSb_2S_4$,[9] and fragments (Sb_3S_7) of such chains in jamesonite, $FePb_4Sb_6S_{14}$.[10] In a very comprehensive classification of the structures of all known complex sulphide minerals[11] the

634

general formula is written in the form $A'_q A''_r (B_m C_n) C_p$ where A' and A'' stand for metals with the following coordination numbers:

$$A' \begin{cases} 2 & Ag, Tl, Hg \\ 3 & Ag, Cu \\ 4 & Ag, Cu, Zn \end{cases} \qquad A'' \begin{cases} 6 & Pb, Fe, Co, Ni, Hg \\ 7 & Pb, Tl \\ 9 & Pb, \end{cases}$$

The atoms B (As, Sb, Bi) and C (S, Se, Te) form the system $B_m C_n$ built of pyramidal BC_3 and/or tetrahedral BC_4 groups, and C_p represents the S (Se, Te) which does not form part of the sulpho-salt complex. The basis of the classification is the nature of the $B_m C_n$ complex which, together with the coordination requirements of the A' and A'' atoms, determines the crystal structure.

Oxysulphides

These compounds appear to be formed by a relatively small number of elements. Those of non-metals are described in other chapters (COS, $P_4 O_6 S_4$). Three types of structure have been found for metallic oxysulphides:

the cubic ZrOS structure,[1]
the $La_2 O_2 S$ structure,[2] also adopted by $Ce_2 O_2 S$ and $Pu_2 O_2 S$,
the BiOCl (PbFCl) structure, of ThOS, PaOS, UOS, NpOS,[3] and
the second form of ZrOS.[4]

(1) AC 1948 **1** 287
(2) AC 1948 **1** 265
(3) AC 1949 **2** 291
(4) ACSc 1962 **16** 791

ZrOS is prepared by passing $H_2 S$ over $Zr(SO_4)_2$ at red heat or over ZrO_2 at $1300°C$; it is dimorphic. In the cubic form the Zr^{4+} ion is 7-coordinated, its neighbours being $3\ O^{2-}$ at $2·13$ Å and $4\ S^{2-}$ at $2·62$ Å. The coordination group is a distorted octahedron consisting of $3\ O + 3\ S$ with the fourth S ion above the centre of the face containing the three oxygen ions. This coordination group is illustrated, in idealized form, in Fig. 3.6(a), p. 66.

The structure of $La_2 O_2 S$ is closely related to that of the high-temperature form of $La_2 O_3$ (the $A-M_2 O_3$ structure of Fig. 12.7). One-third of the oxygen is replaced by sulphur, so that the metal ions have the same 7-coordination group as in cubic ZrOS, here made up of $4\ O + 3\ S$. (For $Ce_2 O_2 S$, Ce–4 O, $2·36$ Å, Ce–3 S, $3·04$ Å.) For further discussion of the $La_2 O_2 S$ structure see p. 1004.

The BiOCl structure has been described and illustrated in Chapter 10. In this structure there is 8-coordination of the metal atom, at the vertices of an antiprism, by $4\ O + 4\ Cl$, but in the isostructural LaOCl a fifth Cl (beyond one square antiprism face) is at the same distance from the metal ion as the other four. In ThOS etc. the positions of the S atoms were not accurately determined, and these atoms were so placed as to give the metal atom 5 equidistant S neighbours, in addition to the 4 O neighbours. (Th–4 O, $2·40$ Å, Th–5 S, $3·00$ Å.)

The compound ZrSiS may be mentioned here since it was at one time mistaken for $Zr_4 S_3$. It has a structure closely related to the PbFCl structure, the main difference being that in the oxysulphides with this structure O–O contacts correspond only to van der Waals bonding whereas in ZrSiS there are Si–Si bonds $(2·51$ Å$)$.[5]

(5) RTC 1964 **83** 776

18

Nitrogen

Introduction

Nitrogen stands apart from the other elements of Group V. It is the most electronegative element in the group, and from the structural standpoint two of its most important characteristics are the following. Only the four orbitals of the L shell are available for bond formation so that nitrogen forms a maximum of four (tetrahedral) bonds, as in NH_4^+, substituted ammonium ions, and amine oxides, R_3N . O. Only three bonds are formed in halides and oxy compounds, the bonds in many of the latter having multiple-bond character. Like the neighbouring members of the first short period, carbon and oxygen, nitrogen has a strong tendency to form multiple bonds. It is the only Group V element which exists as diatomic molecules at ordinary temperatures and the only one which remains in this form ($N\equiv N$) in the liquid and solid states in preference to polymerizing to singly-bonded systems as in the case of phosphorus and arsenic. (For solid N_2 see Chapter 29.) There is an interesting difference between the relative strengths of single and multiple nitrogen–nitrogen, nitrogen–carbon, and carbon–carbon bonds:

C–C	347	C–N	305	N–N	163 kJ mol^{-1}
C=C	611	C=N	615	N=N	418
C≡C	837	C≡N	891	N≡N	946

For the above reasons nitrogen forms many compounds of types not formed by other elements of this group, and for this reason we deal separately with the stereochemistry of this element. For example, the only compounds of N and P which are structurally similar are the molecules in which the elements are 3-covalent and the phosphonium and ammonium ions. There are no nitrogen analogues of the phosphorus pentahalides, and there is little resemblance between the oxygen compounds of the two elements. Monatomic ions of nitrogen and phosphorus are known only in the solid state, in the salt-like nitrides and phosphides of the more electropositive elements. The multiple-bonded azide ion, N_3^-, is peculiar to nitrogen.

The following compounds of nitrogen are described in other chapters:

Ch. 10   Metal nitride halides
Ch. 16   Ions containing N and S
Ch. 19   Phosphorus–nitrogen compounds

The stereochemistry of nitrogen

In all its compounds nitrogen has four pairs of electrons in its valence shell. According to the number of lone pairs there are the five possibilities exemplified by the series

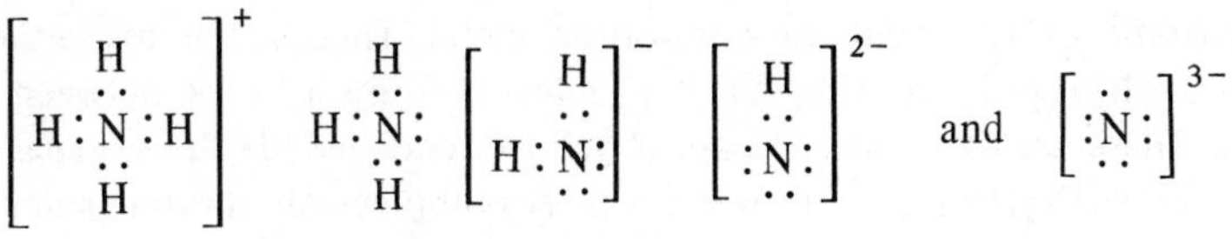

$$\left[\begin{array}{c} H \\ H:N:H \\ H \end{array}\right]^{+} \quad \begin{array}{c} H \\ H:N: \\ H \end{array} \quad \left[\begin{array}{c} H \\ H:N: \end{array}\right]^{-} \quad \left[\begin{array}{c} :N: \end{array}\right]^{2-} \quad \text{and} \quad \left[:N:\right]^{3-}$$

The last three, the NH_2^-, NH^{2-}, and N^{3-} ions, are found in the salt-like amides, imides, and nitrides of the most electropositive metals, but with the exception of the amide ion the stereochemistry of nitrogen is based on N^+ with no lone pairs and N with one lone pair. These two states of the nitrogen atom correspond to the classical trivalent and 'pentavalent' states, now preferably regarded as oxidation rather than valence states.

The stereochemistry of N^+ is similar to that of C, with which it is isoelectronic:

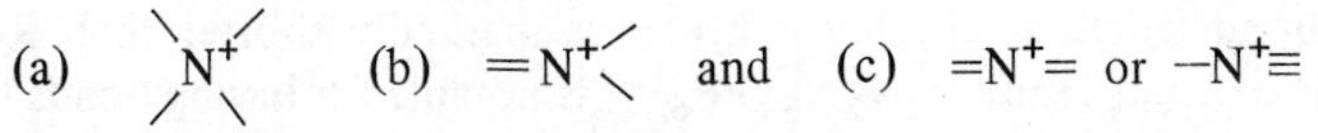

(a) $\diagdown N^+ \diagdown$ (b) $=N^+\diagup$ and   (c) $=N^+=$ or $-N^+\equiv$

tetrahedral   planar   collinear

As in the case of carbon, the valence bond representations (b) and (c) do not generally represent the electron distributions around N in molecules in which this element is forming three coplanar or two collinear bonds. Such bonds often have intermediate bond orders which may be described in terms of resonance between a number of structures with different arrangements of single and multiple bonds or in terms of trigonal (sp^2) or digonal (sp) hybridization with varying amounts of π-bonding.

The central N atom of the N_3^- ion in azides approaches the state $=N^+=$ and diazonium salts[1] illustrate $-N^+\equiv$.

(1) ACSc 1963 **17** 1444

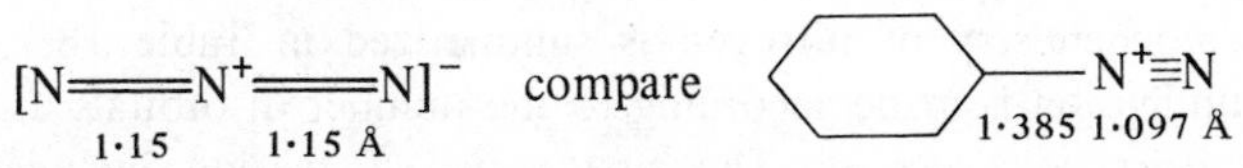

$$[N\!\!=\!\!=\!\!N^+\!\!=\!\!=\!\!N]^- \quad \text{compare}$$

1·15   1·15 Å

$$\text{⬡}\!-\!N^+\!\!\equiv\!\!N$$

1·385 1·097 Å

The bond length N–N in the diazonium ion is the same as in elementary nitrogen (N≡N); the C–N bond is shorter than a normal single bond owing to interaction with the aromatic ring.

Nitrogen

When one of the four orbitals is occupied by a lone pair the possibilities are:

$$(d) \quad :N\!\!\!< \qquad (e) \quad -N\!\!\!\nearrow \qquad or \quad (f) \quad :N\!\!\equiv$$

pyramidal   angular

The unshared pair of electrons in molecules such as NH_3 can be used to form a fourth bond, the coordinate link ($\rightarrow$) of Sidgwick, as in the metal ammines (see Werner coordination compounds, Chapter 27).

An interesting development in the chemistry of nitrogen is the demonstration that the N atoms of neutral N_2 molecules can utilize their lone pairs (like the isoelectronic CO) to bond to a transition metal. There is bonding either through one N or through both, when the N_2 molecule forms a bridge between two metal atoms. These are examples of case (f) above, and include the trigonal pyramidal molecule $CoHN_2(P\phi_3)_3$,[2] in which Co–N is appreciably shorter than in ammines (1·96 Å), the octahedral ion in $[Ru(NH_3)_5N_2]Cl_2$,[3] and the bridged anion

(2) IC 1969 **8** 2719
(3) AC 1968 **B24** 1289

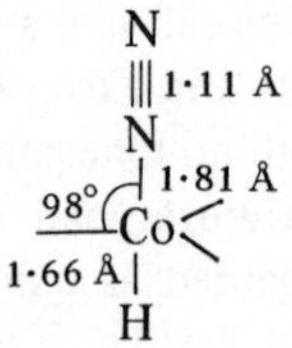

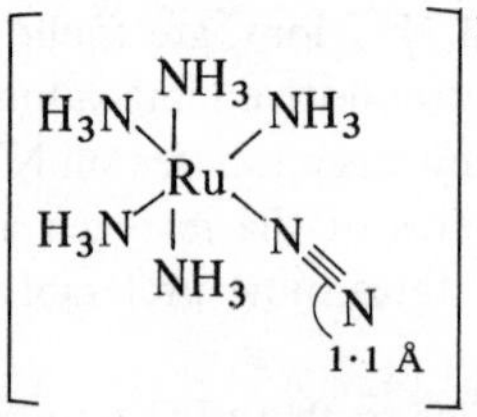

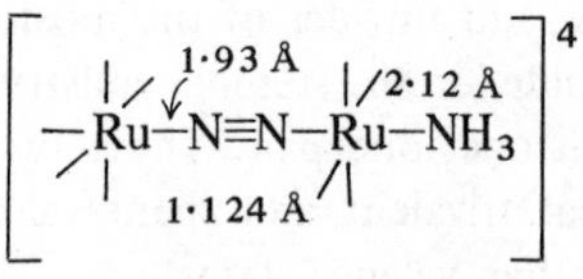

(4) JACS 1969 **91** 6512

(5) IC 1970 **9** 2768

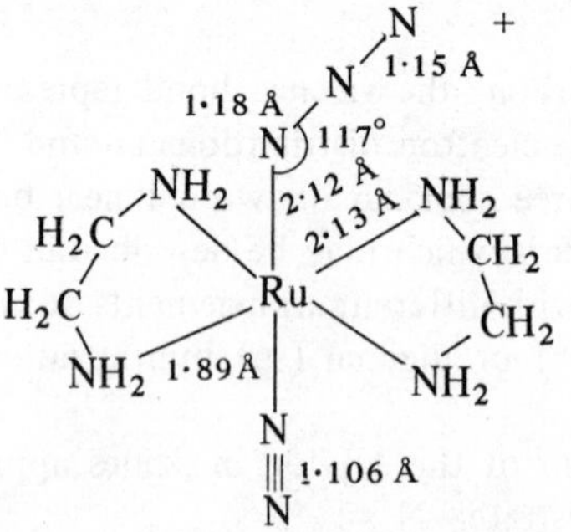

(eclipsed configuration) in $[(NH_3)_5RuN_2Ru(NH_3)_5](BF_4)_4$,[4] in which the Ru–N bond length in the linear bridge is markedly shorter than Ru–NH_3. The cation in the salt $[Ru(en)_2N_2N_3]PF_6$[5] is notable for having bonds from metal to N of three different kinds, to NH_2, N_2 molecules, and N_3^- ligands.

The stereochemistry of nitrogen is summarized in Table 18.1, in which the primary subdivision is made according to the number of orbitals used by bonding and lone pairs of electrons. The first main subdivision includes the simpler hydrogen and halogen compounds, and we shall deal with these first in the order of Table 18.1, except that amides and imides will be included with ammonia. The second and third sections include most of the oxy-compounds, and since it is

638

preferable to deal with oxides and oxy-acids as groups we shall not keep rigidly to the subdivision according to bond arrangement. Certain other special groups, such as azides, sulphides, and nitrides, will also be described separately.

TABLE 18.1
The stereochemistry of nitrogen

Number of σ pairs (bonding and lone pairs)	Type of hybrid	Number of lone pairs	Bond arrangement	Examples
4	sp^3	0	Tetrahedral	NH_4^+, $(CH_3)_3N^+{-}O^-$
		1	Trigonal pyramidal	NH_3, NH_2OH, NF_3, NHF_2, NH_2F, N_2H_4, N_2F_4
		2	Angular	NH_2^-
3	sp^2	0	Trigonal planar	NO_2Cl, $NO_2(OH)$, NO_3^-, N_2O_4 $[ON(NO)SO_3]^-$, $[ON(SO_3)_2]^{2-}$
		1	Angular	$NOCl$, NOF, NO_2^-, N_2F_2
2	sp	0	Collinear	N_2O, NO_2^+, N_3H

Nitrogen forming four tetrahedral bonds

The optical activity of substituted ammonium ions $(Nabcd)^+$ and of more complex ions and molecules (see Chapter 2) provided the earliest proofs of the essentially

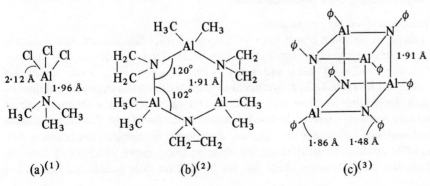

(a)[1]   (b)[2]   (c)[3]

(1) AC 1969 **B25** 377
(2) JACS 1970 **92** 285
(3) AC 1972 **B28** 1619

tetrahedral arrangement of four bonds from a nitrogen atom. Later, this has been confirmed by structural studies of, for example, AlN (wurtzite structure) and BN (wurtzite and zinc-blende structures), of ammonium salts, amine oxides, and the series of molecules (a)–(c) in which N is bonded tetrahedrally to 3 C + 1 Al, 2 C + 2 Al, and 1 C + 3 Al respectively.

Ammonium and related ions

It has not generally proved possible to establish the tetrahedral structure of the NH_4^+ ion in simple ammonium salts by X-ray studies, both because of the small scattering power of H for X-rays and because the ion is rotating or shows orientational disorder in the crystalline salts at ordinary temperatures. In NH_4F, however, rotation of the ions is prevented by the formation of N–H–F bonds, and the tetrahedral disposition of the four nearest F neighbours of a N atom indicates the directions of the hydrogen bonds. It is possible to locate H atoms by neutron diffraction, and detailed studies of several ammonium salts have been made. The N–H bond lengths in ammonium salts have also been determined by n.m.r. and several studies give values close to 1·03 Å.[4] The structures of substituted ammonium ions have been determined in $[N(CH_3)_4]_2SiF_6$, $[N(CH_3)_4]ICl_2$, etc.

The NF_4^+ ion exists in salts such as $(NF_4)(AsF_6)$ and $(NF_4)(SbF_6)$ prepared from NF_3, F_2, and MF_5 heated under pressure. $(NF_4)(AsF_6)$ is a stable, colourless, non-volatile, hygroscopic solid at 25°C, presumably structurally similar to $(PCl_4)(PCl_6)$; n.m.r. shows that all four F atoms attached to N are equivalent.[5]

A neutron diffraction study of $NH_3OH \cdot Cl$ gives N–O, 1·37 Å, and the value 1·41 Å was found in an X-ray study of $NH_3OH \cdot ClO_4$ at −150°C.[7]

Amine Oxides. In the tetrahedral molecule of trimethylamine oxide, $((CH_3)_3NO)$, N–O = 1·40 Å.[8] Trifluoramine oxide, NF_3O, is a stable colourless gas prepared by the action of an electric discharge on a mixture of NF_3 and oxygen. Infrared[9] and n.m.r.[10] data are consistent with a nearly regular tetrahedral shape. Salts such as $(NOF_2)BF_4$ and $(NOF_2)AsF_6$ prepared from NOF_3 and the halide contain planar ions NOF_2^+ similar to the isoelectronic COF_2 (i.r. spectrum).[11]

Nitrogen forming three pyramidal bonds

Ammonia and related compounds

In general three single bonds from a nitrogen atom are directed towards the apices of a trigonal pyramid, as has been demonstrated by spectroscopic and electron diffraction studies of NH_3 and many other molecules NR_3; some results are summarized in Table 18.2. A pyramidal configuration is to be expected whether we regard the orbitals as three p or three of four sp^3 orbitals, the unshared pair of electrons occupying one of the bond positions in the latter case. For pure p orbitals bond angles of 90° would be expected, and mutual repulsion of the H atoms then has to be assumed to account for the observed bond angles, which are close to the tetrahedral value. It seems likely that the bonds have some s character, though less

(4) JCP 1954 **22** 643, 651

(5) IC 1967 **6** 1156
(6) AC 1967 **22** 928
(7) AC 1969 **B25** 1875

(8) AC 1964 **17** 102

(9) IC 1968 **7** 2064
(10) JCP 1967 **46** 2904

(11) IC 1969 **8** 1253

TABLE 18.2

Structural data for molecules NR_3, NR_2X, *and* NRX_2

Molecule	N–H (Å)	N–C (Å)	N–X (Å)	Bond angles	Method	Reference
NH_3	1·015			HNH 106·6°	e.d.	ACSc 1964 **18** 2077
NH_2CH_3	1·011	1·474		HNH 105·9° CNH 112·1°	m.w.	JCP 1957 **27** 343
$NH(CH_3)_2$	1·022	1·466		CNC 111·6° CNH 108·8°	m.w.	JCP 1968 **48** 5058
$N(CH_3)_3$		1·451		CNC 110·9°	m.w.	JCP 1969 **51** 1580
$N(CH_3)_2Cl$			1·75		e.d.	TFS 1944 **40** 164
NH_2Cl				HNCl 102° }	i.r.	JACS 1952 **74** 6076
$NHCl_2$			(1·76)	ClNCl 106° }		
NCl_3		(i.r. and Raman sp. of CCl_4 soln.)				JCP 1968 **49** 3751
NF_3			1·371	FNF 102·2°	e.d. m.w.	JACS 1950 **72** 1182 PR 1950 **79** 513
NHF_2	1·026		1·400	FNF 103° HNF 100°	m.w.	JCP 1963 **38** 456
$N(C_6H_5)_3$		1·42		CNC 116°		JCP 1959 **31** 477

	N–Si	Si–N–Si			
$N(SiH_3)_3$	1·74	120°		e.d.	JACS 1955 **77** 6491
$N(SiH_3)_2H$	1·725	128°		e.d. }	JCS A 1969 1224
$N(SiH_3)_2CH_3$	–·	126°		e.d. }	
$Al[N\{Si(CH_3)_3\}_2]_3$	1·75	118°		X	JCS A 1969 2279

	N–Ge	Ge–N–Ge			
$N(GeH_3)_3$	1·84	120°		e.d.	JCS A 1970 2935

than for sp^3 hybridization, so that the lone-pair electrons have some p character. Since this lone-pair orbital is directed it leads to an atomic dipole, and it is this which is responsible for most of the dipole moment (1·44 D) of NH_3. The importance of the atomic dipole is shown by the fact that the pyramidal NF_3 molecule is almost non-polar (0·2 D), due to cancellation of the moments due to the polar N–F bonds by the lone-pair moment. In contrast to NF_3, NHF_2 has an appreciable moment (1·93 D).

We should expect a molecule N *abc* containing three different atoms or groups to exhibit optical activity, since the molecule is enantiomorphic. However, no such molecule has been resolved in spite of many attempts, and this at one time cast doubt on the non-planar arrangement of three bonds from a neutral N atom. In the case of NH_3 itself it has been suggested that the potential energy curve of the molecule is of the form shown in Fig. 18.1(a), where the energy is plotted against the distance *r* of the N atom from the plane of the H atoms. There are two minima corresponding to the two possible positions of the N atom (on each side of the plane of the H atoms), but they are separated by an energy barrier of only about 25 kJ mol^{-1}. For the resolution of *d* and *l* modifications of a substance to be detectable at room temperature it has been calculated that the energy of activation

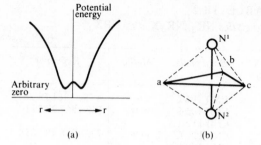

FIG. 18.1. (a) Probable form of the potential energy curve for NH_3. (b) The two possible configurations for the molecule NH_3 or $Nabc$.

(1) JACS 1954 **76** 2645
(2) PRS 1963 **273A** 435, 455
(3) JCS A 1969 2279

for racemization (change of $d \rightleftharpoons l$) must be not less than 84 kJ mol^{-1}. The failure to resolve substituted ammonias may therefore be due to the fact that the molecule can easily turn inside out by movement of the N atom from position N^1 to N^2 (Fig. 18.1(b)). Calculations[1] of the rates of racemization of pyramidal molecules XY_3 with N, P, As, Sb, and S as central atoms, based on a potential energy function derived from known vibrational frequencies and molecular dimensions, suggest that optically active compounds of As, Sb, and S should be stable towards racemization at room temperature, and in the case of P possibly at low temperatures.

Three pyramidal bonds are formed by N in hexamethylenetetramine, $N_4(CH_2)_6$ (Fig. 18.2), which has been studied many times, in the crystalline state by X-ray and neutron diffraction and in the vapour by electron diffraction. In the molecule in the crystal: C–N, 1·476 Å, C–H, 1·088 Å, NCN, 113·6°, and CNC, 107·2°.[2]

Ions and molecules in which π-bonding leads to coplanarity of three bonds from N include a number of molecules containing the silyl group (in the lower part of Table 18.2). The molecule $Al[N\{Si(CH_3)_3\}_2]_3$[3] is also of interest as an example of Al forming three coplanar bonds. The $Si(CH_3)_3$ groups lie out of the AlN_3 plane, the dihedral angle NAlN/AlNSi being 50°. In this connection see also the structures of urea and formamide (Chapter 21).

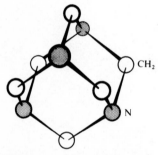

FIG. 18.2. The structure of the molecule of hexamethylene-tetramine, $N_4(CH_2)_6$. N atoms are shaded and H atoms omitted. This molecule, based on a tetrahedral group of N atoms, should be compared with those of $Be_4O(CH_3COO)_6$, P_4O_6, and P_4O_{10}, Figs. 11.1 and 19.7.

(4) JPC 1956 **60** 821

(5) ZaC 1959 **299** 33

Amides and imides. These salts, formed only by the more electropositive metals, contain respectively the NH_2^- and NH^{2-} ions. The alkali, alkaline-earth, and zinc amides are colourless crystalline compounds formed directly by the action of ammonia on the molten metal or in solution in liquid ammonia. A solution of $NaNH_2$ in liquid ammonia is a good conductor of electricity, indicating that the salt is ionized in this solvent.

The high-temperature forms of K, Rb, and Cs amides have the NaCl structure but at ordinary temperatures these compounds have less symmetrical structures. For example, the Cs salt has a (tetragonal) deformed CsCl structure (like γ-NH_4Br). In $NaNH_2$ and $LiNH_2$ at ordinary temperatures there are distorted coordination groups around the NH_2^- ions. In $NaNH_2$[4] there is tetrahedral coordination of both anion and cation, but the arrangement of 4 Na^+ around N is a very distorted tetrahedron, showing that the NH_2^- ions are not rotating. $LiNH_2$ has a slightly distorted zinc-blende structure.[5] The tetrahedral coordination is somewhat irregular (NH_2^- has one Li^+ at 2·35 Å and three Li^+ at 2·15 Å), apparently due to the orientation of the NH_2^- dipoles. With $LiNH_2$ contrast the layer structure of LiOH, in which Li^+ ions occupy a different set of tetrahedral holes in the same arrangement of c.c.p. anions.

Two forms of $Ca(NH_2)_2$ and $Sr(NH_2)_2$ have been described, one having a structure similar to that of anatase[6] and the other a defect NaCl structure.[7] In both polymorphs the NH_2^- ions are apparently in fixed orientations. The complex structures of $Be(NH_2)_2$ and $Mg(NH_2)_2$ have not been determined.[7a]

From the H⋯H distance ($1·63 \pm 0·03$ Å) obtained from the p.m.r. spectrum of KNH_2[8] and the H–N–H bond angle ($104°$) deduced from the i.r. spectrum of $LiNH_2$ the N–H bond length in the amide ion has been estimated to be $1·03$ Å.[9]

The alkali and alkaline-earth imides have typical ionic structures. Li_2NH has the anti-CaF_2 structure,[10] like Li_2O, and the alkaline-earth imides crystallize with the NaCl structure, as do the corresponding oxides, the rotating NH^{2-} ion possessing spherical symmetry. The following radii have been deduced for the amide and imide ions: NH_2^-, $1·73$ Å, and NH^{2-}, $2·00$ Å; compare Cl^-, $1·80$ Å, and SH^-, $2·00$ Å.

Hydroxylamine

In crystalline NH_2OH the N–O bond length is found to be $1·47$ ($0·03$) Å.[11] The H atoms were not located, but their probable positions were deduced from a reasonable hydrogen bonding scheme. We have referred earlier to hydroxylammonium salts.

For the hydroxylamine *N*-sulphonate ion see p. 588.

The trihalides of nitrogen

These compounds present no points of particular interest from the structural standpoint. (No pentahalides are known.) Partially halogenated compounds which have been prepared include NH_2F, NHF_2, NH_2Cl, and $NHCl_2$. Of the trihalides, the pure tribromide has not been prepared, though NBr_3 (and also NH_2Br and $NHBr_2$) has been identified in dilute aqueous solution (resulting from the bromination of NH_3 in buffered solutions) by u.v. absorption and chemical analysis.[12] There is some evidence for the existence of an ammine, $NBr_3 . 6 NH_3$, stable at temperatures below $-70°C$. The so-called tri-iodide made from iodine and ammonia is $NI_3 . NH_3$ and not the simple halide. This ammine does not contain discrete NI_3 molecules but consists of chains of NI_4 tetrahedra sharing two vertices, with one NH_3 attached by a weaker bond to alternate unshared I atoms along the chain.[12a] The trifluoride has been prepared by electrolysis of fused NH_4HF_2 and is quite stable when pure. For structural details of NF_3, NHX_2, and NH_2X see Table 18.2.

Hydrates of ammonia

Ammonia forms two hydrates, $NH_3 . H_2O$ and $2 NH_3 . H_2O$, and the crystal structure of the latter has been studied at $170°K$.[13] The crystal apparently consists of a hydrogen-bonded framework of composition $NH_3 . H_2O$, in which

(6) ZaC 1963 **324** 278
(7) CR 1969 **268** 175
(7a) ZaC 1969 **370** 248

(8) TFS 1956 **52** 802
(9) JPC 1957 **61** 384

(10) ZaC 1951 **266** 325

(11) AC 1958 **11** 511, 512

(12) IC 1965 **4** 899

(12a) ZaC 1968 **357** 225

(13) AC 1954 **7** 194

Nitrogen

NH$_3$ and H$_2$O each form one strong N–H–O bond (2·84 Å) and three weaker ones (3·13, 3·22, and 3·22 Å). The second NH$_3$, which may be rotating, is attached to this framework only by a single N–H–O bond (of length 2·84 Å). In NH$_3$. H$_2$O the H$_2$O molecules are linked by hydrogen bonds into chains which are then cross-connected by further hydrogen bonds to NH$_3$ molecules into a three-dimensional network.[14] There is no hydrogen bonding between one NH$_3$ molecule and another. As in 2 NH$_3$. H$_2$O the unshared pair of electrons on the N atom forms one strong N–H–O bond (2·8 Å) and three weaker ones (3·25 Å). It is interesting to compare with the hydrates the structure of the cubic form of solid NH$_3$.[15] This has a slightly distorted cubic close-packed structure in which each N atom has 6 N neighbours at 3·4 Å and 6 more at 3·9 Å, suggesting that three weak N–H–N bonds are formed by each lone pair of electrons.

Ammines

All the alkali metals and alkaline-earths (but not Be) and also Eu and Yb dissolve in liquid NH$_3$. The alkali-metal solutions are blue, paramagnetic, and conduct electricity. Such solutions also dissolve other metals which are not themselves soluble in liquid NH$_3$, for example, the Na solution dissolves Pb, and in this way a number of intermetallic compounds (Na$_2$Pb, NaPb) have been prepared. From certain of the solutions crystalline ammines can be obtained, including Li(NH$_3$)$_4$ and hexammines of Ca, Sr, Ba, Eu, and Yb. The hexammines are body-centred cubic packings of octahedral molecules M(NH$_3$)$_6$. The Eu and Yb compounds are golden-yellow and are metallic conductors (M–N $\approx$ 3·0 Å), the magnetic and electrical properties suggesting that two electrons occupy a conduction band.[16]

Ammonia also combines with many salts to form ammines in which NH$_3$ utilizes its lone pair of electrons to bond to the metal. The M–NH$_3$ bond is extremely weak in the case of the alkali metals, but stronger to the B subgroup and transition metals. In NaCl . $5\frac{1}{7}$ NH$_3$, for example, Na$^+$ has 5 N neighbours at 2·48 Å and 1 N at 3·39 Å, and the structure is an approximately c.c.p. of NH$_3$ molecules and Cl$^-$ ions.[17] In NH$_4$I . 4 NH$_3$ NH$_4^+$ is hydrogen-bonded to 4 tetrahedral neighbours (N–N, 2·96 Å) and I$^-$ is surrounded by 12 NH$_3$.[18] In contrast to the relatively weak bonding in the ammines of the more electropositive metals NH$_3$ is covalently bonded to metal atoms in numerous molecules and complex ions formed by B subgroup and transition metals, the structures of which are described under the individual element.

Molecules containing the system $\overset{\diagdown}{\diagup}$N–N$\overset{\diagup}{\diagdown}$

The only known symmetrical molecules of the type a$_2$N–Na$_2$ with single N–N bonds are those of hydrazine, N$_2$H$_4$, and dinitrogen tetrafluoride, N$_2$F$_4$. In some molecules in which O atoms are attached to N as, for example, nitramine, (a),

(14) AC 1959 **12** 827

(15) AC 1959 **12** 832

(16) JSSC 1969 **1** 10

(17) AC 1965 **18** 879
(18) ACSc 1960 **14** 1466

substituted nitramines, (b), and dimers of C-nitroso compounds, (c), the co-planarity of the six atoms, the existence of *cis–trans* isomers of the nitroso

```
   H         O          R         O          R         O
    \       /            \       /            \       /
     N—N                  N—N                  N—N
    /       \            /       \            /       \
   H         O          R         O          O         R

      (a)                  (b)                  (c)
```

compounds, and the N–N bond lengths show that the N–N bonds have some double-bond character. Note, however, the extremely *long* N–N bonds in N_2O_3 (p. 651) and N_2O_4 (p. 652).

Hydrazine, N_2H_4

Hydrazine exists in the form of simple N_2H_4 molecules in all states. Hydrogen bonding is responsible for the high b.p. (114°C) and high viscosity of the liquid. Structurally the molecule has one feature in common with H_2O_2, for rotation of the NH_2 groups about the N–N axis is hindered by the lone pair of electrons—the free molecule has a configuration of the *gauche* type (p. 48). For structural details see Table 18.3.

The monohydrate, $N_2H_4 . H_2O$, apparently has a structure of the NaCl type with disordered or rotating N_2H_4 and H_2O molecules.[1]

Hydrazine forms two series of salts, containing the $N_2H_5^+$ and $N_2H_6^{2+}$ ions. The structures of the halides are described in Chapter 8. From their formulae it is evident which type of ion they contain, but this is not true of all oxy-salts. For example, $N_2H_4 . H_2SO_4$ might be $(N_2H_5)HSO_4$ or $(N_2H_6)SO_4$. The type of ion is deduced from the crystal structure (arrangement of hydrogen bonds) or from the p.m.r. spectrum. Of the sulphates, $N_2H_4 . \frac{1}{2}$, 1, and $2 H_2SO_4$, the first is $(N_2H_5)_2SO_4$[2] and the second, $(N_2H_6)SO_4$[3]; the third is presumably $(N_2H_6)(HSO_4)_2$. In $N_2H_4 . H_3PO_4$ there is a 3D H_2PO_4 framework similar to that in KH_2PO_4, and the salt is accordingly $N_2H_5(H_2PO_4)$.[4] On the other hand, $N_2H_4 . 2 H_3PO_4$ is $(N_2H_6)(H_2PO_4)_2$, in which the cation has the staggered configuration.[5]

There is no definite evidence that the N–N bond lengths in the ions $N_2H_5^+$ and $N_2H_6^{2+}$ are appreciably different from that in N_2H_4 (values of 1·43 and 1·42 Å have been recorded) but there is variation in the configuration of the $N_2H_5^+$ ion. In $(N_2H_5)H_2PO_4$ it has the staggered configuration, but in $(N_2H_5)_2SO_4$ there are non-equivalent ions with approximately staggered and eclipsed configurations.

Hydrazine acts as a neutral bridging ligand in salts of the type $Zn(N_2H_4)_2X_2$ (X = Cl,[6] CH_3COO,[7] CNS,[8] etc.), double bridges linking the metal atoms into infinite chains.

Dinitrogen tetrafluoride, N_2F_4

The molecule of this compound, which is prepared by passing NF_3 over various heated metals, has a hydrazine-like structure in which the angle of rotation from the eclipsed configuration is 70°. A Raman spectroscopic study of the liquid (−80

(1) AC 1962 **15** 803

(2) ACSc 1965 **19** 1612
(3) AC 1970 **B26** 536
(4) ACSc 1965 **19** 1629

(5) ACSc 1966 **20** 2483

```
   Cl              Cl
    |               |
   Zn    N—N       Zn
    |         \    /  |
   Cl    N—N       Cl
```

(6) ZK 1962 **117** 241
(7) GCI 1961 **91** 69
(8) AC 1965 **18** 367

to −150°C) indicated an equilibrium mixture of the staggered and *gauche* isomers. N_2F_4 dissociates appreciably to the free radical $\cdot NF_2$ at temperatures above 100°C (compare $N_2O_4 \rightleftharpoons 2\,NO_2$), and this radical can exist indefinitely in the free state; its structure has been determined (Table 18.3). This behaviour of N_2F_4 is in marked

TABLE 18.3
Structural data for hydrazine and related molecules

Molecule	N—N (Å)	N—C(Si) (Å)	C—F (Å)	Angle of rotation from eclipsed	Method	Reference
N_2H_4	1·453			90–95°	i.r.	JCP 1959 **31** 843
	1·449				e.d.	BCSJ 1960 **33** 46
	1·46				X	JCP 1967 **47** 2104
$N_2(SiH_3)_4$	1·46	1·73		82·5°	e.d.	JCS A 1970 318
$N_2H_2(CH_3)_2$	1·45				e.d.	JACS 1948 **70** 2979
$N_2(CF_3)_4$	1·40		1·433	88°	e.d.	IC 1965 **4** 1346
		N—F (Å)	FNF			
N_2F_4	1·53	1·393	104°	70°	e.d.	IC 1967 **6** 304
$\cdot NF_2$		1·363	102·5°		e.d.	IC 1967 **6** 304
N_2F_4 (liquid)	(mixtures of *staggered* and *gauche* isomers)				R	JCP 1968 **48** 3216
			N—O			
$(CH_3)_2N\cdot NO$	1·34	1·46	1·23	C, N, O, all coplanar	e.d.	ACSc 1969 **23** 660
$(CH_3)_2N\cdot NO_2$	1·38	1·46	1·22		e.d.	ACSc 1969 **23** 672

contrast to that of N_2H_4, and is responsible for many reactions in which $-NF_2$ is inserted into molecules, as in the following examples:

$$N_2F_4 \rightleftharpoons \cdot NF_2 \quad + \quad \begin{cases} S_2F_{10} \to SF_5 \cdot NF_2 \\ NO \to ON \cdot NF_2 \\ Cl_2 \to Cl \cdot NF_2 \end{cases}$$

Nitrogen forming two bonds $=N\diagup$

Two bonds formed by a N atom having one or two lone pairs would be expected to be angular with bond angles of 120° (one lone pair, sp^2 bonds) or $109\frac{1}{2}°$ (two lone pairs, sp^3 bonds). (Molecules such as diazirine[1] (cyclic diazomethane) and perfluorodiazirine[2] are special cases like cyclopropane or ethylene oxide.) If the

(1) JACS 1962 **84** 2651
(2) JACS 1967 **89** 5527

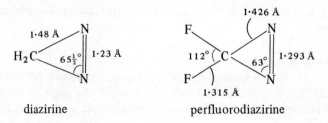

diazirine   perfluorodiazirine

lone-pair orbital has more s character (in the limiting case all s) then angles down to 90° could be expected. Apart from that in N_2H_2 (100°) bond angles lie in the range 110–120° in molecules in which the bonds are formally $=N\diagup$ but they are larger in HN=CO (128°), HN=CS (130°), and CH_3N=CS (142°). This bond arrangement attracted considerable attention in the pre-structural period for it occurs in numerous organic compounds. Early evidence for the non-linearity of one single and one double bond from N came from the resolution into its optical antimers of the oxime

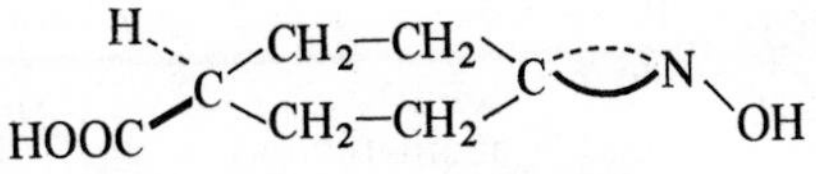

and from the stereoisomerism of unsymmetrical oximes (I).

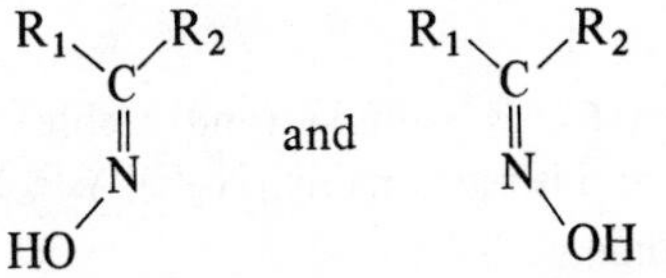

(I) 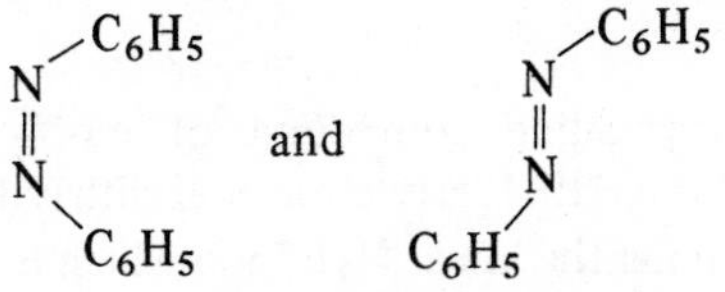   (II)

Among the later demonstrations of this bond arrangement are the crystallographic studies of *cis* and *trans* azobenzene (II) and of dimethylglyoxime. The configuration of the latter molecule changes from the *trans* form (a) in which it exists in the crystalline dioxime to the *cis* form (b) when it forms the well known Ni, Pd, and Cu derivatives.

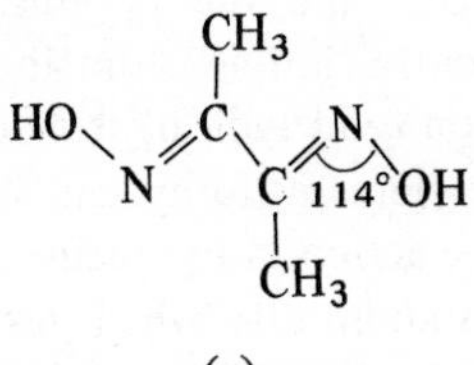

(a) 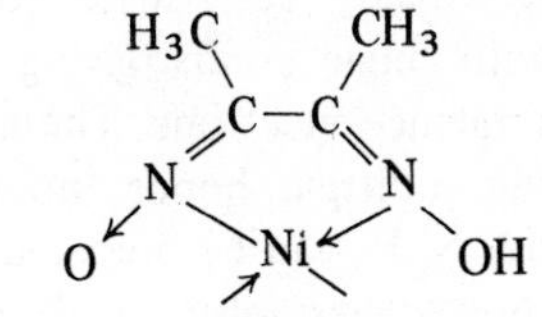   (b)

Di-imide (N_2H_2) *and difluorodiazine* (N_2F_2)

The simplest compound of the type RN=NR, di-imide, has been identified in the mass spectrometer among the solid products of decomposition of N_2H_4 by an electrodeless electrical discharge at 85°K. Both isomers have been identified (i.r.) in inert matrices at low temperatures (as also has imidogen, NH[3]), and the structure (a) has been assigned to the *cis* (planar) isomer [4]

(3) JCP 1965 **43** 507
(4) JCP 1964 **41** 1174

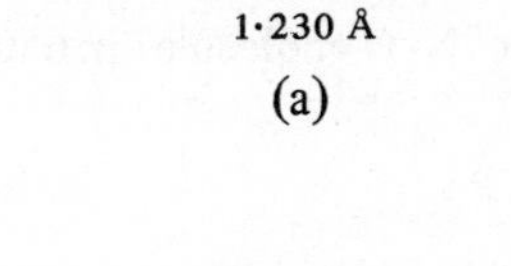

(a) 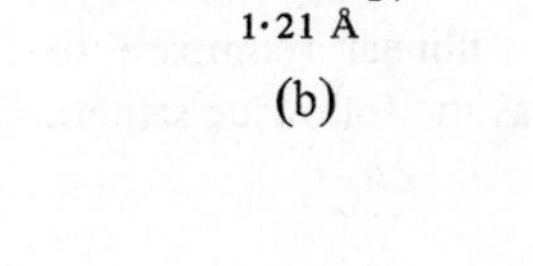   (b)

Nitrogen

Difluorodiazine, N_2F_2, is a colourless gas stable at ordinary temperatures and is prepared by the action of an electric discharge on a stream of NF_3 in the presence of Hg vapour or, with less risk of explosion, by the action of KOH on N,N-difluorourea. The isomer F_2NN is not known, but the two isomers FNNF have been separated by gas chromatography, the *cis* isomer, (b), being the more reactive.[5] The structure of the *trans* isomer,[6] (c), which may be prepared in 45 per cent yield from N_2F_4 and $AlCl_3$, is similar to that of azomethane, (d).

(5) JCP 1963 **39** 1030
(6) IC 1967 **6** 309

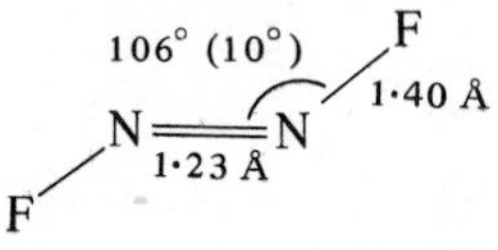

compare azomethane

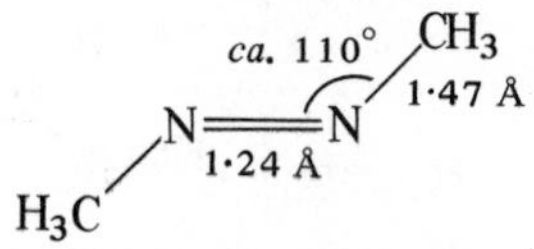

(c)   (d)

(7) JACS 1965 **87** 1889

The crystalline compound of *cis*-N_2F_2 and AsF_5 is soluble (and stable) in anhydrous HF. It is isostructural with $(NO_2)AsF_6$ and is presumably $(N_2F)(AsF_6)$[7] containing the linear N_2F^+ ion which is isoelectronic with NO_2^+.

Compounds containing the system $[N\cdots N\cdots N]$

Apart from organic compounds such as the vicinal triazines, which do not concern us here, this group contains only hydrazoic acid and the azides.

Azides

Reference has already been made to the S_3^{2-} and the I_3^- ions, which may be formulated with single bonds, giving the central iodine atom in the latter ion a group of ten valence electrons. The azide ion (and radical) is of quite a different type, involving multiple bonds; no other element forms an ion of this kind. Hydrazoic acid, N_3H, can be prepared by the action of hydrazine (N_2H_4) on NCl_3 or HNO_2, but it is best obtained from its sodium salt, which results from a very interesting reaction. When nitrous oxide is passed into molten sodamide the following reaction takes place:

$$NH_2^- + N_2O \rightarrow N_3^- + H_2O.$$

When the vapour of hydrazoic acid is passed through a hot tube at low pressure a beautiful blue solid can be frozen out on a finger cooled by liquid air. On warming to $-125°C$ this is converted into NH_4N_3. The blue solid gives no X-ray diffraction pattern and is apparently glassy or amorphous. Although it was at first thought to be $(NH)_x$ the solid is substantially, if not entirely, NH_4N_3, and the colour is probably due to F-centres.[1]

(1) JCP 1959 **30** 349, **31** 564

Numerous structural studies of hydrazoic acid have been made. The three N atoms are collinear (compare the isoelectronic N_2O molecule, p. 650), and the molecule has the following structure:[2]

(2) JCP 1964 **41** 999

648

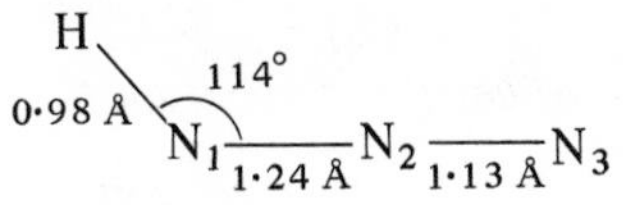

Calculations by the molecular orbital method of the orders of the bonds give $1\cdot65$ for N_1-N_2 and $2\cdot64$ for N_2-N_3.

The azides of the alkali and alkaline-earth metals are colourless crystalline salts which can almost be melted without decomposition taking place. X-ray studies have been made of several ionic azides (Li, Na,[3] K, Sr,[3] Ba[4]); they contain linear symmetrical ions with N—N close to $1\cdot18$ Å. (A number of azides MN_3 are isostructural with the difluorides MHF_2.) Other azides which have been prepared include $B(N_3)_3$, $Al(N_3)_3$, and $Ga(N_3)_3$ from the hydrides, $Be(N_3)_2$ and $Mg(N_3)_2$ from $M(CH_3)_2$ and HN_3 in ether solution, and $Sn(N_3)_4$ from NaN_3 and $SnCl_4$.

Azides of some of the B subgroup metals explode on detonation and are regarded as covalent compounds (e.g. $Pb(N_3)_2$ and AgN_3). In one of the four polymorphs of lead azide there are four non-equivalent N_3 groups, all approximately linear but with a small range of N—N bond lengths ($1\cdot15-1\cdot21$ Å) which may represent real differences.[5] In $Cu(N_3)_2$[6] there are two kinds of non-equivalent N_3 groups bonded in different ways to the Cu atoms. The differences between the results of two studies make it unprofitable to discuss the detailed structures of the N_3 groups. Any asymmetry seems to be less than in the HN_3 molecule.

The azide group can function as a monodentate ligand in coordination compounds such as $[Co(N_3)_2en_2]NO_3$[7] and $[Co(N_3)(NH_3)_5](N_3)_2$,[8] and here again the differences between N—N bond lengths are not too certain. In these complexes the angle between Co—N and the linear N_3 group is around $125°$. The

(3) AC 1968 **B24** 262
(4) AC 1969 **B25** 2638

(5) AC 1969 **B25** 982
(6) ACSc 1967 **21** 2647; AC 1968 **B24** 450

(7) AC 1968 **B24** 1638
(8) AC 1964 **17** 360

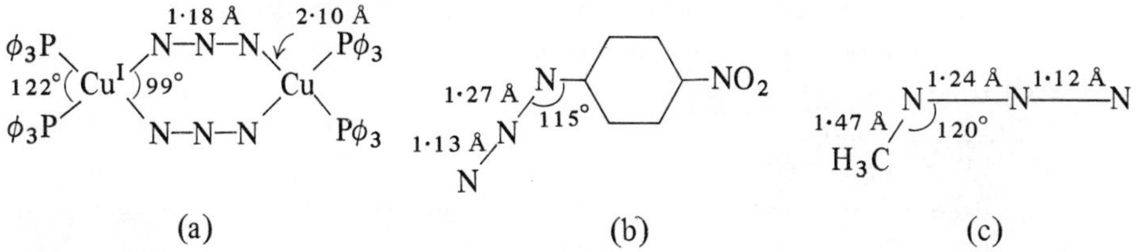

(a)   (b)   (c)

azide group can also act as a bridging ligand (a), when it is symmetrical, with bond lengths similar to those in the ionic azides.[9]

Covalent non-metal azides include the halogen azides N_3X (X = F, Cl, Br, I), silyl azide, H_3SiN_3, and organic azides. Cyanuric[10] and *p*-nitrophenyl azides[11] have been studied in the crystalline state and methyl azide[12] in the vapour state (e.d.). The structures of the last two are shown at (b) and (c).

For comparison with the bond lengths in azides and other compounds standard bond lengths of N—N and C—N bonds are given in Table 18.4.

(9) IC 1971 **10** 1289

(10) PRS 1935A **150** 576
(11) AC 1965 **19** 367
(12) JPC 1960 **64** 756

TABLE 18.4

Observed lengths of N—N *and* C—N *bonds*

(a) 1·86 Å in N_2O_3; 1·75 Å in N_2O_4; 1·53 Å in N_2F_4.

(b) 1·51 Å in amino acids and $(O_2N_2 . CH_2N_2O_2)K_2$.

Bond	Length	Molecule
	(Å)	
N—N(a)	1·46	N_2H_4 and $(CH_3)_2N_2H_2$
N=N	1·23	N_2F_2 and $N_2(CH_3)_2$
N≡N	1·10	N_2
C—N(b)	1·47	CH_3NH_2, $N_4(CH_2)_6$, etc.
C_{ar}—N	≈1·40	C part of aromatic ring, e.g. $C_6H_5 . NH . COCH_3$
C—N	1·33–1·35	Heterocyclic molecules (e.g. pyridine)
C=N	(1·28)	Accurate value not known for a non-resonating molecule
C≡N	1·16	HCN

The oxygen chemistry of nitrogen

Oxides

Five oxides of nitrogen have been known for a long time: N_2O, NO, N_2O_3, N_2O_4, and N_2O_5. A sixth, NO_3, is said to be formed in discharge tubes containing N_2O_4 and O_2 at low pressures. It has at most a short life, like others of this kind which have been described (PO_3, SO, SO_4) and radicals such as OH and CH_3, for the separate existence of which there is spectroscopic evidence. We shall confine our attention to the more stable compounds of ordinary chemical experience.

The trioxide and the pentoxide are the anhydrides of nitrous and nitric acids respectively, but N_2O is not related in this way to hyponitrous acid, $H_2N_2O_2$. Although N_2O is quite soluble in water it does not form the latter acid. Nitrous oxide cannot be directly oxidized to the other oxides although it is easily obtained by the reduction of HNO_3, HNO_2, or moist NO. The oxides NO, N_2O_3, and N_2O_4, on the other hand, are easily interconvertible and are all obtained by reducing nitric acid with As_2O_3, the product depending on the concentration of the acid. The trioxide, N_2O_3, probably exists only in the liquid and solid phases, which are blue in colour, for the vapour is almost completely dissociated into NO and NO_2, as may be seen from its brown colour. The tetroxide, N_2O_4, forms colourless crystals (at $-10°C$), but its vapour at ordinary temperatures consists of a mixture of N_2O_4 and NO_2, and at about $150°C$ the gas is entirely dissociated into NO_2. On further heating decomposition into NO and oxygen takes place, and this is complete at about $600°C$. Since the pentoxide is easily decomposed to N_2O_4 there is the following relation between these oxides:

$$N_2O_5 \rightarrow N_2O_4 \rightleftharpoons \underbrace{NO_2 \rightleftharpoons NO}_{N_2O_3} + O$$

Nitrous oxide, N_2O. This molecule is linear and its electric dipole moment is close to zero (0·17 D). Because of the similar scattering powers of N and O it is not possible to distinguish by electron diffraction between the alternatives NNO and NON, but the former is supported by spectroscopic data. X-ray diffraction data from crystalline N_2O[1] (which is isostructural with CO_2) are explicable only if the NNO molecules are randomly oriented. This disorder would account for the residual entropy of 1·14 e.u. which is close to the theoretical value ($2 \ln 2 = 1·377$ e.u.). The central N atom may be described as forming two sp bonds, the additional π-bonding giving total bond orders of approximately 2·5 (N$\equiv$N) and 1·5 (N$\cdots$O) corresponding to the bond lengths 1·126 and 1·186 Å[2] and force constants 17·88 and 11·39 respectively.[3]

(1) JPC 1961 **65** 1453

Nitric oxide, NO.   Nitric oxide is one of the few simple molecules which contain an odd number of electrons. The N–O bond length is 1·1503 Å[4] and the dipole moment 0·16 D. The monomeric NO molecule in the gas phase is paramagnetic, but the condensed phases are diamagnetic. Evidence for polymerization comes not only from magnetic susceptibility measurements and from the infrared and Raman spectra of the liquid and solid[5] and of NO in a N_2 matrix at 15°K[6] but also from an examination of the structure of the crystalline solid.[7] The variation in degree of association with temperature gives the heat of dissociation of the dimer as $15·52 \pm 0·62$ kJ mol^{-1}. Unfortunately there is disorder in the crystal (resulting in the residual entropy of 1·5 e.u. per mole of dimer), and although the form of the dimer is probably as shown at the right it is not known whether the dimer has the

(2) JCP 1954 **22** 275
(3) JCP 1950 **18** 694

(4) JCP 1955 **23** 57

(5) JACS 1952 **74** 4696
(6) JCP 1969 **50** 3516
(7) AC 1953 **6** 760

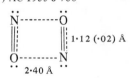

structure $\begin{array}{c} \text{N}\cdots\text{O} \\ | \qquad | \\ \text{O}\cdots\text{N} \end{array}$ or $\begin{array}{c} \text{N}\cdots\text{N} \\ | \qquad | \\ \text{O}\cdots\text{O} \end{array}$ or whether the shape is strictly rectangular. The strength

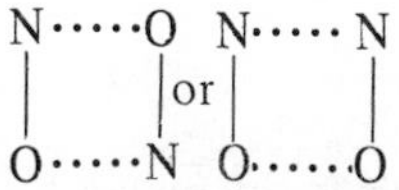

of the bonding between the two molecules of the dimer is certainly much less than that of a single bond.

Nitrogen dioxide, NO_2.   This is also an odd-electron molecule, and some of its reactions resemble those of a free radical, for example, its dimerization, its power of removing hydrogen from saturated hydrocarbons, and its addition reactions with unsaturated and aromatic hydrocarbons. A microwave study[8] gives N–O, 1·197 Å and the O–N–O angle, 134° 15′, in agreement with the results of earlier electron diffraction and infrared studies.

(8) JCP 1956 **25** 1040

Dinitrogen trioxide, N_2O_3.   Pure N_2O_3 cannot be isolated in the gaseous state since it exists in equilibrium with its dissociation products NO and NO_2. However, its m.w. spectrum has been studied[9] at −78°C and 0·1 mm pressure and indicates the planar structure shown. Some of the numerous suggestions as to the nature of the extraordinarily long N–N bond are discussed in ref. (9). The structure of crystalline N_2O_3 (at −115°C) proved too complex to analyse.[10]

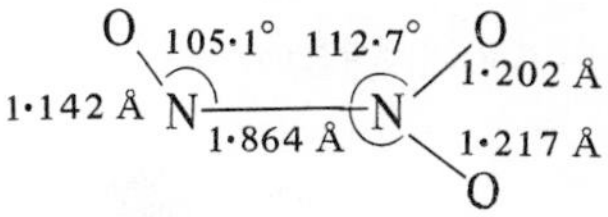

$\begin{array}{c} \text{O} \quad \overset{105\cdot1°}{\frown} \quad \overset{112\cdot7°}{\frown} \quad \text{O} \\ \diagdown \qquad\qquad\qquad \diagup \; 1\cdot202 \text{ Å} \\ 1\cdot142 \text{ Å} \;\; \text{N} \underset{1\cdot864 \text{ Å}}{----} \text{N} \\ \diagdown \; 1\cdot217 \text{ Å} \\ \text{O} \end{array}$

(9) TFS 1969 **65** 1963

(10) AC 1953 **6** 781

Dinitrogen tetroxide, N_2O_4.   The chemistry of this oxide, the dimer of the dioxide, is most interesting, for the molecule can apparently dissociate in three different ways. Thermal dissociation takes place according to the equation

$$N_2O_4 \rightleftharpoons 2\,NO_2.$$

Reactions with covalent molecules are best explained by assuming

$$N_2O_4 \rightleftharpoons NO_2^+ + NO_2^-$$

Nitrogen

(compare the ionization of HNO_3 in H_2SO_4), but in media of high dielectric constant there is ionization

$$N_2O_4 \rightleftharpoons NO^+ + NO_3^-.$$

A number of reactions of liquid N_2O_4 involving nitrosyl compounds and nitrates may be interpreted in terms of this type of ionization. It might appear that these two modes of ionization support the unsymmetrical formula (b) or the bridge formula (c) rather than one of type (a), in which the two NO_2 groups could be

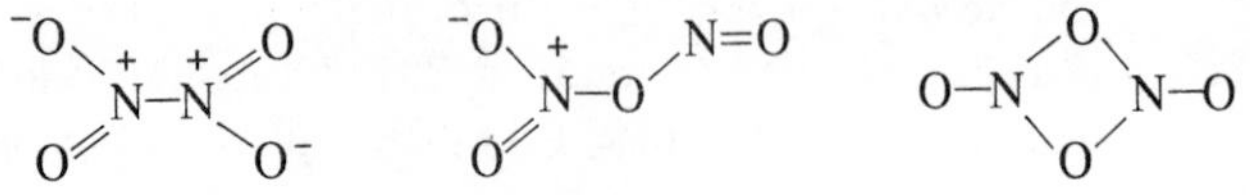

(a)   (b)   (c)

either coplanar or inclined to one another. However, the weakness of this argument is shown by the fact that the isoelectronic oxalate ion, which is known to be symmetrical and of type (a), not only resembles N_2O_4 in its vibration frequencies but also in its chemical reactions:

with $\qquad N_2O_4 \rightarrow NO^+ + NO_3^- \xrightarrow{2\,H^+} NO^+ + H_2NO_3^+ \xrightarrow{H^+} NO^+ + NO_2^+ + H_3O^+$

compare $C_2O_4^{2-} \rightarrow CO + CO_3^{2-} \xrightarrow{2\,H^+} CO + H_2CO_3 \xrightarrow{H^+} CO + CO_2 + H_3O^+.$

The most probable structure of the planar N_2O_4 molecule in the vapour,[11] in the monoclinic form of the solid,[12] and in the crystalline complex with 1 : 4-dioxane[13] is that shown, but somewhat different dimensions were found in the cubic form of the solid (at $-40°C$), namely, N–N, 1·64 Å, and O–N–O, 126°.[14] Note the extraordinary length of the N–N bond compared with the normal single bond length (1·46 Å) in hydrazine, and compare with the long B–B bond in B_2Cl_4 which is also planar. Infrared studies have been made of N_2O_4 trapped in solid argon and oxygen matrices and evidence for various isomers has been given, for example, a twisted non-planar form of type (a) and the isomer (b).[15] The spectrum of this oxide made by the oxidation of NO in liquid ethane–propane mixtures at 80°K has been interpreted as that of $(NO)^+(NO_3)^-$.[16]

Dinitrogen pentoxide, N_2O_5. This oxide consists of N_2O_5 molecules in the gaseous state and also in solution in CCl_4, $CHCl_3$, and $POCl_3$. An early study of the gas by electron diffraction did not lead to a definite configuration of the molecule, and a more precise determination is desirable.[17]

When dissolved in H_2SO_4 or HNO_3 this oxide ionizes

$$N_2O_5 \rightleftharpoons NO_2^+ + NO_3^-$$

and from such solutions nitronium salts can be prepared (see later). The earlier recognition of NO_2^+ and NO_3^- frequencies in the Raman spectrum of crystalline N_2O_5 is confirmed by the determination of the crystal structure. At temperatures down to that of liquid N_2 crystalline N_2O_5 consists of equal numbers of planar NO_3^- ions (N–O $\approx$ 1·24 Å) and linear NO_2^+ ions (N–O $\approx$ 1·15 Å).[18] (The i.r.

(11) JCP 1956 **25** 1282
(12) ACSc 1963 **17** 2419
(13) ACSc 1965 **19** 120
(14) N 1949 **164** 915

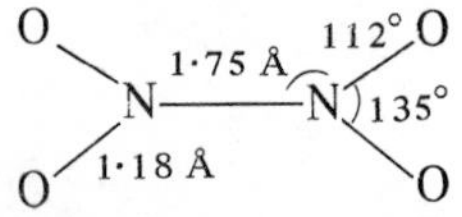

(15) IC 1963 **2** 747; JCP 1965 **42** 85

(16) JCP 1965 **43** 136

(17) JCP 1934 **2** 331

(18) AC 1950 **3** 290

652

spectrum of this oxide suggests that at the temperature of liquid He it is a molecular solid, possibly O_2NONO_2,[19] but this has not been confirmed by a structural study.)

(19) SA 1962 **18** 1641

The reaction of N_2O_5 with SO_3 is noted under nitronium compounds.

Nitrosyl compounds

The loss of an electron by NO should lead to a stable ion $[:N \equiv O:]^+$ since this contains a triple bond compared with a $2\frac{1}{2}$-bond (double + 3-electron bonds) in NO. Hantzsch showed that nitrous acid does not exist as HNO_2 in acid solution but as NO^+ ions:

$$HNO_2 + H^+ \rightleftharpoons NO^+ + H_2O$$

and that the same NO^+ ions are present in solutions of 'nitrosylsulphuric acid', or nitrosyl hydrogen sulphate, $NO^+ . HSO_4^-$. It has been shown by X-ray studies that certain nitrosyl compounds, $NO . ClO_4$, $NO . BF_4$, and $(NO)_2SnCl_6$ (originally written $SnCl_4 . 2 NOCl$), are structurally similar to the corresponding ammonium salts, NH_4ClO_4, etc., and $NO(PtF_6)$ is isostructural with $O_2(PtF_6)$ and $KSbF_6$.[20] These nitrosyl compounds presumably contain the NO^+ ion. The radius of this ion, assumed to be spherical owing to rotation, is 1.40 Å.

(20) IC 1966 **5** 1217

The nitrosyl halide molecules are non-linear (interbond angles, $110°$, $113°$, and $117°$ in NOF,[21] NOCl,[22] and NOBr[23] respectively) and the N—O bond length is close to 1.14 Å, as in NO. This bond length and the fact that the N—X bond lengths (1.52, 1.97, and 2.13 Å) are considerably greater than normal single bond lengths suggest that these molecules are not adequately represented by the simple structure (1) but that there may be resonance with the ionic structure (2):

(21) JCP 1951 **19** 1071
(22) JCS 1961 1322
(23) JACS 1937 **59** 2629
(24) RTC 1943 **62** 289

Support for this polar structure comes from the dipole moments of pure NOCl and NOBr (2.19 D and 1.90 D respectively) which are much greater than the values (around 0.3 D) estimated for structure (1).

Nitroso compounds

Structural studies have been made of the free radical $[(CH_3)_3C]_2NO^{(1)}$ (N—O, 1.28 Å), of nitrosomethane, CH_3NO,[2] (CNO, $113°$ assuming N—O, 1.22 Å), and of F_3CNO,[3] in which N—O is 1.17 Å and angle CNO $121°$. The C—N bond is abnormally long (1.55 Å) as in the nitro compounds, CF_3NO_2, etc. (p. 659). Many

(1) ACSc 1966 **20** 2728
(2) JCP 1968 **49** 591
(3) JPC 1965 **69** 3727

of these compounds dimerize, the example shown being formed by the action of N_2O_3 on ethylene.[4] For the ions $[ON(NO)SO_3]^-$ and $[ON(SO_3)_2]^{2-}$, in both of which N forms three coplanar bonds, see p. 588.

(4) JACS 1969 **91** 1371

Nitrogen

Nitrosyl derivatives of metals

Numerous metallic compounds containing NO have been prepared, most of which are deeply coloured (red to black). Examples include the well known nitroprusside, $K_2[Fe(CN)_5NO]$, ammines such as $[Co(NH_3)_5NO]X_2$, and numerous other complex salts such as $K_3[Ni(S_2O_3)_2NO] \cdot 2\,H_2O$, $K_2[Ni(CN)_3NO]$, and $K_2[RuCl_5NO]$. Many of these compounds can be prepared by the direct action of NO on the appropriate compound of the metal, e.g. nitrosyl halides (e.g. $Co(NO)_2Cl$)[1] from the halides and nitrosyl-carbonyls ($Fe(CO)_2(NO)_2$ and $Co(CO)_3NO$) from the carbonyls, while the action of NO on an ammoniacal solution of a cobaltous salt gives two series (red and black) of salts $[Co(NH_3)_5NO]X_2$. Apparently the red (more stable) form is a dimer, presumably a hyponitrite, $[(NH_3)_5Co-ON=NOCo(NH_3)_5]X_4$, for equivalent conductivity measurements show that it is a 4 : 1 electrolyte.[2] The action of NO under pressure on $Fe_2(CO)_9$ gives $Fe(NO)_2(CO)_2$ or $Fe(NO)_4$ depending on the temperature. Solutions of ferrous salts react with NO in the presence of sulphides to give Roussin's black salts, $M[Fe_4S_3(NO)_7]$, which are converted by alkalis to less stable red salts $M_2[Fe_2S_2(NO)_2]$. If mercaptans are used instead of alkali sulphides esters of the red salt are obtained, for example, $Fe_2S_2(NO)_4(C_2H_5)_2$.

As a monodentate ligand in metal complexes NO behaves in one of two ways: (a) as a 12-electron unit (NO^-) when it forms a single bond to M with M–N–O around 125°, and (b) more usually, as a 10-electron unit (NO^+), when M–N is a multiple bond and M–N–O is approximately linear. Examples of NO behaving as a bridging ligand are given later.

Some of the compounds in Table 18.5 are also of interest in connection with the stereochemistry of the transition-metal atom, for example, the tetragonal pyramidal structures of $[Ir(CO)I(P\phi_3)_2NO]^+$ and the Fe and Co compounds $M(NO)[S_2CN(CH_3)_2]_2$, (c), and $Mo(C_5H_5)_3(NO)$, (d). The Fe molecule of type (c)

(1) ACSc 1967 **21** 1183

(2) IC 1964 **3** 1038

(3) AC 1970 **B26** 1899
(4) JACS 1967 **89** 3645

(a) (b)

(c) (d) (e)

has nearly collinear M–N–O bonds, but the bonding is of type (a) in the Ir and probably also the Co compound, the structure of which is known less accurately. There are approximately collinear M–N–O bonds in the nitroprusside ion, $[Fe(CN)_5NO]^{2-}$, and the similar ions formed by Cr and Mn.

In the molecule (e) NO behaves both as a monodentate ligand of type (b) (angle Cr–N–O, 179°) and also as a bridging ligand. The Cr–Cr distance indicates a metal-metal bond.[3] An even more complex behaviour is exhibited in $(C_5H_5)_3Mn_3(NO)_4$.[4] Here NO is functioning in two ways, as a doubly and as a

TABLE 18.5

Structures of metal nitrosyl compounds

	Angle	M—N	N—O	Reference
Type (a)	M—N—O			
[Co(en)$_2$Cl(NO)]ClO$_4$	124°	1·82 Å	1·11 Å	IC 1970 **9** 2760
[Ir(CO)Cl(Pϕ_3)$_2$NO] BF$_4$	124°	1·97	1·16	JACS 1968 **90** 4486
[Ir(CO)I(Pϕ_3)$_2$NO] BF$_4$.C$_6$H$_6$	125°	1·89	1·17	IC 1969 **8** 1282
Co(NO)[S$_2$CN(CH$_3$)$_2$]$_2$	(≈135°)	(1·7)	(1·1)	JCS 1962 668
Type (b)				
[Cr(CN)$_5$NO][Co(en)$_3$]. 2 H$_2$O	176°	1·71	1·21	IC 1970 **9** 2397
Fe(NO)[S$_2$CN(CH$_3$)$_2$]$_2$	170°	1·72	1·10	JCS A 1970 1275
Na$_2$[Fe(CN)$_5$NO] . 2 H$_2$O	178°	1·63	1·13	IC 1963 **2** 1043
Cr(C$_5$H$_5$)Cl(NO)$_2$	170°	1·71	1·14	JCS A 1966 1095
Cr(C$_5$H$_5$)(NCO)(NO)$_2$	171°	1·72	1·16	JCS A 1970 605, 611
Mo(NO)(C$_5$H$_5$)$_3$	179°	1·75	1·21	JACS 1969 **91** 2528
Mn(CO)$_2$(Pϕ_3)$_2$NO	178°	1·73	1·18	IC 1967 **6** 1575

triply bridging ligand. In Fig. 18.3 the molecule is viewed in a direction perpendicular to the equilateral triangle of Mn atoms (Mn–Mn approx. 2·50 Å) and along the axis of the triply bridging NO, which lies beneath the plane of the paper.

The crystal structure of one of the black Roussin salts, CsFe$_4$S$_3$(NO)$_7$. H$_2$O, has been determined.[5] The nucleus of the ion consists of a tetrahedron of Fe atoms with S atoms above the centres of three of the faces (Fig. 18.4(a)). One iron atom

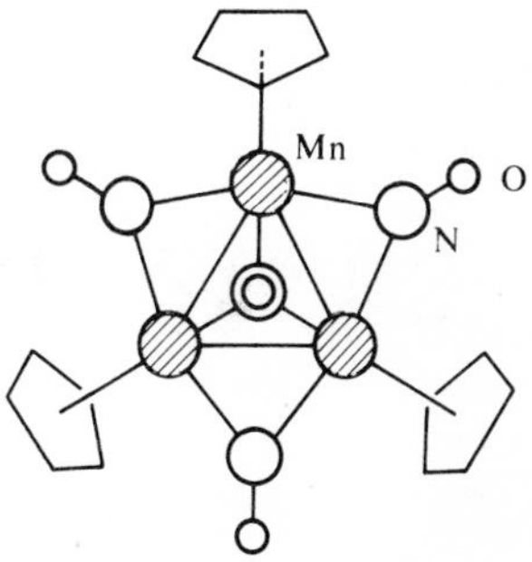

FIG. 18.3. The molecule (C$_5$H$_5$)$_3$-Mn$_3$(NO)$_4$.

(5) AC 1958 **11** 594

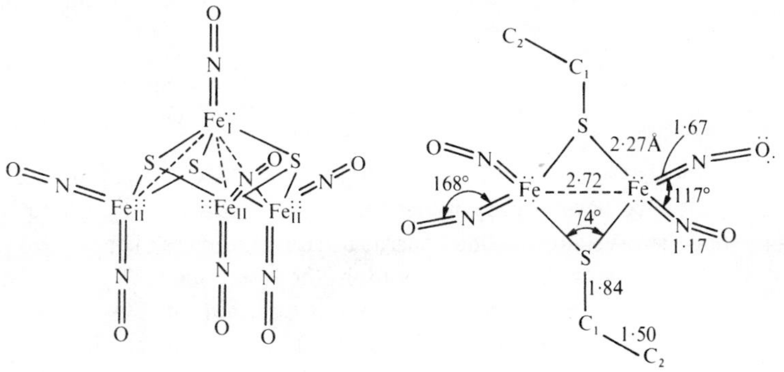

FIG. 18.4. The structures of (a) the Fe$_4$S$_3$(NO)$_7^-$ ion and (b) the Fe$_2$S$_2$(NO)$_4$(C$_2$H$_5$)$_2$ molecule.

(Fe$_I$) is linked to three S atoms and one NO group, and in addition forms weak bonds to the other three iron (Fe$_{II}$) atoms. The Fe$_{II}$ atoms are bonded to two S and two NO and to the Fe$_I$ atom. The length of the Fe$_I$–Fe$_{II}$ bonds (2·70 Å) corresponds to about a 'half-bond', using Pauling's equation (p. 1025), but there are no bonds between Fe$_{II}$ atoms (Fe$_{II}$–Fe$_{II}$, 3·57 Å). The classical valence bond picture (a) gives the iron atoms an effective atomic number of 18 and is consistent with the diamagnetism of the compound, but it implies rather high formal charges on Fe_I (-2) and Fe_{II} (-3) suggesting resonance with structures which will reduce these charges. An alternative model having a four-centre molecular orbital linking the iron atoms has also been suggested. Other bond lengths in this molecule are: Fe–S, 2·23, Fe_I–N, 1·57, Fe_{II}–N, 1·67, and N–O, 1·20 Å.

The molecule of the red ester, $Fe_2S_2(NO)_4(C_2H_5)_2$, has the structure[6] shown in Fig. 18.4(b), in which Fe forms tetrahedral bonds to two S and two NO. The Fe–Fe distance (2·72 Å) is very similar to that in the black salt, and there must be sufficient interaction between these atoms to account for the diamagnetism of this compound. For nitrosyl carbonyls see p. 772.

(6) AC 1958 **11** 599

Nitryl halides and nitronium compounds

The existence of two nitryl halides, NO_2F and NO_2Cl, is well established; the existence of NO_2Br is less certain. The fluoride can be conveniently prepared by the action of fluorine on slightly warm dry sodium nitrite: $NaNO_2 + F_2 \rightarrow NO_2F + NaF$, and is a reactive gas which converts many metals to oxide and fluoride or oxyfluoride, and reacts with many non-metals (Br and Te excepted) to form nitronium salts. Examples of nitronium salts are $(NO_2)BF_4$, $(NO_2)_2SiF_6$, $(NO_2)PF_6$, and $(NO_2)IF_6$.

Nitryl chloride can be prepared from chlorosulphonic acid and anhydrous nitric acid. It combines with halides to form salts such as $(NO_2)SbCl_6$ which in an ionizing solvent (e.g. liquid SO_2) undergo reactions of the type:

$$(NO_2)(SbCl_6) + N(CH_3)_4ClO_4 \rightarrow (NO_2)ClO_4 + N(CH_3)_4SbCl_6.$$

(1) JCS A 1968 1736
(2) JMS 1963 **11** 349

Microwave data for NO_2F[1] and NO_2Cl[2] are consistent with planar molecules:

$$F \xrightarrow{1\cdot47\ Å} N \underset{1\cdot18}{\overset{O}{\diagup}} 136° \qquad Cl \xrightarrow{1\cdot84\ Å} N \underset{1\cdot20}{\overset{O}{\diagup}} 130°$$

(3) AC 1960 **13** 855

Nitronium salts contain the linear NO_2^+ ion (N–O, 1·10 Å), studied in $(NO_2)(ClO_4)$.[3] We have already noted that N_2O_5 ionizes in H_2SO_4 to NO_2^+ and NO_3^- and that the crystalline oxide consists of equal numbers of these two ions. By mixing SO_3 and N_2O_5 in $POCl_3$ solution the compounds $N_2O_5 . 2 SO_3$ and $N_2O_5 . 3 SO_3$ can be obtained, and from a solution of SO_3 in HNO_3 the compounds $HNO_3 . 2 SO_3$ and $HNO_3 . 3 SO_3$ have been isolated. These compounds are in fact $(NO_2)_2S_2O_7$ and $(NO_2)_2S_3O_{10}$, $NO_2(HS_2O_7)$ and $NO_2(HS_3O_{10})$ respectively. X-ray studies show that crystals of $(NO_2)_2S_3O_{10}$[4] and $NO_2(HS_2O_7)$[5] consist of linear $(O-N-O)^+$ ions and $S_3O_{10}^{2-}$ or $HS_2O_7^-$ ions respectively.

(4) AC 1954 7 430
(5) AC 1954 7 402; IC 1971 **10** 2319

Acids and oxy-ions

The existence of three oxy-acids of nitrogen is well established, nitrous, HNO_2, nitric, HNO_3, and hyponitrous, $H_2N_2O_2$. The preparation of other acids or their salts has been described, but we shall comment on these compounds only if structural studies have been made.

Nitrous acid. Only dilute solutions of nitrous acid have been prepared, and these rapidly decompose, evolving oxides of nitrogen. Two structures can be envisaged for the molecule of nitrous acid, the second of which could exist in *cis* and *trans* forms:

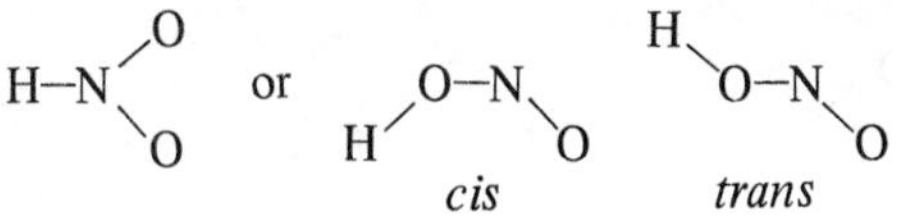

(a) (b)

This tautomerism is suggested by the fact that by the interaction of metal nitrites with alkyl halides two series of compounds are obtained, the nitrites—in which the NO_2 group is attached to carbon through an O atom—and the nitro compounds, in which there is a direct N—C link. From i.r. studies of nitrous acid gas it is concluded[1] that there is no evidence for the existence of form (a) and that the two isomers of (b) exist in comparable amounts, the *trans* isomer being rather more stable. The structure (c) has been assigned to the *trans* isomer.[2]

(1) JCP 1951 **19** 1599

(2) JACS 1966 **88** 5071

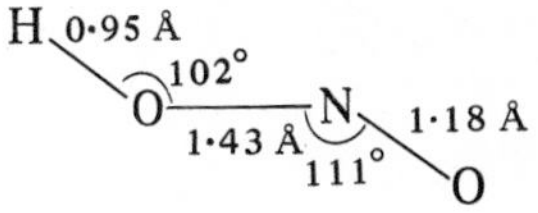

(c)

Metal nitrites and nitrito compounds. Until recently the only stable simple metal nitrites known were those of the alkali and alkaline-earth metals, Zn, Cd, Ag(I), and Hg(I). The alkali nitrites decompose at red heat, the others at progressively lower temperatures. The ionic nitrites of the more electropositive metals contain the NO_2^- ion which in $NaNO_2$[1] has the following configuration: ONO, 115·4°, and N—O, 1·236 Å. It may be of interest to summarize here the structural data for the unique series NO_2^+, NO_2, and NO_2^- with respectively 16, 17, and 18 valence electrons:

(1) AC 1961 **14** 56

	NO_2^+	NO_2	NO_2^-
N—O	1·10	1·19	1·24 Å
ONO angle	180°	134°	115°

(Nothing is known of the structure of the NO_2^{2-} ion which may be present in sodium nitroxylate, Na_2NO_2,[2] formed by the reaction between $NaNO_2$ and Na metal in liquid ammonia as a bright-yellow solid. The paramagnetic susceptibility of this salt is far too small for $(Na^+)_2(NO_2)^{2-}$, and is interpreted[3] as due to about 20 per cent dissociation of the dimeric $N_2O_4^{4-}$ ion.)

The insoluble, more deeply coloured, nitrites of metals such as Hg are probably essentially covalent compounds. Anhydrous nitrites of Ni(II) and Co(II) have been made from NO_2 and a suitable compound of the metal, for example, $Ni(NO_2)_2$ from $Ni(CO)_4$ and NO_2 (each diluted with argon).[4] It is stable up to 260°C. Complex nitrites of transition metals (e.g. $Na_3[Co(NO_2)_6]$ and $K_2Pb[M(NO_2)_6]$ (M = Fe, Co, Ni, Cu)) are well known.

(2) B 1928 **61** 189

(3) ACSc 1958 **12** 578

(4) IC 1963 **2** 228

657

Nitrogen

As a ligand directly bonded to a metal atom the NO_2^- ion can function in the following ways:

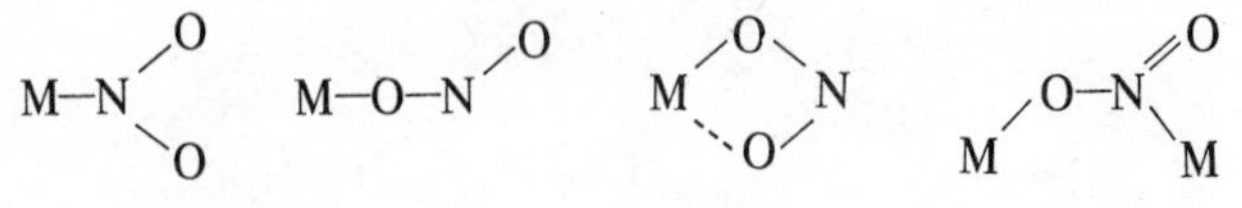

(i)   (ii)   (iii)   (iv)

(5) IC 1966 5 514
(6) AC 1968 **B24** 474
(7) CC 1965 476

(i) Bonded through N as in complex nitrito ions such as $[Co(NO_2)_6]^{3-}$ and $[Cu(NO_2)_6]^{4-}$ (p. 900)—in $K_2Pb[Cu(NO_2)_6]$ the dimensions of NO_2 are the same as for the NO_2^- ion[5]—and in complexes such as $[Co(NH_3)_2(NO_2)_4]^-$ and $[Co(NH_3)_5NO_2]^{2+}$.[6]

(ii) Bonded through O, as in $Ni[(CH_3)_2NCH_2CH_2N(CH_3)_2]_2(ONO)_2$[7]:

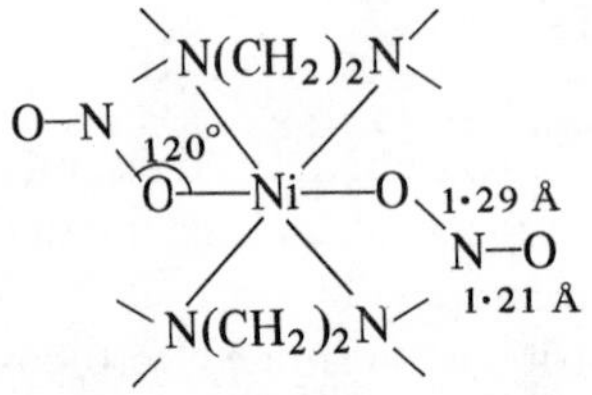

(8) JCS A 1969 1248

(9) JCS A 1969 2081

(iii) As a bidentate ligand. In $[Cu(bipyr)_2(ONO)]NO_3$[8] there is nearly symmetrical bridging. Both Cu—O bonds are weak compared with the normal bond (≈ 2.0 Å), and their mean length is 2.29 Å. In $Cu(bipyr)(ONO)_2$,[9] on the other hand, one bond to each ONO^- is of normal length and the other very long, but the mean (2.24 Å) is close to that in the more symmetrical bridge. Since Cu(II) is by no means a normal element as regards its stereochemistry (see, for example, its nitrato compounds, p. 895) it would be informative to have the structures of compounds of other metals in which NO_2^- is behaving as a bidentate ligand.

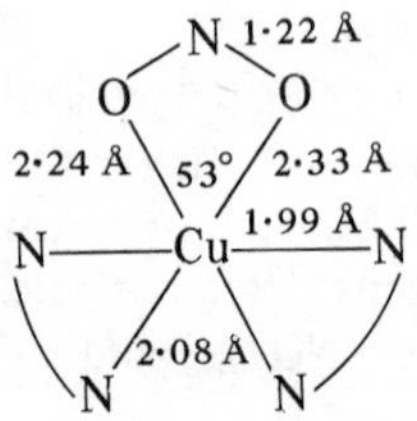

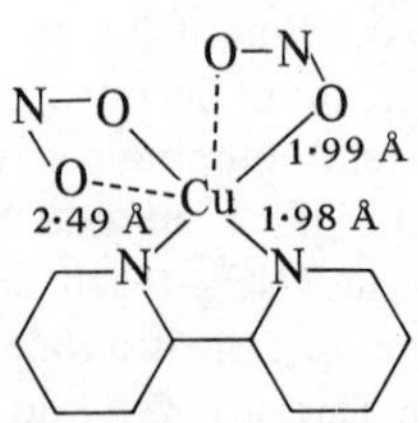

658

(iv) Bridging (unsymmetrically) two M atoms, forming a planar bridge:

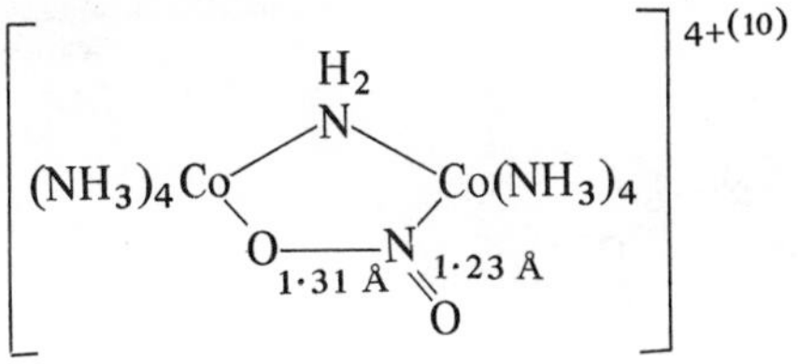

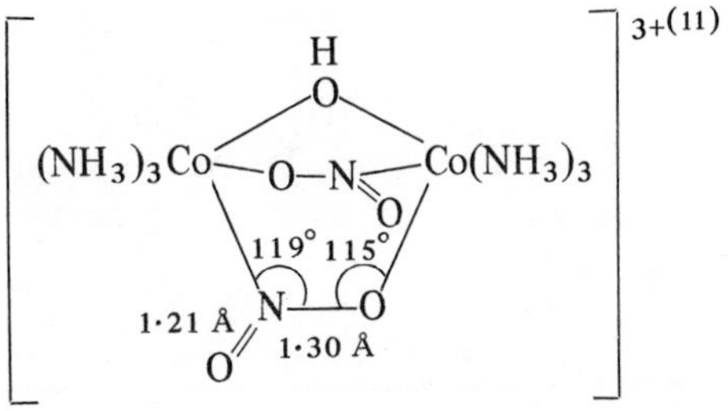

(10) IC 1970 **9** 1604
(11) AC 1970 **B26** 81

Organic nitro compounds.   The NO_2 group is known to be symmetrical in organic nitro compounds, with N–O, 1·21 Å, and ONO 125–130° (nitromethane, CH_3NO_2,[1] and various crystalline nitro compounds). Rather larger angles (132° and 134°) have been found in CF_3NO_2 and CBr_3NO_2.[2] In these molecules the C–F, C–Br, and N–O bond lengths are normal, 1·33, 1·92, and 1·21 Å respectively, but C–N is unusually long (1·56 and 1·59 Å), and the system

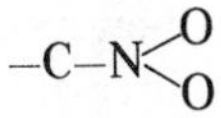

(1) JCP 1956 **25** 42
(2) JCP 1962 **36** 1969

non planar.

Hyponitrous acid.   In contrast to nitrous and nitric acids, hyponitrous acid crystallizes from ether as colourless crystals which easily decompose, explosively if heated. The detailed molecular structure of this acid has not been determined, but it is known that the molecular weights of the free acid and its esters correspond to the double formula, $H_2N_2O_2$, that it is decomposed by sulphuric acid to N_2O, and that it can be reduced to hydrazine, H_2N–NH_2. Infrared and Raman studies show conclusively that the hyponitrite ion has the *trans* configuration (a), but the N–N frequency suggests that the central bond has an order of rather less than two.[1]

(1) JACS 1956 **78** 1820

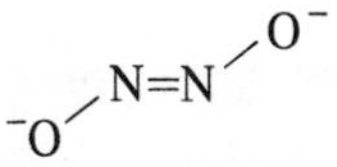

(a) 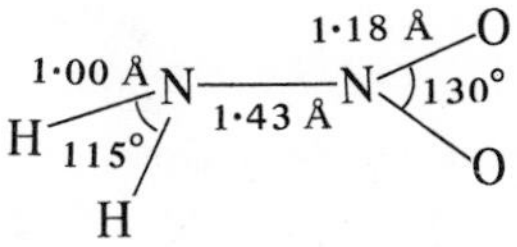   (b)

The isomer nitramide, $H_2N . NO_2$, which decomposes in alkaline solution to N_2O and which like $H_2N_2O_2$ may be reduced to hydrazine, has the structure (b), in which the angle between the NH_2 plane and the NNO_2 plane is 52°.[2]

(2) JMS 1963 **11** 39

Oxyhyponitrite ion.   The salt $Na_2N_2O_3$ (Angeli's salt) is prepared by the action of sodium methoxide on hydroxylamine and butyl nitrate at 0°C:

$$C_4H_9ONO_2 + H_2NOH + 2\,NaOCH_3 \rightarrow Na_2(ONNO_2) + C_4H_9OH + 2\,CH_3OH.$$

(1) IC 1962 **1** 938; ibid, 1964 **3** 900; ibid, 1969 **8** 693

Two forms of the salt are known, and although their structures have not been determined, i.r. and Raman spectroscopic and other indirect evidence[1] are said to support the formulae

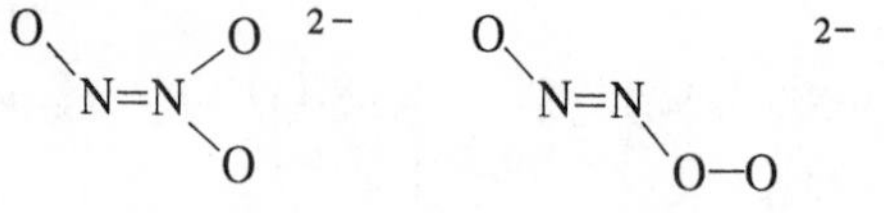

(α)   (β)

(1) JINC 1964 **26** 453; JCS A 1968 450; JCS A 1970 925

Peroxynitrite ion. This ion, produced by oxidizing azide solutions by O_3 or NH_2Cl by O_2, apparently exists in methanol solutions for weeks. Its presence has been demonstrated by its u.v. absorption spectrum, but no solid salts have been obtained. The structure $(O=N-O-O)^-$ has been suggested.[1]

Nitric acid and nitrate ion. Pure nitric acid can be made by freezing dilute solutions (since the freezing point of HNO_3 is about $-40°C$), but it is not stable. Dissociation into N_2O_5 and H_2O takes place, the former decomposes, and the acid becomes more dilute. If strong solutions of the acid are kept in the dark, however, decomposition of the N_2O_5 is retarded and the concentration of nitric acid remains constant. The anhydrous acid and also the crystalline mono- and tri-hydrates have been studied in the crystalline state and the structures of the hydrates in solution

(1) ACSc 1954 **8** 374
(2) JCP 1965 **42** 3106
(3) AC 1951 **4** 120
(4) AC 1951 **4** 239

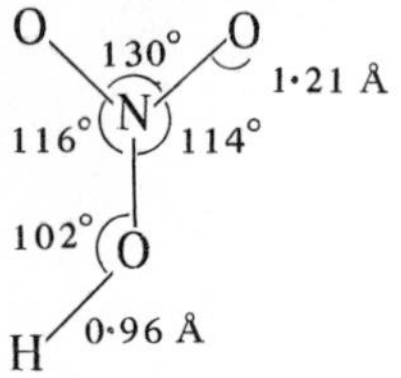

(5) AC 1953 **6** 157

have also been studied.[1] In the vapour state HNO_3 exists as planar molecules with the structure shown.[2] Reliable data on the structure of the molecule in the crystalline anhydrous acid were not obtained owing to submicroscopic twinning of the crystals.[3] In the monohydrate[4] there are puckered layers of the kind shown in Fig. 15.25(b) (p. 563), where the shaded circles represent O atoms of water molecules. N.m.r. studies indicate groups of three protons at the corners of equilateral triangles, showing that this hydrate is $(H_3O)^+(NO_3)^-$ and contains H_3O^+ ions, as do the monohydrates of certain other acids (p. 562). In $HNO_3 . 3 H_2O$[5] there is a complex system of hydrogen bonds linking the NO_3 groups and H_2O molecules into a 3-dimensional framework. In the NO_3 groups the bond lengths are approximately equal (1·24 Å) and the angles are close to 120°, so that here also the crystal appears to contain NO_3^- ions rather than HNO_3 molecules.

Simple ionic nitrates (see later) are quite different structurally from salts of the type $A_m(XO_3)_n$ formed by other Group V elements. They contain the planar equilateral NO_3^- ion, geometrically similar to the BO_3^{3-} and CO_3^{2-} ions, in contrast to the pyramidal SO_3^{2-} and ClO_3^- ions of the elements of the second short Period and the polymeric $(PO_3)_n^{n-}$ ions in metaphosphates. The N–O bond length in NO_3^- is close to 1·22 Å in a number of nitrates,[6] though the value 1·27 Å was found in a n.d. study of $Pb(NO_3)_2$[7a] and 1·26 Å in $AgNO_3$.[7b]

(6) AC 1957 **10** 567
(7a) AC 1957 **10** 103
(7b) JCS A 1971 2058

In addition to normal nitrates salts $M[H(NO_3)_2]$ and $M[H_2(NO_3)_3]$ are known in which M is a large ion such as K^+, Rb^+, Cs^+, or ammonium or arsonium ions. Two quite different structures have been found for the $H(NO_3)_2^-$ ion; (a) a slightly distorted tetrahedral group in $[Rh(pyr)_4Br_2]H(NO_3)_2$,[8] in which the positions

(8) CC 1967 62

of the H atoms are not obvious, and (b) a coplanar pair of NO_3^- ions linked by a short hydrogen bond in $[As\phi_4]H(NO_3)_2$.[9] In $NH_4H_2(NO_3)_3$[10] the $H_2(NO_3)_3^-$ system consists of a central nitrate ion hydrogen-bonded to two HNO_3 molecules which lie in a plane perpendicular to that of the NO_3^- ion, (c).

(9) CC 1967 1211
(10) AC 1970 **B26** 1117

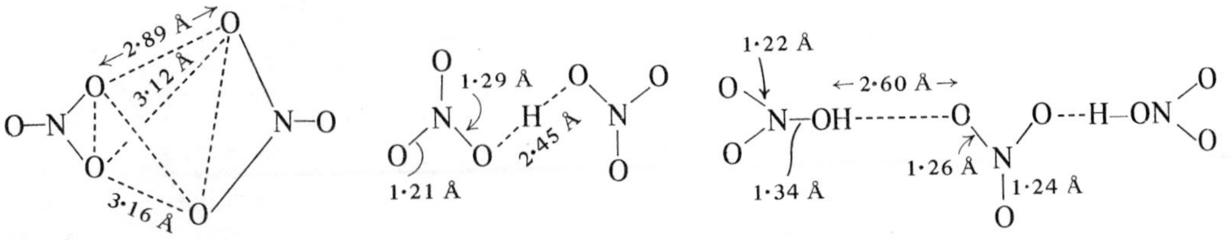

(a)   (b)   (c)

Metal nitrates and nitrato complexes.   There are several groups of compounds in which the NO_3 group is directly bonded to one or more metal atoms by bonds which presumably have considerable covalent character, namely:

(i) volatile metal nitrates, $M(NO_3)_n$,
(ii) nitrato complexes, $[M(NO_3)_n]^{m-}$,
(iii) oxy-nitrato compounds, $MO_x(NO_3)_y$, and
(iv) coordination complexes containing $-NO_3$ as a ligand.

Examples of compounds of classes (i)–(iii) are given in Table 18.6.

TABLE 18.6
Metal nitrates and nitrato complexes

Class (i)			
$Cu(NO_3)_2$[1] (also Be, Zn, Mn, Hg, Pd)	$Al(NO_3)_3$ (also Cr, Fe, Au, In, Co[2]	$Ti(NO_3)_4$[3] (also Sn,[4] Zr)	
Class (ii)			
$[Au(NO_3)_4]K$[5]		$[Al^{III}(NO_3)_4]NO$ $[In^{III}(NO_3)_4](NEt_4)$ $[Co^{II}(NO_3)_4](As\phi_4)_2$[6] $[Mn^{II}(NO_3)_4](As\phi_4)_2$[7] $[Fe^{III}(NO_3)_4](As\phi_4)$[8]	$[Ce^{IV}(NO_3)_6]\cdot(NH_4)_2$[9] $[Th^{IV}(NO_3)_6]Mg\cdot8\,H_2O$[10] $[Ce^{III}(NO_3)_6]_2Mg_3\cdot24\,H_2O$[11]
Class (iii)			
$Fe^{III}O(NO_3)$ $Ti^{IV}O(NO_3)_2$ $Nb^VO(NO_3)_3$	$V^VO_2NO_3$ $Cr^{VI}O_2(NO_3)_2$ $UO_2(NO_3)_2$ $[UO_2(NO_3)_3]Rb$[12]	$Re^{VII}O_3NO_3$	$Be_4O(NO_3)_6$

(1) JACS 1963 **85** 3597. (2) CC 1968 871. (3) JCS A 1966 1496. (4) JCS A 1967 1949. (5) JCS A 1970 3092. (6) IC 1966 **5** 1208. (7) JCS A 1970 226. (8) CC 1971 554. (9) IC 1968 **7** 715. (10) AC 1965 **18** 698. (11) JCP 1963 **39** 2881. (12) AC 1965 **19** 205.

In these various compounds the NO_3 group behaves in one of the three following ways:

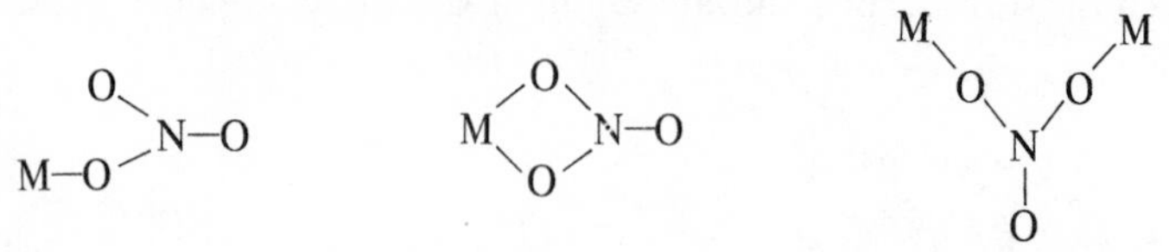

(a) monodentate   (b) bidentate   (c) bridging

In (b) and (c) it appears that the NO_3 group can bond symmetrically or unsymmetrically, that is, the two bonds to the metal atom(s) have the same or different lengths. This is discussed further on p. 896 in connection with $Cu(NO_3)_2$ and related compounds. Owing to lack of sufficient structural information it is not yet possible to say whether all these possibilities are realized in each of the above four classes, but examples can certainly be given of the following:

Class (i)		(b)	(c)
(ii)		(b)	
(iii)		(b)	(c)
(iv)	(a)	(b)	

We now deal briefly with the four groups of compounds listed above.

(i) *Anhydrous metal nitrates.*   Anhydrous *ionic* nitrates include the familiar water-soluble compounds of the alkali and alkaline-earth metals and those of a few B subgroup metals (e.g. $AgNO_3$, $Cd(NO_3)_2$, $Pb(NO_3)_2$) and transition metals ($Co(NO_3)_2$, $Ni(NO_3)_2$). The nitrates of Mg and many 3d metals are normally obtained as hydrates, and while the hydrates are normal ionic crystals the anhydrous salts are in many cases surprisingly volatile and are less readily prepared. Heating of many hydrated nitrates produces hydroxy-salts or the oxide, but the volatile anhydrous nitrates of many transition metals and of the less electropositive metals (Be, Mg, Zn, Hg, Sn^{IV}, Ti^{IV}) may be prepared by the action of liquid N_2O_5 or of a mixture of N_2O_4 and ethyl acetate on the metal or a compound of the metal. For example, if metallic Cu is treated with N_2O_4 and dried ethyl acetate at room temperature for 4 hours the addition of further N_2O_4 then precipitates $Cu(NO_3)_2 . N_2O_4$. On heating this is converted into anhydrous $Cu(NO_3)_2$, which may be purified by sublimation. Some of these anhydrous nitrates (e.g. $Sn(NO_3)_4$ and $Ti(NO_3)_4$) are powerful oxidizing agents and explode or inflame with organic compounds, whereas others (of Zn, Ni^{II}, Mn^{II}) are inert.

There are interesting differences in series such as Zn, Cd, and Hg nitrates. The Zn and Hg compounds are volatile, in contrast to $Cd(NO_3)_2$ which is a typical ionic nitrate.

Examples of volatile nitrates which have been studied in the crystalline state include $Cu(NO_3)_2$, which forms molecules of type (a) in the vapour—for the structure of the solid see p. 895, $Co(NO_3)_3$, and $Ti(NO_3)_4$, which exist as molecules in the crystal. In all these molecules NO_3 is behaving as a bidentate ligand. The relative orientation of the two NO_3 groups in the vapour molecule $Cu(NO_3)_2$ was not determined. The molecule of $Co(NO_3)_3$ is essentially octahedral, allowing for the smaller angle subtended at the metal by the bidentate ligand; the bond angles not given in Fig. 18.5(a) are close to $100°$. The $Ti(NO_3)_4$ molecule is a flattened tetrahedral grouping of 4 NO_3 around Ti, the 8 O atoms bonded to the metal atom forming a flattened dodecahedral group (Fig. 18.5(b)).

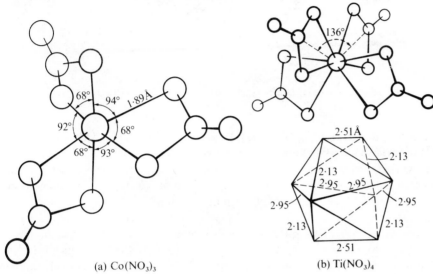

      (a) $Co(NO_3)_3$ (b) $Ti(NO_3)_4$

FIG. 18.5. The molecules (a) $Co(NO_3)_3$, (b) $Ti(NO_3)_4$.

There is apparently some asymmetry of the NO_3 group, (b). The molecule of

$Sn(NO_3)_4$ is similar. In all these molecules there is symmetrical bonding of the NO_3 group to the metal, that is, equal M–O bonds.

(ii) *Complex nitrato ions.* Salts containing these ions are prepared either from aqueous solution (hydrated compounds) or by special methods such as reacting a complex chloride with liquid N_2O_5 or N_2O_4 in CH_3CN. In this way $Cs_2[Sn(NO_3)_6]$ is prepared from Cs_2SnCl_6.

In the $[Au(NO_3)_4]^-$ ion Au forms four coplanar bonds to monodentate NO_3 ligands (Fig. 18.6(a)). The four Au–O distances of $2·85$ Å are too long to indicate appreciable interaction between the Au and O atoms. The structures of all

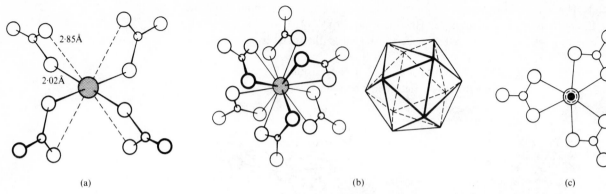

FIG. 18.6. The ions (a) $[Au(NO_3)_4]^-$, (b) $[Ce(NO_3)_6]^{3-}$, (c) $[UO_2(NO_3)_3]^-$.

the ions $[Mn(NO_3)_4]^{2-}$, $[Fe(NO_3)_4]^-$, and $[Co(NO_3)_4]^{2-}$ (studied in their tetraphenylarsonium salts) are of the same general type as the $Ti(NO_3)_4$ molecule (i.e. symmetry close to D_{2d}). However, in the first two ions there is symmetrical bidentate coordination of the NO_3 groups but in $[Co(NO_3)_4]^{2-}$ unsymmetrical coordination of NO_3 to the metal, though the two different Co–O bond lengths are apparently 2·03 and 2·36 Å for two NO_3 groups as compared with 2·11 and 2·54 Å for the other two. This difference in behaviour of the NO_3 ligand in these tetranitrato ions is attributed to the lower symmetry of Co(II), d^7(high spin) as compared with Ti(IV), d^0, Mn(II) and Fe(III), d^5 (high spin), and Sn(IV), d^{10}.

Several hexanitrato ions have been studied. There is nearly regular icosahedral coordination of the metal atom in $Mg[Th(NO_3)_6] . 8 H_2O$ and $Mg_3[Ce(NO_3)_6]_2 . 24 H_2O$. Since Mg^{2+} is completely hydrated in both these crystals they are preferably formulated $[Mg(H_2O)_6][Th(NO_3)_6] . 2 H_2O$ and $[Mg(H_2O)_6]_3[Ce(NO_3)_6]_2 . 6 H_2O$. For an ion of this type see Fig. 18.6(b). The anion in $(NH_4)_2[Ce(NO_3)_6]$ also has six bidentate NO_3 groups bonded to the metal atom, but here they are arranged with approximate T_h symmetry. There is marked asymmetry of the NO_3 group (N–O, 1·235 and 1·282 Å) but rather less than in $Ti(NO_3)_4$ (see above).

(iii) *Oxynitrato compounds.* Transition-metal compounds $MO_x(NO_3)_y$ are prepared by methods such as the action of liquid N_2O_5 on the metal oxide or hydrated nitrate to give addition compounds with N_2O_5 which on heating *in vacuo* give in some cases the anhydrous nitrate and in others an oxynitrato compound. The volatile $Be_4O(NO_3)_6$, formed by heating $Be(NO_3)_2$ *in vacuo*, presumably has a tetrahedral structure similar to that of $Be_4O(CH_3COO)_6$.

Compounds containing oxynitrato ions include $Rb[UO_2(NO_3)_3]$. In the anion in this salt (Fig. 18.6(c)) the axis of the linear UO_2 group is normal to the plane of the paper and three bidentate NO_3 groups complete the 8-coordination group around the metal atom.

(iv) *Coordination complexes containing* NO_3 *groups.* There are numerous molecules in which NO_3 behaves either as a monodentate or as a bidentate ligand, for example:

Monodentate:	$Cu(en)_2(ONO_2)_2$	AC 1964 **17** 1145
	$Cu(C_5H_5NO)_2(NO_3)_2$	AC 1969 **B25** 2046
Bidentate:	$Cu^I(P\phi_3)_2NO_3$	IC 1969 8 2750
	$Co(Me_3PO)_2(NO_3)_2$	JACS 1963 **85** 2402
	$Th(NO_3)_4(OP\phi_3)_2$	IC 1971 **10** 115
	$La(NO_3)_3(bipyridyl)_2$	JACS 1968 **90** 6548

In the *trans* octahedral molecule of $Cu(en)_2(NO_3)_2$, (a), Cu^{II} forms two long bonds to O atoms; in $Cu(P\phi_3)_2NO_3$ there is distorted tetrahedral coordination of Cu^I, (b). The structure of $Co(Me_3PO)_2(NO_3)_2$ may also be described as tetrahedral if the bidentate NO_3 is regarded as a single ligand, for the optical and magnetic properties of this molecule (and also the ion $Co(NO_3)_4^{2-}$) suggest that the metal atom is situated in a tetrahedral ligand field.

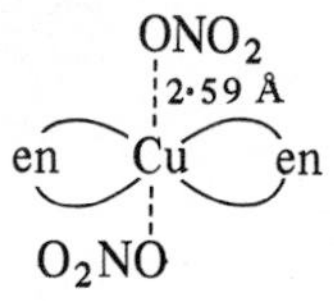

(a) 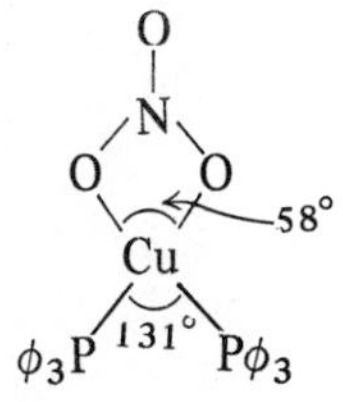 (b) 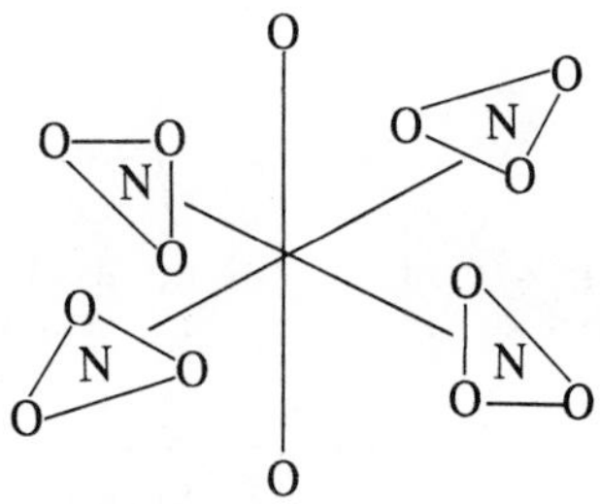 (c)

The molecules $Th(NO_3)_4(OP\phi_3)_2$ (and the isostructural Ce compound) and $La(NO_3)_3(bipyridyl)_2$ are interesting examples of 10-coordination. In the former, (c), the 'pseudo-octahedral' coordination group is composed of four bidentate NO_3 groups and two phosphoryl O atoms. (There is 11-coordination of Th in $Th_2(OH)_2(NO_3)_6(H_2O)_6$ and $Th(NO_3)_4 . 5 H_2O$, in which the coordination groups around Th are $(OH)_2(H_2O)_3(NO_3)_3$ and $(H_2O)_3(NO_3)_4$ respectively.)

Covalent nitrates. Apart from organic nitrates covalent nitrates of non-metals are limited to those of H, F, and Cl. ($ClNO_3$ has been prepared from anhydrous HNO_3 and ClF as a liquid stable at $-40°C$ in glass or stainless steel vessels.[1]) Electron diffraction studies have been made of the explosive gas FNO_3[2] and of the (planar) methyl nitrate molecule.[3] A refinement of the crystal structure of pentaerythritol nitrate, $C(CH_2ONO_2)_4$,[4] shows that the nitrate group has the same structure, (d), as in nitric acid.

(1) IC 1967 6 1938
(2) JACS 1935 **57** 2693; ZE 1941 47 152
(3) JCP 1961 **35** 191
(4) AC 1963 **16** 698

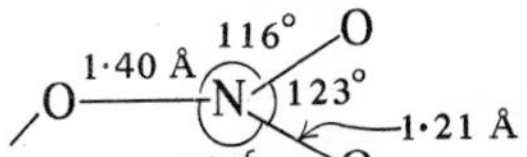

(d)

The sulphides of nitrogen and related compounds

These compounds have little in common with the oxides of nitrogen, either as regards chemical properties or structure, and only two sulphides, N_2S_4 and N_2S_5, have formulae similar to those of oxides. The structure of N_2S_5 is not known; for N_2S_4 see later. This group of compounds contains molecules with 4-, 5-, 6-, 7-, and 8-membered rings. We shall include here the simpler compounds NSF and NSF_3 because of their close relation to $(NSF)_4$.

The simplest ring system is that of N_2S_2. If N_4S_4 (see later) is sublimed through a plug of silver wool at a pressure of 0·01 mm at 300°C and the products are cooled rapidly, a volatile white solid is obtained which has a molecular weight corresponding to the formula N_2S_2. This compound is too unstable for structural studies above about −80°C, for it readily polymerizes to the blue-black $(NS)_x$, but its structure has been established in the crystalline adduct $N_2S_2(SbCl_5)_2$, (a).[1] In the planar N_2S_2 ring the four bonds are equal in length (1·62 Å, as in N_4S_4). The Sb—N bonds are appreciably weaker than in $N_4S_4 . SbCl_5$ (see later), and one

(1) IC 1969 **8** 2426

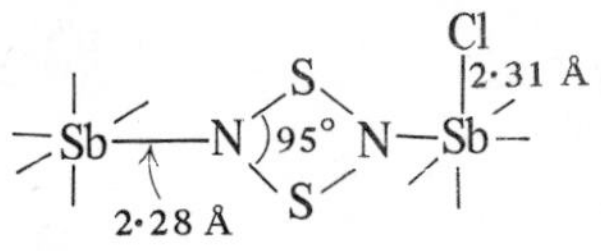

(a) 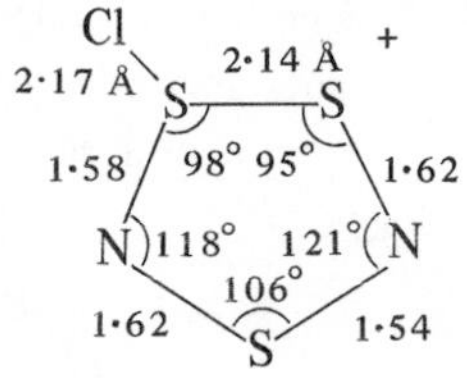   (b)

$SbCl_5$ is easily removed, leaving $N_2S_2 . SbCl_5$. There would appear to be a correlation between the Sb—N and Sb—Cl bond lengths in the adducts of $SbCl_5$:

	$N_2S_2 . SbCl_5$	$CH_3CN . SbCl_5$	$N_4S_4 . SbCl_5$
Sb—N	2·28	2·23	2·17 Å
Sb—Cl	2·31	2·36	2·39 Å

By refluxing a suspension of NH_4Cl in S_2Cl_2 (or from the interaction of N_4S_4 or $(NSCl)_3$ with S_2Cl_2) orange crystals of $N_2S_3Cl_2$ are obtained. This is a salt, $(N_2S_3Cl)^+Cl^-$, containing the very puckered 5-ring (b).[2]

(2) IC 1966 **5** 1767

(3) AC 1966 **20** 192

(4) AC 1966 **20** 186

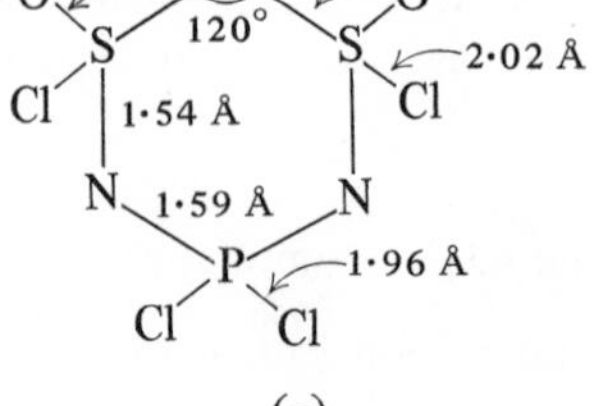

The N_3S_3 ring, with alternate N and S atoms, is chair-shaped in $(NSCl)_3$,[3] formed by chlorination of N_4S_4, and also in α-sulphanuric chloride, $N_3S_3O_3Cl_3$.[4] All the N—S bonds are of equal length (1·61 and 1·57 Å respectively) in these two compounds (Fig. 18.7(a) and (b)), but there is a small

(c)

(possibly real) difference between the S—Cl bond lengths. In the very similar compound, $N_3S_2PO_2Cl_4$,[5] containing one P atom in the ring, (c), all bond lengths are similar to those in $N_3S_3O_3Cl_3$, but here the ring is nearly planar (compare the planar $(PNCl_2)_3$ ring). On refluxing with a mixture of S_2Cl_2 and CCl_4 $N_2S_3Cl_2$ is converted quantitatively into N_3S_4Cl, and this in turn can be converted into $(N_3S_4)NO_3$ by dissolving in concentrated HNO_3. In this salt[6] the cation consists of the planar 7-membered ring shown in Fig. 18.7(c). The S—S bond is apparently a single bond but the approximate equality of the S—N bond lengths and the well-defined u.v. absorption spectrum suggest a 10-electron π system.

The parent compound from which many nitrogen–sulphur compounds can be derived is N_4S_4. It is readily prepared by the interaction of sulphur with NH_3 in the absence of water or preferably by passing S_2Cl_2 vapour through hot pellets of NH_4Cl. It forms yellow crystals and has remarkable chemical properties. The N_4S_4[7] molecule has the structure shown in Fig. 18.7(d). The N_4S_4 ring is bent into an extreme cradle form, giving two 'non-bonded' S—S distances of 2·58 Å. These distances, intermediate between a single S—S bond length (2·08 Å) and the normal van der Waals distance of 3·3 Å, suggest some kind of weak covalent bonding. (The corresponding distance in N_4Se_4[8] is 2·75 Å, intermediate between 2·34 Å and 4·0 Å.)

(5) AC 1969 **B25** 651

(6) IC 1965 **4** 681

(7) AC 1963 **16** 891; IC 1966 **5** 906

(8) AC 1966 **21** 571

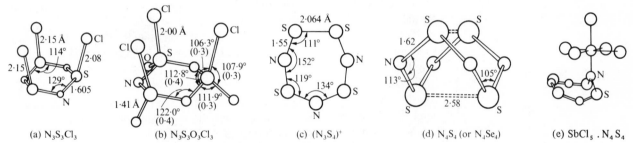

FIG. 18.7. Cyclic nitrogen–sulphur systems: (a) $N_3S_3Cl_3$, (b) $N_3S_3O_3Cl_3$, (c) $(N_3S_4)^+$, (d) N_4S_4, (e) $N_4S_4 . SbCl_5$.

In two adducts, $N_4S_4 . BF_3$[9] and $N_4S_4 . SbCl_5$[10] (Fig. 18.7(e)), the N_4S_4 ring has the cradle form but with the four S atoms coplanar instead of the four N atoms as in N_4S_4. The halide molecule is attached to one N atom of the ring in $N_4S_4 . BF_3$, with N—B, 1·58 Å and N—S, 1·66 Å. (S$\cdots$S, 3·8 Å).

Mild fluorination by AgF_2 in an inert solvent converts N_4S_4 into $(NSF)_4$.[11] In this puckered 8-membered ring (boat configuration) the F atoms are attached to S atoms, and there is a definite alternation of longer and shorter bonds, (d). More vigorous fluorination of N_4S_4 yields NSF, (e),[12] and NSF_3, (f),[13] in both of which F is bonded to S. Hydrolysis of NSF (or the action of NH_3 on $SOCl_2$) gives HNSO,[14] *cis*-thionyl imide, with the structure shown at (g).

We now come to the imides S_7NH, $S_6(NH)_2$, $S_5(NH)_3$, and $S_4(NH)_4$. The first three result from the action of NH_3 on S_2Cl_2 in dimethylformamide at low

(9) IC 1967 **6** 1906
(10) ZaC 1960 **303** 28

(11) AC 1963 **16** 152

(12) JACS 1963 **85** 1726
(13) JACS 1962 **84** 334

(14) JACS 1969 **91** 2437

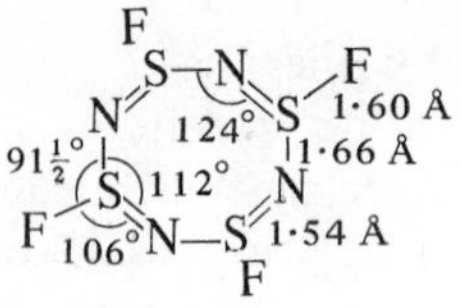

(e)

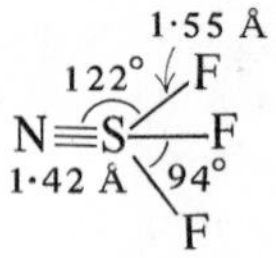

(f)

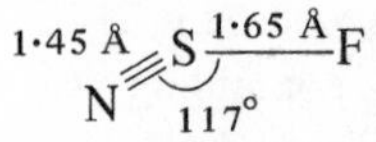

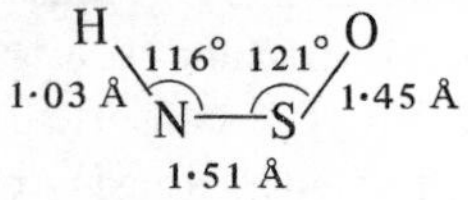

(g)

(d)

(15) AC 1971 **B27** 2480
(16) AC 1969 **B25** 611

(17) AC 1967 **23** 574

(h)

(18) JCS A 1971 136

(i)

(19) JINC 1958 7 421

temperatures, and are separated chromatographically. $S_4(NH)_4$ is made by reducing a solution of N_4S_4 in benzene by alcoholic $SnCl_2$. Three of the four possible structural isomers of $S_6(NH)_2$, namely, $1:3$,[15] $1:4$,[16] and $1:5$[16] have been characterized, and the crystal structures of several of these compounds have been studied. All are structurally similar to the S_8 molecule of rhombic sulphur, that is, they have the crown configuration, though in $N_4S_4H_4$[17] the distance separating the two parallel planes in which the 4 S and 4 N atoms lie is only 0·57 Å (i.e. the ring is nearly planar). Neutron diffraction confirms that H is attached to N and the N bonds are coplanar (mean bond angles in ring, SNS, 128°, NSN, 110°, and S–N, 1·65 Å). In $1:4$ $S_6(NH)_2$[16] the bond lengths are S–S, 2·05 Å, and N–S, 1·72 Å, and both enantiomorphic forms of the molecule are present in the crystal.

The N–S bond lengths in this family of compounds suggest the following standard values: S–N, 1·72 Å (compare 1·73 Å in O_3SNH_3) and S=N, 1·54 Å; lengths of formal triple bonds in NSF and NSF_3 are 1·45 Å and 1·42 Å.

If N_4S_4 is heated in solvents N_2S_5 is produced, while by the action of S in CS_2 solution under pressure N_4S_4 is converted into N_2S_4, a dark-red, rather unstable solid. Solutions of N_2S_4 in organic solvents are diamagnetic, showing that this sulphide is not dissociated, in contrast to N_2O_4. The detailed structure of the N_2S_4 molecule is not known, but a review of its spectra (i.r., Raman, u.v., and mass spectra) indicates the structure (h).[18]

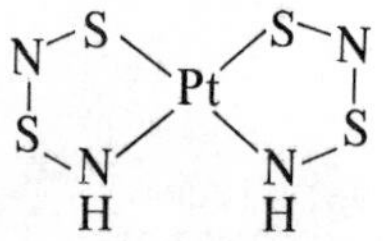

Thionitrosyl complexes of a number of metals have been prepared, for example, $Ni(NS)_4$ and the similar Co and Fe compounds from N_4S_4 and the carbonyls, and $Pt(NS)_4$ from the action of N_4S_4 on H_2PtCl_6 in solution in dimethylformamide. An incomplete X-ray study of the last compound indicates a planar molecule with the *cis* configuration (i).[19]

Nitrides

Most elements combine with nitrogen, and the nitrides include some of the most stable compounds known. The following metals are *not* known to form nitrides: (Na?), K, Rb, Cs, Au, Ru, Rh, Pd, Os, Ir, and Pt. We can make a rough division into four groups, but it should be emphasized that there is no sharp dividing line, particularly in the case of the IIA–IVA subgroup metals. In some cases an element forms nitrides of quite different types, and behaviour as both an ionic compound

668

and as a metallic conductor is a feature of a number of nitrides (for example, HfN, below). Thus Ca forms Ca_3N_2, one form of which (α) has the anti-Mn_2O_3 structure,[1] but there are also black and yellow forms of the same compound. Moreover Ca also forms Ca_2N, Ca_3N_4, and $Ca_{11}N_8$. Ca_2N forms lustrous green-black crystals with the anti-$CdCl_2$ (layer) structure,[2] in which the Ca–Ca distances within the layer (3·23 and 3·64 Å) are much shorter than in metallic Ca (3·88 and 3·95 Å). Nevertheless this is apparently an ionic compound, for the action of water produces $NH_3 + H_2$. The lustre and semiconductivity suggest formulation as $Ca_2^{2+}N^{3-}(e)$. There are even shorter Ca–Ca distances in $Ca_{11}N_8$ (3·11 Å),[3] a compound which is decomposed by water vapour. Zirconium forms not only the brown Zr_3N_4 but also blue Zr_xN (defect NaCl structure)[4a] and the metallic yellow ZrN. Hafnium forms a nitride HfN which has a defect NaCl

(1) AC 1968 **B24** 494

(2) IC 1968 7 1757

(3) AC 1969 **B25** 199
(4a) ZaC 1964 **329** 136

Li_3N Be_3N_2 Mg_3N_2 Ca_3N_2 etc.	Transition-metal nitrides Interstitial and more complex structures	BN Si_3N_4 AlN GaN Ge_3N_4 InN Hg_3N_2	Molecular nitrides of non-metals
ionic	*semi-metallic*	*covalent or ionic–covalent*	*covalent*

structure,[4b] the density indicating that one-eighth of the sites are unoccupied. When dissolved in HF all the nitrogen is converted into NH_3 (NH_4^+) and the compound is a metallic conductor ($Hf^{4+}N^{3-}(e)$?).

(4b) ZaC 1967 **353** 329

In the first and third of our groups of nitrides the formulae *usually* correspond to the normal valences of the elements and this is also true of some transition-metal compounds (e.g. Sc, Y, and 4f nitrides MN, Zr_3N_4) and of some of the molecular nitrides (e.g. P_3N_5);[5] others, such as N_4S_4, have formulae and structures which are not understood. The molecular nitrides are few in number, and if their structures are known they are described under the structural chemistry of the appropriate element.

(5) JACS 1957 **79** 1765

In addition to the binary nitrides described here, some ternary nitrides have been prepared, for example, Li_5TiN_3, Li_7NbN_4, and Li_9CrN_5, all with the fluorite type of structure (superstructures in most cases),[6] and alkaline-earth compounds with Re, Os, Mo, or W (e.g. $Sr_9Re_3N_{10}$, Ca_5MoN_5).[7]

(6) ZaC 1961 **309** 276
(7) IC 1970 **9** 1849

Ionic nitrides

Nitrides of the alkali metals other than Li do not appear to be known in a pure state. The simple structure assigned to Li_3N[8] is peculiar to this compound. The unusual coordination of Li^+ is of course related to the very small size of this ion compared with N^{3-}, but we noted in Chapter 6 that a structure with tetrahedral coordination of Li^+ is geometrically impossible. One-third of the Li^+ ions have 2 nearest N neighbours (at 1·94 Å) and the remainder 3 (at 2·11 Å). The N^{3-} ion is surrounded by 2 Li^+ at 1·94 Å and 6 more at 2·11 Å. We have noted the ReO_3 structure of Cu_3N (and the perovskite structure of $GaNCr_3$) in Chapter 6.

(8) ZE 1935 **41** 102

Nitrogen

(9) ZaC 1969 **369** 108

The nitrides M_3N_2 (M = Be, Mg, Zn, Cd), and α-Ca_3N_2 have the anti-Mn_2O_3 structure, in which the metal ions have 4 tetrahedral neighbours and N has two types of distorted octahedral coordination groups. (Be_3N_2 also has a close-packed high-temperature form in which Be is tetrahedrally coordinated;[9] it is described on p. 139.) The anti-Mn_2O_3 structure of α-Ca_3N_2 is remarkable not only for the tetrahedral coordination of Ca^{2+} but also because this leads to a very open packing of the N^{3-} ions. The Ca—N distance (2·46 Å) is about the value expected for 6-coordinated Ca^{2+}, but the N—N distances are very large. In this structure there are empty N_4 tetrahedra with all edges 4·4 Å, while the edges of a CaN_4 tetrahedron are 3·76 Å (four) and 4·4 Å (two). All these distances are large compared with twice the radius of N^{3-} (about 3 Å). Presumably the abnormally low c.n. of Ca^{2+} (and the resulting low density of the structure) is due either to the fact that there is no alternative structure of higher coordination (say, 6 : 9 or 8 : 12) or to the fact that the structure is determined by the octahedral coordination of N^{3-}, to which the 4-coordination of Ca^{2+} is incidental.

The radius of the N^{3-} ion in these nitrides would appear to be close to 1·5 Å, larger than most metal ions. These compounds are the analogues of the oxides, and form transparent colourless crystals. They are formed by direct union of the metal with nitrogen, or in some cases by heating the amide. Compare, for example:

$$3 \, Ba(NH_2)_2 \longrightarrow Ba_3N_2 + 4 \, NH_3$$
$$\text{with} \qquad Ba(OH)_2 \longrightarrow BaO + H_2O.$$

The nitrides have high melting points, ranging from 2200°C (Be_3N_2) to 900°C (Ca_3N_2) and are hydrolysed by water to the hydroxides with liberation of ammonia.

The structures of compounds of the type M_3N_2 are summarized in Table 18.7, where *A* represents the anti-Mn_2O_3, *B* the Zn_3P_2, and *C* the anti-La_2O_3 structure. The Zn_3P_2 structure is closely related to the anti-Mn_2O_3 structure. Zinc atoms are surrounded tetrahedrally by four P and P by six Zn at six of the corners of a distorted cube. The idealized (cubic) Zn_3P_2 structure is similar to that of Bi_2O_3 illustrated in Fig. 20.3 (p. 711).

TABLE 18.7

Crystal structures of compounds M_3N_2, etc.

	Be	Mg	Zn	Cd	Ca
N	*A*	*A*	*A*	*A*	*A*
P	*A*	*A*	*B*	*B*	
As		*A*	*B**	*B**	
Sb		*C*			
Bi		*C*			

(10) AC 1956 9 685
(11) AC 1968 **B24** 1062

* Neither of these compounds has the simple Zn_3P_2 structure but closely related structures (Zn_3As_2,[10] Cd_3As_2[11]).

670

Covalent nitrides

In contrast to the ionic nitrides of Zn or Cd, the chocolate-coloured explosive Hg_3N_2 made from HgI_2 and KNH_2 in liquid ammonia, is presumably a covalent compound. For the nitrides of B, Al, Ga, In, and Tl the geometrical possibilities are the same as for carbon, there being an average of 4 valence electrons per atom in MN.

Boron nitride has been known for a long time as a white powder of great chemical stability and high melting point with a graphite-like layer structure. It has also been prepared with the zinc-blende structure[12] by subjecting the ordinary form to a pressure of 60 kbar at 1400–1800°C. This form, 'borazon', is very much more dense, 3·47 as compared with 2·25g/cc for the ordinary form. In contrast, BP can be prepared by heating red P with B at 1100°C under a pressure of only 2 atm; it also has the zinc-blende structure.[13] The same structure is adopted by BAs.[14a] The structures of the compounds of Group III elements with N, P, etc. are summarized below. While BN is presumably a covalent compound, the bonds in AlN, GaN, and InN probably have appreciable ionic character, and the compounds with the heavier Group V elements are essentially metallic compounds. (For InBi see p. 218.) The structure of $BeSiN_2$ is a superstructure of the wurtzite structure of AlN etc.[14b]

(12) JCP 1957 **26** 956

(13) N 1957 **179** 1075
(14a) AC 1958 **11** 310

(14b) ZaC 1967 **353** 225

	B	Al	Ga	In
N	ZL	W	W	W
P	Z	Z	Z	Z
As	Z	Z	Z	Z
Sb	–	Z	Z	Z

L = BN layer structure; Z = zinc-blende; W = wurtzite

The structures of several nitrides M_3N_4 are known. Both Si_3N_4 and Ge_3N_4 have two polymorphic forms.[15] The structures of both are closely related to that of Be_2SiO_4 (phenacite); M has 4 tetrahedral neighbours and N forms coplanar bonds to 3 M atoms. There is probably appreciable ionic character in the bonds in Ge_3N_4 and also in Th_3N_4. In Th_3N_4[16] there is closest packing of the metal atoms of the same kind as in metallic Sm (*chh*). One-half of the N atoms occupy tetrahedral and the other half octahedral holes in such a way that one-third of the metal atoms have 6 N and the remainder 7 N neighbours. (The compound originally described as Th_2N_3 is Th_2N_2O, in which N occupies tetrahedral and O octahedral holes in a h.c.p. arrangement of Th atoms; all Th atoms have 4 N + 3 O neighbours.[16]). For an alternative description of this (Ce_2O_2S) structure see p. 1004.

(15) AC 1958 **11** 465

(16) AC 1966 **21** 838

Nitrides of transition metals

Some of these compounds have already been mentioned. They are extremely numerous, more than one nitride being formed by many transition metals. For example, five distinct phases (apart from the solid solution in the metal) have been recognized in the Nb–N system up to the composition NbN.[17] Of the great variety of structures adopted by these compounds we shall mention only some of the simpler ones.

(17) ZaC 1961 **309** 151

Nitrogen

In the so-called 'interstitial' nitrides the metal atoms are approximately, or in some cases exactly, close-packed (as in ScN, YN, TiN, ZrN, VN, and the rare-earth nitrides with the NaCl structure), but the arrangement of metal atoms in these compounds is generally *not* the same as in the pure metal (see Table 29.13, p. 1054). Since these interstitial nitrides have much in common with carbides, and to a smaller extent with borides, both as regards physical properties and structure, it is convenient to deal with all these compounds in Chapter 29.

The nitrides of Mn, Fe, Co, and Ni form a group of less stable compounds of greater complexity than those of the earlier Periodic Groups—compare the formulae of the nitrides of the metals of the first transition series:

ScN	Ti_2N	V_3N	Cr_2N	MnN (?)	Fe_2N	CoN	Ni_3N	Cu_3N
	TiN	VN	CrN	Mn_6N_5	Fe_3N	Co_2N	Ni_4N	
				Mn_3N_2	Fe_4N	Co_3N		
				Mn_2N	Fe_8N	Co_4N		
				Mn_4N				
				Mn_xN (δ)				

(18) AC 1952 5 404

Although the N atoms in these compounds are usually few in number compared with the metal atoms they seem to exert great influence, and it has been suggested[18] that the tendency of the N atoms to order themselves is great enough to cause rearrangement of the metal atoms in, for example, the change from ϵ-Fe_2N to ζ-Fe_2N.

(19) ACSc 1954 8 204

Interstitial structures are not possible for the larger P and As atoms, and apart from a few cases such as LaP, PrP, and GeP with the NaCl structure there is usually little similarity between nitrides and phosphides (or arsenides). Compare, for example, the formulae of the nitrides and phosphides of Mo and W: Mo_2N, W_2N, MoN, and WN, but Mo_3P, W_3P, MoP, WP, MoP_2, and WP_2.[19]

A short section on the structures of metal phosphides is included in Chapter 19.

Table 18.8 summarizes the coordination numbers of metal and N atoms in a number of nitrides.

TABLE 18.8

Coordination numbers in crystalline nitrides

(20) ACSc 1962 16 1255

Nitride	Structure	C.N. of M	C.N. of N
Cu_3N	Anti-ReO_3	2	6
Ti_2N	Anti-rutile[20]	3	6
Co_2N	Anti-$CdCl_2$	3	6
Ca_3N_2	Anti-Mn_2O_3	4	6
ScN etc.	NaCl	6	6
Th_3N_4		6, 7	4, 6
Ge_3N_4	Phenacite	4	3
AlN	Wurtzite	4	4
Ca_3N_2	Anti-Mn_2O_3	4	6
$AlLi_3N_2$	Anti-CaF_2	4	8

Phosphorus

The stereochemistry of phosphorus

Phosphorus, like nitrogen, has five valence electrons. These are in the third quantum shell, in which there are d orbitals in addition to s and p orbitals. Phosphorus forms up to six separate bonds with other atoms, but covalencies greater than four are *usually* exhibited only in combination with halogens (in the pentahalides and PX_6^- ions) and with groups such as phenyl, C_6H_5. (For an example of 5 coordination by O see p. 684.) The d orbitals are used for σ bonding if more than four bonds are formed, but they are also used for π bonding in tetrahedral oxy-ions and molecules. A very simple summary of the stereochemistry of P may be

TABLE 19.1

The stereochemistry of phosphorus

No. of σ pairs	Type of hybrid	No. of lone pairs	Bond arrangement	Examples
4	sp^3	0	Tetrahedral	PH_4^+, PCl_4^+ $POCl_3$, PO_4^{3-}, $PO_2F_2^-$
		1	Trigonal pyramidal	PCl_3, $P(CN)_3$, $P(C_6H_5)_3$
5	$sp^3d_{z^2}$	0	Trigonal bipyramidal	PCl_5, PF_3Cl_2, $P(C_6H_5)_5$
6	$sp^3d_\gamma^2$	0	Octahedral	PCl_6^-

given if it is assumed that the arrangement of bonds is determined by the number of σ electron pairs which are used for either σ bonds or lone pairs (Table 19.1). For PX_3 molecules this picture is valid if appreciable hybridization occurs. In PH_3 the bonds are likely to be essentially p^3 bonds, and this may also be true in the trihalides, but the interpretation of bond angles around $100°$ is not simple. The bond angles in the P_3 rings in P_4 and P_4S_3 are to be treated as special cases.

Elementary phosphorus

Phosphorus crystallizes in at least five polymorphic forms. The white form is metastable and is prepared by condensing the vapour. There are apparently two closely related modifications of white P, with a transition point at $-77°C$. The

Phosphorus

(1) JCP 1935 **3** 699
(2) N 1952 **170** 629

(3) AC 1965 **19** 684

molecular weight in various solvents shows that white P exists in solution as P_4 molecules, presumably similar to those in the vapour of white P, the configuration of which was determined in an early e.d. study.[1] The same tetrahedral molecules exist in crystalline white P,[2] but a complete study of the crystal structure has not been made. In the P_4 molecule each P atom is bonded to three others, the interbond angles being 60° and the bond length 2·21 Å, similar to that of many other single P–P bonds. There has been much discussion of the nature of the bonds in the P_4 molecule. Although the use of pd^2 hybrids (theoretical bond angle 66°26′) would result in the least strain it seems that the promotion of two 3p electrons to 3d orbitals is unlikely on energetic grounds, and that the bonds are largely p in character. In the P_2 molecule in the vapour of red P the P≡P bond length is 1·895 Å. (In PN, P≡N is 1·49 Å.)

Black P has a layer structure[3] in which each atom is bonded to three others. A layer of this structure is shown, in idealized form, in Fig. 19.1. In spite of its appearance it is simply a puckered form of the simple hexagonal net, the interbond

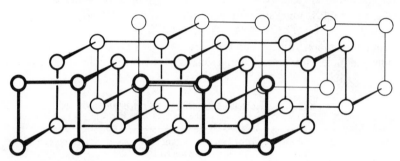

FIG. 19.1. The crystal structure of black phosphorus—portion of one layer (idealized).

angles being two of 102° and one of $96\frac{1}{2}$°. (See also p. 88 for the relation of this structure to the simple cubic lattice.) Although black P is the most stable crystalline form of the element (see the figures for heats of solution given below) it had until comparatively recently been made only under high pressure. It can in fact be made by heating white P to 220-370°C for eight days with Hg on Cu as catalyst in the presence of a seed of black P, though it then contains Hg which can only be removed with difficulty.[3a]

(3a) JAP 1963 **34** 3630

	Density (g/cc)	Heat of solution in Br_2
White P	1·83	249 kJ mol⁻¹
Red	2·31	178
Black	2·69	160

Hittorf's red P may be made by dissolving white P in thirty times its weight of molten lead, cooling slowly, and dissolving away the lead electrolytically. It has an

674

extraordinarily complex structure.[4] Complex chains with 21 atoms in the repeat unit are linked into layers by bonds to similar chains which lie at right angles to the first set but in a parallel plane. Two such systems of cross-linked chains form a complex layer (Fig. 19.2) in which there are no P–P bonds between the atoms of one half of the layer (broken lines) and the other (full lines). The whole crystal is an assembly of composite layers of the type shown in Fig. 19.2(b) held together by van der Waals bonds. Within a layer the mean P–P bond length is 2·22 Å and the angle P–P–P, 101°.

(4) AC 1969 **B25** 125

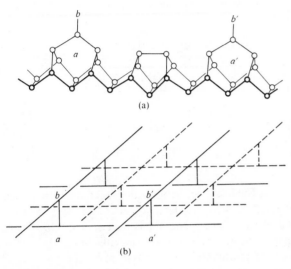

FIG. 19.2. The crystal structure of red phosphorus: (a) chain sub-unit, (b) layer formed from cross-linked chains (diagrammatic).

It has been suggested that a brown form of P, formed by condensing the vapour of white P at 1000°C or the vapour of red P at 350°C on to a cold finger at −196°C, may consist of P_2 molecules.[5] A vitreous form of P has been prepared by heating white P with Hg at 380°C; at 450°C black P is formed.[6] The conversion of P under pressure to forms with the As structure (at 80 kbar) and a (metallic) simple cubic structure (at 110 kbar) has been reported, but details are not available.[7]

(5) JACS 1953 **75** 2003
(6) IC 1963 **2** 22

(7) Sc 1963 **139** 1291

Phosphides of metals

This is a very large class of compounds. Many metals, particularly transition metals, form more than one phosphide and many of the compounds are polymorphic. As regards bonding they range from compounds in which the bonds are probably essentially ionic (e.g. Mg_3P_2 with the anti-Mn_2O_3 structure) containing P^{3-} ions of radius about 1·9 Å, through 'covalent' compounds (e.g. PdP_2) to essentially metallic compounds. (With increasing metallic character of the metalloid the phases become much more complex than the nitrides and phosphides. For example, in the K–As system we find K_3As, K_5As_4, and KAs; K and Sb form similar compounds and also KSb_2, while Cs and Sb form not only compounds of these four types but also Cs_5Sb_2, Cs_2Sb, and Cs_3Sb_7.) Somewhat surprisingly some of the alkali-metal compounds are apparently not simple ionic crystals, judging by their formulae —their structures are not known—for Li forms LiP, in addition to the expected Li_3P, while the other alkali metals form compounds such as Na_2P_5 (and Na_3P).

The structures of the alkal-metal phosphides, arsenides, etc. are shown in Table 19.2, where N stands for the Na_3As structure and L for the Li_3Bi structure (p. 1035). The Na_3As structure is somewhat reminiscent of an alloy structure, As having 11 Na neighbours at distances from 2·94 to 3·30 Å, and that of Li_3Bi is a typical intermetallic structure. With the Group VB metalloids Li also forms 1 : 1

TABLE 19.2

Crystal structures of phosphides, M_3P, *etc.*

	Li	Na	K	Rb	Cs
P	N	N	N		
As	N	N	N	N	
Sb	α, N β, L	N	N	NL	L
Bi	α, ? β, L	N	N	L	L

compounds. In LiAs (with which NaSb is isostructural) the As atoms form helical chains (As—As, 2·46 Å, compare 2·51 Å in the element), and Li has six neighbours arranged approximately octahedrally. These compounds have metallic lustre and electrical conductivity and there is clearly some degree of metallic bonding.

The formulae and structures of phosphides are complicated by the presence of metal-metal and P—P bonds, as already noted in Chapter 1 for PdP_2 and CdP_4. The extent of the linking of P atoms by P—P bonds provides one way of classifying these compounds, but it has little direct relation to the nature of the bonds. Many phosphides have geometrically simple structures (Table 19.3), but except in cases such as Mg_3P_2 or Zn_3P_2 the structure does not indicate the type of bonds between metal and P atoms. Thus LaP has the NaCl structure, but so also does ZrP, and indeed also ZrC, ZrN, ZrO, and ZrS. The metallic lustre of compounds such as ScP and YP indicates that these are not simple ionic compounds. Table 19.3 shows that in some phosphides there are discrete P atoms (that is, P bonded only to metal nearest neighbours), whereas others contain systems of linked P atoms of all the following types:

P—P groups in compounds with the pyrites or marcasite structures,
   discrete P_4 rings in IrP_3 etc. ($CoAs_3$ structure),
   chains in PdP_2,
   layers in CdP_4.

In all these compounds the P—P bond length is close to 2·2 Å (single bond). In some of the transition-metal phosphides P—P has rather larger values, 2·4 Å (NiP), and 2·6–2·7 Å (CrP, MnP, FeP, CoP).

TABLE 19.3
Structures of metallic phosphides

Structure	System of linked P atoms	Examples
Zinc-blende NaCl NiAs TiP[1] MnP Anti-CaF$_2$ Anti-PbCl$_2$ Anti-Mn$_2$O$_3$ Zn$_3$P$_2$	Isolated P atoms	AlP, GaP, InP LaP, SmP, ThP, UP, ZrP, etc. VP HfP, β-ZrP FeP, CoP, WP Ir$_2$P, Rh$_2$P Co$_2$P, Ru$_2$P Be$_3$P$_2$, Mg$_3$P$_2$ Cd$_3$P$_2$
Pyrites or Marcasite	P$_2$ groups	PtP$_2$,[2a] NiP$_2$ (h.p.)[2] FeP$_2$,[2a] OsP$_2$, RuP$_2$
CoAs$_3$	P$_4$ rings	RhP$_3$, IrP$_3$, PdP$_3$[2a]
	P chains double chains	PdP$_2$,[3] NiP$_2$, CdP$_2$,[4] BaP$_3$[4a] ZnPbP$_{14}$[5]
	P layers	CdP$_4$[6]

(1) ACSc 1967 **21** 1773
(2) IC 1968 7 998
(2a) ACSc 1969 **23** 2677
(3) AC 1963 **16** 1253
(4) AC 1970 **B26** 1883
(4a) NW 1971 **58** 623
(5) ZaC 1958 **294** 257
(6) ZaC 1956 **285** 15

Examples of relatively simple structures include

α-CdP$_2$	CdP$_4$	Th$_3$P$_4$
Cd—4 P (tetrahedral)	Cd—6 P (octahedral)	Th—8 P (dodecahedral)

P—$\begin{cases} 2\,P \\ 2\,Cd \end{cases}$ (tetrahedral)   P—$\begin{cases} 3\,P \\ 1\,Cd \end{cases}$ or $\begin{cases} 2\,P \\ 2\,Cd \end{cases}$ (tetrahedral) P—6 Th (octahedral)

The form of the chains of P atoms in BaP$_3$ is shown in Fig. 19.3, which also shows a projection of the structure of CdP$_4$. The structure of TiP, in which P occupies

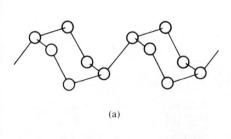

(a)

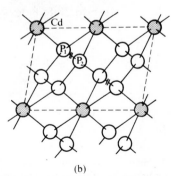

(b)

FIG. 19.3. (a) Chain of P atoms in BaP$_3$; (b) projection of structure of CdP$_4$ showing layers of linked P atoms which are arranged around screw axes perpendicular to the plane of the paper.

677

Phosphorus

octahedral and trigonal prism interstices in a non-close-packed sequence of Ti layers, is mentioned on p. 129. The structure of PdP_2 is noted in Chapter 1 (p. 17), that of Th_3P_4 in Chapter 5 (p. 160), and the relation of the $CoAs_3$ structure to the ReO_3 structure is described in Chapter 6 (p. 216). Recent studies of phosphides with this structure show that the P_4 ring is rectangular rather than exactly square (P–P, 2·23 and 2·31 Å); in $CoAs_3$ the As_4 group has a similar rectangular shape (As–As, 2·46 and 2·57 Å).[1]

We have emphasized here the existence of P–P bonds in the phosphorus-rich compounds. The existence of metal–metal bonds in the 'lower' phosphides is also of interest, since this has a direct bearing on their magnetic properties.[2]

The structures of simple molecules

Molecules PX_3

If p orbitals are used in simple molecules PX_3 interbond angles not much greater than $90°$ are expected. A value of $93·8°$ is found in PH_3 and angles around $100°$ in many pyramidal molecules (Table 19.4). It is of interest that PF_3 behaves like CO in forming complexes:

$Ni(PF_3)_4$	compare	$Ni(CO)_4$
$Pt(PF_3)_2Cl_2$		$Pt(CO)_2Cl_2$
$F_3P . BH_3$		$OC . BH_3$

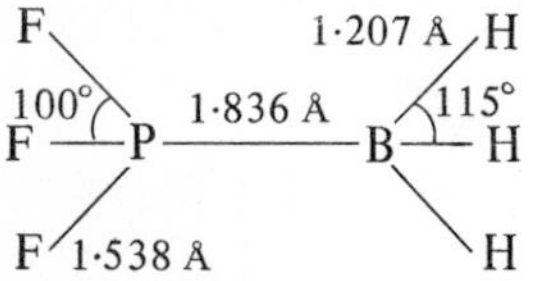

The $d\pi$ bonding postulated in the Ni and Pt compounds (p. 984) cannot explain the shortness of the P–B bond in $F_3P . BH_3$ compared with that in BP (zinc-blende structure), 1·96 Å, or $H_3P . BH_3$, 1·93 Å (ref. p. 850):

HCP and the PH_2^- ion

Many simple molecules containing P (e.g. the hydrides) are much less stable than the corresponding N compounds. This is also true of methinophosphine, HCP, the analogue of HCN. Prepared by passing PH_3 through a specially designed carbon arc, this reactive gas readily polymerizes; it is converted by HCl into $H_3C . PCl_2$ (H–C 1·07, C≡P, 1·54 Å).[1]

The PH_2^- ion, the effective radius of which has been estimated as 2·12 Å, occurs in KPH_2 and $RbPH_2$, both of which have distorted NaCl structures.[2]

Hydrides and molecules P_2X_4 and P_3X_5

In addition to PH_3 there are a number of lower hydrides.[1] P_2H_4 decomposes in the presence of water to a solid $P_{12}H_6$, but in the dry state to P_9H_4 or non-stoichiometric hydrides. P_3H_5 has been prepared by the photolysis of P_2H_4.[2] The substituted diphosphines $P_2(CH_3)_4$ and $P_2(CF_3)_4$ have been prepared, and structural studies have been made of compounds $P_2R_4Y_2$ where R is an organic radical (CH_3, C_6H_5, etc.) and Y is S, BH_3, or $Fe(CO)_4$, for example;

(1) AC 1971 **B27** 2288

(2) JCP 1968 **48** 263

(1) JCP 1964 **40** 1170

(2) AC 1962 **15** 420

(1) JACS 1956 **78** 5726

(2) JACS 1968 **90** 6062

TABLE 19.4

Structural data for molecules PX$_3$, POX$_3$ *and* PSX$_3$, *and* PX$_5$

Molecules PX$_3$ etc.

Bond	Length (Å)	X–P–X	Method	Reference
P–H	1·4206 ⎱ 1·437 ⎰	93·8°	m.w. e.d.	⎱ JCP 1959 **31** 449 ⎰
P–F	1·570	97·8°	e.d.	IC 1969 **8** 867
HFP–F	1·58	99°	m.w.	JACS 1968 **90** 1705
P–Cl	2·04	100·3°	m.w.	JCP 1962 **36** 589
P–Br	2·20	101·0°	e.d./m.w.	IC 1971 **10** 2584
P–I	2·43	102°	e.d.	AC 1950 **3** 46
P–CH$_3$	1·843	98·9°	e.d.	⎱ JCP 1960 **32** 512, 832 ⎰
H$_2$P–CH$_3$	1·858	–	e.d.	
P–C$_6$H$_5$	1·828	–	X	JCS 1964 3799
P–SiH$_3$	2·248	96·5°	e.d.	JCS A 1968 3002

Molecules POX$_3$ and PSX$_3$

	P–X	P–O (S)	X–P–X	Method	Reference
POF$_3$	1·524	1·436	101·3°		For references see:
POCl$_3$	1·993	1·449	103·3°	e.d.	IC 1971 **10** 344
PSF$_3$	1·53	1·87	100·3°		
PSCl$_3$	2·011	1·885	101·8°		
POCl$_3$ (cryst.)	1·98	1·46	(105°)	X	AC 1971 **B27** 1459
POBr$_3$ (cryst.)	2·14	1·44	(105½°)	X	AC 1969 **B25** 974

Molecules PX$_5$

P–F	1·577 (axial) 1·534 (equatorial)	e.d.	IC 1965 **4** 1775
P–Cl	2·14 (axial) 2·02 (equatorial)	e.d.	See: IC 1971 **10** 344

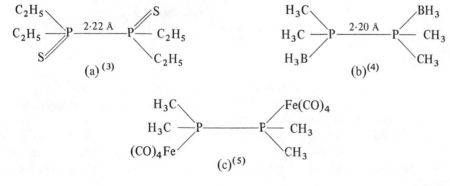

(a)[3] (b)[4] (c)[5]

(3) AC 1961 **14** 178
(4) AC 1968 **B24** 699
(5) JCS A 1968 622

Phosphorus

Bond lengths include: P–C, 1·83 Å, P–S, 1·94 Å, P–B, 1·95 Å, and P–P, 2·20 Å. In contrast to the symmetrical structure of $P_2S_2(CH_3)_4$ the analogous As compound has the unsymmetrical structure $(CH_3)_2S . As . S . As(CH_3)_2$ (p. 724). Infrared and ^{31}P n.m.r. suggest the symmetrical structure for $P_2I_4S_2$, which is made from the elements in CS_2 solution.[6]

Other compounds containing the system $>$P–P$<$ include the dihalides P_2X_4, hypophosphoric and diphosphorous acids and their ions (see later), and the infinite chain in $CuBr[Et_2P–PEt_2]$ (p. 882).

Three dihalides are known, P_2F_4, P_2Cl_4, and P_2I_4. The fluoride is made from PF_2I and Hg, the chloride by subjecting a mixture of PCl_3 and H_2 to an electric discharge, and the iodide directly from the elements in CS_2 solution. The P_2I_4 molecule in the crystalline iodide is centrosymmetrical, with the structure shown in Fig. 19.4.[7] The bond lengths are: P–P, 2·21, P–I, 2·48 Å, and mean angle I–P–P, 94°.

(6) IC 1964 **3** 780

(7) JPC 1956 **60** 539
For further references to these compounds see: IC 1969 **8** 2086, 2797

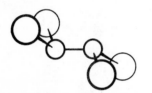

FIG. 19.4. The structure of the P_2I_4 molecule.

Phosphoryl and thiophosphoryl halides

Results of electron and X-ray diffraction studies of these molecules are listed in Table 19.4. All these molecules are tetrahedral, but since the X–P–X angles range from 100° to 108° the tetrahedra are not regular. The P–O bond length is close to 1·45 Å. In the Raman spectrum of solid $POBr_3$ there is an extra vibrational line indicating a lowering of the molecular symmetry, which is C_{3v} for the free molecule. This is attributed to weak charge-transfer bonds between O and Br atoms of different molecules, as indicated by the distance 3·08 Å in the infinite chains. All other intermolecular Br–O distances are greater than the sum of the van der Waals radii (3·35 Å). The intermolecular Cl···O distance in the chains in crystalline $POCl_3$ (3·05 Å) may indicate an interaction slightly stronger than van der Waals bonds. In

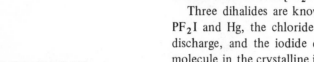

the thiophosphoryl halides the P–S bond length (1·87 Å) is, like the P–O bond length in the phosphoryl halides, close to that expected for a double bond. For other phosphorus thiohalides see p. 696.

$POCl_3$ and certain other molecules mentioned below react with some tetra- and penta-halides to form molecules in which an octahedral group around the metal atom has been completed by one or two O atoms of $POCl_3$ (or other) molecules.

680

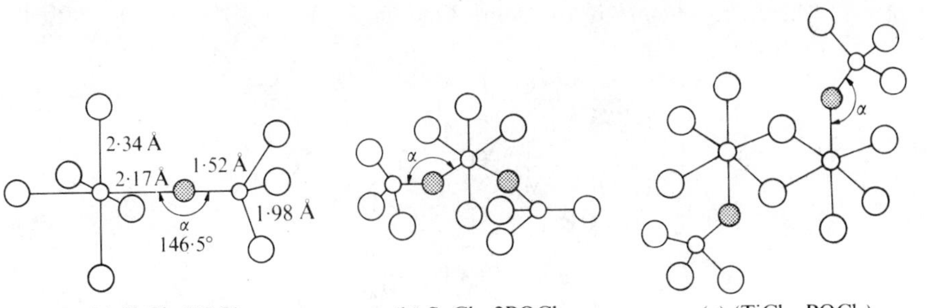

(a) $SbCl_5.POCl_3$ (b) $SnCl_4. 2POCl_3$ (c) $(TiCl_4. POCl_3)_2$

FIG. 19.5. Molecular structures: (a) $SbCl_5 . POCl_3$, (b) $SnCl_4 . 2 POCl_3$, (c) $(TiCl_4 . POCl_3)_2$ (O atoms shaded).

Figure 19.5 shows examples of three molecules of this general type:

(a) $MCl_5 . POCl_3$ (M = Sb, Nb, Ta),[1] and $SbCl_5 . OP(CH_3)_3$[1]
(b) $MCl_4 . 2 POCl_3$ (M = Sn,[2] Ti[3])
(c) $(MCl_4 . POCl_3)_2$ (M = Ti[4]).

(1) ACSc 1963 **17** 353
(2) ACSc 1963 **17** 759
(3) ACSc 1962 **16** 1806
(4) ACSc 1960 **14** 726

In these adducts there is an increase in the c.n. of the acceptor atom while the structure of the donor molecule is essentially unchanged. Note, however, the remarkable difference between the Sb—O bond lengths in the molecules $SbCl_5. POCl_3$ and $SbCl_5. OP(CH_3)_3$, 2·17 and 1·94 Å respectively (Sb—Cl, 2·34 Å). In the $POCl_3$ adducts the O bond angle lies in the range 140–152°. It is interesting that $SnCl_4. 2 POCl_3$ has, like $SnCl_4. 2 SeOCl_2$, the *cis* configuration, in contrast to molecules such as *trans*-$GeCl_4$ (pyridine)$_2$.

The O-bridged molecule (a) has been identified by its ^{31}P n.m.r. spectrum.[5]

(5) IC 1964 **3** 280

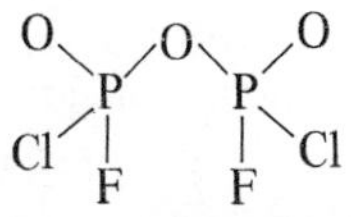

(a)

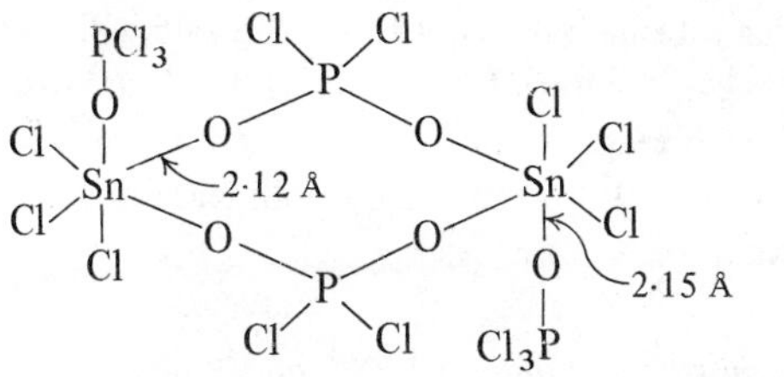

(b)

In the molecules of Fig. 19.5 $POCl_3$ is attached to a metal atom through the single O atom. The action of Cl_2O on a solution of $SnCl_4$ in $POCl_3$ gives a compound with the formula $(SnO_3P_2Cl_8)_2$. In addition to $POCl_3$ ligands on the metal atoms the latter are also bridged by tetrahedral O_2PCl_2 groups, (b), so that a (puckered) 8-membered ring is formed.[6] Bridges of the same kind occur in the compound $Mn(PO_2Cl_2)_2(CH_3COOC_2H_5)_2$[7] formed by the action on MnO of $POCl_3$ dissolved in ethyl acetate. The bridging PO_2Cl_2 groups lead to infinite chains (c) of octahedral coordination groups in which the $CH_3COOC_2H_5$ molecules occupy *cis* positions and successive 8-membered rings are in nearly perpendicular

(6) AC 1969 **B25** 1720, 1726
(7) ACSc 1963 **17** 1971

(8) ACSc 1970 **24** 59

planes. There are chains of the same kind in $Mg(PO_2Cl_2)_2(POCl_3)_2$ with two $OPCl_3$ groups on each metal atom.[8]

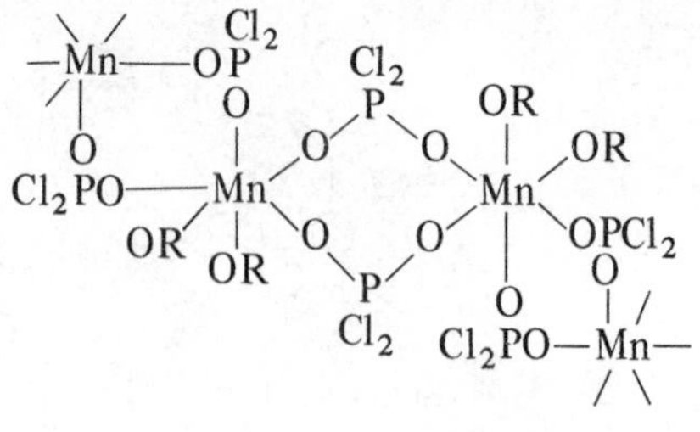

(c)

(1) IC 1968 7 2582
(2) ACSc 1965 **19** 879
(3) JCP 1967 **46** 357
(4a) AC 1969 **B25** 617
(4b) JCS **A** 1969 1804

Other tetrahedral molecules and ions

Further examples of tetrahedral molecules are shown at (a)[1] and (b)[2]. Others which have been studied structurally include $F_3P \cdot BH_3$,[3] $PS(OCH_3)(C_6H_5)_2$,[4a] $OP(NH_2)_3$,[4b] and many oxy- and thio-ions described later in this chapter.

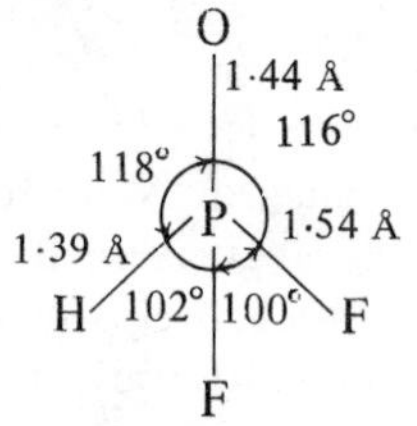

(a) 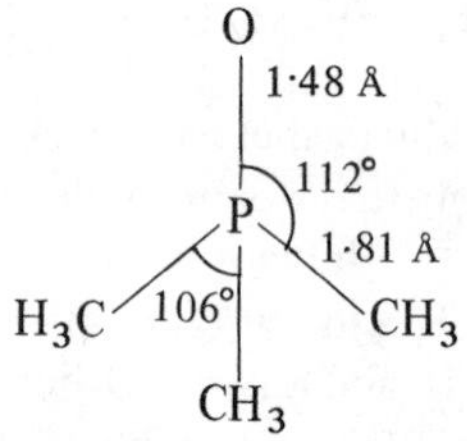   (b)

(5) JCP 1967 **47** 1818

The phosphorus analogue of the ammonium ion, the phosphonium ion, has a regular tetrahedral structure in the salts $(PH_4)X$. A n.d. study of PH_4I gives P–H, 1·414 Å and shows that the P–H bonds are directed towards I^- ions. The anion has 8 H neighbours at the corners of a distorted cube, 4 at 2·87 Å and 4 at 3·35 Å; the atoms P–H–I are practically collinear (172°).[5] Ions PX_4^+ are included in the next section.

Phosphorus pentahalides, PX_4^+ *and* PX_6^- *ions*

The stability of the pentahalides decreases rapidly with increasing atomic weight of the halogen. The pentafluoride is stable up to high temperatures; PCl_5 is about half dissociated at 200°C; PBr_5 is less stable, and PI_5 is not known. In the vapour state PF_5 and PCl_5 have been shown to exist as trigonal bipyramidal molecules, the stereochemistry of which has been discussed in Chapter 7. The axial bonds are longer than the equatorial ones as in other molecules of this type (see below). Ionization presumably takes place in nitrobenzene, for solutions of PCl_5 in this solvent have appreciable electrical conductivity, and crystalline PCl_5 is built of tetrahedral PCl_4^+ and octahedral PCl_6^- ions, which are packed together in much the same way as the ions in CsCl.[1] The PCl_4^+ ion also exists in the crystalline compounds PCl_6I,[2] $(PCl_4)FeCl_4$,[3] $(PCl_4)(Ti_2Cl_9)$, and $(PCl_4)_2(Ti_2Cl_{10})$; for

(1) JCS 1942 642
(2) JACS 1952 74 6151
(3) IC 1968 7 2150

682

the last two compounds see p. 392. PCl_6I, which is prepared by direct union of PCl_5 and ICl in CS_2 or CCl_4, ionizes in polar solvents and consists of PCl_4^+ and linear ICl_2^- ions. Crystalline $(PCl_4)FeCl_4$ is an aggregate of PCl_4^+ ions (P–Cl, 1·91 Å) and tetrahedral $FeCl_4^-$ ions (Fe–Cl, 2·19 Å).

In contrast to the pentachloride PBr_5 crystallizes as $(PBr_4)Br$, the P–Br bond length in the PBr_4^+ ion being 2·15 Å.[4] The same cation occurs in PBr_7,[5] formed by the action of Br_2 on PBr_5, in which the anion is the linear (unsymmetrical) Br_3^- ion, and also presumably in compounds such as PBr_6I and PBr_5ICl.

Of metallic salts $M(PX_6)_n$ only hexafluorophosphates appear to be known, and of these the salts with large cations such as NH_4^+, K^+, Cs^+, Ba^{2+}, and $Co(NH_3)_6^{3+}$, are the most stable in the solid state. The first three have been shown to crystallize with a NaCl-like packing of M^+ and octahedral PF_6^- ions.[6] The crystal structure of $HPF_6 . 6 H_2O$ is described on p. 544. In the PF_6^- ion in this crystal P–F was found to be 1·73 Å, but in $NaPF_6$[7] the length of this bond is apparently only 1·58 Å. Moreover, in $NaPF_6 . H_2O$[8] the ion was found to be distorted, with four (equatorial) P–F bonds of length 1·58 Å (these F atoms have a Na^+ ion as nearest external neighbour) and two of 1·73 Å (these F atoms having a H_2O neighbour at 2·90 Å).

(4) AC 1970 **B26** 443
(5) AC 1967 **23** 467

(6) ZaC 1951 **265** 229

(7) ZaC 1952 **268** 20
(8) AC 1956 **9** 825

Molecules PR_5, $PR_{5-n}X_n$, and mixed halides

Molecules in which five ligands are attached to P have trigonal bipyramidal configurations, and in PR_5 and PX_5 the axial bonds are longer than the equatorial ones. For example, in $P(C_6H_5)_5$[1] the length of the axial bonds is 1·99 Å as compared with 1·85 Å for the equatorial bonds; the latter have the same length as in $P(C_6H_5)_3$. Assuming that the more significant steric interactions are those between groups at 90° to one another and that the magnitude of ligand repulsions increases in the order F–F < F–R < R–R it is expected that in molecules PF_nR_{5-n} the groups R will preferentially occupy equatorial positions. This has been confirmed (e.d.) for CH_3PF_4 and $(CH_3)_2PF_3$[2] and for CF_3PF_4.[3] The first two of these compounds have structures very similar to those of the corresponding molecules formed by S and Cl with respectively one and two lone pairs (Fig. 19.6). An equatorial position is occupied by H in PHF_4,[4] but in many molecules of this general type there is apparently rapid interchange between axial and equatorial

(1) JCS 1964 2206

(2) IC 1965 **4** 1777
(3) JCP 1968 **49** 1307

(4) JCP 1968 **48** 2118

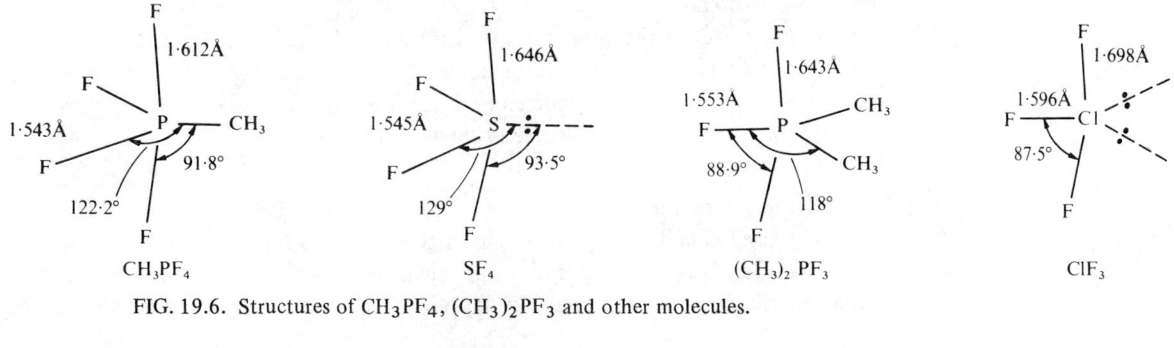

FIG. 19.6. Structures of CH_3PF_4, $(CH_3)_2PF_3$ and other molecules.

(5) QRCS 1966 **20** 245

positions at ordinary temperatures. An extensive review of 5-coordination is available.[5] The molecule (a), with three isopropoxy groups and one phenanthrene

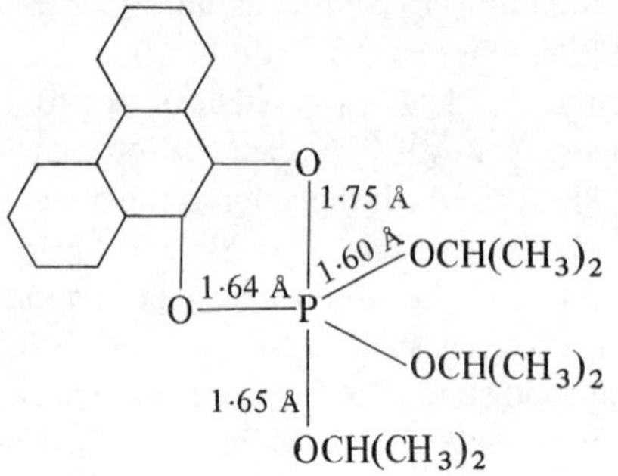

(a)

(6) JACS 1967 **89** 2268, 2272

quinone molecule bonded to P, is of special interest as an example of a trigonal bipyramidal arrangement of 5 O atoms around P. There is very little distortion of the bond angles from the ideal values (90° and 120°).[6]

All the mixed halides PCl_nF_{5-n} have been prepared. They have low-temperature forms in which they exist as trigonal bipyramidal molecules and these rearrange slowly at room temperature to form ionic crystals. For example, PCl_2F_3 is a gas at room temperature while the ionic form $(PCl_4)^+PF_6^-$ is a hygroscopic salt which sublimes, with some decomposition, at 135°C. At temperatures above 70°C

(7) ZaC 1957 **293** 147; IC 1965 **4** 738

it changes into $PF_5 + PCl_4F$, and the latter exists both as a non-polar liquid and also as an ionic solid, $(PCl_4)F$.[7]

The oxides and oxysulphide

In addition to the trioxide and pentoxide there are crystalline phases with compositions in the range $PO_{2.0}$–$PO_{2.25}$ which behave chemically as compounds containing P(III) and P(V). For example, they hydrolyse to mixtures of phosphorous and phosphoric acids. It is convenient to discuss these phases after the trioxide and pentoxide.

Phosphorus trioxide

This oxide exists in the vapour state as tetrahedral molecules, P_4O_6, at moderate temperatures; the structure of crystalline P_2O_3 is not known. The results of e.d. studies of the vapour of this oxide and related molecules are given in Table 19.5, and the structure of the P_4O_6 molecule is illustrated in Fig. 19.7.

Phosphorus pentoxide

(1) JCP 1961 **35** 1271

The structural chemistry of this oxide in the solid and liquid states is complex. In addition to a high-pressure form[1] and a glass there are three polymorphs stable at atmospheric pressure. As noted in Chapter 3 these three crystalline forms represent

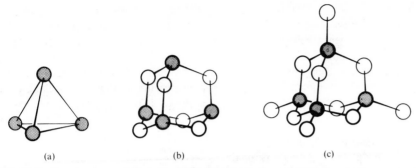

FIG. 19.7. The structures of the molecules (a) P_4, (b) P_4O_6, and (c) P_4O_{10} in the vapour state (diagrammatic). Shaded circles represent P atoms.

TABLE 19.5

Observed interatomic distances and interbond angles

Molecule	M—O (Å)	M—O′ or M—S (Å)	Angle		Reference
			M—O—M	O—M—O	
P_4O_6	1·65		128°	99°	JACS 1938 **60** 1814
P_4O_{10}	1·60	1·40	124°	OPO 102° / OPO′ 117°	K 1959 **4** 360
As_4O_6	1·80		126°	100°	JCS 1961 5486
$P_4O_6S_4$	1·61	1·85	124°	OPO 102° / OPS 117°	JACS 1939 **61** 1130

different ways of linking tetrahedral PO_4 groups through *three* vertices to form (a) finite P_4O_{10} molecules[2] of the same form as those in the vapour (Fig. 19.7(c)), (b) a layer structure[3] based on the simplest planar 3-connected net, and (c) a 3-dimensional framework[4] built of rings of ten PO_4 groups based on one of the two simplest 3D 3-connected networks. The accurate data for the layer structure (P–O, 1·56 Å, P–O′, 1·49 Å and P–O–P 145°) may be compared with those for the P_4O_{10} molecule given in Table 19.5. These crystal structures throw light on the physical and chemical properties of the three polymorphs. The metastable form, consisting of discrete P_4O_{10} molecules, is the most volatile and hygroscopic; it is formed by condensation of the vapour. The rearrangement of the PO_4 tetrahedra required to form the infinite networks of the other two forms does not take place if the vapour is rapidly condensed; it occurs only if the molten oxide is maintained at a high temperature for a considerable time. These forms are much less volatile and less rapidly attacked by water because the extended systems of linked tetrahedra must be broken down when the crystal melts, vaporizes, or reacts with water.

The P_4O_{10} molecule has been studied in the vapour by e.d. with the results summarized in Table 19.5. The unshared (outer) O atoms are distinguished as O′; the (multiple) bond P–O′ is the shortest recorded phosphorus–oxygen bond.

(2) AC 1964 **17** 677
(3) AC 1964 **17** 679
(4) RTC 1941 **60** 413

Phosphorus

P(III)P(V) *oxides.* Oxides with compositions in the range $PO_{2.02}$–$PO_{2.25}$ (α oxide) form a series of solid solutions containing molecules P_4O_8 and P_4O_9 which have the same general shape as the P_4O_{10} molecules but with either two or one of the outer O' atoms removed.[5] Crystals of composition $PO_{2.25}$ consist entirely of P_4O_9 molecules, but apparently up to 90 per cent of these molecules may be replaced by P_4O_8 molecules without any change in the mode of packing of the molecules. The β phase, stable in the range $PO_{2.0}$–$PO_{1.93}$ apparently consists of a statistical arrangement of P_4O_8 and P_4O_7 molecules.[6]

Phosphorus oxysulphide

The oxysulphide $P_4O_6S_4$, formed by heating P_4O_6 with sulphur, forms a tetrahedral molecule similar to that of P_4O_{10}. The P–S bond length (1·85 Å) is similar to that in the thiophosphoryl halides.

Molecules of the same geometrical type as P_4O_{10}

It is convenient to mention here two molecules which are structurally similar to P_4O_{10}. The basal ring of P_4O_{10} can be replaced by a cyclohexane ring, as in the phosphoric ester and its S analogue[1] which is illustrated in Fig. 19.8. The action of excess $Ni(CO)_4$ on P_4O_6 yields $P_4O_6[Ni(CO)_3]_4$, in which a $Ni(CO)_3$ group is attached to each P.[2]

The oxy-acids of phosphorus and their salts

The number and stability of the oxy-acids of phosphorus are in marked contrast to those of nitrogen. The formulae and basicities of the acids are set out in Table 19.6.

(5) AC 1964 **17** 1593

(6) AC 1966 **21** 34

(1) ACSc 1960 **14** 829

(2) AC 1966 **21** 288

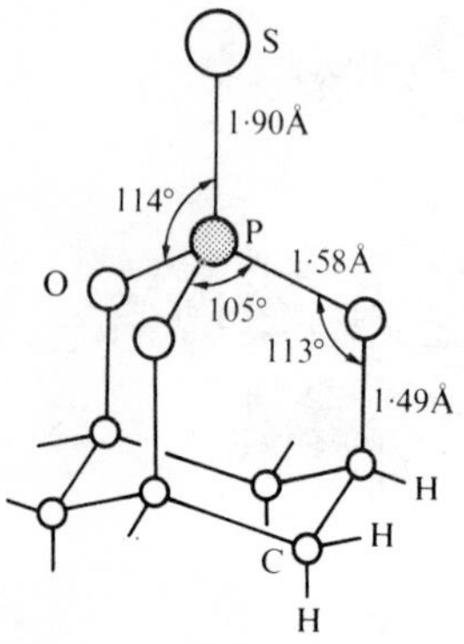

FIG. 19.8. The molecular structure of $C_6H_9O_3PS$.

TABLE 19.6
Oxy-acids of phosphorus

Acid	Formula	Basicity	Ion	Reference
Orthophosphorous	H_3PO_3	2	HPO_3^{2-}	AC 1969 **B25** 227
Hypophosphorous	H_3PO_2	1	$H_2PO_2^-$	AC 1969 **B25** 1932
Pyrophosphorous	$H_4P_2O_5$	2	$H_2P_2O_5^{2-}$	JACS 1957 **79** 2719
Diphosphorous	$H_4P_2O_5$	3	$HP_2O_5^{3-}$	JACS 1957 **79** 2719
	$H_5P_3O_8$	5	$P_3O_8^{5-}$	AC 1969 **B25** 1077
	$H_6P_6O_{12}$	6	$P_6O_{12}^{6-}$	ZaC 1960 **306** 30
Hypophosphoric	$H_4P_2O_6$	2	$H_2P_2O_6^{2-}$	AC 1964 **17** 1352
	$H_4P_2O_6 \cdot 2\,H_2O$	2	$(H_2P_2O_6)^{2-}$	AC 1971 **B27** 1520
Isohypophosphoric	$H_4P_2O_6$	3	$HP_2O_6^{3-}$	IC 1967 **6** 1137
Orthophosphoric	H_3PO_4	3	PO_4^{3-}	
Metaphosphoric	HPO_3	1	$(PO_3)_n^{n-}$	see text
Pyrophosphoric	$H_4P_2O_7$	4	$P_2O_7^{4-}$	

The so-called 'per-acids', H_3PO_5 and $H_4P_2O_8$, are omitted as nothing is known of their structures. The latter is presumably analogous to $H_2S_2O_8$. The structures, and indeed the existence, of meta- and pyro-phosphorous acids must still be regarded as doubtful. Crystalline anhydrous pyrophosphites are well known, for example, $Na_2(H_2P_2O_5)$, which is prepared by heating Na_2HPO_3. By analogy with orthophosphorous acid the pyro-acid would be

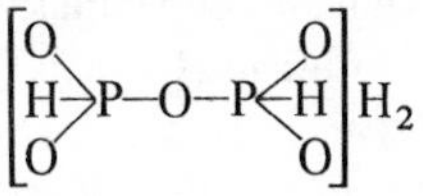

$$\left[\begin{array}{c}O\\H{>}P{-}O{-}P{<}H\\OO\end{array}\right]H_2$$

a formulation which is supported by the n.m.r. spectrum of $Na_2(H_2P_2O_5)$, which shows that the ion is symmetrical and that there is one H attached to each P atom.

Apart from metaphosphoric acid, which is usually obtained as a glass, the other acids in Table 19.6 can be obtained crystalline at ordinary temperatures. Nevertheless, the crystal structures of only two of the anhydrous acids, H_3PO_3 and H_3PO_4, have been studied.

Orthophosphorous acid

A striking feature of the phosphorous acids is their abnormal basicities. We might have expected H_3PO_3 to form salts containing pyramidal PO_3^{3-} ions analogous to the pyramidal SO_3^{2-} and ClO_3^- ions, but the acid is in fact dibasic. Normal salts are of the type $Na_2(HPO_3)$ and $Ba(HPO_3)$, though acid salts such as $NaH(HPO_3)$ have also been prepared.

In crystalline phosphorous acid the P and three O atoms form a pyramidal group, and the fourth tetrahedral position is occupied by a H atom. Unlike the other two H atoms this third one does not take part in the hydrogen-bonding scheme. The H atoms were not located in the X-ray study, but their positions were inferred from the short intermolecular contacts $O\cdots H\cdots O$ (around 2·60 Å); they have been confirmed by a n.d. study. Unlike the phosphite ion in $MgHPO_3 . 6 H_2O$ (with P–O, 1·51 Å, and O–P–O angles of $110°$)[1] the $H_2(HPO_3)$ molecule does not possess trigonal symmetry. The 'normal' phosphite ion is accordingly (a) and the hypophosphite ion (b):

(1) AC 1956 **9** 991

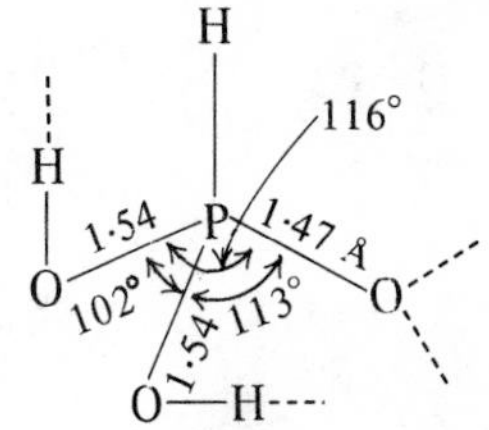

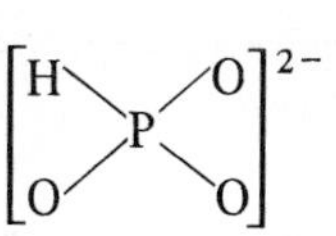

$$\left[\begin{array}{c}H\\O\end{array}{>}P{<}\begin{array}{c}O\\O\end{array}\right]^{2-}\qquad\left[\begin{array}{c}H\\H\end{array}{>}P{<}\begin{array}{c}O\\O\end{array}\right]^{-}$$

(a) 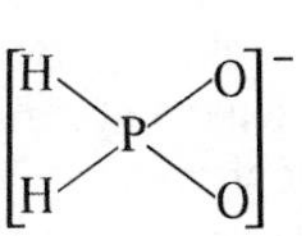   (b)

We give in Table 19.6 the reference to $CuHPO_3 . 2 H_2O$, which includes references to work on H_3PO_3.

Hypophosphorous acid

The structure of crystalline H_3PO_2 is not known, but the structures of several salts have been studied. These salts MH_2PO_2 are not acid salts, in which there would be PO_2^{3-} ions joined together by hydrogen bonds, for the closest approach of O atoms attached to different P atoms in $NH_4H_2PO_2$ is 3·45 Å. Instead the salts contain tetrahedral $H_2PO_2^-$ ions, though the H atoms were not directly located in the X-ray studies. The mean dimensions of the ion are as shown.

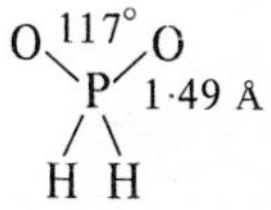

We come now to a group of acids in which there are two or more P atoms directly bonded.

$2-$

$3-$

$5-$

$6-$

The $P_6O_{12}^{6-}$ ion

Phosphorus

Hypophosphoric acid

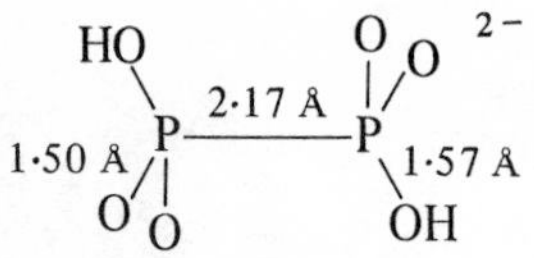

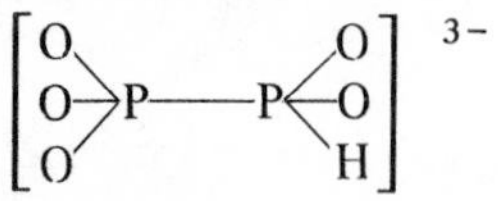

Cryoscopic measurements show that the molecular weights of the ethyl ester and of the free acid correspond to the formulae $(C_2H_5)_4P_2O_6$ and $H_4P_2O_6$ respectively. The magnetic evidence is conclusive on this point. The molecule H_2PO_3 would be paramagnetic since there would be an unpaired electron, but the Na and Ag salts are diamagnetic. An X-ray study of $(NH_4)_2H_2P_2O_6$ shows that the ion has the staggered configuration. The bond lengths P–OH and P–O are very similar to those in KH_2PO_4 (1·58 and 1·51 Å) and H_3PO_4 (1·57 and 1·52 Å); the ion has a similar structure in the 'dihydrate' of the acid, which is $(H_3O)_2^+(H_2P_2O_6)^{2-}$. For isohypophosphoric acid see later.

Diphosphorous acid

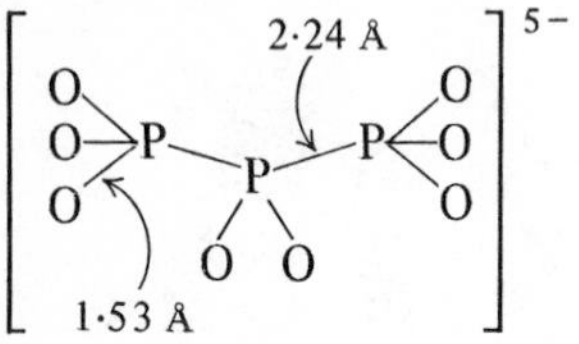

A salt described as a diphosphite, $Na_3HP_2O_5$. 12 H_2O, has been prepared by hydrolysing PBr_3 with ice-cold aqueous $NaHCO_3$. Its n.m.r. spectrum is consistent with the structural formula shown.

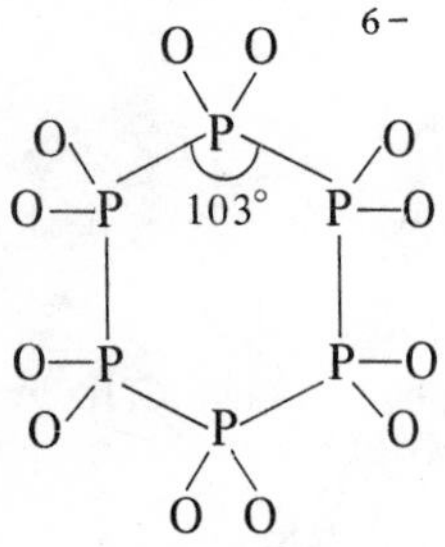

$H_5P_3O_8$ *and* $H_6P_6O_{12}$. The ion $P_3O_8^{5-}$ has been studied in $Na_5P_3O_8$. 14 H_2O. The P–P bond has the length of a normal single bond in this ion, in which P atoms are in the formal oxidation states III, IV, and III.

Salts of an acid $H_6P_6O_{12}$ have been prepared by treating red P suspended in alkali hydroxide solution with hypochlorite. An X-ray study of the Cs salt shows the ion to contain a chair-shaped ring of six directly bonded P atoms in which P–P is approximately 2·2 Å.

Isohypophosphoric acid

The hypophosphoric acid of an earlier paragraph results from the oxidation of P by moist air. The mild alkaline hydrolysis of PCl_3 gives a complex mixture of products which have been separated chromatographically. One component has been identified as isohypophosphoric acid, and the free acid has been prepared from PCl_3 and H_3PO_4. The trisodium salt, $Na_3(HP_2O_6)$. 4 H_2O, results from heating a mixture of Na_2HPO_4 and NaH_2PO_3. The n.m.r. spectrum is consistent with the structures (a) and (b) for the ion and the acid.

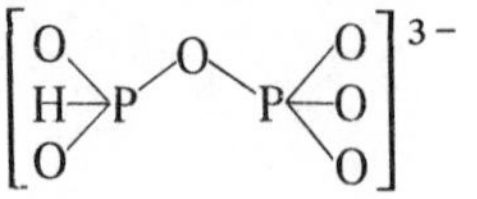

(a) 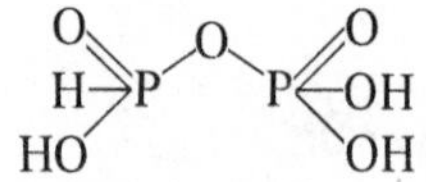   (b)

Phosphoric acid and phosphates

We now come to the extensive oxygen chemistry of P(v) based on discrete PO_4^{3-} ions or on tetrahedral PO_4 groups sharing one or two O atoms; sharing of three O atoms by all PO_4 tetrahedra leads to the neutral oxide P_2O_5. Discrete PO_4^{3-} ions

688

are found in the orthophosphates and systems of linked PO_4 groups in pyro-, poly-, and metaphosphates.

Orthophosphoric acid and orthophosphates

Discrete PO_4^{3-} ions exist in normal orthophosphates. In dihydrogen phosphates such as KH_2PO_4 and $(N_2H_5)H_2PO_4$ $PO_2(OH)_2$ units are hydrogen-bonded to form extended anions—see p. 318. Several X-ray and n.d. studies[1] give P–O close to 1·51 Å and P–OH 1·55–1·58 Å. The length of the hydrogen bonds is around 2·50 Å. In the (layer) structure of H_3PO_4[2] there are also hydrogen bonds of length 2·84 Å; this structure is described in Chapter 8. The P–O bond lengths found in H_3PO_4 and its hemihydrate[3] are P–O, 1·52 and 1·49 Å, and P–OH, 1·57 and 1·55 Å respectively.

(1) ACSc 1965 **19** 1629

(2) ACSc 1955 **9** 1557

(3) AC 1969 **B25** 776

The structures of a number of orthophosphates are similar to those of forms of silica or of silicates. For example, $AlPO_4$ crystallizes with all the three normal silica structures[4] and also undergoes a transition from a low- to a high-temperature form in each case; YPO_4 and $YAsO_4$ crystallize with the zircon ($ZrSiO_4$) structure.

(4) ZK 1967 **125** 134

Pyrophosphates

Pyrophosphates $M^{IV}P_2O_7$, $M_2^{II}P_2O_7$, and $M_4^{I}P_2O_7$ (hydrated) have been studied in some detail, particular interest being centred in the bond angle at the bridging O atom. Salts of divalent metals often have low-(α) and high-temperature (β) forms in which the configuration of the pyrophosphate ion varies somewhat with the size of M^{2+}. For small metal ions the configuration is approximately staggered, and for large ions nearly eclipsed.[1] There is considerable variation in the O bond angle, from values of $134°$ ($Na_4P_2O_7 \cdot 10 H_2O$),[2] $139°$ in one form of SiP_2O_7,[3] to $156°$ in $\alpha\text{-}Mg_2P_2O_7$. The apparent collinearity of the O bonds in the high-temperature forms of certain salts $M_2P_2O_7$ is now generally attributed either to positional disorder (random arrangement throughout the crystal of O_b atoms around but off

(1) JSSC 1970 **1** 120
(2) AC 1964 **17** 672
(3) AC 1970 **B26** 233

the P–P line, P-----P) or to highly anisostropic motion of O_b; compare high-cristobalite. An early study of the cubic polymorph of ZrP_2O_7 (with which the Si, Ge, Sn, Pb, Ti, Hf, Ce, and U compounds are isostructural) indicated collinear bonds in the P_2O_7 ion. It has now been shown, by a careful study of the high-temperature form of SiP_2O_7[4] that the true unit cell has 27 times the volume of the cell to which the earlier structure was referred, and that the mean P–O–P angle is $150°$, though there are still a few apparently collinear O bonds in the structure. There appears to be a correlation between the O bond angle and the difference in length between the terminal and bridging P–O bonds: P–O–P, $130°$, $P\text{-}O_b$, 1·61, $P\text{-}O_t$, 1·52 Å, ($Na_4P_2O_7 \cdot 10 H_2O$); P–O–P, $156°$, $P\text{-}O_b$, 1·58, $P\text{-}O_t$, 1·55 Å ($\alpha\text{-}Mg_2P_2O_7$).

(4) JSSC 1973 **7** 69

Linear polyphosphates

These contain 'hybrid' ions intermediate between the pyrophosphate ion and the infinite linear metaphosphate ion, in which the terminal PO_4 groups share one O and the intermediate groups two O atoms. The normal sodium triphosphate,

Phosphorus

$Na_5P_3O_{10}$, is known anhydrous and as the hexahydrate; $Na_2H_3P_3O_{10}$ and $Na_3H_2P_3O_{10}$ have also been prepared. The triphosphate, which is used as a detergent in mixtures with soaps and sulphonates, can be prepared in various ways, for example, by heating together

$$2\,Na_2HPO_4 + NaH_2PO_4 \rightarrow Na_5P_3O_{10} + 2\,H_2O.$$

(1) AC 1964 **17** 674

X-ray studies of the two crystalline forms of $Na_5P_3O_{10}$[1] show the structure of the ion to be

The linear tetraphosphate ion results from alkaline hydrolysis of the cyclic metaphosphate, and the salts of large ions such as Pb^{2+}, Ba^{2+}, and $N(CH_3)_4^+$ can be crystallized from the acidified solution; they tend to remain in alkaline solution, forming viscous liquids. Material marketed as 'hexasodium tetraphosphate', $Na_6P_4O_{13}$, is not a simple chemical individual but is a mixture of $Na_5P_3O_{10}$ and $NaPO_3$. The acids $H_5P_3O_{10}$ and $H_6P_4O_{13}$ have been obtained by passing solutions of their tetramethylammonium salts through a cation exchange resin. The polyphosphates containing more than four P atoms are not phase-diagram entities,[2] and only Ca hexaphosphate can be prepared relatively easily. However, gram quantities of pure polyphosphates containing 4–8 P atoms have been prepared chromatographically. The basic starting material is a polyphosphoric acid with average chain length around five which results from heating 85 per cent aqueous H_3PO_4 in a gold dish at 400°C for 12 hours.

(2) JACS 1967 **89** 2884;
ZaC 1964 **330** 78

(3) AC 1965 **19** 363

The salt $K_4(P_3O_9NH_2)\cdot H_2O$ contains an ion of the same type as $P_3O_{10}^{5-}$ in which one O has been replaced by NH_2.[3]

Metaphosphates

In these compounds each PO_4 group shares two O atoms to form rings or chains of composition $(PO_3)_n^{n-}$ analogous to those in metasilicates. Metaphosphoric acid itself has not been obtained crystalline, but as a glass or syrup; it is obviously highly polymerized.

Some metaphosphates are rubbery or horn-like masses and some are mixtures of various metaphosphates. Thus the solid of empirical composition $NaPO_3$ obtained by heating $NaNH_4HPO_4$ or NaH_2PO_4 is not homogeneous, part being soluble and the remainder insoluble in water, but under controlled conditions numerous well defined crystalline salts can be prepared. Also, by rapidly cooling molten $NaPO_3$ to below 200°C a brittle, transparent, glassy form is obtained, called Graham's salt. A radial distribution function derived from X-ray data for a $NaPO_3$ glass shows that it consists largely of long chains of PO_4 tetrahedra linked by sharing two vertices and held together by O–Na–O bonds. However, these phosphate glasses also contain small quantities of cyclic metaphosphates. The penta- and hexa-metaphosphates have been isolated from Graham's salt, and the presence of the 7- and 8-ring ions proved by paper chromatography. Of these higher cyclic ions the 6-ring is most resistant to hydrolysis.

The two simplest water-soluble (cyclic) sodium metaphosphates, which are important water-softening agents, are $Na_3P_3O_9$, formed by heating NaH_2PO_4 to 550–600°C, and $Na_4P_4O_{12}$, conveniently prepared by treating the soluble form of P_2O_5 with a cold suspension of $NaHCO_3$. The trimetaphosphates contain cyclic $P_3O_9^{3-}$ ions with the chair configuration and the dimensions shown at (a); salts studied include $Na_3P_3O_9$ and its monohydrate[1] and $LiK_2P_3O_9 . H_2O$.[2] An unusual feature of $Na_3P_3O_9 . H_2O$ is that its crystal structure is almost identical

(1) AC 1965 **18** 226
(2) AC 1962 **15** 1280

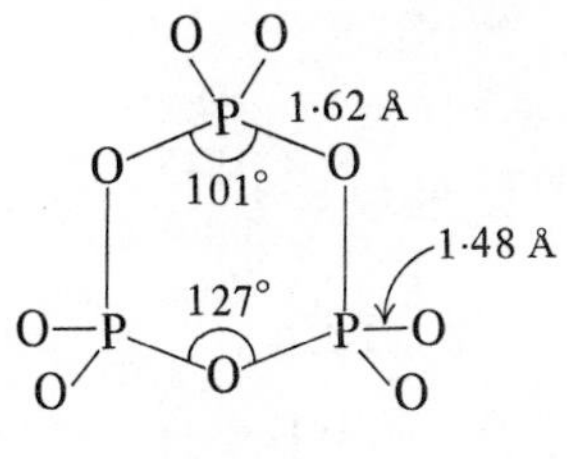

(a)

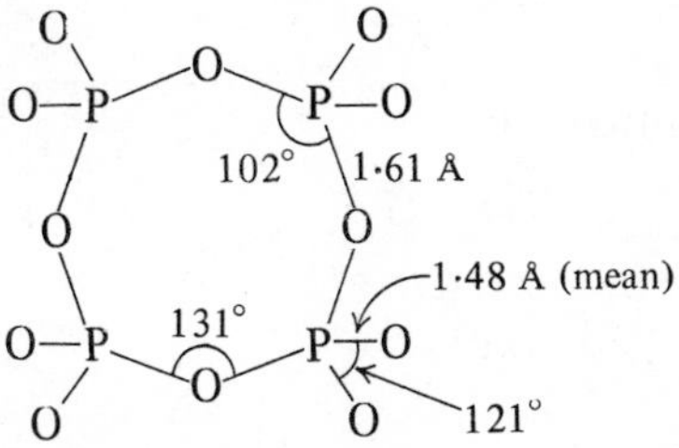

(b)

with that of the anhydrous salt. In the structure of the latter there is almost sufficient room for the H_2O molecule, and only a small expansion of the structure is necessary to accommodate it. The chair-shaped tetrametaphosphate ring, (b), has been studied in the anhydrous NH_4[3a] and Al salts, and in two crystalline forms of $Na_4P_4O_{12} . 4 H_2O$.[3b] The structure of the hexametaphosphate ion has been determined in $Na_6P_6O_{18} . 6 H_2O$.[4] The 6 P atoms of the ring are coplanar, and bond lengths are similar to those in the $P_3O_9^{3-}$ and $P_4O_{12}^{4-}$ ions.

(3a) JCS A 1970 435
(3b) AC 1961 **14** 555; 1964 **17** 1139
(4) AC 1965 **19** 555

The insoluble metaphosphates contain infinite chain ions with the mean inter-bond angles and bond lengths shown:

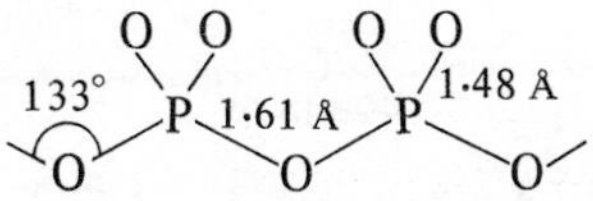

Some of these compounds exist in more than one form with different configurations of the $(PO_3)_n^{n-}$ chain, as in the so-called Maddrell and Kurrol salts[5] (two forms of $NaPO_3$). The nature of the chains in Na, K, Rb,[6] Ag, and Pb metaphosphates and in $LiAsO_3$ and $NaAsO_3$ is illustrated in the discussion of tetrahedral structures in Chapter 23. A complex configuration of the chain, with 8 tetrahedra in the repeat unit, is found in $[Na_3H(PO_3)_4]_n$.[7]

(5) AC 1968 **B24** 1621
(6) AC 1964 **17** 681

(7) AC 1968 **B24** 992

Mono- and di-fluorophosphoric acids
Intermediate between the neutral molecule POF_3 and the ion PO_4^{3-} are the ions $PO_2F_2^-$ and PO_3F^{2-}. Salts of both mono- and di-fluorophosphoric acids have been

Phosphorus

prepared, for example, the NH_4 and K salts and several others. Anhydrous monofluorophosphoric acid, which can be prepared in quantitative yield from anhydrous metaphosphoric acid and liquid anhydrous HF, is an oily liquid which sets to a glass at the temperature of solid CO_2 and shows a strong resemblance to H_2SO_4.

Both $BaPO_3F$ and KPO_2F_2 are structurally similar to $BaSO_4$. Accurate data are available for the PO_3F^{2-} ion in the Ca and NH_4[1] salts (a), and for the $PO_2F_2^-$ ion (b), in its K,[2] Cs, and NH_4 salts.[3] The $PO_2F_2^-$ is appreciably distorted from regular tetrahedral shape and is extremely similar in structure to the isoelectronic SO_2F_2 molecule (c). The acid HPS_2F_2 and many of its salts have been prepared.

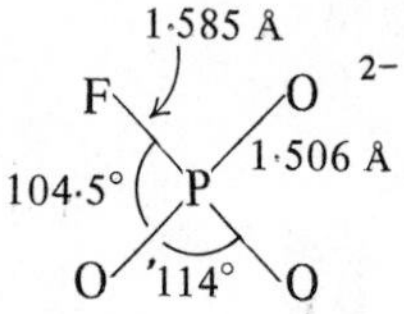

(1) AC 1972 **B28** 2183, 2191
(2) JCS A 1966 1775
(3) JCS A 1969 1783

(a) 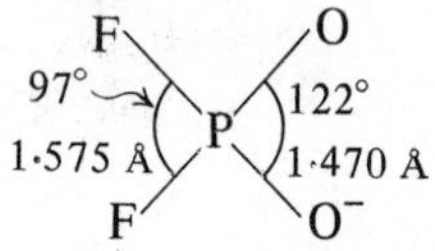   (b)   compare   (c)

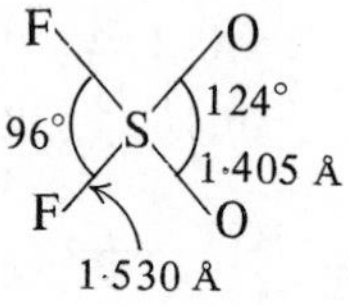

(4) AC 1964 **17** 671

Phosphoramidates

In $NaHPO_3(NH_2)$[4] there are zwitterions $^+H_3N \cdot PO_3^{2-}$, (d), analogous to the isoelectronic sulphamic acid, (e), $^+H_3N \cdot SO_3^-$ (p. 586). The N–H–O bonds link the ions into a 3-dimensional framework, in the interstices of which are located the Na^+ ions. Replacement of one O by S in $PO_3NH_2^{2-}$ gives the ion $[PO_2S(NH_2)]^{2-}$, in which the bond lengths shown at (f) come from a study of the di-ammonium salt.[5a] There are similar bond lengths in the diamidothiophosphate ion, studied in $NH_4[POS(NH_2)_2]$.[5b]

(5a) AC 1969 **B25** 1256
(5b) ZaC 1968 **358** 282

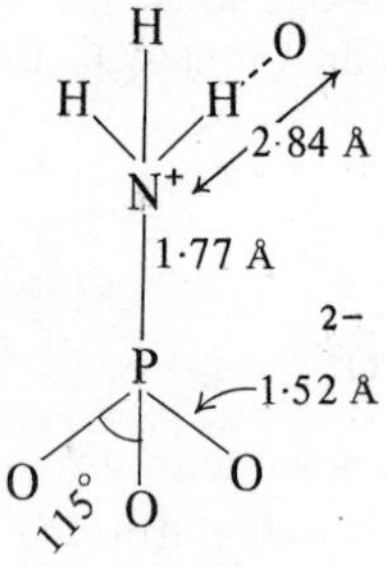

(d)   compare   (e)

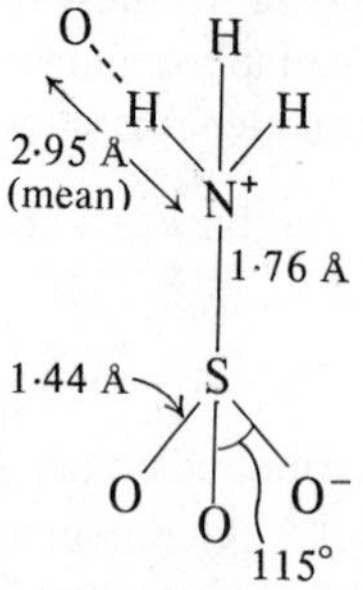

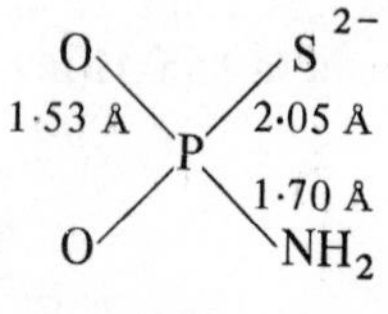

(f)

Phosphorothioates

Sulphur can replace O in the PO_4^{3-} ion to give the whole range of phosphorothioate ions from PSO_3^{3-} to PS_4^{3-}. The sodium salts are all hydrated and generally hygroscopic. Thiophosphates have been made from $Na_2S-P_2S_5$ melts with S : P up to $3\frac{1}{2}$: 1 as glasses which are very unstable to water.

In potassium O-O-dimethylphosphorodithioate the ion has the structure (g).[6]

(6) AC 1962 **15** 765

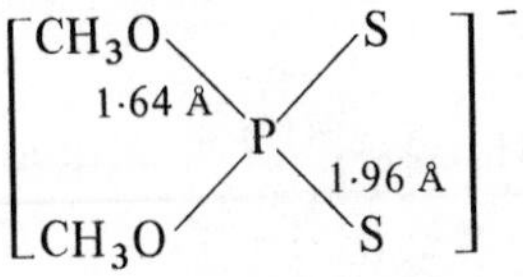

(g)

Other substituted phosphoric acids, etc.

Replacement of H in H_3PO_4 by a radical R gives substituted phosphoric acids $OP(OH)_2OR$ and $OP(OH)(OR)_2$ and finally the phosphoric ester, $OP(OR)_3$. The structure of dibenzylphosphoric acid, $HPO_2(OCH_2C_6H_5)_2$,[7] was described in Chapter 8; the molecules are linked in pairs by O–H–O bonds as in dimers of carboxylic acids. Examples of mono- and di-substituted ions are the phenyl-phosphate ion, (h), studied in the K salt,[8] and the diethylphosphate ion, (i), in the Ba salt,[9] and of an ester, triphenylphosphate, (j).[10]

(7) AC 1956 **9** 327

(8) IC 1967 **6** 1998
(9) AC 1966 **21** 49
(10) AC 1965 **19** 645

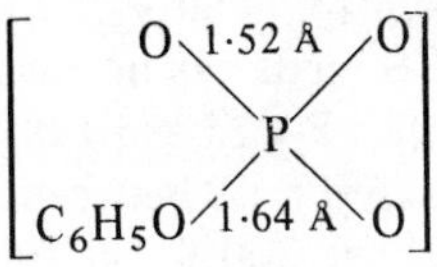

(h)

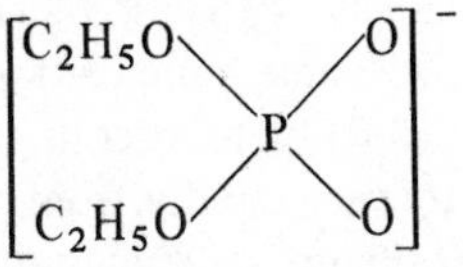

(i)

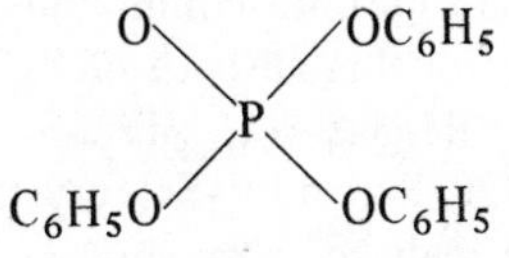

(j)

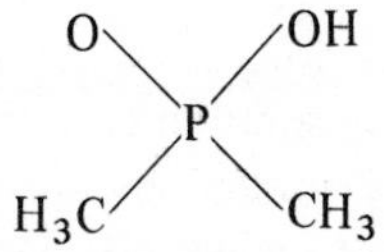

(k)

Replacement of OH in H_3PO_4 by R gives $RPO(OH)_2$ $R_2PO(OH)$, and finally the phosphine oxide, R_3PO. Structural studies have been made of dimethylphosphinic acid, (k),[11] (see also p. 317), and of its salts.[12] Some of the salts $M(R_2PO_2)_2$, of Be, Zn, and other divalent metals, have interesting properties. They range from high-melting crystalline salts to compounds which can be melted and drawn into fibres and others which are waxy solids. The linear chain in the Zn compound, with alternate single and triple $-OP(R_2)O-$ bridges, was mentioned in Chapter 3 as an interesting type of chain structure.

(11) AC 1967 **22** 678
(12) JCS A 1968 757, 763

Phosphorus

Since groups R_2PO_2 or R_2PS_2 can act either as bridging ligands or as bidentate ligands numerous types of polymeric system can be formed, of which the following are examples. The first two are dimers, the others infinite linear molecules.

(13) IC 1969 **8** 2410
(14) IC 1965 **4** 99

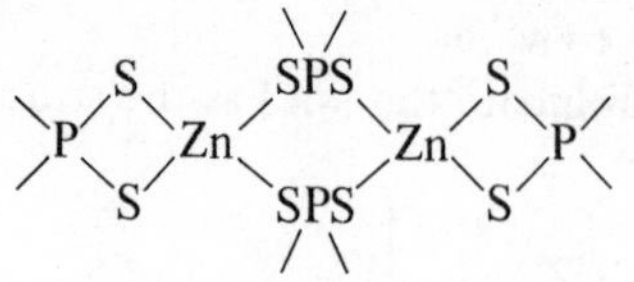

$\{Zn[PS_2(i\text{-}C_3H_7)_2]_2\}_2$ [(13)] 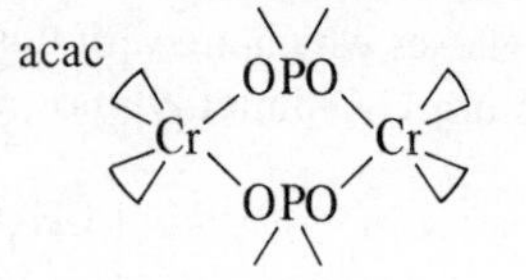 $[(acac)_2Cr \cdot PO_2\phi_2]_2$ [(14)]

(15) AC 1969 **B25** 2303
(16) AC 1969 **B25** 1057

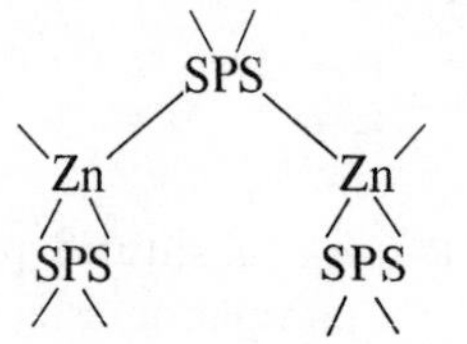

$\{Zn[PS_2(C_2H_5)_2]_2\}_n$ [(15)] $\{Zn[PO_2\phi(C_4H_9)]_2\}_n$ [(16)]

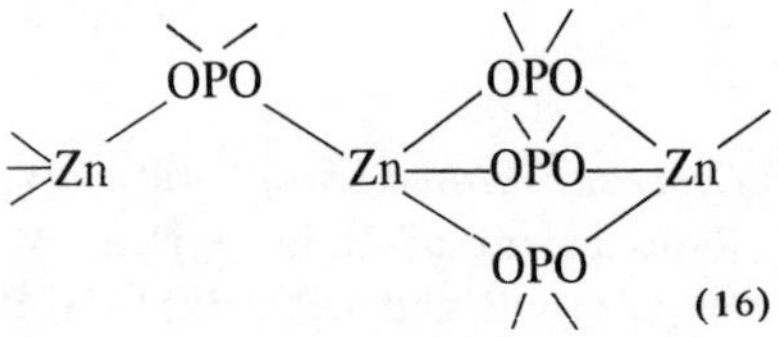

Phosphorus sulphides

Although many sulphides of phosphorus have been described it appears that only five are definite compounds, namely, P_4S_3, P_4S_5, P_4S_7, P_4S_9, and P_4S_{10}. Note the surprising absence of P_4S_6. At some compositions P–S melts can be quenched to brittle glasses (S : P ratio 3·5–3·0) or to viscous gums (S : P 1·25 and 1·00), but P_4S_3, P_4S_7, and P_4S_{10} crystallize very rapidly from melts. For S : P between 2 and 3 the melt viscosity shows a maximum at temperatures above 300°C like sulphur, suggesting that the incorporation of P into molten S leads first to branching and cross-linking and then to gradual breakdown of the polymeric S structures owing to the formation of small cage molecules.

All the phosphorus sulphides are yellow crystalline solids, all have molecular weights (determined either in CS_2 solution or in the vapour state) corresponding to the formulae given, and all are formed by direct union of the elements under various conditions. There are interesting relations between these compounds. For example, P_4S_3 readily combines with sulphur to give P_4S_5 or P_4S_7, and on heating, P_4S_5 forms P_4S_3 and P_4S_7. In spite of the fact that all the molecules are of the type P_4S_n structural studies show that there is no P_4 unit common to these molecules. The molecules of P_4S_3, P_4S_5, and P_4S_7 are illustrated in Fig. 19.9. In all three of these molecules there are bonds between P atoms. In the first two there are single bonds of about the same length as in white P; in P_4S_7, however, the only P–P bond is a weak one, of length 2·33 Å. In sulphur-deficient (β) P_4S_7 apparently some of the external S atoms are missing, and the unique P–P bond has a more normal length (2·26 Å). The P–S bond lengths are generally close to one of two values, 2·10 Å if S is forming two bonds, and 1·93 Å to an 'external' S attached to only

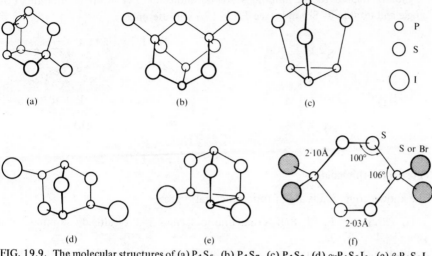

FIG. 19.9. The molecular structures of (a) P_4S_5, (b) P_4S_7, (c) P_4S_3, (d) α-$P_4S_3I_2$, (e) β-$P_4S_3I_2$, (f) $P_2S_6Br_2$.

one P atom. Similar bond lengths are found in P_4S_{10}, the structure of which is similar to that of the P_4O_{10} molecule (Fig. 19.7, p. 685). Although the S analogue of P_4O_6 is not known there is a sulphide P_4S_9 analogous to P_4O_9. Its structure is derived from that of P_4S_{10} by removing one of the terminal S atoms; it is therefore $P^{III} P_3^V S_9$. Interbond angles lie for the most part in the range 100–$115°$ except those in the P_3 ring of P_4S_3, in the quadrilateral P_3S ring of P_4S_5 (mean angle, $87°$), and an exterior angle of $125°$ in P_4S_5. References are included in Table 19.7.

TABLE 19.7

Phosphorus sulphides, thiohalides, and related molecules

Molecule	P—P (Å)	P—S (Å)	P=S (Å)	*Reference*
P_4S_3	2·235	2·090		AC 1957 **10** 574
P_4S_5	2·25	2·11 (mean)	1·94	AC 1965 **19** 864
α-P_4S_7	2·33	2·10 (mean)	1·92	AC 1965 **19** 864
β-P_4S_7	2·36			AC 1965 **18** 221
P_4S_{9-x}				ZaC 1969 **366** 152
P_4S_9		2·11	1·93	AC 1969 **B25** 1229
P_4S_{10}		2·10	1·91	AC 1965 **19** 864
$P_2S_6Br_2$			1·98	IC 1965 **4** 186
α-$P_4S_3I_2$	2·20	2·10		AC 1959 **12** 455
β-$P_4S_3I_2$	2·22	2·12	(P—I, 2·49)	JCS A 1971 1100
$P_2S_4(CH_3)_2$		2·14	1·95	JCS 1964 4065
$P_2S_4(i$-$C_3H_7O)_4$		2·07	1·91	IC 1970 **9** 2269
P–S melts				JINC 1963 **25** 683
		P—Se	P—I	
$P_4Se_3I_2$	2·22	2·24	2·47	AC 1970 **B26** 2092

Phosphorus

Closely related to the sulphides are the molecules (a) and (b) containing both single and double P–S bonds (see Table 19.7 for references).

(a)

(b)

Phosphorus thiohalides

The known compounds are of four structural types:

(i) PSX_3 (X = F, Cl, Br): tetrahedral molecules $S=PX_3$ already summarized in Table 19.4 (p. 679).

(ii) $P_2S_2I_4$: structure not known but presumably a diphosphine (p. 678), $I_2SP–PSI_2$.

(iii) $P_2S_5Br_4$ and $P_2S_6Br_2$: The structure of $P_2S_5Br_4$ is not known. In crystalline $P_2S_6Br_2$ (formed by the action of Br_2 on P_4S_6) the molecule consists of a P_2S_4 ring, with the skew boat configuration, and to each P are attached two atoms which are either Br (P–Br, 2·07 Å) or S (P–S, 1·98 Å). There is apparently complete disorder as regards the choice of these atoms. The very flexible skew boat configuration is apparently the ring configuration which minimizes repulsions between non-bonded atoms consistent with S dihedral angles close to 100°. The dihedral angles for P are small (40° and 42°)–Fig. 19.9(f) and Table 19.7.

(iv) $P_4S_3I_2$. This compound exists in two forms. The α form is prepared in CS_2 solution from the elements or by treating the β form with excess iodine. The molecule (Fig. 19.9(d)) consists of one 6- and two 5-membered rings, but has different relative arrangements of the P and S atoms as compared with P_4S_3. This compound illustrates an interesting feature of the phosphorus–sulphur compounds, the facile rearrangement of the P and S atoms. The reaction of P_4S_3 with I_2 under mild conditions gives β-$P_4S_3I_2$. The structure of this form (Fig. 19.9(e)) is much more closely related to P_4S_3, one P–P bond having been broken and two P–I bonds formed. The $P_4Se_3I_2$ molecule has the same structure.

Cyclic phosphorus compounds

Compounds containing P_n rings

We have mentioned that connected systems of P atoms occur in some metal phosphides and that there are pairs of directly bonded P atoms in P_2H_4 and substituted diphosphines, in P_2Cl_4 and P_2I_4, in certain phosphite ions, and also in molecules such as $R_2SP–PSR_2$. The systems P–P and P–P–P form parts of the ring systems in phosphorus sulphides, while the P_4 molecule is built of P_3 rings. There

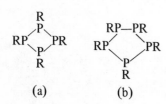

(a)   (b)

are also molecules based on larger P_n rings; the acid $[PO(OH)]_6$ has already been mentioned.

The cyclic molecules (a) and (b), in which R is $-CF_3$, are prepared by the action of Hg on $F_3C.PI_2$. Both the P_4[1] and P_5[2] rings are non-planar, with P–P–P bond angles of 85° and 101° (mean) respectively and normal bond lengths P–P (close to 2·2 Å). Compounds originally thought to be the analogues of azo compounds, that is, RP=PR, and now known to be cyclic polymers, include $(CH_3P)_5$ and the phosphobenzenes $(PC_6H_5)_5$ and $(PC_6H_5)_6$ (three polymorphs). In the pentamer the P_5 ring, (c), is non-planar with dihedral angles ranging from 3°

(1) AC 1962 **15** 564
(2) AC 1961 **14** 250; AC 1962 **15** 509

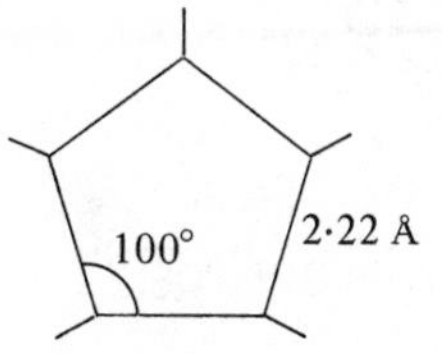

(c) 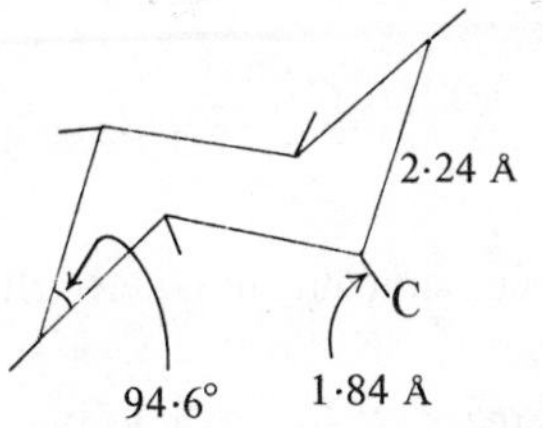   (d) 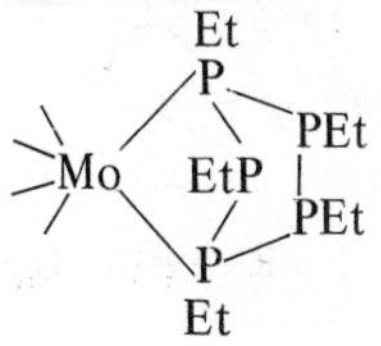   (e)

to 61° (mean 38°),[3] and in the hexamer the 6-ring, (d), has the chair configuration[4] with the six phenyl groups in equatorial positions and dihedral angles 85° (compare 69° in $(CsPO_2)_6$, p. 688, and 89° in $(AsC_6H_5)_6$). For a comparison of the structures of these cyclic molecules see reference.[5] The reaction between $(PC_2H_5)_4$, a molecule of type (a), and $Mo(CO)_6$ is of interest as converting a P_4 ring into a P_5 ring, the product being $(CO)_4Mo(PC_2H_5)_5$, (e).[6]

(3) JCS 1964 6147
(4) JCS 1965 4789

(5) AC 1962 **15** 708

(6) JCS A 1968 1221

Compounds containing $(PN)_n$ rings

An example of the smallest ring of this kind is found in $(CH_3NPCl_3)_2$,[1] Fig. 19.10(a), prepared from PCl_5 and $(CH_3NH_3)Cl$. This reaction also produces another compound of the same general type, namely, $P_4(NCH_3)_6Cl_8$,[2] with the structure

(1) JCS A 1966 1023; ZaC 1966 **342** 240
(2) ZaC 1967 **351** 152

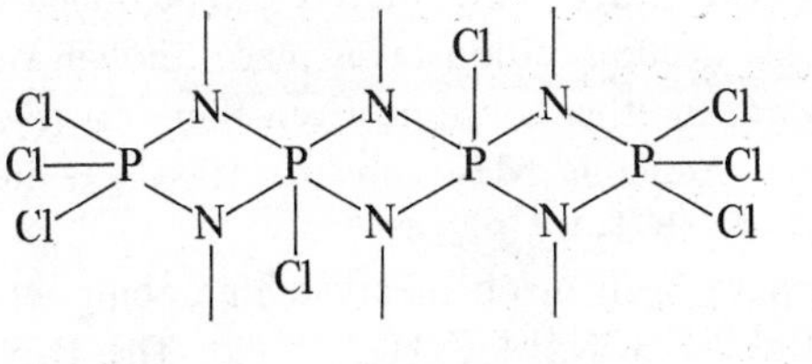

in which there is trigonal bipyramidal coordination of P as in the simpler molecule of Fig. 19.10(a). The $P_2(NCH_3)_2Cl_6$ molecule can be regarded as built from two PCl_5 molecules, replacing one equatorial and one axial Cl by N and then joining to form a 4-membered ring. The corresponding angles and bond lengths are very similar to those in PCl_5.

A number of larger $(PN)_n$ rings have been studied in the cyclophosphazenes. The phosphonitrile chlorides, $(PNCl_2)_n$, may be prepared by heating together PCl_5 and NH_4Cl, for example, in tetrachloroethane. The extract in light petroleum yields

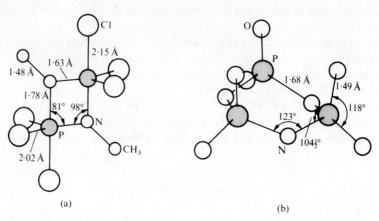

FIG. 19.10. The molecular structures of (a) $(CH_3NPCl_3)_2$, (b) $(HN.PO_2)_3^{3-}$.

(3) JCS 1960 2542

crystalline cyclic polymers $(PNCl_2)_n$, $n = 3$ to 8; the residue is an oil which contains polymers up to at least $n = 17$.[3] The fraction insoluble in the petroleum is a viscous oil of composition $(PNCl_2).PCl_5$ probably consisting of linear polymers $Cl.(PNCl_2)_nPCl_4$. This material is converted by boiling in tetrachloroethane into the rubbery $(PNCl_2)_n$, presumably a linear polymer

$$\diagup N\!\!=\!\!\underset{Cl_2}{P}\!\!=\!\!N\!\!=\!\!\underset{Cl_2}{P}\!\!=\!\!N\diagdown$$

The fluorides have been prepared up to $(PNF_2)_{17}$. For example, both $(PNF_2)_3$ and $(PNF_2)_4$ are volatile solids stable up to $300°C$, above which temperature they are converted into colourless liquid polymers.

The $(PN)_n$ rings have considerable stability. For example, $(PNCl_2)_4$ can be heated to $250°C$ before further polymerization takes place, and depolymerization takes place only above $350°C$. This chloride can be boiled with water, acid, or alkali without appreciable decomposition takes place, though slow hydrolysis occurs if it is shaken with water in ether solution. Even then, however, the ring system is not broken (see later). Aqueous NH_3 converts $(PNCl_2)_3$ into $P_3N_3Cl_4(NH_2)_2$ and liquid NH_3 gives $P_3N_3(NH_2)_6$.

(4a) JCS 1963 3211
(4b) JCS A 1971 1450
(4c) AC 1970 **B26** 1812
(5) IC 1970 **9** 1656
(6) AC 1969 **B25** 316, 2040

X-ray studies have been made of crystalline compounds $(PNX_2)_3$ where X is F,[4a] Cl,[4b] Br,[4c] SCN,[5] C_6H_5,[6] etc. The P_3N_3 ring in $(PNF_2)_3$ and $(PNCl_2)_3$ is essentially planar with angles of $120°$ and six equal P–N bonds of length $1·56$ Å (compare $1·78$ Å expected for a single bond), and is clearly an aromatic system in which P uses its d orbitals. In all these compounds both halogen atoms are attached to P, and the lengths of P–F ($1·51$ Å) and P–Cl ($1·99$ Å) correspond approximately to those of single bonds.

(7) JCS 1961 4777

The 8-membered P_4N_4 rings show considerable variation in shape in compounds $P_4N_4X_8$. It is planar in $(PNF_2)_4$[7] with a rather shorter P–N bond ($1·51$ Å) and angles P–N–P, $147°$, and N–P–N, $123°$, but in the stable form of

$P_4N_4Cl_8$[8] the ring has the chair conformation and in the less stable (K) form[9] the ring is boat-shaped. The molecule $[NP(CH_3)_2]_4$ is also boat-shaped,[10] and the same conformation is retained in $[(NPMe_2)_4H]CuCl_3$.[11] By reacting $[NPMe_2]_4$ with certain transition-metal halides in methyl ethyl ketone compounds such as $[(NPMe_2)_4H]CuCl_3$ and $[(NPMe_2)_4H]_2CoCl_4$ are produced. In the former a $CuCl_3$ group is attached to one N of the ring and H to the opposite N atom, as shown at (a). The bond arrangement around Cu(II) is distorted square planar, the bond angles being $143°$, $134°$, and four of $97°$ (mean). In the Co compound[12] there is simply one H attached to the ring, (b), and the crystal contains discrete tetrahedral $(CoCl_4)^{2-}$ ions. An interesting feature of both structures is the formation of short $N-H\cdots Cl$ bonds (3·20 Å). In the Co compound there are rings with both the boat and approximately the saddle conformations.

(8) AC 1968 **B24** 707
(9) AC 1962 **15** 539
(10) JCS 1961 5471
(11) JCS A 1970 455

(12) JCS A 1970 460

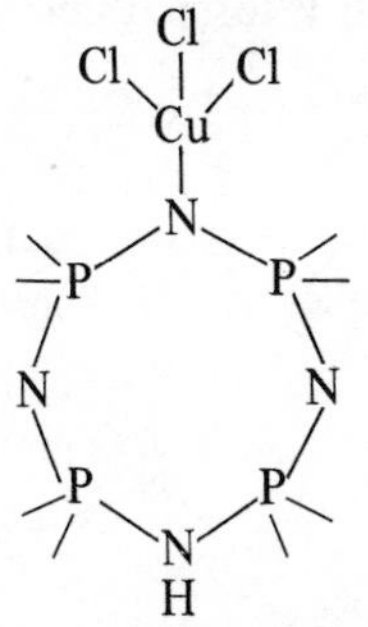

(a) 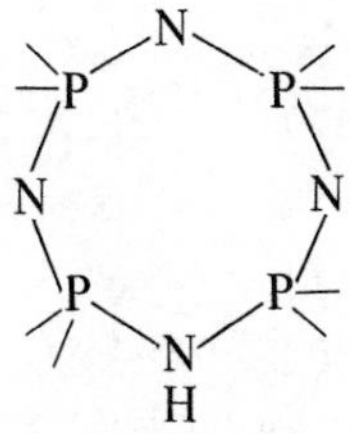 (b)

X-ray studies have also been made of molecules containing 10-, 12-, and 16-membered rings. $(PNCl_2)_5$[13] contains an approximately planar ring with no suggestion of alternating bond type (mean P–N, 1·52 Å). The 12-membered ring in $P_6N_6(NMe_2)_{12}$[14] is highly puckered with all bonds in the ring of length 1·56 Å (*exo* P–N bonds, 1·67 Å) and angles in the ring, $148°$ (at N) and $120°$ (at P). The 16-membered ring in $P_8N_8(OMe)_{16}$[15] consists of two approximately planar 6-segments in two parallel planes joined at a 'step'; the P–N bond length is the same as in the 12-membered ring.

(13) JCS A 1968 2317

(14) AC 1968 **B24** 1423

(15) JCS A 1968 2227

Hydrolysis of the halides yields the corresponding cyclic metaphosphimic acids or their salts, such as $Na_3(PO_2NH)_3 . 4 H_2O$. The carbon and sulphur analogues of these acids are the polymerized forms of cyanic acid and sulphimide:

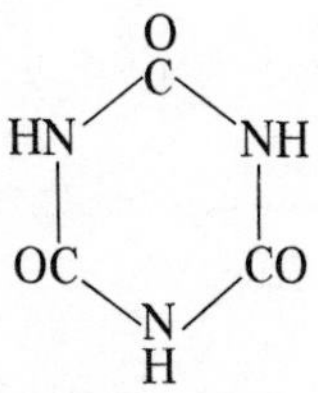

Cyanuric acid 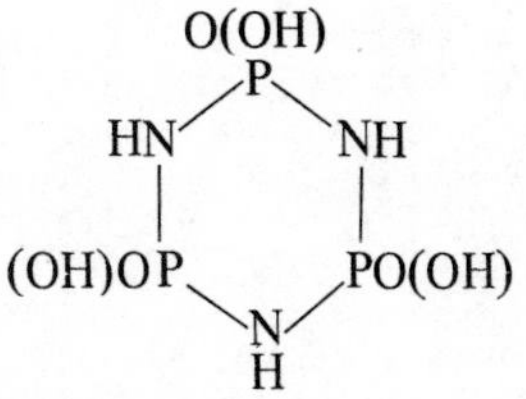   Trimetaphosphimic acid 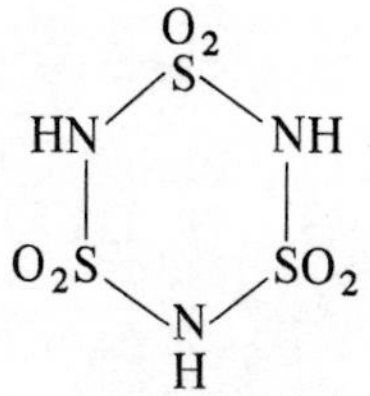   Trisulphimide

699

(16) AC 1965 **19** 596

(17) JCS A 1968 3026

(18) AC 1964 **19** 603
(19) AC 1971 **B27** 740

(20) AC 1969 **B25** 651

(21) IC 1971 **10** 2591

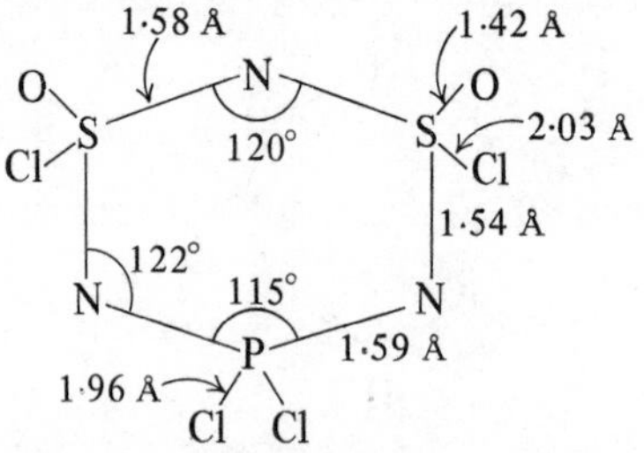

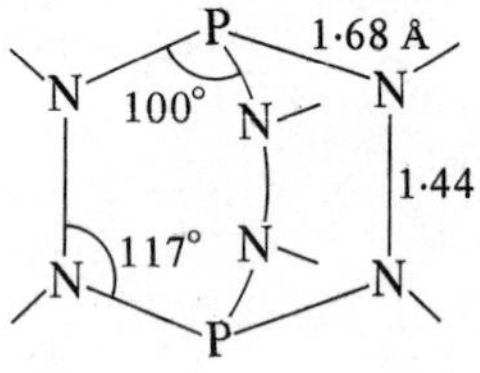

In contrast to the planar 6-ring in the halides and the isothiocyanate the ring of the trimetaphosphimate ion in the sodium salt has the chair conformation (Fig. 19.10(b))[16] and the P—N bond is appreciably longer (1·68 Å) than in $(PNX_2)_4$. It is confirmed that one H atom is attached to each N atom as shown above for the acid. In the fully methylated compound $(NMe)_3(PO_2Me)_3$[17] P—N is 1·66 Å, close to the value in the trimetaphosphimate ion, and the ring has a distorted boat shape. The tetrametaphosphimate ring shows considerable variation in shape like the 8-ring in $(PNX_2)_4$. It has the boat conformation in $(NH_4)_4[P_4N_4H_4O_8] . 2 H_2O$,[18] the chair conformation in the K salt (tetrahydrate),[19] and the saddle (cradle) conformation in the Cs salt (hexahydrate).[19] The P—O bond lengths are close to 1·50 Å and P—N, 1·66–1·68 Å.

An interesting ring containing N, P, and S is that in $N_3S_2PO_2Cl_4$[20] (chair conformation). A multiple ring system in which P forms three single (pyramidal) P—N bonds is the cage-like molecule $P_2N_6(CH_3)_6$.[21]

20

Arsenic, Antimony, and Bismuth

Elementary arsenic, antimony, and bismuth

Arsenic is apparently trimorphic and antimony dimorphic. The yellow (non-metallic) forms are metastable, and are prepared by condensing the vapour at very low temperatures. Arsenic also forms a polymorph isostructural with black P; it is prepared by heating amorphous As at 100–175°C in the presence of Hg.[1] The yellow forms revert to the metallic forms on heating or on exposure to light; they probably consist of tetrameric molecules, but owing to their instability their structures have not been studied. The As_4 molecule in arsenic vapour has the same tetrahedral configuration as the P_4 molecule, with As–As, 2·44 Å. Mass spectrometric studies show that the molecules As_4, Sb_4, and Bi_4 and all combinations of these atoms exist in the vapours from liquid mixtures of the elements.

(1) ZaC 1956 **283** 263

The metallic forms of As, Sb, and Bi are isostructural, with a layer structure (p. 59) in which each atom has three equidistant nearest neighbours, the next set of three neighbours being at a greater distance. The distinction between these two sets of neighbours grows less going from As to Bi, as seen from the following figures:

	3 at	3 at	M-M-M
As	2·51 Å	3·15 Å	97°
Sb[1a]	2·91	3·36	96°
Bi	3·10	3·47	94°

(1a) AC 1963 **16** 451

The structural chemistry of As, Sb, and Bi

The series P, As, Sb, and Bi show a gradation of properties from non-metal to metal. We shall be concerned here chiefly with the stereochemistry of As and Sb, for the Bi analogues of many of the simple molecules containing As and Sb are either much less stable or have not been prepared. Simple molecules containing multiple bonds are not formed by any of these elements. For example, arseno-methane, $(AsCH_3)_n$, is not structurally similar to azomethane, $H_3C–N=N–CH_3$. Arsenomethane exists in two forms, yellow and red. Crystals of the yellow form consist of molecules $(AsCH_3)_5$ having the form of puckered 5-membered rings (mean As bond angle, 102°, As–As, 2·43 Å, and As–C, 1·95 Å).[2] Arseno-benzene, on the other hand, consists of 6-membered rings $[As(C_6H_5)]_6$ which are

(2) JACS 1957 **79** 859

701

(3) AC 1961 **14** 369
(4) JACS 1966 **88** 378
(5) ZaC 1967 **350** 9

(6) JACS 1969 **91** 5631, 5633

chair-shaped, with As–As, 2·46 Å and As bond angle, 91°.[3] The cyclic $(CF_3As)_4$ has been characterized by its i.r. spectrum,[4] and $As_4(NMe)_6$,[5] prepared from $AsCl_3$ and CH_3NH_2, is structurally similar to $N_4(CH_2)_6$. (For the $CoAs_3$ structure, which contains As_4 rings, see p. 216.) Among the few examples of molecules containing As–As bonds which are appreciably shorter than that in As_4 (2·44 Å) are the molecules (a) and (b)[6] which contain tetrahedral nuclei related to the As_4 molecule. An interesting difference between P and As is that only one pure arsenyl

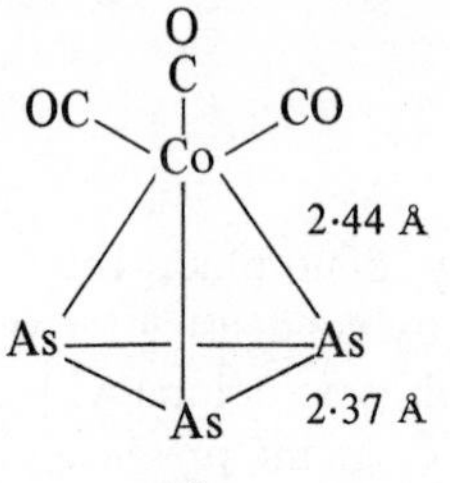

(a) 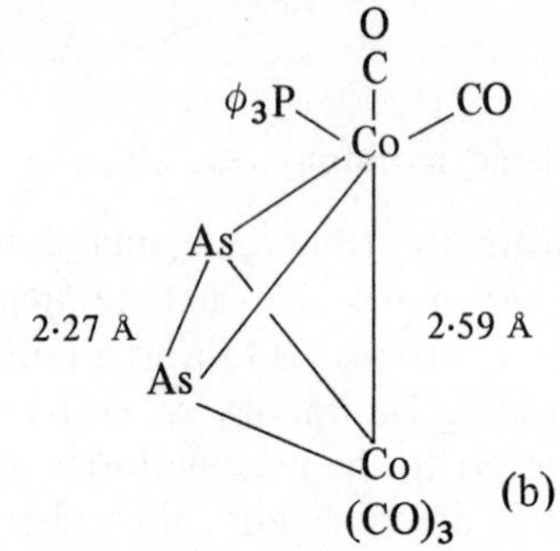 (b)

halide is known; this is $AsOF_3$. All the elements As, Sb, and Bi form a hydride MH_3. In the preparation of AsH_3 and SbH_3 (by reduction of the trichlorides in aqueous HCl by sodium hydroborate) As_2H_4 and Sb_2H_4 are formed as secondary products, but no higher homologues have been observed.

In addition to the usual formal oxidation states III and V Bi appears to form metal–metal complexes in molten halides and in crystalline compounds obtained from such systems. The 'sub-halide' Bi_6Cl_7 is noted later in this Chapter; it contains $BiCl_5^{2-}$, $Bi_2Cl_8^{2-}$, and Bi_9^{5+} ions. From a solution of Bi in molten $NaAlCl_4$ the compounds Bi_4AlCl_4 and $Bi_5(AlCl_4)_3$ have been obtained, possibly containing ions Bi_5^{3+} and Bi_8^{2+}.[7] In this connection we may also mention the $Bi_6(OH)_{12}^{6+}$ ion formed when solutions of Bi_2O_3 in $HClO_4$ are hydrolysed. There is presumably some interaction between the Bi atoms in the octahedral Bi_6 nucleus, although the Bi–Bi distance (3·7 Å) is appreciably greater than in elementary Bi (3·10Å).[8]

(7) IC 1968 **7** 198

(8) IC 1968 **7** 183

TABLE 20.1

The stereochemistry of As, Sb, *and* Bi

Total number of electron pairs	Bond type	Lone pairs	Bond arrangement	Proved for
4	sp^3	0	Tetrahedral	As^+
		1	Pyramidal	As, Sb, Bi
5	sp^3d	0	Trigonal bipyramidal	Sb
		1	See text	Sb
6	sp^3d^2	0	Octahedral	As^-, Sb^-, (Bi)
		1	Square pyramidal	Sb
7		1	Octahedral	Sb

The bond arrangements of Table 20.1 have been established by structural studies of simple ions or covalent molecules for the elements indicated. The presence of a lone pair implies M^{III}; otherwise M^V. The paucity of information about Bi is due to the more metallic character of this element, which does not form many of the simple covalent molecules formed by As and Sb. The octahedral bonds formed by Bi in, for example, the crystalline pentafluoride have considerable ionic character. For 7-coordination (sp^3d^3) a pentagonal bipyramidal bond arrangement would be expected. We noted in Chapter 7 that $SbBr_6^{3-}$ should form a distorted octahedron since there is a valence group of 14 electrons including one lone pair. In fact the ion shows very little distortion from a regular octahedron. (see p. 706).

Molecules MX_3: *valence group* 2, 6

The trigonal pyramidal shape of many molecules MX_3 formed by As, Sb, and Bi has been demonstrated by e.d. or m.w. studies of the vapours or by X-ray studies of the solids (Table 20.2).

TABLE 20.2

Structural data for molecules MX_3 and halides MX_3

Molecule	M—X (Å)	X—M—X	Method	Reference
AsH_3	1·519	91·83°	m.w.	PR 1955 **97** 684
$As(CH_3)_3$	1·96	96°	m.w.	SA 1959 **15** 473
$As(CF_3)_3$	2·053	100°	e.d.	TFS 1954 **50** 463
AsF_3	1·706	96·2°	m.w.	
$AsCl_3$	2·161	98·7°	m.w.	BCSJ 1966 **39** 71
$AsBr_3$	2;33	99·7°	e.d.	IC 1970 **9** 805
AsI_3	2·55	100·2°	e.d.	
	2·56	102°	X	ZK 1965 **121** 81
$As(SiH_3)_3$	2·355	94°	e.d.	JCS A 1968 3006
$As(CH_3)I_2$	2·54	104°	X	AC 1963 **16** 922
$AsBr(C_6H_5)_2$	2·40	C—As—C 106°	X	JCS 1962 2567
		C—As—Br 95°		
$As(CN)_3$		90·5°	X	AC 1966 **20** 777
$As(CN)_2CH_3$		94°	X	
SbF_3	1·92	87·3°	X	JCS A 1970 2751
SbH_3	1·707	91·3°	m.w.	PR 1955 **97** 680
$Sb(CF_3)_3$	2·202	100·0°	e.d.	TFS 1954 **50** 463
$SbCl_3$	2·325	99·5°	m.w.	JCP 1954 **22** 86
	2·36	95·2°	X	JINC 1956 **2** 345
$SbBr_3$	2·51	97°	e.d.	AC 1950 **3** 46
α (cryst.)	2·50	95°	X	JCS 1964 4162
β	2·49	95°	X	JCS 1962 2218
SbI_3	2·719	99·1°	e.d.	ACSc 1963 **17** 2573
	See text		X	ZK 1966 **123** 67
BiF_3	See p.		X	ACSc 1955 **9** 1206, 1209
$BiCl_3$	2·50	84°, 94° (2)	X	AC 1971 **B27** 2298
$BiCl_3$	2·48	100°	e.d.	TFS 1940 **36** 681
$BiBr_3$	2·63	100°	e.d.	
BiI_3	See text		X	ZK 1966 **123** 67

For the structures of crystalline SbX_3 and BiX_3 see p. 706.

Tetrahedral ions MX_4^+: *valence group* 8

There are no arsenic or antimony analogues of the simple ammonium or phosphonium halides. (Of the latter, PH_4Cl, PH_4Br, and PH_4I are known, the latter forming brilliant cubes which can be sublimed like an ammonium salt.) The tetrahedral configuration of substituted arsonium ions has been demonstrated in crystalline $[As(CH_3)_4]Br$[1] and $[As(C_6H_5)_4]I_3$.

(1) JCS 1963 4051

There are many compounds in which As^V and Sb^V form tetrahedral bonds (valence group 8) as, for example, GaAs, GaSb, InAs, and InSb which, like the corresponding phosphides, crystallize with the zinc-blende structure. Tetrahedral bonds are formed by As^V in the AsO_4^{3-} ion in orthoarsenates (and arsine oxides) but note the quite different behaviour of Sb^V in its oxy-compounds (see later).

Ions MX_4^-: *valence group* 2, 8

The formation of 4 bonds by M^{III} would result in the valence group 2, 8, for which there are two closely related bond arrangements, (a) and (b). The structure of a

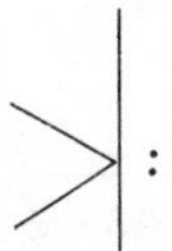

(a) 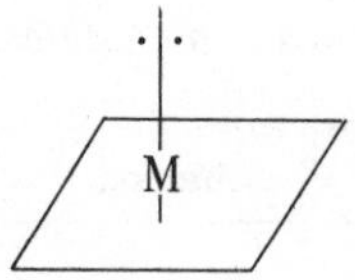 (b) 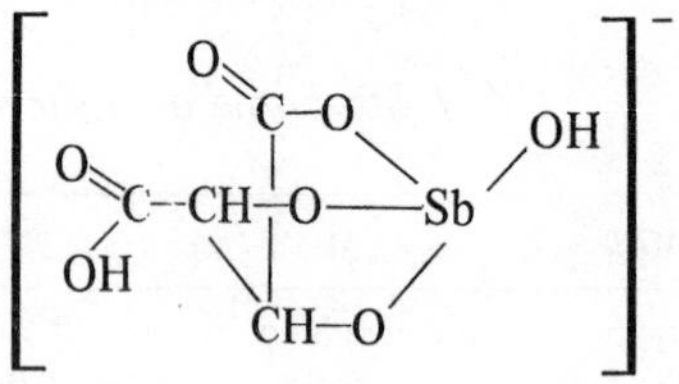 (c)

simple ion MX_4^- is not yet known; finite ions $(MX_4)^-$ containing As, Sb, or Bi probably do not exist. Tetrachloroarsenites (for example, $(CH_3)_4N.AsCl_4$) have been prepared but they presumably do not contain finite $(AsCl_4)^-$ ions but chain ions analogous to $(SbCl_4)_n^{n-}$ etc. described on p. 708.

An example of this unusual valence group (2, 8) is found for Sb(III) in antimonyl tartrates. In the potassium salt, $KSbC_4H_4O_7.\frac{1}{2}H_2O$,[1] the bond arrangement around Sb appears to be close to (a), while in the ammonium salt[2] it is closer to (b). The ion is shown at (c).

(1) AC 1965 **19** 197
(2) DAN 1964 **155** 545

Pentahalides and molecules MX_5: *valence group* 10

The following pentahalides are known:

$$AsF_5 \quad SbF_5 \quad BiF_5$$
$$SbCl_5$$

The trigonal bipyramidal configuration of AsF_5 and $SbCl_5$ molecules in the vapour state has been demonstrated by e.d., and the same stereochemistry has been confirmed for $SbCl_5$ in the crystalline state (Table 20.3). Other molecules shown to have this shape in the crystalline state include $As(C_6H_5)_5$[1] $Sb(C_6H_5)_4OH$,[2] $Sb(C_6H_5)_4OCH_3$ and $Sb(C_6H_5)_3(OCH_3)_2$,[3] $Sb(CH_3)_3Cl_2$,[4] and $Bi(C_6H_5)_3Cl_2$.[5] However, in contrast to the analogous P and As compounds $Sb(C_6H_5)_5$[6] has a tetragonal pyramidal configuration with Sb approximately

(1) JCS 1964 2206
(2) JACS 1969 **91** 297
(3) JACS 1968 **90** 1718
(4) ZK 1938 **99** 367
(5) JCS A 1968 2539
(6) JACS 1968 **90** 6675

TABLE 20.3

Structural data for pentahalides and adducts

Molecule	M—X (Å)	Sb—O (N) (Å)	Reference
AsF_5	1·711 (axial)	–	IC 1970 **9** 805
	1·656 (equat.)	–	
$SbCl_5$ (cryst.)	2·34 (axial)		JACS 1959 811
	2·29 (equat.)		
(liquid)	–		JPC 1958 **62** 364
(vapour)	2·43 (axial)		
	2·31 (equat.)		APL 1940 **14** 78
$SbF_5 . SO_2$	1·85	2·13	JACS 1968 **90** 1358
$SbCl_5 . POCl_3$	2·33	2·17	ACSc 1963 **17** 353
$SbCl_5 . OC(H) . N(CH_3)_2$	2·34	2·05	AC 1966 **20** 749

0·5 Å above the base of the pyramid and an angle of 102° between the axial and equatorial bonds. At present this remains the only example of a Group V molecule of this kind (10-electron valence group) which is not trigonal bipyramidal.

BiF_5 has the α-UF_5 structure (p. 994) in which there is octahedral coordination of the metal atom.[7] (An early e.d. study established the trigonal bipyramidal configuration of the molecules $NbCl_5$, $NbBr_5$, $TaCl_5$, and $TaBr_5$.[8]

(7) JACS 1959 **81** 6375
(8) TFS 1940 **36** 668

Formation of octahedral bonds by As^V, Sb^V, *and* Bi^V: *valence group* 12
Octahedral bonds are formed by As in $KAsF_6$ and $NO(AsF_6)$, presumably also in salts such as $Na(AsF_5OH)$, and possibly in the trimeric ion (a) formed by heating this salt. Resolution of the tricatechol derivative (b) into its optical antimers is evidence for the octahedral arrangement of the six As—O bonds. The octahedral

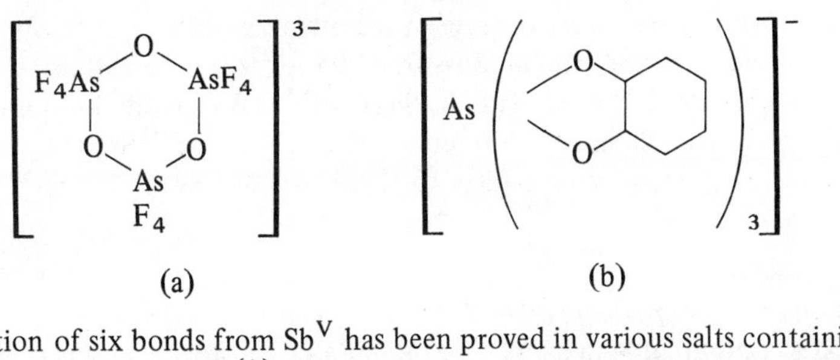

(a)   (b)

configuration of six bonds from Sb^V has been proved in various salts containing the ions SbF_6^- (Sb—F, 1·88 Å)[1] and $SbCl_6^-$, and also in the ions $Sb(OH)_6^-$ and $SbF_4(OH)_2^-$ to which we refer later. For the $(Sb_2F_{11})^-$ ion in $(XeF)(Sb_2F_{11})$ and $(BrF_4)(Sb_2F_{11})$ see pp. 321 and 335 respectively, and for $(Br_2)(Sb_3F_{16})$ see p. 334. No metallic salts containing the BiF_6^- ion appear to be known; the octahedral coordination of Bi in BiF_5 has already been noted.

(1) AC 1962 **15** 1098

The valence group 10 in $SbCl_5$ is readily expanded to the octahedral group 12 not only in the $SbCl_6^-$ ion but also in numerous adducts in which an O or N atom occupies the sixth bond position. Crystalline $SbCl_5 . POCl_3$ consists of molecules

705

The diagrams (a) and (b) show molecular structures with bond lengths:

(a): Sb center with Cl atoms, bond lengths 2.17 Å, 2.33 Å, angle 147°; O at 1.52 Å; P–Cl with Cl at 1.98 Å.

(b): Sb center with F atoms, bond length 2.13 Å, angle 139°; S with O at 1.45 Å and O at 1.38 Å, angle 119°.

(a)   (b)

(a), and adducts are also formed with $PO(CH_3)_3$, $(CH_3)_2SO_2$, etc. (Table 20.3). The molecule $SbF_5 . SO_2$ is shown at (b). In $SbCl_5 . N_4S_4$ the sixth octahedral bond is formed to a N atom of the cyclic N_4S_4 molecule (Sb–N, 2·17Å, Sb–Cl, 2·39 Å). In the compound with ICl_3, $ISbCl_8$ (p. 332) there are recognizable $SbCl_6^-$ ions, though the bond lengths suggest that the system may be intermediate between the states $(ICl_2)^+(SbCl_6)^-$ and $(ICl_4)^-(SbCl_4)^+$.

Formation of octahedral bonds by Sb^{III}: *valence group 2, 12*

We noted in Chapter 16 that not only does the $TeCl_6$ molecule, in which Te has the valence group 12, have the expected octahedral configuration but also the ion $TeCl_6^{2-}$ (valence group 2, 12) has a regular octahedral configuration in salts such as K_2TeCl_6 (cubic K_2PtCl_6 structure). Ions $Sb^{III}X_6^{3-}$ are isoelectronic with the corresponding $Te^{IV}X_6^{2-}$ ions. The jet-black salt with the empirical composition $(NH_4)_2SbBr_6$ is in fact $(NH_4)_4(Sb^{III}Br_6)(Sb^VBr_6)$.[2] Its structure is a super-structure of the K_2PtCl_6 type, the octahedral ions containing Sb^{III} and Sb^V alternating as shown in Fig. 20.1 so that the unit cell has twice the c dimension of the simple K_2PtCl_6 structure. In spite of the presence of the lone pair on Sb^{III} the $SbBr_6^{3-}$ octahedron is *un*distorted (Sb–Br, 2·795 Å); there is slight distortion of the $SbBr_6^-$ ions (Sb–Br, 2·564 Å). Salts $R_6Sb_4Br_{24}$[3] (R = pyridinium) also contain Sb^{III} and Sb^V, and should be formulated $(C_5H_5NH)_6(Sb^{III}Br_6)(Sb^VBr_6)_3$. The Sb–Br bond lengths are very similar to those just quoted.

Formation of square pyramidal bonds by Sb^{III} and Bi^{III}: *valence group 2, 10*

If the valence group of M in a molecule or ion MX_5 consists of six electron pairs the lone pair is expected to occupy the sixth octahedral position, giving a square pyramidal arrangement of bonds. Some complex halides of Sb^{III} are of interest in this connection (p. 708).

The crystalline trihalides of As, Sb, and Bi

The trigonal pyramidal shape of the molecules in the vapours has already been noted; Table 20.2 also includes literature references to those halides which have

(2) JACS 1966 **88** 616

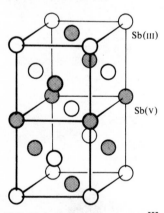

FIG. 20.1. Arrangement of Sb^{III} and Sb^V atoms in $(NH_4)_4(Sb^{III}Br_6)(Sb^VBr_6)$.

(3) IC 1971 **10** 701

706

been studied in the crystalline state, namely:

$$AsI_3$$
$$SbF_3 \quad SbCl_3 \quad SbBr_3 \quad SbI_3$$
$$BiF_3 \quad BiCl_3 \quad \quad \quad BiI_3$$

In crystals of AsI_3 and all the Sb trihalides there are 3 closer and 3 more distant neighbours, and it is interesting that the distance to the latter in SbI_3 is *less* than the corresponding distances in $SbCl_3$, in both polymorphs of $SbBr_3$, or in AsI_3 (see below):

		SbF_3	$SbCl_3$	α-$SbBr_3$	β-$SbBr_3$	SbI_3
Sb–X in gas molecule		?	2·33	2·51	2·51	2·72 Å
Crystal	Sb–3 X	1·92	2·36	2·50	2·49	2·87 Å
	Angle X—Sb—X	87°	95°	96°	95°	96°
	Sb–3 X	2·61	≥ 3·5	≥ 3·75	≥ 3·6	3·32 Å

The Bi halides are quite different. The ionic (YF_3) structure of BiF_3 is noted on p. 357.

The chlorides $SbCl_3$ and $BiCl_3$ and the two polymorphs of $SbBr_3$ form a group of related structures in which there are well defined molecules MX_3. However, $BiCl_3$ is abnormal in having all its cell dimensions smaller than those of $SbCl_3$ in spite of the longer M–X bonds (Sb–Cl, 2·37 Å, Bi–Cl, 2·50 Å). There appears to be a slight but real asymmetry in the molecule, and *five* more Cl neighbours (at 3·22–3·45 Å) complete a bicapped trigonal prism coordination group. This tendency towards a structure of 8-coordination results in the denser packing in this crystal, and is somewhat reminiscent of the structure of $PbCl_2$, another 'inert pair' halide.

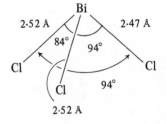

In the iodides, MI_3, M occupies octahedral interstices between h.c.p. I atoms, but M is progressively further from the centre of the I_6 octahedron in the series BiI_3, SbI_3, and AsI_3:

$$Bi{-}6\ I \quad 3{\cdot}1\ Å \qquad Sb\begin{cases} 3\ I & 2{\cdot}87\ Å \\ 3\ I & 3{\cdot}32 \end{cases} \qquad As\begin{cases} 3\ I & 2{\cdot}56\ Å \\ 3\ I & 3{\cdot}50 \end{cases}$$

M–I in MI_3
molecule
in vapour

?   2·72   2·55

These figures suggest that SbI_3 is intermediate between the (polarized) octahedral layer structure of BiI_3 and the molecular structure of AsI_3.

Complex halides of trivalent Sb *and* Bi

Complex fluorides of the following types are formed with alkali metal or similar ions: $MSbF_4$, M_2SbF_5, MSb_2F_7, MSb_3F_{10}, and MSb_4F_{13}, but apparently salts

707

(1) AK 1951 **3** 461
(2) AK 1952 **4** 175

(3a) AK 1953 **6** 77

(3b) IC 1971 **10** 1757

(3c) IC 1971 **10** 2793

(3d) AK 1951 **3** 17

M_3SbF_6 are not known. K_2SbF_5[1] contains SbF_5^{2-} ions with the same tetragonal pyramidal configuration as $SbCl_5^{2-}$ in $(NH_4)_2SbCl_5$ (see below). In $KSbF_4$[2] similar SbF_5 groups are joined by sharing two F atoms to form cyclic $Sb_4F_{16}^{4-}$ ions, while in $NaSbF_4$[3a] there are apparently infinite chains formed from SbF_5 groups linked in a similar way. In these complexes the Sb atom lies about 0.2–0.3 Å *below* the square base of the pyramid, and this is also true of the 5-coordinated Sb atoms in Sb_2S_3 (see later) and certain complex sulphides (e.g. $FeSb_2S_4$). The type of Sb–F complex in salts MSb_2F_7 depends on the nature of the cation. In KSb_2F_7[3b] there are distorted trigonal bipyramidal SbF_4^- ions, (a), and SbF_3 molecules (Sb–F, 1.94 Å) very similar to those in crystalline SbF_3. However, whereas in crystalline SbF_3 3 F at 2.61 Å complete a distorted octahedral group (or capped octahedron if the lone pair is included) in KSb_2F_7 there are only 2 additional (F_a) neighbours (at 2.41 Å and 2.57 Å) which, *together with the lone pair*, complete a distorted octahedron. The structure of $CsSb_2F_7$[3c] is quite different. There are well defined $Sb_2F_7^-$ ions, (b), formed from two distorted trigonal bipyramidal SbF_4 groups sharing axial F atoms, with long bridge bonds. (The next shortest Sb–F distance is 2.77 Å). Note that the anion in KSb_2F_7 could be described as an infinite chain of $SbF_2(F_a)_2$(2e) groups sharing axial F atoms with pseudo-octahedral $SbF_3(F_a)_2$(2e) groups but with longer (unsymmetrical bridges of 2.41 Å and 2.57 Å. The nature of the complex ion in KSb_4F_{13} is not clear.[3d]

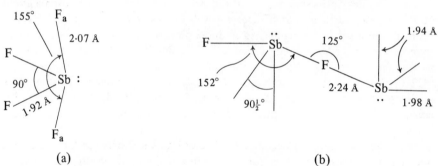

(a)　　　　　　　　　　　　　　　　(b)

(4) JCS A 1970 1356

(5) JCS A 1971 298

Structural studies have been made of complex chlorides or bromides containing ions of the following types: $(SbX_4)^-$, $(SbX_5)^{2-}$, $(SbX_6)^{3-}$, and the enneahalide ion $(Sb_2X_9)^{3-}$. The anion in $(SbCl_4)(C_5H_5NH)$[4] is an inifinite chain of very distorted octahedra sharing 'skew' edges (p. 174). All Cl–Sb–Cl angles are close to 90°, but the Sb–Cl bond lengths are: 2 of 2.38 Å, 2 of 2.64 Å, and 2 of 3.12 Å, the four latter being involved in the unsymmetrical bridges between the $SbCl_6$ groups, as shown in Fig. 20.2(a). In $(NH_4)_2SbCl_5$[5] the NH_4^+ and Cl^- ions form a distorted closest packing with vacant Cl^- sites in every third layer, and the Sb atoms occupy what would be octahedral holes in a normal closest packing. Owing to the absence of some Cl^- ions the nearest neighbours of Sb are 5 Cl arranged at five of the vertices of an octahedron, the bond lengths being: axial, 2.36 Å, equatorial 2.58 Å (two), and 2.69 Å (two); the lone pair presumably occupies the sixth bond position, Fig. 20.2(b).

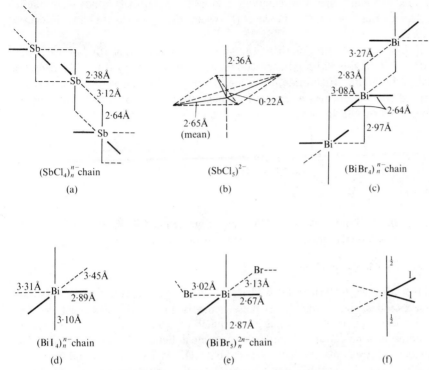

FIG. 20.2. Environments of Sb or Bi in complex halides (see text).

The $(SbBr_6)^{3-}$ ion has been studied in halides which also contain the $(SbBr_6)^-$ ion, as noted in an earlier section. The enneabromide ion occurs in the pyridinium salt $(C_5H_5NH)_5(Sb_2Br_9)Br_2$.[6a] It consists of two octahedral $SbBr_6$ groups sharing a face, with Sb—Br: terminal, 2·63 Å, and bridging, 3·00 Å.

Complex fluorides of Bi are, like BiF_3, ionic compounds. For example, NH_4BiF_4[6b] has an ionic structure in which the infinite $(BiF_4)_n^{n-}$ layer is built of BiF_9 coordination groups (Bi—F, 2·19–2·86 Å) of the same type as in orthorhombic BiF_3 (p. 357). The ions $(SbI_4)_n^{n-}$, $(BiBr_4)_n^{n-}$, and $(BiI_4)_n^{n-}$ in the 2-picolinium salts[7] consist of infinite 'skew' octahedral chains, Fig. 20.2(c) and (d), very similar in structure to the $(SbCl_4)^-$ chains shown at (a), with three pairs of bonds of similar length. In the bispiperidinium salt $(C_5H_{10}NH_2)_2BiBr_5$,[8] with which the Sb compound is isostructural, there are chains of octahedral MX_6 groups sharing *cis* vertices, and here also there is the same pattern of bond lengths, Fig. 20.2(e). If we regard the Bi—Br bond in the $BiBr_3$ molecule (2·63 Å) as a single bond then the two next shortest bonds in (c) and (e) correspond to bond order $\frac{1}{2}$, using Pauling's equation, $d_n = d_1 - 0.6 \log n$. Similarly, if the six octahedral bonds in crystalline BiI_3 (3·07 Å) are regarded as having bond order $\frac{1}{2}$ then the shortest bonds in (d) are single bonds. Thus in each case the total bond order is close to 3. It may be noted that the four stronger bonds could alternatively be regarded as distorted trigonal

(6a) JCS A 1970 1359

(6b) ACSc 1964 **18** 1554

(7) JPC 1967 **71** 3531

(8) JPC 1968 **72** 532

709

bipyramidal bonds, when the remaining two weak (equatorial) bonds would be in the plane of the lone pair as shown at (f), where the approximate bond orders are indicated.

An example of Bi forming tetragonal pyramidal bonds is provided by $Bi_{12}Cl_{14}$.[9] In this extraordinary compound, which is formed by dissolving the metal in the molten trichloride, there are groups of 9 Bi atoms (at the vertices of a tricapped trigonal prism), tetragonal pyramidal $BiCl_5^{2-}$ groups, and $Bi_2Cl_8^{2-}$ groups formed from two $BiCl_5^{2-}$ ions sharing two Cl atoms:

(9) IC 1963 **2** 979

The structural formula of this halide may be written $Bi_9^{5+}(BiCl_5)_2^{2-}(Bi_2Cl_8)_{\frac{1}{2}}^{2-}$ to indicate the relative numbers of the three kinds of structural unit.

The oxygen chemistry of trivalent As, Sb, and Bi

In this section we shall summarize what is known of the structures of the trioxides, complex oxides, and oxyhalides of these elements. In the trioxides and in the one complex oxide of Sb^{III} which has been studied the stereochemistry of the Group VB atoms is simple; they form three pyramidal bonds like P^{III} in P_4O_6. The behaviour of Sb in its oxyhalides and Sb_3O_6OH is less simple, and moreover different from that of Bi. For this reason the oxyhalides of Sb and Bi are considered separately. Structurally related to the oxyhalides of bismuth are a number of complex oxides which will also be mentioned.

The trioxides of As, Sb, *and* Bi

Each of the trioxides exists in at least two polymorphic modifications (Table 20.4). Vaporization of As and Sb trioxides gives molecules As_4O_6 and Sb_4O_6 which break down to the simpler As_2O_3 and Sb_2O_3 molecules only at high temperatures. The

TABLE 20.4

Polymorphic forms of As_2O_3, Sb_2O_3, *and* Bi_2O_3

	As_2O_3	Sb_2O_3	Bi_2O_3
Low-temperature form	Monoclinic (claudetite)	Orthorhombic (valentinite)	(α) Monoclinic
High-temperature form	Cubic (arsenolite)	Cubic (senarmontite)	(β) Tetragonal
			(γ) Cubic?
Transition temperature	110°C	606°C	710°C ($\alpha - \beta$)

(1) AKMG 1942 **15B** No. 22

structure of As_4O_6 in the vapour state has been shown to be similar to that of P_4O_6 (see Table 19.5, p. 685). In the cubic forms of As_4O_6 and Sb_4O_6[1] there exist molecules of the same type as in the vapour of As_4O_6.

710

The mineral claudetite, As_2O_3,[2] has the simplest possible type of layer structure for a compound with formula A_2X_3, namely, a hexagonal net of As atoms joined through O atoms. The layers are puckered because the bond angles O–As–O are approximately 100° and the O bond angles, 123 (3)°. The As–O bond length is 1·80 (0·02) Å, as in arsenolite. (Compare the structure of orpiment, As_2S_3, p. 723.)

(2) AM 1951 **36** 833; ZaC 1951 **266** 293

The second form of antimony trioxide, valentinite, has a different type of structure.[3] Instead of finite molecules Sb_4O_6 there are infinite double chains of the type shown below, with Sb–O, 2·00 Å, O bond angles of 116° and 132°, and Sb bond angles of 81°, 93°, and 99°.

(3) ZK 1937 **98** 1

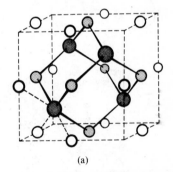

A rather complex (and unique) structure has been proposed[4] for the monoclinic form of Bi_2O_3, in which there is irregular coordination of Bi atoms of two kinds in a 3D framework of Bi and O atoms. The Bi^I atoms have 5 O nearest neighbours at five of the vertices of a distorted octahedron at distances from 2·08 to 2·63 Å (and a sixth at 3·25 Å), while Bi^{II} has an irregular octahedral arrangement of 6 O at distances from 2·14 to 2·80 Å. Three of these are appreciably closer (2·14–2·29 Å) than the other three (2·48–2·80 Å). There still appears to be some doubt about the number and structures of the high-temperature forms of this oxide. Some of the confusion is undoubtedly due to the fact that this compound is very easily contaminated. For example, fusion for a short time in a porcelain crucible gives a body-centred phase with the composition $SiBi_{12}O_{20}$; there are isostructural compounds of Ge and Ti (see later). If Bi_2O_3 is fused for a long time in a porcelain crucible a simple cubic phase is obtained which probably has the structure shown in Fig. 20.3(a) stabilized by a small amount of impurity. This structure is similar to that of Zn_3P_2 (p. 670) and it has been claimed that this is in fact the structure of high-temperature (pure) Bi_2O_3.[5] Each Bi atom has 6 O neighbours (at 2·40 Å) arranged at six of the eight vertices of a cube, those at two diagonally opposite vertices being missing. This structure is simply related geometrically to the fluorite structure which is shown in the same orientation in Fig. 20.3(b). Removal of one-quarter of the anions (those shown as dotted circles) from the unit cell of the latter leaves the arrangement shown in (a). If the metal atoms and the shaded O atoms are drawn in a little towards the centre of the cube the structure becomes that of As_4O_6 or Sb_4O_6 containing discrete molecules. The remaining O atoms then form parts of neighbouring molecules. The tetragonal (pseudo-cubic) phase has $a = 10·93$ and $c = 5·62$ Å, corresponding to four unit cells of the cubic structure ($a = 5·52$ Å), with atomic positions slightly different from those of the idealized structure of Fig. 20.3(a).

(4) ACSc 1970 **24** 384

(5) ZaC 1962 **318** 176

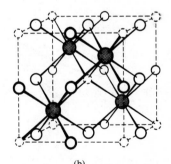

(a)

(b)

FIG. 20.3. The structures of (a) Bi_2O_3, and (b) CaF_2. Metal atoms are shaded.

Arsenic, Antimony, and Bismuth

In the $M^{IV}Bi_{12}O_{20}$ phase (ref. Table 20.5) Bi has a very unusual coordination by oxygen. In many Bi^{III} oxy-compounds the metal atom has 5 or 6 neighbours at distances from about 2·1 Å to 2·7 Å (mean close to 2·4 Å) and then a small number

TABLE 20.5

Coordination of Bi^{III} in some oxy-compounds

Compound	Total c.n.	Bi to 5 or 6 O atoms	Further O to	Mean Bi to 5 or 6 O	Reference
$Bi_4Si_3O_{12}$	9	2·15–2·62 Å	3·55 Å	2·39 Å	ZK 1966 **123** 73
Bi_2GeO_5	7	2·13–2·66	3·18	2·35	ACSc 1964 **18** 1555
$Bi(OH)CrO_4$ (monoclinic)	9	2·23–2·58	3·33	2·38 ⎫	ACSc 1964 **18** 1937
$Bi(OH)CrO_4$ (orthorhombic)	9	2·19–2·68	3·09	2·38 ⎬	
$Bi_2O_2SO_4 . H_2O$	10	{2·22–2·57 / 2·10–2·59}	3·45 / 3·35	2·38 / 2·32 ⎫	ACSc 1964 **18** 2375
$Bi(OH)SeO_4 . H_2O$	9	2·19–2·64	3·16	2·40 ⎭	JCP 1967 **47** 4034
$Bi_{12}GeO_{20}$	7	2·08–2·64	3·17	2·36	

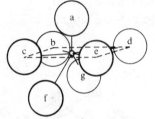

FIG. 20.4. Coordination of Bi in $GeBi_{12}O_{20}$.

of additional neighbours. In some cases, for example, $GeBi_{12}O_{20}$, the coordination group consists of three very close O atoms, a, b, and c, (Fig. 20.4) at 2·08, 2·22, and 2·23 Å, arranged pyramidally, then d and e at the same distance (2·63 Å) and finally f and g at 3·08 and 3·17 Å respectively. Regarded as a group of 5 neighbours the atoms a–e form a square pyramidal group, with mean Bi–O 2·36 Å as in many other B^{III} oxy-compounds (see Table 20.5).

Meta-arsenites

Incorrect formulae have been given to some of these compounds. For example, a salt sometimes formulated as Na_2HAsO_3 is actually the polymeta-arsenite $(NaAsO_2)_n$, and an X-ray study shows that it consists of chains of pyramidal AsO_3 groups held together by Na^+ ions.[1]

(1) AC 1958 **11** 742

(2) AKMG 1943 **17B** No. 5

(3) AK 1955 8 245

Complex oxides of trivalent As *and* Sb

Little is known of the compounds of this group, structural studies having been made only of $ZnSb_2O_4$,[2] with which a number of other complex oxides are isostructural. This compound may be prepared by heating Sb_2O_3 with ZnO *in vacuo* to 500°C. It has the same structure as Pb_3O_4 (p. 462), Sb replacing Pb^{II} and Zn, Pb^{IV}. The Zn atoms are octahedrally coordinated by 6 O, and assuming Zn–O 2·05 Å, Sb–O was determined as 2·01 (2) and 1·87 Å (1). The Sb bond angles are: two of of 101° and one of 95° (pyramidal). The corresponding compounds of Mg, Mn^{II}, Fe^{II}, Co^{II}, and Ni^{II}, and also $NiAs_2O_4$, are all isostructural with $ZnSb_2O_4$.

Some crystalline compounds formed by As_2O_3 with alkali and ammonium halides, for example, $NH_4Cl . As_2O_3 . \frac{1}{2} H_2O$, $KCl . 2 As_2O_3$, etc., are closely related structurally to arsenious oxide. In the former,[3] there are hexagonal As_2O_3 sheets, with all O atoms directed to one side of the sheet, arranged in pairs with the

O atoms turned toward each other and with water molecules enclosed between the layers. Between these pairs of As_2O_3 layers there are NH_4^+ and Cl^- ions.

Complex oxides of trivalent Bi

Bismuth forms with other metallic oxides an extraordinary variety of complex oxides, in some of which O can be replaced by F.

The systems Ca(Sr, Ba, Cd, Pb)O–Bi_2O_3.   Studies have been made of some of the phases formed when Bi_2O_3 is fused with one of the oxides CaO.[1] SrO,[1] BaO,[1] CdO,[2] or PbO.[3] The phases studied show variation in composition over rather wide ranges. For example, in the system BaO–Bi_2O_3 there is a rhombohedral phase ($x = 0 \cdot 10$–$0\cdot22$) and a tetragonal phase ($x = 0\cdot22$–$0\cdot50$), where x is the atomic fraction of Ba in $Ba_{2x}Bi_{2-2x}O_{3-x}$. Structures have been proposed for a number of such phases, in which it is concluded that the metal content of the unit cell is constant but the oxygen content variable.

X-ray studies have also been made of a series of complex oxides with tetragonal or pseudotetragonal symmetry which all have the *a* dimension of the unit cell close to 3·8 Å, and are based on sequences of Bi_2O_2 layers of the same type as in Fig. 20.6, interleaved with portions of perovskite-like structure. They are summarized in Table 20.5. and illustrated in Fig. 20.5. The simplest member of the family is at present represented only by compounds in which part of the O is replaced by F.

(1) AKMG 1943 **16A** No. 17

(2) ZPC 1941 **B49** 27
(3) ZK 1939 **101** 483

TABLE 20.6

Complex oxides and oxyhalides of bismuth

Number of perovskite layers	Formula	c	Reference
		(Å)	
1	Bi_2NbO_5F, Bi_2TaO_5F, $Bi_2TiO_4F_2$	~16·5	AK 1952 **5** 39
2	$PbBi_2Nb_2O_9$, Bi_3NbTiO_9	25·5	AK 1949 **1** 463
3	$Bi_4Ti_3O_{12}$	32·8	AK 1949 **1** 499
4	$BaBi_4Ti_4O_{15}$	41·7	AK 1950 **2** 519

From the structural standpoint these compounds are closely related to the complex bismuth oxyhalides (p. 716) in which the square Bi_2O_2 layers are interleaved with halogen layers. A great variety of compounds can be prepared, for example, $(MBi)_6R_4O_{18}$, in which M can be Na, K, Ca, Sr, Ba, or Pb, and R is Ti, Nb, or Ta. Note that compounds of this family such as Bi_2TaO_5F are quite unrelated to those described on p. 720, for example, $BiTa_2O_6F$.

The oxyhalides of trivalent Sb

Structural data are available for only one oxyfluoride. Four forms of SbOF have been prepared thermally from SbF_3 and Sb_2O_3. In one form the coordination group around Sb^{III} consists of 2 O and 2 F at four of the vertices of a trigonal

713

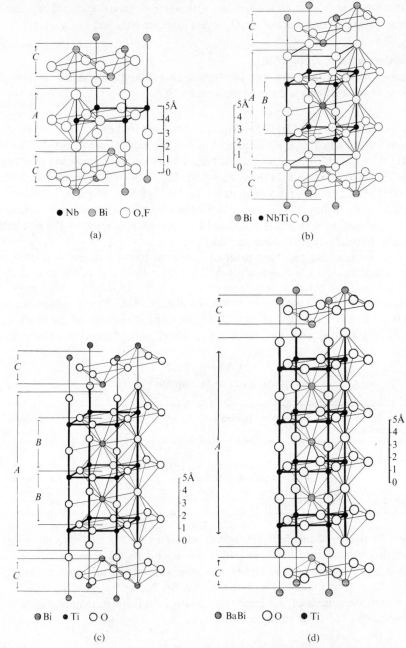

FIG. 20.5. The crystal structures of (a) Bi_2NbO_5F, (b) Bi_3NbTiO_9, (c) $Bi_4Ti_3O_{12}$, and (d) $BaBi_4Ti_4O_{15}$, showing one-half of each unit cell. In each diagram C represents the Bi_2O_2 layer (in (a), $Bi_2(O, F)_2$, in (d) $(Ba, Bi)_2O_2$). In (a) A denotes the layer of $Nb(O, F)_6$ octahedra and in (b), (c), and (d) the perovskite portions of the structure. The regions B in (b) and (c) mark off unit cells of the hypothetical perovskite structures $BaNb_{0.5}Ti_{0.5}O_3$ and $BaTiO_3$ respectively.

714

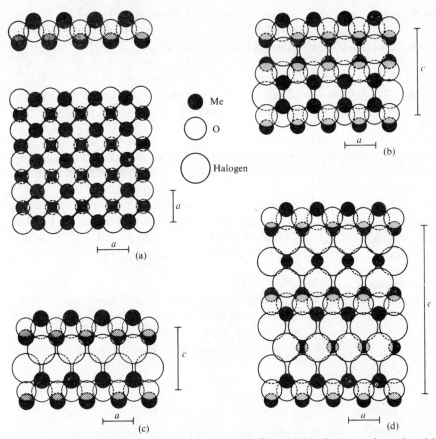

FIG. 20.6. The structures of complex bismuth oxyhalides. (a) Metal–oxygen sheet viewed in directions parallel and perpendicular to the sheet. (b), (c), (d) Elevations of structures built from (b) single halogen layers X_1, (c) double halogen layers X_2, and (d) triple halogen layers X_3 (Sillén)

bipyramid, the lone pair occupying an equatorial bond position, as in Sb_2O_4 (Fig. 20.8(c)). These groups share edges to form infinite chains.[1a]

Other oxyhalides, mostly oxychlorides and oxybromides, result from the controlled hydrolysis of the trihalides, and are of interest for two main reasons. First, they are quite unrelated to the oxyhalides of bismuth. Although both antimony and bismuth form compounds MOX the structures of the antimony compounds are quite different from those of the compounds BiOX, which have been described on p. 408. The more complex oxyhalides of Sb have no analogues among Bi compounds. Second, a feature of the published structures of the antimony oxyhalides is the coordination of Sb by either three or four O atoms. It should perhaps be remarked here that the investigation of the structures of these complex compounds is difficult, and the precise positions of the O atoms are by no means certain. However, it appears that a feature of these compounds is the formation of extended Sb–O systems, generally layers, interleaved with halogen

(1a) ACSc 1971 **25** 1519

(1b) AK 1953 6 89
(2) ACSc 1947 1 178
(3) AK 1955 8 257
(4) AK 1955 8 279

ions. Structural studies have been made of the following compounds, listed in order of decreasing $X:Sb$ ratio: $SbOCl$,[1b] $Sb_4O_5Cl_2$ and $Sb_4O_5Br_2$,[2] $[Sb_8O_8(OH)_4][(OH)_{2-x}(H_2O)_{1+x}]Cl_{2+x}$,[3] and $Sb_8O_{10}(OH)_2X_2$[4] ($X = Cl$, Br, or I).

SbOCl is built of puckered sheets in which two-thirds of the Sb atoms are linked to $2 O + 1 Cl$ and the remainder to $4 O$, the sheets $(Sb_6O_6Cl_4)_n^{2n+}$ being held together by Cl^- ions. (The distinction between the two types of Cl is based on the much greater Sb–Cl distance (2·9 Å) between Sb and interlayer Cl than Sb–Cl within a layer (2·3–2·5 Å).

$Sb_4O_5Cl_2$. Again there are infinite layer ions, here of composition $(Sb_4O_5)_n^{2n+}$, in which one half of the Sb atoms are 3- and the others 4-coordinated, with Cl^- ions between the layers.

In the two more complex oxyhalides the characteristic structural feature is apparently a double chain, in which all the Sb atoms are 4-coordinated, which is joined through OH groups to form infinite layers $[Sb_8O_8(OH)_4]_n^{4n+}$ in the first compound and complex cylindrical chain ions $[Sb_4O_5(OH)]_n^{n+}$ in the second.

Oxyfluorides of Bi

We have seen that BiF_3 is structurally different from all other trihalides of As, Sb, and Bi. Similarly, Bi oxyfluorides, with the exception of BiOF, are not related structurally to the compounds formed by Bi (or Sb) with the other halogens. Oxidation of BiF_3 gives the following oxyfluorides, the structures of which are described elsewhere as indicated:

$BiO_{0.1}F_{2.8}$: tysonite (LaF_3) structure (p. 356): ACSc 1955 **9** 1206
$BiO_{0.5}F_2$ (Bi_2OF_4): cubic (p. 358): ARPC 1952 **3** 369
BiOF: BiOCl structure (p. 408): ACSc 1964 **18** 1823, 1851

Complex oxyhalides of Bi with Li, Na, Ca, Sr, Ba, Cd, and Pb

By fusing the bismuth oxyhalides with excess of one of the halides of the above metals, Sillén and co-workers have prepared a large number of complex oxyhalides in which the halogen is Cl, Br, or less usually I. They all form tetragonal leaflets with an a axis of about 4 Å, but c axes up to 50 Å are found. These compounds are closely related to the simple bismuth oxyhalides and like those compounds contain characteristic oxygen layers. The metal–oxygen sheet is shown in Fig. 20.6(a). The simplest possibility is that sheets of this type alternate with single, double, or triple halogen layers, forming the X_1, X_2, or X_3 structures of Fig. 20.6(b), (c), and (d). The simple bismuth oxyhalides are, of course, of Sillén's X_2 structure type. More complex structures arise if halogen layers of more than one kind occur in the same structure, as in $SrBi_3O_4Br_3$ (X_1X_2 structure). In the X_1 compounds, $M^{II}BiO_2X$ or $M^IBi_3O_4Cl_2$, and the X_1X_2 compounds $M^{II}Bi_3O_4Cl_3$, all the metal atoms are associated with the metal–oxygen layer (a), apparently arranged at random in the available metal positions, but in the X_3 structures there are also metal atoms between the halogen layers (d). Not all these latter positions are occupied, however, so that structures containing X_3 layers are apparently always defect structures, as shown by the formulae of the cadmium bismuth oxyhalides in Table 20.7. A further structure type (O_1X_2) has been suggested for a barium bismuth oxyiodide in which the characteristic metal–oxygen layers are separated alternately by double halogen layers and oxygen layers.

Other compounds also related structurally to these oxyhalides include $Bi_{24}O_{31}Cl_{10}$ and the corresponding Br compound. The existence of Bi oxyhalides other than BiOX has been disputed, but according to Sillén the compounds $Bi_{24}O_{31}X_{10}$ (originally formulated $Bi_7O_9X_3$) do exist, and the chloride, for example, can be prepared by heating BiOCl to temperatures above 600°C. A structure has been proposed for these compounds. The various structure types proposed by Sillén for these oxyhalides are shown diagrammatically in Fig. 20.6, and some representative compounds are listed in Table 20.7. It should perhaps be remarked that certain features of these structures cannot be regarded as wholly satisfactory, though there seems little doubt about the general scheme according to which they are built. The general geometrical analogies with the layer-type hydroxyhalides will be obvious.

(a) ZaC 1942 **250** 173; ZK 1942 **104** 178; AKMG 1947 **25** 49.

TABLE 20.7
Structure types of bismuth oxyhalides[a]

Structure type	Examples
X_1	$BaBiO_2Cl$, $CdBiO_2Br$, $LiBi_3O_4Cl_2$
X_2	$BiOCl$, $LaOCl$
X_3	$Ca_{1.25}Bi_{1.5}O_2Cl_3$
X_1X_2	$SrBi_3O_4Cl_3$
X_2X_3	$Ca_{2-3x}Bi_{3+2x}O_4Cl_5$
$X_1X_1X_2$	$SrBi_2O_3Br_2$
$X_1X_2X_3$	$Cd_{2-3x}Bi_{5+2x}O_6Cl_7$

(1) AC 1968 **B24** 987
(2) AC 1966 **21** 808

The oxygen chemistry of pentavalent arsenic and antimony

The oxy-compounds of pentavalent arsenic

The pentoxides of As, Sb, and Bi cannot be prepared, like P_2O_5, by direct oxidation of the element, for they decompose at high temperatures with loss of oxygen. As_2O_5 is prepared by dehydration of its hydrates:

$$As_2O_5 \cdot 7H_2O \xrightarrow{-30°} As_2O_5 \cdot 4H_2O \xrightarrow{36°C} As_2O_5 \cdot \tfrac{5}{3}H_2O \xrightarrow{170°C} As_2O_5$$
$$(H_3AsO_4 \cdot 2H_2O) \quad (H_3AsO_4 \cdot \tfrac{1}{2}H_2O) \quad (H_5As_3O_{10})$$

The second of these compounds. $H_3AsO_4 \cdot \tfrac{1}{2}H_2O$, is a hydrate of H_3AsO_4 for it consists of a hydrogen-bonded assembly of tetrahedral $AsO(OH)_3$ molecules and H_2O molecules.[1] The third compound, $H_5As_3O_{10}$, consists of infinite chains of octahedral and tetrahedral groups sharing vertices as shown in Fig. 20.7.[2] The H

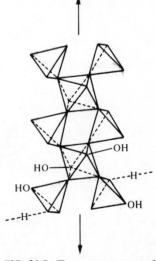

FIG. 20.7. The structure of $H_5As_3O_{10}$.

atoms were not directly located, their probable positions being deduced from the O—O separations between the chains; a n.d. study is called for.

(3) NBS Circular 539, Vol. 10 (1960)

Sb_2O_5 is formed by dissolving Sb in HCl, precipitating with HNO_3, and heating at 780°C for 3 minutes. It is a well defined cubic compound,[3] but its structure is not known. Bi_2O_5, prepared by oxidizing Bi_2O_3 in solution or by fusion with $KClO_3$, is a rather ill defined compound, often containing alkali and/or water; it is not certain that pure Bi_2O_5 has been prepared.

(4a) JINC 1963 **25** 79, 87, 93

The tetroxides As_2O_4[4a] and Sb_2O_4 (p. 720) are definite compounds.

Orthoarsenates are well known, and pyro- and meta-salts are formed by heating mono- and di-hydrogen arsenates (compare phosphates). However, it does not appear possible to prepare anhydrous H_3AsO_4, or the pyro- or meta-acids by dehydrating hydrates of the ortho-acid. The structures of several orthoarsenates have been determined. YPO_4 and $YAsO_4$ are isostructural with $ZrSiO_4$, $AlPO_4$ crystallizes with all three normal silica structures, and KH_2AsO_4 is isostructural

(4b) AC 1970 **B26** 1889
(4c) AC 1970 **B26** 1584
(5) AC 1956 9 87, 811

with KH_2PO_4. The $(AsO_3OH)^{2-}$ ion has been studied in $(NH_4)_2HAsO_4$[4b] and $Na_2HAsO_4 . 7 H_2O$;[4c] bond lengths are close to: As—O, 1·67 Å, As—OH, 1·74 Å. X-ray studies of $LiAsO_3$ and $NaAsO_3$[5] show that they contain infinite chain ions similar to those in diopside:

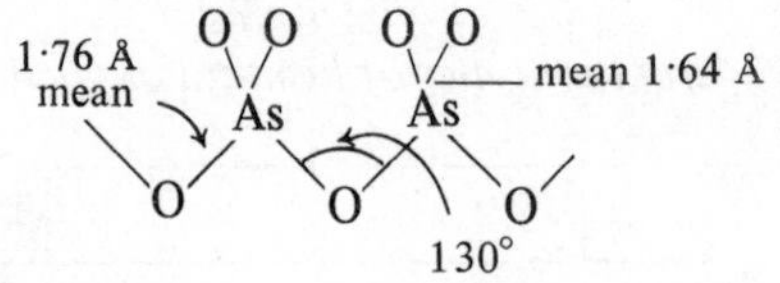

(6) JCS 1965 4466

There is also tetrahedral coordination of As^V in cacodylic acid,[6] which form hydrogen-bonded dimers in the crystal like carboxylic acids;

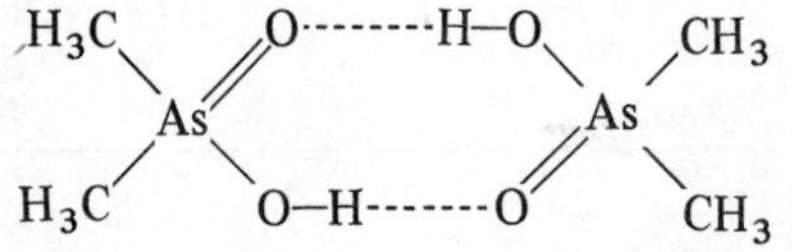

(7) BCSJ 1962 **35** 1600

and in arsonic acids, $R . AsO(OH)_2$.[7]

Octahedral coordination of As^V by O in the catechol derivative and in $H_5As_3O_{10}$ has already been mentioned.

The oxy-compounds of pentavalent antimony

The oxygen chemistry of pentavalent antimony was formerly very perplexing, and many compounds—some anhydrous, some hydrated—were described as ortho-, meta-, and pyro-antimonates and were assigned formulae analogous to those of phosphates. As a result of the study of the structures of many of these crystalline compounds it is found that their structural chemistry is very simple and quite different from that of phosphates, arsenates, and vanadates, being based not on tetrahedral but on octahedral coordination of Sb^V by oxygen. Many of the old formulae require revision. The compounds in question fall into two main groups:

(a) salts containing $Sb(OH)_6^-$ ions,

(b) complex oxides of the following (and possibly other) types, all based on octahedral SbO_6 coordination groups: $MSbO_3$, $M^{III}SbO_4$, $M^{II}Sb_2O_6$, $M^{II}Sb_2O_7$.

The free oxy-acids of antimony are even less stable than those of arsenic. Various hydrated forms of the trioxide and pentoxide have been described as antimonious and antimonic acids, but their structures are unknown. Finite oxy-ions SbO_4^{3-}, SbO_3^-, and $Sb_2O_7^{4-}$ do not exist. Compounds formerly described as ortho-, meta-, and pyro-antimonates all contain Sb^V octahedrally coordinated by oxygen, either as $Sb(OH)_6^-$ ions, in compounds which were once described as hydrated meta- or pyro-antimonates, or as SbO_6 groups in the compounds of type (b), which are best described as complex oxides, and not as antimonates, for reasons given later.

(a) *Salts containing* $Sb(OH)_6^-$ *ions.* Typical of the old formulae are $LiSbO_3.3H_2O$ and $Na_2H_2Sb_2O_7.5H_2O$ for hydrated meta- and pyro-antimonates. In fact no meta- or pyro-ions exist in these salts, which were formulated thus to bring them into line with the phosphorus compounds. In no case is there any structural similarity between the phosphorus and antimony compounds. A study of the crystal structure of sodium 'pyroantimonate', $Na_2H_2Sb_2O_7.5H_2O$, shows that its correct structural formula is $NaSb(OH)_6$, and the isostructural silver salt is $AgSb(OH)_6$[1]. In the pseudo-cubic sodium salt the Na and Sb atoms are arranged in the same way as the Na^+ and Cl^- ions in NaCl, and each Sb atom is the centre of an $Sb(OH)_6^-$ ion. By the action of 40 per cent HF, $Na[Sb(OH)_6]$ is converted into $Na(SbF_6)$ which has a quite similar structure. If a solution of sodium 'pyroantimonate' is added to one of a magnesium salt the precipitate has the empirical composition $Mg(SbO_3)_2.12H_2O$ and this was considered to be a hydrated meta-antimonate. Its formulation as $[Mg(H_2O)_6][Sb(OH)_6]_2$, resulting from an X-ray study of the crystals[2], shows that it is a hydrated antimonate of exactly the same type as sodium pyroantimonate. The old and new formulae of such antimonates are summarized below.

(1) ZaC 1938 **238** 241

(2) RTC 1937 **56** 931

Old formula	Correct structural formula
$Na_2H_2Sb_2O_7.5H_2O$	$Na[Sb(OH)_6]$
$Mg(SbO_3)_2.12H_2O$	$Mg[Sb(OH)_6]_2.6H_2O$ or $[Mg(H_2O)_6][Sb(OH)_6]_2$
(Co and Ni salts similar)	
$Cu(NH_3)_3(SbO_3)_2.9H_2O$	$[Cu(NH_3)_3(H_2O)_3][Sb(OH)_6]_2$
$LiSbO_3.3H_2O$	$Li[Sb(OH)_6]$

The partial hydrolysis of $NaSbF_6$ leads to the formation of a salt originally formulated as $NaF.SbOF_3.H_2O$. Its correct formula is $Na[SbF_4(OH)_2]$, showing that it is of the same type as $NaSbF_6$ and $NaSb(OH)_6$.

(b) *Complex oxides based on* SbO_6 *coordination groups.* In all these compounds antimony is octahedrally coordinated by oxygen, and in some cases the structures are similar to those of simple ionic crystals viz. $NaSbO_3$, ilmenite structure ($FeTiO_3$, p. 216), $FeSbO_4$, random rutile structure (p. 488), and $ZnSb_2O_6$, trirutile structure. It is therefore preferable to describe these compounds as complex oxides rather than as 'antimonates' of various types.

M^ISbO_3. The salt $NaSbO_3$ may be prepared by heating $NaSb(OH)_6$ in air or by fusing Sb_2O_3 with sodium carbonate—there are similar silver and lithium salts. The sodium salt has the ilmenite structure; the compound thought to be a polymorphic modification is actually $Na_2Sb_2O_5(OH)_2$ (see later). The structure of $LiSbO_3$[1] is related to that of $NaSbO_3$ in the way described in Chapter 4. In both structures M and Sb each occupy one-third of the octahedral holes in a h.c.p. assembly of O atoms, but the choice of holes occupied is different in the two structures (p. 145).

$KSbO_3$ is prepared by heating $KSb(OH)_6$ and is trimorphic. One form has the ilmenite structure, and the other a cubic structure[2] illustrated in Fig. 5.33 (p. 183). In the latter structure octahedral SbO_6 groups share edges and vertices, and this is also true of $K_3Sb_5O_{14}$.[3] There is apparently no compound $K_2Sb_2O_5$.

The compound originally formulated $CaSb_2O_5$ is actually $CaNaSb_2O_6OH$. The first product of heating $NaSb(OH)_6$ is $Na_2Sb_2O_5(OH)_2$, or possibly $NaSbO_3 . xH_2O$, which finally decomposes to $NaSbO_3$.[4]

All the following compounds have very similar structures: $Na_2Sb_2O_5(OH)_2$, the minerals pyrochlore, $CaNaNb_2O_6F$, and romeite, $CaNaSb_2O_6(OH)$, $Sb_3O_6(OH)$, $BiTa_2O_6F$, and $AgSbO_3$. The compound $Sb_3O_6(OH)$ is of interest in connection with the structure of Sb_2O_4. When hydrated Sb_2O_5 is heated (at about 800°C) it is not directly converted into Sb_2O_4, as was at one time supposed, but into $Sb_3O_6(OH)$. This latter compound is converted into the tetroxide only after prolonged heating at 900°C, and $Sb_3O_6(OH)$ had been confused with the tetroxide, which is also formed by the oxidation of Sb_2O_3 in air.

The compound $BiTa_2O_6F$ is prepared by heating BiOF with Ta_2O_5. The structures of the above-mentioned compounds are related in the following way. In the unit cell of the pyrochlore structure (see also p. 209) there are eight formula weights, that is, $8[CaNaNb_2O_6F]$. The 8 Ca + 8 Na may be replaced by 8 Sb(III), the Nb by Sb(V), and the F by OH, giving $Sb^{III}Sb_2^VO_6(OH)$. Alternatively the 8 Ca + 8 Na may be replaced by 16 Ag and the 8 OH omitted, giving $AgSbO_3$. These relationships may be summarized as follows:

Pyrochlore	$AgSbO_3$	$Sb_3O_6(OH)$	$BiTa_2O_6F$
8 Ca + 8 Na	16 Ag	8 Sb (III)	8 Bi (III)
16 Nb	16 Sb (V)	16 Sb (V)	16 Ta
48 O	48 O	48 O	48 O
8 F	—	8 OH	8 F

Sb_2O_4. The normal (orthorhombic) α form, the mineral cervantite, is isostructural with $SbNbO_4$ and $SbTaO_4$, and there is also a monoclinic β form made by heating the α form in air or oxygen. The structures of the two polymorphs are very similar, one being a 'sheared' version of the other. Both consist of corrugated

(1) ACSc 1954 **8** 1021

(2) AKMG 1940 **14A** 1

(3) AK 1966 **25** 505

(4) CJC 1970 **48** 1323

sheets formed from Sb^VO_6 octahedra sharing all their equatorial vertices (as in the plane layer in K_2NiF_4) and the Sb^{III} atoms lie between the layers in positions of rather irregular pyramidal 4-coordination. The structure of β-Sb_2O_4[5a] is illustrated in Fig. 20.8. The compound studied in the α series was $Sb^{III}Nb^VO_4$,[5b] in which the coordination of Sb is similar to that of Fig. 20.8(c), with Sb–O distances 2·01, 2·06, 2·13 and 2·33 Å and angles 95° and 151°, but the environment of Sb^{III} in α-Sb_2O_4 itself may be slightly different. $Bi^{III}Sb^VO_4$ has the same structure.[5c]

(5a) PCS 1964 400

(5b) CC 1965 611

(5c) AK 1951 3 153

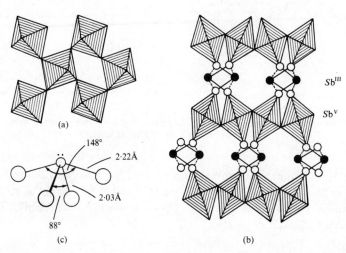

FIG. 20.8. (a) Layer of composition SbO_4 formed from Sb^VO_6 octahedra sharing equatorial vertices. (b) Showing Sb^{III} atoms situated between the SbO_4 layers, which are perpendicular to the plane of the paper. Dotted lines represent the longer Sb—O bonds. (c) The pyramidal coordination of Sb^{III}.

$M^{II}Sb_2O_6$.   Three types of structure have been found for these compounds. The small ions of Mg and certain 3d metals form the trirutile structure (p. 203), sometimes described as the tapiolite structure, after the mineral of that name, $FeTa_2O_6$. Compounds with this structure include:

$MgSb_2O_6$	$FeSb_2O_6$	$MgTa_2O_6$	$FeTa_2O_6$
Zn	Co		Co
	Ni		Ni

The somewhat larger Mn^{2+} ion prefers the niobite structure (p. 498). These structures are not possible for larger M^{2+} ions, and a hexagonal structure is adopted if the ionic radius of M is about 1 Å or more.[6] In this structure (Fig. 20.9) the M^{2+} and Sb^{5+} ions are segregated in layers so that ions varying from Ca^{2+} to Ba^{2+} can be accommodated by pushing the Sb–O layers further apart.

(6) AKMG 1941 **15B** No. 3

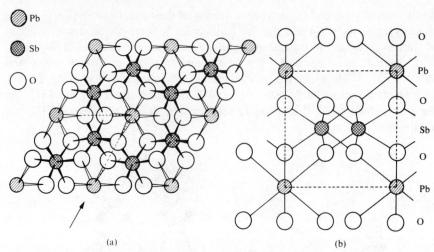

FIG. 20.9. The crystal structure of $PbSb_2O_6$: (a) plan, and (b) elevation (viewed in the direction of the arrow) on a slightly larger scale.

$M^{III}SbO_4$.   A number of compounds of this type have been shown to have a statistical rutile structure, that is, the ions M^{3+} and Sb^{5+} occupy at random the Ti^{4+} positions in the rutile structure. They include $FeSbO_4$, $AlSbO_4$, $CrSbO_4$, $RhSbO_4$, and $GaSbO_4$.[7]

$M_2^{II}Sb_2O_7$.   Two structures have been found for compounds of this type,[8] viz. the 'weberite' structure for $Ca_2Sb_2O_7$, $Sr_2Sb_2O_7$, and $Cd_2Sb_2O_7$, and the pyrochlore structure for $Pb_2Sb_2O_7$, $Ca_2Ta_2O_7$, $Cd_2Ta_2O_7$, and $Cd_2Nb_2O_7$.[9] These two closely related structures are adopted by a number of complex oxides and fluorides. In the weberite structure (of Na_2MgAlF_7)[10] we can distinguish a framework of linked MgF_6 and AlF_6 octahedra in which all vertices of each MgF_6 and four vertices of each AlF_6 octahedron are shared, so that the framework has the composition $(MgAlF_7)$ or more generally M_2X_7 when all atoms in the octahedra are of the same kind. (There is no essential difference in bond type between the Mg–F or Al–F and the Na–F bonds but the Na^+ ions lie in the interstices in positions of 8-, or strictly 4 + 4- or 2 + 4 + 2-coordination, and when comparing the two structures it is convenient to describe the Mg–Al–F frameworks.) In the pyrochlore framework all octahedral MX_6 groups share each X with another group, so that the composition of the framework is $MX_3(M_2X_6)$. There is room for a seventh X atom but this is not essential to the stability of the framework, and while compounds such as ralstonite, $Na_xMg_xAl_{2-x}(F, OH)_6 . H_2O$, $Ca_2Ta_2O_7$, etc. have this structure so also has $AgSbO_3$, already mentioned on p. 720. Since it is difficult to illustrate clearly complicated packings of octahedra we show in Fig. 20.10 only the Mg and Al atoms of these M_2X_7 and M_2X_6 frameworks. Each line joining two M atoms represents M–X–M, that is, the sharing of an octahedron vertex, though in general the X atoms do not lie actually along these lines. For the pyrochlore structure see also Chapters 6 and 13; the structure is illustrated in Fig. 7.4(c).

(7) AKMG 1943 **17A** No. 15
(8) AKMG 1944 **18A** No. 21

(9) PR 1955 **98** 903

(10) AKMG 1944 **18B** No. 10

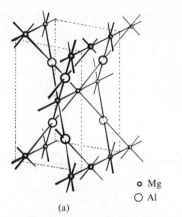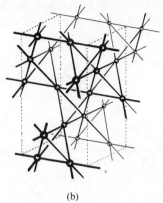

o Mg
O Al

(a)   (b)

FIG. 20.10. (a) The M_2X_7 framework in weberite, $Na_2(MgAlF_7)$, (b) the M_2X_6 (pyrochlore) framework in ralstonite, $Na_xMg_xAl_{2-x}(F, OH)_6H_2O$.

The sulphides of arsenic, antimony, and bismuth

These elements form the following sulphides:

$$As_4S_3$$
$$As_4S_4$$
$$As_2S_3 \qquad Sb_2S_3 \qquad Bi_2S_3$$
$$As_2S_5 \qquad Sb_2S_5$$

Bismuth does not form a pentasulphide; the structures of As_2S_5 and Sb_2S_5 are not known.

The sulphide As_4S_3 occurs as a mineral (dimorphite) and is dimorphic; the structure of one form (α) has been determined.[1] It consists of molecules very similar to those of P_4S_3 (Fig. 19.9(c)) with As–S, 2·21 Å, As–As, 2·45 Å, S–As–S, 98°, As–As–As, 60°, As–As–S, 102°, and As–S–As, 106°.

Realgar, As_4S_4, is a molecular crystal which sublimes readily *in vacuo*, and at temperatures up to about 550°C the vapour density corresponds to the formula As_4S_4. At higher temperatures dissociation to As_2S_2 takes place, and is complete at 1000°C. The structure of the As_2S_2 molecule is not known, but the As_4S_4 molecule has been studied by electron diffraction[2] and shown to have the configuration of Fig. 20.11, with As–S, 2·23 (0·02), As–As, 2·49 (0·04) Å, As–S–As, 101 (4)°, S–As–S, 93°, and S–As–As, 100°. This molecule has conventional bond arrangements and bond lengths. The crystalline material is built of the same molecules,[3] though As–As was found to be somewhat larger (2·59 Å) than in the vapour.

On exposure to light realgar changes very easily to orpiment, As_2S_3. Crystalline orpiment has a layer structure[4] (Fig. 20.12) of essentially the same kind as the monoclinic form of As_2O_3 (p. 710), but vaporizes to As_4S_6 molecules with the same configuration as As_4O_6 (p. 685). In the As_4S_6 molecule As–S is 2·25 Å,

(1) JCS A 1970 1800

(2) JACS 1944 **66** 818

As

FIG. 20.11. The As_4S_4 molecule.

(3) AC 1952 **5** 775

(4) MJ 1954 **1** 160

As–S–As, 100°, and S–As–S, 114°. In crystalline orpiment the mean As–S bond length is 2·24 Å and the mean bond angles at both S and As are 99°.

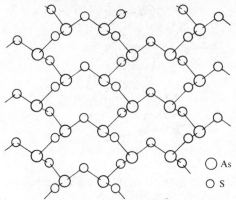

FIG. 20.12. A layer of the crystal structure of orpiment, As_2S_3.

○ As

○ S

(5) JCS 1964 219

The molecule of 'cacodyl disulphide' is of interest as containing As^{III} (pyramidal) and As^V (tetrahedral);[5]

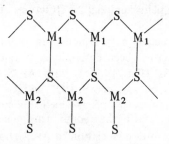

A simple chain with the composition M_2S_3 for an element M forming three bonds is

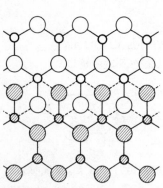

FIG. 20.13. The double chains in Sb_2S_3 (diagrammatic). The larger circles represent S atoms. All the atoms of one chain are shaded, and the weak secondary Sb–S bonds are shown as broken lines.

(6) AC 1957 **10** 99

and chains of this type occur in the structure of Sb_2S_3 and of the isostructural Bi_2S_3. The interatomic distances indicate, however, that these chains are associated in pairs as shown diagrammatically in Fig. 20.13, for whereas the atoms M_1 (see above) have three S neighbours at 2·50 Å and no others nearer than 3·14 Å the atoms M_2 have S neighbours as follows: one at 2·38, two at 2·67, and two more at 2·83 Å in addition to others at greater distances. The M_2–S bonds of length 2·83 Å are shown as dotted lines in Fig. 20.13. It is clear that no simple bond picture can be given for crystalline Sb_2S_3, but the bonds from the two kinds of non-equivalent Sb atoms can be given fractional bond numbers using Pauling's equation (p. 1025). This structure for Sb_2S_3 has been confirmed by a later study of the isostructural Sb_2Se_3.[6]

724

21

Carbon

Introduction

We shall be concerned here with only some of the simpler compounds of carbon, for the very extensive chemistry of this element forms the subject-matter of organic chemistry. Carbon is unique as regards the number and variety of its compounds based on skeletons of similar atoms directly linked together. There are also compounds based on mixed C–N, C–O, and C–N–O skeletons, including cyclic systems, to some of which we shall refer later. Two major classes of organic compounds are recognized: aliphatic compounds, based on the tetrahedral carbon atom, and aromatic compounds, based on the hexagonal C_6 ring, in which some of the C atoms may be replaced by N, etc. Corresponding to these two types of carbon skeleton in finite molecules are the two polymorphic forms of crystalline carbon, diamond–in which each C atom is linked by tetrahedral sp^3 bonds to its four neighbours, and graphite–in which each atom forms three coplanar sp^2 bonds leading to the arrangement of the atoms in sheets.

$$\overset{\diagup}{\underset{\diagdown}{C}}\diagdown \qquad \text{(diamond)–aliphatic compounds}$$

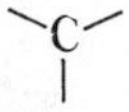

$$\diagdown\underset{|}{C}\diagup \qquad \text{(graphite)–aromatic compounds}$$

There are also 'mixed' compounds containing C_6 rings and, for example, aliphatic side chains.

In contrast to the chemistry of living organisms based on systems of linked C atoms there is the chemistry of silicates based on Si atoms linked via O atoms. We shall note some of the more important differences between C and Si at the beginning of Chapter 23. Here we note only one other feature of carbon chemistry which differentiates this element from silicon, namely, the presence of multiple bonds in many simple molecules, a characteristic also of nitrogen and oxygen.

The stereochemistry of carbon

The structural formulae of carbon compounds were originally based on the four bond pictures:

$$\text{(a)} \quad \overset{\diagup}{\underset{\diagdown}{C}}\diagdown \qquad \text{(b)} \quad \overset{\parallel}{\underset{\diagdown}{C}}\diagdown \qquad \text{(c)} \quad =C= \quad \text{and} \quad \text{(d)} \quad \equiv C-.$$

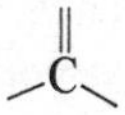

Carbon

It is now known, however, that (b) is not an adequate representation of the behaviour of a carbon atom in many molecules in which it is attached to three other atoms. Similarly (c) and (d) are unsatisfactory for many molecules containing a carbon atom forming two collinear bonds because there is resonance (π bonding) leading to non-integral bond orders, when neither $=C=$ nor $\equiv C-$ accurately represents the electronic structure. Their lengths and physical properties indicate that many carbon–carbon bonds are intermediate between C–C, C=C, and C≡C. The stereochemistry of this element may accordingly be discussed under the following heads:

(a) *The tetrahedral carbon atom*—forming four sp^3 bonds,

(b) *3-covalent carbon* —forming sp^2 bonds. With the proviso that intermediate types of bond are to be expected, it will simplify our treatment if we recognize three cases, to which a number of molecules approximate fairly closely:

   (i) the simple system of one double and two single bonds already noted,
   (ii) one (single) and two equivalent bonds, and
   (iii) three equivalent bonds.

   (i) $=C\!\!<$: $(H_3C)_2C=CH_2, H_2C=O, X_2C=O,$

   (ii) $-C\!\!\lesssim$: carboxylate ions, benzene, and cyclic C–N systems,

   (iii) $---C\!\!\lesssim$: carbonate, guanidinium, etc. ions, urea, graphite.

(c) *and* (d) *2-covalent carbon* (sp bonds). The compounds to be considered include CO_2, COS, and CS_2 (with which the monoxide and suboxide are conveniently included), acetylene, cyanogen, cyanates, and thiocyanates. (There are also relatively unstable species containing 2-covalent carbon such as CF_2,[1] for which m.w. data give C–F, 1·30 Å and bond angle $105°$ (ground state) or $135°$ (excited state), and the CO_2^- ion. This ion, detected by its e.s.r. spectrum, is formed when CO_2 is adsorbed on u.v.-irradiated MgO and in γ-irradiated sodium formate. The ion persists for months if formed in a KBr matrix; the bond angle has been estimated as $127 \pm 8°$.[2])

(1) JCP 1966 **45** 1067, 1068

(2) JPC 1965 **69** 2182; JCP 1966 **44** 1913

Although we shall adopt the above rough classification according to the type of bonds formed by carbon, it will not be convenient to adhere strictly to this order. For example, cyanogen derivatives polymerize to cyclic molecules in which C forms three bonds; these cyclic C–N compounds are logically included after the simple cyanogen derivatives. The following groups of compounds are included in Chapter 22: metal cyanides, carbides, and carbonyls, and certain classes of organo-metallic compound.

The tetrahedral carbon atom

Diamond: saturated organic compounds
The earliest demonstration of the regular tetrahedral arrangement of carbon bonds was provided by the analysis of the crystal structure of diamond. In this crystal

each C atom is linked to four equidistant neighbours and accordingly the linking shown in Fig. 21.1 extends throughout the whole crystal. The bond angles and the C–C bond length (1·54 Å) are the same as in a simple molecule such as $C(CH_3)_4$ or in any saturated hydrocarbon chain. The hexagonal form of diamond, related to cubic diamond in the same way as wurtzite to zinc-blende, has been produced at high temperature and pressure.[1] The molecule of symmetrical tricyclo-decane ('adamantane'), $C_{10}H_{16}$, has the same configuration as $N_4(CH_2)_6$ (Fig. 18.2, p. 642), and the molecule may be regarded as a portion of the diamond structure which has been hydrogenated. A larger portion of the diamond network is the carbon skeleton of $C_{14}H_{20}$.[2] The relation of $C_{10}H_{16}$ and $C_{14}H_{20}$ to diamond may be compared with that of large planar aromatic hydrocarbons such as coronene to graphite. Another interesting molecule is that of 'cubane', C_8H_8, with a cubic framework of carbon atoms (C–C, 1·552 Å).[3] The octaphenyl compound, $C_8(C_6H_5)_8$, is, however, a derivative of cyclooctatetraene (p. 730), the lengths of alternate bonds in the 8-membered ring being 1·343 Å and 1·493 Å.[4]

Many simple carbon compounds have been studied by diffraction or spectroscopic methods, and in all molecules Ca_4 the interbond angles have the regular tetrahedral value (109°28′) to within the experimental error. Some reliable values of bond lengths in simple molecules of carbon and other Group IV elements are listed in Table 21.1.

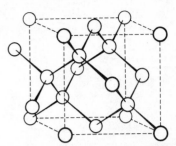

FIG. 21.1. The crystal structure of diamond.

(1) JCP 1967 **46** 3437

(2) JACS 1965 87 918

(3) JACS 1964 **86** 3889

(4) JCS 1965 3136

TABLE 21.1

Bond lengths in molecules—Group IV elements

Bond	Bond length (Å)	Molecule	Reference
C–H	1·06	C_2H_2	PRS A 1956 **234** 306
	1·10	C_2H_4	JCP 1965 **42** 2683
	1·11	C_2H_6 ⎱	JCP 1968 **49** 4456
C–C	1·53–	C_2H_6 ⎰	
	1·56	C_2Cl_6	ACSc 1964 **18** 603
C=C	1·34	C_2H_4	JCP 1965 **42** 2683
C≡C	1·20₅	C_2H_2	PRS A 1956 **234** 306
C–N	See Table 18.4, p. 650		
C–O	See Table 21.5, p. 739		
C–S	See Table 21.6, p. 739		
C–F	1·32–	CF_4	ARPC 1954 **5** 395
	1·39	CH_3F, CCl_3F	JCP 1960 **33** 508
C–Cl	1·75₅–	CF_3Cl	JCP 1962 **36** 2808
	1·78	CH_3Cl	JCP 1952 **20** 1420
C–Br	1·91–	CF_3Br	TFS 1954 **50** 444
	1·94	CH_3Br, CBr_4	JCP 1952 **20** 1112
C–I	2·14	CF_3I, CH_3I	JCP 1958 **28** 1010
Si–H	1·45₅–	$H_3SiC≡CH$	JCP 1963 **39** 1181
	1·48	SiH_4	JCP 1955 **23** 922
Si–C	1·83–	$H_3SiC≡CH$	JCP 1963 **39** 1181
	1·87	$(CH_3)_3SiH$	JCP 1960 **33** 907
Si–Si	2·33	Si_2H_5F	JCP 1966 **44** 2619
	2·36	$Si[Si(CH_3)_3]_4$	IC 1970 **9** 2436
Si–O	1·64	$Si(OCH_3)_4$	JCP 1950 **18** 1414

T A B L E 21.1–*continued*

Bond	Bond length (Å)	Molecule	Reference
Si—F	1·56–	$SiBrF_3$	JCP 1951 **19** 965
	1·59	SiH_3F	PR 1950 **78** 64
Si—Cl	1·98–	SiF_3Cl	JCP 1951 **19** 965
	2·05	SiH_3Cl	ACSc 1954 **8** 367
Si—Br	2·15–	SiF_3Br	JCP 1951 **19** 965
	2·21	SiH_3Br	PR 1949 **76** 1419
Si—I	2·39–	SiF_3I	JCP 1967 **47** 1314
	2·43	SiH_3I	N 1953 **171** 87
Si—N	1·72	$(CH_3)_3Si(NHCH_3)$	TFS 1962 **58** 1686
Ti—Cl	2·19	$TiCl_4$	
Ti—Br	2·32	$TiBr_4$	BCSJ 1956 **29** 95
Zr—Cl	2·33	$ZrCl_4$	
Th—Cl	2·61	$ThCl_4$	TFS 1941 **37** 393
Ge—H	1·52–	GeH_4	JCP 1954 **22** 1723
	1·54	GeH_3F–GeH_3Br	JCP 1965 **43** 333
Ge—C	1·90–	$H_3GeC{\equiv}CH$	JCP 1966 **44** 2602
	1·95	$GeH_2(CH_3)_2$	JCP 1969 **50** 3512
Ge—Ge	2·41	Ge_2H_6, Ge_3H_8	JACS 1938 **60** 1605
Ge—Si	2·36	$H_3Ge . SiH_3$	JCP 1967 **46** 2007
Ge—F	1·69–	GeF_3Cl	PR 1951 **81** 819
	1·73$_5$	GeH_3F	JCP 1965 **43** 333
Ge—Cl	2·07–	GeF_3Cl	PR 1951 **81** 819
	2·15	GeH_3Cl	JCP 1965 **43** 333
Ge—Br	2·31	GeH_3Br	
Ge—I	2·50	GeI_4	TFS 1941 **37** 393
Sn—H	1·70	SnH_4	JCP 1956 **25** 784
Sn—C	2·14	SnH_3CH_3	JCP 1952 **20** 1761
Sn—Cl	2·31–	$SnCl_4$	JCP 1959 **30** 339
	2·37	CH_3SnCl_3–$(CH_3)_3SnCl$	TFS 1944 **40** 164
Sn—Br	2·44–	$SnBr_4$	TFS 1941 **37** 393
	2·49	CH_3SnBr_3–$(CH_3)_3SnBr$	TFS 1944 **40** 164
Sn—I	2·64–	SnI_4	TFS 1941 **37** 393
	2·72	CH_3SnI_3–$(CH_3)_3SnI$	TFS 1944 **40** 164
Pb—C	2·20	$Pb(CH_3)_4$	JCP 1958 **28** 1007
Pb—Cl	2·43	$PbCl_4$	TFS 1941 **37** 393
Pb—Pb	2·88	$Pb_2(CH_3)_6$	TFS 1940 **36** 1209

Carbon fluorides (*fluorocarbons*)

The product of the decomposition of $(CF)_n$ (see later) or the direct fluorination of carbon is a mixture of fluorides of carbon from which the following have been isolated in quantity and studied: CF_4, C_2F_6, C_3F_8, C_4F_{10} (two isomers), C_2F_4, C_5F_{10}, C_6F_{12} and C_7F_{14}. All are extremely inert chemically and are notable for the very low attractive forces between the molecules, approaching the atoms of the noble gases in this respect. There is a great difference between the boiling points of the fluorocarbons and the hydrocarbons, as shown in Fig. 21.2.

Mixed fluoro-chloro compounds have been prepared by fluorinating the appropriate chloro compounds; they include $CHClF_2$, CCl_2F_2, $CHCl_2F$, CCl_3F, and $CCl_2F . CCl_2F$. The unsaturated compound tetrafluoroethylene, C_2F_4, can be polymerized under pressure to a solid polymer which is, like the simple fluorocarbons, very inert chemically. Polytetrafluoroethylene, $(CF_2)_n$, consists of chain molecules $-CF_2-CF_2-$ but in contrast to the planar zigzag $(CH_2)_n$ chains in

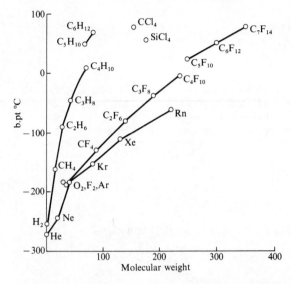

FIG. 21.2. Boiling-points of hydrocarbons, fluorocarbons, and noble gases.

polyethylene they have a helical configuration. The solid carbon monofluoride $(CF)_n$ is closely related to graphite, the layers of which have been pushed apart by the introduction of F atoms (see later).

Carbon forming three bonds

On p. 726 we recognized three main groups, (i)–(iii), of compounds in which carbon forms three bonds. We shall now give examples of each group, and for comparison with observed bond lengths we give in Table 21.2 standard values for single, double, and triple bonds. It is possible to select examples illustrating a simple correlation between types of bond and interbond angles, in which the logical sequence would be:

125°	120°	115°
C 110°	C 120°	C 130°
125°	120°	115°
(i)	(iii)	(ii)
$COCl_2$	CO_3^{2-}	$(R\,.\,COO)^-$

but we shall see that there are no clear-cut dividing lines between these classes, particularly between (ii) and (iii); cyclic molecules such as C_6H_6 illustrate this point. As we proceed it will become evident that the bonds in many molecules

729

Carbon

(1) AC 1967 **23** 410

cannot be described in any simple way; a good example is the molecule of ethyl carbamate, in which there are three different C—O bond lengths:[1]

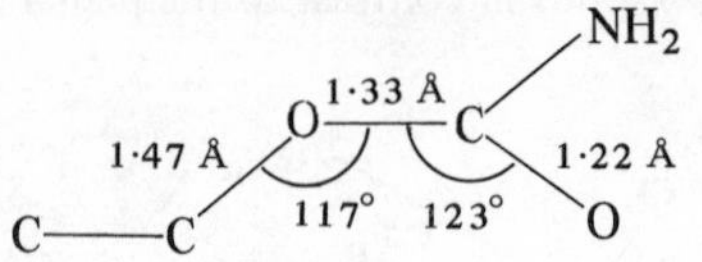

TABLE 21.2
Observed bond lengths (Å)

	Single	Double	Triple
C—C	1·54	1·34	1·20
C—N	1·47	[1·28]	1·16
C—O	1·41	1·22	1·13

Bond arrangement =C<

For carbon forming one double and two single bonds the bond angles

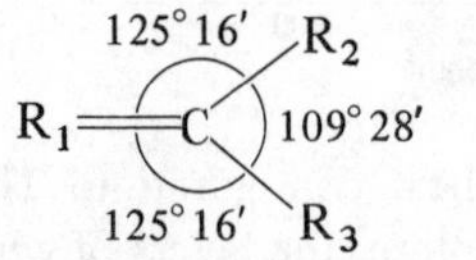

should be as shown and the lengths of the bonds C=R_1, C—R_2, and C—R_3 should be those expected for one double and two single bonds respectively. The most symmetrical molecules of this type include isobutene, $(CH_3)_2C=CH_2$, formaldehyde, $H_2C=O$, ketones, $R_2C=O$, carbonyl halides, $X_2C=O$, and their sulphur analogues, $X_2C=S$. The free molecule $(OH)_2C=O$ is not known, but the reaction between a suspension of Na_2CO_3 in dimethyl ether and HCl at $-33°C$ is said to give the etherate, $OC(OH)_2 . O(CH_3)_2$.[2] Less symmetrical molecules approximating to this bond arrangement include the boat-shaped molecule of cyclooctatetraene, (a),[3] the bond lengths in which indicate that there is comparatively little interaction between the bonds in the ring, formyl fluoride, (b),[4] and carboxylic acids.

(2) ZaC 1968 **357** 78

(3) JCP 1958 **28** 512
(4) JCP 1961 **34** 1847

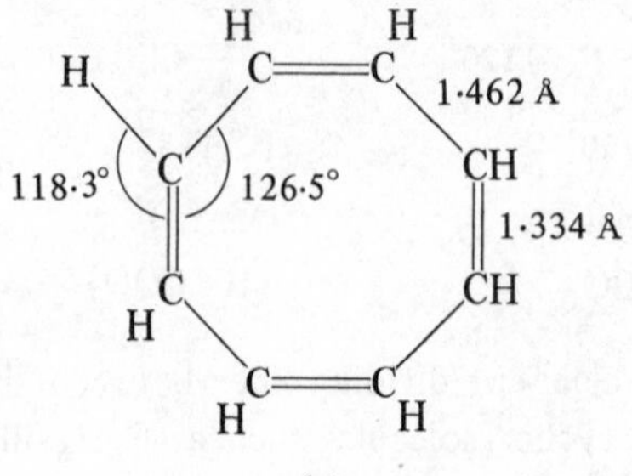

(a)

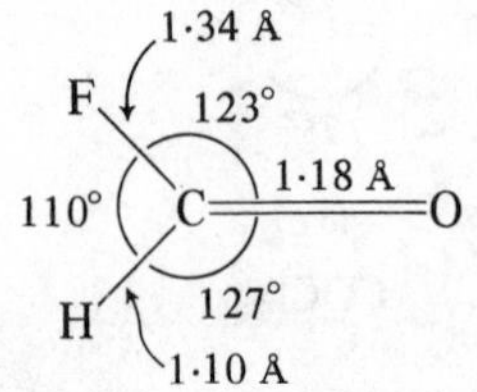

(b)

Carbonyl halides and thiocarbonyl halides. All the compounds COF_2,[5] $COCl_2$,[6] and $COBr_2$,[7] have been studied in the vapour state by e.d. and/or m.w., and $COCl_2$ (phosgene) has also been studied in the crystalline state.[8] The C–X bond lengths are close to the values expected for single bonds, while the C=O

(5) JCP 1962 **37** 2995
(6) JCP 1953 **21** 1741
(7) JACS 1933 **55** 4126
(8) AC 1952 **5** 833

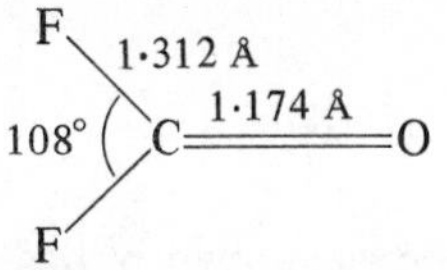

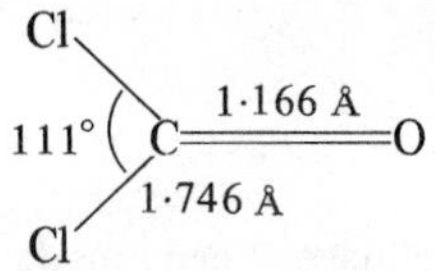

bonds are slightly shorter than 'normal' double bonds. There is no sign of disorder in crystalline phosgene, so that the residual entropy (1·63 e.u.) calculated from calorimetric and spectroscopic data is unexplained.

An early electron diffraction study of thiophosgene, $CSCl_2$, assumed C–S, 1·63 Å, which could profitably be checked. The dimer, produced by photolysis of $CSCl_2$, has the structure (c),[9] and the compounds with the empirical formulae $(COCl_2)_2$ and $(COCl_2)_3$ are (d) and (e).[10]

(9) ZaC 1969 **365** 199

(10) JCS 1957 618

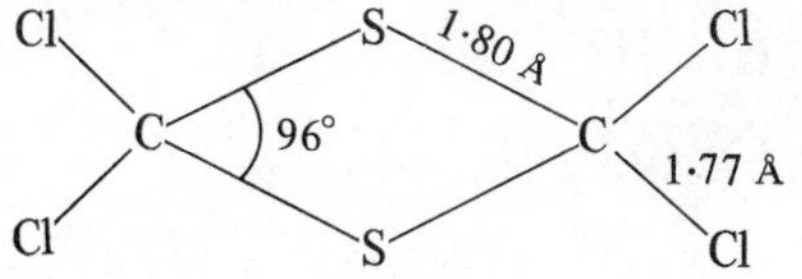

Carboxylic acids and related compounds. Acids such as formic and acetic dimerize in the vapour state. The structures of the monomer and dimer of formic acid are shown at (f) and (g)[11] as determined by e.d. studies of the vapour; the

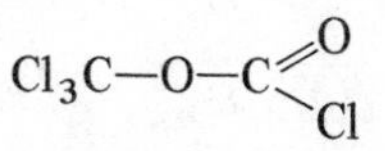

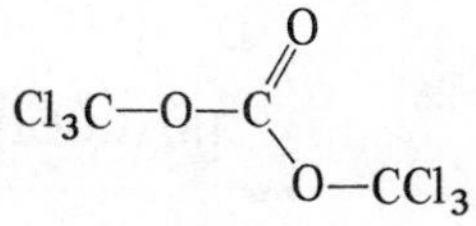

(11) ACSc 1969 **23** 2848

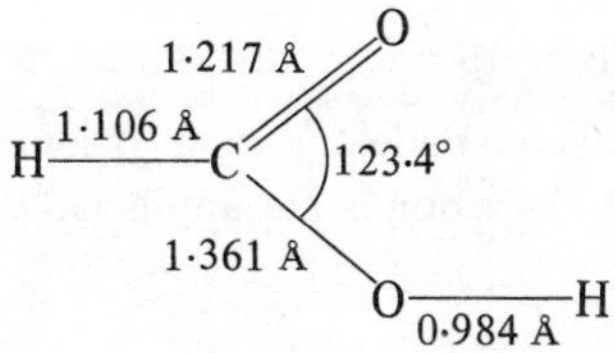

acid has also been studied in the crystalline state.[12] The dimers of these acids are very nearly planar molecules;[13] it may be supposed that sp^2 hybridization of the C atom requires also sp^2 hybridization of the carbonyl O atoms. Consequently the two lone pairs of the latter are extended at 120° to the C=O bond and are coplanar with it and with the other C–O bond. The whole structure is therefore essentially planar. An electron diffraction study of thioacetic acid[14] gave the

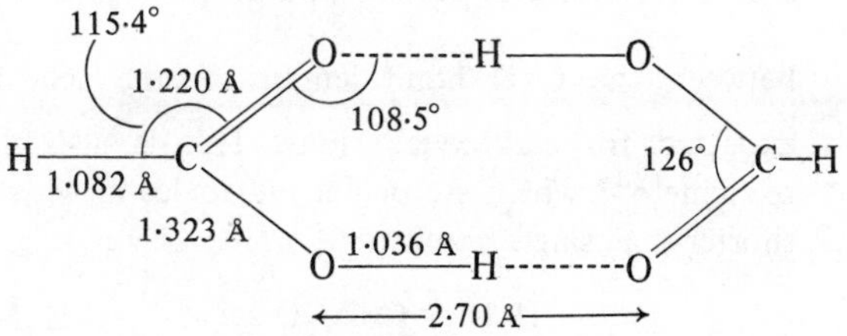

(12) AC 1953 **6** 127
(13) AC 1963 **16** 430

(14) JCP 1946 **14** 560

731

following result, values enclosed in brackets being assumed:

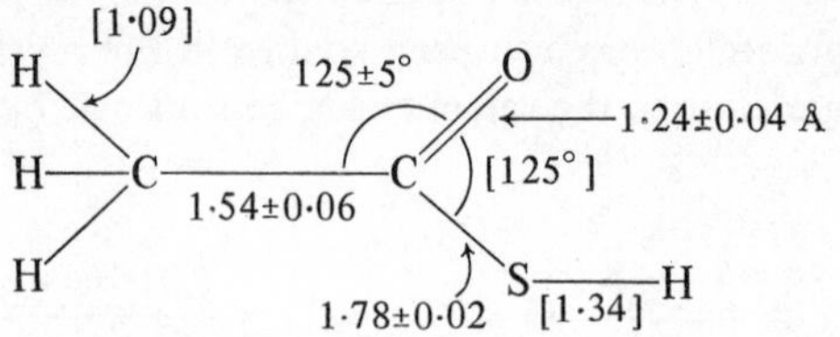

(15) JCS 1952 4854
(16) AC 1965 **18** 410
(17) AC 1969 **B25** 469
(18) ACSc 1968 **22** 2953

(19) AC 1969 **B25** 2423

Many studies have been made of oxalic acid, which has two anhydrous polymorphs[15] and of oxalates, $(NH_4)_2C_2O_4 . H_2O$,[16] $K_2C_2O_4 . H_2O$,[17] KHC_2O_4[18] etc. Although the dihydrate of the acid is not an oxonium salt there is comparatively little difference between the dimensions of the $(COOH)_2$ molecule in $(COOH)_2 . 2 H_2O$,[19] (h), and of the ion in alkali-metal salts, (i). Points of

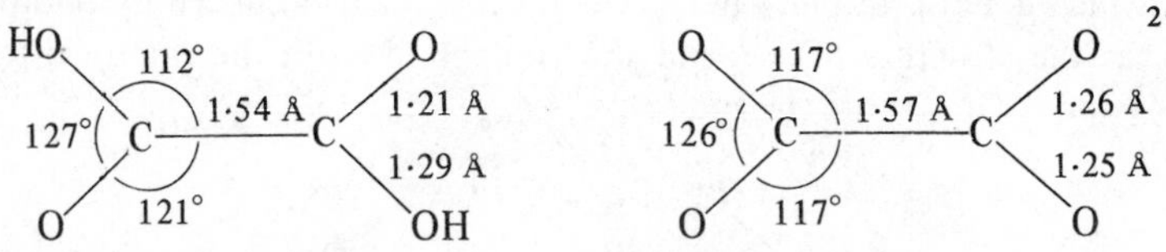

(h) $(COOH)_2 . 2 H_2O$ (i) $C_2O_4^{2-}$ ion

interest are the planarity of the oxalic acid molecule, in which the C–C bond is at least as long as a single bond, and the variation in conformation of the oxalate ion, which is planar in $K_2C_2O_4 . H_2O$ but non-planar in KHC_2O_4 and $(NH_4)_2C_2O_4 . H_2O$. In the last two crystals the angles between the planes of the COO groups are $13°$ and $27°$ respectively. The non-planarity is attributed to strong hydrogen bonding, which also affects slightly the C–O bond lengths in a number of oxalates.

Evidently, although formic acid is close to $O=C\langle^H_{OH}$, with a marked difference

between the C–O bond lengths, oxalic acid is closer to $-C\langle$, the structure expected for carboxylate ions. This is also true of oxamide[20] and dithio-oxamide[21] which are planar molecules in which the C–N bonds are appreciably shorter than single bonds:

(20) AC 1954 **7** 588
(21) JCS 1965 396

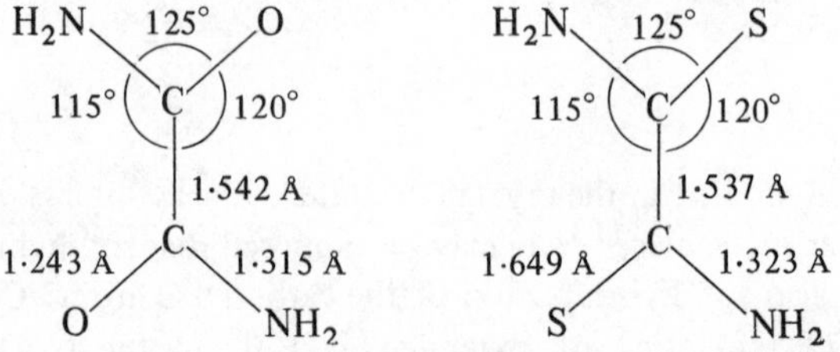

Two cyclic ions are conveniently included here. They are the square planar $C_4O_4^{2-}$ ion in the potassium salt of 3:4-diketocyclobutene diol, $K_2C_4O_4$.

H_2O,[22] and the croconate ion, $C_5O_5^{2-}$, studied in $(NH_4)_2C_5O_5$.[23] The latter has almost perfect pentagonal symmetry. In both ions the bond lengths are C–C, 1·46 Å, C=O, 1·26 Å.

(22) JCP 1964 **40** 3563
(23) JACS 1964 **86** 3250

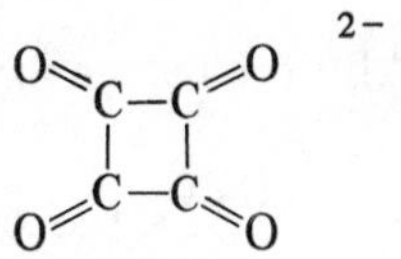

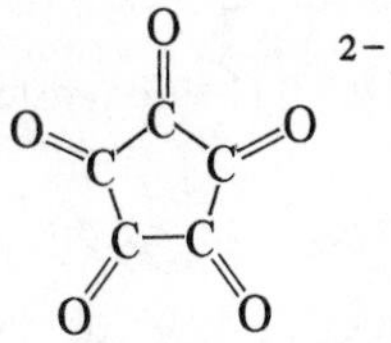

Bond arrangement —C⟨

Carboxylate ions. Ions R–COO⁻ of carboxylic acids are symmetrical with two equal C–O bonds (1·26 Å) and O–C–O bond angle in the range 125–130°. In sodium formate,[1] for example, the ion is reported to have C–O, 1·27 Å, and O–C–O, 124 (4)°, and bond lengths close to 1·25 Å have been found in alkaline-earth formates.

(1) JACS 1940 **62** 1011

Benzene. A number of studies give values for C–C close to 1·395 Å.[2]

(2) PRS A 1958 **247** 1

Bond arrangement ---C⟨

A number of symmetrical ions (and one free radical) have been shown to be planar, or approximately so, in the crystalline state. They are listed in Table 21.3, which also gives the other ion in the salt studied.

TABLE 21.3
Planar ions CX_3^{2-}, CX_3^{-}, *and* CX_3^{+}

Ion	C–X	Reference
CO_3^{2-} : Ca^{2+} ⎱	1·29 Å (C–O)	JCP 1967 **47** 3297
: K_2^{+}.3H_2O ⎰		
CS_3^{2-} : K_2^{+}.H_2O	1·71 (C–S)	AC 1970 **B26** 877
$C(CN)_3^{-}$: NH_4^{+}	1·40 (C–C)	ACSc 1967 **21** 1530
: K^{+}	1·39 (C–C)	AC 1971 **B27** 1835
: Ag^{+}	–	IC 1966 **5** 1193
$C(SO_2CH_3)_3^{-}$: NH_4^{+}	1·70 (C–S)	AC 1965 **19** 651
$C(p\text{-}C_6H_4NO_2)_3$	1·47 (C–C)	ACSc 1967 **21** 2599
$C(NH_2)_3^{+}$: NO_3^{-}, Cl^{-}	1·32 (C–N)	AC 1965 **19** 676
$C(N_3)_3^{+}$: $SbCl_6^{-}$	1·34 (C–N)	AC 1970 **B26** 1671
$C(C_6H_5)_3^{+}$: ClO_4^{-}, BF_4^{-}	1·45 (C–C)	AC 1965 **18** 437
$C(C_6H_4NH_2)_3^{+}$: ClO_4^{-}	1·45 (C–C)	AC 1971 **B27** 1405

733

Carbon

Carbonate ion. Redetermination of the C—O bond length in calcite and in $K_2CO_3 \cdot 3\,H_2O$ gives values close to 1·29 Å, though the lower value 1·25 Å was found in sodium sesquicarbonate (p. 314) which has been studied by both X-ray and neutron diffraction.

Triazidocarbonium ion. The dimensions found for this ion are shown at (a)

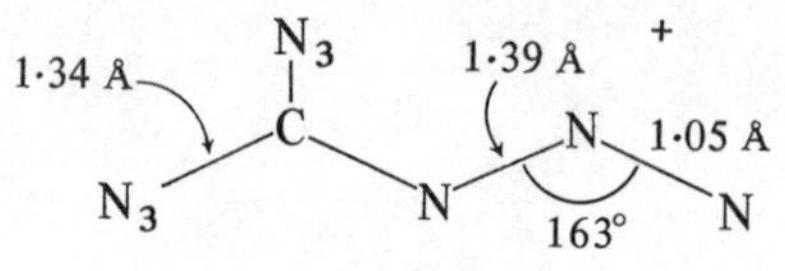

(a) 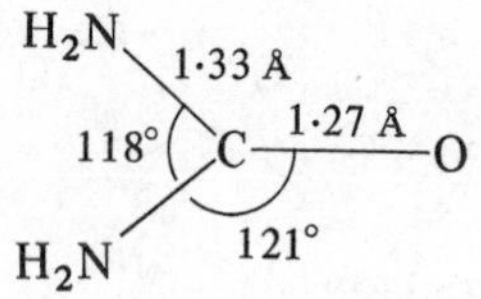   (b)

Tricyanomethanide ion. The very small deviations from planarity of this ion observed in certain salts have been attributed to crystal interactions. This ion can form interesting polymeric structures with metal ions. In the argentous compound the group acts as a 3-connected unit in layers based on the simplest 3-connected plane net, and these layers occur as interwoven pairs as described on p. 90.

Urea. This molecule, (b), is intermediate between CO_3^{2-} and $C(NH_2)_3^+$ (guanidinium ion).

A number of studies have shown that the H atoms are coplanar with the C, N, and O atoms;[1] other molecules in which N forms three coplanar bonds include formamide, $H_2N \cdot CHO$, and $N(SiH_3)_3$. Bond lengths in thiourea are C—N, 1·33 Å, and C—S, 1·71 Å.[2]

(1) AC 1969 **B25** 404

(2) JCS 1958 2551

Graphite. A portion of the crystal structure of graphite is illustrated in Fig. 21.3. The large separation of the layers (3·35 Å) compared with the C—C bond length of 1·42 Å in the layers indicates relatively feeble binding between the atoms of different layers, which can therefore slide over one another, conferring on graphite its valuable lubricating properties. (The interlayer spacing varies slightly with the degree of crystallinity, from 3·354 Å in highly crystalline graphite to 3·44 Å in poorly crystalline carbons.[1]) In the graphite structure (Fig. 21.3) the carbon atoms of alternate layers fall vertically above one another, and the structure can be described in terms of a hexagonal unit cell with $a = 2·456$ Å and $c = 6·696$ Å. The structure of many graphites is rather more complex.[2] In Fig. 21.4 the heavy and light full lines represent alternate layers in the normal graphite structure, and it is seen that there is a third possible type of layer (broken lines) symmetrically related to the other two. A second graphite structure is therefore possible in which these three types of layer alternate, a structure with $a = 2·456$ Å and $c = \frac{3}{2}\,(6·696) = 10·044$ Å. Several natural and artificial graphites have been found to have partly this new structure and partly the normal structure. It is interesting to note that in the normal graphite structure the carbon atoms are not all crystallographically equivalent, whereas they are in the second form of Fig. 21.4.

(1) AC 1951 4 253, 558

(2) PRS A 1942 **181** 101; ZK 1956 **107** 337

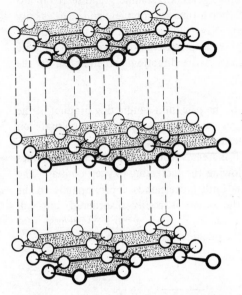

FIG. 21.3. The crystal structure of graphite.

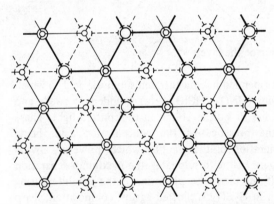

FIG. 21.4. Relative positions of atoms in successive layers of the two forms of graphite (see text).

Also, this new form is closely related to diamond, from which it could be derived by making coplanar the atoms in the puckered layers parallel to (111) and lengthening the bonds between these layers. Graphite can be converted into diamond under a pressure greater than 125 kbar at about 3000°K.[3]

(3) JCP 1967 **46** 3437

The great difference between the bonding within and between the layers of graphite is responsible for the formation of numerous compounds in which atoms or molecules are introduced between the layers. Some of these compounds will now be described.

(4) QRCS 1960 **14** 1

Derivatives of graphite[4]

Charcoal is essentially graphitic in character but with a low degree of order, and because of its extensive internal surfaces and unsatisfied valences possesses the property of adsorbing gases and vapours to a remarkable extent and of catalysing

735

reactions by bringing together the reacting gases. Crystalline graphite is far less reactive in this way since the layers of carbon atoms are complete and there is only a weak residual attraction between them. Nevertheless it undergoes a number of reactions in which atoms or radicals are taken up between the layers. The graphite crystal does not distintegrate but expands in the direction perpendicular to the layers.

Graphitic oxide.　This name is applied to the pale, non-conducting product of the action of strong oxidizing agents such as nitric acid or potassium chlorate on graphite. The graphite swells in one direction, and measurements of the inter-layer distance in these compounds show that it increases from the value 3·35 Å to values between 6 and 11 Å, this expansion following the increasing oxygen content. The composition is not very definite but the limit corresponds approximately to the formula $C_4O(OH)$. The structure of the oxide is highly disordered, but e.d. results[5] on the dehydrated material (inter-layer spacing ~6·2 Å) suggest that oxygen may be attached in two ways to the puckered graphite-like sheets:

(5) AC 1963 **16** 531

There is no regularity in the stacking of the layers.[6]

(6) ZaC 1969 **369** 327

Graphitic 'salts'.　If graphite is suspended in concentrated sulphuric acid to which a small quantity of oxidizing agent (HNO_3, $HClO_4$, etc) is added, it swells and develops a steel-blue lustre and blue or purple colour in transmitted light. These products are stable only in the presence of the concentrated acid, being decomposed by water with regeneration of the graphite which, however, always contains some strongly held oxygen. In the 'bisulphate', the maximum separation of layers is approximately 8 Å. If the electrode potential of the graphite during oxidation or subsequent reduction in H_2SO_4 is plotted against composition, sharp breaks are observed in the curve corresponding to the ratios C_{96}^+, C_{48}^+, and C_{24}^+ to HSO_4^-.[7] Other graphitic 'salts', perchlorate, selenate, nitrate, etc., have been described.

(7) PRS A 1960 **258** 329, 339

Carbon monofluoride.　When graphite, previously degassed, is heated in fluorine to temperatures around 400–450°C a grey solid is formed with the approximate composition $(CF)_n$. The expansion of the inter-layer separation to 8·17 Å shows that the fluorine is contained between the layers. Here again the characteristic lustre and conductivity of graphite are absent, the material having a specific resistance about 10^5 times that of graphite. On heating to higher temperatures no further fluorine enters the structure, which eventually decomposes into a mixture of various carbon fluorides together with soot—compare the final oxidation of 'graphitic acid' to mellitic acid, $C_6(COOH)_6$. A compound $(C_4F)_n$ has also been described and a structure has been suggested, but there does not appear to be general agreement as to the positions of the F atoms in either $(C_4F)_n$ or $(CF)_n$.[8]

(8) SR 1947–8 **11** 212–217

Compounds of graphite with alkali metals and bromine. On treatment with the molten metal or its vapour graphite forms the following compounds C_nM with the alkali metals:[9]

(9) JCP 1968 49 434

		n				
Li	6	12	18			
Na		only		64		
K $\rbrace$						
Rb $\rbrace$	8	10	24	36	48	60
Cs $\rbrace$						

(10) PRS A 1965 **283** 179

Absorption of bromine gives C_8Br and $C_{28}Br$,[10] and since it seems that C_8K and C_8Br are similar from the geometrical standpoint, having alternate layers of C and K (Br), these compounds are conveniently discussed together. The marked temperature-dependent diamagnetism of graphite, due to the large π-orbitals, is changed to a weak temperature-independent diamagnetism in C_8Br (comparable with that of metallic Sn or Zr), and a comparatively strong temperature-independent parmagnetism in C_8K (comparable with Ca), stronger than that of metallic K itself. Both C_8K and C_8Br are better conductors of electricity than graphite. The interlamellar separation of $3 \cdot 35$ Å in graphite becomes $7 \cdot 76$ Å in C_8K and $7 \cdot 05$ Å in C_8Br. It would seem that if these structures were ionic in character there would be a larger separation of the carbon layers in the bromine compound (since the radius of the Br^- ion is $1 \cdot 95$ Å while that of K^+ is $1 \cdot 38$ Å), whereas the metallic radius of K is $2 \cdot 23$ Å as compared with the covalent radius $1 \cdot 19$ Å of Br. It has therefore been suggested that there are 'metallic' bonds between K or Br and the graphite layers, the empty electron band in graphite being partly filled in the K compound and the full band being partly emptied in the Br compound. The fact that Na does not form a similar compound, while Rb and Cs do, may be due to the fact that the ionization energy of Na is higher than that of K.

Graphite complexes with metal halides. In addition to the well-known compounds with $FeCl_3$ and $AlCl_3$ complexes can be formed with many other halides, for example, $CuCl_2$, $CrCl_3$, 4f trihalides, $MoCl_5$, and also chromyl fluoride and chloride. (Crystalline BN also forms similar compounds with $AlCl_3$ and $FeCl_3$.) An electron diffraction study[11] has been made of the graphite–$FeCl_3$ system. The best characterized phase in the graphite–$MoCl_5$ system consists of layers of Mo_2Cl_{10} molecules between sets of four graphite layers.[12] Within a layer the Mo_2Cl_{10} molecules are close-packed.

(11) AC 1956 9 421

(12) AC 1967 **23** 770

737

Carbon

The oxides and sulphides of carbon

Carbon monoxide

In this molecule, which has a very small (possibly zero) dipole moment, the bond length is 1·131 Å. The length, force constant, and bond energy (Table 21.4) show that it is best represented C≡O.

TABLE 21.4

Properties of carbon-oxygen bonds

Bond	Length	Force constant	Bond energy
	(Å)	N m^{-1}	(kJ mol^{-1})
C=O in H.CHO	1·209	0·0123	686
O=C=O	1·163	0·0155	803
C≡O	1·131	0·0186	1075

Carbon dioxide and disulphide: carbonyl sulphide

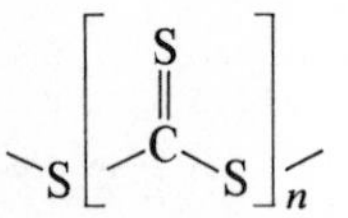

The molecules OCO, SCS, and OCS are linear. Bond lengths are given in Tables 21.5 and 21.6. In carbon dioxide the carbon-oxygen bond is intermediate in length between a double and triple bond. The studies of CO_2 and CS_2 by high resolution infrared spectroscopy provide an example of the use of this method for molecules which cannot be studied by the microwave method because they have no permanent dipole moment. The structure of the CS_2 molecule has also been studied in the crystalline state. (C–S, 1·56 Å).[1] Under a pressure of 30 kbar CS_2 polymerizes to a black solid for which a chain structure has been suggested.[2]

(1) JCP 1968 **48** 2974
(2) CJC 1960 **38** 2105

Carbon suboxide

The third oxide of carbon, the so-called suboxide C_3O_2, is a gas at ordinary temperatures (it boils at +6°C) and may be prepared by heating malonic acid or its diethyl ester with a large excess of phosphorus pentoxide at 300°C. Its preparation from, and conversion into, compounds containing the –C–C–C– nucleus are consistent with its structure, O–C–C–C–O. The action of water regenerates malonic acid, ammonia yields malonamide, and HCl malonyl chloride:

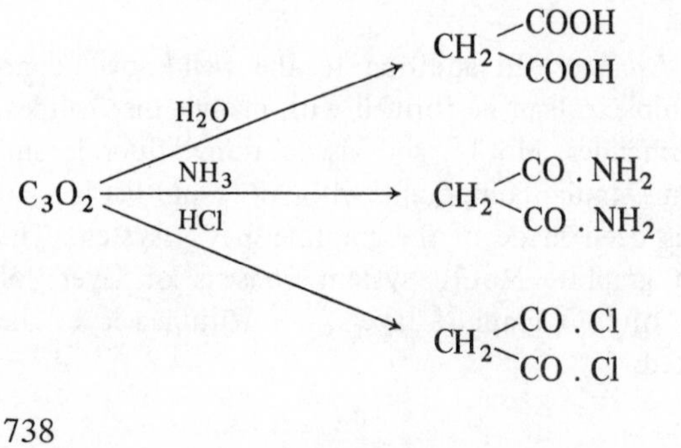

C_3O_2 →
H₂O → CH₂⟨COOH / COOH⟩
NH₃ → CH₂⟨CO.NH₂ / CO.NH₂⟩
HCl → CH₂⟨CO.Cl / CO.Cl⟩

all of the form

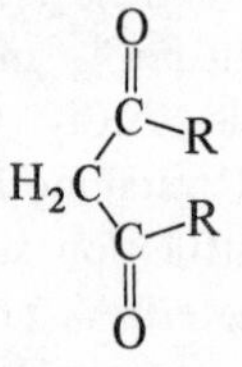

H_2C⟨C(=O)–R / C(=O)–R⟩

where R = OH, NH₂, or Cl

738

Carbon suboxide polymerizes readily at temperatures above −78°C. An e.d. study indicates a linear molecule with C–C, 1·28 Å, and C–O, 1·16 Å,[3] but the possibility of small deviations from linearity could not be excluded.

The oxides C_2O[4] and CO_3[5] have been identified in low-temperature matrices. The latter is an unstable species formed by the reaction of O atoms with CO_2; its i.r. spectrum has been studied in an inert matrix and the cyclic structure (a) suggested.

(3) JACS 1959 **81** 285

(4) JCP 1965 **43** 3734
(5) JCP 1966 **45** 4469

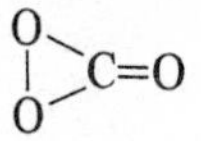

(a)

TABLE 21.5
Lengths of C—O *bonds*

Molecule	Bond length	Method	Reference
CO	1·131 Å	m.w.	PR 1950 **78** 140
$COCl_2$	1·166	m.w.	JCP 1953 **21** 1741
COSe	1·159	m.w.	PR 1949 **75** 827
CO_2	1·163	i.r.	JRNBS 1955 **55** 183
COS	1·164	m.w.	PR 1949 **75** 270
HNCO	1·184	X	AC 1955 **8** 646
H_2CO	1·209	e.d.	BCSJ 1969 **42** 2148
CH_3OH	1·427	m.w.	JCP 1955 **23** 1200
HCOO—CH_3	1·437	m.w.	JCP 1959 **30** 1529

TABLE 21.6
Lengths of C—S *bonds*

Molecule	Bond length	Method	Reference
HNCS	1·56 Å	m.w.	JCP 1953 **21** 1416
OCS	1·56	m.w.	PR 1949 **75** 270
TeCS	1·56	m.w.	PR 1954 **95** 385
SCS	1·55	i.r.	JCP 1958 **28** 682
$(NH_2.CS)_2$	1·65	X	JCS 1965 396
$SC(NH_2)_2$	1·71	X	JCS 1958 2551
CH_3SH	1·81	m.w.	PR 1953 **91** 464
$(CF_3)_2S$	1·83 ⎫	e.d.	TFS 1954 **50** 452
$(CF_3)_2S_3$	1·85 ⎭		

Two molecules with some structural resemblance to CO_2 and CS_2:

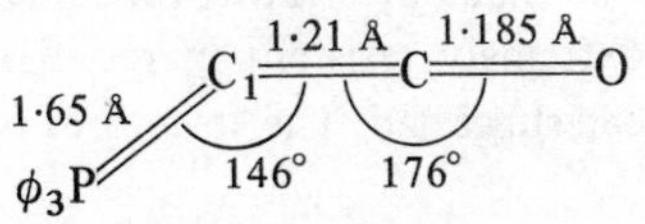

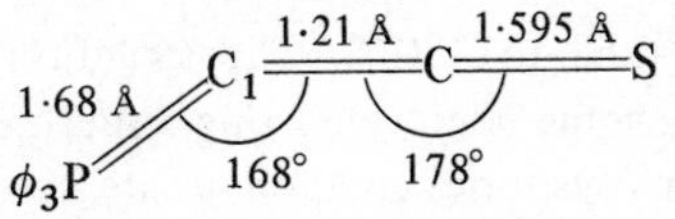

are remarkable for the non-linear bonds formed by C_1.[6],[7] The C_1—C bond length corresponds to a triple bond and P—C to a double bond, the length of a single P—C bond being at least 1·85 Å.

(6) JCS A 1966 1703
(7) JCS A 1967 1913

Carbon

Acetylene and derivatives

A literature reference for acetylene is given in Table 21.1 (p. 727). All molecules R–C≡C–R are linear, and in all cases the length of the C≡C bond is close to the value in acetylene itself (1·205 Å). In compounds containing the system –C≡C–C≡C– the central bond is only slightly longer (1·37 Å) than the double bond in ethylene (1·34 Å). Also, the terminal bond in molecules X–C≡C–X is appreciably shorter than a normal single bond. For example, in the molecules $H_3C . C≡CX$ (X = Cl, Br, or I) H_3C–C is approximately 1·46 Å and C–X is 1·64,[1] 1·79,[2] and 1·99 Å[2] as compared with the single-bond lengths, C–C, 1·54, C–Cl, 1·77, C–Br, 1·94, and C–I, 2·14 Å.

Very short C–X bonds are also found in HCCF and HCCCl:[3]

$$ F\underset{1\cdot279}{\text{——}}C\underset{1\cdot198}{\text{——}}C\underset{1\cdot053\,A}{\text{——}}H \qquad Cl\underset{1\cdot637}{\text{——}}C\underset{1\cdot204}{\text{——}}C\underset{1\cdot055\,A}{\text{——}}H $$

these C–X bond lengths being very similar to those in cyanogen halides.

Cyanogen and related compounds

Cyanogen

For the cyanogen molecule, which is linear, the simplest formulation is N≡C–C≡N. The length of the central bond, 1·38 Å, however, is the same as in diacetylene, HC≡C–C≡CH, showing that there is considerable π bonding. The C≡N bond length is the same as in HCN. Bond lengths in cyanogen and related compounds are given in Table 21.7.

TABLE 21.7

Structural data for cyanogen and related compounds (Å)

	C≡N	C–C	C≡C	*Reference*
N≡C–C≡N	1·15	1·38	–	JCP 1965 **43** 3193
N≡C–C≡CH	1·157	1·382	1·203	JACS 1950 **72** 199
N≡C–C≡C–C≡N	1·14	1·37	1·19	AC 1953 **6** 350

Cyanogen and its derivatives are of special interest because of the ease with which they polymerize. When cyanogen is prepared by heating (to 450°C) mercuric cyanide, preferably with the chloride, there is also obtained a quantity of a solid polymer, paracyanogen $(CN)_n$, which may also be made by heating oxamide in a sealed tube to 270°C. The structure of this extremely resistant material is not known; some conjugated ring structures have been suggested. The structures of the trimeric (cyanuric) compounds are noted later.

HCN: *cyanides and isocyanides of non-metals*

Only one isomer of HCN is known, but organic derivatives are of two kinds, cyanides RCN and isocyanides RNC. The latter have been formulated R–N≡C, but

(1) JCP 1955 **23** 2037
(2) JCP 1952 **20** 735
(3) TFS 1963 **59** 2661

in fact the C—N bond length in CH_3NC is only slightly greater (1·167 Å) than in CH_3CN (1·158 Å), which is close to that in HCN itself (1·155 Å). The force constants of the C—N bonds in HCN, CH_3CN, and the cyanogen halides, XCN, are in all cases close to 1800 N m^{-1}, and therefore cyanides are satisfactorily represented R—C≡N.

The structures in the vapour state of HCN and of a number of covalent cyanides and isocyanides containing the linear systems R(M)—C—N and R(M)—N—C respectively are summarized in Table 21.8. There are two crystalline forms of HCN,

TABLE 21.8
Structures of covalent cyanides and isocyanides

Molecule	C—N	Other data		Reference
HCN	1·155 Å	C—H	1·063 Å	TFS 1963 **59** 2661
CH_3CN	1·158	C—C	1·460	PR 1950 **79** 54
				JACS 1955 **77** 2944
CH_3NC	1·167	H_3C—N	1·426	RMP 1948 **20** 668
CF_3CN	1·15	C—C	1·475	JACS 1955 **77** 2944
		C—F	1·300	JCP 1952 **20** 591
	C—M—C	M—C	M—C≡N	
$S(CN)_2$	96°	1·73 Å	177°	AC 1966 **21** 970
$P(CN)_3$	93°	1·78	172°	AC 1964 **17** 1134
$As(CN)_3$	92°	1·90	171°	AC 1963 **16** 113
H_3SiCN	–	1·847	–	JCP 1960 **32** 1577

the low-temperature orthorhombic form changing reversibly into a tetragonal polymorph at −102·8°C. The structures of both forms are essentially the same, the molecules being arranged in endless chains:[1]

(1) AC 1951 4 330

$$\cdots N≡C-H\cdots\cdots N≡C-H\cdots$$
$$\leftarrow 3·18\ \text{Å} \rightarrow$$

The dipole moment of HCN is 2·6 D in benzene solution and 2·93 D in the vapour. The abnormally high dielectric constant of liquid HCN is attributed to polymerization, and there is i.r. evidence for dimerization in an inert matrix at low temperatures.[2]

(2) JCP 1968 **48** 1685

HCN polymerizes to a dark-red solid which on recrystallization yields almost colourless crystals of $(HCN)_4$. This 'tetramer' is actually diaminomaleonitrile:[3]

(3) AC 1961 **14** 589

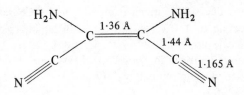

Carbon

The approximately collinear configuration of M–C≡N has been confirmed in a number of molecular cyanides (Table 21.8). In some of these cyanides there are intermolecular M···N distances less than the sums of the van der Waals radii, and in $S(CN)_2$, for example, the two next nearest neighbours of S are coplanar with the intramolecular C atoms, suggesting that intermolecular interactions may be responsible for the small deviations of the M–C≡N systems from exact collinearity.

The structures of metal cyanides are described in the next chapter.

Cyanogen halides

All four cyanogen halides are known. The fluoride is a gas at ordinary temperatures and is prepared by pyrolysis of $(FCN)_3$ which in turn is made from $(ClCN)_3$ and NaF in tetramethylene sulphone. The chloride is a liquid boiling at $14 \cdot 5°C$ which readily polymerizes to cyanuric chloride, $Cl_3C_3N_3$, and is prepared by the action of chlorine on metal cyanides or HCN. The bromide and iodide, also prepared from halogens and alkali cyanides, form colourless crystals. All four halides have been studied in the vapour state and all except FCN also in the crystalline state (Table 21.9).

TABLE 21.9

Structures of cyanogen halides

	C–X	C–N	C–X in CX$_4$	X···N in crystal	Method	*Reference*
FCN	1·26 Å	1·16 Å	1·32 Å	–	m.w.	TFS 1963 **59** 2661
ClCN	1·63	1·16	1·77	–	m.w.	JCP 1965 **43** 2063
				3·01 Å	X	AC 1956 **9** 889
BrCN	1·79	1·16	1·94		m.w.	PR 1948 **74** 1113
				2·87	X	JCP 1955 **23** 779
ICN	2·00	(1·16)	2·14		m.w.	PR 1948 **74** 1113
				2·8	X	RTC 1939 **58** 448

The molecules are linear, and in the crystals are arranged in chains

$$----X{-}C{\equiv}N----X{-}C{\equiv}N----$$

There appears to be some sort of weak covalent bonding between X and N atoms of adjacent molecules, the effect being most marked in the iodide, where the intermolecular I···N distance is only $2 \cdot 8$ Å, far less than the sum of the van der Waals radii (see p. 232). The C–X bond lengths within the molecules are appreciably less than the single-bond values in the molecules CX_4 and are the same as in $H_3C{-}C{\equiv}C{-}X$, but C≡N is very close to the value in HCN.

Cyanamide, $H_2N \cdot CN$

This colourless crystalline solid formed from NH_3 and ClCN readily polymerizes, for example, by heating to $150°C$, to the trimer melamine (see later). The bond lengths in the molecule are:[1] N–C, $1 \cdot 15$ Å, and C–NH$_2$, $1 \cdot 31$ Å. The group N–C–N also is unsymmetrical in (covalent) heavy metal derivatives such as Pb(NCN)[2] (bond lengths approximately $1 \cdot 17$ and $1 \cdot 25$ Å) presumably because it

(1) SPC 1961 **6** 147

(2) AC 1964 **17** 1452

is more strongly bonded to the metal through one N. Derivatives of more electro-positive metals are salts containing the linear symmetrical NCN^{2-} ion (studied in $CaNCN$);[3] compare N–C, 1·22 Å, with N–N, 1·17 Å, in the isoelectronic N_3^- ion. (Only approximate bond lengths were determined in $Sr(NCN)$.)[4]

(3) ACSc 1962 **16** 2263

(4) ACSc 1966 **20** 1064

Dicyandiamide, $NCNC(NH_2)_2$

The dimer of cyanamide is formed when calcium cyanamide is boiled with water. This also is a colourless crystalline compound, the molecule of which has the structure:[5]

(5) JACS 1940 **62** 1258

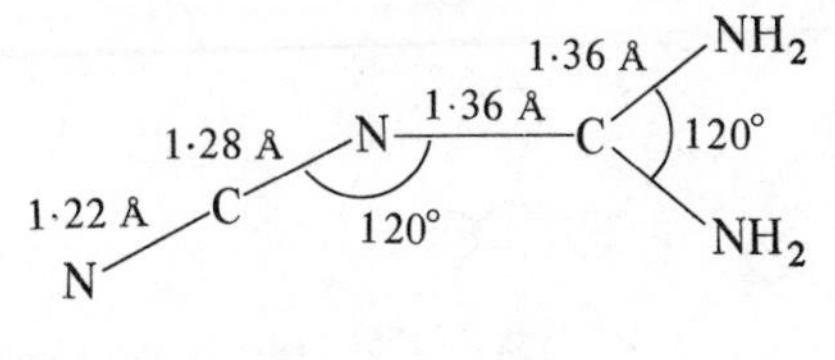

Cyanuric compounds

The cyanogen halides polymerize on standing to the trimers, the cyanuric halides. Cyanamide is converted into the corresponding trimer, melamine, on heating to about 150°C. Isocyanic acid, however, polymerizes far more readily. If urea is distilled, isocyanic acid is formed but polymerizes to cyanuric acid, $(NCOH)_3$, a crystalline solid, the vapour of which, on rapid cooling, yields isocyanic acid as a liquid which, above 0°C, polymerizes explosively to cyamelide, a white porcelain-like solid. This latter material is converted into salts of cyanuric acid by boiling with alkalis. These reactions are summarized in Chart 21.1. Cyanuric derivatives

CHART 21.1

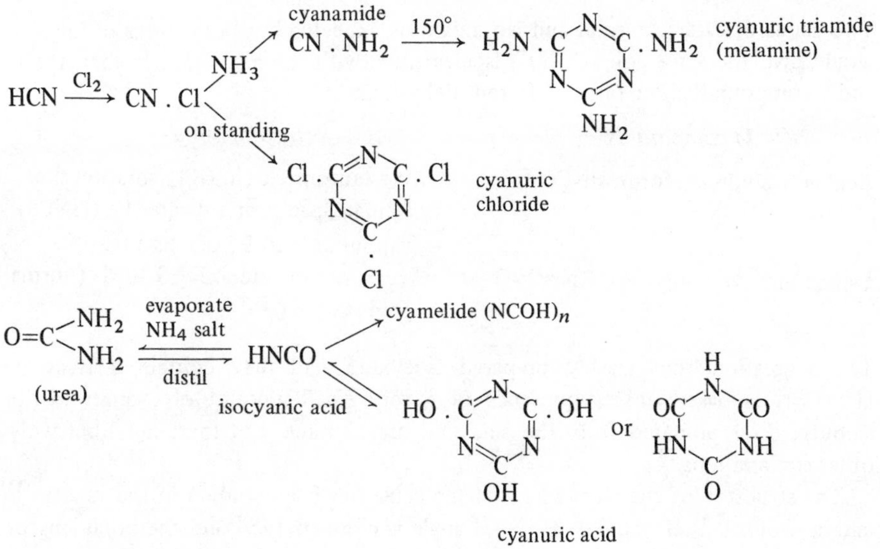

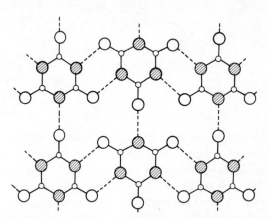

Carbon

may be interconverted. For example, if cyanuric bromide is warmed with water, cyanuric acid is formed. Cyanuric chloride is converted by the action of NaF in tetramethylene sulphone to the fluoride, $(FCN)_3$, pyrolysis of which provides pure FCN in good yield.

The structure of the cyanuric ring has been studied in several derivatives, for example, the triazide, triamide (melamine), and cyanuric acid (Fig. 21.5). The ring is a nearly regular hexagon with six equal C–N bonds of length 1·34 Å. In crystalline cyanuric acid the molecules are arranged in layers,[6] six N–H–O bonds linking each molecule to four neighbours as shown in Fig. 21.6.

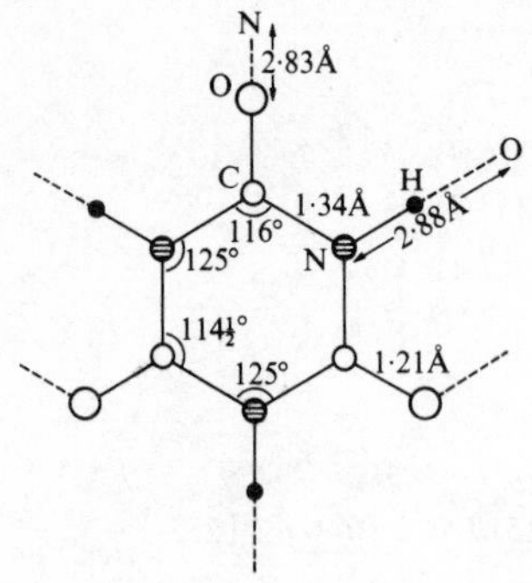

FIG. 21.5. The configuration of the molecule in crystalline cyanuric acid. Broken lines indicate N–H····O bonds to adjacent molecules.

(6) JACS 1952 **74** 6156

FIG. 21.6. Layer of hydrogen-bonded molecules in crystalline cyanuric acid.

Isocyanic acid and isocyanates

In the acids HNCO and HNCS there is the possibility of tautomerism:

$$H–O–C\equiv N \quad \text{or} \quad H–N=C=O \quad (iso \text{ form})$$
$$(S) \qquad\qquad\qquad (S)$$

Two series of esters, normal and iso, might be expected but both forms of the acid would give the same ion $(OCN)^-$, isoelectronic with the azide ion. In fact, the O and S compounds show rather different behaviours:

O *compounds*	S *compounds*
Acid entirely in *iso* form, HNCO	Acid tautomeric (in CCl_4 solution the Raman spectrum indicates *iso* (HNCS); vapour at least 95 per cent *iso*)
Esters: one type only, *iso*, RN=C=O	Two series of esters: R–S–C≡N (normal) R–N=C=S (*iso*)

The suggestion that freshly prepared isocyanic acid may contain a trace of HO–C≡N is based on the presence of a very small amount of cyanate in the isobutyl ester prepared from the acid and diazobutane, but there are admittedly other explanations.[1]

The structure of the isocyanic acid molecule has been studied in the vapour[2] and crystalline[3] states. The H–N–C angle is close to 128° and the bond lengths

(1) ACSc 1965 **19** 1768
(2) JCP 1950 **18** 990
(3) AC 1955 **8** 646

744

are: N–H, 0·99 Å, N–C, 1·20 Å, and C–O, 1·18 Å. Values of the R–N–C bond angle in esters include 126° in an early e.d. study of $H_3C . NCO$[4] and 150 ± 3° in $(CH_3)_3Si . NCO$.[5] The value 180° found in $Si(NCO)_4$ should perhaps be accepted with reserve.

Few studies have been made of ionic (iso)cyanates. The fact that the sodium salt is isostructural with the azide confirms the linear structure of the ion, but the bond lengths N–C, 1·21 Å, and C–O, 1·13 Å, were not claimed to be highly accurate.[6]

In AgNCO[7] the isocyanate groups are linked by Ag atoms which form two collinear bonds (Ag–N, 2·12 Å). (The silver–oxygen distances, 3·00 Å, are too long for normal bonds; compare 2·05 Å in Ag_2O.)

(4) JACS 1940 **62** 3236
(5) JACS 1966 **88** 416
(6) MSCE 1943 **30** 30
(7) AC 1965 **18** 424

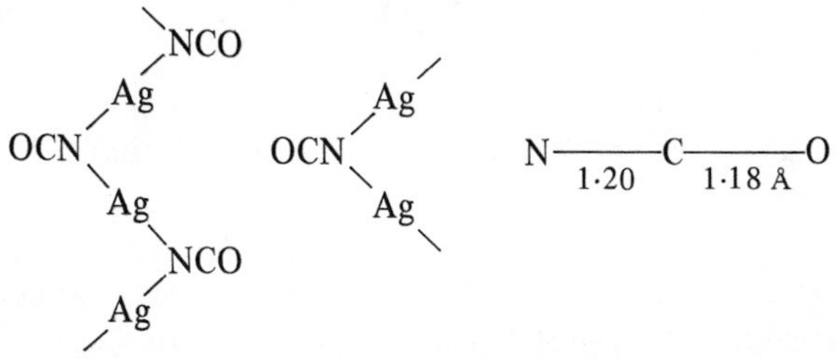

(There are two crystalline forms of the isomeric fulminate, AgCNO, one containing chains somewhat like those in the isocyanate and the other rings of six Ag atoms

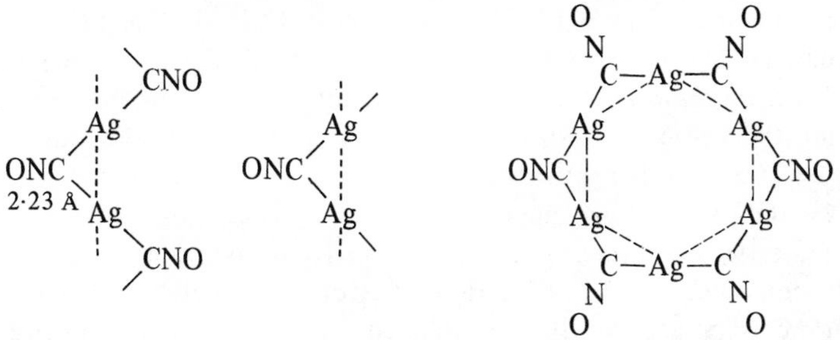

linked through CNO groups.)[8] In the chains the Ag–Ag distance, 2·93 Å, is only slightly longer than in the metal (2·89 Å), and in the rings it is 2·83 Å.

(8) AC 1965 **19** 662

Isothiocyanic acid, thiocyanates, and isothiocyanates
The free acid, prepared by heating KSCN with $KHSO_4$, is in the iso form to at least 95 per cent in the vapour state, and the molecule has the structure (a).[1] A

(1) JMS 1963 **10** 418

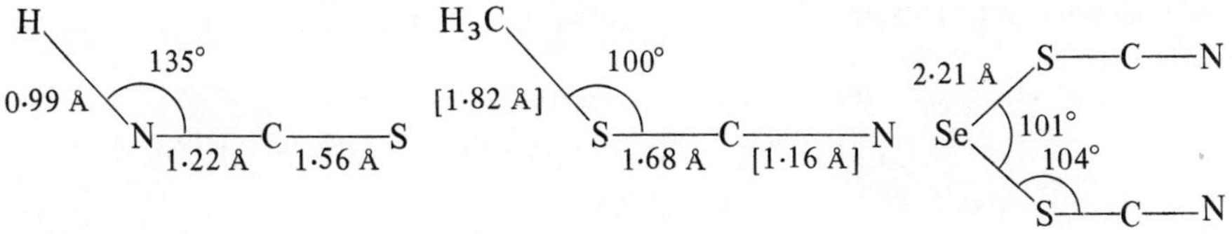

(a)   (b)   (c)

(2) JCP 1965 **43** 3583
(3) JACS 1954 **76** 2649

(4) JACS 1949 **71** 927
(5) JACS 1966 **88** 416
(6) TFS 1962 **58** 1284
(7) SA 1962 **18** 1529 (but see comments in ref. 5)

microwave study of methyl thiocyanate[2] gives the structure (b), values in brackets being assumed. In crystalline $Se(SCN)_2$,[3] in which the system S–C–N was assumed to be linear, the configuration (c) was found, with dihedral angle SSeS/SeSC 79°, but the S–C and C–N bond lengths were not accurately determined.

In molecular isothiocyanates the angle R–N–C has been determined as 142° in an early m.w. study of $H_3C . NCS$,[4] 154° in $(CH_3)_3Si . NCS$,[5] but 180° in $H_3Si·NCS$[6] and $Si(NCS)_4$.[7]

Metal thiocyanates and isothiocyanates

As in the case of metallic cyanides (Chapter 22) we may distinguish three main types:

(i) *ionic crystals containing the SCN⁻ ion,*

(ii) *compounds containing NCS bonded to one metal atom only, either through S (thiocyanate) or through N (isothiocyanate), and*

(iii) *compounds in which —NCS— acts as a bridge between two metal atoms.*

We shall give an example of a compound of type (ii) which is both a thiocyanate and isothiocyanate; $Ni(NH_3)_3(SCN)_2$ may represent a fourth class in which there are both bridging and singly attached NCS groups in the same crystal. In (i) and (iii) there is no distinction between thiocyanate and isothiocyanate.

(1) AC 1968 **B24** 1125
(2) IC 1965 **4** 499

(i) *Ionic thiocyanates.* The dimensions of the (linear) SCN⁻ ion have been accurately determined in KSCN:[1] S–C, 1·69 Å, C–N, 1·15 Å. (The approximate bond lengths in SeCN⁻ in KSeCN are:[2] Se–C, 1·83 Å, C–N, 1·12 Å.)

(ii) *Covalent thiocyanates and isothiocyanates.* Numerous ions and molecules containing the —SCN or —NCS ligands are formed by transition and B subgroup metals (e.g. Cd, Hg). It appears that in general metals of the first transition series except Cu(II) form isothiocyanates, —M–N–C–S, while those of the second half of the second series and those of the third series form thiocyanates, —M–S–C–N, or bridged compounds of class (iii). However, for some metals (e.g. Pd) the choice between M–NCS and M–SCN is affected by the nature of the other ligands attached to M. The evidence comes from i.r. spectroscopy, the stretching frequency of the S–C bond apparently making possible a decision between thiocyanate and isothiocyanate; for example, $Pd(P\phi_3)_2(NCS)_2$ but $Pd(Sb\phi_3)_2(SCN)_2$. The claim that both isomers of $Pd(As\phi_3)_2(SCN)_2$ have been isolated has not been confirmed by a structural study.[3] Nevertheless there is both a thiocyanato and an isothiocyanato ligand in the remarkable square planar Pd complex (a).[4]

(3) IC 1966 **5** 1632
(4) IC 1970 **9** 2754

(5) For refs. see JCS 1961 4590; JACS 1967 **89** 6131

X-ray crystallographic studies have been made of many compounds, of which the following are examples:[5]

Isothiocyanates   tetrahedral: $[Co(NCS)_4]^{2-}$
octahedral: $[Cr(NH_3)_2(NCS)_4]^-$, $Ni(NH_3)_4(NCS)_2$

Thiocyanates   planar: $[Pt(SCN)_4]^{2-}$
tetrahedral: $[Hg(SCN)_4]^{2-}$
octahedral: $[Rh(SCN)_6]^{3-}$, $Cu(en)_2(SCN)_2$.

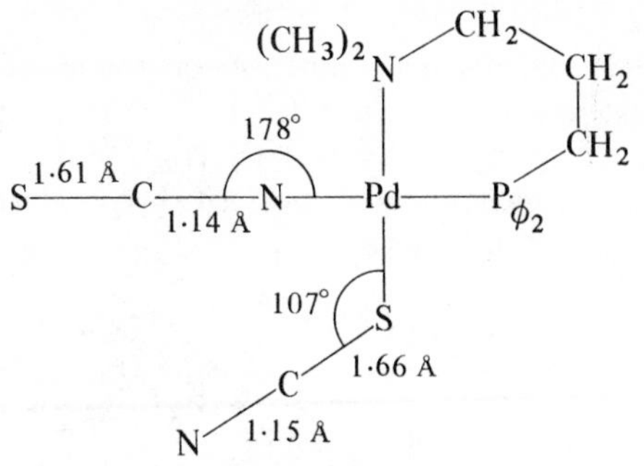

(a)

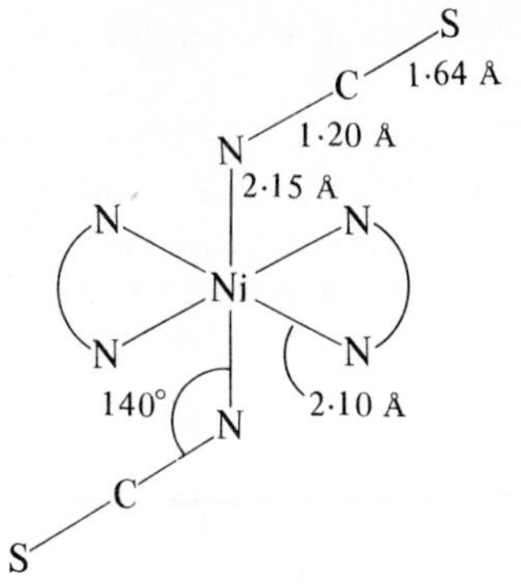

(b)

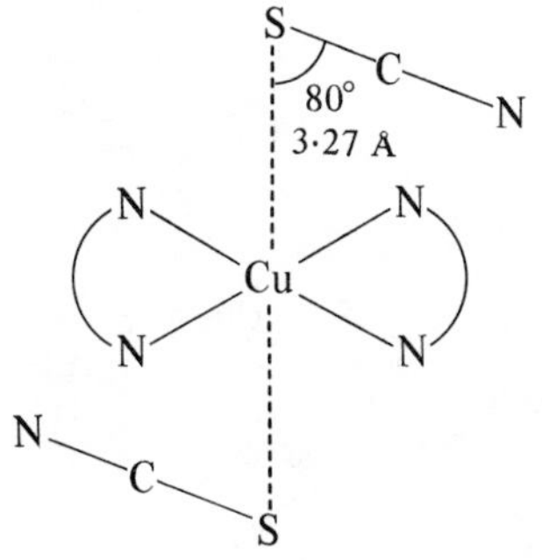

(c)

A feature of the isothiocyanate complexes is the large variation in the angle M–N–C, from 111° in $[Co(NCS)_4]^{2-}$ and 140° in $Ni(en)_2(NCS)_2$[6], (b), to 180° in $[Cr(NH_3)_2(NCS)_4]^-$. The correlation of bond angle with bond lengths, to be expected if the system approximates to $M–N^+ \equiv C–S^-$ or $_M \nearrow N{=}C{=}S$ would not appear to be justified by the data at present available. Bond angles M–S–C found in metal thiocyanato complexes include 80° in $Cu(en)_2(SCN)_2$[7], (c), and 107° in the Pd compound (a), above. Bond lengths in a number of complex thiocyanates and isothiocyanates are close to the values, C–N, 1·16 Å, and C–S, 1·62 Å. Both –S–C–N and –N–C–S are linear to within the experimental errors.

(iii) *Thiocyanates containing bridging –S–C–N– groups.* A study of crystalline AgSCN showed that the crystal is built of infinite chains which are bent at the S atoms and apparently also at the Ag atoms (where the bond angle was determined as

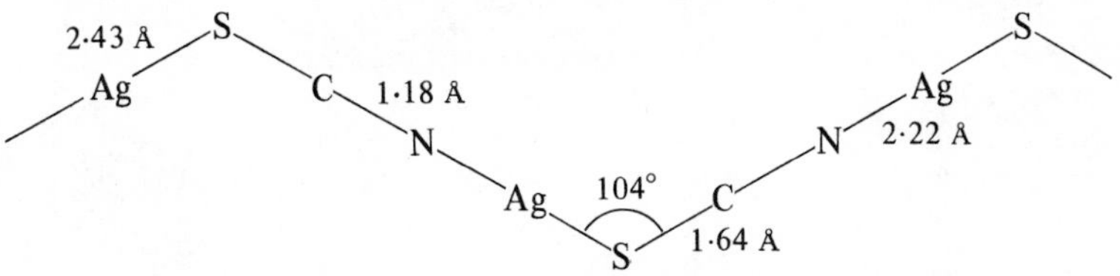

(6) AC 1963 **16** 753

(7) AC 1964 **17** 254

Carbon

(8) AC 1957 **10** 29

(9) JCS 1961 1416 (see p. 979)
(10) AC 1960 **13** 125

165°), though the positions of the lighter atoms are not very certain.[8] On the subject of the Ag bond angle see also p. 879.

Bridging of metal atoms by —S—C—N— occurs in a number of compounds of Cd, Hg, Pb, Co, Ni, Cu, Pd, and Pt. An example of a finite molecule is the Pt complex (a),[9] while infinite linear molecules are formed in $Cd(etu)_2(SCN)_2$[10], (b), and the isostructural Pb compound. (etu = ethylenethiourea, $SC(NHCH_2)_2$). Mercury provides examples of 2D and 3D complexes (Chapter 26).

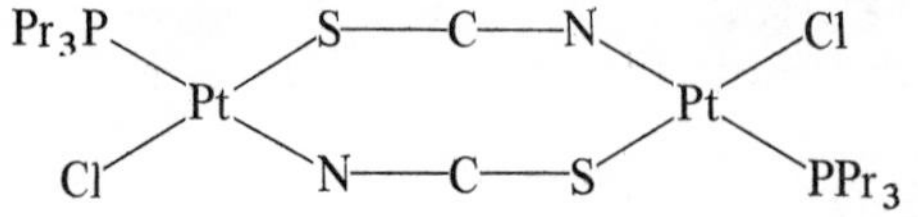

(a)

$(Pr = C_3H_7)$

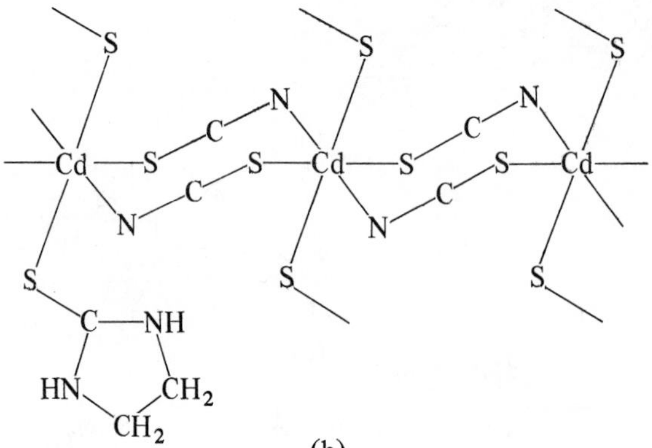

(b)

22

Metal Cyanides, Carbides, Carbonyls, and Alkyls

In this chapter we deal first with compounds formed by metals containing the isoelectronic groups CN^-, C_2^{2-}, and CO. (HCN and non-metal cyanides, and also cyanates and thiocyanates are included in Chapter 21.) We then describe briefly compounds of metals with hydrocarbon radicals or molecules other than olefine compounds of Pd and Pt (Chapter 27) and certain compounds of Cu and Ag (Chapter 25).

Metal cyanides

We may recognize three classes of metal cyanides corresponding to the behaviour of the cyanide ion or group:

(a) *simple ionic cyanides, containing the ion* CN^-,
(b) *molecules or complex ions in which* $-CN$ *is attached to one metal atom only* (*through the* C *atom*), *and*
(c) *compounds in which the* $-CN-$ *group is bonded to metal atoms through both* C *and* N, *so performing the same bridging function as halogen atoms, oxygen, etc.*

In a few complex cyanides CN groups of types (b) and (c) are found, for example, $KCu(CN)_2$ and $K_2Cu(CN)_3 . H_2O$. The structures of these salts are described in Chapter 25.

(a) *Simple ionic cyanides*

At ordinary temperatures NaCN, KCN, and RbCN crystallize with the NaCl structure, and CsCN and TlCN with the CsCl structure. These structures would be consistent with either free rotation of the CN^- ion, when it would behave as a sphere of radius about 1·9 Å (intermediate in size between Cl^- (1·80 Å) and Br^- (1·95 Å)), or with random orientations of CN^- ions along directions [111] consistent with cubic symmetry. A neutron diffraction study of KCN at room temperature indicates free rotation of the ion (C–N, 1·16 Å).[1]

All the above alkali cyanides exist in less symmetrical 'low-temperature' forms, in which the CN^- ion behaves as an ellipsoid of rotation, with long axis about 4·3 Å

(1) AC 1961 **14** 1018

749

and maximum diameter about 3·6 Å. Their structures are closely related to the NaCl or CsCl structures of the high-temperature forms (Table 22.1). (There is also a

TABLE 22.1

Crystal structures of the alkali cyanides

	Low-temp. form	High-temp. form	Transition point	Reference
NaCN	Orthorhombic (Fig. 22.1)	NaCl	+15·3°C	ZK 1938 **100** 201
KCN			−108°	RTC 1940 **59** 908
RbCN	Monoclinic?		Between −100° and −180°	SR 1942–4 **9** 138–142
CsCN	Rhombohedral (Fig. 22.2)	CsCl	−55°?	
NH₄CN	Tetragonal (Fig. 22.3(b)	?	Continuous change	RTC 1944 **63** 39
LiCN	Orthorhombic (Fig. 22.3(a)	−	No transition	RTC 1942 **61** 244

(2) JCP 1968 **48** 1018
(3) AC 1962 **15** 601
(4) JCP 1949 **17** 1146

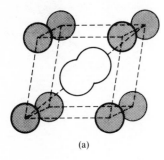

(a)

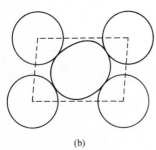

(b)

FIG. 22.2. The crystal structure of the low-temperature (rhombohedral) form of CsCN: (a) unit cell (Cs$^+$ ions shaded); (b) section through the structure showing the packing of the Cs$^+$ and CN$^-$ ions.

metastable rhombohedral form of NaCN made by quenching the cubic form from 200–300°C and also a second low-temperature form.[2] A second low-temperature form of KCN is stable only in the range −108° to −115°C.[3])

The structure of the (orthorhombic) low-temperature form of NaCN (and the isostructural KCN) is illustrated in Fig. 22.1; its close relation to the NaCl structure is evident. (The portion of structure shown is enclosed by (110) planes.) All the CN$^-$ ions are parallel and there is 6-coordination of both anion and cation.

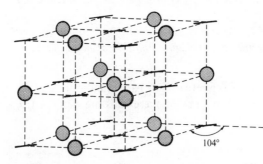

FIG. 22.1. The structure of the low-temperature form of NaCN. The CN$^-$ ions are indicated by short straight lines.

Low-temperature RbCN also has a deformed NaCl structure, but the details of the structure are not known with certainty. The metastable rhombohedral form of NaCN is closely related structurally to the high-temperature form (for $\alpha = 53\cdot2°$) and not, like low-temperature CsCN, to the CsCl structure.[4] The low-temperature form of CsCN is rhombohedral, having a slightly deformed CsCl structure, illustrated in Fig. 22.2, in which the CN$^-$ ions lie along the trigonal axis. The coordination is 8-fold as in CsCl. Lithium cyanide differs from the other alkali cyanides in many ways. It has a low melting point (160°C, compare 550° for NaCN and 620° for KCN), a low density (1·025 g/c.c.), and a loosely-packed crystal structure in which each ion is only 4-coordinated (Fig. 22.3(a)). Also, it is possible

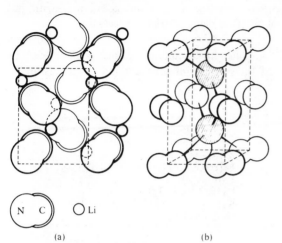

FIG. 22.3. The crystal structures of (a) LiCN, projected on (100), and (b) NH_4CN. In (a) the heavier circles represent atoms lying in a plane 1·86 Å above the others. In (b) the shaded circles represent NH_4^+ ions.

N C   ◯ Li

(a)   (b)

to distinguish the C and N atoms in the CN^- ions, which is not possible in the other alkali cyanides. From an electron count in a Fourier projection it appears that the negative charge resides on the N atom, and that the tetrahedral coordination group around Li^+ consists of one C and three N, while a CN^- ion is surrounded by four Li^+, one near the C and three around the N. The change from $8 \to 6 \to 4$-coordination in CsCN, NaCN, and LiCN is probably not a simple size effect, for the small Li^+ ion appears to polarize the CN^- ion appreciably.

The structure of NH_4CN, which shows no polymorphism in the temperature range +35° to −80°C, also has some interesting features. The structure is very simply related to that of CsCl, the unit cell being doubled in one direction owing to the way in which the CN^- ions are oriented (Fig. 22.3(b)). Although each NH_4^+ ion is surrounded by eight ellipsoidal CN^- ions, owing to the orientations of the latter there are strictly only four in contact with NH_4^+ (at 3·02 Å), the remaining four being farther away (3·56 Å).

(b) Covalent cyanides containing −CN

It should be noted that in X-ray studies of the heavier metal cyanides it is not possible to distinguish between C and N atoms. However, from neutron diffraction studies of $K_3Co(CN)_6$[1] and $K_2Zn(CN)_4$[2] (Zn−C, 2·024 Å, C−N, 1·157 Å) it is concluded that the C atom is bonded to the metal, as has always been assumed.

Only one metal cyanide, $Hg(CN)_2$, is known to have a molecular structure of the type we are considering. The crystal is built of nearly linear molecules[2a]

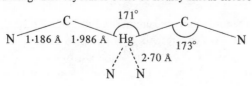

(1) AC 1959 **12** 674
(2) AC 1966 **20** 910

(2a) ACSc 1958 **12** 1568

The two weaker bonds from Hg to N atoms in a plane perpendicular to the C–Hg–C plane may account for the departure of the C–Hg–C bond angle from $180°$. In the ions to be discussed now the C–N bond is collinear with the metal–carbon bond; the system M–C–O in metal carbonyls is also linear.

Many heavy metal cyanides redissolve in the presence of excess CN^- ions to form complex ions $[M(CN)_n]^{m-}$. Since these ions are primarily of interest as illustrating the stereochemistry of the metal atoms, details and references are in most cases given in other chapters.

Bond arrangement	Examples
2 collinear	$K[NC-Ag-CN]$ (also Au)
4 coplanar	$Ba[Pd(CN)_4] \cdot 4\,H_2O$ (also Ni, Pt)
4 tetrahedral	$K_3[Cu(CN)_4]$
6 octahedral	$K_4[Fe(CN)_6] \cdot 3\,H_2O$
8 dodecahedral	$K_4[Mo(CN)_8] \cdot 2\,H_2O$ (also W, Re)
8 antiprism	$Na_3[Mo(CN)_8] \cdot 4\,H_2O$ (also W)

The octahedral configuration of a number of $M^{III}(CN)_6^{3-}$ ions has been established by X-ray studies of, for example, $K_3Co(CN)_6$ (above), $[Co(NH_3)_6][Cr(CN)_6]$, and $[Co(NH_3)_5H_2O][Fe(CN)_6]$, the last two crystallizing with the $[Ni(H_2O)_6][SnCl_6]$ structure. The ferrocyanide ion has been studied in

(3) AC 1969 **B25** 1685

$K_4Fe(CN)_6 \cdot 3\,H_2O$ and in the free acid, $H_4Fe(CN)_6$ (Fe–C, 1.89 Å).[3]

Octacyanide ions are formed by Mo(IV) and (V), W(IV) and (V), and Re (V) and (VI). Since the original proof of the dodecahedral (D_{2d}) configuration of the anion in $K_4Mo(CN)_8 \cdot 2\,H_2O$ (with which the W compound is isostructural) there has been much interest in the configuration of these octa-coordinate ions because n.m.r., e.s.r., magnetic and other measurements suggested antiprismatic (D_{4d}) geometry. X-ray studies show that (i) the same ion, for example, $Mo(CN)_8^{3-}$, may have different configurations in different salts, and (ii) the same configuration may be adopted by ions such as $W(CN)_8^{3-}$ and $W(CN)_8^{4-}$ in spite of the presence of the extra electron in the latter. There is only a slight, though possibly significant, difference between the shapes of the antiprismatic coordination groups in the last two cases. Evidently there is not a very large energy difference between the D_{2d} and D_{4d} configurations, the detailed geometry of which is given in Chapter 3. It appears that the geometry of these ions cannot be deduced from factors such as ligand–ligand repulsions, π-bonding, etc., but that lattice energy is the most important factor. An extreme example of the importance of lattice energy is the series of fluorotantalates Na_3TaF_8, K_2TaF_7, and $CsTaF_6$, where the size of the cation actually determines the *composition* of compounds crystallizing from solution.

Structural data for these ions are summarized in Table 22.2.

A complex cyanide which presents a problem in bonding is the salt with the empirical formula $K_2Ni(CN)_3$, originally supposed to contain Ni(I). In fact the salt is diamagnetic both in solution and in the solid state; Ni(I) would contain one

752

TABLE 22.2

Structures of octacyanide ions

Dodecahedral (D_{2d})	Antiprismatic (D_{4d})	Reference
$Mo(CN)_8^{3-}[(n\text{-}C_4H_9)_4N]_3$		IC 1970 **9** 356
	$Mo(CN)_8^{3-}Na_3 \cdot 4 H_2O$ $W(CN)_8^{3-}Na_3 \cdot 4 H_2O$ }	AC 1970 **B26** 684
$Mo(CN)_8^{4-}K_4 \cdot 2 H_2O$		JACS 1968 **90** 3177
	$W(CN)_8^{4-}H_4 \cdot 6 H_2O$	AC 1970 **B26** 1209

unpaired electron. The complex ion is apparently dimeric, but the nature of the bridge between the metal atoms is not known.[4] Reduction of this salt yields the yellow $K_4Ni(CN)_4$, presumably containing a tetrahedral ion $Ni^0(CN)_4^{4-}$ iso-electronic with the tetrahedral $Ni(CO)_4$ molecule.

(4) JACS 1956 **78** 702

(c) *Covalent cyanides containing* –CN–

In some metal cyanides the CN group acts as a bridge between pairs of metal atoms by forming bonds from both C and N, the atoms M–C–N–M being collinear or approximately so. The simplest finite molecule of this kind which has been studied is $(NH_3)_5Co \cdot NC \cdot Co(CN)_5$ (a), in its monohydrate.[1] A more complex example

(1) IC 1971 **10** 1492

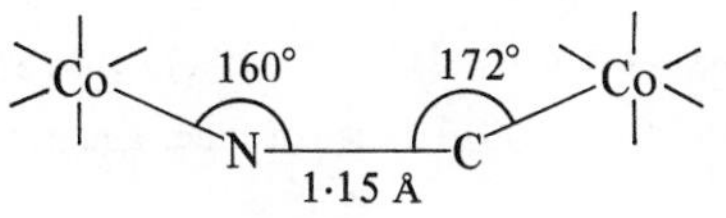

(a)

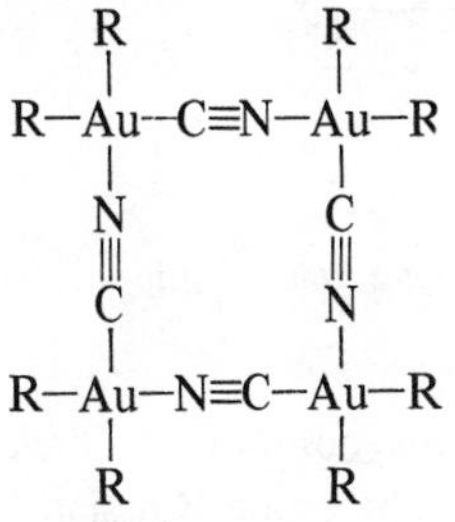

(b)

(2) PRS A 1939 **173** 147

is the cyclic molecule $[Au(n\text{-}C_3H_7)_2 \cdot CN]_4$ (b).[2] Examples of the other main types of structure include:

(3) JCS 1958 3412
(4) APURSS 1945 **20** 247
(5) CRURSS 1941 **31** 350

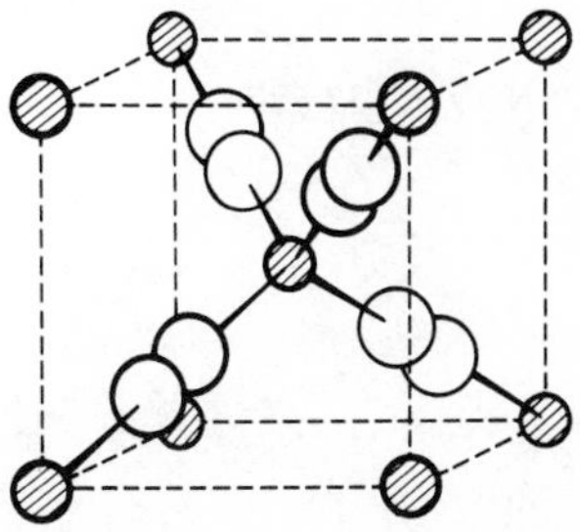

chain structures: AgCN and AuCN, $Ag_3CN(NO_3)_2$ (for which see Chapter 25),
layer structures: not known for simple cyanides, but found in $Ni(CN)_2 \cdot NH_3 \cdot C_6H_6$[3] (p. 101), and
3-dimensional frameworks: $Zn(CN)_2$[4] and $Cd(CN)_2$[5] with the anti-Cu_2O structure (p. 106), and Prussian blue and related compounds which are discussed in the next section. The anti-cuprite structure is shown in Fig. 22.4. As in Cu_2O the structure consists of two identical interpenetrating nets which are not cross-connected by any Cu–C or Cu–N bonds (see Chapter 3).

FIG. 22.4. Structure proposed for $Cd(CN)_2$ and the isostructural $Zn(CN)_2$.

Prussian blue and related compounds

When cyanides of certain metals, for example, Fe, Co, Mn, Cr, are redissolved by the addition of excess alkali cyanide solution, complex ions $M(CN)_6$ are formed.

The alkali and alkaline-earth salts of these complexes are fairly soluble in water and crystallize well. The ferrocyanides and chromocyanides (for example, $K_4Fe(CN)_6$ and $K_4Cr(CN)_6$) are pale-yellow in colour. The ferricyanides and manganicyanides, for example, $K_3Fe(CN)_6$ and $K_3Mn(CN)_6$, are deeper in colour and are made by oxidation of the $M(CN)_6^{4-}$ salts. Solutions of the manganese and cobalt $M(CN)_6^{4-}$ ions oxidize spontaneously in the air to the $M(CN)_6^{3-}$ salts. When solutions of these complex cyanides are added to solutions of salts of transition metals or Cu^{II}, insoluble compounds are precipitated in many cases. Thus addition of potassium ferrocyanide to ferric salt solutions gives a precipitate of Prussian blue, and cupric salts produce brown cupric ferrocyanide. These precipitates are usually flocculent and may form colloidal solutions on washing. When dried they are very stable towards acids, whereas alkali ferrocyanides yield the acid $H_4Fe(CN)_6$ on treatment with dilute HCl. Such compounds are obviously quite different in structure from the simple soluble ferrocyanides.

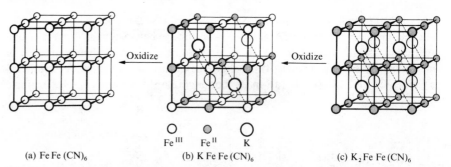

FIG. 22.5. The relationship between (a) Berlin green, (b) Prussian blue, and (c) potassium ferrous ferrocyanide.

The iron compounds have been known for several centuries and valued as pigments. The name Prussian (Berlin) blue is given to the compound formed from a solution of a ferric salt and a ferrocyanide, while that obtained from a ferrous salt and a ferricyanide has been known as Turnbull's blue. It appears from Mössbauer studies that both of these compounds are (hydrated) $Fe_4^{III}[Fe^{II}(CN)_6]_3$, with high-spin Fe^{III} and low-spin Fe^{II} in the ratio 4 : 3.[1] There is also a 'soluble Prussian blue' with the composition $KFe[Fe(CN)_6]$ which is also a ferrocyanide. Most, if not all, of the compounds we are discussing are hydrated; we consider the water of hydration later.

An early X-ray study[2] of Prussian blue and some related compounds showed that in ferric ferricyanide (Berlin green), $FeFe(CN)_6$, Prussian blue, $KFeFe(CN)_6$, and the white insoluble $K_2FeFe(CN)_6$, there is the same arrangement of Fe atoms on a cubic face-centred lattice. In Fig. 22.5 ferrous atoms are distinguished as shaded and ferric as open circles. In (a) all the iron atoms are in the ferric state; in (b) one-half the atoms are Fe^{II} and the others Fe^{III}, and alkali atoms maintain electrical neutrality. These are at the centres of alternate small cubes, and it was supposed that in hydrated compounds water molecules could also be accommodated in the interstices of the main framework. Lithium and caesium, forming

(1) JCP 1968 **48** 3597

(2) N 1936 **137** 577

very small and very large ions respectively, do not form compounds with this structure. In (c), with all the iron in the ferrous state, every small cube contains an alkali-metal atom. The CN groups were placed between the metal atoms along all the full lines of Fig. 22.5, so that each transition metal atom is at the centre of an octahedral group of 6 C or 6 N atoms. The whole system, of composition $M'M''(CN)_6$ thus forms a simple 6-connected 3D framework. Alkali-metal compounds with the Prussian blue structure include $KCuFe(CN)_6$, $KMnFe(CN)_6$, $KCoFe(CN)_6$, $KNiFe(CN)_6$, and $KFeRu(CN)_6$ (ruthenium purple).

Many ferro- and ferri-cyanides of multivalent metals also have cubic structures of the same general type. The unit cell of Fig. 22.5 contains 4 M', 4 M'', and 24 CN, where M' is the atom at the origin and face-centres and M'' at the body-centre and mid-point of each edge. For a complete 3D framework we therefore require a formula of the type $M'M''(CN)_6$, but in many of these complex cyanides, including Prussian blue itself, the ratio of $M' : M''$ is not $1 : 1$. There are two possibilities for

TABLE 22.3

Prussian blue and related compounds

Number of atoms for 24 CN per cell		Example	Number of M'' for 4 M'	Number of CN per cell
M'	M''			
[C] 3	4	$Ti_3[Fe(CN)_6]_4$	$5\frac{1}{3}$	(32) [D]
4	4	$FeFe(CN)_6$	4	24
$5\frac{1}{3}$	4	$Fe_4[Fe(CN)_6]_3$	3	18
[A] 6	4	$Co_3[Co(CN)_6]_2$	$2\frac{2}{3}$	16 [B]
8	4	$Cu_2[Fe(CN)_6]$	2	12

the structure of these compounds. If the metal–cyanide framework is to be maintained intact (requiring 24 CN per cell) there would be excess M' atoms in some of these structures, the group A in the left-hand column of Table 22.3. Alternatively there could be a deficit of M'' atoms (which implies fewer than 24 CN per cell) as at B in the right-hand column. Density measurements distinguish between A and B for $Co_3^{II}[Co^{III}(CN)_6]_2 \cdot 12 H_2O$.[3] (and for the isostructural Cd and Mn[4] cobalticyanides) and confirm B. The unit cell contains $1\frac{1}{3}$ formula weights, that is, $Co_4^{II}Co_{2\frac{2}{3}}^{III}(CN)_{16} \cdot 16 H_2O$, and therefore one-third of the Co^{III} positions are unoccupied and at the same time one-third of the CN positions are occupied by 8 H_2O molecules. The remaining 8 H_2O are situated at the centres of the octants of the cell (K^+ positions in Fig. 22.5(c)). The average environment of Co^{II} is $N_4(H_2O)_2$. In Prussian blue, $Fe_4^{III}[Fe(CN)_6]_3 \cdot 15 H_2O$ (made by slowly diluting a solution of soluble Prussian blue in concentrated HCl) there is 1 formula weight per cell (i.e. 4 Fe^{III}), and replacement of one-quarter of the CN by H_2O, giving the mean environment of Fe^{III} as $N_{4.5}(H_2O)_{1.5}$.[5] For a compound such as $Ti_3[Fe(CN)_6]_4$ the converse situation arises, as at D in Table 22.3, and since the

(3) HCA 1968 **51** 2006
(4) IC 1970 **9** 2224

(5) C 1969 **23** 194

total number of CN per cell cannot exceed 24 the possibility C would presumably be realized if the same general type of structure is adopted, but this point has not yet been verified.

Miscellaneous cyanide and isocyanide complexes

In addition to complex cyanides containing the simple CN group bonded to metal (through C) there are also metal complexes in which cyanide molecules R . CN or isocyanides R . NC are bonded to metal atoms. In the former case the bond to the metal is necessarily through N, as in $[Fe(NCH)_6](FeCl_4)_2$, a salt made by dissolving $FeCl_3$ in anhydrous HCN (Fe–Cl, 1·89 Å, Fe–N, 2·16 Å, angle Fe–N–C, 171°).[1] Isocyanides must bond to metal through C and numerous compounds of this type are formed by transition metals. In these isocyanide complexes CN . R behaves like CO—compare $Cr(CN . C_6H_5)_6$ with $Cr(CO)_6$, and $Fe(NO)_2(CNR)_2$ with $Fe(NO)_2(CO)_2$. Like CO isocyanides stabilize low oxidation states of metals, as in the diamagnetic $[Mn^I(CNR)_6]I$ and the yellow paramagnetic $[Co^I(CN . CH_3)_5]ClO_4$. The cation in this salt has a trigonal bipyramidal configuration with essentially linear Co–C–N–C and Co–C, 1·87 Å (compare 2·15 Å for a single bond).[2] Another example is the salt $[Fe(CN . CH_3)_6]Cl_2 . 3 H_2O$, in which there is a regular octahedral arrangement of six CN . CH_3 molecules around Fe, with Fe–C, 1·85 Å, and C–N, 1·18 Å.[3] There are also mixed coordination groups in molecules such as the *cis* and *trans* isomers of $Fe(CN)_2(CN . CH_3)_4$.[4]

Organic dicyanides NC . R . CN can act as bridging ligands by coordinating to metal through both N atoms, and examples of structures of this kind were noted in Chapter 3.

For the nitroprusside ion, $[Fe(CN)_5NO]^{2-}$ see p. 654.

Metal carbides

Many metals form carbides M_xC_y which are prepared either by direct union of the elements, by heating the metal in the vapour of a suitable hydrocarbon, or by heating the oxide or other compound of the metal with carbon. Their chemical and physical properties suggest that they may be divided into four main groups:

(1) the salt-like carbides of metals of the earlier Periodic Groups,
(2) the carbides of 4f and 5f elements,
(3) the interstitial carbides of transition metals, particularly of the IVth, Vth, and VIth Periodic Groups, and
(4) a less well-defined group of carbides comprising those of metals of Groups VII and VIII and also some of those of elements of Group VI.

Although carbides MC_2 in groups (1) and (2) are similar structurally they have very different properties, so that it is not convenient to adopt a purely structural classification. Moreover the dividing lines in Table 22.4 are not entirely clear-cut. For example, Tb_2C and Ho_2C (like Y_2C) have the anti-$CdCl_2$ structure, and above 900°C yttrium carbide has a range of composition (YC_x, $0·3 < x < 0·7$) and a defect NaCl structure.[1] These c.p. structures are characteristic of the interstitial

(1) JSSC 1970 **2** 421

(2) IC 1965 **4** 318

(3) JCS 1945 799
(4) JCS 1957 719

(1) JCP 1969 **51** 3863, 3872

carbides of class 3. On the other hand, UC has the NaCl structure like many interstitial carbides but unlike them belongs to the group of reactive heavy metal carbides (PuC, Pu_2C_3, CeC_2, UC_2, etc.) which have metallic properties but are readily hydrolysed. Also, whereas WC does not have a c.p. structure, WC_{1-x} has the NaCl structure with c.p. metal atoms.

TABLE 22.4

Examples of metal carbides

Be_2C	Al_4C_3	Ti_2C	V_2C	$Cr_{23}C_6$	Mn_4C	Fe_3C	Co_3C	Ni_3C
		TiC	V_4C_3	Cr_7C_3	$Mn_{23}C_6$	$Fe_{20}C_9$	Co_2C	
CaC_2	*Class 1*		V_6C_5	Cr_3C_2	Mn_3C			
SrC_2			V_8C_7		Mn_3C_2.			
BaC_2					Mn_7C_3	*Class 4*		

4f and 5f carbides				W_2C		
M_2C_3, MC_2	Y_2C			WC_{1-x}		
ThC_2	Tb_2C			WC	RuC	OsC
UC	Ho_2C	*Class 3*				
Class 2						

This rough classification is illustrated by the examples of Table 22.4. Little appears to be known of the structures of carbides of B subgroup elements or of the carbides M_2C_2 formed by the alkali metals.[2]

Class 1 The carbides of the more electropositive elements have many of the properties associated with ionic crystals. They form colourless, transparent crystals which at ordinary temperatures do not conduct electricity. They are decomposed by water or dilute acids, and since the negative ions are unstable, hydrocarbons are evolved. According to the type of C_x^{n-} ion present in the crystal we may divide these carbides into two main groups:

(a) Those containing discrete C atoms (or C^{4-} ions).
(b) Those in which C_2^{2-} ions may be distinguished.
(There is no proof yet of more complex carbon anions in ionic carbides, but the fact that Mg_2C_3 on hydrolysis yields chiefly allylene, $CH_3-C\equiv CH$, has been taken to suggest the presence of C_3^{4-} ions in this crystal. It is interesting that MgC_2, which may be prepared by the action of acetylene on $Mg(C_2H_5)_2$ and yields acetylene on hydrolysis, is unstable at high temperatures and breaks down into Mg_2C_3 and carbon.)

Carbides of type (a) yield CH_4 on hydrolysis. Examples are Be_2C, with the anti-fluorite structure, and Al_4C_3. The structure of the latter is rather more complex and its details do not concern us here.[3] It is sufficient to note that each carbon atom is surrounded by Al atoms at distances from 1·90 to 2·22 Å, the shortest C–C distance being 3·16 Å. As in Be_2C therefore there are discrete C atoms, accounting for the hydrolysis to CH_4. The alkaline-earth carbides, type (b), crystallize at room temperature with the CaC_2 structure (Fig. 22.6). (There is some

(2) For Li_2C_2 see AC 1962 **15** 1042
(3) AC 1963 **16** 559

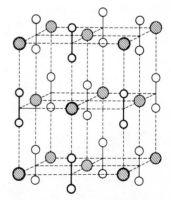

FIG. 22.6. The crystal structure of CaC_2.

(4) JACS 1943 **65** 602, 1482

doubt about the structure of MgC_2.[4]) The CaC_2 structure is a NaCl-like arrangement of M^{2+} and C_2^{2-} ions, the symmetry having dropped from cubic to tetragonal owing to the parallel alignment of the anions. Carbides of type (b) yield acetylene on hydrolysis; compare the hydrolysis of BaO_2 (with the same structure) to H_2O_2.

(5) JACS 1958 **80** 4499

Class 2 The 4f elements from La to Ho form carbides of three types[5]: M_3C, in which the C atoms occupy at random one-third of the octahedral holes in a NaCl-like structure, M_2C_3 with the Pu_2C_3 structure, and MC_2 with the CaC_2 structure. In the Pu_2C_3 structure[6] all the C atoms are present as C_2 groups; there are 6 C_2 groups in the unit cell, which contains 4 Pu_2C_3.

(6) AC 1952 **5** 17

The structure of ThC_2 is very similar to the CaC_2 structure but of lower symmetry (monoclinic). The similarity to the NaCl structure is not readily seen from the conventional diagram showing a unit cell of the structure, but can be seen from Fig. 22.7 where the relation to the monoclinic axes is indicated.

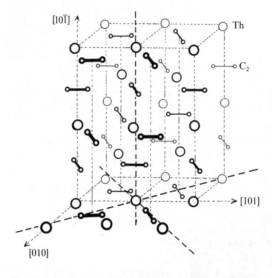

FIG. 22.7. The crystal structure of ThC_2. The heavy broken lines are edges of an orthogonal NaCl-type cell with dimensions 11·7, 11·7, and 10·3 Å.

Although the 4f and 5f carbides MC_2 crystallize with the CaC_2 or ThC_2 structures these are not salt-like carbides like those of the alkaline-earths. For example, CaC_2 is an insulator but LaC_2, ThC_2, and UC_2 have metallic lustre and electrical conductivities similar to those of the metals. Also, whereas CaC_2 gives largely C_2H_2 on hydrolysis, ThC_2 gives almost exclusively CH_4, as also does UC at temperatures between 25° and 100°C. The fact that there is no direct relation between the structures of heavy metal carbides and their hydrolysis products is clearly shown by the behaviour of LaC_2 and CeC_2.[7] At room temperature the product, though a complex mixture of hydrocarbons, consists predominantly of acetylene, but at 200°C contains no acetylene.

(7) IC 1962 **1** 345, 683

The differences in physical and chemical properties between these carbides MC_2 and M_2C_3 appear to be related to the C–C bond lengths in the C_2^{2-} ions; a number of these have now been determined by neutron diffraction (Table 22.5). The value in CaC_2 is that expected for the $(C\equiv C)^{2-}$ ion, while those in the 4f and 5f carbides, approaching the value for a double bond, suggest that one or more electrons have entered the conduction bond, leading to metallic conduction. (The cell dimensions of EuC_2 suggest that this compound is probably $Eu^{II}C_2$, but the C–C bond length is not known.[8])

(8) IC 1964 **3** 335

Most, if not all, of the dicarbides of Classes 1 and 2 have one or more high-temperature forms in which the fixed alignment of the C_2^{2-} ions is relaxed. Compounds such as LaC_2 and UC_2, which at ordinary temperatures have the tetragonal CaC_2 structure, become cubic, the C_2^{2-} ions being randomly oriented along [111] axes (random pyrites, or high-KCN structure). On the other hand, ThC_2 changes first to a tetragonal structure in which there is random orientation of the anions in the basal (001) plane and at 1480°C to a cubic form in which cations and anions are centred at the sites of the NaCl structure. As noted under KCN it is difficult to ascertain whether there is completely random orientation of the ions in

TABLE 22.5
Bond lengths in carbides

C–C	Crystal	Reference
1·19 Å	CaC_2	JCP 1961 **35** 1950
		ACSc 1962 **16** 1212
1·24	La_2C_3, Pr_2C_3, Tb_2C_3	JCP 1961 **35** 1960
1·276	Ce_2C_3	JCP 1967 **46** 4148
1·28–1·29	MC_2 (M = Y, La, Ce, etc.)	JCP 1967 **46** 1891
1·295	U_2C_3	AC 1959 **12** 159
1·34	UC_2 (cubic)	AC 1965 **18** 291
1·35	UC_2 (tetragonal)	JCP 1967 **47** 1188
1·30–1·35	ThC_2	AC 1968 **B24** 1121

a high-temperature cubic form, and if one ion is linear orientation along one of the [111] directions is initially more likely.

Class 3   The interstitial carbides are refractory materials with certain of the characteristic properties of metals (lustre, metallic conductivity) and in addition extraordinary hardness and infusibility. They derive their name from the fact that the metal atoms are close-packed, the C atoms occupying octahedral interstices, but it should be noted that the arrangement of metal atoms is not always the same as in the metal itself. In such cases a carbide MC is not to be regarded as the limit of solid solution of carbon in the metal structure; on the contrary the presence of the C atoms has resulted in the rearrangement of the metal atoms to form the NaCl structure. Because the C atoms occupy octahedral holes in a c.p. metal structure there is a lower limit (around 1·3 Å) to the radius of the metal atom that can form interstitial carbides; metals with smaller radii form carbides with less simple structures. The radii (for c.n. 12) of a number of metals are set out in the accompanying table. The metals to the left of the vertical line, with the exception of Th (which forms ThC_2) form interstitial carbides; those to the right form carbides in which the metal atoms are not close-packed (Class 4).

759

Ti	1·47 Å	V	1·35 Å			Cr	1·29 Å
Zr	1·60	Nb	1·47	Mo	1·40 Å	Mn	1·37[a]
Hf	1·59	Ta	1·47	W	1·41	Fe	1·26
Th	1·80					Co	1·25
						Ni	1·25

(a) The crystal structures of the forms of Mn are complex (see p. 1017).

The main structural features of the transition-metal interstitial carbides are as follows. The maximum carbon content depends on the c.p. layer sequence. The two octahedral interstices on either side of an *h* layer are located directly above one another, and only one of these is ever occupied. This restriction gives the following limiting formulae:

Layer sequence	*h*	*hcc*	*hhc*	*hhcc*	*c*
Limiting formula	M_2C	M_3C_2	M_3C_2	M_4C_3	MC
Example	See below	Mo_3C_2	$(Ta_2V)C_2$	V_4C_3	See below

Examples of structures based on the two simplest c.p. sequences are numerous. The h.c.p. structures are examples of the various ways of filling one-half of the octahedral interstices, and the simpler patterns of sites have been illustrated in Chapter 4. Examples are given in Table 22.6; a review of these structures is available.[9] (The literature of these compounds is confusing. Many have different

(9) AC 1970 **B26** 153

TABLE 22.6

Structures of interstitial carbides

Hexagonal c.p. metal atoms Octahedral sites occupied	Structure	Examples
One-half of all sites at random	–	β-V_2C, γ-Nb_2C
All sites between alternate pairs of layers	anti-CdI_2	α-Ta_2C, α-W_2C
One-half between each pair of c.p. layers	⎧ anti-$CaCl_2$ ⎨ ζ-Fe_2N ⎩ ζ-Nb_2C	Co_2C α-Mo_2C ζ-Nb_2C
One-third and two-thirds between successive pairs of c.p. layers	ϵ-FeN	ϵ-V_2C, ϵ-Nb_2C, ϵ-W_2C

Cubic c.p. metal atoms	Examples	Reference
All octahedral sites occupied (NaCl structure)	HfC, TiC, ZrC, VC, NbC, TaC, β-MoC_{1-x}	
Ordered defect NaCl structures	⎧ V_8C_7 ⎪ V_6C_5 ⎨ Sc_2C, Ti_2C, Zr_2C ⎪ anti-$CdCl_2$ (Ho_2C ⎫ ⎩ Y_2C, etc.) ⎭	AC 1970 **B26** 1882 PM 1968 **18** 177 AC 1970 **B26** 153

structures at different temperatures and the forms have been designated by Greek letters which have not been used consistently by different authors.) The c.c.p. structures include not only the carbides MC with the NaCl structure but also a number in which there is an ordered arrangement of vacancies; a feature of certain of these structures is the helical arrangement of the vacancies (V_6C_5 and V_8C_7). As remarked earlier, the arrangement of C atoms in the carbides of the heavier metals cannot be regarded as firmly established unless neutron diffraction studies have been made, as in the examples given in Table 22.6.

Class 4 The carbides of this class do not possess the extreme properties of the interstitial carbides. For example, whereas titanium carbide is not attacked by water or HCl even at $600°C$, these carbides are decomposed by dilute acids (Fe_3C and Ni_3C) or even water (Mn_3C). Although the carbon is present as discrete C atoms the products include, in addition to hydrogen, complex mixtures of hydrocarbons.

We shall not attempt here to survey the structures of these carbides, which are often complex. At low carbon contents there are marked similarities to borides, but when there is more non-metal the similarities are much less marked, for C does not show the same tendency as B to form extended systems of non-metal–non-metal bonds. There are groups of isostructural carbides and borides (with different formulae types) in all of which the non-metal atom has very similar environments, often a trigonal prismatic arrangement with from one to three additional neighbours beyond the vertical prism faces. Both $(6 + 1)$- and $(6 + 2)$-coordination of C occur in Cr_3C_2 (and the isostructural Hf_3P_2),[10] and tricapped trigonal prism coordination in the following group of closely related structures:

(10) ACSc 1969 **23** 1191

(11) AC 1965 **19** 463

(12) AC 1962 **15** 878

Fe_3C (cementite),[11] Mn_3C, Co_3B, Pd_3P, Al_3Ni
Mn_5C_2 and Pd_5B_2
Cr_7C_3, Re_7B_3, Th_7Fe_3[12]

It will be noticed that some of these structures are adopted by metallic phases in addition to carbides and borides. The WC structure, illustrated in Fig. 4.8(a), p.128, provides another example of a structure that is not peculiar to carbides, for it is adopted by MoP and WN in addition to WC, RuC, and OsC.[13]

(13) AC 1961 **14** 200

In addition to the binary carbides noted above there are ternary carbides and also carbonitrides and oxide–carbides. The behaviour of metals in metal–carbon–nitrogen systems ranges from that of Al, which forms a 'homologous' series of compounds with definite compositions and structures intermediate between those of AlN and Al_4C_3 (namely, $Al_8C_3N_4$, $Al_7C_3N_3$, $Al_6C_3N_2$, and Al_5C_3N)[14] and that of Th, in which there is a complete series of solid solutions between ThC and ThN; there are also the compounds ThC_2 and Th_3N_4. The compound Yb_2OC is mentioned in Chapter 12 as originally mistaken for the monoxide. A number of the 4f elements form compounds of the type MO_2C_2[15] or M_4O_3C.[16] The conductivity of the silver-grey Nd_4O_3C, which has a NaCl-like structure, is similar to that of Nd metal; it hydrolyses largely to methane, and could be formulated $Nd_4^{3+}O_3^{2-}C^{4-}(e)_2$. The structure of Al_4O_4C has been studied.[17] It consists of a 3D framework formed from AlO_3C tetrahedra, an unusual feature of which is the association of certain of the tetrahedra in edge-sharing pairs, the shared edges being $O\cdots O$ edges.

(14) AC 1966 **20** 538

(15) IC 1966 **5** 1567
(16) JACS 1968 **90** 1715

(17) AC 1963 **16** 177

Metal carbonyls

Preparation and properties

The compounds of CO with the alkali metals and the alkaline-earths are quite different in structure from those of the transition metals. The compound $K_2(CO)_2$ formed by the action of CO on the metal dissolved in liquid NH_3 is a salt containing the linear acetylenediolate ion $(OC≡CO)^{2-}$, in which the bond lengths are C–C, 1·21 Å, and C–O, 1·28 Å.[1] This salt is also formed from CO and the molten metal at low temperatures; at higher temperatures (above 180°C) the product is predominantly $C_6(OK)_6$.[2] The so-called carbonyls of the alkaline-earths made from the metals and CO in liquid NH_3 are mixtures of the metal acetylenediolates and methoxides with ammonium carbonate.[3]

(1) HCA 1963 **46** 1121

(2) HCA 1964 **47** 1415

(3) HCA 1966 **49** 907

TABLE 22.7

Carbonyls, carbonyl hydrides, and carbonyl halides

Periodic Group	Element	Carbonyls	Carbonyl hydrides	Carbonyl halides
V	V	$V(CO)_6$		
VI	Cr, Mo, W	$M(CO)_6$	$Cr(CO)_5H_2$	
VII	Mn	$Mn_2(CO)_{10}$	$Mn(CO)_5H$	$Mn(CO)_5I$
				$Mn_2(CO)_8I_2$
	Tc, Re	$M_2(CO)_{10}$		$M(CO)_5I$
		$Re_3(CO)_{12}$	$Re(CO)_5H$	$M_2(CO)_8I_2$
VIII	Fe	$Fe(CO)_5$	$Fe(CO)_4H_2$	$Fe(CO)_2X_2$
		$Fe_2(CO)_9$	$Fe_3(CO)_{11}H_2$	$Fe(CO)_4X_2$
		$Fe_3(CO)_{12}$	$Fe_4(CO)_{13}H_2$	$Fe(CO)_5X_2$
				$Fe_3(CO)_9X_6$
	Co	$Co_2(CO)_8$	$Co(CO)_4H$	$Co(CO)X_2$
		$Co_4(CO)_{12}$		
		$Co_6(CO)_{16}$		
	Ni	$Ni(CO)_4$	$Ni_2(CO)_6H_2$	

Certain of the salts of the I B subgroup metals combine with CO forming, for example, $CuCl(CO) \cdot 2H_2O$, $Ag_2SO_4 \cdot CO$, and $AuCl(CO)$, of which the last is a comparatively stable volatile compound. The structures of these compounds are not known, and we shall confine our attention here to the carbonyls and related compounds of transition metals.

Compounds of transition metals containing CO include carbonyls, carbonyl hydrides (hydrocarbonyls), and carbonyl halides, of which examples are set out in Table 22.7, a few nitrosyl carbonyls, and numerous more complex compounds containing a variety of ligands. In Group V only vanadium is known to form a simple carbonyl, $V(CO)_6$—its structure is not known—though compounds $R \cdot HgTa(CO)_6$ ($R = CH_3$, C_2H_5, C_6H_5) have been described. The pentacarbonyls of Ru and Os are presumably structurally similar to $Fe(CO)_5$, but $Ru_3(CO)_{12}$ and $Os_3(CO)_{12}$ have different molecular structures from $Fe_3(CO)_{12}$. The structure of $Os_2(CO)_9$, which has recently been prepared, is not known. Rhodium and iridium form $M_4(CO)_{12}$ and $M_6(CO)_{16}$ like Co, but the existence of $Rh_2(CO)_8$ and $Ir_2(CO)_8$ is doubtful.[4] Ru and Os form $M(CO)_4H_2$ like Fe, while Rh and Ir form $M(CO)_4H$, like Co. No simple carbonyls of Pd or Pt have yet been prepared. With the exception of $Ni(CO)_4$ and the pentacarbonyls of Fe, Ru, and Os, all the carbonyls of Table 22.7 are crystalline solids at ordinary temperatures.

The carbonyls are in general volatile compounds with an extensive chemistry which presents many problems as regards valence and stereochemistry. Some are reactive and form a variety of derivatives, as shown in Chart 22.1 for the iron compounds, while others are relatively inert, as for example, $Cr(CO)_6$ etc. and $Re_2(CO)_{10}$. This rhenium compound, although converted to the carbonyl halides by gaseous halogens, is stable to alkalis and to concentrated mineral acids. A few carbonyls may be prepared by the direct action of CO on the metal, either at atmospheric pressure ($Ni(CO)_4$) or under pressure at elevated temperatures ($Fe(CO)_5$, $Co_4(CO)_{12}$). Others are prepared from halides or, in the case of Os and Re, from the highest oxide. The polynuclear carbonyls are prepared photosynthetically, by heating the simple carbonyls, or by other indirect methods.

The carbonyl hydrides are prepared either from the carbonyls, as shown in the chart for $Fe(CO)_4H_2$, by reduction with Na in liquid NH_3, for $Ni_2(CO)_6H_2$, or directly from the metal by the action of a mixture of CO and H_2. Cobalt in this way gives $Co(CO)_4H$.

The heavy metal derivatives of $H_2Fe(CO)_4$ and $HCo(CO)_4$, the so-called 'mixed metal carbonyls', show very considerable differences in properties. $HgFe(CO)_4$ is a stable yellow solid, insoluble in both polar and non-polar solvents; these properties suggest a polymeric structure for example, of type (a), whereas the derivatives of $HCo(CO)_4$, $M[Co(CO)_4]_2$, where M = Zn, Cd, Hg, Sn, or Pb, resemble the polynuclear carbonyls, being soluble in non-polar solvents, insoluble in water, and subliming without decomposition; they consist of simple linear molecules in the crystalline state (see later).

The formation of carbonyl halides does not coincide with that of carbonyls, as shown in the accompanying table, where full lines enclose the elements which form simple carbonyls and broken lines those which form carbonyl halides:

V	Cr	Mn	Fe	Co	Ni	Cu	Zn
Nb	Mo	Tc	Ru	Rh	Pd	Ag	Cd
Ta	W	Re	Os	Ir	Pt	Au	Hg

These compounds may be formed directly from the metal halides by the action of CO (for example, CO and anhydrous CoI_2 give $Co(CO)I_2$) or, in the case of iron indirectly from the carbonyls or carbonyl hydrides. The only carbonyl fluoride which has been well characterized has the empirical formula $Mo(CO)_2F_4$;[5] a dimeric structure with two bridging F atoms has been suggested. The compounds $Pt(CO)_2F_8$ and $Rh(CO)_2F_3$ have been described as the products of the action of

(4) IC 1969 **8** 1206

(5) IC 1970 **9** 2611

$$\left[\begin{array}{c} OC \quad\ \ CO \\ {-}Fe{-}Hg{-} \\ OC \quad\ \ CO \end{array} \right]_n$$

(a)

CO on the tetrafluorides. Although simple carbonyls of Pd and Pt are not known the dichlorides combine with CO to form compounds with the empirical formulae $PdCl_2 . CO$, $PtCl_2 . CO$, and $PtCl_2(CO)_2$. The compound $PtCl_2(CO)_2$ is monomeric and a non-electrolyte, and the dipole moment (4·85 D) shows that it has the *cis* configuration,[6] but $PtCl_2 . CO$, which is a solid melting at 195°C is dimeric and non-polar and analogous to the compounds of $PtCl_2$ with arsines and phosphines, of the general type

(6) JACS 1954 **76** 4271

$$\begin{array}{c} R \diagdown \quad \diagup Cl \diagdown \quad \diagup Cl \\ \quad Pt \quad \quad Pt \\ Cl \diagup \quad \diagdown Cl \diagup \quad \diagdown R \end{array}$$

All the carbonyl halides are decomposed by water, but their stability towards water increases from the chloride to the iodide.

The nitrosyl carbonyls result from the action of NO on the polynuclear carbonyls of iron and cobalt. Like the carbonyls the nitrosyl carbonyls react with neutral molecules such as organic nitrogen compounds and with halogens, and in such reactions CO, and not NO, is replaced. Thus $Fe(CO)_2(NO)_2$ with iodine gives

CHART 22.1

Some reactions of iron carbonyls

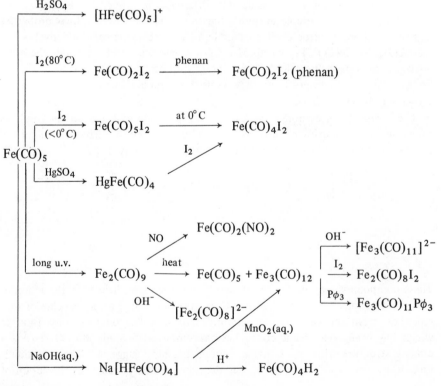

764

$Fe(NO)_2I$. No reaction has yet been detected between NO and the much more stable hexacarbonyls of Cr, Mo, and W. Whereas Mn, Fe, and Co form respectively $Mn(NO)_3CO$, $Fe(NO)_2(CO)_2$, and $Co(NO)(CO)_3$, Ni does not form a nitrosyl carbonyl of this type (see later) and reacts with NO only in the presence of water. A blue compound, $Ni(NO)OH$, is then formed which is soluble in water and possesses reducing properties. It is regarded as a compound of Ni(I) and is obviously quite different in type from the nitrosyl carbonyls of Co and Fe.

Chart 22.1 summarizes some of the reactions of the iron carbonyls.

The structures of carbonyls and related compounds

The main features are as follows.

1. The CO molecule is bonded to the metal atom(s) through C; this has been proved for $Fe(CO)_5$ and $Cr(CO)_6$ and is assumed to be true in all other molecules.

2. The CO molecule behaves as a monodentate ligand, (a), as a bridge between two metal atoms, (b), or (less frequently) between three metal atoms, (c).

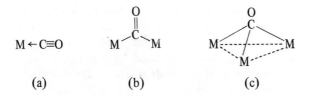

(a) (b) (c)

3. The formulae of most simple carbonyls and related compounds are consistent with the view that in many of these compounds the metal atom acquires a share in sufficient electrons to attain a noble gas configuration. Assuming that CO bonded as at (a) contributes 2 electrons, and NO 3 electrons, then in all the following compounds the metal would have the Kr configuration:

$Ni(CO)_4$

$Co(CO)_3NO$
$Fe(CO)_5$ $Fe(CO)_2(NO)_2$
$Mn(CO)(NO)_3$

$Cr(CO)_6$

In $Fe_2(CO)_9$ the same electronic configuration would be reached by the formation of a metal–metal bond; metal–metal bonds are an important feature of the polynuclear carbonyls. This simple electron counting is consistent with the formation of mononuclear carbonyls by the even-numbered elements, $Cr(CO)_6$, $Fe(CO)_5$, and $Ni(CO)_4$, and of polynuclear carbonyls by Mn and Co; it is not, however, consistent with the formation of $V(CO)_6$.

4. It is supposed that the single σ M–C bond in (a) above, formed by the lone pair of electrons of the ligand, is supplemented by π-bonding from the filled metal d orbitals to a vacant orbital of the ligand, as shown diagrammatically in Fig. 22.8.

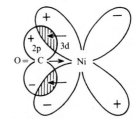

FIG. 22.8. Dative π-bond in $Ni(CO)_4$.

This 'back-bonding' reduces the negative formal charge on M implied by the formulation $M \leftarrow C \equiv O$ and accounts for the fact that CO and other π-bonding ligands stabilize low oxidation states of transition metals. In $Ni(CO)_4$ the formal oxidation state of Ni is zero, so that it is forming $4s4p^3$ (tetrahedral) bonds. This type of bonding, which also operates in the case of NO, isocyanides, substituted phosphines, sulphides, etc., is to be distinguished on structural grounds from that operative in the pure π-complexes formed by $C_5H_5^-$, C_6H_6, etc. which are discussed later. The nodal planes of the π orbitals used to supplement the σ bonding in the carbonyls must include the axis of the σ bond, requiring that the system M—C—O must be linear. In π-complexes in which the bonding involves only the π orbitals of the ligand the metal atoms lie out of the molecular plane of the C_5H_5 or other ligand.

The evidence for the multiple character of the M—C bond in these compounds is indirect. It cannot be deduced directly from the M—C bond length since, for example, the length of a single Ni—C bond formed by zerovalent Ni is not known, for the good reason that such a bond does not exist. An indirect approach is to compare with a bond in a similar molecule which can have no π character owing to the non-availability of ligand orbitals for the π-bonding. For example, in $Mo(CO)_3(dien)$ there are no such N orbitals available, and since the radius of N (sp^3) is similar to that of C (sp) the shortness of Mo—C as compared with Mo—N suggests multiple character of the former bond. Unequivocal spectroscopic evidence for the multiple character of M—C bonds is not readily obtained since the M—C stretching frequencies lie in a region of the spectrum which makes their correct assignment difficult. Assuming that strengthening of the M—C bond should be accompanied by weakening of the C—O bond the nature of M—C can be deduced from the properties of the C—O bond. The length is not a suitable criterion since the C—O bond length is obviously not sensitive to bond order in the region in which we are interested (varying only from $1 \cdot 13$ Å in CO to $1 \cdot 17$ Å or so in carbonyls), but there are marked reductions in C—O stretching frequencies in carbonyls which support the idea of multiple bonding.

The carbonyls $M(CO)_n$ have the highly symmetrical structures listed in Table 22.8.

Closely related to $Fe(CO)_5$ are the trigonal bipyramidal molecules $Fe(CO)_4PH\phi_2$[1] (phosphine in axial position) and $Fe(CO)_4$ (acrylonitrile),[2]

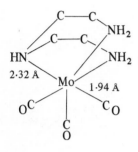

2·32 Å

1·94 Å

(1) JCS A 1969 1906
(2) AC 1962 **15** 1117

TABLE 22.8

Structures of carbonyls $M(CO)_n$

Molecule	Shape	M—C	C—O	Reference
$Ni(CO)_4$	Regular tetrahedral	1·83 Å	1·15 Å	AC 1952 **5** 795
$Fe(CO)_5$	Trigonal bipyramidal	1·806 (ax.)	1·145	AC 1969 **B25** 737
		1·833 (eq.)		ACSc 1969 **23** 2245
$Cr(CO)_6$		1·916	1·171	AC 1967 **23** 977
$Mo(CO)_6$	Regular octahedral	2·063	1·145	ACSc 1966 **20** 2711
$W(CO)_6$		2·058	1·148	

where there is one equatorial bond to the double bond of $H_2C=CH \cdot CN$. Surprisingly long axial bonds (1·99 Å) were found (compare 1·77 Å for the equatorial bonds) and 2·10 Å to the ethylenic C atoms. Also related to $Ni(CO)_4$ and $Fe(CO)_5$ are the bridged molecules (a)[3] and (b),[4] in which the bond arrangements around Ni and Fe are the same as in the simple carbonyls. In (b) the axial Fe–C bond appears to be significantly shorter (1·71 Å) than the equatorial bonds (1·79 Å) and C–O longer (1·29 Å as compared with 1·16 Å). In many formulae in the following pages we shall indicate –CO simply by a stroke (–).

(3) JCS A 1967 1744
(4) JCS A 1968 622

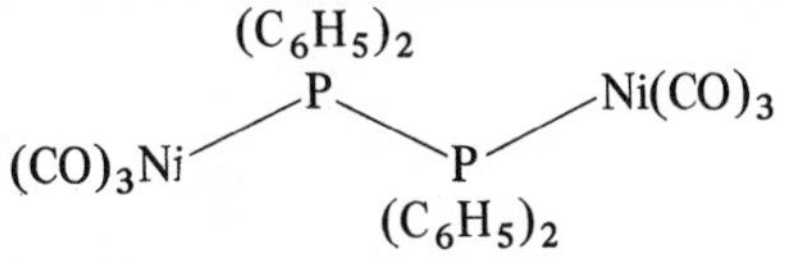

(a)

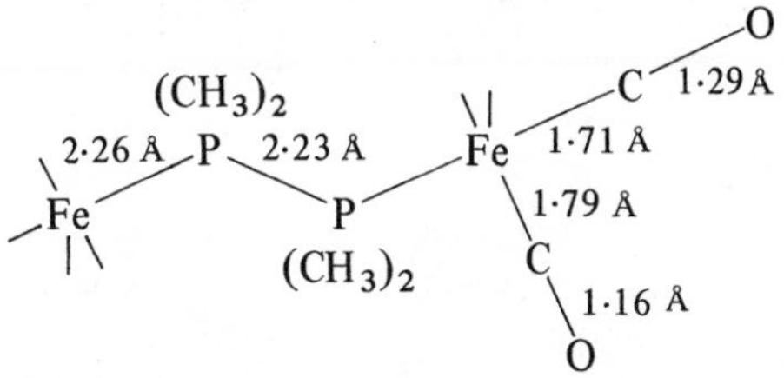

(b)

An early X-ray study of $Fe_2(CO)_9$[5] showed that the molecule has the structure of Fig. 22.9(a) with three bridging CO groups. The accuracy of location of the light atoms does not justify detailed comparison of the two different Fe–C and C–O bond lengths (see $Co_2(CO)_8$, later) but indicates appreciably lower bond orders in the bridging bonds than in the terminal ones. The Fe–Fe distance (2·46 Å), equal to twice the 'metallic' radius of Fe for 8-coordination, shows that there is a Fe–Fe bond, accounting for the diamagnetism and giving Fe the Kr closed-shell configuration.

(5) JCS 1939 286

There are many compounds related more or less closely to $Fe_2(CO)_9$ including (c) with a triple $(CH_3)_2Ge$ bridge,[6] compounds of type (d) with double bridges, where the bridging group is, for example, NH_2[7] or SC_2H_5 (see p. 774), and singly bridged compounds, (e), $(CO)_4Fe \cdot R \cdot Fe(CO)_4$, where R is $P(CH_3)_2$, $Ge(C_6H_5)_2$, etc. The Ge compounds are of interest as containing 3-membered $GeFe_2$ rings—compare the 5-membered rings in the compound noted on p. 778.

(6) JACS 1968 90 3587
(7) JACS 1968 90 5422

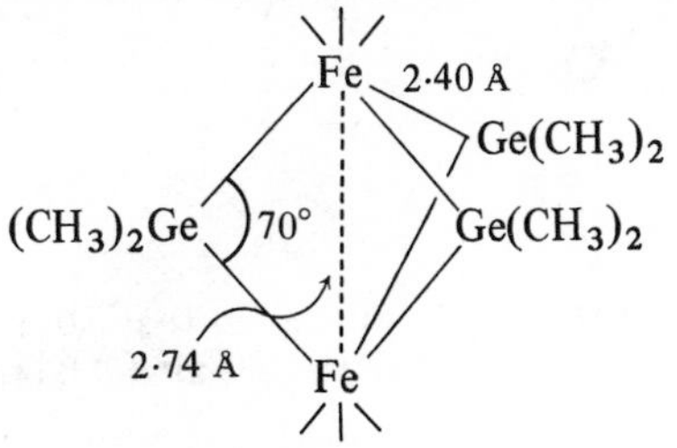

(c)

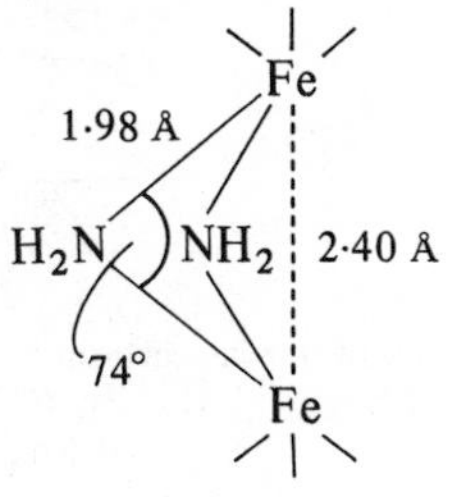

(d)

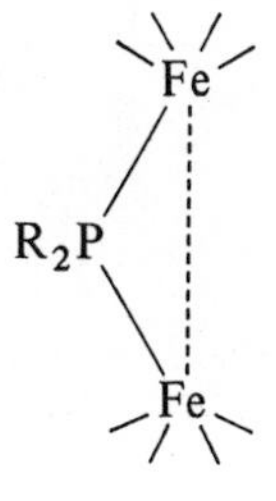

(e)

The $Co_2(CO)_8$ molecule,[8] Fig. 22.9(b), has essentially the same structure as $Fe_2(CO)_9$ with one of the bridging CO groups removed. By forming a metal–metal bond (2·52 Å) Co acquires the same closed-shell configuration as Fe in $Fe_2(CO)_9$.

(8) AC 1964 17 732

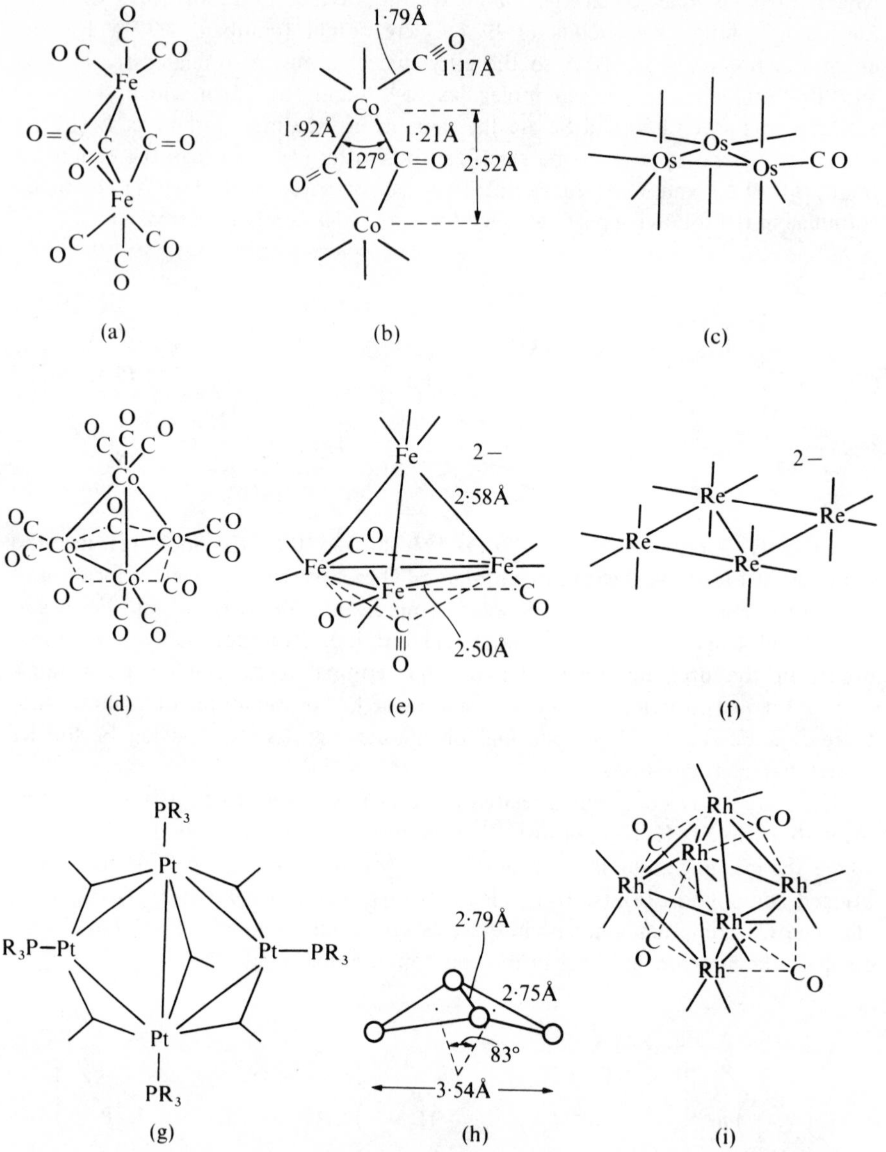

FIG. 22.9. The structures of metal carbonyls: (a) $Fe_2(CO)_9$, (b) $Co_2(CO)_8$, (c) $Os_3(CO)_{12}$, (d) $Co_4(CO)_{12}$, (e) $[Fe_4(CO)_{13}]^{2-}$, (f) $[Re_4(CO)_{16}]^{2-}$, (g) and (h) $Pt_4(CO)_5[P(C_2H_4CN)_3]_4$, (i) $Rh_6(CO)_{16}$.

(9) AC 1963 **16** 419
(10) IC 1965 **4** 1140

In $Mn_2(CO)_{10}$[9] and $Re_2(CO)_{10}$[10] (and the isostructural $Tc_2(CO)_{10}$) there are no bridging CO groups, the molecule consisting of two $-M(CO)_5$ joined by a metal–metal bond. Each metal atom forms six octahedral bonds (to 5 CO and M) and the two octahedra are rotated through about 45° from the eclipsed position.

768

The long M–M bonds (Mn–Mn, 2·92 Å, Re–Re, 3·02 Å, and Tc–Tc, 3·04 Å) are attributed to the large negative formal charges and to repulsions between the CO groups. Carbonyl ions $[M_2(CO)_{10}]^{2-}$ exist in the Na salts (M = Cr, Mo, W) prepared by reducing the hexacarbonyls with $NaBH_4$, and ions $[(CO)_5M–M'(CO)_5]^-$ have been isolated in $[N(C_2H_5)_4]^+$ salts (M = Mn, Re; M' = Cr, Mo, W).[10a] The red diamagnetic form of $[Co(CN.CH_3)_5]_2(ClO_4)_2$ contains dimeric ions of the same structural type as $Mn_2(CO)_{10}$, with Co–Co, 2·74 Å.[11]

The molecule of $Fe_3(CO)_{12}$ in the solid can be described as resulting from the replacement of one of the bridging CO groups in $Fe_2(CO)_9$ by the unit $Fe(CO)_4$, the third Fe atom being equidistant from the other two.[12] The three Fe atoms are situated at the corners of an isosceles triangle (f), and the twelve CO are arranged at the vertices of an icosahedron (as is also the case in $Co_4(CO)_{12}$, below). The apparent incompatibility of the i.r. spectrum of $Fe_3(CO)_{12}$ in solution with

(10a) JACS 1967 **89** 539

(11) PCS 1964 175

(12) JACS 1969 **91** 1351

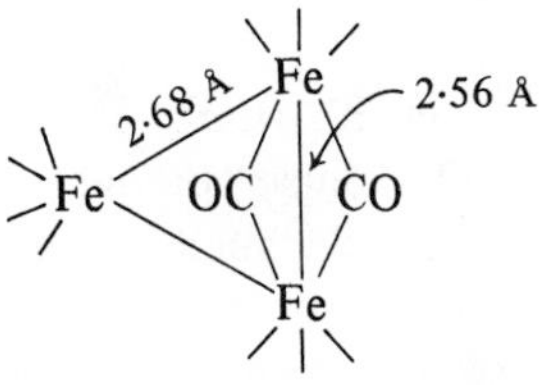

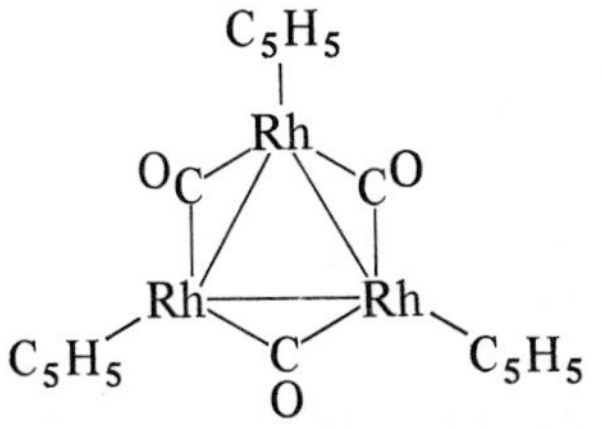

the model (f) may be due to rearrangement in solution, a view supported by the fact that one isomer of $Rh_3(CO)_3(C_5H_5)_3$ has a structure similar to (f) and the other has the symmetrical structure (g).[12a]

The anion in $[(C_2H_5)_3NH][Fe_3(CO)_{11}H]$ [13] apparently has the same type of structure as $Fe_3(CO)_{12}$. The H atom was not located, but presumably replaces one bridging CO group; there is the same isosceles triangle of Fe atoms (edges 2·58 and 2·69 Å (two)). In $Fe_3(CO)_{11}P\phi_3$ one terminal CO is replaced by the phosphine. The crystal contains two structural isomers, (h) and (i), and the bridges are slightly unsymmetrical,[14] with *a* and *b* approximately 2·04 Å and 1·87 Å (means).

(12a) JOC 1967 **10** P3, 331

(13) IC 1965 **4** 1373

(14) JACS 1968 **90** 5106

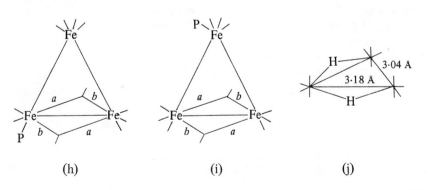

(h)　　　　(i)　　　　(j)

(15) JCS A 1968 778
(16) IC 1962 **1** 521

In contrast to $Fe_3(CO)_{12}$ the molecules $Ru_3(CO)_{12}$[15] and $Os_3(CO)_{12}$[16] (originally formulated as enneacarbonyls) have the symmetrical structure of Fig. 22.9(c), with 4 CO bonded to each metal atom. In the equilateral triangle of metal atoms the bond lengths are: Ru–Ru, 2·85 Å, and Os–Os, 2·88 Å, and each metal atom forms four normal and two 'bent' octahedral bonds—compare the similar 'bent' bonds in $Co_2(CO)_8$, $Rh_2(CO)_4Cl_2$, etc.

(17) JACS 1968 **90** 7135

The anion in $[Re_3(CO)_{12}H_2][As(C_6H_5)_4]$ has a somewhat similar structure, (j).[17] The H atoms were not located but presumably bridge the two longer edges of the Re_3 triangle. As in $Ru_3(CO)_{12}$ there are no bridging CO groups.

Molecules and ions containing 4 metal atoms are of two main types, tetrahedral and quadrilateral (planar or dihedral). In $Ir_4(CO)_{12}$ the metal atoms form a regular tetrahedron of edge 2·68 Å,[18] with 3 CO bonded to each Ir. The anion in $[Re_4(CO)_{12}H_6][As(C_6H_5)_4]$ also has a regular tetrahedral shape,[19] and the H atoms (not located) are logically placed along (or close to) the edges. The mean Re–Re distance (3·16 Å) is close to the larger value in $[Re_3(CO)_{12}H_2]^-$ and appreciably larger than the value (around 3·0 Å) for non-bridged Re–Re distances in, for example, $Re_2(CO)_{10}$.

(18) IC 1967 **8** 2011
(19) JACS 1969 **91** 1021

(20) IC 1969 **8** 2384

The tetrahedral molecules $Co_4(CO)_{12}$[20] and $Rh_4(CO)_{12}$[20] are less symmetrical (Fig. 22.9(d)), with three bridging CO groups. One metal atom forms octahedral bonds to three others and to 3 CO, while the bond arrangement around the basal metal atoms is approximately pentagonal bipyramidal. Owing to crystallographic complications (disorder and twinning) the bond lengths are not known accurately, but mean metal–metal distances are respectively 2·49 Å and 2·73 Å.

(21) JACS 1966 **88** 4847

The carbonyl anion $[Fe_4(CO)_{13}]^{2-}$ has a tetrahedral structure (Fig. 22.9(e)) of a quite different kind.[21] An apical $Fe(CO)_3$ unit is bonded to the base of the tetrahedron only by Fe–Fe bonds, and a feature of the molecule is the triply bonded CO below the base. (There is very weak bonding between the C atom of each CO in the basal plane and a second metal atom.)

(22) IC 1968 **7** 2606

The ion $[Re_4(CO)_{16}]^{2-}$ (Fig. 22.9(f)) has an entirely different shape. The four metal atoms are coplanar, all Re–Re distances are close to 3·0 Å, and the arrangement of bonds around Re is octahedral.[22]

We noted in Chapter 3 some families of tetrahedral molecules, including molecules $M_4X_4Y_4$ and $M_4X_6Y_4$. The molecule $Fe_4(CO)_4(\pi\text{-}C_5H_5)_4$ is tetrahedral with a triply bridging CO on each face, while $Ni_4(CO)_6[P(C_2H_4CN)_3]_4$ has a bridging CO along each edge and a phosphine molecule attached to each vertex (Ni).[23] Removal of one of the edge-bridging CO molecules from a molecule of this kind leads to opening-up of the tetrahedral to a dihedral ('butterfly') structure, as in $Pt_4(CO)_5[P(CH_3)_2C_6H_5]_4$,[24] as shown diagrammatically in Fig. 22.9(g) and (h). The dihedral angle is $83°$ instead of $109\frac{1}{2}°$, the value in the tetrahedron. Removal of one bridging CO and of one metal–metal bond reduces the electron count to 16 for the 5-covalent metal atoms, leaving the two 7-coordinated atoms with a closed shell configuration.

(23) JACS 1967 **89** 5366

(24) JACS 1969 **91** 1574

The black crystals of $Rh_6(CO)_{16}$, previously formulated $Rh_4(CO)_{11}$, are composed of octahedral molecules (Fig. 22.9(i)) in which there are two CO attached to each Rh by normal Rh—C bonds and also four bridging CO groups.[25] The latter are situated above four of the faces of the Rh_6 octahedron, and each is bonded to three metal atoms. The Rh—Rh distance is 2·78 Å, and since the electron count for a Rh atom is 56 this compound is an exception to the 'noble gas rule'.

(25) JACS 1963 **85** 1202

Carbonyl hydrides

Early electron diffraction studies[26] of $Fe(CO)_4H_2$ and $Co(CO)_4H$ indicated a tetrahedral arrangement of 4 CO around the metal atoms, but the H atoms were not located. No O—H frequency is observed in the i.r. spectra. The structures of these molecules are still in doubt.

(26) TFS 1939 **35** 681

It has been shown that in a number of transition-metal hydrido complexes H atoms are directly bonded to M and occupy definite coordination positions (as in K_2ReH_9, $HPtBr[P(C_2H_5)_3]_2$, etc., for which see Chapter 27). This would also appear to be true in certain carbonyl hydrides, though the H atoms have been directly located by n.d. in only one form of $Mn(CO)_5H$. In $Mn(CO)_5H$ five CO are situated at five of the vertices of an octahedron (Mn—C, 1·84 Å) and the H atom is situated at the sixth vertex, with Mn—H, 1·60 Å.[27] (C—O, 1·134 Å).

(27) IC 1969 **8** 1928

The ion $HCr_2(CO)_{10}^-$ has been studied in the $[N(C_2H_5)_4]^+$ salt, and here the logical position for the H atom would be at the bridging position, (a), where it would complete each octahedral coordination group.[28] The Cr—Cr distance

(28) JACS 1966 **88** 366

(a)   (b)

(3·41 Å) is too large for a normal Cr—Cr bond, but 1·70 Å is a reasonable value for Cr—H (compare Re—H, 1·68 Å, in K_2ReH_9). A similar position seems likely for the H atom in $HRe_2Mn(CO)_{14}$,[29] (b), in which Re—Re is 3·39 Å (giving Re—H, 1·70 Å) as compared with Re—Mn, 2·96 Å. Infrared and n.m.r. studies of $Os(CO)_4H_2$[30] indicate a *cis* octahedral configuration.

(29) IC 1967 **6** 2086

(30) IC 1967 **6** 2092

Carbonyl halides

Structural studies have been made of several of these compounds. $Ru^{II}(CO)_4I_2$[31] forms octahedral molecules with the I atoms in *cis* positions (Ru—C, 2·01 Å, Ru—I, 2·72 Å), and a similar configuration has been assigned to $Fe(CO)_4I_2$ on the basis of i.r. evidence.[32] The structure of $[Rh(CO)_2Cl]_2$ is more complex.[33] This compound is dimeric in solution and also in the crystal, in which the dimers are bonded into infinite chains by Rh—Rh bonds as shown in Fig. 22.10(a). To account for the diamagnetism it has been suggested that bent metal–metal bonds are formed by overlap of d^2sp^3 orbitals at an angle of 56°.

(31) AC 1962 **15** 946

(32) ZaC 1956 **287** 223
(33) JACS 1961 **83** 1761

(34) AC 1963 **16** 611

(34a) AC 1968 **B24** 424

$Mn_2(CO)_8Br_2$[34] forms a symmetrical bridged molecule (Fig. 22.10(b)) with Mn–Br, 2·53 Å, and Mn–C, 1·81 Å. Molecules $M_2(CO)_8X_2$ are formed by Mn, Tc, and Re in which X = Cl, Br, or I. The closely related molecule $Ru_2Br_4(CO)_6$[34a] is shown in Fig. 22.10(c).

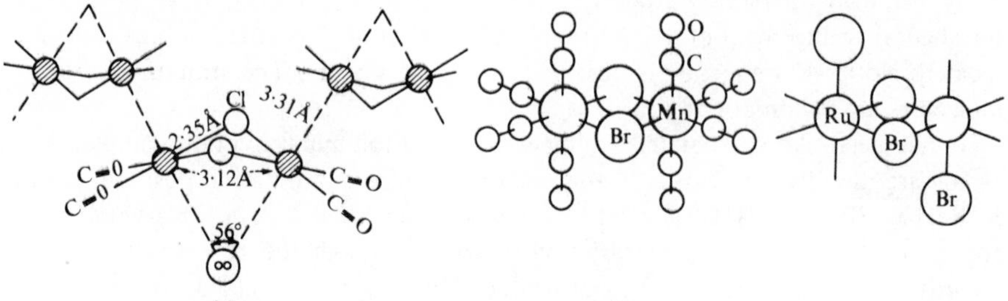

FIG. 22.10. The structures of (a) $[Rh(CO)_2Cl]_2$, (b) $[Mn(CO)_4Br]_2$, (c) $[RuBr_2(CO)_3]_2$.

Nitrosyl carbonyls

Early e.d. studies[35] of $Fe(CO)_2(NO)_2$ and $Co(CO)_3NO$ indicated a tetrahedral arrangement of the four ligands with

(35) TFS 1937 **33** 1233

$$Fe\underset{1·84}{\rule{1.5cm}{0.4pt}}C\underset{1·15}{\rule{1.5cm}{0.4pt}}O \quad \text{and} \quad Fe\underset{1·77}{\rule{1.5cm}{0.4pt}}N\underset{1·12\ \text{Å}}{\rule{1.5cm}{0.4pt}}O$$

and very similar bond lengths in the Co compound, though it is not likely that the bond lengths are of high accuracy. These studies did not prove the linearity of the –M–N–O system, but this has since been demonstrated by m.w. studies of molecules such as $(C_5H_5)NiNO$.[36] Manganese forms $Mn(NO)_3CO$ and $Mn(CO)_4NO$, and on irradiation the latter is converted into $Mn_2(CO)_7(NO)_2$—compare the conversion of $Fe(CO)_5$ to $Fe_2(CO)_9$. The molecule $Mn(CO)_4NO$ has a trigonal bipyramidal structure, like the isoelectronic $Fe(CO)_5$, with NO at one equatorial position and linear M–N–O bonds.[37] Bond lengths are Mn–C, 1·89 Å (axial), and 1·85 Å (equatorial), Mn–N, 1·80 Å.

(36) N 1958 **181** 1157

(37) IC 1969 8 1288

(38) IC 1972 **11** 382

The Ru and Os compounds of the type $M_3(CO)_{10}(NO)_2$[38] are interesting for they possess double NO bridges. The two shorter sides of the isosceles triangle correspond to normal metal–metal bonds (2·87 Å) (a).

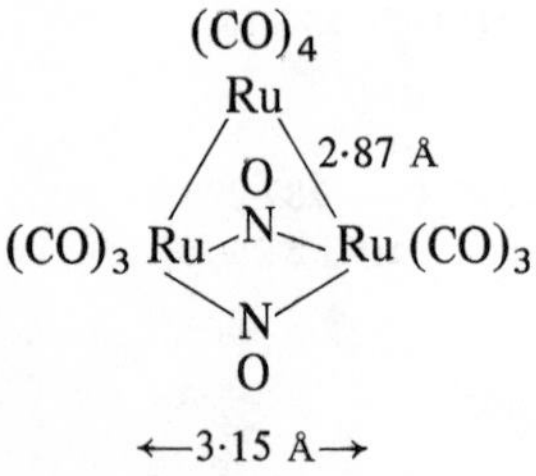

Mixed metal carbonyls

Many carbonyls have been prepared containing atoms of two or more different elements, some with only CO ligands and others with ligands of various types. The special interest lies in the metal–metal bonds and in the stereochemistry of the metal atoms. We give in Table 22.9 a few simple examples of typical molecules:

772

TABLE 22.9

Structures of mixed metal carbonyls

Molecule	M—M	Metal stereochemistry	Reference
$(CH_3)_3Sn—Mn(CO)_5$	2·674 Å	Sn tetr.; Mn oct.	JCS A 1968 696
$(CO)_5Mn—Fe(CO)_4—Mn(CO)_5$	2·815	Mn and Fe octahedral	AC 1967 **23** 1079
$(CO)_4Co—Zn—Co(CO)_4$	2·305	Co trig. bipyr.; Zn linear	JACS 1967 **89** 6362
$(CO)_4Co—Hg—Co(CO)_4$	2·50	Co trig. bipyr.; Hg linear	JCS A 1968 1005

Miscellaneous carbonyl derivatives of metals

Molecules containing CO and other ligands are very numerous, and range from simple molecules and ions in which all the ligands form a normal coordination group around the transition-metal atom to those in which there are more complex cyclic and polyhedral systems of metal atoms. Compounds containing unsaturated hydrocarbon ligands are described in the following pages, and we mention here only a few molecules which present points of special interest.

The compound $Fe_5(CO)_{15}C$ is produced in small quantity when $Fe_3(CO)_{12}$ is heated in petroleum ether with methylphenylacetylene. It has the tetragonal pyramidal structure shown in Fig. 22.11(a), in which C is located just below the

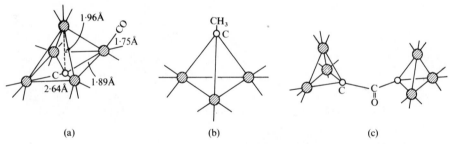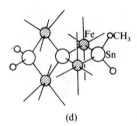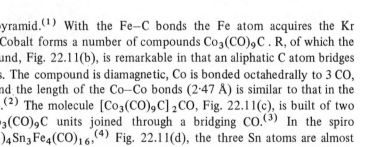

(a) (b) (c) (d)

FIG. 22.11. The molecular structures of (a) $Fe_5(CO)_{15}C$, (b) $Co_3(CO)_9C(CH_3)$, (c) $[Co_3(CO)_9C]_2CO$, (d) $(CH_3)_4Sn_3Fe_4(CO)_{16}$.

base of the pyramid.[1] With the Fe—C bonds the Fe atom acquires the Kr configuration. Cobalt forms a number of compounds $Co_3(CO)_9C$. R, of which the methyl compound, Fig. 22.11(b), is remarkable in that an aliphatic C atom bridges three Co atoms. The compound is diamagnetic, Co is bonded octahedrally to 3 CO, 2 Co, and C, and the length of the Co—Co bonds (2·47 Å) is similar to that in the metal (2·51 Å).[2] The molecule $[Co_3(CO)_9C]_2CO$, Fig. 22.11(c), is built of two tetrahedral $Co_3(CO)_9C$ units joined through a bridging CO.[3] In the spiro molecule $(CH_3)_4Sn_3Fe_4(CO)_{16}$,[4] Fig. 22.11(d), the three Sn atoms are almost exactly collinear and are bridged by $Fe(CO)_4$ groups. There are two different Sn—Fe bond lengths, terminal 2·63 Å and central 2·75 Å. The bond arrangement around both types of Sn atom is tetrahedral: Fe—C, 1·75 Å, Sn—C, 2·22 Å.

(1) JACS 1962 **84** 4633

(2) JACS 1967 **89** 261

(3) AC 1969 **B25** 107

(4) IC 1967 **6** 749

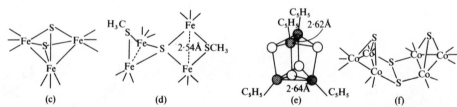

There are many complex carbonyls containing S or—SR ligands, some of which are structurally similar to molecules already described. For example, $[(C_2H_5S)Fe(CO)_3]_2$,[5] Fig. 22.12(a), is of type (d), p. 767, there being no bond between the S atoms (S—S, 2·93 Å), while in the closely related $[SFe(CO)_3]_2$,[6] Fig. 22.12(b), the S atoms are bonded (S—S, 2·01 Å). In $S_2Fe_3(CO)_9$ there is the more complex nucleus (c), which has been studied in the 1 : 1 molecular compound $[S_2Fe_2(CO)_6][S_2Fe_3(CO)_9]$.[7] The molecule $[CH_3SFe_2(CO)_6]_2S$,[8] (d), is of a different kind. There is a central S atom bonded tetrahedrally to 4 $Fe(CO)_3$ units which are also bridged in pairs by SCH_3 ligands. It is convenient to include here also $(C_5H_5)_4Fe_4S_4$,[9] (e). The nucleus of the molecule is an elongated tetrahedron of Fe atoms (of which only two edges correspond to Fe—Fe bonds, the others being much longer (3·37 Å)), above the faces of which are arranged the S atoms, each bridging 3 Fe. The lengths of the Fe—S bonds in this and the other molecules of Fig. 22.12 are very close to that in pyrites (2·26 Å).

(5) IC 1963 **2** 328
(6) IC 1965 **4** 1

(7) IC 1965 **4** 493
(8) IC 1967 **6** 1236

(9) IC 1966 **5** 892

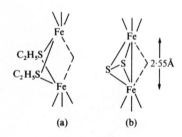

FIG. 22.12. The molecular structures of (a) $[(C_2H_5S)Fe(CO)_3]_2$, (b) $[SFe(CO)_3]_2$, (c) $S_2Fe_3(CO)_9$, (d) $[CH_3SFe_2(CO)_6]_2S$, (e) $(C_5H_5)_4Fe_4S_4$, (f) $[SCo_3(CO)_7]_2S_2$.

(10) IC 1967 **6** 1229

(11) JACS 1967 **89** 3727

(12) JACS 1968 **90** 3960, 3969, 3977

Of the numerous sulphur-containing Co carbonyls we give only a few examples. The molecule $SCo_3(CO)_9$[10] is of the same general type as $CH_3C . Co_3(CO)_9$, Fig. 22.11(b), and $[SCo_3(CO)_7]_2S_2$[11] consists of two identical tetrahedral units $SCo_3(CO)_7$ bridged by an S_2 group, as shown in Fig. 22.12(f). The following molecules[12] all contain bridging —SC_2H_5 groups: $Co_3(CO)_4(SC_2H_5)_5$, $Co_5(CO)_{10}(SC_2H_5)_5$, and $SCo_6(CO)_{11}(SC_2H_5)_4$, the latter also containing a Co_3S unit resembling that in Fig. 22.12(f).

Compounds of metals with hydrocarbons

Compounds of metals containing hydrocarbon radicals or molecules are numerous. The simplest contain only metal and hydrocarbon radical (metal alkyls and aryls) or unsaturated hydrocarbon molecules (for example, $Cr(C_6H_6)_2$), while in others there are also ligands such as halogens or CO. Because they are most closely related to the metal carbonyls we deal first with transition-metal compounds containing unsaturated hydrocarbons and CO ligands and later with the alkyls of non-transition metals. The formulae of many of the simpler transition-metal compounds are consistent with the view that the metal gains the following numbers of electrons from the ligands:

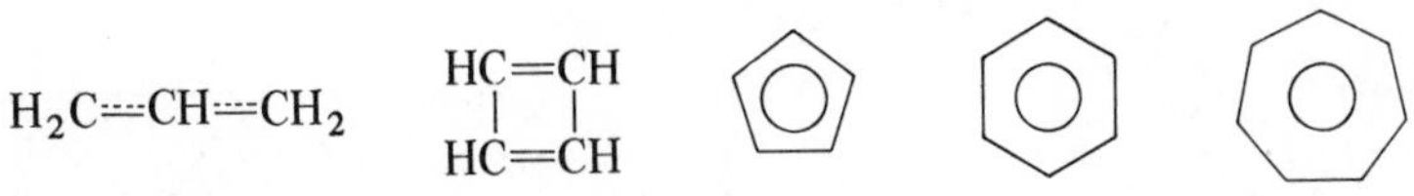

$$H_2C \cdots CH \cdots CH_2 \qquad \begin{matrix} HC=CH \\ | \quad | \\ HC=CH \end{matrix}$$

| 3 | 4 | 5 | 6 | 7 |

and thereby reaches the electronic structure of a noble gas—compare the carbonyls and nitrosyl carbonyls. In all the following diamagnetic compounds of the 3d elements the metal has the Kr configuration:

A.N.

23	$V(CO)_4(C_5H_5)$		
24		$Cr(CO)_3(C_6H_6)$	$Cr(C_6H_6)_2$
25	$Mn(CO)_3(C_5H_5)$		
26		$Fe(CO)_3(C_4H_4)$	$Fe(C_5H_5)_2$ $Fe(CO)(C_3H_5)(C_5H_5)$
27	$Co(CO)_2(C_5H_5)$		
28			$Ni(C_5H_5)NO$

while a molecule such as $Ni(C_5H_5)_2$ clearly has 2 electrons in excess of the noble gas number. The molecule of cyclooctatetraene, C_8H_8 (COT), functions as a 4-electron donor in $(CO)_3Fe(COT)$ but as two (cyclobutadiene) halves in $(CO)_3Fe(C_8H_8)Fe(CO)_3$ (see later).

We have already noted the geometrical difference between the bond formed by a transition metal to CO, which is a true π bond (overlap of metal d orbitals with the π orbital of CO in the nodal plane) and the bond to an unsaturated hydrocarbon in which the metal orbital is perpendicular to the nodal plane of the π orbital. The term μ-bond has been suggested for this latter type of bond.

Acetylene complexes

The reactions of acetylene with metal carbonyls can be grouped into two main classes: (i) those in which no new C—C bonds are formed, and the acetylene is bonded to the metal atom(s) by μ-bonds, 'bent' σ bonds, or by a combination of μ and σ bonds, and (ii) those in which new C—C bonds are formed by combination of some of the CO with the acetylene to form a cyclic hydroxy-ligand or a lactone ring.

(i) Acetylenes replace two CO in $Co_2(CO)_8$, and a study of crystalline $Co_2(CO)_6$(diphenylacetylene) shows that the molecule has the configuration of Fig. 22.13(a). The metal atoms form six bonds, arranged octahedrally, one to the other metal atom, two to the acetylene carbon atoms, and three to CO molecules; the C—C bond in the acetylene has lengthened to about 1·46 Å.[1] In $Co_4(CO)_{10}C_2H_5C \cdot C \cdot C_2H_5$[2] the acetylene forms two σ bonds to two Co atoms and μ bonds to the other two metal atoms (Fig. 22.13(b)).

(1) JACS 1959 **81** 18
(2) JACS 1962 **84** 2451

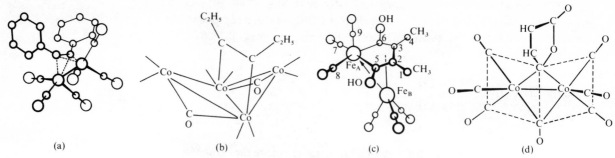

FIG. 22.13. The molecular structures of:
(a) (diphenylacetylene)$Co_2(CO)_6$, (b) $Co_4(CO)_{10}C_2H_5CCC_2H_5$,
(c) $(CO)_3Fe(C_6H_8O_2)Fe(CO)_3$, (d) $(CO)_3Co(CO)(C_4H_2O_2)Co(CO)_3$.

(3) AC 1961 **14** 139

(4) PCS 1959 156

(ii) An interesting reaction is the displacing of two CO groups from two molecules of iron or cobalt carbonyl hydrides by acetylenes or other unsaturated molecules. A study of the crystal structure[3] of the but-2-yne complex from $Fe(CO)_4H_2$, which has the formula $(OC)_3Fe(C_6H_8O_2)Fe(CO)_3$, shows that the molecule has the structure shown in Fig. 22.13(c). All the atoms C^1 to C^8 are coplanar, and the group C^9O is perpendicular to this plane. The three CO groups of Fe_B are arranged with three-fold symmetry around an axis which is perpendicular to the plane of $C^6C^3C^2C^5$. There is a covalent Fe—Fe bond, and the π bonding between Fe_B and the ring of carbon atoms is apparently similar to that in cyclopentadienyl compounds.

The action of CO under pressure on $Co_2(CO)_6C_2H_2$ yields a compound with the empirical formula $Co_2(CO)_9C_2H_2$. In this molecule, Fig. 22.13(d), the two Co atoms are bridged by one CO and a C atom of a lactone ring.[4] Each Co is bonded to five C atoms in a square pyramidal configuration, and the two square pyramids are joined along a basal edge about which the molecule is folded until the Co—Co separation has the value 2.5 Å. The molecular structure is very similar to that of $Co_2(CO)_8$.

Cyclopentadienyl complexes

Cyclopentadiene, C_5H_6, is a colourless liquid which readily forms a monosodium derivative. This in turn reacts with anhydrous transition-metal halides to form derivatives $M(C_5H_5)_n$, some of which may be made direct from the hydrocarbon and the metal carbonyls at about 300°C. Some of these compounds, including $Fe(C_5H_5)_2$ ('ferrocene') can be oxidized to cations. The following are examples of cyclopentadienyl compounds:

$(C_5H_5)_2M$:	M = Fe, Co, Ni, Ru, Rh, Ir, Cr, V, Mn, Mg;
$[(C_5H_5)_2M]X$:	M = Fe, Co, Ni, Ru, Rh, Ir;
$[(C_5H_5)_2M]X_2$:	M = Ti, Zr, V;
carbonyls:	$(C_5H_5)_2Mo_2(CO)_6$, $(C_5H_5)_2Fe_2(CO)_4$

776

The interaction of $Fe(CO)_5$ with excess of cyclopentadiene gives first the bridged compound I which decomposes at 200°C to ferrocene (IIa). (Fig. 22.14). The compound I can be isolated if the temperature is kept in the range 100–200°C.

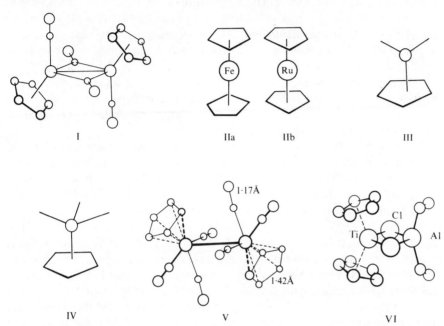

I   IIa   IIb   III

IV   V   VI

FIG. 22.14. The structures of cyclopentadienyl derivatives of metals: (I) $Fe_2(CO)_4(C_5H_5)_2$, (IIa) $Fe(C_5H_5)_2$, (IIb) $Ru(C_5H_5)_2$, (III) $M(C_5H_5)(CO)_2$, (IV) $M(C_5H_5)(CO)_3$, (V) $Mo_2(CO)_6(C_5H_5)_2$, (VI) $(C_5H_5)_2TiCl_2Al(C_2H_5)_2$.

The bond lengths, derived from the crystal structure,[1] are shown at (a). The Fe–Fe bond length is very close to that in $Fe_2(CO)_9$, and the general resemblance of the two molecules is evident.

(1) AC 1958 **11** 620

(a)

Ferrocene has been examined by electron[2] and X-ray diffraction.[3] The cyclopentadienyl rings are parallel, and in the crystal the carbon atoms are situated

(2) JCP 1955 **23** 1966
(3) AC 1956 **9** 373

at the vertices of a pentagonal antiprism. There is probably free rotation of the rings in the vapour at 400°C. The perpendicular distance between the rings is 3·25 Å—compare 3·35 Å in graphite. Bond lengths are: Fe–C, 2·05 Å, C–C, 1·40 Å, assuming C–H, 1·09 Å. The length of the Fe–C bond is about the value to be expected for a single bond, in contrast to the value 1·84 Å in $Fe(CO)_5$, and here also Fe has the Kr configuration. The ferricinium ion has been studied in $(C_5H_5)_2Fe(I_3)$,[4] but disorder in the crystal made it impossible to determine the detailed structure of the ion.

Magnetic measurements indicate respectively 1 and 2 unpaired electrons in $Ni(C_5H_5)_2^+$ and $Ni(C_5H_5)_2$. The cyclopentadienyl derivatives of Co, Ni, Cr, V, and Mg are isostructural with ferrocene, but in $Ru(C_5H_5)_2$ the rings are in the eclipsed positions (IIb in Fig. 22.14), with Ru–C, 2·21 Å.[5]

In the mononuclear carbonyl compounds a symmetrical arrangement of the CO groups around the axis in pyramidal structures III and IV is consistent with the i.r. spectra (Fig. 22.14) and has been confirmed by an X-ray study of crystalline $C_5H_5Mn(CO)_3$.[6] Bond lengths found were: Mn–CO, 1·80 Å, Mn–C_5H_5, 2·165 Å, and distance of Mn from centre of ring, 1·80 Å.

An X-ray study[7] of $C_5H_5Mo(CO)_6MoC_5H_5$ gives the structure V in Fig. 22.14. All the atoms drawn in heavily lie in one plane. The arrangement of bonds around a metal atom is such that if the Mo atom is imagined to be at the centre of a cube then a C_5H_5 ring lies centrally in one face and the other four bonds (to Mo and three C) are directed towards the vertices of the opposite cube face. In this molecule Mo has a completed outer shell of eighteen electrons and the compound is therefore diamagnetic. Bond lengths include: Mo–C, 2·34 Å, Mo–Mo, 3·22 Å, and the bond angle Mo–C–O is approximately 175°.

In the bridged molecule $(C_5H_5)_2TiCl_2Al(C_2H_5)_2$, shown in Fig. 22.14 (VI), there is an approximately square bridge of side 2·5 Å between the two metal atoms.[8] The bonds from both Ti and Al are arranged tetrahedrally, so that the general shape of the molecule is similar to that of the Al_2X_6 molecules. The compound $(C_5H_5)_3UCl$ provides an example of a molecule in which three C_5H_5 rings are bonded to the same metal atom. The bond arrangement is tetrahedral.[9]

The molecule (a) in Fig. 22.15 is included here as an interesting example of a 5-membered ring of metal atoms of three different kinds.[10a]

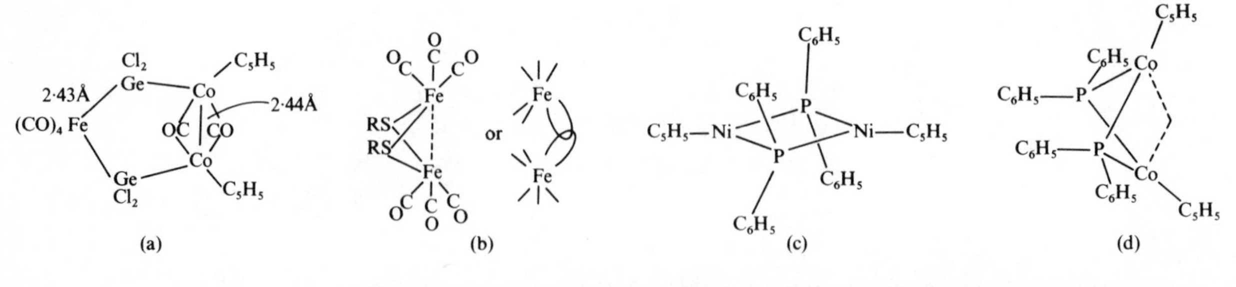

(4) AC 1968 **B24** 1640

(5) AC 1959 **12** 28

(6) AC 1963 **16** 118

(7) JCP 1957 **27** 809

(8) JACS 1958 **80** 755

(9) AC 1965 **18** 340

(10a) JACS 1970 **92** 208

FIG. 22.15. The molecular structures of (a) $(CO)_4Fe(GeCl_2)_2[Co(CO)(C_5H_5)]_2$, (b) $Fe_2(RS)_2(CO)_6$, (c) $[(\phi_2P)Ni(C_5H_5)]_2$, (d) $[(\phi_2P)Co(C_5H_5)]_2$.

We referred earlier (p. 767) to compounds related to $Fe_2(CO)_9$ of type (b) (Fig. 22.15) involving a metal-metal bond. The reality of this bond (which may alternatively be represented as a 'bent' bond) is shown by a comparison of the structures of the dimeric $[(\phi_2P)Ni(C_5H_5)]_2$ and $[(\phi_2P)Co(C_5H_5)]_2$.[10b] The Ni compound, (c), has a planar bridge system, but in the Co compound, (d), the two CoP_2 planes are inclined to one another owing to the formation of the Co–Co bond (Co–Co, 2·56 Å, compare Ni–Ni, 3·36 Å). An electron count, assuming that the two $P\phi_2$ ligands provide three, C_5H_5 five, and the metal-metal bond one electron to each metal atom, shows that in all these molecules a noble-gas configuration is attained: Fe, 26 + 9 + 1, Co, 27 + 8 + 1, and Ni, 28 + 8.

(10b) JACS 1967 **89** 542

The structures of a number of non-transition metal cyclopentadienyls have been studied. The molecule of $Be(C_5H_5)_2$ in the vapour[11] has the staggered configuration of Fig. 22.14 IIa, but unlike the Fe compound the molecule is unsymmetrical as regards the position of the metal atom, which is 1·485 Å from one ring and 1·980 Å from the other (perpendicular distances). In the vapour state InC_5H_5[12] and TlC_5H_5[13] form simple pyramidal molecules (In–C, 2·62 Å, Tl–C, 2·71 Å). In the crystalline compounds,[14] on the other hand, there are chains (Fig. 22.16(a)) in which In is equidistant from two C_5H_5 rings (at 3·2 Å) and the angle at the In atom is 137°. The shortest In–In distance between chains is 4·0 Å. There are rather similar chains in crystalline $Pb(C_5H_5)_2$[15] with an additional C_5H_5 bonded to each Pb (Fig. 22.16(b)). The latter is rather more strongly bonded to the metal (Pb–C, 2·76 Å) than are the bridging C_5H_5 groups (Pb–C, 3·06 Å). The angles at the Pb atoms are approximately 120° and Pb–Pb in the chain is 5·64 Å.

(11) JCP 1964 **40** 3434

(12) JCP 1964 **41** 717
(13) N 1959 **183** 1182
(14) N 1963 **199** 1087

(15) AC 1966 **21** 823

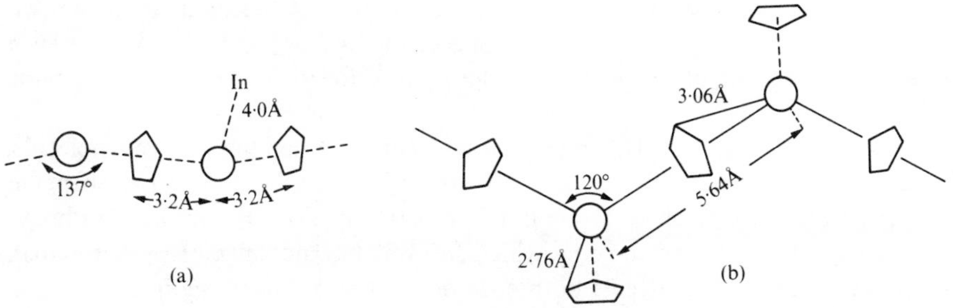

FIG. 22.16. The structures of (a) $In(C_5H_5)$, (b) $Pb(C_5H_5)_2$.

Complexes containing benzene or cyclooctatetraene
We give here only a few of the simpler complexes of these types. Closely related to the cyclopentadienyl compounds is the compound $Cr(C_6H_6)_2$, prepared by heating a mixture of $CrCl_3$, $AlCl_3$, Al powder, and benzene at 150°C. This is a crystalline compound (m.p. 284°C) in which molecules of the type shown in Fig. 22.17(a) are arranged in cubic closest packing. The Cr–C distance is 2·15 Å and the perpendicular distance between the rings is 3·23 Å.[1]

It is interesting to compare with the molecule $(CO)_3Cr(C_6H_6)$[2] (Fig. 22.17(b))

(1) ACSc 1965 **19** 41
(2) JACS 1959 **81** 5510

779

Metal Cyanides, Carbides, Carbonyls, and Alkyls

(3) JACS 1966 88 1877

the environment of Cu^+ in C_6H_6 . $CuAlCl_4$,[3] which is shown in Fig. 22.17(c). This compound contains tetrahedral $AlCl_4^-$ ions, and the neighbours of Cu are a C_6H_6 ring and 3 Cl atoms of $AlCl_4^-$ ions.

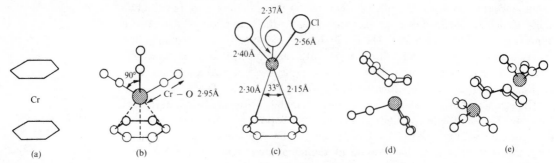

FIG. 22.17. The molecular structures of (a) $Cr(C_6H_6)_2$, (b) $(CO)_3Cr(C_6H_6)$, (c) $(C_6H_6)CuAlCl_4$, (d) $(CO)_3Fe(C_8H_8)$, (e) $(CO)_3Fe(C_8H_8)Fe(CO)_3$.

(4) JACS 1958 80 5075

(5) JCP 1959 63 845

In crystalline $AgClO_4$. C_6H_6[4] there are columns of C_6H_6 molecules and Ag^+ ions, two of the latter being associated with each benzene ring, and each Ag^+ is distant 2·50 and 2·63 Å from the two C atoms of the nearest C—C bond. The energy of the 'charge-transfer' bonds has been estimated at about 66 kJ/bond mole. In $AgNO_3$. C_8H_8[5] also there are columns of Ag^+ ions and hydrocarbon molecules, between which the anions are situated, and as in the benzene compound each Ag^+ is closely associated with one bond of the organic ring, here with a double bond. These shortest Ag—C bonds are again close to 2·5 Å in length. There are two weaker bonds from Ag^+ to two other atoms of the ring (Ag—C, c. 2·8 Å). The even longer Ag—C bonds, of length around 3·2 Å, link together the Ag^+ . C_8H_8 units into columns.

(6) JCP 1962 37 2084

In $AgNO_3$. C_8H_8 the C_8H_8 ring has the boat configuration like the free molecule, but two other shapes of this ring are found,[6] namely, a dihedral shape in $(CO)_3Fe(C_8H_8)$ and a 'chair' shape in $(CO)_3Fe(C_8H_8)Fe(CO)_3$. In the tricarbonyl Fe is attached to part of the ring which behaves like butadiene (Fig. 22.17(d)), whereas in the second molecule both ends of the ring are behaving in this way, as shown at (e) in Fig. 22.17. There are two configurations of the hydrocarbon ring also in $Ti_2(C_8H_8)_3$. The molecule has a double sandwich structure, C_8H_8

(7) AC 1968 B24 58

$-Ti-C_8H_8$ $-Ti-C_8H_8$,[7] in which the outer rings are planar and the central one boat-shaped.

Metal alkyls

Simple alkyls $M(alkyl)_n$ are formed by two groups of metals:

(i) the alkali metals, Be, Mg, the alkaline-earths, and Al, and

(ii) the B subgroup metals; little appears to be known of alkyls of the intervening transition metals. (For compounds $[Pt(CH_3)_3X]_4$, one of which, the hydroxide,

780

was earlier thought to be $Pt(CH_3)_4$, see p. 982.) It is convenient to deal first with group (ii) because with the possible exception of alkyls of Cu, Ag, and Au (of which the structures are not known) the B subgroup alkyls are normal covalent compounds like those of B, C, Si, etc. in which a C atom of the alkyl group forms a single bond to M.

Alkyls of B subgroup metals

Alkyls of IB metals include $CuCH_3$, $AgCH_3$, AgC_2H_5, and trimethyl gold. The last compound has been prepared at low temperatures in ether solution, in which it probably exists as $(CH_3)_3Au \cdot O(C_2H_5)_2$. This etherate reacts with ethereal HCl to give dimethyl auric chloride, presumably a bridged molecule, $(CH_3)_2AuCl_2Au(CH_3)_2$.

The IIB alkyls are normal covalent compounds, the chemical reactivity of which decreases from that of the violently reactive $Zn(CH_3)_2$ through $Cd(CH_3)_2$ to the relatively inert $Hg(CH_3)_2$. The high resolution Raman spectra of these compounds show that the molecules are linear; the bond lengths are respectively 1·93, 2·11, and 2·09 Å.[1]

The alkyls of Ga, In, and Tl form plane molecules $Ga(CH_3)_3$ etc.—contrast Ga_2Cl_6, Ga_2H_6, and the dimeric methyl gallium hydrides—as shown by a spectroscopic study of $Ga(CH_3)_3$,[2] and an early e.d. study of $In(CH_3)_3$ in the vapour state,[3] and later by crystallographic studies of $In(CH_3)_3$[4] and $Tl(CH_3)_3$.[5] Since these are planar molecules a metal atom necessarily has neighbours in directions approximately normal to the plane of $M(CH_3)_3$, and these complete a distorted trigonal bipyramidal arrangement of nearest and next nearest neighbours. The intra- and shortest inter-molecular bond lengths are:

In—C, 2·16 Å (three), 3·11 Å (one), and 3·59 Å (one);

Tl—C, 2·29 Å (three), 3·16 Å (one), and 3·31 Å (one).

Alkyls of groups I and II metals and Al

The alkyls of Na, K, Rb, and Cs are colourless amorphous solids which are insoluble in all solvents except the liquid zinc dialkyls, with which they react to form compounds such as $NaZn(C_2H_5)_3$. The Li compounds are quite different, for they are soluble to varying extents in solvents such as benzene, in which they are polymeric, and both lithium methyl and ethyl exist as colourless crystalline solids at ordinary temperatures.

Group II provides the crystalline $Be(CH_3)_2$, the very reactive magnesium dialkyls, and compounds of the alkaline-earths such as the dimethyls of Ca, Sr, and Ba.[6]

A number of aluminium alkyls have been prepared: $Al(CH_3)_3$ by refluxing Al with CH_3I in a nitrogen atmosphere, and the ethyl, n- and iso-propyl compounds by heating Al with the appropriate mercury dialkyl at 110°C for 30 hours. At ordinary temperatures they are water-white liquids (the trimethyl melts at 15°C), very reactive and spontaneously inflammable in air. The trimethyl is dimeric in benzene solution and in the vapour state, the ethyl and n-propyl dimers show

(1) CJP 1960 **38** 1516

(2) JCS 1964 3353
(3) JACS 1941 **63** 480
(4) JACS 1958 **80** 4141
(5) JCS A 1970 28

(6) JACS 1958 **80** 5324

(7) JACS 1946 **68** 2204

measurable dissociation, while the isopropyl compound is monomeric.[7] There is a striking similarity between $Al(CH_3)_3$ and $AlCl_3$. Both exist as dimers in the vapour state, and the energy of association is of the same order, viz. 84 and 121 kJ mol^{-1} respectively. Both combine with amines and ethers to give coordination compounds.

Since the structures of the compounds of the heavier alkali metals and of the alkaline-earths are not known our discussion is restricted to the alkyls of Li, Be, Mg, and Al. All are electron-deficient compounds in which some or all of the CH_3 groups form two or three bonds to metal atoms. These are described as 3- or 4-centre bonds formed by overlap of an sp^3 orbital of C with appropriate orbitals of the metal atoms.

Crystalline methyl lithium consists of a b.c.c. packing of molecules $Li_4(CH_3)_4$. The Li atoms are arranged at the vertices of a regular tetrahedron of edge 2·56 Å and a CH_3 group is situated above the centre of each face (Li–C, 2·28 Å). Since all Li and C atoms lie on body-diagonals of the unit cell Li has another CH_3 belonging to a neighbouring molecule as its next nearest neighbour (at 2·52 Å). The

(8) JOC 1964 **2** 197

intramolecular contacts are Li–3 Li (2·56 Å) and 3 CH_3 (2·28 Å).[8] In crystalline

(9) AC 1963 **16** 681

$Li_4(C_2H_5)_4$ the tetrahedral molecule is apparently less symmetrical, Li–Li ranging from 2·42 to 2·63 Å[9] and instead of three equidistant C at 2·28 Å Li has 3 C at 2·19, 2·25, and 2·47 Å, this last bond length being very close to the distance to the nearest C of an adjacent molecule (2·53 Å). These bond lengths are extremely difficult to understand, particularly in view of the strong bonding between the Li atoms—compare Li–Li, 2·67 Å in Li_2 and 3·04 Å in the metal.

Dimethyl beryllium is a white solid which sublimes at 200°C and reacts violently with air and water. It is isostructural with SiS_2, the crystal being built of infinite chains (a) in which Be forms tetrahedral bonds (Be–C, 1·93 Å) to four methyl

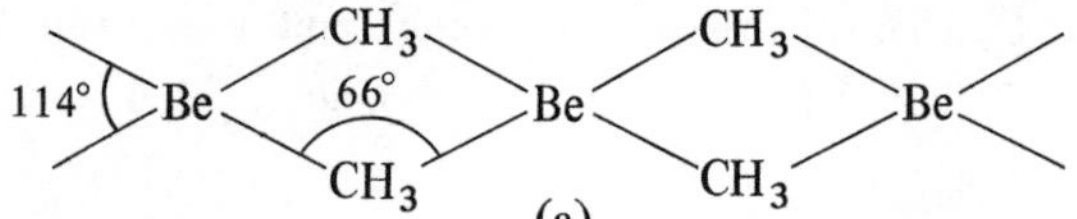

(a)

(10) AC 1951 **4** 348

groups.[10] $Mg(CH_3)_2$ has the same structure (Table 22.10) and the same type of chain is found in $LiAl(C_2H_5)_4$ where the metal atoms along the chain are alternately Li and Al and CH_3 is replaced by C_2H_5 (Table 22.10).

In Group III we find a third type of structure for the metal alkyl, namely, the bridged dimer (b) of $Al_2(CH_3)_6$ which is structurally similar to the Al_2X_6 molecules. There are also less symmetrical molecules with the same type of structure, for example, $(CH_3)_5Al_2N(C_6H_5)_2$, where the bridge consists of one CH_3

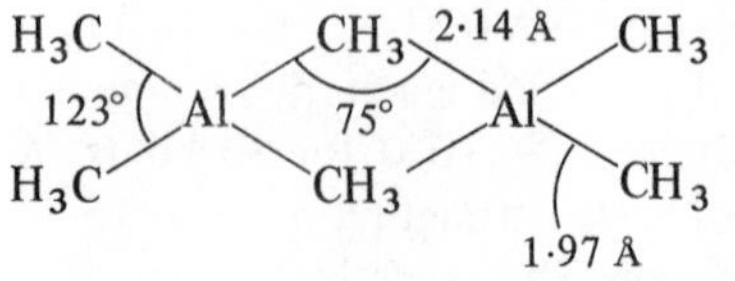

(b)

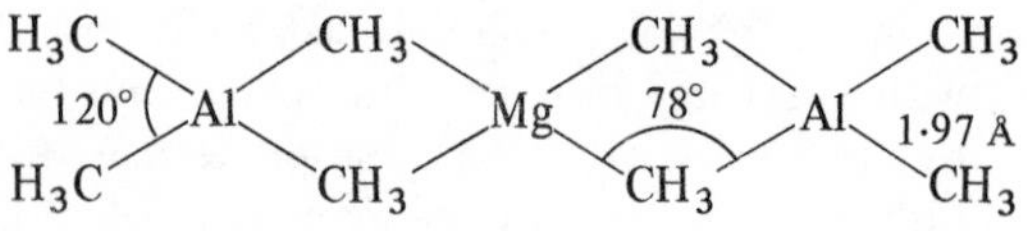

(c)

(Al–C–Al, 79°) and the $N(C_6H_5)_2$ group (Al–N–Al, 86°) and the central 4-ring is not planar as in (b). The mixed alkyl $MgAl_2(CH_3)_8$, made by dissolving $Mg(CH_3)_2$ in $Al(CH_3)_3$, has the structure (c) intermediate between those of the simple Al and Mg methyls. Some details of the structures of these compounds are summarized in Table 22.10. The fluoro-diethyl, $AlF(C_2H_5)_2$,[11] is a very viscous liquid soluble in

(11) IC 1966 **5** 503

TABLE 22.10
Structures of metal alkyls

	M–M	M–C (bridge)	M–C–M	Reference
$[Mg(CH_3)_2]_\infty$	2·72 Å	2·24 Å	75°	JACS 1964 **86** 4825; JOC 1964 **2** 314
$[LiAl(C_2H_5)_4]_\infty$	2·71	2·02 (Al) 2·30 (Li)	77°	IC 1964 **3** 872
$Al_2(CH_3)_6$	2·60	2·14	75°	JACS 1967 **89** 3121
$MgAl_2(CH_3)_8$	2·70	2·10 (Al) 2·21 (Mg)	78°	JACS 1969 **91** 2538
$Al_2(CH_3)_5N(C_6H_5)_2$	2·72	2·14	79°	JACS 1969 **91** 2544

benzene, in which it is tetrameric; a F-bridged cyclic structure has been suggested but not confirmed by a structural study. This is not, like the simple alkyls, an electron-deficient molecule. From an early e.d. study it was concluded that in the dimeric dimethyl monochloro compound the Al atoms were bridged by the Cl atoms,[12] but the Raman spectra of $(CH_3)_4Al_2X_2$ and $(CH_3)_2Al_2X_4(X = Cl, Br,$ or I) indicate that there are bridging methyl groups, as in $Al_2(CH_3)_6$.[13] Other mixed metal alkyls include $Li_2Zr(CH_3)_6$,[14] prepared from $LiCH_3$ and $ZrCl_4$ at a low temperature.

(12) JACS 1941 **63** 3287
(13) BSCB 1956 **65** 362
(14) ZaC 1970 **372** 292

23

Silicon

Introduction

Although the silicon atom has the same outer electronic structure as carbon its chemistry shows very little resemblance to that of carbon. It is true that elementary silicon has the same crystal structure as one of the forms of carbon (diamond) and that some of its simpler compounds have formulae like those of carbon compounds, but there is seldom much similarity in chemical or physical properties. Since it is more electro-positive than carbon it forms compounds with many metals which have typical alloy structures (see the silicides, p. 789) and some of these have the same structures as the corresponding borides. In fact, silicon in many ways resembles boron more closely than carbon, though the formulae of the compounds are usually quite different. Some of these resemblances are mentioned at the beginning of the next chapter. Silicides have few properties in common with carbides but many with borides, for example, the formation of extended networks of linked Si (B) atoms, though on the other hand few silicides are actually isostructural with borides because Si is appreciably larger than B and does not form some of the polyhedral complexes which are peculiar to boron and are one of the least understood features of boron chemistry.

Silicon is compared with germanium in a short section on the latter element in Chapter 26. Here we shall note some points of difference from and resemblance to carbon. The outstanding feature of carbon chemistry is the unlimited number of molecules that can be formed by direct bonding of carbon to carbon. Simple molecules containing Si–Si bonds are few in number; they include the silanes, molecules X_3Si–SiX_3, some oxyhalides, and compounds such as $[Si(C_6H_5)_2]_5$,[1] which contains a ring of five Si atoms. Many silicides contain systems of linked Si atoms, but apart from these cases the chemistry of complex compounds of silicon is largely based on linking of Si atoms through O atoms.

In its simple molecules and ions Si does not exhibit a covalency of less than four except (possibly) in the silyl ion (see later). Unlike carbon it does not form a small number of multiple bonds, as does carbon in CO, CO_2, –CN, etc. Only one oxide of Si is stable at ordinary temperatures, and this exists in a number of crystalline forms, in all of which—with the exception of stishovite (p. 804)—there is a tetrahedral arrangement of four bonds from each Si atom, and in every case the Si and O atoms form an infinite 3D network. The lower oxide SiO has nothing in common with the gaseous CO. It is produced by heating SiO_2 with Si at

(1) JACS 1963 **85** 1016

temperatures above 1250°C and disproportionates on slow cooling. It has been claimed that if it is quenched it can be obtained at room temperature, but apparently not in crystalline form.[2] The reaction $Si + SiCl_4 \rightarrow 2SiCl_2$ is said to take place at temperatures above 1100°C.[3] Certainly SiF_4 reacts with Si at 1150°C to form gaseous SiF_2, a species with a half-life at room temperature of 150 seconds, (compare 1 second for CF_2 and a few hundredths of a second for CH_2); see also under silicon halides.[4] There is no stable sulphide CS, but material with the composition SiS can be sublimed from a mixture of $SiS_2 + Si$. At room temperature the coloured sublimates are not crystalline and have probably disproportionated to $SiS_2 + Si$.[5] The structure of the crystalline SiS_2 bears no relation to that of the simple CS_2 molecule; the normal form of SiS_2 consists of infinite chains of SiS_4 tetrahedra sharing pairs of opposite edges. (There is also a high-pressure polymorph with a compressed cristobalite-like structure, M–S–M, $109\frac{1}{2}°$.[6])

Many of the differences between Si and C, like those between P and N, S and O, or Cl and F, are attributable to the fact that the elements of the second short Period can utilize d orbitals which are not available in the case of the first row elements. We find covalences greater than 4 in compounds of Si, P, and S with the more electronegative elements. There are no carbon analogues of the fluorosilicates and there is no nitrogen analogue of the PF_6^- ion. Although the silyl compounds, containing $-SiH_3$, are formally analogous to methyl compounds, they behave in many ways very differently from their carbon analogues. The abnormal properties of many silyl compounds are probably connected with the powerful electron-accepting properties of $-SiH_3$. Although Si is less electronegative than C, $-SiH_3$ is a stronger electronic-acceptor than $-CH_3$ because π bonding takes place between a p_π orbital on the atom attached to Si by a σ bond and a vacant d_π orbital of Si; there are no 3d orbitals on the C atom. (Usually π bonding is stronger the *more* electronegative is the donor atom.) This characteristic of the silyl group shows itself in, for example, the planarity of $N(SiH_3)_3$ (p. 642), and the weaker coordinating power of silyl compounds. For example, $N(SiH_3)_3$ forms no quaternary salts $[N(SiH_3)_4]X$ or $[N(SiH_3)_3H]X$, and the addition compound with BF_3 is much less stable than the compound with $N(CH_3)_3$.

The last point we have to mention is the great difference in stability towards oxygen and water between pairs of carbon and silicon compounds. In contrast to the stable CCl_4 and CS_2 the corresponding silicon compounds are decomposed by water to form 'silicic acid' (and, of course, HCl and H_2S respectively), a term applied to the colloidal silica which may be obtained hydrated to varying degrees. (The preparation of well defined di-, tri-, and tetra-silicic acids has been claimed[7] by the hydrolysis and disproportionation of ortho silicic esters.) Silicon forms a number of hydrides (silanes), and although these are stable towards pure water they are hydrolysed by dilute alkali, which has no action at all on the corresponding hydrocarbons. In view of the ease with which silicon compounds are converted into the oxide or other oxy-compounds, it is not surprising that silicon occurs in nature exclusively as oxy-compounds, and we shall devote much of this chapter to a review of the extensive chemistry of the silicates.

(2) JPC 1959 **63** 1119; ZaC 1967 **355** 265
(3) ZaC 1953 **274** 250

(4) JACS 1965 **87** 2824

(5) JACS 1955 **77** 904

(6) Sc 1965 **149** 535

(7) ZaC 1954 **275** 176; **276** 33.

Silicon

The stereochemistry of silicon

In the crystalline element and in the great majority of its compounds stable under ordinary conditions Si forms four tetrahedral bonds; numerous examples are given in this and other chapters. The next most important coordination number is six, and there is apparently no tendency to exceed this number. Thus with acetylacetone Si forms $[Si(acac)_3]AuCl_4$ and $Si(acac)_3Cl.HCl$ (a compound which has been resolved into its optical antimers),[1] but it does not, like Ti and Zr, form $M(acac)_4$; K_3SiF_7, formed by heating K_2SiF_6 in dry air, is $K_3(SiF_6)F$.[2] The octahedral configuration has been established for finite complexes such as the SiF_6^{2-} ion—for example, K_2SiF_6 has the K_2PtCl_6 structure—, the ion (a) in its pyridinium salt,[3] and the neutral molecule (b),[4] and it is probable for the

(1) JACS 1959 **81** 6372

(2) AC 1962 **15** 186

(3) JACS 1969 **91** 5756

(4) AC 1969 **B25** 156

(a)   (b)

(5) JCS A 1969 363

(6a) IC 1962 **1** 236; IC 1966 **5** 1979

(6b) JSSC 1973 **7** 69

(6c) AC 1971 **B27** 594

(7) AC 1970 **B26** 233

(8) ACSc 1967 **21** 1374

trisoxalato ions in salts $(NR_4)_2[M^{IV}(C_2O_4)_3]$, (M = Si, Ge, Sn, Ti),[5] and derivatives $PcSiX_2$ of the phthalocyanine.[6a] Octahedral coordination by oxygen occurs in all forms of SiP_2O_7[6b] stable under atmospheric pressure, in $[Ca_3Si(OH)_6.12H_2O]SO_4.CO_3$[6c] and in a number of high-pressure oxide phases, including one form of SiO_2 itself (stishovite), synthetic $KAlSi_3O_8$, and a garnet.[7] There is also octahedral coordination of Si in the high-pressure forms of $SiAs_2$ and SiP_2. In the usual form of SiP_2 the coordination numbers of Si and P are 4 and 3, but in the high-pressure form (pyrites structure) they are 6 (octahedral) and 4 (3 Si + P, tetrahedral).[8] The persistence of octahedral coordination in the *high-temperature* (cubic) form of SiP_2O_7 is noteworthy, as also is the fact that this structure is based on the same (6,4)-connected 3D net as the *high-pressure* form of SiP_2, from which it is derived (topologically) by inserting an O atom along each link (Si-P or P–P) of the SiP_2 (pyrites) structure.

(9) IC 1969 **8** 450

(10) JACS 1968 **90** 6973

(11) JACS 1968 **90** 4026

(12) JACS 1968 **90** 5102

There are several compounds in which Si is 5-coordinated. The most straightforward case would be an ion SiX_5^- such as occurs in salts $(NR_4)(SiF_5)$,[9] in which R must be larger than CH_3, but their structures are not yet known. Structural studies have been made of various chelate species, for example, the anion (c) in its tetramethylammonium salt,[10] and the molecules (d)[11] and (e).[12] (There are some unexpected bond lengths in (c)—a very short C–C, 1·29 Å, in the phenylene

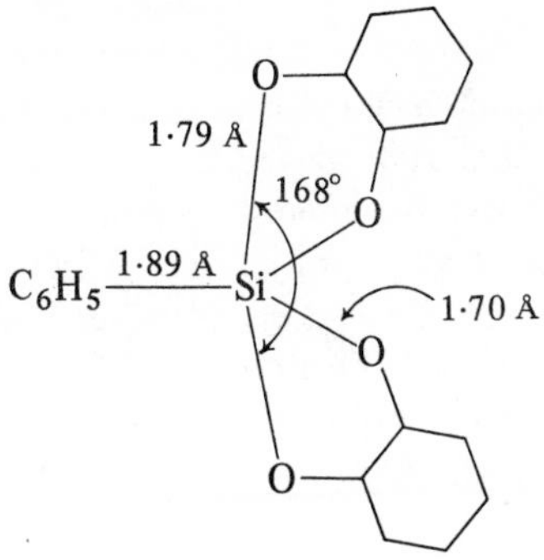

(c)

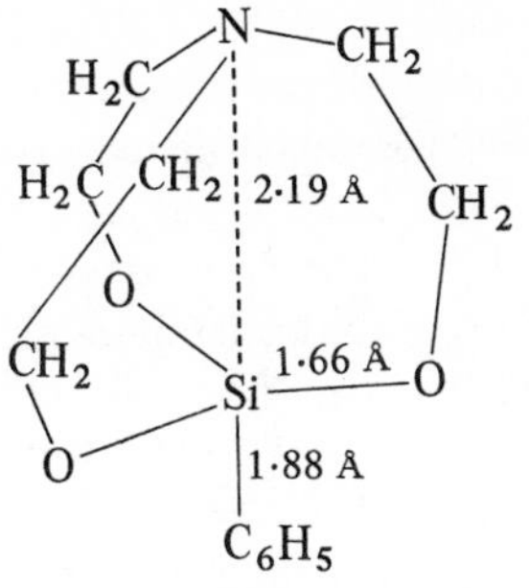

(d)

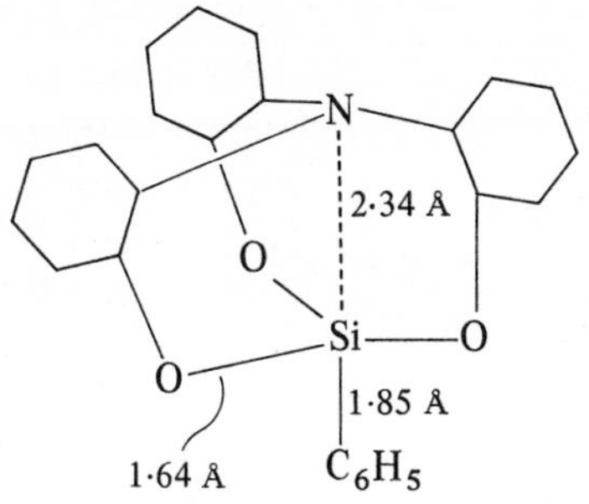

(e)

rings, and C–O, 1·34 Å.) A notable feature of (d) and (e) both of which are approximately trigonal bipyramidal in shape, is the extreme length of the Si–N bond (compare 1·74 Å for tetrahedral coordination). In the cyclic molecule [(CH$_3$)$_2$N . SiH$_3$]$_5$,[13] (f), there is presumably trigonal bipyramidal coordination

(13) JACS 1967 89 5157

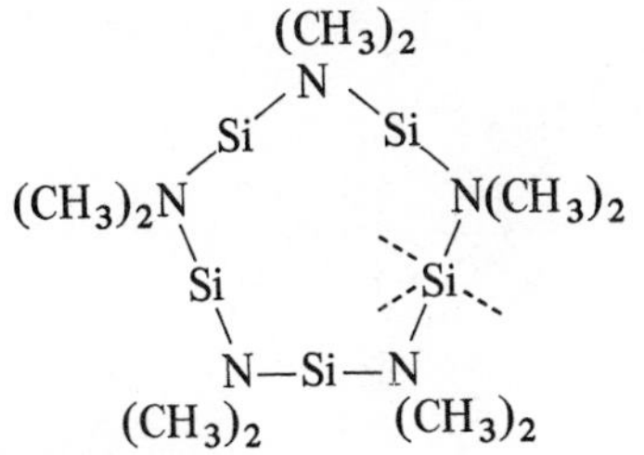

(f)

of Si by 3 H (equatorial) and 2 N; the H atoms were not located. Here the Si–N bond has the same length as in the 6-coordinated molecule (b).

The action of metallic K on SiH$_4$ or Si$_2$H$_6$ in aprotic solvents such as 1 : 2-dimethoxyethane produces KSiH$_3$, which forms colourless crystals with the NaCl structure.[14] The H atoms were not located, but the SiH$_3^-$ ion would contain 3-covalent Si. For the unstable SiF$_2$ and SiCl$_2$ species see p. 785.

(14) JACS 1961 83 802

Elementary silicon and carborundum

We have mentioned in the previous chapter the two crystalline forms of carbon: diamond and graphite. The latter structure is unique among the elements, but a number of other elements of Group IV crystallize with the diamond structure; silicon, germanium, and tin (grey modification). In the B subgroup only lead has a typical metallic structure (cubic close-packed), whereas all the elements of the A subgroup crystallize with close-packed structures. In elementary silicon Si–Si = 2·35 Å.

Silicon

(1) AC 1964 **17** 752

Under moderate pressure both Si and Ge transform to more dense (about 10 per cent) modifications with different 4-connected nets, both different from the diamond structure.[1] There is little change in bond lengths, but appreciable angular distortion—in Si to three angles each of 99° and 108° and in Ge to angles ranging from 88° to 135°. The Ge structure is particularly interesting since the same network is the basis of the structure of one of the high-pressure forms of SiO_2 (keatite), and also of ice III. These 4-connected nets have been described in Chapter 3. At still higher pressures Ge transforms to the white tin structure.

Carborundum, SiC crystallizes in a large number of forms. The cubic form (β-SiC) has the zinc-blende structure (p. 102), and in addition there are numerous crystalline modifications all with structures closely related to the zinc-blende and wurtzite structures, which are collectively referred to as α-SiC. The atomic positions

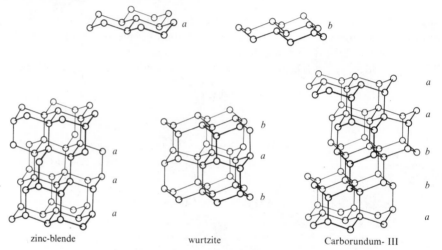

FIG. 23.1. The geometrical relationship between the structures of zinc-blende, wurtzite, and carborundum III.

in zinc-blende are the same as in diamond, alternate atoms being Zn and S. The structures of zinc-blende and wurtzite are not layer structures in the sense in which we use this term, but we may dissect these structures into layers of types (a) and (b) as shown in Fig. 23.1 (where alternate circles represent Si and C or Zn and S atoms as the case may be). It will be seen that the zinc-blende structure is built up of layers of type (a) only, while in wurtzite (a) and (b) layers alternate. More complex sequences are found in the various forms of α-SiC, for example:

				Layer sequence	*Alternative symbols*	
Zinc-blende	.	.	.	a		
Wurtzite	.	.	.	a b		2H
Carborundum III	.	.	a a b b	2 2	4H
Carborundum I	.	.	a a a b b	3 2	15R
Carborundum II	.	.	a a a b b b	3 3	6H

788

The first three structures are illustrated in Fig. 23.1. We should, perhaps, point out that the *crystallographic* repeat unit is, in general, a multiple of the layer sequence listed in the table because there we have disregarded the relative translations of the layers. Thus in zinc-blende the unit cell contains three (a) layers and in carborundum I 15 layers (that is, three sets of *aaabb*). A convenient symbol shows the number of layers in the repeat unit followed by H if the symmetry is hexagonal and by R if it is rhombohedral. The carborundum structures set out above are the most abundant forms in the ordinary material and are relatively simple examples of a very large group of structures with increasingly complex layer sequences which in some cases repeat only after several hundred layers. The layer sequence has been determined in more than seventy of these structures, which are described as *polytypes* to distinguish them from normal polymorphs. Carborundum is apparently formed by a sublimation process in the electric furnace, and the explanation of these regular sequences must lie in the mode of growth of the crystals. It is possible that more than one mechanism may be operative, but growth by screw dislocations probably provides the best explanation of the long range order.[2]

(2) AC 1969 **B25** 477

Silicides

Silicides form a large group of compounds the formulae of many of which are not explicable in terms of the normal valences of the elements. For example, the compounds of the alkali metals include $Li_{15}Si_4$ and Li_2Si, KSi and KSi_6, $CsSi$ and $CsSi_8$. Bonding ranges from essentially metallic or ionic to covalent, and in many silicides there are bonds of more than one kind. We shall therefore adopt here only a very general geometrical classification, since a knowledge of the structure must necessarily precede any discussion of the nature of the bonding in a particular compound.

At one extreme lie silicides containing isolated Si atoms as, for example, Ca_2Si[1] with the anti-$PbCl_2$ structure, Mg_2Si, which has the antifluorite structure (like Mg_2Ge, Mg_2Sn, and Mg_2Pb), and a variety of compounds with typically metallic structures—V_3Si, Cr_3Si, and Mo_3Si (β-W structure,), Cu_5Si (β-Mn structure), Mn_3Si (random b.c.c.), and Fe_3Si (Fe_3Al superstructure). The interpretation of interatomic distances in these structures is not simple (compare intermetallic compounds, Chapter 29); for example, in $TiSi_2$ (see later) Si has 2 Si and 2 Ti at 2·54 Å and 3 Ti and 3 Si at approximately 2·75 Å. In other silicides there are Si—Si distances close to that in elementary silicon (2·35 Å). A uniquely short Si—Si distance, 2·18 Å, has been found in $IrSi_3$.[2a] The metal atoms lie, in positions of 9-coordination, between plane 4-connected nets of Si atoms of the kind shown in Fig. 29.13(b). If we recognize distances up to 2·5 Å or so as indicating appreciable interaction between Si atoms we may distinguish systems of

(1) ZaC 1955 **280** 321

(2a) IC 1971 **10** 1934

linked Si atoms of the following kinds:

(2b) ZaC 1961 **313** 90

Type of Si *complex*	Examples
Si_2 groups	U_3Si_2 (Si—Si, 2·30 Å)
Si_4 groups	$KSi^{(2b)}$
Chains	USi (Si—Si, 2·36 Å)
	CaSi (Si—Si, 2·47 Å)
Plane hexagonal net	β-USi_2
Puckered hexagonal net	$CaSi_2$
3D framework	$SrSi_2$, α-USi_2 ($ThSi_2$)

(3) 1949 **2** 94

It is interesting that the silicides of U provide examples of all the main types of silicide structure,[3] for in addition to those listed there is U_3Si which has a distorted version of the Cu_3Au structure (p. 842), in which there are discrete Si atoms entirely surrounded by U atoms. They also show that some silicides adopt the same structures as certain borides which contain extended systems of linked B atoms, though because of the greater size of Si and its different electronic structure borides and silicides are not generally isostructural.

(4) AC 1966 **20** 572

(5) IC 1967 **6** 842

The FeB structure of USi and other 4f and 5f monosilicides[4] (and also TiSi and ZrSi) with infinite silicon chains, and the AlB_2 structure of β-USi_2 (and other 4f and 5f disilicides)[5] with plane hexagonal Si layers, are described and illustrated in the next chapter. Here, therefore, we shall first note the structures of some silicides with alloy-like structures containing discrete Si atoms and then describe the structure of $CaSi_2$, which contains puckered Si layers, and finally the structures of $SrSi_2$ and α-$ThSi_2$ as examples of 3D framework structures.

$TiSi_2$, $CrSi_2$, *and* $MoSi_2$. The structures of $TiSi_2$ (orthorhombic), the isostructural $CrSi_2$, VSi_2, $NbSi_2$, and $TaSi_2$ (hexagonal), and the isostructural $MoSi_2$, WSi_2, and $ReSi_2$ (tetragonal) are related in a very simple way, which is unexpected in view of their quite different symmetries. TiS_2 is built of close-packed layers of the type shown in Fig. 23.2(a), with the Si : Ti ratio 2 : 1. The layers are not, however, superposed in such a way as to make the structure as a whole close-packed (that is, to give each atom twelve equidistant nearest neighbours) but, instead, the Ti atoms of successive layers fall above the points *B, C,* and *D* respectively (Fig. 23.2(a)). Each atom has therefore ten nearest neighbours, Ti being approximately equidistant from 10 Si and each Si approximately equidistant from 5 Ti and 5 Si. Crystals of $CrSi_2$ contain the same type of close-packed layers, and, moreover, the layers are superposed in a similar way, giving each atom ten nearest neighbours (Cr to 10 Si, Si to 5 Cr + 5 Si). In this (hexagonal) structure the repeat unit comprises only three layers, the Cr atoms of successive layers falling above the points *B* and *D* only (Fig. 23.2(b)). From the geometrical standpoint the structure of $MoSi_2$ is related to that of CaC_2, as may be seen by comparing Fig. 23.2(c) with Fig. 22.6, p. 757, though the Si atoms are not associated in pairs as are the C atoms in CaC_2. (Both structures are tetragonal, CaC_2 having $a = 3·89$, $c = 6·38$ Å, and the parameter of the C atom $(00z)$ $z_C = 0·406$; $MoSi_2$ has $a = 3·20$, $c = 7·86$ Å and $z_{Si} \approx 0·33$. The conventional body-centred unit

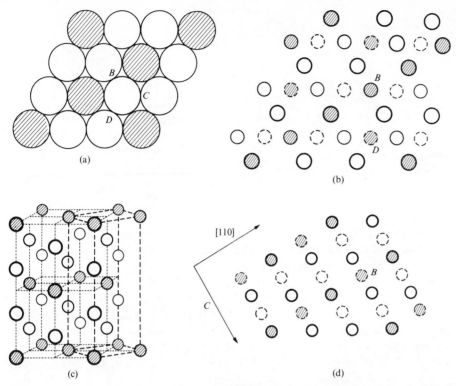

FIG. 23.2. (a) Close-packed MSi$_2$ layer in some disilicides. (b) Three successive layers in CrSi$_2$. (c) The crystal structure of MoSi$_2$. (d) Projection of the structure of MoSi$_2$ on (110) showing the relative positions of atoms in two adjacent layers (full and dotted circles). Metal atoms are represented by shaded circles.

cell is indicated by broken lines in Fig. 23.2(c).) However, the projection of the structure on (110), Fig. 23.2(d), shows that MoSi$_2$ is very closely related to TiSi$_2$ and CrSi$_2$. In the (110) plane the atoms are arranged at the points of a hexagonal net of exactly the same type as in the latter structures, and here also successive layers are superposed in such a way that each type of atom has the same arrangement of ten nearest neighbours. In MoSi$_2$ atoms of *alternate* layers fall vertically above one another, adjacent layers being displaced with respect to one another, so that a Mo atom of an adjacent layer falls over the point B in Fig. 23.2(d). These three structures thus form an interesting series, all being built from the same type of close-packed layers superposed in essentially the same way, to give 10-instead of 12-coordination, but differing in the number of layers in the repeat unit, which is four layers in TiSi$_2$, three in CrSi$_2$, and two in MoSi$_2$. It is found that crystals with compositions intermediate between TiSi$_2$ and MoSi$_2$ adopt the intermediate (3-layer) CrSi$_2$ structure, and the same is true in the TiSi$_2$–ReSi$_2$ series.[6]

(6) AC 1964 **17** 450

CaSi$_2$.   In this silicide the Ca and Si atoms are arranged in alternate layers, as shown in Fig. 23.3. A silicon layer might be described as a puckered graphite layer

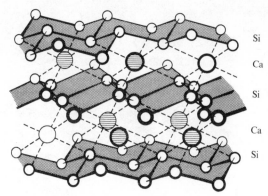

Si

Ca

Si

Ca

Si

FIG. 23.3. The crystal structure of CaSi₂.

(7) ZaC 1927 **160** 152

(8) AC 1972 **B28** 2326

or a section of the diamond structure parallel to the octahedron plane (111), in which each Si has three neighbours at 2·48 Å, the bond angles having the normal tetrahedral value. The Ca atoms lie between these layers in positions of (6 + 1)-coordination (capped octahedron) with Ca–Si close to 3 Å.[7] The hydrolysis of this compound is mentioned on p. 794.

SrSi₂. The structure of this compound[8] consists of a 3D framework of Si atoms each bonded to 3 others (Si–Si, 2·39 Å), in the interstices of which are situated the Sr atoms (ions) in positions of 8-coordination (approximately a hexagonal bipyramid, Sr–6 Si, 3·24 Å, Sr–2 Si (axial), 3·36 Å). This unusual arrangement of 8 neighbours may be compared with that of the A atoms in the pyrochlore structure (Fig. 6.12). A Si atom has 4 close Sr neighbours (in addition to 3 Si of the framework), so that the structure may be described as an AX₂ structure of 8 : 4 coordination. A reasonable interpretation of the structure is to suppose that the 3-connected Si framework carries a charge of approximately –1 on each Si atom. The framework is the 3D net of Fig. 3.28(a) slightly distorted from its most symmetrical configuration (Si bond angles 117·7° instead of 120°), as noted on p. 98; the structure is therefore enantiomorphic. It is noteworthy that the second simplest 3-connected 3D net, also built of rings of 10 Si atoms, is found in α-ThSi₂.

α-ThSi₂. Many 4f and 5f disilicides adopt the α-ThSi₂ structure (and/or a less symmetrical variant of this structure) and/or the AlB₂ structure.[5] In α-ThSi₂ the Si atoms form a very open, 3-coordinated, 3D network in which the Si–Si bond length is 2·39 Å, close to the value in elementary Si. This 3D network may be compared with the 3-coordinated boron layer in the AlB₂ structure (p. 842); it is emphasized in Fig. 23.4. In the interstices are the Th atoms, each bonded to 12 Si neighbours (at 3·16 Å). The next nearest neighbours of a Si atom are 6 Th at this distance. The close relation between the ThSi₂ and AlB₂ structures is emphasized by the facts that the disilicides of Th, U, and Pu crystallize with both structures, and the disilicides of the earlier 4f metals (La–Ho) crystallize with the ThSi₂ structure (in some cases distorted) while those of Er, Tm, Yb, and Lu adopt the AlB₂ structure.

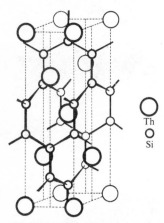

○ Th

○ Si

FIG. 23.4. The crystal structure of ThSi₂.

Silanes

Hydrides from SiH_4 to Si_7H_{16} have been prepared by the action of 20 per cent HCl on magnesium silicide in an atmosphere of hydrogen, or preferably by the action of NH_4Br on the same compound in liquid NH_3. The mixture of gases evolved is cooled in liquid nitrogen and fractionated as in the case of the boron hydrides. All the above hydrides are obtained in this way in regularly decreasing amounts. The two isomers of Si_4H_{10} have been separated by gas chromatography and identified by their i.r. and n.m.r. spectra.[1] Mixed Si—Ge hydrides (e.g. Si_3GeH_{10} and Si_5GeH_{14}) have been prepared by hydrolysis of Mg—Ge—Si alloys and by pyrolysis of mixtures of silanes and germanes.[2] The thermal stability of the silanes decreases with increasing molecular weight and the higher members break down on heating into mixtures of the lower hydrides—compare the 'cracking' of hydrocarbons, which is a similar process. The silicon hydrides are not attacked by water but are hydrolysed by dilute alkali solutions, hydrogen being evolved and a silicate formed. Like the hydrocarbons they may be halogenated, and the hydrolysis of the products is described later.

(1) IC 1964 **3** 946

(2) JCS 1964 1467; JCS A 1969 2937

A convenient general method of preparing Group IV hydrides, simple and substituted, is by the action of lithium aluminium hydride on the appropriate halide in ether solution:

$$4\,MR_yX_{4-y} + (4-y)LiAlH_4 \rightarrow 4\,MR_yH_{4-y} + (4-y)LiX + (4-y)AlX_3,$$

where R represents alkyl or aryl groups and X a halogen. In this way, $SiCl_4 \rightarrow SiH_4$, $Si_2Cl_6 \rightarrow Si_2H_6$, $Et_2SiCl_2 \rightarrow Et_2SiH_2$, etc. (The same method has been used to prepare GeH_4, SnH_4, and the three methyl stannanes.)

The silanes are very easily oxidized, Si_2H_6, for example, being spontaneously inflammable in air. In marked contrast to carbon, silicon appears to form no simple unsaturated hydrides corresponding to the olefines, acetylenes, and aromatic hydrocarbons. Solid unsaturated hydrides such as $(SiH_2)_n$, of unknown constitution, have been described, though in a study of the higher silanes no solid products were found at any stage. Structurally the silanes are presumably similar to the saturated hydrocarbons. Many substituted monosilanes have been prepared, for example:

$$[Si(CH_3)_4] \quad \begin{array}{l} Si(CH_3)_3Cl \\ Si(CH_3)_2Cl_2 \\ Si(CH_3)Cl_3 \end{array} \quad [SiCl_4] \quad \begin{array}{l} SiCl_3H \\ SiCl_2H_2 \\ SiClH_3 \end{array} \quad [SiH_4]$$

Halogenated and alkylated disilanes are also known, for example, F_3Si—SiF_3, Cl_3Si—$SiCl_3$, $(CH_3)_3Si$—$Si(CH_3)_3$, etc.

Various linear and cyclic fully methylated polysilanes can be made by catalytic rearrangement of $Si_6(CH_3)_{14}$, and the cyclic compounds $[Si(CH_3)_2]_n$ ($n = 5-7$) have been prepared by the action of Li on $Si(CH_3)_2Cl_2$ in tetrahydrofuran.[3]

(3) JACS 1969 **91** 5440

Structural data for some of these compounds are included in Table 21.1 (p. 727).

Silicon halides

(1) JCP 1967 **47** 5031
(2) JACS 1969 **91** 2536

The lower halides SiF_2 and $SiCl_2$ have already been mentioned. In SiF_2, $Si-F = 1.591$ Å and the angle $F-Si-F = 101°$[1]; this compound has also been studied in solid Ne and Ar matrices.[2] Condensation of gaseous SiF_2 at low temperatures yields a plastic polymer $(SiF_2)_n$ which burns spontaneously in moist air and on heating generates all the perfluorosilanes from SiF_4 to $Si_{14}F_{30}$. With BF_3 in a trap at the temperature of liquid nitrogen SiF_2 forms compounds such as Si_2BF_7, Si_3BF_9, etc. which have been identified mass-spectrometrically. These are presumably linear molecules $F_3Si-(SiF_2)_n-BF_2$.[3]

(3) JACS 1965 **87** 2824, 3819

Electron diffraction and spectroscopic studies show that in the gaseous state the halides SiX_4 form tetrahedral molecules; the simple molecular structure of SiF_4 in the crystalline state has been confirmed by an X-ray study at $-145°C$.[4]

(4) AC 1954 7 597

Higher chlorosilanes have been prepared, some conveniently by methods such as the disproportionation of Si_2Cl_6 in the presence of trimethylamine, when $5\ Si_2Cl_6 \rightarrow 4\ SiCl_4 + Si_6Cl_{14}$.[5] Controlled hydrolysis of these compounds is discussed later (p. 795); complete hydrolysis of these halides and of other chloro-silicon compounds gives oxy-compounds which are in many cases formally analogous to carbon compounds. The silicon compounds bear little rememblance, however, as regards physical or chemical properties to their carbon analogues, the final products of hydrolysis being highly polymerized solids. All chlorosilanes are readily hydrolysed, for example SiH_2Cl_2 gives $(H_2SiO)_n$, 'prosiloxane', which appears in many polymeric forms depending on the mode of hydrolysis. The final product of the hydrolysis of Si_2Cl_6 is a white powdery solid of empirical composition $(SiO_2H)_n$, called silico-oxalic acid. In spite of its name and formula this compound bears no chemical resemblance to oxalic acid, for it is insoluble in water and possesses no acidic properties. No salts or esters are known, and it dissolves in alkalis with evolution of hydrogen. It is obviously a complex polymer. The 'silico-mesoxalic acid', $(Si_3O_6H_4)_n$, formed by the action of moist air on Si_3Cl_8, has rather similar properties. Another polymerized product, also of unknown constitution, $(Si_2H_2O)_n$, is produced by the action of alcoholic hydrochloric acid on $CaSi_2$ (p. 791). It is spontaneously inflammable in air; it is a reducing agent, and is rapidly decomposed by hot water. It forms flakes of the same shape as the calcium silicide from which it was formed, which is reminiscent of the graphite derivatives described in Chapter 21.

(5) JINC 1964 **26** 421, 427, 435

Oxyhalides and thiohalides

The oxyhalide $SiOCl_2$, the analogue of phosgene, is not known; it would imply the formation by silicon of a double bond to oxygen. The simplest oxychloride known is the liquid Si_2OCl_6, formed by passing a mixture of air and $SiCl_4$ vapour through a red-hot tube. An electron diffraction study of this compound gives the following data: $Si-O$, 1.64 (0.04), $Si-Cl$, 2.02 (0.03) Å, $Cl-Si-Cl$, 109 $(2)°$. The $Si-O-Si$ angle was not determined but is probably close to $130°$.[1]

(1) JCP 1950 **18** 1414

The linear oxychlorides $Si_nO_{n-1}Cl_{2n+2}$ have been prepared by passing a

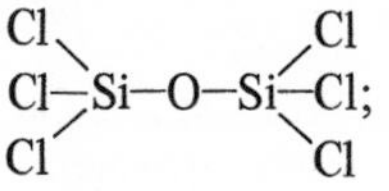

$$Cl \diagdown \atop Cl - Si - O - Si \diagdown Cl \atop Cl \diagup \diagdown Cl$$
$$Si_2OCl_6$$

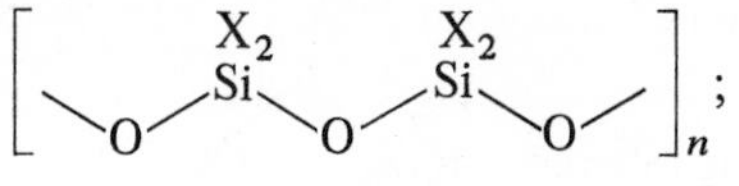

$$\left[\diagdown O \diagdown Si \diagup X_2 \diagdown O \diagdown Si \diagup X_2 \diagdown O \diagdown \right]_n ;$$
$$Si_nO_{n-1}X_{2n+2}$$

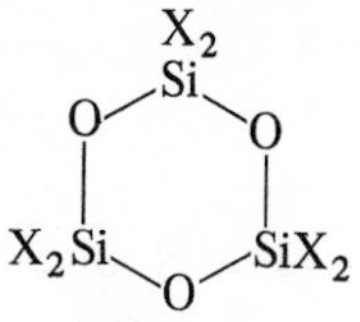

$$\begin{array}{c} X_2 \\ Si \\ O \diagup \diagdown O \\ X_2Si \diagdown \diagup SiX_2 \\ O \end{array} \quad etc.$$
$$(SiOX_2)_n$$

mixture of $O_2 + 2\,Cl_2$ over heated silicon and also by controlled hydrolysis of $SiCl_4$ in ether solution at $-78°C$ followed by fractional distillation. All up to and including $Si_7O_6Cl_{16}$ have been prepared and characterized.[2] They are colourless, oily, liquids, which are incombustible and miscible with solvents such as chloroform, carbon tetrachloride, and carbon disulphide. They are all hydrolysable, the lower ones most rapidly, and with alcohol they form esters, for example, $Si_6O_5(OC_2H_5)_{14}$. The cyclic compounds $(SiOCl_2)_n$, $n = 3$, 4, or 5, have been prepared.[3] The corresponding linear oxybromides are known up to $Si_6O_5Br_{14}$, and the cyclic $(SiOBr_2)_4$ has also been prepared.[4] Mixed fluoro-chloro compounds which have been prepared include $Si_2OF_3Cl_3$ and $Si_2OF_4Cl_2$.[5]

Whereas controlled hydrolysis of $SiCl_4$ at $-78°C$ gives the linear oxychlorides $Si_nO_{n-1}Cl_{2n+2}$, a similar treatment of Si_2Cl_6 gives another series $Si_{2n+2}O_nCl_{4n+6}$ containing pairs of linked Si atoms bonded into chains by O atoms:

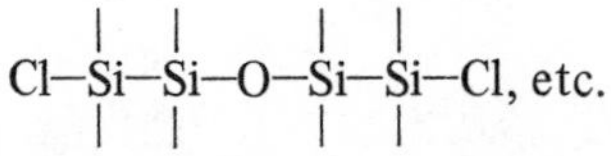

$$Cl - \overset{|}{Si} - \overset{|}{Si} - O - \overset{|}{Si} - \overset{|}{Si} - Cl, \ etc.$$

In this way Si_4OCl_{10}, $Si_6O_2Cl_{14}$, and $Si_8O_3Cl_{18}$ have been obtained as colourless liquids.[6]

Thiochlorides that have been prepared include Si_2SCl_6 (I), the cyclic compound (II).

$$Cl_3Si \diagdown S \diagup SiCl_3$$

$$Cl_2Si \overset{S}{\diagdown \diagup} \overset{\diagup}{\diagdown} SiCl_2 \atop S$$

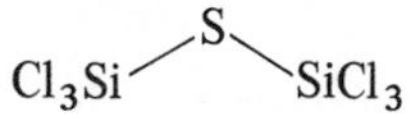

I 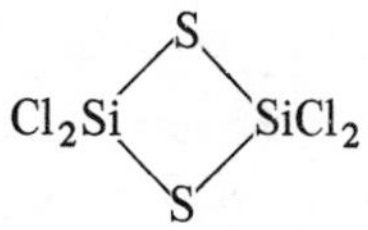 II

and the crystalline $(SiSCl_2)_4$.

Cyclic silthianes

Electron diffraction studies[1] have been made of tetramethyl cyclodisilthiane (a) and hexamethylcyclotrisilthiane (b):

$$(CH_3)_2 \atop Si$$
$$S \overset{105°}{\diagup \diagdown} S \atop Si \atop (CH_3)_2$$
$$2·18\ (·03)\ Å$$

$$(CH_3)_2 \atop Si$$
$$S \diagup \diagdown S$$
$$110°$$
$$(CH_3)_2Si \qquad 115° \quad Si(CH_3)_2$$
$$2·15\ (·03)\ Å$$
$$S$$

(a) (b)

(2) JACS 1941 **63** 2753

(3) JCS 1960 5088
(4) JACS 1937 **59** 261
(5) JACS 1945 **67** 1092

(6) JACS 1954 **76** 2091

(1) JACS, 1955 **77** 4484

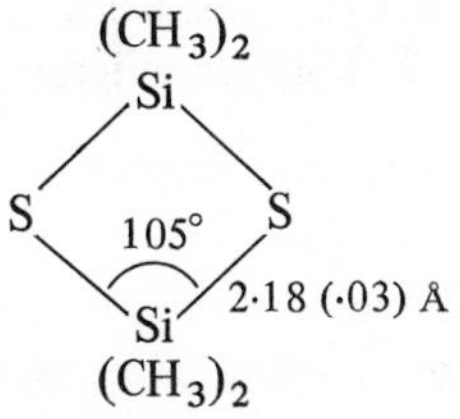

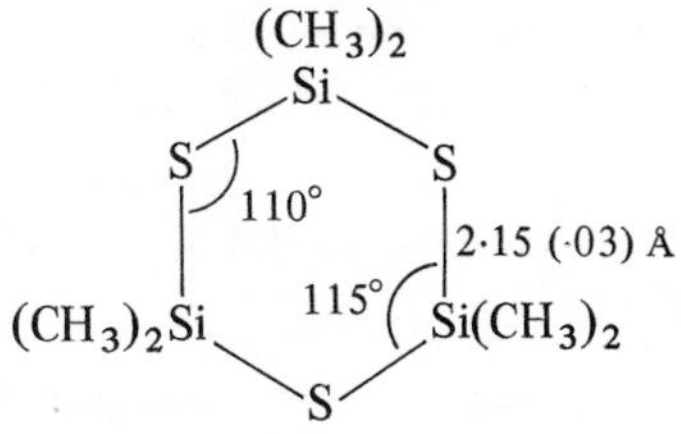

Silicon

(C–H was assumed to be 1·09 Å, C–Si–C, 110°, and H–C–H, tetrahedral.) With these data may be compared those for SiS_2, namely, Si–S, 2·14 Å, S–Si–S, 100°,

and Si–S–Si, 80° in 4-membered rings $Si\diagdown\diagup{S}\diagup\diagdown{S}Si$. It was not possible to distinguish

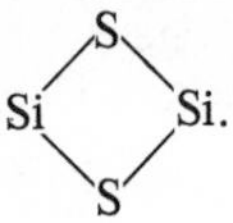

between the chair and boat forms for the trisilthiane ring, but it was stated that the ring is definitely not planar like the oxygen analogue.

Some silicon–nitrogen compounds

We mention here a few compounds which present points of particular interest. The planarity of $N(SiH_3)_3$ has been noted; coplanar N bonds are also formed in the cyclodisilazane (a).[1] More complex Si–N ring systems include the cyclotetrasilazane (b),[2] which is of interest because both chair and cradle forms of the

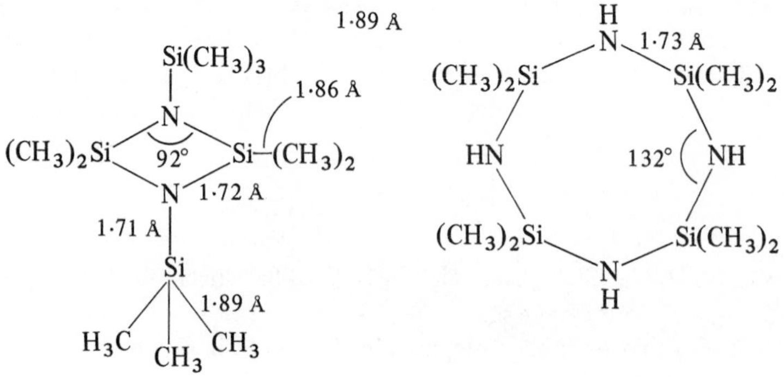

(1) JCS 1962 1721

(2) AC 1963 **16** 1015

(a)

(b)

(3) ACSc 1964 **18** 1879

(4) AC 1969 **B25** 2157

molecule exist in the same crystal. The refractory Si_2N_2O,[3] made by heating to 1450°C a mixture of Si + SiO_2 in a stream of argon containing 5 per cent N_2, is related structurally to both SiO_2 and Si_3N_4.[4] The latter has two crystalline forms, both with phenacite-like structures (p. 77) in which Si has 4 tetrahedral neighbours and N is bonded to 3 Si (Si–N, 1·74 Å). The oxynitride is built of SiN_3O tetrahedra which are linked to form a 3-dimensional framework. Puckered hexagonal nets consisting of equal numbers of Si and N atoms are linked together through O atoms which complete the tetrahedra around the Si atoms. One half of the Si atoms are linked to the layer above and the remainder to the layer below. In the resulting structure every O atom is connected to 2 Si (as in SiO_2), and N and Si are respectively 3- and 4- connected, as in Si_3N_4 (Fig. 3.16(a)); see Figure 23.5.

Organo-silicon compounds and silicon polymers

Silicon forms a great variety of compounds containing carbon. In spite of the formal resemblance between these and purely organic compounds, there is little similarity in chemical properties except in the case of compounds such as the

796

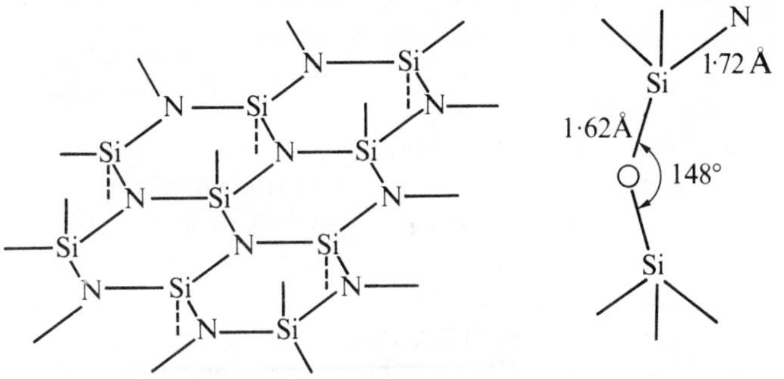

FIG. 23.5. The structure of Si_2N_2O (diagrammatic).

aliphatic quaternary silanes SiR_4. Here the resemblance is presumably due to the fact that the silicon atom is completely shielded by the organic groups. A few organo-silicon compounds are extremely stable. For example, $Si(C_6H_5)_4$ may be distilled unchanged in air at 425°C, and although it is unwise to generalize about the stability of silicon compounds we may note two differences between the silicon and carbon compounds:

(1) Silicon has a much greater affinity than carbon for oxygen. For example, $SiCl_4$ is violently hydrolysed by water, whereas CCl_4 is stable towards water and is only slowly hydrolysed by aqueous alkali. The silicon hydrides are hydrolysed to silicic acid, and disilane is spontaneously inflammable in air—contrast the stability of the corresponding hydrocarbons.

(2) No silicon compounds containing multiple bonds to silicon are known. The hydrolysis of halogen compounds of silicon (discussed later) does not yield the silicon analogues of ketones and carboxylic acids by elimination of water from each molecule. Instead, water is eliminated from two or more molecules leading to polymerized products.

The following are some typical methods of synthesizing organo-silicon compounds:

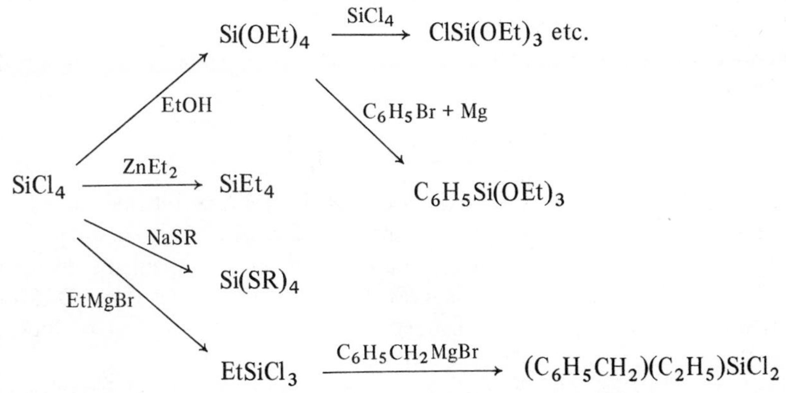

(Et = C_2H_5; R = organic radical)

By successive reactions with different Grignard reagents ($RMgBr$) compounds such as

$(C_6H_5CH_2)(C_2H_5)(C_3H_7)Si(CH_2C_6H_4SO_3H)$ have been prepared. Here we shall be chiefly interested in the hydrolysis of substituted chlorosilanes.

Substituted chlorosilanes, silanols, and siloxanes

Chlorosilanes were formerly obtained by the action of Grignard compounds on $SiCl_4$ but it is also possible to prepare them directly from the alkyl or aryl halide and silicon. The vapour of the halide is passed over sintered pellets of silicon containing some 15 per cent of silver or copper as catalyst, the products being condensed and then fractionated. Most important technically are the methyl chlorosilanes, CH_3SiCl_3, $(CH_3)_2SiCl_2$, and $(CH_3)_3SiCl$. (For structural data see Table 21.1, p. 727.) The following vinyl, allyl, and phenyl compounds are among those which have been prepared in this way:

$$\begin{array}{lll} C_2H_3SiCl_3 & (C_3H_5)HSiCl_2 & C_6H_5SiCl_3 \\ (C_2H_3)_2SiCl_2 & C_3H_5SiCl_3 & (C_6H_5)_2SiCl_2 \\ & (C_3H_5)_2SiCl_2 & (C_6H_5)_3SiCl. \end{array}$$

Intensive studies of the products of hydrolysis of substituted chloro-silanes and esters have led to an important new branch of chemistry. Many organo-silicon compounds were prepared by Kipping and his school in the early years of this century and it was observed that glue-like products often resulted from the hydrolysis of these compounds, as well as intractable solids of the kinds already mentioned. No study was made of these polymeric products which are now the basis of *silicone chemistry*.

By analogy with carbon we might expect mono-, di-, and tri-chloro silanes to hydrolyse to the silicon analogues of alcohols, $R_3Si . OH$, ketones $R_2Si=O$ (by loss of water from $R_2Si(OH)_2$), and carboxylic acids, $RSi\underset{\diagdown OH}{\overset{\diagup O}{}}$ (by loss of water from the trihydroxy compound). In fact compounds containing $Si=O$ are never formed, though the generic name *silicone* (by analogy with ketone) is applied to the various hydrolysis products of dichlorosilanes.

Silanols $R_3Si(OH)$ were obtained in 1911. These compounds tend to condense to the disiloxane:

$$2\,R_3Si(OH) \rightarrow R_3Si{-}O{-}SiR_3$$

at rates depending on the nature of the group R. Dicyclohexylphenyl silanol was isolated by Kipping, and $(C_6H_5)_3Si . OH$ may be distilled under reduced pressure without forming the disiloxane, but the trialkyl silanols are not isolable. The term *disiloxane* is applied to all compounds $R_3Si{-}O{-}SiR_3$ and includes $H_3Si{-}O{-}SiH_3$, $(C_2H_5)_3Si{-}O{-}Si(C_2H_5)_3$, and oxyhalides, $X_3Si{-}O{-}SiX_3$. The first of these is made by the hydrolysis of SiH_3Cl by water at $30°C$.

Silanediols $R_2Si(OH)_2$. In the early studies the only diols isolated were those in which R is an aromatic radical, but even these compounds condense to linear and cyclic polysiloxanes.

798

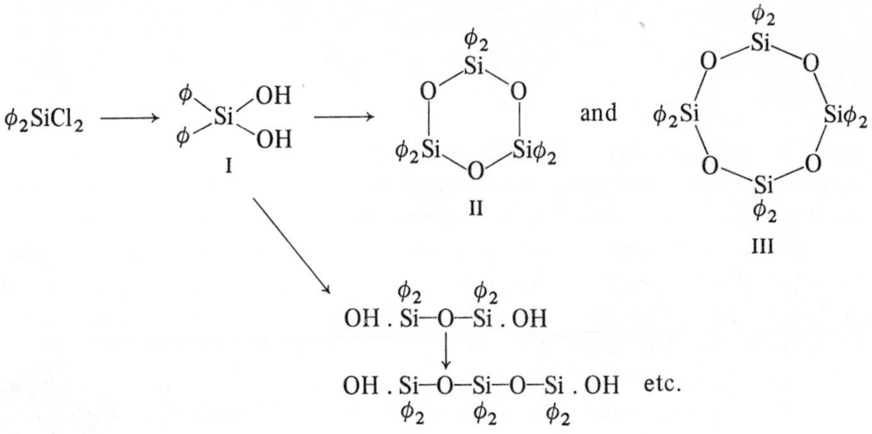

$$\phi_2\text{SiCl}_2 \longrightarrow \quad \begin{matrix} \phi \\ \phi \end{matrix}\text{Si}\begin{matrix} \text{OH} \\ \text{OH} \end{matrix}$$

I

(followed by cyclic structures)

II

and

III

$$\text{OH}\cdot\underset{\phi_2}{\text{Si}}-\text{O}-\underset{\phi_2}{\text{Si}}\cdot\text{OH}$$

$$\text{OH}\cdot\underset{\phi_2}{\text{Si}}-\text{O}-\underset{\phi_2}{\text{Si}}-\text{O}-\underset{\phi_2}{\text{Si}}\cdot\text{OH}\quad \text{etc.}$$

$$[\phi = \text{C}_6\text{H}_5]$$

All the compounds I, II, and III were isolated during the years 1912– 1914, the last two being the first cyclic silicon polymers to be characterized. The first dialkylsilanediol was prepared in 1946 by the hydrolysis, by weak alkali, of $(\text{C}_2\text{H}_5)_2\text{SiCl}_2$: it is a stable crystalline solid. Later, $(\text{CH}_3)_2\text{Si(OH)}_2$ was prepared by the hydrolysis of $(\text{CH}_3)_2\text{Si(OCH}_3)_2$ as a crystalline solid extremely sensitive to acid or alkali, which cause condensation to polysiloxanes, mostly the cyclic tri- and tetra-compounds. Other diols, prepared include the n- and iso-propyl and butyl, and tert-butyl compounds, in addition to tetramethyldisiloxane-1,3-diol (IV). No silanetriol RSi(OH)_3 has been isolated.

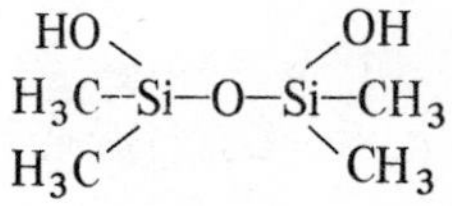

$$\begin{matrix} \text{HO} & & \text{OH} \\ \text{H}_3\text{C}-\text{Si}-\text{O}-\text{Si}-\text{CH}_3 \\ \text{H}_3\text{C} & & \text{CH}_3 \end{matrix}$$

IV

The structures of silanols and siloxanes

Although the stereochemistry of silicon in these compounds presents no points of special interest, the bond arrangement being the normal tetrahedral one, the siloxanes include some interesting ring systems.

Silanols. Diethylsilanediol, $\text{Si(C}_2\text{H}_5)_2(\text{OH})_2$, is a simple tetrahedral molecule:

O—Si—O, 110°, C—Si—C, 111°, Si—OH, 1·63 Å, and Si—C, 1·90 Å.[1]

For the crystal structure of $\text{OHSi(CH}_3)_2\cdot\text{C}_6\text{H}_4\cdot\text{Si(CH}_3)_2\text{OH}$[2] see p. 95.

Linear polysiloxanes. Several studies have been made of the simplest member of this family, $\text{O(SiH}_3)_2$. The most reliable estimate of the Si—O—Si bond angle[3] may be compared with the much smaller value in the sulphur analogue:[4]

(H₃Si, 144°, SiH₃, O 1·634 Å, 1·486 Å) compare (H₃Si, 100°, SiH₃, S)

(1) JCP 1953 **21** 167

(2) JPC 1967 **71** 4298

(3) ACSc 1963 **17** 2455
(4) JCP 1958 **29** 921

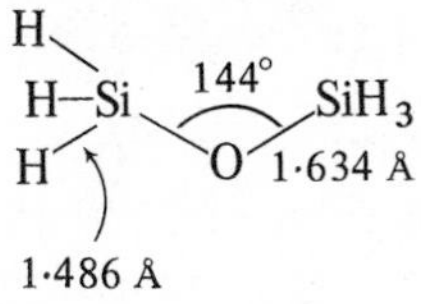

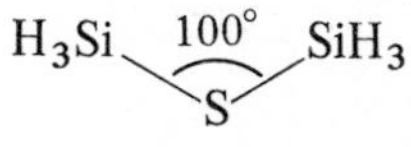

Silicon

Two studies have been made of hexamethyldisiloxane

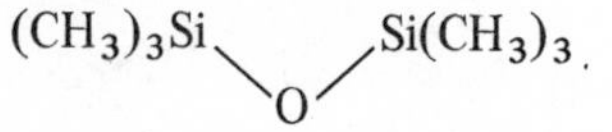

$$(CH_3)_3Si\diagdown_O\diagup Si(CH_3)_3.$$

(5) JACS 1946 **68** 1794

(6) JCP 1950 **18** 1414

From the dipole moment[5] the Si–O–Si angle was estimated as 160 (15)°, but a much lower value, 130 (10)°, was deduced from an electron diffraction study[6] which gave Si–O, 1·63 (0·03) Å, Si–C, 1·88 (0·03) Å, and C–Si–C, 111 (4)°.

Cyclic polysiloxanes. Particular interest attaches to the configurations of these molecules, for the related carbon compounds paraldehyde and metaldehyde have puckered rings. The true silicon analogue of paraldehyde, tri-methylcyclo-trisiloxane, has been prepared but its structure is not known.

(7) JCP 1950 **18** 42

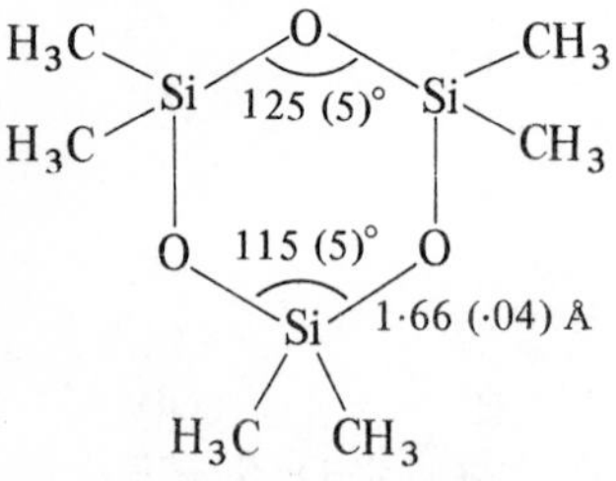

A detailed electron diffraction study[7] of hexamethylcyclotrisiloxane established the planarity of the Si_3O_3 ring but indicated considerable thermal motion of the methyl groups. To maintain planarity in the Si_4O_4 ring the oxygen bond angle would have to increase to about 160°. In fact, the Si_4O_4 ring is puckered, with Si–O–Si, 142·5° (compare 144° in α-quartz and 147° in α-cristobalite), Si–O, 1·65 Å, and Si–C, 1·92 Å. These data come from an X-ray study of the low-temperature form of $[(CH_3)_2SiO]_4$.[8] The conformation of the 8-ring of $[(CH_3)_2SiO]_4$ is not one of the four more symmetrical ones (see footnote, p. 571); contrast the occurrence of both 'chair' and 'cradle' forms in crystalline $[(CH_3)_2SiNH]_4$, p. 796.

(Si–C = 1·88 (·04) Å)

(8) AC 1955 **8** 420

'Spiro' compounds are also known, and the crystal structure of octamethylspiro (5.5) pentasiloxane has been determined.[9] The planar rings lie in two mutually perpendicular planes. There is evidence that the methyl groups 'librate' about the Si–O bonds as if the Si atoms to which they are attached were free to move in a 'ball-and-socket' joint. Data obtained include: Si–O, 1·64 ± 0·03 Å, Si–C, 1·88 ± 0·03 Å, Si bond angles tetrahedral, Si–O–Si angles 130°.

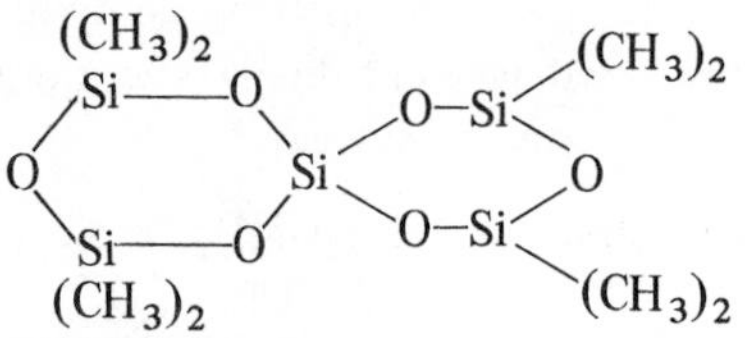

(9) AC 1948 **1** 34

Silicone chemistry

The basis of silicone chemistry is the use of four types of unit. The monofunctional unit (*M*) results, for example, from the hydrolysis of $(CH_3)_3SiCl$ or $(CH_3)_3Si(OC_2H_5)$. These units alone can only give disiloxanes, and in a mixture (see later) an *M* unit ends a chain.

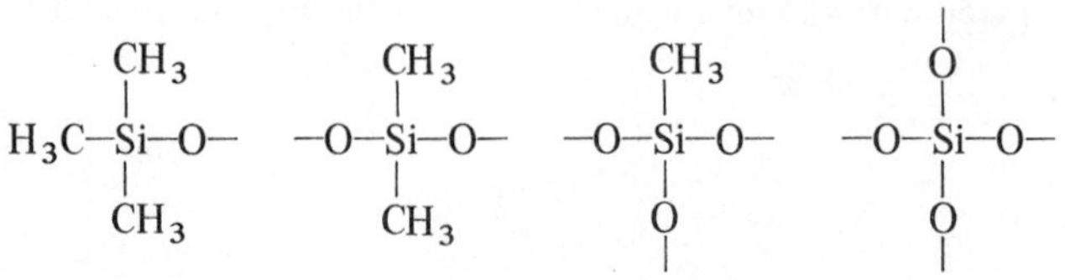

(In these formulae O strictly represents half an oxygen atom, since on condensation O is shared between two Si atoms.)

The difunctional unit (*D*) is formed from $(CH_3)_2SiCl_2$ or $(CH_3)_2Si(OC_2H_5)_2$ and *D* units, alone, form cyclic polysiloxanes. The trifunctional unit (*T*) is formed from *mixtures* containing CH_3SiCl_3; complete hydrolysis of methyltrichlorosilane alone gives an infusible white powder (see later). The quaternary unit (*Q*) may be introduced by using $SiCl_4$. The polysiloxanes are the analogues of the silicon-oxygen ions found in silicates. The infinite chain polymer (a) is the analogue of the pyroxene chain ion (b) (see p. 816):

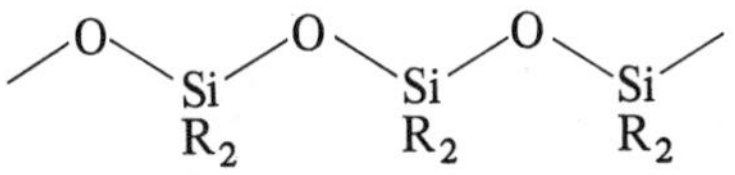

$$\text{(a)} \qquad\qquad\qquad \text{(b)}$$

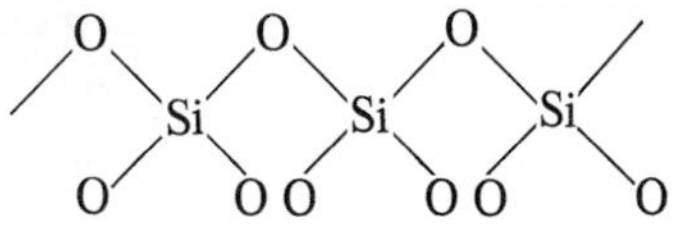

and the ring polymers are the analogues of the $Si_3O_9^{6-}$ and $Si_6O_{18}^{12-}$ ions found in the silicates benitoite and beryl (p. 815)

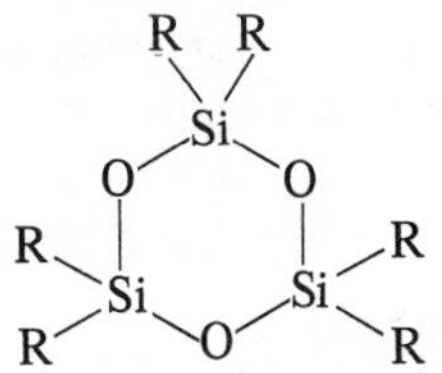

compare

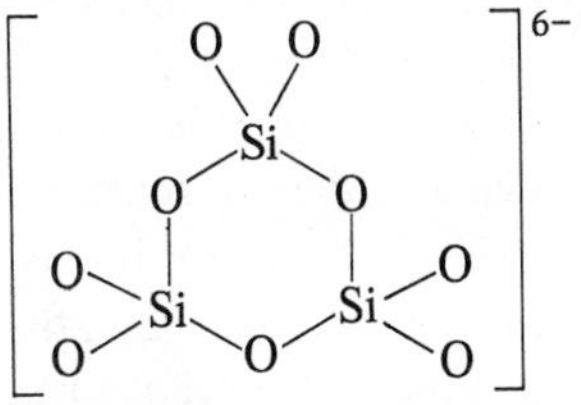

The next type of polymer would be the infinite 2-dimensional polymer containing Si atoms joined to three others via O atoms, the fourth bond being that to the organic group R (or halogen). The layer would have the composition $(Si_2O_3R_2)_n$ and would correspond to the petalite or apophyllite layers of Fig. 23.13, p. 818. Polymers intermediate between infinite chains and layers would correspond to the amphibole chain, for example, having the composition $(Si_4O_5R_6)_n$. Although these 'pure' types, apart from the chains and rings, are not known, 2- and 3-dimensional linking by means of *T* and *Q* groups presumably occurs in the gels and resins. Polymers have also been made from silicon compounds containing unsaturated organic groups. The vinyl and allyl chlorosilanes already mentioned can be hydrolysed to alkylene polysiloxanes which can polymerize further through the unsaturated organic groups.

Technically useful silicones arise by co-hydrolysis of suitable mixtures of chlorosilanes or esters yielding mixtures of two or more of the basic *M, D, T,* or *Q* units. The product of hydrolysis of $(CH_3)_2SiCl_2$ is a colourless non-volatile oil, but this is not stable because the chains end in *D* groups and further condensation, to rings or longer chains, can take place. To stabilize the product it is necessary to introduce some *M* units, by cohydrolyzing $(CH_3)_3Si(OC_2H_5)$ and $(CH_3)_2Si(OC_2H_5)_2$ or $(CH_3)_2SiCl_2$. However, if the proportion of the latter is increased more cyclic compounds D_n are formed, and to obtain the desired mixtures of longer linear polysiloxanes (for oils and greases of various viscosities) a device known as *catalytic equilibration* is used. A liquid methylpolysiloxane mixture, the cyclic tetrasiloxane, or a mixture of $(CH_3)_3Si.O.Si(CH_3)_3$ and any source of D units is shaken with sulphuric acid for a few hours, when rearrangement takes place to a mixture of MD_xM

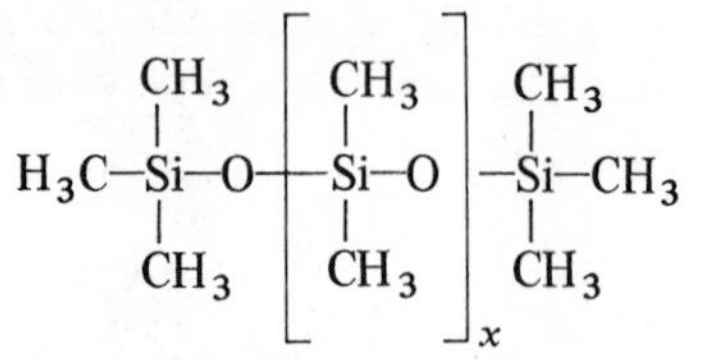

$$H_3C-\underset{\underset{CH_3}{|}}{\overset{\overset{CH_3}{|}}{Si}}-O\left[\underset{\underset{CH_3}{|}}{\overset{\overset{CH_3}{|}}{Si}}-O\right]_x-\underset{\underset{CH_3}{|}}{\overset{\overset{CH_3}{|}}{Si}}-CH_3$$

$$M \qquad D_x \qquad M$$

These equilibrated silicone oils have valuable properties: they are very stable to heat and oxidation, they have good electrical properties, and are water-repellent. They find application as vacuum oils and greases, water-proofing agents, and paint media.

From these oils the compounds M_2D_5 to M_2D_9 have been obtained by fractional distillation under reduced pressure. All the following compounds have been prepared: linear M_2, M_2D to M_2D_9, cyclic D_3 to D_9. The last can be prepared by the hydrolysis of $(CH_3)_2SiCl_2$ or by the depolymerization of dimethylsiloxane high polymers. Rearrangement to low cyclic polymers takes place on heating the latter to 230–240°C under reduced pressure with NaOH as catalyst, and these can be separated by fractional distillation under reduced pressure.

Other types of polysiloxane obtainable by co-hydrolysis include TM_3 and QD_n. The co-hydrolysis product from $SiCl_4$ and $(CH_3)_2SiCl_2$ undergoes rearrangement in nitrogen at 400–600°C to give spirosiloxanes such as $Si_5O_6(CH_3)_8$, QD_4, described on p. 800. Combinations of appropriate proportions of Q and T give silicone resins which can be hardened by heat treatment. We noted earlier that total hydrolysis of a trifunctional silane $RSiX_3$ such as $CH_3Si(OR)_3$ or CH_3SiCl_3 gives an infusible solid ('T-gel') of unknown structure, but partial hydrolysis, for example of $CH_3Si(OC_2H_5)_3$ in benzene solution, gives solid mixtures from which well defined, high-melting solids have been obtained by fractional sublimation. Structures of the type

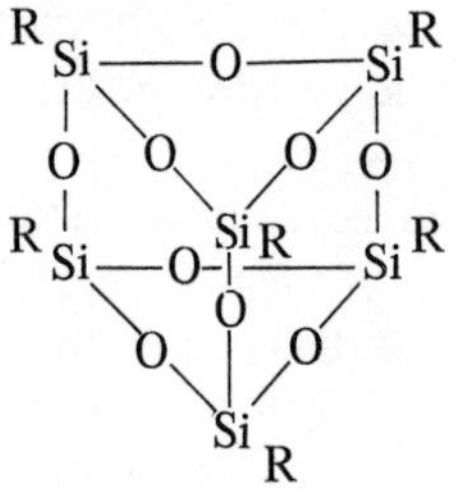

and

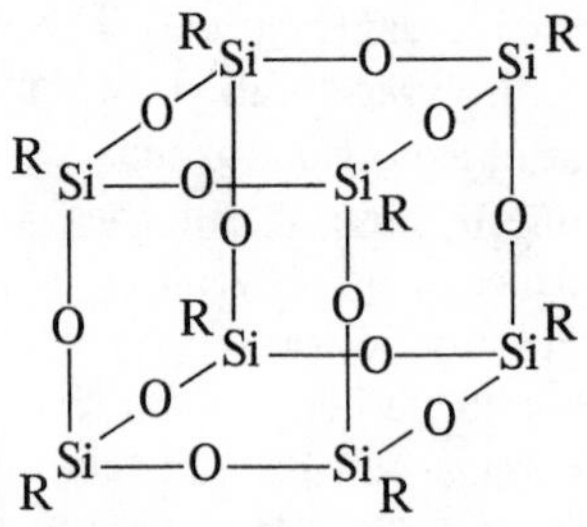

for hexa-and octa- (ethylsilsesquioxane) have been suggested.

In siloxanes the silicon atoms are linked, as in complex silicates, through the oxygen atoms. It is also possible to link them through carbon atoms, and some molecules containing silicon–carbon chains have been prepared, for example

$$(CH_3)_3Si\diagdown_{CH_2}\diagup Si(CH_3)_3, \qquad\qquad Cl_3Si\diagdown_{CH_2}\diagup SiCl_3,$$

$$(CH_3)_3Si\diagdown_{CH_2}\diagup \overset{\displaystyle (CH_3)_2}{\underset{}{Si}}\diagdown_{CH_2}\diagup Si(CH_3)_3.$$

The methyl compounds are prepared by reactions of the type

$$(CH_3)_3Si . CH_2 . MgCl + (CH_3)_3SiCl \rightarrow (CH_3)_3Si . CH_2 . Si(CH_3)_3$$

and the chloro compound by passing methylene chloride over silicon-copper (compare the preparation of the substituted chlorosilanes, p. 798).

The Si–O skeletons in mineral silicates can be recognized in the siloxanes prepared from them. The ground mineral is treated with a mixture of cold concentrated HCl, isopropyl alcohol, and hexamethyldisiloxane. The siloxane layer is separated and analysed chromatographically. The trimethylsilyl derivatives correspond to the type of Si–O complex in the original silicate. For example, from olivine (an orthosilicate) 70 per cent of the Si is recovered as $Si[OSi(CH_3)_3]_4$, from hemimorphite (pyrosilicate) 78 per cent as $(Me_3SiO)_3SiOSi(OSiMe_3)_3$ and $(Me_3SiO)_3SiOSi(OH)(OSiMe_3)_2$; there is some cleavage of the original Si–O–Si bonds. Differences are found between the products from framework alumino-silicates depending on the sequence of Si and Al atoms in the framework.[1]

(1) IC 1964 **3** 574

The crystalline forms of silica

At atmospheric pressure silica exists in three crystalline forms, which are stable in the temperature ranges indicated below:

$$\underset{Quartz}{\xrightarrow{\hspace{1.5cm}}} 870° \underset{Tridymite}{\xleftrightarrow{\hspace{1.5cm}}} 1470° \underset{Cristobalite}{\xleftrightarrow{\hspace{1.5cm}}} 1710°C \text{ (m.p.)}$$

Crystallization of the liquid is difficult, and it usually solidifies to a glass which has great value on account of its extremely small coefficient of thermal expansion and high softening point (in the region of 1500°C). Silica in this glassy form is found in nature as the mineral obsidian. These three polymorphic forms of silica are not readily interconvertible as is shown by the fact that all three are found as minerals, though tridymite and cristobalite are rare in comparison with quartz. Furthermore, each of the three forms exists in a low- and high-temperature modification (α and β respectively) with the following transition points: α-β quartz, 573°, α-β tridymite, 110–180°, and α-β cristobalite, 218° ± 2°C. The fact that in the last two cases these transitions can be studied at temperatures at which the particular polymorphic forms are metastable is a further indication of the difficulty of changing one of the three varieties of silica into another. The general structural relations between these crystalline forms of silica have been already noted in. our discussion of nets in Chapter 4 and are as follows. The three main forms, quartz, tridymite, and cristobalite, are all built of SiO_4 tetrahedra linked together so that every oxygen atom is common to two tetrahedra (thus giving the composition SiO_2), but the arrangement of the linked tetrahedra is quite different in the three crystals. These structures are illustrated, as packings of tetrahedra, in Fig. 23.20, between pp. 826 and 827. The difference between, for example, tridymite and cristobalite is like that between wurtzite and zinc-blende, the two forms of ZnS. On the other hand, the α and β forms of one of the three varieties differ only in detail, there being, for example, slight rotations of the tetrahedra relative to one another without any alteration in the general way in which they are linked together. The change from quartz to tridymite thus involves the breaking of Si—O—Si bonds and the linking together of the tetrahedra in a different way, and accordingly is a very sluggish process, whereas the conversion of $\alpha \rightarrow \beta$ quartz merely involves a small distortion of the structure without any radical rearrangement and is an easy and reversible process. Quartz is optically active, in contrast to the other forms, and the fact that a crystal of left-handed α-quartz remains left-handed after conversion into the β form is further evidence of the relatively small change in structure at the transition point.

The high-temperature forms, tridymite and cristobalite, differ from quartz in having much more open structures, as shown by their densities: quartz, 2·655, tridymite, 2·30, and cristobalite, 2·27 g/c.c. There are also high-pressure forms of silica, coesite,[1] formed by heating dry sodium metasilicate with $(NH_4)_2HPO_4$ at 700°C under a pressure of 40,000 atm, keatite,[2] which is the stable form in the pressure range between those for quartz and coesite, and stishovite, formed at very high pressures. The last form has the 6-coordinated rutile structure, (Si—4 O, 1·76 Å, Si—2 O, 1·81 Å)[3] while coesite and keatite are based on 4-connected frameworks in which the smallest rings are respectively 4- and 8- membered (coesite) and 5-, 7-, and 8-membered (keatite). For the basic keatite framework see p. 111.

In β-quartz, the structure of which is illustrated in Fig. 23.6, there may be

(1) ZK 1959 **111** 129

(2) ZK 1959 **112** 409

(3) AC 1971 **B27** 2133

FIG. 23.6. Plan of the structure of β-quartz. Small black circles represent Si atoms. The oxygen atoms lie at different heights above the plane of the paper, those nearest the reader being drawn with heaviest lines. Each atom is repeated at a certain distance above (and below) the plane of the paper along the normal to that plane so that the Si_3O_3 rings in the plan represent helical chains.

distinguished helices of linked tetrahedra, and in a given crystal these are all either left- or right-handed, the crystal being laevo- or dextro-rotatory. Some idea of the relation between the α and β forms of quartz may be obtained from Fig. 23.7, which shows, in plan, the arrangement of the silicon atoms in the two crystals. It will be seen that the α form is a distorted version of the β form, but that the general scheme according to which the tetrahedra are linked together is the same in the two forms.

In Fig. 23.8 the structures of β-tridymite and β-cristobalite are shown with the O atoms midway between, and on the straight lines joining, the pairs of Si atoms. In fact, the O atoms lie off the lines joining pairs of nearest Si atoms, giving

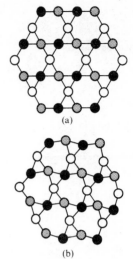

FIG. 23.7. The arrangement of the Si atoms (in plan) in (a) β-quartz and (b) α-quartz.

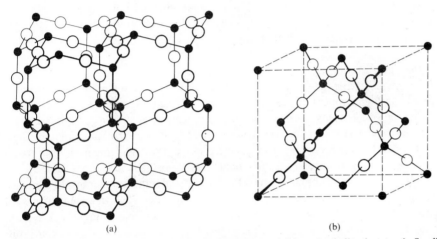

FIG. 23.8. The idealized structures of (a) β-tridymite, and (b) β-cristobalite (see text). Small black circles represent Si atoms.

Si–O = 1·61 Å and an O bond angle close to 144°. A detailed study of the geometry of the α-(low) and β-(high) quartz structures shows that in neither structure are the SiO_4 tetrahedra strictly regular. In α-quartz there are two different Si–O bond lengths (1·597 and 1·617 Å) and the O bond angle is 144°.[4] (In various felspars values of this angle range from 125–160° with a mean value close to 143°.) In tetragonal (α) cristobalite Si–O = 1·61 Å and O bond angle = 147°.[5]

Tridymite presents a more complex picture. The less dense structures of tridymite and cristobalite are presumably maintained at high temperatures by the thermal movements of the atoms, and it is possible that they are stabilized at lower temperatures by the inclusion of foreign atoms.

Natural tridymite and cristobalite often, if not invariably, contain appreciable concentrations of foreign ions, particularly of the alkali and alkaline-earth metals,[6] suggesting that they may have been deposited as the stable phases at temperatures well below the respective transition points. The variable composition of these minerals is presumably the reason for wide ranges of temperatures recorded for the α-β transition points. The origin of the high-temperature tridymite studied by Gibbs is not known but it appears that the monoclinic form stable at room

(4) AC 1963 **16** 462, 542

(5) ZK 1965 **121** 369

(6) AM 1954 **39** 600

temperature undergoes some sort of structural change at approximately 107°C. Above 180°C it transforms to an orthorhombic structure in which the apparent Si–O distance is 1·56 Å, but corrected for the very large thermal motion this becomes 1·61 Å.[7]

(7) AC 1967 **23** 617

There is a remarkable silica-containing mineral called *melanophlogite* which occurs in the Sicilian sulphur deposits in limestone saturated with liquid bituminous matter. It contains sulphur and organic matter, including long-chain hydrocarbons, and its density is only 2·05 g/cc; it has cubic symmetry and remarkable optical properties. Preliminary analysis of the X-ray data[8] indicates that the structure is probably like that of the 12 Å gas hydrate (p. 545), in which case it provides another example of the close analogy between SiO_2 frameworks and those formed by water molecules. From the structural standpoint melanophlogite has obvious analogies with 'β-quinol' and the urea–hydrocarbon clathrates, the structures of which have been noted in earlier chapters.

(8) Sc 1965 **148** 232

Stuffed silica structures

This name is given to structures in which the mode of linking of the tetrahedra in one of the high-temperature forms of silica is retained, but some Si is replaced by Al and cations (usually alkali-metal ions) are accommodated in the interstices to maintain electrical neutrality. The holes in the high-quartz structure are only large enough for Li^+ (and possibly Be^{2+}), but those in the high-temperature forms of tridymite and cristobalite can accommodate the large alkali-metal ions (Table 23.1). We note later that the normal form of $LiAlSi_2O_6$ (the mineral spodumene) contains metasilicate chains; it is one of the monoclinic pyroxenes. Heat treatment of glasses with this composition gives crystalline forms with stuffed β-quartz and keatite structures. Spodumene is therefore preferably formulated $LiAl(Si_2O_6)$, having both Li^+ and Al^{3+} in positions of octahedral coordination between the infinite linear anions, while the stuffed silica polymorphs are $Li(AlSi_2O_6)$, indicating that Al and Si are tetrahedrally coordinated in a silica-like framework.

TABLE 23.1

Stuffed silica structures

Silica polymorph	Stuffed silica structure	Reference
β-quartz	High-LiAlSiO$_4$ (eucryptite)	AC 1948 **1** 27
β-tridymite	KAlSiO$_4$ (kalsilite)	AC 1948 **1** 42
	KNa$_3$Al$_4$Si$_4$O$_{16}$ (nepheline)	
	BaAl$_2$O$_4$ (?), KLiSO$_4$	
β-cristobalite	NaAlSiO$_4$ (high-carnegieite)	DAN 1956 **108** 1077
	Na$_2$BeSi$_2$O$_6$	
keatite and β-quartz	LiAlSi$_2$O$_6$	AC 1971 **B27** 1132

Silicates

The earth's crust is composed almost entirely of silicates and silica, which constitute the bulk of all rocks and of soils, clays, and sands, the breakdown products of rocks. All inorganic building materials, ranging from natural rocks such as granite to artificial products such as bricks, cement, and mortar, are silicates, as also are ceramics and glasses. Metallic ores and other mineral deposits form an insignificant proportion of the mass of the earth's crust. Table 23.2, due to

TABLE 23.2

Compositions of the lithosphere and hydrosphere

	Lithosphere				Hydrosphere	
Element	% (weight)	% (atomic)	Radius (Å)[a]	% (volume)	Element	% (weight)
O	46·59	62·46	1·32	91·77	O	85·89
Si	27·72	21·01	0·39	0·80	H	10·80
Al	8·13	6·44	0·57	0·76	Cl	1·93
Fe	5·01	1·93	0·82	0·68	Na	1·07
Mg	2·09	1·84	0·78	0·56	Mg	0·13
Ca	3·63	1·93	1·06	1·48	S	0·09
Na	2·85	2·66	0·98	1·60	Ca	0·04
K	2·60	1·43	1·33	2·14	K	0·04
Ti	0·63	0·28	0·64	0·22		
All other elements are present in smaller amounts					All others less than 0·01%	

(a) These ionic radii differ slightly from those given in Chapter 7, since Goldschmidt based his values on O^{2-} 1·32 Å.

Goldschmidt, gives the average composition of the lithosphere; that of the hydrosphere is given for comparison. It is seen that more than nine-tenths of the volume of the earth's crust is occupied by oxygen. In many silicates the oxygen atoms are close-packed, or approximately so, and the more electropositive ions, which are nearly all smaller than oxygen, are found in the interstices between the close-packed oxygen atoms. We have already discussed the relation between radius ratio and coordination number in Chapter 7. Here we need only remind the reader of the usual oxygen coordination numbers of the ions commonly found in silicate minerals:

Be	4
Li	4 and 6
Si	4
Al	4, 6 (rarely 5)
Mg, Ti, Fe	6
Na	6 (8)
Ca	8
K	6—12

Since there are a number of ions of similar size which can occupy, for example, octahedral holes (i.e. coordination number 6), and since atoms of many kinds were present in the melts or solutions from which the minerals crystallized, isomorphous substitution is very common in the silicates, which rarely have the 'ideal' composition of a simple chemical compound. The analysis of a typical specimen of hornblende illustrates this point well. Hornblende belongs to the group of silicate minerals called amphiboles (see later), and their ideal composition is represented by the following formula: $(OH)_2Ca_2Mg_5(Si_4O_{11})_2$. The numbers of atoms present for every twenty-four oxygen atoms, that is, $O_{22} + (OH)_2$, are:

$$
\begin{array}{lll}
\left.\begin{array}{ll} \text{Si} & 5{\cdot}98 \\ \text{Al} & 2{\cdot}32 \end{array}\right\} 8{\cdot}30 &
\left.\begin{array}{ll} \text{Ti} & 0{\cdot}13 \\ \text{Fe}^{3+} & 0{\cdot}90 \\ \text{Mg} & 0{\cdot}46 \\ \text{Fe}^{2+} & 3{\cdot}16 \\ \text{Mn}^{2+} & 0{\cdot}10 \end{array}\right\} 4{\cdot}75 &
\left.\begin{array}{ll} \text{Na} & 0{\cdot}37 \\ \text{K} & 0{\cdot}66 \\ \text{Ca} & 1{\cdot}84 \end{array}\right\} 2{\cdot}87
\end{array}
$$

First, some of the silicon in tetrahedral positions is replaced by Al, for this atom may occupy either tetrahedral or octahedral holes. Some, however, of the Al atoms (0·30) must be in octahedral holes since in this particular structure there is room for only eight atoms in tetrahedral positions. The realization that aluminium can play this dual role in silicates was an important result of X-ray studies, for previously it was very difficult to interpret correctly analytical figures which gave only total Al without distinguishing between that which was replacing Si in the 'acid radical' part of the formula and that which was behaving like Fe^{3+}, Mg^{2+}, etc., as simple positive ion. Secondly, we see that in this particular hornblende most of the magnesium in the above formula is replaced by divalent iron. From the list of elements in octahedral positions (second group) it will be seen that size is more important than charge in determining which elements may replace others by isomorphous substitution. Any substitution of an ion of one charge by one of another must, of course, be accompanied by some other adjustment in the structure to maintain electrical neutrality. The replacement of silicon by aluminium, which is very commonly found, requires the presence of extra positive ions, usually alkali or alkaline-earth, in the structure, and many examples of aluminosilicates are given later in this chapter.

In our discussion of structures based on the closest packing of anions (Chapter 4) we saw that the choice of tetrahedral sites occupied determines whether there are discrete MX_4 ions in a structure or more extended systems of such tetrahedral groups sharing vertices (X atoms) or edges (pairs of X atoms). In silicates usually only sharing of vertices between SiO_4 (or AlO_4) groups occurs, and therefore silicates are conveniently classified according to the way in which these groups are linked together. Discrete SiO_4^{4-} ions occur in orthosilicates. Higher O : Si ratios than 4 : 1 are possible if there are separate O^{2-} ions in addition to the SiO_4^{4-} ions; the structure of Sr_3SiO_5, for example, is of this type—it is very similar to that of Cs_3CoCl_5. Lower O : Si ratios result if there is sharing of O atoms between the SiO_4 tetrahedra. The simplest structures arise if all the SiO_4 groups are topologically equivalent, that is, they all share the same number of vertices:

(a) One O atom of each SiO_4 is shared with another SiO_4 giving the pyrosilicate ion $Si_2O_7^{6-}$.

808

(b) Two O atoms of every SiO_4 are shared with other SiO_4 tetrahedra, resulting in closed rings or infinite chains, as in

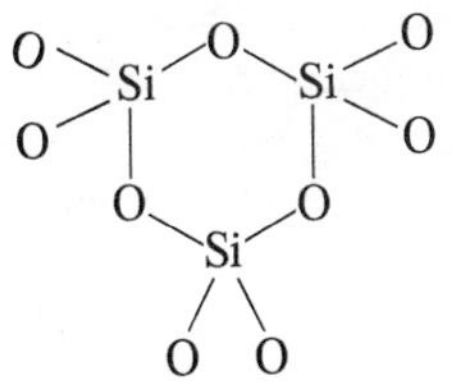

and larger rings, and

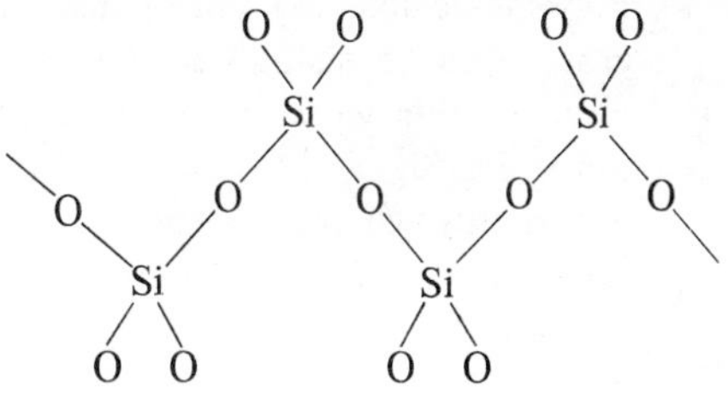

These complexes are all of the type $(SiO_3)_n^{2n-}$.

(c) Three O atoms of each SiO_4 are shared with other SiO_4 tetrahedra to give systems of composition $(Si_2O_5)_n^{2n-}$. These may be finite or infinite in one, two, or three dimensions, as explained in Chapter 3 for 3-connected systems and in particular in Table 5.2 (p. 161) for tetrahedra sharing three vertices.

(d) Sharing of all vertices of each SiO_4 tetrahedron leads to infinite 3D frameworks, as in the various forms of silica, or to double layers. If some of the Si is replaced by Al the framework—always of composition $(Si, Al)O_2$—is negatively charged, and positive ions must be accommodated in the holes in the structure. We shall mention later the felspars and zeolites as the most important classes of aluminosilicates with 3D framework structures. An alternative way of linking $(Si, Al)O_4$ tetrahedra so that all vertices are shared is to form double layers, as found in the hexagonal forms of $CaAl_2Si_2O_8$ and $BaAl_2Si_2O_8$. Replacement of Si by Al to different extents is illustrated by the formulae of the following minerals:

	Si : Al ratio
The various forms of silica: SiO_2	∞
Orthoclase $K(AlSi_3O_8)$	3:1
Analcite $Na(AlSi_2O_6) . H_2O$	2:1
Natrolite $Na_2(Al_2Si_3O_{10}) . 2 H_2O$	3:2
Anorthite $Ca(Al_2Si_2O_8)$	1:1

In the groups (a)–(d) above we have considered only those Si_xO_y complexes in which there are the same numbers of shared and unshared oxygen atoms around every Si atom, viz.

3 unshared, 1 shared	Si_2O_7
2 unshared, 2 shared	SiO_3
1 unshared, 3 shared	Si_2O_5
4 shared	SiO_2

There are other possible complexes in which the environment of every Si atom, in terms of shared and unshared O atoms, is not the same. This topological non-equivalence of SiO_4 tetrahedra could arise in complexes of the following kinds:

(i)   polynuclear ions intermediate between $Si_2O_7^{6-}$ and $(SiO_3)_n^{2n-}$ chains,

(ii) ions formed by union of chain ions (multiple chains),

(iii) complex layers, or

(iv) frameworks in which SiO_4 tetrahedra share different numbers of vertices.

(1a) AC 1968 **B24** 845

(1b) NW 1972 **59** 35

(2) NW 1965 **52** 512

Polysilicate ions, (i), analogous to $P_3O_{10}^{5-}$ and $S_3O_{10}^{2-}$, are extremely rare: Si_3O_{10} groups can be distinguished in a rare arsenic vanadium silicate, ardennite[1a] and also, together with SiO_4 groups, in $Ho_4(Si_3O_{10})(SiO_4)$.[1b] The linear $(Si_6O_{16})^{8-}$ ion in $Ba_4Si_6O_{16}$[2] consists of three pairs of *edge*-sharing tetrahedral SiO_4 groups joined through their vertices, as shown below. The more complex chains (ii) are described later, as also is one example of a more complex layer, (iii), in which

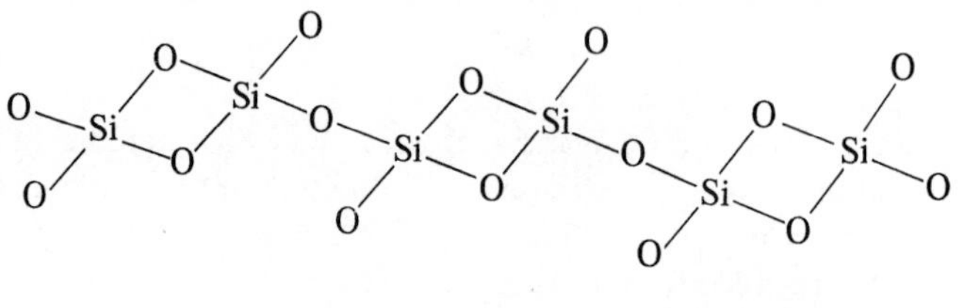

two-thirds of the SiO_4 tetrahedra share 3 and the remainder 4 vertices, giving the composition Si_3O_7 (p. 814). Neptunite (p. 824) provides an example of (iv).

We shall now describe some of the more important silicate structures, dealing in turn with the following groups:

1. Orthosilicates, containing discrete SiO_4^{4-} ions,
2. Silicates with $Si_2O_7^{6-}$ ions,
3. Silicates with cyclic $(SiO_3)_n^{2n-}$ ions,
4. Silicates with chain structures—pyroxenes and amphiboles,
5. Silicates with layer structures—clay minerals and micas,
6. Silicates with framework structures—felspars and zeolites.

It should be remarked that although for convenience we adopt this classification in terms of the silicon–oxygen complex we are in fact classifying according to the (Al, Si)–O complex in many cases, particularly layer structures, where replacement of some Si by Al is common, and in framework structures, in which replacement of some Si by Al is essential to form a charged framework. For this reason it is more logical to recognize not only Al but other tetrahedrally coordinated ions as forming part of the complex. These include the ions of

$$\text{Li} \quad \text{Be} \quad \text{B}$$
$$(\text{Na} \quad \text{Mg}) \quad \text{Al} \quad \text{Zn} \quad \text{Ga} \quad \text{Ge}$$

(3) ZK 1967 **124** 115

Thus we find $AlPO_4$ and BPO_4 with silica-type structures, and phenacite, Be_2SiO_4, is more logically described as a 3D framework M_3O_4 than as an orthosilicate, as shown by the fact that Ge_3N_4 has the same structure. Hemimorphite (p. 814) contains Si_2O_7 groups, but tetrahedrally coordinated Zn and Si together form a 3D framework. Again, Zn plays a role similar to that of Si in $Pb(ZnSiO_4)$, larsenite,[3] though here the framework is more complex than in the silica structures. In the melilite family (p. 814) there are layers in which XO_4 tetrahedra share 4 and YO_4

tetrahedra share 3 vertices, giving a layer of composition XY_2O_7, as in $Ca_2(Si_2MgO_7)$ or $Ca_2(Al_2SiO_7)$. Finally, there is the same type of X_2O_5 layer in gillespite, $BaFe(Si_4O_{10})$, datolite (p. 860), $Ca(BSiO_4OH)$, herderite, $Ca(BePO_4F)$, and gadolinite, $FeY_2(Be_2Si_2O_{10})$.[4]

(4) K 1959 **4** 324

It should also be noted that the type of Si–O complex is not necessarily the most important feature of a crystalline silicate in certain contexts. In the transformation of a single crystal of one silicate into another of related structure (e.g. by heating) there is certainly rearrangement of the Si–O complex rather than disturbance of the cation–oxygen system.[5] Thus in the conversion of rhodonite (p. 817) to β-$CaSiO_3$ there is movement of Si from the 5T chain to form a 3T chain which lies in a different direction from the original chain. This involves far less movement of the larger ions (particularly O^{2-}) than would a rearrangement of the 5T chain by twisting about its own axis. A similar case is the 'dehydration' of xonotlite to wollastonite; for these structures see p. 817.

(5) AC 1961 **14** 8, 18

Orthosilicates

These contain SiO_4^{4-} ions, the cations occupying interstices in which they are surrounded by a number of O^{2-} ions appropriate to their size:

	C.N. of M
Be_2SiO_4 (phenacite)	4
$(Mg, Fe)_2SiO_4$ (olivine)	6
$ZrSiO_4$ (zircon)	8

The phenacite structure has been mentioned in the discussion of (3,4)-connected nets in Chapter 3.

The garnets are a group of orthosilicates of the general formula $R_3^{II}R_2^{III}(SiO_4)_3$, where R^{II} is Ca, Mg, or Fe^{2+}, and R^{III} is Al, Cr, or Fe^{3+}, as in grossular, $Ca_3Al_2(SiO_4)_3$, uvarovite, $Ca_3Cr_2(SiO_4)_3$, and andradite, $Ca_3Fe_2(SiO_4)_3$. The garnet structure, in which R^{II} and R^{III} are respectively 8- and 6-coordinated by oxygen, has been described in Chapter 13. The synthetic garnet pyrope, a high-pressure phase of composition $Mg_3Al_2Si_3O_{12}$, is of interest as an example of 8-coordinated Mg^{2+} (distorted cube, 4 O at 2·20, at 2·34 Å).[1]

(1) AM 1965 **50** 2023

It is not necessary that all the oxygen atoms in an orthosilicate are bonded to silicon. We have already mentioned Sr_3SiO_5[2] as containing O^{2-} ions which are not bonded to silicon. The olivine structure may be described either as an assembly of SiO_4^{4-} and Mg^{2+} ions or as an array of approximately close-packed O atoms with Si in tetrahedral and Mg (or Fe) in octahedral holes. The idealized (h.c.p.) structure is illustrated, in plan, in Fig. 23.9(a). The plan of a layer of the $Mg(OH)_2$ structure is shown at (b). Owing to the similar packing of the oxygen atoms in these two compounds portions of the two structures fit together perfectly, giving rise to a series of minerals of general formula $m[Mg_2SiO_4] . n[Mg(OH)_2]$—the so-called chondrodite series. If we compare the compositions of the horizontal rows of atoms in the plans of Fig. 23.9(a) and (b) we see that they are of two kinds, x and y, the former containing Mg and O and the latter Mg, O, and Si. In brucite, $Mg(OH)_2$, the

(2) AC 1965 **18** 453

Silicon

rows are all of the type x; in olivine x and y alternately. Other sequences are possible and those found in the chondrodite minerals are as follows:

Repeat unit		
xxy	Norbergite	$Mg_2SiO_4 . Mg(OH)_2$
xy, xxy	Chondrodite	$2 Mg_2SiO_4 . Mg(OH)_2$
xy, xy, xxy	Humite	$3 Mg_2SiO_4 . Mg(OH)_2$
xy, xy, xy, xxy	Clinohumite	$4 Mg_2SiO_4 . Mg(OH)_2$

The compound Al_2SiO_5 exists in three polymorphic forms, cyanite, sillimanite, and andalusite. In each of the three forms of Al_2SiO_5 one-half of the Al atoms are 6-coordinated and the other half are in positions of 4-, 5-, and 6-coordination in sillimanite, andalusite, and cyanite respectively. The latter is accordingly the most compact of the three forms, the oxygen atoms being cubic close-packed. Just as the

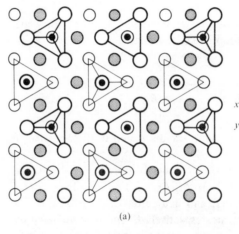

(a)

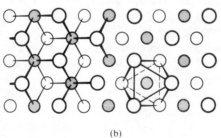

(b)

FIG. 23.9. Plans of the structures of (a) Mg_2SiO_4, and (b) $Mg(OH)_2$. Small black circles represent Si, shaded circles Mg, and open circles O atoms. In (a) light and heavy lines are used to distinguish between SiO_4 tetrahedra at different heights. To the left in (b) the Mg–OH bonds are shown, and to the right an octahedral coordination group is outlined.

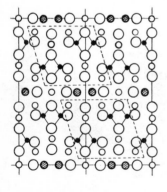

Si   Al   Fe   O

FIG. 23.10. Relation between the structures of staurolite, $HFe_2Al_9Si_4O_{24}$, and cyanite, Al_2SiO_5.

structures of the minerals of the chondrodite series are closely related to those of olivine and brucite, so we find a more complex mineral, staurolite—with the ideal composition $HFe_2Al_9Si_4O_{24}$—related in a rather similar way to cyanite. Two unit cells of the staurolite structure are shown, in plan, in Fig. 23.10, and the broken lines outline cyanite unit cells. It is seen that the cyanite structure is interrupted by the insertion of Fe and Al atoms.[3]

A much higher O : Si ratio is found in Fe_7SiO_{10}, extracted from the reheating furnace of a steel plant. In a c.c.p. assembly of O atoms Si occupies tetrahedral holes, 5 Fe^{2+} octahedral holes, and 2 Fe^{3+} is equally distributed between tetrahedral and octahedral holes.[4]

Portland cement. Portland cement may conveniently be mentioned here since orthosilicates are important constituents. It is formed by burning a mixture of finely ground limestone or chalk with clay or shale, using coal, oil, or gas fuel. The main constituents are CaO, 60–67 per cent, SiO_2, 17–25 per cent, Al_2O_3, 3–8 per cent, and some minor ingredients which may have an appreciable effect on its properties. The more important compounds formed are Ca_2SiO_4, Ca_3SiO_5, $Ca_3Al_2O_6$, and Ca_2AlFeO_5. Phase-rule studies show that tricalcium silicate is stable only between 1250° and 1900°C, outside which range it decomposes into CaO + Ca_2SiO_4, though it will remain indefinitely in a metastable state when quenched to room temperature. Ca_2SiO_4 has four polymorphs, of which the two high-temperature forms (α' and α) are isostructural with β- and α-K_2SO_4 respectively. This polymorphism is important because the different forms do not all hydrate equally readily. The γ form (olivine structure)[5] is very inert towards water and therefore valueless as a component of cement, which owes its strength to the inter-locking crystals of the hydration products. The β form, on the other hand, which is metastable at ordinary temperatures, sets to a hard mass when finely ground and mixed with water. (The β form has a monoclinic structure closely related to that of the α' form.[6])

Little is yet known of the structures of the hydration products of Portland cement, which include hydrated silicates, 'aluminates', and more complex compounds containing SO_4^{2-} ions arising from the gypsum which is added to control the setting time. In two compounds of this family there are apparently discrete tetrahedral SiO_3OH^{3-} groups having OH directly attached to Si. They are afwillite, $Ca_3(SiO_3OH)_2 . 2 H_2O$[7] and dicalcium silicate α-hydrate, $Ca_2(SiO_3OH)OH$.[8]

There are numerous hydrated silicates of the alkali and alkaline-earth metals. Of the sodium compounds, with empirical formulae $Na_2O . SiO_2$. 5, 6, 8, and 9 H_2O, the first and last have been studied. Their structural formulae are $Na_2[SiO_2(OH)_2] . 4$[9] and 8 H_2O.[10] They apparently contain ions $[SiO_2(OH)_2]^{2-}$ in which Si–O = 1·59 Å and Si–OH = 1·67 Å.

Silicates containing $Si_2O_7^{6-}$ ions

We noted earlier that from the structural standpoint it is logical to group together all the tetrahedrally coordinated atoms in certain structures rather than single out only the Si_xO_y grouping as a particular type of silicate ion. The latter view regards as simple pyrosilicates only the rare minerals thortveitite, $Sc_2Si_2O_7$,[1] and barysilite, $MnPb_8(Si_2O_7)_3$,[2] and the 4f pyrosilicates, $M_2Si_2O_7$.[3] There has been considerable interest in pyrosilicates because of the variation in the Si–O–Si bond angle. The 4f pyrosilicates adopt one or more of four structures according to

(3) AC 1958 **11** 862

(4) AC 1969 **B25** 1251

(5) AC 1965 **18** 787

(6) AC 1952 **5** 307

(7) AC 1952 **5** 477
(8) AC 1952 **5** 724

(9) AC 1966 **21** 583
(10) AC 1966 **20** 688

(1) AC 1962 **15** 491;
K 1972 **17** 859
(2) AC 1966 **20** 357
(3) AC 1970 **B26** 484

the size of M^{3+}:

	(La) Ce ...... Eu	Gd Tb Dy	(Y) Ho Er	Tm Yb Lu Sc
c.n. of M	←———— 8 ————→		←— 6 —→	←——— 6 ———→
		←————————— 7 —————————→		(thortveitite structure)

and these compounds provide examples of the whole range of Si–O–Si angles: $180°$ (Sc, Yb), $159°$ (Gd), and $133°$ (Nd). It is evident that the structure of the $Si_2O_7^{6-}$ ion is very dependent on its environment. There are also minerals which contain both SiO_4^{4-} and $Si_2O_7^{6-}$ ions:

(4) ZK 1931 **78** 422

(5) AC 1954 **7** 53

Vesuvianite	$Ca_{10}Al_4(Mg, Fe)_2(Si_2O_7)_2(SiO_4)_5(OH)_4$ [4]
Epidote	$Ca_2FeAl_2(SiO_4)(Si_2O_7)O(OH)$ [5]

In the following crystals we may regard the Si_2O_7 groups as part of more extensive tetrahedral systems in which the larger cations are 8-coordinated (melilite, danburite):

		Nature of tetrahedral complex
Melilite type	$Ca_2(MgSi_2O_7)$	A_3X_7 layer
Hemimorphite	$[Zn_4Si_2O_7(OH)_2]H_2O$	A_2X_3 3D framework
Danburite	$Ca(B_2Si_2O_8)$	AX_2 3D framework

Certain members of the melilite family are formally pyrosilicates since they contain Si_2O_7 units, for example, akermanite, $Ca_2MgSi_2O_7$, and hardystonite, $Ca_2ZnSi_2O_7$; the mineral melilite itself is $(Ca, Na)_2(Mg, Al)(Si, Al)_2O_7$. However, there is complete solid solution formation between $Ca_2MgSi_2O_7$ and $Ca_2Al_2SiO_7$ (gehlenite), and the basic feature of the structure is the XY_2O_7 layer built from XO_4 and YO_4 tetrahedra sharing respectively 4 and 3 O atoms, as described on p. 163.

Hemimorphite, an important zinc ore, is of interest as an example of the way in which X-ray studies lead to the assignment of correct structural formulae. It was formerly written $H_2Zn_2SiO_5$, but one-half of the H in this empirical formula is in the form of H_2O molecules which may be removed by moderate heating of the crystal, whereby its form and transparency are not destroyed. The removal of further H_2O, at temperatures above $500°C$, results in destruction of the crystal. These facts are consistent with the crystal structure, which shows that the composition is $Zn_4(OH)_2Si_2O_7 \cdot H_2O$. [6] The structure may be regarded as built from $ZnO_3(OH)$ and SiO_4 tetrahedra: these share three vertices to form layers of

(6) ZK 1967 **124** 180

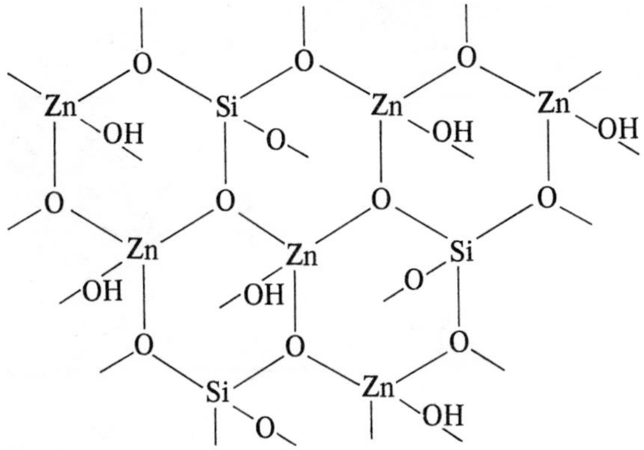

the type shown which are then joined through Si–O–Si or Zn–OH–Zn links to form a 3D framework. (The Si–O–Si angle is $150°$.) There is a strong resemblance to the structure of Si_2N_2O, as may be seen by writing the formulae

$$Si_4 \quad Si_2 \quad N_6 \quad O \quad O_2$$
$$Zn_4 \quad Si_2 \quad O_6 \quad O \quad (OH)_2$$

and comparing with Fig. 23.5 (p. 797). The H_2O molecules are accommodated in the tunnels in the framework. We refer to danburite on p. 826.

Silicates containing cyclic $(SiO_3)_n^{2n-}$ *ions*

The existence of the $Si_3O_9^{6-}$ and $Si_6O_{18}^{12-}$ ions in minerals has long been established. The former occurs in benitoite, $BaTiSi_3O_9$, and catapleite, $Na_2ZrSi_3O_9 . 2 H_2O$, and the latter in beryl (emerald), $Be_3Al_2Si_6O_{18}$, dioptase, $Cu_6Si_6O_{18} . 6 H_2O$[1] (originally formulated as an orthosilicate, CuH_2SiO_4), and tourmaline, $(Na, Ca)(Li, Al)_3Al_6(OH)_4(BO_3)_3Si_6O_{18}$.[2] The structures of beni- toite and beryl are very similar in general type, as can be seen from the plans in Fig. 23.11. In each structure the cyclic ions are arranged in layers with their planes parallel, but the structures are not layer structures, as might appear from these plans, for the metal ions lie between the layers of anions and bind together the anions at different levels. Both the Ti^{4+} and Ba^{2+} ions in $BaTiSi_3O_9$ are 6-coordinated (the Ba–O distance being greater than the Ti–O), but in beryl the Be^{2+} ions are in tetrahedral and the Al^{3+} in octahedral holes. The low-temperature form of cordierite, $Al_3Mg_2(Si_5Al)O_{18}$, has a similar structure to beryl.

The synthetic metasilicate $K_4(HSiO_3)_4$[3] contains the silicon analogue of the cyclic $P_4O_{12}^{4-}$ ion. The $Si_4O_{12}^{8-}$ ion also exists in axinite, a complex borosilicate, but only the approximate structure is known;[4] the same ion may exist in the mineral baotite.[5] A recent addition to the crystal chemistry of silicon has been the discovery of the $Si_8O_{24}^{16-}$ cyclic ion in the mineral murite,[6] $Ba_{10}(Ca, Mn, Ti)_4Si_8O_{24}(Cl, OH, O)_{12} . 4 H_2O$; cyclic metasilicate ions are now known con- taining 3, 4, 6, and 8 Si atoms.

(1) AC 1955 8 425

(2) AC 1969 **B25** 1524

(3) AC 1964 17 1063; AC 1965 18 574

(4) AC 1952 5 202

(5) K 1960 5 544

(6) Sc 1971 **173** 916

Silicon

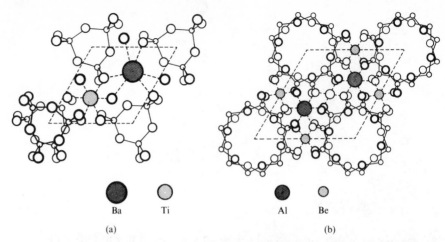

Ba   Ti   Al   Be

(a)   (b)

FIG. 23.11. Plans of the structures of (a) benitoite, $BaTiSi_3O_9$, and (b) beryl, $Be_3Al_2Si_6O_{18}$. Atoms belonging to complex ions lying at different heights above the plane of the paper are shown as heavy and light circles in order to show the nature of the coordination groups around the metal ions. Only one of the upper complex ions is shown in (a) and many oxygen atoms are omitted from (b) for the sake of clarity.

Silicates containing chain ions

Single chains formed from tetrahedral MX_4 groups sharing two vertices adopt various configurations in crystals. Although there are different numbers of tetrahedra in the repeat unit, namely, 1, 2, and 3 for the chains (a), (b), and (c) of Fig. 23.12 (described as 1T, 2T, 3T, etc. chains) the composition is in all cases MX_3

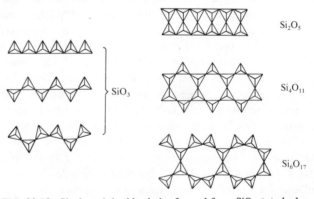

SiO_3

Si_2O_5

Si_4O_{11}

Si_6O_{17}

FIG. 23.12. Single and double chains formed from SiO_4 tetrahedra.

since each tetrahedron shares two vertices. In the double chains, (d)–(g), there are different proportions of tetrahedra sharing 2 and 3 vertices, Examples of the various chains are summarized in Table 23.3. The most important single chain is (b) which occurs in synthetic metasilicates of Li and Na and in the pyroxene group of minerals, a group which includes enstatite, $MgSiO_3$, diopside, $CaMg(SiO_3)_2$, jadeite,

816

$NaAl(SiO_3)_2$, and spodumene, $LiAl(SiO_3)_2$. Details of the chain in Na_2SiO_3 are shown inset; in the various crystals the parallel chains pack so as to provide suitable environments for the cations–6-coordination of Mg^{2+}, 8-coordination of Ca^{2+} (in diopside).

Silicates $M_nSi_2O_5$ prefer the hexagonal layer to the double chain ion (see next section). Sillimanite is strictly an orthosilicate, for it contains discrete SiO_4 groups. However, since one-half of the Al atoms are 4-coordinated it may alternatively be regarded as an aluminosilicate in which the Si atoms and these Al atoms form double chains of type (d). The octahedrally coordinated Al atoms lie between these chains, and the compound may be formulated $Al(AlSiO_5)$. In the mineral world the most important compounds containing double chains are the amphiboles, of which

TABLE 23.3

Crystals containing single and double chain ions

Single chain $(MO_3)_n$	Double chain
1T $CuGeO_3$, K_2CuCl_3	M_2O_5 $Al(AlSiO_5)$ (sillimanite)
2T Pyroxenes, Li_2SiO_3, Na_2SiO_3[1]	M_4O_{11} (e) amphiboles
$RbPO_3$,[2] $LiAsO_3$,[3] SO_3	(f) $Na_2ZrSi_4O_{11}$ (vlasovite)[12]
3T β-Wollastonite,[4] high-$NaPO_3$	M_6O_{17}, $Ca_6Si_6O_{17}(OH)_2$[13] (xonotlite)
(Maddrell salt),[5] $Ca_2NaHSi_3O_9$	
(pectolite)[6]	
4T Linear: $Pb(PO_3)_2$[7]	
Helical: $NaPO_3$ (Kurrol salt A),[8] $AgPO_3$[8]	
5T $CaMn_4Si_5O_{15}$[9] (rhodonite)	
7T $(Ca, Mg)(Mn, Fe)_6Si_7O_{21}$[10] (pyroxmangite)	
8T $Na_3H(PO_3)_4$[11]	

(1) AC 1967 **22** 37. (2) AC 1964 **17** 681. (3) AC 1956 **9** 87. (4) PNAS 1961 **47** 1884. (5) AC 1955 **8** 752. (6) ZK 1967 **125** 298. (7) AC 1964 **17** 1539. (8) AC 1961 **14** 844. (9) AC 1959 **12** 182. (10) AC 1959 **12** 177. (11) AC 1968 **B24** 992. (12) CRURSS 1961 **141** 958. (13) AC 1956 **9** 1002.

tremolite, $(OH)_2Ca_2Mg_5(Si_4O_{11})_2$, is a typical member. The term *asbestos* was originally reserved for the more fibrous amphiboles, for example, tremolites, actinolites,[1] and crocidolites.[2] The name is also, however, applied to fibrous varieties of serpentine which are not amphiboles but minerals more closely related to talc and the other clay minerals. In fact, some 50 per cent of commercial asbestos is chrysotile, for the structure of which see p. 821. Synthetic amphiboles (NaMg, NaCo, NaNi) have been prepared hydrothermally as long, flexible, fibrous crystals.[3]

(1) AC 1955 **8** 301
(2) AC 1960 **13** 291

(3) IC 1964 **3** 1001

A different type of chain, also with the composition Si_4O_{11}, has been found in the mineral vlasovite, $Na_2Zr(Si_4O_{11})$. This consists of rings of four tetrahedra (Fig. 23.12(f)) and like the amphibole chain has equal numbers of tetrahedra sharing two and three vertices. An example of the Si_6O_{17} chain is included in Table 23.3. Also included in the Table are some other meta-salts containing tetrahedral chain ions. Chains differ not only in the number of tetrahedra in the repeat unit along the chain (1T, 2T, etc.) but also in the relative orientations of the tetrahedra within a

Silicon

repeat unit. In Table 23.3 we distinguish two types of 4T chain, one essentially linear (two pairs of tetrahedra) and the other helical.

Silicates with layer structures

In a layer formed from SiO_4 tetrahedra linked to three others by sharing vertices the Si atoms lie at the points of a 3-connected net, which may be plane or puckered. Silicate structures based on four different nets are known (Fig. 23.13), of

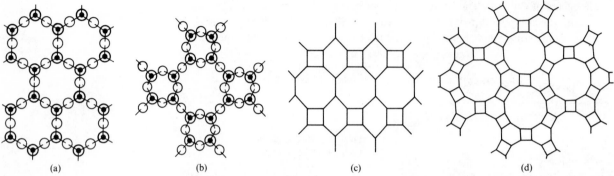

(a)   (b)   (c)   (d)

FIG. 23.13. Layers of SiO_4 tetrahedra sharing 3 vertices which are found in silicates. In (a) and (b) the small black circles represent Si atoms and the open circles O atoms. Oxygen atoms lying above Si atoms are drawn more heavily. The O atoms are omitted from the more complex nets (c) and (d).

which the two simplest are the most frequently encountered. Three main classes of layer structure may be recognized according to the way in which the basic layer is incorporated in the structure.

   (i) There are single layers, of composition Si_2O_5 or $(Si, Al)_2O_5$, held together by cations. Structures of this type are rare in the mineral world but a few examples are known, and this type of structure is also found in some synthetic disilicates:

(a) 6 net: $Li_2Si_2O_5$ and the isostructural $H_2Si_2O_5$,[1] $Na_2Si_2O_5$,[2] $Ag_2Si_2O_5$,[3] $LiAlSi_4O_{10}$ (petalite),[4]

(b) 4:8 net: $BaFeSi_4O_{10}$ (gillespite),[5] $CaCuSi_4O_{10}$ (Egyptian blue),[6] $Ca_4Si_8O_{20} . KF . 8 H_2O$ (apophyllite),[7]

(c) 4:6:8 net: $K_{1.7}Na_{0.3}ZrSi_6O_{15}$ (dalyite),[8]

(d) 4:6:12 net: $(Mn_{12}Fe_3Mg)Si_{12}O_{30}(OH, Cl)_{20}$ (manganpyrosmalite).[9]

The hexagonal layers in $Li_2Si_2O_5$ are very puckered, as described shortly; the conformation of Si_2O_5 layers has been reviewed.[10] An interesting 'topochemical' reaction is the conversion of a single crystal of α-$Na_2Si_2O_5$ into one of $Ag_2Si_2O_5$ by heating in molten $AgNO_3$. Acids $H_2Si_2O_5$ have been prepared which are

(1) ZK 1964 **120** 427

(2) AC 1968 **B24** 13, 1077

(3) AC 1961 **14** 537

(4) AC 1961 **14** 399

(5) AM 1943 **28** 372

(6) AC 1959 **12** 733

(7) AM 1971 **56** 1222, 1234, 1243

(8) ZK 1965 **121** 349

(9) AC 1968 **B24** 690

(10) AC 1968 **B24** 690

818

structurally similar to two of the alkali disilicates, and gillespite from which the metal ions have been leached by 5N HCl gives sufficient X-ray diffraction to show that the layer structure is retained in the crystals (of composition $H_4Si_4O_{10} \cdot \frac{1}{2} H_2O$), which are colourless and have a density of 2·1 g/cc—compare 3·4 g/cc for the red crystals of gillespite.[11]

(11) AM 1958 **43** 970

All the layers of Fig. 23.13 are *simple* layers in the sense that they can be represented on a plane as 3-connected nets without any links overlapping or intersecting. More complex layers can be envisaged which have some extension in a direction perpendicular to the layer, as in the double layer described in (ii). An intermediate possibility is realized in $Na_2Si_3O_7$.[11a] We may represent the end-on view of an idealized metasilicate chain as in Fig. 23.14(a). In (b) each pair of vertex-sharing tetrahedra represents a chain perpendicular to the plane of the paper,

(11a) N 1967 **214** 794

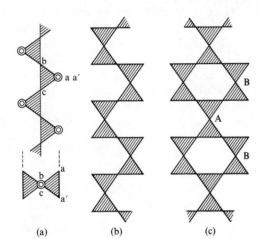

(a)   (b)   (c)

FIG. 23.14. (a) Metasilicate chain ion $(SiO_3)_n^{2n-}$ and end-on view. (b) Layer of composition Si_2O_5 built of SiO_3 chains perpendicular to the plane of the paper. (c) Layer of composition Si_3O_7 similarly represented.

so that (b) represents a buckled layer in which each SiO_4 group shares 3 vertices. This is approximately the configuration of the Si_2O_5 layer in $Li_2Si_2O_5$. In (c) the tetrahedra in the B chains share 3 vertices and those in the A chains all 4 vertices, these layers being members of the series Si_mO_{2m+1}:

SiO_3	Si_2O_5	Si_3O_7	...	SiO_2
chain	layers			3D framework

All the layers are in fact slices of the cristobalite structure, which is the end-member of the series, and the Si_3O_7 layer is the anion in $Na_2Si_3O_7$. This A_3X_7 layer, in which all circuits are rings of six tetrahedra, may be compared with the simple A_3X_7 layer in melilite and related compounds which is composed entirely of rings of five tetrahedra. We referred to this layer under pyrosilicates; it is illustrated in Fig. 5.7(c), p. 163. In both A_3X_7 layers two-thirds of the tetrahedra share 3 vertices and the remainder 4 vertices.

(ii) If the (unshared) vertices of all the tetrahedra of a plane layer point to the

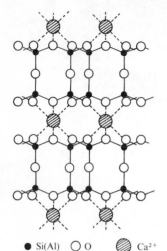

● Si(Al)   ○ O   ◍ Ca²⁺

FIG. 23.15. Elevation of the structure of hexagonal $CaAl_2Si_2O_8$ showing the double layers interleaved with Ca^{2+} ions.

(12) AC 1959 **12** 465

(13) MJ 1958 **2** 311

same side of the layer, two layers can be combined to form a double layer which has the composition $(Si, Al)O_2$, since all the vertices of each tetrahedron are now shared. There must be replacement of some of the Si by Al (or possibly Be) since otherwise the layer would be neutral, of composition SiO_2. The hexagonal forms of $CaAl_2Si_2O_8$[12] and $BaAl_2Si_2O_8$[13] contain double layers of this kind (Fig. 23.15) formed from two 6-gon layers of the type of Fig. 23.13(a).

The simplest plane 3-connected net is found as a single layer of composition Si_2O_5 in (i) and forms the double layer in the structures just mentioned in (ii). There is a third way in which this net plays a part in silicate structures which is due to the approximate dimensional correspondence between a plane hexagonal Si_2O_5 layer and the $Al(OH)_3$ or $Mg(OH)_2$ layers.

(iii) The third, and by far the most important, family of layer structures is based on composite layers built of one or two silicon–oxygen layers combined with layers of hydroxyl groups bound to them by Mg or Al atoms. These composite layers are the units of structure, and we first consider their internal structure before seeing how they are packed together in crystals. Fig. 23.16(a) shows a portion of a sheet of linked SiO_4 tetrahedra, each sharing three vertices, with all the unshared vertices pointing to the same side of the layer. Fig. 23.16(b) shows the arrangement of the OH groups in a layer of the $Mg(OH)_2$ or $Al(OH)_3$ structures. The distance between the upper O atoms in (a), drawn with heavy full lines, is approximately the same as that between the corresponding OH groups in (b). It is therefore possible by

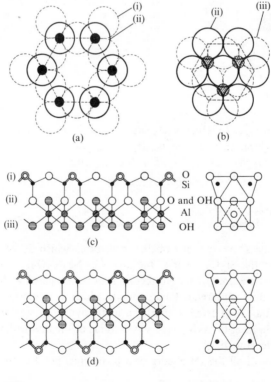

FIG. 23.16. The formation of composite silicon–aluminium–oxygen or silicon–magnesium–oxygen layers (see text).

inverting (a) and placing it on top of (b) to form a composite layer having these O atoms in common. The composite layer is shown in elevation in Fig. 23.16(c); it has been illustrated in Fig. 5.44 (p. 191) as an example of a structure built from tetrahedral and octahedral coordination groups. It is possible to repeat this process of condensation on the other side of the (b) layer giving the more complex layer shown at (d). In illustrating structures based on these layers we shall use the diagrammatic forms shown to the right of (c) and (d) to represent layers of these types extending indefinitely in two dimensions, a simplification suggested by Pauling. Just as the $Mg(OH)_2$ and $Al(OH)_3$ layers have the same arrangement of OH groups and differ, from the geometrical point of view, only in the number of octahedral holes occupied by metal atoms, so in these composite layers there may be either Mg or Al ions or, of course, other di- or tri-valent ions of appropriate size. The ideal compositions of these layers in the extreme cases are:

$$\text{(c)} \quad Mg_3(OH)_4Si_2O_5 \quad \text{or} \quad Al_2(OH)_4Si_2O_5$$
$$\text{(d)} \quad Mg_3(OH)_2Si_4O_{10} \quad \text{or} \quad Al_2(OH)_2Si_4O_{10}$$

according to whether the octahedral holes are occupied by Mg or Al. The structural chemistry of these layer minerals is further complicated by the fact that Si can be partly replaced by Al (in tetrahedral coordination) giving rise to charged layers. Such layers are either interleaved with positively charged alkali or alkaline-earth ions, as in the micas, by layers of hydrated ions in montmorillonite, or by possitively charged $(Mg, Al)(OH)_2$ layers in the chlorites. We may distinguish five main structural types, remembering that in each of the five classes the composition may range from Mg_3 to Al_2 as already indicated.

Layers of type (c) *only.* The pure Mg layer occurs in chrysotile, $(OH)_4Mg_3Si_2O_5$, in which the structural unit is a kaolin-like layer of this composition, instead of $(OH)_4Al_2Si_2O_5$ as in Fig. 23.16. Since the dimensions of the brucite $(Mg(OH)_2)$ part of the composite layer do not exactly match those of the Si_2O_5 sheet, the layer curls up, the larger (brucite) portion being on the outside. The fibres are built of curled ribbons forming cylinders several thousand Å long with a dozen or so layers in their walls and overall diameters of several hundred Å;[14] the detailed structures of chrysotiles are still not known, nor is the nature of the material filling the tubes.

We referred earlier to the silylation of silicates to give trimethylsilyl derivatives corresponding to the Si—O—Si systems in the original minerals. A development of this idea is to treat chrysotile with a mixture of HCl and $ClSi(CH_3)_3$ when the Mg and OH ions are stripped off the outside of the layer and replaced by $-OSi(CH_3)_3$, giving a gel which on treatment with water forms a fibrous mass of ribbons. When dry these curl up to form fibres similar in shape to those of the original chrysotile.[15]

The $(OH)_4Al_2Si_2O_5$ layer is the structural unit in the kaolin (china-clay) minerals (Fig. 23.17). The three minerals kaolinite, dickite,[16] and nacrite all have the composition $Al_2(OH)_4Si_2O_5$ and their structures differ only in the number of

(14) AC 1956 9 855, 862, 865

(15) IC 1967 6 1693

(16) AC 1956 9 759

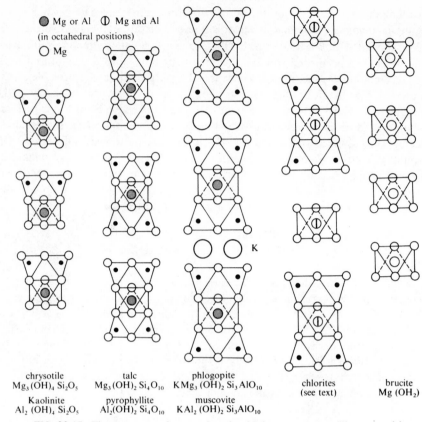

● Mg or Al   ⏻ Mg and Al
(in octahedral positions)
○ Mg

K

chrysotile	talc	phlogopite	chlorites	brucite
$Mg_3(OH)_4Si_2O_5$	$Mg_3(OH)_2Si_4O_{10}$	$KMg_3(OH)_2Si_3AlO_{10}$	(see text)	$Mg(OH_2)$
Kaolinite	pyrophyllite	muscovite		
$Al_2(OH)_4Si_2O_5$	$Al_2(OH)_2Si_4O_{10}$	$KAl_2(OH)_2Si_3AlO_{10}$		

FIG. 23.17. The structures of some minerals with layer structures (diagrammatic).

kaolin layers (1, 2, and 6 respectively) in the repeat unit. In halloysite, the commonest of the kaolin minerals, there is an irregular sequence of layers; in amesite, with the ideal formula $(Mg_2Al)(SiAlO_5)(OH)_4$, the replacement of part of the Mg_3 by Al is balanced by substitution of Al for half of the Si.[17]

(17) AC 1951 **4** 552

Layers of type (d) *only.* The two extremes are talc, $Mg_3(OH)_2Si_4O_{10}$, and pyrophyllite, $Al_2(OH)_2Si_4O_{10}$ (Fig. 23.17). As in the kaolins the layers are electrically neutral, and there are only feeble attractive forces between neighbouring layers. These minerals are therefore soft and cleave very easily, and talc, for example, finds applications as a lubricant (french chalk).

Charged layers of type (d) *interleaved with ions.* Replacement of one-quarter of the Si in talc and pyrophyllite layers gives negatively-charged layers which are interleaved with K^+ ions in the *micas* phlogopite and muscovite (Fig. 23.17):

(18) AC 1966 **20** 638

(19) AC 1960 **13** 919

$$KMg_3(OH)_2Si_3AlO_{10} \quad \text{phlogopite}^{[18]}$$
$$KAl_2(OH)_2Si_3AlO_{10} \quad \text{muscovite}^{[19]}$$

The potassium ions occupy large holes between twelve oxygen atoms so that the K–O electrostatic bond strength is only one-twelfth. These bonds are easily broken and the micas accordingly possess very perfect cleavage parallel to the layers. In the so-called brittle micas, typified by margarite, $CaAl_2(OH)_2Si_2Al_2O_{10}$, further substitution of Al for Si has taken place so that the negative charge on the layers is twice as great as in phlogopite and muscovite. The layers are therefore held together by divalent ions such as Ca^{2+}. The increased strength of the interlayer Ca–O bonds, with electrostatic bond strength now equal to one-sixth, results in greater hardness, as shown by the figures for the hardnesses on the Mohs scale:

	Hardness
Talc, pyrophyllite	1–2
Micas	2–3
Brittle micas	$3\frac{1}{2}$–5

Charged layers of type (d) *interleaved with hydrated ions.* In addition to the micas, which are anhydrous, there are some very important minerals, sometimes called 'hydrated micas', which are built of layers with smaller charges per unit area than in the micas, interleaved with layers of hydrated alkali or Mg^{2+} ions. Such feebly charged layers can arise by replacing part of the Al_2 in a pyrophyllite layer by Mg as in

$$\underbrace{[Mg_{0\cdot33}Al_{1\cdot67}Si_4O_{10}(OH)_2]}_{\text{total } 2\cdot00}{}^{-0\cdot33}Na_{0\cdot33} \quad \text{(montmorillonite)}[20]$$

(20) TFS 1948 **44** 306, 349

(21) AM 1954 **39** 231

or there can be substitution in a talc layer not only in the octahedral positions but also in the tetrahedral positions, as in vermiculite:

$$[(Mg_{2\cdot36}Fe^{3+}_{0\cdot48}Al_{0\cdot16})(Si_{2\cdot72}Al_{1\cdot28})O_{10}(OH)_2]^{-0\cdot64}[Mg_{0\cdot32}(H_2O)_{4\cdot32}]^{+0\cdot64}$$

As might be expected, this mineral dehydrates to a talc-like structure. A section through the structure[21] is shown in Fig. 23.18, but only a proportion of the interlamellar Mg^{2+} and H_2O positions are occupied.

Clay minerals. It is customary to group under this general heading a number of groups of minerals with certain characteristic properties. They are soft, easily hydrated, and exhibit cation exchange with simple ions and also with many organic ions. They include the kaolin group, the montmorillonite-beidellite group, and the illite clays, related to vermiculite. Apart from their fundamental importance as constituents of soils they find many industrial applications. Kaolins are used in pottery and ceramics, as rubber fillers, and for filling and coating paper. Montmorillonite (bentonite), an important constituent of fuller's earth, finds many applications owing to its property of forming thick gelatinous suspensions at concentrations of only a few per cent. For example, sodium bentonite swells readily in water and is used as a binder for foundry sand and as a thickener in

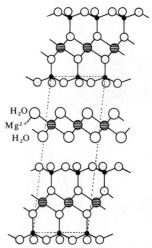

FIG. 23.18. Elevation of the structure of vermiculite.

oil-well drilling muds, while the non-swelling calcium bentonite is converted by acids into adsorbents used in oil refining. Vermiculite has many uses, ranging from soil conditioner to a porous filler in light-weight plastics and concretes.

Charged layers of two kinds. In all these layer structures so far described all the layers in a given crystal are of the same kind. There are also more complicated structures with layers of two kinds, and the chlorite minerals are in this class. In kaolin, talc, and pyrophyllite the uncharged layers are held together by comparatively weak forces, and in the micas the negatively-charged layers are cemented together by positive ions. In the chlorites another possibility is realized, the alternation of negatively-charged layers of the mica type with positively-charged layers, that is the structures contain infinite 2-dimensional ions of opposite charge. The mica-like layers have compositions ranging from $[Mg_3(AlSi_3O_{10})(OH)_2]^-$ to $[Mg_2Al(Al_2Si_2O_{10})(OH)_2]^-$. The positive layers result from the replacement by Al of one-third of the Mg in a brucite $(Mg(OH)_2)$ layer, giving the composition $Mg_2Al(OH)_6^+$. The sequence of layers in the chlorites is compared with that in $Mg(OH)_2$ and phlogopite in Fig. 23.17. Since there are many ways of stacking the layers in the chlorites, as in other layer structures, polytypes are numerous.[22]

(22) AC 1969 **B25** 632

Silicates with framework structures

The possible types of 3D structure have been noted earlier in this chapter, namely:

(a) 3 vertices of each SiO_4 tetrahedron shared: composition $(Si_2O_5)_n^{2n-}$ or more highly charged if some Si is replaced by Al (or Be),

(b) 3 or 4 vertices shared.

(c) 4 vertices shared: composition SiO_2 or $(Si, Al)O_2$, the charge on the framework in the latter case depending on the extent to which Al replaces Si.

We have already described the forms of SiO_2 and noted some minerals with closely related structures in which some Si is replaced by Al or Be. In each class we could imagine additional tetrahedra inserted between the 3- or 4-connected ones and having the effect of simply extending the links in the basic 3-, (3,4)-, or 4-connected net. Little is known of structures of classes (a) or (b), but there is an interesting example of a framework based on tetrahedra sharing 3 vertices some of which are linked through 2-connected tetrahedra. In neptunite, $LiNa_2K(Fe,Mg,Mn)_2(TiO)_2Si_8O_{22}$,[1] there are equal numbers of SiO_4 tetrahedra sharing 2 and 3 vertices. They form two interpenetrating 3-connected nets of the $ThSi_2$ type (Fig. 3.15(e), p. 76) in which there are two additional tetrahedra (sharing two vertices) inserted along one-third of the links of each net. Such nets have the same O : Si ratio as the amphibole chain: $OSi(O_{\frac{1}{2}})_2(Si_2O_7)_{\frac{1}{2}}\equiv Si_4O_{11}$. They account for all except two of the O atoms in the above formula; these are not bonded to Si but, together with the O atoms of the frameworks, complete octahedral coordination groups around the Ti atoms.

We shall be concerned here exclusively with frameworks in which tetrahedra share all four vertices. In all these structures some (often about one-half) of the tetrahedral positions are occupied by Al (rarely by Be), and positive ions are present to neutralize the negative charge of the $(Si,Al)O_2$ framework. These

(1) AC 1966 **21** 200

framework structures, and some of their physical properties, are more easily understood if they are subdivided according to whether or not there are obvious polyhedral cavities or tunnels in the structure. We shall deal here with three groups of minerals (and some related synthetic compounds), the felspars, zeolites, and ultramarines. The felspar structures are relatively compact, but in the other two groups there are polyhedral cavities or tunnels in which are accommodated water molecules (in the zeolites) or finite anions (Cl^-, SO_4^{2-}, S^{2-}, etc. in the ultramarines) in addition, of course, to the necessary numbers of cations. Between the felspars and the framework structures which may be described as space-filling arrangements of polyhedra lie structures of an intermediate nature. For example, milarite, $K_2Ca_4(Be_4Al_2Si_{24}O_{60}) \cdot H_2O$[2] contains tightly knit hexagonal prisms built from 12 tetrahedra which are linked through single tetrahedra into a 3D network (Fig. 23.19). The prisms have the composition $(Be_{0.10}Si_{0.90})_{12}O_{30}$, while the separate tetrahedra have the mean composition $(Be_{0.27}Al_{0.37}Si_{0.40})O_4$.

(2) AC 1952 **5** 209

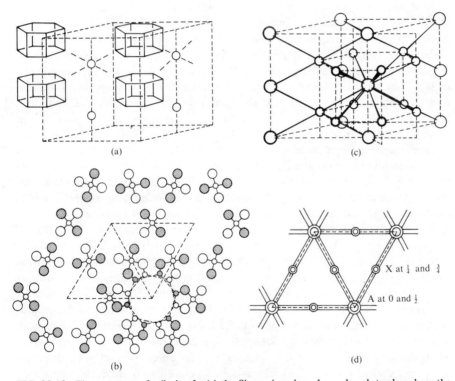

(a)

(b)

(c)

(d)

X at $\frac{1}{4}$ and $\frac{3}{4}$

A at 0 and $\frac{1}{2}$

FIG. 23.19. The structure of milarite. In (a) the Si_{12} prisms have been shrunk to show how the Be atoms link them together to form the 3D framework. The actual environment of Be is a tetrahedral group of O atoms belonging to four different $Si_{12}O_{30}$ prisms, as shown in the projection (b). The shaded O atoms belong to the lower 6-rings of $Si_{12}O_{30}$ groups and the unshaded ones to the upper 6-rings of prisms below. If the Si_{12} prisms are shrunk to become 12-connected points the basic framework is the (12,4)-connected AX_3 net shown at (c) and in plan at (d).

Silicon

Felspars. These are the most important rock-forming minerals, comprising some two-thirds of the igneous rocks. Granite, for example, is composed of quartz, felspars, and micas. The felspars are subdivided into two groups according to the symmetry of their structures, typical members of the groups being:

(3) AC 1961 **14** 443

(1) Orthoclase $KAlSi_3O_8$[3]
   Celsian $\quad BaAl_2Si_2O_8$.

(4) AC 1962 **15** 1005, 1017

(2) The plagioclase felspars; Albite $\quad NaAlSi_3O_8$
   Anorthite $CaAl_2Si_2O_8$.[4]

We have already noted that there are hexagonal forms of both $BaAl_2Si_2O_8$ and $CaAl_2Si_2O_8$ with double-layer structures. It will be noticed that in the first example in each group one-quarter of the tetrahedral positions are occupied by Al, requiring a monovalent ion to balance the charge on the $AlSi_3O_8$ framework. In celsian and anorthite further replacement of Si by Al necessitates the introduction of divalent ions. The plagioclase felspars, albite and anorthite, are of peculiar interest since they are practically isomorphous and many minerals of intermediate composition are known. The composition of labradorite, for example, ranges from AbAn to AbAn₃, where Ab = albite and An = anorthite. This isomorphous replacement of (K + Si) by (Ba + Al) or (Na + Si) by (Ca + Al) is characteristic of felspars and other framework silicates. The felspars of the first group contain the large K^+ and Ba^{2+} ions (radii 1·38 and 1·36 Å respectively), whereas the plagioclase felspars contain the smaller Na^+ and Ca^{2+} ions (radii 1·02 and 1·00 Å respectively). It may be noted here that only the comparatively large positive ions are found in felspars. The smaller ions of Fe, Cr, Mn, etc., so commonly found in chain and layer silicates do not occur in these minerals, presumably because the frameworks cannot close in around these smaller ions. The difference between the symmetries of the two groups of felspars is associated with this difference in size of the positive ions, for the framework is essentially the same in all the above felspars but contracts slightly in the second group around the smaller sodium and calcium ions.

(5) AC 1953 **6** 613

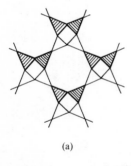

(a)

(b)

FIG. 23.22. Layers of tetrahedra which, by further linking through O atoms, form the framework structures of (a) paracelsian and (b) the common felspars. Open and shaded triangles represent tetrahedra the fourth vertices of which project either above or below the plane of the paper.

The $(Si,Al)O_2$ frameworks of felspars may be regarded as built from layers of tetrahedra placed at the points of the plane 4 : 8 net of Fig. 23.13(b), p. 818, the fourth vertex of each tetrahedron pointing either upwards or downwards out of the plane of the paper. These layers are then joined through the projecting vertices so that adjacent layers are related by planes of symmetry. Different 3-dimensional frameworks arise according to the relative arrangement of upward- and downward-pointing tetrahedra. The layers of Fig. 23.22(a) give the framework of paracelsian,[5] one of the forms of $Ba(Al_2Si_2O_8)$ (and the isostructural danburite, $Ca(B_2Si_2O_8)$), while that of the common felspars arises by joining up the more complex layers of Fig. 23.22(b). The idealized paracelsian framework is illustrated in Fig. 23.21(a). The compound $Ba(Al_2Si_2O_8)$ provides one of the best examples of polymorphism. Not only are there the two felspar structures of celsian and paracelsian with different 3-dimensional frameworks based on the nets of Fig. 23.22 but there is also the hexagonal form with the double-layer structure described on p. 820). Moreover, the latter has low- and high-temperature

FIG. 23.20 Stereoscopic photographs showing the structures of the three forms of silica, (A) cristobalite, (B) quartz, and (C) tridymite, as systems of linked SiO$_4$ tetrahedra.

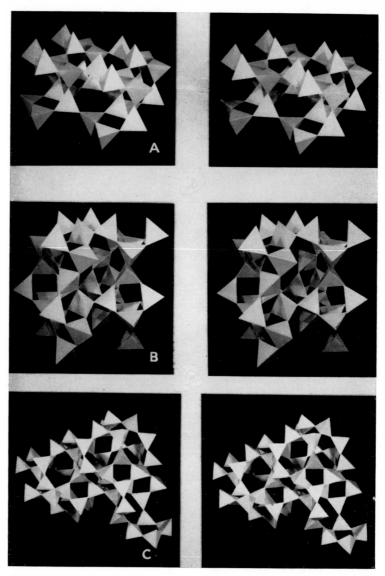

FIG. 23.21 Stereoscopic photographs showing the structures of (A) paracelsian, $Ba(Al_2Si_2O_8)$, (B) a fibrous zeolite, $Na(AlSi_2O_6).H_2O$, and (C) ultramarine, $Na_8(Al_6Si_6O_{24})S_2$, as systems of linked (Al, Si)O_4 tetrahedra.

modifications which differ in the precise configuration of the hexagonal layers, with different coordination numbers of the Ba^{2+} ions between the layers.

It seems likely that the finer structural differences between felspars of similar types may be connected with the degree of ordering of Si and Al in the tetrahedra of the frameworks, for there is an appreciable difference between the distances Al–O and Si–O. Estimates have been made of the percentage of Si and Al occupying a particular set of tetrahedral sites in a number of structures, though there has been some discussion of the precise values of Si–O and Al–O to be adopted as standards.[6] For felspars values close to 1·605 Å for Si–O and 1·76 Å for Al–O are probably appropriate. The difference between the low- and high-temperature forms of albite, $NaAlSi_3O_8$, is apparently a question of the ordering of Si and Al in the tetrahedral sites. In the low-temperature form there is an ordered arrangement of Al in one-quarter of the tetrahedra and Si in the remaining sites, but there is a random arrangement in the high-temperature form (Al(Si)–O, 1·64 Å).[7] A similar value was found for both of the two types of non-equivalent sites in sanidine, corresponding to a statistical arrangement of $\frac{1}{4}$ Al + $\frac{3}{4}$ Si in each type of tetrahedron.[8]

We noted earlier the change of a synthetic sanidine ($KAlSi_3O_8$) under high pressure (120 kbar, 900°C) to the hollandite structure, in which Si is 6-coordinated.[9]

(6) AC 1968 **B24** 355

(7) AC 1969 **B25** 1503

(8) AC 1949 **2** 280

(9) AC 1967 **23** 1093

Zeolites. The zeolites, like the felspars, consist of $(Si,Al)_nO_{2n}$ frameworks, substitution of Al for some of the Si in SiO_2 giving the framework a negative charge which is balanced by positive ions in the cavities. The structures are much more open than those of the felspars, and a characteristic property of the zeolites is the ease with which they take up and lose water, which is loosely held in the structure. In addition to water, a variety of other substances can be absorbed, including gases like CO_2 and NH_3, alcohol, and even mercury. Also the positive ions may be interchanged for others merely by soaking the crystal in a solution of the appropriate salt. The 'permutites' used for water-softening are sodium-containing zeolites which take out the calcium from the hard water and replace it by sodium, thereby rendering it 'soft'. The permutite is regenerated by treatment with brine, when the reverse process—replacement of Ca by Na—takes place, making the permutite ready for further use. In this interchange of positive ions the actual number of ions in the crystal may be altered provided the total charge on them remains the same. Thus one-half of the Ca^{2+} ions in the zeolite thomsonite, $NaCa_2(Al_5Si_5O_{20}) \cdot 6 H_2O$, may be replaced by $2 Na^+$, thereby increasing the total number of positive ions from three to four, there being sufficient unoccupied positions in the structure for the additional sodium ions.

A characteristic feature of zeolites is the existence of tunnels (or systems of interconnected polyhedral cavities) through the structures, and we may therefore expect three main types in which the tunnels are parallel to (a) one line, giving the crystals a fibrous character, (b) two lines and arranged in planes, so that the crystals have a lamellar nature, or (c) three non-coplanar lines, such as cubic axes, when the crystals have no pronounced fibrous or lamellar structure. The most symmetrical frameworks have cubic symmetry, and in a number of zeolites the framework corresponds to the edges in a space-filling arrangement of polyhedra.

(a) The structure of a typical fibrous zeolite, edingtonite, $Ba(Al_2Si_3O_{10}) . 4 H_2O$, is illustrated in Fig. 23.23. The structure may be described in terms of a characteristic chain formed by the regular repetition of a group of five tetrahedra. These chains are linked to other similar chains through the projecting O atoms. These atoms, shown as shaded circles in Fig. 23.23(b), are at heights $3c/8$ and $5c/8$ (c being the repeat distance along the chain) above the Si atom which lies

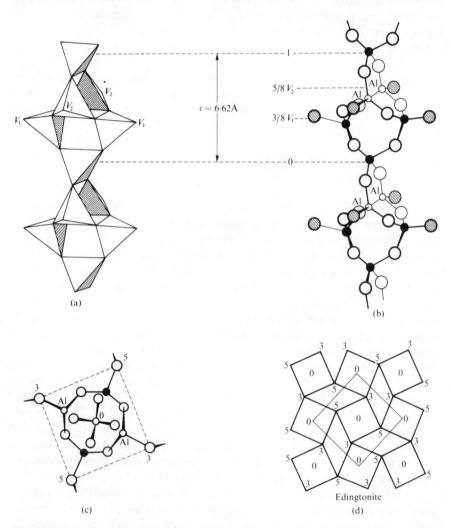

(a)

(b)

(c)

(d)

Edingtonite

FIG. 23.23. (a) The common structural feature of the fibrous zeolites, the chain of linked tetrahedra. It is attached to neighbouring chains by the vertices V_1 and V_2. (b) the same chain, showing the silicon–oxygen arrangement. The vertices V_1 and V_2 are at heights $\frac{3}{8}$ and $\frac{5}{8}$ in the repeat of the pattern. (c) The chain viewed along its length. (d) Diagrammatic projection of the structure of edingtonite, with numbers 3, 5, to indicate the heights of attachment.

on the axis of the chain. Although the resulting structure contains a 3D framework there is a marked concentration of atoms in the chains, giving the crystal its fibrous character. The structures of thomsonite, $NaCa_2(Al_5Si_5O_{20}) . 6 H_2O$, and natrolite, $Na_2(Al_2Si_3O_{10}) . 2 H_2O$,[10] are closely related to that of edingtonite.

(10) ZK 1960 **113** 430

The structure of a fibrous zeolite is illustrated in Fig. 23.21 (between pp. 826 and 827) as a packing of tetrahedra.

(b) Lamellar zeolites are frequently important constituents of sedimentary rocks. An example is phillipsite, $(K,Na)_5Si_{11}Al_5O_{32} . 10 H_2O$.[11] Each triangle in Fig. 23.24(b) represents a pair of tetrahedra sharing a common vertex as shown at (a). These pairs are perpendicular to the plane of the paper in (b), so that one of the portions of (b) drawn with heavier lines represents a non-linear double chain of tetrahedra. In planes above and below that of the paper these double chains (light lines) are displaced, and the whole structure therefore consists of a 3D framework

(11) AC 1962 **15** 644

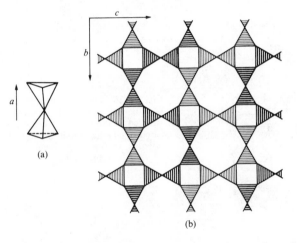

FIG. 23.24. The structure of a lamellar zeolite (phillipsite). The double tetrahedral unit shown at (a) projects in the plan (b) as a single (equilateral) triangle.

of tetrahedra each of which is linked to four others through its vertices. There are tunnels parallel to the a axis (perpendicular to the paper in (b)) of cross-sectional area 12 Å^2 and parallel to the b axis of approximately 9 Å^2 cross-sectional area, but none parallel to the c axis.

Not all zeolites fit neatly into any simple classification. In yugawaralite, $Ca_2Al_4Si_{12}O_{32} . 8 H_2O$, there is a 3D framework consisting of 4-, 5-, and 8-membered rings in which the widest tunnels form a 2D system (being parallel to the a and c axes) but there are also smaller channels parallel to the b axis.[12]

(12) AC 1969 **B25** 1183

(c) We have remarked that the edges in certain space-filling arrangements of polyhedra represent the basic (4-connected) frameworks in some aluminosilicates. We shall refer to the simplest of these, the space-filling by truncated octahedra, in connection with the ultramarines. If the square faces of the truncated octahedra are separated by inserting cubes as shown in Fig. 23.25 we have another space-filling, by truncated octahedra (t.o.), cubes, and truncated cuboctahedra (t.c.o.). Examination of Fig. 23.25 shows that this structure can be described in two other ways. The t.c.o.'s form a continuous system in three dimensions, since each shares its

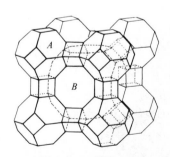

FIG. 23.25. Space-filling arrangement of truncated octahedra cubes, and truncated cuboctahedra.

octagonal faces with six other t.c.o.'s; there are accordingly tunnels parallel to the three cubic axial directions through the larger cavities B. A third way of describing the same structure is illustrated in Fig. 23.27(a) later.

(13) AC 1960 **13** 737; IC 1966 **5** 1537, 1539

The framework of Fig. 23.25 is that of the synthetic zeolite 'Linde A', which can be prepared with different Si : Al ratios[13] as in $Na_{12}(Al_{12}Si_{12}O_{48})$. $27 H_2O$ or $Na_9(Al_9Si_{15}O_{48})$. $27 H_2O$. The ion-exchange properties of zeolites have long been known and utilized, but more recent commercial interest in these compounds has been due to their possible use as 'molecular sieves'. The type of gas molecule that can pass through a tunnel is determined by the diameter of the tunnel, so that if this has a critical value the zeolite can separate mixtures of gases by allowing the passage of only those with molecular diameters smaller than that of the tunnel. The cubes in Fig. 23.25 do not represent cavities in the actual aluminosilicate framework for each consists of a tightly knit group of eight

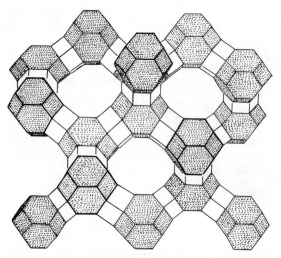

FIG. 23.26. The crystal structure of faujasite. The Si (Al) atoms are situated at the apices of the truncated octahedra, which are joined together to form a 3-dimensional framework.

tetrahedra, but water can enter both the small holes (A) and the larger holes (B), whereas O_2 can apparently enter only the larger (B) holes (compare the hydrate $CHCl_3$. $2 H_2S$. $17 H_2O$, p. 546).

In the space-filling arrangement of truncated octahedra each of the polyhedra is in contact with 14 others. If hexagonal prisms are placed in contact with alternate hexagonal faces, which are parallel to the faces of a tetrahedron, an open packing (not a space-filling) of these two kinds of polyhedron may be built (Fig. 23.26) which is equivalent to placing truncated octahedra at the points of the diamond net and joining through four of the eight hexagonal faces. This is the basic (Si,Al) framework in the natural zeolite, faujasite, $NaCa_{0.5}(Al_2Si_5O_{14})$. $10 H_2O$.[14]

(14) NJM 1958 **9** 193

Corresponding to these structures which result from joining t.o.'s through cubes and hexagonal prisms there is another family derived from the t.c.o., each being joined to neighbouring ones through (a) 12 cubes (on square faces), (b) 8 hexagonal prisms (on hexagonal faces), or (c) 6 octagonal prisms (on 8-gon faces). These are illustrated in Fig. 23.27. The first, (a), is simply a third way of describing the 'Linde

A' structure. The second, (b), is the framework of the synthetic zeolite ZK-5, with composition approximating to $Na_{30}(Al_{30}Si_{66}O_{192}) . 98 H_2O.$[15] There are two sets of (interpenetrating) 3D channel systems through the approximately planar 8-rings, with free diameter of about 3·8 Å, as compared with only one 3D system of channels in the 'Linde A' zeolite. The third, Fig. 23.27(c), is not known as a zeolite structure, but is of interest because the polyhedra of Fig. 23.27(c) fill one-half of space. There are accordingly two sets of interpenetrating 3D systems of tunnels like those shown in Fig. 23.27(c).

(15) ZK 1965 **121** 211

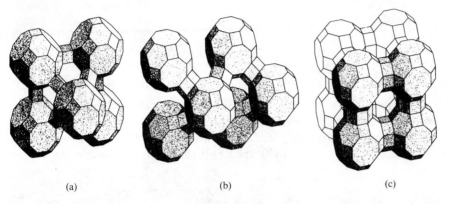

(a) (b) (c)

FIG. 23.27. Aluminosilicate frameworks of zeolites: (a) Linde A, (b) zeolite ZK 5, (c) hypothetical structure.

Polyhedral cavities of less symmetrical types occur in the closely related minerals gmelinite and chabazite.[16] A Ca-chabazite has the approximate composition $Ca(Al_2Si_4)O_{12} . 6 H_2O$, but the Ca may be partly replaced by Na. These structures contain hexagonal prisms (as do millarite and the zeolite ZK-5) and larger cavities. Their structures may be built up by a process analogous to the closest packing of equal spheres, using hexagonal prisms as the units instead of spheres.[17] In Fig. 23.28(a) and (b) each hexagonal ring represents a hexagonal prism. The layer sequence AB . . . (h.c.p.) corresponds to the structure of gmelinite and the sequence ABC . . . to chabazite. The resulting polyhedral cavities are illustrated in Fig. 23.28(c) and (d); (e) shows how the 3D framework is formed.

(16) AC 1964 **17** 374

(17) N 1964 **203** 621

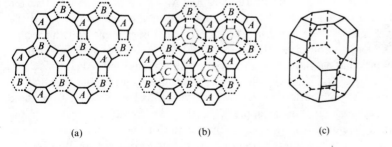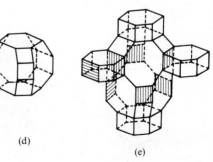

(a) (b) (c) (d) (e)

FIG. 23.28. The structure of chabazite (see text).

Silicon

Ultramarines. The last group of framework silicates that we shall mention includes the materials called ultramarines, the coloured silicates which have been manufactured for use as pigments. The mineral lapis lazuli is of the same type, and since a number of colourless minerals such as sodalite are closely similar in structure, we shall for simplicity refer to all these silicates as ultramarines. Like the other framework silicates they are based on $(Si,Al)O_2$ frameworks with positive ions in the interstices, but a characteristic of the crystals of this group is that they also contain negative ions such as Cl^-, SO_4^{2-}, or S^{2-}. Like the felspars and in contrast to the zeolites the ultramarines are anhydrous. Formulae of representative members of the group are:

Ultramarine	$Na_8Al_6Si_6O_{24} . S_2$
Sodalite	$Na_8Al_6Si_6O_{24} . Cl_2$
Noselite	$Na_8Al_6Si_6O_{24} . SO_4$
Helvite	$(Mn, Fe)_8Be_6Si_6O_{24} . S_2$

The last-mentioned is of interest as a silicate in which one-half of the tetrahedral positions are occupied by beryllium.

All of these compounds contain essentially the same framework of linked tetrahedra (Figs. 23.21 and 23.29), the Si(Al) atoms being situated at the vertices

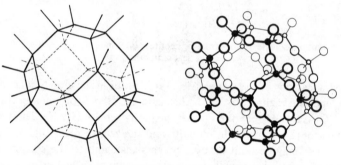

FIG. 23.29. The basket-like framework of linked SiO_4 tetrahedra which is the basis of the structure of the ultramarines. The silicon atoms are arranged at the vertices of the polyhedron shown on the left and the Si–O framework extends indefinitely in three dimensions.

of Fedorov's space-filling array of truncated octahedra. The positive and negative ions are situated in the numerous cavities in the framework. A refinement of the structure of sodalite[18] gives Si–O, 1·628 Å, and Al–O, 1·728 Å, showing that the Si and Al atoms are arranged in an ordered way in the framework. As in the zeolites, exchange of these ions for others is possible. For example, sodalite, which contains Cl^- ions, is converted into noselite by heating in fused sodium sulphate. Also, many ultramarines have been prepared containing ions such as Li^+, Tl^+, Ca^{2+}, and Ag^+ instead of Na^+ and with Se or Te replacing S. Considerable variations in colour, from nearly colourless to yellow, red, violet, and blue, result from these various replacements. It will be appreciated that the formulae assigned to the compounds in the above list are 'ideal' formulae and that considerable variation in composition is possible, subject always to the balancing of the total positive and negative charges.

(18) AC 1967 **23** 434

24

Boron

Introduction

Boron has little in common with the other elements of Group III. Being the most electronegative element of the group it resembles in general the non-metals, particularly silicon, far more than aluminium and the metals of the A and B subgroups. Boron is the only element of Group III which forms an extended series of hydrides. These bear no resemblance to the solid ionic hydrides of the alkali and alkaline-earth metals but are volatile compounds like the molecular hydrides of the non-metals of the later groups and form a remarkable group of electron-deficient compounds. Unlike aluminium, boron forms no stable normal oxy-salts (nitrate, sulphate, etc.). An acid sulphate is said to be produced by the action of SO_3 on boric acid, and tetraacetyl diborate, $(CH_3COO)_2BOB(CH_3COO)_2$, has been made (but no normal acetate). The salt $(CH_3)_4N[B(NO_3)_4]$ is formed as a white crystalline powder, stable at room temperature and insoluble in water, by the action of excess N_2O_4 at $-78°C$ on solid $(CH_3)_4N(BCl_4)$. The compounds $B(ClO_4)_3$, $B(ClO_4)_2Cl$, and $B(ClO_4)Cl_2$ have all been prepared from BCl_3 and anhydrous $HClO_4$ at $-78°C$ as low-melting salts which are sensitive to moisture and decompose on warming. On the other hand, boron forms numerous oxy-salts in which B is bonded to three or four O atoms, and it also forms the silica-like BPO_4 and $BAsO_4$.

The halides BX_3 are monomeric in the vapour, in contrast to the Al_2X_6 molecules in the vapours of $AlCl_3$, $AlBr_3$, and AlI_3, and these molecules persist in the crystalline halides (contrast $AlCl_3$). Boron also forms the halides B_2X_4, B_4X_4, and B_8X_8, the last two of which are of a type peculiar to this element. The monofluoride, BF, can be prepared in 85 per cent yield by passing BF_3 over B at $2000°C$; its life is even shorter than that of SiF_2. In Group III the extensive stereochemistry based on the formation of three coplanar bonds is characteristic of B alone; there are also many compounds in which four tetrahedral bonds are formed. For example, $B(CH_3)_3$ unites with NH_3 to form a stable solid $B(CH_3)_3 . NH_3$ which can be distilled without decomposition, and the halides form similar compounds. In this respect the aluminium halides behave similarly with organic oxy-compounds; $AlCl_3$ reacts with many ketones to give $R_2C=O . AlCl_3$, and also forms volatile esters like boron, for example, $Al(OC_2H_5)_3$.

The chemical, though not structural, resemblance of boron to silicon is marked. Its halides, like those of Si but unlike those of C, are easily hydrolysed, boric acid being formed. The acid HBF_4 is formed (together with boric acid) when BF_3 reacts with water, and $NaBF_4$ may be prepared from solutions of boric acid and $NaHF_2$. In the case of Si there is, of course, a change from SiF_4 to SiF_6^{2-}. Salts $M(BCl_4)$ can be prepared in anhydrous solvents and are stable only if M is a fairly large ion, for example, K^+, Rb^+, Cs^+, $N(CH_3)_4^+$, or PCl_4^+. The salts $N(C_2H_5)_4BBr_4$ and (pyridinium)BI_4 have been prepared in the liquid hydrogen halides and isolated as very hygroscopic white solids. Boron rivals silicon in the number and complexity of its oxy-salts. In these compounds B forms three coplanar or four tetrahedral bonds to O; in many borates there is both planar and tetrahedral coordination in the same anion. Since various borate ions are in equilibrium in borate solutions it is not possible to predict the formula of the borate which will crystallize from a solution or be formed by double decomposition. Anhydrous borates with a particular composition are therefore made by fusion of a mixture of the metal oxide (or carbonate) with B_2O_3 in the appropriate proportions. Here again borates resemble silicates, for many borates made in this way supercool to glasses (like B_2O_3 itself). However, the formation of series of borates containing OH groups is not paralleled by silicon. We deal later with boric acids and borates.

Boron monoxide is prepared in a pure state by heating the element with the trioxide to $1350°C$. At ordinary temperatures it forms an amorphous amber-coloured glass which reacts vigorously with water and alcohol. The molecular species in the vapour has been shown to be B_2O_2.[1] There is a second form of the monoxide which is a soft white hygroscopic solid (see Chart 24.1, p. 846). Unstable boron-oxygen species include B_2O (p. 847), BO_2[2] (formed by heating $ZnO + B_2O_3$ in a Knudsen cell), and the BO_2^- ion. The latter has been studied by introducing some KBH_4 into a pressed KBr disc and heating, when it is formed together with $B_3O_6^{3-}$ by oxidation due to included oxygen. The i.r. spectrum indicates a linear structure. The ion is not stable, its concentration decreasing appreciably after a few months.[3]

The sulphide B_2S_3 is formed by the decomposition *in vacuo* at $300°C$ of HBS_2 which in turn is produced by the action of H_2S on B at $700°C$. Mass spectrometric study of the vapour of B_2S_3 shows the presence of many polymeric species.[4] Like B_2O_3 it usually forms a glass, but it has been crystallized though its complex structure has not yet been determined.[5]

There is an extensive organo-boron chemistry. Among the simpler types of molecule are:

(1) JCP 1956 **25** 498; JPC 1958 **62** 490

(2) CJP 1961 **39** 1738

(3) IC 1964 **3** 168

(4) JACS 1962 **84** 3598

(5) IC 1970 **9** 1776

$RB(OH)_2$, $R_2B(OH)$, $\overset{R}{\underset{R}{>}}B-O-B\overset{<R}{\underset{R}{}}$, $(OH)_2B-\bigcirc-B(OH)_2$, and

Compounds $RB(OH)_2$ polymerize to cyclic trimers only, in contrast to $R_2Si(OH)_2$ which give polymeric linear siloxanes; compare the formation of only one cyclic metaborate ion, $B_3O_6^{3-}$. The formation of 6-membered rings is a prominent feature of boron chemistry, as in boroxine (p. 862) and the numerous substituted boroxines (boroxoles), (a), B_3N_3 rings, (b), and B_3S_3 rings, (c):

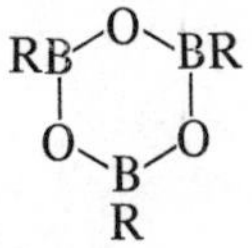

(a) R = H, F, OH, CH_3, OCH_3 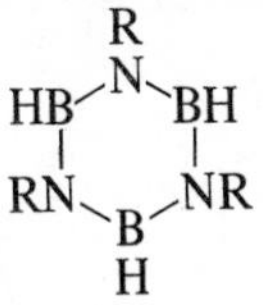   (b) 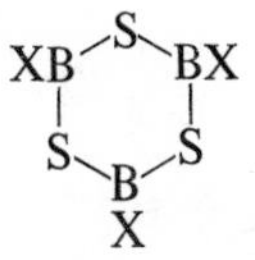   (c) X = Cl, Br, SH

(d)

The 6-membered B_3O_3 ring also occurs in many more complex borate anions. Cyclic boron–nitrogen molecules are considered later in this chapter. The cyclic molecule (c), X = SH, is the predominant species in the vapour of HBS_2 at temperatures below $100°C$.[6] In the crystalline tribromo compound of type (c), $(BrBS)_3$, the molecular structure (d) was found.[7]

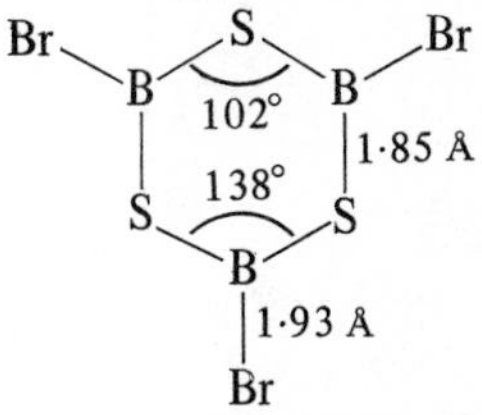

(6) JACS 1966 **88** 2935
(7) SPC 1959 **3** 569

The stereochemistry of boron

The stereochemistry of boron is simple in many of its compounds with the halogens (but note B_8Cl_8), oxygen, nitrogen, and phosphorus, three coplanar or four tetrahedral bonds being formed. The more complex stereochemistry of the element in electron-deficient systems (elementary B, some borides, and the boranes) is dealt with separately. The planar arrangement of three bonds from a B atom has been demonstrated in many molecules BX_3 and BR_3 (Table 24.1), in cyclic molecules of the types noted in the previous section, in many oxy-ions (see later section), and in crystals such as the graphite-like form of BN (p. 847) and AlB_2 (p. 842).

TABLE 24.1

The structures of planar molecules BX_3 and BR_3

Molecule	B—X	Other data	Reference
BF_3	1·31 Å		JCP 1968 **48** 1571
BCl_3	1·74		BCSJ 1966 **39** 1134
BBr_3	1·87		JACS 1937 **59** 2085
BI_3	2·10		IC 1962 **1** 109 (crystal)
HBF_2	1·31	B–H, 1·19 Å, F–B–F, 118°	JCP 1968 **48** 1
$B(CH_3)_3$	–	B–C, 1·58 Å	JCP 1965 **42** 3076
$B(CH_3)_2F$	1·29		JACS 1942 **64** 2686
$B(OCH_3)_3$	–	B–O, 1·38 Å	JACS 1941 **63** 1394

The tetrahedral arrangement of four bonds from B was established by the methods of classical stereochemistry. A molecule such as that of borosalicylic acid (I) would possess a plane of symmetry if the boron bonds were coplanar. The optical activity of this molecule was demonstrated by its resolution into optical antimers using strychnine as the active base. This really only eliminates a strictly coplanar configuration of the complex (or other configuration possessing $\bar{n}$ symmetry, as explained in Chapter 2). More recently the optically active cation (II)

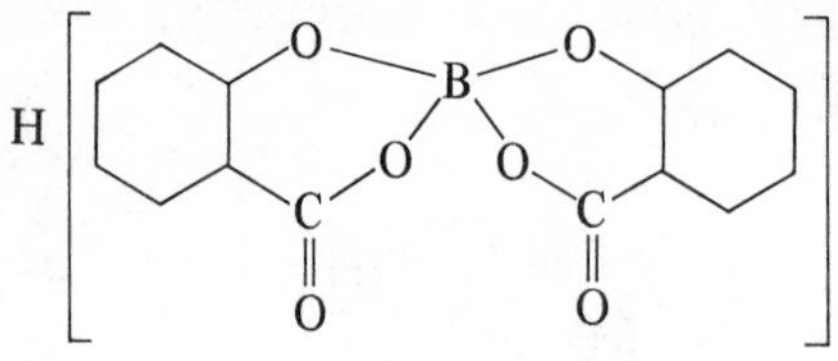

(I) 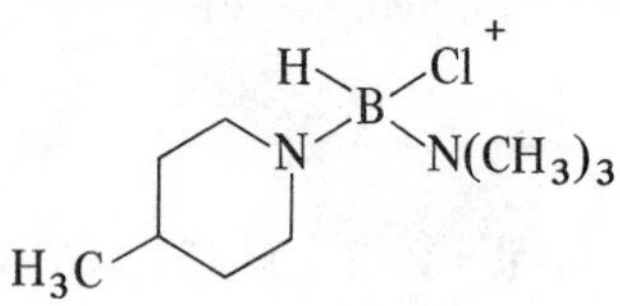   (II)

(1) IC 1971 **10** 667

was resolved by hand selection of crystals, and the detailed structure of the analogous bromo cation has been determined in the crystalline hexafluoro-phosphate.[1]

The simplest groups containing B forming four tetrahedral bonds are the ions BX_4^-. Some fluoroborates are isostructural with oxy-salts containing tetrahedral ions, for example, $NaBF_4$ with $CaSO_4$, $CsBF_4$ and $TlBF_4$ with $BaSO_4$, and

(2) ZaC 1966 **344** 279
(3) AC 1971 **B27** 1102

$(NO)BF_4$ with $(NO)ClO_4$. The B–F bond length is 1·41 Å in $NaBF_4$[2] and 1·406 Å in NH_4BF_4[3]; it is interesting that there is no hydrogen bonding in the ammonium salt, the shortest N–H–F distance being 2·93 Å (compare 2·69 Å in NH_4F). The ion BF_3OH^- occurs in $H(BF_3OH)$ and $(H_3O)(BF_3OH)$, originally

(4) AC 1964 **17** 742
(5) AC 1957 **10** 199

respectively formulated as $BF_3 . H_2O$ and $BF_3 . 2 H_2O$,[4] and the tetrahedral ion BOF_3^{2-} (B–F, 1·43 Å, B–O, 1·435 Å) in salts such as $BaBOF_3$.[5] The $B(OH)_4^-$ ion and many examples of tetrahedral BO_4 coordination groups are mentioned under boron–oxygen compounds later in this chapter.

Simple tetrahedral molecules include $H_3B . N(CH_3)_3$, borine carbonyl,

(6) PR 1951 **78** 1482
(7) IC 1969 **8** 836

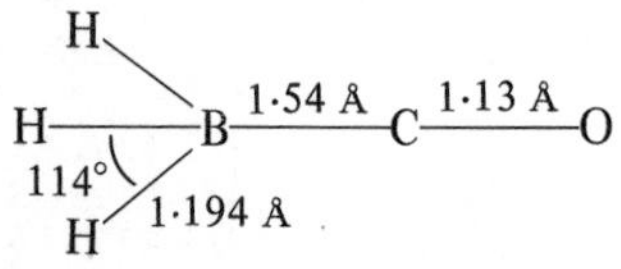

$H_3B . CO$,[6] (a), the numerous adducts of BF_3, and the remarkable molecule $B_4F_6 . PF_3$,[7] (b). This compound is formed as a very reactive colourless solid when the high-temperature species BF is condensed on to a cold surface with PF_3.

(a) 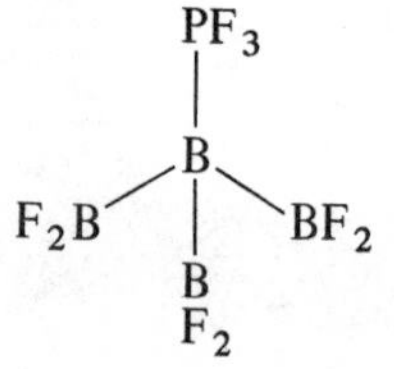   (b)

This molecule is of special interest as containing one B atom bonded tetrahedrally to 1 P (B–P, 1·83 Å) and to 3 B (B–B, 1·68 Å), these B atoms forming three coplanar bonds (B–F, 1·31 Å, P–F, 1·51 Å).

Many molecules and crystals containing equal numbers of B and N (or P) atoms form structures similar to those of the corresponding compounds of Group IV elements. Examples include BN and BP with the diamond structure, the silica-like structures of BPO_4 already mentioned, and analogues of substituted cyclohexanes such as $[H_2B . N(CH_3)_2]_3$ with the same chair configuration as C_6H_{12}.

Elementary boron and related borides

Boron is notable for the structural complexity of its polymorphs, some of which are almost as hard as diamond. Decomposition of boranes or BI_3 between 800–1200°C yields α rhombohedral boron (B_{12}); the subscript number is the number of atoms in the unit cell. This form is considered to be thermodynamically unstable relative to the β rhombohedral B_{105} which crystallizes from any fairly pure molten boron or by recrystallizing the element at any temperature above 1500°C and is the one polymorph that is readily available. The reaction between BBr_3 and H_2 over a Ta filament at 1200–1400° gives α tetragonal B_{50}. Massive (glassy) boron and the thin filaments which have high tensile strength are apparently nearly 'amorphous', giving only two diffuse X-ray diffraction rings. All three known boron structures and also those of a number of boron-rich borides contain icosahedral B_{12} groups with, in some cases, additional B atoms which form an essential part of the framework.

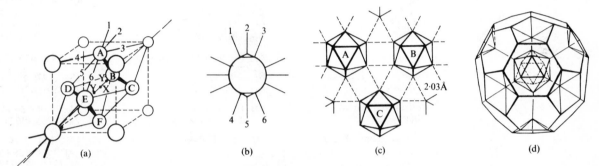

FIG. 24.1. The crystal structures of α-B_{12} and β-B_{105}. (a) and (b), packing of B_{12} units in α-B_{12}. (c) Section through the structure of α-B_{12} perpendicular to a 3-fold axis showing the 3-centre bonds. (d) The B_{84} unit of β-B_{105} showing the central icosahedron and six only of the 'half-icosahedra', one of which is attached to each B atom of the central B_{12} group.

In α B_{12} all the B atoms belong to icosahedral groups which are arranged in approximately cubic closest packing. This type of packing may be referred to a rhombohedral unit cell with interaxial angle close to 60°. If B_{12} icosahedra are placed at all the points of the rhombohedral lattice of Fig. 24.1(a) 6 B atoms of each (those of two opposite faces) can bond to a B atom of another icosahedron by a B–B bond of length 1·71 Å. These bonds lie along the lattice directions (cell edges) and result in a 3D framework in which each icosahedron is connected to six others; the directions of these bonds correspond ideally to certain of the 5-fold axes of the icosahedron (Fig. 24.1(b)). The remaining six B atoms of an icosahedron, which form a buckled hexagon around its equator, cannot bond in this way, the shortest contacts in the equatorial plane ABC (shaded) being 2·03 Å. The bonding in this plane is by means of '3-centre' bonds, as shown in (c).

This framework can also serve as the basis of the structures of certain borides. Groups of atoms YXY can be placed along the body-diagonal of the cell of Fig. 24.1(a), for at the centre of the cell there is an octahedral hole surrounded by the

837

icosahedra ABCDEF. In B_4C these are linear C–C–C groups, and the 3-centre bonding in the equatorial plane is replaced by bonds between the six equatorial B atoms of the B_{12} groups and the terminal C atoms of the chains. One-half of the B atoms therefore form bonds to 6 B (as in α-B_{12}) and the remainder are bonded to 5 B + 1 C. Later work suggests that the only congruently melting borocarbide has the composition $B_{13}C_2$, that is B_{12}. CBC, and it seems probable also that B_4C is $(B_{11}C)CBC$ rather than $(B_{12})C_3$; this field is still being actively studied. In AlC_4B_{40} there are apparently linear CBC and CBB chains and also non-linear C(Al)B chains. (References to these compounds are given in Table 24.2.)

TABLE 24.2

Polymorphs of boron; icosahedral borides

Structure	Units	Reference
α-B_{12} (rhombohedral)	B_{12} (c.c.p.) only	AC 1959 **12** 503
B_4C	$B_{11}C$; chains CBC	CR 1963 **257** 3927
$B_{13}P_2$	B_{12}; chains PBP	AC 1962 **15** 1048
$B_{2.89}Si$	$(B_{10.3}Si_{1.7})(Si_2)_3$	ASc 1962 **16** 449
AlC_4B_{40}	B_{12}; chains CBC, CBB, CAlB	AC 1970 **B26** 315
$Al_{2.1}C_8B_{51}$	B_{12} (h.c.p.); CBC (CAlC) chains	AC 1969 **B25** 1223
NaB_{15}	B_{12}, 3 B, Na	JSSC 1970 **1** 150
$MgAlB_{14}$	B_{12}, 2 B, Mg, Al	AC 1970 **B26** 616
α-B_{50} (tetragonal)	$(B_{12})_4$, 2 B	JACS 1958 **80** 4507
BeB_{12}	B_{12}, Be	ZaC 1960 **306** 266
β-B_{105} (rhombohedral)	$(B_{84})(B_{10}B . B_{10})$	JSSC 1970 **1** 268; AC 1970 **B26** 1800
YB_{66}		AC 1969 **B25** 237

There are many compounds with the same general type of structure as B_4C and rhombohedral B_{12} though the details of their structures are not in all cases certain. The C_3 or CBC chains may be replaced by S in $B_{12}S$, by P–B–P in $B_{13}P_2$ (B–P in chain, 1·87 Å, P–$B_{icos.}$, 2·03 Å), O–B–O in $B_{13}O$, or by 2 Si in $B_{2.89}Si$. Since there would appear to be room for only 2 Si replacing 3 C there is apparently some (statistical) replacement of B by Si in the icosahedra, giving a composition close to $B_{31}Si_{11}$ (i.e. $B_{31}Si_5$. Si_6 or $B_{2.89}Si$).

There is a closely related (orthorhombic) phase $Al_{2.1}C_8B_{51}$ (variously described as AlB_{10}, AlB_{12}, or AlC_4B_{24}) in which there is distorted *hexagonal* closest packing of B_{12} icosahedra and linear CBC chains, in some of which Al replaces B. As in B_4C each B_{12} group is directly bonded to six others via B–B bonds (1·77 Å) so that six B atoms of each icosahedron form 6 B–B bonds and the remainder are bonded to 5 B + C. The bond lengths are similar to those in B_4C:

Bond	$Al_{2.1}C_8B_{51}$	B_4C
B–B (intra-icosahedral)	1·816 Å	1·789 Å
B–B (inter-icosahedral)	1·774	1·718
$B_{icos.}$–C	1·623	1·604
B_{chain}–C	1·467	1·435

The close relation between these two structures is shown by the fact that the Al compound undergoes a topotactic transition by loss of Al at 2000°C to form B_4C, a process involving a change from h.c.p. to c.c.p. B_{12} icosahedra.

In the isostructural NaB_{15} and $MgAlB_{14}$ the B_{12} icosahedra are packed in layers which are stacked vertically above one another, so that only two direct pentagonal pyramidal contacts between icosahedra result (perpendicular to the layers). In the layers each B_{12} is bonded to four others at the same distance (1·75 Å) but there is also bonding through Mg, Al, and additional B atoms. The interstitial atoms obviously play an essential part in stabilizing this mode of packing of the icosahedra.

A feature of the polymorphs of boron is the large proportion of atoms which form a sixth bond (pentagonal pyramidal coordination) directed outwards along the 5-fold axis of the icosahedron. The fractions are:

Rhombohedral B_{12}	50%
Rhombohedral B_{105}	80%
Tetragonal B_{50}	64%

If all the B atoms of each icosahedron were to form a sixth bond to a B atom of another icosahedron there would be formed a radiating icosahedral packing which is not periodic (Chapter 4). The structure of rhombohedral B_{105} may be visualized as built of units $(B_{84})(B_{10}BB_{10})$. The B_{84} unit consists of an icosahedron bonded to 12 half-icosahedra, one attached, like an umbrella, to each vertex (Fig. 24.1(d)). These units are placed at the points of the rhombohedral lattice of Fig. 24.1(a) so that six of the half-icosahedra of each B_{84} meet at the mid-points of the edges of the cell forming further icosahedra. The basic framework therefore consists of icosahedra at the points of the lattice of Fig. 24.1(a) each connected to six others situated at the mid-points of the cell edges. These icosahedra are directly bonded only to two others. The remaining six half-icosahedra of a B_{84} unit do not juxtapose to form icosahedra but together with the B_{10} units provide further bonding in the equatorial plane—compare the 3-centre bonds in B_{12} and the bonding through C atoms in B_4C. Along the diagonal of the cell, replacing the CBC chain in B_4C, are $B_{10}BB_{10}$ groups, there being a unique B atom at the body-centre of the cell.

It is convenient to mention here the structure of YB_{66} which is extremely complex, with 1584 B and 24 Y atoms in the unit cell, but has high (cubic) symmetry. The majority of the B atoms (1248) are contained in units consisting of 13 icosahedra (156 atoms), twelve around a central one, which are further linked by B–B bonds (1·62–1·82 Å) into a 3D framework. The grouping of twelve icosahedra around a central one represents the beginning of a radiating packing which, as already noted, would not be periodic in three dimensions. The packing of the B_{156} units already leaves large channels in which the remaining B and the Y atoms are accommodated.

The structure of tetragonal B_{50} is comparatively simple. There are 50 atoms in the unit cell of which 48 form four nearly regular icosahedral groups which are linked together both directly and through the remaining B atoms. The latter form tetrahedral bonds, but the atoms of the icosahedra form six bonds, five to their neighbours in the B_{12} group and a sixth to an atom in another B_{12} or to an 'odd' B atom. Although all the icosahedral B atoms form 'pentagonal pyramidal' bonds, for 8 of the 12 B atoms there is considerable deviation $(20°)$ of the external bond from the 5-fold axis of the icosahedron. The bond lengths in this structure are $1·60$ Å for the 4-coordinated B, $1·68$ Å for bonds between B_{12} groups and $1·81$Å within the B_{12} groups. Several borides have been prepared which have the same basic structure as tetragonal B_{50}, for example, NiB_{25} and BeB_{12}. The metal atoms clearly occupy some of the numerous holes in the B_{50} structure.

Metal borides and borocarbides

We deal here with borides and borocarbides other than those containing icosahedral B_{12} groups. Like carbides and silicides they can be prepared by direct union of the elements at a sufficiently high temperature; alternatively some borides and silicides result from reduction of borates or silicates by excess metal. As in the case of silicides it is not possible to account for the formulae of borides in terms of the normal valences of the metals. Thus we have NaB_6, CaB_6, and YB_6; yttrium also forms YB_2, YB_4, YB_{12}, and YB_{66}, and chromium, for example, forms CrB, Cr_3B_4, CrB_2, and CrB_4.

We have described the structures of some boron-rich borides in which there are extensive 3D systems of B—B bonds. At the other extreme there are crystalline borides with low boron content in which there are isolated B atoms, that is, B atoms surrounded entirely by metal atoms as nearest neighbours. With increasing boron content the B atoms link together to form first B_2 units, then chains, layers, or 3D frameworks extending throughout the whole crystal (Table 24.3).

Borides in which there are discrete B atoms include Be_4B and Ni_3B (and Co_3B) with the cementite (Fe_3C) structure. In this structure the metal atoms form slightly corrugated layers of the kind shown in Fig. 24.2(a). The B atoms lie between the layers with six nearest neighbours (B—Ni, $2·04$ Å) at the apices of a trigonal prism. These metal layers are closely related to the layers (parallel to $10\bar{1}1$) in hexagonal close-packing, as shown in the figure. Next we have Be_2B with the fluorite structure and the numerous borides M_2B with the $CuAl_2$ structure described and illustrated on p. 1046. In this structure the metal atoms form layers of the type of Fig. 24.2(b). Successive layers are rotated through $45°$ relative to one another and the B atoms are at the centres of antiprisms formed by two square groups in adjacent layers. In Fe_2B, for example, B has eight Fe neighbours at $2·18$ Å, and the two nearest B atoms are at $2·12$ Å; there are no B—B bonds.

Isolated pairs of B atoms occur in the structures of certain double borides $M_2'M''B_2$ in which M' is a large atom (Mo, Ti, Al, or a mixture of these) and M'' is a small atom (Cr, Fe, Ni). Here again the large atoms form the 5-connected net of Fig. 24.2(b) and the M'' and B atoms lie between the layers, M'' at the centres of tetragonal prisms and B at the centres of pairs of adjacent trigonal prisms (Fig. 24.2(c)). In the B_2 groups the B—B bond length is $1·73$ Å.

TABLE 24.3
The crystal structures of metal borides

Nature of B complex	Structure	Examples	Reference
Isolated B atoms	Be_4B		ZaC 1962 **318** 304
	Fe_3C	Ni_3B, Co_3B	ACSc 1958 **12** 658
	$CuAl_2$	$Mn_2B, Fe_2B, Co_2B, Ni_2B,$	ACSc 1949 **3** 603;
		Mo_2B, Ta_2B	ACSc 1950 **4** 146
	CaF_2	Be_2B	DAN 1955 **101** 97
	Rh_2B		AC 1954 **7** 49
B_2 groups	$M_2'M''B_2$	$(Mo, Ti, Al)CrB_2$	AC 1958 **11** 607
Single chains	CrB	NbB, TaB, VB	AC 1965 **19** 214
	FeB	CoB, MnB, TiB, HfB	AC 1966 **20** 572
	MoB	WB	ACSc 1947 **1** 893
	Ni_4B_3		ACSc 1959 **13** 1193
	$Ru_{11}B_8$		ACSc 1960 **14** 2169
Double chains	Ta_3B_4	$Cr_3B_4, Mn_3B_4, Nb_3B_4$	ACSc 1961 **15** 1178
Layers	AlB_2	M = Ti, Zr, Nb, Ta, V, Cr, Mo, U, Mg	AC 1953 **6** 870; ACSc 1959 **13** 1193
	ϵ—Mo—B	ϵ—W—B phase	ACSc 1947 **1** 893
3D frameworks	CrB_4		ACSc 1968 **22** 3103
	UB_4	M = Y, La, 4f, Th	AC 1953 **6** 269; JACS 1958 **80** 3479
	CaB_6	M = Sr, Ba, Y, La, 4f, Th, U	IC 1963 **2** 430; JSSC 1970 **2** 332
	UB_{12}	M = Y, Zr, 4f	AC 1965 **19** 1056

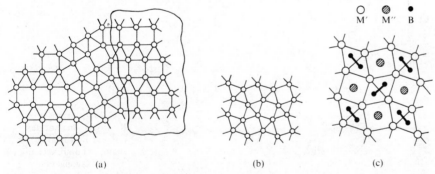

FIG. 24.2. (a) Layers of Ni atoms in Ni_3B. At the right is shown a portion of a layer parallel to the plane (10$\bar{1}$1) in hexagonal close-packing. (b) Layer of metal atoms in borides with the $CuAl_2$ structure. (c) The structure of borides $M_2'M''B_2$.

The two polymorphs of Ni_4B_3 form a link between structures such as Ni_3B with isolated B atoms and those with chains of B atoms. In one form of Ni_4B_3 two-thirds of the B atoms form zigzag chains, the remainder being isolated B atoms, while in the other polymorph all the B atoms are in chains.

Next come three closely related structures (FeB, CrB, and MoB) in which there are zigzag chains of boron atoms. In FeB (and the isostructural CoB) each B atom is surrounded by 6 Fe at the vertices of a trigonal prism (Fe—B, 2·12–2·18 Å), but its nearest neighbours are two B atoms (at 1·77 Å) as shown in Fig. 24.3(a). The B atoms therefore form infinite chains.

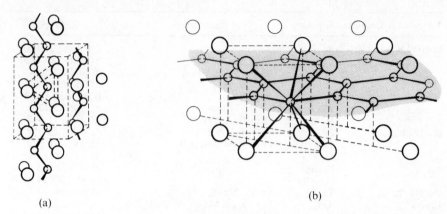

(a)

(b)

FIG. 24.3. The crystal structures of (a) FeB, and (b) AlB_2. The smaller circles represent B atoms.

In borides of the type M_3B_4 the boron atoms form double chains in which the central bonds are apparently much stronger than those along the length of the chain (Fig. 24.4). The Ta_3B_4 structure is in this way intermediate between the chain and layer borides.

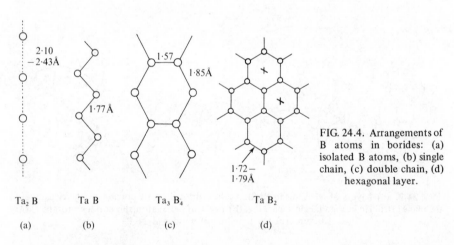

2·10
— 2·43Å

1·57

1·85Å

1·77Å

1·72 —
1·79Å

Ta_2B

$Ta B$

$Ta_3 B_4$

$Ta B_2$

(a)

(b)

(c)

(d)

FIG. 24.4. Arrangements of B atoms in borides: (a) isolated B atoms, (b) single chain, (c) double chain, (d) hexagonal layer.

In AlB_2 the B atoms are arranged in layers with layers of Al interleaved between them. The structure of a B layer is the same as that of a layer in the graphite structure (p. 734). Each B is here equidistant from three other B atoms (at 1·73 Å), the next nearest neighbours being a set of six Al at the vertices of a trigonal prism (Fig. 24.3(b)). Most of the 'layer' borides have the AlB_2 structure, but in the Mo—B and W—B systems there are also phases with the ideal composition M_2B_5, apparently with additional B atoms at the points X of the hexagonal nets of Fig. 24.4(d). The tantalum–boron system provides examples of four types of boride structure (Fig. 24.4); compare the silicon complexes in metallic silicides.

With higher boron–metal ratios some extremely interesting 3-dimensional boron frameworks are found. In CrB_4 the B atoms form a very simple 4-connected framework. Square groups of B atoms (B–B, 1·68 Å) at different heights are linked by somewhat longer bonds (1·91 Å) as shown in Fig. 24.5. Each B forms two bonds of each type to other B atoms, while Cr has 2 Cr and 12 B as nearest neighbours. In CaB_6 (and numerous isostructural compounds) the boron network consists of octahedral B_6 groups joined together as shown in Fig. 24.6. Alternatively this is a 5-connected net the links of which are the edges of a space-filling array of octahedra and truncated cubes. There are large holes surrounded by 24 B atoms which accommodate the metal atoms (Ca–B, 3·05 Å). The B–B distances of about 1·7 Å are similar to those in the FeB and AlB_2 structures.

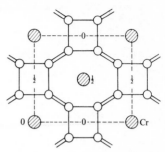

FIG. 24.5. The crystal structure of CrB_4.

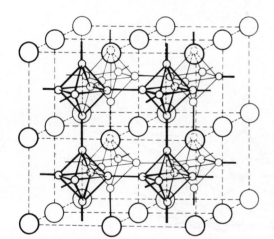

FIG. 24.6. The crystal structure of CaB_6.

The UB_4 structure is of interest because of the geometrical relationships with the structures of borides MB_2 and MB_6. In the AlB_2 structure there are close-packed metal layers in which each M has six equidistant neighbours, and these layers are superposed directly above one another so that the B atoms in the holes between the layers are surrounded by six M at the vertices of a trigonal prism (Figs. 24.3(b) and 24.7(a)). In CaB_6 there is simple cubic packing of the M atoms (Fig. 24.7(c)) so that there are holes between groups of eight M atoms situated at the vertices of a cube. The CaB_6 structure can therefore be described as a simple cubic M lattice expanded by the introduction of the octahedral B_6 groups, though since these are joined to neighbouring B_6 groups there is a continuous 3-dimensional B framework. In UB_4 the metal atoms are in layers in which every M has five neighbours, and between the layers there are both 'trigonal prism' and 'cubic' holes, as may be seen from Fig. 24.7(b). In the holes of the first type there are single B atoms and in the latter B_6 groups, and since there are pairs of adjacent holes of the former type the actual B network is that shown in Fig. 24.7(d), with B_6 groups joined through pairs of B atoms. In this structure some B atoms are linked to three B, others to five B, and the B—B distances are in the range $1 \cdot 7 - 1 \cdot 8$ Å. Although the B—B distances in the rigid boron framework structures are consistent with $r_B = 0 \cdot 86$ Å, the metal–boron distances suggest larger effective radii—from $0 \cdot 86$ to $1 \cdot 10$ Å. Yet another way of describing the CaB_6 structure is as a CsCl-like packing of M atoms and B_6 groups. We noted in Chapter 22 that the bonding is not

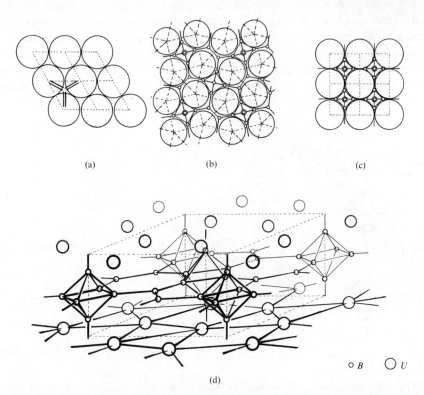

(a) (b) (c)

(d)

∘ B ◯ U

FIG. 24.7. The crystal structures of metallic borides: (a) close-packed metal layer in UB_2 (AlB_2) with B in trigonal prism holes between layers, (b) metal layer in UB_4 with positions of 6- and 8-coordination between layers, (c) metal layer in CaB_6 (ThB_6) structure, (d) the UB_4 structure.

of the same type in all carbides MC_2; this is also true in borides MB_6. These have been prepared containing metals in various oxidation states, namely, 1 (KB_6), 2 (Ca, Sr, Yb), and >2 (La, Th). The electronic properties of KB_6 have not yet been studied, but compounds of the second group are semiconductors and those of the third are metallic conductors.

The boron framework in UB_{12} is illustrated in Fig. 3.47(d) (p. 116) as a space-filling arrangement of cuboctahedra, truncated octahedra, and truncated tetrahedra, the edges of which form a 5-connected 3D net. The U atoms are situated at the centres of the largest cavities in positions of 24-coordination—compare the

24-coordination, truncated cubic, of the metal atoms in the CaB$_6$ structure. Alternatively the structure may be described as a NaCl-like packing of B$_{12}$ groups and U atoms. The cuboctahedral B$_{12}$ groups may be contrasted with the icosahedral groups in the polymorphs of elementary B and related borides. The B–B bonds in the B$_{12}$ groups are appreciably shorter (1·68 Å) than those linking the groups (1·78 Å in ZrB$_{12}$, 1·80 Å in UB$_{12}$, 1·81 Å in YB$_{12}$).

Numerous borocarbides have been prepared, for example, those of Al already described and compounds such as Mo$_2$BC,[1] in which C is octahedrally co-ordinated by Mo and B has trigonal prismatic coordination and forms chains as in CrB. Of special interest are the borocarbides MB$_2$C$_2$ formed by Sc and the 4f elements. The metal atoms are situated between plane 3-connected layers which in the 4f compounds are the 4 : 8 net (with alternate B and C atoms) but in ScB$_2$C$_2$ are the 5 : 7 net, necessarily with pairs of adjacent B and C atoms. The Sc atoms are in pentagonal prism holes between the layers, but there are also Sc–Sc distances very nearly the same as in h.c.p. Sc metal, so that Sc has (14 + 5)-coordination (C–C, 1·45, B–B, 1·59, B–C, 1·52–1·61 Å).[2]

(1) AC 1969 **B25** 698

(2) AC 1965 **19** 668

Lower halides and diboron compounds

Other compounds in which there are B–B bonds include the halides B$_2$X$_4$ (all four of which have been prepared), B$_4$Cl$_4$ and B$_8$Cl$_8$, and diboron compounds. Salts containing the B$_2$Cl$_6^{2-}$ ion include [(CH$_3$)$_4$N]$_2$B$_2$Cl$_6$, prepared in liquid an-hydrous HCl, and (PCl$_4$)$_2$B$_2$Cl$_6$. Methods of preparation and relations between some of the diboron compounds are indicated in Chart 24.1.

Halides B$_2$X$_4$
The fluoride, B$_2$F$_4$ is an explosive gas. The structure of the planar, centro-symmetrical molecule, (a), has been determined in the crystal at −120°C.[1] The chloride B$_2$Cl$_4$ is formed by passing BCl$_3$ vapour through an electric discharge

(1) JCP 1958 **28** 54

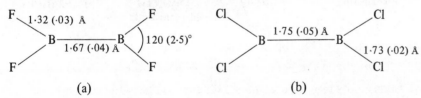

(a)   (b)

between mercury electrodes. It is a colourless liquid which decomposes at temperatures above 0°C into a mixture of the yellow crystalline B$_4$Cl$_4$, red B$_8$Cl$_8$, and the paramagnetic B$_{12}$Cl$_{11}$.[2] Among its derivatives are esters B$_2$(OR)$_4$ and the ethylene compound Cl$_2$B . C$_2$H$_4$. BCl$_2$ mentioned later. The compounds B$_2$Br$_4$ and B$_2$I$_4$ (a yellow crystalline compound) are formed by methods similar to that used for the chloride.

(2) IC 1963 **2** 405

The crystal structure of B$_2$Cl$_4$ has been determined (at −165°C) and it is found that the molecule is planar with bond angles close to 120°, (b).[3] The B–Cl bond is normal but B–B is long compared with that expected for a single bond (around 1·6 Å)—compare the long bond in N$_2$O$_4$ and the normal single bond in P$_2$I$_4$. In the vapour, however, the B$_2$Cl$_4$ molecule has the 'staggered' configuration, the planes

(3) JCP 1957 **27** 196

(4) JCP 1969 **50** 4986

(5) JCP 1965 **43** 503

of the two halves being approximately perpendicular (B–B, 1·70, B–Cl, 1·75 Å, Cl–B–Cl, 119°).[4] The eclipsed configuration in the crystal may result from better packing of the molecules, for m.o. calculations show that the staggered form is more stable than the eclipsed by at least 4 kJmol^{-1}.[5]

CHART 24.1

Preparation of diboron compounds †

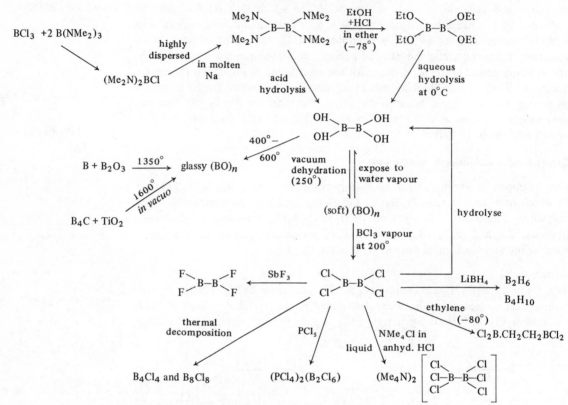

† See, for example: JACS 1954 **76** 5293; JACS 1960, **82**, 6242, 6245; JACS 1961 **83** 1766, 4750: PCS 1964 242.

(6) AC 1956 **9** 668

The molecule $Cl_2B . C_2H_4 . BCl_2$ is also planar to within the limits of experimental error, though the bond lengths are only approximate:[6]

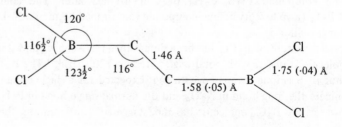

The B–C bond length may be compared with the (approximate) value 1·52 (0·07) Å found in $C_6H_5 . BCl_2$.[7]

Halides B_4X_4 and B_8X_8

Unlike halides B_2X_4 these halides are electron-deficient. The structures of the B_4Cl_4 and B_8Cl_8 molecules in the crystalline state have been determined. In B_4Cl_4[8] a nearly regular tetrahedron B_4 is surrounded by a tetrahedral group of 4 Cl, also nearly regular (Fig. 24.8(a)). The B–Cl bonds (1·70 Å) are single, but the length of the B–B bonds (also 1·70 Å) corresponds to a bond order of about $\frac{2}{3}$, assuming a boron radius of 0·8 Å and using Pauling's equation (p. 1025). Assuming that one electron of each B is used for a single B–Cl bond there remain 8 electrons (4 electron pairs) for the 6 B–B bonds, as shown diagrammatically in the figure. The molecule B_8Cl_8 also consists of a polyhedral boron nucleus to each B atom of which is bonded a Cl atom. The polyhedron (Fig. 24.8(b)) is closer to a dodecahedron (bisdisphenoid) than to a square antiprism, but there is a considerable range of B–B bond lengths. The four longer B–B bonds (a) range from 1·93 to 2·05 Å and the remainder from 1·68 to 1·84 Å. The mean B–Cl bond length is 1·74 Å.[9] Since there are 8 electron pairs available for B–B bonds in this polyhedron with 18 edges, of which four are appreciably longer than the other fourteen, no simple description of the bonding is possible.

(7) JPC 1955 **59** 193
(8) AC 1953 **6** 547
(9) AC 1966 **20** 631

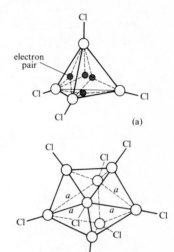

FIG. 24.8. The molecular structures of (a) B_4Cl_4, (b) B_8Cl_8.

Boron–nitrogen compounds

Boron nitride

Boron nitride, BN, is made by the action of nitrogen or ammonia on boron at white heat and in other ways. In the crystalline state it is very inert chemically, though it can be decomposed by heating with acids, and is described as melting under pressure at about 3000°C. The crystal structure of one form is very closely related to that of graphite (Fig. 21.3, p. 735) being built of hexagonal layers of the same kind, but in BN these are arranged so that atoms of one layer fall vertically above those of the layer below (Fig. 24.9). The B–N bond length is 1·446 Å.[1] In spite of the structural resemblance to graphite the physical properties of BN are very different from those of graphite. It is white, a very good insulator, and its diamagnetic susceptibility is very much smaller than that of graphite. Like graphite BN can be prepared with a turbostratic (unordered layer) structure, which can be converted to the ordered hexagonal structure by suitable heat treatment.[2] BN also crystallizes with the zinc-blende and wurtzite[3] structures. In cubic BN the length of the B–N (single) bond is 1·57 Å.[4]

Also isoelectronic with graphite is B_2O, prepared by reducing B_2O_3 with B or Li at high temperatures under high pressure. (In the 'tetrahedral anvil' pressures of 50–75 kbar and temperatures of 1200–1800°C are reached.) The B and O atoms were not definitely located in the graphite-like structure of B_2O.[5]

(1) AC 1952 **5** 356

(2) JACS 1962 **84** 4619
(3) JCP 1963 **38** 1144
(4) JCP 1960 **32** 1569

(5) IC 1965 **4** 1213

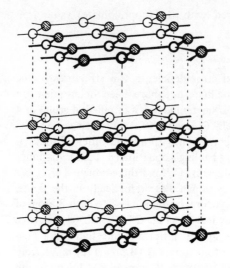

FIG. 24.9. The crystal structure of boron nitride, BN.

Boron—nitrogen analogues of carbon compounds

The simplest compounds of this kind are molecules

$$>N\!\!\overset{1\cdot40\ \text{Å}}{=\!=\!=}B< \qquad \text{and} \qquad >N\!\!\overset{1\cdot60\ \text{Å}}{-\!\!-}B<$$

(a)   (b)

which correspond to substituted ethylenes and ethanes. In (a) both atoms form three coplanar bonds and the B—N bond length is $1 \cdot 40 \pm 0 \cdot 02$ Å. This length is close to values in one group of cyclic molecules to which we refer shortly and corresponds to a double bond. In (b) both atoms form tetrahedral bonds and the B—N bond is a single bond of length close to $1 \cdot 60$ Å. Examples include:

		B—N	*Reference*
(a)	$(CH_3)_2N—BCl_2$	$1 \cdot 38$ Å	IC 1970 9 2439
	$(CH_3)_2N—B(CH_3)_2$	$1 \cdot 42$	JCS A 1970 992
(b)	$(CH_3)_3N—BX_3$	$1 \cdot 58 - 1 \cdot 64$	AC 1969 **B25** 2338
			IC 1971 **10** 200

(All four halides of type (b) have been studied, but since there are differences of as much as $0 \cdot 05$ Å between the B—N bond length found in crystal and vapour molecule detailed discussion of bond lengths is probably premature.)

Cyclic molecules with alternate B and N atoms include 'aromatic' 4-, 6-, and 8-membered rings and saturated molecules with 4- and 6-membered rings. In the former B—N is close to $1 \cdot 42$ Å and in the latter, $1 \cdot 60$ Å. Examples are given in Fig. 24.10. In the molecule (a) the 4-ring is planar and the three bonds from each N are coplanar, but the shortness of the *exo*cyclic B—N bond ($1 \cdot 44$ Å, the same as in the ring) is not consistent with the fact that the planes NSi_2 are approximately

848

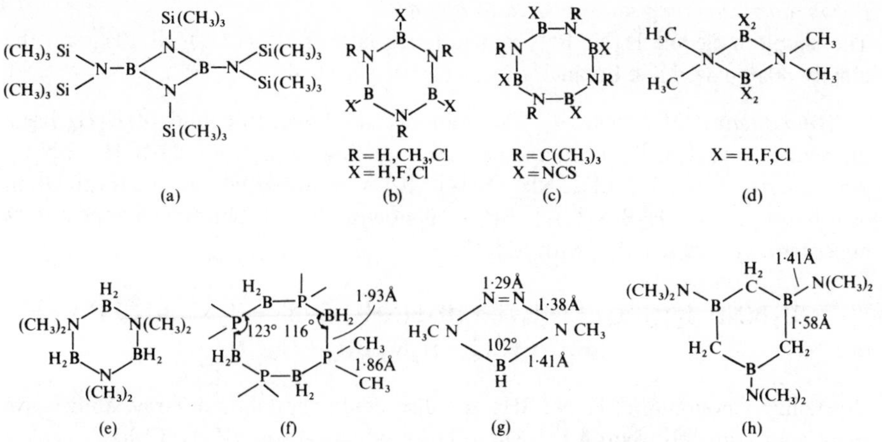

FIG. 24.10. Cyclic boron–nitrogen, boron–phosphorus, and boron–carbon molecules.

perpendicular to that of the 4-ring. (B–N, 1·45, N–Si, 1·75, Si–C, 1·87 Å).[1] Molecules of type (b) which have been studied include the symmetrical benzene-like molecule of borazine itself, $B_3N_3H_6$ (B–N, 1·44 Å),[2a] $B_3N_3Cl_6$[2b] (B–Cl, 1·76 Å, N–Cl, 1·73 Å), B-monoaminoborazine, $B_3N_4H_7$ (B–N_{ring}, 1·42, B–NH_2, 1·50 Å),[3] and B-trichloroborazine (B–N, 1·41, B–Cl, 1·75 Å).[4] (This compound is made by heating together BCl_3 and NH_4Cl at temperatures above 110°C—compare $(PNCl_2)_n$.) For references to numerous cyclic B–N compounds see reference (10). Compounds $(RN . BX)_4$ of type (c) have been prepared from BCl_3 and primary alkylamines and a number of derivatives have been made. In crystalline $[(CH_3)_3C . NB . NCS]_4$ both B and N form three coplanar bonds and there is a slight alternation in the B–N bond lengths in the boat-shaped ring:[5]

(1) AC 1969 **B25** 2342

(2a) IC 1969 8 1683
(2b) AC 1971 **B27** 1997
(3) JACS 1969 **91** 551
(4) JACS 1952 **74** 1742

(5) JCS 1965 6421

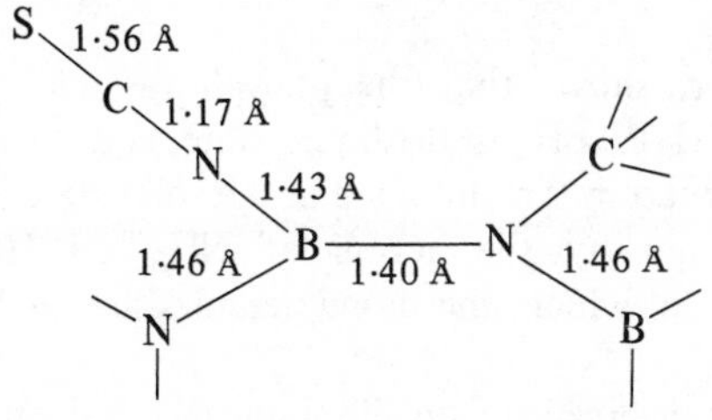

The ethylene-like molecules $R_2N . BX_2$ readily polymerize, and the dimers and trimers are the analogues of substituted cyclobutanes and cyclohexanes. Molecules of type (d) have planar 4-membered rings[6],[7] with B–N close to 1·60 Å, and the molecule (e) has the chair configuration.[8] The phosphorus analogue of (e) has been shown to have a similar configuration with B–P, 1·94 Å,[9a] and the molecule $[H_2B . P(CH_3)_2]_4$ has the structure (f).[9b] (The As compounds $[(CH_3)_2As . BH_2]_n$ (n = 3 and 4) have been prepared.)

Miscellaneous cyclic boron compounds include (g), with a 5-membered N_4B ring,[10] and (h), with a B_3C_3 ring and double bonds to the $N(CH_3)_2$ groups.[11]

(6) AC 1970 **B26** 1905
(7) JCS A 1966 1392
(8) AC 1961 **14** 273
(9a) AC 1955 8 199
(9b) JACS 1962 **84** 2457

(10) IC 1969 8 1677
(11) AC 1969 **B25** 2334

Boron

Boron–nitrogen compounds related to boranes

The simple molecule $H_3N \cdot BH_3$ is mentioned later. Both $(CH_3)_3N \cdot BH_3$ and the closely related aziridine borane[1] (a), have been studied.

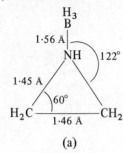

(a)

(1) JCP 1967 **46** 357

'Ammoniates' of boranes. The ammonia addition compound of B_2H_6 has a molecular weight in liquid ammonia corresponding to the formula $B_2H_6 \cdot 2\,NH_3$. An ionic structure $NH_4^+[BH_3 \cdot NH_2 \cdot BH_3]^-$ has been suggested, but the formulation as a borohydride, $[H_2B(NH_3)_2]^+(BH_4)^-$, is now preferred. This would appear to be consistent with its reaction with NH_4Cl:

$$[H_2B(NH_3)_2](BH_4) + NH_4Cl \rightarrow [H_2B(NH_3)_2]^+Cl^- + H_3N \cdot BH_3 + H_2$$

or

$$BH_4^- + NH_4^+ \rightarrow H_3N \cdot BH_3 + H_2$$

(2) JACS 1956 **78** 502, 503

(3) JACS 1959 **81** 3551

Crystalline (monomeric) $H_3N \cdot BH_3$ is a disordered crystal, and X-ray studies give B–N approximately 1·60 Å.[2] The other product, $[H_2B(NH_3)_2]Cl$ has a typical ionic crystal structure,[3] consisting of layers of $[H_2B(NH_3)_2]^+$ ions (b) interleaved with Cl^- ions.

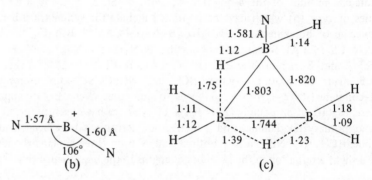

(b)   (c)

(4) IC 1966 **5** 723

(5) AC 1960 **13** 535

(High resolution n.m.r. shows that the phosphorus analogue of $N_3N \cdot BH_3$ is certainly (monomeric) $H_3P \cdot BH_3$ in the liquid state, and the i.r. and Raman spectra show that the same structure is maintained in the solid state.[4] The closely related $H_3B \cdot P(NH_2)_3$, prepared by the action of NH_3 on $H_3B \cdot PF_3$, consists of tetrahedral molecules in which the bond lengths are P–N, 1·65 Å, and P–B, 1·89 Å.)[5]

(6) JACS 1959 **81** 3538

The action of sodium amalgam on diborane in ethyl ether gives a product of composition NaB_2H_6 which is actually a mixture of $NaBH_4$ and NaB_3H_8. The latter reacts with NH_4Cl in ether at $25^\circ C$ to give $H_3N \cdot B_3H_7$. This is a white crystalline solid which has a disordered structure at temperatures above about $25^\circ C$; the structure of the molecule has been studied in the low-temperature form.[6] The framework of the molecule (c) consists of a triangle of B atoms to one of which the N is attached. The B_3H_7 portion may be regarded as a distorted fragment of tetraborane resulting from symmetrical cleavage of the double bridge. (The estimated accuracy of the B–H bond lengths is ±0·05 Å and of the other bond lengths ±0·005 Å.) For '$B_4H_{10} \cdot 2\,NH_3$' see under $B_3H_8^-$ ion.

850

Aminodiboranes. These compounds are derived from B_2H_6 by replacing one H of the bridge by NH_2. Electron diffraction studies of $H_2N \cdot B_2H_5$ and $(CH_3)_2N \cdot B_2H_5$ [7] give the following data:

(7) JACS 1952 **74** 954

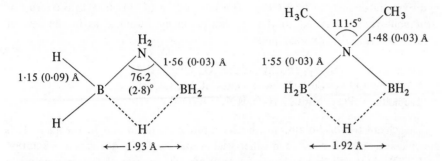

(In $H_2N \cdot B_2H_5$ the following distances and angles were assumed: N–H, 1·02, B–H$_{bridge}$, 1·35 Å, angles HNH, $109\frac{1}{2}°$, and HBH, 120°.) The B–N bonds are similar in length to those in the addition compounds of BF_3 described later, but the B–B distance is very much increased over the value (1·77 Å) in B_2H_6 owing to the interposition of the N atom.

The oxygen chemistry of boron

Boron, like silicon, occurs in nature exclusively as oxy-compounds, particularly hydroxyborates of calcium and sodium. In borates based exclusively on BO_3 coordination groups there would be a simple relation between the O : B ratio and the number of O atoms shared by each BO_3 group, assuming these to be equivalent and each bonded to 2 B atoms:

O : B ratio		Number of O atoms shared
3	Orthoborates: discrete BO_3^{3-} ions	0
$2\frac{1}{2}$	Pyroborates: discrete $B_2O_5^{4-}$ ions	1
2	Metaborates: cyclic or chain ions	2
$1\frac{1}{2}$	Boron trioxide:	3

Intermediate ratios (e.g. $1\frac{3}{4}$) would correspond to the sharing of different numbers of O atoms by different BO_3 groups, as in the hypothetical $(B_4O_7)_n^{2n-}$ chain analogous to the amphibole chain formed from SiO_4 groups. All of the above four possibilities are realized in compounds in which B is exclusively 3-coordinated, but there are two factors which complicate the oxygen chemistry of boron. First, there is tetrahedral coordination of B in many oxy-compounds, either exclusively or admixed with 3-coordination in the same compound. There is, therefore, no simple relation between O : B ratios and the structures of borates; we shall see later that a ratio such as 7 : 4 can be realized in a number of ways which, incidentally, do not include that mentioned above. Second, there are many hydroxyborates containing

Boron

OH bonded to B as part of a 3- or 4-coordination group; this is in marked contrast to the rarity of hydroxysilicates. Examples are known of all 4-coordination groups from BO_4 through $BO_3(OH)$ and $BO_2(OH)_2$, both of which occur in $CaB_3O_4(OH)_3 . H_2O$, and $BO(OH)_3$ (in $Mg[B_2O(OH)_6]$), to $B(OH)_4$ in tetrahydroxyborates such as $NaB(OH)_4$. The recognition of this fact, as the result of structural studies, has led to the revision of many formulae as in the case of antimonates (p. 719), for example:

Colemanite, $Ca_2B_6O_{11} . 5 H_2O$, is $CaB_3O_4(OH)_3 . H_2O$,
Bandylite, $CuCl_2 . CuB_2O_4 . 4 H_2O$, is $CuClB(OH)_4$, and
Teepleite, $NaBO_2 . NaCl . 2 H_2O$, is $Na_2ClB(OH)_4$.

The chemistry of borates is complex both in solution and in the melt. It is concluded from ^{11}B n.m.r. and other studies that $Na_3B_3O_6$ dissociates in solution to $B(OH)_4^-$ ions and that at low concentrations tetraborates dissociate completely into $B(OH)_3$ and $B(OH)_4^-$, but that in more concentrated solutions of borates various polyborate ions coexist in equilibrium with one another. From melts extensive series of borates are obtained, the product depending on the composition of the melt, for example:

	Li_3BO_3	$Li_4B_2O_5$	$Li_6B_4O_9$	$LiBO_2$	LiB_3O_5	$Li_2B_8O_{13}$	LiB_5O_8
O:B ratio	3	2·5	2·25	2	1·66	1·625	1·6

Little is yet known of the structures of metal-rich compounds of these types, but the structures of a number of anhydrous borates with O : B ratios between 1·75 and 1·55 are described later.

Boron trioxide

This compound was known only in the vitreous state until as late as 1937, but it may be crystallized by dehydrating metaboric acid under carefully controlled conditions or by cooling the molten oxide under a pressure of 10–15 kbar. The normal form has a density of 2·56 g/cc and consists of a 3D network of BO_3 groups joined through their O atoms,[1] in which B–O = 1·38 Å. This net, an assembly of $B_{10}O_{10}$ rings has been described in Chapter 3. Under a pressure of 35 kbar at 525°C B_2O_3-II is formed (density, 3·11 g/cc).[2] This much more dense polymorph is built of irregular tetrahedra, three vertices of which are common to 3 BO_4 groups and one to 2 BO_4 groups; the structure may be described in terms of vertex-sharing pairs of tetrahedra (details shown at (a)). Mass spectrometric studies show that liquid B_2O_3 vaporizes predominantly as B_2O_3 molecules. These are polar, eliminating the trigonal bipyramidal model, but the structure of the molecule is still uncertain;[3] a V-shaped molecule, (b), has been tentatively suggested.

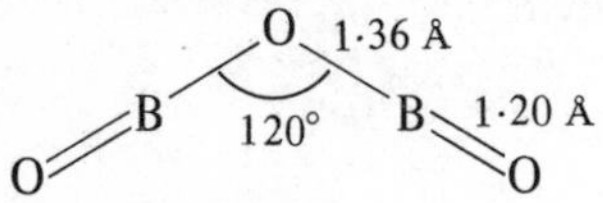

Orthoboric acid and orthoborates

In H_3BO_3 planar $B(OH)_3$ molecules (B–O, 1·36 Å) are linked into plane layers by O–H–O bonds of length 2·72 Å, the angle between which (at a given O atom) is

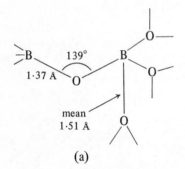

(a)

(1) AC 1971 **B27** 1662

(2) AC 1968 **B24** 869

(b)

(3) JCP 1968 **48** 3339

852

114° (Fig. 24.11). The H atoms were located by X-ray diffraction at positions about one Å from the O atoms to which they are bonded, a result confirmed by a n.d. study of D_3BO_3 (O–D, 0.97 Å along a linear O–D···O bond of length 2.71 Å).[1]

(1) AC 1966 **20** 214

Not very much is known about the structures of orthoborates of the alkali metals, and indeed not many appear to have been prepared. In α-Li_3BO_3 Li^+ has a distorted tetrahedral arrangement of 4 O neighbours at approximately 2 Å and a fifth at 2.5 Å.[2] Na_3BO_3 is formed from $Na_3B_3O_6$ at temperatures above 680°C but has not been obtained pure, there being an equilibrium: $Na_3B_3O_6 \rightleftharpoons Na_3BO_3 + B_2O_3$. There is reason to suppose that some of the alkali orthoborates cannot exist (see Chapter 7). Known compounds $M_3(BO_3)_2$ include salts of Mg, Ca, Ba, Cd, and Co,[3] and there are complex borates such as $NaCaBO_3$ and $CaSn(BO_3)_2$, the latter being isostructural with dolomite, $CaMg(CO_3)_2$. From the structural standpoint the

(2) AC 1971 **B27** 704

(3) ACSc 1949 **3** 660

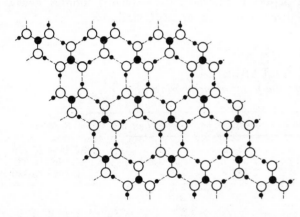

FIG. 24.11. Portion of a layer of H_3BO_3. Broken lines indicate O–H···O bonds.

simplest orthoborates are $M^{III}BO_3$, which furnish examples of crystals isostructural with all three polymorphs of $CaCO_3$: $ScBO_3$ and $InBO_3$ (calcite structure), YBO_3 and $LaBO_3$ (aragonite structure), and $SmBO_3$ (vaterite structure). Some recent references are given to these compounds.[4]

(4) AM 1961 **46** 1030; AC 1966 **20** 283; JCP 1969 **51** 3624

Pyroborates

Two rather different configurations of the pyroborate ion have been found, in the magnesium and cobalt salts. In $Co_2B_2O_5$,[1] (a), the planes of the BO_3 groups make angles of 7° with the plane of the central B–O–B system and are twisted in opposite directions (mean B–O approximately 1.30 Å). In $Mg_2B_2O_5$,[2] (b), the

(1) ACSc 1950 **4** 1054

(2) AC 1952 **5** 574

$$
\begin{array}{cc}
\text{(a) } 153° & \text{(b) } 131\tfrac{1}{2}°
\end{array}
$$

angle between the planes of the BO_3 groups is found to be 22° 19′, and the mean B–O bond length is 1·36 Å. The oxygen bond angle in (b) is close to that found in metaborates, viz. 130° in CaB_2O_4, and $126\frac{1}{2}$° in the $B_3O_6^{3-}$ and $B_5O_{10}^{5-}$ ions. No difference was found between the lengths of the central and terminal B–O bonds.

For the $(OH)_3B-O-B(OH)_3^{2-}$ ion see p. 860.

Metaboric acid and metaborates

Metaboric acid is known in three crystalline modifications, which provide a good example of monotropism (Fig. 24.12) and of the increase in density with change from 3- to 4-coordination of B (Table 24.4). The monoclinic form is readily prepared by the dehydration of H_3BO_3 in an open vessel at 140°C; quenching of the molten material gives a glass which later recrystallizes as the orthorhombic form. If the melt is held at 175°C the most stable (cubic) form is slowly precipitated, while complete dehydration at about 230°C yields B_2O_3.

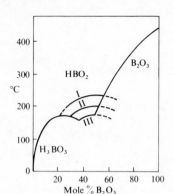

FIG. 24.12. Phase diagram for the $B_2O_3{-}H_2O$ system.

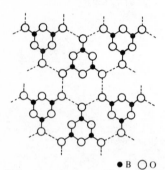

● B ○ O

FIG. 24.13. Arrangement of $B_3O_3(OH)_3$ molecules in a layer of one form of crystalline metaboric acid.

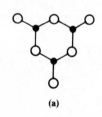

(a)

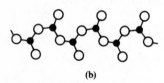

(b)

FIG. 24.14. Metaborate ions: (a) cyclic $(B_3O_6)^{3-}$, (b) infinite $(BO_2)_n^{n-}$ chain ion in CaB_2O_4 and $LiBO_2$-I. Small black circles represent B atoms.

TABLE 24.4
Crystalline forms of metaboric acid

	M.P.	Density	C.N. of B	Reference
Orthorhombic	176°C	1·784 g/cc	3	AC 1964 **17** 229
Monoclinic	201°	2·045	3 and 4	AC 1963 **16** 385
Cubic	236°	2·487	4	AC 1963 **16** 380

Orthorhombic metaboric acid is built of molecules $B_3O_3(OH)_3$ which are linked into layers by O–H–O bonds (Fig. 24.13). Monoclinic metaboric acid is apparently built of chains of composition $[B_3O_4OH(OH_2)]$, recognition of the OH group and the H_2O molecule resting on the location of the H atoms. The direct bonding of a water molecule to B is most unexpected. Cubic HBO_2 has a framework structure built from tetrahedral BO_4 groups with hydrogen bonds between certain pairs of O atoms.

Metaborates are anhydrous compounds $M_x(BO_2)_y$; certain compounds originally formulated as hydrated metaborates contain $B(OH)_4^-$ ions; for example, $NaBO_2 \cdot 4\,H_2O$ is $NaB(OH)_4 \cdot 2\,H_2O$. The normal forms of these salts stable under atmospheric pressure contain either the cyclic $B_3O_6^{3-}$ ion (Fig. 24.14(a)), as in $Na_3B_3O_6$, $K_3B_3O_6$, and $Ba_3(B_3O_6)_2$, or the infinite linear $(BO_2)_n^{n-}$ ion of Fig. 24.14(b), as in $LiBO_2$, CaB_2O_4, and SrB_2O_4. At higher pressures some of these compounds undergo changes to forms (designated by Roman numerals II, III, etc.) in which some or all of the B atoms become 4-coordinated (Table 24.5). For example, $LiBO_2$-II is a superstructure of the zinc-blende type, in which both Li and B are surrounded tetrahedrally by 4 O, at 1·96 and 1·48 Å respectively.

A surprising difference is found between the structures of the $B_3O_6^{3-}$ ions in $Na_3B_3O_6$ and $K_3B_3O_6$, the structures of which have been carefully refined:

$(Na_3B_3O_6)$ 1·280 Å   1·433 Å $(K_3B_3O_6)$ 1·331 Å   1·398 Å

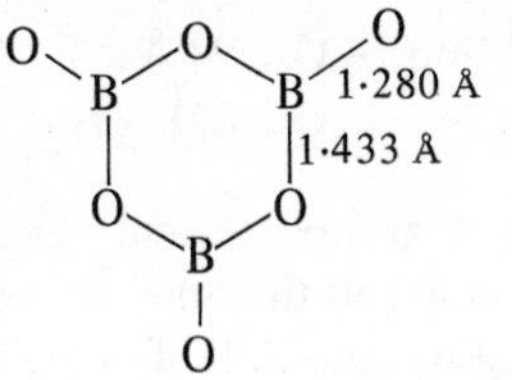

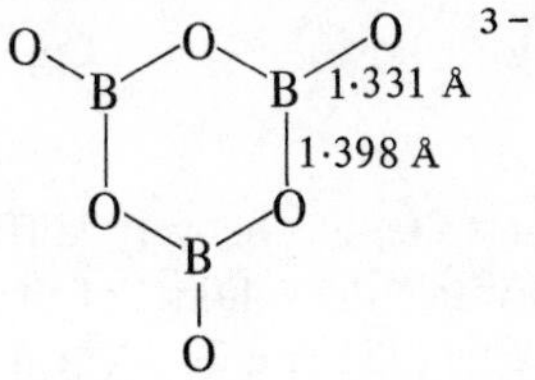

Since these compounds are isostructural these data are not consistent with Pauling's rules if the same relation between bond strength and length is assumed for both compounds.

TABLE 24.5

Crystalline forms of metaborates

	Pressure (kbar)	Density (g/cc)	C.N. of B	C.N. of Ca	Reference
$Ca(BO_2)_2$-I	–	2·702	All 3	8	AC 1963 **16** 390
$Ca(BO_2)_2$-II	12–15	2·885	Half 3 Half 4	8	AC 1967 **23** 44
$Ca(BO_2)_2$-III	15–25	3·052	One-third 3 two-thirds 4	8 and 10	AC 1969 **B25** 955
$Ca(BO_2)_2$-IV	25–40	3·426	All 4	(9 + 3) and 12	AC 1969 **B25** 965

	Type of structure	Reference
$Sr(BO_2)_2$-I	Infinite chain ion	AC 1964 **17** 314
$Sr(BO_2)_2$-III $Sr(BO_2)_2$-IV	Isostructural with $Ca(BO_2)_2$	AC 1969 **B25** 1001
$LiBO_2$-I	Infinite chain ion	AC 1964 **17** 749
$LiBO_2$-II	Zinc-blende superstructure	JCP 1966 **44** 3348
$Na_3B_3O_6$ $K_3B_3O_6$	Cyclic $B_3O_6^{3-}$ ions	AC 1963 **16** 594 AC 1970 **B26** 1189
$Ba(BO_2)_2$		AC 1966 **20** 819
$Cu(BO_2)_2$	3D tetrahedral framework	AC 1971 **B27** 677

The structures of CaB_2O_4-IV and CuB_2O_4 are of special interest because the oxy-ion is a 3D framework formed from tetrahedral BO_4 groups sharing all vertices—compare silica structures. Since the framework contains planar B_3O_3 rings it may alternatively be described as built from rings of three tetrahedra similar to the S_3O_9 molecule or $Si_3O_9^{6-}$ ion which share the extra-annular O atoms with those of six similar rings. In CaB_2O_4-IV the framework forms around Ca^{2+} ions which are either 12- or (9 + 3)-coordinated:

$$Ca_I - 6 \ O \ 2·70 \ \text{Å} \qquad Ca_{II} - 3 \ O \ 2·39 \ \text{Å}$$
$$6 \ O \ 2·67 \qquad\qquad 3 \ O \ 2·48$$
$$3 \ O \ 2·60$$
$$3 \ O \ 3·14$$

Boron

In CuB_2O_4 the framework accomodates (or is cross-linked by) Cu^{2+} ions which have only 4 coplanar nearest neighbours:

$$Cu_I-4\ O\ 2{\cdot}00\ \text{Å} \qquad Cu_{II}-4\ O\ 1{\cdot}94\ \text{Å}$$
$$(O\ 3{\cdot}07)$$

In spite of the very different coordination groups around the cations in these crystals the volumes of the unit cells, one cubic and the other tetragonal, are almost equal (731 and 741 Å^3). These structures raise a question of nomenclature.

The prefixes ortho, pyro, and meta applied to acids and salts of Si, P, and As refer to *tetrahedral* ions (or molecules in the case of the acids) which share respectively 0, 1, and 2 O atoms, that is, to compounds containing MO_4, M_2O_7, or $(MO_3)_n$ groups. As applied to borates they normally refer to acids and ions in which B is 3-coordinated, that is, ortho, BO_3^{3-}, pyro, $B_2O_5^{4-}$, and meta, $(BO_2)_n^{n-}$ (cyclic or linear). The tetrahedral B ions would be the (unknown)

$$\text{ortho, } BO_4^{5-}, \text{ pyro, } B_2O_7^{8-}, \text{ and meta, } (BO_3)_n^{3n-},$$

and in addition there would be oxy-ions in which all BO_4 groups share 3 or 4 O atoms, namely,

$$(B_2O_5)_n^{4n-}, \text{ layers or 3D frameworks,}$$
$$(BO_2)_n^{n-}, \text{ double layers or 3D frameworks.}$$

Sharing of two opposite edges of each BO_4 would give a chain $(BO_2)_n^{n-}$ structurally similar to the SiS_2 chain. Three of the ions of the tetrahedral family have formulae the same as those of the trigonal family. At present the only known type of borate oxy-ion in which all B atoms are tetrahedrally coordinated and all shared O atoms bonded to 2 B, as we have assumed above, is the $(BO_2)_n^{n-}$ framework ion in CaB_2O_4-IV and CuB_2O_4. These compounds are called metaborates, and indeed CaB_2O_4-IV is a high-pressure polymorph of the compound which in its normal form contains the metaborate ion formed from BO_3 groups. If we retain the term metaborate for all compounds $M(BO_2)_x$ it loses its earlier structural significance, for it includes not only the two extreme types of structure (all 3-coordinated B or all 4-coordinated B) but also the intermediate structures (e.g. CaB_2O_4-II and III) in which there is coordination of both types.

Hydroxyborates and anhydrous polyborates
We discuss these compounds together because their anions either consist of or are built from simple cyclic units of the kind shown in Fig. 24.15. The number of possible anions is large because

(i) there are different basic ring systems, three of which, (a), (b), and (c), are illustrated,

(ii) the number of extra-annular O atoms can in principle increase until all the B atoms are 4-coordinated, as in the series $(a_3)-(a_6)$,

(iii) the extra-annular O atoms can be O of OH groups or O atoms shared

856

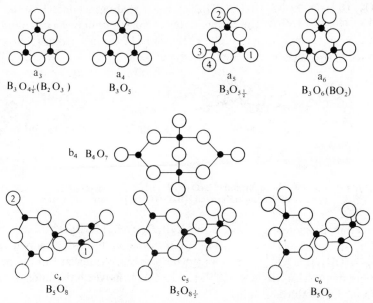

FIG. 24.15. Cyclic boron-oxygen systems in hydroxyborates and/or polyborates. The subscript is the number of extra-annular O atoms. The formula shows the composition of the 3D anion formed if all of these are shared with other similar units.

between two units. Some of the finite units of Fig. 24.15 in which all the extra-annular O atoms belong to OH groups exist in hydroxyborates (Table 24.6), but various numbers of these O atoms can be shared to form ions extending indefinitely in one, two, or three dimensions. The *minimum* numbers of shared O atoms are obviously 2 for a chain and 3 for a layer or 3D system; the known layers are built from 4- or 5-connected units. If some OH groups remain the result is a hydroxy-anion, as found in many borates crystallized from aqueous solution; if all the extra-annular O atoms are shared the anion is of the type $B_x O_y$, characteristic of anhydrous polyborates prepared from the melt.

(iv) A particular borate anion may be built of units all of the same kind (for example, a_4) or it may be built from units of two or more kinds (for example, a_4 and c_4). We now amplify (ii)–(iv).

Of the fully 'hydroxylated' units of type (a), a_3 is the cyclic $B_3 O_3 (OH)_3$ molecule, a_4 and a_6 are not known, but a_5 occurs in a series of hydrated calcium hydroxyborates[1] which includes the minerals meyerhofferite, $Ca[B_3 O_3 (OH)_5]$. $H_2 O$, and inyoite, the tetrahydrate. Sharing of the two O atoms 1 and 2 of a_5 gives the infinite chain ion in $Ca[B_3 O_4 (OH)_3] \cdot H_2 O$, colemanite, (Fig. 24.16(a)), and sharing of all the atoms 1–4 gives the 2D ion in $Ca[B_3 O_5 (OH)]$, a layer based on the simplest plane 4-connected net. (The 3D structure of $CaB_2 O_4$-IV may be described as built of units a_6 sharing all extra-annular O atoms.)

The finite hydroxy-ion b_4 is the anion in $K_2[B_4 O_5 (OH)_4] \cdot 2 H_2 O$ and also in borax,[2] a fact necessitating the revision of a familiar chemical formula,

(1) JINC 1964 **26** 73

(2) MJ 1956 **2** 1

$Na_2B_4O_7 . 10 H_2O$, to $Na_2[B_4O_5(OH)_4] . 8 H_2O$. Hydrogen bonds link the units into chains (Fig. 24.16(b)). The cation–water complex in borax was mentioned in Chapter 5 as an example of an infinite chain formed from octahedral $[Na(H_2O)_6]$ groups sharing two edges to give the $H_2O : Na$ of $4 : 1$.

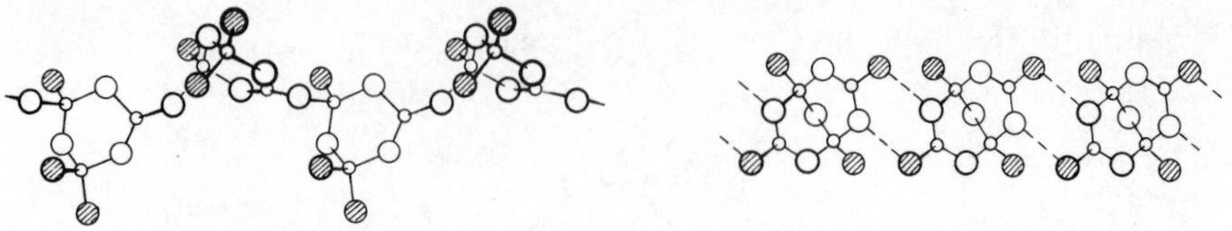

(a)   (b)

FIG. 24.16. (a) The infinite-chain ion $[B_3O_4(OH)_3]_n^{2n-}$ in $CaB_3O_4(OH)_3 . H_2O$; (b) the system of hydrogen-bonded $[B_4O_5(OH)_4]^{2-}$ ions in borax.

(3) AC 1963 **16** 376

The tetrahydroxy-ion c_4 occurs in $K[B_5O_6(OH)_4] . 2 H_2O$,[3] and chains formed by sharing the O atoms 1 and 2 form the anion in the mineral larderellite,

(4) AC 1969 **B25** 2264

(4a) Sc 1971 **171** 377

$NH_4[B_5O_7(OH)_2] . H_2O$.[4] An intermediate possibility is realized in the mineral ammonioborite, $(NH_4)_3[B_{15}O_{20}(OH)_8] . 4 H_2O$,[4a] in which the anion is a finite group formed from three c_4 units:

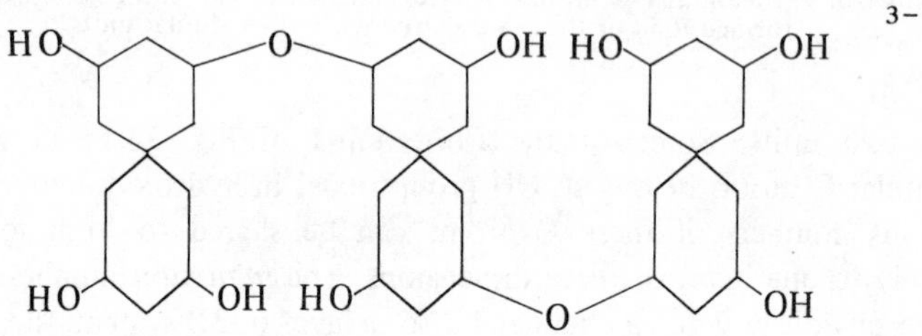

The formation of 3D framework anions requires the sharing of *at least* 3 O by each sub-unit, but usually 4 or 5 are shared. The simplest structures arise from the units a_4, b_4, and c_4 which have their four extra-annular O atoms disposed at the vertices of (irregular) tetrahedra. These units can therefore link up to form 3D frameworks based on the diamond net, as noted in Chapter 3. There are a number of points of special interest. In CsB_3O_5[5] the units form one framework, but the

(5) AC 1960 **13** 889

(6) AC 1968 **B24** 179

(7) AC 1965 **18** 1088

(8) AC 1965 **18** 77; AC 1969 **B25** 2153

(9) AC 1965 **19** 297

(10) AC 1967 **23** 427

more bulky b_4 and c_4 sub-units form two interpenetrating identical frameworks (in $Li_2B_4O_7$[6] and KB_5O_8[7] respectively). This is also true in $Ag_2B_8O_{13}$,[8] where there is the additional complication that each framework is composed of alternate units of two kinds (a_4 and c_4): $B_3O_5 + B_5O_8 = B_8O_{13}$. The units a_5 and c_5 have 5 extra-annular O atoms. In BaB_4O_7[9] alternate units of these types form a 3D 5-connected framework by sharing all these O atoms: $B_3O_{5\frac{1}{2}} + B_5O_{8\frac{1}{2}} = 2(B_4O_7)$. An even more complex system is the anion in CsB_9O_{14}[10] which consists of two interpenetrating 3D nets each built of two kinds of sub-unit. These are the a_3 and a_4 units of Fig. 24.15, which are present in the ratio $2a_3 : 1a_4$, so that the composition is $2(B_3O_{4\frac{1}{2}}) + B_3O_5 = B_9O_{14}$.

These framework ions usually contain no OH groups, all the extra-annular atoms being bridging O atoms. An exception is the anion in $K_2[B_5O_8(OH)] \cdot 2\,H_2O$,[11] formed from c_5 units sharing only four of the extra-annular O atoms. This ion is also notable as the first 3D anion in a hydroxyborate crystallized from solution. It is, however, made under extreme conditions, namely, by evaporating a very viscous supersaturated solution made from $5H_3BO_3 + 2\,KOH$ at $90°C$, conditions approximating the anhydrous melts from which anhydrous polyborates are crystallized.

(11) AC 1969 **B25** 1787

TABLE 24.6
Hydroxyborates and polyborates

Number of O atoms shared	Cyclic unit of Fig. 24.15		
	a_5	b_4	c_4
0	$[B_3O_3(OH)_5]Ca \cdot \begin{array}{c}H_2O\\2\,H_2O\\4\,H_2O\end{array}$	$[B_4O_5(OH)_4]Na_2 \cdot 8\,H_2O$	$[B_5O_6(OH)_4]K \cdot 2H_2O$
2	$[B_3O_4(OH)_3]Ca \cdot H_2O$	$[B_4O_6(OH)_2]^{2-}$	$[B_5O_7(OH)_2]NH_4 \cdot H_2O$
4	$[B_3O_5(OH)]Ca†$	$[B_4O_7]Li_2*$	$[B_5O_8]K*$

† Layer structure. * Two interpenetrating 3D frameworks.

The hexahydroxy-ion c_6 occurs in ulexite, $NaCaB_5O_6(OH)_6 \cdot 5\,H_2O$.[12] Table 24.6 summarizes the formulae of anions formed from the sub-units a_5, b_4, and c_4.

(12) Sc 1964 **145** 1295

We have noted two ways of constructing anions $(B_4O_7)_n^{2n-}$, in $Li_2B_4O_7$ and BaB_4O_7. (The framework in CdB_4O_7[13] is of the same general type as in $Li_2B_4O_7$.) A third possibility is realized in SrB_4O_7[14] which has a 3D anion in which all the B atoms are tetrahedrally coordinated and $2/7$ of the O atoms are 3-coordinated. We thus have three quite different ways of attaining the O : B ratio 7 : 4:

(13) AC 1966 **20** 132

(14) AC 1966 **20** 274

	C.N. of B	C.N. of O
$CdB_4O_7 \; (Li_2B_4O_7)$		
BaB_4O_7	3 and 4	2
$SrB_4O_7 \; (PbB_4O_7)$	4	2 and 3

We conclude this section with examples of structures based on a more complex tricyclic unit with the composition $B_6O_7(OH)_6^{2-}$ (Fig. 24.17(a)). This unit is planar, apart from the OH groups attached to the tetrahedral B atoms, and is of interest as containing a central O atom bonded to three B atoms. It is found as a discrete anion in $Mg_2[B_6O_7(OH)_6]_2 \cdot 9\,H_2O$.[15] Sharing of the O atoms shown as shaded circles in Fig. 24.17(b) results in a layer of composition $[B_6O_9(OH)_2]_n^{2n-}$ based on the simplest 4-connected plane net (Chapter 3); this is the anion in the mineral tunellite, $SrB_6O_9(OH)_2 \cdot 3\,H_2O$.[16] In $(Ca,Sr)_2B_{14}O_{20}(OH)_6 \cdot 5\,H_2O$[17] this tricyclic unit is found as the sub-unit in a much more complex layer. The

(15) AANL 1969 **47** 1

(16) AM 1964 **49** 1549

(17) AM 1970 **55** 1911

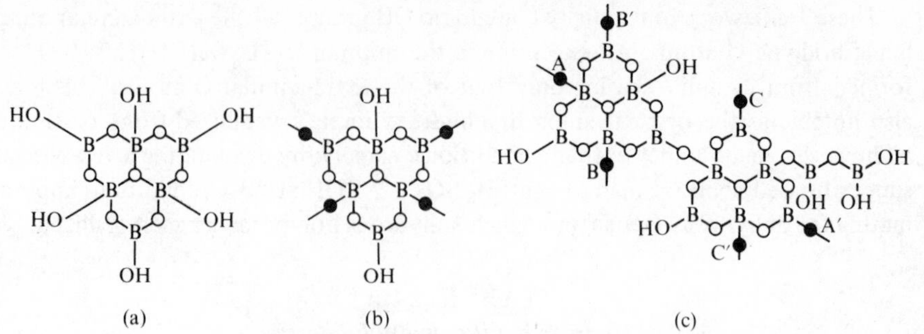

FIG. 24.17. Tricyclic boron–oxygen unit in hydroxyborates (see text).

multiple unit consists of two B_6 units to one of which is attached a B_2 side chain, and this B_{14} complex is linked into layers by sharing six O atoms with other similar units (Fig. 24.17(c)).

Other borate structures containing tetrahedrally coordinated boron
Coordination groups ranging from BO_4 to $B(OH)_4$ are found in some borates, and we shall note here a number of the more interesting structures.

The simplest structures containing BO_4 coordination groups are those of BPO_4 and $BAsO_4$, with silica-like structures. Further similarity to silicon is shown by the isomorphism of $TaBO_4$ and $ZrSiO_4$ and by the structure of $Zn_4B_6O_{13}$.[1] The B atoms in the 3D framework of this crystal are situated at the vertices of Fedorov's packing of truncated octahedra, the BO_4 tetrahedra being linked in the same way as the SiO_4 tetrahedra in, for example, sodalite, $Na_4Si_3Al_3O_{12}Cl$. One O atom does not form part of the B_6O_{12} framework, so that the compound may be formulated $Zn_4O(B_6O_{12})$.

Examples of crystals containing $BO_3(OH)$ coordination groups include the minerals datolite,[2] $CaBSiO_4(OH)$, and colemanite, $CaB_3O_4(OH)_3 . H_2O$. The structure of the latter has been mentioned in the previous section. Datolite consists of apophyllite-like sheets of tetrahedra (p. 818) held together by Ca^{2+} ions. The tetrahedral groups composing the sheets are alternately SiO_4 and BO_3OH, and each tetrahedron shares three vertices with tetrahedra of the other kind, the unshared vertices being O of SiO_4 and OH of BO_3OH groups. The composition of the layer is therefore $BO_{\frac{3}{2}}OH . SiO_{\frac{5}{2}} = BSiO_4OH$ (Fig. 24.18(a)).

The mineral pinnoite,[3] originally formulated $MgB_2O_4 . 3 H_2O$, contains ions

$$\left[(OH)_3B \overset{\overset{O}{\diagup \diagdown}}{\underset{124°}{\quad\quad}} B(OH)_3 \right]^{2-}$$

(1) ZK 1961 **115** 460

(2) ZaC 1967 **125** 286

(3) AC 1967 **23** 500

(3a) ZaC 1966 **342** 188
(4) AC 1963 **16** 1233
(5) AC 1969 **B25** 1811

in which B forms tetrahedral bonds.
Tetrahedral $B(OH)_4^-$ ions are found in salts such as $LiB(OH)_4$[3a] (B–OH, 1·48 Å), $NaB(OH)_4 . 2H_2O$[4] and $Ba[B(OH)_4]_2 . H_2O$[5] and in the minerals

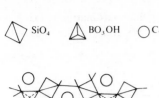

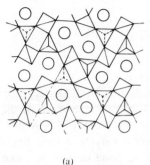
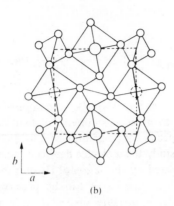
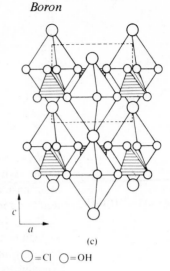

$\bigcirc$ = Cl $\bigcirc$ = OH

FIG. 24.18. (a) The $BSiO_4OH$ layer in datolite, $CaBSiO_4(OH)$, (b) mode of linking of distorted octahedral $[Cu(OH)_4Cl_2]$ and tetrahedral $[B(OH)_4]$ coordination groups in bandylite, $CuClB(OH)_4$. (c) elevation of the structure of bandylite showing the layers (b) held together by long Cu—Cl bonds.

teepleite,[6] $Na_2ClB(OH)_4$, and bandylite,[7] $CuClB(OH)_4$. The latter is also of interest in connection with the stereochemistry of the cupric ion (p. 905). The structure may be dissected into puckered layers composed of tetrahedral $B(OH)_4$ groups connected by Cu^{II} atoms which are thereby surrounded by four coplanar OH groups at the corners of a square (Fig. 24.18(b)). The layers are held together by long Cu—Cl bonds between Cu atoms of adjacent layers and Cl atoms situated between them (Fig. 24.18(c)), the coordination group around Cu being a distorted octahedron (Cu–4 OH, 1·98 Å, Cu–2 Cl, 2·80 Å). The B—OH bond length is 1·42 Å).

The mineral hambergite[8], $Be_2(BO_3)OH$, is a hydroxy-orthoborate containing planar BO_3^{3-} and OH^- ions, not tetrahedral $(BO_3OH)^{4-}$ ions. Similarly fluoborite,[9] $Mg_3(OH,F)_3BO_3$, consists of a c.p. assembly of O^{2-}, OH^-, and F^- ions with B in positions of triangular coordination and Mg in octahedral holes. In general, O : B ratios > 4 : 1 do *not necessarily* imply tetrahedral coordination of B since all the O atoms are not necessarily bonded to B. For example, we may have an assembly of O atoms in which B atoms occupy some positions of 3-coordination and the metal atoms octahedral interstices, as in fluoborite. This is the case in the 'boroferrites' such as $Mg_2Fe_2^{III}B_2O_8$[10] and warwickite,[11] $Mg_3TiB_2O_8$ (p. 497) and in $Co_4Fe_2^{III}B_2O_{10}$. Such compounds may be regarded as intermediate between ortho-borates and oxides. As an example of a compound of this type in which there *is* tetrahedral coordination of B we may quote Fe_3BO_6,[12] which is isostructural with $Mg_3SiO_4(OH,F)_2$.

(6) RS 1951 **21** No. 7
(7) AC 1951 **4** 204

(8) AC 1963 **16** 1144
(9) AC 1950 **3** 208

(10) AC 1950 **3** 473
(11) AC 1950 **3** 98
(12) AC 1965 **19** 1060

The lengths of B—O *bonds*

Observed lengths of B—O bonds range from 1·20 Å for B=O in the gaseous B_2O_3

861

molecule to around 1·55 Å. The mean value for triangular coordination is 1·365 Å and for tetrahedral coordination 1·475 Å, but there are considerable ranges of lengths for both types of coordination:

$$\begin{array}{ccccc} B{=}O & & BO_3 & & BO_4 \\ 1{\cdot}20\ \text{Å} & 1{\cdot}28 \longleftrightarrow 1{\cdot}43 & & 1{\cdot}43 \longleftrightarrow 1{\cdot}55\ \text{Å} \\ & 1{\cdot}365 & & 1{\cdot}475\ \text{Å} \end{array}$$

Cyclic $H_2B_2O_3$, *boroxine,* $H_3B_3O_3$, *and boranocarbonates*

Cyclic $H_2B_2O_3$ is formed as an intermediate in the oxidation of B_5H_9, B_4H_{10}, etc. as an unstable species with a half-life of only 2–3 days at room temperature. As the result of a m.w. study the molecule has been assigned the structure (a).[1] Boroxine, $H_3B_3O_3$, is prepared by the action of H_2 on a mixture of $B + B_2O_3$. Although it is a 'high-temperature' species it can be preserved for an hour or two at room temperature under a pressure of 1–2 torr in the presence of excess argon. It decomposes to $B_2O_3 + B_2H_6$ but can be oxidized to cyclic $H_2B_2O_3$. The structure (b) has been assigned to the molecule.[2]

(1) JCP 1967 **47** 4186; AC 1969 **B25** 807

(2) IC 1969 **8** 1689

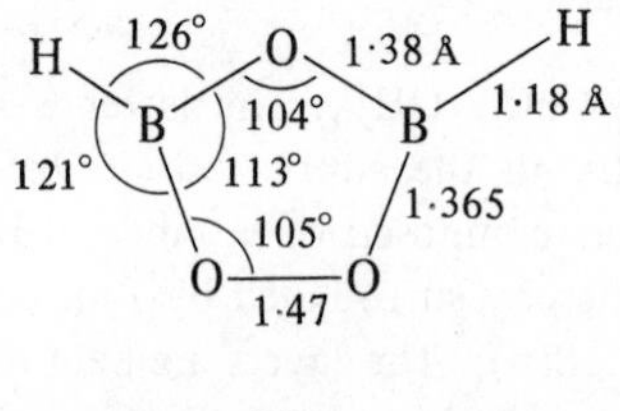

(3) IC 1967 **6** 817

(a) 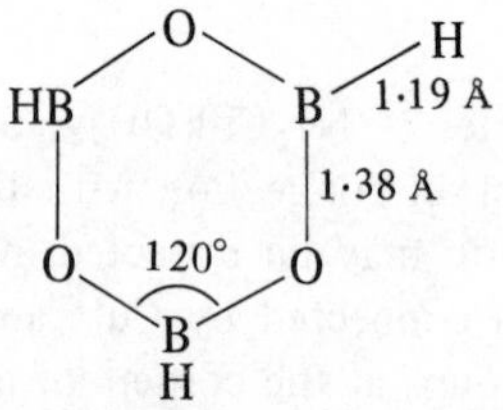   (b)

A BH_3-substituted carbonate ion has been made by reacting $H_3B \cdot CO$ with KOH to give the boranocarbonate, $K_2(H_3B \cdot CO_2)$.[3]

Boranes and related compounds

Preparation and properties

The boron hydrides (boranes) were originally prepared by the action of 10 per cent HCl, or preferably 8N-phosphoric acid, on magnesium boride. The chief product of this method was B_4H_{10}, mixed with small quantities of B_5H_9, B_6H_{10}, and $B_{10}H_{14}$. This mixture was separated into its components by fractional distillation. B_2H_6 had to be obtained indirectly (with B_5H_9 and $B_{10}H_{14}$) by heating B_4H_{10} at 100°C. It is interesting to note that the action of acid on Mg_2Si gives the silanes from SiH_4 to Si_6H_{14} in decreasing amounts. A later method of preparing boron hydrides was to pass the vapour of BCl_3 with hydrogen in a rapid stream at low pressure through an electric discharge between copper electrodes. The main boron-containing product of this reaction is B_2H_5Cl, which decomposes when kept at 0°C into B_2H_6 and BCl_3. These and other methods of preparing diborane were adequate while the compound was only of theoretical interest. Diborane is now a useful intermediate and reagent; for example, it converts metal alkyls to boro-hydrides, reduces aldehydes and ketones to alcohols, and decomposes to pure boron at high temperatures. Moreover, it appeared at one time to have a future as a rocket fuel, and efforts were therefore made to find better ways of preparing B_2H_6 on a large scale from readily accessible starting materials. Methyl borate can be prepared in over 90 per cent yield and in a very pure state by heating B_2O_3 with an excess of methanol and removing the methanol from the $CH_3OH-B(OCH_3)_3$ azeotrope so formed with LiCl. Methyl borate is then converted into $NaBH(OCH_3)_3$ by refluxing with NaH at 68°C, and the following reaction gives a nearly theoretical yield of diborane:

$$8\,(C_2H_5)_2O \cdot BF_3 + 6\,NaBH(OCH_3)_3 \rightarrow$$

$$B_2H_6 + 6\,NaBF_4 + 8\,(C_2H_5)_2O + 6\,B(OCH_3)_3$$

Diborane may alternatively be made directly from B_2O_3 by heating the oxide with $Al + AlCl_3$ at 175°C under a pressure of 750 atm of H_2.

An outstanding feature of borane chemistry is the large number of reactions which result in the conversion of one or more boranes into others. These reactions make it possible to develop the whole of borane chemistry from one simple starting material, diborane. This point is emphasized in Chart 24.2, which shows only a small part of the very complex chemistry of boranes. Pyrolysis of B_2H_6 yields a number of the lower boranes including, for example, the thermally unstable B_5H_{11}, but this compound is preferably prepared by utilizing the equilibrium

$$B_2H_6 + 2\,B_4H_{10} \rightleftharpoons 2\,B_5H_{11} + 2\,H_2.$$

This borane can be converted catalytically into B_6H_{10}:

$$2\,B_5H_{11} \rightarrow B_6H_{10} + 2\,B_2H_6$$

or reacted with B_4H_{10} to give B_5H_9:

$$B_5H_{11} + B_4H_{10} \rightarrow B_5H_9 + 2\,B_2H_6.$$

Recycling of the products of a particular reaction may give a satisfactory yield of a desired product (for example, $B_{10}H_{14}$ from the pyrolysis of B_2H_6), and the use of the silent electric discharge (with or without the addition of hydrogen) figures prominently in the preparation of boranes.

In some respects the hydrogen chemistry of boron resembles that of carbon and in others that of silicon. For example, boranes undergo many substitution reactions, H being replaced by halogen, CN, and organic ligands; derivatives of $B_{10}H_{14}$ include $B_{10}H_{13}I$ and $B_{10}H_{12}I_2$ and $B_{10}H_{12}L_2$, where L is R_2S, RCN, R_3N, etc. Mono-iododiborane reacts with sodium in exactly the same way as does ethyl iodide in the Wurtz reaction:

$$2\,B_2H_5I + 2\,Na \rightarrow B_4H_{10} + 2\,NaI$$

CHART 24.2
Reactions of diborane

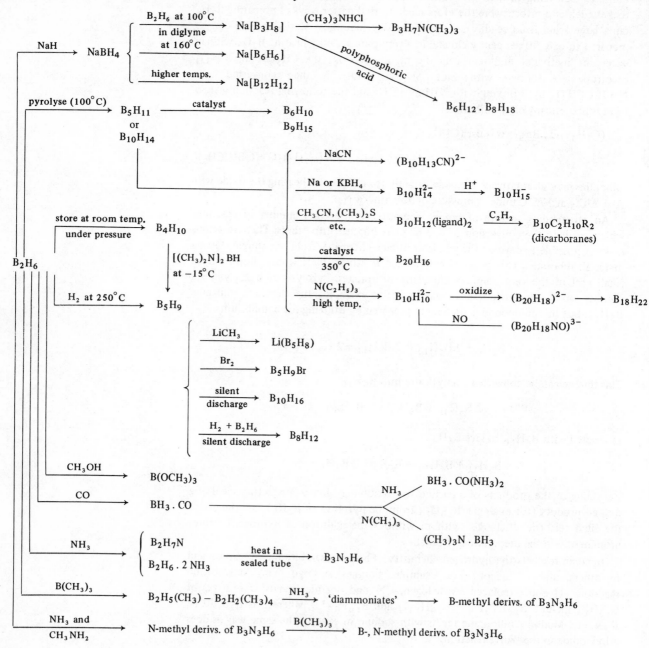

compare

$$2\,C_2H_5I + 2\,Na \rightarrow C_4H_{10} + 2\,NaI$$

In their reactions with halogens and hydrogen halides, however, the boranes behave quite differently from hydrocarbons. Diborane, for example, reacts with HCl to give a chloro derivative and hydrogen, and $B_{10}H_{14}$—which from its formula would appear to be an unsaturated compound—forms with a halogen a substituted derivative and not an addition product as is the case with C_2H_4 and other unsaturated hydrocarbons. There are large differences in thermal stability between boranes such as B_5H_9 and the very unstable B_5H_{11}, but generally in their low stability and vigorous reaction with water the boranes resemble silanes rather than hydrocarbons.

In addition to the neutral boranes many borohydride ions have been prepared, ranging from BH_4^- to $B_{20}H_{18}^{2-}$ and including a remarkable series of polyhedral ions $B_nH_n^{2-}$ (n from 6–12). Unlike the boranes these ions have highly symmetrical structures and appear to be the boron analogues of the aromatic carbon compounds. There is extensive delocalization of a small number of electrons, in contrast to the localized 3-centre bonds in the neutral boranes, and the alkali-metal salts of $B_{10}H_{10}^{2-}$ and $B_{12}H_{12}^{2-}$ are extremely stable compounds. Borohydride ion chemistry is complicated by the fact that not only are there polyhedral ions $B_nH_n^{2-}$ but there are also (a) many substituted ions (for example, $B_{12}H_{11}OH^{2-}$), (b) ions with the same composition but different charge (for example, $B_8H_8^{2-}$ and the paramagnetic $B_8H_8^-$), and (c) ions with the same boron, or boron–carbon, skeleton but different numbers of H atoms and different charges (for example, $B_{11}H_{13}^{2-}$ and $B_{11}H_{14}^-$, which are reversibly interconvertible). Fully halogenated ions include $B_{12}Br_{12}^{2-}$ and $B_{10}Cl_{10}^{2-}$; the latter has been isolated in the free acid, $(H_3O)_2B_{10}Cl_{10} \cdot 5\,H_2O$. Furthermore, oxidation of $B_{10}H_{10}^{2-}$ gives the $B_{20}H_{18}^{2-}$ ion, which consists of two B_{10} units joined by two 3-centre bonds (contrast the borane $B_{20}H_{16}$). There are also ions in which the two 'halves' are linked by NO or a metal atom in place of the 3-centre bonds, for example, $B_{20}H_{18}NO^{3-}$ and $Fe(\pi\text{-}B_9C_2H_{11})_2^{2-}$.

Isoelectronic with ions $B_nH_n^{2-}$ are the dicarboranes, $B_nC_2H_{n+2}$, prepared from appropriate boranes and acetylene, which also have aromatic character. These also form composite π-bonded ions which form salts such as $[(C_2H_5)_4N]_2 [Cu(C_2B_9H_{11})_2]$.

We have therefore three main groups of structures to summarize:

(a) the boranes and their derivatives, including the higher members formed from two simpler units joined at a common vertex ($B_{10}H_{16}$), a common edge ($B_{16}H_{20}$, $B_{18}H_{22}$), or a common face ($B_{20}H_{16}$). In Table 24.7 an asterisk distinguishes boranes that were prepared by Stock; all of these, and most of the others, fall into one of two families, B_nH_{n+4} and B_nH_{n+6}, the former being generally far more stable than the latter.

(b) borohydride ions and metal borohydrides, and

(c) the polyhedral ions $B_nH_n^{2-}$ and the isoelectronic carboranes $B_nC_2H_{n+2}$ and their derivatives, including the composite ions mentioned above.

Table 24.7 lists the boranes and some simple derivatives, with references to structural studies.

TABLE 24.7
Boranes and simple derivatives

B_nH_{n+4}	B_nH_{n+6}	Others	Reference
$B_2H_6^*$			JCP 1968 **49** 4456 (vapour)
			JCP 1965 **43** 1060 (crystal)
	$B_4H_{10}^*$		JACS 1953 **75** 4116 (vapour)
			JCP 1957 **27** 209 (crystal)
$B_5H_9^*$			JCP 1954 **22** 262 (vapour)
			AC 1952 **5** 260 (crystal)
	$B_5H_{11}^*$		JCP 1957 **27** 209
$B_6H_{10}^*$			JCP 1958 **28** 56
	B_6H_{12}		
B_8H_{12}			AC 1966 **20** 631
		B_8H_{16}	
		B_8H_{18}	
	B_9H_{15}		JCP 1961 **35** 1340
	i-B_9H_{15}		
$B_{10}H_{14}^*$			IC 1969 **8** 464 (n.d.)
	$B_{10}H_{16}$		JCP 1962 **37** 2872
		$B_{10}H_{18}$	
$B_{16}H_{20}$			IC 1970 **9** 1452
n-$B_{18}H_{22}$			AC 1966 **20** 631
i-$B_{18}H_{22}$			JCP 1963 **39** 2339
		$B_{20}H_{16}$	JCP 1964 **40** 866
	$B_2H_2(CH_3)_4$		IC 1968 **7** 219
	B_5H_8I		AC 1965 **19** 658
	$B_5H_7(CH_3)_2$		IC 1966 **5** 1752
	$B_{10}H_{13}I$		IC 1967 **6** 1281
	$B_{10}H_{12}I_2$		JACS 1957 **79** 2726
	$B_{10}H_{12}[S(CH_3)_2]_2$		AC 1962 **15** 410

The molecular structures of the boranes

Because of the number and complexity of the boranes and their derivatives we shall not attempt to describe all their structures in detail. Apart from B_2H_6 most of the boranes have boron skeletons which are usually described as icosahedral fragments. (They could equally well be related to the more recently discovered carboranes $B_nC_2H_{n+2}$ since the boron–carbon skeletons in these compounds are highly symmetrical triangulated polyhedra (Table 24.8) which include the icosahedron.) The great theoretical interest of the boranes stems from the fact that they are electron-deficient molecules, that is, there are not sufficient valence electrons to bond together all the atoms by normal electron-pair bonds. In these molecules the number of atomic orbitals (1 for each H and 4 for each B) is greater than the total number of valence electrons:

	Total number of atomic orbitals	Total number of valence electrons	Number of 3-centre bonds
B_2H_6	14	12	2
B_4H_{10}	26	22	4
B_5H_9	29	24	5
$B_{10}H_{14}$	54	44	10
$B_{18}H_{22}$	94	76	18

Structural studies show that in molecules of boranes some H atoms (on the periphery of the molecule) are attached to a single B (B–$H_{terminal}$, 1·2 Å) while others are situated between B atoms forming B–H–B bridges. These bridges are usually symmetrical (B–H_{bridge}, 1·34 Å); an exception is the slightly unsymmetrical bridge in $B_{10}H_{14}$, with B–H_b, 1·30 and 1·35 Å. The B atoms are at the vertices of triangulated polyhedra or fragments of such polyhedra, with B–B usually 1·70–1·84 Å but in rare cases smaller (1·60 Å for the basal B–B in B_6H_{10}) or larger (1·97 Å for two B–B bonds in $B_{10}H_{14}$). In general each B atom is bonded to one or more terminal H atoms, but in some boranes which consist of two 'halves' the B atoms involved in the junctions are bonded only to B atoms or to B atoms and bridging H atoms:

$$B_{10}H_{16}: \text{ two B bonded to 5 B only}$$
$$i\text{-}B_{18}H_{22}: \text{ one B} \quad\quad 5 B + 2 H_b$$
$$\text{one B} \quad\quad 7 B \text{ only}$$
$$n\text{-}B_{18}H_{22}: \text{ two B} \quad\quad 6 B + 1 H_b.$$

In order to account for the molecular structures, retaining two-electron bonds and at the same time utilizing the excessive numbers of orbitals available it has been supposed that three or more atomic orbitals combine to form only one bonding orbital. The structures of the simpler boranes may be formulated with 3-centre bonds of three kinds. (Fig. 24.19). In the central (closed) bond, (a), the three B atoms use hybrid orbitals and are situated at the corners of an equilateral triangle.

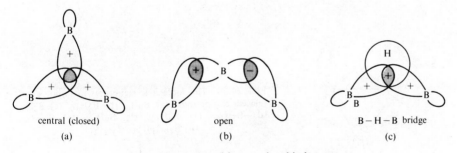

central (closed)	open	B – H – B bridge
(a)	(b)	(c)

FIG. 24.19. Types of 3-centre bond in boranes.

In the open 3-centre bond, (b), they form an obtuse-angled triangle, and the intermediate B atom uses p orbitals, while in (c) the two B atoms are bridged by a H atom, and the orbitals used are 1s of H and hybrid orbitals of B. In some of the more complex systems, such as the polyhedral ions, complete delocalization of a number of electrons (implying the use of a larger number of orbitals) appears to be necessary to account for the 'aromatic' character—compare $C_5H_5^-$, C_6H_6, and $C_7H_7^+$.

Boron

Diborane, B_2H_6.   The results of the latest electron diffraction (sector, micro-photometer) study of this (diamagnetic) molecule are:

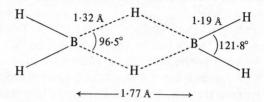

The plane of the central BH_2B system is perpendicular to those of the terminal BH_2 groups (Fig. 24.20(a)). An X-ray study of the crystalline β form (at 4·2°K)

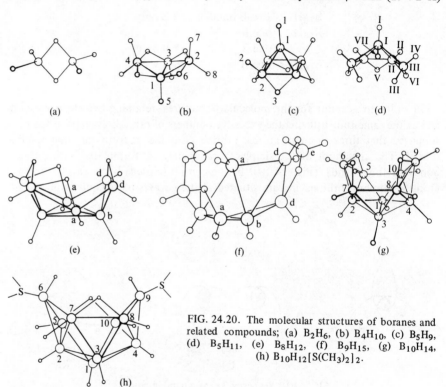

FIG. 24.20. The molecular structures of boranes and related compounds; (a) B_2H_6, (b) B_4H_{10}, (c) B_5H_9, (d) B_5H_{11}, (e) B_8H_{12}, (f) B_9H_{15}, (g) $B_{10}H_{14}$, (h) $B_{10}H_{12}[S(CH_3)_2]_2$.

gives a similar B—B distance but shorter B—H bond lengths (B—H_t, 1·09 Å, and B—H_b, 1·24 Å) and angles H_tBH_t, 124°, and H_bBH_b, 90°. The absence of a direct B—B bond in diborane accounts for reactions such as

$$B_2H_6 + 2\,CO \rightarrow 2\,BH_3 . CO$$
$$B_2H_6 + 2\,N(CH_3)_3 \rightarrow 2\,BH_3 . N(CH_3)_3$$

and

$$B_2H_6 + 2\,NH_3 \rightarrow NH_4(H_3B . NH_2 . BH_3)$$

Of the six H atoms in B_2H_6 only four can be replaced by CH_3 and in these methylated compounds there are never more than two CH_3 groups on a particular B atom. Also, $B(CH_3)_3$ is known, but not $BH(CH_3)_2$ or $BH_2(CH_3)$, which could obviously condense to $B_2H_2(CH_3)_4$ and $B_2H_4(CH_3)_2$ respectively. In the molecule $(CH_3)_2BH_2B(CH_3)_2$ the following distances were determined: B–H, 1·36 Å, B–C, 1·59 Å, and B–B, 1·84 Å.

Tetraborane (10), B_4H_{10}. The configuration of Fig. 24.20(b) has been established by e.d. and by a study of the crystal structure. The B atoms lie at the corners of two triangles hinged at the line B_1B_3 with dihedral angle $B_1B_3B_2/B_1B_3B_4 = 124\frac{1}{2}°$ and angle $B_2B_1B_4 = 98°$. This second angle would be $90°$ or $108°$ respectively if the group of four B atoms was a fragment of an octahedron or icosahedron.

Pentaborane (9), B_5H_9. The boron framework has the form of a tetragonal pyramid (Fig. 24.20(c))–compare the octahedral B_6 groups in CaB_6 and $B_6H_6^{2-}$ (see later). The mean B–B bond lengths are close to 1·80 Å in the base (B_2B_2) and to 1·70 Å for B_1–B_2. A very similar B_5 skeleton is found in B_5H_8I (I attached to apical B_1) and in $B_5H_7(CH_3)_2$ (CH_3 bonded to two adjacent basal, B_2, atoms).

Pentaborane (11), B_5H_{11}. In this hydride the boron skeleton (Fig. 24.20(d)) may be regarded as a fragment of the icosahedron-like arrangement in $B_{10}H_{14}$ or as related to the tetragonal pyramid of B_5H_9 by opening up one of the basal B–B bonds. The bond lengths are:

B_I–B_{II} = 1·72 Å (mean) B–H_t = 1·10 Å (mean)
B_{II}–B_{III}= 1·76 (mean) B–H_b = 1·22
B_{II}–B_{II} = 1·77 B_{III}–H_{VII} = 1·72 (mean)
B_I–B_{III} = 1·87

A feature of this molecule is the unique H_{VII} which is bonded to B_1 (1·09 Å) but is also at a distance of 1·72 Å from the two B_{III} atoms.

Hexaborane (10), B_6H_{10}. An X-ray study of B_6H_{10} showed that in this hydride the B atoms are arranged at the apices of a pentagonal pyramid, so that here also the boron skeleton is a portion of an icosahedron. Bond lengths are shown in Fig. 24.21, the estimated standard deviations being about 0·05 Å for B–H and 0·01 Å for B–B. Note the unsymmetrical nature of the base of the pyramidal molecule, with one short B–B bond.

Octaborane (12), B_8H_{12}. This thermally unstable borane is produced in very small amounts by the action of a silent electric discharge on a mixture of B_5H_9, B_2H_6, and H_2. The molecule is closely related to B_9H_{15}, the doubly bridged BH_2 marked e in Fig. 24.20(f) being replaced by a bridging H atom. Certain of the B atoms in Fig. 24.20(e) and (f) are distinguished by letters to facilitate comparison of the molecular structures.

Enneaborane (15), B_9H_{15}. The much less regular skeleton of this hydride (Fig. 24.20(f)) can be derived from an icosahedron by removing three connected B atoms

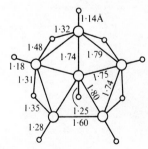

FIG. 24.21. Bond lengths in B_6H_{10}.

Boron

which do not form a triangular group and then opening out the structure around the 'hole' so formed.

Decaborane (14), $B_{10}H_{14}$.   This (solid) hydride is one of the most stable boranes. The B atoms occupy ten of the twelve vertices of a distorted icosahedron, and ten of the H atoms project outwards approximately along the 5-fold axes of the icosahedron so that the outer surface of the molecule consists entirely of H atoms. We have already noted the longer B–B bonds (1·97 Å) in this molecule (they are the bonds B_7–B_8 and B_5–B_{10} in Fig. 24.20(g)) and the slight asymmetry of the hydrogen bridges. Two isomers of $B_{10}H_{13}I$ have similar structures to $B_{10}H_{14}$ and in $B_{10}H_{12}I_2$ the iodine atoms are attached to B_2 and B_4. In $B_{10}H_{12}[S(CH_3)_2]_2$, on the other hand, the substituents are bonded to B_6 and B_9 and there is rearrangement of the hydrogen bridges (Fig. 24.20(h)).

Decaborane (16), $B_{10}H_{16}$.   This borane is produced directly from B_5H_9 (electric discharge) with loss of two H atoms. The molecule (Fig. 24.22(a)) consists of two B_5 units as in B_5H_9 joined by a single B–B bond.

Boranes $B_{16}H_{20}$, $B_{18}H_{22}$, *and* $B_{20}H_{16}$. A number of higher boranes consist of two icosahedral fragments joined by sharing an edge ($B_{16}H_{20}$ and the two isomers of $B_{18}H_{22}$) or of two icosahedra sharing two faces ($B_{20}H_{16}$). Their molecular structures are illustrated in Fig. 24.22(b)–(d).

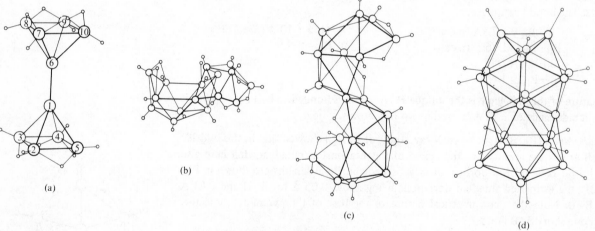

FIG. 24.22. Structures of boranes: (a) $B_{10}H_{16}$, (b) $B_{16}H_{20}$, (c) $B_{18}H_{22}$, (d) $B_{20}H_{16}$.

Borohydride ions and carboranes

Metal derivatives of boranes and related compounds are of at least three types:

  (a) salts containing ions, which range from BH_4^- and $B_3H_8^-$ to polyhedral ions,
  (b) covalent molecules and ions in which BH_4 and B_3H_8 groups are bonded to metal atoms by hydrogen bridges, and
  (c) π-type sandwich structures formed by carboranes.

We shall not adhere strictly to this order.

870

The BH$_4^-$ *ion.* Numerous borohydrides M(BH$_4$)$_n$ have been prepared by the action of metal halides on NaBH$_4$, obtained from B(OCH$_3$)$_3$ and NaH at 250°C, and in other ways. In the series NaBH$_4$, LiBH$_4$, Be(BH$_4$)$_2$, and Al(BH$_4$)$_3$ there is a gradation in chemical and physical properties, for example, an increase in volatility and decreased stability towards air and water. The sodium compound is a crystalline solid stable *in vacuo* up to 400°C, though very hygroscopic. At room temperature it has the NaCl structure[1] (like the K, Rb, and Cs salts), but below −83°C there is a lowering of symmetry to body-centred tetragonal[2] (compare the ammonium halides). The lithium salt melts at 275°C and apparently has the zinc-blende or wurtzite structure.[3] These salts presumably contain tetrahedral BH$_4^-$ ions similar to the tetrahedral BH$_4$ groups in the covalent compounds to be described shortly.

Al(BH$_4$)$_3$ is a volatile liquid boiling at 44·5°C and the borohydrides M(BH$_4$)$_4$ of U, Th, Hf, and Zr are the most volatile compounds known of these metals.

In the more covalent metal borohydrides BH$_4$ is bonded to the metal by hydrogen bridges, as shown in Fig. 24.23 for Be(BH$_4$)$_2$,[4] Al(BH$_4$)$_3$,[5] [(C$_6$H$_5$)$_3$P]$_2$.CuBH$_4$,[6] and Al(BH$_4$)$_3$.N(CH$_3$)$_3$.[7a] Note the quite different mode of bonding of BH$_4$ in the Be and Al compounds, and the trigonal prismatic arrangement of the 6 H atoms around the metal atom in the latter. It was not possible to locate the H atoms in the room temperature form of Al(BH$_4$)$_3$.

(1) JACS 1947 **69** 987
(2) JCP 1954 **22** 434

(3) JACS 1947 **69** 1231

(4) ACSc 1968 **22** 859
(5) ACSc 1968 **22** 328
(6) IC 1967 **6** 2223
(7a) IC 1968 **7** 1575

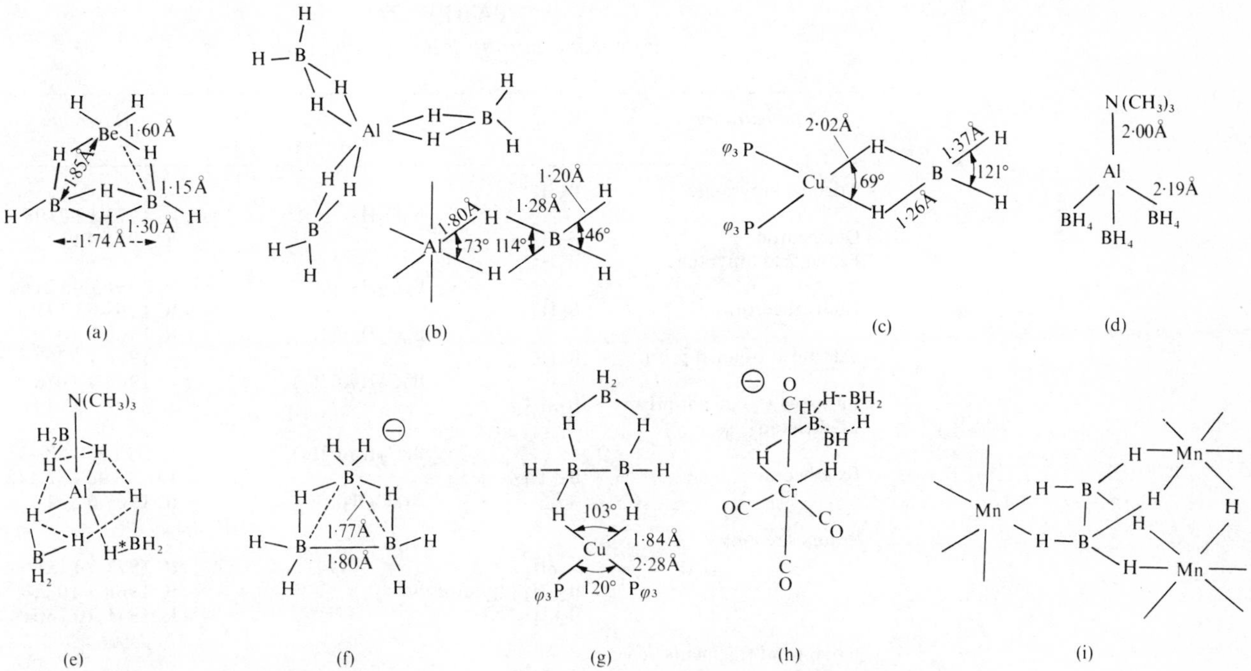

FIG. 24.23. Molecular structures of borohydrides and related compounds: (a) Be(BH$_4$)$_2$, (b) Al(BH$_4$)$_3$, (c) [(C$_6$H$_5$)$_3$P]$_2$Cu(BH$_4$), (d) and (e) Al(BH$_4$)$_3$.N(CH$_3$)$_3$, (f) (B$_3$H$_8$)$^-$, (g) [(C$_6$H$_5$)$_3$P]$_2$Cu(B$_3$H$_8$), (h) [Cr(CO)$_4$B$_3$H$_8$]$^-$, (i) HMn$_3$(CO)$_{10}$(BH$_3$)$_2$.

$N(CH_3)_3$, in which the gross structure of the molecule is tetrahedral, (d), but in the low temperature form the three BH_4 groups are not equivalent. One is rotated so that one H (H*) occupies an apical position of a pentagonal bipyramidal group around Al, (e). In the molecule $Zr(BH_4)_4$ the BH_4 groups are arranged tetrahedrally around the metal atom with triple bridges:[7b]

(7b) IC 1971 **10** 590

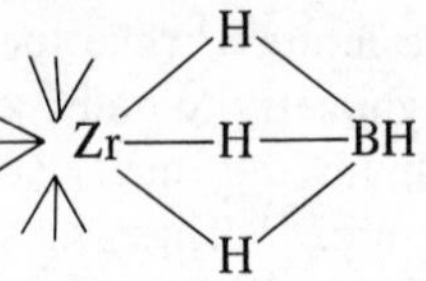

The $B_3H_8^-$ *ion.* The structure of this ion, Fig. 24.23(f), has been determined in $[H_2B(NH_3)_2](B_3H_8)$[8] a compound originally formulated $B_4H_{10} . 2 NH_3$ which is prepared from $B_4H_{10} + NH_3$ in ether at $-78°C$.

Examples of covalent molecules in which B_3H_8 is attached to the metal by hydrogen bridges include $[(C_6H_5)_3P]_2 . Cu(B_3H_8)$[9] and $[Cr(CO)_4B_3H_8]$ $N(CH_3)_4$,[10] Fig. 24.23(g) and (h). All H atoms were located in an X-ray study of $HMn_3(CO)_{10}(BH_3)_2$, (i), the molecule of which contains not only metal–H–boron bridges but also Mn–H–Mn bridges.[11]

(8) JACS 1960 **82** 5758

(9) IC 1969 **8** 2755
(10) IC 1970 **9** 367

(11) JACS 1965 **87** 2753

TABLE 24.8

Polyhedral borohydride ions and carboranes

Configuration	$B_nH_n^{2-}$	$B_{n-2}C_2H_n$ or derivative	Reference
Trigonal bipyramid	$B_5H_5^{2-}$		
		$B_3C_2H_5$	IC 1973 **12** 2108
Octahedron	$B_6H_6^{2-}$		IC 1965 **4** 917
Pentagonal bipyramid	$B_7H_7^{2-}$		
		$B_5C_2H_7$	JCP 1965 **43** 2166
Dodecahedron	$B_8H_8^{2-}$		IC 1969 **8** 2771
		$B_6C_2H_6(CH_3)_2$	IC 1968 **7** 1070
Tricapped trigonal prism	$B_9H_9^{2-}$		IC 1968 **7** 2260
		$B_7C_2H_7(CH_3)_2$	IC 1968 **7** 1076
Bicapped square antiprism	$B_{10}H_{10}^{2-}$		JCP 1962 **37** 1779
11-skeleton			
		$B_9C_2H_9(CH_3)_2$	JACS 1966 **88** 4513
Icosahedron	$B_{12}H_{12}^{2-}$		JACS 1960 **82** 4427
		$B_{10}C_2H_{10}Br_2$	IC 1967 **6** 874
Composite ions			
	$B_{20}H_{18}^{2-}$		IC 1971 **10** 151
	$B_{20}H_{18}^{2-}$ (photoisomer)		IC 1968 **7** 1085
	$B_{20}H_{18}NO^{3-}$		IC 1971 **10** 160
Icosahedral fragments			
	$B_{11}H_{13}^{2-}$		IC 1967 **6** 1199
		$B_7C_2H_{11}(CH_3)_2$	IC 1967 **6** 113
		$B_4C_2H_8$	IC 1964 **3** 1666

Polyhedral ions. From the structural standpoint the simplest polyhedral ions are the family $B_nH_n^{2-}$ which are known for $n = 6$ to 12 inclusive and are conveniently grouped with the isoelectronic carboranes $B_{n-2}C_2H_n$. These ions are usually isolated in alkali metal or substituted ammonium salts. The boron (or boron–carbon) skeletons are the highly symmetrical triangulated polyhedra listed in the upper part of Table 24.8, as shown by structural studies of ions, carboranes, or substituted carboranes for which references are given. One terminal H is attached to each vertex of the polyhedral B_n or $B_{n-2}C_2$ nucleus; there are no bridging H atoms.

In addition to ions $B_nH_n^{2-}$ numerous ions containing more H atoms have been prepared, and the structures of some have been studied. Such ions include: $B_5H_8^-$, $B_6H_9^-$, $B_9H_{14}^-$, $B_{10}H_{12}^{2-}$, $B_{10}H_{13}^-$ $B_{10}H_{14}^{2-}$, $B_{10}H_{15}^-$, and $B_{11}H_{13}^{2-}$. In ions of this type the excess of H atoms, over the number required for one terminal H on each B, are available for bridge formation, which results in rearrangement of the boron skeleton. The same situation arises in the carboranes. Thus $B_7C_2H_7(CH_3)_2$ is a substituted derivative of a borane of the $B_{n-2}C_2H_n$ family, and the B_7C_2 polyhedron is the tricapped trigonal prism in which C atoms cap two of the prism faces (Fig. 24.24(a)). On the other hand, $B_7C_2H_{11}(CH_3)_2$ is a derivative of $B_7C_2H_{13}$ and the skeleton is an 'opened out' icosahedral fragment with two H bridges and one H on each B and C atom (Fig. 24.24(b)). Similarly, $B_4C_2H_6(CH_3)_2$ has the form of a pentagonal pyramid (not an octahedron, like $B_6H_6^{2-}$), while $B_{11}H_{13}^{2-}$ is strictly an icosahedron less one vertex (two H bridges), Fig. 24.24(c), in

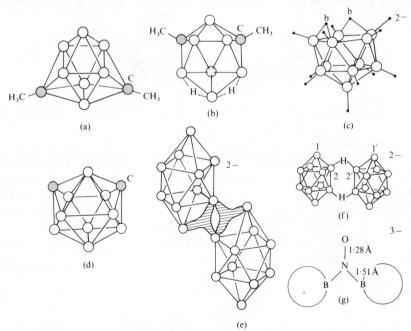

FIG. 24.24. Molecular structures of (a) $B_7C_2H_7(CH_3)_2$, (b) $B_7C_2H_{11}(CH_3)_2$, (c) $B_{11}H_{13}^{2-}$, (d) $B_9C_2H_{11}$, (e) and (f) $B_{20}H_{18}^{2-}$, (g) $(B_{20}H_{18}NO)^{3-}$.

contrast to $B_9C_2H_{11}$ which has the more symmetrical shape (d) formed by adding one B atom to the skeleton of the $B_{10}H_{14}$ type.

Composite ions include the two isomers of $B_{20}H_{18}^{2-}$, (e) and (f), and the $B_{20}H_{18}NO^{3-}$ ion, (g). The parent ion $B_{20}H_{18}^{2-}$, (e) is formed by oxidation of $B_{10}H_{10}^{2-}$ by ferric or ceric ion; action of u.v. light in acetonitrile solution gives the photoisomer (f). The ion (g) is formed by the action of NO on $B_{20}H_{18}^{2-}$.

Metal derivatives of carboranes. The last group of metal compounds to be mentioned here are those in which one or more carborane ions are π-bonded to a transition-metal atom to form either a composite ion or a neutral molecule. Examples are shown in the self-explanatory Fig. 24.25, namely, the ions $[Co(B_9C_2H_{11})_2]^{-}$[1] and $[B_9C_2H_{11}Re(CO)_3]^{-}$,[2] both studied in their Cs salts,

(1) IC 1967 **6** 1911
(2) IC 1966 **5** 1189

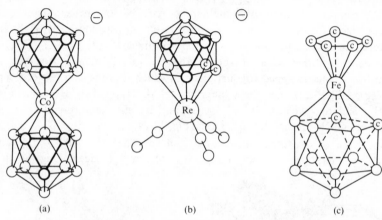

(a)   (b)   (c)

FIG. 24.25. The molecular structures of (a) $[Co(B_9C_2H_{11})_2]^{-}$, (b) $[B_9C_2H_{11}Re(CO)_3]^{-}$, (c) $Fe(C_5H_5)(B_9C_2H_{11})$. H atoms are omitted.

(3) JACS 1965 **87** 3988

(4) JACS 1970 **92** 1173
(5) JACS 1968 **90** 4828
(6) JACS 1970 **92** 1187

and the 'sandwich' molecule $Fe(\pi\text{-}C_5H_5)(\pi\text{-}B_9C_2H_{11})$.[3] A number of molecules and ions of the type of Fig. 24.25(a) have been studied containing Fe, Co, Ni, Cu, and Au in various oxidation states and in some cases with substituents in place of some of the H atoms, for example, $Ni^{IV}(B_9C_2H_{11})_2$,[4] $[Cu^{III}(B_9C_2H_{11})_2]^{-}$,[5] and $[Ni^{II}(B_9C_2H_{11})_2]^{2-}$.[6] Some of these complexes have the symmetrical staggered configuration of Fig. 24.25(a) while others have a less symmetrical 'sheared' structure; the difference may be associated with the number of d electrons on the 3d transition-metal atom. In the anion in $Cs_2[(B_9C_2H_{11})Co(B_8C_2H_{10})Co(B_9C_2H_{11})].H_2O$ the metal atoms are bridged by a 10-atom icosahedral fragment and also bonded to 11-atom fragments.[7]

(7) IC 1969 **8** 2080

25

Copper, Silver, and Gold

Valence states

Each of these elements follows a transition metal (Ni, Pd, and Pt respectively) with a completed d shell. They might be expected to behave as non-transition metals and to form ions M^+ by loss of the single electron in the outermost shell or to use the s and p orbitals of that shell to form collinear sp or tetrahedral sp^3 bonds. In fact, both Cu and Ag form ions M^+ and bonds of both these types, but Au^+ is not known and Au(I) shows a marked preference for 2- as opposed to 4-coordination. Moreover, all these elements exhibit higher valences as a result either of losing one or more d electrons (e.g. Cu^{2+}) or of utilizing d orbitals of the penultimate shell in combination with the s and p orbitals of the valence shell. In these higher valence states these elements have some of the characteristic properties of transition metals, for example, the formation of coloured paramagnetic ions. This dual behaviour greatly complicates the structural chemistry of these metals. Since not very much is known of the structural chemistry of these elements in certain oxidation states we shall deal separately and in more detail with Cu(I), Ag(I), and Au(I), Cu(II), and Au(III), and include what is known about Cu(III), Ag(II), and Ag(III) in our preliminary survey.

In spite of the general similarity in the electronic structures of their atoms, Cu, Ag, and Au differ very considerably in their chemical behaviour. First, the valences exhibited in their common compounds are: Cu, 1 and 2, Ag, 1, and Au, 1 and 3. In addition, Cu and Ag may be oxidized to the states Cu(III) and Ag(II) and Ag(III) respectively, but no simple compounds of Au(II) are known. Most crystalline compounds apparently containing Au(II), for example, $CsAuCl_3$ and $(C_6H_5CH_2)_2S . AuBr_2$,[1] actually contain equal numbers of Au(I) and Au(III) atoms. The first fully characterized paramagnetic compound of Au(II) is $[Au^{II}(mnt)_2] [(n-C_4H_9)_4N]_2$,[2] which is stable in the absence of air but oxidizes rapidly in solution; according to an e.s.r. study the phthalocyanin is a derivative of Au(II).[3] Second, the more stable ion of Cu is the (hydrated) Cu^{2+} ion, whereas that of silver is Ag^+. In contrast to copper and silver, there is no ionic chemistry of gold in aqueous solution, for the Au^+ and Au^{3+} ions do not exist in aqueous solutions of gold salts, at least in any appreciable concentration. The only water-soluble compounds of gold, aurous or auric, contain the metal in the form of a complex ion as, for example, in solutions of $K[Au(CN)_2]$ or $Na_3[Au(S_2O_3)_2]$. $2 H_2O$. Coordination compounds of Au are considerably more stable than the

(1) JCS 1952 3686

(2) JACS 1965 87 3534

(3) JACS 1965 87 2496

mnt = maleonitrile dithiolate

corresponding simple salts. For example, AuCl is readily decomposed by hot water, which does not affect $[Au(etu)_2]Cl . H_2O$. Similarly, aurous nitrate has not been made but $[Au(etu)_2]NO_3$ is a quite stable compound. Auric nitrate can be prepared under anhydrous conditions and complex auric nitrates such as $K[Au(NO_3)_4]$ are known.

As already noted, the more stable ion of copper in aqueous solution is the cupric ion. Cuprous oxy-salts such as Cu_2SO_4 are decomposed by water, $2 Cu^+ \rightarrow Cu + Cu^{2+}$, and $CuNO_3$ and CuF are not known. Although anhydrous Cu_2SO_3 is not known, the pale-yellow $Cu_2SO_3 . \frac{1}{2} H_2O$ can be prepared, and also NH_4CuSO_3 and $NaCuSO_3 . 6 H_2O$.[4] We refer later to salts containing both Cu(I) and Cu(II). The stable cuprous compounds are the insoluble ones, in which the bonds have appreciable covalent character (the halides, Cu_2O, Cu_2S), and the halides and the cyanide are actually more stable in the presence of water than the cupric compounds. Thus CuI_2 and $Cu(CN)_2$ decompose in solution to the cuprous compounds. The cuprous state is, however, stabilized by coordination, and derivatives such as $[Cu(etu)_4]NO_3$ and $[Cu(etu)_3]_2SO_4$ are much more stable than the simple oxy-salts (etu = ethylene-thiourea). In pyridine the equilibrium $2 Cu^+ \rightleftharpoons Cu^{2+} + Cu$ is strongly in favour of Cu^+. In general the cupric salts of only the stronger acids are stable, $Cu(NO_3)_2$ and $CuSO_4$ for example; those of weaker acids are unstable, and only 'basic salts' are generally known, as in the case of the carbonate, nitrite, etc. However, if a coordinated ion such as $Cu(en)_2^{2+}$ is formed, then stable compounds result, for example, $[Cu(en)_2](NO_2)_2$, $[Cu(en)_2]SO_3$ etc. If methyl sulphide is added to a solution of a cupric salt, the cup*rous* compound is precipitated, while if ethylene diamine is added to a solution of cup*rous* chloride in KCl (in absence of air) the cup*ric* ion $Cu(en)_2^{2+}$ is formed with precipitation of Cu. From these facts it is clear that the relative stabilities of Cu^+ and Cu^{2+} cannot be discussed without reference to the environment of the ions, that is, the neighbouring atoms in the crystal, solvent molecules or coordinating ligands if complex ions are formed. For the reaction

$$2 Cu^+ (g) \rightarrow Cu^{2+} (g) + Cu (s)$$

ΔH = +870 kJ mol^{-1}, corresponding to a large absorption of energy. If, however, we wish to compare the stabilities of the two ions in the crystalline or dissolved states this ΔH will be altered by a (large) amount corresponding to the difference between the interactions of Cu^+ and Cu^{2+} with their surroundings, as represented by lattice energies or solvation energies. These will be much greater for Cu^{2+} than for Cu^+, so that $\Delta H'$ may become negative, that is, ionic cuprous salts are less stable than cupric. With increasing covalent character of the Cu–X bonds, $\Delta H'$ again becomes positive, and in the case of the iodide and the cyanide it is the cuprous compound which is more stable. The actual configuration of a coordinating molecule may be important in determining the relative stabilities of cuprous and cupric compounds, as shown by the following figures for $(Cu^{II})/(Cu^I)^2$ in the presence of various diamines.

(4) JINC 1964 **26** 1122

$$(Cu^{II})/(Cu^{I})^2$$

Ethylene diamine	$\sim 10^5$
Trimethylene diamine	$\sim 10^4$
Pentamethylene diamine	3×10^{-2}
(cf. Ammonia	2×10^{-2})

Whereas ammonia stabilizes Cu^I in the reaction

$$2\,Cu(NH_3)_2^+ \rightleftharpoons Cu(NH_3)_4^{2+} + Cu,$$

the first two diamines stabilize the cupric state. These compounds can form chelate complexes with Cu^{II} but apparently not with Cu^I, while pentamethylene diamine presumably can be attached by only one NH_2 to either Cu^{II} or Cu^I and therefore behaves like ammonia.

Compounds of Cu(III)

Very few compounds of Cu(III) are known. The hydrated periodates $M_nH_{7-n}Cu(IO_6)_2$ (M is an alkali metal) and also $KCuO_2$[1] are diamagnetic, from which it was concluded that the metal atoms are forming four coplanar (dsp^2) bonds as in the diamagnetic planar 4-coordinated complexes of Au(III). This bond arrangement has been demonstrated for (a), which is diamagnetic and isostructural with the Au(III) compound.[2] Both Cu^{III}–Br (2·31 Å) and Cu^{III}–S (2·19 Å) are shorter than the corresponding bonds to Cu(II). In crystalline $Na_3KH_3[Cu(IO_6)_2]\cdot 14\,H_2O$[3] there are complex anions, (b), consisting of two octahedral IO_6 groups linked by a Cu atom (Cu–4 O, 1·9 Å) which forms a fifth bond to a water molecule (Cu–O, 2·7 Å). The complex fluoride K_3CuF_6,[4] on the other hand, is paramagnetic with μ_{eff} corresponding to 2 unpaired electrons, and Cu(III) has

(1) ZaC 1952 **270** 69

(2) IC 1968 **7** 810

(3) NW 1960 **47** 377

(4) AnC 1950 **62** 339

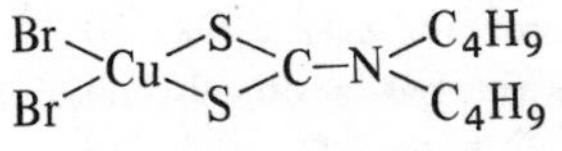

(a)

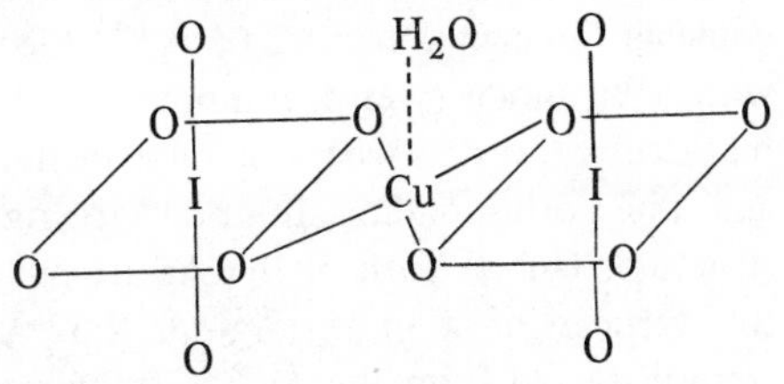

(b)

been provisionally assigned an octahedral ($4s4p^3 4d^2$) configuration in this compound.

Higher oxidation states of Ag

The simple argentic ion has been produced in concentrated nitric acid solution, but apart from the fluoride, AgF_2, compounds of Ag(II) can be prepared only in the

(1) JPCS 1971 **32** 543

presence of molecules which will coordinate to the metal and form complex ions. The structure of AgF_2,[1] which is quite different from that of CuF_2, is described on p. 223. The nitrate is formed by anodic oxidation of $AgNO_3$ solution in the presence of pyridine and isolated as $[Ag(py)_4](NO_3)_2$ and the persulphate is formed by double decomposition and isolated as $[Ag(py)_4]S_2O_8$. Many other coordination compounds of Ag(II) have been prepared, and their magnetic moments (around 2 BM) correspond to 1 unpaired electron. An incomplete X-ray

(2) JCS 1936 775

study[2] of the isostructural cupric and argentic salts of picolinic acid, (a), provides the only evidence for the coplanar arrangement of four bonds formed by Ag(II) in a

(a)

finite complex. A preliminary study of $AgL_2 . H_2O$ (L = pyridine-2,6-dicarboxylate) indicates a distorted octahedral structure in which the two ligands coordinate to the metal with different Ag—O and Ag—N bond lengths (Ag—O, 2·20 and 2·54 Å, Ag—N, 2·08 and 2·20 Å).[3]

(3) JACS 1969 **91** 7769

The oxide AgO is prepared by slow addition of $AgNO_3$ solution to an alkali persulphate solution, and is commercially available. It is a black crystalline powder which is a semiconductor and is diamagnetic. It is $Ag^IAg^{III}O_2$, and its structure is a distorted form of the PtS structure, another variant of which is the structure of CuO (tenorite). In CuO all Cu atoms have 4 coplanar neighbours, but in AgO Ag(I) has two collinear O neighbours (Ag^I—O, 2·18 Å), the other two atoms of the original square planar coordination group being at 2·66 Å, while Ag^{III} has 4

(4) JES 1961 **108** 819

coplanar O neighbours at 2·05 Å.[4] Electrolysis of an aqueous solution of $AgNO_3$ with a Pt anode gives a compound with the empirical composition Ag_7NO_{11} as

(5) AC 1965 **19** 180

black cubic crystals with a metallic lustre. This is an oxynitrate, $Ag(Ag_6O_8)NO_3$[5] and the corresponding fluoride, $Ag(Ag_6O_8)F$ apparently has a closely related structure. One-seventh of the Ag atoms (presumably Ag^+) have 8 O neighbours at the vertices of a cube (Ag—O, 2·52 Å as in $AgClO_3$). The remainder are all equivalent and form the Ag_6O_8 framework; they have a square coordination group (Ag—O, 2·05 Å). There is clearly interchange of electrons between Ag atoms in higher oxidation states, to give a neutral framework of composition Ag_6O_8, leading to the colour and semiconductivity (compare the bronzes, p. 505). This compound, like AgO, contains the metal in more than one oxidation state. Diamagnetic compounds containing exclusively Ag(III) presumably include the yellow salts $KAgF_4$ and $CsAgF_4$ (which are readily decomposed by moisture), salts such as $K_6H[Ag(IO_6)_2] . 10 H_2O$ and $Na_6H_3[Ag(TeO_6)_2] . 18 H_2O$ which are probably structurally similar to the Cu(III) compounds, and the red salts (sulphate, nitrate, etc.) of the very stable ethylenebiguanide complex (b).

(b)

The structural chemistry of Cu(I), Ag(I), and Au(I)

The simplest possibilities are the formation of two collinear (sp) or four tetrahedral (sp^3) bonds. In addition, Cu(I) and Ag(I) form three bonds in a number of crystals, and we shall give examples of this bond arrangement after dealing with the two simpler ones.

The formation of two collinear bonds by Cu(I), Ag(I), *and* Au(I)

The formation of only two collinear bonds by Cu(I) is very rare, and is observed only with the electronegative O atom, in Cu_2O (with which Ag_2O is isostructural), $CuFeO_2$ and $CuCrO_2$ (p. 478) and also apparently in $KCuO$, with which $KAgO$, $CsAgO$, and $CsAuO$ are isostructural.[1a] Although two collinear bonds are formed to N atoms in the diazoaminobenzene compound (a), (of length 1·92 Å) the Cu–Cu distance (2·45 Å) is less than in metallic Cu (2·56 Å) and presumably indicates some interaction between the metal atoms.[1b] The formation of metal–metal bonds in addition to 2, 3, or 4 bonds to non-metals seems to be a feature of the structural chemistry of Cu(I) and Ag(I). The reluctance of Cu(I) to form only two collinear bonds leads to many differences between the chemistry of Cu(I) on the one hand and Ag(I) and Au(I) on the other. For example, whereas Ag(I) and Au(I) form two collinear bonds in AgCN and AuCN and also in the $M(CN)_2^-$ ions, CuCN has a much more complex (unknown) structure with 36 CuCN in the unit cell,[2] and in $KCu(CN)_2$ Cu(I) forms three bonds (p. 884). With long-chain amines cuprous halides form 1:1 and 1:2 complexes. The former are tetrameric and presumably similar structurally to $[CuI . As(C_2H_5)_3]_4$ (p. 883), while the 1:2 complexes are dimeric in benzene solution and may be bridged molecules, $(RNH_2)_2Cu . X Cu(NH_2R)_2$. The compounds formed by Cu(I) and Au(I) of the type P(N)–M–C, where P(N) is a phosphine (amine) and C an acetylene, have entirely different structures. The gold compounds form linear molecules, for example,

(1a) ZaC 1968 **360** 113

(1b) AC 1961 **14** 480

$$(\text{i-}C_3H_7)H_2N \underset{2\cdot03\ \text{Å}}{\rule{2.5cm}{0.4pt}} Au \underset{1\cdot94\ \text{Å}}{\rule{2.5cm}{0.4pt}} C{\equiv}C(C_6H_5)^{[3]}$$

$$C_6H_5N{-}Cu{-}NC_6H_5$$
$$N \qquad\qquad N$$
$$C_6H_5N{-}Cu{-}NC_6H_5$$

(a)

(2) AIC 1957 **28** 316

(3) AC 1967 **23** 156

whereas the phosphine $(CH_3)_3P{-}Cu{-}C{\equiv}C(C_6H_5)$ is tetrameric (p. 883), though with a quite different structure from $(R_3PCuI)_4$.

There are numerous examples of Ag(I) forming two collinear bonds. In AgCN there are infinite linear chains of metal atoms linked through CN groups (Ag–Ag, 5·26 Å),[4] and very similar chains exist in $Ag_3CN(NO_3)_2$, where the NO_3^- ions and the other Ag^+ ions lie between the chains.[5] There are also linear anions as in $K[Ag(CN)_2]$ (Ag–C, 2·13 Å, C–N, 1·15 Å)[6] and linear cations as in $[Ag(NH_3)_2]_2SO_4$. Although two bonds from Ag(I) are usually collinear this is apparently not always so. In crystalline AgSCN (p. 747) a bond angle of 165° was found, and in the complex sulphides proustite, Ag_3AsS_3, and pyrargyrite, Ag_3SbS_3, the same value was found for the S–Ag–S bond angle.[7] In the complex with pyrazine, $AgNO_3 . N_2C_4H_4$,[8] Ag forms two bonds to N but also four weak bonds to O (two of length 2·72 Å, two of 2·94 Å), (b), while in $KAgCO_3$[9]

(4) ZK 1935 **90** 555
(5) AC 1965 **19** 815
(6) ZK 1933 **84** 231

(7) JCP 1936 **4** 381
(8) IC 1966 **5** 1020
(9) JCS 1963 2807

(10) ZaC 1959 **299** 328

there are infinite zigzag chains of Ag atoms bridged by CO_3 groups, (c). In the remarkable compound $(Ag_3S)NO_3$,[10] formed from CS_2 and concentrated $AgNO_3$ solution, the NO_3^- ions are situated in the interstices of a framework built from SAg_6 groups (with a configuration intermediate between an octahedron and a trigonal prism). Here also there are non-linear bonds from the Ag atoms (S–Ag–S, 157°).

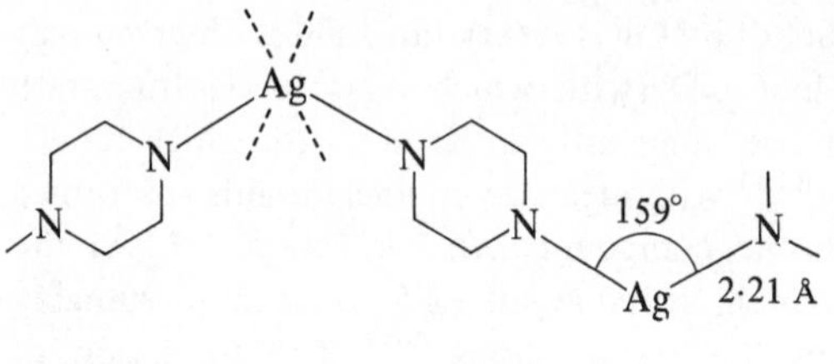

(b)

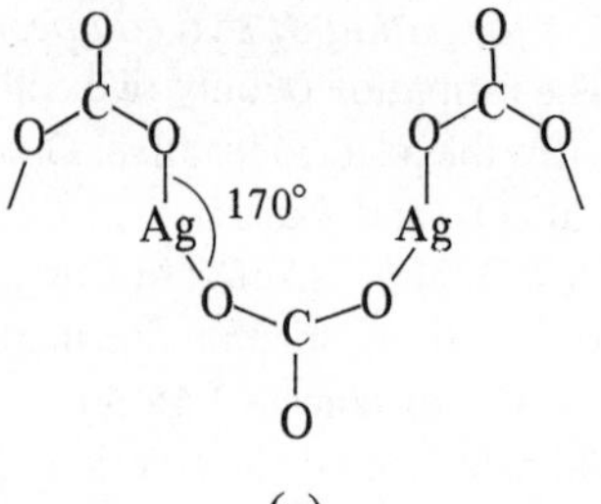

(c)

(11) AC 1959 **12** 709

For Au(I) two is the preferred coordination number. AuCN and $K[Au(CN)_2]$[11] are isostructural with the Ag compounds; only approximate bond lengths were determined. The most stable amino derivatives of AuCl are $H_3N \cdot AuCl$ and $(H_3N-Au-NH_3)Cl$, and whereas with ligands such as thioacetamide Cu and Ag form salts of type (d), Au forms only the rather unstable salt (e).

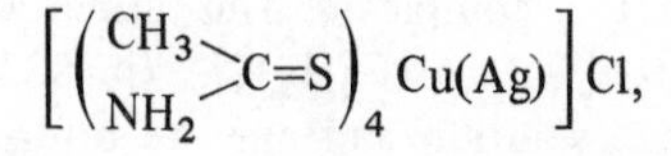

(d)

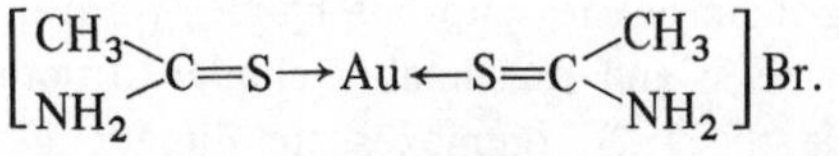

(e)

With trialkyl phosphines and arsines cuprous and argentous halides form the tetrameric molecules $[R_3P(As) \cdot Cu(Ag)X]_4$, but Au forms only $R_3P(As) \cdot AuX$. In addition to the cases already noted, X-ray studies have demonstrated the formation of two collinear bonds by Au(I) in AuI,[12] which consists of chains of the type

(12) ZK 1959 **112** 80

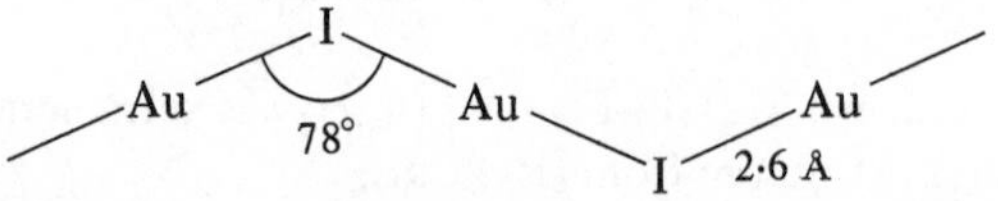

(13) RTC 1962 **81** 307
(14) JACS 1938 **60** 1846

in $Cl-Au-PCl_3$,[13] and in $Cs_2(Au^{I}Cl_2)(Au^{III}Cl_4)$[14] (p. 393).

The formation of four tetrahedral bonds by Cu(I) *and* Ag(I)

No example of Au(I) forming four tetrahedral bonds has yet been established by a structural study, but tetrahedral bonds are formed in numerous compounds of Cu(I) and Ag(I), the simplest examples being the cuprous halides and AgI with the zinc-blende structure (p. 349).

Finite complexes of Cu(I) include the $Cu(CN)_4^{3-}$ ion (mean Cu–C, 2·00 Å),[1] the thioacetamide complex $[Cu\{SC(NH_2)(CH_3)\}_4]Cl$[2] (with which the Ag compound is isostructural), and the assortment of molecules shown in Fig. 25.1.

(1) AC 1968 **B24** 269
(2) JCS 1962 1748

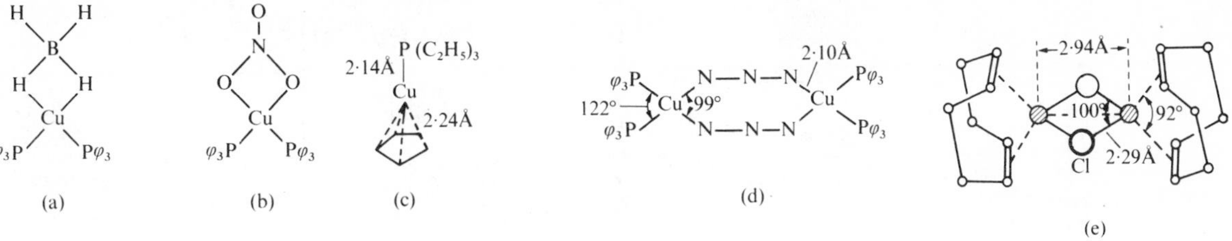

FIG. 25.1. Some molecules formed by Cu(I).

References (Fig. 25.1):
(a) JACS 1967 **89** 3929
(b) IC 1969 **8** 2750
(c) AC 1970 **B26** 515
(d) JACS 1970 **92** 738
(e) JACS 1963 **85** 1009

There is considerable distortion of the tetrahedral bond angles in the dimeric azido complex, in which the N_3 group is apparently symmetrical (mean N–N, $1·18$ Å), as compared with the normal unsymmetrical form (N–N, $1·15$ and $1·21$ Å) in transition-metal complexes in which it is attached to a metal through one N atom only. We include the cyclopentadienyl compound (c) since here C_5H_5 contributes 5 electrons and Cu acquires the Kr configuration as in the tetrahedral complexes. In C_6H_6 . $CuAlCl_4$[3] Cu has 3 Cl neighbours of different $AlCl_4^-$ ions and the C_6H_6 ring, which is distorted towards a cyclohexatriene structure, (a). We return to polymeric finite complexes later.

(3) JACS 1963 **85** 4046

In tris-thiourea Cu(I) chloride, $[Cu\{SC(NH_2)_2\}_3]Cl$[4] there are infinite chain ions, (b), in which one of the three thiourea molecules forms a bridge between two Cu atoms:

(4) AC 1964 **17** 928

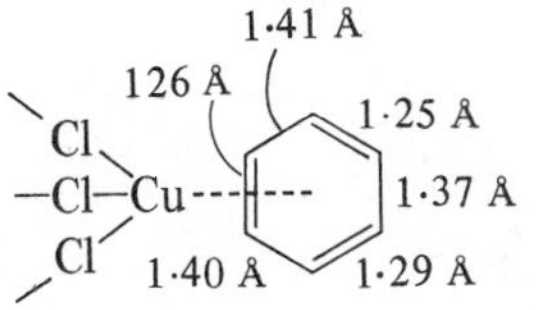

(a)

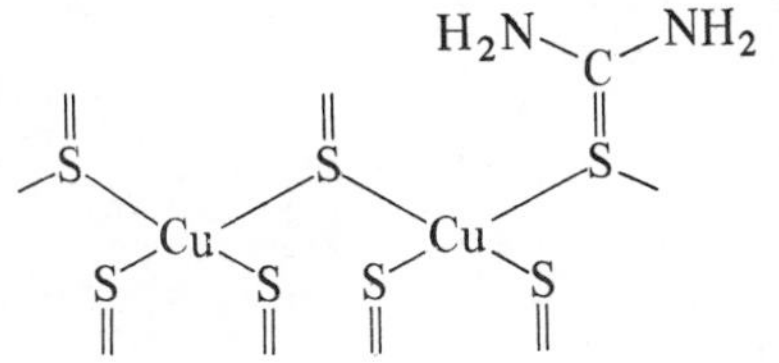

(b)

Chains built from CuX_4 tetrahedra form the infinite anions in a number of complex halides of Cu(I) and Ag(I)–Table 25.1. The two simplest are the vertex-sharing MX_3 (pyroxene) chain and the MX_2 chain in which the tetrahedral groups share opposite edges (as in $BeCl_2$ and SiS_2). There is some angular distortion of the $CuCl_4$ tetrahedra in the chain anion in the 'paraquat' salt, $pq^{2+}(CuCl_2)_2$, where pq^{2+} represents N,N'-dimethyl-4,4'-bipyridylium ion. The cation is an electron-acceptor, and the black colour, semiconductivity, and paramagnetism of this salt suggest that there is some interchange of electrons between Cu(I) and Cu(II) in the chains. The M_2X_3 chain found in $CsCu_2Cl_3$ and $CsAg_2I_3$ consists of double rows of tetrahedra in which each MX_4 group shares three edges and two vertices with other similar groups (Fig. 25.2(a)). Chains of a similar type form the anion in the diazonium salt $(C_6H_5N_2)(Cu_2Br_3)$, an intermediate in the Sandmeyer reaction. The more complex M_3X_5 chain of Fig. 25.2(b) is found in $(Cu_3^I Cl_5)(Cu^{II}L_2)$, where L is N-benzoylhydrazine. Within the chain distorted $CuCl_4$ tetrahedra share a vertex and either two or three edges. Each Cu(I) is bonded (through Cl) to 3 or 4

TABLE 25.1

Complex halides of Cu(I) *and* Ag(I)

Nature of M–X complex	Examples	Reference
MX_3 chain	K_2CuCl_3, Cs_2AgCl_3, Cs_2AgI_3, $(NH_4)_2CuCl_3$, $(NH_4)_2CuBr_3$, K_2AgI_3, Rb_2AgI_3, $(NH_4)_2AgI_3$	AC 1949 **2** 158; AC 1952 **5** 506; AC 1952 **5** 433
MX_2 chain	$[Ni(en)_2](AgBr_2)_2$	ACSc 1969 **23** 3498
	$(pq)(CuCl_2)_2$	JCS A 1969 1520
MXL' chain	$CuI . CH_3NC$	JCS 1960 2303
MXL'' chain	$CuBr[(C_2H_5)_2P\!-\!P(C_2H_5)_2]$	ZaC 1970 **372** 150
M_2X_3 chain (Fig. 25.2 (a))	$CsCu_2Cl_3$. $CsAg_2I_3$. $(C_6H_5N_2) . (Cu_2Br_3)$	AC 1954 **7** 176 CC 1965 299
M_3X_5 chain (Fig. 25.2 (b))	$(Cu_3^ICl_5)(L_2Cu^{II})$	IC 1971 **10** 138
M_2X_2L'' layer	$Cu_2Cl_2 . N_2(CH_3)_2$	AC 1960 **13** 28

other Cu(I) atoms in the chain, and certain of the Cl atoms (shaded) complete the (4 + 2)-coordination around the Cu(II) atoms.

The chain of Fig. 25.2(c) represents the structure of CuI . CH_3NC, in which the centrally situated Cu atoms are bonded to 4 I as in CuI and the outer ones to 2 I and 2 CH_3NC molecules. The composition of the chain is MXL', where L' is a

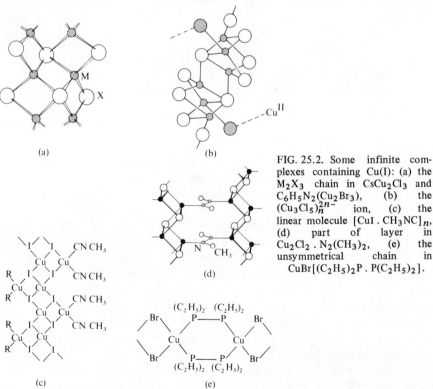

(a)

(b)

(c)

(d)

(e)

FIG. 25.2. Some infinite complexes containing Cu(I): (a) the M_2X_3 chain in $CsCu_2Cl_3$ and $C_6H_5N_2(Cu_2Br_3)$, (b) the $(Cu_3Cl_5)_n^{2n-}$ ion, (c) the linear molecule $[CuI . CH_3NC]_n$, (d) part of layer in $Cu_2Cl_2 . N_2(CH_3)_2$, (e) the unsymmetrical chain in $CuBr[(C_2H_5)_2P . P(C_2H_5)_2]$.

ligand forming a bond to one metal atom only. A ligand L″ capable of bonding at both ends can link together chains of type (a) to form layers of composition $Cu_2X_2L″$, as in $Cu_2Cl_2 . N_2(CH_3)_2$, Fig. 25.2(d). In this 'chain' the bonds shown as broken lines are appreciably weaker (2·55 Å) than those shown as full lines (2·35 Å); Cu–N, 1·99 Å. There are unsymmetrical chains of a novel type (Fig. 25.2(e)) in $CuBr(Et_2P–PEt_2)$.

We noted earlier the formation of one metal–metal bond by Cu(I), in addition to two bonds to N, in $[Cu(\phi NNN\phi)]_2$. There are other polymeric complexes in which Cu(I) or Ag(I) has close metal neighbours in addition to 3 or 4 non-metal atoms at normal distances. The tetrameric molecule $Cu_4I_4(AsEt_3)_4$ consists of a central tetrahedron of Cu atoms surrounded by tetrahedra of I atoms and triethylarsine molecules (Fig. 25.3(a)). In this molecule Cu has, in addition to 3 I and 1 As at

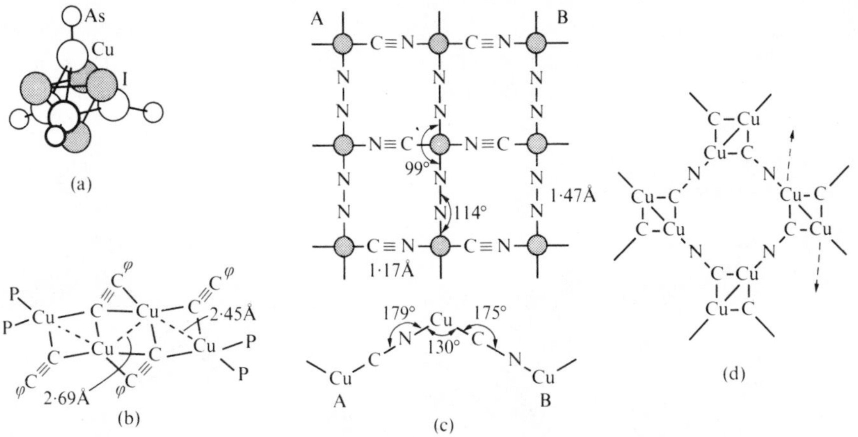

FIG. 25.3. (a) The molecule $Cu_4I_4[As(CH_3)_3]_4$, (b) $[(CH_3)_3PCuC\equiv C(C_6H_5)]_4$, (c) layer in $CuCN . N_2H_4$, (d) layer in $CuCN . NH_3$ (NH_3 omitted).

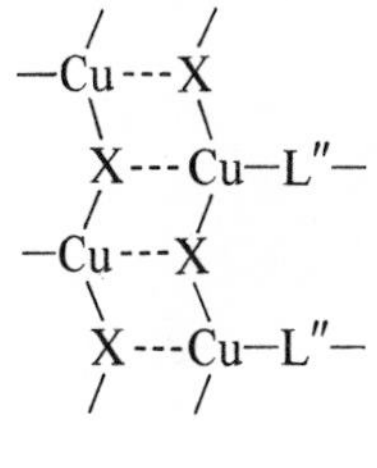

normal single bond distances, 3 Cu at 2·6 Å, similar to the interatomic distance in metallic Cu (2·56 Å).[5] The compound $(CH_3)_3P . Cu . C\equiv C(C_6H_5)$ forms tetrameric molecules of a quite different kind. The bonding system is obviously complex, involving Cu–C bonds (1·96–2·22 Å) and Cu–Cu bonds of two different lengths (Fig. 25.3(b)).[6]

Several cuprous compounds form layer structures based on the simplest plane 4-connected net. In the puckered layers in $CuCN . N_2H_4$,[7] Fig. 25.3(c), the bond lengths are: Cu–C, 1·93 Å, Cu–N, 1·95 Å, and Cu–2 N (of N_2H_4), 2·17 Å; the Cu(I) bond angles being 99° (three), 113° (two), and 130° (one). Nitriles of aliphatic dibasic acids can function as bridging ligands, $–NC–(CH_2)_n–CN–$ giving compounds of the type $(CuR_2)X$ where R is the nitrile and X, for example, NO_3^- or ClO_4^-. All the salts $Cu(succinonitrile)_2ClO_4$,[8] $Cu(glutaronitrile)_2NO_3$,[9] and $Ag(adiponitrile)_2ClO_4$[10] form layers based on the planar 4-gon net. The structure of $Cu(adiponitrile)_2NO_3$,[11] which consists of six interpenetrating identical 3D nets of the cristobalite type, was mentioned in Chapter 3.

(5) ZK 1936 **94** 447

(6) AC 1966 **21** 957

(7) AC 1966 **20** 279

(8) AC 1969 **B25** 1518
(9) BCSJ 1959 **32** 1216
(10) IC 1969 8 2768
(11) BCSJ 1959 **32** 1221

Copper, Silver, and Gold

(12) AC 1965 **19** 192

The ammine $CuCN \cdot NH_3$ is built of a unique kind of CuCN layer, Fig. 25.3(d), to each Cu of which one NH_3 is attached, the $Cu-NH_3$ bond being approximately perpendicular to the plane of the layer.[12] The distinction between C and N atoms is not certain, but the arrangement shown appears most likely. Within the planar C_2Cu_2 units (which are inclined to the plane of the figure) the $Cu-Cu$ distance (2·42 Å) is shorter than in the metal, and in addition to this bond Cu has four neighbours in a somewhat distorted tetrahedral arrangement ($Cu-N$, 1·98 Å, $Cu-NH_3$, 2·07 Å, $Cu-C$, 2·09 and 2·13 Å). The $Cu-C$ bonds are abnormally long, and the nature of the bonding is not clear.

Some further examples of tetrahedral Cu(I) are included in the section on compounds of Cu(I) and Cu(II).

The formation of three bonds by Cu(I) *and* Ag(I)

(1) JPC 1957 **61** 1388

As already noted, the structure of $KCu(CN)_2$[1] is quite different from that of the isostructural $KAg(CN)_2$ and $KAu(CN)_2$, which contain linear $(NC-M-CN)^-$ ions. The complex ions consist of helical chains (Fig. 25.4(a)) bonded by K^+ ions lying

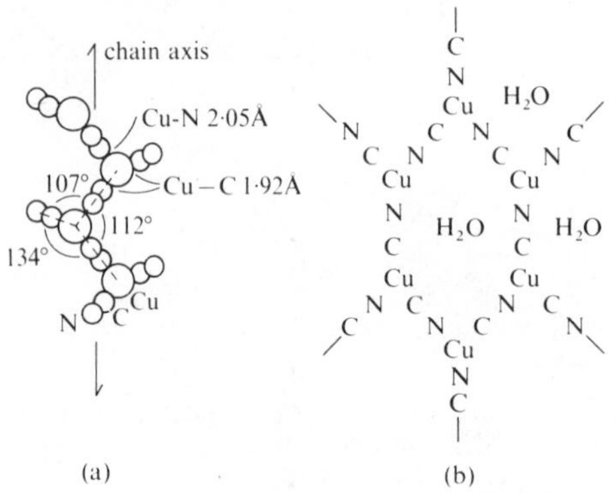

FIG. 25.4. (a) The helical chain ion in $KCu(CN)_2$, (b) the layer anion in $KCu_2(CN)_3 \cdot H_2O$.

between them. The interpretation of the bonding in the chains is not, however, clear-cut. The two $Cu-C$ bonds are equal in length (1·92 Å), but the $Cu-N$ bond in the chain is considerably longer (2·05 Å) than expected for a single bond ($\approx 1·85$ Å). We may therefore describe the Cu as forming three approximately coplanar bonds (possibly sp^2, two orbitals overlapping sp orbitals of C atoms and one overlapping the s orbital of N), or alternatively the stronger $Cu-C$ bonds form non-linear $NC-Cu-CN$ units which are then bonded weakly together by $Cu \cdots N$ bonds to form the chains. Whereas Ag(I) and Au(I) form only $KM(CN)_2$, Cu(I) also

(2) AC 1962 **15** 397

forms $KCu_2(CN)_3 \cdot H_2O$ and $K_3Cu(CN)_4$. The latter contains tetrahedral $Cu(CN)_4^{3-}$ ions (p. 880). In $KCu_2(CN)_3 \cdot H_2O$[2] each Cu(I) forms three nearly coplanar bonds, as in $KCu(CN)_2$, (interbond angles, two of 112° and one of 134°) in puckered layers of the simplest possible type (Fig. 25.4(b)). The lengths of the $Cu-C$ and $Cu-N$ bonds are very similar to those in $KCu(CN)_2$, and the shortest $Cu-Cu$ distance is 2·95 Å, as compared with 2·84 Å in $KCu(CN)_2$. The H_2O

884

molecules are not bonded to the metal atoms, but occupy holes in the layers, the latter being held together by the K^+ ions. We have noted the quite different behaviour of Cu(I) in $CuCN \cdot N_2H_4$ and $CuCN \cdot NH_3$ in the previous section.

Cuprous chloride forms crystalline complexes with numerous unsaturated hydrocarbons, of which one has already been mentioned. In several of these Cu(I) forms two bonds to Cl and a third to a multiple bond of the hydrocarbon, the three bonds from Cu being coplanar. In the 1,5-cyclooctatetraene complex, $CuCl \cdot C_8H_8$,[3] the Cu and Cl atoms form an infinite chain (Fig. 25.5(a)), while in the complexes with norbornadiene (C_7H_8)[4] and 2-butyne (C_4H_6)[5] there are Cu_4Cl_4 rings (Fig. 25.5(b)).

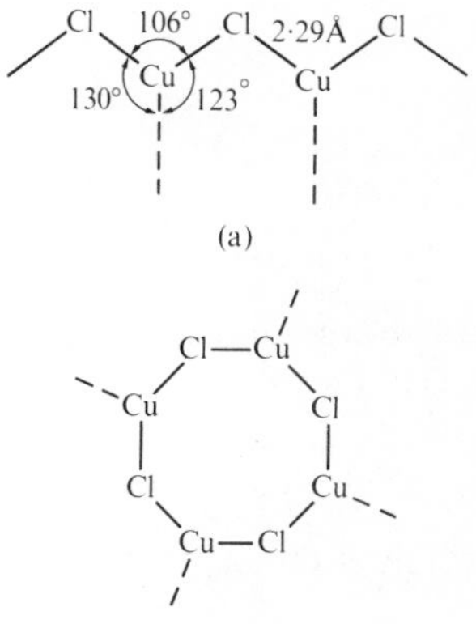

In polymeric complexes with certain S-containing ligands Cu(I) and Ag(I) form bonds to 3 S atoms and there are also 3 or 4 metal–metal bonds. The nucleus of the molecule of $Cu_4[(i-C_3H_7O)_2PS_2]_4$[6] is a tetrahedral Cu_4 group having four shorter (2·74 Å) and two longer (2·95 Å) edges to which are bonded four ligand molecules as shown in Fig. 25.6(a). Each Cu has 3 S and 3 Cu nearest neighbours. There is also a tetrahedral nucleus of metal atoms in $[(C_2H_5)_2NCS_2Cu]_4$.[7a] The type of polymer appears to be determined by the geometry of the ligand, for both $[(C_3H_7)_2NCOSCu]_6$[7b] and $[(C_3H_7)_2NCS_2Ag]_6$[7c] are hexameric. In the Cu_6 octahedron in the former Cu–Cu ranges from 2·70–3·06 Å (mean 2·88 Å). The hexameric molecule of silver dipropyldithiocarbamate, $[(C_3H_7)_2NCS_2Ag]_6$, contains a distorted octahedral Ag_6 nucleus which has six short edges (2·9–3·2 Å) and six longer ones (3·45–4·0 Å), the latter being the edges of a pair of opposite faces. The ligands behave in very much the same way as in the tetramer (a), one S being bonded to one Ag and the other to two Ag atoms. As a result each Ag is bonded to 3 S (Ag–S, 2·43–2·56 Å) and also has 2 Ag neighbours (Ag–Ag, 2·9, 3·0 Å) and 2 more distant Ag (3·45, 4·0 Å)–Fig. 25.6(b). The anion in salts $(Cu_8L_6)(As\phi_4)_4$,[8] in which L is the ligand (a), consists of a cubic cluster of 8 Cu atoms of edge 2·83 Å surrounded by 12 S atoms of the six ligands which are situated at the mid-points of the edges of a larger cube surrounding the Cu_8 group (Fig. 25.6(c)). Each Cu has 3 S and 3 Cu atoms as nearest neighbours at distances very similar to those in the tetramer described above (Cu–S, 2·25 Å).

FIG. 25.5. Cu–Cl complexes in (a) $CuCl \cdot C_8H_8$, (b) $CuCl \cdot C_4H_6$ and $CuCl \cdot C_7H_8$. Dotted lines indicate bonds from Cu to multiple bonds of hydrocarbon molecules.

(3) IC 1964 **3** 1529
(4) IC 1964 **3** 1535
(5) AC 1957 **10** 801

(6) IC 1972 **11** 612

(7a) AK 1963 **20** 481
(7b) ACSc 1970 **24** 1355
(7c) ACSc 1969 **23** 825

(8) JACS 1968 **90** 7357

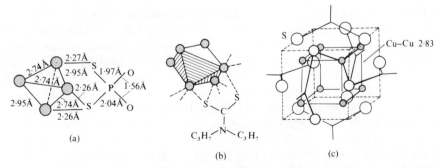

FIG. 25.6. Details of some polymeric Cu(I) and Ag(I) complexes (see text). In (b) the long Ag–Ag bonds are the edges of the unshaded face and of the opposite face of the Ag_6 octahedron.

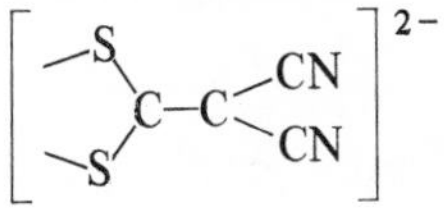

(9) IC 1966 **5** 1193

The unique structure adopted by silver tricyanomethanide, $Ag[C(CN)_3]$,[9] is illustrated in Chapter 3 (p. 90). It consists of pairs of interwoven hexagonal layers in which Ag is approximately 0·5 Å out of the plane of its three N neighbours. Presumably single plane layers would have too open a structure—compare $KCu_2(CN)_3 . H_2O$, in which there are K^+ ions and H_2O molecules to occupy the large holes in the sheets.

Three bonds are apparently formed by Cu(I) in certain sulphides, the bonding in which is discussed in a later section of this chapter, and also by Ag(I) in a number of S-containing compounds, but here also there is often no simple interpretation of the bonding. In Ag_3AsS_3 there is both planar and pyramidal 3-coordination of Ag,[10] while in $Ag(thiourea)_2Cl$[11] Ag has three approximately coplanar S neighbours but also one Cl at a rather large distance (2·95 Å).

(10) AC 1968 **B24** 77
(11) IC 1968 **7** 1351

Although the stereochemistry of Ag(I) is simpler than that of Cu(I) in some compounds the reverse is true in many complexes with organic compounds. For example, the structure of $C_6H_6 . AgAlCl_4$[12] is much less simple than that of the Cu(I) compound (p. 780), Ag forming four bonds to Cl (2·59, 2·77, 2·80, and 3·03 Å) and a fifth to one bond of the benzene molecule. In $(C_6H_5C{\equiv}C)Ag$ $[P(CH_3)_3]$[13] there are two types of Ag atom with entirely different environments. One type forms 2 collinear bonds to terminal C atoms of $C{\equiv}C . C_6H_5$ and the other two bonds to $P(CH_3)_3$ molecules (P–Ag–P, 118°) and two bonds to triple bonds of the hydrocarbon.

(12) JACS 1966 **88** 3243

(13) AC 1966 **20** 502

Certain silver oxy-salts, in particular the perchlorate, are notable for forming crystalline complexes with organic compounds. For $AgClO_4 . C_6H_6$ and $AgNO_3 . C_8H_8$ see p. 780. In the compound with dioxane, $AgClO_4 . 3 C_4H_8O_2$,[14] the Ag atoms are at the points of a simple cubic lattice, the ClO_4^- at the body-centre, and the dioxane molecules along the edges. The metal ions are therefore surrounded octahedrally by six O atoms (at about 2·46 Å), presumably bonded by some sort of ion-dipole bonds. Both the perchlorate ions and the dioxane molecules are rotating, probably quite freely.

(14) AC 1956 **9** 741

Salts containing Cu(I) and Cu(II)

We comment later in this chapter on the difficulty of assigning oxidation numbers to Cu atoms in the sulphides, and in Chapter 17 we noted the structure of KCu_4S_3. Apart from these semi-metallic compounds there are normal coloured salts containing Cu(I) and Cu(II) each with an entirely different stereochemistry. In the red salt $Cu_2^ISO_3 . Cu^{II}SO_3 . 2 H_2O$[1] Cu(I) has a tetrahedral environment (though Cu–S, 2·14 Å, is remarkably similar to Cu–O, 2·11–2·14 Å), and Cu(II) has a typical (4 + 2)-octahedral coordination group. In the pale-violet $Na_4[Cu^{II}(NH_3)_4]$ $[Cu^I(S_2O_3)_2]_2$[2] the $Cu(NH_3)_4^{2+}$ ions are planar (with no other close neighbours of Cu(II)) and the anion consists of chains (a), in which Cu(I) has four tetrahedral S neighbours at 2·36 Å. In contrast to the pale-yellow α and β CuNCS and the light-blue $Cu(NCS)_2(NH_3)_2$ the compound $Cu_2(NCS)_3(NH_3)_3$ is greenish-black.[3] There is tetrahedral coordination of Cu(I), by 2 N and 2 S of NCS groups, and

(1) ACSc 1965 **19** 2189

(2) AC 1966 **21** 605

(3) IC 1969 **8** 304

tetragonal pyramidal coordination of Cu(II), (b). The Cu(I) and Cu(II) atoms presumably interact through the S atoms, (c).

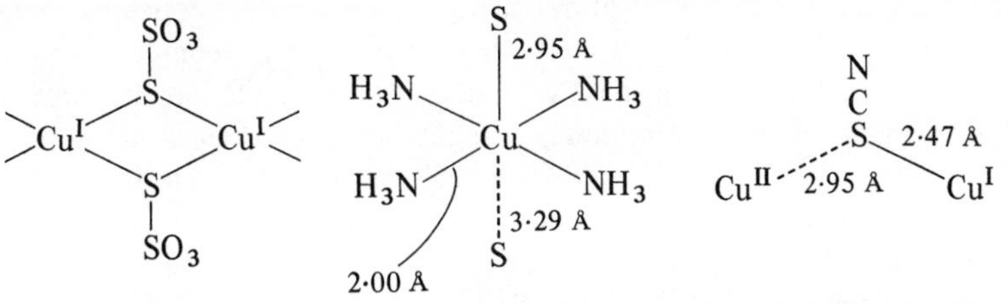

(a)   (b)   (c)

The structural chemistry of cupric compounds

This subject is of quite outstanding interest, for in one oxidation state this element shows a greater diversity in its stereochemical behaviour than any other element. Before commencing this survey it is interesting to note that a number of simple cupric compounds are either difficult to prepare or are not known. Although crystalline $Cu(OH)_2$ can be prepared by special methods (for example, by treating $Cu_4(OH)_6SO_4$ with aqueous NaOH, dissolving the hydroxide in concentrated NH_4OH and removing the NH_3 slowly *in vacuo* over H_2SO_4), the ready decomposition of the freshly precipitated compound is well known. (For the structure of $Cu(OH)_2$ see p. 521.) Two hydroxycarbonates, azurite, $Cu_3(OH)_2(CO_3)_2$, and malachite, $Cu_2(OH)_2CO_3$, are well-known minerals, but the normal carbonate, $CuCO_3$, is not known, though a rhombohedral substance which may have been $CuCO_3$ has been produced in admixture with malachite by hydrothermal treatment of a precipitated hydroxycarbonate.[1] A number of oxy-salts (e.g. nitrate and perchlorate) crystallize from aqueous solution as hydrates which cannot be dehydrated to the anhydrous salts, but both $Cu(NO_3)_2$ and $Cu(ClO_4)_2$ have been prepared by other methods (p. 662). They can both be volatilized without decomposition and dissolve in a number of polar organic solvents. The structures of some of the numerous hydroxy-salts are described in other chapters (Cu(OH)Cl and $Cu_2(OH)_3Cl$ in Chapter 10, and $Cu(OH)IO_3$ and $Cu_2(OH)_3NO_3$ in Chapter 14). (See also p. 143 for the relationship of atacamite, $Cu_2(OH)_3Cl$, to the NaCl structure.) The compounds CuI_2, $Cu(CN)_2$, and $Cu(SCN)_2$ are extremely unstable, and CuS is not a simple cupric compound (see later).

At one time it appeared that the characteristic behaviour of Cu(II) was the formation of four coplanar bonds, supposedly dsp^2 as in the case of Ni(II). This implied that the ninth 3d electron was placed in a 4p orbital:

(1) E 1960 **16** 447

		3s	3p	3d	4s	4p
2	8	2	6	8 + 2	2	4 + 1

shared

887

and a similar assumption was made for Ag(II). In view of the resistance of Cu(II) to further oxidation the existence of a single unpaired 4p electron seems unlikely, and it is incompatible with the results of paramagnetic resonance and optical studies. It should be emphasized that most of our structural information concerns compounds in which the atoms bonded to Cu(II) are N, O, F, or Cl; comparatively little is known about the bond arrangement when the ligands are the less electronegative S, I, CN, etc., many of these compounds being much less stable (or unknown). Since therefore most of the Cu(II) bonds we shall be discussing are likely to have a considerable degree of ionic character it is now usual to relate the complex stereochemistry of Cu(II) to the asymmetry of the $Cu^{2+}(d^9)$ ion. Furthermore the above simple treatment does not account for the quite characteristic formation of one, or more often two, weaker bonds in addition to the four coplanar bonds, or for the fact that no cupric compounds apart from the phthalocyanin and similar complexes are isostructural with those of Ni(II) and Pd(II). The closest structural resemblances of Cu(II) to Pd(II) compounds are those of CuO to PdO and of $CuCl_2$ to $PdCl_2$, to both of which we refer later.

In fact, the only simple salts known to be isostructural with those of Cu(II), d^9, are certain compounds of Cr(II), d^4, as might be expected from ligand field considerations. This similarity does not, however, extend to all compounds of these elements in this oxidation state. For example, while $CrBr_2$ and CrI_2 have structures similar to $CuCl_2$ and $CuBr_2$, $CrCl_2$ has a distorted rutile-type structure (but distorted in a different way from CrF_2 and the isostructural CuF_2) in contrast to the CdI_2-like structure of $CuCl_2$. In contrast to the complex structure of CuS, CrS has the type of structure that might have been expected for CuS (and CuO); this is a structure in which Cr has four approximately equidistant coplanar neighbours and two more completing a distorted octahedron, that is, the (4 + 2)-coordination characteristic of many Cu(II) compounds. (The oxide CrO is not known.) Examples of pairs of isostructural Cu(II) and Cr(II) compounds include the difluorides and dibromides, $KCuF_3$ and $KCrF_3$, and the dimeric acetates $[M(CH_3COO)_2 . H_2O]_2$ (p. 897).

The stereochemistry of Cu(II) is summarized in Fig. 25.7. Four or five bonds are formed of length corresponding to normal single bonds. Four bonds are coplanar or directed towards the vertices of a flattened tetrahedron; regular tetrahedral

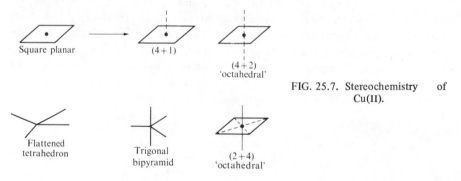

Square planar   (4+1)   (4+2) 'octahedral'

Flattened tetrahedron   Trigonal bipyramid   (2+4) 'octahedral'

FIG. 25.7. Stereochemistry of Cu(II).

coordination is not observed. Five bonds of approximately single bond length are directed towards the vertices of a trigonal bipyramid; this bond arrangement is rare. The most frequently observed coordination group of Cu(II) is a group of 4 coplanar neighbours with one or two more distant neighbours completing a distorted tetragonal pyramid or distorted octahedron. In view of the Jahn–Teller distortion (usually to (4 + 2)-coordination) regular octahedral coordination is not to be expected; on this point see p. 273. The coordination of Cu^{2+} by 6 F^- in $KCuF_3$ and K_2CuF_4 is not far removed from regular octahedral. In these compounds there are two close and four more distant neighbours, but the difference between the two Cu–F distances is much smaller than in, for example, CuF_2, where there is (4 + 2)-coordination.

From the geometrical standpoint the square coplanar arrangement may be regarded as the limit of the extension of an octahedron along a 4-fold axis; it may alternatively be regarded as the limit of compression of a tetrahedron along a 2-fold axis (normal to an edge). In the former case the six equal *bonds* become two long and four short, and in the latter case the six equal *angles* of a regular tetrahedral bond arrangement ($109\frac{1}{2}°$) become two larger (in the limit $180°$) and four smaller (in the limit $90°$):

$$\text{six of } 109\tfrac{1}{2}° \longrightarrow \begin{cases} \text{two equal angles} \longrightarrow 180° \\ (> 109\tfrac{1}{2}°) \\ \text{four equal angles} \longrightarrow 90° \\ (< 109\tfrac{1}{2}°) \end{cases}$$

regular tetrahedron (flattened tetrahedron) square planar

We show in Table 25.2 examples of these intermediate configurations which are described later in the text; for $CuCr_2O_4$ see p. 493.

TABLE 25.2

Configurations intermediate between square coplanar and regular tetrahedral in Cu(II) *compounds*

Compound	Interbond angles	
	4 of	2 of
(Square coplanar)	90°	180°
Chelates	93°	154°
	96°	141°
Cu(imidazole)$_2$	97°	140°
$(CuCl_4)^{2-}$	100°	130°
$CuCr_2O_4$	103°	123°
(Regular tetrahedron)	6 of $109\frac{1}{2}°$	

The available evidence appears to indicate that having formed four strong coplanar bonds Cu(II) forms if possible one or two additional weaker bonds perpendicular to the plane of the four coplanar bonds. Naturally there must always be additional neighbours of some kind in directions normal to this plane, and it is necessary to consider their distances from the metal atom in order to decide whether they should be regarded as bonded to Cu(II). An alternative to (4 + 1)- or (4 + 2)-coordination is to rearrange the four close neighbours to form a flattened tetrahedral group. We now examine a number of structures in which Cu(II) has 4-, (4 + 1)-, or (4 + 2)-coordination.

Crystalline CuO[2a] (the mineral tenorite) provides the simplest example of Cu(II) forming four coplanar bonds. The structure is a distorted version of the PdO (PtS) structure in which the O–Cu–O angles are two of $84\frac{1}{2}°$ and two of $95\frac{1}{2}°$; Cu has 4 O′ neighbours at 1·96 Å, the next nearest neighbours being two O″ at 2·78 Å. The ratio of these distances is much greater than for the usual distorted octahedral coordination of Cu(II), and the line O″–Cu–O″ is inclined at 17° to the normal to the Cu(O′)$_4$ plane. The shortest distance Cu–Cu is 2·90 Å. Presumably, alternative structures such as a distorted NaCl structure with (4 + 2)-coordination for Cu(II), implying also this type of coordination of O, would be less stable than the observed structure with planar coordination of Cu(II) and tetrahedral coordination of O, but the relevant lattice energy calculations have not been made. There is also apparently square planar coordination of Cu(II) in CaCu$_2$O$_3$ and Sr$_2$CuO$_3$ the structures of which have not been determined in detail.[2b] The crystalline derivative of imidazole, Cu(C$_3$N$_2$H$_3$)$_2$,[3] is a 3D network of Cu atoms linked by imidazole ligands (a), each N forming a bond to a metal atom. The Cu atoms are of two kinds, one-half having 4 coplanar N neighbours (interbond angles close to 90°) and no other close neighbours, and the remainder a flattened tetrahedral arrangement of 4 neighbours (Table 25.2). It is evidently not possible to bring additional N atoms closer to the Cu(II) atoms owing to the way in which the imidazole rings are packed.

The structural similarity of the PdO and CuO structures suggests comparison of cupric with palladous halides. In contrast to the normal rutile structure of PdF$_2$, CuF$_2$ has a distorted variant of that structure with (4 + 2)-coordination of Cu^{2+}, the four stronger bonds linking Cu and F into layers (Fig. 6.6, p. 202). The bond lengths Pd^{II}–Cl and Cu^{II}–Cl are the same (2·30 Å) and planar chains of exactly the same type are found in CuCl$_2$ and in one form of PdCl$_2$:

$$\text{>M}\underset{\text{Cl}}{\overset{\text{Cl}}{<}}\text{M}\underset{\text{Cl}}{\overset{\text{Cl}}{<}}\text{M<}$$

However, the chains pack together quite differently in the two halides (Fig. 25.8). In PdCl$_2$ the next nearest neighbours of Pd are 4 more Cl at 3·85 Å arranged at the corners of a rectangle, the plane of which is perpendicular to that of the chain; in CuCl$_2$ the chains are packed side-by-side to form a CdI$_2$-like layer in which Cu(II) has 2 additional Cl neighbours at 2·95 Å completing a very distorted octahedral coordination group (Fig. 25.9).

(2a) AC 1970 **B26** 8

(2b) ZaC 1969 **370** 134
(3) AC 1960 **13** 1027

(a)

890

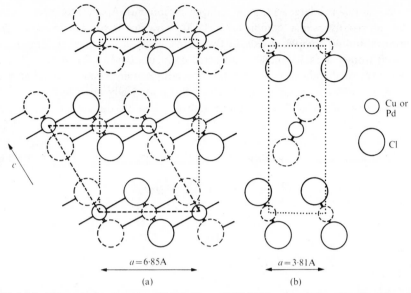

FIG. 25.8. The crystal structures of (a) CuCl$_2$, and (b) PdCl$_2$ viewed along the direction of the chains showing how these are packed differently in the two crystals. Atoms at $y = 0$ and $y = \frac{1}{2}$ are represented by full and dotted circles respectively. In (a) the broken lines enclose the monoclinic cell, and the dotted lines indicate the pseudorhombic unit cell.

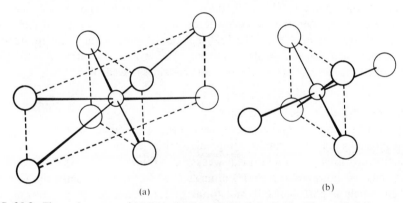

FIG. 25.9. The environment of (a) Pd in PdCl$_2$, and (b) Cu in CuCl$_2$. The small circles represent metal atoms.

Finite complexes containing only monodentate ligands range from cations through neutral molecules to anions:

$$\left[\begin{array}{c} L \\ L \end{array} \!\! Cu \!\! \begin{array}{c} L \\ L \end{array} \right]^{2+} \quad \begin{array}{c} L \\ X \end{array} \!\! Cu \!\! \begin{array}{c} X \\ L \end{array} \quad \left[\begin{array}{ccc} X & L & X \\ X & Cu & L & Cu & X \\ X & X & X \end{array} \right]^{2-} \quad \left[\begin{array}{c} X \\ X \end{array} \!\! Cu \!\! \begin{array}{c} X \\ X \end{array} \right]^{2-}$$

891

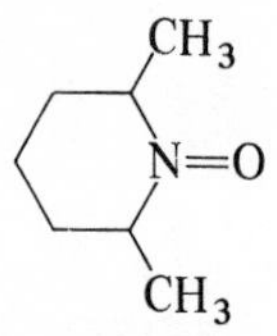

and others noted in a later section on halogen compounds. Apparently *cis* and *trans* isomers of strictly planar complexes CuL_2X_2 have not been characterized. The dichloro dilutidine complex exists in two forms, a green *trans* planar form and a yellow *cis* isomer with a somewhat distorted configuration.[4] The state of Cu(II) in complexes of the above types in solution is not generally known—completion of the octahedron by solvation is likely where this is sterically possible—certainly in solids the formation of *only* four bonds is rare and then it is due to the geometry of the complex. Thus in $[Cu(\text{pyridine oxide})_4](BF_4)_2$[5] and the perchlorate[6] Cu has 4 coplanar O neighbours at approximately 1·95 Å but no more neighbours closer than 3·35 Å. In the bridged cation in the compound (b)[7] Cu is bonded to

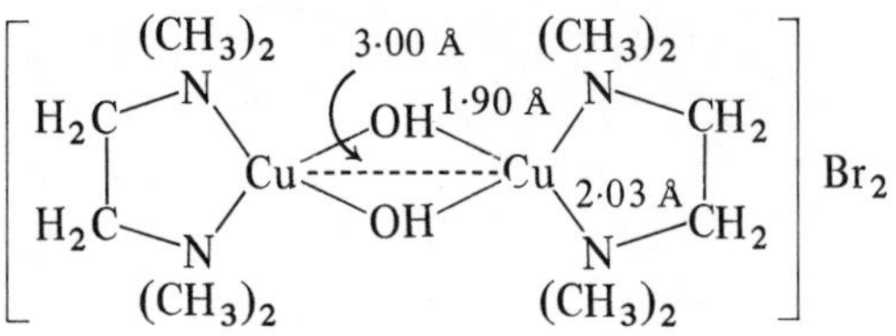

(b)

only 4 coplanar neighbours, the next nearest neighbours being at 4·78 Å. On the other hand, $Cu(\text{pyridine oxide})_2(NO_3)_2$[8] and $Cu(\text{pyridine oxide})_2Cl_2$[9] form dimers in their crystals, each Cu forming a weaker fifth bond to the O atom of a pyridine oxide molecule (Fig. 25.10(a)). For suitable ligands X and suitable geometry of the monomer this process leads to the formation of the infinite chain of Fig. 25.10(b). In some cases the bridging 'bonds' are too long to be regarded as much more than van der Waals bonds (e.g. 3·05 Å in $Cu(\text{pyridine})_2Cl_2$).[10] A review of complexes of Cu(II) with aromatic N-oxides is available.[11]

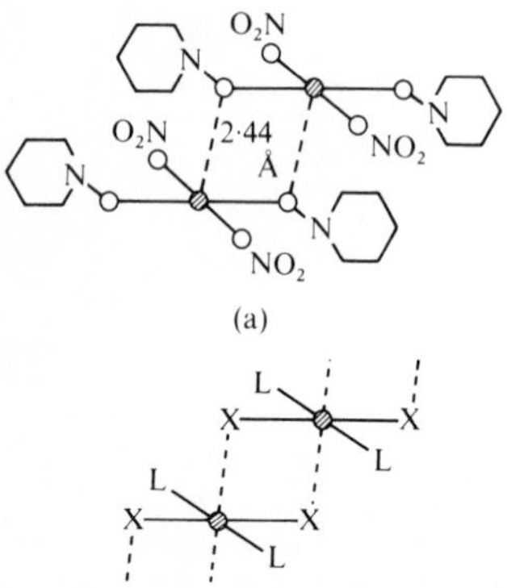

Structures of chelate cupric compounds

Many chelate cupric compounds have been made, particularly with bidentate and tetradentate ligands. In a simple compound such as $Cu(\text{en})_2(BF_4)_2$[1] there is no difficulty in forming two additional weaker Cu–F bonds (2·56 Å) to F atoms of the anions. If, however, the Cu(II) atom is at the centre of a large tetradentate molecule such as phthalocyanin and forms four coplanar bonds to atoms of that molecule, or if the four atoms bonded to Cu(II) belong to two large coplanar molecules as in many chelates CuR_2, then no other bonds from Cu shorter than about 3·4 Å are possible since this is the minimum van der Waals separation of adjacent parallel molecules (Fig. 25.11(a)). Only four coplanar bonds are formed by Cu(II) in the bis-salicylaldiminato complex (a),[2] but such cases are rare, and a number of interesting possibilities are realized in crystals consisting of molecules of this general type.

(i) If parts of the ligands can be bent out of the plane of the central Cu atom and its four coplanar neighbours it may be possible to have two additional Cu–N or Cu–O bonds of, say $2\frac{1}{2}$ Å while maintaining the necessary $3\frac{1}{2}$ Å between successive

(lutidine)

(4) IC 1969 8 308
(5) AC 1969 **B25** 1595
(6) AC 1969 **B25** 1378

(7) AC 1970 **B26** 2096

(8) AC 1969 **B25** 2046
(9) IC 1967 **6** 951

FIG. 25.10. (a) Dimers in $Cu(\text{pyridine oxide})_2(NO_3)_2$, (b) chains CuL_2X_2.

(10) AC 1957 **10** 307
(11) IC 1969 8 1879

(1) AC 1968 **B24** 730

(2) JCS A 1966 680

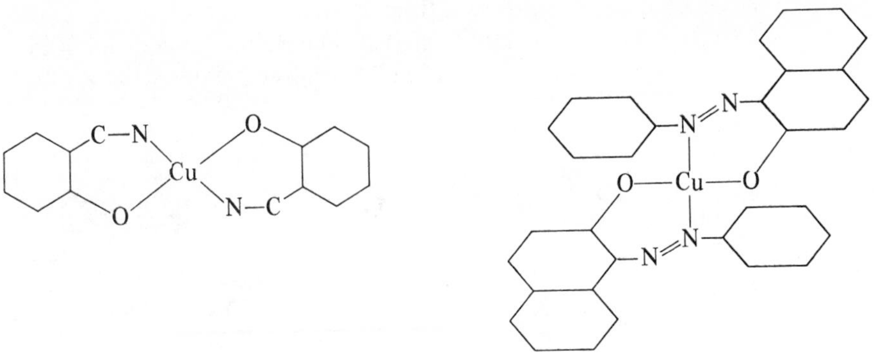

(a)

(b)

molecules (Fig. 25.11(b)). This occurs in the cupric derivative of benzene azo-β-naphthol, (b).[3]

(ii) Alternatively there may be other atoms in the chelating molecules to which Cu can form the two additional bonds, when plane molecules may be stacked as in Fig. 25.11(c) with a sideways shift of each successive molecule. The salicylaldoxime compound[4] (c), has this type of structure. Such additional O atoms are present in the molecule of dimethylglyoxime (DMG), but here the presence of the bulky methyl groups leads, for the Ni compound, to stacking of the molecules directly above one another, alternate molecules being rotated through 90° to facilitate

(3) AC 1961 **14** 961

(4) AC 1964 **17** 1109, 1113

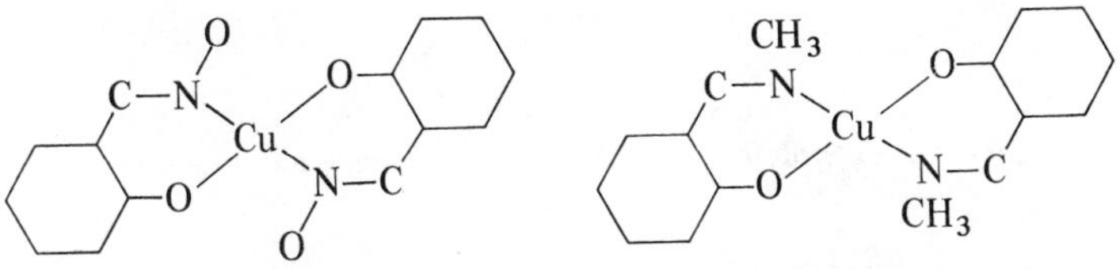

(c)

(d)

packing of the CH$_3$ groups (Fig. 25.11(d)). The only intermolecular bonds formed by Ni are those to Ni atoms of neighbouring molecules at 3·25 Å. This type of packing is adopted in one crystalline form of bis-(N-methylsalicylaldiminato)Cu(II), (d),[5] in which Cu has, in addition to its four coplanar neighbours within the molecule, no other neighbours nearer than 2 Cu at 3·33 Å.

(iii) Cu(DMG)$_2$[6] is not isostructural with the Ni compound although the structure of Fig. 25.11(d) would be geometrically possible. Stacking of the molecules to form infinite columns as in Fig. 25.11(c) is not possible because of the CH$_3$ groups, but this type of relation is possible between a *pair* of molecules which

(5) AC 1961 **14** 1222

(6) JCS A 1970 218

are bent to increase the clearance between them. Crystalline $Cu(DMG)_2$ consists of dimers (Fig. 25.11(e)). The two molecules of DMG bonded to a given Cu atom are inclined at an angle of $23°$ to one another, and the metal atom is displaced slightly

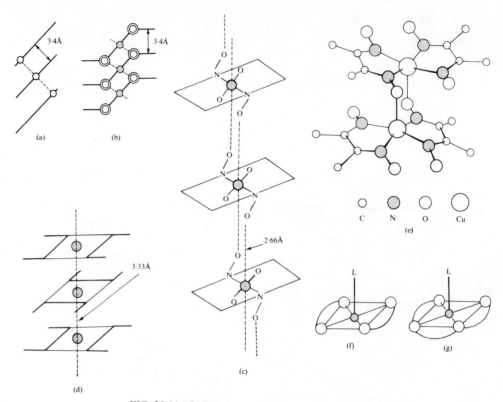

FIG. 25.11. Structures of chelated cupric compounds (see text).

(by $0·3$ Å) out of the plane of its four N atoms towards an O atom of the other half of the dimer. The Cu–N bonds are of normal length ($1·95$ Å) but the fifth bond (Cu–O, $2·30$ Å) is much longer than a normal one ($1·9$–$2·0$ Å).

(iv) Dimers of the same type are formed in the chelates formed with 8-hydroxy-quinoline[7] and NN′-ethylene salicylideneimine, but the corresponding propylene compound forms a monohydrate (Fig. 25.11(f)) in which the fifth bond is formed to a water molecule (Cu–OH_2, $2·53$ Å).[8] Other square pyramidal monomers of this type include the chelates with o-hydroxyacetophenone (L = 4-methyl pyridine)[9] and trien (L = SCN) in [Cu(trien)SCN]NCS.[10] (Fig. 25.11(g)). In all these molecules the fifth bond is appreciably longer than a normal single bond.

(v) If large alkyl groups are substituted for H on N in the salicylaldiminato ligand in chelates CuL_2, (e), there is slight buckling of the molecule and the bond

(7) AC 1967 **22** 476

(8) JCS 1960 2639

(9) AC 1969 **B25** 2245
(10) IC 1969 8 2763

arrangement becomes 'flattened tetrahedral':

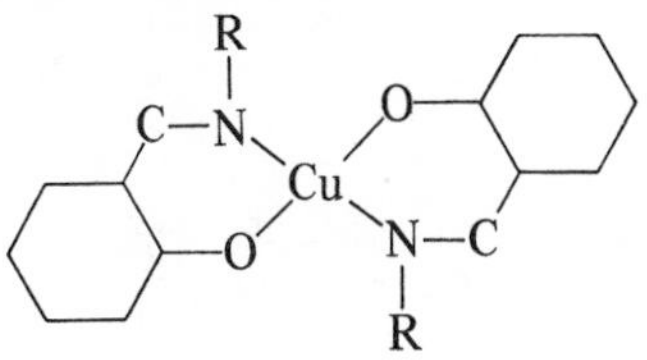

(e) 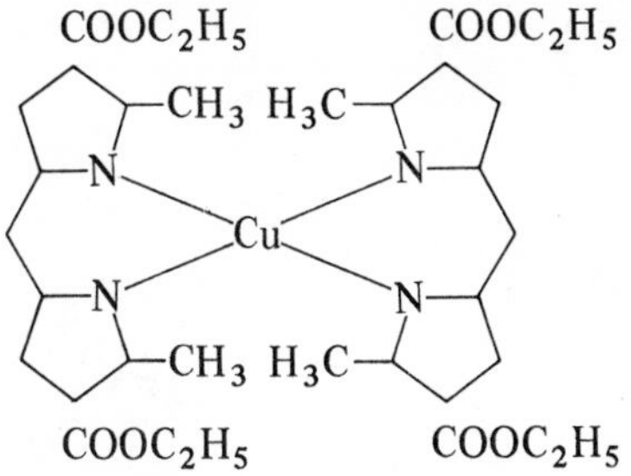   (f)

$R = C_2H_5$[11] bond angles four of 93° and two of 154°, $R = t\text{-}C_4H_9$,[12] bond angles four of 96° and two of 141°. In the dipyrromethene complex (f) interference between the methyl groups of the two ligands leads to a very non-planar Cu coordination. The ligands remain planar, but the angle between their planes is 66°.[13]

(11) AC 1967 **23** 537

(12) JCS A 1966 685

(13) JCS A 1969 2556

The structures of oxy-salts

Crystalline oxy-salts provide many examples of (4 + 1)- and (4 + 2)-coordination in 3D structures.

The structures of one of the two crystalline forms of anhydrous $Cu(NO_3)_2$ and those of a number of complexes formed with organic molecules are of interest not only in connection with the stereochemistry of Cu(II) but also as examples of the behaviour of the NO_3^- ligand. (The structure of the vapour molecule $Cu(NO_3)_2$ is described on p. 662.) The molecule $Cu(en)_2(NO_3)_2$ (p. 899) is an example of NO_3^- behaving as a monodentate ligand, it being bonded to Cu(II) through two long bonds (2·59 Å) in a (4 + 2)-coordination group. The basic framework of $\alpha\text{-}Cu(NO_3)_2$[1] is a corrugated layer (4-gon net) in which each NO_3 bridges two Cu atoms as shown diagrammatically in Fig. 25.12(a). One-half of the anions (N^x) also form two weak bonds, both from the same O atom. The longer bond (2·68 Å) is to one of the Cu atoms in its own layer, the other (2·43 Å) to Cu of an adjacent layer. Each Cu thus forms 4 stronger coplanar bonds (Cu–O, 1·98 Å) and 2 weaker ones (2·43 and 2·68 Å) (Fig. 25.12(b)). The same type of corrugated layer occurs in $Cu(NO_3)_2 \cdot CH_3NO_2$.[2] As in $\alpha\text{-}Cu(NO_3)_2$ all NO_3^- are bridging ions, and to each Cu is bonded a CH_3NO_2 molecule through an O atom (Cu–O, 2·31 Å), giving (4 + 1)-coordination (Fig. 25.12(c)). (There are also some still weaker interactions, Cu–N, 2·74 Å, and Cu–O, 2·75 Å.)

In $Cu(NO_3)_2 \cdot 2\frac{1}{2}H_2O$[3] the four strong bonds from Cu define groups $Cu(NO_3)_2(H_2O)_2$ (Fig. 25.12(d)) and as in the anhydrous nitrate there are two kinds of non-equivalent NO_3^- ions. Here both types of NO_3^- ion are behaving as 'pseudo-bidentate' ligands, and a second bond from O of one NO_3^- links the square coplanar groups into infinite chains. The environment of Cu(II) is very similar to that in the nitromethane complex, namely, Cu–4 O, 1·97 Å, 1 O, 2·39 Å, and 2 O,

(1) JCS 1965 2925

(2) AC 1966 **20** 210

(3) AC 1970 **B26** 1203

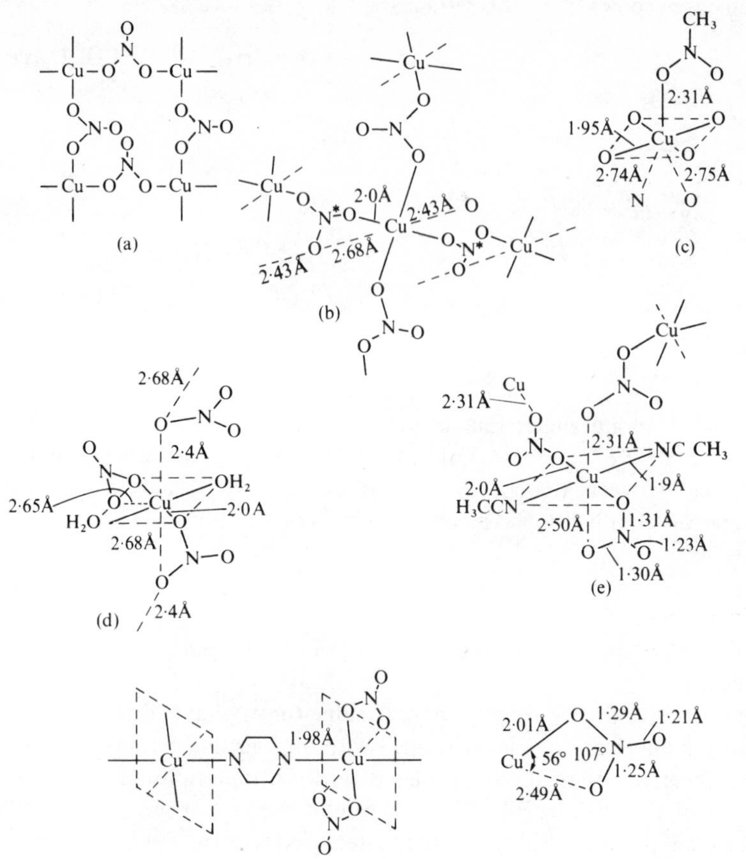

FIG. 25.12. Structures of cupric nitrato complexes: (a) and (b), α-$Cu(NO_3)_2$, (c) $Cu(NO_3)_2 . CH_3NO_2$, (d) $Cu(NO_3)_2 . 2\frac{1}{2} H_2O$, (e) $Cu(NO_3)_2 . 2 CH_3CN$, (f) $Cu(NO_3)_2 . C_4N_2H_4$.

2·65 and 2·68 Å. One-fifth of the H_2O molecules are not bonded to Cu but are hydrogen bonded to O atoms of NO_3^- and H_2O.

(4) AC 1968 **B24** 396

In $Cu(NO_3)_2 . 2 CH_3CN$[4] also there are two types of NO_3^- ion, pseudo-bidentate and bridging. The bridging is here unsymmetrical (Fig. 25.12(e)) and links the tetragonal pyramidal groups into chains. A simpler behaviour of the NO_3^- ligand is found in the chain structure of the complex with pyrazine, $Cu(NO_3)_2 . C_4N_2H_4$[5] (Fig. 25.12(f)). Here all these groups are behaving as unsymmetrical bidentate ligands. Owing to the dimensions of the NO_3^- ion the two weaker Cu–O bonds are inclined at $34°$ to the normal to the plane of the four strong bonds.

(5) AC 1970 **B26** 979

We noted in Chapter 18 the behaviour of NO_3^- as a monodentate, symmetrical bidentate, and bridging ligand. We summarize in Fig. 25.13 the types of NO_3^- ligand observed in the compounds of Fig. 25.12. There appears to be a correlation

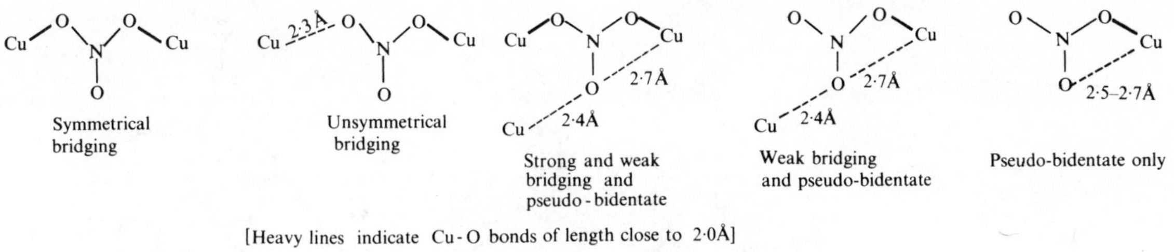

Symmetrical bridging   Unsymmetrical bridging   Strong and weak bridging and pseudo-bidentate   Weak bridging and pseudo-bidentate   Pseudo-bidentate only

[Heavy lines indicate Cu–O bonds of length close to 2·0Å]

FIG. 25.13. Behaviour of NO_3^- ligand in cupric nitrato complexes.

between the dimensions of the ion and its behaviour as a ligand, though it is not certain that the accuracy of location of the light atoms in all these structures justifies detailed discussion at this time.[6]

Cupric salts of carboxylic acids illustrate some of the many types of complex that can be formed by bridging –OC(R)O– groups. In anhydrous cupric formate[7] there is a 3D network of Cu atoms linked by –O . CH . O– groups to give square coordination around Cu(II), but these planar CuO_4 groups are drawn together in pairs so that each Cu forms a fifth bond (2·40 Å) as in $Cu(DMG)_2$. In the tetrahydrate[8] there are infinite chains of composition $Cu(HCOO)_2(H_2O)_2$, in which Cu has 4 O at 2·00 Å and 2 H_2O at 2·36 Å, joined together by hydrogen bonding through the other two water molecules as shown in Fig. 25.14(a). In the monohydrate of the acetate, on the other hand, there are dimers $Cu_2(CH_3COO)_4 . 2 H_2O$ of the kind shown in Fig. 25.14(b). The distorted octahedral coordination group around Cu is composed of 4 coplanar O atoms (of acetate groups) at 1·97 Å, one H_2O at 2·20 Å, and the other Cu atom of the dimer at 2·65 Å.[9] This last distance is close to Cu–Cu in metallic copper. The corresponding Cr(II)[10] and Rh(II)[11] salts have the same structure, and the same type of dimer is formed in $Cu_2(HCOO)_4[OC(NH_2)_2]_2$[12] and in both crystalline forms of $Cu_2(CH_3COO)_4 (C_5H_5N)_2$,[13][14] urea or pyridine molecules replacing water in Fig. 25.14(b). As in the case of the Cu–H_2O bond, the Cu–N bond (2·19 Å) is appreciably longer than a normal single bond. In these molecules Cu is displaced slightly (about 0·2 Å) out of the plane of the 4 O atoms and away from the other Cu atom.

In the dihydrate of cupric succinate[15] one-half of the water is present as isolated molecules of water of crystallization and the remainder is present in dimeric complexes like those of $Cu_2(ac)_4(H_2O)_2$. The succinate ions form not only the bridges in the dimers but also link these units into infinite chains (Fig. 25.14(c)).

In the trihydrate of cupric benzoate[16] there are chains CuX_2B of the type illustrated in Fig. 25.16(c), p. 903, in which X is H_2O and B is the benzoate group, –OC(C_6H_5)O–; the remaining H_2O and $C_6H_5COO^-$ ions are accommodated between the chains.

It is interesting to note that (4 + 2)-coordination of Cu(II) is found in α-$Cu(NO_3)_2$, $CuSO_4$[17] (Cu–4 O, 1·9–2·0, 2 O, 2·4 Å), and $CuSO_4 . 5 H_2O$ (p. 557), but (4 + 1)-coordination in $Cu(NH_3)_4SO_4 . H_2O$.[18] (Cu–4 N, 2·05 Å, Cu–OH_2, 2·59 Å).

(6) QRCS 1971 **25** 289

(7) JCS 1961 3289

(8) AC 1954 7 482

(9) AC 1953 6 227
(10) AC 1953 6 501
(11) DAN 1962 **146** 1102
(12) IC 1970 9 1626
(13) JCS 1961 5244
(14) AC 1964 17 633

(15) AC 1966 **20** 824

(16) IC 1965 4 626

(17) AC 1961 14 321

(18) AC 1955 8 137

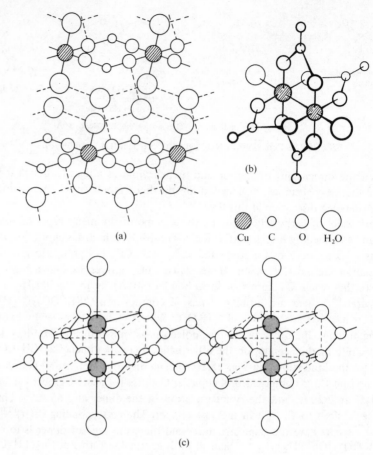

FIG. 25.14. (a) Crystal structure of $Cu(HCOO)_2 \cdot 4 H_2O$, (b) dimeric units in cupric acetate monohydrate, (c) chain structure of cupric succinate dihydrate.

(1) IC 1968 **7** 1111

(2) ACSi 1966 **32** 162
(3) JCS 1963 5691
(4) JACS 1967 **89** 6131
(5) AC 1968 **B24** 595

(6) IC 1971 **10** 1061

The formation of trigonal bipyramidal bonds by $Cu(II)$

The simplest example is the anion in $[Cr(NH_3)_6](CuCl_5)$.[1] The fact that the equatorial bonds (2·39 Å) are longer than the axial ones (2·30 Å) indicates that a simple electrostatic treatment of ligand electron-pair repulsions is inadequate and that account must also be taken of the interaction of the asymmetric d electrons with the ligand electron pairs. Another simple example of this arrangement of five monodentate ligands is the anion in $Ag[Cu(NH_3)_2(NCS)_3]$[2] (Fig. 25.15(a)). Ions or molecules containing polydentate ligands include iodo-bis-(2,2′-dipyridyl)Cu(II) iodide,[3] (b), $[Cu(tren)NCS]SCN$,[4] (c), and $[Cu(Me_6 tren)Br]Br$.[5] In the latter case this stereochemistry is not peculiar to Cu(II) for there are similar compounds of Ni and Co. Note the square pyramidal configuration of $[Cu(trien)SCN]^-$ (p. 894). In contrast to $[Cr(NH_3)_6]CuCl_5$ there are bridged $Cu_2Cl_8^{4-}$ ions, (d) in $Co(en)_3CuCl_5 \cdot H_2O$,[6] and, of course, additional Cl^- ions. This compound should be formulated $[Co(en)_3]_2 (Cu_2Cl_8)Cl_2 \cdot 2 H_2O$.

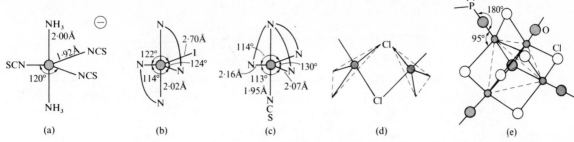

FIG. 25.15. Trigonal bipyramidal coordination of Cu(II) (see text).

There is trigonal bipyramidal arrangement of bonds from Cu(II) in the tetra-hedral molecules $Cu_4OCl_6L_4$ (p. 81) in which L is, for example, pyridine, 2-methyl pyridine,[7] or $(C_6H_5)_3PO$ (Fig. 25.15(e)).[8] The central O is bonded tetrahedrally to 4 Cu (Cu–O, 1·90 Å) and each pair of Cu atoms is bridged by Cl (Cu–Cl, 2·38 Å). The $Cu_4OCl_6(\phi_3PO)_4$ molecule is notable as one of the comparatively rare examples of O forming two collinear bonds. The Cu–Cu distances in these molecules (around 3·1 Å) are too large for metal–metal inter-actions, and the magnetic moments (2·2 BM) are larger than the usual range (1·8–1·9 BM) observed for trigonal bipyramidal Cu(II).[9] A similar moment is observed for the ion $(Cu_4OCl_{10})^{4-}$ in its $[N(CH_3)_4]^+$ salt,[10] which has the same type of structure as Fig. 25.15(e) with L = Cl, but the moment is lower (1·84 BM) for the cubane-like complex Cu_4L_4[11] formed with the tridentate Schiff base (a) in which also there is trigonal bipyramidal coordination of Cu(II). These magnetic moments are discussed in reference 10.

In crystalline $Cu_2O(SO_4)$,[12] which is formed as brown crystals by heating $CuSO_4 . 5 H_2O$ to 650°C, one half of the metal atoms have octahedral (4 + 2)-coordination and the remainder trigonal bipyramidal coordination; these are presumably essentially ionic bonds. There is also trigonal bipyramidal coordination of Cu(II) in Cu_3WO_6[13] and very distorted octahedral coordination of W (3 O, 1·79 Å, 3 O, 2·09 Å).

Octahedral complexes

Special interest attaches to octahedral complexes because of the expected Jahn–Teller distortion. Complexes of type (a) naturally lend themselves to this type of distorted octahedral structure, and this has been confirmed in a number of cases such as $Cu(en)_2X_2$[1] (X = SCN, NO_3, etc.) Complexes CuL_6, (b), still present problems, in particular salts containing ions CuL_6^{2+} or CuL_6^{2-} which have cubic symmetry. The salts $Cu(NH_3)_6X_2$ are tetragonal (X = Cl, Br, or I) with distorted octahedral cations (Cu–4 N, 2·1 Å and Cu–2 N, 2·6 Å approximately).[2] The cubic phases reported for the bromide and iodide may be deficient in ammonia, but the BF_4^- and ClO_4^- salts are also reported to be cubic. This symmetry would presumably be consistent with distorted octahedral cations only if there is assumed to be either restricted rotation or statistical orientation of a statically deformed

(7) IC 1970 9 1619
(8) IC 1967 6 495

(9) TFS 1963 59 1055

(10) IC 1969 8 1982

(11) CC 1968 1329

(12) AC 1963 16 1009
(13) ACSc 1969 23 221

(a)

(1) AC 1964 17 254, 1145
(2) IC 1967 6 126

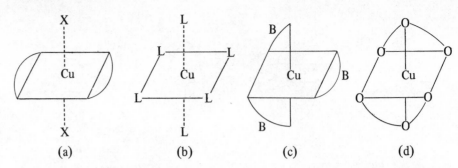

(a)   (b)   (c)   (d)

group or some type of 'dynamic' tetragonal distortion. There is a similar difficulty with salts $K_2Pb[M(NO_2)_6]$ (M = Fe, Co, Ni, Cu) which are cubic at room temperature.[3] It is also considered possible that the interactions of the Pb^{2+} ion with the O atoms of the NO_2 groups might counteract the tendency of Cu(II) to have a static unsymmetrical environment. The Cu–N bond (2·11 Å) is appreciably longer than the normal value (2·00 Å) for a distorted octahedral environment. For $K_2Ba[Cu(NO_2)_6]$ spectroscopic data (polarized electronic and e.s.r.) suggest a tetragonally distorted octahedral environment of the Cu(II) atom.[4]

A rhombic distortion of $Cu(H_2O)_6^{2+}$ groups has been found in two salts. The structure of the cupric Tutton's salt, $Cu(NH_4)_2(SO_4)_2 . 6 H_2O$,[5] is generally similar to those of the Mg and Zn compounds (in which there is little, if any, distortion of the $M(H_2O)_6$ octahedra), but there is definitely a non-tetragonal distortion of $Cu(H_2O)_6^{2+}$, and a similar effect is found in $Cu(ClO_4)_2 . 6 H_2O$:[6]

	2 O at	2 O at	2 O at
$Cu(NH_4)_2(SO_4)_2 . 6 H_2O$:	1·96	2·10	2·22 Å
$Cu(ClO_4)_2 . 6 H_2O$	2·09	2·16	2·28 Å

Complexes of type (c) are found at positions with site symmetry 32 (D_3) in $Cu(en)_3SO_4$[7] (Cu–N, 2·15 Å) and in $CuB_3(ClO_4)_2$[8] where B represents the ligand (e). The neutral molecule $CuB_2(ClO_4)_2$[9] has the expected structure of type (a) with long bonds (2·55 Å) to O atoms of the ClO_4 ligands, but in $CuB_3(ClO_4)_2$, which is isostructural with the Co and Mg compounds, Cu has six

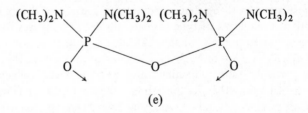

(e)

equidistant O neighbours (Cu–O, 2·07 Å). In an unsymmetrical complex such as $[Cu(bipyr)_2NO_2]NO_3$[10] there is necessarily unsymmetrical octahedral co-ordination because of the different dimensions of the ligands, (f); but note the different Cu–N bond lengths and the two long bonds to O atoms of NO_2.

(3) IC 1971 **10** 1264

(4) JCS A 1969 386, 1845

(5) AC 1966 **20** 659

(6) ZK 1961 **115** 97

(7) IC 1970 **9** 1858
(8) JACS 1968 **90** 5623
(9) IC 1970 **9** 162

(10) JCS A 1969 1248

There would appear to be a definite Jahn–Teller effect in Cu(bipyridine) (11) JACS 1969 **91** 1859 $(F_3C . CO . CH . CO . CF_3)_2$,[11] in which there are unequal Cu–O bond lengths and moreover distortion of the normally symmetrical ligand, (g).

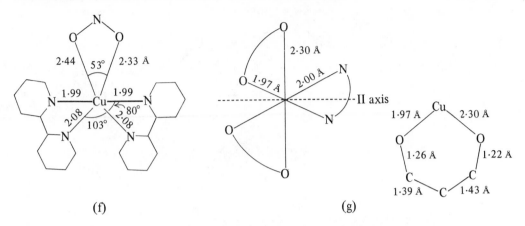

(f)   (g)

In Cu(diethylenetriammine)$_2$NO$_3$, of type (d), there is tetragonal distortion, with two shorter *trans* Cu–N bonds (2·01 Å) and four longer bonds (mean 2·22 Å).[12] (12) JCS A 1969 883

Cupric halides–simple and complex

Some details of the structure of these compounds are given in Table 25.3. The structure types (second column) are the 'ideal' structures with approximately regular octahedral coordination; they have all been described in earlier chapters, and earlier in this chapter we have referred to the cupric halides. The following points amplify the Table.

(i) The distortion of the coordination group around Cu^{2+} in KCuF$_3$ and K$_2$CuF$_4$ is in the opposite sense to that in CuF$_2$ and Na$_2$CuF$_4$, but the shorter Cu–F bond length is the same in all three compounds. The weighted mean of the six Cu–F bond lengths is the same for the two types of distortion (about 2·03 Å), but the difference between the two Cu–F bond lengths is very different. For (4 + 2)-coordination the difference is 0·34 Å in CuF$_2$ (0·5 Å in CuF$_2$. 2 H$_2$O) but for (2 + 4)-coordination it is only 0·11 Å in KCuF$_3$ and 0·13 Å in K$_2$CuF$_4$.

(ii) Both K$_2$CuF$_4$ and (NH$_4$)$_2$CuCl$_4$ crystallize with distorted forms of the K$_2$NiF$_4$ structure (Fig. 5.16, p. 172) but whereas in the former compound the distortion is not sufficient to justify distinguishing linear CuF$_2$ molecules, in the latter case the difference between the two Cu–Cl bond lengths (0·48 Å) is perhaps sufficient to justify description in terms of planar CuCl$_4^{2-}$ ions.

(iii) The structure of the CuCl$_4^{2-}$ ion is very sensitive to its surroundings. We have just mentioned (NH$_4$)$_2$CuCl$_4$. In (C$_2$H$_5$NH$_3$)$_2$CuCl$_4$ the longer bonds in the (4 + 2)-coordination group, 2·98 Å, can hardly correspond to more than van der Waals bonding. In the bright-yellow Cs salt and in the N,N′-dimethyl-4,4′-bipyridinium (DMBP) salt the ion has a flattened tetrahedral shape, with Cl–Cu–Cl

TABLE 25.3

Crystal structures of cupric halogen compounds

Compound	Octahedral structure type	Coordination group of Cu (or Cr)		Reference
CuF_2	Rutile	4 F 1·93 Å	2 F 2·27 Å	JACS 1957 **79** 1049
CrF_2		4 F 2·00	2 F 2·43	PCS 1957 232
Na_2CuF_4	Na_2MnCl_4	4 F 1·91	2 F 2·37	ZaC 1965 **336** 200
$CuF_2 . 2 H_2O$		2 F 1·90 2 O 1·94	2 F 2·47	JCP 1962 **36** 50 (n.d.)
$KCuF_3$	Perovskite	2 F 1·96	4 F 2·07	JCS 1959 4126
$KCrF_3$		2 F 2·00	4 F 2·14	AC 1961 **14** 19 (n.d.)
K_2CuF_4	K_2NiF_4	2 F 1·95	4 F 2·08	JCP 1959 **30** 991
		4 X	**2 X**	
$CsCuCl_3$	$CsNiCl_3$	2·28, 2·36	2·78	IC 1966 **5** 277
$[(CH_3)_2NH_2]CuCl_3$		2·25–2·35	2·73 (one)	JCP 1966 **44** 39
$(NH_4)_2CuCl_4$	K_2NiF_4	2·31	2·79	JCP 1964 **41** 2243
$LiCuCl_3 . 2 H_2O$		2·33	O 2·60, Cl 2·90	AC 1963 **16** 1037; JCP 1963 **39** 2923 (n.d.)
$CuCl_2(en)_2 . H_2O$		4 N 2·00	O 2·62, Cl 2·81	JCS A 1967 1435
$CuCl_2 . 2 H_2O$	$CoCl_2 . 2 H_2O$	2 Cl 2·29 2 O 1·96	2·94	ACSc 1970 **24** 3510
$K_2CuCl_4 . 2 H_2O$		2 Cl 2·29 2 O 1·97	2·90	AC 1970 **B26** 827 (n.d.)
$CuCl_2$	CdI_2	2·30	2·95	JCS 1947 1670
$(C_2H_5NH_3)_2CuCl_4$		2·28	2·98	ICA 1970 **4** 367
$KCuCl_3$	NH_4CdCl_3	2·29	3·03	JCP 1963 **38** 2429
$Cu_3Cl_6(C_6H_7NO)_2 . 2 H_2O$		See text		IC 1968 **7** 2035
$CuCl_2(pyr)_2$	α-$CoCl_2(pyr)_2$	2 N 2·02 2 Cl 2·28	3·05	AC 1957 **10** 307
$CuBr_2$	CdI_2	2·40	3·18	JACS 1947 **69** 886
Cs_2CuCl_4		2·20		JPC 1961 **65** 50
Cs_2CuBr_4		2·39 See text		AC 1960 **13** 807
$(DMBP)CuCl_4$		2·25		AC 1969 **B25** 1691
$Cs_3Cu_2Cl_7 . 2 H_2O$		See text		AC 1971 **B27** 1528

bond angles in the ranges 124–130° (two) and 103–100° (four). Quite similar bond angles are found in the distorted spinel structure of $CuCr_2O_4$.

The description of the anions in the chloro compounds depends on the interpretation placed on the two Cu–Cl bond lengths. The description in terms of distorted octahedral coordination groups brings together a number of structures based on octahedral chains of the types described in Chapter 5. The simple edge-sharing chain has the composition $(AX_4)_n^{2n-}$ if all the larger circles in Fig. 25.16(a) represent halogen atoms, AX_2L_2, (b), if the shaded circles represent neutral ligands such as H_2O, NH_3 or pyridine, or AX_2B, (c), if B is a ligand capable of bridging two Cu atoms. The simple octahedral chain ion $(CuCl_4)_n^{2n-}$ has not yet been found in a complex cupric chloride, but the $CuCl_2 . L_2$ chain is the structural unit in $CuCl_2 . 2 H_2O$ and other compounds listed in Table 25.4. Examples of the $CuCl_2B$ chain of Fig. 25.16(c) include the complexes with 1 : 2 : 4-triazole and

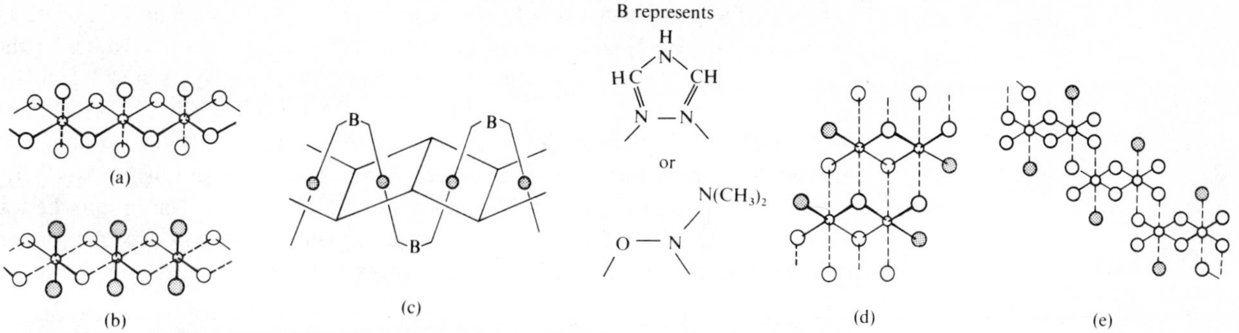

B represents

FIG. 25.16. Infinite linear ions and molecules with octahedral coordination of metal atoms: (a) AX_4, (b) AX_2L_2 in $CuCl_2 . 2 H_2O$, (c) AX_2B in $CuCl_2 . C_2N_3H_3$ or $CuCl_2 . ONN(CH_3)_2$, (d) $(CuCl_3)_n^{n-}$ in $KCuCl_3$ or in $CuCl_2 . NCCH_3$ (if shaded circles represent N), (e) $[CuCl_3(H_2O)]_n^{n-}$ in $LiCuCl_3 . 2 H_2O$ (shaded circles represent H_2O molecules).

TABLE 25.4

A family of octahedral chain structures

Type of complex	Regular octahedral coordination	Distorted octahedral coordination	Reference to Cu compound
Single chain Fig. 25.16			
(a) AX_4^{2-}	Na_2MnCl_4	Na_2CuF_4	ZaC 1965 **336** 200
(b) AX_2L_2	$CoCl_2 . 2 H_2O$	$CuCl_2 . 2 H_2O$ $\left.\vphantom{\begin{matrix}a\\b\end{matrix}}\right\}$ $CuF_2 . 2 H_2O$	See Table 25.3
	$CdCl_2(NH_3)_2$	α-$CuBr_2(NH_3)_2$	AC 1959 **12** 739
	α-$CoCl_2(pyr)_2$	$CuCl_2(pyr)_2$	
(c) AX_2B	—	$CuCl_2$ (triazole)	AC 1962 **15** 964
		$CuCl_2(Me_2NNO)$	AC 1969 **B25** 2460
Double chain			
(d) AX_3^-	NH_4CdCl_3	$KCuCl_3$	See Table 25.3
AX_2L		$CuCl_2(NCCH_3)$	JCP 1964 **40** 838
(e) AX_4^{2-}	$TcCl_4$		
AX_3L^-		$Li[CuCl_3(H_2O)] . H_2O$	See Table 25.3

$(CH_3)_2NNO$. In the latter compound the bond lengths found were: Cu–4 Cl, 2·29 Å, Cu–O, 2·29 Å, and Cu–N, 3·00 Å, suggesting $(4 + 1)$- rather than $(4 + 2)$-coordination.

The double chain ion $(CuCl_3)_n^{n-}$ of Fig. 25.16(d) is the anion in $KCuCl_3$, with which $KCuBr_3$ is isostructural; it also represents the linear molecule $CuCl_2 . NC(CH_3)$. In the AX_4 chain of Fig. 25.16(a) the AX_6 octahedra share two opposite edges. In the chain of Fig. 25.16(e), necessarily of the same composition, a different pair of edges is shared, giving a staggered arrangement of pairs of octahedra. This is the anion in $LiCuCl_3 . 2 H_2O$, the shaded circles representing H_2O molecules. The composition of the chain is $CuCl_3(H_2O)$; the second H_2O and the Li^+ ions are situated between the chains. These structures are summarized in Table 25.4.

By describing the structures of these compounds in this way we have emphasized the octahedral (4 + 2)-coordination of Cu(II). Alternatively we may describe the systems formed by the four stronger Cu–X bonds. These are layers in CuF_2, chains in $CsCuCl_3$ and $CuCl_2$, ions $Cu_2Cl_6^{2-}$ in $KCuCl_3$ and $LiCuCl_3 . 2 H_2O$, and $CuCl_2(H_2O)_2$ molecules in $CuCl_2 . 2 H_2O$ and $K_2CuCl_4 . 2 H_2O$. The structure of this last compound may therefore be described as an aggregate of $CuCl_2(H_2O)_2$ molecules, K^+, and Cl^- ions; it does not contain $CuCl_4^{2-}$ ions. The infinite helical chains $(CuCl_3)_n^{n-}$ in $CsCuCl_3$ are of the type (a) while those in $CuCl_2$ are formed from planar $CuCl_4$ groups sharing opposite edges, (b).

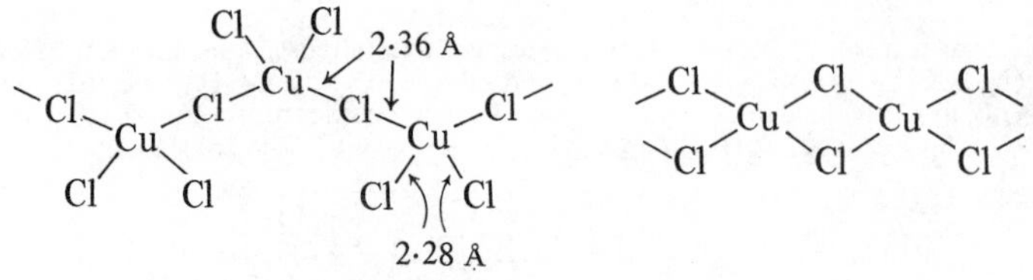

(a)　　　　　　　　　　　　　　　(b)

Two structures are of some interest in this connection. In $[(CH_3)_2NH_2]CuCl_3$ planar $Cu_2Cl_6^{2-}$ ions are linked into chains by a fifth 'bond' (2·73 Å) from each Cu atom. The sixth octahedral position is occupied by a methyl group (Cu–CH_3, 3·78 Å), so that Cu(II) is here exhibiting 5-coordination (tetragonal pyramidal) (c). In a 2-picoline-N oxide complex, $Cu_3Cl_6(C_6H_7NO)_2 . 2 H_2O$ dimers $Cu_2Cl_4L_2$ are bridged by $CuCl_2(H_2O)_2$ molecules through Cu–Cl bonds of two lengths, 2·65 Å and 2·96 Å, to form chains in which there is both 5- and 6-coordination of Cu(II) atoms (d).

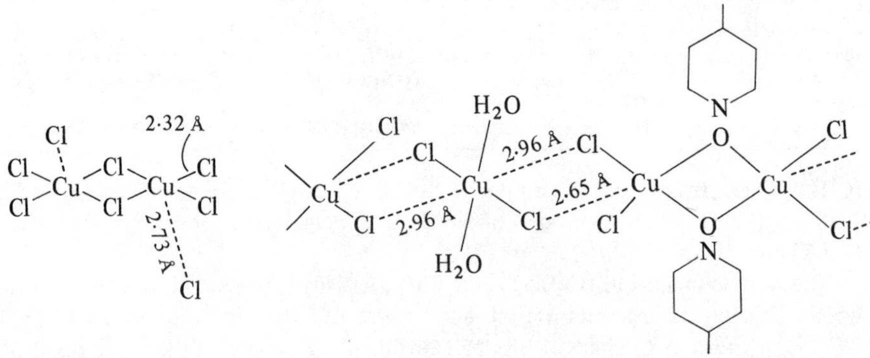

(c)　　　　　　　　　　　　　　　(d)

The $Cu_2Cl_6^{2-}$ ion is simply a portion of the $CuCl_2$ chain, and there are some compounds of $CuCl_2$ with organic molecules which have structures very closely related to these two systems; they are prepared by dissolving anhydrous $CuCl_2$ in hot solvents such as acetonitrile or n-propanol and crystallizing the resultant

$$H_3C . CN \diagdown Cu \diagup Cl \diagdown Cu \diagup Cl \diagdown NC . CH_3$$

$$(e)$$

$$H_3C . CN \diagdown Cu \diagup Cl \diagdown Cu \diagup Cl \diagdown Cu \diagup Cl \diagdown NC . CH_3$$

$$(f)$$

solutions. The yellow-red $CuCl_2 . CH_3CN$ consists of dimeric molecules (e) which are packed in the same way as the $Cu_2Cl_6^{2-}$ ions in $KCuCl_3$. The longer Cu–Cl bonds link each molecule to *two* others to form chains (Table 25.4), the CH_3CN molecules projecting from both sides of the double octahedral chain as indicated by the shaded circles (N atoms) in Fig. 25.16(d). The packing of the longer molecules $Cu_3Cl_6(CH_3CN)_2$ (f), and $Cu_5Cl_{10}(C_3H_7OH)_2$ is different, the longer Cu–Cl bonds linking atoms of one molecule to atoms of *four* adjacent ones, so that layer structures are formed. The structures of the layers are different in these two compounds; the layer in the former is shown in Fig. 25.17.

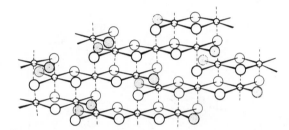

FIG. 25.17. Portion of layer of $Cu_3Cl_6(CH_3CN)_2$ showing how each molecule is bonded to four others by long Cu–Cl bonds (dotted). Only N atoms (shaded) of CH_3CN molecules are shown.

We note the bridged $Cu_2Cl_8^{4-}$ ion (p. 898) as an example of Cu(II) forming trigonal bipyramidal bonds. There are apparently bridged ions $[Cu_2Cl_5(H_2O)_2]^-$ in the salt $Cs_3Cu_2Cl_7 . 2 H_2O$. The (4 + 2)-coordination is unusually unsymmetrical (g), but the range of Cu–Cl bond lengths (2·09–2·41 Å) suggests that further study of this structure may be desirable.

$$(g)$$

Cupric hydroxy-salts

Copper(II) forms an extraordinary variety of hydroxy-oxysalts,[1] many of which occur as (secondary) minerals. Examples include $Cu(OH)IO_3$, $Cu_2(OH)_3NO_3$, $Cu_2(OH)_2CO_3$ (malachite),[2] $Cu_2(OH)PO_4$ (libethenite), $Cu_3(OH)_3PO_4$, $Cu_5(OH)_4(PO_4)_2$, and $Cu_4(OH)_6SO_4$ (brochantite). The structures of $Cu(OH)IO_3$ and $Cu_2(OH)_3NO_3$ have been described in Chapter 14, where the general features of the structures of some hydroxy-salts are noted, $CuB(OH)_4Cl$ (bandylite) in Chapter 24, and the (layer) structure of $CuHg(OH)_2(NO_3)_2(H_2O)_2$[3] on p. 102. In most of these compounds Cu(II) has 4 coplanar O and 2 more distant ones

(1) For earlier references see: FM 1961 **39** 59; AC 1963 **16** 124

(2) AC 1967 **22** 146, 359

(3) AC 1969 **B25** 800

(4) AC 1965 **18** 777

completing a distorted octahedral coordination group, but in some cases the sixth O is so far away from Cu that the coordination is preferably described as tetragonal pyramidal (for example, in $Cu_3(OH)_3AsO_4$, clinoclase).[4]

The structures of two forms of $Cu_2(OH)_3Cl$, atacamite and botallackite, and of $Cu_2(OH)_3Br$ have been described under 'Hydroxyhalides' (p. 410), where references are given. Atacamite has a framework structure which, in its idealized form, is derivable from the NaCl structure, as shown in Fig. 4.22 (p. 143). Botallackite and $Cu_2(OH)_3Br$ have layer structures of the CdI_2 type. In all these structures there is the characteristic distorted (4 + 2)-coordination.

$$Cu_2(OH)_3Cl \qquad\qquad Cu_2(OH)_3Br$$

$$Cu \begin{cases} \dfrac{4\,OH\ \ 2{\cdot}02\ Å}{2\,Cl\ \ 2{\cdot}76} & or & \dfrac{4\,OH\ \ 2{\cdot}00\ Å}{\substack{OH\ \ 2{\cdot}36 \\ Cl\ \ 2{\cdot}76}} \end{cases} \qquad Cu \begin{cases} \dfrac{4\,OH\ \ 2{\cdot}0\ Å}{2\,Br\ \ 3{\cdot}0} & or & \dfrac{4\,OH\ \ 2{\cdot}0\ Å}{\substack{OH\ \ 2{\cdot}3 \\ Br\ \ 2{\cdot}8}} \end{cases}$$

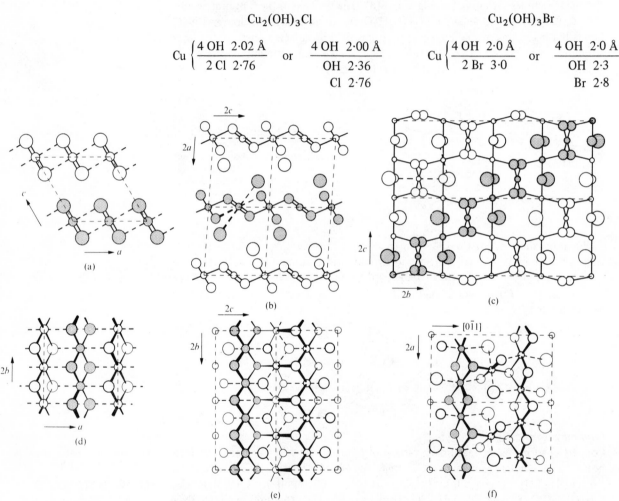

FIG. 25.18. (a), (b), (c). Elevations of the crystal structures of $CuCl_2$, $Cu_2(OH)_3Br$, and $Cu_2(OH)_3Cl$ (atacamite). (d), (e), (f). Projections of portions of these structures (the atoms shaded in (a)–(c)) showing a layer of metal atoms (smallest circles) in the plane of the paper, and close-packed layers of halogen atoms (or halogen atoms and OH groups) above and below the plane of the paper. The largest circles represent halogen atoms. Full lines indicate short Cu–OH or Cu–X bonds, and the heavier broken lines long Cu–OH or Cu–X bonds. (See text).

The metal atoms do not all have the same environment in these hydroxyhalides, one half being surrounded by 4 OH + 2 X and the remainder by 5 OH + X. These coordination groups in a compound M_2A_3B are a consequence of the relative numbers of A and B atoms and of the coordination numbers of M (6), A (3), and B (3). It is not possible for all the M atoms in such a structure to have the same environment, and the fact that the M atoms are (crystallographically) of two kinds has no chemical significance.

In Fig. 25.18 are shown elevations and plans of portions of the structures of $CuCl_2$, $Cu_2(OH)_3Br$, and $Cu_2(OH)_3Cl$. Although atacamite does not have a layer structure it can be dissected into approximately close-packed layers of the same type as occur in the hydroxybromide, as described on p. 410 and illustrated in Fig. 25.18(e) and (f).

The sulphides of copper

The Cu–S system is remarkably complex, and the structures of CuS and Cu_2S are not consistent with their formulation as cupric and cuprous sulphides. The phases recognized, with their mineral names, are set out in Table 25.5.

TABLE 25.5

The sulphides of copper

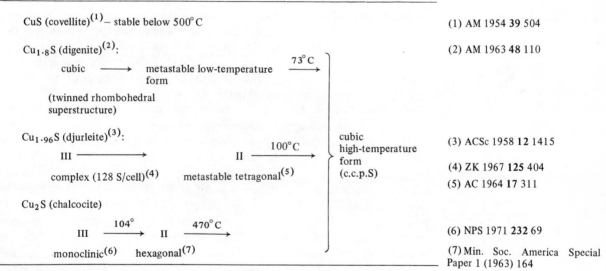

CuS (covellite)[1] – stable below 500° C

$Cu_{1.8}S$ (digenite)[2]:

cubic ⟶ metastable low-temperature form ⟶ 73° C

(twinned rhombohedral superstructure)

$Cu_{1.96}S$ (djurleite)[3]:

III ⟶ complex (128 S/cell)[4]   II ⟶ 100°C   metastable tetragonal[5]

Cu_2S (chalcocite)

III ⟶ 104° II ⟶ 470°C

monoclinic[6]   hexagonal[7]

cubic high-temperature form (c.c.p.S)

(1) AM 1954 **39** 504

(2) AM 1963 **48** 110

(3) ACSc 1958 **12** 1415

(4) ZK 1967 **125** 404

(5) AC 1964 **17** 311

(6) NPS 1971 **232** 69

(7) Min. Soc. America Special Paper 1 (1963) 164

Covellite, which can be ground to a blue powder, is a moderate conductor of electricity and has an extraordinary and unique structure. One-third of the metal atoms have three S neighbours (at 2·19 Å) at the corners of a triangle, and the remainder have four S neighbours arranged tetrahedrally (at 2·32 Å). Moreover, two-thirds of the S atoms are present as S_2 groups like those in pyrites, so that if we regard this as a normal covalent structure it would be represented $Cu_4^I Cu_2^{II}(S_2)_2 S_2$.

The three distinct species with formulae near to or equal to Cu_2S all have a high-temperature form based on c.c.p. S atoms in which there is statistical arrangement of the Cu atoms, probably in tetrahedral holes, this $Cu_{2-x}S$ phase being stable over the composition range $x = 0.0-0.2$. In high-digenite there is on the average 9/10 of a Cu atom within each S_4 tetrahedron but not centrally situated. In the intermediate metastable form some of the Cu atoms have moved so that now some S_4 tetrahedra are unoccupied and in the occupied tetrahedra Cu is statistically distributed over 4 equivalent (off-centre) positions.

The extremely complex structure of $Cu_{1.96}S$-III is not known, but in the metastable tetragonal form there is almost perfect cubic closest packing of the S atoms in which Cu occupies trigonal holes (Cu–3 S, 2.31 Å); S is surrounded by 6 Cu at the vertices of a trigonal prism. There is very slight displacement of the Cu atom towards a tetrahedral hole, and in addition to its three S neighbours Cu has 2 Cu at 2.64 Å and 2 more at 2.97 Å. In hexagonal chalcocite-II the Cu atoms are disordered in a h.c.p. array of S atoms, being statistically distributed in three types of site. In the unit cell containing 2 S atoms there is the following average occupancy of sites: 1.74 Cu in sites of 3-coordination (Cu–S, 2.28 Å), 1.42 Cu in sites of tetrahedral coordination with Cu non-central (Cu–3 S, 2.59 Å, Cu–1 S, 2.15 Å), and 0.84 Cu in sites of 2-coordination, on edges of S_4 tetrahedra (Cu–2 S, 2.06 Å). In addition, there are separations of only 2.59 Å between the 3- and 4-coordinated Cu atoms.

A very careful study of low-chalcocite, which was originally referred to an orthorhombic cell containing 96 Cu_2S, has shown that the true symmetry is certainly not higher than monoclinic. In the proposed structure, which has 48 Cu_2S in the unit cell, all Cu atoms are 3-coordinated, mostly in or near positions of planar coordination in a h.c.p. array of S atoms. Certain features of the structure, notably two very long Cu–S distances (2.88 Å, compared with 2.33 Å, mean), suggest that although this structure must be very near to the true one, there may still be complications due to submicroscopic twinning.

The physical properties of these compounds, which include metallic lustre, intrinsic semiconductivity, and in some cases ductility, the close contacts between Cu atoms in some of the phases, and the variety of arrangements of the bonds from the Cu atoms, indicate that these are semi-metallic phases, as also are some of the numerous complex sulphides. On the grounds that the chalcopyrite (zinc-blende superstructure) structure of $CuFeS_2$ is adopted by $CuAlS_2$ and $CuGaS_2$ we could formulate this compound as $Cu^IFe^{III}S_2$, with tetrahedral coordination of Cu^I, but in stromeyerite, AgCuS,[8] there is trigonal coordination of Cu^I. It does not seem possible to assign oxidation numbers to Cu in the numerous complex sulphides (e.g. $CuFe_2S_3$, cubanite (p. 633), $Cu_3Fe_4S_6$, Cu_5FeS_4, bornite,[9] and Cu_5FeS_6) in such a way as to give a consistent stereochemistry of Cu(I) and Cu(II).

The structural chemistry of Au(III)

In this oxidation state gold forms four coplanar covalent bonds. When forming these bonds the effective atomic number of Au(III) is 84, five shared electrons

(8) ZK 1955 **106** 299

(9) AC 1964 **17** 351

being added to the 79 of the free atom, as is seen from the bond pictures

$$\left[\begin{array}{c} | \\ -Au- \\ | \end{array} \right]^{-} \text{ in } K[AuBr_4] \text{ or } \begin{array}{c} | \\ -Au\leftarrow \\ | \end{array} \text{ in } Au_2Cl_6$$
$$\text{(see below).}$$

The electronic structure of the 4-covalent Au^{III} atom is thus

				5s	5p	5d	6s	6p
2	8	18	32	2	6	8 + 2	2	4

shared

so that four coplanar dsp^2 bonds are formed.

Information about the structural chemistry of Au(III) comes from (a) auric compounds and (b) compounds containing both Au(I) and Au(III). The square coplanar structure has been confirmed for the anions in $K(AuF_4)$,[1] (Au–F, 1·95 Å), $K[Au(NO_3)_4]$, (p. 663) and $K[Au(CN)_4] \cdot H_2O$,[2] (Au–C, 1·98 Å), and for the $AuCl_4^-$ ion in $Cs_2Au^IAu^{III}Cl_6$ (p. 393). The structures of the auric halides themselves are less simple. Crystalline AuF_3 has a unique helical (chain) structure formed from planar AuF_4 groups sharing *cis* F atoms (Fig. 25.19).[2a] The lengths of the Au–F bonds in the chains are 1·91 Å and 2·04 Å (bridging), and the F bond angle is 116°. Weaker bonds of length 2·69 Å complete octahedral groups around the Au(III) atoms. The vapour density of the chloride between 150° and 260°C indicates Au_2Cl_6 molecules, and the elevation of the b.p. of bromine by the bromide shows that this compound exists in solution as Au_2Br_6 molecules. An X-ray study[3] showed that crystalline $AuCl_3$ consists of planar dimers:

(1) JCS A 1969 1936

(2) AC 1970 **B26** 422

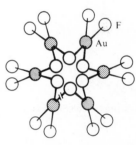

FIG. 25.19. The helical AuF_3 molecule viewed along its length.

(2a) JCS A 1967 478

(3) AC 1958 **11** 284

(4) JCS 1937 1690

(5) JCS 1946 438

The compound $Au_2Br_2(C_2H_5)_4$[4] consists of bridged dimeric molecules, and the coplanar configuration of four bonds from Au(III) has also been confirmed in $AuBr_3P(CH_3)_3$[5] and in the compounds (a)–(c):

(6) IC 1968 7 810

(7) IC 1968 7 2636

(8) IC 1969 8 1661

There appears to be weak additional bonding in (c) (as in the Ni and Pt compounds, where M–I distances are 3·21 Å and 3·50 Å). The sum of the 'covalent radii' gives Au–I = 2·73 Å for a single bond.

The action of AgCN on $Au_2Br_2(C_2H_5)_4$ gives the cyanide, which has a 4-fold molecular weight. The bridged structure noted above for the bromo compound is possible because the angle between two bonds from a halogen atom can be approximately 90°, but a similar structure is not possible for the cyano compound since the atoms linked by the CN group must be collinear with the C and N atoms. Accordingly this compound forms a square molecule of type (d), the structure having been established for the case R = n-C_3H_7.[9]

(9) PRS 1939 A **173** 147

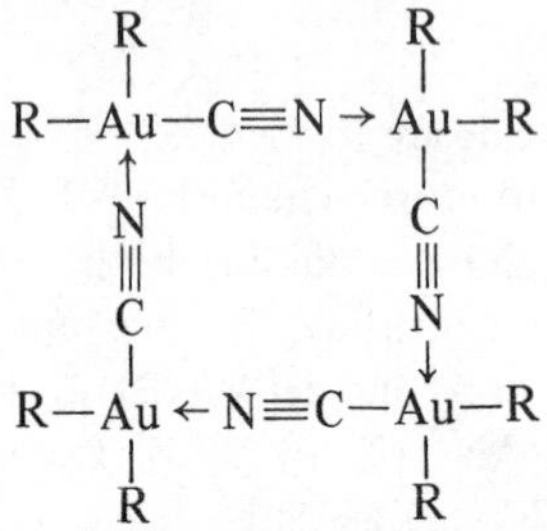

(d)

Dimethyl auric hydroxide, $[(CH_3)_2AuOH]_4$ is also a cyclic tetramer, (e), but there is apparently a strange asymmetry and the interbond angles at Au(III) range from 81 to 106°; the range of Au–O bond lengths, 1·87–2·40 Å, is inexplicably large.[10]

(10) JACS 1968 **90** 1131

Examples of compounds containing Au(I) and Au(III) include $Cs_2(AuCl_2)$ $(AuCl_4)$ and $(C_6H_5CH_2)_2S . AuBr_2$, which have already been noted, and the compound (f).[11] All these compounds contain Au(I) forming 2 collinear and Au(III) forming 4 coplanar bonds. In contrast to the black $Cs_2(AuCl_2)(AuCl_4)$

(11) IC 1968 **7** 805

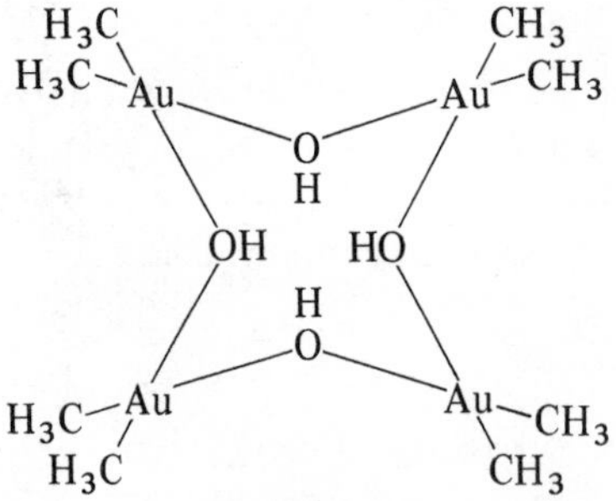

(e)

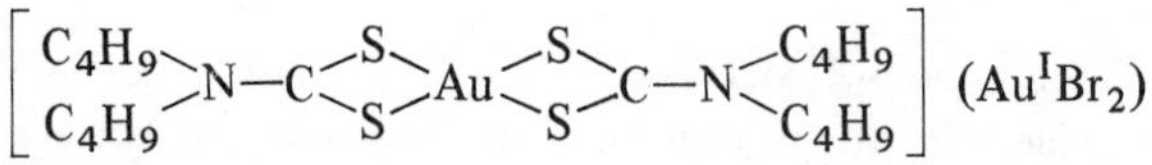

(f)

there is no interaction between Au(I) and Au(III) in the compound (f), which forms yellow needles. In crystalline $[Au^{III}(dimethylglyoxime)_2](Au^ICl_2)$ Au(III) forms two weak bonds in addition to the four coplanar bonds. The crystal consists of planar $[Au^{III}(DMG)_2]^+$ and linear $(Au^ICl_2)^-$ ions so arranged that there are chains of alternate Au(I) and Au(III) atoms in which the metal–metal distance is 3·26 Å.[12]

(12) JACS 1954 **76** 3101

910

26

The Elements of Subgroups IIB, IIIB, and IVB

Introduction

Before the structural chemistry of these elements is considered in more detail attention should be drawn to one feature common to a number of them. We have seen that the elements Cu, Ag, and Au can make use of d electrons of the penultimate quantum group and in the case of Cu can lose one or two of the 3d electrons to form ions Cu^{2+} and Cu^{3+}. Some elements of the later B subgroups show a quite different behaviour. In addition to forming the normal ion M^{N+} by loss of all the N electrons of the outermost shell (N being the number of the Periodic Group) there is loss of only the p electron(s), the pair of s electrons remaining associated with the core—the so-called 'inert pair'. For a monatomic ion this implies that M must have at least 3 electrons in the valence shell, and we have therefore to examine the evidence for the existence of ions M^+ in Group IIIB and ions M^{2+} in Group IVB. Mercury retains the structure 78 (2) in its monatomic vapour and would have the same effective atomic number in $(Hg–Hg)^{2+}$ if the free ion exists; this point is mentioned later. The very small degree of ionization of mercuric halides was regarded by Sidgwick as evidence for the inertness of the pair 6s electrons of Hg, but the Hg^{2+} ion obviously exists in crystalline HgF_2 (fluorite structure). Evidence for the existence of ions may be adduced from the properties of compounds in solution or in the fused state, or from the nature of their crystal structures.

Ga^+, In^+, and Tl^+

All the monohalides of Ga are known as vapour species, being prepared by methods such as heating Ga + CaF_2 in a Knudsen cell or by vaporizing $GaCl_2$, but only GaI has been prepared as a solid (actually $GaI_{1.06}$);[1] its structure is not known. Reduction of the dibromide by metal proceeds only as far as $GaBr_{1.30}$.[2] Both Ga_2O and Ga_2S have been prepared, but their structures also await examination. However, the Ga^+ ion exists in the crystalline 'dichloride', $Ga^+(GaCl_4)^-$,[3] and the fused salt (at 190°C) also has this ionic structure for its Raman spectrum is very similar to that of the $GaCl_4^-$ ion formed in a solution of $GaCl_3$ containing excess Cl^- ions. The fused dibromide is similar.[4] The Ga^+ ion can be introduced into the β-alumina structure (q.v.) and then replaced by Na^+ in molten NaCl (810°C) to form the original compound.[5]

All the crystalline monohalides of In exist except InF, which is known only as a

(1) JACS 1955 77 4217
(2) JACS 1958 80 1530
(3) JINC 1957 4 84
(4) JCS 1958 1505
(5) IC 1969 8 994

vapour species, though these compounds are much less stable than the thallous compounds. The In^+ ion is present in crystalline InBr and InI (see Chapter 9), but it is not stable in water, in which both InCl and $InCl_2$ form $In^{3+} + In$. Solid $InCl_2$ is diamagnetic and is presumably $In(InCl_4)$; In also forms the chlorides In_4Cl_5, In_4Cl_7, and In_2Cl_3, the latter presumably being $In_3^I(In^{III}Cl_6)$. The Tl^+ ion is certainly stable, for the soluble TlOH is a strong base like KOH, and the halides have ionic crystal structures; note, however, the unsymmetrical structures of TlF and yellow TlI (Chapter 9).

For compounds apparently containing divalent Ga, In, or Tl, there are two simple possibilities: equal numbers of M^+ and M^{3+} (or M^I and M^{III} in a covalent compound) or ions $(M–M)^{4+}$ analogous to Hg_2^{2+}. The former alternative has been confirmed in a number of compounds; there is no evidence for binuclear ions. The diamagnetic GaS[6] is not an ionic crystal but contains Ga bonded to four tetrahedral neighbours (1 Ga + 3 S) to form a layer structure (Fig. 26.1). With the equivalence of all the Ga atoms in GaS compare the presence of equal numbers of Tl(I) and Tl(III) in TlS (p. 928).

InS[7] may be described as built of puckered 3-connected layers similar to those in GeS and SnS), but the layers are translated relative to their positions in GeS so that there are close In–In contacts between the layers (2·80 Å). The structures are compared in Fig. 26.2. As a result In has a distorted tetrahedral arrangement of 4 nearest neighbours (3 S, 2·57 Å, 1 In, 2·80 Å). A S atom has 3 In neighbours and also 1 S at the surprisingly short distance of 3·09 Å (in addition to 4 S at 3·71 Å and 2 S at 3·94 Å); confirmation of this distance would seem desirable.

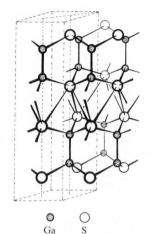

FIG. 26.1. The crystal structure of GaS.

Ga S

(6) ZaC 1955 **278** 340

(7) NW 1954 **41** 448

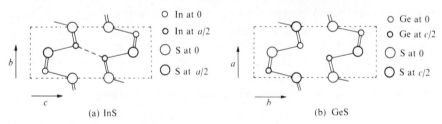

	In at 0		Ge at 0
	In at $a/2$		Ge at $c/2$
	S at 0		S at 0
	S at $a/2$		S at $c/2$

(a) InS   (b) GeS

FIG. 26.2. Projections of the structures of (a) InS, (b) GeS (layers perpendicular to plane of paper), showing one In–In bond in (a) as a broken line.

Ge^{2+}, Sn^{2+}, and Pb^{2+}

Compounds of divalent Ge are well known. There is no evidence that GeO is a stable phase at temperatures below 1000°K (compare SiO), but compounds stable at ordinary temperatures include GeS, all four dihalides, and complex halides such as $MGeCl_3$. The Ge^{2+} ion is not stable in water, and its crystal structure shows that GeF_2 is not a simple ionic crystal (p. 929). There seem to be no simple ionic crystalline stannous compounds. In solution Sn^{2+} presumably exists as complexes; its easy conversion into Sn^{4+} gives stannous compounds their reducing properties. Lead presents a quite different picture. The stable ion is Pb^{2+}. This has no reducing properties, and there is no evidence that Pb^{4+} can exist in aqueous solution, but it certainly exists in crystalline compounds such as PbO_2 (rutile structure).

The structural chemistry of zinc

The structures of many of the simple compounds of the IIB elements have been described in earlier chapters. Cadmium, like zinc, has only one valence state in its normal chemistry and its structural chemistry presents no points of special interest. We therefore confine our attention here to zinc and mercury.

From the geometrical standpoint the structural chemistry of zinc is comparatively simple. There is only one valence state to consider (Zn^{II}) and in most molecules and crystals the metal forms 4 tetrahedral or 6 octahedral bonds; two collinear bonds are formed in the gaseous ZnX_2 molecules and presumably in $Zn(CH_3)_2$, and some examples of 5-coordination are noted later.

A comparison of the structures of compounds of Be, Mg, Zn, and Cd reveals an interesting point; we exclude Hg from this series because there is practically no resemblance between the structural chemistries of Zn and Hg apart from the fact that one form of HgS has the zinc-blende structure. In the following group of compounds italic type indicates tetrahedral coordination of the metal atom in the crystal; in other cases the metal has 6 octahedral neighbours.

BeO	MgO	*ZnO*	CdO
BeS	MgS	*ZnS*	*CdS*
BeF$_2$	MgF$_2$	ZnF$_2$	CdF$_2$
BeCl$_2$	MgCl$_2$	*ZnCl*$_2$	CdCl$_2$
β-*Be(OH)*$_2$	Mg(OH)$_2$	*Zn(OH)*$_2$	Cd(OH)$_2$

Apart from the halides (for which see Chapter 9) the Be and Zn compounds are isostructural. Octahedral coordination of Zn is found in ZnF_2 and also in the following compounds, all of which are isostructural with the corresponding Mg compounds: $ZnCO_3$, $ZnWO_4$, $ZnSb_2O_6$, $Zn(ClO_4)_2 . 6 H_2O$, and $ZnSO_4 . 7 H_2O$. Comparison with the Be and Cd compounds is not possible for all these salts either because Be does not form the analogous compound or because their structures are not known. Compounds in which Zn, like Be, has 4 tetrahedral oxygen neighbours include ZnO, $Zn(OH)_2$, Zn_2SiO_4 (contrast the 6-coordination of Mg in Mg_2SiO_4), and complex oxides such as K_2ZnO_2 (chains of edge-sharing ZnO_4 tetrahedra)[1] and $SrZnO_2$ (layers of vertex-sharing tetrahedra).[2] These differences in oxygen coordination number are not due to the relative sizes of the ions Be^{2+}, Mg^{2+}, and Zn^{2+}, for the last two have similar radii; moreover, there is 6-coordination of Mg^{2+} by Cl^- in $MgCl_2$ but 4-coordination of the metal in all three crystalline forms of $ZnCl_2$. The tetrahedral coordination by Cl persists in $ZnCl_2 . 1\frac{1}{3} H_2O$[3] and in $ZnCl_2 . \frac{1}{2} HCl . H_2O$.[4] In the former all the Cl is associated with two-thirds of the Zn atoms in tetrahedral $ZnCl_4$ groups which share two vertices to form chains $(ZnCl_3)_n^{n-}$. The remaining Zn atoms lie between the chains surrounded octahedrally by 2 Cl and 4 H_2O; note that the more electronegative O (of H_2O) belongs to the octahedral coordination groups. The structural formula could be written $(Zn^{tetr} . Cl_3)_2 [Zn^{oct} . (H_2O)_4]$ though this does not show the actual coordination numbers of the two kinds of Zn atom (ion). In the second compound $ZnCl_4$ tetrahedra each share 3 vertices to form a 3D framework of composition

(1) ZaC 1968 **360** 7
(2) ZaC 1961 **312** 87

(3) AC 1970 **B26** 1679
(4) AC 1970 **B26** 1544

$(Zn_2Cl_5)_n^{n-}$ which forms around $(H_5O_2)^+$ ions, i.e. $(Zn_2Cl_5)^-(H_5O_2)^+$—see Chapter 15.

It would seem that the tetrahedral Zn–O bonds have less ionic character than octahedral ones. Evidently the character of a Zn–O bond will depend on the environment of the *oxygen* atom, though a satisfactory discussion of this question is not yet possible. In ZnO the O atom is forming four equivalent bonds (a), in Zn_2SiO_4 bonds to 2 Zn and 1 Si (b), and in $ZnCO_3$ bonds to 2 Zn and 1 C of a CO_3^{2-} ion, within which the C–O bonds are covalent in character, (c).

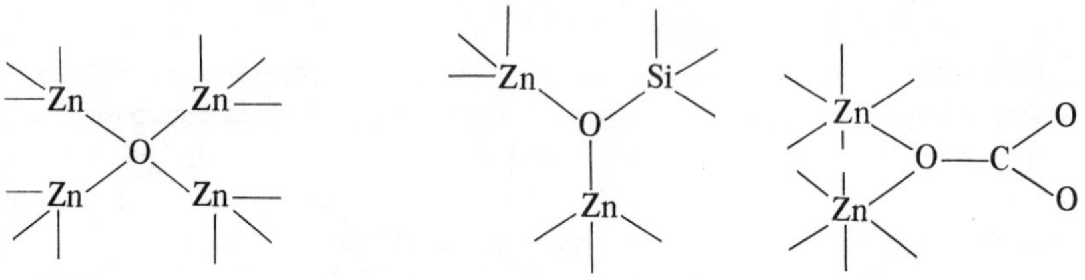

(a) (b) (c)

Both tetrahedrally and octahedrally coordinated Zn occur in a number of crystalline oxy-compounds, including $Zn_2Mo_3O_8$, $Zn_2(OH)_2SO_4$, γ-$Zn_3(PO_4)_2$, $ZnMn_3O_7 . 3\,H_2O$, $Zn_5(OH)_8Cl_2 . H_2O$, and $Zn_5(OH)_6(CO_3)_2$. The structures of the hydroxy-salts are described in other chapters. In some crystals there appears to be a clear-cut difference in length between the tetrahedral and octahedral Zn–O bonds, for example, 2·02 and 2·16 Å respectively in $Zn_5(OH)_8Cl_2 . H_2O$ and 1·95 and 2·10 Å in the hydroxycarbonate, while in $ZnMn_3O_7 . 3\,H_2O$ the octahedral coordination group is made up of 3 O of H_2O molecules (Zn–O, $\approx$2·15 Å) and 3 O of the Mn_3O_7 layer (Zn–O, $\approx$1·95 Å), and in the inverse spinel $Zn(Sb_{0.67}Zn_{1.33})O_4$ there is apparently little difference in length between the two types of Zn–O bond (all close to 2·05 Å).[5] It is probably not profitable to discuss this subject in more detail until more precise information on bond lengths is available.

A similar difference in M–O bond character presumably exists in oxy-compounds of divalent lead. Many anhydrous oxy-salts of Pb^{II} are isostructural with those of Ba (and sometimes Sr and Ca):

(5) AC 1963 **16** 836

	C.N. of M
$PbSO_4$, $BaSO_4$	12
$Pb(NO_3)_2$, $Ba(NO_3)_2$	6 + 6
$PbCO_3$, $BaCO_3$	9

in which it is reasonable to conclude that the Pb is present as (colourless) Pb^{2+} ions with radius close to those of Ba^{2+} and Sr^{2+}. With these compounds compare the coloured PbO (two forms) in which Pb has only 4 nearest neighbours, in contrast to the NaCl structure of the colourless BaO and SrO.

914

5-*covalent* Zn

Monomeric chelate complexes with a trigonal bipyramidal bond arrangement include [Zn(tren)NCS] SCN[1] and $ZnCl_2$(terpyridyl),[2] Fig. 26.3(a). The stereochemistry of such complexes is strongly influenced by the structure of the polydentate ligand, but there are other compounds which, like the cupric compounds discussed earlier, emphasize the importance of considering the structure as a whole. Some chelates such as the 8-hydroxyquinolinate[3] form dihydrates which consist of octahedral molecules (Fig. 26.3(b)), while others form monohydrates in which the configuration of the molecule is intermediate between tetragonal pyramidal and trigonal bipyramidal. In the monoaquo bis-(acetylacetonate) molecule,[4] Fig. 26.3(c), the configuration is nearer the former, with all Zn—O = 2·02 Å and the Zn atom 0·4 Å above the base of the pyramid, but we saw in Chapter 3 that the choice of description for this type of configuration is

(1) JACS 1968 **90** 519
(2) AC 1966 **20** 924

(3) AC 1964 **17** 696

(4) AC 1963 **16** 748

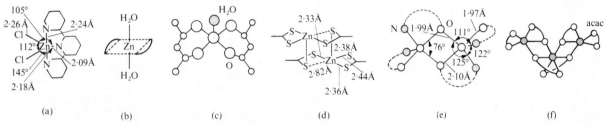

FIG. 26.3. Zn forming 5 or 6 bonds in molecules.

somewhat arbitrary. The configuration of the monohydrated complex with NN′-disalicylidene ethylene diamine is of the same general type, with Zn 0·34 Å above the plane of the O_2N_2 square.[5] The reason for the formation by these chelates of a monohydrate rather than a dihydrate is presumably associated with the packing of the molecules in the crystal. In crystals of the last compound there is strong hydrogen bonding of H_2O to two O atoms of an adjacent molecule (O—H···O ≈2·5 Å), and this may be a decisive factor in determining the composition and structure of the crystalline compound.

In the foregoing examples Zn achieves 5-coordination in a monomeric complex. A fifth bond is formed in some compounds by dimerization, as in Fig. 26.3(d), a type of structure already noted for a number of Cu^{II} complexes. In the diethyldithiocarbamate[6] apparently one of the four short bonds is formed to a S atom of the second molecule of the dimer. A further possibility is realized in bis-(N-methylsalicylaldiminato) zinc, Fig. 26.3(e), in which the five bonds from each Zn are formed within a simple bridged dimer;[7] the Mn and Co compounds are isostructural. A rather surprising structure is adopted by anhydrous $Zn(acac)_2$. We note elsewhere the polymeric structures of $[Co(acac)_2]_4$ and $[Ni(acac)_2]_3$ formed from respectively 4 and 3 octahedral coordination groups. The Zn compound is trimeric but structurally quite unlike the linear Ni compound. The central Zn is octahedrally coordinated, but the terminal Zn atoms are 5-coordinated (Fig. 26.3(f)), the bond arrangement being described as approximately trigonal bipyramidal.[8]

(5) JCS A 1966 1822

(6) AC 1965 **19** 898

(7) IC 1966 **5** 400

(8) AC 1968 **B24** 904

Coordination compounds with tetrahedral or octahedral Zn bonds

We note here a few compounds presenting particular points of interest.

(1) AC 1966 **21** 536

Zinc dimethyl dithiocarbamate[1] forms dimeric molecules, Fig. 26.4(a), in which the ligand behaves in two different ways, Zn forming somewhat irregular tetrahedral bonds. The diethylthiophosphinate forms dimers of the same kind, $S_2P(C_2H_5)_2$ replacing $S_2CN(CH_3)_2$.[2] By way of contrast there is the much simpler behaviour of a similar ligand in zinc ethylxanthate, $Zn(S_2COC_2H_5)_2$.[3] Here each ligand forms a single bridge between tetrahedrally coordinated Zn atoms to give a layer based on the simple plane 4-gon net (Fig. 26.4(b)).

(2) JCS A 1970 714
(3) AC 1966 **21** 919

In salts $Zn(N_2H_4)_2X_2$ the hydrazine molecules link the Zn atoms into infinite chains, Fig. 26.4(c), and the two X ligands complete the octahedral coordination

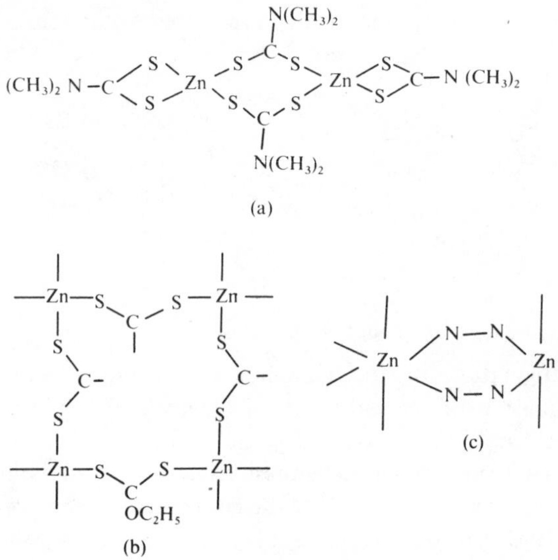

FIG. 26.4. Zn forming tetrahedral bonds in (a) $Zn[S_2CN(CH_3)_2]_2$ and (b) $Zn(S_2COC_2H_5)_2$, and octahedral bonds in (c) $Zn(N_2H_4)_2Cl_2$.

(4) AC 1963 **16** 498

group of Zn. In the chloride,[4] the Zn–Cl distances are surprisingly large (2·58 Å) compared with values around 2·25 Å for tetrahedral $ZnCl_4^{2-}$ in $(NH_4)_3ZnCl_4$. Cl etc. A similarly long Zn–Cl bond (2·53 Å) is found in the *trans* octahedral molecules of the biuret complex,[5] in which H_2N . CO . NH . CO . NH_2 behaves as a bidentate ligand like acac in $Zn(acac)_2(H_2O)_2$, Fig. 26.3(b). In the dihydrazine isothiocyanate[6] all six Zn–N bonds are said to be equal in length (2·17 Å).

(5) AC 1963 **16** 343

(6) AC 1965 **18** 367

The structural chemistry of mercury

Mercury forms two series of compounds, mercurous and mercuric, but the former are not compounds of monovalent Hg in the sense that cuprous compounds, for example, are derivatives of monovalent Cu. A number of elements form compounds in which there are metal–metal bonds but mercury is unique in forming, in addition to Hg^{2+} and the normal mercuric compounds, a series of compounds based on the grouping (–Hg–Hg–). (Evidence for the formation of the Cd_2^{2+} ion in molten $Cd_2(AlCl_4)_2$ at 250°C is limited to the observation of one Raman line.[1])

(1) IC 1962 1 700

916

Mercurous compounds

The fact that the 'mercurous' ion in aqueous solution is not Hg^+ but Hg_2^{2+} (presumably, $(H_2O–Hg–Hg–OH_2)^{2+}$) was demonstrated by investigating the equilibrium between Ag^+ and mercurous ions in the presence of excess liquid mercury and by measuring the e.m.f. of a cell containing mercurous nitrate solution at two different concentrations with mercury electrodes. As far as is known all mercurous compounds contain pairs of Hg atoms in linear groups X–Hg–Hg–X in which the bonds formed by Hg are two collinear sp bonds as in molecules of mercuric compounds X–Hg–X. Even Hg_2F_2 has the same linear molecular structure as the other halides, though the structure has an extraordinarily large F–F separation (3·85 Å) between different molecules, and the (hydrated) nitrate and perchlorate do not contain $Hg–Hg^{2+}$ ions but linear groups $H_2O–Hg–Hg–OH_2^{2+}$. Since Hg in these compounds has only two nearest neighbours its coordination group is necessarily completed by a number of more distant neighbours. In the halides Hg_2X_2 these complete a distorted octahedral group, but the number and arrangement of next nearest neighbours of Hg in crystalline mercurous compounds in variable. In the bromate, which consists of linear molecules $O_3Br–Hg–Hg–BrO_3$, two more O atoms complete a very distorted tetrahedral group. The sulphate consists of infinite linear molecules and the next nearest neighbours of Hg are three

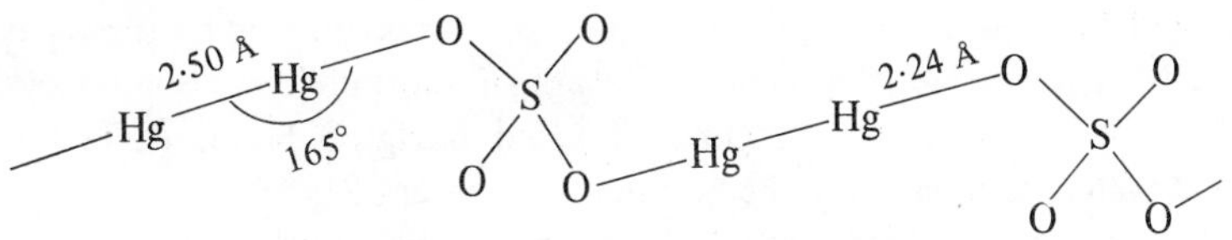

O atoms of adjacent chains at 2·50, 2·72, and 2·93 Å.

The latest data on mercurous halides (Table 26.1) do not confirm the earlier conclusion that the length of the Hg–Hg bond increases with decreasing electronegativity of the halogen. Organic mercurous compounds containing the system –C–Hg–Hg–C– are not known. 'Mercurous oxide' is apparently a mixture of HgO and Hg.[2]

(2) ZaC 1933 **211** 233

TABLE 26.1

Interatomic distances in mercurous compounds

	Hg–Hg	Hg–X	Hg–4 X	Reference
Hg_2F_2	2·51 Å	2·14 Å	2·72 Å	
Hg_2Cl_2	2·53	2·43	3·21	JCS 1956 1316;
Hg_2Br_2	2·49	2·71	3·32	CC 1971 466
Hg_2I_2	[2·72	2·68	3·51]	(Under investigation)
$Hg_2(NO_3)_2 . 2 H_2O$	2·53			JCS 1956 1312
$Hg_2(ClO_4)_2 . 4 H_2O$	2·50	Hg–O 2·12 (0·05)		ACSc 1966 **20** 553
$Hg_2(BrO_3)_2$	2·51			ACSc 1967 **21** 2834
Hg_2SO_4	2·50			ACSc 1969 **23** 1607

Mercuric compounds

In most of its compounds Hg(II) forms either two collinear or four tetrahedral bonds which are presumably essentially covalent and may be described as using sp or sp^3 hybrids. There are no examples as yet of the formation of six equivalent octahedral bonds, though this may possibly occur in compounds such as $[Hg(C_5H_5NO)_6](ClO_4)_2$.[1] A few examples of 3-covalent Hg(II) are noted later. The only simple ionic compound is HgF_2 (fluorite structure), in which Hg^{2+} has 8 equidistant F^- neighbours. Even oxygen is not sufficiently electronegative to ionize Hg, and we find Hg forming two short collinear bonds to O in both forms of HgO and also in crystals such as $HgSO_4 . H_2O$,[2a] where Hg has a distorted octahedral coordination group of 5 O (of anions) and 1 H_2O:

(1) IC 1962 **1** 182

(2a) AC 1964 **17** 933

(2b) IC 1971 **10** 2331

$$O \quad\quad O \quad\quad (4\ at\ 2{\cdot}50\ \text{Å})$$
$$O \xrightarrow{\ 2{\cdot}17\ \text{Å}\ } Hg \xrightarrow{\ 2{\cdot}24\ \text{Å}\ } OH_2$$
$$169°$$
$$O \quad\quad O$$

In $HgCN(NO_3)$[2b] linear chains —Hg—CN—Hg—CN— are stacked with their axes parallel, and the coordination group around Hg(II) is completed by three pairs of O atoms of NO_3^- ions. Here the metal atom forms 2 short collinear bonds (2·06 Å) to C or N (which were not distinguishable) and 6 Hg—O bonds (2·73 Å) in the equatorial plane (compare uranyl compounds, Chapter 28).

In all crystals in which Hg(II) forms only two short collinear bonds the coordination group necessarily includes a number of other atoms, as in the case of mercurous compounds. These additional neighbours often complete a distorted octahedral group, so that an analogy may be drawn between this (2 + 4)-coordination of Hg(II) and the (4 + 2)- or less common (2 + 4)-coordination of Cu(II) described in the previous chapter. In other cases there are only two next nearest neighbours forming, with the two close neighbours, a distorted tetrahedral group. This type of coordination is associated with departures from collinearity of the two strong bonds. Before dealing with selected groups of compounds we mention some examples of Hg(II) forming 2 collinear or 4 tetrahedral bonds.

Molecules (other than the dihalides, for which see later) which have been shown to be linear include H_3C—Hg—CH_3, F_3C—Hg—CF_3, mercaptides RS—Hg—SR,[3] C_2H_5S—Hg—Cl, Cl—Hg—SCN, and Br—Hg—SCN. From the bond lengths H_3C—Hg—Cl and H_3C—Hg—Br,[4] namely, Hg—C, 2·07 Å, Hg—Cl, 2·28 Å, and Hg—Br, 2·41 Å, the radius of 2-covalent Hg has been deduced as 1·30 Å. (The value 1·48 Å, on the Pauling scale, has been suggested for tetrahedral Hg(II).)

(3) JCP 1964 **40** 2258

(4) JCP 1954 **22** 92

(5) ACSc 1958 **12** 1568

In several crystalline compounds X—Hg—X or X—Hg—Y in which one or both of the bonds are to C or S, the bonds are not collinear. A combined X-ray and neutron diffraction study of $Hg(CN)_2$[5] shows that there are two additional weak Hg—N bonds in a plane perpendicular to that of the two Hg—C bonds, (a), the angle between the latter being 171(±2)°. Smaller interbond angles are found in the

918

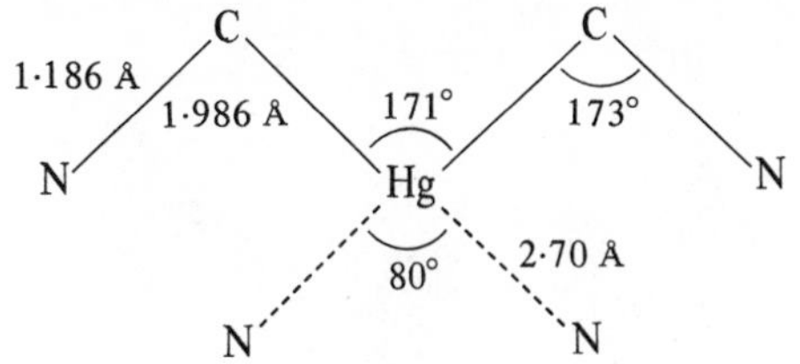

The structures at the top show two diagrams labelled (a) and (b).

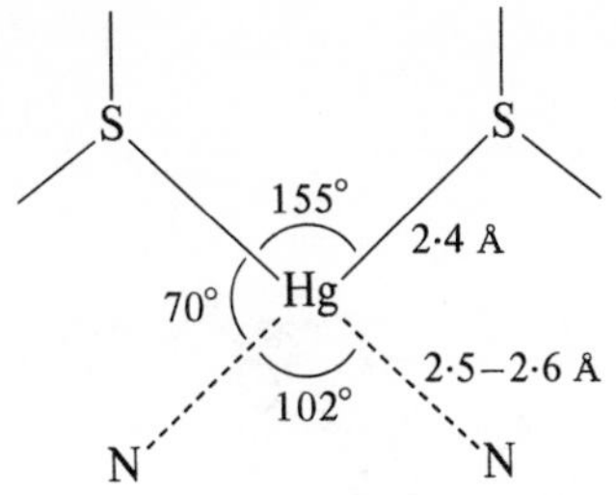

(a): 1·186 Å, 1·986 Å, 171°, 173°, 80°, 2·70 Å

(b): 155°, 2·4 Å, 70°, 102°, 2·5–2·6 Å

dithizone complex, $Hg(C_{13}H_{11}N_4S)_2 . 2 C_5H_5N$, (b), which crystallizes with two pyridine molecules of crystallization.[6] This very distorted tetrahedral arrangement with one large angle between bonds of normal length and a small angle (80–100°) between the two weaker bonds occurs in a number of crystals and also in the molecule $HgCl_2[OAs(C_6H_5)_3]_2$,[7] shown at the right, in which the Hg–Cl bonds have the normal length but Hg–O is considerably longer than in, for example, HgO (2·05 Å). In $KI . Hg(CN)_2$,[8] on the other hand, Hg forms two approximately collinear bonds in $Hg(CN)_2$ molecules but $4 I^-$ ions complete an octahedral co-ordination group (Hg-I, 3·38 Å). It seems likely that the dipole moments (in solution) of molecules HgX_2 (dihalides, diphenyl, etc.) are due to interaction with the solvent, since such molecules are linear in the solid and vapour states.[9]

Some examples of the formation of four tetrahedral bonds by Hg(II) are given on the next few pages. We note here some complexes, finite, 1-, 2-, and 3-dimensional, in which Hg forms bonds to 4 S atoms which in three of the compounds belong to SCN groups. There is a tetrahedral anion in $[Hg(SCN)_4]$ $[Cu(en)_2]$;[10] the thioxan complex (a), is a tetrahedral molecule.[11] In the infinite chain (b) the trithiane molecules act as bridges between the $HgCl_2$ groups.[12] The layer in $CoHg_2(SCN)_6 . C_6H_6$[13] was noted in Chapter 3 as a

(6) JCS 1958 4136

(7) ACSc 1963 **17** 1363

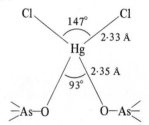

(8) AC 1963 **16** 105
(9) JCP 1964 **41** 3276

(10) AC 1953 **6** 651
(11) JCS A 1967 271
(12) JCS A 1966 1190
(13) HCA 1964 **47** 1889

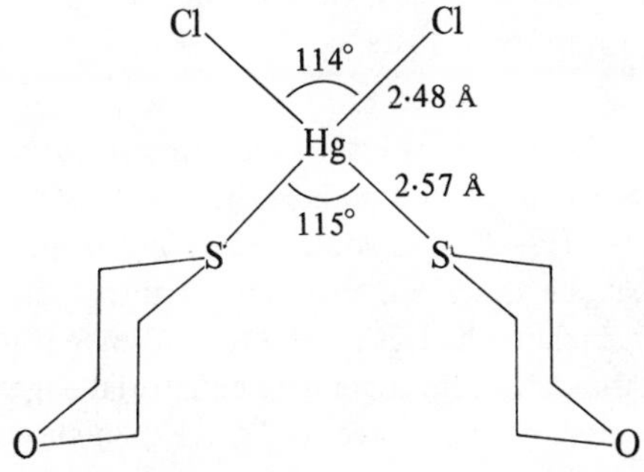

The lower structures show diagrams labelled (a) and (b).

(a): Cl, Cl, 114°, 2·48 Å, 2·57 Å, 115°, S, S, O, O

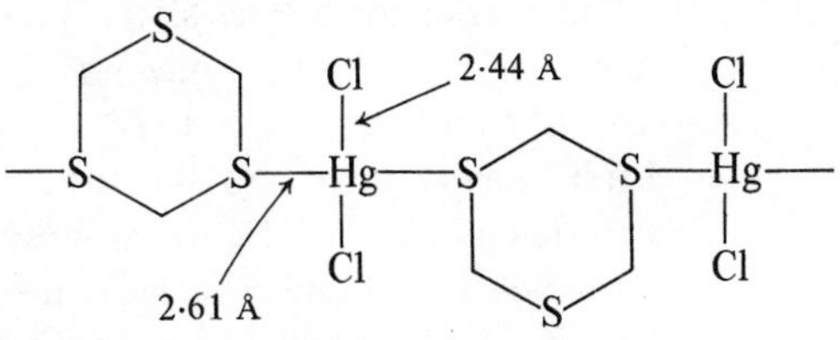

(b): S, Cl, 2·44 Å, Cl, 2·61 Å, Cl, Cl

rather complex example of a layer based on the plane 4-gon net. Octahedral $Co(SCN)_6$ groups are linked through their terminal S atoms by pairs of Hg atoms which are bridged by two S atoms (Fig. 26.5); the C_6H_6 molecules are

(14) AC 1968 **B24** 653

accommodated between the layers. In the diamond-like structure of $CoHg(SCN)_4^{(14)}$ the —SCN— ligands link Co and Hg into a 3D framework in which Co has four tetrahedral N and Hg four tetrahedral S neighbours.

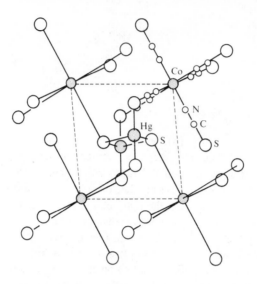

FIG. 26.5. Portion of layer in $CoHg_2(SCN)_6 . C_6H_6$.

Mercuric halides—simple and complex. In contrast to HgF_2, which crystallizes with the fluorite structure, $HgCl_2$, $HgBr_2$, and the yellow form of HgI_2 crystallize as linear molecules X—Hg—X; four halogen atoms of other molecules complete very distorted octahedral coordination groups around Hg(II). (Alternatively, the $HgBr_2$ structure may be described as built of very distorted CdI_2-type layers.) The red form of HgI_2 has a quite different layer structure in which each Hg atom has four equidistant I neighbours arranged tetrahedrally. In the remarkable structure assigned to an orange form of HgI_2 multiple groups of four tetrahedra, similar to the P_4O_{10} molecule, replace the single tetrahedra in the layer of red HgI_2. The molecules $HgCl_2$, $HgBr_2$, and HgI_2 have been shown by electron diffraction to form linear molecules in the vapour state. Interatomic distances in these halides are included in Table 26.2.

Many complex halides $MHgCl_3$ and M_2HgX_4 are known; the structures of fluoro compounds are not yet known. In a number of chloro compounds there is the same (2 + 4)-coordination of Hg(II) as in $HgCl_2$ (Hg—2 Cl at about 2·3 Å and 4 more distant neighbours), and the structures may be described in terms of these very distorted octahedral $HgCl_6$ groups. They include $K_2HgCl_4 . H_2O$ built of edge-sharing chains (a), and NH_4HgCl_3 in which the octahedra share four equatorial edges to form layers in which the Hg—Cl bonds are so much longer (2·96 Å) than those perpendicular to the layers (2·34 Å), that this crystal is an aggregate of $HgCl_2$ molecules, Cl^-, and NH_4^+ ions (Fig. 26.6). The salt $CsHgCl_3$ has a deformed perovskite structure in which Hg has 2 Cl neighbours at 2·29 Å and 4 Cl at 2·70 Å.

On the other hand, in salts of very large cations there is trigonal coordination of

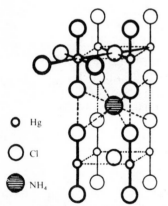

O   Hg

◯   Cl

⊜ NH_4

FIG. 26.6. The crystal structure of NH_4HgCl_3.

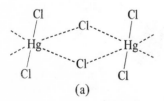

(a)

TABLE 26.2

Interatomic distances in mercuric compounds

Compound	Numbers of X atoms around Hg *in crystal*			Hg—X *in vapour molecule*[a]
	2 at	2 at	2 at	
HgF$_2$		8 at 2·40 Å		
HgCl$_2$[b]	2·25	3·34	3·63	2·34 Å
HgBr$_2$[c]	2·48	3·23	3·23	2·44
HgI$_2$ (yellow)[d]	2·62	4 at 3·51		2·61
HgI$_2$ (red)[d]		4 at 2·78		
HgI$_2$ (orange)[h]		4 at 2·68 Å		
NH$_4$HgCl$_3$[e]	2·34	2·96	2·96	
K$_2$HgCl$_4$. H$_2$O[f]	2·29	2·8	3·15	
CsHgCl$_3$[f]	2·29	4 at 2·70		
(Bu$_3$As)$_2$(HgBr$_2$)$_3$[g]	2·25	4 at 3·1–3·2		

(*a*) TFS 1937 **33** 852. (*b*) AK 1950 **22** 14. (*c*) ZK 1931 **77** 122. (*d*) IC 1967 **6** 396. (*e*) ZK 1938 **100** 208. (*f*) DAN 1955 **102** 1115. (*g*) JCS 1940 1209. (*h*) ZK 1968 **128** 97.

Hg(II) as in, for example, $(CH_3)_4N(HgCl_3)$ and the isostructural bromide[1] and iodide. In the bromide, which was studied in detail, approximately planar $HgBr_3^-$ groups (Hg–Br, 2·52 Å) are linked into chains by a fourth bond from each Hg (Hg–Br, 2·90 Å), a system intermediate between discrete $HgBr_3^-$ ions and chains formed from $HgBr_4$ tetrahedra sharing two vertices as in a metasilicate chain. In $(CH_3)_3SHgI_3$[2] there is also approximately planar coordination of Hg (Hg–3 I, 2·71 Å) but in this case the HgI_3 groups are linked into chains by very much weaker bonds (Hg–I, 3·52 and 3·69 Å), the five bonds forming very distorted trigonal bipyramidal coordination groups around the metal atoms.

Tetrahedral coordination of Hg(II) by halogen atoms occurs in only one of the dihalides (red HgI_2) but it is a feature also of complex halides such as Ag_2HgI_4 and Cu_2HgI_4 (p. 630). However, the tetrahedral ions HgX_4^{2-} (including HgF_4^{2-}) are stable in combination with large cations, as in the mercurichloride of the alkaloid perloline (Hg–Cl, 2·50 Å),[3] in $[N(CH_3)_4]_2HgBr_4$, and $[(CH_3)_3S]_2HgI_4$. Note the large difference between Hg–Cl when two collinear bonds are formed (2·3 Å) and the value 2·5 Å for tetrahedral bonds. Both types of bond occur in $C_8H_{12}S_2 . Hg_2Cl_4$,[4] in which some of the Hg atoms form two approximately collinear bonds (Hg–Cl, 2·30 Å, Cl–Hg–Cl, 168°) and the remainder form four tetrahedral bonds (Hg–2 Cl, 2·51 Å, Hg–2 S, 2·53 Å).

Mercuric halides combine with tertiary phosphines and arsines (e.g. $As(C_4H_9)_3$) to form crystalline products $(R_3As)_m(MX_2)_n$. The compound $[(C_4H_9)_3As]_2$ $[HgBr_2]_2$ consists of molecules in which the two Hg atoms are joined by halogen atoms:

(1) AC 1963 **16** 397

(2) AC 1966 **20** 20

(3) PCS 1963 171

(4) JCS 1965 5988

$(C_4H_9)_3As$ ⟍ Br ⟍ Br
     Hg   Hg
 Br ⟍ Br ⟍ $As(C_4H_9)_3$

921

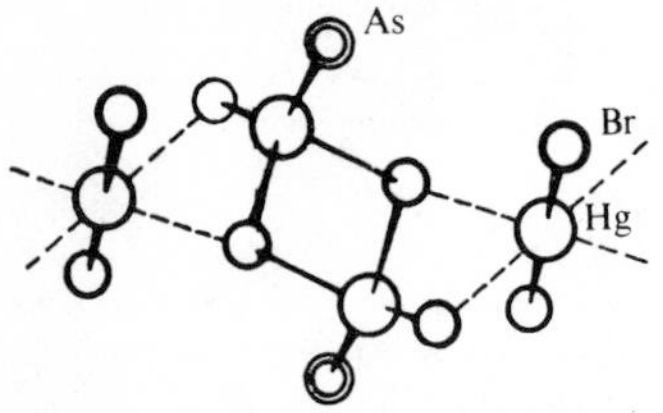

FIG. 26.7. The environment of a mercury atom of a $HgBr_2$ molecule in $[(C_4H_9)_3As]_2 Hg_2 Br_4 . HgBr_2$.

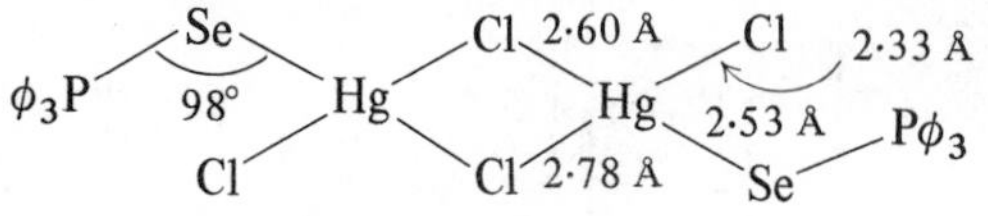

(5) JCS A 1969 2501
For other references see Table 26.2

(1) ACSc 1968 **22** 2529

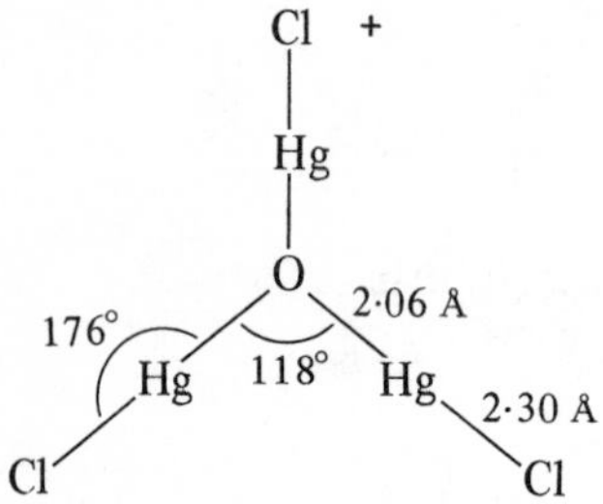

(2) AK 1964 **22** 517, 537
(3) ACSc 1964 **18** 1305

This compound is therefore similar in general type to the gold compounds already described (p. 909) and to the bridged palladium compounds mentioned later (p. 978), except that the bonds from the metal atoms are disposed tetrahedrally instead of being coplanar. The product richer in mercuric bromide and having the empirical formula $[(C_4H_9)_3As]_2[HgBr_2]_3$ is found to be a mixed crystal of $[(C_4H_9)_3As]_2 Hg_2 Br_4$, the above bridged molecule, and $HgBr_2$. In this crystal the environment of a Hg atom of a $HgBr_2$ molecule is very similar to that in $HgBr_2$ itself. There are two near Br neighbours and four more distant ones which belong to bridged molecules, as shown in Fig. 26.7. An unsymmetrical bridge is found in $Hg_2Cl_4(Se . P\phi_3)_2$[5]:

Oxychlorides and related compounds. Like many other B subgroup elements mercury forms numerous oxy- and thio-salts. The oxidation state of the metal in all of these compounds is not obvious, as for example in the complex structure of $2 HgO . Hg_2 Cl_2$.[1] The existence of the following mercuric compounds has been confirmed: Hg_3OCl_4 (colourless), $Hg_3O_2Cl_2$ (black), $Hg_4O_3Cl_2$, (yellow, the mineral kleinite), and $Hg_5O_4Cl_2$ (red). The structure of Hg_3OCl_4 shows it to be an ionic compound, trichloromercury oxonium chloride, for the crystal is an aggregate of Cl^- and $OHg_3Cl_3^+$ ions. The latter have the form of flat pyramids with the configuration shown.[2]

The structures of crystalline compounds of B subgroup elements in which the metal forms a small number of stronger bonds can be described in terms of the molecules or ions delineated by the stronger bonds. With the description of NH_4HgCl_3 as $HgCl_2$ molecules, NH_4^+, and Cl^- ions, compare the structure of $2 HgO . NaI$,[3] which contains infinite zigzag chains like those in orthorhombic HgO embedded in a mixture of Na^+ and I^- ions. Three I^- neighbours complete a

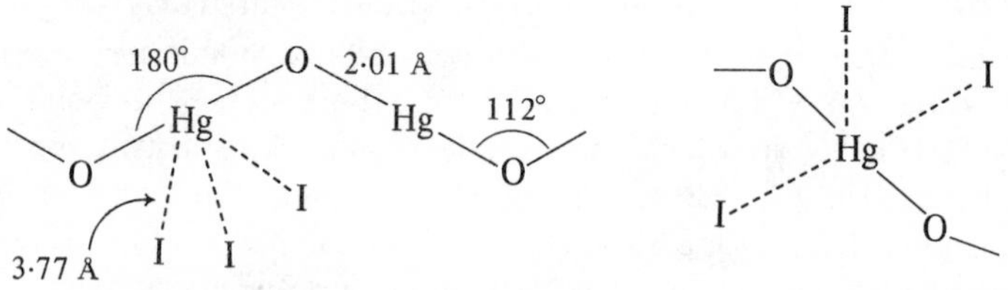

tetragonal pyramidal group around Hg(II). The structure of $NH_4Pb_2Br_5$ will be noted later in this chapter as a further example of a compound of this type—$NH_4^+(PbBr_2)_2Br^-$.

The structure of the black $Hg_3O_2Cl_2$ which, like Hg_3OCl_4, may be prepared by the action of marble chips on aqueous $HgCl_2$, is complex. There are two kinds of Hg(II) atoms with quite different environments. The structure is not known

922

sufficiently accurately to warrant detailed description[4] but appears to contain HgO chains, Hg^{2+}, and Cl^- ions. It has also been described as built of HgOCl layers and mercuric ions.[5]

(4) AK 1964 **23** 205

(5) AC 1955 **8** 379

The structures of two of the three polymorphs of the corresponding sulphur compound, $Hg_3S_2Cl_2$, are accurately known. In both there are continuous frameworks $(Hg_3S_2)_n^{2n-}$ built from pyramidal SHg_3 groups sharing Hg atoms (S–Hg–S, $166°$, Hg–S–Hg, $92°$ (α form), $95°$ (γ form)). In the α form this is a 3D network (see p. 95) and in the γ form a 2D system; Cl^- ions complete distorted octahedral groups around Hg (Hg–2 S, $\approx 2 \cdot 4$ Å, Hg–4 Cl at distances from $2 \cdot 7$–$3 \cdot 5$ Å).[6]

(6) AC 1968 **B24** 156, 1661

In $HgO \cdot Hg(CN)_2$ there appear to be molecules $NC \cdot HgOHg \cdot CN$, but the two different Hg bond angles ($180°$ and $156°$)[4] suggest that the light atoms may not have been accurately located. The structural chemistry of these compounds will be known to greater accuracy when neutron diffraction studies have been made, as in the case of Hg_3OCl_4 and HgO.

Mercuric oxide and sulphide. The two well-known forms of HgO are the stable orthorhombic and the metastable hexagonal forms. The former is made, for example, by heating the nitrate in air and the latter by the prolonged action of NaOH on K_2HgI_4 solution at $50°C$. Both form red crystals which grind to a yellow powder like the precipitated form. The former is built of infinite planar zigzag chains:[1]

(1) ACSc 1964 **18** 1305

and the next nearest neighbours of Hg are 4 O atoms of different chains at $2 \cdot 81$–$2 \cdot 85$ Å. The hexagonal form[2] is isostructural with cinnabar (HgS) and is built of helical chains in which the bond angles and bond lengths are the same as in the planar chains of the orthorhombic form. Two further O atoms at $2 \cdot 79$ Å and two at $2 \cdot 90$ Å complete a very distorted octahedral coordination group around Hg. The determination of these structures provides a good example of the use of both X-ray and neutron diffraction. It is not possible to locate the O atoms by X-ray diffraction but it is possible using neutrons, for which the relative scattering amplitudes are much more closely similar and moreover independent of the angle of scattering. Apparently at least two other forms of HgO can be prepared hydrothermally.

(2) ACSc 1958 **12** 1297

Mercuric sulphide, HgS, is dimorphic. Metacinnabarite, a rare mineral, crystallizes with the zinc-blende structure, in which Hg(II) forms tetrahedral bonds (Hg–S, $2 \cdot 53$ Å).[3] The more common form, cinnabar, has a structure[4] which is unique among those of monosulphides, for the crystal is built of helical chains in which Hg has two nearest neighbours (at $2 \cdot 36$ Å), two more at $3 \cdot 10$ Å, and two at $3 \cdot 30$ Å. Interbond angles are S–Hg–S, $172°$ and Hg–S–Hg, $105°$. Cinnabar is notable for its extraordinarily large optical rotatory power which is, of course, a property of the solid only.

(3) ACSc 1965 **19** 522

(4) ACSc 1950 **4** 1413

(5) ZaC 1964 **329** 110

(6) AK 1959 **13** 515

Two other oxy-compounds may be mentioned here, though the details of their structures are not established. Alkali-metal compounds of the type Na_2HgO_2[5] apparently contain linear $(O-Hg-O)^{2-}$ ions $(Hg-O \approx 1.95$ Å$)$, and one of the two forms of HgO_2[6] (prepared from HgO and H_2O_2 at $-15°C$) consists of chains

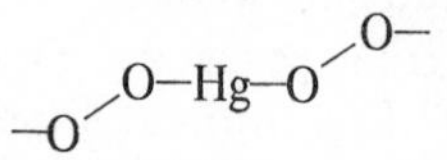

The relation of the structure of β-HgO_2 to PdS_2 and AgF_2 is described on p. 223.

Mercury–nitrogen compounds. By the interaction of mercuric salts with ammonia in aqueous solution a number of compounds can be obtained, depending on the conditions. They include:

$Hg(NH_3)_2X_2$, diamminohalides (X=Cl, Br)
$HgNH_2X$, aminohalides (X=F,† Cl, Br)
Hg_2NHX_2, iminodihalides (X=Cl, Br)

(1) ZaC 1956 **287** 24

and Hg_2NX, Millon's base $(Hg_2N . OH . xH_2O)$ and salts (X=Cl, Br, I, NO_3, and ClO_4). The chlorides $Hg(NH_3)_2Cl_2$ and $HgNH_2Cl$ have long been known as fusible and infusible white precipitates respectively.

The structural feature common to all these compounds is the presence of complexes containing Hg^{II} forming two collinear (sp) bonds and N forming four tetrahedral bonds, though in the iminodihalide Hg_2NHBr_2 there is also an unusual $(3 + 2)$-coordination of one-quarter of the Hg atoms.

The diamminohalide $Hg(NH_3)_2Cl_2$ (with which the dibromide is isostructural) has a very simple cubic structure[2] (Fig. 26.8(a)) in which the unit cell contains $\frac{1}{2}[Hg(NH_3)_2Cl_2]$, so that only one-sixth of the Hg positions are occupied (statistically). Each Hg position has two equidistant NH_3 neighbours and every NH_3 will have, on the average, one Hg neighbour. Accordingly there are isolated

(2) ZK 1936 **94** 231

$(H_3N . Hg . NH_3)^{2+}$ ions randomly arranged. In $HgNH_2Br$[3] the positions available for Hg, N, and Br are the same as in $Hg(NH_3)_2Br_2$ but there are twice as many Hg atoms, so that each shaded circle in Fig. 26.8(a) now represents Hg/3. Since any

(3) ZaC 1952 **270** 145

particular Hg atom has two equidistant N atoms as nearest neighbours and since the number of Hg and N atoms is the same, each N atom will have an average of two Hg neighbours. There must therefore be chains

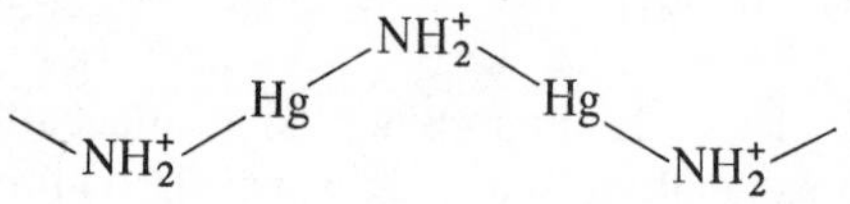

but these are not straight and parallel over long distances but bent at each N atom and irregularly arranged in the crystal.

† This compound is apparently not structurally similar to the chloride and bromide.[1] It may be a 3-dimensional framework of the Millon's base type with NH_4^+ and F^- ions in the interstices, that is $(Hg_2N)F . NH_4F$.

924

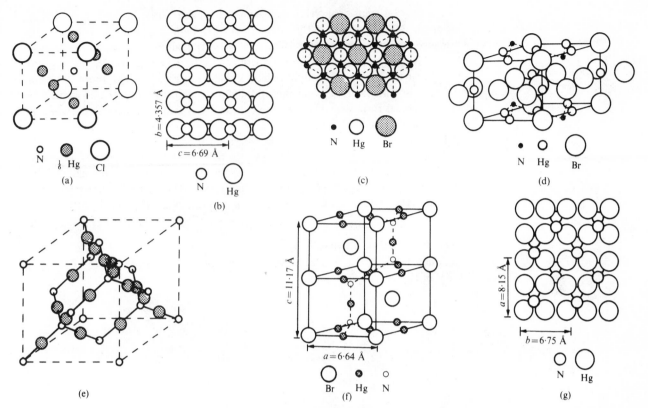

FIG. 26.8. Structures of mercury–nitrogen compounds: (a) $Hg(NH_3)_2Cl_2$, (b) $-Hg-NH_2-$ chains in orthorhombic $HgNH_2Cl$, (c) $[Hg_3(NH)_2]$ layer in Hg_2NHBr_2, (d) unit cell of Hg_2NHBr_2, (e) NHg_2NO_3, (f) NHg_2Br, (g) layer in $Hg_2N_2H_2Cl_2$.

Independent X-ray studies of $HgNH_2Cl$[4] and $HgNH_2Br$[5] led to the assignment of an orthorhombic structure (Fig. 26.8(b)) also containing infinite chain ions bound together by halide ions. In this (regular) structure the zigzag chains are arranged in parallel array, but it appears[6] that this structure is not characteristic of pure $HgNH_2X$, being formed only when there is a small amount of halide HgX_2 in solid solution. (The nuclear resonance spectrum of 'infusible white precipitate' is consistent with the presence of NH_2 groups[7] and eliminates older formulae involving NH_4^+ or NH_3.)

With increasing $Hg:N$ ratio more extended complexes become possible. In Hg_2NHBr_2[8] three-quarters of the Hg atoms form hexagonal layers of composition $(NH)_2Hg_3$ in which N is attached to three Hg and one H (Fig. 26.8(c)), the remainder of the Hg atoms being situated between the layers surrounded by three coplanar Br atoms at 2.6 ± 0.1 Å and two more at 3.08 Å completing a trigonal bipyramidal coordination group. A unit cell of the structure is shown in Fig. 26.8(d).

(4) AC 1951 **4** 266
(5) AC 1952 **5** 604

(6) ZaC 1954 **275** 141

(7) JCS 1954 3697

(8) AC 1955 **8** 723

(9) ZaC 1953 **274** 323; AC 1951 **4** 156; AC 1954 **7** 103

Finally, 3-dimensional NHg_2 frameworks similar to those in the forms of SiO_2 occur in Millon's base, $NHg_2OH \cdot xH_2O$ ($x = 1$ or 2), and its salts.[9] The framework in the nitrate, NHg_2NO_3, is of the cristobalite type, and other salts also have this cubic structure if prepared from the nitrate (Fig. 26.8(e)), but it seems likely that the hexagonal (tridymite-like) structure of Fig. 26.8(f) is more stable for the halides. The base itself and the bromide and iodide can be prepared initially in this hexagonal form or by heating the cubic modifications. In these frameworks N forms four tetrahedral bonds and Hg two collinear bonds (Hg–N $\approx 2 \cdot 07$ Å). The OH^-, X^- or other ions occupy the interstices in the frameworks.

Many heavy-metal salts are reduced to the metal by hydrazine, and this is true of mercuric oxy-salts. From the chloride and bromide, however, compounds of the types $[Hg(N_2H_4)]X_2$, $[Hg(N_2H_4)_2]X_2$ and $(Hg_2N_2H_2)X_2$ may be prepared. The chloride $(Hg_2N_2H_2)Cl_2$, is closely related structurally to $HgNH_2Cl$, but whereas $H_2N{<}$ can form only chains the unit ${>}HN{-}NH{<}$ forms layers[10] (Fig. 26.8(g)), between which are situated the Cl^- ions.

(10) ZaC 1956 **285** 5; ZaC 1957 **290** 24

Summarizing, these mercury–nitrogen compounds provide examples of all four types of Hg–N complex, namely:

$$\text{finite } (H_3N{-}Hg{-}NH_3)^{2+} \text{ ions in } Hg(NH_3)_2X_2,$$
$$\text{chains } {-}NH_2{-}Hg{-}NH_2{-}Hg{-} \text{ in } HgNH_2X,$$
$$\text{layers } Hg_3(NH)_2 \text{ in } Hg_2NHBr_2,$$
$$Hg_2N_2H_2 \text{ in } (Hg_2N_2H_2)Cl_2,$$

and 3-dimensional frameworks in Hg_2NX.

The structural chemistry of gallium and indium

To our earlier remarks on the lower valence states of these elements we add here a summary of their structural chemistry in the trivalent state. Many Ga(III) and In(III) (and a few Tl(III)) compounds have similar structures, and there is a strong resemblance to Al, which forms compounds structurally similar to all those of Table 26.3. (The structures of most of these compounds are described in other

TABLE 26.3

Environment of Ga(III) *and* In(III) *in simple compounds*

Tetrahedral coordination	Octahedral coordination
$Li(GaH_4)$	GaF_3
Ga_2Cl_6 and In_2Cl_6 (vapour molecules)	$InCl_3$ (and $TlCl_3$)
$GaPO_4$ (cristobalite structure)	$GaSbO_4$ (statistical rutile structure)
GaN, InN (wurtzite structure)	α-Ga_2O_3 (corundum structure but
GaP, InP (zinc-blende structure)	also C-M_2O_3 structure like In_2O_3)
	$In(OH)_3$
Ga_2S_3	
Both tetrahedral and octahedral coordination in β-Ga_2O_3, In_2S_3	

926

chapters.) Like Al, Ga exhibits both tetrahedral and octahedral coordination by Cl and O; in β-Ga_2O_3 there are both kinds of coordination in the same crystal. For $Ga(CH_3)_3$ and $In(CH_3)_3$ see p. 781.

More complex examples of tetrahedrally bonded Ga(III) include $Na\{Ga[OSi(CH_3)_3]_4\}$[1] and the cyclic molecule $Ga_4(OH)_4(CH_3)_8$,[2] which has a structure similar to that of the corresponding gold compound but with tetrahedral instead of coplanar bonds from the metal atoms.

(1) JCS 1963 3200
(2) JACS 1959 **81** 3907

In $[N(C_2H_5)_4]_2$ $InCl_5$[3] In forms five bonds directed towards the vertices of a square pyramid, the metal atom being 0·6 Å above the base; compare the similar structure of $Sb(C_6H_5)_5$. The configuration of the $InCl_5^{2-}$ ion is close to that calculated for minimum Cl–Cl repulsions using a simple inverse square law. Contrast the trigonal bipyramidal structure of the isoelectronic $SnCl_5^-$ ion.

8-coordinated In presumably occurs in the anion in $[N(C_2H_5)_4]$ $[In(NO_3)_4]$[4] which is isoelectronic with $Sn(NO_3)_4$.

(3) IC 1969 **8** 14

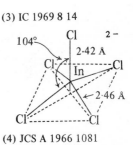

(4) JCS A 1966 1081

The structural chemistry of thallium

The structures of some simple thallous and thallic compounds have been noted in earlier chapters. The thallous halides show a remarkable resemblance to those of Ag(I) in their colours and solubilities, though their crystal structures are different owing to the different sizes and polarizabilities of the cations. TlF, like AgF, is soluble in water and the other halides are very insoluble. TlOH is a strong base, soluble in water like the alkali hydroxides, and Tl_2SO_4 is isostructural with K_2SO_4.

Compounds which apparently contain Tl(II) in fact contain Tl(I) and Tl(III). For example, both $TlCl_2$ and Tl_2Cl_3 are diamagnetic, whereas Tl(II) would have one unpaired electron. The former is presumably similar to $TlBr_2$[1] which is isostructural with Ga_2Cl_4, that is, it is $Tl^I(Tl^{III}Br_4)$ and contains the same tetrahedral ion as $CsTlBr_4$ (Tl–Br, 2·51 Å). The complex structure of Tl_2Cl_3 has not been determined; the compound is presumably $Tl_3^I(Tl^{III}Cl_6)$ containing the same octahedral ions as $K_3TlCl_6 . 2 H_2O$ and similar salts. The compound TlI_3 is not thallic iodide but is isostructural with NH_4I_3 and is therefore $Tl^+I_3^-$.[2] The halogen complexes of Tl(III) have been reviewed.[3a]

Sulphides of thallium which have been described include Tl_2S (anti-CdI_2 structure), TlS, Tl_4S_3, and TlS_2. The details of the (distorted) anti-CdI_2 structure

(1) JCS 1963 3459

(2) AC 1963 **16** 71
(3a) IC 1965 **4** 502

(3b) ZK 1939 **101** 367

of $Tl_2S^{(3b)}$ do not appear to be known. In crystalline TlS the metal atoms are of two kinds, one set having four (tetrahedral) and the other eight nearest neighbours. The former are linked by the S atoms into chains of tetrahedra sharing opposite edges:

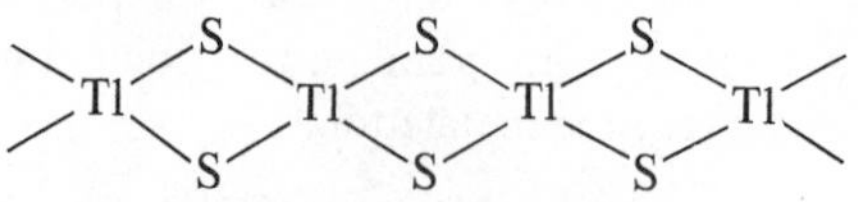

(4) ZaC 1949 **260** 110

whereas the latter lie between the chains in position of 8-coordination. The large distance between these latter Tl atoms and their eight S neighbours (3·32 Å) compared with the Tl–S distance of only 2·60 Å in the chains suggests the formulation of the compound as an ionic thallous thallic sulphide, $Tl^I(Tl^{III}S_2)$.[4]

(5) JCS 1965 6107

In addition to the tetrahedral and octahedral coordination of Tl(III) or Tl^{3+} in halide complexes and other compounds there may be 5-coordination in compounds such as $(C_6H_5)_2L_2Tl(NO_3)$, where L is $OP(C_6H_5)_3$, but their structures are not yet known.[5]

In its covalent compounds thallium shows no reluctance to utilizing the two 6s electrons for bond formation. Indeed monoalkyl derivatives Tl(Alk) in which the valence group would be (2, 2) are not known, whereas trialkyls (valence group 6) are known and the most stable alkyl derivatives are the dialkyl halides such as $[Tl(CH_3)_2]I$. These are ionic compounds—$[Tl(CH_3)_2]$ OH being a strong base—and in the $Tl(Alk)_2^+$ ions the thallium atom has the same outer electronic structure as mercury in CH_3–Hg–CH_3, viz. (4). Accordingly the $(CH_3$–Tl–$CH_3)^+$ ion is

(6) ZK 1934 87 370

linear, as shown by the determination of the crystal structure of $Tl(CH_3)_2I$.[6] In molecules such as $Tl(Alk)_2A$, where A represents a molecule of a β-diketone, Tl apparently has the valence group (8).

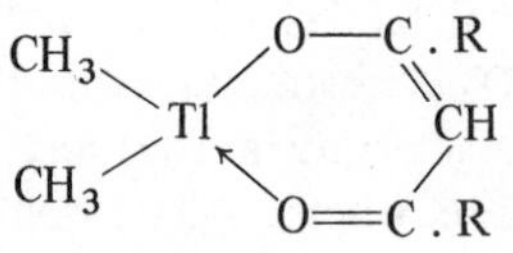

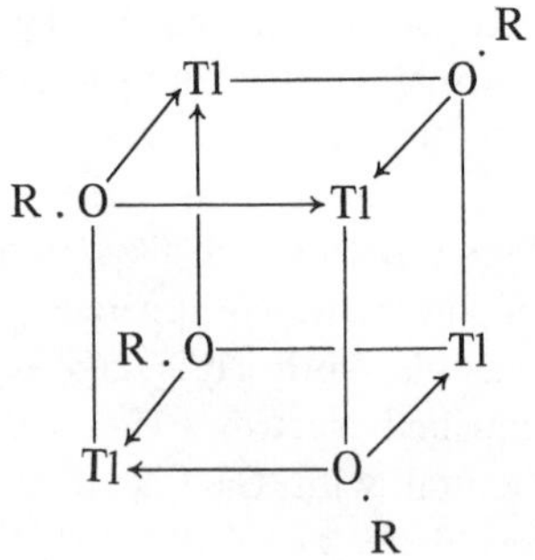

(7) JINC 1962 **24** 357

Thallous alkoxides exist as tetramers in benzene solution, and an X-ray study of $(TlOCH_3)_4$[7] shows a tetrahedral arrangement of the metal atoms as in the idealized structure shown above, but the alkoxyl groups were not located. In such molecules Tl forms pyramidal bonds, with valence group (2, 6).

928

The structural chemistry of germanium

The more important characteristics of this element are perhaps most easily seen by comparing it with silicon. The resemblance of Ge(IV) to silicon is very marked. Not only are the ordinary forms of elementary Ge and Si isostructural but so also are GeI_4 and SiI_4, hexagonal GeO_2[1] and SiO_2 (high-quartz), tetragonal GeO_2 and the rutile (high-pressure) form of SiO_2, and many oxy-compounds of the two elements. The Ge analogues of all the major types of silicates and aluminosilicates have been prepared, ranging from those containing finite ions to chain, layer, and 3D framework structures.[2] Examples of germanates containing various types of complex ion which are isostructural with the corresponding silicates include: Be_2GeO_4 and Zn_2GeO_4 with the phenacite and willemite structures respectively, $Sc_2Ge_2O_7$ and $Sc_2Si_2O_7$, $BaTiGe_3O_9$ with the same type of cyclic ion as in benitoite, and $CaMg(GeO_3)_2$ with a chain ion similar to that in diopside. The extent of this structural resemblance to Si is seen from the facts that two crystalline forms of Ca_2GeO_4 are isostructural with two forms of Ca_2SiO_4 while Ca_3GeO_5 crystallizes with no fewer than four of the structures of Ca_3SiO_5. In contrast to the close similarity of the oxygen chemistry of Ge to that of Si, the structures of the normal forms of SiS_2 and GeS_2 are entirely different, though there is tetrahedral coordination of Si and Ge. The normal form of SiS_2 has a simple chain structure whereas GeS_2 has a unique 3D structure in which the GeS_4 terahedra share all vertices as in silica-like structures. However, both SiS_2 and GeS_2 have the same structure at high pressures, a 'compressed' cristobalite-like structure (p. 612) in which the S bond angle is close to the regular tetrahedral value. Silanes up to Si_7H_{16} have been prepared, and germanes up to Ge_9H_{20} have been characterized. (Reduction of an aqueous germanate solution by KBH_4 gives a good yield of GeH_4, from which the higher members are produced by the action of a spark discharge.) In all the above compounds Si or Ge form four tetrahedral bonds.

The major differences between these elements may be summarized: (i) the much greater stability of Ge(II) than Si(II), (ii) the greater tendency of Ge(IV) to form 6 bonds, and (iii) the formation of salts by Ge(IV). We deal briefly with these three points.

(i) The divalent state does not enter into normal silicon chemistry, though it may be important at high temperatures (see SiO, p. 784), but as already noted a number of compounds of Ge(II) are stable at ordinary temperatures. Six octahedral bonds are formed by Ge(II) in GeI_2 (CdI_2 structure) and in $CsGeCl_3$[3] (high-temperature form, perovskite structure, low-temperature form, deformed perovskite structure), and in GeS. In GeS, however, three S atoms are appreciably closer than the others (p. 937) as in SnS, and this preference for 3-coordination is more pronounced in GeF_2. This halide (made from GeF_4 + Ge at $300°C$) has a unique structure in which trigonal pyramidal GeF_3 coordination groups share two F atoms (F bond angle $157°$) to form infinite chains[4] (Fig. 26.9). The bond to the unshared F atom is appreciably shorter ($1\cdot79$ Å) than those in the chain ($1\cdot91$, $2\cdot09$ Å), and weaker bonds ($2\cdot57$ Å) link the chains into a 3D structure. This is obviously not an ionic crystal, and the stereochemistry of Ge(II) is very similar to

(1) AC 1964 **17** 842

(2) FM 1966 **43** 230

(3) ACSc 1965 **19** 421

(4) JCS A 1966 30

929

that of Sn(II) in $NaSn_2F_5$:

		GeF_2	$(Sn_2F_5)^-$		GeF_2	$(Sn_2F_5)^-$
(For the labelling	a	1·79 Å	2·08 Å	ab	92°	89°
of the bonds see	b	1·91 Å	2·07 Å	cd	163°	142°
Fig. 26.9).	c	2·09 Å	2·22 Å	ac	86°	81°
	d	2·57 Å	2·53 Å	bc	85°	84°

Regarding Ge as forming three pyramidal bonds the valence group is (2, $\underline{6}$); if the much more distant fourth F is included the valence group would be (2, $\underline{8}$), derived from a trigonal bipyramid, as shown at (b) in Fig. 26.9. (Gaseous GeF_2 exists only at high temperatures in the presence of excess metal, and the vapour contains polymeric species up to $Ge_3F_5^+$; SnF_2 behaves similarly.)[5]

(5) IC 1968 **7** 608

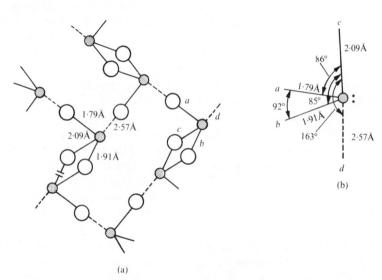

FIG. 26.9. The crystal structure of GeF_2: (a) projection along the chains, (b) environment of Ge.

(ii) We noted at the beginning of Chapter 23 that the number of molecules, ions, or crystals in which Si forms 6 bonds is very small. Although the structural chemistry of Ge(IV) is very largely based on 4-coordinated (or 4-covalent) Ge examples of octahedral coordination are more numerous than in the case of Si. For example, there is 6-coordination in the stable (tetragonal) form of GeO_2 (Ge–4 O, 1·87 Å, Ge–2 O, 1·90 Å)[5a] in contrast to 4-coordination in the soluble quartz-like form. Examples of 6-coordinated Ge(IV) include molecules such as *trans*-$GeCl_4(pyr)_2$[6] (Ge–Cl, 2·27 Å), the GeF_6^{2-} and $GeCl_6^{2-}$ ions, and $FeGe(OH)_6$,[7] which is isostructural with $FeSn(OH)_6$ and $NaSb(OH)_6$ (Ge–6 OH, 1·96 Å). There is both tetrahedral and octahedral coordination of Ge in $K_3HGe_7O_{16} . 4 H_2O$,[2] which has an elegant framework structure built of GeO_4 and GeO_6 groups (Fig. 5.45, p. 191) and possesses zeolitic properties, and in $Na_4Ge_9O_{20}$.[8]

(5a) AC 1971 **B27** 2133

(6) JCS 1960 366

(7) AC 1961 **14** 205

(8) ACSc 1963 **17** 617

(iii) The more metallic character of Ge is shown in the formation of salts such as $Ge(SO_4)_2$, $Ge(ClO_4)_4$, and $GeH_2(C_2O_4)_3$, which have no silicon analogues.[9]

(9) ACSc 1963 17 597

The structural chemistry of tin and lead

A comparison of the structural chemistry of these two elements reveals some interesting resemblances and also some remarkable differences. The atoms of each element have the same outer electronic structure, two s and two p electrons, and each has oxidation states of 2 and 4. The more metallic nature of lead is shown by the difference between the structures of the elements (Chapter 29) and by many differences between stannous and plumbous compounds. It will be convenient to deal first with Sn(IV) and Pb(IV) since the structural chemistry is more straightforward for the higher oxidation state. Few compounds of Sn(II) and Pb(II) are isostructural and compounds containing these metals in both oxidation states are different for the two elements (e.g. Sn_2S_3 and Pb_3O_4).

Stannic and plumbic compounds

Bond arrangements found are tetrahedral, trigonal bipyramidal, and octahedral for completely shared valence groups of 8, 10, and 12 electrons respectively. Higher coordination numbers are exhibited by Sn(IV), 7, and 8, and by Pb(IV), 8, in certain complexes formed by chelating ligands, most or all of the bonds being formed to O atoms.

Tetrahedral coordination. The simplest examples are molecules MX_4 and $MX_{4-x}Y_x$, a number of which have been shown to be tetrahedral either in the vapour or crystalline states:

$SnCl_4$, $SnBr_4$, SnI_4, $Sn(C_6H_5)_4$[1a] (crystal); $(CH_3)_3SnX$ etc., $PbCl_4$,

(1a) JCS A 1970 911

$Pb(CH_3)_4$ and $Pb_2(CH_3)_6$ in the vapour state (references in Table 21.1, p. 727). In contrast to the tetrahedral structure of $(CH_3)_3SnCl$ in the vapour state there is considerable distortion of the molecule in the crystal, weak bridging Sn–Cl bonds suggesting a tendency to polymerize to an octahedral chain structure:[1b]

(1b) JCS A 1970 2862

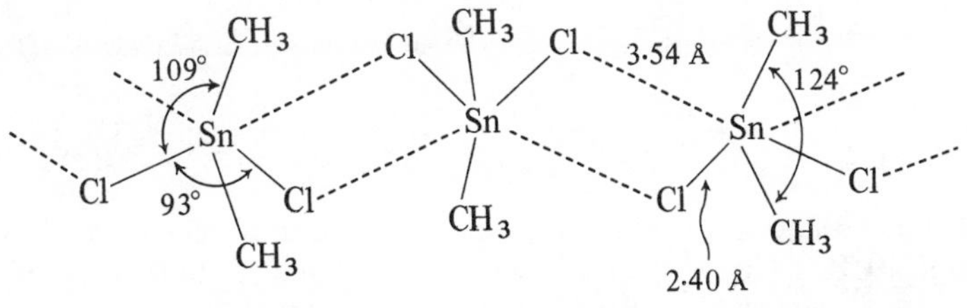

Some compounds $(CH_3)_3SnX$ and $(CH_3)_2SnX_2$ show a clear-cut change from a tetrahedral monomer to a polymeric structure with 5- or 6-coordination of Sn (see later).

Monomeric $Sn^{II}R_2$ compounds generally polymerize to Sn^{IV} compounds, either linear (for example, $H[Sn(C_6H_5)_2]_6H$) or cyclic molecules. An example of the latter is $[Sn(C_6H_5)_2]_6$, Fig. 26.10(a), which has been studied in the crystalline

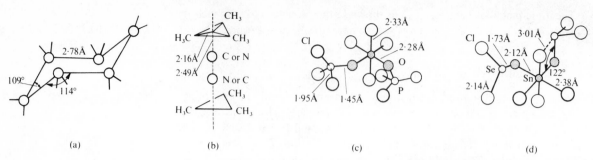

FIG. 26.10. Stereochemistry of Sn(IV): (a) cyclic $[Sn(C_6H_5)_2]_6$ molecule (phenyl groups omitted), (b) infinite chains in $(CH_3)_3SnCN$, (c) $SnCl_4 . 2 POCl_3$, (d) $SnCl_4 . 2 SeOCl_2$.

(2) IC 1963 **2** 1310

complex with *m*-xylene.[2] The Sn–Sn bond length in the ring (2·78 Å) is the same as in grey tin (2·80 Å). There is also tetrahedral coordination of Sn in complexes containing the ligand $-SnCl_3$ such as $(C_8H_{12})_3Pt_3(SnCl_3)_2$ (see p. 372). In these complexes Sn is bonded to 3 Cl and 1 Pt atom.

There are numerous molecules in which Sn is bonded to other metal atoms to which are attached ligands such as CO, C_5H_5, and phosphines. Four bonds from Sn in such cases are arranged tetrahedrally with various degrees of distortion from regular tetrahedral bond angles. For example, in $Sn[Fe(CO)_4]_4$[3] there is appreciable distortion to bring together one pair of Fe atoms (Fe–Fe, 2·87 Å). The special interest of these compounds lies in the metal–metal bonding.

(3) JCS A 1967 382

(4) JACS 1964 **86** 733

Trigonal bipyramidal coordination. The simplest example is the $(SnCl_5)^-$ ion[4] which has been studied in the salt of the substituted cyclobutenium ion (a). Both 5- and 6-coordination of Sn(IV) occur in crystalline $[(CH_3)_2SnCl(terpyridyl)]^+$ $[(CH_3)_2SnCl_3]^-$, which is an assembly of the ions (b) and (c).[5]

(5) JCS A 1968 3019

(6) IC 1966 **5** 511

Crystalline compounds $(CH_3)_3SnX$ do not consist of tetrahedral molecules (like, for example, $(CH_3)_3GeCN$[6]) or of ions $Sn(CH_3)_3^+$ and X^- In $(CH_3)_3SnF$ there are chains in which $Sn(CH_3)_3$ groups are linked through F atoms. Owing to disorder in the crystals it is not possible to interpret the X-ray data unambiguously, but Sn is certainly 5-coordinated and the bridging $-F-$ bonds are non-linear.[7] Crystalline $(CH_3)_3SnCN$[8a] consists of infinite chains of planar $Sn(CH_3)_3$ and CN groups (Fig. 26.10(b)) in which Sn forms three short bonds (Sn–C, 2·16 Å) to CH_3 and two longer ones (2·49 Å) to C or N (not distinguishable in the X-ray study). The structure of $(CH_3)_3SnOH$ is apparently similar. In $Sn(CH_3)_3$ (dicyanamide)[8b] the

(7) JCS 1964 2332
(8a) IC 1966 **5** 507

(8b) IC 1971 **10** 1938

932

(angular) NCN . CN group behaves like CN in $Sn(CH_3)_3CN$, $Sn(IV)$ forming trigonal bipyramidal bonds to 3 equatorial CH_3 groups and to 2 axial N atoms. (In $Sn(CH_3)_2(NCNCN)_2$ there is *octahedral* coordination of Sn, as in $Sn(CH_3)_2F_2$ (see next section). There is a square net of Sn atoms which are bridged by $-NCNCN-$ groups, the octahedron being completed by two axial CH_3 groups. The structure of the dicyanamide ligand is worthy of note: it is shown at the right.

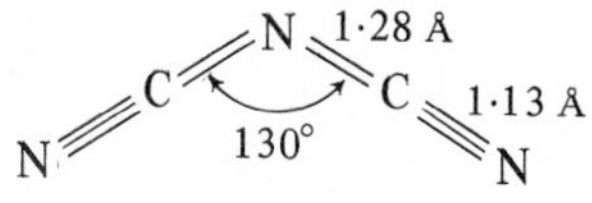

Infrared studies of other salts $(CH_3)_3SnX$ ($X = ClO_4$, BF_4, NO_3) suggest strong interaction between $(CH_3)_3Sn$ and the anion.[9] The perchlorate and nitrate react with ammonia to form the ion $(CH_3)_3Sn(NH_3)_2^+$ in which the valence group of 10 shared electrons would be the same as in $(pyr)(CH_3)_3SnCl$.[10] In this trigonal bipyramidal molecule the methyl groups occupy the equatorial positions; no example of this bond arrangement around $Pb(IV)$ has yet been established.

For the unusual square pyramidal coordination of Sn^{IV} (and Pb^{IV}) in $K_2M^{IV}O_3$ see p. 463.

Octahedral coordination. The structures of crystalline SnF_4[11] and $SnF_2(CH_3)_2$[12] have been described in Chapter 5. Octahedral SnF_6 and $SnF_4(CH_3)_2$ groups respectively share four equatorial vertices (F atoms) to form infinite layers:

	Sn—F (bridging)	Sn—F (terminal)	Sn—C
SnF_4	2·02 Å	1·88 Å	
$SnF_2(CH_3)_2$	2·12 Å		2·08 Å

Numerous octahedral ions MX_6^{2-} are formed by these elements. For example, Rb_2SnCl_6 and Rb_2PbCl_6 both crystallize with the K_2PtCl_6 structure; ions intermediate between SnF_6^{2-} and $Sn(OH)_6^{2-}$ have been studied in solution[13] and isolated in salts such as $M_2SnF_5(OH)$ and $M_2SnF_4(OH)_2$.[14]

In the crystalline compounds $SnCl_4 . A_2$, where A is, for example, $POCl_3$[15] or $SeOCl_2$[16], two O atoms of the A molecules complete octahedral coordination groups around $Sn(IV)$ with the *cis* configuration (Fig. 26.10(c) and (d)). The compound with C_4H_8S (tetrahydrothiophene)[17] is exceptional not only in having the *trans* configuration, attributable to the bulky ligand, but also in having a high dipole moment in solution (5 D) for which there is no obvious explanation. Other examples of octahedral $Sn(IV)$ complexes include the molecule of dimethyl Sn bis-(8-hydroxyquinolinate), (a),[18] and $SnCl_4[NC(CH_2)_3CN]$,[19] (b), in

(9) IC 1963 **2** 1020

(10) JCS 1963 1524

(11) NW 1962 **49** 254
(12) IC 1966 **5** 995

(13) PCS 1964 407
(14) ZaC 1963 **325** 442
(15) ACSc 1963 **17** 759
(16) AC 1960 **13** 656

(17) JCS 1965 1581

(18) IC 1967 **6** 2012
(19) IC 1968 **7** 1135

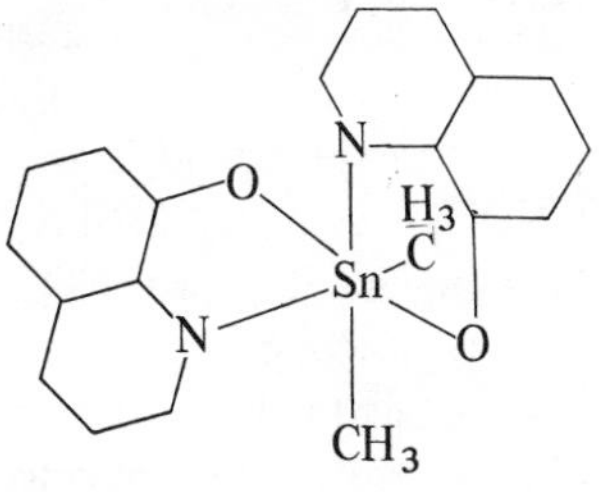

(a)

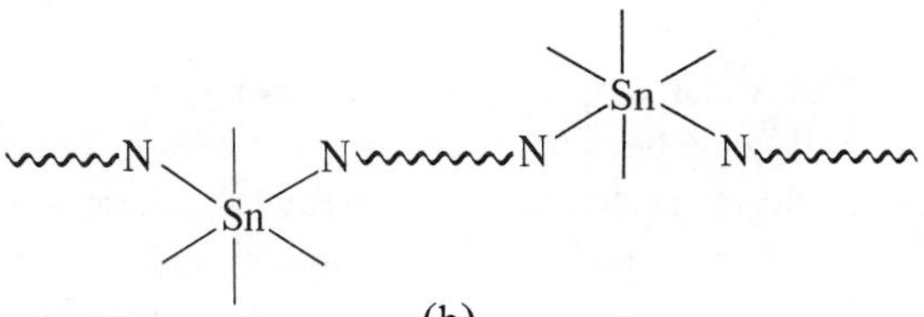

(b)

which the glutaronitrile molecules link the Sn atoms into infinite chains (note the *cis* N atoms in the coordination group of Sn).

There is octahedral coordination of the metal atoms in the dioxides of Sn and Pb; these are essentially ionic crystals. The structure of $Sn(OH)_4$ is not known. The gel-like $Sn(OH)_4$, precipitated from $SnCl_4$ solution by NH_4OH, dries to constant weight at $110°C$ with the composition SnO_3H_2. Dehydration to SnO_2 above $600°C$ apparently takes place through a number of definite crystalline phases with the compositions $Sn_2O_5H_2$, $Sn_4O_9H_3$, and $Sn_8O_{16}H_2$.[20]

In SnS_2 (CdI_2 structure) Sn has six octahedral neighbours and S forms three pyramidal bonds. If this is regarded as a covalent crystal Sn has a valence group of 12 shared electrons. Lead forms no disulphide under atmospheric pressure, but two high-pressure forms of PbS_2 have been made,[21] one of which probably has the CdI_2 structure, like SnS_2. The Sn(IV) atoms have six octahedral neighbours in $Sn^{II}Sn^{IV}S_3$, which is isostructural with NH_4CdCl_3.[22] For references to octahedral Sn(IV) complexes, see ref. (23).

7- and 8-coordinated Sn(IV); *8-coordinated* Pb(IV). Examples of structural studies which definitely establish c.n.'s greater than 6 for Sn(IV) or Pb(IV) are few in number, and all involve coordination wholly or largely by pairs of O atoms in special chelating ligands.

In the monohydrate $SnY(H_2O)$,[24] where $H_4Y = $ edta, Sn(IV) is bonded to three pairs of O atoms of the ligand and to one H_2O molecule. There is no simple description of the geometry of the 7-coordination polyhedron, which is very similar to that in $Mn^{II}HY(H_2O)$—illustrated in Fig. 27.3(d). With the tropolonato (T) ligand, (a), the metal forms chelate complexes SnT_3Cl and $SnT_3(OH)$ the configuration

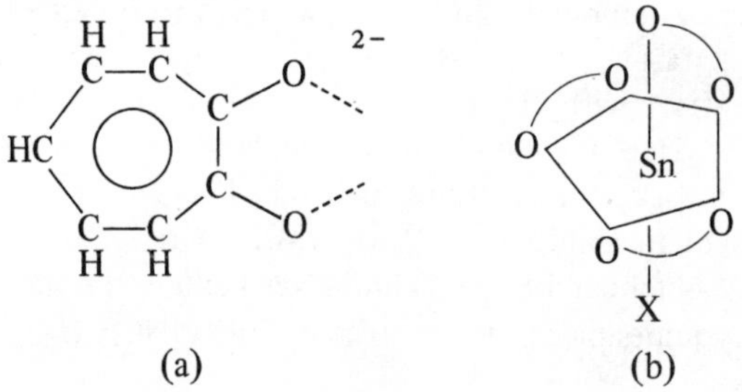

(a)   (b)

of which approximates to pentagonal bipyramidal, (b).[25] In the ligand T the distance between the O atoms is 2·56 Å. In the bidentate NO_3^- ion it is only 2·14 Å, and in the $Sn(NO_3)_4$ molecule[26] there is a tetrahedral arrangement of four NO_3^- groups around the 8-coordinated Sn(IV) atom. This special type of 8-coordination is also possible with the acetato ligand, in which O–O is similar to that in NO_3^-; 8-coordination of Pb(IV) is found in $Pb(ac)_4$, the coordination polyhedron having the form of a flattened triangulated dodecahedron.[27]

(20) IC 1967 6 1294

(21) IC 1966 5 2067

(22) AC 1962 15 913
(23) JCS A 1970 1257

(24) IC 1971 10 2313

(25) JACS 1970 92 3636

(26) JCS A 1967 1949

(27) AC 1963 16 A34

Stannous and plumbous compounds

The structural chemistry of Sn(II) is complex, and it is evident that the structures of even the stannous halides are not those of simple ionic compounds. Details of the distorted rutile-like structure of SnF_2[1] are not yet known. Although the structure of $SnCl_2$[2] could be described as a distorted $PbCl_2$ structure the range covered by the Sn–Cl distances is so large that it is more reasonable to describe crystalline $SnCl_2$ as formed from pyramidal $SnCl_3$ groups sharing 2 Cl atoms (Sn–1 Cl, 2·66 Å, Sn–2 Cl, 2·78 Å, interbond angles 80° (two) 105° (one)). The next nearest neighbours are at 3·06 Å (2), 3·22 Å (1), 3·30 Å (1), and 3·86 Å (2). The Raman spectrum of molten $SnCl_2$[3] is very similar to that of the solid, suggesting that the liquid also consists of chains of linked $SnCl_3$ groups like the solid. The compound originally formulated as $SnCl_2 . 2 H_2O$ is in fact $[SnCl_2 (H_2O)] H_2O$,[4] consisting of pyramidal groups (Fig. 26.12(a)) and separate H_2O

(1) AC 1962 **15** 509
(2) AC 1962 **15** 1051

(3) JCP 1967 **47** 1823

(4) JCS 1961 3954

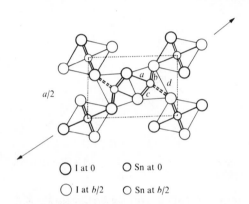

$a/2$

FIG. 26.11. Projection of one-half of a unit cell of the crystal structure of SnI_2. The chains containing the two kinds of Sn atom are perpendicular to the paper.

○ I at 0   ○ Sn at 0

○ I at $b/2$ ○ Sn at $b/2$

molecules. Similarly, the salt $K_2(SnCl_3)Cl . H_2O$ was formerly thought to be isostructural with $K_2HgCl_4 . H_2O$ and to contain $(MX_4)_n^{2n-}$ ions formed from octahedral MX_6 groups sharing opposite edges. In fact Sn has only 3 nearest neighbours which form a pyramidal $SnCl_3^-$ ion in which the mean Cl–Sn–Cl bond angle is 85° and mean Sn–Cl, 2·59 Å.[5a] There are similar pyramidal $SnCl_3^-$ ions in $CsSnCl_3$[5b] (3 Cl at 2·52 Å, 3 Cl at 3·21 Å) and pyramidal $Sn(O_2CH)_3^-$ ions (Sn–O, 2·16 Å, O–Sn–O, 82°) in $KSn(HCOO)_3$.[5c]

  The structure of $SnBr_2$ is at present being studied; that of SnI_2[5d] is a unique AX_2 structure of unexpected complexity which is shown in projection in Fig. 26.11. One-third of the Sn atoms (Sn′) are in positions of nearly regular octahedral coordination in rutile-like chains which are cross-linked by double chains containing the remaining metal atoms. The latter (Sn″) form 5 shorter and 2 longer bonds to I atoms:

(5a) JINC 1962 **24** 1039
(5b) ACSc 1970 **24** 150
(5c) ACSc 1969 **23** 3071
(5d) AC 1972 **B28** 2965

$$Sn''\text{--}I \begin{cases} a & 3\cdot00\ \text{Å} \\ 2b & 3\cdot20 \\ 2c & 3\cdot25 \end{cases} \begin{matrix} \text{mean} \\ 3\cdot18\ \text{Å} \end{matrix} \qquad Sn'\text{--}I \qquad 6\ \text{at}\ 3\cdot16\ \text{Å}$$

$$\begin{matrix} 2d & 3\cdot72 \end{matrix}$$

The bonds b link the double chains to the rutile-like chains to form layers in the direction of the arrows in Fig. 26.11. These layers are linked by the weaker d bonds into a 3D structure. All five of the shorter bonds lie to one side of Sn″. The coordination of Sn″ may be described as monocapped trigonal prismatic, the two longer bonds going to atoms of one edge; with this $(5 + 2)$-coordination compare the $(7 + 2)$-coordination of Pb in $PbCl_2$. The two longer bonds from Sn″, with electrostatic bond strength 1/6 as compared with 1/3 for the other five, can be regarded as replacing the sixth octahedral bond formed by Sn.

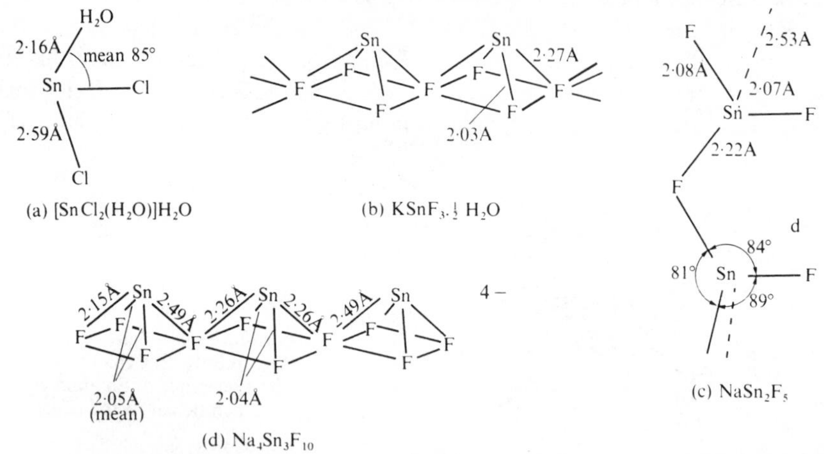

FIG. 26.12. Halogen–metal complexes in stannous compounds: (a) $[SnCl_2(H_2O)]\,H_2O$, (b) $KSnF_3 . \frac{1}{2}H_2O$, (c) $NaSn_2F_5$, (d) $Na_4Sn_3F_{10}$.

In the system $NaF\text{-}SnF_2 - H_2O$ the only compounds formed are $NaSn_2F_5$ and $Na_4Sn_3F_{10}$, while KF forms $KSnF_3 . \frac{1}{2}H_2O$; note the absence of M_2SnF_4. In $KSnF_3 . \frac{1}{2}H_2O$ the infinite anion consists of square pyramidal SnF_4 groups sharing opposite vertices (Fig. 26.12(b)), and the water molecules are not bonded to the Sn atoms.[6] In $NaSn_2F_5$[7] pairs of pyramidal SnF_3 groups share a F atom to form $Sn_2F_5^-$ ions which are then linked through weaker bonds (2·53 Å) to form chains (Fig. 26.12(c)). The terminal and bridging bonds are very similar in length to those in $KSnF_3 . \frac{1}{2}H_2O$. In $Na_4Sn_3F_{10}$[8] there are complex anions formed from three square pyramidal SnF_4 groups, but there is a large range of bond lengths (Fig. 26.12(d)). In particular the outer bridging bonds are very long, suggesting a tendency to split into a central SnF_4^{2-} ion and two SnF_3^- ions. Clearly it is not easy to give a satisfactory description of the bonding in these complex fluorides.

Simple dihydroxides $M(OH)_2$ do not appear to be formed by Sn or Pb; instead they form $M_6O_4(OH)_4$[9] containing M_6O_8 groups similar to the Mo_6Cl_8 groups described in Chapter 9. By careful heating the Sn compound may be converted into a metastable (red) form of SnO. The usual blue-black form of SnO is isostructural with tetragonal PbO. In contrast to the 4-coordination by O in the tetragonal forms of SnO and PbO there is 3-coordination of the metal in rhombic PbO and also in $SnSO_4$,[10] in which Sn has 3 close O (at a mean distance of 2·26 Å bond angle, 78°) and no other neighbours closer than 2·95 Å.

(6) AC 1968 **B24** 803

(7) AC 1964 **17** 1104

(8) AC 1970 **B26** 19

(9) N 1968 **219** 372

(10) AC 1972 **B28** 864

936

The structure of GeS and SnS is a layer structure similar to that of black P (see pp. 674 and 912) which may also be described as a very deformed version of the NaCl structure in which Sn(II) has 3 pyramidal neighbours; compare the environments of a metal atom in GeS, SnS, and PbS:

$$
Ge \begin{cases} 1\,S & 2{\cdot}47\,\text{Å} \\ 2\,S & 2{\cdot}64 \\ 1\,S & 2{\cdot}91 \\ 2\,S & 3{\cdot}00 \end{cases}
\qquad
Sn \begin{cases} 1\,S & 2{\cdot}62\,\text{Å} \\ 2\,S & 2{\cdot}68 \\ \overline{2\,S \quad 3{\cdot}27} \\ 1\,S & 3{\cdot}39 \end{cases}
\qquad
Pb \begin{cases} 6 \text{ equidistant} \\ S \text{ atoms at } 2{\cdot}97\,\text{Å}. \end{cases}
$$

The simplest way of describing the structural chemistry of many compounds of Sn(II) and Pb(II) is to regard the bonds as covalent and to suppose that the preferred bond arrangements are trigonal pyramidal or tetragonal pyramidal with valence groups (2, 6) and (2, 8) respectively, the lone pair occupying the remaining bond position in each case. If we show the origin of the electrons by using the symbols (−) and (→) to indicate normal and coordinate covalent bonds we then have the following bond pictures, of which the pyramidal arrangement (a) is similar to that for Sb(III):

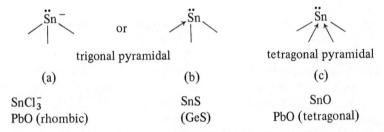

	trigonal pyramidal		tetragonal pyramidal
(a)	(b)		(c)
SnCl$_3^-$	SnS		SnO
PbO (rhombic)	(GeS)		PbO (tetragonal)

Plumbous compounds fall into two groups as regards their structural chemistry. In combination with the more electronegative halogens Pb forms the colourless Pb^{2+} ion, rather larger than Sr^{2+}, which is 8-coordinated in the high-temperature form of PbF_2 and also in $Pb_2Cu(OH)_4Cl_2$. In low-PbF_2 and also in $PbCl_2$ and $PbBr_2$ there is the less regular 7 (+2)-coordination described in Chapter 6. In PbI_2 and in many complex halides Pb(II) is octahedrally coordinated, and the compounds are often isostructural with Cd salts:

Cs_4PbX_6 (X = Cl, Br, I)	K_4CdCl_6 structure with discrete PbX_6^{4-} ions
$CsPbX_3$	perovskite-type structures, or NH_4CdCl_3 structure for the yellow low-temperature form of $CsPbI_3$.

There is apparently appreciable distortion of the PbI_6 octahedra in $CsPbI_3$, in which there are three Pb–I distances, $3{\cdot}01$, $3{\cdot}25$, and $3{\cdot}42$ Å.

In essentially covalent molecules Pb(II) forms 2, 3, or 4 bonds. In *molecules* of the dihalides we should have the valence group (2, 4), assuming simple electron-pair bonds, and such p^2 bonds would be mutually perpendicular. The structures of the

937

(11) JCS 1937 119

free molecules $PbCl_2$, $PbBr_2$, and PbI_2, and $SnCl_2$, $SnBr_2$, and SnI_2 have been studied by the electron diffraction method; they are non-linear. The salt $NH_4Pb_2Br_5$ consists of $PbBr_2$ molecules, NH_4^+, and Br^- ions.[11] The nearest neighbours of a Pb atom are 2 Br at 2·89 Å, 2 Br at 3·16 Å, and 4 Br at 3·35 Å, and the angle between the bonds from Pb to its two nearest Br neighbours is $85\frac{1}{2}°$.

(12) ZaC 1966 **343** 315

(13) K 1956 1 514

(14) AC 1966 **21** 350

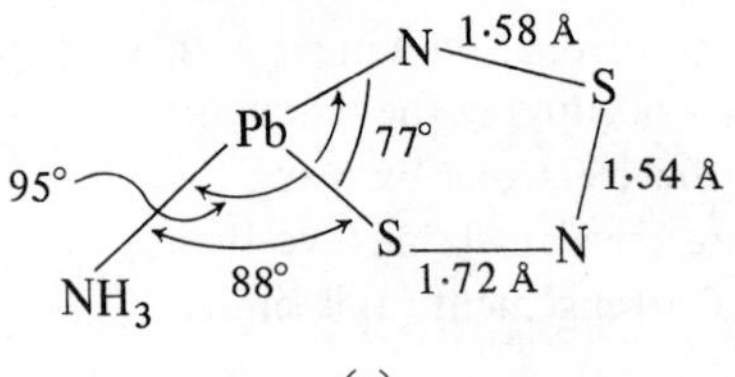

A rather irregular pyramidal arrangement of three bonds is formed by Pb in $Pb(N_2S_2)NH_3$,[12] in which the system PbN_2S_2 is planar, Pb–S, 2·73, Pb–NH_3, 2·24, and Pb–N_{ring}, 2·29 Å. The square pyramidal bond arrangement is found in molecules of the diethyl thiocarbamate,[13] (b), and the closely related ethyl-xanthate.[14] Presumably an appreciable degree of ionic character is responsible for

(a) 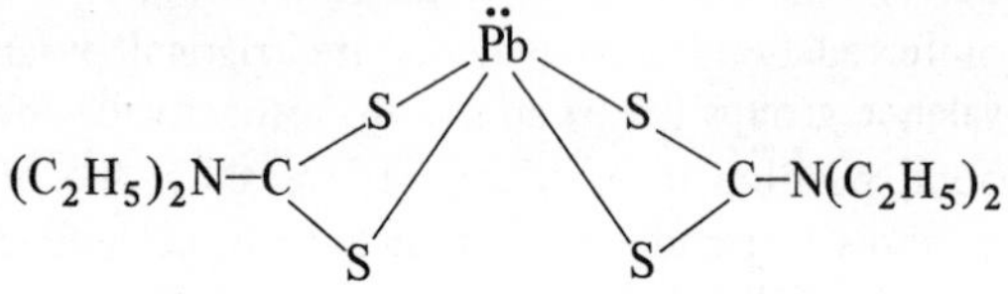   (b)

(15) AC 1959 **12** 727

(16) AC 1960 **13** 898

the much less regular arrangements of 7 or 8 neighbours found in compounds such as $PbCl_2(thiourea)_2$[15] and Pb(thiourea) acetate, $Pb[SC(NH_2)_2](C_2H_3O_2)_2$.[16]

In contrast to the colourless PbF_2, $PbCl_2$, and oxy-salts such as the nitrate, sulphate, etc. containing Pb^{2+} ions, the oxides are highly coloured. (Note also that whereas $PbSO_4$ and $PbMoO_4$ are colourless and $PbCrO_4$ pale-yellow, coprecipitated solid solutions of these compounds with appropriate $CrO_4 : SO_4 : MoO_4$ ratios have intense scarlet colour, the scarlet chrome pigments.) In tetragonal PbO and

(17) ACSc 1950 **4** 613

(18) AC 1961 **14** 747

Ag_2PbO_2[17] a Pb atom is bonded to 4 O atoms at the basal vertices of a square pyramid, while in Pb_3O_4, Pb_2TiO_4, $PbCu(OH)_2SO_4$[18] Pb^{II} is 3-coordinated, and an analogy may be drawn with Sb(III). Instead of emphasizing the $Pb^{IV}O_6$ octahedra in Fig. 12.15, p. 462, we could distinguish $(Pb_2^{II}O_4)_n^{4n-}$ chains in which Pb^{II} forms three pyramidal bonds, like Sb^{III} in $MgSb_2O_4$. We have already noted the NaCl structure of PbS, but the physical properties of galena, for example, its opacity and brilliant metallic lustre, are not those of simple ionic crystals.

For $Pb(C_5H_5)_2$ see p. 779; for $Pb_4(OH)_4^{4+}$ and $OPb_6(OH)_6^{4+}$, p. 517, and for $Pb_6O_4(OH)_4$ and $Sn_6O_4(OH)_4$, pp. 319 and 936.

938

27

Group VIII and Other Transition Metals

Introduction

This chapter is largely concerned with the structural chemistry of Fe, Co, Ni, Pd, and Pt, for the most part in finite complexes. The structures of the simpler compounds of these and other transition metals have been described under Halides, Oxides, etc. in the appropriate chapters. Other groups of compounds described in earlier chapters include hydrido compounds, oxo-, peroxo-, and superoxo-compounds, carbonyls, and nitrosyls, in Chapters 11, 12, and 18, respectively. We note here a few general points.

The formation of complexes with certain ligands or with particular types or arrangements of bonds seems to be characteristic of small numbers of transition metals. For example, there are metal–metal bonds in compounds of many elements (Chapter 7) but Nb, Ta, Mo, W, Re (and Tc) are notable for forming metal-cluster compounds with the halogens (p. 364); some other groups of complexes containing metal–metal bonds are summarized in Table 7.8 (p. 253). The formation of complexes containing, in addition to other ligands, a small number of O atoms strongly bonded to the metal is an important feature of the chemistry of V(IV) (see pp. 425 and 943 for vanadyl compounds), Mo(V), and Mo(VI) (p. 425). Complexes containing neutral molecules such as CO, NO, NH_3, etc. have been known for a long time, and more recently complexes have been prepared containing N_2 (Co, Ru) and SO_2 (Ir, Ru). It appears that certain ligands bond to transition metals in two quite different ways, for example, SO_2 (p. 581), NO (p. 654), and SCN^- (p. 746). There are interesting differences between compounds formed with molecular N_2 and O_2. In the N_2 compounds (p. 638) there is end-on coordination of the N_2 molecule, which retains the same structure as the free molecule, but there is a radical rearrangement of O_2 when it bonds to Ir. Moreover, this rearrangement differs in $IrCl(CO)O_2(P\phi_3)_2$, which is formed reversibly from $IrCl(CO)(P\phi_3)_2$ in benzene solution, and the analogous I compound, which is formed irreversibly (p. 418).

Of special interest are the multiple bonds to N studied in some compounds of Os and Re. In the anion in $K_2[Os^{VI}NCl_5]$,[1] (a), the bond to N is a formal triple bond, as in the 5-coordinated complex $Re^V NCl_2(P\phi_3)_2$,[2] which has a configuration intermediate between tetragonal pyramidal and trigonal bipyramidal. In the molecule (b)[3] the Re^V—N bond is longer (double ?) and in (c)[4] longer still; this unexpected result suggests that steric factors may affect the metal–nitrogen bond length. Note the lengthening of the M—Cl bond *trans* to M—N in both (a) and (c).

(1) IC 1969 8 709
(2) IC 1967 6 204

(3) IC 1969 8 703
(4) IC 1967 6 197

$$
\begin{array}{c}
N \\
\| \; 1.61 \text{ Å} \\
Cl - Os - Cl \quad 2.36 \text{ Å} \\
Cl \quad 2.61 \text{ Å} \quad Cl \\
Cl
\end{array}
$$

(a)

$$
\begin{array}{c}
CH_3 \\
N \\
\| \; 1.685 \text{ Å} \\
Cl - Re - P\phi_2Et \\
Et\phi_2P \quad 2.41 \text{ Å} \quad Cl \\
Cl
\end{array}
$$

(b)

$$
\begin{array}{c}
N \\
| \; 1.79 \text{ Å} \\
Et_2\phi P - Re - Cl \quad 2.45 \text{ Å} \\
Et_2\phi P \quad 2.56 \text{ Å} \quad P\phi Et_2 \\
Cl
\end{array}
$$

(c)

Certain bond arrangements are favoured by particular ligands in finite complexes. The unusual trigonal prismatic coordination of the metal by 6 S in the disulphides of Nb, Ta, Mo, W, and Re is also found for a number of these metals in chelate complexes with certain S-containing ligands. Examples include $Mo(SCHCHS)_3$ and the isostructural W compound,[5] $Re(S_2C_2\phi_2)_3$,[6a] $V(S_2C_2\phi_2)_3$, and the isostructural Cr compound.[6b] The dodecahedral bond arrangement has been found so far for very few metals in covalent complexes; see, for example, the octacyano complexes of Mo and W (p. 752). The trigonal bipyramidal arrangement of five bonds has been found in a number of transition-metal complexes, including the following d^8 complexes of the heavier Group VIII elements: $Os^0(CO)_3(P\phi_3)_2$,[7] $Rh^IH(CO)(P\phi_3)_3$,[8] and $Ir^ICl(CO)_2$-$(P\phi_3)_2$.[9] The ligands in these complexes are necessarily those which stabilize low oxidation states, as in the case of the tetrahedral $Ni(CO)_4$ and $Pt^0CO(P\phi_3)_3$;[10] the last molecule is of special interest since Pt does not form a simple carbonyl.

The formal oxidation state may be defined as the charge left on the metal atom when the attached ligands are removed in their closed-shell configurations, for example, NH_3 molecule, Cl^-, NO^+, or H^- (if directly bonded to the metal). The oxidation states recognized for Mn then cover the whole range from $+7$ (d^0), in Mn_2O_7, to -3 (d^{10}) in $Mn(NO)_3CO$; they are nearly as numerous for Cr and Fe and fall off rapidly on each side of these elements. All the 3d elements from Ti to Ni exhibit 'zero valence', but the low oxidation states are observed only with special ligands which are of two types. The π-acceptor (π-acid) ligands (e.g. CO, CN^-, RNC, PR_3, AsR_3) have lone pairs which form a σ bond to the metal and they also have vacant orbitals of low energy to which the metal can 'back-donate' some of the negative charge acquired in the formation of the σ bonds. Less important numerically are the π-complexes formed by unsaturated organic molecules and ions (e.g. C_5H_5, C_6H_6) in which all bonding between metal and ligand involves ligand π orbitals; their structures are described in Chapter 22 and later in this chapter (Pd

(5) JACS 1965 87 5798
(6a) IC 1966 5 411
(6b) IC 1967 6 1844

(7) IC 1969 8 419
(8) AC 1965 18 511
(9) IC 1969 8 2714
(10) IC 1969 8 2109

and Pt compounds). The highest oxidation states are exhibited in combination with the most electronegative elements O and F, but two points are worth noting in this connection. First, the highest oxidation state is not in all cases exhibited in combination with F, the most electronegative element; for example, Mn and Tc exhibit their 'group valence' in Mn_2O_7 and Tc_2O_7 but they do not form heptafluorides, and Ru and Os form tetroxides but not octafluorides (Table 27.1). Second, several elements exhibit higher oxidation states in complex fluoro- or

TABLE 27.1

Highest oxidation states in binary fluorides and oxides

Fluorides					*Oxides*				
Groups	Mn	Fe	Co	Ni	*Groups*	Mn	Fe	Co	Ni
III–VI	4	3	3	2	III–VI	7	3	2	2
3–6	Tc	Ru	Rh	Pd	3–6	Tc	Ru	Rh	Pd
	6	6	6	4		7	8	3	2
	Re	Os	Ir	Pt		Re	Os	Ir	Pt
	7	6	6	6		7	8	4	4

oxy-ions than in simple fluorides or oxides. Compare $Cs_2Co^{IV}F_6$ and $Cs_2Ni^{IV}F_6$ with CoF_3 and NiF_2, the highest fluorides at present known; a somewhat similar difference is found in the oxy-compounds. Iron rises to Fe(VI) in $(FeO_4)^{2-}$ as compared with Fe(III) in the highest oxide. The highest simple oxide of Co is CoO (there is no evidence for the existence of pure Co_2O_3), but Co^{3+} exists in Co_3O_4 and Co(IV) may exist in Ba_2CoO_4, though this is not certain. (The only Co^{3+} salts known are CoF_3 (anhydrous and hydrated), $Co_2(SO_4)_3 . 18 H_2O$ and alums, and $Co(NO_3)_3$. The last salt has been prepared as dark green hygroscopic crystals by the action of N_2O_5 on CoF_3. It is reduced by water.[11] For its structure see p. 663). Similarly, there is no evidence for the existence of pure anhydrous Ni_2O_3 but Ni^{3+} is present in $LiNiO_2$. (In Table 27.1 we have disregarded the compound $IrO_{2.7}$ described as formed by fusing Ir with Na_2O_2.)

(11) JCS A 1969 2699

Following brief sections on finite complexes of Ti(IV) and V(IV) and on complexes in which certain metals exhibit unusually high coordination numbers, we deal in some detail with the structural chemistry of Fe, Co, Ni, Pd, and Pt.

The stereochemistry of Ti(IV) in some finite complexes

Ti(IV) exhibits a variety of coordination numbers, including 4 in $TiBr_4$ (and TiI_4) and in $Ti_2Cl_4O(C_5H_5)_2$,[1] Fig. 27.1(a), and the cyclic $[TiOCl(C_5H_5)]_4$,[1a] if C_5H_5 is regarded as one ligand, 5 in certain complexes noted later, and 7 (but see the remarks below). The preferred coordination number, however, particularly with O or oxygen-containing ligands is 6. Thus, the tetrahalides readily form adducts with molecules such as $POCl_3$ and ethyl acetate in which octahedral coordination is attained, and the alkoxides form tetramers in which the metal is octahedrally coordinated. With acetylacetone Ti forms the octahedral molecule $Ti(acac)_2Cl_2$

(1) JACS 1959 **81** 5510
(1a) AC 1970 **B26** 716

941

and, like Si, the ion $[Ti(acac)_3]^+$, but does not, like Zr, form $M(acac)_4$. Only with special ligands such as *diarsine* and the bidentate nitrate ion is Ti known to exhibit 8-coordination in a finite complex. With diarsine the presumably octahedral complex $TiCl_4(diarsine)$ is formed and also $TiCl_4(diarsine)_2$, which has a dodecahedral structure, Fig. 27.1(b), like the $Mo(CN)_8^{4-}$ ion.[2] Similar complexes are formed by Zr, Hf, V, and Nb. The As atoms occupy the vertices of type A in Fig. 3.7(a) and the Cl atoms the vertices of type B, with Ti–Cl, 2·46 Å, and Ti–As, 2·71 Å.

(2) JCS 1962 2462

Examples of 5-coordination include the square pyramidal anion (c) in $(TiCl_4O)(NEt_4)_2$,[3] which is isostructural with the corresponding V compound, and the bridged molecule (d),[4] in which the coordination is described as close to

(3) AC 1968 **B24** 282
(4) AC 1968 **B24** 281

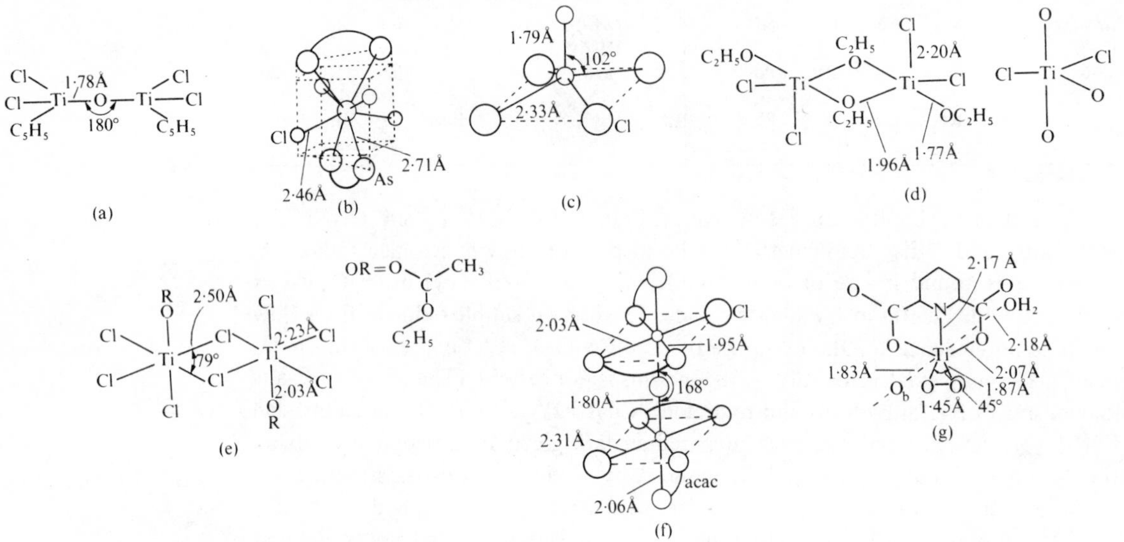

FIG. 27.1 Structures of ions and molecules containing Ti (see text).

trigonal bipyramidal. In the related (red) phenoxy-compound the (planar) bridge was found to be unsymmetrical, with Ti–O, 1·91 Å and 2·12 Å.[5]

(5) IC 1966 **5** 1782
(6) AC 1968 **B24** 1107

The tetrameric molecule of an alkoxide such as $[Ti(OCH_3)_4]_4$[6] has been illustrated in Fig. 5.8(h), p. 165, as an assembly of four octahedra. In this tetramer there are two pairs of non-equivalent Ti atoms in coordination octahedra which share 2 or 3 edges, and the Ti–O bond lengths range from 1·8–2·2 Å; some O atoms are bonded to one Ti only, others to two or three. These compounds are colourless, in contrast to the yellow or red compounds of type (d). Another type of bridged molecule is that of $Ti_2Cl_8(CH_3COOC_2H_5)_2$,[7] (e), in which an ethyl acetate ion is bonded to each Ti atom and two Cl atoms form the bridge between the two edge-sharing octahedral coordination groups.

(7) AC 1966 **20** 739

We noted above the acetylacetone complexes $Ti(acac)_2Cl_2$ and $[Ti(acac)_3]^+$. In the orange-yellow $Ti_2Cl_2O(acac)_4$,[8] which crystallizes with one molecule of

(8) IC 1967 **6** 963

942

CHCl$_3$ as solvent of crystallization, there is an approximately linear Ti—O—Ti bridge and short bridging Ti—O bonds (Fig. 27.1(f)). The Ti—O—Ti bridge in the red dipicolinic acid salt K$_2$[Ti$_2$O$_5$(C$_7$H$_3$O$_4$N)$_2$] . 5 H$_2$O[9] is strictly linear (O$_b$ lying on a 2-fold axis), and the coordination group around Ti, (g), is similar to that in K$_2$W$_2$O$_{11}$. 4 H$_2$O (p. 423) which also contains peroxo-groups bonded to the metal atoms. In Fig. 27.1(g) the broken lines represent bonds perpendicular to the plane of the chelate ligand and the peroxo-group, so that the coordination around Ti may be described as either pentagonal bipyramidal or as octahedral if the O$_2$ group is counted as one ligand.

(9) IC 1970 **9** 2391

The stereochemistry of V(IV) in some finite complexes

The structural chemistry of V in various oxidation states (II–V) in binary compounds has been noted in previous chapters, in particular the unusual coordination of V(V) by oxygen in oxides and oxy-salts, where there is very irregular square pyramidal or octahedral coordination (in addition to tetrahedral coordination). Our chief concern here is with compounds of V(IV), but we shall mention incidentally several complexes of V(III) and V(V).

A feature of the 5- and 6-coordinated compounds of V(V) is the formation of bonds with lengths in the ranges 1·6–2·0 Å or 1·6–2·3 Å respectively, with one or sometimes two bonds much shorter than the others. This irregular 6-coordination is also found in methyl vanadate, VO(OCH$_3$)$_3$,[1] which consists of chains of octahedral coordination groups each sharing two (non-opposite) edges. These groups are very distorted, there being five different V—O bond lengths, the two shortest (1·54 Å and 1·74 Å) to unshared O atoms (Fig. 27.2(a)). The V—O—V

(1) IC 1966 **5** 2131

bridges are alternately symmetrical (all four V—O approximately 2·02 Å) and unsymmetrical (V—O, 1·85 Å and 2·27 Å), and the six V—O bonds from any V(V) atom have the lengths: 1·54, 1·74, 1·85, 2·02 (two), and 2·27 Å. With these figures compare the distances to the *five* near neighbours in vanadyl compounds (Table 27.2), which include one very short bond and four longer ones which in some cases appear to fall into two groups (2 + 2). The mean distance to the four more distant

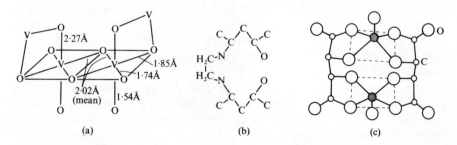

(a)   (b)   (c)

FIG. 27.2. (a) Portion of chain in VO(OCH$_3$)$_3$, (b) the ligand 'acen', (c) the divanadyl-*dl*-tartrate ion.

(2) AC 1969 **B25** 1354

O neighbours in these complexes is similar to V–O (1·98 Å) in $V^{III}(acac)_3$,[2] a compound that is readily oxidized to $VO(acac)_2$. We have included in the Table examples of complexes containing mono-, bi-, and quadri-dentate ligands; the skeleton of the ligand acen is shown in Fig. 27.2(b). The divanadyl-*dl*-tartrate ion has an interesting structure containing two V atoms, both of which form square pyramidal bonds (Fig. 27.2(c)).

TABLE 27.2

Bond lengths in vanadyl complexes

	V–O bond lengths			Reference
	one	two	two	
$[VO(NCS)_4](NH_4)_2 .5 H_2O$	1·62 Å	2·05 (V–N)		JCS 1963 5745
$VO(acac)_2$	1·57	1·97		JCP 1965 **43** 3111
$VO(bzac)_2$	1·61	1·97		JCP 1965 **43** 1323
$VO(acen)$	1·59	1·96 (V–O)		IC 1970 **9** 130
		2·05 (V–N)		
$Na_4[VO\text{-}dl\text{-}tartrate] . 12 H_2O$	1·62	1·91	2·00	IC 1968 **7** 356
$(NH_4)_4[VO\text{-}dl\text{-}tartrate] .2 H_2O$	1·60	1·86	2·02	JCS A 1967 1312

The characteristic bond arrangement in the vanadyl complexes of Table 27.2 (and also in $VOSO_4$ and $VOMoO_4$, p. 513) is square pyramidal, the V atom lying slightly above the equatorial plane, with the short V–O bond apical. On the other hand, the molecule $VOCl_2(NMe_3)_2$[3] has the trigonal bipyramidal structure (a), in

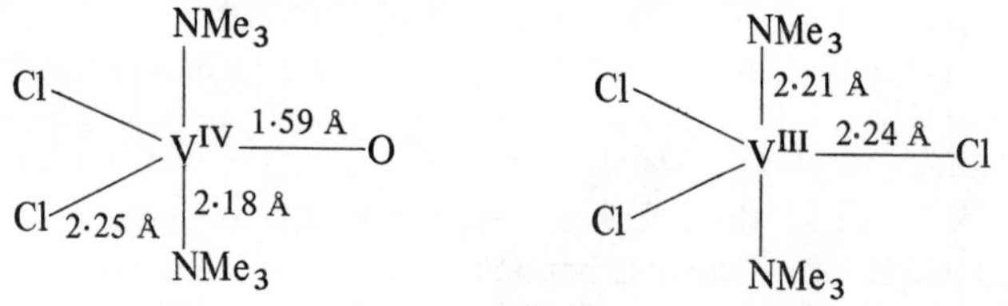

(3) JCS A 1968 1000

(a)   (b)

which the equatorial bond angles are 120° to within the experimental error. It seems that this configuration is favoured by the shape of the tertiary amine, for the methyl groups are in the staggered positions relative to the equatorial ligands, a view supported by the fact that the V(III) molecule $VCl_3(NMe_3)_2$,[4] (b), has almost exactly the same structure.

(4) JCS A 1969 1621

We comment elsewhere on the rarity of trigonal prism coordination, which has been observed in complexes $M(S_2C_2\phi_2)_3$ in which M is V, Cr, Mo, or Re. In the V complex[5] V–S = 2·34 Å and S–V–S = 82° (mean), and the shortest distances between S atoms of different ligands are close to 3·1 Å; the dimensions of the Re and Mo complexes are almost identical to those of the V complex. It has been suggested that some type of weak interaction between the S atoms stabilizes this

(5) IC 1967 **6** 1844

configuration, which is not found for all chelate sulphur-containing ligands. In the maleonitrile dithiolate ion in $[V(mnt)_3](NMe_4)_2$[6] the 6-coordination around V is irregular (V–S, 2·36 Å), with 2·93 Å as the shortest *inter*-ligand S–S distance. The coordination group is intermediate between octahedral and trigonal prismatic, as shown by the angle S–V–S for pairs of S atoms that are farthest apart. For a regular octahedron this angle would be 180° or 173° for the nearest approach to a regular octahedron taking into account the constraints due to the rigid bidentate ligand. In $V(S_2C_2\phi_2)_3$ this is 136° (trigonal prismatic coordination), and in the dithiolate ion it is 159°. We have included these complexes with compounds of V(IV) though the formal oxidation state of the metal could be regarded as zero, depending on the degree of delocalization of the four electrons on the three chelate ring systems.

(6) JACS 1967 89 3353

The structural chemistry of Cr(IV), Cr(V), and Cr(VI)

Compounds of Cr(IV)

In addition to the well known ferromagnetic CrO_2 (rutile structure) CrF_4 and $CrBr_4$ have been synthesized; their structures are at present unknown. The complex fluoride K_2CrF_6 contains octahedral CrF_6^{2+} ions.

The olive-green Ba_2CrO_4, with a magnetic moment of 2·82 BM, corresponding

TABLE 27.3

Compounds of Cr *in higher oxidation states*

Cr(IV)	Cr–O (Å)		Reference
CrO_2	2 at 1·88, 4 at 1·92		JAP 1962 33 1193
Sr_2CrO_4	4 at 1·66–1·95, mean 1·82		AK 1966 26 157
K_2CrF_6			ZaC 1956 286 136
Cr(V)			
$Ca_5OH(CrO_4)_3$	4 at 1·66		ACSc 1965 19 177
$Ca_2Cl(CrO_4)$	4 at 1·70 (mean)		
	Angles 105° (4), 119° (2)		AC 1967 23 166
Cr(VI)	Cr–O (bridge)	Cr–O (term.)	Cr–O–Cr
CrO_3	1·75	1·60	143°	AC 1970 B26 222
$K_2Cr_2O_7$	1·79	1·63	126°	CJC 1968 46 933
$(NH_4)_2CrO_4$	–	1·66	–	AC 1970 B26 437
Cr(III) and Cr(VI)	Cr^{III}–6 O	Cr^{VI}–4 O		
Cr_5O_{12}	1·97	1·65		ACSc 1965 19 165
KCr_3O_8	1·97	1·60		ACSc 1958 12 1965
$LiCr_3O_8$	2·05	1·66		AK 1966 26 131
$CsCr_3O_8$	1·96	1·63		AK 1966 26 141
Cr(IV) and Cr(VI)	Cr^{IV}–6 O	Cr^{VI}–4 O		
$K_2Cr_3O_9$	1·97	1·65		ACSc 1969 23 1074

For the preparation of high-pressure oxides see ACSc 1968 22 2565.

to $Cr(IV)$, is isostructural with Ba_2FeO_4 and Ba_2TiO_4 (notable as the only example of Ti^{4+} tetrahedrally coordinated by oxygen) and like the green Ba_3CrO_5 is a compound of $Cr(IV)$. In Sr_2CrO_4 rather distorted CrO_4^{4-} groups were found, which may be compared with the nearly regular octahedral coordination in CrO_2 (Table 27.3). Oxides $MCrO_3$ (M = Ca, Sr, Ba, Pb) have been prepared under pressure.

Compounds of $Cr(V)$

These include a number of compounds isostructural with silicates or phosphates, for example:

$NdCrO_4$, $YCrO_4$: (zircon structure),
$Ca_5OH(CrO_4)_3$: apatite structure, $Cr^V–O$, $1·66$ Å,
$Ca_2Cl(CrO_4)$: isostructural with $Ca_2Cl(PO_4)$.

$Cr(V)$ is tetrahedrally coordinated by oxygen and has a similar radius to $Cr(VI)$. The complete structure determination of $Ca_2Cl(CrO_4)$ shows that all the CrO_4^{3-} ions are equivalent, i.e. this is not a compound containing $Cr(IV)$ and $Cr(VI)$. The magnetic moment has the expected value, $1·7$ BM. The CrO_4^{3-} tetrahedra are slightly distorted.

Compounds of $Cr(VI)$

The structural chemistry of $Cr(VI)$ is very similar to that of $S(VI)$, and accurate structural data are available for several simple compounds.

The crystalline trioxide is built of infinite chains formed by the linking of CrO_4 tetrahedra by two vertices:

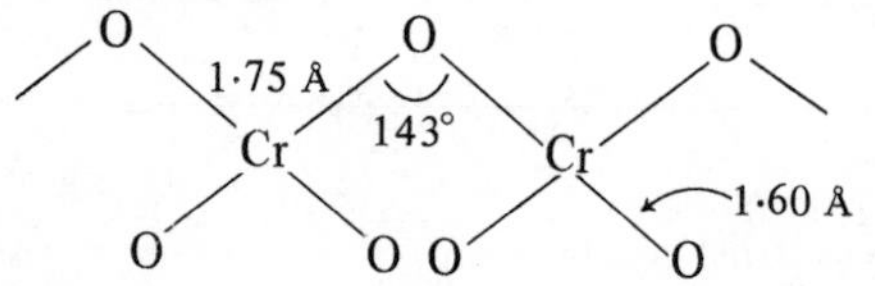

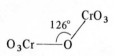

There are only van der Waals forces between the chains, consistent with the comparatively low melting point ($197°C$).

In simple crystalline chromates there are tetrahedral CrO_4^{2-} ions in which Cr–O is $1·66$ Å (in the ammonium salt). The tetrahedral CrO_3F^- ion (Cr–O, $1·58$ Å) is the analogue of SO_3F^-. The sizes of O and F being very similar, $KCrO_3F$ crystallizes with the scheelite ($CaWO_4$) structure with random arrangement of O and F. $KCrO_3Cl$ has a distorted form of the same structure due to the much greater difference between Cr–O and Cr–Cl ($2·12$ Å). Polychromates result from the sharing of vertices between limited numbers of CrO_4 tetrahedra. In $K_2Cr_2O_7$ the anion has the structure shown. The bridging Cr–O bonds are much longer ($1·79$ Å) than the terminal ones (mean, $1·63$ Å) as in the disulphate ion.

In the tetrahedral CrO_2Cl_2 molecule the bond lengths are: Cr–O, $1·57$ Å, and Cr–Cl, $2·12$ Å. (See Table 10.12, p. 406).

Compounds containing Cr *in two oxidation states*

The structure of Cr_2F_5, a compound of Cr(II) and Cr(III), has been described in Chapter 5. We note here some compounds which contain Cr(III) and Cr(VI) or Cr(IV) and Cr(VI).

A number of oxides have been prepared which are intermediate between CrO_2 and CrO_3. They include Cr_2O_5 and Cr_3O_8, resulting from the thermal decomposition of CrO_3 under oxygen pressures up to 1 kbar in the temperature range $200-400°C$, and an oxide Cr_5O_{12} which forms under oxygen pressures up to 3 kbar. This last oxide is a compound of Cr(III) and Cr(VI), $Cr_2^{III}(Cr^{VI}O_4)_3$, in which Cr(III) atoms occupy positions of octahedral coordination (Cr–O, 1·97 Å) and Cr(VI) atoms positions of tetrahedral coordination (Cr–O, 1·65 Å) in an approximately c.c.p. assembly of oxygen atoms. The black metallic-looking KCr_3O_8 prepared by heating $K_2Cr_2O_7$ with CrO_3 in oxygen under pressure is also a compound of Cr(III) and Cr(VI), i.e. $KCr^{III}(Cr^{VI}O_4)_2$. Octahedral $Cr^{III}O_6$ and tetrahedral $Cr^{VI}O_4$ groups are linked to form layers of composition Cr_3O_8 which are held together by the potassium ions. The corresponding lithium salt has a quite different structure which is essentially the same as that of a number of chromates $MCr^{VI}O_4$ (M = Co, Cu, Zn, Cd), phosphates (Cr, In, Tl), and sulphates (Mn, Ni, Mg). The octahedral holes between O atoms of CrO_4 groups are occupied *at random* by Li^+ and Cr^{3+}. (This structure may also be described as an approximately cubic closest packing of O atoms with 1/8 of the tetrahedral and 1/4 of the octahedral holes occupied.

The compounds $M_2Cr_3O_9$ (M = Na, K, Rb) have been prepared under pressure as red crystals insoluble in water. The structure consists of rutile-type chains of $Cr^{IV}O_6$ octahedra which are cross-linked by $Cr^{VI}O_4$ tetrahedra.

Higher coordination numbers of metals in finite complexes

We have collected together in this section some examples of finite complexes formed with polydentate ligands in which transition metals exhibit coordination numbers from 7 to 10. For chelates of 4f metals see p. 67.

We noted in Chapter 10 two rather symmetrical arrangements of 7 nearest neighbours found in oxy- and fluoro-compounds of transition metals, namely, pentagonal bipyramidal (ZrF_7^{3-}), and monocapped trigonal prism (NbF_7^{2-}). The ions formed from ethylenediamine tetracetic acid and the closely related diamino cyclohexane acid are potentially sexadentate ligands, for there are 4 O atoms and

2 N atoms which can coordinate to a metal atom. In the formulae in this section and in Fig. 27.3 we use the abbreviations edta and dcta to mean the ions (4 —)

formed from these acids. Three different types of complex formed with Ni, Co, and Fe are illustrated in Fig. 27.3 (a)–(c). The Ni(II) and Co(III) complexes are octahedral, but in the former, (a), the ligand forms only five bonds to Ni, the metal atom completing its coordination group with a water molecule.[1] (Although only 2 H of edta are replaced by Ni only one acetate group is unattached and presumably carries one H as OH. The location of the other H atom is not clear, but it has been provisionally allocated as shown, on the basis of the Ni–O bond lengths.) In the (diamagnetic) Co(III) complex,[2] (b), edta is behaving as a sexadentate ligand. On the other hand, in $Rb[Fe^{III}(H_2O)edta].H_2O$[3] the anion is a 7-coordinated complex, Fig. 27.3(c), with one H_2O in addition to the 6 bonds from edta. The bond arrangement is very approximately pentagonal bipyramidal, two O atoms and two N atoms of edta and one H_2O molecule forming the

(1) JACS 1959 **81** 556

(2) JACS 1959 **81** 549
(3) IC 1964 **3** 34

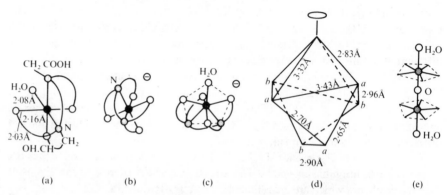

FIG. 27.3. Structures of (a) $[Ni(edta)H_2(H_2O)]$, (b) $[Co(edta)]^-$, (c) $[Fe(edta)(H_2O)]^-$, (d) coordination polyhedron of $[MnH(edta)(H_2O)]^-$, (e) $[(H_2O)LFe-O-FeL(H_2O)]^{4+}$, where the pentagon represents the ligand L shown in the margin.

(4) IC 1964 **3** 27

(5) JACS 1966 **88** 3228

equatorial group of five ligands. In contrast to Ni(II), Mn(II) forms a salt of a hydrated aquo-complex, $Mn^{2+}[Mn^{II}(H_2O)H(edta)]_2.8 H_2O$,[4] with replacement of only 3 H atoms of edta. The Mn^{2+} ions in the crystal have normal octahedral coordination (by 4 H_2O and 2 O atoms of anions), but within the complex Mn(II) is 7-coordinated. The bond arrangement is, however, different from that in the iron complex (c) and may be regarded as a skew form of the NbF_7^{2-} coordination group. A small clockwise rotation of the face *aaa* and an equivalent anticlockwise rotation of *bbb*, Fig. 27.3(d), would convert this coordination group into a monocapped trigonal prism. The Fe(III) complex with dcta has a configuration very similar to that of the Mn complex (d); it was studied in $Ca[Fe^{III}(H_2O)dcta]_2.8 H_2O$.[5]

It is evident that the geometries of these 7-coordinated complexes are very much dependent on the constraints due to the shapes of the polydentate ligands. A more extreme example is the pentagonal bipyramidal coordination in complexes formed by the pentadentate ligand L, such as the cations in the salts $[FeL(NCS)_2]ClO_4$, (where the NCS groups occupy the apical positions with Fe–N, 2·01 Å as compared

(L)

with 2·23 Å for the equatorial bonds), and $[(H_2O)LFe–O–FeL(H_2O)](lCO_4)_4$.[6] (6) JACS 1967 **89** 720
The cation in this salt, Fig. 27.3(e), is also of interest as an example of a bridged
oxo compound of Fe(III) with a linear Fe–O–Fe bridge.

With bidentate ligands of the type R . CO . CH . CO . R yttrium and the smaller
4f ions form a number of 7-coordinated complexes in which the seventh ligand is a
water molecule. These are noted in Chapter 3 in our discussion of 7-coordination.

Although Zr is 6-coordinated in ions such as $ZrCl_6^{2-}$, with the smaller ligands O
and F there is a marked preference for higher coordination numbers. In a number
of cases this is 7-coordination as in $(NH_4)_3ZrF_7$, in monoclinic ZrO_2, and in
ZrOS. The Zr^{4+} ion is obviously a borderline case, for in contrast to the above
examples there is 8-coordination (dodecahedral) in K_2ZrF_6, in the high-
temperature forms of ZrO_2, and in numerous hydrated and hydroxy-salts (p. 531).
From the energy standpoint there is clearly very little to choose between the two
predominant types of 8-coordination, namely antiprismatic and dodecahedral. In a
3D structure the crystal energy can be minimized by adjusting the shapes of all the
various coordination groups, and in extreme cases these geometrical requirements
actually determine the *composition* of the salt that crystallizes from solution, as in
the salts Na_3TaF_8, K_2TaF_7, and $CsTaF_6$, or $(NH_4)_3ZrF_7$ and K_2ZrF_6. In *finite*
complexes containing chelate ligands the choice between the two coordination
polyhedra is probably determined by 'non-bonded' interactions between atoms of
the ligands, and we find

dodecahedral: $Na_4[Zr(C_2O_4)_4] . 3 H_2O$ IC 1963 **2** 250
                  $K_2[Zr(NTA)_2] . H_2O^{(a)}$ JACS 1965 **87** 1610

antiprismatic: $Zr(acac)_4$ IC 1963 **2** 243
                  $[Zr_4(OH)_8(H_2O)_{16}] . Cl_8 . 12 H_2O$ See p. 532

(a) NTA is the nitrilo triacetate ion $N(CH_2COO)_3^{3-}$.

The 9-coordination of La^{3+} and the larger 4f ions in both finite complexes and
3D structures has been noted in previous chapters. We have seen that 6-fold
coordination by the ligand edta is possible for small ions M^{2+} and M^{3+}. Even the
smallest 4f ion is too small for 'octahedral' coordination by edta, and with the
larger ions the ligand occupies only part of the coordination sphere, the remainder
of which can be filled with, for example, H_2O molecules. The very symmetrical
9-coordination group found in a number of 4f compounds is not consistent with
the geometry of some polydentate ligands, and we find coordination groups more
closely related to the dodecahedral 8-coordination group but with additional
ligands giving 9-coordination or the much less common 10-coordination. Thus in
the 9-coordinated anion in $K[La(H_2O)_3edta] . 5 H_2O$ the 4 O and 2 N atoms of
edta and one H_2O molecule occupy seven of the vertices of the triangulated
dodecahedron characteristic of 8-coordination, and there are isostructural salts in
which the 4f metal is Nd, Gd, Tb, and Er.[7] The arrangement of seven of the (7) JACS 1965 **87** 1612
ligand atoms around La is similar in $[LaH(edta)(H_2O)_4] . 3 H_2O$,[8] where there are (8) JACS 1965 **87** 1611
4 H_2O molecules completing a 10-coordination group. It should, however, be
remarked that this polyhedron approximates closely to a bicapped square
antiprism.

High coordination numbers are also possible if a group such as NO_3^- or CH_3COO^- behaves as a bidentate ligand. In Chapter 18 we noted that 8 O atoms (four NO_3^- ions) surround the metal in $[Co(NO_3)_4]^{2-}$ and $Ti(NO_3)_4$ and 12 O (six bidentate NO_3^- ions) in nitrato ions of Ce(III), Ce(IV), and Th(IV). Three bidentate nitrate ions and two bipyridyl molecules give La 10-coordination in $La(NO_3)_3$-(bipyridyl)$_2$ in an arrangement that has been described as a bicapped dodecahedron. A similar description could be given to the 'pseudo-octahedral' 10-coordination group in $Th(NO_3)_4(OP\phi_3)_2$ (p. 665); for references to these compounds see the section on metal nitrates and nitrato complexes in Chapter 18.

The structural chemistry of iron

Although all formal oxidation states from and including −2 to +6 are recognized, only Fe(II), and Fe(III) are of practical importance. Oxidation states lower than +2 involve π-type ligands and the states −2, −1, and +1 apply to compounds such as $Fe(CO)_2(NO)_2$ and $Fe(NO)_2X$ and the ion $[Fe(H_2O)_5NO]^{2+}$ where the meaning of the term oxidation state is somewhat dubious. The diamagnetic compound $Fe(NO)_2I$ exists as dimers, (a),[1] very similar in structure to (b)[2] except for the larger Fe—Fe distance. Unless there is some sort of superexchange interaction involving the I atoms in (a) it would seem necessary to account for the diamagnetism by postulating metal-metal interaction of some kind, and to suppose that the large Fe—Fe distance, as compared with that in (b), is due to the size of I. The bond arrangement around Fe is tetrahedral. Crystalline $Co(NO)_2I$ has a

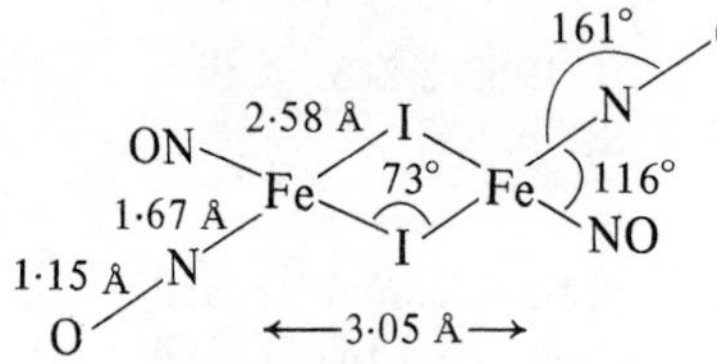

(1) JACS 1969 **91** 1653
(2) AC 1958 **11** 599

(a)

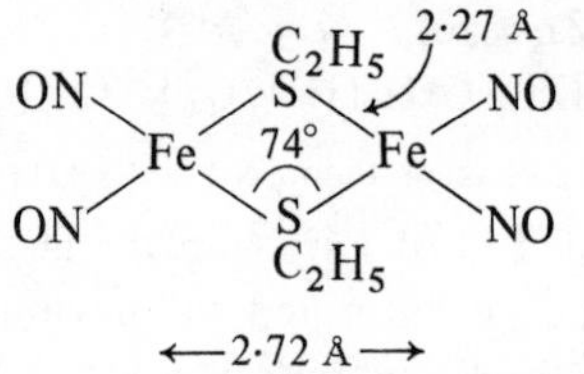

(b)

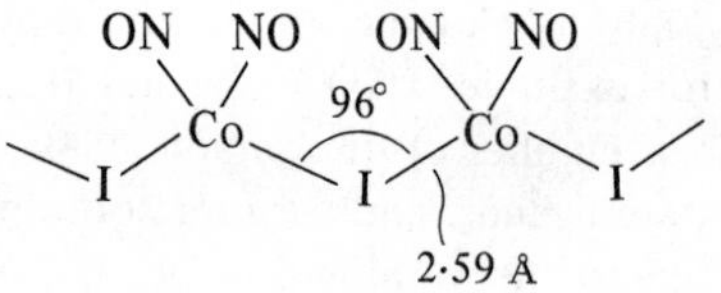

(c)

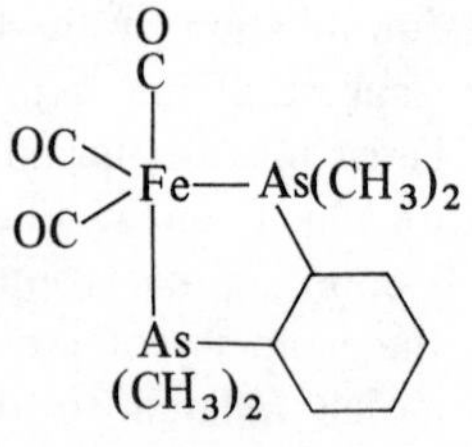

(d)

tetrahedral chain structure (c).[1] The small bond angle at I in (a) seems to be a feature of metal-metal bonded systems; compare

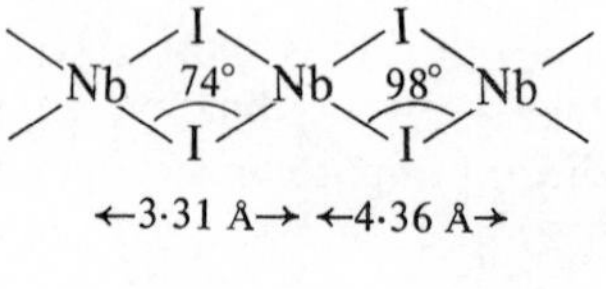

The only bond arrangement established for Fe(0) is the trigonal bipyramidal arrangement of five bonds, as in $Fe(CO)_5$, and $Fe(CO)_3$(diarsine)[3] (d).

(3) AC 1967 **22** 296

The normal coordination of Fe(II) is octahedral. Many of the ionic compounds in which Fe^{2+} is surrounded by 6 X or 6 O are isostructural with those of other metals forming similarly sized ions:

MO (NaCl structure): Mg, Mn, Fe, Co, Ni

MF_2 (rutile structure)

$MSO_4 . 7 H_2O$ ⎫

$K_2M(SO_4)_2 . 6 H_2O$ ⎬ M = Mg, Mn, Fe, Co, Ni, Zn

$M_3Bi_2(NO_3)_{12} . 24 H_2O$ ⎭

The structural similarity between these M(II) compounds does not, however, extend to the more covalent compounds. For example, the complex cyanides formed by Fe, Co, and Ni are of quite different kinds:

$K_4[Fe(CN)_6]$, $K_3[Co(CN)_5]$, $K_2[Ni(CN)_4]$, and $K_4Ni_2(CN)_6$ (p. 970).

(The Co compound was incorrectly described in the older literature as $K_4[Co(CN)_6]$.) The complex thiocyanates are:

$Na_4[Fe(NCS)_6] . 12 H_2O$, $Na_2[Co(NCS)_4] . 4 H_2O$, and $Na_4[Ni(NCS)_6] . 12 H_2O$, of which the Ni compound was earlier assigned the formula $Na_2[Ni(NCS)_4] . 8 H_2O$.

The octahedral complexes of Fe(II) are high-spin ($\mu \approx 5$ BM) except for those containing strong-field ligands such as $Fe(CN)_6^{4-}$ and $Fe(CNR)_6^{2+}$, which are diamagnetic. Neutral molecules include *trans*-$Fe(NCS)_2(pyr)_4$[4], (e).

(4) ACSc 1967 **21** 2028

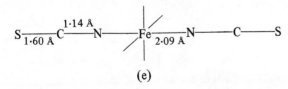

(e)

Tetrahedral coordination of Fe(II) is comparatively rare, and in fact has not yet been directly established by a structure determination. It has been deduced from the existence of compounds isostructural with analogous compounds of Mn, Co, or Ni:

$(\phi_3MeAs)_2 (MCl_4)$, M = Mn, Fe, Co, Ni; JCS 1961 3512

$(Me_4N)_2 [M(NCS)_4]$, M = Fe, Co; JCS 1965 268

$M(\phi_3AsO)_2Cl_2$; M = Fe, Co; JCS 1965 454

Like $(NiCl_4)^{2-}$ the $(FeCl_4)^{2-}$ ion is stable only in the presence of large cations; high-spin Fe(II), d^6, has the expected moment of approximately 5 BM corresponding to 4 unpaired electrons.

951

Square coplanar coordination of Fe(II) is observed in the phthalocyanin ($\mu = 3 \cdot 96$ BM) and also in gillespite (p. 818). In this crystal Fe^{2+} has 4 O at $1 \cdot 97$ Å (compare Fe–6 O (octahedral), $2 \cdot 14$ Å) and no other neighbours closer than $3 \cdot 98$ Å. The visible and i.r. absorption spectra have been discussed in terms of high-spin Fe^{2+}.[5]

(5) IC 1966 **5** 1268

The structural chemistry of Fe(III) is largely that of octahedral coordination, in crystalline halides, Fe_2O_3, and numerous finite complexes ($Fe(en)_3^{3+}$, $Fe(acac)_3$, $Fe(C_2O_4)_3^{3-}$, etc.), but tetrahedral coordination is less rare than for Fe(II). Thus there is octahedral coordination of the metal in crystalline $FeCl_3$ but tetrahedral coordination in the vapour dimer Fe_2Cl_6 and in the $FeCl_4^-$ ion, and there is both tetrahedral and octahedral coordination of Fe^{3+} in Fe_3O_4 (inversed spinel structure) and also in the complex of $FeCl_3$ with dimethyl sulphoxide (L), $FeCl_3L_2$. This compound consists of equal numbers of tetrahedral $(FeCl_4)^-$ ions (Fe–Cl, $2 \cdot 16$ Å) and octahedral *trans* $(FeCl_2L_4)^+$ ions (Fe–Cl, $2 \cdot 34$ Å, Fe–O, $2 \cdot 01$ Å).[6] The complex (f), containing the ligands thio-*p*-toluoyl disulphide and dithio-*p*-toluate, is of interest as containing 4-membered rings (FeS_2C) and 5-membered rings (FeS_3C).[7] As in the case of Fe(II) compounds the magnetic moments of Fe(III) compounds correspond to high spin (d^5, $5 \cdot 9$ BM) except for complexes such as $[Fe(CN)_6]^{3-}$ and $[Fe(dipyr)_3]^{3+}$ containing strong-field ligands.

(6) AC 1967 **23** 581

(7) JACS 1968 **90** 3281

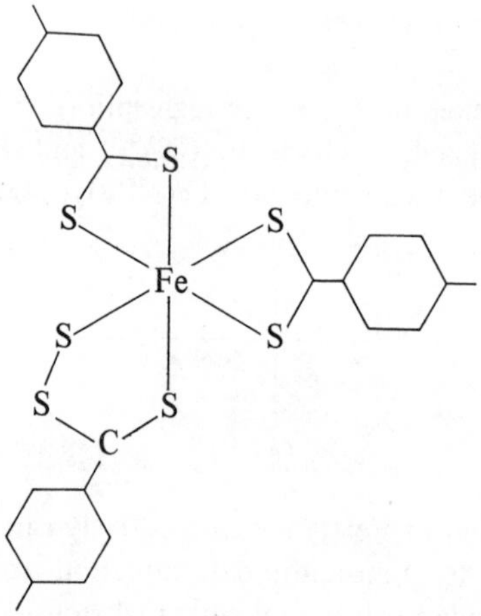

(f)

(8) IC 1970 **9** 447

The square pyramidal bond arrangement is found in a number of finite complexes. Apart from $FeH(SiCl_3)_2(CO)C_5H_5$[8] they are mostly monomers containing two bidentate ligands, (g), or dimers in which a fifth weaker bond is formed as shown at (h). The metal atom is situated about $\frac{1}{2}$ Å above the centre of the square base of the pyramid. Dimerization of a molecule of type (g) gives a complex of type (i) in which the metal atom has distorted octahedral coordination.

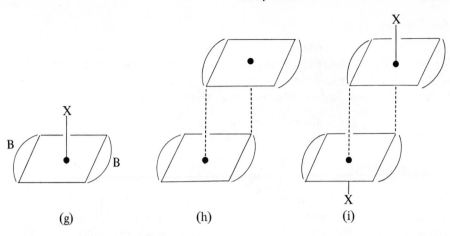

(g)   (h)   (i)

Examples include the following:

	Bond lengths (Å)	Reference
(g) $Fe[S_2CN(C_2H_5)_2]_2Cl$	4 S 2·32, Cl 2·27	IC 1967 **6** 712
Fe(salen)Cl	2 O1·88, 2 N 2·08	JCS A 1967 1598
	Cl 2·24	
(h) $\{Fe[S_2C_2(CN)_2]_2\}_2$ (*n*-Bu$_4$N)$_2$	4 S 2·23, S 2·46	IC 1967 **6** 2003
(i) $[Fe(salen)Cl]_2$	2 N 2·10, Cl 2·29	JCS A 1967 1900
	3 O 1·90, 1·98, 2·18	

Salen = NN′-*bis*-salicylideneiminato.

Structural studies of the (very few) compounds containing Fe in higher oxidation states appear to be limited to the following.

Fe(IV): Ba_2FeO_4—isostructural with Ba_2TiO_4 (tetrahedral coordination of Fe(IV). (ZaC 1956 **383** 330).
    $BaFeO_{2·47-2·92}$—hexagonal $BaTiO_3$ structure (JPC 1964 **68** 3786)

The black hygroscopic compound $[Fe(diarsine)_2Cl_2](BF_4)_2$ may contain an octahedral complex of Fe(IV), but its structure is not known.[9] The magnetic moment indicates 2 unpaired electrons which would be consistent with the low-spin d^4 configuration.

(9) JCS A 1966 162

Fe(VI): K_2FeO_4—isostructural with K_2SO_4 (ZaC 1950 **263** 175)
    $BaFeO_4$—isostructural with $BaSO_4$

The magnetic moments of these compounds are close to the value (2·83 BM) expected for 2 unpaired electrons (d^2). For the reduction of the oxidation state of Fe by pressure, for example, $SrFeO_3 \rightarrow SrFeO_{2·86}$, see JCP 1969 **51** 3305, 4353.

The structural chemistry of cobalt

The stereochemistry of cobalt (II)–d^7

We have to discuss the spatial arrangement of 4, 5, or 6 bonds. At one time many deductions of bond arrangements were made from magnetic moments, which would be expected to have the following (spin-only) values:

$$\begin{array}{lll} & & \mu_{\text{eff}} \\ \text{low spin} & \left.\begin{array}{l}\text{planar}\\\text{octahedral}\end{array}\right\} & 1\cdot73\ \text{BM} \\[2ex] \text{high spin} & \left.\begin{array}{l}\text{tetrahedral}\\\text{octahedral}\end{array}\right\} & 3\cdot87 \end{array}$$

However, the magnetochemistry of Co(II) is complicated by spin–orbital interactions. Observed moments are almost invariably higher than the above values, and for some time it was supposed that typical ranges for the two kinds of complex were 1·8–2·1 and 4·3–5·6 BM. More recently many intermediate values have been recorded for octahedral complexes (e.g. 2·63 BM for Co(terpyr)$_2$Br$_2$. H$_2$O) and in fact these now cover the entire range between 2 and 4 BM. The moment is evidently very sensitive to the environment of the Co(II) atom and can no longer be regarded as a reliable indication of the stereochemistry. With the discovery of high-spin square planar complexes with moments around 4 BM magnetic moments do not even distinguish between planar and tetrahedral bonds. Since it is necessary also to be sure of the coordination number of the metal, which cannot always be correctly deduced from the chemical formula, we shall confine our examples to those for which diffraction studies have been made.

Co(II) *forming 4 bonds.* The preferred arrangement of four bonds is the tetrahedral one, which is found in the CoX$_4^{2-}$ ions (X = Cl, Br, I), which are stable in the salts of large cations (Cs$_2$CoCl$_4$, (NBu$_4$)$_2$CoI$_4$, and Cs$_3$(CoCl$_4$)Cl), in the [Co(SCN)$_4$]$^{2-}$ ion in, for example, K$_2$Co(SCN)$_4$. 4 H$_2$O,[1] and in the blue forms of compounds with organic amines such as CoCl$_2$(aniline)$_2$.[2] In the pink or violet forms of these compounds, for example, CoCl$_2$(pyridine)$_2$,[3] there are infinite chains of octahedral coordination groups which share a pair of opposite edges.

There is distortion of the tetrahedral coordination group if the ligands are bidentate, as in [Co(O$_2$C . CF$_3$)$_4$] (Asϕ_4)$_2$[4] and [Co(NO$_3$)$_4$] (Asϕ_4)$_2$.[5] In complexes of this general type there are 8 O atoms (of four bidentate groups) around the metal atom and these form two groups, 4 A + 4 B, which separately define two interpenetrating (non-regular) tetrahedra (Fig. 27.4(a)). The 8 atoms lie at the vertices of a dodecahedron (Fig. 27.4(b)), and in Ti(NO$_3$)$_4$ the two sets of M–O distances (to A and B atoms) are equal. In [Co(NO$_3$)$_4$]$^{2-}$ the distances to A and B atoms are appreciably different (2·07 Å and 2·45 Å) so that Co is forming 4 stronger tetrahedral bonds (i.e. (4 + 4) distorted 8-coordination). In the trifluoroacetato complex the distortion from dodecahedral coordination is much greater, the two Co–O distances being 2·00 Å and 3·11 Å. The four B atoms are coplanar with Co, and the angle O$_A$–Co–O$_A$ is 97°, so that the coordination is much closer to regular tetrahedral. The magnetic moments of both complexes are close to 4·6 BM,

(1) ZFK 1950 **24** 1339

(2) K 1956 **1** 49

(3) AC 1957 **10** 307

(4) IC 1966 **5** 1420

(5) IC 1966 **5** 1208

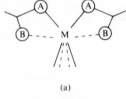

(a)

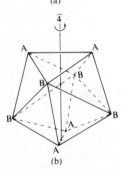

(b)

FIG. 27.4. Coordination of metal by four bidentate ligands (see text).

typical of Co(II) high-spin complexes. (The arrangement of P atoms around Co in CoH(PF$_3$)$_4$ is distorted tetrahedral,[6] with bond angles ranging from 102–118°, but since H is presumably bonded to Co this is an example of 5-covalent Co. The H atom was not located in the X-ray study.)

(6) IC 1970 9 2403

Square coplanar bonds are formed only in complexes with certain polydentate ligands or combinations of ligands such that steric factors prevent a tetrahedral configuration. They include the phthalocyanine (low-spin), the bis-salicylaldehyde diimine complex (a)[7] (in which the maximum deviation of any atom from the mean plane of the complex is 0·6 Å and Co–O(N) is 1·85 Å), and the ion (b) in the salt [Co(mnt)$_2$] [N(C$_4$H$_9$)$_4$]$_2$.[8] The magnetic moment of this salt (3·92 BM) corresponds to 3 unpaired electrons (d^7). The Ni((II) compound is isostructural. Ions [M(mnt)$_2$]$^-$ are formed by Ni(III) and Cu(III).[9] The similar ion formed by

(7) AC 1969 B25 1675

(8) IC 1964 3 1500

(9) IC 1964 3 1507

(a)   (b)

(c)

Co(III) and the ligand (c) has a magnetic moment of 3·18 BM, slightly higher than the spin-only value for 2 unpaired electrons (d^6); it also is planar.[10]

The planar molecule *trans*-Co(PEt$_2$'ϕ]$_2$(mesityl)$_2$[11] is an example of a spin-paired d^7 complex with monodentate ligands.

(10) IC 1968 7 741

(11) JCS 1963 3411

Co(II) *forming 5 bonds*.   We shall, somewhat arbitrarily, divide the examples into three groups: (i) those approximating to trigonal bipyramidal complexes, with monodentate or 'tripod-like' tetradentate ligands, (ii) those with a fairly characteristic square pyramidal structure, with the metal atom 0·3–0·6 Å above the centre of the base, and (iii) dimeric complexes formed from square planar complexes ML$_2$ in which each M atom forms a fifth (longer) bond to an atom in the other half of the dimer. Many complexes have configurations intermediate between trigonal bipyramidal and square pyramidal, and in some cases the 5-coordination is to be regarded as indicating a failure to attain 6-coordination owing to the bulky nature of the ligand. We shall include these intermediate structures in classes (i) or (ii).

(i) The two simple types to which we have referred are illustrated in Fig. 27.5(a) and (b). Examples of (a) include Co(2-picoline N oxide)$_5$ (ClO$_4$)$_2$,[1] with axial bonds (2·10 Å) longer than the equatorial ones (1·98 Å), and the less symmetrical Co(CO)$_4$SiCl$_3$.[2] In [CoBr(Me$_6$tren)] Br,[3] Fig. 27.5(b), Co lies 0·3 Å below the equatorial plane, giving N^1–Co–N^2 81° and N^2–Co–N^3 118°, and the axial Co–N^1, 2·15 Å, slightly longer than the equatorial Co–N^2 (2·08 Å).

(1) IC 1970 9 767

(2) IC 1967 6 1208

(3) IC 1967 6 955

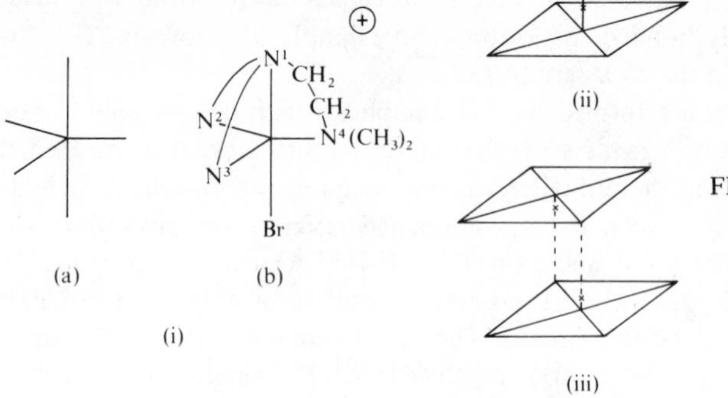

(a)   (b)

(i)

(ii)

(iii)

FIG. 27.5. Co(II) forming five bonds (see text).

(4) CC 1967 763

(5) IC 1966 **5** 879

(6) IC 1967 **6** 483
(7) IC 1969 **8** 2729

These distortions are presumably associated with the geometry of the ligand. There is more distortion in Co(QP)Cl,[4] where the equatorial bond angles are 109°, 113°, and 138° (QP is the P analogue of QAS, p.981), and still more in $CoBr_2(HP\phi_2)_3$,[5] where the equatorial bond angles are 98°, 126°, and 136° and there is a marked difference between the lengths of the two Co–Br bonds (both equatorial, 2·33 Å and 2·54 Å). (The H atom is omitted from the formula throughout ref. (5).) Intermediate configurations are also found with bulky amine ligands, as in the high-spin complexes $Co(Et_4dien)Cl_2$[6] and $Co(Me_5dien)Cl_2$,[7] in which the ligands are:

$$HN{\Large<}^{(CH_2)_2NEt_2}_{(CH_2)_2NEt_2} \quad (Et_4dien); \qquad MeN{\Large<}^{(CH_2)_2NMe_2}_{(CH_2)_2NMe_2} \quad (Me_5dien)$$

(1) JCS A 1966 1317

(ii) Square pyramidal monomeric complexes, Fig. 27.5 (ii), include $CoCl_2(L_3)$,[1] where L_3 is the tridentate ligand (a), the structure of which is very similar to that of $ZnCl_2(terpyr)$, p. 915. The metal atom lies above the base of the pyramid (0·4 Å).

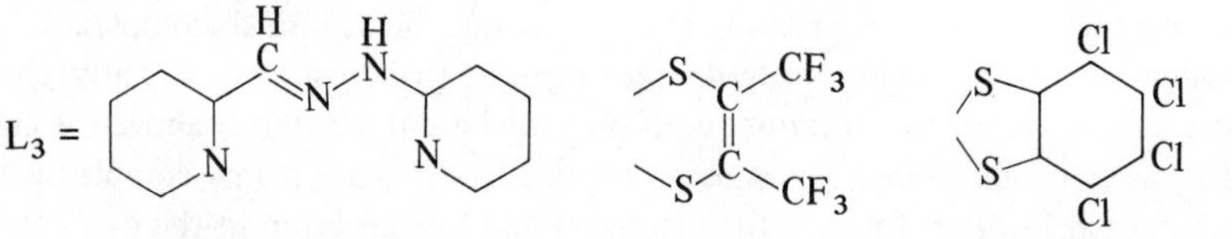

$$L_3 = $$

(a)   (b)   (c)

(2) IC 1965 **4** 1729
(3) JACS 1968 **90** 4253

(iii) Dimers of the type shown in Fig. 27.5 (iii) exist in $[Co(b)_2]_2$[2] and $[Co(c)_2]_2(NBu_4)_2$,[3] in which (b) and (c) are the ligands shown above. In both complexes the metal atom lies slightly above the plane of the four close S ligands (0·37 Å and 0·26 Å respectively), and the vertical bonds to the fifth S are longer than the four basal bonds (approximately 2·40 Å and 2·18 Å respectively).

Co(II) *forming 6 bonds.* With one exception, noted below, the bond arrangement is octahedral, and this is found in 3D complexes in crystalline binary compounds

(CoO, CoX$_2$, etc.), in layer structures such as K$_2$CoF$_4$ and BaCoF$_4$, in chain structures of the type CoCl$_2$(pyr)$_2$, and in numerous finite complexes which include *trans*-Co(NC$_5$H$_5$)$_4$Cl$_2$, Co(dimethylglyoxime)$_2$Cl$_2$, and Co(acac)$_2$-(pyr)$_2$.[1] In its complexes with acetylacetone Co(II) consistently achieves 6-coordination. The acetylacetonate sublimes as the tetramer, Co$_4$(acac)$_8$,[2] but from solution in inert solvents containing a little water various polymeric species can be obtained: Co$_3$(acac)$_6$. H$_2$O,[3] Co$_2$(acac)$_4$(H$_2$O)$_2$,[4] and finally the simple monomeric Co(acac)$_2$(H$_2$O)$_2$.[5] These correspond to splitting the tetrameric [Co(acac)$_2$]$_4$, Fig. 27.6(a), at a shared face (*A*), giving (b), or at both (*A*) and (*B*), leaving Co$_2$(acac)$_2$(H$_2$O)$_2$, (c); an unoccupied bond position is filled with a H$_2$O molecule.

In the red diamagnetic form of [Co$_2$(CNCH$_3$)$_{10}$](ClO$_4$)$_4$[6] one of the octahedral bonds from each Co is to a second metal atom. In this ion Co(II) is isoelectronic with Mn(0) and forms the same type of structure as Mn$_2$(CO)$_{10}$ (staggered configuration, Co–Co, 2·74 Å). The blue solution of this compound and the blue crystalline form presumably contain [Co(CNCH$_3$)$_5$]$^{2+}$, the paramagnetism corresponding to one unpaired electron.

(1) IC 1968 **7** 1117
(2) IC 1965 **4** 1145

(3) JACS 1968 **90** 38
(4) IC 1966 **5** 423
(5) AC 1959 **12** 703

(6) IC 1964 **3** 1495

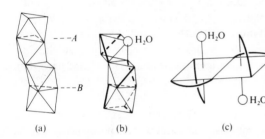

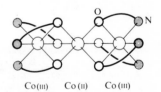

FIG. 27.6. The molecular structures of (a) Co$_4$(acac)$_8$, (b) Co$_3$(acac)$_6$. H$_2$O, (c) Co$_2$(acac)$_4$(H$_2$O)$_2$.

(a)   (b)   (c)

The very irregular 6-coordination of Co(II) in Co(NO$_3$)$_2$[OP(CH$_3$)$_3$]$_2$[7] is noted in Chapter 18.

The ion [Co$_3$(OCH$_2$CH$_2$NH$_2$)$_6$]$^{2+}$ is the sole example of trigonal prism coordination of Co(II) by oxygen. In this ion (Fig. 27.7) there is octahedral coordination of the terminal Co(III) atoms but trigonal prism coordination of the central Co(II).[8] It is possible that this unusual coordination is connected with the geometry of the ligands, there being apparently better packing of the CH$_2$ groups (which are not shown in Fig. 27.7) with trigonal prismatic than with octahedral coordination around the central Co atom.

(7) JACS 1963 **85** 2402

Co(III)   Co(II)   Co(III)

FIG. 27.7. The structure of the ion [Co$_3$(OCH$_2$CH$_2$NH$_2$)$_6$]$^{2+}$ (diagrammatic).

(8) JACS 1969 **91** 2394

Cobaltammines

Introductory. The discovery that CoCl$_3$ could form stable compounds with 3, 4, 5, or 6 molecules of NH$_3$ led to Werner's suggestion that in 'molecular compounds' such as CoCl$_3$. 6 NH$_3$ the six NH$_3$ molecules are arranged in the first 'coordination sphere' around the metal atom and the three Cl atoms in an outer sphere. This was consistent with the chemical behaviour of the compounds, with the numbers of ions formed from one 'molecule', and with the numbers of isomers of mixed coordination groups such as $[Co(NH_3)_4Cl_2]^+$. Some or all of the NH_3 molecules in $[Co(NH_3)_6]^{3+}$ may be replaced by other ligands such as Cl^-, H_2O, NO_2^-, CO_3^{2-}, $C_2O_4^{2-}$, $NH_2.CH_2.CH_2.NH_2$ (en), etc. Of the ammines of $CoCl_3$ containing 3, 4, 5, and $6 NH_3$ the first is a non-electrolyte while the others ionize to give respectively 2, 3, and 4 ions from one formula-weight, so that they are $Co(NH_3)_3Cl_3$, $[Co(NH_3)_4Cl_2]Cl$, $[Co(NH_3)_5Cl]Cl_2$, and $[Co(NH_3)_6]Cl_3$. Owing to the large number of combinations of atoms and groups which may be attached to Co these compounds are very numerous. The only comparable group of compounds is that formed by Cr(III), but these have been much less studied in recent years.

Many cobaltammines are easily prepared by oxidizing ammoniacal solutions of Co(II) salts by a current of air. For example, if the solution contains $CoCl_2$ and NO_2^- or CO_3^{2-} ions the brown Erdmann's salt, $[Co(NH_3)_2(NO_2)_4]NH_4$, or the red $[Co(NH_3)_4CO_3]Cl$ may be crystallized out. Crystals of the bridged salt $[(NH_3)_5Co.O_2.Co(NH_3)_5](SCN)_4$ are formed in a strongly ammoniacal solution of $Co(SCN)_2$ which is left exposed to the atmosphere. In such preparations mixtures of cobaltammine ions are formed, both mononuclear and polynuclear, solubility differences determining the nature of the product obtained in a particular preparation. As an illustration of the stability of certain cobaltammines we may quote the action of concentrated H_2SO_4 on $[Co(NH_3)_6]Cl_3$ to form the sulphate of the complex, which is not disrupted in the process. The aquo-pentammine, $[Co(NH_3)_5H_2O]Cl_3$, retains its water up to a temperature of $100°C$, and even then the complex does not break up but a rearrangement takes place to form $[Co(NH_3)_5Cl]Cl_2$.

Cobaltammines, including $Co(NH_3)_3F_3$ (but not K_3CoF_6, which has $\mu_{eff.} = 5.3$ BM), are diamagnetic except in special cases where an unpaired electron is introduced with a ligand such as O_2^- (see later). Their stability is presumably associated with the fact that with 12 bonding electrons added to the d^6 configuration of Co^{III} there are 18 electrons to fill the 9 available orbitals; for Cr^{III}

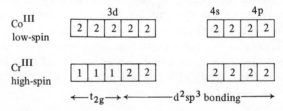

there is a single electron in each of the t_{2g} orbitals. (For Cr^{III} there is no difference between the magnetic moment for high-spin and low-spin complexes.)

The isomerism of cobaltammines. The octahedral arrangement of the six ligands has now been established by X-ray studies of many crystalline cobaltammines, but when it was first suggested it was not by any means generally accepted. However, it was consistent with the existence and numbers of isomers of many complexes. For example, there should be only two isomers (*cis* and *trans*) of an octahedral complex CoA_4B_2 but three for a plane hexagonal or trigonal prism model, and the fact that only two isomers of the numerous complexes of this type have ever been isolated is strong presumptive evidence in favour of the octahedral model. The demonstration of the optical activity of $K_3Co(C_2O_4)_3$ provided striking confirmation of the octahedral arrangement of the bonds from the Co atom. It is now well known that any octahedral complex MA_3 in which A represents a bidentate group such as $C_2O_4^{2-}$ or 'en' is enantiomorphic, but it will be appreciated that the resolution of such a compound appeared surprising, for it was previously formulated as a double salt, $3\,K_2C_2O_4\,.\,Co_2(C_2O_4)_3$. Nevertheless, it is a curious fact that the resolution of salts like $K_3[Co(C_2O_4)_3]$ and $[Co(NH_2.CH_2.CH_2.NH_2)_3]Br_3$ into their optical antimers was not considered by some chemists to prove Werner's hypothesis. There still lingered a belief that the optical activity was in some way dependent on the presence of an organic radical in the complex. It was not until Werner resolved a purely inorganic cobaltammine that the last doubts were removed. The compound in question has the formula $[Co_4(OH)_6(NH_3)_{12}]Cl_6$; the probable structure of the complex ion is illustrated in Fig. 27.8(e) (p. 962).

Some idea of the complexity of the chemistry of the cobaltammines may be gathered from the fact that no fewer than nine compounds with the empirical formula $Co(NH_3)_3(NO_2)_3$ are said to have been prepared. There are many possible types of isomerism and polymerism, as the following examples will show.

Isomers have the same molecular weight but differ in physical and/or chemical properties. The following kinds of isomerism may be distinguished.

(i) *Stereoisomerism.* The isomers contain exactly the same set of bonds, that is, the same pairs of atoms are linked together in the isomers.

(a) The positions of all the atoms relative to one another are the same. The only difference between the isomers is that they are related as object and mirror image (*optical isomerism*).

(b) The constituent atoms are arranged differently in space relative to one another (*geometrical isomerism*).

Optically active complexes which have been resolved include $[Co(C_2O_4)_3]^{3-}$, $[Co(en)_3]^{3+}$, and $[Co(en)_2Cl_2]^+$. A more complex type of isomerism can arise when the chelate group is unsymmetrical. For a symmetrical group occupying two coordination positions (e.g. C_2O_4), which we may represent $A-A$, there is only the one stereoisomer (I) which, of course, exists in right- and left-handed modifications. If, however, the group is unsymmetrical there are two possible forms (II and III) exactly analogous to the two stereoisomers of a complex CoA_3B_3, and each form is

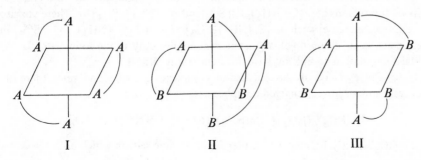

$\qquad\qquad$ I $\qquad\qquad\qquad\qquad$ II $\qquad\qquad\qquad\qquad$ III

enantiomorphic. Two such forms of the tri-glycine complex $Co(NH_2CH_2COO)_3$ have been prepared.

Some bridged complexes provide interesting examples of enantiomorphism. For example, $[(en)_2Co(NH_2)(NO_2)Co(en)_2]Br_4$ exists in *d*- and *l*-forms and also in a meso form:

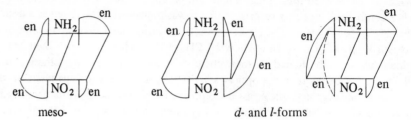

meso-   *d*- and *l*-forms

The active forms of this compound were obtained by Werner by resolution with *d*-bromcamphorsulphonic acid.

Many pairs of geometrical isomers have been made, such as the 'violeo' and 'praeseo' series, *cis* and *trans* $[Co(NH_3)_4Cl_2]Cl$.

(ii) *Structural isomerism.* This includes all cases of isomerism where different pairs of atoms are linked together in the various isomers. These may therefore possess very different chemical properties. We may distinguish three simple kinds of structural isomerism.

(a) A number of cobaltammines containing groups such as NO_2 or SCN exist in two forms, the isomerism being due to the possibility of linking the group to the metal atom in the two ways, $M-NO_2$ or $M-ONO$ and $M-SCN$ or $M-NCS$ respectively. When aquo-pentammine cobaltic chloride $[Co(NH_3)_5H_2O]Cl_3$ is treated with nitrous acid under the correct conditions a red compound is formed which evolves nitrous fumes if treated with dilute mineral acids and which at ordinary temperatures is slowly (and at 60°C rapidly) converted into a brown isomer, The latter is quite stable towards dilute acids. The isomers are formulated $[Co(NH_3)_5ONO]Cl_2$ and $[Co(NH_3)_5NO_2]Cl_2$. Pairs of 'linkage' isomers of which structural studies have been made include $[Co(NH_3)_5SCN]Cl_2 . H_2O$ and $[Co(NH_3)_5NCS]Cl_2$, $[Co(NH_3)_5NCCo(CN)_5] . H_2O$ and $[Co(NH_3)_5CNCo(CN)_5] . H_2O$ (see p. 753 and Table 27.4).

(b) The term 'ionization isomerism' has been applied to the isomerism of pairs of compounds like $[Co(NH_3)_5Br]SO_4$ and $[Co(NH_3)_5SO_4]Br$. There are also more interesting cases such as $[Co(NH_3)_5SO_3]NO_3$ and $[Co(NH_3)_5NO_2]SO_4$. In these pairs, as in those in (c), the isomerism applies only to the solid state, for the actual coordination complexes have different compositions.

(c) Since many of the octahedral complexes are ions and since there are positively and negatively charged complexes, as in the series

$$Co(NH_3)_4(NO_2)_2^+, \quad Co(NH_3)_3(NO_2)_3, \quad Co(NH_3)_2(NO_2)_4^-,$$

it should obviously be possible to prepare crystalline salts in which both the cation

960

TABLE 27.4

References to the structures of cobaltammines

Compound	Reference
$[Co(NH_3)_6]I_3$	AC 1969 **B25** 168
$[Co(CN)_6][Cr(en)_3] \cdot 6\,H_2O$	IC 1968 7 2333
$Co(N_3\phi_2)_3$	JACS 1967 **89** 1530
$[Co(NH_3)_5NO_2]Br_2$	AC 1968 **B24** 474
$[Co(NH_3)_5NO_2]Cl_2$	ACSc 1968 **22** 2890
$[Co(NH_3)_5CO_3]Br \cdot H_2O$	JCS 1965 3194
$[Co(NH_3)_5N_3](N_3)_2$	AC 1964 17 360
$[Co(NH_3)_5SCN]Cl_2 \cdot H_2O$ and $[Co(NH_3)_5NCS]Cl_2$	AC 1972 **B28** 1908
$[Co(en)_2(N_3)_2]NO_3$	AC 1968 **B24** 1638
$[Co(en)_2Cl_2]Cl \cdot HCl \cdot 2\,H_2O$	BCSJ 1952 **25** 331
$[Co(NH_3)_4CO_3]Br$	JCS 1962 586;
	ACSc 1963 **17** 1630
$Co(NH_3)_3(NO_2)_3$	IC 1971 **10** 1057
$Co(NH_3)_3(NO_2)_2Cl$	BCSJ 1953 **26** 420
$[Co(NH_3)_3(H_2O)Cl_2]Cl$	BCSJ 1952 **25** 328
$[Co(NH_3)_2(NO_2)_4]Ag$	ZK 1936 **95** 74
$[(NH_3)_5NCCo(CN)_5] \cdot H_2O$	IC 1971 **10** 1492
$[(NH_3)_5Co \cdot NH_2 \cdot Co(NH_3)_5](NO_3)_5$	AC 1968 **B24** 283
$\left[(NH_3)_4Co \underset{OH}{\overset{OH}{<}} Co(NH_3)_4\right]Cl_4 \cdot 4\,H_2O$	JCS 1962 4429; ACSc 1963 **17** 85
$\left[(NH_3)_4Co \underset{NH_2}{\overset{Cl}{<}} Co(NH_3)_4\right]Cl_4 \cdot 4\,H_2O$	IC 1970 9 2131
$\left[(NH_3)_3Co \underset{OH}{\overset{OH}{\underset{}{-}}} OH - Co(NH_3)_3\right]Br_3$	ACSc 1967 **21** 243
$[(en)_2Co(NH_2)(SO_4)Co(en)_2]Br_3$	AC 1971 **B27** 1744
$[Co_3(NH_3)_8(OH)_2(NO_2)_2(CN)_2](ClO_4)_3 \cdot NaClO_4 \cdot 2\,H_2O$	AC 1970 **B26** 1709

and the anion are octahedral complexes. We could then have pairs of salts with the same composition of which the following are the simplest types: $(M^1A_6)(M^2B_6)$ and $(M^1B_6)(M^2A_6)$ containing two different metals. The pairs of salts $[Co(NH_3)_6][Cr(CN)_6]$, $[Cr(NH_3)_6][Co(CN)_6]$ and $[Co(NH_3)_4(H_2O)_2] \cdot [Cr(CN)_6]$, $[Cr(NH_3)_4(H_2O)_2][Co(CN)_6]$ have been prepared.

(iii) *Polymerism.* Consider a salt of the type $(MA_nB_{6-n})(MA_{6-n}B_n)$ in which the central metal atom is the same in both complexes. Crystals of the compound consist of equal numbers of octahedral complexes of two kinds, so that the simplest structural formula is that given above. The empirical formula is, however, MA_3B_3, which corresponds to an entirely different compound, in which all the structural units are identical. The compounds $Co(NH_3)_3(NO_2)_3$ and $[Co(NH_3)_6][Co(NO_2)_6]$ are related in this way, and $[(NH_3)_3Co(OH)_3 \cdot (Co(NH_3)_3]Cl_3$ has the same composition as $[Co\{(OH)_2Co(NH_3)_4\}_3]Cl_6$.

The structures of cobaltammines. The structures of many cobaltammines have been established by X-ray crystallographic studies. The special interest attaching to a particular compound may lie in the bond lengths, in the mode of attachment of the ligands to the metal, in their relative arrangement (geometrical isomerism), or in the actual gross stereochemistry in the case of more complex (bridged) complexes. For example, in $[Co(NH_3)_6]I_3$ the length of the bond $Co^{III}-N$ is 1·94 Å, as compared with $Co^{II}-N$, 2·11 Å in $[Co(NH_3)_6]Cl_2$ (and planar $Co^{II}-N$, 1·85 Å).

In $[Co(NH_3)_4CO_3]Br$ there is a bidentate CO_3^{2-} ligand, Fig. 27.8(a); the lengthening of the Co—N bonds *trans* to the bidentate ligand is probably real. Contrast the monodentate CO_3^{2-} group in $[Co(NH_3)_5CO_3]Br . H_2O$, where there is apparently hydrogen bonding between one O and one NH_3 (b).

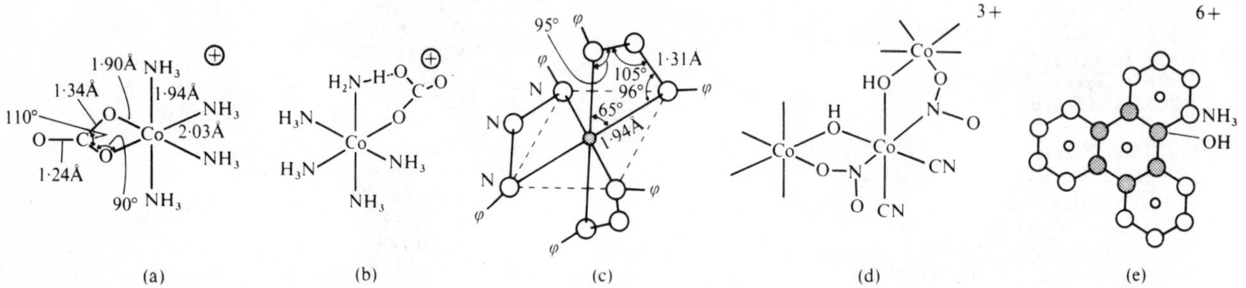

FIG. 27.8. The structures of cobaltammines: (a) $[Co(NH_3)_4CO_3]^+$, (b) $[Co(NH_3)_5CO_3]^+$, (c) $Co(N_3\phi_2)_3$, (d) $[Co_3(NH_3)_8(OH)_2(NO_2)_2(CN)_2]^{3+}$, (e) probable structure of $[Co_4(OH)_6(NH_3)_{12}]^{6+}$.

The structure of the diazoaminobenzene (1,3-diphenyltriazene) complex $Co(N_3\phi_2)_3$, (c), is of interest since the dimeric structure of the cuprous derivative (p. 879) might have been attributed to the difficulty of forming a 4-ring. In fact, 4-membered rings occur in the Co(III) compound and also in all complexes containing bidentate oxy-ions such as CO_3^{2-} (above), NO_3^-, NO_2^-, and $OOC . CH_3^-$ which are noted elsewhere. The azide group N_3^- is a linear ligand which has been studied in the $[Co(NH_3)_5N_3]^{2+}$ and $[Co(en)_2(N_3)_2]^+$ ions:

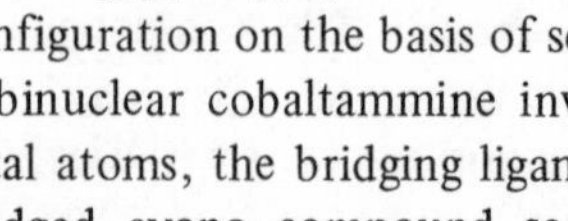

In complexes CoA_4B_2 (and more complex types) the structural study is necessary to establish unambiguously the relative arrangement of the ligands, e.g. the *trans* arrangement of the Cl atoms in $[Co(NH_3)_3(H_2O)Cl_2]^+$ and of the two NH_3 molecules in $[Co(NH_3)_2(NO_2)_4]^-$. The latter ion, of Erdmann's salt, had been assigned the *cis* configuration on the basis of self-consistent chemical evidence.

The three types of binuclear cobaltammine involve a single, double, or triple bridge between the metal atoms, the bridging ligands being Cl, OH, NH_2, or O_2 in most cases. For a bridged cyano compound see p. 753 and Table 27.4. The structures of the red diamagnetic peroxo- and the green paramagnetic superoxo-compounds are described in Chapter 11. Examples of bridged complexes are

962

included in Table 27.4; some details include:

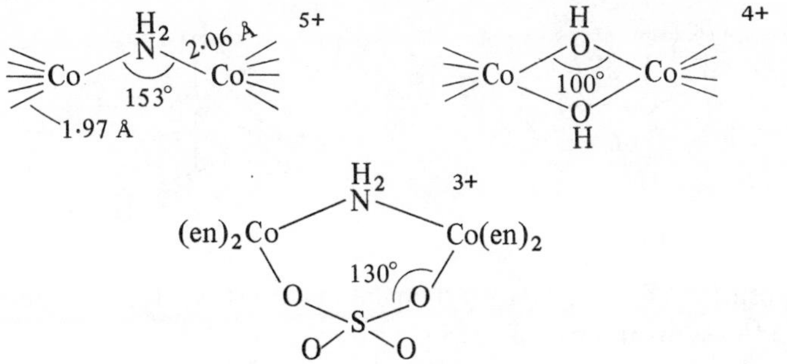

The ion $[Co_3(NH_3)_8(OH)_2(NO_2)_2(CN)_2]^{3+}$ is of special interest as containing not only bridging OH groups but also NO_2^- ions bridging through O and N atoms (Fig. 27.8(d)); in $[(en)_2Co(NH_2)(SO_4)Co(en_2)]^{3+}$ there is a bridging SO_4 group (see above).

Bridging can proceed further as in complexes such as

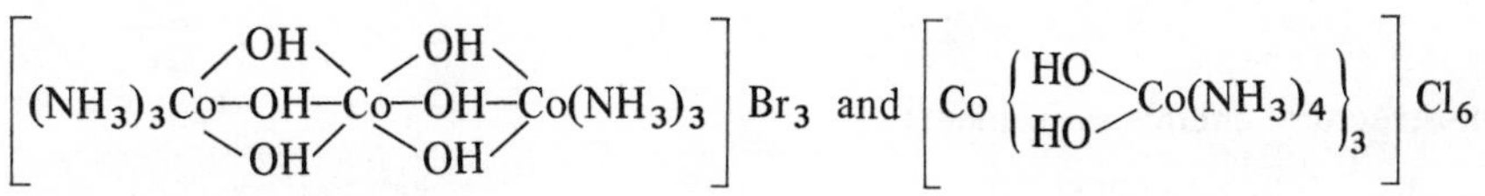

the second of which, Fig. 27.8(e), has already been mentioned in connection with the optical activity of cobaltammines. It is noteworthy that the closely related ion $[Cr_4(OH)_6(en)_6]^{6+}$ has the different structure illustrated in Fig. 5.10 (p. 167).

Other oxidation states of Co

The formal oxidation state -1 may be recognized in the tetrahedral molecule $Co(CO)_3NO$ (p. 772). Little appears to be known about the structures of compounds of Co(0).

Co(I). Structural studies do not appear to have been made of complexes in which the metal is forming 4 or 6 bonds, in ions such as $[Co(CN)_3CO]^{2-}$ and $[Co(dipyr)_3]^+$. On the other hand, several structural studies have established the arrangement of 5 bonds. The tetragonal pyramidal structure of the molecule $Co[S_2CN(CH_3)_2]_2NO$ is described on p. 654, and X-ray studies have established the trigonal bipyramidal structure of the following complexes. The molecule $CoH(P\phi_3)_3N_2$[1a] is of special interest as an example of the coordination of molecular nitrogen to a metal atom, (a). The bond length in the N_2 ligand is similar to that in the free molecule, and the length of the Co–N bond suggests some multiple bond character (single Co–N, 1·95 Å). There is apparently similar coordination of N_2 to the metal atom in the $[Ru^{II}(NH_3)_5N_2]^{2+}$ ion,[1b] but owing to disorder in the crystal less accurate data were obtained. In our second example, $Co(P\phi_3)(CO)_3(CF_2 \cdot CHF_2)$,[2] (b), the three CO ligands occupy the

(1a) IC 1969 8 2719

(1b) AC 1968 **B24** 1289

(2) JCS A 1967 2092

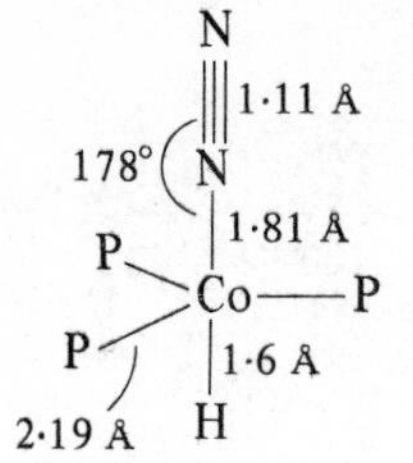

(a) 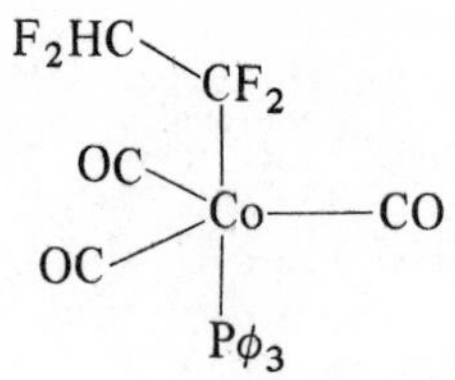   (b)

equatorial positions. The trigonal bipyramidal ion $[Co(CNCH_3)_5]^+$, studied in the perchlorate,[3] was mentioned on p. 756.

(3) IC 1965 **4** 318

Co(III). With very few exceptions, such as the heteropolyacid ion $(CoW_{12}O_{40})^{5-}$ (p. 436), the bond arrangement is octahedral. We deal separately with examples of the extremely numerous cobaltammines. High-spin Co(III), d^6, is rare; it occurs in K_3CoF_6 but not in $Co(NH_3)_3F_3$ or most cobaltammines.

Co(IV). The pale-blue compound originally thought to be K_3CoF_7 was later shown to be K_3CoF_6. However, the brownish-yellow Cs_2CoF_6 has the K_2PtCl_6 structure,[4] and e.s.t. and optical spectra suggest the presence of Co(IV) in α-Al_2O_3 doped with Mg and Co.[5]

(4) ZaC 1961 **308** 179
(5) TFS 1965 **61** 2597

The structural chemistry of nickel

The stereochemistry of Ni(II)–d^8

The following bond arrangements have been established by structural studies of finite ions or molecules:

C.N.	Low spin (diamagnetic)	High spin (paramagnetic)
4	Coplanar	Tetrahedral
5	Trigonal bipyramidal	Trigonal bipyramidal
	Tetragonal pyramidal	Tetragonal pyramidal
6	(Distorted octahedral)	Octahedral

Much confusion in the literature on this subject has been due to the placing of too much reliance on magnetic data of dubious quality and to incorrect deductions of coordination numbers from chemical formulae. There are regrettably few examples of compounds of which both thorough magnetic *and* structural studies have been made. The following facts have been established:

(a) Diamagnetism is indicative of planar as opposed to tetrahedral coordination if the c.n. is known to be 4,

(b) 5-coordinated complexes may be either diamagnetic or paramagnetic,

(c) Paramagnetism is exhibited by 4-, 5-, and 6-coordinated Ni(II). The spin-only moment for d^8 is 2·83 BM, but spin-orbital interactions in *magnetically dilute* systems always lead to higher values. It is probably safe to interpret values of $\mu > 3·4$ BM as indicative of tetrahedral coordination and values $< 3·2$ BM as associated with octahedral coordination; the interpretation of values in the intermediate range is not always clear. Examples of magnetic moments of compounds whose structures have been established include:

		μ
Tetrahedral:	$(NiCl_4)[N(C_2H_5)_4]_2$	3·90 BM
Trigonal bipyramidal:	$NiCl_3(H_2O)[N(C_2H_4)_3NCH_3]$	3·7
Octahedral:	$Ni(pyr)_4(ClO_4)_2$	3·24

The coordination number of the metal atom in a complex cannot in general be deduced simply from the chemical formula because a ligand may be shared between two (or more) metal atoms in a polymeric (finite or infinite) grouping, or a particular ligand may behave in different ways, as illustrated by the behaviour of acac in Pt compounds (p. 983). For example, the (paramagnetic) nickel acetylacetonate is not a tetrahedral molecule but a trimer, $[Ni(acac)_2]_3$, containing octahedrally coordinated Ni, and compounds MX_2L_2 may be either finite (that is, planar or tetrahedral) or an infinite chain of edge-sharing octahedral groups. The stereochemistry of Ni is further complicated by the fact that there is evidently little difference in stability between the following pairs of bond arrangements for certain types of complex:

(i) square planar (4) and tetrahedral (4),
(ii) trigonal bipyramidal (5) and square pyramidal (5),
(iii) square planar (4) and octahedral (6).

(i) Many compounds NiX_2L_2 exist in two forms. For example, $NiBr_2(PEt\phi_2)_2$ crystallizes as a dark-green paramagnetic form from polar solvents and as a brown diamagnetic form from CS_2. Both forms are monomeric in solution and presumably tetrahedral and planar complexes respectively. (The blue and yellow forms of $Ni(quinoline)_2Cl_2$ are both paramagnetic and may be examples of finite tetrahedral and octahedral chain structures.) Square planar–tetrahedral isomerism seems to have been observed only with certain series of Ni complexes, and pure specimens of both isomers have been isolated only with diphenyl alkyl ligands. In series of compounds $NiX_2(PR_3)_2$, with X changing from Cl–Br–I, and R_3 from (alkyl)$_3$ to (phenyl)$_3$, the change in configuration usually occurs around $NiBr_2(PR\phi_2)_2$, that is, the Cl compounds are mostly planar, the I compounds mostly tetrahedral, and the Br compounds often exist in both forms.[1] In one form of $NiBr_2[P\phi_2(benzyl)]_2$ the crystal contains both planar and tetrahedral isomers. This compound crystallizes in a red diamagnetic form and also in a green paramagnetic form (moment 2·7 BM). In the latter, one-third of the molecules are square planar and two-thirds tetrahedral; the name *interallogon* compound has been suggested for a crystal of this type containing two geometrical isomers with different geometries.[2] Examples of complexes with bond arrangements intermediate between square planar and tetrahedral are noted later.

(ii) Numerous 5-covalent diamagnetic complexes have been studied, containing both mono- and poly-dentate ligands (examples are given later) and here

(1) IC 1965 4 1701

(2) JCS A 1970 1688

again two configurations of a complex have been found in the same crystal. The $Ni(CN)_5^{3-}$ ion is stable only in the presence of large cations—at low temperatures K^+ is sufficiently large—and its configuration is dependent on the nature of the cation. In $[Cr(tn)_3] [Ni(CN)_5] . 2 H_2O$ (tn = 1,3-propane diamine) all the anions are square pyramidal (a), but in $[Cr(en)_3] [Ni(CN)_5] . 1\frac{1}{2} H_2O$ there are ions of both types (a) and (b).[3]

(3) IC 1970 9 2415

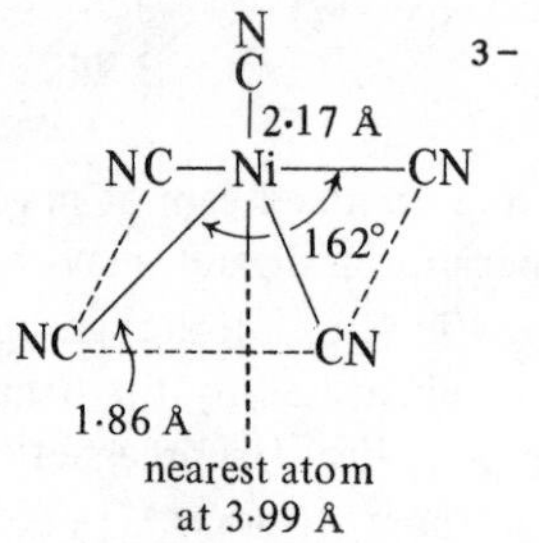

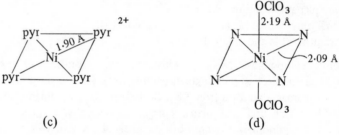

(a) 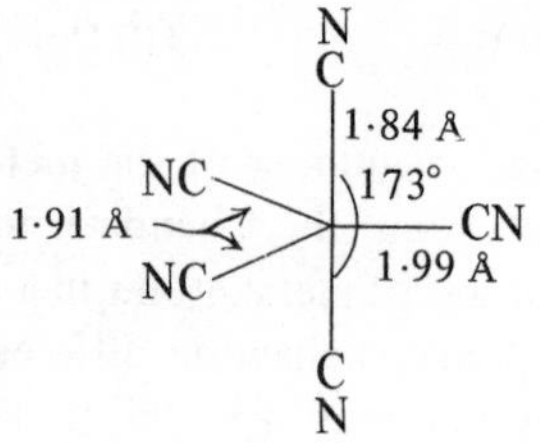   (b)

(iii) Compounds $Ni(pyr)_4(ClO_4)_2$, in which 'pyr' represents pyridine or a substituted pyridine, exist in yellow diamagnetic and blue paramagnetic forms. The former contain planar ions, (c), the next nearest neighbours of Ni being O atoms at

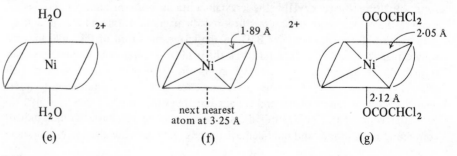

(c)   (d)

3·34 Å, while the latter consist of octahedral molecules, (d).[4] The compounds studied were the 3,5-dimethylpyridine compound, (c), and the 3,4-dimethyl-pyridine compound, (d). The compounds containing the bidentate ligand meso-stilbene diamine, $C_6H_5 . CH(NH_2)–CH(NH_2) . C_6H_5$, are still more interesting.[5] The blue and yellow forms are interconvertible in solution, and while the hydrated blue form ($\mu = 3·16$ BM) of the dichloroacetate contains octahedral ions, (e), the yellow form ($\mu = 2·58$ BM) contains planar ions of type (f) and octahedral

(4) AC 1968 **B24** 745, 754

(5) IC 1964 3 468

(e)   (f)   (g)

966

molecules (g) in the ratio 1 : 2. This is the same ratio of diamagnetic to paramagnetic complexes as in $NiBr_2(P\phi_2Bz)_2$ mentioned above, and results in a similar intermediate value of μ_{eff}.

We referred in Chapter 3 to the clathrate compounds based on layers of the composition $Ni(CN)_2 \cdot NH_3$. In the clathrates containing C_6H_6 or $C_6H_5NH_2$ the layers are directly superposed, with NH_3 molecules of different layers directed towards one another, so that the interlayer spacing is 8·3 Å and there are large cavities between the layers (Fig. 1.9(b)).[6] In the hydrate and in the anhydrous compound the layers are packed much more closely (interlayer spacing 4·4 Å), the NH_3 molecules of one layer projecting towards the centres of the rings of adjacent layers (pseudo body-centred).[7]

Ni(II) *forming 4 coplanar bonds*. Planar diamagnetic complexes include the $Ni(CN)_4^{2-}$ ion, which has been studied in the K[1] and other salts[2] (Ni–C, 1·85 Å), and molecules such as (a)[3] and (b).[4] The numerous complexes with bidentate ligands include the thio-oxalate ion, (c),[5] molecules of the types (d)[6] and (e),[7] and the glyoximes, (f),[8] notable for the short intramolecular hydrogen bonds. Many divalent metals form phthalocyanins,[9] M(II) replacing 2 H in $C_{32}H_{18}N_8$ (Fig. 27.9).

(6) JCS 1958 3412

(7) ZK 1966 **123** 391

(1) ACSc 1964 **18** 2385
(2) ACSc 1969 **23** 14, 61; AC 1970 **B26** 361
(3) AC 1968 **B24** 108
(4) JCS A 1967 1750
(5) JCS 1935 1475
(6) IC 1968 7 2140, 2625; AC 1969 **B25** 909, 1294, 1939
(7) ZaC 1968 **363** 159
(8) AC 1967 **22** 468
(9) JCS 1937 219

(a)

(b)

(c)

(d)

(e)

(f)

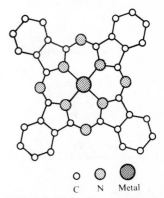

FIG. 27.9. Molecule of a metal phthalocyanine, $C_{32}H_{16}N_8M$.

Polymeric species include finite molecules and chains. In Ni_3L_4 (L = $H_2N \cdot CH_2 \cdot CH_2 \cdot S$),[10] (g), the three Ni atoms are collinear with a planar arrangement of bonds around each. The whole molecule is twisted owing to the

(10) IC 1970 9 1878

symmetrical arrangement of the chelates around the central Ni atom and the *cis* configuration around the terminal Ni atoms. The Ni—Ni distance in this complex is 2·73 Å.

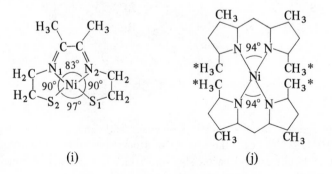

(g)

(11) JACS 1965 **87** 5251

Nickel mercaptides are mixtures of insoluble polymeric materials and hexamers. The molecule of $[Ni(SC_2H_5)_2]_6^{(11)}$ is shown diagrammatically in Fig. 27.10. The six Ni atoms are coplanar and are bridged by 12 mercaptan groups. The bond arrangement around Ni is square planar, with all angles in the Ni_2S_2 rings equal to 83°, Ni—S, 2·20 Å, and Ni—Ni, 2·92 Å (mean). The dimethylpyrazine complex, $NiBr_2L,^{(12)}$ (h), is an example of a chain in which Ni is 4-coordinated by 2 Br and by 2 N atoms of the bridging ligands:

(12) IC 1964 **3** 1303

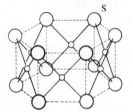

FIG. 27.10. The molecule $[Ni(SC_2H_5)_2]_6$ (diagrammatic).

(13) IC 1965 **4** 1726

(14) IC 1970 **9** 783

Departures from exact coplanarity of Ni and its four bonded atoms may occur in unsymmetrical complexes, for example, those containing polydentate ligands. In the (diamagnetic) complex (i),$^{(13)}$ there is slight distortion (tetrahedral rather than square pyramidal), the angles S_1NiN_1 and S_2NiN_2 being 173°. Large deviations from planarity, leading to paramagnetic tetrahedral complexes, can result if there is steric interference between atoms forming parts of independent ligands. In the tetramethyl dipyrromethenato complex (j)$^{(14)}$ there is interference between the CH_3* groups, and the angle between the planes of the two ligands is 76°; it is not obvious why this angle is not 90°. The 3-substituted bis(N-isopropylsalicylaldi-

(i)   (j)

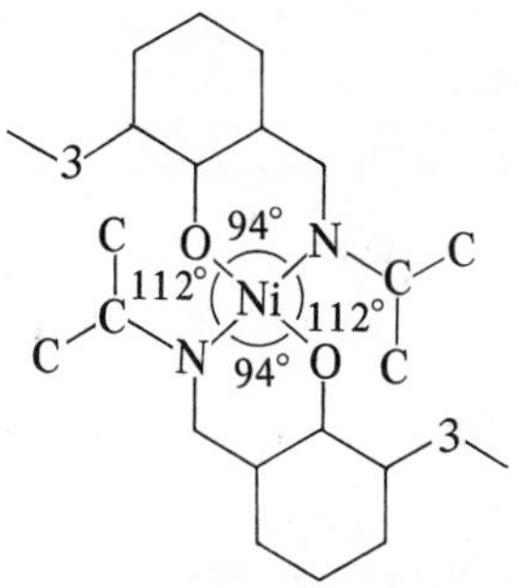

(k)

minato) Ni complexes, (k),[15] are of interest in this connection, for they fall into two groups; tetrahedral ($\mu \approx 3\cdot3$ BM) or planar diamagnetic according to the nature of the substituent in the 3 position. The detailed stereochemistry of these compounds is not readily understandable, for the 3-CH_3 compound is planar while the 3-H and 3-C_2H_5 compounds are tetrahedral.

Ni(II) *forming 4 tetrahedral bonds.* The simplest examples are the complexes NiX_4^{2-}, NiX_3L^-, NiX_2L_2, and NiL_2', where L is a monodentate and L' is a bidentate ligand. The failure to recognize, until fairly recently, the existence of halide ions NiX_4^{2-} was due, not to their intrinisic instability, but to their instability in aqueous solution, where X is displaced by the more strongly coordinating H_2O molecule. The $NiCl_4^{2-}$ ion can be formed in melts and studied in solid solutions in, for example, Cs_2ZnCl_4, but Cs_2NiCl_4 dissociates on cooling. On the other hand, Cs_3NiCl_5 (analogous to $Cs_3(CoCl_4)Cl$) can be prepared from the molten salts and quenched to room temperature, but on slow cooling it breaks down to a mixture of CsCl and $CsNiCl_3$ (in which there is octahedral coordination of Ni). By working in alcoholic solution salts of large organic cations can be prepared, for example, $(NEt_4)_2NiCl_4$ (blue, $\mu_{eff.}$, 3·9 BM) and $(As\phi_3Me)_2NiI_4$ (red, $\mu_{eff.}$, 3·5 BM). There is no Jahn–Teller distortion in the $NiCl_4^{2-}$ ion, which has a regular tetrahedral shape in $(As\phi_3Me)_2NiCl_4$[1] but is slightly flattened (two angles of 107° and four of 111°) in the (NEt_4) salt,[2] presumably the result of crystal packing forces. Ions NiX_4^{2-} have been prepared in which X is Cl, Br, I, NCS, or NCO.[3] Ions NiX_3L^- which have been studied include those in the salts $[NiBr_3(quinoline)](As\phi_4)$[4] and $[NiI_3(P\phi_3)][N(n\text{-}C_4H_9)_4]$.[5] In molecules NiX_2L_2 there may be considerable distortion from regular tetrahedral bond angles as, for example, Br–Ni–Br, 126° in $NiBr_2(P\phi_3)_2$,[6] and Cl–Ni–Cl, 123° and P–Ni–P, 117°, in $NiCl_2(P\phi_3)_2$.[7] We have already commented on the planar–tetrahedral isomerism of molecules NiX_2L_2 and of molecules NiL_2' containing certain bidentate ligands.

Some more exotic examples of molecules in which Ni forms tetrahedral bonds are shown at (a)–(c).

(15) AC 1967 **22** 780

(1) IC 1966 **5** 1498
(2) AC 1967 **23** 1064
(3) JINC 1964 **26** 2035
(4) IC 1968 **7** 2303
(5) IC 1968 **7** 2629

(6) JCS A 1968 1473
(7) JCS 1963 3625

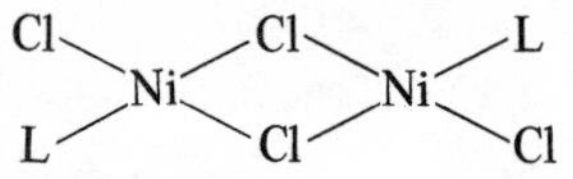

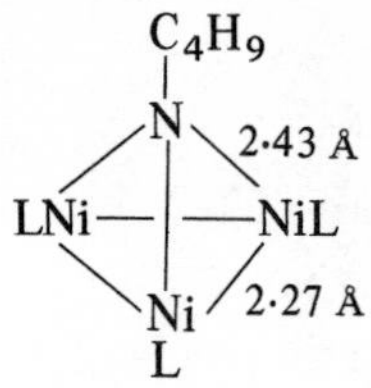

C_4H_9

N 2·43 Å

LNi——NiL

Ni 2·27 Å

L

(8) HCA 1962 **45** 647
(9) IC 1968 **7** 261

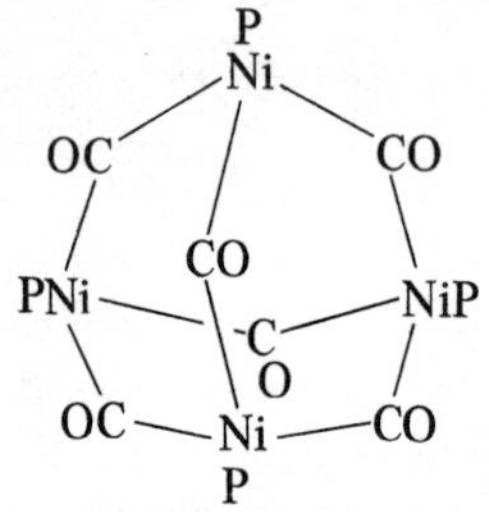

L = cyclobutadiene[8]

L = cyclopentadiene[9]

(a)

(b)

P
Ni

OC   CO

CO

PNi——   ——NiP

C
O

OC——Ni——CO
P

(10) JACS 1967 **89** 5366

$P = P(C_2H_4CN)_3$[10]

(c)

In $Ni_4(CO)_6[P(C_2H_4CN)_3]_4$ Ni acquires a closed shell configuration if the three metal–metal bonds (2·57 Å, compare 2·49 Å in the metal) are included with those to the four nearest neighbours (Ni–3 C, 1·89 Å, Ni–P, 2·16 Å), and for this reason Ni should perhaps be regarded as 7- rather than 4-coordinated in this compound.

(11) NW 1952 **39** 300;
JACS 1956 **78** 702

The bridged structure assigned to the anion in the red diamagnetic $K_4Ni_2(CN)_6$[11] does not appear to have been elucidated in detail.

Ni(II) *forming* 5 *bonds.* The two most symmetrical configurations, trigonal bipyramidal and square (tetragonal) pyramidal, are closely related geometrically, so that descriptions of configurations intermediate between the two extremes are sometimes rather arbitrary. However, complexes with geometries very near to both configurations are found for both diamagnetic and paramagnetic compounds of $Ni(II)$, and in one case (p. 966) two distinct configurations of the $Ni(CN)_5^{3-}$ ion are found in the same crystal. The choice of configuration is influenced by the geometry of the ligand if this is polydentate. Apart from the $Ni(CN)_5^{3-}$ ion nothing is known of the structures of ions NiX_5^{3-}. In salts such as $Rb_3Ni(NO_2)_5$ there may be bridging NO_2 groups in an octahedral chain, as in $[Ni(en)_2(NO_2)]^+$. Complexes with monodentate ligands include the diamagnetic molecule (a) with a nearly ideal trigonal bipyramidal configuration (though with $P\phi(OEt)_2$ ligands an intermediate configuration is adopted[1]) and the paramagnetic complex (b) ($\mu = 3\cdot7$ BM).[2]

(1) IC 1969 **8** 1084, 1090
(2) IC 1969 **8** 2734

Complexes with polydentate ligands include the types (c)–(f). We shall not discuss all these cases individually since the configuration often depends on the detailed structure of the (organic) ligand. For example, in derivatives of Schiff bases $R \cdot C_6H_3(OH)CH=N \cdot CH_2CH_2NEt_2$ (type (d)) the compound NiL_2 may be

970

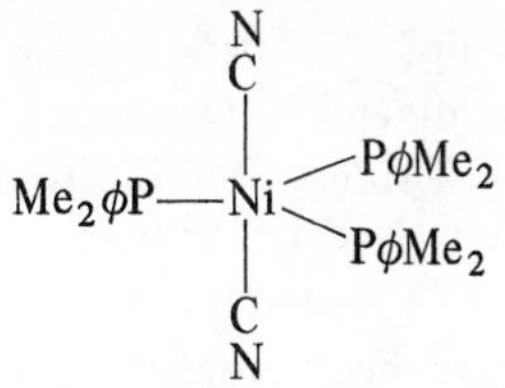

(a)

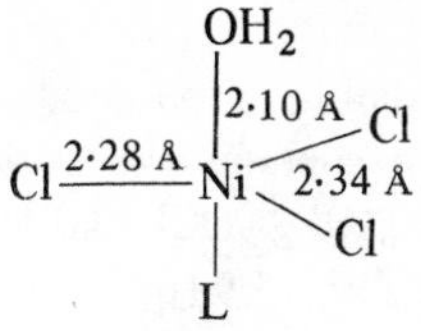

(b)

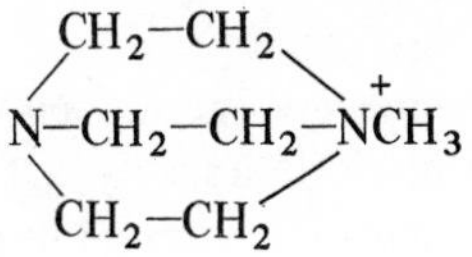

$$N-CH_2-CH_2-\overset{+}{N}CH_3$$
with CH_2-CH_2 bridges

(L)

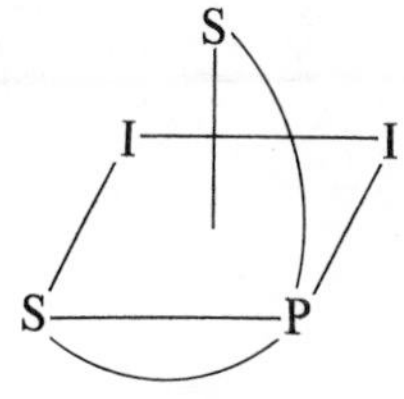

(c)

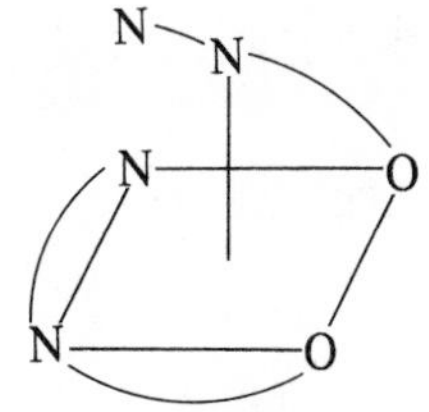

(d)

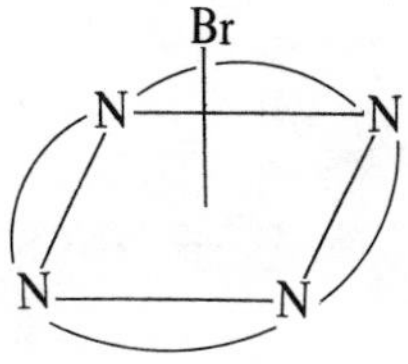

(e)

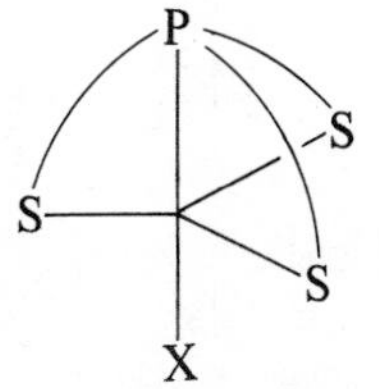

(f)

diamagnetic planar, paramagnetic 5-coordinated, or paramagnetic octahedral, depending on the nature of the substituent R.[3] The 'tripod-like' tetradentate ligands such as TSP and TAP tend to favour the trigonal bipyramidal configuration,

(TSP)

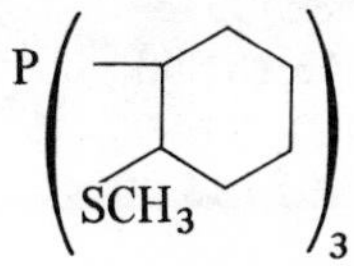

and $P[-(CH_2)_3As(CH_3)_2]_3$

(TAP)

as in the diamagnetic compounds [Ni(TSP)Cl] ClO$_4$[4] and [Ni(TAP)CN] ClO$_4$.[5] References to a number of square pyramidal complexes are given in reference (6).

The structure of NiBr$_2$(TAS),[7] in which TAS is the tridentate ligand

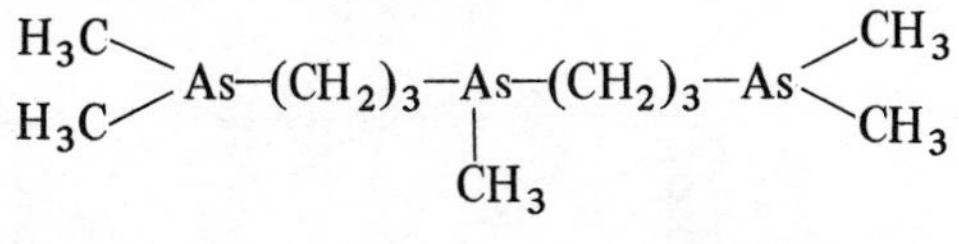

(TAS)

is described on p. 980, where it is compared with a 5-coordinated Pd complex with somewhat similar geometry.

(3) JACS 1965 87 2059

(4) IC 1969 8 1072
(5) JACS 1967 89 3424
(6) IC 1969 8 1915
(7) PCS 1960 415

971

Ni(II) *forming 6 octahedral bonds.* The simplest examples of octahedrally coordinated Ni^{2+} are the crystalline monoxide and the dihalides. Octahedral complexes mentioned elsewhere include the cation in $[Ni(H_2O)_6]SnCl_6$, the trimeric $[Ni(acac)_2]_3$[1a] and $Ni_6(CF_3COCHCOCH_3)_{10}(OH)_2(H_2O)_2$[1b] both of which are illustrated in Chapter 5 (Fig. 5.9 (p. 166) and Fig. 5.11 (p. 167)).

(1a) IC 1965 **4** 456
(1b) IC 1969 **8** 1304

Many octahedral paramagnetic complexes of Ni(II) have been studied, and

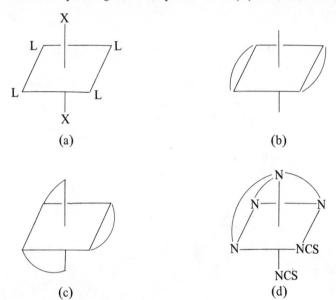

(a)　　　　　　　　　　(b)

(c)　　　　　　　　　　(d)

examples of the general types (a)–(d) include:

(a) Ni(pyrazole)$_4$Cl$_2$	AC 1969 **B25** 595
(b) Ni(acac)$_2$(pyr)$_2$	IC 1968 **7** 2316
Ni(en)$_2$(NCS)$_2$	AC 1963 **16** 753
(c) [Ni(en)$_3$](NO$_3$)$_2$	AC 1960 **13** 639
(d) Ni(tren)(NCS)$_2$	ACSc 1959 **13** 2009

In contrast to Ni(tu)$_4$Cl$_2$, which forms finite molecules in which there is an (unexplained) asymmetry (the lengths of the two *trans* Ni–Cl bonds being 2·40 and 2·52 Å),[2] the thiocyanate Ni(tu)$_2$(NCS)$_2$ consists of infinite chains of edge-sharing octahedra, (e);[3] in contrast to Ni(en)$_2$(NCS)$_2$, type (b), the nitrito compound Ni(en)$_2$ONO is built of chains of vertex-sharing octahedra, (f).[4]

(2) JCS 1963 1309
(3) AC 1966 **20** 349
(4) JCS 1962 3845

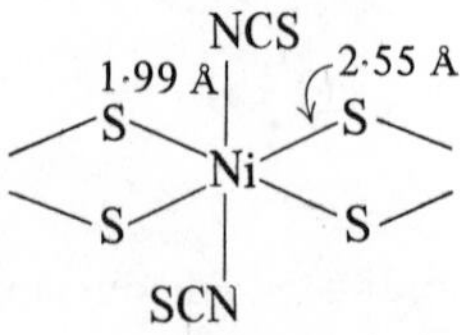

NCS
1·99 Å　　2·55 Å
S　　　S
　Ni
S　　　S
SCN

(e) 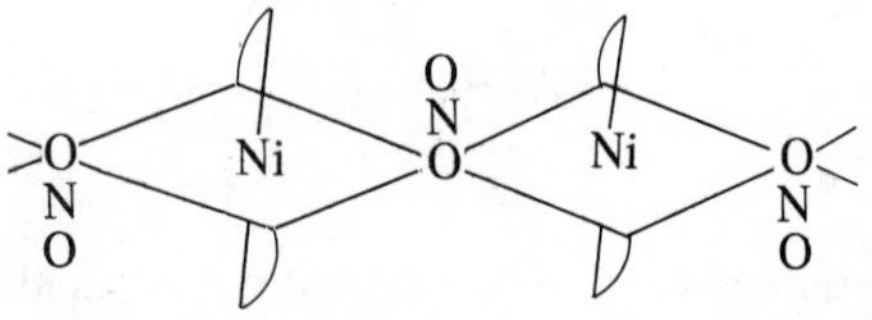　　　　　　　　　　(f)

The situation with regard to diamagnetic octahedral complexes is still unsatisfactory. Halogen compounds Ni(diarsine)$_2$X$_2$ behave in solution in CH$_3$NO$_2$ as uni-univalent electrolytes, but in crystals of the brown form of Ni(diarsine)$_2$I$_2$ there are octahedral molecules of type (b) in which Ni–As is 2·29 Å but Ni–I, 3·21 Å.[5] This Ni–I bond length may be compared with 2·55 Å in [NiI$_3$(Pϕ_3)]$^-$ (tetrahedral) and 2·54 Å in NiI$_2$[(C$_6$H$_5$)P(C$_6$H$_4$.SCH$_3$)$_2$] (square pyramidal, ref. (6) of previous section). The nature of the Ni–I bonds is uncertain, for the close approach of I to Ni is prevented by the methyl groups which project above and below the equatorial plane. Structural studies have apparently not been made of compounds such as [Ni(diarsine)$_3$](ClO$_4$)$_2$ which are often quoted as examples of octahedral spin-paired Ni(II). (The electronic structure of the paramagnetic [Ni(diarsine)$_2$Cl$_2$]Cl, containing one unpaired electron, is uncertain.[6])

(5) AC 1964 **17** 592

(6) JACS 1968 **90** 1067

Other oxidation states of Ni

The oxidation states other than II which are firmly established are 0, III, and IV.

Ni(0). Only tetrahedral complexes are known; they include Ni(CO)$_4$, Ni(PF$_3$)$_4$, Ni(CNR)$_4$, and the Ni(CN)$_4^{4-}$ ion in the yellow potassium salt. Structural information regarding these complexes is summarized in Chapter 22.

Ni(III). Compounds of Ni(III) may include NiBr$_3$(PEt$_3$)$_2$, formed by oxidation of *trans*-NiBr$_2$(PEt$_3$)$_2$, and Ni(diarsine)$_2$Cl$_3$, which results from the oxidation of the dichloro compound by O$_2$ in the presence of excess Cl$^-$ ions. Both these compounds have magnetic moments corresponding to 1 unpaired electron, but their structures are not yet known. For the former a trigonal bipyramidal configuration would be consistent with its dipole moment,[1] and this structure is likely in view of the structure of the compound Ni^{III}Br$_3$.P$_2'$.0·5 (Ni^{II}Br$_2$P$_2'$).C$_6$H$_6$,[2] in which P$'$ = P(C$_6$H$_5$)(CH$_3$)$_2$. This compound consists of *trans* planar molecules NiBr$_2$P$_2'$ with twice as many trigonal bipyramidal molecules NiBr$_3$P$_2'$ in which P occupies axial positions (Ni–P, 2·27 Å, Ni–Br, 2·35 Å). (There is a slight distortion of the latter molecules, with one equatorial bond about 0·03 Å longer than the other two and an angle of 133° opposite this longer bond. The Ni–P bonds are not significantly different in length from those in the square Ni(II) complex; the Ni–Br bonds are approximately 0·05 Å longer.) The diarsine complex is presumably [Ni(diarsine)$_2$Cl$_2$]Cl containing Ni(III) forming octahedral bonds, though there is some doubt as to whether such octahedral complexes should be regarded as compounds of Ni(III) since e.s.r. measurements suggest that they should perhaps be regarded as Ni(II) compounds in which the unpaired electron spends a large part of its time on the As atom.[3]

(1) ACSc 1963 **17** 1126
(2) IC 1970 **9** 453

(3) JACS 1968 **90** 1067

(4) ZaC 1961 **308** 179

(5) JSSC 1971 **3** 582

(6) JAP 1969 **40** 434

Structural information about simple Ni(III) compounds is limited to those containing F or O. K$_3$NiF$_6$ has been assigned the K$_3$FeF$_6$ structure and has a moment of 2·5 BM, intermediate between the values for low- and high-spin.[4] It contains octahedrally coordinated Ni(III). Oxides LnNiO$_3$ formed with the 4f elements have the perovskite structure,[5] while NiCrO$_3$ has a statistical corundum structure; it is concluded from the magnetic properties that this compound contains high-spin Ni(III).[6]

(7) ZaC 1949 **258** 221
(8) ZaC 1956 **286** 136
(9) AC 1951 **4** 148

Ni(IV). The only definite structural information relates to the red salts K_2NiF_6[7] and Cs_2NiF_6.[8] They are diamagnetic and have the K_2SiF_6 structure, with octahedral coordination of Ni(IV). The oxide $BaNiO_3$ has been quoted for many years as an example of a Ni(IV) compound,[9] but there is still doubt about its composition ($BaNiO_{2.5}$?) and the interpretation of its magnetic properties.

The diarsine cation mentioned above can be further oxidized to $[Ni(diarsine)_2Cl_2]^{2+}$ which is isolated as the deep-blue perchlorate, possibly containing Ni(IV).

The structural chemistry of Pd and Pt

Planar complexes of Pd(II) *and* Pt(II)

As long ago as 1893 Werner suggested that the four atoms or groups attached to divalent Pd or Pt atoms in certain molecules were coplanar with the metal atom, this suggestion being made to account for the existence of compounds such as $Pt(NH_3)_2Cl_2$ in two isomeric forms. For coplanar bonds there would be *cis* and

trans forms whereas if the bonds were arranged tetrahedrally there would be only one form of such a molecule. It should be noted that no amount of evidence from chemical reactions, *cis–trans* isomerism, or optical activity can prove conclusively the coplanar arrangement of four *bonds* from a metal atom. For example, even if the approximate coplanarity of the four ligands is admitted, this does not prove that the central metal atom lies in the same plane; it could be situated at the vertex of a tetragonal pyramid. In pre-structural days the only way of attempting to prove this unusual bond arrangement was to study geometrical and optical isomerism, and much ingenuity was used in making appropriate compounds. In addition to the *cis–trans* isomers of compounds PtX_2R_2 three isomers of a planar complex Mabcd are possible as compared with only one form of a tetrahedral complex, and all three isomers of $[Pt(NH_3)(NH_2OH)(NO_2)(C_5H_5N)]NO_2$ have been reported. [Some confusion arose in cases where two forms of a compound have the same empirical formula but different structures, for example, $Pt(NH_3)_2Cl_2$ and $[Pt(NH_3)_4](PtCl_4)$.] More complex examples of *cis–trans* isomers include the compounds of Pd and Pt with two molecules of glycine or unsymmetrical glyoximes. A molecule such as $Pd(glycine)_2$ (a), with unsymmetrical rings, would

(a)

be enantiomorphic if the bonds from the metal atom are disposed tetrahedrally, whereas if these bonds are coplanar with the metal atom only *cis–trans* isomerism is possible. Attempts to resolve certain compounds of this type have been made and have been unsuccessful, but the failure to effect a resolution is clearly of doubtful value as evidence for the disposition of the bonds from the metal atom. This difficulty was overcome in an ingenious way by Mills and Quibell (1935) who prepared a compound which would be optically active if the bonds from the central metal atom were coplanar and non-resolvable if these bonds were arranged tetrahedrally. The following sketches show that if the Pt bonds are coplanar, (b), the ion possesses neither a plane nor a centre of symmetry, whereas if the Pt bonds are tetrahedral, (c), then the planes of the two rings are perpendicular to one another and the ion possesses a plane of symmetry.

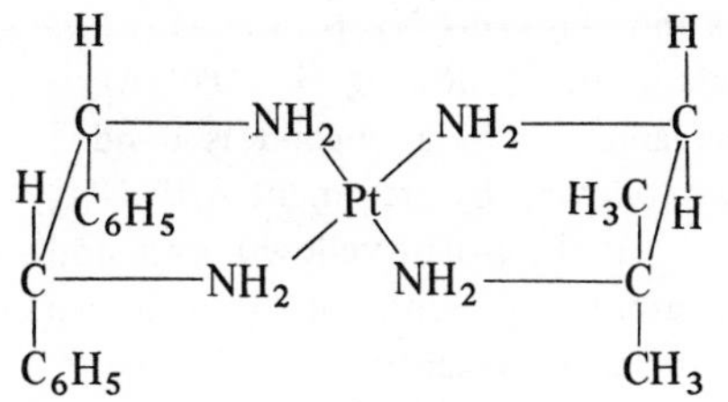

(b) 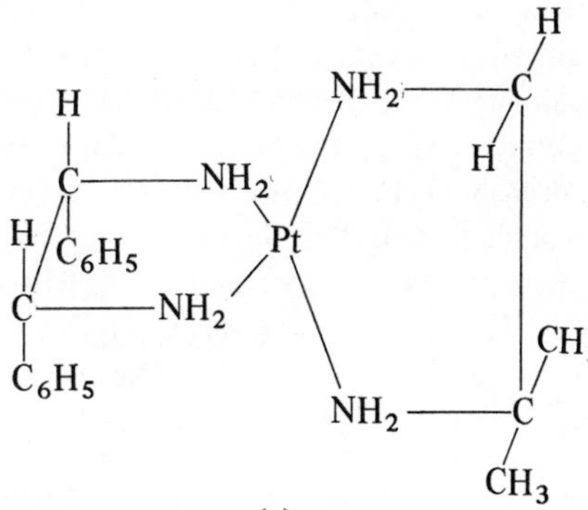   (c)

The compound,

$$[Pt(NH_2 . CH . C_6H_5 . CH . C_6H_5 . NH_2)(NH_2 . C(CH_3)_2CH_2NH_2)] Cl_2,$$

was resolved into highly stable optical antimers. The corresponding ion containing Pd was also shown, in the same way, to have coplanar rings. Here again it should be noted that the resolution of these compounds is not conclusive proof of the coplanar arrangement of the four Pt or Pd bonds. The metal atom could lie above the plane of the four NH_2 groups, and the molecule would still be optically active. Nevertheless, as was pointed out by Mills and Quibell, certain simpler complexes should be resolvable if the metal bonds were pyramidal.

Dipole moment measurements make possible a distinction between *cis* and *trans* planar molecules or ions, the *cis* isomer having a larger dipole moment and the *trans* a smaller or zero moment, depending on the nature of the ligands, for example:

	μ		
trans $PtBr_2(Et_3P)_2$	0	*cis* $PtBr_2(Et_3P)_2$	11·2 D
$PtCl_2(Pr_2S)_2$	2·35 D	$PtCl_2(Pr_2S)_2$	9·5

(In the phosphine compounds the P bonds are tetrahedral, so that the resultant moment of the R_3P group is directed along the Pt–P bond, whereas the S bonds in the dialkyl sulphide groups are pyramidal and the resultant moment of the ligand is not directed along the Pt–S bond.) Note that the existence of only one isomer with a high dipole moment is not proof of a planar configuration, for a tetrahedral molecule can also have a high moment. (Nickel compounds NiX_2L_2 do not exhibit *cis-trans* isomerism, but some exhibit planar–tetrahedral isomerism, as already noted.)

975

In 1922 the planar structure of the ions $PdCl_4^{2-}$ and $PtCl_4^{2-}$ was proved by the determination of the structures of the potassium salts. The coplanar arrangement of four bonds from Pd(II) and Pt(II) has since been demonstrated in many crystals. Binary compounds (chlorides, oxides, sulphides, etc.) are dealt with elsewhere; here we give examples of finite molecules and ions.

Pd(II) *compounds*

The structures of square planar mononuclear complexes have been established by diffraction studies of compounds such as K_2PdCl_4, $[Pd(NH_3)_4]Cl_2 . H_2O$, and numerous complex cyanides (many of which are isostructural with the Ni and Pt analogues), for example, $Ca[Pd(CN)_4] . 5 H_2O$ and $Na_2[Pd(CN)_4] . 3 H_2O$. Compounds having points of special interest include the following. In $[Pd(en)_2]$-$[Pd(en)S_2O_3]$[1] both ions are square planar, and the S_2O_3 ligand is bonded through S, (a). The dissolution of Pd in HNO_3 followed by treatment with NH_3 does not give the expected $[Pd(NH_3)_4](NO_3)_2$ but the bright-yellow compound $[Pd(NH_3)_3NO_2]_2[Pd(NH_3)_4](NO_3)_4$ which contains two kinds of planar ion in the proportion of $2 : 1$.[2] The n-propyl mercaptide is hexameric,[3] like the Ni compound illustrated in Fig. 27.10, and the molecule (b) is 'basin-shaped'.[4] In the

(1) AC 1970 **B26** 1698

(2) IC 1971 **10** 651
(3) AC 1968 **B24** 1623;
AC 1971 **B27** 2292
(4) AC 1969 **B25** 1659

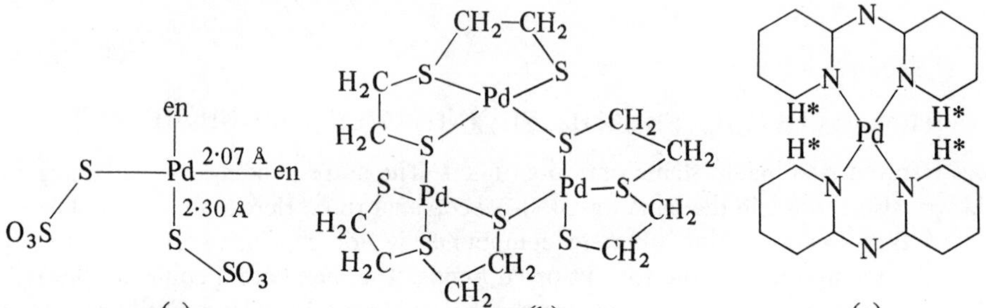

(5) AC 1965 **18** 845

(a)   (b)   (c)

2,2'-dipyridyl imine complex (c),[5] in which there is interference between the H^x atoms, the bond arrangement around Pd remains square planar and the ligands distort, in contrast to the Ni methenato complex mentioned earlier (p. 968).

Palladium forms complexes such as the dithio-oxalate ion, $Pd(N_2S_2H)_2$, glyoximes, and phthalocyanine which are structurally similar to the compounds of Ni and Pt, and also numerous bridged compounds. Early X-ray studies established the planar *trans* configuration of (d) and (e):

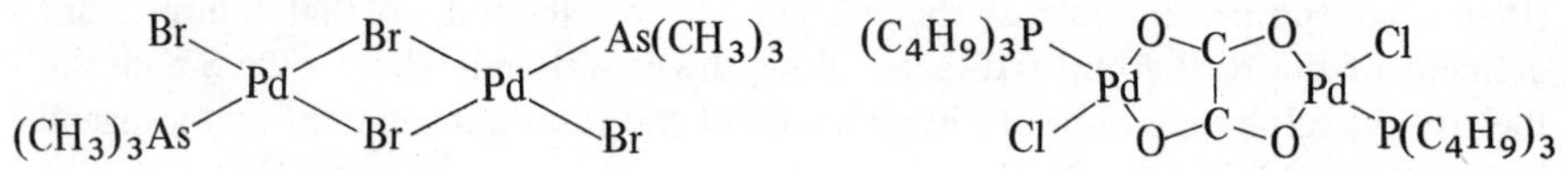

(d)   (e)

but there have been more recent studies of bridged compounds of Pt to which we refer shortly.

976

Examples of the rare compounds in which Pd(II) does *not* form square coplanar bonds include $[PdAlCl_4(C_6H_6)]_2$ and $[PdAl_2Cl_7(C_6H_6)]_2$ which are formed when $AlCl_3$, Al metal, and $PdCl_2$ are reacted together in boiling benzene.[6] The molecules have the structures sketched at (f) and (g). These molecules are notable

(6) JACS 1970 **92** 289

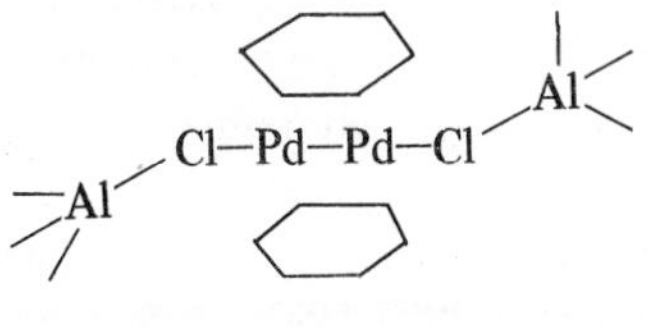

(f)

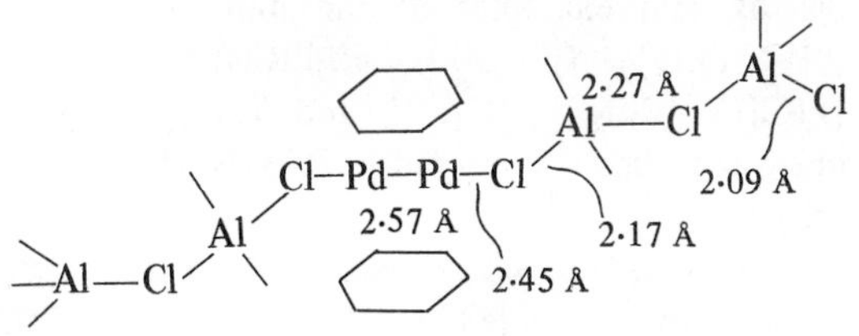

(g)

for the short Pd—Pd bonds, which are shorter than in the metal (2·75 Å), and long Pd—Cl bonds.

Compounds of Pt(II)

There is no evidence for the formation of more than four bonds (which are coplanar) except in the cases noted on pp. 980 and 981. As in the case of Pt(IV) (p. 983) the formulae of certain compounds suggest other coordination numbers, but structural studies have confirmed 4-coordination. In $[Pt(acac)_2Cl]K$,[1] (a), 5-coordination is avoided by coordination to C, and an X-ray study of the compound originally formulated as $[Pt(NH_3)_4(CH_3CN)_2]Cl_2 . H_2O$ shows that

(1) JCS A 1969 485

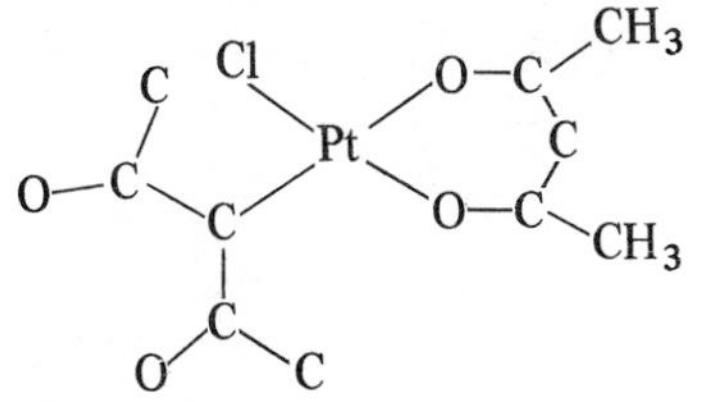

(a)

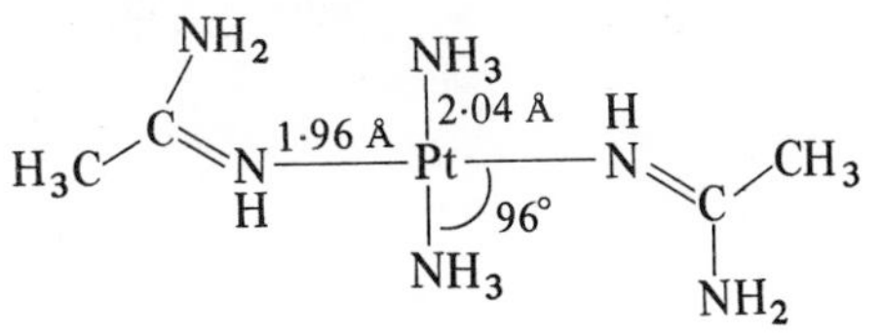

(b)

(2) JINC 1962 **24** 801

this is not an example of 6-coordinated Pt(II) but is $\{Pt(NH_3)_2[CH_3 . C(NH_2) . NH]_2\}Cl_2 . H_2O$.[2] There is planar 4-coordination of the metal by two NH_3 and two acetamidine molecules, (b).

(3) JINC 1962 **24** 791, 797; AC 1964 **17** 1517

The compounds of Ni, Pd, and Pt of the type $M(diarsine)_2X_2$ appear to behave in solution as uni-univalent electrolytes, indicating a close association of one halogen atom with the metal, $[M(diarsine)_2X]^+$. In the crystalline state[3] they form very distorted octahedral units with Ni—As, 2·3 Å, and Pd(Pt)—As, 2·4 Å, but M—X very much longer than a normal covalent bond. In $Pt(diarsine)_2Cl_2$ the length of the Pt—Cl bond is 4·16 Å, suggesting Cl^- ions resting on the four CH_3 groups projecting out of the equatorial plane of the two diarsine molecules. The *shorter* M—I distances (3·2, 3·4, and 3·5 Å for the Ni, Pd, and Pt compounds) may be due to considerable polarization of the I^- ions; they are intermediate between the covalent

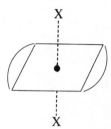

radius sums and the sums of the van der Waals (or ionic) radii. The Ni–I distance has been compared with the much smaller Ni–I distances in paramagnetic tetrahedral and square pyramidal molecules on p. 973.

Like Pd(II), Pt(II) forms numerous square planar complexes with monodentate ligands, and examples of the more recent structural studies include those of PtHBr(PEt$_3$)$_2$ (p. 300), *cis*-PtCl$_2$(PMe$_3$)$_2$,[4] *cis*- and *trans*-PtCl$_2$(NH$_3$)$_2$,[5] [PtCl(CO)(PEt$_3$)$_2$] BF$_4$,[6] and (PtCl$_3$NH$_3$)K . H$_2$O.[7] Complexes containing bidentate ligands include Pt(S$_2$N$_2$H)$_2$,[8] the dimethyl glyoxime,[9] and diglycine.[10]

In certain complexes there is observed a lengthening of the Pt–X bonds which are *trans* to a π-bonded ligand ('*trans*-effect'). This is seen in PtCl$_2$(PMe$_3$)$_2$, (c), and also in the bridged molecule (d),[11] where it leads to an unsymmetrical bridge (and three different Pt–Cl bond lengths). A similar effect is observed in Pt$_2$Cl$_4$(PPr$_3$)$_2$.[12]

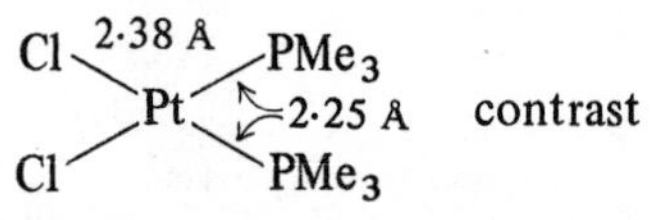

(4) IC 1967 6 725
(5) JCS A 1966 1609
(6) JACS 1967 89 3360
(7) IC 1970 9 778
(8) JINC 1958 7 421
(9) AC 1959 12 1027
(10) AC 1969 B25 1203
(11) JCS A 1970 168
(12) AC 1969 B25 1760

(c)

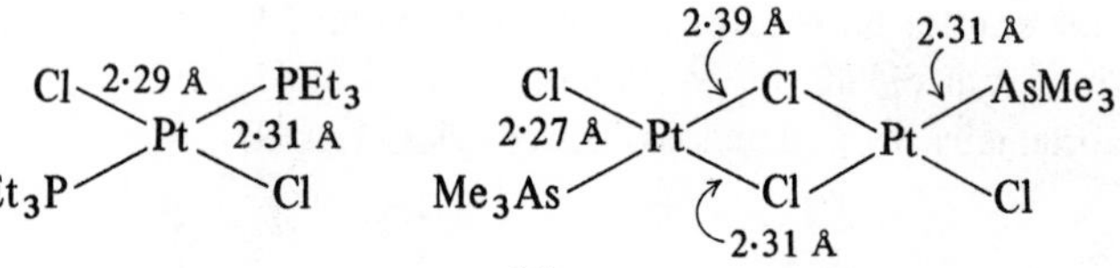

(d)

Bridged compounds of Pd *and* Pt

Solutions containing PdX$_4^{2-}$ or PtX$_4^{2-}$ ions react with tetramethyl ammonium salts to give compounds with empirical formulae of the type (NR$_4$)PdX$_3$ or (NR$_4$)PtX$_3$ in which X is Br or Cl. An X-ray study of [N(C$_2$H$_5$)$_4$]PtBr$_3$[1] showed that it contains a bridged ion and accordingly salts of this kind should be formulated

(1) AC 1964 17 587

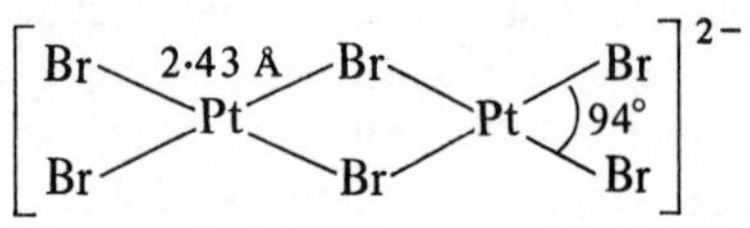

(NR$_4$)$_2$(Pt$_2$X$_6$). By the reaction of PtCl$_2$ with melted L$_2$PtCl$_2$ or of *cis* L$_2$PtCl$_2$ with K$_2$PtCl$_4$ in solution bridged compounds Pt$_2$Cl$_4$L$_2$ are obtained, L being one of the neutral ligands CO, C$_2$H$_4$, NH$_3$, P(OR)$_3$, PR$_3$, AsR$_3$, SR$_2$, etc. We have already noted that Pd forms similar compounds, and that the molecule Pd$_2$Br$_4$-

$[As(CH_3)_3]_2$ is planar, apart from the methyl groups, with the *trans* configuration I:

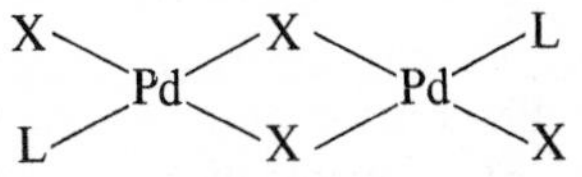

I

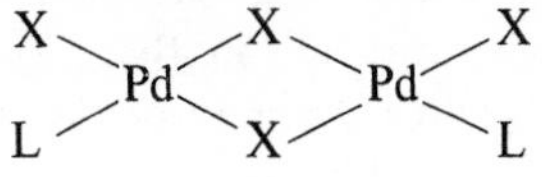

II

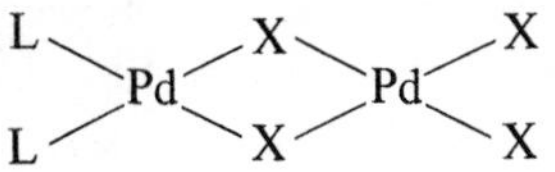

III

Although three isomers are theoretically possible, all the compounds in which L is one of the ligands listed above and X is a halogen are known in one form only, the *trans* isomer I, as shown by dipole moments.

The bridge between the metal atoms may also be formed by

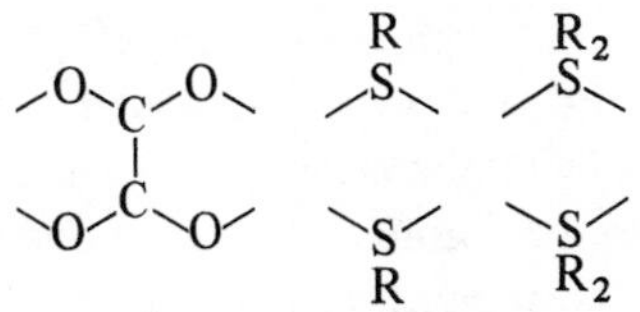

or

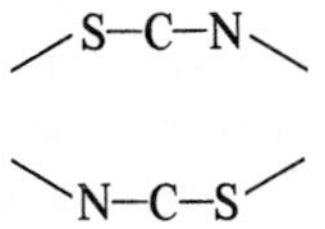

Two isomers of the ethyl-thiol compound $(Pr_3P)_2Pt_2(SR)_2Cl_2$ have been isolated, with dipole moments 10·3 D (stable form) and zero, from which it follows that the stable form is the *cis* isomer IV and the labile isomer V:[(2)]

(2) JCS 1953 2363

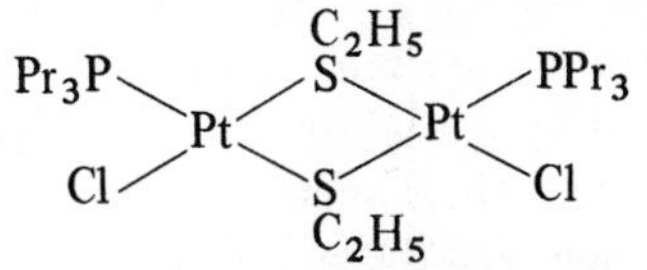

IV

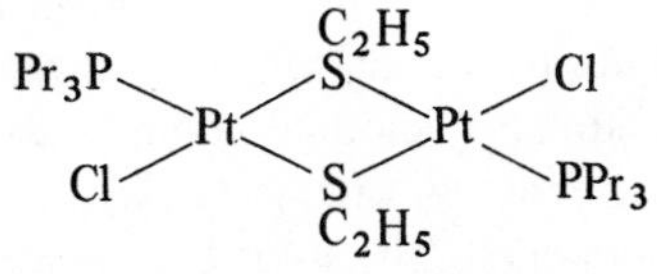

V

It is interesting that in $Pt_2Br_4(SEt_2)_2$ the bridge consists of two $S(C_2H_5)_2$ groups whereas in the closely related $Pd_2Br_4(SMe_2)_2$ the two Br atoms form the bridge:[(3)]

(3) JCS A 1968 1852

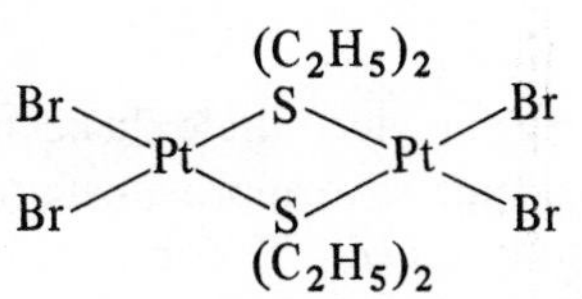

contrast

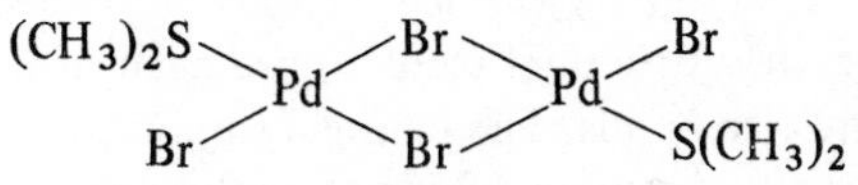

X-ray examination of the two isomers of $Pt_2Cl_2(SCN)_2(Pr_3P)_2$[(4)] shows that they have the structures VI and VII:

(4) JCS A 1970 2770

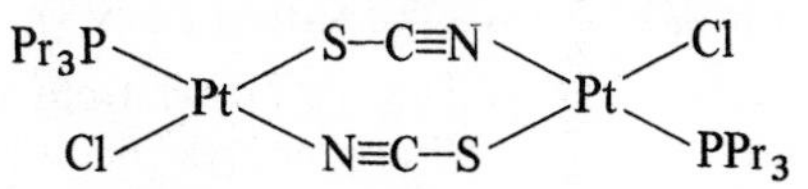

VI

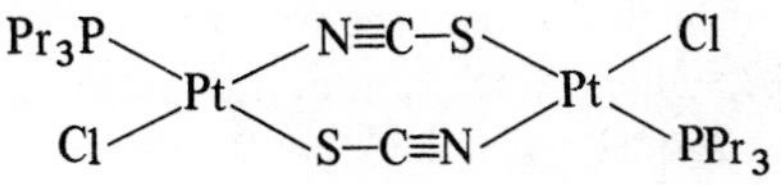

VII

Some highly-coloured compounds of Pt

A number of crystalline salts consist of alternate planar ions of two kinds stacked vertically above one another. The anions and cations of 'Magnus's green salt', $[Pt(NH_3)_4]PtCl_4$, are respectively red and colourless in solution, and the green colour and abnormal dichroism of the crystals are associated with the metal–metal bonds, which are of the same length (3·25 Å) as Ni–Ni in Ni dimethylglyoxime. Many salts with the same structure are pink, for example, $[Pd(NH_3)_4]PdCl_4$ and $[Pd(NH_3)_4]PtCl_4$; the unusual optical properties arise only when both ions contain Pt(II) and when Pt–Pt is short. Thus $[Pt(NH_3)_4]PtCl_4$ and $[Pt(CH_3NH_2)_4]PtCl_4$ are green but $[Pt(NH_2C_2H_5)_4]PtCl_4$ is pink. In the last compound all atoms of the cation are not coplanar, the planes PtN_4 and $PtCl_4$ being inclined at 29° and the Pt–Pt distance is 3·62 Å.[1] Of interest in this connection are the colours of hydrated platinocyanides such as $K_2Pt(CN)_4 . 3 H_2O$ (yellow) and $MgPt(CN)_4 . 7 H_2O$ (red-purple, green surface reflection from certain faces) which crystallize from colourless aqueous solutions.

In another type of salt intense colour results from interaction between Pt(II) and Pt(IV). The deeply-coloured salt with the composition $Pt(NH_3)_2Br_3$ contains planar $Pt^{II}(NH_3)_2Br_2$ and octahedral $Pt^{IV}(NH_3)_2Br_4$ molecules.[2] These are arranged in columns in the crystal (Fig. 27.11), the planar and octahedral molecules alternating. The Pt(II) atom appears to be forming two additional weak bonds (3·03 Å) to Br atoms of $Pt^{IV}(NH_3)_2Br_4$ molecules; compare Pt-Br within the complexes, 2·50 Å. Other salts of this kind include $[Pt^{II}(en)Br_2][Pt^{IV}(en)Br_4]$ [3] and Wolfram's red salt, $[Pt^{II}(C_2H_5NH_2)_4]^{2+}[Pt^{IV}(C_2H_5NH_2)_4Cl_2]^{2+}$-$Cl_4 . 4 H_2O$.[4] In the latter the Cl^- ions (and H_2O molecules) lie between the chains, which are of the same general type as those of Fig. 27.11. The structure of the (anhydrous) Br analogue of Wolfram's salt has also been determined.[5]

There are also coloured compounds formed by partial oxidation of $K_2Pt(CN)_4$ or partial reduction of $K_2[Pt(CN)_4X_2]$, in which the formal oxidation number of Pt is non-integral. The Pt atoms are all equivalent and the complexes are stacked in columns with very short Pt–Pt distances (around 2·85 Å), not very much greater than in the metal (2·78 Å). Examples are $[K_2Pt(CN)_4] . Cl_{0.32} . 2·6 H_2O$ and $Mg_{0.82}[Pt(C_2O_4)_2] . 5·3 H_2O$.[6]

Pd(II) *and* Pt(II) *forming 5 bonds*

Very few structural studies have been made of complexes in which these elements form five bonds. The configuration of the Pd complexes appears to approximate to tetragonal pyramidal but with one abnormally long bond. The molecule $PdBr_2L_3$,[1] in which L is the phosphine shown in Fig. 27.12(a), has a configuration close to tetragonal pyramidal but with the axial Pd–Br bond much longer than the equatorial ones. In addition, Br^2 lies below the equatorial plane, and in this respect the structure resembles that of the Ni compound $NiBr_2(TAS)$[2] (b), where TAS is the tridentate ligand shown on p. 971. The distortion from tetragonal pyramidal configuration is much greater in the Ni compound and is tending towards trigonal bipyramidal (when the angles 95°, 111°, and 154° would all be 120°). There is apparently no distortion of this type in the ion

(1) AC 1971 **B27** 480

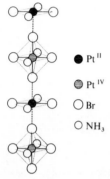

● Pt^{II}

⊖ Pt^{IV}

○ Br

○ NH$_3$

FIG. 27.11. Portion of the infinite chain in $Pt(NH_3)_2Br_2 . Pt(NH_3)_2-Br_4$.

(2) AC 1958 **11** 624

(3) JACS 1961 **83** 2814

(4) AC 1961 **14** 475

(5) AC 1966 **21** 177

(6) ZaC 1968 **358** 67, 97

(1) JCS 1964 1803

(2) PCS 1960 415

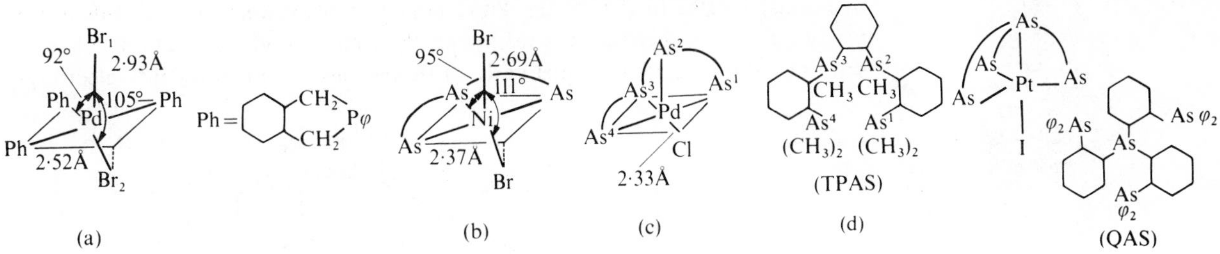

FIG. 27.12. 5-coordinated complexes of Pd, Ni, and Pt (see text).

[Pd(TPAS)Cl]$^+$,[3] (c), containing the tetradentate ligand TPAS, but here also one of the five bonds (Pd–As^2, 2·86 Å) is much longer than normal (mean of other three, 2·37 Å).

(3) JCS 1967 1650

Another example of the formation by Pd(II) of four stronger and one weaker bond is provided by crystalline *trans* PdI$_2$(PMe$_2\phi$)$_2$.[4] Here the complex is a square planar one (Pd–P, 2·34 Å, Pd–I, 2·63 Å), and the molecules are arranged so that on one side Pd has an I atom of another molecule as its next nearest neighbour (at 3·28 Å); on the other side of the square planar coordination group Pd has two H atoms of different C$_6$H$_5$ groups as next nearest neighbours. This case is, of course, not strictly comparable with (a) and (c) where Pd forms five bonds within each finite molecule or ion.

(4) CC 1965 237

Pt(II) shows as much reluctance as Pd(II) to form five bonds, and the only examples of 5-covalent Pt(II) are both special cases. One is the ion (d) of Fig. 27.12 in [Pt(QAS)I](Bϕ_4),[5] of which no details are available; it is described as trigonal bipyramidal, with normal single covalent bonds. Here the 3-fold symmetry of the tetradentate ligand might be expected to favour the trigonal bipyramidal structure. The other compound of which a preliminary study has been made is (ϕ_3PMe)$_3$ [Pt(SnCl$_3$)$_5$],[6] in which the five ligands are attached to Pt by Pt–Sn bonds. Here also the configuration is described as trigonal bipyramidal, but no details are available.

(5) PCS 1961 170

(6) JACS 1965 87 658

Pd(II) and Pt(II) forming 6 octahedral bonds

The reluctance of Ni(II) to form low-spin octahedral complexes, which has already been noted, is also shown by Pd(II) and Pt(II). There is octahedral coordination of Pd(II) in PdF$_2$ (rutile structure) and in Pd^{II}(Pd^{IV}F$_6$), but the formation of six covalent bonds would require the use of a d orbital of the outermost shell in addition to one $(n-1)$d, one ns, and three np, possibly as four coplanar dsp^2 and two pd hybrids. Certainly the two bonds completing octahedral coordination of low-spin complexes of these elements are much longer than the normal single bonds. Known examples have 4 As coordinated to the metal, and it has been suggested that with a planar arrangement of 4 As around M back donation of electrons from the metal (increasing its positive charge) enables the metal atom to bond to more highly polarizable anions such as I$^-$. With only 2 As bonded to the metal there is not sufficient back donation, and we find in (a)[1] only four bonds

(1) AC 1970 **B26** 1655

(2) JINC 1962 **24** 797

of normal length. In (b),[2] the Pd–I bonds are very weak, though this may be partly due to the fact that I is resting on the four methyl groups. In nitrobenzene solution $Pt(diarsine)_2I_2$ apparently behaves as a uni-univalent electrolyte,

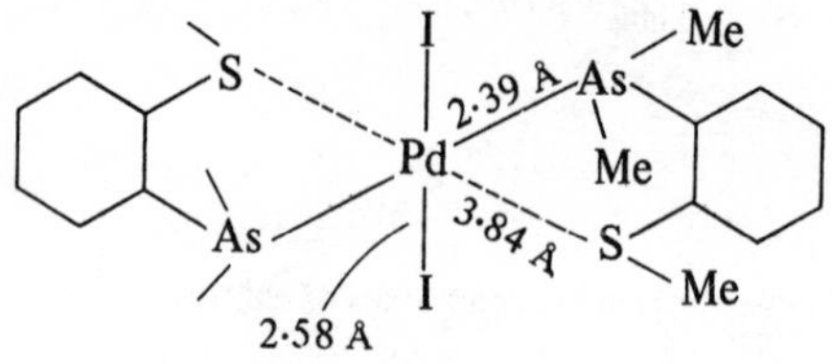

(a)

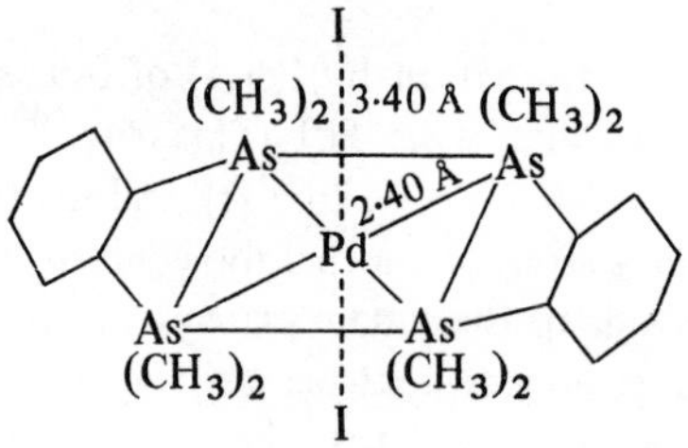

(b)

(3) JINC 1962 **24** 791
(4) AC 1964 **17** 1517

$[Pt(diarsine)_2I]$ I. Its structure is very similar to that of the Pd compound, with Pt–I, 3·50 Å.[3] However, in the dichloro compound, $Pt(diarsine)_2Cl_2$,[4] the Pt–Cl distance is 4·16 Å, which can only be interpreted as indicating an ionic bond. As noted earlier, the Ni compounds present a similar picture

Trimethyl platinum chloride and related compounds

These compounds are of quite unusual interest because their empirical formulae suggest coordination numbers 4, 5, and 7 for Pt^{IV}. Platinum forms a number of compounds $Pt(CH_3)_3X$, in which X is Cl, I, OH, or SH, which exist as tetrameric molecules (Fig. 27.13(a)) in which Pt and X atoms occupy alternate vertices of a (distorted) cube. These compounds are notable for the facts that the hydroxide was

(1) AC 1968 **B24** 287
(2) AC 1968 **B24** 157
(3a) JACS 1947 **69** 1561
(3b) JCS A 1971 90

(4) IC 1968 **7** 2165

formerly described as the tetramethyl compound[1] and the iodide as $Pt_2(CH_3)_6$.[2] In the early study of $[Pt(CH_3)_3Cl]_4$ Pt–Cl was found to be 2·48 Å;[3a] the value, 2·58 Å determined in $[Pt(C_2H_5)_3Cl]_4$[3b] should be more reliable. The following figures for the hydroxide show that there is some distortion from the highest possible cubic symmetry:[4]

O–Pt–O	78°	Pt–C	2·04 Å
C–Pt–C	87°	Pt–O	2·22
Pt–O–Pt	101°	Pt–Pt	3·43

982

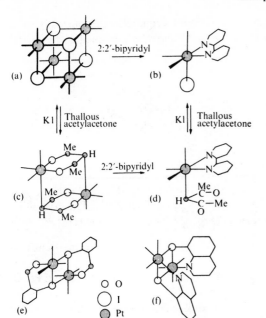

FIG. 27.13. Trimethylplatinum halides and related compounds.

○ O

○ I

● Pt

· C

A neutron diffraction study of this compound has also been made.[5] Since OH or Cl together provide 5 electrons (one normal and two dative covalent bonds) Pt acquires the same electronic structure as in $PtCl_6^{2-}$.

The bipyridyl derivative (Fig. 27.13(b)) is a normal monomeric octahedral Pt^{IV} molecule, but the empirical formulae of the β-diketone compounds $Pt(CH_3)_3(R . CO . CH . CO . R)$, (c),[6] and $Pt(CH_3)_3(bipyr)(R . CO . CH . CO . R)$, (d),[7] would suggest 5- and 7-coordination of the metal respectively, since β-diketones normally chelate through both O atoms. In fact (c) is a dimer in which each diketone molecule chelates to a Pt atom and is also attached to the other Pt atom through the active methylene carbon atom. This Pt—CH bond is very much longer (2·4 Å) than the Pt—CH$_3$ bonds (2·02 Å). (The 6-rings are actually boat-shaped.) In the molecule (d) use of the O atoms of the diketone would give 7-coordination of Pt^{IV}; instead, the metal is bonded only to the CH group, so that there is octahedral bonding by Pt as in (a), (b), and (c). The bond to the methylene carbon is similar to that in (c), namely, 2·36 Å. In $[(CH_3)_3Pt(acac)]_2(en)$[8] the ethylenediamine bridge completes the octahedral coordination group of each Pt atom, and acac coordinates to each Pt through its O atoms in the normal way:

(5) JOC 1968 **14** 447

(6) PRS A 1969 **254**, 205, 218; JCS A 1969 2282 (n.d.)
(7) PRS A 1962 **266** 527

(8) JCS 1965 630

acac

H_3C — Pt — NH_2 — CH_2 — CH_2 — NH_2 — Pt — CH_3

H_3C — Pt — CH_3

CH_3 CH_3

acac

The molecules of salicylaldehyde, I, and 8-hydroxyquinoline, II, normally form chelate compounds in which two O atoms or O and N are bonded to the metal atom:

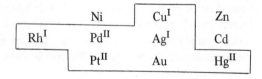

(I)   (II)

(9) JCS A 1967 1955
(10) JCS 1965 6899

so that the compounds with the empirical formulae $(C_7H_5O_2)Pt(CH_3)_3$ and $(C_9H_6NO)Pt(CH_3)_3$ would appear to contain 5-coordinated Pt^{IV}. The crystalline compounds are composed of dimers, (Fig. 27.13(e)),[9] and (f),[10] both containing $Pt{<}^O_O{>}Pt$ bridges in which an O atom of the ligand forms bonds to both Pt atoms. In (f) the bond lengths are: Pt–O, 2·24 Å, Pt–N, 2·13 Å, and Pt–C, 2·06 Å.

In all these compounds there is octahedral coordination of Pt^{IV} which is achieved in a variety of ways in compounds which present the possibility of coordination numbers of 4, 5, and 7.

Olefine compounds

It was originally thought that in coordination compounds a bond is formed between an atom with one or more lone pairs of electrons and a metal atom requiring electrons to complete a stable group of valence electrons. This does not, however, account for the formation of compounds such as, for example, $(C_2H_4PtCl_3)K$, by olefines (which have no lone pairs) or for the fact that basicity does not run parallel with donor properties in series of ligands, as it might be expected to do. For example, basicity falls rather rapidly in the series NH_3, PH_3, AsH_3, SbH_3, but PR_3 and AsR_3 form complexes more readily than NR_3 (R is an organic radical).

The elements forming reasonably stable olefine compounds are those with filled d orbitals having energies close to those of the s and p orbitals of the valence shell:

	Ni	Cu^I	Zn
Rh^I	Pd^{II}	Ag^I	Cd
	Pt^{II}	Au	Hg^{II}

Examples include: $C_2H_4Pt(NH_3)Cl_2$, C_2H_4CuCl, $C_6H_{10}Ag^+ \cdot$ aq., $C_2H_4Hg^{2+} \cdot$ aq., and bridged compounds $(diene)_2Rh_2^ICl_2$ where diene indicates, for example, cycloocta-1:5-diene. It is unlikely that the bond between C_2H_4 and Pt in $(C_2H_4PtCl_3)K$ is formed by donation of π electrons to Pt, for C_2H_4 does not combine even at low temperatures with an acceptor molecule such as $B(CH_3)_3$, whereas amines readily do so. Similarly, the P atom in PF_3 forms no compound with the very strong acceptor BF_3, showing that its lone pair is very inert, yet it combines with Pt to form the stable volatile carbonyl-like compounds $(PF_3)_2PtCl_2$ and $(PF_3PtCl_2)_2$. Also, $Ni(PF_3)_4$ has been prepared as a volatile liquid resembling $Ni(CO)_4$. These facts suggest the possibility of a similar type of bonding in compounds in which C_2H_4, CO, PF_3, and possibly other ligands are attached to Pt and similar metals, and suggest that an essential part is played by d orbitals.

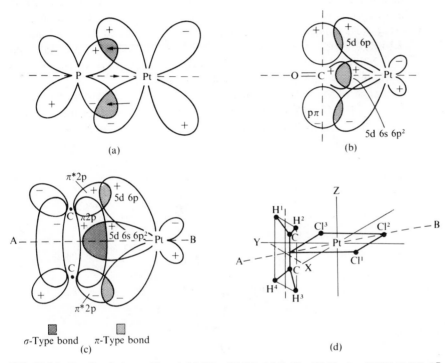

FIG. 27.14. Bonding between Pt and (a) PF_3, (b) CO, (c) C_2H_4. (d) The ion $[Pt(C_2H_4)Cl_3]^-$.

In the free PF_3 molecule the inertness of the lone pair is due to the withdrawal of electrons by the very electronegative F. It is supposed that there is interaction between the filled d orbitals of Pt and the d orbitals of P (Fig. 27.14(a)), as suggested earlier by Pauling for the interaction of CO with Pt (Fig. 27.14(b)), this d_π bonding being assisted by the electronegative F atoms attached to P. At the same time this process would tend to make the lone pair of electrons on the P atom more readily available for σ bond formation. The Pt—CO bond may be described as a combination of the ordinary dative (σ) bond, due to overlap of the carbon orbital containing the lone pair and an empty $5d6s6p^2$ orbital, with a π bond resulting from overlap of a *filled* orbital of Pt and the empty p-molecular orbital of CO. As regards the nature of the filled orbital used for the π bonding, it is thought likely that some sort of hybrid 5d6p orbital will be preferred to a simple 5d orbital.

985

A rather similar combination of σ and π bonds may be postulated for the C_2H_4–Pt bond, except that here both the orbitals of C_2H_4 are π orbitals, namely, the $2p\pi$ 'bonding' orbital containing the two π electrons and the (empty) $2p\pi^*$ 'antibonding' orbital (Fig. 27.14(c)). A feature of this bond is that the bonding molecular orbital of the olefine is still present. The infra-red spectra of $K[Pt(C_3H_6)Cl_3]$ and $(C_3H_6)_2Pt_2Cl_4$ show the C=C bond is still present, though its stretching frequency is lowered by 140 cm^{-1} as compared with the free olefine, the bond being weakened by use of some of its electrons in the bond to the Pt atom. In the spectra of $K[Pt(C_2H_4)Cl_3]$ and $(C_2H_4)_2Pt_2Cl_4$ the C=C absorption band is very weak, indicating symmetrical bonding of C_2H_4 to Pt as indicated in Fig. 27.14(d). From the dipole moments of molecules such as *trans* $C_2H_4PtCl_2 \cdot NH_2C_6H_4 \cdot CH_3$ and *trans* $C_2H_4PtCl_2 \cdot NH_2C_6H_4Cl$ it is estimated that the Pt–C_2H_4 bond has about one-third double-bond character.

The existence of very stable complexes of PtX_2 with cyclooctatetraene (a) and CH_2=CH.CH_2CH_2.CH=CH_2 (b) is consistent with this view of the metal–olefine bond because the π bonds in these *cis* molecules must use two *different* d orbitals

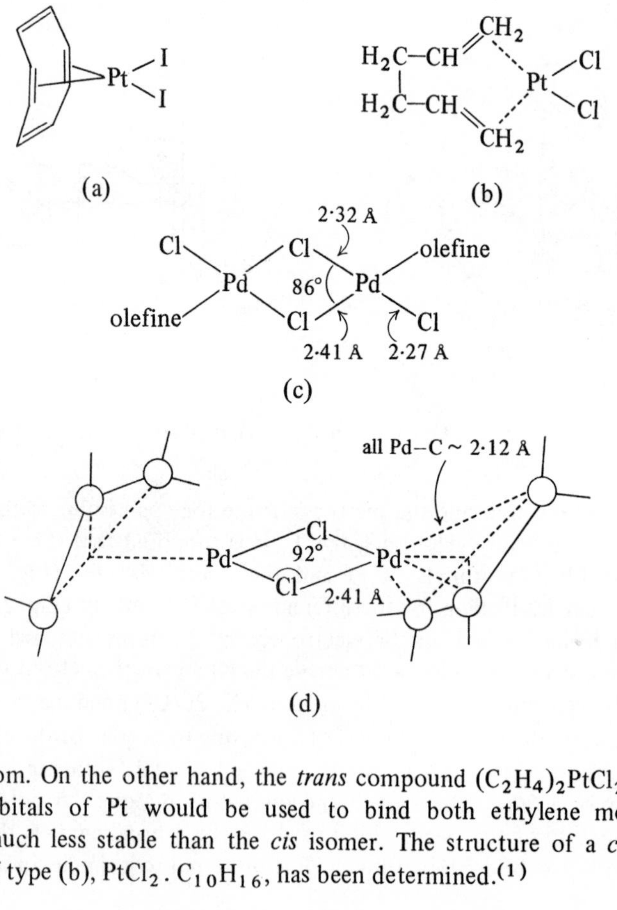

(a)

(b)

(c)

(d)

of the Pt atom. On the other hand, the *trans* compound $(C_2H_4)_2PtCl_2$, in which the same orbitals of Pt would be used to bind both ethylene molecules, is apparently much less stable than the *cis* isomer. The structure of a *cis* dichloro compound of type (b), $PtCl_2 \cdot C_{10}H_{16}$, has been determined.[1]

(1) AC 1965 18 237

The hypothesis that olefines are linked to Pt or Pd by interaction between the π orbital of the double bond and a dsp^2 orbital of the metal is supported by the crystal structures of the (dimeric) styrene–PdCl$_2$[2] and ethylene–PdCl$_2$[3] complexes. Both these bridged molecules have the *trans* configuration (c), and the plane of the ethylene molecule or of H$_2$C=CH–C in the case of the styrene compound, [(C$_6$H$_5$.CH=CH$_2$)PdCl$_2$]$_2$, is perpendicular (or approximately so) to the plane containing Pd, the two ethylene C atoms and Cl. An interesting point is that the bridge system is unsymmetrical in both the ethylene and styrene compounds, the Pd–Cl bonds opposite to the Pd–olefine bonds being longer than the other two. In the symmetrical molecule Pd$_2$Cl$_2$(C$_3$H$_5$)$_2$[4] (d), all the Pd–Cl (bridge) bonds are long (2·41 Å).

(2) JACS 1955 77 4987
(3) JACS 1955 77 4984

(4) AC 1965 18 331

(5) AC 1960 13 149
(6) AC 1971 B27 366

The structures of the molecule PtCl$_2$(C$_2$H$_4$)NH(CH$_3$)$_2$[5] and the ion in Zeise's salt, K(PtCl$_3$.C$_2$H$_4$).H$_2$O,[6] (see also Fig. 27.14(d)), are shown at (e) and (f). The structure of Zeise's salt is very similar geometrically to that of

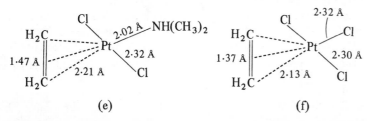

(e)   (f)

K(PtCl$_3$NH$_3$).H$_2$O. Within the double layer there are ionic bonds between anions and cations and also H$_2$O–Cl bonds (3·28 Å), and between the layers van der Waals bonds between C$_2$H$_4$ molecules; in the ammine there are N–H---Cl bonds between the layers. In the ion (f) C=C is very nearly perpendicular to the PtCl$_3$ plane, its centre being approximately 0·2 Å above that plane. There is very little difference between the lengths of the Pt–Cl bonds *cis* and *trans* to C$_2$H$_4$; the latter may be slightly longer, but the difference is certainly very small and possibly within the limits of accuracy of the structure determination.

28

The Lanthanides and Actinides

The crystal chemistry of the lanthanides (rare-earths)

In the fourteen elements (Ce to Lu) which follow La the 4f shell is being filled while the outer structure of the atoms is either $5d^1 6s^2$ or $5d^0 6s^2$. Since many of the compounds of La are structurally similar to those of the rare-earths we shall include this element in our discussion.

The characteristic valence of the lanthanides is three, but there are some interesting differences between the rare-earths. Ce, Pr, and Tb exhibit tetravalence, and Sm, Eu, Yb, and possibly Tm, divalence, that is, from the elements following La and Gd it is possible to remove two 4f electrons (in addition to two 6s

		4	4	?					4	?					
(La)	Ce	Pr	Nd	Pm	Sm	Eu	(Gd)	Tb	Dy	Ho	Er	Tm	Yb	(Lu)
			?	2	2						?	2	

electrons), and the elements immediately preceding Gd and Lu can form ions M^{2+} in addition to M^{3+}. The radii of the 4f ions M^{3+} fall steadily throughout the series (p. 260), but this *lanthanide contraction* is not shown by Eu and Yb in the metals or in compounds such as the hexaborides, where the lower valence of these two elements is evident from the larger radii compared with those of neighbouring elements. The rare-earths therefore fall into two series, with a break at Gd, showing the same sequence of valences in each. In Gd there is one electron in each of the seven 4f orbitals, so that further electrons have to be paired with those already present, and we find a marked resemblance between atoms having the same number of doubly occupied orbitals and those with singly occupied orbitals—with Gd compare Mn, the middle member of the 3d series of elements.

Trivalent lanthanides

A feature of the crystal chemistry of 4f compounds is that numerous structures are found for each group of compounds (for example, halides) owing to the variation in ionic (or atomic) radius throughout the series. In the compounds of the earlier 4f elements there is often a high c.n. of the metal ion which falls to lower values as the ionic radius decreases. For example, there are at least seven structures for trihalides, with (somewhat irregular) 9-coordination of M in the LaF_3 and YF_3 structures, 9- or (8 + 1)-coordination in other halides of the earlier 4f elements, and 6-coordination (YCl_3 or BiI_3 structures) in the compounds of the later elements. (For a

988

summary see Table 9.19, p. 358.) A number of 4f trihydroxides have the same 9-coordinated (UCl_3) structure as the trichlorides of the earlier rare-earths (p. 358). Oxyhalides are described in Chapter 10.

In oxides M_2O_3 there is 6- and/or 7-coordination of M, the La_2O_3 structure being notable for the unusual (monocapped octahedral) coordination of M (Fig. 12.7, p. 452). This structure is also described in a different way on p. 1004 in connection with the structures of compounds such as U_2N_2S and U_2N_2Te; it is also the structure of Ce_2O_2S and other 4f oxysulphides. Sesquisulphides adopt one or other of three structures, in which the c.n. of M falls from 8 to 6:

	Ce_2S_3	α-Gd_2S_3	Ho_2S_3	Yb_2S_3
c.n. of M	8	8 and 7	7 and 6	6

Other compounds are described in other chapters:

borides: p. 840;	hydrides: p. 294;
carbides: p. 756;	nitrides: p. 668.

Tetravalent lanthanides

The oxidation state IV is firmly established only for Ce, Pr, and Tb, though the preparation of complex fluorides containing Nd(IV) and Dy(IV) has been claimed. The only lanthanide ion M^{4+} stable in aqueous solution is Ce^{4+}, and this is probably always present in complexes, as in the nitrato complex in $(NH_4)_2[Ce(NO_3)_6]$ (p. 664). The only known binary solid compounds of these elements in this oxidation state are:

CeO_2	PrO_2	TbO_2	(fluorite structure)
CeF_4	–	TbF_4	(UF_4 structure).

Some complex fluorides and complex oxides have been prepared, for example, alkaline-earth compounds $MCeO_3$ with the perovskite structure (p. 484). The metal–oxygen systems are complex, there being various ordered phases intermediate between the sesquioxide (p. 483) and the dioxide. In the Tb–O system[1] there are $TbO_{1.715}$ (Tb_7O_{12}), $TbO_{1.81}$, and $TbO_{1.83}$, and one or more of these phases occur in the Ce–O[2] and Pr–O[3] systems. For Tb_7O_{12} and Pr_6O_{11} see p. 449.

(1) JACS 1961 **83** 2219

(2) JINC 1955 **1** 49
(3) JACS 1954 **76** 5239

Divalent lanthanides

In the lanthanides the oxidation state II is most stable for Eu, appreciably less so for Yb and Sm (in that order), and extremely unstable for Tm and Nd. It appears that Eu is probably the only lanthanide which forms a monoxide, the compounds previously described as monoxides of Sm and Yb being Sm_2ON and Yb_2OC. Of the many 4f compounds MX which crystallize with the NaCl structure only EuO, SmS, EuS, and YbS contain M(II); all other compounds MS, for example, the bright yellow CeS (and also LaS) are apparently of the type $M^{3+}S^{2-}$ (e). The cell dimensions of the 4f compounds MS decrease steadily with increasing atomic number except those of SmS, EuS, and YbS, the points for which fall far above the

(1) AC 1965 **19** 214

curve;[1] all compounds MN, including those of Sm, Eu, and Yb, are normal compounds of M(III). Europium forms an oxide Eu_3O_4 ($Eu^{II}Eu_2^{III}O_4$) which is isostructural with $SrEu_2O_4$ (p. 496).

All dihalides of Eu, Sm, and Yb are known, but less is known of the dihalides of Nd and Tm. Of the structures of the Yb dihalides only that of YbI_2(CdI_2 structure) appears to be known. The structures of Eu and Sm dihalides are summarized in Table 9.15 (p. 353) and the accompanying text, where the similarity to the alkaline-earth compounds is stressed. Compounds MI_2 formed from the metal and

(2) IC 1965 4 88

MI_3 by La, Ce, Pr, and Gd[2] are not compounds of M(II), but have metallic properties and may be formulated $M^{III}I_2$ (e).

The ionic radius of Yb^{2+} is close to that of Ca^{2+} and those of Sm^{2+} and Eu^{2+} are practically identical to that of Sr^{2+}. Accordingly many compounds of these three 4f elements are isostructural with the corresponding alkaline-earth compounds, for example, dihalides, EuO and MS with the alkaline-earth sulphides (NaCl structure), $EuSO_4$ and $SmSO_4$ with $SrSO_4$ (and $BaSO_4$).

The actinides

Introduction

Just as the lanthanides form a series of closely related elements following La in which the characteristic ions M^{3+} have from 1 to 14 4f electrons, so the actinide series might be expected to include the 14 elements following the prototype Ac (which, like La, is a true member of Group III), with from 1 to 14 electrons entering the 5f in preference to the 6d shell. In fact there are probably no 5f electrons in Th and the number in Pa is uncertain, and these elements are much more characteristically members of Groups IV and V respectively than are the corresponding lanthanides Ce and Pr. Thus the chemistry of Th is essentially that of Th(IV), whereas there is an extensive chemistry of Ce(III) but only two solid binary compounds of Ce(IV), namely, the oxide and fluoride. In contrast to Pa, the most stable oxidation state of which is V, there are no compounds of Pr(V).

Owing to the comparable energies of the 5f, 6d, 7s, and 7p levels, the chemistry of the actinides is much more complex than the predominantly ionic chemistry of the trivalent lanthanides. The oxidation state III is common to all actinides, though it is observed only in the solid state for Th and it is unimportant for Pa. Except for these two elements the behaviour of 4f and 5f elements in this oxidation state is similar, both as regards their solution chemistry and the structures of compounds in the solid state—witness the trifluorides AcF_3 and UF_3—CmF_3 with the LaF_3 structure and the corresponding trichlorides with the same structure as $LaCl_3$. (Compounds MO and MS formed by U, Np, and Pu are not regarded as compounds of M(II), for compounds MC and MN have the same (NaCl) structure—see p. 992.) In contrast to the exceptional II and IV oxidation states of certain 4f elements the actinides as a group exhibit all oxidation states from III to VI, and there is no parallel in the 4f series either to the elements Th and Pa or to the closely related group consisting of U, Np, Pu, and Am, all of which exhibit oxidation numbers from III to VI inclusive (Table 28.1). Other differences between the lanthanides

TABLE 28.1

The oxidation states of the actinide elements

Ac	Th	Pa	U	Np	Pu	Am	Cm	Bk	Cf
3	3	3	3	3	3	3	**3**	**3**	**3**
	4	4	4	4	**4**	4	4	4	
		5	5	5	5	5			
			6	6	6	6			

(Heavy type indicates the most stable oxidation state.)

and actinides include the formation of ions MO_2^{2+} (by U, Np, Pu, and Am) and MO_2^+ (by Np and Am; the U and Pu ions are very unstable), the greater tendency of the actinides to form complexes, and the more complex structures of the metals themselves; Am is the first actinide to crystallize with a close-packed structure like the majority of the lanthanides.

The radii of actinide ions M^{3+} and M^{4+} decrease with increasing atomic number in much the same way as do those of the lanthanides, the radius of M^{3+} being about 0·10 Å greater than that of M^{4+} (e.g. U^{3+}, 1·03 Å, U^{4+}, 0·93 Å). The screening effect of the f electrons does not entirely compensate for the increased nuclear charge, and since the outermost electronic structure remains the same the radii decrease ('lanthanide' and 'actinide' contractions). The effect is clearly seen in series of isostructural compounds such as the dioxides (fluorite structure) or trifluorides (LaF_3 structure).

The crystal chemistry of thorium

Chemically thorium is essentially a member of Group IV of the Periodic Table. It is tetravalent in most of its compounds, it is the most electropositive of the tetravalent elements, and its chemistry strongly resembles that of Hf. Its ionic radius (Th^{4+}) is, however, closer to that of Ce^{4+} than to that of Hf^{4+}, and in its crystal chemistry it has much in common not only with U^{IV} but also with Ce^{IV}. For example, like Ce^{IV} Th forms no normal carbonate but complex carbonates of the type $(NH_4)_2M(CO_3)_3 . 6 H_2O$, and both elements form complex sulphates and nitrates, respectively $K_4M(SO_4)_4$ and $K_2M(NO_3)_6$. Again, many compounds of Th are isostructural with compounds of U^{IV} and the transuranium elements, as shown in Table 28.2 for some of the simpler compounds of these elements. On the one hand, ThF_4 is isostructural with CeF_4, ZrF_4, and HfF_4, and on the other with PaF_4, UF_4, NpF_4, PuF_4, and AmF_4. Some complex fluorides M_2ThF_6 are structurally similar to U(IV) compounds, but other complex fluorides M_nThF_{4+n} have distinctive structures, some of which are noted in a later section. The remarkable framework in compounds $M_7^I M_6^{IV} F_{31}$ (p. 397) is formed not only by Zr and Th but also by Pr, Pa, U, Np, Pu, Am, and Cm. Compounds of Th mentioned in other chapters include ThC_2 (p. 758) and Th_3N_4 (p. 671). There is a note on the sulphides on p. 1006.

(*a*) JACS 1953 **75** 1236
(*b*) JACS 1953 **75** 4560
(*c*) JACS 1954 **76** 2019

(*d*) AC 1949 **2** 291
(*e*) AC 1949 **2** 57

(*f*) AC 1948 **1** 265

(1) JCS 1964 5450
(2) RJIC 1966 **11** 1318
(3) CR 1968 **266** 1056
(4) MRB 1969 **4** 443

TABLE 28.2

Structures of compounds of actinide elements

	Ac	Th	Pa	U	Np(*a*)	Pu	Am(*b*),(*c*)	Cm	Structure
MC		N		N		N			N = NaCl
MN		N		N	N	N			
MO		N	N	N	N	N	N	N	
MS(*d*)		N		N		N			
M_2O_3	A					A/C	A/C	A/C	A or C—M_2O_3
M_2S_3	D(*e*)	S(*d*)		S(*d*)	S(*d*)	D(*e*)	D(*e*)		D = Ce_2S_3 (p. 621)
									S = Sb_2S_3
MO_2		F	F	F	F	F	F	F	F = CaF_2
MS_2		P(*d*)							P = $PbCl_2$
MC_2				C	C				C = CaC_2 (p. 757)
MSi_2		Si		Si	Si				Si = $ThSi_2$ (p. 792)
MOS		B(*d*)	B	B(*d*)	B(*d*)				
MOCl	B					B	B		B = BiOCl(PbFCl)
MF_3	L			L	L	L	L	L	L = LaF_3
MF_4		U	U	U	U	U	U(*b*)	U	U = UF_4
MCl_3	V(*f*)			V(*f*)	V(*f*)	V(*f*)	V(*f*)	V	V = UCl_3
MCl_4		W	W	W	W				W = UCl_4

As noted earlier Th(III) is known only in the solid state, in $ThCl_3$, $ThBr_3$, and ThI_3,[1][2] Th_2S_3 (Table 28.2), and oxyfluorides. The latter include ThOF (statistical fluorite structure)[3] and $ThO_{0.5}F_{2.5}$[4] (superstructure of LaF_3 type), this compound presumably containing equal numbers of Th^{3+} and Th^{4+} ions. The properties of ThI_2, the structure of which is described on p. 352, suggest formulation as $Th^{IV}I_2(e)_2$.

The crystal chemistry of protoactinium

Although the most stable oxidation state of this element is v no examples are yet known of crystalline compounds of Pa(v) which are isostructural with those of Nb and Ta. Either the compounds are isostructural with those of U(v) or they have structures peculiar to this element, as shown by the examples of Table 28.3. The c.n. of Pa^{5+} is generally higher than that of Nb^{5+} (or Ta^{5+}), consistent with the larger radius of that ion (about 0·9 Å) as compared with about 0·7 Å for the group V ions. In view of the fact that recent work on Nb_2O_5 and Ta_2O_5 has shown that the structural chemistry of these oxides is extremely complex, it is likely that earlier statements on the similarity of Pa_2O_5 to these oxides should be checked. Certainly there is no resemblance between the complex oxides formed by Pa(v) with 4f elements and those formed by Nb and Ta. For example, Pa does not form the analogues of MNb_3O_9 and MTa_3O_9 but instead forms extensive solid solutions M_2O_3–Pa_2O_5 with fluorite-like structures.[1]

Compounds of Pa(IV) are structurally similar to those of other actinide (IV) compounds. For example, all the compounds PaF_4, $PaCl_4$, $LiPaF_5$, and $(NH_4)_4PaF_8$ are isostructural with the corresponding compounds of U(IV);[2] see also Table 28.2 for PaO_2 and PaOS.

(1) AC 1967 **23** 740

(2) IC 1967 **6** 544
For references to complex halides of Pa(V) see Table 28.5 (p. 998).

TABLE 28.3

Comparison of halides of Nb (*and* Ta), Pa(V), *and* U(V)

C.N. of M		*C.N. of* M			*C.N. of* M	
NbF_5 TaF_5	6	PaF_5	7	$UF_5\ \alpha$	6 (octahedral)	
Tetramer M_4F_{20} (octahedral)		Isostructural with β-UF_5		$UF_5\ \beta$	7 (Fig. 28.2)	
$NbCl_5$ $TaCl_5$	6	$PaCl_5$	7	UCl_5	6	
Dimer M_2Cl_{10} (octahedral)		Pentagonal bipyramidal chain		U_2Cl_{10} dimer		
$RbNbF_6$ $RbTaF_6$	6	$RbPaF_6$	8 and	$RbUF_6$	8	
Octahedral MF_6^-		Dodecahedral chain ion				
K_2NbF_7 K_2TaF_7	7	K_2PaF_7	9	K_2UF_7	?	
Monocapped trigonal prism MF_7^{2-}		Tricapped trigonal prism chain			?	
Na_3TaF_8	8	Na_3PaF_8	8 and	Na_3UF_8	8	
Antiprism MF_8^{3-}		Cubic coordination (fluorite superstructure)				

The crystal chemistry of uranium

The structures of the metal, hydrides, and carbides are described in other chapters, as also are the halides MX_3, MX_4, and MX_5. Here we devote sections to certain halide structures peculiar to U, complex fluorides of Th and U, oxides of U, uranyl compounds and uranates, nitrides and related compounds, and conclude with a note on the sulphides of U, Th, and Ce.

Halides of uranium. Our present knowledge of the structures of the majority of the 5f halides is summarized (with references) under Trihalides, Tetrahalides, etc. in Chapter 9. The halides of U include:

	C.N. of U		*C.N. of* U
UF_3 : LaF_3 structure	7 + 2	UF_4 : ZrF_4 structure	8
UCl_3 } UCl_3 structure UBr_3 }	9	UCl_4 : $ThCl_4$ structure	8
UI_3 : $PuBr_3$ structure	8 + 1	$UF_5\ {}^{\alpha}_{\beta}$ see text	6 7
		UCl_5 : U_2Cl_{10} dimers	6
		UF_6 } molecular UCl_6 }	6

We comment here on the two forms of UF_5[1] and on U_2F_9:[2]

(1) AC 1949 **2** 296

(2) AC 1949 **2** 390

The two crystalline forms of the pentafluoride are of some interest as illustrating two ways in which a ratio of 5 F : 1 U can be attained. In α-UF_5 (Fig. 28.1) there are infinite chains formed from octahedral UF_6 groups sharing a pair of opposite vertices. In β-UF_5 each U atom is surrounded by seven F, of which three are

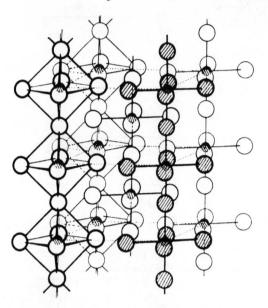

FIG. 28.1. The crystal structure of α-UF_5. (Small circles represent U atoms).

bonded to one U only and the other four are shared with a second U atom: $(3 \times 1) + (4 \times \frac{1}{2}) = 5$. A projection of this structure is shown in Fig. 28.2 to show the way in which the 7-coordination groups are linked together. Interatomic distances in these essentially ionic structures are: in α-UF_5, U-2 F, 2·23 Å, U-4 F, 2·18 Å: in β-UF_5, U-7 F, 2·23 Å (mean).

In the black U_2F_9 all the U atoms have nine F neighbours (at 2·31 Å), and since all the U atoms are crystallographically equivalent there is presumably interchange of electrons between ions with different charges (e.g. U^{4+} and U^{5+}), accounting for the colour. The same metal and fluorine positions are occupied by Th^{4+} and F^- in $NaTh_2F_9^{(2)}$ but in addition Na^+ ions occupy two-thirds of the available octahedral holes between six F^- ions (Na–6 F, 2·34 Å, Th–9 F, 2·40 Å, very similar to the corresponding distances in β_2-Na_2ThF_6).

Complex fluorides of the 5f elements. As already pointed out in Chapter 10 the closest relation between the crystal structures of simple and complex halides is found for certain of the latter which have the structures of simple halides but with statistical distribution of two or more kinds of metal ions in the cation positions. Examples include:

$$\left.\begin{array}{l} KLaF_4 \\ \alpha\text{-}K_2UF_6 \end{array}\right\} \text{disordered } CaF_2 \text{ structure;} \qquad \left.\begin{array}{l} BaThF_6 \\ BaUF_6 \end{array}\right\} \text{disordered } LaF_3 \text{ structure.}$$

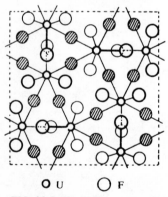

○ U   ◯ F

FIG. 28.2. Projection of the crystal structure of β-UF_5. The fluorine ions which belong to the coordination groups of two uranium ions are shown as shaded circles.

The statistical fluorite structure is to be expected for a compound $A_m BF_n$ when both A and B are large enough for 8-coordination by F. For smaller A ions other structures are found, as for example, $NaLaF_4$ and $Na_2 ThF_6$ with the β_2-$Na_2 ThF_6$ structure, which is described later, in which there is 6-coordination of A and 9-coordination of B ions. Other ways in which a more complex structure may be related to an $A_m X_n$ structure are exemplified by γ-$Na_2 UF_6$ and $Na_3 UF_8$ (fluorite superstructures), $Na_3 UF_7$ (anion defective fluorite structure), and $Na_2 UF_8$ (cation defective fluorite structure).

The γ-$Na_2 UF_6$ structure is a slightly deformed fluorite structure in which both types of positive ion are 8-coordinated. It is illustrated in Fig. 28.3.

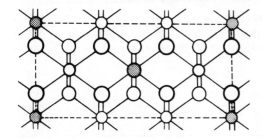

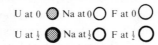

FIG. 28.3. Projection of the crystal structure of γ-$Na_2 UF_6$. The heights of the ions above the plane of the paper, are given as fractions of the length of the b axis.

The $Na_3 UF_7$ structure is also closely related to the fluorite structure. It is tetragonal, with $a = 5 \cdot 45$ Å and $c = 10 \cdot 896$ Å. A projection of the structure (Fig. 28.4) shows that the doubling of one axis is due to the regular arrangement of U in one-quarter of the cation positions. One-eighth of the F positions of the fluorite structure are unoccupied, as shown at the left in Fig. 28.4, with the result that both

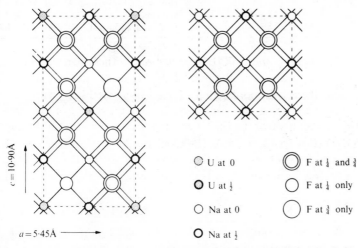

FIG. 28.4. Projection on (010) of the (tetragonal) crystal structure of $Na_3 UF_7$ (left), to which the key refers. The heights are fractions of b (5·45 Å) above the plane of the paper. At the right is shown a projection of the (cubic) fluorite structure (compare Fig. 6.9, p. 204), in which the heights of the atoms are indicated in the same way as for $Na_3 UF_7$.

995

Na^+ and U^{4+} are 7-coordinated. Summarizing, all the following complex fluorides of U have structures related to the fluorite structure:

Statistical	Superstructures	Anion defective	Cation defective
High-K_2UF_6	Na_2UF_6 Na_3UF_8	Na_3UF_7	Na_2UF_8

In the complex fluorides we are describing there are high c.n.'s of the 4f and 5f elements, and of special interest are (i) the persistence of these high c.n.'s down to low F : M ratios, and (ii) the unexpected variations in c.n. in certain series of compounds, notably some complex fluorides of thorium. The first point is illustrated by the structures of compounds of U(IV), some of which are set out in Table 28.4. The same type of 9-coordination (tricapped trigonal prism) is found in $LiUF_5$ and in two forms of K_2UF_6.

TABLE 28.4

Coordination of U(IV) *in some complex fluorides*

Compound	C.N. of U(IV)	Coordination polyhedron	Reference
$LiUF_5$	9	t.c.t.p.	AC 1966 **21** 814
Rb_2UF_6	8	Dodecahedron	IC 1969 8 33
β_2-Na_2UF_6 β_1-K_2UF_6	9	t.c.t.p.	AC 1948 **1** 265; AC 1969 **B25** 2163
Na_3UF_7	7	Fig. 28.4	AC 1948 **1** 265
K_3UF_7	7	Pentagonal bipyramid	AC 1954 7 792
Li_4UF_8	8 (+1)	See text	JINC 1967 **29** 1631
$(NH_4)_4UF_8$	8	Distorted antiprism	AC 1970 **B26** 38

The β_2-Na_2ThF_6 and β_1-K_2UF_6 structures. These two structures are very closely related. Projections on the base of the hexagonal unit cell are given in Fig. 28.5. In both there is 9-coordination of the heavy metal ions, with fluorine ions at the apices of a trigonal prism and three beyond the centres of the vertical prism faces. The arrangement of Th (or U) and F is the same as that of Sr and H_2O in $SrCl_2 . 6 H_2O$ (Fig. 15.14, p. 556). The alkali-metal ions are situated between these chains in positions of 6- or 9-coordination in β_2-Na_2ThF_6 and β_1-K_2UF_6 respectively. The former structure is therefore generally preferred by sodium salts

β_2-Na_2ThF_6 structure	β_1-K_2UF_6 structure
β_2-Na_2ThF_6	β_1-K_2ThF_6
β_2-Na_2UF_6	β_1-K_2UF_6
β_2-K_2UF_6	
	β_1-$KLaF_4$
$NaPuF_4$	β_1-$KCeF_4$

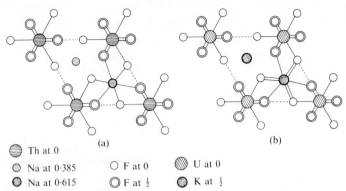

Th at 0
Na at 0·385   F at 0   U at 0
Na at 0·615   F at $\frac{1}{2}$ K at $\frac{1}{2}$

FIG. 28.5. The crystal structures of (a) β_2-Na$_2$ThF$_6$, and (b) β_1-K$_2$UF$_6$. The heights of atoms are given as fractions of the lengths of the hexagonal c axes, which are normal to the plane of the paper.

and the latter by potassium salts. In the isostructural compounds ABX$_4$ there is A$_{\frac{3}{2}}$B$_{\frac{3}{2}}$X$_6$ in the unit cell instead of, for example, Na$_2$ThF$_6$ or K$_2$UF$_6$, that is, the same total number of cations. Of these one B ion occupies the Th (or U) position and $\frac{3}{2}$ A + $\frac{1}{2}$ B the Na (or K) positions, statistically. (Compare, for example, certain of the spinels, p. 490.) The structure of NaNdF$_4$ (and isostructural Ce and La salts) is similar to but different from the β_2-Na$_2$ThF$_6$ structure originally assigned; there is 9-coordination (t.c.t.p.) of the 4f ion.[1]

(1) IC 1965 4 881

The structures of compounds below the line in Table 28.4 differ from those above the line in containing discrete UF$_7$ or UF$_8$ groups. The structure of Na$_3$UF$_7$ has already been noted. One form of K$_3$UF$_7$ is isostructural with K$_3$UO$_2$F$_5$; the other is similar structurally to K$_3$ZrF$_7$ and (NH$_4$)$_3$ZrF$_7$ (p. 399). In both polymorphs the 7 F atoms are arranged at the vertices of a pentagonal bipyramid (U–F, 2·26 Å), as in Fig. 28.6.

A number of 8-coordination polyhedra are found for U(IV)– contrast the cubic coordination of U(V) and U(VI) in Na$_3$UF$_8$ and Na$_2$UF$_8$ respectively. It should be remarked that whereas in LiUF$_5$ there is t.c.t.p. coordination of U with all U–F distances between 2·26 and 2·59 Å, the coordination polyhedron of U in Li$_4$UF$_8$ is strictly a bicapped trigonal prism (8 F at 2·21–2·39 Å). There is a ninth neighbour (at 3·3Å) beyond the third prism face, and if this is included in the coordination group the structure no longer contains discrete UF$_8^{4-}$ ions, since this more distant F$^-$ is necessarily shared between UF$_9$ groups.

Our second point concerns the complex fluorides of Th, which are remarkable in several respects (Table 28.5). Both (NH$_4$)$_3$ThF$_7$ and (NH$_4$)$_4$ThF$_8$ contain infinite linear ions of composition (ThF$_7$)$_n^{3n-}$ formed from tricapped trigonal prism groups sharing two edges, though the details of the chains are different in the two crystals. The eighth F in (NH$_4$)$_4$ThF$_8$ is a discrete F$^-$ ion, not bonded to Th. For (NH$_4$)$_4$ThF$_8$ there are two types of structure more obvious than the one adopted, namely, discrete ThF$_8^{4-}$ ions (which actually occur in (NH$_4$)$_5$ThF$_9$) or a chain of tricapped trigonal prisms sharing two *vertices*. Instead of either of these structures

FIG. 28.6. The configuration of the UF$_7^{3-}$ or UO$_2$F$_5^{3-}$ ions.

TABLE 28.5

Coordination of metal in some complex fluorides of Th^{4+} *and* Pa^{5+}

Compound	C.N. of M	Coordination polyhedron	Reference
$RbTh_3F_{13}$	9	t.c.t.p. sharing edges and vertices	AC 1971 **B27** 1823
β_2-Na_2ThF_6 β_1-K_2ThF_6	9	t.c.t.p. sharing two faces	See Table 28.4
$(NH_4)_3ThF_7$	9	t.c.t.p. sharing two edges	AC 1971 **B27** 2279
$(NH_4)_4ThF_8$	9	t.c.t.p. sharing two edges	AC 1969 **B25** 1958
$(NH_4)_5ThF_9$	8	Dodecahedron	AC 1971 **B27** 829
$K_7Th_6F_{31}$	8	Antiprism	AC 1971 **B27** 2290
$RbPaF_6$ $KPaF_6$ NH_4PaF_6	8	Dodecahedra sharing 2 edges	AC 1968 **B24** 1675 IC 1966 5 659
K_2PaF_7	9	t.c.t.p. sharing 2 edges	JCS A 1967 1429
Na_3PaF_8	8	Cube	JCS A 1969 1161

there are chains of composition $(ThF_7)_n^{3n-}$ similar to those in K_2PaF_7 (tricapped trigonal prisms sharing two *edges*) plus additional F^- ions. More remarkable still is $(NH_4)_5ThF_9$, which might have been expected to contain ThF_9^{5-} ions or F^- ions plus $(ThF_8)_n^{4n-}$ chains built from ThF_9 groups sharing two vertices (thus retaining 9-coordination). In fact this crystal contains discrete dodecahedral ThF_8^{4-} plus F^- ions, that is, the c.n. of Th is lower in $(NH_4)_5ThF_9$ than in some compounds with lower $F:Th$ ratios (for example, $(NH_4)_4ThF_8$, $(NH_4)_3ThF_7$, and $RbTh_3F_{13}$, in all of which Th is 9-coordinated) but the same (8) as in $K_7Th_6F_{31}$ with an intermediate $F:Th$ ratio. Note that the 8-coordination of Th is dodecahedral in $(NH_4)_5ThF_9$ but antiprismatic in $K_7Th_6F_{31}$, which is isostructural with $Na_7Zr_6F_{31}$ (p. 397).

Oxides of uranium. The oxygen chemistry of this element is extraordinarily complex, for although there are oxides with simple formulae such as UO (NaCl structure), UO_2, and UO_3, the last two have ranges of composition which depend on the temperature and there are numerous intermediate phases with ordered structures. Moreover UO_3 is polymorphic, as also is U_3O_8. We summarize here what is known of oxides in the range UO_2–UO_3; references to structural studies are given in Table 28.6.

At ordinary temperatures UO_2 has the normal fluorite structure. At higher temperatures some of the O atoms move into the large interstices at $(\frac{1}{2}\frac{1}{2}\frac{1}{2})$ etc. and UO_2 takes up oxygen to the composition $UO_{2\cdot25}$. A n.d. study of the disordered solid solution at the intermediate composition $UO_{2\cdot13}$ shows three types of O atom, O_I close to the original fluorite positions (but moved slightly towards the large interstitial holes), and O_{II} and O_{III} in positions about 1 Å from these large holes. At the composition $UO_{2\cdot25}$ (U_4O_9) these three types of O atom are arranged in an orderly way forming a superstructure referable to a much larger unit cell (64 times the volume of the UO_2 cell).

The phase $UO_{2\cdot6}$, both forms of $UO_{2\cdot67}$ (U_3O_8), and the orthorhombic $\alpha\text{-}UO_3$ (see later) all appear to have related structures, but not all of these are known in detail.

TABLE 28.6

The higher oxides of uranium

Oxide	Polymorph	Coordination of U	Reference
$UO_{2\cdot13}$	–		N 1963 **197** 755
U_4O_9	–	See text	AC 1972 **B28** 785
$UO_{2\cdot6}$	–		JINC 1964 **26** 1829
U_3O_8	α	All 7 (pentagonal bipyramidal)	AC 1964 **17** 651
	β	2 U, 7 (pentagonal bipyramidal)	AC 1969 **B25** 2505;
		1 U, 6 (distorted octahedral)	AC 1970 **B26** 656
UO_3	α	Not known	RTC 1966 **85** 135;
			JINC 1964 **26** 1829
	β	All 6 or 7 (three kinds of U atom)	AC 1966 **21** 589
	γ	All 6 (distorted octahedral)	AC 1963 **16** 993
	δ ($UO_{2\cdot8}$)	6 (regular octahedral, 2·07 Å, ReO$_3$ structure)	JINC 1955 **1** 309
	High pressure	All 7 (pentagonal bipyramidal)	AC 1966 **20** 292

Two forms of U_3O_8 are known, the α form being usually encountered. Their structures are closely related, but differ as follows. In $\alpha\text{-}U_3O_8$ all U atoms are 7-coordinated, with 6 O in the range 2·07–2·23 Å but with the seventh at 2·44 Å for one-third of the U atoms and at 2·71 Å for the remainder. For a projection of this structure see Fig. 12.8(c), p. 454. In the β form (prepared by heating α to 1350°C in air and cooling slowly to room temperature) there is both distorted octahedral and 7-coordination (pentagonal bipyramidal). In contrast to the α form one-third of the U atoms (U^{VI} ?) have two close O neighbours (1·89 Å), but the remaining U atoms have quite different environments:

$$U_I : \quad 1\cdot89\,(2) \qquad 2\cdot11\,(1) \qquad 2\cdot30\,(2) \qquad 2\cdot37\,\text{Å}\,(2)$$
$$U_{II} : \quad 2\cdot09\,(4) \qquad 2\cdot28\,\text{Å}\,(2)$$
$$U_{III} : \quad 2\cdot02\,(1) \qquad 2\cdot08\,(2) \qquad 2\cdot29\,(2) \qquad 2\cdot40\,\text{Å}\,(2).$$

(The numbers in brackets are the numbers of neighbours at the particular distance.)

At atmospheric pressure UO_3 appears to have six polymorphs, and a seventh has been prepared under high pressure. The original 'hexagonal $\alpha\text{-}UO_3$' studied by Zachariasen is apparently a highly twinned form of the orthorhombic phase referred to as $UO_{2\cdot9}$ or orthorhombic UO_3. The experimental densities of both 'forms' of $\alpha\text{-}UO_3$ (7·25 and 7·20 g/cc respectively) are not reconcilable with the calculated (X-ray) density (8·4 g/cc for both unit cells), and the structures are not known. The δ form has the cubic ReO$_3$ structure, with 6 equidistant octahedral neighbours at 2·07 Å; it was determined for a crystal with the composition $UO_{2\cdot8}$. In the β and γ forms U has 6 and/or 7 nearest O neighbours (at distances in the approximate range 1·8–2·4 Å), the coordination being irregular; only in the high-pressure form are there well defined UO_2 groups with two short collinear bonds (1·83 Å) and five U–O 2·20–2·56 Å.

Uranyl compounds. An interesting feature of the chemistry of U and certain other 5f elements is the formation of ions MO_2^+ and MO_2^{2+}, the former by Pu(v) and Am(v) and the latter by U(vi), Np(vi), Pu(vi), and Am(vi). Uranyl compounds contain a linear group O–U–O with short U–O bonds (about 1·8 Å) and an equatorial group of 4, 5, or 6 O atoms completing 6-, 7-, or 8-coordination of U (Table 28.7). Different total coordination numbers are found even in different polymorphs of a compound (e.g. $UO_2(OH)_2$).

TABLE 28.7

Coordination of U in uranyl compounds

Compound	U–2 O in UO_2 group	Additional neighbours	Reference
α-$UO_2(OH)_2$	1·71 Å	6 OH, 2·48 Å	AC 1971 **B27** 1088
β-$UO_2(OH)_2$	1·81	4 OH, 2·30	AC 1970 **B26** 1775;
			AC 1971 **B27** 2018
γ-$UO_2(OH)_2$			JINC 1964 **26** 1671, 1829
$UO_2(OH)_2 . H_2O$			AM 1960 **45** 1026
UO_2F_2	1·74	6 F, 2·43	AC 1970 **B26** 1540 (n.d.)
UO_2Cl_2			
$UO_2Cl_2 . H_2O$	See text		AC 1968 **B24** 400
$UO_2Cl_2 . 3 H_2O$			
$(UO_2)_2(OH)_2Cl_2(H_2O)_4$	1·79	2 OH, 2 H_2O, Cl	ACSc 1969 **23** 791
$K_3UO_2F_5$	1·76	5 F, 2·24 Å	AC 1954 **7** 783
$(NH_4)_3UO_2F_5$	(1·9)	(5 F, 2·2 Å)	AC 1969 **B25** 67
$Cs_2UO_2Cl_4$	1·81	4 Cl, 2·62 Å	AC 1966 **20** 160
$Cs_xUO_2OCl_x$	1·84	3 Cl, 2 O	AC 1964 **17** 41
UO_2CO_3	(1·7)	6 O of CO_3^{2-}, 2·49 Å	AC 1955 **8** 847
$UO_2(NO_3)_2 . 6 H_2O$	1·76	2 H_2O, 4 O of NO_3^-	AC 1965 **19** 536 (n.d.)
$UO_2(NO_3)_2 . 2 H_2O$	1·76	2 H_2O, 4 O of NO_3^-	IC 1971 **10** 323 (n.d.)
$Rb(UO_2)(NO_3)_3$	1·78	6 O of NO_3^-, 2·48 Å	AC 1965 **19** 205
$Na(UO_2)(OOC . CH_3)_3$	1·71	6 O, 2·49 Å	AC 1959 **12** 526
$Cs_2(UO_2)(SO_4)_3$	(1·74)	5 O, 2·37–2·47 Å	JINC 1960 **15** 338
$[UO_2(CH_3COO)_2(\phi_3PO)]_2$	(1·78)	5 O, 2·36 Å	IC 1969 **8** 320
NpO_2F_2			AC 1949 **2** 388
$KAmO_2F_2$			JACS 1954 **76** 5235
$KAmO_2CO_3$			IC 1964 **3** 1231

'Hydrates' of UO_3 include $UO_3 . \frac{1}{2} H_2O$, $UO_3 . H_2O$, and $UO_3 . 2 H_2O$. The structure of the 'hemihydrate' is not known. There appear to be three forms of the 'monohydrate', which is actually $UO_2(OH)_2$, and possibly two of the 'dihydrate' $(UO_2(OH)_2 . H_2O)$. The latter has been studied as the mineral schoepite; it consists of $UO_2(OH)_2$ layers with H_2O molecules interleaved between them. The type of layer is similar to that in α-$UO_2(OH)_2$. The structures of two polymorphs of $UO_2(OH)_2$ have been determined. The β form consists of layers of octahedral $UO_2(OH)_4$ groups which share all four equatorial OH groups to form a layer based on the simplest 4-connected plane net (Fig. 28.7(a)). The H atoms have been located by neutron diffraction in both forms of $UO_2(OH)_2$, confirming the formation of O—H—O bonds (2·76 and 2·80 Å in α and β respectively) between the layers.

The β form of $UO_2(OH)_2$ is very readily converted by slight pressure to the α form (with increase in density from 5·73 to 6·73 g/cc), in which U has 6 OH

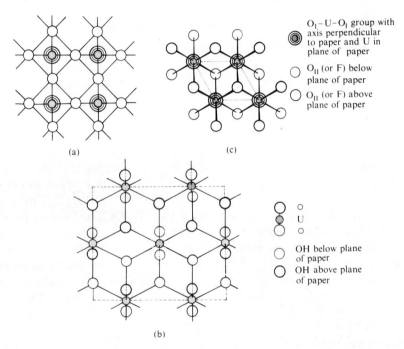

(a)   (c)

◉ O_I–U–O_I group with axis perpendicular to paper and U in plane of paper

○ O_{II} (or F) below plane of paper

○ O_{II} (or F) above plane of paper

(b)

○ O
◉ U
○ o

○ OH below plane of paper

○ OH above plane of paper

FIG. 28.7. Forms of $UO_2(O_2)$ layers: (a) tetragonal layer in β-$UO_2(OH)_2$ and $BaUO_4$, (b) hexagonal layer in α-$UO_2(OH)_2$, (c) hexagonal layer in UO_2F_2 and $CaUO_4$. In (a) and (c) O–U–O is perpendicular to the plane containing the U atoms; in (b) it is inclined at 70° to the plane of the U atoms.

neighbours in a plane approximately perpendicular to the UO_2 groups, the layer having the form of Fig. 28.7(b)). The 'tetragonal' layer of Fig. 28.7(a) and the 'hexagonal' layer of Fig. 28.7(b), or its more symmetrical variant (c), form the basis of the structures of a number of uranyl compounds and uranates.

Crystalline UO_2F_2 consists of layers of the type shown in Fig. 28.7(c), in which linear UO_2 groups are perpendicular to the plane of the paper and the U atoms also have 6 rather more distant neighbours forming a puckered hexagon (or flattened octahedron). There are no primary bonds between the layers, and a characteristic feature of this structure is the presence of stacking faults, which have been studied in a later n.d. investigation (see Table 28.7). NpO_2F_2 has a similar structure. If

The Lanthanides and Actinides

U(vı) in this layer is replaced by Am(v) we have the 2D ion in $K(AmO_2F_2)$, in which K^+ ions are accomodated between the layers. From X-ray powder photographic studies of UO_2Cl_2 and its mono- and tri-hydrate it has been concluded that the UO_2 group has 5 additional neighbours in the equatorial plane, but the interatomic distances found in certain of these structures suggest that these compounds require further study. The dihydrate of the hydroxychloride forms very interesting dimeric molecules $[UO_2Cl(OH)(H_2O)_2]_2$ (a), consisting of two pentagonal bipyramidal groups sharing an edge (2 OH).

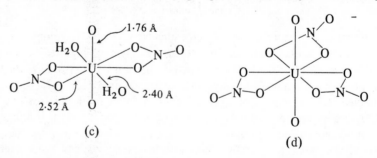

(a)   (b)

Complex ions of various kinds are found in salts containing U oxyhalide ions. The pentagonal bipyramidal structure of $UO_2F_5^{3-}$ in $K_3UO_2F_5$ (and also in the ammonium salt) has been illustrated in Fig. 28.6. In $Cs_2(UO_2Cl_4)$ there are octahedral ions (b), but in the oxychloride $Cs_x(UO_2)OCl_x$ ($x \approx 0.9$) there is pentagonal bipyramidal coordination of U. The coordination groups share O atoms to form infinite double chains as shown diagrammatically in Fig. 28.8.

Two studies of $UO_2(NO_3)_2 . 6 H_2O$ give essentially similar structures showing that this is a tetrahydrate of the complex (c), in which two H_2O molecules and two bidentate NO_3^- ions form a planar equatorial group of 6 O around U. This is the structural unit in the dihydrate, in which very similar U–O distances were found (U–2 O, 1.76 Å, U–OH_2, 2.45 Å, U–ONO_2, 2.50 Å). A somewhat similar coordination of U, but with three bidentate NO_3^- ions, is found in $Rb(UO_2)(NO_3)_3$, (d). A n.d. redetermination of the structure of $UO_2(NO_3)_2 . 2 H_2O$ gives the same U–O bond length (1.76 Å) in the uranyl group as the n.d. study of the hexahydrate.

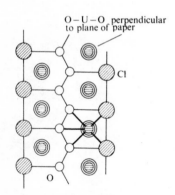

O–U–O perpendicular to plane of paper

Cl

O

FIG. 28.8. The infinite chain ion in $Cs_x(UO_2)OCl_x$.

(c)   (d)

The acetate ion, $H_3C . COO^-$ (ac), can behave like NO_3^- as a bidentate or bridging ligand. Thus we find $UO_2(ac)_2L_2$ with the octahedral structure (c), in which H_2O has been replaced by L (ϕ_3PO or ϕ_3AsO), and $[UO_2(ac)_2L]_2$ with the bridged structure of type (e), containing both bridging and bidentate acetate

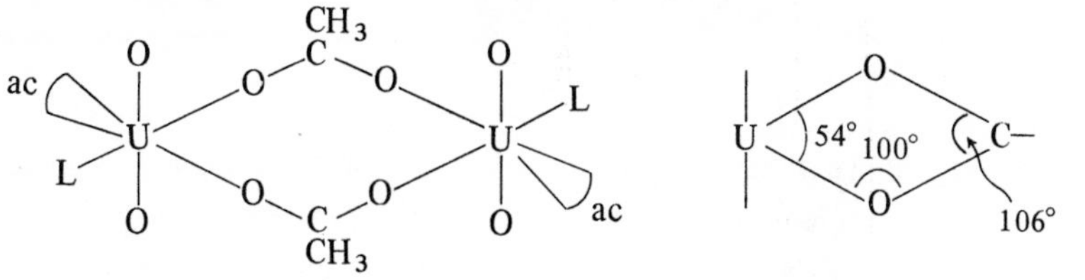

(e)

groups. The geometry of the bidentate CH_3COO^- group is very similar to that of the bidentate nitrate ion (p. 658).

Uranates and complex oxides of uranium. From their empirical formulae uranates would appear to be normal ortho- and pyro-salts containing UO_4^{2-} or $U_2O_7^{2-}$ ions. However, all the known uranates (alkali, alkaline-earth, Ag) are insoluble in water, including $Na_2U_2O_7 \cdot 6\,H_2O$—contrast the soluble alkali chromates, molybdates, and tungstates. In fact the alkali and alkaline-earth uranates $M_2^IUO_4$ and $M^{II}UO_4$ contain infinite 2D ions of the same general types as the tetragonal or hexagonal layers of Fig. 28.7:

hexagonal layer: Na_2UO_4, K_2UO_4,[1] $CaUO_4$[2] (and rhombohedral $SrUO_4$)

tetragonal layer: $SrUO_4$ (rhombic), $BaUO_4$.[2]

(1) AC 1948 **1** 281
(2) AC 1969 **B25** 787

In these 'uranates' there are well defined UO_2 groups, but this is not a feature of all complex oxides of U^{VI}, as may be seen from the following figures which show (in brackets) the number of O neighbours of U at the particular distance:

$CaUO_4$: 1·96 (2)	2·30 (6)	Ca_2UO_5: U_I 2·02 (4), 2·25 (2)
		 U_{II} 1·95 (2), 2·13 (2), 2·21 (2)
$SrUO_4$: 1·87 (2)	2·20 (4)	
$BaUO_4$: 1·89 (2)	2·20 (4)	Ca_3UO_6: 2·03 (3), 2·11 (2), 2·18 (1)

The distinction between two close and the remaining neighbours can still be detected for one-half of the U atoms in Ca_2UO_5 but it has disappeared entirely in Ca_3UO_6.

In contrast to the 2D ions in $SrUO_4$ and $BaUO_4$ there is linking of quite similar distorted UO_6 octahedra into infinite chains in the rutile-type superstructure of $MgUO_4$: U–2 O, 1·92 Å, U–4 O, 2·18 Å.[3a] This formation of two stronger bonds by U in the chains leads to a similar, and unusual, unsymmetrical environment of Mg between the chains: Mg–2 O, 1·98 Å, Mg–4 O, 2·19 Å.

The 3D framework of $UO_2(O_4)$ octahedra which forms the anion in BaU_2O_7[3b] has been described and illustrated in Chapter 5 and Fig. 5.28.

(3a) AC 1954 **7** 788

(3b) JINC 1965 **27** 1521

In all the uranyl compounds and in some of the uranates U has two nearest O neighbours and from four to six next nearest neighbours:

4 in $BaUO_4$, $MgUO_4$
5 in $K_3UO_2F_5$, $Cs_2(UO_2)(SO_4)_3$
6 in $CaUO_4$, UO_2F_2.

Values of the short U–O distance range from around 1·7 Å to 1·96 Å, and the longer U–O distances are usually in the range 2·2–2·5 Å. A bond strength (number) s may be assigned to each U–O bond such that Σs = valence (i.e. 2 for Ba, 6 for U, and so on), the bond strength being derived from an equation of the type suggested by Pauling for metals: $D_1 - D_s = 2\,k\log s$, where s is the bond strength, D_1 the single bond length, D_s the bond length, and k a constant (here 0·45).[4] Bond strengths estimated in this way are, for example, 1·75 Å ($s = 2$), 2·05 Å ($s = 1$, and 2·45 Å ($s = 0·33$). Although these bond strengths correspond to electrostatic bond strengths in ionic crystals it is not implied that the bonds are simple ionic bonds. Where U has two close O neighbours and a group of 4, 5, or 6 additional equatorial neighbours the geometry of the coordination groups suggests that the bonds have some covalent character. The six next nearest neighbours of U in UO_2F_2 and $CaUO_4$ are only 0·5 Å above or below the equatorial plane (perpendicular to the axis of the UO_2 group); in $Rb(UO_2)(NO_3)_3$ they are actually coplanar, being three pairs of O atoms of bidentate NO_3^- ions. A satisfactory description of the bonds in these crystals cannot yet be given, but the possibility of f hybridization in the formation of 2 + 6 bonds has been discussed.[5]

Nitrides and related compounds of Th *and* U; *the* La_2O_3 *and* Ce_2O_2S *structures*
Both Th and U form 'interstitial' nitrides MN (and also MC and MO, all with the NaCl structure). In addition Th forms the typical Group IV nitride Th_3N_4, while U forms UN_2 (fluorite structure) and U_2N_3 (Mn_2O_3 structure).[1] Closely related to the nitrides are compounds such as Th_2N_2O and U_2N_2S which are members of a large group of compounds M_2Y_2X formed by 4f and 5f metals; some examples are set out in Table 28.8.

(4) AC 1954 7 795

(5) JCS 1956 3650

(1) IC 1965 4 115

TABLE 28.8
Compounds with the La_2O_3 (Ce_2O_2S) *and* U_2N_2X *structures*

	$Ce_2O_2S^{(1)}$ Se	La_2O_2O S	$Th_2N_2(NH)^{(6)}$ $Th_2(N, O)_2P^{(3)}$ As	$Th_2N_2O^{(5)}$ S Se	$U_2N_2P^{(3)}$ As	U_2N_2S Se
$Ce_2O_2Sb^{(2)}$ Bi	Ce_2O_2Te	La_2O_2Te Nd_2O_2Te	$Th_2N_2Sb^{(4)}$ Bi	Th_2N_2Te	$U_2N_2Sb^{(4)}$ Bi	U_2N_2Te

(1) AC 1949 2 60. (2) AC 1971 **B27** 853. (3) AC 1969 **B25** 294. (4) AC 1970 **B26** 823. (5) AC 1966 **21** 838. (6) ZaC 1968 **363** 245.

The compounds in which X = O, S, Se, P, or As have the A-M_2O_3 or La_2O_3 structure, while those containing larger X atoms (Te, Sb, or Bi) have a different

structure, which is also that of $(Na_{\frac{1}{4}}Bi_{\frac{3}{4}})_2O_2Cl$. In both structures there is a rigid
layer of composition MY (M_2Y_2) built of YM_4 tetrahedra, which in the La_2O_3
structure share 3 edges (Fig. 28.9(a)) and in the U_2N_2X structure, (b), share 4
edges. The latter is the same layer as in tetragonal PbO or LiOH. In the La_2O_3
structure the third O (X) is situated between the layers surrounded octahedrally by
6 M, while the M atom (on the surface of the 'layer') has 4 O (Y) neighbours
within the layer and 3 O (X) neighbours between the layers, as shown at (c). The
O (Y) atoms within the layers have 4 tetrahedral M (La) neighbours. This La_2O_3
(Ce_2O_2S) structure of 7 : $\frac{3}{4}$ coordination is adopted by a number of 4f sesquioxides

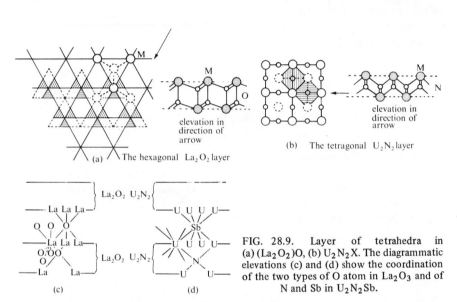

(a) The hexagonal La_2O_2 layer

(b) The tetragonal U_2N_2 layer

(c)

(d)

FIG. 28.9. Layer of tetrahedra in (a) $(La_2O_2)O$, (b) U_2N_2X. The diagrammatic elevations (c) and (d) show the coordination of the two types of O atom in La_2O_3 and of N and Sb in U_2N_2Sb.

M_2O_3 and oxysulphides M_2O_2S, by Ac_2O_3, Pu_2O_3, β-Am_2O_3, Pu_2O_2S, and also
by β-Al_2S_3. In La_2O_3 there is a difference of about 0·3 Å between the distance
La–4 O within the tetrahedral layer and La–3 O between the layers; the
corresponding difference is, of course, much greater in the U compounds, where Y
= N and X = P, As, S, or Se, and in Ce_2O_2S, where Y = O and X = S:

M_2Y_2X	M—4 Y	M—3 X
La_2O_2O	2·38-2·45 Å	2·72 Å
U_2N_2S	2·28	2·87
U_2N_2As	2·27	2·97
Ce_2O_2S	2·36	3·04

In the U_2N_2X structure the tetragonal layers (Fig. 28.9(b)) stack so as to provide
positions of 8-coordination (cubic) for the larger X atoms (Sb, Bi, Te) between

1005

them, and the coordination groups of the atoms are:

$$
\begin{array}{ll}
\text{U:} & \text{4 N + 4 Sb (cubic)} \\
\text{N:} & \text{4 U (tetrahedral)} \\
\text{Sb:} & \text{8 U (cubic)}
\end{array}
$$

as shown diagrammatically in Fig. 28.9(d).

Sulphides of U, Th, and Ce

Most of these compounds are very refractory materials melting at temperatures of 1800°C or higher (CeS melts at $2450 \pm 100°$), and some are highly coloured: CeS, brass-yellow, Ce_2S_3, red, but Ce_3S_4, black; the uranium compounds are all metallic-grey to black in colour. The following compounds are known:

$$
\begin{array}{lll}
CeS & ThS & US \\
Ce_2S_3 & Th_2S_3 & U_2S_3 \\
& ThS_2 & US_2 \\
Ce_3S_4 & Th_7S_{12} & \\
& (ThS_{1.71-1.76}) &
\end{array}
$$

Although the monosulphides all have the NaCl structure they are evidently not simple ionic crystals. ThS is, like all the thorium sulphides, diamagnetic, but the susceptibilities of the paramagnetic CeS and US indicate respectively 1 and 2 unpaired electrons. Assuming that electrons not used for bonding would remain unpaired in the orbitals available and that, judging from the magnetic properties of the transition metals, d electrons will pair to form metallic bonds, the paramagnetism may be attributed to unpaired f electrons. We then have the following picture:

CeS	ThS	US	Ce_2S_3	Th_2S_3	U_2S_3	
1	2	2	–	1	2	d electrons used in metallic bonding
1	–	2	1	–	1	unpaired f electrons

Calculated lattice energies (using estimated ionization potentials where necessary) increase in the order BaS, CeS, ThS, that is, as the number of d electrons available for metallic bonding increases from 0 to 1 to 2. The bond lengths M–S show a decrease in the same order in both MS and M_2S_3:

	BaS	CeS	ThS	(US)
M–S	3·17	2·88	2·84	2·74 Å

	(La_2S_3)	Ce_2S_3	Th_2S_3	U_2S_3
	(3·01)	2·98	2·90	2·82 Å

(1) For references see Table 28.2 (p. 992)

whereas Th is normally larger than Ce. The comparison is perhaps less justifiable for the sesquisulphides because they do not all have the same structure.[1] Those of

Th, U, and Np have the Sb_2S_3 structure, but the sesquisulphides of La, Ce, Ac, Pu, and Am have a defect Th_3P_4 structure. The Th_3P_4 structure is a 3D array of Th and P atoms in which Th is bonded to 8 equidistant P and P to 6 Th. In these sesquisulphides $10\frac{2}{3}$ metal atoms are distributed statistically over the 12 metal sites occupied in Th_3P_4; in Ce_3S_4 all the metal sites are occupied.

The sulphide Th_7S_{12} has a disordered structure[2] in which Th is surrounded by either 8 or 9 S, and ThS_2 has the $PbCl_2$ structure. The structure of US_2 is not known.

(2) AC 1949 2 288

29

Metals and Alloys

The structures of the elements

We have already mentioned the structures of a number of non-metals in earlier chapters. In this concluding chapter we shall be concerned chiefly with the structures of the metallic elements and intermetallic compounds. We shall confine our attention to metals in the solid state and generally to the forms stable under atmospheric pressure. High-pressure polymorphs, which are numerous, will be mentioned only if they are of special interest. Not very much is known of the structures of liquid metals; in the few cases in which diffraction studies have been made the structural information is limited to the average numbers of neighbours within particular ranges of distance, and typical numbers of nearest neighbours are: liquid K (70°C) 8, Li (200°C), 9·8, Hg and Al, 8–9.[1] The existence of diatomic molecules in the vapours of a number of metals has been demonstrated and their dissociation energies determined; they are noted in the discussion of metal–metal bonds in Chapter 7 (p. 251). Diatomic molecules have also been detected in the vapours above certain metal solutions, for example, GeCr and GeNi.[2] Before proceeding to the intermetallic systems we shall review briefly the structures of all the elements in the solid state as far as they are known. For this purpose it is convenient to consider them in groups as indicated below.

(1) AC 1965 **19** 807

(2) JCP 1968 **49** 3579

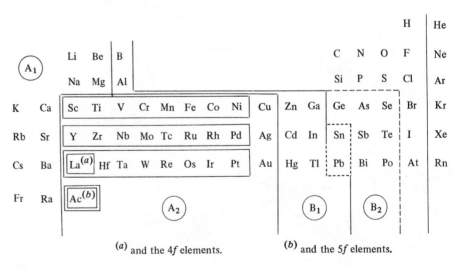

(a) and the 4f elements.   (b) and the 5f elements.

The groups are:

(1) The noble gases.
(2) Hydrogen, the non-metals of the first two short Periods (excluding B) and the later B subgroup elements.
(3) Boron, aluminium, the elements of subgroups IIB and IIIB, Sn, and Pb.
(4) The transition elements and those of IB.
(5) The typical and A subgroup elements of Groups I and II.

We shall discuss the structures of intermetallic compounds in terms of the groups A_1, A_2, B_1, and B_2 (see above), of which A_1 and A_2 comprise the A subgroup elements together with the Group VIII triads and Cu, Ag, and Au, while B_1 and B_2 include the more metallic B subgroup elements apart from those of IB. The reason for placing these latter elements with the transition metals rather than with the B subgroup elements will be apparent later.

(1) *The noble gases*

All these elements except radon have been studied in the crystalline state; they are monatomic in all states of aggregation. Details are given in Table 29.1. Helium is unique among the elements in that it forms a true solid only under pressure, a minimum of 25 atm being necessary. The predominance of the c.c.p. structure is not reconcilable with calculations of lattice energies, which suggest that h.c.p. should be preferred.

TABLE 29.1

Crystal structures of the noble gases, H_2, N_2, O_2, and F_2

Element	Crystalline forms	Reference
He	Three polymorphs of each isotope (^{3}He and ^{4}He): b.c.c. → h.c.p. → c.c.p. with increasing pressure.	PRL 1962 **8** 469
Ne	c.c.p.	
Ar	c.c.p. and h.c.p. (metastable)	JCP 1964 **41** 1078
Kr	c.c.p.	
Xe	c.c.p.	
For calculations of lattice energies see: JCP 1964 **40** 2744		
H_2	Normally h.c.p. but also f.c.c. under special conditions of crystallization	JCP 1966 **45** 834
N_2	α cubic $\big\}$ see text β hexagonal	AC 1972 **B28** 984 JCP 1964 **41** 756
O_2	α monoclinic	JCP 1967 **47** 592
	β rhombohedral	AC 1962 **15** 845
	γ cubic	AC 1964 **17** 777
	Also α′ and amorphous	AC 1969 **B25** 2515
F_2	α monoclinic (below 45·6°K)	JSSC 1970 **2** 225
	β cubic	JCP 1964 **41** 760

(2) *Non-metals and the later B subgroup elements*

Crystalline hydrogen, oxygen, and nitrogen consist of diatomic molecules. Solid H_2 has a h.c.p. structure in which the rotation of the molecules probably persists at temperatures down to the absolute zero. The (isostructural) high-temperature forms of O_2 and F_2 have an interesting cubic structure with 8 molecules in the unit cell, of which 2 are spherically disordered while the remaining 6 behave as oblate spheroids. A spherically disordered molecule (A in Fig. 29.1) has 12 F_2 neighbours at 3·7 Å, while a molecule of type B has 2 neighbours at 3·3 Å, 4 at 3·7 Å, and 8 at 4·1 Å. Van der Waals distances between (stationary) F_2 molecules would be expected to be 2·7 Å for lateral contacts and approximately 4·1 Å for end-to-end contacts. The shortest intermolecular contacts are within 'chains' of molecules running parallel to the cubic axes. This structure is clearly not consistent with any form of dimer, as at one time suggested for $\gamma\text{-}O_2$.

The low-temperature (α) form of F_2 apparently consists of a distorted closest packing of molecules in which each F *atom* has 12 nearest neighbours in addition to the other atom of its own molecule (at 1·49 Å). The molecular axes are nearly perpendicular to the plane of a layer and the interatomic distances are greater

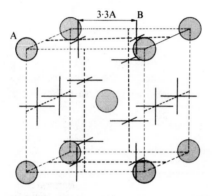

FIG. 29.1. The crystal structure of the high-temperature form of F_2. The two molecules at (000) and ($\frac{1}{2}\frac{1}{2}\frac{1}{2}$) are spherically disordered. The other six, at ($\frac{1}{4}\frac{1}{2}0$) etc., show an oblate spheroidal distribution of electron density. These molecules are represented by their major axes, the minor axes lying along the broken lines.

within a molecular layer (F—9 F, 3·24 Å) than between the layers (F—3 F, 2·84 Å). Contrast the isostructural Cl_2, Br_2, and I_2 (p. 329).

As noted in Chapter 3 many of the elements of this group crystallize at atmospheric pressure with structures in which an atom has $8 - N$ nearest neighbours, N being the ordinal number of the Periodic Group:

VII diatomic molecules F_2, etc.,
 VI rings (S_6, S_8, S_{12}, Se_8) or chains (fibrous S, 'metallic' Se and Te),
  V tetrahedral molecules in white P and presumably in the metastable yellow forms of As and Sb, or layer structures (P, red and black, As, Sb, and Bi),
 IV diamond structure (C, Si, Ge, grey Sn).

The structures of these elements are described in earlier chapters. The graphite structure is obviously exceptional, and the heavier elements in the subgroups IVB–VIB become increasingly metallic. In IVB Sn also forms the more metallic white polymorph, and Pb is a metal. In VB there is a progressive decrease in the difference between the distances to the 3 nearest and 3 next nearest neighbours in the layer structures of As, Sb, and Bi, while in VIB the interatomic distances between the chains in Se and Te indicate interactions between the chains stronger than normal van der Waals bonding (p. 232). The last member of the VIB subgroup is a metal. An X-ray study of metallic Po (in which a powder photograph was obtained from 100 μg) showed this metal to be dimorphic, one form being simple cubic and the other rhombohedral.[1] In each form Po has only 6 (equidistant) nearest neighbours (at 3·35 Å).

(1) JCP 1949 **17** 1293

(3) Boron, aluminium, the elements of subgroups IIB *and* IIIB, *tin and lead*
These elements are grouped together because (i) certain of them have peculiar structures which do not conform with the principles governing those of the elements in the adjacent groups in our classification, and (ii) those which have typical metallic structures show certain abnormalities which make it convenient to distinguish them from the 'true' metals of our fourth and fifth groups. Of the IIB subgroup elements zinc and cadmium crystallize with a distorted form of hexagonal closest packing, the axial ratio being nearly 1·9 instead of 1·63 as in perfect hexagonal closest packing. Each atom has six nearest neighbours in its own plane, the other six (three above and three below) being at a rather greater distance. The distances to these two sets of neighbours are 2·659 and 2·906 Å in Zn and 2·973 and 3·286 Å in Cd. It is not satisfactory simply to regard the atoms in these structures as packing like prolate ellipsoids, for in the close-packed ϵ phase of Cu–Zn (p. 1044) the axial ratio is *less* than 1·63, that is, we should have to assume that the atoms are behaving at some compositions as prolate and at others as oblate spheroids. It has been suggested that there is some tendency to form covalent bonds (dp^2 hybrids) in the basal planes, so accounting for the smaller interatomic distances in the close-packed planes. The striking decrease in the axial ratio of $MgCd_3$ from 1·72 to 1·63 on forming the ordered superstructure at 80°C (this superstructure consists of identical close-packed layers of type (d), Fig. 4.15) has been attributed to the decrease in the number of Cd–Cd contacts on formation of the ordered structure. (There is presumably also some kind of directional bonding in Sc, Y, and the hexagonal rare-earths with axial ratio *less* than 1·63, for example, $c : a \approx 1·57$ for Y, Ho, Er, and Tm.)

The normal (α) form of crystalline Hg has a quite different (rhombohedral) structure which may be derived from a simple cubic packing by distorting it so that the interaxial angle changes from 90° to $70\frac{1}{2}$°. Each atom has 6 nearest neighbours at 2·99 Å and 6 more at 3·47 Å. Alternatively the structure may be described as resulting from compression of c.c.p. along a body-diagonal, so that it is referable to either of the rhombohedral cells (compare Fig. 6.3, p. 197):

$$Z = 1: a = 2·99 \text{ Å}, \alpha = 70\tfrac{1}{2}°,$$

or

$$Z = 4: a = 4·58 \text{ Å}, \alpha = 98°.$$

A second (β) form is produced under pressure and after formation is the stable form below 79°K. It is b.c. tetragonal and results from compressing c.c.p. along a cube edge, giving each atom 2 nearest neighbours (2·83 Å) and 8 more distant ones (3·16 Å).[1] This structure is also adopted by Cd–Hg alloys containing 37–74 atomic per cent Hg.

(1) JCP 1959 **31** 1628

The structures of the typical and B subgroup elements of the third Periodic Group present many problems. The unexpectedly complex structures of boron are described in Chapter 24. Aluminium and thallium have c.p. structures which we shall discuss later. Gallium normally crystallizes with a rather complex structure in which an atom has the following set of neighbours: 1 at 2·44, 2 at 2·70, 2 at 2·73, and 2 at 2·79 Å. In a second (β) metastable form the two nearest neighbours define chains, but the differences between the distances to the various sets of neighbours are comparatively small, there being two at each of the following distances: 2·68, 2·77, 2·87, and 2·92 Å.[2] Gallium is notable for its low melting point (about 30°C), and the association of the atoms in pairs in the normal form of the element may persist in the liquid; it is reported that the X-ray diffraction pattern of liquid gallium is different from that of a simple close-packed liquid metal such as mercury. Indium crystallizes with a structure which is only a slightly distorted form of cubic closest packing, each atom having four neighbours at 3·24 and eight more at 3·36 Å. Thallium is dimorphic, having cubic (β) and hexagonal (α) modifications, and lead, which we include in this group, has the cubic close-packed structure.

We have said that the elements Al, In, Tl, and Pb, which crystallize with close-packed or approximately close-packed structures, are abnormal in certain respects. In the case of Al there is evidence that the atoms are not completely ionized in the pure metal and that the atomic diameter of the fully ionized Al^{3+} is nearer 2·70 than 2·86 Å, the closest distance of approach of atoms in aluminium itself. The melting point of the metal is only 8°C higher than that of Mg, in contrast to the rise in melting point with increase in valence in the first row and A subgroup elements:

(2) AC 1969 **B25** 995

Li	Be	B	C
178°	1285°	2300°	3500°C
Na	Mg	Al	Si
97°	650°	658°	1410°
K	Ca	Sc	Ti
64°	845°	1540°	1675°

Indium, thallium, and lead form a group of close-packed metals isolated from the rest of the close-packed metals. As may be seen from Table 29.2, the interatomic distances in these metals are very large compared with those in the neighbouring elements. Larger distances are to be expected on account of the much higher coordination number, but the distances are much greater than in the close-packed silver and gold, whereas there is very little variation in interatomic distance throughout the whole series of elements from Cu to As or from Ag to Sb, excluding In and white Sn. The figure 2·80 Å given for tin in Table 29.2 is the interatomic distance in grey tin (diamond structure). The normal form of this element is white tin, which has a less symmetrical structure than the grey variety. In white tin (Fig. 29.2) each atom has 4 nearest neighbours at 3·02 Å, forming a very flattened tetrahedron around it, and 2 more at 3·18 Å. Although the four nearest neighbours

TABLE 29.2

Interatomic distances in the elements (*to nearest neighbours*)

Cu	Zn	Ga	Ge	As
2·55	2·66	2·70(a)	2·44	2·51 Å
Ag	Cd	In	Sn	Sb
2·88	2·97	3·24	2·80(b)	2·90
			3·02(c)	
Au	Hg	Tl	Pb	Bi
2·88	2·99	3·40	3·49	3·10

(*a*) Weighted mean for 7 nearest neighbours. (*b*) Grey tin. (*c*) White tin.

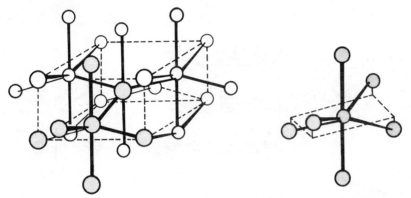

FIG. 29.2. The crystal structure of white tin, showing the deformed octahedral group of neighbours around each atom.

are at a greater distance than the four nearest neighbours in grey Sn (at 2·80 Å) there are also 4 Sn at 3·77 Å and 8 Sn at 4·41 Å, as compared with the 12 next nearest neighbours at 4·59 Å in grey Sn. White Sn is accordingly much more dense (7·31 g/cc) than grey Sn (5·75 g/cc), in spite of the fact that it is the high-temperature form. For the relation of white to grey Sn[3] see also p. 103. The suggestion that Sn is in different valence states in its two polymorphs is supported by the fact that if white and grey Sn are dissolved in concentrated HCl the compounds isolated from the resulting solutions are $SnCl_2 \cdot 2 H_2O$ and $SnCl_4 \cdot 5 H_2O$ respectively.[4]

Under high pressure (120 kbar) Ge, which is a semiconductor, is converted into a highly conducting form with the white tin structure.[5] On decompression at room temperature this changes into the high-pressure (tetragonal) form illustrated in Fig. 3.41(d), p. 110, which is some 11 per cent more dense than the normal cubic form (5·91 as compared with 5·33 g/cc). The bond length (2·48 Å) in this form is very similar to that in cubic Ge, but there is a very considerable range of bond angles (88–135°).[6]

(3) PRS 1963 A **272** 503

(4) JCP 1956 **24** 1009

(5) Sc 1963 **139** 762

(6) AC 1964 **17** 752

It is considered probable that in Tl and Pb, and possibly also in In, the atoms are only partly ionized, for example, to the stages Tl^+ and Pb^{2+} in the former elements. The peculiar stability of the pair of s electrons in the outer shells of these elements has already been noted in Chapter 26.

(4) *The transition elements and those of subgroup* IB

This group of more than 50 elements, including the 4f and 5f elements, comprises most of the metals. The structure of at least one form of most of these metals is known (Table 29.3), and with few exceptions they crystallize with one or more of three structures. These are the hexagonal and cubic close-packed structures and the body-centred cubic structure.

The two simplest forms of closest packing, hexagonal and cubic, have already been described in Chapter 4. In Fig. 29.3 we show these in their conventional orientations together with the body-centred cubic structure. Sufficient atoms of adjacent unit cells are depicted to show the full set of nearest neighbours of one

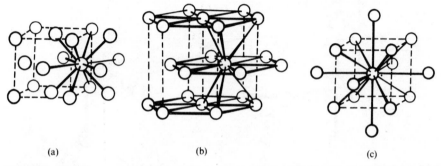

(a)   (b)   (c)

FIG. 29.3. The three commonest metallic structures: (a) cubic close-packing, (b) hexagonal close-packing, and (c) body-centred cubic.

atom. The body-centred cubic structure is slightly less closely packed than the others, the effective volume per atom (of radius a) being $6.16a^3$ as compared with $5.66a^3$ in close-packed structures. In the latter an atom has twelve equidistant nearest neighbours, the next set of neighbours being 41 per cent farther away.

The body-centred cubic structure is commonly the high-temperature form of metals which are close-packed at lower temperatures. The transition temperature is above room temperature for many metals (Ca, Sr, La, Ce, etc., Tl, Ti, Zr, Th, Mn) but below room temperature for Li and Na ($78°K$ and $36°K$ respectively). In this structure an atom has 8 equidistant nearest neighbours, but the six next nearest neighbours are only 15 per cent farther away. Neglecting this comparatively small difference, or alternatively, adopting the more logical definition of coordination number based on the *polyhedral domain* of an atom (p. 61), the c.n. in this structure is 14. Note that if this structure is compressed or extended along one of the 4-fold axes to become body-centred tetragonal with axial ratio $c : a \sqrt{\frac{2}{3}}$ or $\sqrt{2}$ respectively, the number of nearest neighbours becomes ten or twelve. (Compare the structure of Pa, p. 1018.) The latter case corresponds to the face-centred cubic structure.

TABLE 29.3
The crystal structures of metals

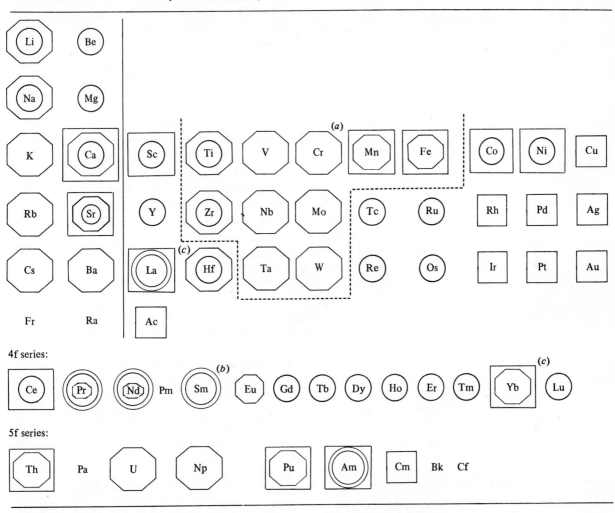

The primary object of this Table is to emphasize the widespread occurrence of the b.c.c. and the c.p. structures:

close-packed structures

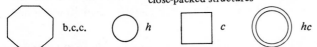

Certain metals show a more complex behaviour and are further discussed in the text, for example, Mn, W, 4f and 5f metals (Table 29.4, p. 1019).

(*a*) The electrodeposited h.c.p. structure is now known to be that of the hydride CrH (p. 298).

(*b*) Normally close-packed *chh* type.

(*c*) A c.c.p. form of La has been described and an intermediate h.c.p. form of Yb (260°–720°C) is apparently stabilized by impurities. For references to b.c.c. and c.p. metals see: JACS 1963 **85** 1238.

Metals and Alloys

There is an interesting qualitative difference in properties between cubic and hexagonal close-packed metals. The ductility, malleability, and softness of pure metals depend, at least to some extent, on the ease with which adjacent planes and rows of atoms can glide over one another. (These properties are also very much influenced by the numbers of flaws and crystal boundaries in the material—a single crystal of a metal like cadmium being very soft and easily deformed, whereas the polycrystalline material is quite hard and brittle—and also by the presence of minute amounts of impurity.) Gliding takes place on the most closely packed planes, and the direction of gliding in these planes follows the lines of most closely packed atoms. In a cubic close-packed metal there are four equivalent sets of parallel closest-packed planes, perpendicular to the four 3-fold axes which are the characteristic symmetry elements of a cubic crystal, as compared with only one set of such planes in a hexagonal close-packed crystal. Owing to the greater possibility of gliding in a cubic close-packed crystal, such crystals are in general more malleable and ductile than those with hexagonal close-packed structures. Thus Cu, Ag, Au, and γ-Fe are fairly soft, ductile, and malleable in comparison with such metals as Cr, V, and Mo which have hexagonal close-packed and/or body-centred structures. Iron is a particularly important metal which crystallizes with either the cubic close-packed or the body-centred cubic structure, depending on the heat treatment it has received. We shall deal briefly with iron and steel in the section on interstitial compounds.

There is one feature peculiar to the polymorphism of certain of these transition elements. In the case of other elements a particular polymorphic form is, under a given pressure, stable over one temperature range. (At atmospheric pressure one form may, of course, always be more stable than the other.) Tin, for example, has a transition temperature of 13·2°C, below which α (grey) tin is the stable form and above which β (white) tin is the stable form. Iron, on the other hand, crystallizes at temperatures below 906°C with the body-centred cubic structure and from 906° to 1401°C with the cubic close-packed structure, but from 1401°C to the melting point (1530°C) the body-centred cubic structure again becomes the stable form. We have therefore the face-centred structure stable over a certain temperature range, on both sides of which the body-centred structure is stable. Cobalt exhibits a more complex behaviour.[1] At temperatures above 500°C the structure is cubic close-packed but on cooling it becomes partly hexagonal. The sequence of layers is not, however, $ABAB\ldots$ or $ABCABC\ldots$ throughout but a random sequence. Thus in a typical specimen of 'hexagonal close-packed' cobalt at room temperature the probability that alternate planes are similar is less than one; a value 0·90 was found in one specimen.

The pure hexagonal structure of Ni is observed only in very thin films deposited electrolytically on hexagonal Co. It has the same interatomic distance (2·50 Å) as in c.c.p. Ni and cell dimensions, a = 2·50 Å, and c = 3·98 Å. A hexagonal phase with Ni–Ni 2·63 Å (a = 2·63 Å, c = 4·31 Å) is the carbide Ni_3C which can exist, if the particle diameter is less than 200 Å, with as little as 1 atomic per cent carbon.[2]

A second point of interest in connection with the structures of the transition metals is the sequence of structures adopted by the elements preceding the metals

(1) PRS A 1942 **180** 268, 277

(2) AC 1966 **21** 1001

1016

Cu, Ag, and Au. Table 29.3 shows that a similar sequence occurs in each of the series, Ti–Cu, Zr–Ag, and Hf–Au. It has been suggested that these changes in structure are associated with the extent to which the d-shell is filled. In the IB metals the d-shells are full and in these metals the closed d-shells are in contact, resulting in intense repulsive forces and hence in close-packed structures. With the decrease in number of electrons in these shells a more open structure becomes more stable. It should, however, be noted that close-packed structures again occur for the elements of the earlier Periodic Groups (see Table 29.3), so that we may alternatively regard the elements enclosed by the broken lines (which, except for Ti and Zr, are not close-packed at ordinary temperatures) as constituting a group of exceptional elements surrounded by those with close-packed structures.

Manganese.   The normal (α) form has a complex cubic structure ($Z = 58$) which is further discussed on p. 1039. In β-Mn (stable between 800° and 1100°C, but can be quenched to room temperature) there are two kinds of atom with different environments, each set having 12 nearest neighbours at distances from 2·36 Å to 2·67 Å. At around 1100°C the γ form appears. Quenched material has tetragonal symmetry corresponding to a distorted form of the c.c.p. structure which it possesses at the high temperature. At 1134°C there is a further transformation to a b.c.c. structure, this form persisting up to the melting point (1245°C).

Tungsten.   This element is normally encountered in the b.c. cubic (α) form, but a second (β) form, described as being stable at temperatures below about 650°C, has long been recognized. This form has to be prepared at temperatures below 650°, above which it is irreversibly, though rather sluggishly, converted into α-W, by chemical methods such as reduction of an oxide. The claim that 'β-W' is in fact an oxide W_3O[3] and not a polymorph of the metal (see p. 473) is not universally

(3) AC 1954 **7** 351; N 1955 **175** 131

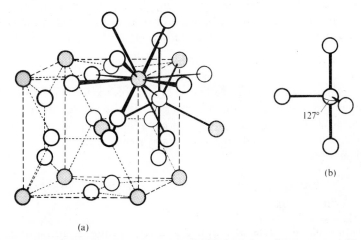

(a)

(b)

127°

FIG. 29.4. (a) The 'β-tungsten' structure. As regards environment the atoms are of two types. The twelve equidistant neighbours of an atom of the first type (shaded circles) and the six nearest neighbours of an atom of the second type (open circles) are shown. (b) The arrangement of the four nearest neighbours of an atom in metallic uranium (α form).

(4) JPC 1956 **60** 1148
(5) ZaC 1957 **293** 241

accepted,[4] for it has been stated[5] that β-W can be prepared with less than 0·01 O atoms per atom of W. In any case it is convenient to refer to the structure of Fig. 29.4(a) as the β-W structure. It is adopted by a number of intermetallic compounds and silicides, including NiV_3, GeV_3, SiV_3, and $SiCr_3$.

The 4f *(lanthanide or rare-earth) metals.*   The abnormal chemical properties of the elements immediately preceding Gd and Lu have already been noted in our account of the 4f elements (Chapter 28). In the elementary state also Eu and Yb are abnormal. The majority of the 4f metals crystallize with the h.c.p. structure, though the *hc* type of closest packing is found in Pr and Nd (also La) and the 9-layer *chh* sequence is the normal structure of Sm.[6] On the other hand Eu is b.c.c. and Yb c.c.p. The metallic radii of the other 4f elements fall steadily from about 1·82 Å (Pr) to 1·72 Å (Lu), but Eu (2·06 Å) and Yb (1·94 Å) have much larger radii, presumably due to the withdrawal of one electron into an inner orbital. The behaviour of Yb under pressure is extremely interesting.[7] At 20 kbar the metal becomes semiconducting, and at 40 kbar it again becomes a metallic conductor. This latter change is accompanied by a structural change (f.c.c. → b.c.c.), that is, the c.n. has altered from 12 to 8 + 6 with a decrease in volume of 3·2 per cent at the transition point. The radius of Yb in the b.c.c. structure (1·75 Å, corresponding to 1·82 Å for 12-coordination) is considerably less than that of Yb(II) in the normal form (1·94 Å) and lies on the curve for other 4f metals M(III), showing that the transition at the higher pressure corresponds to a change from Yb(II) to Yb(III). This transformation requires higher pressure as the temperature is reduced. With this behaviour contrast that of Ce, which under high pressure changes to another f.c.c. phase, but this change takes place at *lower* pressures as the temperature is reduced—at atmospheric pressure at 109°K. The volume decrease of 16·5 per cent accompanying this change is attributed to the promotion of a 4f electron to a bonding orbital. Both f.c.c. phases can coexist at low temperatures, the cell dimensions at 90°K being 5·14 Å and 4·82 Å.[8]

(6) AC 1970 **B26** 1043

(7) IC 1963 **2** 618

(8) JCP 1950 **18** 145

The 5f *(or actinide) metals.*   The elements between Th (c.c.p.) and Am (close-packed *ABAC* . . .) show a very complex behaviour in the elementary state. Not only is polymorphism common, Pu having as many as six forms, but a number of the structures are peculiar to the one element. This is true of the body-centred tetragonal structure of Pa, in which an atom has ten practically equidistant neighbours, and of the α-U, β-U, α-Np, β-Np, and γ-Pu structures.

Uranium is trimorphic:

$$\alpha\text{-U} \quad (660°C) \quad \beta\text{-U} \quad (760°C) \quad \gamma\text{-U}$$
$$\text{orthorhombic} \quad\quad\quad \text{tetragonal} \quad\quad\quad \text{b.c.c.}$$

The structure of α-U could be regarded as a very deformed version of hexagonal closest packing in which each atom has only four nearest neighbours, two at 2·76 Å and two at 2·85 Å. The arrangement of these neighbours is very unusual, for they lie in two perpendicular planes at four of the five vertices of a trigonal bipyramid

1018

(Fig. 29.4(b)); for further remarks on this structure see p. 1025. β-U apparently has a typical metallic structure with high c.n. (12 or more), but there is not yet agreement as to the details of this structure.

Neptunium shows some resemblance to uranium:

$$\alpha\text{-Np} \quad (278°C) \quad \beta\text{-Np} \quad (540°C) \quad \gamma\text{-Np}$$

$$\text{orthorhombic} \qquad \text{tetragonal} \qquad \text{b.c.c.(?)}$$

In the α form each atom has only four nearest neighbours. For one-half of the atoms these are arranged as in α-U, but for the others they are arranged at one polar and three equatorial vertices of a trigonal bipyramid. In the β form also each atom forms four strong bonds, but here one-half of the atoms have their nearest neighbours at the vertices of a deformed tetrahedron while for the others they lie at the basal vertices of a very flat tetragonal pyramid.

As already mentioned, Pu is notable for having six polymorphs. The α form has a complex monoclinic structure in which there are 8 kinds of non-equivalent atom, each of which forms a number of short bonds (2·57–2·78 Å) with the remainder in the range 3·19–3·71 Å (compare α-U, α-Np, and β-Np), giving total c.n.'s of 12–16. Three-quarters of the atoms form 4 short and 10 long, one-eighth form 3 short and 13 long, and the remainder 5 short and 7 long bonds. The β form also has a complex monoclinic structure ($Z = 34$, 7 kinds of non-equivalent atom). Co-ordination numbers range from 12 to 14, and there is no obvious relation to the α or γ structures. In the γ form (Fig. 29.5) Pu has 10 nearest neighbours: 4 at 3·026 Å, 4 at 3·288 Å, and 2 at 3·159 Å. The three high-temperature forms have much simpler structures, (Table 29.4), the δ' structure being a slightly distorted version of δ.

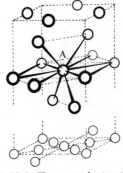

FIG. 29.5. The crystal structure of γ-plutonium showing the atom A and its ten nearest neighbours projected on to the base of the unit cell.

TABLE 29.4
Crystal structures of 5f metals

Element		Crystal structure	Reference
Pa		Body-centred tetragonal: c.n. 10, 8 at 3·21 Å, 2 at 3·33 Å	AC 1952 **5** 19
U	α	Orthorhombic (A 20 structure) } see text	AC 1970 **B26** 129
	β	Tetragonal ($Z = 30$)	AC 1971 **B27** 1740
	γ	Body-centred cubic (U—U, 3·01 Å)	
Np	α	Orthorhombic	
	β	Tetragonal } See text	AC 1952 **5** 660, 664
	γ	Cubic (b.c.?)	
Pu	α	Monoclinic (c.n. 12–16)	AC 1963 **16** 777
	β	Monoclinic (c.n. 12–14)	AC 1963 **16** 369
	γ	F.c. orthorhombic (c.n. 10)	AC 1955 **8** 431
	δ	F.c. cubic (Pu–Pu, 3·28 Å) }	JCP 1955 **23** 365
	ϵ	B.c. cubic (Pu–Pu, 3·15 Å)	
	δ'	B.c. tetragonal	TAIME 1956 **206** 1256
Am		F.c.c. and *hc* (density 13·67 g/cc, $r = 1·73$ Å)	JINC 1962 **24** 1025
		hc (density 11·87 g/cc, $r = 1·82$ Å).	JACS 1956 **78** 2340
Cm		F.c.c. (density 19·2 g/cc, $r = 1·55$ Å)	JCP 1969 **50** 5066
		hc (density 13·5 g/cc, $r = 1·74$ Å)	JINC 1964 **26** 271

Metals and Alloys

In two studies of Am metal the *hc* form was found but with rather different densities and cell dimensions (Table 29.4). Although the less dense form was not reproduced in the later study it is conceivable that there are two forms of the metal with the same structure, as in the case of Ce. In the case of Cm also the *hc* structure reported earlier was not found in a later study; here the difference between the densities of two (close-packed) forms is more than a little surprising.

(5) *The typical and* A *subgroup elements of groups* I *and* II
The metals of this group call for little discussion. The alkali metals adopt the b.c.c. structure under ordinary conditions, but under high pressure Cs has been obtained in a close-packed form and at low temperatures both Li and Na, but not K, Rb, or Cs, transform to close-packed structures.[1] These are essentially hexagonal close-packed, but Na exhibits stacking faults and in the case of Li cold-working induces a transformation to cubic closest packing. As shown in Table 29.3 Ca and Sr are trimorphic, the sequence of structures with rising temperature being the same for the two metals:

$$\text{Ca: f.c.c. } (250^\circ\text{C}) \text{ h.c.p. } (450^\circ\text{C}) \text{ b.c.c.}$$
$$\text{Sr: f.c.c. } (215^\circ\text{C}) \text{ h.c.p. } (605^\circ\text{C}) \text{ b.c.c.}$$

though a later paper describes only the f.c.c. and b.c.c. forms of Ca.[2]

Interatomic distances in metals: metallic radii

Apart from the intrinsic interest of the interatomic distances in metals, it is useful to have a set of radii to refer to when discussing the structures of alloys. Since the c.n. 12 is the most common in metals, it is usual to draw up a standard set of radii for this coordination number. For the metals with ideal close-packed structures the radii are simply one-half the distances between an atom and its twelve equidistant nearest neighbours. Many structures, however, deviate slightly from ideal hexagonal closest packing in such a way that six of the neighbours are slightly farther away than the other six, for example,

	Axial ratio	6 at	6 at	Mean
Be	1·5848	2·2235	2·2679	2·25 Å
Y	1·588	3·595	3·663	3·63
α-Zr	1·589	3·166	3·223	3·19

In such cases the mean of the two distances is taken. (We are not here referring to zinc and cadmium which show a very much larger deviation from closest packing, with axial ratios 1·856 and 1·885 respectively.) For metals which crystallize with structures of lower coordination the radii for 12-coordination have to be derived in other ways. From a study of the interatomic distances in many metals and alloys Goldschmidt found that the apparent radius of a metal atom varies with the coordination number in the following way. The relative radii for different

(1) AC 1956 **9** 671

(2) AC 1959 **12** 419

1020

coordination numbers are:

C.N.	
12	1·00
8	0·97
6	0·96
(4	0·88)

For the alkali metals, for example, the radius for c.n. 12 is derived from one-half the observed interatomic distance in the b.c.c. structure by multiplying by 1·03. This ratio, which is equal to $6^{\frac{1}{2}}2^{\frac{1}{3}}/3$, results from assuming that no change of volume accompanies the transition from 8- to 12-coordination (see also p. 1018). For some of the B subgroup elements which crystallize with structures of very low coordination (for example, Ge, Sn) the problem is more difficult. The use of a ratio such as that given above is difficult to justify for such a large change in c.n., since the bonds in structures such as the diamond structure are probably nearer to covalent than to purely metallic bonds. There are alternative ways of deriving radii for c.n. 12 for such elements. Many of these elements form alloys with true metals in which both elements exhibit high coordination numbers. The radius of the B subgroup element can be found from the interatomic distance in such an alloy, for example, that of Sb from the Ag–Sb distance in the hexagonal close-packed Ag_3Sb or that of Ge from the Cu–Ge distance in the hexagonal close-packed Cu_3Ge. This method is open to objection since there is probably not true metallic binding in such compounds and it is necessary, of course, to assume that the radius of Ag, for example, in Ag_3Sb is the same as in pure silver. The second method is to derive the radius of the B subgroup element from the variation in cell dimensions of the solid solution it forms in a true metal. By extrapolation, the cell dimension of the hypothetical form of the B subgroup element with c.n. 12, and hence the interatomic distance therein, may be obtained, assuming a linear relation between cell dimension and concentration of solute atoms. Unfortunately this relation is not always linear, thus necessitating the use of empirical relations. Also, it is not certain that the radius of the solute element is not affected by the nature of the solvent metal, for there may be alterations in the state of ionization of an atom in different alloys. Nevertheless it is possible to derive a set of radii for many elements for c.n. 12 which are valuable when considering the structures of alloys. Such a set of radii is given in Table 29.5.

The actual interatomic distances in the elements are plotted against atomic number in Fig. 29.6, which shows a number of points of interest. The general periodic variation is apparent, as also is the fact that there is no general increase in the size of atoms with increasing atomic number. For example, Th (a.n. 90) is not much larger than Li (a.n. 3) and actually smaller than Na (a.n. 11), owing to the tighter binding of the electrons in the heavier elements with increased nuclear charge. A second point is that although the transition and B subgroup elements of the second long period are larger than those of the first, there is no similar increase in size from the second to the third long period. Compare, for example, the atomic diameters of Cu, 2·551 Å, Ag, 2·883 Å, and Au, 2·877 Å. We have already referred to the lanthanide contraction and the very close resemblance in chemical properties

TABLE 29.5

Metallic radii for 12-coordination (Å)

Li	Be
1·57	1·12

Na	Mg	Al
1·91	1·60	1·43

K	Ca	Sc	Ti	V	Cr	Mn	Fe	Co	Ni	Cu	Zn	Ga	Ge	
2·35	1·97	1·64	1·47	1·35	1·29	1·37	1·26	1·25	1·25	1·28	1·37	1·53	1·39	

Rb	Sr	Y	Zr	Nb	Mo	Tc	Ru	Rh	Pd	Ag	Cd	In	Sn	Sb
2·50	2·15	1·82	1·60	1·47	1·40	1·35	1·34	1·34	1·37	1·44	1·52	1·67	1·58	1·61

Cs	Ba	La	Hf	Ta	W	Re	Os	Ir	Pt	Au	Hg	Tl	Pb	Bi
2·72	2·24	1·88	1·59	1·47	1·41	1·37	1·35	1·36	1·39	1·44	1·55	1·71	1·75	1·82

4f elements: Ce (1·82)–Lu (1·72) but Eu, 2·06, Yb, 1·94

5f elements:

	Th	Pa	U	Np	Pu	Am	Cm
	1·80	1·63	1·56	1·56	1·64	See Table 29.4	

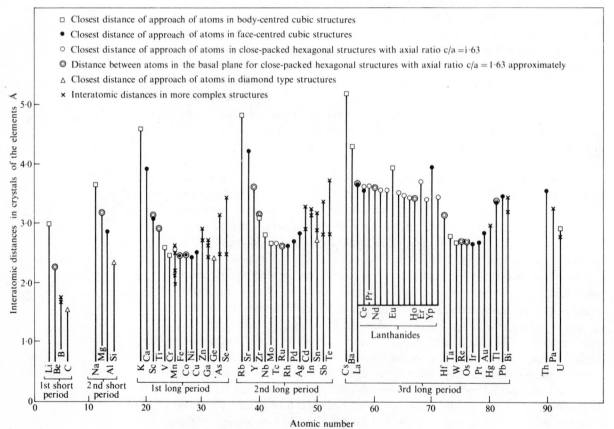

FIG. 29.6. Interatomic distances in metals.

between such pairs of elements as Zr and Hf or Nb and Ta. The radii of the intervening 4f elements fall from 1·82 Å (Ce) to 1·72 Å (Lu) while the 4f sub-shell is being filled, except that Eu and Yb have much larger radii, as mentioned on p. 1018. The variation of atomic radii in the long periods, with minima in the neighbourhood of the Group VIII triads, corresponds to maxima or minima in a number of physical properties—compressibility, tensile strength, hardness, and melting point. Pauling has suggested that if an element such as iron can utilize its 3d orbitals as well as those of the valence shell in its carbonyls and in the ferrocyanide and other complex ions, this may also be the case in the metal. Assuming that for the elements of the first long period 3d, 4s, and 4p orbitals (nine in number) are available for bond formation, then the maximum number of bonds could be formed when there are nine electrons available per atom. This occurs at Co, Rh, and Ir, for the first, second, and third long periods respectively. In the preceding elements there are fewer available electrons and in the later ones more electrons than orbitals available. Pauling has considered in some detail the nature of the bonds in many metals and some alloys in relation to the bond lengths in the crystals, and his theory of resonance in metallic structures will be briefly outlined.

Theories of metallic bonding

In the conventional theory of the metallic state, which has developed from the early electron-gas theories of Drude and Lorentz, the methods of wave mechanics are applied to the behaviour of electrons in a 3-dimensional periodic field, the periodicity of which is that of the crystal lattice. The possible states of the electrons are described in terms of permitted energy bands (zones) which are separated by ranges of forbidden energies, and such a theory gives a satisfactory picture of normal conductors, semiconductors, and insulators. It has had considerable success in calculating properties such as lattice spacings and energies and the compressibilities of a number of metals, assuming only the type of crystal structure (for example, f.c.c.). It does *not* offer an explanation of supra-conductivity or, of course, of mechanical properties outside the elastic range, the latter being due to the secondary structure (mosaic structure, dislocations, etc.). We propose to say no more about the wave-mechanical theory of metals here but only emphasize that in the band theory the assembly of electrons is treated as a whole, first in a simple periodic 3-dimensional field, and then later in a field which is modified by supposing that in the vicinity of atoms the wave functions have symmetry characteristics resembling those (the s, p, and d functions) of free atoms. The wave function of each electron extends over the whole crystal, and in this sense the band theory of metals may be compared with the molecular orbital treatment of molecules, though as is well known it is not in practice possible to deal with genuine molecular orbitals. The analogue of the valence-bond treatment would be a theory of metals in which the bonds from a metal atom to its neighbours are considered in terms of the electronic structure of the atom, regarding the whole crystal as an assembly of atoms linked by localized bonds.

1023

(1) JACS 1947 **69** 542; PRS A 1949 **196** 343

Pauling[1] adopts the viewpoint that there is no essential difference between metallic bonds and ordinary covalent bonds, a view originally put forward by V. M. Goldschmidt in 1928. However, there are usually very high coordination numbers in metallic crystals as compared with normal covalent crystals, and, moreover, in a metal such as sodium only four orbitals (one s and three p) are available for forming the 8 + 6 bonds of the b.c.c. structure. Pauling therefore assumes that all or most of the outer electrons of the atom, including the d electrons in the case of the transition metals, take part in bond formation, and that there is resonance of a new type, to which we refer shortly. These concepts imply fractional bond orders and valences. The drop in atomic diameter from K through Ca, Sc, and Ti to V (and similarly from Rb to Nb and from Cs to Ta) and the approximate constancy of size from the element of Group V through the Group VIII triad in each series of transition metals is interpreted to mean that on passing from K to V the numbers of bonding electrons increase from 1 to 5 per atom, with a regular increase in the number of covalent bonds between which resonance can occur and hence a steady drop in interatomic distance. It is then supposed that in the elements from Cr to Ni not all of the nine available orbitals (s, three p, and five d) are used in bonding, but that only 5·78 are stable spd *bonding* orbitals, 2·44 being *atomic non-bonding* d orbitals and the remaining 0·78 (the *metallic* orbital) being necessary to allow non-synchronized resonance between the individual valence bonds. These numbers are derived from the saturation moments of (ferromagnetic) iron, cobalt, and nickel. Pauling's electronic structures for a number of metals are shown in Table 29.6

TABLE 29.6
Electronic structures in the metallic state (Pauling)

Metal	Atomic d orbitals		Saturation moment (BM)		Total number of electrons in bonding 3d orbitals	Total number of 3d and 4s electrons
	+	−	Assumed	Observed		
Cr	0·22	0	0·22	—	5·78	6
Mn	1·22	0	1·22	—	5·78	7
Fe	2·22 3·22	0	2·22	2·22	5·78	8
Co	2·44 4·22	0·78	1·66	1·61	5·78	9
Ni	2·44	1·78	0·66	0·61	5·78	10

For Cr, Mn, and Fe the number of electrons assigned to atomic d orbitals is less than the number of orbitals so that there is no pairing of spins, but in Co, for example, there are 3·22 electrons to be accommodated in 2·44 orbitals, and therefore 0·78 are paired and the moment drops to 2·44−0·78 = 1·66 BM.

This picture of a gradual building up of a core of 3d electrons on passing from Cr to Ni might account for the difference (leading, for example, to superstructure formation in solid solutions) between elements apparently so similar as Fe and Co or Co and Ni. With this may be compared the conventional explanation of the magnetic properties of, for example, Ni, that on the average 9·39 of the ten outer electrons occupy the 3d shell and that the remaining 0·61 electron occupies the 4s shell and is responsible for the cohesive forces, the 'hole' of 0·61 electron in the 3d level giving rise to the magnetic moment of this value (in Bohr magnetons). For Co the cohesion would be due to 0·61 s electron compared with 0·22 for Fe and 1·00 for Cu, corresponding to magnetic moments: Fe, 2·22, Co, 1·61, and Cu, diamagnetic. The difference between normal and metallic resonance may be illustrated by considering grey and white tin. In grey tin the atoms are forming normal tetrahedral sp^3 bonds, using only four of the nine ($d^5 sp^3$) orbitals. There is no 'metallic orbital' available so that the crystals are essentially non-metallic. In white tin it is supposed that the fourteen outer electrons of Sn are in eight of the nine orbitals—six pairs in six ($d^5 s$) hybrids and two electrons in two other orbitals, leaving one 'spare' orbital. Two bonds are then supposed to resonate among six bond positions but not necessarily synchronously because of the existence of the metallic orbital. The resonance involves neutral 2-covalent Sn, 3-covalent Sn^-, and 1-covalent Sn^+, and it is supposed that the crystal is stabilized by this resonance and also by the energy resulting from interchange of one electron between p and s orbitals.

The figures in Table 29.6 imply valences of 5·78 for the transition metals Cr, Mn, Fe, Co, and Ni, and the same is assumed for the series Mo–Pd, W–Pt, and U–Cm. It is less easy to justify the valences of 5·44, 4·44, and 3·44 respectively for the elements of Groups IB, IIB, and IIIB, which follow from the interpretation of interatomic distances in terms of an essentially empirical equation. This relates the metallic radius $R(n)$ for a bond of order n (i.e. one involving n electron pairs) to the single-bond radius $R(1)$:

$$R(1) - R(n) = 0.300 \log_{10} n \qquad (1)$$

and values of $R(1)$ for many metals have been tabulated (Pauling, ref. (1)). The values of the bond-order n are in general non-integral. For a close-packed structure, in which each atom has twelve equidistant neighbours, $n = v/12$, where v is the valence. For example, for the hexagonal close-packed form of Zr, $n = \frac{1}{3}$, and for the body-centred cubic form, $n = \frac{1}{2}$ (neglecting the six more distant neighbours). For more complex structures in which there are various numbers of bonds of different lengths the bonding powers of the atoms are divided up among the different bonds in accordance with the interatomic distances, so that $\Sigma Nn = v$. The application of these ideas to uranium will illustrate the method.

One form of metallic uranium has the body-centred cubic structure, from which $R(1)$ is found to be 1·421 Å, assuming $v = 5.78$. The interatomic distances in the orthorhombic α form then correspond to the following bond orders:

	N		n	
	2 at	2·76 Å	1·36	
U	2	2·85	0·96	$\Sigma Nn = v = 5.96$
	4	3·27	0·19	
	4	3·36	0·14	

(2) AC 1948 1 212

This treatment has also been applied to FeSi,[2] with which CrSi, MnSi, ReSi, and CoSi are isostructural. In the FeSi structure each atom has 7 neighbours of the other kind at the distances 2·29 (1), 2·34 (3), and 2·52 (3), and 6 further neighbours of its own kind at 2·75 Å (for Fe) or 2·78 Å (for Si). Application of equation (1) using the $R(1)$ values from elementary iron and silicon gives, however, a calculated valence of 6·85 ($\Sigma\, Nn$) for Si, whereas the value 4 is to be expected. In order to retain a valence of 4 for Si it is necessary to neglect the bonding of the Si to 6 Si at 2·78 Å (to which bond equation (1) assigns a bond-order of 0·19), to assume for Fe a valence of 6 instead of the 5·78 previously assumed, and to use different single-bond radii for Fe^{VI} for the Fe—Si and Fe—Fe bonds on the grounds that these (hybrid) bonds have different amounts of d character. More recently it has been pointed out[3] that the Pauling equation is not consistent with the interatomic distances for 12- and 8-coordination in most of the metals which show a transition from a c.p. to the b.c.c. structure, the distances being in the ratio 1·03 : 1, and that it is questionable whether the equation should be used to discuss interatomic distances in intermetallic compounds.

(3) JACS 1963 85 1238

In later developments of the Pauling theory attempts are made to interpret the fractional valences as averages of integral valences, and in addition the hypothesis was advanced that a special stability is to be associated with bond numbers that are simple fractions ($\frac{1}{2}$, $\frac{1}{3}$, etc.). The criticism has been made that the Pauling theory is essentially a discussion of known experimental data in terms of empirical assumptions for which no independent evidence is available, that a large number of arbitrary assumptions have been made, and that no useful generalizations have yet resulted. The need is not for further discussions of structures but for the calculation of physical properties by fundamental methods which do not assume the values of the properties concerned, or, alternatively, the generalization of facts to a number substantially greater than that of the assumptions involved.

Other authors have attempted to develop hybrid bond theories to account for the apparent directional characteristics of metallic bonding. Not only has the concept of varying amounts of d orbitals been used but also the concept of partial occupation of hybrid orbitals.[4] Four tetragonal pyramidal d^4 bonds might be regarded as forming eight cubic 'delocalized' bonds in an electron-deficient system comparable with the π-electron systems of conjugated molecules. Alternatively eight equivalent bonds directed towards the vertices of a cube may arise from tetrahedral sd^3 and tetragonal pyramidal d^4 hybrids, and six longer bonds in the b.c.c. structure from delocalized d^3 trigonal prism hybrids.

(4) PRS A 1957 240 145

Crystal structure and physical properties

In this section we shall describe the crystal structures of ideal crystals, that is, the structures which would extend without interruption throughout ideal crystals but which in the normal material, with a mosaic and moreover a polycrystalline structure, persist only over comparatively small volumes. In the case of metals and alloys the imperfections in the crystals are of the utmost importance in determining the properties of the material, and it is necessary to remember that the properties of the greatest value in metals cannot be explained in terms of the ideal structures we shall be describing. For the details in the structure of an alloy to be seen under the optical microscope they must be of the order of magnitude of 10^{-4} cm, that is, the polycrystalline structure or regions of different composition in a non-homogeneous alloy must be on this scale. Atomic arrangement as determined by the X-ray method is on the scale of 10^{-8} cm, interatomic distances lying in most cases between 2 and 3×10^{-8} cm and cell dimensions between 2 and 12×10^{-8} cm. The structures which are found in the intermediate range, from 10^{-4} to 10^{-8} cm, are probably of as much importance in determining the mechanical properties of a metal as the structures visible under the optical microscope. Most metals and alloys which have valuable mechanical and physical properties are neither in true equilibrium nor are they structurally homogeneous or perfect crystals.

A pure metal such as copper or iron, with a close-packed structure, is extremely soft, being an aggregate of ions held together not by direct bonds between the atoms as in a homopolar or ionic crystal but by the 'atmosphere' of electrons. As far as we know, the elastic limit of a perfect single crystal of a metal, that is, without impurity atoms or flaws, may tend to zero as perfect purity and regularity of structure are approached. Useful properties arise as the result of the introduction of impurities, often in comparatively small concentration, or after cold-working (rolling or drawing) or age-hardening. The effect of cold-working is to break up the single crystal into an aggregate of small crystals with dimensions of the order of 10^{-4}-10^{-5} cm, and the properties of the resultant material depend on the average size, the state of strain, and the relative orientations of these minute portions of the polycrystalline mass. The physical properties of a large number of commercial alloys are enhanced by heat-treatment processes referred to as age-hardening and precipitation-hardening. In the simple case of a binary alloy the hardening constituent is melted with the solvent metal and the alloy is quenched. Since the solubility of the former metal is lower at lower temperatures the cold solid solution is supersaturated, but the rapid cooling prevents rearrangement of the atoms. The alloy is then subjected to the 'ageing' treatment, which consists in keeping it at a temperature far below its melting point for a considerable time so that the supersaturated solid solution begins to break down. The Cu—Al alloy mentioned below can be fully hardened at a temperature of 140°C over a period of 22 hours. The alteration in physical properties takes place during this ageing process, but the alloy may attain maximum hardness before any precipitation visible under the microscope has taken place. Apparently the separation of solute from the super-saturated solution has taken place to different extents in different alloys when maximum hardness is reached. For example, in a silver-copper alloy (7·5 per cent Cu) of maximum hardness, precipitation is practically complete, that is, copper has separated out from the silver lattice. On the other hand, in an aluminium-copper alloy containing 4·8 per cent Cu, maximum hardness is reached before such precipitation has taken place, the separation of the solute having proceeded only as far as the segregation of Cu atoms into 'islands' on the Al lattice. As our last example we may take an alloy with the approximate composition $FeCu_4Ni_3$. This is cubic close-packed above 800°C and if quenched

1027

retains this structure as a random substitutional solid solution. If annealed for a week at 650°C it breaks up into two phases of approximate compositions $FeCu_{1.3}Ni_6$ and $FeCuNi_2$, both of which are still cubic close-packed but have slightly different cell dimensions. On slow cooling, which is equivalent to a treatment intermediate between quenching and quenching followed by complete annealing, the alloy apparently attempts to remain a single phase but cannot quite achieve this because of the difference in cell dimensions of the two phases mentioned above. The result is apparently a lamellar structure with layers alternately rich in copper and iron, some hundreds of atoms wide and about fifty atoms thick.

Solid solutions

A characteristic property of metals is that if two (or more) are melted together in suitable proportions a homogeneous solution often results. When cooled this is called a solid solution because, as in the case of a liquid solution, the solute and solvent atoms (applying the term solvent to the metal which is in excess) are arranged at random. Random arrangement of the two kinds of metal atom is always found if the alloy is cooled rapidly (quenched). In certain solid solutions with particular concentrations of solute a regular atomic arrangement develops on slow cooling or appropriate subsequent heat treatment, and we shall describe shortly some of the features of this *superstructure* formation. It was necessary to qualify the first statement in this paragraph because (i) solid solutions are not formed by all pairs of metals, and (ii) when they are, the range of composition over which solid solutions are formed varies from the one extreme, complete miscibility, to the other, immiscibility, depending on the metals concerned. In what follows we are referring to substitutional solid solutions, that is, those in which the atoms of the solute replace some of those of the solvent in the structure of the latter. There is another kind of solid phase, the interstitial solid solution, in which the small atoms of some of the lighter non-metals occupy the interstices between the atoms in metal structures. We deal with these interstitial solid solutions in a later section. It is possible to have a combination of substitutional and interstitial solid solution in ternary or higher systems. For example, an austenitic manganese steel is a substitutional solid solution of manganese in iron and also an interstitial solution of carbon in the (Fe, Mn) structure.

The conditions determining the formation of solid solutions are as follows:

(1) The tendency to form solid solutions is small if the metals are chemically dissimilar. In general we may say that extended solid solution formation is common between metals of our classes A_1 and A_2, subject to the further restrictions discussed below, and between metals of the same subgroup of the Periodic Table. Thus the following pairs of metals form continuous series of solid solutions: K–Rb, Ag–Au, Cu–Au, As–Sb, Mo–W, and Ni–Pd. For elements of greatly differing electronegativity, for example, an A subgroup metal and a member of one of the later B subgroups, not only are the structures of the pure elements quite different but, owing to the difference in electronegativity, compound formation is likely to

occur rather than solid solution. For this reason we divided the elements into groups A_1, A_2, B_1, and B_2, and we shall later consider in turn the types of intermetallic compounds formed between pairs of elements from different groups.

(2) For two elements in the same group in our classification the range of composition over which solid solutions are formed depends on the relative sizes of the two atoms. This is to be expected, since if some of the atoms in a structure are replaced (at random) by others of a different size, distortion of the structure must occur and the cell dimensions alter as the concentration of the solute increases. To a first approximation they vary linearly with the atomic percentage of the solute (Végard's law), though in many cases this law is not exactly obeyed. If the difference between the radii of the metals is greater than about 15 per cent (of the radius of the solvent atom) there is no extensive formation of solid solutions. There is rather more tolerance at high temperatures, but then precipitation-hardening generally occurs on quenching. For metals crystallizing with very different types of structure the application of this 'relative size' criterion is complicated by the difference in coordination number in the structures of the two metals. We need not, however, go into this point as we shall only be considering solid solutions of metals with typical metallic structures of high coordination.

(3) The mutual solubilities of metals are not reciprocal. Hume-Rothery observed that, other things being equal, a metal of lower valence is likely to dissolve more of one of higher valence than vice versa—the 'relative valence effect'. Though there may be exceptions to this generalization, it is nevertheless, a useful one. The following figures, taken from the much more extensive data quoted by Hume-Rothery, give some idea of the striking differences in solubilities in cases where the size factor is favourable.

Solubility of:	Zn in Ag	37·8% atomic (Zn)
	Ag in Zn	6·3% atomic (Ag)
	Zn in Cu	38·4% atomic (Zn)
	Cu in Zn	2·3% atomic (Cu)

We shall now describe some of the changes in structure which take place in certain substitutional solid solutions on cooling slowly or annealing.

Order–disorder phenomena and superstructures

Some alloys which are random solid solutions when quenched from the molten state undergo rearrangement when given a suitable heat treatment or in some cases simply when cooled sufficiently slowly. Such a structural change, which begins at a temperature often hundreds of degrees below the melting point, results in a change from the random arrangement to one in which there is regular alternation of atoms of different kinds throughout the structure. We may illustrate such a change by showing the structure of a portion of the alloy CuAu (a) as a random solid solution and (b) as the ordered superstructure (Fig. 29.7).

The ordered structure has lower (tetragonal) symmetry than the disordered, being a slightly deformed cubic closest packing of the atoms. The structures of Fig. 29.7 correspond to temperatures (a) above 420°C and (b) below 380°C. At

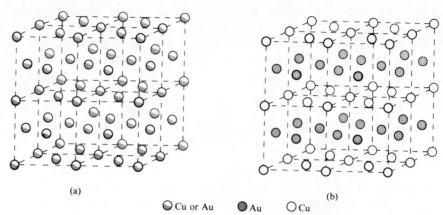

(a)   (b)

◔ Cu or Au   ● Au   ○ Cu

FIG. 29.7. The crystal structure of the alloy CuAu in (a) the disordered, and (b) the ordered state.

intermediate temperatures the structure is more complex, being a superstructure with a 10-fold unit cell. Moreover the period of the superstructure can be altered by adding a third element of different valence, that is, it is affected by the electron : atom ratio. The possibility of rearrangement to form a superstructure is determined by the thermal energy of the atoms, the difference in potential energies of the ordered and disordered states, and the magnitude of the energy barrier that has to be surmounted before two atoms can change places. The formation of a superstructure is a co-operative phenomenon like the loss of ferromagnetism, which occurs at the Curie point, or the onset of rotation of ions or molecules in crystals. A characteristic of such processes is that the behaviour of a particular atom (or molecule) is affected by that of its neighbours. Once the process has started it speeds up, giving a curve like that shown in Fig. 29.8(a) for the specific heat of

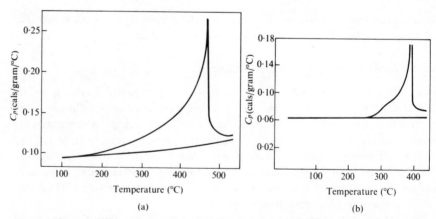

(a)   (b)

FIG. 29.8. The variation with temperature, on heating, of the specific heat of (a) β-brass, and (b) Cu_3Au (after cooling at 30°C per hour). In each case the upper curve is drawn through the experimental values and the lower curve is that to be expected in the absence of a transformation.

1030

β-brass (see below). The rotation of one methane molecule, for example, in a crystal of that compound loosens the attachments of the others so that not only is the process facilitated for each succeeding molecule but also the thermal energy of the molecules is increasing all the time since the temperature is rising. Although order–disorder transformations may sometimes be loosely compared with melting, since a regular arrangement of atoms gives place to a random one and movement of the atoms relative to one another takes place, it is clear that they differ from the process of melting in certain important respects. There is no general collapse of the whole structure and the transformation takes place over a wide range of temperature, though in this latter respect the melting of a glass is similar. In a liquid, however, the arrangement of the atomic centres is less regular than in a solid and, moreover, is constantly changing, whereas in both the disordered and the ordered phases of an alloy the positions of the atomic centres are the same, or the same to within very close limits.

Starting with an ordered alloy at a low temperature two types of order–disorder transformation may be recognized:

(a) Order decreases continuously, passing through all intermediate states, and finally disappears completely at the critical temperature. The specific heat shows an abnormally great increase with rising temperature and at the critical temperature falls suddenly to a value only slightly above the theoretical value.

(b) The behaviour is similar to that described in (a), except that there is still some order left when the critical temperature is reached so that the last stage of the rearrangement takes place suddenly. In this case there is also a latent heat in addition to the specific heat anomaly.

We shall now give some examples of order–disorder transformations observed in alloys of two types, XY and X_3Y.

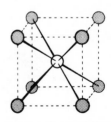

β-Brass

The structures of the ordered and disordered phases are illustrated in Fig. 29.9. The atoms are arranged on a body-centred cubic lattice, in one case in a regular way and in the other at random. Instead of drawing a large portion of the structure to show the random arrangement it is more convenient to use a circle with intermediate shading to represent an atom of either kind. The probability of such an atom being either Cu or Zn is one-half. In this alloy the transformation takes place over a temperature range of some 300°C. Below 470°C it takes place continuously and reversibly, there being no sudden change in the degree of order at the transition point and therefore no latent heat. The specific heat/temperature curve for a change of this sort is shown (for CuZn) in Fig. 29.8(a). To account for the fact that the specific heat is still abnormal above the transition point it has been suggested that although the long-distance order has disappeared there is still a tendency for atoms to prefer unlike atoms as nearest neighbours.

 Cu or Zn Cu Zn

FIG. 29.9. The structures of the disordered and ordered forms of β-brass.

Alloys X_3Y

Order–disorder transformations have been observed in the following alloys: Cu_3Au, Cu_3Pd, Cu_3Pt, Ni_3Fe, and Fe_3Al. In the first four cases the atoms occupy the positions of a face-centred cubic structure, and the relation of the ordered to the disordered structure is that shown in Fig. 29.10(a). These transformations are sluggish and inhibited by quenching from high temperatures, and annealing for several months is necessary to produce a high degree of order in Ni_3Fe. The specific heat/temperature curve for Cu_3Au is shown in Fig. 29.8(b). Investigation of the variation with temperature of the electrical resistance of Cu_3Au shows that a marked degree of order sets in at the critical temperature, for on cooling through this temperature a sudden drop of 20 per cent occurs. Accordingly there is a latent heat associated with the order–disorder transformation in Cu_3Au. The absence of the expected latent heat in the case of Ni_3Fe is probably due to the extreme

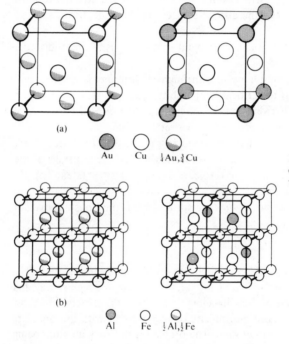

(a)

Au   Cu $\frac{1}{4}$Au,$\frac{3}{4}$Cu

(b)

Al   Fe $\frac{1}{4}$Al,$\frac{1}{2}$Fe

FIG. 29.10. The crystal structures of (a) Cu_3Au, and (b) Fe_3Al in the disordered and ordered states.

slowness of the rearrangement. The transformation in Fe_3Al is illustrated in Fig. 29.10(b).

Certain other alloys X_3Y provide interesting examples of superstructures and also of close-packed structures containing atoms of two types. The structures of Al_3Ti and Al_3Zr are superstructures derived from the cubic close-packed structure of Al. A unit cell of the (tetragonal) structure of Al_3Ti is shown in Fig. 29.11 together with an equivalent portion of the structure of Al. One-quarter of the Al atoms are replaced by Ti in a regular manner, and it will be seen that the plane corresponding to (111) of Al is now a close-packed X_3Y layer of the type shown in Fig. 4.15(c). The more complex structure of Al_3Zr is closely related to that of Al_3Ti but contains both types of X_3Y layer (Fig. 4.15(c) and (d)). It will be appreciated that the relation of Al_3Ti to Al is similar to that of Fe_3Al (and the isostructural Fe_3Si) to Fe. Al_3Ti arises by the regular replacement (by Ti) of one-quarter of the Al in a face-centred cubic structure and Fe_3Al by the regular replacement of one-quarter of the Fe by Al in the body-centred cubic structure (Fig. 29.10(b)). Another group of alloys have structures which are superstructures derived from *hexagonal* or more complex types of closest packing. In Ni_3Sn (and the isostructural Mg_3Cd and Cd_3Mg) and Ni_3Ti the layers are of the type of Fig. 4.15(d) but the structures differ in the sequence of layers (h in Ni_3Sn, hc in Ni_3Ti).

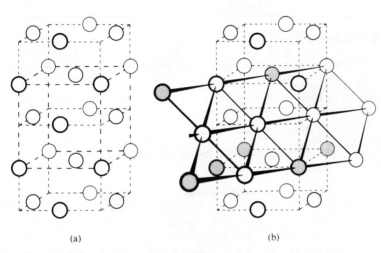

(a) (b)

FIG. 29.11. The crystal structures of (a) Al (two unit cells), and (b) Al_3Ti (one unit cell). A portion of one close-packed Al_3Ti layer is emphasized in (b) where the shaded circles represent Ti atoms.

Examples of intermetallic phases X_3Y based on more complex c.p. layer sequences are included in Table 4.4 (p. 134). It is tempting to speculate on the reasons for the choice of different layer sequences in these compounds. The cubic Cu_3Au structure implies that all interatomic distances are the same, and this structure is adopted by, for example, the compounds Al_3M formed by the smaller 4f metals (e.g. Er). In the hexagonal or rhombohedral structures adjustment is possible to give two different interatomic distances as required by the larger 4f metals and Y:[1]

(1) AC 1967 **23** 729

	Al_3Er	Low-Al_3Y	High-Al_3Y
Layer sequence	c	h	chh
	Er – 12 Al 2·98 Å	Y – 12 Al 3·11 Å	Y – 12 Al 3·07 Å
	$Al\begin{cases}4\ Er \\ 8\ Al\end{cases}$ 2·98	$Al\begin{cases}4\ Y & 3·11 \\ 8\ Al & 2·96\end{cases}$	$Al\begin{cases}4\ Y & 3·07 \\ 8\ Al & 2·98\end{cases}$

(2) JNM 1965 **15** 1, 57; AC 1965 **19** 184

The polymorphism of Al_3Pu[2] suggests that the relative sizes of the atoms are more important at lower temperatures. This compound has at least three polymorphs with different layer sequences:

$$\text{———}1027°\text{C———}1210°\text{C———}1400°\text{C}$$

| 9-layer | 6-layer | 3-layer |
| *chh* | *cch* | *c* |

(3) AC 1955 8 349

The compound WAl_5 is an interesting example of a more complex formula arising in a close-packed structure by the alternation of layers of composition WAl_2 and Al_3 in the sequence $ABAC\ldots.$[3]

The structures of alloys

We shall now describe briefly some of the structures adopted by alloys. We shall limit our survey to binary systems. Adopting the classification of the elements indicated on p. 1008, we have to consider three main classes of alloys:

I. Alloys of two A metals, AA

II. Alloys of an A and a B subgroup metal, AB $\begin{cases} A_1B_1 \\ A_2B_1 \\ A_1B_2 \\ A_2B_2 \end{cases}$

III. Alloys of two B subgroup metals, BB.

Since we subdivide the A metals into two groups A_1 and A_2 and also distinguish the earlier from the later B subgroup metals as B_1 and B_2 respectively we make further subdivisions in class II as shown; we shall not deal systematically with the three subdivisions in each of the classes I and III. We have dealt with several systems of type AA in our discussion of solid solutions and superstructures, and the only other examples we shall mention of class I alloys are some phases formed by transition metals (group (c), p. 1038).

It is not easy to draw a hard-and-fast line between truly metallic alloys and homopolar compounds in some cases, particularly those involving elements of the later B subgroups (As, Sb, Se, Te), and it is also not possible to adhere rigidly to the classification of elements into the groups A_1, A_2, B_1, and B_2. This classification is used in the Chart 29.1, which shows the groups of alloys we shall consider, but in (b), for example (alloys XY_5, X_2Y_{17}, XY_{11}, etc.), X is an atom taken from our A_1 group extended to include Ga, La, Ce, and Th, while Y is taken from the group Fe, Co, Ni, Cu, Be, Zn, and Cd. It is convenient to make our classification somewhat flexible so as to allow us to bring together families of structures with a common structural theme. Size factors play an important part in determining the appearance of the so-called Laves phases with the closely related $MgZn_2$, $MgCu_2$, and $MgNi_2$ structures. From the structural standpoint these phases are probably most closely related to the σ phases formed by transition elements, so that although the Laves phases are not typically combinations of transition (A_2) metals they are conveniently mentioned in connection with the σ phases.

(a) *The NaTl and related structures*

Two structures commonly found for phases of composition $A_1 B_1$ are the caesium chloride and 'sodium thallide' structures. Among the alloys crystallizing with the former structure are LiHg, LiAl, LiTl, MgTl, CaTl, and SrTl. The structure needs no description here, but we shall refer to these alloys in section (d). In the NaTl structure (which is also that of LiZn, LiCd, LiGa, LiIn, and NaIn) the coordinates of the atoms are the same as in the CsCl structure (that is, those of a body-centred cubic lattice), but the arrangement of the two kinds of atoms is such as to give each atom four nearest neighbours of its own kind and four of the other kind (Fig. 29.12). The interatomic distances in alloys with the NaTl structure are very much shorter (some 13 per cent for NaTl itself) than the sums of the radii derived from the structures of the pure metals.

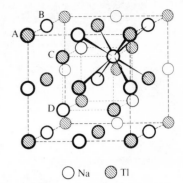

○ Na   ◎ Tl

FIG. 29.12. The crystal structure of NaTl.

CHART 29.1

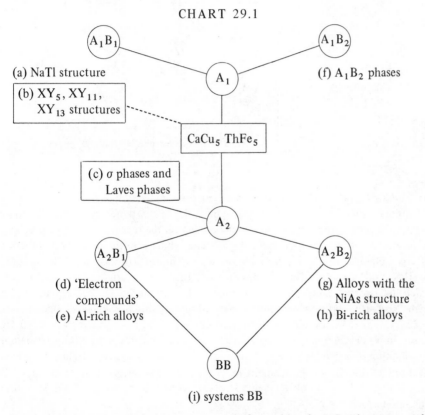

(a) NaTl structure

(b) XY_5, XY_{11}, XY_{13} structures

(c) σ phases and Laves phases

$A_1 B_1$

A_1

$A_1 B_2$

(f) $A_1 B_2$ phases

CaCu₅ ThFe₅

A_2

$A_2 B_1$

(d) 'Electron compounds'
(e) Al-rich alloys

$A_2 B_2$

(g) Alloys with the NiAs structure
(h) Bi-rich alloys

BB

(i) systems BB

It is convenient to include here compounds formed by those B subgroup metals which lie on the borderline between B_1 and B_2 elements in our classification on p. 1008. The same atomic positions as those of the NaTl structure (that is, those of the b.c.c. lattice) are occupied in some phases $X_3 Y$, but the arrangements of the X and Y atoms lead to quite different structures. $Li_3 Bi$, $Li_3 Sb$, and $Li_3 Pb$ have the same structure as the ordered form of $Fe_3 Al$ (Fig. 29.10(b)). In $Cs_3 Sb$, which is of interest on account of its photoemissivity, Cs atoms occupy the sites A and C of Fig. 29.12 while equal numbers of Cs and Sb atoms occupy at random the sites B and D. Since the combination of B and D sites corresponds to a diamond-like arrangement it has been suggested that ionization takes place: $3 \, Cs + Sb \rightarrow 2 \, Cs^+ + Cs^{3-} + Sb^+$, and that Cs^{3-} and Sb^+ form tetrahedral $6s6p^3$ and $5s5p^3$ bonds respectively, the Cs^+ ions occupying the interstices in the diamond network.[1] These structures are summarized in Table 29.7, and for completeness we add two other compounds already mentioned in earlier chapters (Bi_2OF_4, p. 358, and NaY_3F_{10}, p. 357).

(1) PRS A 1957 **239** 46

TABLE 29.7

Structures of the CsCl–NaTl *family*

	4-fold positions of Fig. 29.12			
	A	B	C	D
CsCl	Cs	Cs	Cl	Cl
NaTl	Na	Tl	Na	Tl
Cs_3Sb	Cs	Cs, Sb	Cs	Cs, Sb
Li_3Bi	Bi	Li	Li	Li
Fe_3Al (ordered)	Al	Fe	Fe	Fe
Fe_3Al (disordered)	Fe	Fe	Fe, Al	Fe, Al
Bi_2OF_4	4 Bi	$\longleftarrow$	2 O + 8 F	$\longrightarrow$
NaY_3Fe_{10}	Na, 3 Y (Fig. 9.7)	$\longleftarrow$	10 F	$\longrightarrow$

(b) The XY_5, XY_{11}, XY_{13}, *and related structures*

A feature of many compounds of the type A_1B_1 is the tendency of the B_1 atoms to group together, the type of group depending on the relative numbers of A_1 and B_1 atoms. For example, in the Na–Hg system, which is notable for the number of different phases formed, three of the phases are (in order of increasing Hg content) Na_3Hg_2, NaHg, and $NaHg_2$.[1] In the first there are nearly square Hg_4 groups which are isolated units entirely surrounded by Na atoms. In NaHg pairs of Hg atoms (Hg–Hg, 3·05 Å) are linked into infinite chains by slightly longer bonds (3·22 Å), while $NaHg_2$ has the AlB_2 structure (p. 842) with layers of linked Hg atoms. In Na_2Tl there are tetrahedral Tl_4 groups which form part of the icosahedral coordination group of 9 Na + 3 Tl around Tl.[2] There are also tetrahedral groups of atoms of the Group IV element in $NaSi_2$ and $BaSi_2$, KGe, KSn, and NaPb.[3] With NaHg and the complex NaPb structure contrast the simple NaTl structure illustrated in Fig. 29.12.

With higher $B_1 : A_1$ ratios 3-dimensional networks of B_1 atoms are found, the more compressible A_1 atoms occupying the interstices. (In the XY_5 structure we shall now describe Y is not strictly a B-type atom in our classification of p. 1008, as already noted.) The $CaCu_5$ structure[4] is built from alternate layers of the types shown in Fig. 29.13(a) and (b), so that each Ca is surrounded by six Cu in one plane (at 2·94 Å) and by two more sets of six in adjacent planes (at 3·27 Å),

(1) AC 1954 **7** 277

(2) AC 1967 **22** 836
(3) AC 1953 **6** 197

(4) ZaC 1940 **244** 17

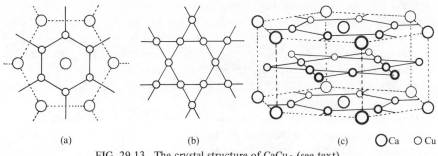

(a)   (b)   (c)   ●Ca ○Cu

FIG. 29.13. The crystal structure of $CaCu_5$ (see text).

making a total c.n. of 18, For this structure R_A should be less than $1\cdot6\,R_B$; it is in fact found for values of R_A between $1\cdot37$ and $1\cdot58\,R_B$. Examples of phases with this structure include $CaNi_5$, $CaZn_5$, $LaZn_5$, $ThZn_5$, $CeCo_5$, and $ThCo_5$. Sr and Ba are near the upper size limit, and variants of this structure are found for $SrZn_5$ and $BaZn_5$.[5] By replacing in a regular manner certain of the X atoms in XY_5 by pairs of Y atoms there arise the related structures of Th_2Fe_{17},[6] $ThMn_{12}$, and $TiBe_{12}$.[7]

In the XY_{11}[8] and XY_{13} structures the X atoms occupy holes in which they are surrounded by 22 and 24 Y atoms respectively. For a given Y atom the smaller X metals form the XY_{11} phase and the larger ones the XY_{13} phase, while if X is too large even the XY_{13} phase is not formed (Table 29.8). This XY_{13} structure is of particular interest on account of the coordination polyhedra. The X atom has 24 Y neighbours at the vertices of a nearly regular snub cube, which is one of the less familiar Archimedean solids and has 6 square faces parallel to those of a cube and

(5) AC 1956 **9** 361

(6) AC 1969 **B25** 464

(7) AC 1952 **5** 85

(8) AC 1953 **6** 627

TABLE 29.8
Phases with the XY_{11} and XY_{13} structures

Radius (Å)			Phases formed		
Cs	2·72				$CsCd_{13}$
Rb	2·50			$RbZn_{13}$	$RbCd_{13}$
K	2·35			KZn_{13}	KCd_{13}
Ba	2·24			$BaZn_{13}$	$BaCd_{11}$
Sr	2·15		$SrBe_{13}$	$SrZn_{13}$	$SrCd_{11}$
Ca	1·97		$CaBe_{13}$	$CaZn_{13}$	
Na	1·91		–	$NaZn_{13}$	
La	1·88		$LaBe_{13}$	$LaZn_{11}$	
U	1·56		UBe_{13}		

32 equilateral triangular faces. The unique Y atom at the origin has a nearly regular icosahedral coordination group of 12 Y atoms, while the coordination group of the other twelve Y atoms is much less regular, as shown by the figures for $RbZn_{13}$:[9]

(9) AC 1971 **B27** 862

$$Rb - 24\,Zn_{II} \quad 3\cdot62\,\text{Å}$$
$$Zn_I - 12\,Zn_{II} \quad 2\cdot68 \qquad \text{(compare the shortest Zn–Zn}$$
$$Zn_{II} - \begin{cases} 10\,Zn_{I,II} & 2\cdot61\text{-}2\cdot96 \\ 2\,Rb & 3\cdot62 \end{cases} \quad \text{in Zn metal, } 2\cdot66\,\text{Å.)}$$

(c) *Transition-metal σ phases and Laves phases*

Between electron compounds and 'normal valence' compounds, which are characterized by relatively small numbers of structures, there lie large groups of intermetallic compounds in which the structural principles are less clear. In the structures to be described in this section the importance of geometrical factors is becoming evident, a special feature of these structures being the high coordination numbers which range from 12 to 16.

In a sphere packing the smallest hole enclosed within a polyhedral group of spheres is obviously a tetrahedral hole, and therefore the closest packing is achieved if the number of such holes is the maximum possible. In a 3-dimensional closest-packing of N equal spheres there are $2N$ tetrahedral holes but also N octahedral holes; more dense packings with only tetrahedral holes are possible, but these extend indefinitely in only one or two dimensions.[1] A local density higher than that of the ordinary closest packings is achieved, for example, by packing twenty spheres above the faces of an icosahedron, and this grouping of atoms is in fact found in some of the structures we shall describe. However, this type of packing of equal spheres cannot be extended to fill space, for holes appear which reduce the mean density of the packing to a value below that of closest packing. If, on the other hand, there are larger spheres to fill the holes (surrounded by 14–16 spheres) between the icosahedral groups a very efficient packing results, and this suggests at least a partial explanation of the occurrence of the structures we shall describe.

From the fact that the smallest kind of interstice between spheres in contact is a tetrahedral hole it follows that we should expect to find coordination polyhedra with only triangular faces, in contrast to those in, for example, cubic closest packing which have square in addition to triangular faces. Moreover, it seems likely that the preferred coordination polyhedra will be those in which five or six triangular faces (and hence five or six edges) meet at each vertex, since the faces are then most nearly equilateral. It follows from Euler's relation (p. 61) that for such a polyhedron, $v_5 + 0v_6 = 12$, where v_5 and v_6 are the numbers of vertices at which five or six edges meet, so that starting from the icosahedron ($v_5 = 12$) we may add 6-fold vertices to form polyhedra with more than twelve vertices.

It can readily be shown that there is no polyhedron of this family with $Z = 13$, only one for $Z = 14$, one for $Z = 15$, and two for $Z = 16$ (one of which has a pair of adjacent 6-fold vertices). Polyhedra with $Z > 16$ have at least one pair of adjacent 6-fold vertices. It is found that the three coordination polyhedra for c.n. >12 in a number of alloy structures are in fact those for $Z = 14$, 15, and 16 which have no adjacent 6-fold vertices; they are illustrated in Fig. 29.14. The icosahedron (a) may be derived from a pentagonal antiprism by adding atoms above the mid-points of the opposite pentagonal faces; the 14-hedron (b) is similarly derived from a hexagonal antiprism. Since the distance from the centre to a vertex of a regular icosahedron is about 0·95 of the edge length this group may be regarded as a coordination polyhedron for radius ratio 0·90. In the 14-hedron (b) the twelve spheres of the antiprism are equal and different in radius from the central sphere, but also different in size from the remaining two which cap the hexagonal faces.

(1) PRR 1952 7 303

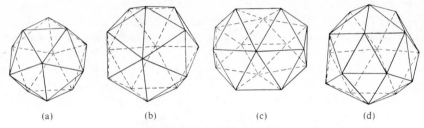

(a)   (b)   (c)   (d)

FIG. 29.14. Coordination polyhedra in transition metal structures.

The 15-hedron (c) may be described as a hexagonal antiprism with one atom beyond one hexagonal face and two above the other. If the polyhedron (d) for c.n. 16 consists of a truncated tetrahedral group of 12 B atoms around a central A atom plus a tetrahedral group of 4 A atoms which lie above the centres of the hexagonal faces it is sometimes referred to as the Friauf polyhedron, after the 'Friauf phases' $MgCu_2$ and $MgZn_2$, though these together with $MgNi_2$ are also called the 'Laves phases'. We discuss later the description of this coordination group as one corresponding to the radius ratio 1·23.

In the 14-hedron the lines joining the centre of the polyhedron to the 6-fold vertices are collinear, in the 15-hedron they are coplanar (at 120°), and in the 16-hedron they are directed towards the vertices of a tetrahedron; compare the stereochemistry of carbon. Frank and Kasper have pointed out that the basic geometry of certain structures in which the coordination numbers of all the atoms are 12, 14, 15 and/or 16 can be described in a very elegant way.[2] They describe sites with c.n. >12 as *major sites*, the lines joining major sites which have six neighbours in common as *major ligands*, and the system of major ligands as the *major skeleton*. Since a major ligand connects two points of c.n. 14, 15, or 16 and passes through a 6-fold vertex there is a ring of six atoms around every major ligand, and the major skeleton generally defines the complete structure uniquely, apart from small displacements and distortions. If the only sites in a structure with c.n. > 12 are 14-coordinated then the major skeleton is simply a set of non-intersecting lines, because there are only two major ligands from any point, but in structures with sites of 15- or 16-coordination the major skeleton may be a planar or 3-dimensional net. In the σ phase to be described shortly the major skeleton formed by joining the 15-coordinated sites is a planar net, and in the Laves phases the skeletons formed by linking the 16-coordinated sites correspond to the diamond, wurtzite, and carborundum-III structures.

In α-Mn there are four kinds of crystallographically nonequivalent atoms with the c.n.'s shown in Table 29.9 and Mn—Mn distances ranging from 2·26 to 2·93 Å. It has been supposed for a long time that the complexity of this structure is due to the presence of atoms in different valence states, but there has been no generally accepted interpretation of the structure. In the complex σ phases formed by a number of transition metals there are no fewer than five crystallographically different kinds of atom with c.n.'s ranging from 12 to 15. Of the 30 atoms in the unit cell 10 have the icosahedral coordination group of Fig. 29.14(a), 16 the

(2) AC 1958 **11** 184; AC 1959 **12** 483

V Cr Mn Fe Co Ni

Mo

(a)   (b)

14-group of (b), and 4 the c.n. 15 (c). Neutron diffraction studies of Ni–V σ phases (Ni_9V_{21}, $Ni_{11}V_{19}$, $Ni_{13}V_{17}$) and σ-FeV, and X-ray studies of σ-MnMo and σ-FeMo show that there is a definite segregation of certain atoms into sites of highest c.n. and of others into sites of low c.n. If we distinguish between atoms to the left and the right of Mn as (a) and (b), we find (a) atoms with c.n. 15, both (a) and (b) in positions of c.n. 14, and only (b) in (icosahedral) 12-coordination.

In the μ phases Fe_7W_6, Fe_7Mo_6, and the corresponding Co compounds, Fe or Co is found in positions of 12-coordination (icosahedral) while Mo or W is found with higher c.n.'s (14–16) for which the coordination polyhedra are also those of Fig. 29.14. Further examples of structures of high c.n. are included in Table 29.9, which includes references to some of the more recent work. It is interesting that of the three phases in the Mg–Al system the one with the simplest formula (Mg_2Al_3) has the most complex structure, with which may be compared the equally complex structure of Cu_4Cd_3. We have noted the description of these structures in terms of the 'major skeleton'. An alternative, for structures in which there is a reasonable proportion of 16-coordinated atoms, is to describe them in terms of the linking of the Friauf polyhedra, since by sharing some or all of their hexagonal faces these form a rigid framework of centred truncated tetrahedra. For references to such descriptions see Table 29.9, in which Z is the number of atoms in the unit cell.

TABLE 29.9

Coordination numbers in some intermetallic phases

Compound	Z	No. of atoms with c.n.					Major ligand network	Reference
		12	13	14	15	16		
Cu_4Cd_3	1124	736	–	120	144	124	For description in terms of Friauf polyhedra see refs. to $Mg_{23}Al_{30}$ and Cu_4Cd_3	AC 1967 **23** 586
$Mg_2Al_3(\beta)^{(a)}$	1168	672				252		AC 1965 **19** 401
ϵ-$Mg_{23}Al_{30}{}^{(b)}$	53	24	2	13	–	8		AC 1968 **B24** 1004
R-phase	53	27	–	12	6	8		AC 1960 **13** 575
γ-$Mg_{17}Al_{12}$								
χ-phase	58	24	24	–	–	10		JM 1956 **8** 265
α-Mn								AC 1970 **B26** 1499
δ-phase (Mo–Ni)	56	24	–	20	8	4	4 interpenetrating (3,4)-connected nets	AC 1963 **16** 997
P-phase								AC 1957 **10** 1
σ-phase (Ni–V)	30	10	–	16	4	–	2 D + 1 D	AC 1956 **9** 289
Zr_4Al_3	7	3	–	2	2	–		AC 1960 **13** 56
$MgZn_2$ (C 14)	12	8	–	–	–	4	3 D 4-connected nets	AC 1968 **B24** 7, 1415
$MgCu_2$ (C 15)	24	16	–	–	–	8		
$MgNi_2$ (C 36)	24	16	–	–	–	8		
μ-Fe_7W_6	13	7	–	2	2	2	3 D + 2 D	AC 1962 **15** 543
$Mg_{32}(Al, Zn)_{49}$	162	98	–	12	12	40	See text	AC 1957 **10** 254

(*a*) Also 244 atoms with miscellaneous types of coordination (c.n.'s 10–16).
(*b*) Also 6 atoms with c.n. 11.

The structure of $Mg_{32}(Al, Zn)_{49}$ provides a beautiful illustration of the importance of geometrical factors in determining a crystal structure, for no well-defined Brillouin zone could be found to account for its stability. Starting from one atom surrounded by an icosahedral group of 12 others, Fig. 29.15(a), 20 more are placed beyond the mid-points of the icosahedron faces at the vertices of a pentagonal dodecahedron, (b). Beyond the faces of the latter 12 more atoms are placed forming a larger icosahedron (c). With the previous 20 these 32 atoms lie at the vertices of a rhombic triacontahedron. This figure has 30 rhombus-shaped faces. Atoms are now placed beyond the centres of the 60 triangular half-rhombs of the triacontahedron; these lie at the vertices of the truncated icosahedron (d). The addition of 12 more atoms gives an outer group of 72 atoms which lie on the faces of a truncated octahedron (e). Such groups can now be packed to fill space by sharing the atoms on their surfaces (f). Within each truncated octahedron there is a nucleus

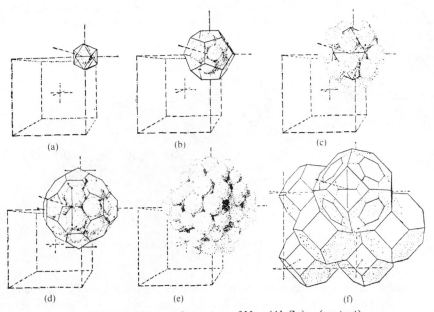

(a)   (b)   (c)

(d)   (e)   (f)

FIG. 29.15. The crystal structure of $Mg_{32}(Al, Zn)_{49}$ (see text).

of 45 atoms, and since each atom of the outer group of 72 is shared with another similar group the number of atoms per unit cell is $2\{45 + \frac{1}{2}(72)\} = 162$. In the resulting structure

$$
\left.
\begin{array}{ll}
\text{98 atoms have c.n. 12} & \text{Zn, Al} \\
\text{12 atoms have c.n. 14} \\
\text{12 atoms have c.n. 15} \left.\right\} \text{Mg} \\
\text{40 atoms have c.n. 16}
\end{array}
\right\} Mg_{32}(Al, Zn)_{49}
$$

and

In the Laves phases AB_2 typified by $MgZn_2$, $MgCu_2$, and $MgNi_2$, with three closely related structures, the coordination of A by 12 B + 4 A is consistent with the view that the structures are determined *primarily* by size factors. These phases are formed by a great variety of elements from many Periodic Groups, and the same element may be A or B in different compounds. In some ternary systems two or all of the structures occur at different compositions, the change from one structure to another being connected with electron–atom ratios, suggesting that factors of more than one kind may be important in determining the choice of one of a number of closely related structures.

The structures of certain crystals are derived from close-packed assemblies from which a proportion of the atoms have been removed in a regular way, as in the ReO_3 structure in which O atoms occupy three-quarters of the positions of cubic closest packing. It is of interest to note here some ways of removing one-half of the atoms from different kinds of closest packings. From a close-packed layer A we remove one-quarter of the atoms as shown in Fig. 29.16(a), and from the layer B above we remove three-quarters of the atoms. This leaves a set of isolated

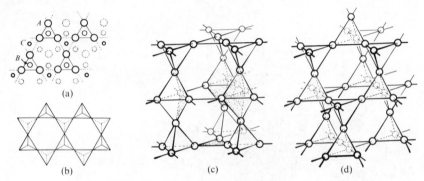

FIG. 29.16. The structures of the Laves phases (see text).

tetrahedra. Below the A layer we could place either another B layer (leaving pairs of tetrahedra base to base) or a layer in position C, in which case we are left with a layer of tetrahedra with vertices pointing alternately up and down (Fig. 29.16(b)).

Certain sequences of layers of this kind form space-filling assemblies of tetrahedra and truncated tetrahedra, as shown in Fig. 29.16(c) and (d). That of Fig. 29.16(d) is of particular interest as a system of tetrahedra linked entirely through vertices, and it is alternatively derived by placing atoms at the mid-points of the bonds in the diamond structure and joining them to form tetrahedra around the points of that structure. The centres of the truncated tetrahedral holes in (d) also correspond to the positions of the atoms of the diamond structure; in (c) they correspond to the wurtzite structure. The number of these large holes is equal to one-half the number of atoms forming the tetrahedral network, so that if we place an atom A in each hole (that is, replacing a tetrahedral group of four B atoms in the original close-packed assembly) we have a compound of formula AB_2 in which A is surrounded by twelve B at the vertices of a truncated tetrahedron and is also linked tetrahedrally to four A neighbours lying beyond the hexagonal faces of that

1042

polyhedron. These are the structures of the Laves phases, $MgCu_2$, $MgZn_2$, and $MgNi_2$, which are related as follows:

	Sequence of close-packed layers		Network formed by A atoms
$MgCu_2$	$ABC\ldots$	(c)	Diamond
$MgZn_2$	$ABAC\ldots$	(hc)	Wurtzite
$MgNi_2$	$ABC\ BAC\ BC\ldots$	$(cchc)$	Carborundum-III

These structures provide an elegant example of the interrelations of nets, open packings of polyhedra, space-filling arrangements of polyhedra, and the closest packing of equal spheres.

It has been remarked[3] that the description of the Laves phases as suitable for atoms with radius ratio $r_A : r_B = 1\cdot225$ is an over-simplification, for the observed range of the ratio is $1\cdot05$–$1\cdot67$. In the simplest of these structures, the cubic $MgCu_2$ structure, the relative interatomic distances are the following fractions of the cell edge a:

(3) AC 1968 **B24** 7, 1415

$$\text{A–A}\ \ 0\cdot433, \quad \text{B–B}\ \ 0\cdot354, \quad \text{and} \quad \text{A–B}\ \ 0\cdot414$$

from which it is evident that all three types of contact are not simultaneously possible for rigid spheres. Radii r_A and r_B may be chosen consistent with any two of the three types of contact:

$r_A : r_B$ 1·09	1·225	1·34
r_A 0·216	0·216	0·237
r_B 0·198	0·177	0·177
Contacts: A–A	A–A	A–B
A–B	B–B	B–B
B compressed	No A–B contacts	A compressed

The majority of compounds with the $MgCu_2$ structure have $r_A : r_B > 1\cdot225$, suggesting that it is important to ensure A–B contacts even at the expense of compressing the A atoms, and the largest radius ratios are found for the largest electronegativity difference between A and B.

A similar difficulty arises in the β-W ($A\ 15$) structure (Fig. 29.4(a)) for a compound A_3B (i.e. atom B at the origin), since the environments of the atoms are:

$$\text{A} \begin{cases} 2\,\text{A}\ \ 0\cdot500a \\ 4\,\text{B}\ \ 0\cdot559 \\ 8\,\text{A}\ \ 0\cdot613 \end{cases} \quad \text{and B–12 A } 0\cdot559a$$

where a is the length of the edge of the cubic unit cell. There is evidently compression of the A atoms in chains parallel to the cubic axes, and in fact the stability and widespread adoption of this structure by binary intermetallic phases may be due to special interactions between pairs of A atoms, leading to interatomic distances which are not reconcilable with rigid spherical atoms.

(d) Electron compounds

In systems A_2B_1, containing a transition metal or Cu, Ag, or Au and a metal of one of the earlier B subgroups, one or more of three characteristic structures are generally found. These are termed the β, γ, and ϵ structures, the solid solution at one end of the phase diagram being designated the α phase. These β, γ, and ϵ phases are not necessarily stable down to room temperature. In the Cu–Al system, for example, the β phase is not stable at temperatures below about 540°C. The structures of these phases are:

β body-centred cubic
γ complex cubic structure containing 52 atoms in the unit cell.
ϵ hexagonal close-packed.

Although we shall assign formulae to these phases, which are often termed *electron compounds* for reasons that will be apparent later, they may appear over considerable ranges of composition. Moreover, although we might expect the β (body-centred) structure to contain at least approximately equal numbers of atoms of the two kinds, it sometimes appears with a composition very different from this. Thus, although the coordinates of the positions occupied in these phases are always the same, the distribution of a particular kind of atom over these positions is variable. In the Ag–Cd system the β phase is homogeneous at 50 per cent Cd and has the CsCl structure, but in the Cu–Sn and Cu–Al systems it appears with the approximate compositions Cu_5Sn and Cu_3Al respectively. In these cases there is random arrangement of the two types of atom in the body-centred structure. It is interesting that whereas Cu_3Al has the β structure, Ag_3Al and Au_3Al crystallize with the β-Mn structure (p. 1017), and the same difference is found between Cu_5Sn (β structure) and Cu_5Si (β-Mn structure). In the case of alloys with the γ structure, the rather complex formulae such as Ag_5Cd_8, Cu_9Al_4, Fe_5Zn_{21}, and $Cu_{31}Sn_8$ are based on the number of atoms (52) in the unit cell. It will be seen that the total numbers of atoms in the formulae quoted are respectively 13, 13, 26, and 39. In the Ag–Cd system the γ phase is stable over the range 57–65 atomic per cent Cd, allowing a considerable choice of formulae, that chosen (Ag_5Cd_8) being consistent with the crystal structure. A selection of phases with the β, γ, and ϵ structures is given in Table 29.10.

A striking feature of this table is the variety of formulae of alloys with a particular structure. Hume–Rothery first pointed out that these formulae could be accounted for if we assume that the appearance of a particular structure is determined by the ratio of valence electrons to atoms. Thus for all the formulae in the first two columns we have an electron : atom ratio of 3 : 2, for the third column 21 : 13, and for the fourth 7 : 4, if we assume the normal numbers of valence electrons for all the atoms except the triads in Group VIII of the Periodic Table. These fit into the scheme only if we assume that they contribute no valence electrons, as may be seen from the following examples:

CuBe	$(1 + 2)/2$	Cu_5Zn_8	$(5 + 16)/13$	$CuZn_3$	$(1 + 6)/4$
Cu_3Al	$(3 + 3)/4$ $\left.\right\}\frac{3}{2}$	Cu_9Al_4	$(9 + 12)/13$ $\left.\right\}\frac{21}{13}$	Cu_3Sn	$(3 + 4)/4$ $\left.\right\}\frac{7}{4}$
Cu_5Sn	$(5 + 4)/6$	Fe_5Zn_{21}	$(0 + 42)/26$	Ag_5Al_3	$(5 + 9)/8$
CoAl	$(0 + 3)/2$	$Na_{31}Pb_8$	$(31 + 32)/39$		

TABLE 29.10

Relation between electron : atom ratio and crystal structure

Electron : atom ratio 3 : 2		Electron : atom ratio 21 : 13	Electron : atom ratio 7 : 4
β b-c structure	β Mn cubic structure	'γ brass' structure	ϵ close-packed hexagonal structure
CuBe	Ag_3Al	Cu_5Zn_8	$CuZn_3$
CuZn	Au_3Al	Cu_9Al_4	Cu_3Sn
Cu_3Al	Cu_5Si	Fe_5Zn_{21}	$AgZn_3$
Cu_5Sn	$CoZn_3$	Ni_5Cd_{21}	Ag_5Al_3
CoAl		$Cu_{31}Sn_8$	Au_3Sn
(for MgTl, etc., see later)		$Na_{31}Pb_8$	

Certain other alloys also with the body-centred cubic structure, LiHg, MgTl, etc., which have already been mentioned, have sometimes been regarded as exceptions to the Hume-Rothery rules. They fall into line with the other β structures only if we assume that Hg or Tl provides one valence electron. Since, however, the radii of the metal atoms in these alloys, as in those with the NaTl structure, are smaller than the normal values, it is probably preferable not to regard these as β electron compounds.

Although the alloys with compositions giving the above comparatively simple electron : atom ratios generally fall within the range of homogeneity of the particular phases, it now appears that the precise values of those ratios, 3/2, 21/13, and 7/4, have no special significance. By applying wave mechanics to determine the possible energy states of electrons in metals it has been found possible to derive theoretical values for the electron : atom ratios at the boundaries of the α, β, and γ phases, the α phase being the solid solution with the close-packed structure of one of the pure metals. In comparing these values with the electron : atom ratios found experimentally we have to remember that phase boundaries may change with temperature, that is, the tie-lines separating the regions of stability of different phases on the phase diagram are not necessarily parallel to the temperature axis.

TABLE 29.11

Experimental electron : atom ratios

System	Electron : atom ratios for		
	Maximum solubility in α phase	*β phase boundary with smallest electron concentration*	*γ phase boundaries*
Cu—Al	1·408	1·48	1·63–1·77
Cu—Zn	1·384	1·48	1·58–1·66
Cu—Sn	1·270	1·49	1·67–1·67
Ag—Cd	1·425	1·50	1·59–1·63
Theoretical	1·362	1·480	1·538

This is invariably the case with β phases, the range of composition over which the phase is stable decreasing at lower temperatures. In Table 29.11 are given the experimental values of the electron : atom ratios for four systems in which both β and γ phases occur. The second column gives the electron : atom ratio for maximum solubility in the α phase, the third for the β phase boundary with smallest electron concentration, and the fourth for the boundaries of the γ phase. It will be seen that there is general agreement with the theoretical values, particularly in the second and third columns, though all the values for the γ phase boundaries exceed the theoretical electron : atom ratio.

(e) *Some aluminium-rich alloys* A_2B_1

We have already noted that we have yet little understanding of the principles determining the structures of many phases formed by transition metals. Here we shall simply indicate some of the features of a number of structures which have been discussed in more detail elsewhere.[1]

(1) See, for example: AC 1955 **8** 175; AcM 1954 **2** 684; AcM 1956 **4** 172

Some of these compounds have simple structures with 8-coordination of the transition metal, for example,

CsCl structure	FeAl, CoAl, NiAl
Ni_2Al_3 structure	Pd_2Al_3 (also Ni_2In_3, Pt_2Ga_3, etc.)
CaF_2 structure	$PtAl_2$, $AuAl_2$, etc.

[The Ni_2Al_3 structure is a distorted CsCl structure in which one-third of the body-centring (Ni) positions are unoccupied.]

Certain other structures can be dissected into the 5-connected net of Fig. 29.17(a) and/or simple square nets. Depending on the relative orientations and translations of successive nets of these kinds (composed of Al atoms) there are formed polyhedral holes between eight or nine Al atoms in which the transition-metal atoms are found. A simple example is the $CuAl_2$-θ structure, in which alternate layers of type (a) are related by the translation *ac*; the Cu atoms are situated between the layers surrounded by eight Al at the vertices of the square antiprism of Fig. 29.17(b). In Co_2Al_9 (Fig. 29.17(c)) the sequence: square net, type (a) net, square net rotated and translated relative to the first, provides positions of 9-fold coordination for the Co atoms.

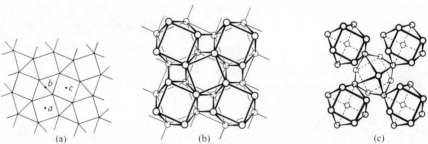

(a)	(b)	(c)

FIG. 29.17. The $CuAl_2$ (θ) and Co_2Al_9 structures: (a) Al layer of the $CuAl_2$ structure, (b) 8-coordination of Cu in $CuAl_2$ between two layers of type (a), (c) 9-coordination of Co in Co_2Al_9.

The structures of Co_2Al_5 and $FeAl_3$ may be interpreted as close-packed structures which have been modified so as to permit 9- and 10- (instead of 12-) coordination of the transition metal. In Co_2Al_5 (Fig. 29.18) close-packed layers at heights $c = \frac{1}{4}$ and $\frac{3}{4}$ are split apart by Co atoms X and Y (which have nine Al neighbours) with the result that layers are formed at $c \approx 0$ and $c \approx \frac{1}{2}$ (the broken lines in Fig. 29.18(b)) which are of the type of Fig. 29.18(c). The Co atoms in the original close-packed layers are 10-coordinated. The very complex structure of $FeAl_3$ may be dissected into layers of two kinds, flat layers as in Fig. 29.19(a) alternating with puckered layers of type (b). The flat layers are made up of close-packed regions and 'misfit' regions, and the triangles of the (b) layers lie approximately above and below the Fe atoms of an (a) layer. These Fe atoms have ten Al neighbours, whereas those in a (b) layer have nine Al neighbours.

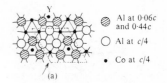

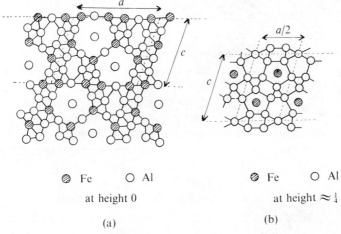

⊘ Fe   ○ Al   ⊘ Fe   ○ Al

at height 0   at height $\approx \frac{1}{4}$

(a)   (b)

FIG. 29.18. The structure of Co_2Al_5: (a) projection along c-axis of atoms at height $\approx \frac{1}{4}$, (b) elevation of one unit cell viewed in direction of arrow in (a)—X and Y are atoms which split such planes into two, (c) projection as in (a) showing atoms at height ≈ 0.

FIG. 29.19. The crystal structure of $FeAl_3$ showing the two kinds of layer into which the structure may be dissected.

(f) Systems A_1B_2

The systems A_1B_2 and A_2B_2, containing elements of the later B subgroups, call for little discussion here since arsenides have been dealt with to some extent in Chapter 20, and sulphides, selenides, and tellurides in Chapter 17. The phase diagrams of systems A_1B_2 are generally very simple, showing very restricted solid solution and only one, usually very stable, compound with a formula conforming to the ordinary valences of the elements (Mg_2Ge, Mg_3As_2, $MgSe$, etc.). These intermetallic compounds have simple structures which are similar to those of simple salts and they are electrical insulators. The structures of some of these compounds are set out below.

$$\left.\begin{matrix} Mg_2Si \\ Mg_2Ge \\ Mg_2Sn \\ Mg_2Pb \end{matrix}\right\} \text{anti-}CaF_2 \qquad \left.\begin{matrix} Mg_3P_2 \\ Mg_3As_2 \end{matrix}\right\} \text{anti-}Mn_2O_3 \qquad \left.\begin{matrix} MgS \\ MgSe \end{matrix}\right\} NaCl$$
$$\left.\begin{matrix} Mg_3Sb_2 \\ Mg_3Bi_2 \end{matrix}\right\} \text{anti-}La_2O_3 \qquad MgTe \text{ wurtzite}$$

1047

The fact that compounds such as Mg_2Si to Mg_2Pb have such high resistances and crystallize with the antifluorite structure does not mean that they are ionic crystals. Wave-mechanical calculations show that in these crystals the number of energy states of an electron is equal to the ratio of valence electrons : atoms (8/3) so that, as in other insulators, the electrons cannot become free (that is, reach the conduction band) and so conduct electricity. That the high resistance is characteristic only of the crystalline material and is not due to ionic bonds between the atoms is confirmed by the fact that the conductivity of molten Mg_2Sn, for example, is about the same as that of molten tin.

As indicated on p. 1034, the dividing line between B_1 and B_2 metals is somewhat uncertain. We have included here some compounds of Pb, Sb, and Bi; certain other phases containing these metals were included with the A_1B_1 structures on p. 1035.

As examples of structures found in systems A_2B_2 we shall mention first the NiAs structure and then the structures of some Bi-rich phases formed by certain transition elements.

(g) Phases A_2B_2 with the nickel arsenide structure

The A_2 metals and the elements of the earlier B subgroups (B_1 metals) form the electron compounds already discussed. With the metals of the later B subgroups the A_2 metals, like the A_1, tend to form intermetallic phases more akin to simple homopolar compounds, with structures quite different from those of the pure metals. The nickel arsenide structure has, like typical alloys, the property of taking up in solid solution a considerable excess of the transition metal. From Table 29.12

TABLE 29.12

Compounds crystallizing with the NiAs structure

	Cu	Au	Cr	Mn	Fe	Co	Ni	Pd	Pt
Sn	CuSn	AuSn			FeSn		NiSn	PdSn	PtSn
Pb									PtPb
As				MnAs			NiAs		
Sb			CrSb	MnSb	FeSb	CoSb	NiSb	PdSb	PtSb
Bi				MnBi			NiBi		PtBi
Se			CrSe		FeSe	CoSe	NiSe		
Te			CrTe	MnTe	FeTe	CoTe	NiTe	PdTe	PtTe

it will be seen that many A_2B_2 compounds crystallize with this structure, which has been illustrated on p. 609, where it is discussed in more detail. The following compounds crystallize with the pyrites structure:

$$PtP_2$$
$$PdAs_2 \qquad PtAs_2$$
$$AuSb_2 \qquad PdSb_2 \qquad PtSb_2$$
$$\alpha\text{-}PtBi_2$$

(h) *Some bismuth-rich alloys* A_2B_2

The structural principles in a group of alloys of Bi with Ni, Pd, and Rh are not simple, for although there is in some cases an obvious attempt by Bi to form three strongest pyramidal bonds (as in metallic bismuth itself) there is also a tendency to attain the much higher coordination numbers characteristic of the metallic state. For example, RhBi has a deformed NiAs structure in which the c.n. of Rh is raised to 8 and that of Bi to 12. An interesting feature of a number of these alloys is that they may be described in terms of the packing of monocapped trigonal prisms (Fig. 29.20(a)). The transition-metal atoms occupy the centres of these polyhedra but

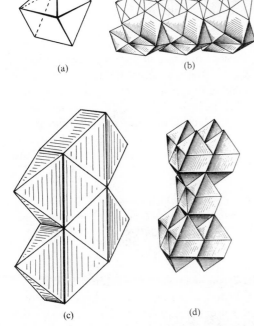

(a)   (b)

(c)   (d)

FIG. 29.20. Structures of some bismuth-rich alloys: (a) coordination polyhedron around transition metal atom, (b)–(d) packing of these coordination polyhedra in α-Bi_2Pd, Bi_3Ni, and BiPd.

are also bonded to similar atoms in neighbouring polyhedra. For example, in Bi_3Ni (Fig. 29.20(c)) these polyhedra are linked into columns, and a Ni atom is bonded to two Ni in neighbouring polyhedra (at 2·53 Å) as well as to seven Bi, making a total c.n. of 9. Fig. 29.20(b) and (d) show how these Bi_7 polyhedra are packed in α-Bi_2Pd and BiPd respectively; in all these packings there are (empty) tetrahedral and octahedral holes between the Bi_7 polyhedra, which do not, of course, form space-filling assemblies. In α-Bi_4Rh there is 8-coordination of Rh by Bi (square antiprism), so that the characteristic c.n.'s of the transition-metal atoms in a number of these alloys are 8 or 9 (7 Bi + 2 transition metal) while those of Bi are high, 11–13.

1049

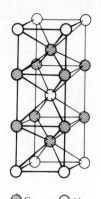

◍ Cr   ◯ Al

FIG. 29.21. The crystal structure
of Cr_2Al.

Metals and Alloys

In contrast to α-Bi_2Pd the high-temperature β form has a more symmetrical (tetragonal) structure which is a superstructure of the f.c.c. structure, of the same general type as Cr_2Al (Fig. 29.21), which is the analogous b.c.c. superstructure. Note that although both Cr_2Al and β-Bi_2Pd have the same space group ($F/4mmm$) and the same positions are occupied, namely Al(Pd) in (000, $\frac{1}{2}\frac{1}{2}\frac{1}{2}$) and Cr(Bi) in $\pm(00z, \frac{1}{2}\frac{1}{2}\frac{1}{2} + z)$, for Cr_2Al $c:a \approx 3$ ($z = 0.32$), while β-Bi_2Pd has $c:a \approx 3\sqrt{2}$ ($z = 0.36$). These compounds are not isostructural; see the discussion of this topic in Chapter 6.

(i) *Systems* BB

Alloys containing elements of only the earlier B subgroups have typically metallic properties. Solid solutions are formed to an appreciable extent only by elements of the same subgroup and, of course, the relative size criterion applies as in other systems. Thus Cd and Hg form solid solutions over quite large ranges of composition and Cd and Zn over smaller ranges; compare the radii: Zn, 1·37, Cd, 1·52, and Hg, 1·55 Å. Cadmium and tin, on the other hand, are practically immiscible. When both the metals belong to later B subgroups a 1 : 1 compound with the NaCl structure occurs in a number of cases, for example, SnSe, SnTe, PbSe, and PbTe. Finally the zinc-blende or wurtzite structure is generally found for 1:1 compounds in which the average number of valence electrons per atom is four, that is, when the atoms belong to the Nth and $(8 - N)$th subgroups. Examples of compounds with these structures are the sulphides, selenides, and tellurides of Zn, Cd, and Hg, and GaAs, GaSb, and InSb. (Compounds such as BeS and AlP also crystallize with the zinc-blende structure, which is not restricted to elements of the B subgroups.)

In contrast to InSb, TlSb (and also TlBi) has the CsCl structure; InBi is also exceptional in having the B 10 structure (p. 218) in which In has four Bi neighbours arranged tetrahedrally, but Bi has four In neighbours on one side at the basal vertices of a square pyramid.

The formulae of alloys

Substitutional solid solutions can have any composition within the range of miscibility of the metals concerned, and there is random arrangement of the atoms over the sites of the structure of the solvent metal. At particular ratios of the numbers of atoms superstructures may be formed, and an alloy with either of the two extreme structures, the ordered and disordered, but with the same composition in each case, can possess markedly different physical properties. Composition therefore does not completely specify such an alloy. Interstitial solid solutions also have compositions variable within certain ranges. The upper limit to the number of interstitial atoms is set by the number of holes of suitable size, but this limit is not necessarily reached, as we shall see later. When a symmetrical arrangement is possible for a particular ratio of interstitial to parent lattice atoms this is adopted. In intermediate cases the arrangement of the interstitial atoms is random.

When we come to alloys which are described as intermetallic compounds as opposed to solid solutions, we find that in some cases the ratios of the numbers of atoms of different kinds which we should expect to find after examination of the crystal structure are never attained in practice. When dealing with electron compounds we noted many cases of alloys with quite different types of formulae which crystallize with the same structure (e.g. CuZn, Cu_3Al, and Cu_5Sn all with the β structure). We know that in cases of this kind the compositions are determined by the electron : atom ratios. At an appropriate temperature an electron compound, like a solid solution, is stable over a range of composition, and the particular formulae adopted were selected to conform with the numbers of equivalent positions in the crystal structure (e.g. Cu_9Al_4 rather than, say, Cu_9Al_5 which also lies within the range of stability of this phase) and/or Hume-Rothery's simple electron : atom ratios. We have seen that the precise values of these ratios, 3/2, 21/13, and 7/4, have no theoretical significance. An alloy such as Ag_3Al is completely disordered, there being a total of 20 atoms in the unit cell in two sets of 12-fold and 8-fold equivalent positions. There are, however, alloys other than those with the β, γ, and ϵ structures the compositions of which are never those of the ideal structures. In the Cu–Al system, for example, there is a θ phase with ideal formula $CuAl_2$, the structure of which has been illustrated in Fig. 29.17. In this (tetragonal) structure there are, in one unit cell, four Cu and eight Al atoms, the symmetry requiring these numbers of the two kinds of atom. If, however, an alloy is made up with the composition $CuAl_2$ it is not homogeneous but consists of a mixture of alloys with the $CuAl_2$ and CuAl structures. In other words, the $CuAl_2$ structure is not stable with the Cu : Al ratio equal to 1 : 2 but prefers a rather greater proportion of aluminium. Thus, although this θ phase is stable over a certain small range of composition, the alloy $CuAl_2$ lies outside this range. The laws applicable to conventional chemical compounds (see, however, FeS, p. 610) do not hold in metal systems, and these facts concerning $CuAl_2$ are not surprising when we remember that the approximate compositions of the phases in this system with the β and γ structures are Cu_3Al and Cu_9Al_4 respectively. The β phase in the Cr–Al system provides another example of the same phenomenon. The body-centred solid solution of Al in Cr is stable at high temperatures up to some 30 per cent Al, that is, beyond the composition Cr_2Al. If, however, an alloy containing about 25 per cent Al is cooled slowly, the body-centred cubic structure changes to the tetragonal β structure illustrated in Fig. 29.21. As may be seen from the diagram, this structure is closely related to the body-centred cubic structure and its ideal composition is clearly Cr_2Al. Although this β structure is stable over a considerable range of composition, the alloy Cr_2Al lies outside this range. The slow cooling of an alloy with the exact composition Cr_2Al therefore gives an inhomogeneous mixture which actually consists of an Al-rich Cr_2Al component and some body-centred cubic solid solution.

Interstitial carbides and nitrides

We saw in Chapter 4 that from the geometrical standpoint the structures of many inorganic compounds, particularly halides and chalconides, may be regarded as assemblies of close-packed non-metal atoms (ions) in which the metal atoms occupy tetrahedral or octahedral interstices between 4 or 6 c.p. non-metal atoms. The numbers of such interstices are respectively $2N$ and N in an assembly of N c.p. spheres, and occupation of some or all of them in hexagonal or cubic closest packing gives rise to the following simple structures:

Interstices occupied	h.c.p.	c.c.p.	Formula
All tetrahedral	–	antifluorite	A_2X
$\frac{1}{2}$ tetrahedral	wurtzite	zinc-blende	AX
All octahedral	NiAs	NaCl	AX
$\frac{1}{2}$ octahedral	$\left\{\begin{array}{l} CaCl_2 \text{ (rutile)} \\ CdI_2 \end{array}\right.$	$\left\{\begin{array}{l} \text{atacamite} \\ CdCl_2 \end{array}\right.$	AX_2

The description of these structures in terms of the closest packing of the halide or chalconide ions is both convenient and realistic, because in most cases these ions are appreciably larger than the metal ions. At the other extreme there are many compounds of metals with the smaller non-metals in which the non-metal atoms occupy interstices between c.p. metal atoms. For structural reasons it is preferable to deal separately with hydrides (p. 291) and borides (p. 840). In the latter compounds B–B bonds are an important feature of many of the structures, and their formulae and structures are generally quite different from those of carbides and nitrides. The structures of carbides of the type MC_2 have been described in Chapter 22. In the LaC_2 and ThC_2 structures the carbon atoms are in pairs as C_2^{2-} ions. Although these structures may be regarded as derived from c.c.p. metal structures with the C_2^{2-} ions in octahedral interstices, there is considerable distortion from cubic symmetry owing to the large size and non-spherical shape of these ions, and these carbides are therefore not to be classed with the interstitial compounds. Certain oxides are sometimes included with the interstitial carbides and nitrides; these will be mentioned later. It is convenient to deal separately with the carbides and nitrides of Fe, which are much more chemically reactive than, and structurally different from, the compounds to be considered here.

Interstitial carbides and nitrides have many of the properties characteristic of intermetallic compounds, opacity (contrast the transparent salt-like carbides of Ca, etc.), electrical conductivity, and metallic lustre. In contrast to pure metals, however, these compounds are mostly very hard and they melt at very high temperatures. Compounds of the type MX are generally derived from cubic close packing; those of the type M_2X from hexagonal close packing. The melting points and hardnesses of some interstitial compounds are set out below. These compounds may be prepared by heating the finely divided metal with carbon, or in a stream of ammonia, to temperatures of the order of 2200°C for carbides, or 1100–1200°C for nitrides. Alternatively, the metal, in the form of a wire, may be heated in an atmosphere of a hydrocarbon or of nitrogen. The solid solution of composition $4\,TaC + ZrC$ melts at the extraordinarily high temperature 4215°K. These compounds are very inert chemically except towards oxidizing agents. Their electrical conductivities are high and decrease with rising temperature as in the case of metals, and some exhibit supraconductivity. (The hardness is according to Mohs' scale, on which that of diamond is 10.)

	M.P. (°K)	Hardness		M.P. (°K)	Hardness
TiC	3410	8–9	TiN	3220	8–9
HfC	4160		ZrN	3255	8
W_2C	3130	9–10	TaN	3360	
NbC	3770				

The term 'interstitial compound' or 'interstitial solid solution' was originally given to these compounds because it was thought that they were formed by the interpenetration of the non-metal atoms into the interstices of the metal structure, implying that no gross rearrangement of the metal atoms accompanied the formation of the interstitial phase. This view of their structures was apparently supported by their metallic conductivity, by the variable composition of many of these phases, and by the fact that there is an upper limit to the size of the 'interstitial' atom compared with that of the metal atom. While it is certainly true that the C or N atoms occupy interstices (usually octahedral) in an array of (usually) c.p. metal atoms, it is now known that the arrangement of the metal atoms in the interstitial compound is generally different from that in the metal from which it is formed, although initially the metal structure may be retained if a solid solution is formed. Thus Ti dissolves nitrogen to the stage $TiN_{0.20}$, this phase being a solid solution of N in the h.c.p. structure of α-Ti. At the composition $TiN_{0.50}$ the ε phase has the anti-rutile structure, with distorted h.c.p. Ti, but at $TiN_{0.60}$ (at 900°C) the arrangement of metal atoms becomes c.c.p. (defect NaCl structure).[1] The high-temperature form of the metal (above 880°C) is b.c.c. All the metals V, Nb, Cr, Mo, and W have the b.c.c. structure, but V_2C and Nb_2C have h.c.p. metal atoms in which C atoms occupy (in various sets of sites) one-half of the octahedral holes, while VC, VN, VO, and NbC have the NaCl (f.c.c.) structure. The Nb–N system is more complex, and NbO has a unique structure derived from the NaCl structure by omitting one-quarter of the atoms of each kind (p. 193). The b.c.c. Cr, Mo, and W form the following assortment of compounds, in none of which is the arrangement of metal atoms the same as in the metal itself: Cr_2N (hexagonal) and CrN (NaCl structure), low-Mo_2N (f.c.c.), W_2N (f.c.c.), and WN (hexagonal).

In table 29.13 are listed the structures of compounds MC and MN with metallic character and known crystal structure. Of all the metals in Table 29.13 forming a carbide MC or a nitride MN with the NaCl structure only four have the cubic close-packed structure. For all the other compounds the arrangement of metal atoms in the compound MX is *different* from that in the metal itself. (This is also true of many hydrides—see p. 294.) Also, although many of these compounds exhibit variable composition, some do not, for example, UC, UN, and UO. In any case, variable composition is not confined to interstitial compounds—see, for example, the note on non-stoichiometric compounds, p. 5.

(1) ACSc 1962 **16** 1255

1053

TABLE 29.13

The structures of metals and of interstitial compounds MX

Metal	Structure	Carbide	Nitride	Oxide
Sc	A 1, A 3	–	B 1	–
La	A 1, A 3	–	B 1	–
Ce	A 1, A 3	–	B 1	–
Pr	A 3	–	B 1	–
Nd	A 3	–	B 1	–
Ti	A 2, A 3	B 1	B 1	B 1 (p. 465)
Zr	A 2, A 3	B 1	B 1	B 1
Hf	A 2, A 3	B 1	B 1 (?)	–
Th	A 1	B 1	B 1	B 1
V	A 2	B 1	B 1	B 1
Nb	A 2	B 1	(1) (2)	See text
Ta	A 2	B 1	(2)	–
Cr	A 2	Hex. ?	B 1	–
Mo	A 2, A 3	Hex.	Hex.	–
W	A 2	Hex.	Hex.	–
U(γ)	A 2 (a)	B 1	B 1	B 1

A 1 indicates cubic close-packed, A 2, body-centred cubic, A 3, hexagonal close-packed, B 1, NaCl structure.
(a) Also α and β forms of lower symmetry.
(1) B 1 structure for composition $NbN_{0.9}O_{0.1}$.
(2) For Nb and Ta nitrides see text (pp. 671 and 1055) and Table 29.14 (p. 1056).

(2) AC 1948 **1** 180

It would seem that the salient characteristics of these compounds are: (a) adoption in most cases of the NaCl structure irrespective of the structure of the metal, (b) high melting point and hardness, and (c) electrical conductivity. Rundle therefore suggested[2] that these properties indicate metal–non-metal bonds of considerable strength, the bonds from the non-metal atoms being directed octahedrally but not localized (to account for the conductivity). The compounds are regarded as electron-deficient compounds in which the non-metal atoms form six octahedral bonds either (1) by using three 2p orbitals (for three electron pairs), the 2s orbital being occupied by an electron pair, or (2) by using two hybrid sp orbitals (bond angle 180°) and two p orbitals, which are perpendicular to the hybrid orbitals and to each other. The six (octahedral) bonds would then become equivalent by resonance. In (1) the bonds would be '$\frac{1}{2}$-bonds'; in (2) they would be '$\frac{2}{3}$-bonds', using Pauling's nomenclature (p. 1024). Use of the 2s orbital by an unshared pair of electrons would be expected if the non-metal is sufficiently electronegative compared with the metal (as in 'suboxides', MO), or possibly in the nitrides of the more electropositive metals. On this view the total number of valence electrons is used for the metal–non-metal bonds in the Group IIIA nitrides and the Group IVA carbides, so that the metal–metal distances would be the result of the M–X bonding. For Group IV nitrides and Group V carbides there is one electron per metal atom available for metal–metal bonds, which would therefore have bond-number 1/12 in the NaCl structure. In the Group V nitrides and Group VI carbides there are two electrons per metal atom for M–M bonds which would accordingly be $\frac{1}{6}$-bonds. These bonds are now sufficiently strong to influence the stability of the NaCl structure. Accordingly the Nb–N and Ta–N systems show a complex behaviour, and for the carbides and nitrides of Cr, Mo, and W we also find structures other than $B\,1$ for the compounds MX. (UC and UN are exceptional, possibly because U is not hexavalent in these compounds.)

An interesting feature of this treatment of these compounds is that it offers some explanation of the fact that they are limited to compounds MC and MN (and sometimes MO) of the elements of Groups IIIA, IVA, VA, and VIA, owing to the requirements that (1) one element (M) must have more stable bond orbitals than valence electrons, and must therefore generally be a metal, (2) the second element must have relatively few bond orbitals, and is therefore generally a non-metal, and (3) the electronegativities of the two elements must not differ so much that the bond is essentially ionic (hence all fluorides and some oxides are excluded). Boron, for example, is classified with the metals in this scheme, so that borides are structurally different from the carbides and nitrides. Only metals having more than four stable bond orbitals will require C, N, and O to use a single orbital for more than one bond; these are the A subgroup metals. In the B subgroup metals the d levels below the valence group are filled, as for example in Ga, In, and Tl, which have tetrahedral (sp^3) orbitals and therefore form normal, as opposed to interstitial, mononitrides. Of the A subgroup metals, the alkalis and alkaline-earths are too electropositive to form essentially covalent bonds with C, N, and O; hence, the interstitial compounds begin in Group III.

A comparison of the V–C[3] and Ta–N[4] systems is instructive. Vanadium metal (b.c.c.) dissolves very little carbon, probably not more than 1 atomic per cent at 1000°C. It does however, form V_2C (with composition range from $VC_{0.37}$ to $VC_{0.50}$) with C in one-half of the octahedral interstices of a hexagonal close-packed assembly of V atoms, and VC with the NaCl structure (that is, c.c.p. V). Although C and N are not very different in size and Ta has the same structure as V, there are considerable differences between the V–C and Ta–N systems. Following the distorted body-centred cubic β phase, containing about 5 atomic per cent N, there is a hexagonal close-packed γ phase of approximate composition Ta_2N (actually $TaN_{0.40}$–$TaN_{0.45}$) isostructural with V_2C, but after that there is no further similarity between the two systems. The δ phase, $TaN_{0.8-0.9}$, has the WC structure with a simple hexagonal sequence of metal layers (p. 128), and ϵ-TaN has a new structure ($B\,35$) illustrated in Fig. 29.22. In the hexagonal unit cell of this structure there are metal–metal contacts much shorter relative to Ta–N than would occur in the NaCl structure (where $M–12\,M = \sqrt{2}(M–N)$):

(3) ACSc 1954 **8** 624
(4) ACSc 1954 **8** 199

Ta_I	2 Ta	2.908 Å		Ta_{II}	2 Ta	2.908 Å
	12 Ta	3.402			3 Ta	2.994
	6 N	2.593			6 Ta	3.402
					6 N	2.204

(compare $\sqrt{2}(Ta_I–N) = 3.67$ Å) $\qquad$ (compare $\sqrt{2}(Ta_{II}–N) = 3.11$ Å)

The metal atom sequences in a number of nitrides and carbides are summarized in Table 29.14.

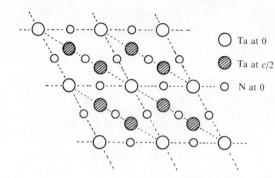

FIG. 29.22. Projection of four unit cells of the TaN (B 35) structure.

TABLE 29.14

Structures of some interstitial compounds

Metal layer sequence	Nitrides	Carbides
$A\,A\ldots$	γ-NbN$_{0.8}$, δ-TaN$_{0.8-0.9}$, MoN, WN (WC structure)[a]	γ-MoC, WC
$A\,B\ldots$	δ-NbN$_{0.95}$ (anti-NiAs structure)	Mo$_2$C, W$_2$C
$A\,B\,C\ldots$	VN, TiN, ZrN (NaCl structure)	TiC, ZrC
$A\,A\,B\,B\ldots$	ϵ-NbN	γ'-MoC
$A\,B\,A\,C\ldots$	Ta$_3$MnN$_4$	

(a) For the relation of this structure to that of NiAs see p. 128.

Iron and steel

A simple steel consists of iron containing a small amount of carbon. Many steels now used for particular purposes also contain one or more of a number of metals (Cr, Mo, W, V, Ni, Mn) which modify the properties of the steel. We shall restrict our remarks to the simple Fe–C system. The properties of the various kinds of 'iron' and of steels which make these materials so valuable are dependent on the amount of carbon and the way in which it is distributed throughout the metal. The Fe–C system is, for two reasons, more complicated than the metal–non-metal systems giving the interstitial compounds just described. Firstly, iron is dimorphic. The form stable at ordinary temperatures is called α-iron. It has the body-centred cubic structure and is ferromagnetic. The body-centred structure is stable up to 960°C and again from 1401° to 1530°C, the melting point. Over the intermediate range of temperature, 906–1401°C, the structure is face-centred cubic (γ-iron), in which form the metal is non-magnetic. The Curie point (766°C) is lower than the α-γ transition point, and the term β-iron is applied to iron in the temperature range 766–906°C. Since there is no change in atomic arrangement at the Curie point we shall not refer to β-iron in what follows. The second complicating factor is that the C : Fe radius ratio (0·60) lies near the critical limit for the formation of interstitial solid solutions. Accordingly, in addition to the latter a carbide Fe$_3$C is formed. Thus according to the carbon content and heat treatment the carbon may be present either in the free state as graphite, in solid solution, or as cementite (Fe$_3$C).

Iron is obtained by smelting oxy-ores with coke, so that in the melt there is an excess of carbon. Molten iron dissolves up to 4·3 per cent of carbon, the eutectic mixture solidifying at 1150°C. Pig-iron (cast-iron) contains therefore about 4 per cent C. There is also up to 2 per cent Si from the clays associated with the ores. The carbon in cast-iron may be in the form of cementite (white cast-iron) or graphite (grey cast-iron), depending on the silicon content. The presence of silicon favours the decomposition of cementite into graphite, and in general some of the carbon in cast-iron is present in both forms. The material which solidifies at 1150°C containing 4·3 per cent C is a mixture: the maximum solubility of C in γ-Fe at that temperature, that is, in a homogeneous solid solution, is 1·9 per cent. This solubility falls to 0·9 per cent at 690°C, the lowest temperature at which γ-Fe is rendered stable by the presence of carbon. (Iron can be retained in the non-magnetic γ form at ordinary temperatures by adding elements such as Mn and Ni which form solid solutions with γ- but not with α-iron.) Removal of all but the last traces of impurity from pig-iron gives wrought iron, the purest commercial iron. Steels contain up to 1·5 per cent C; mild steels from about 0·1 to 0·5 per cent. The production of steels therefore involves either the controlled reduction of the amount of carbon if they are made from pig-iron or the controlled addition of carbon if they are made from wrought iron (the process of cementation). We may summarize the processes which take place in the production of steels as follows.

Above 906°C the steel is in the form of a (non-magnetic) solid solution of carbon in γ-iron (austenite). This is a simple interstitial solid solution in which the carbon atoms are arranged at random since there are not sufficient to form a regular structure. It is fairly certain that in austenite the carbon atoms occupy octahedral holes in the γ-Fe lattice. When austenite is cooled slowly, the first process which takes place is the separation of the excess carbon as cementite, since the solubility of carbon falls to 0·9 per cent at 690°C. Below this temperature γ-Fe is no longer stable, and the solid solution of C in γ-Fe changes at 690° into a eutectoid mixture of ferrite and cementite. Ferrite is nearly pure α-Fe; it contains some 0·06 per cent of C in interstitial solid solution. The remaining carbon goes into the cementite. Pearlite, which is the name given to this eutectoid mixture, has a fine-grained banded structure with a pearly lustre and is very soft. The other extreme form of heat treatment is to quench the austenite to a temperature below 150°C, when martensite is formed. This is a super-saturated solid solution of C in α-Fe and may contain up to 1·6 per cent C. It is very hard, extreme hardness being a characteristic of quenched steels. (The original γ solid solution, austenite, can be preserved after quenching only if other metals are present, as mentioned above.) The hard, brittle quenched steels are converted into more useful steels by the process of tempering. This consists in reheating the martensite to temperatures from 200 to 300°C. The object of tempering is the controlled conversion of the quenched solid solution into ferrite and cementite. The mixture produced by tempering has a coarser texture than pearlite and is termed sorbite. The tempering reduces the hardness but increases the toughness of the steel. Sorbite is thus an intermediate product in the sequence: austenite-martensite-sorbite-pearlite. These forms of heat treatment are summarized on p. 1058.

It is interesting that the structure of cementite is not related in any simple way to those of α- or γ-Fe. In the γ structure each Fe atom has twelve equidistant nearest neighbours (at 2·52 Å, the value obtained by extrapolation to room temperature), and in the α structure eight, at 2·48 Å. In cementite some of the Fe atoms have twelve neighbours at distances ranging from 2·52 to 2·68 Å and the others eleven at distances from 2·49 to 2·68 Å. The most interesting feature of the structure is the environment of the carbon atoms. The six nearest neighbours lie at the apices of a distorted trigonal prism, but the range of Fe–C distances is considerable, 1·89–2·15 Å, and two further neighbours at 2·31 Å might also be included in the coordination group of the carbon atoms. The reason for this unsymmetrical environment—contrast the octahedral environment in austenite—is not known.

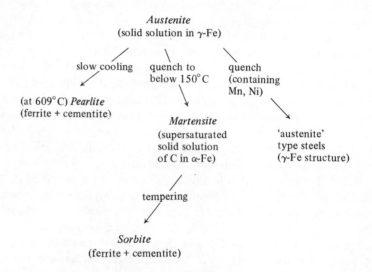

The process of case-hardening of steel provides an interesting example of the formation of interstitial compounds. In one method both C and N are introduced into the surface of steel either by immersion in a molten mixture of NaCN, Na_2CO_3, and NaCl at about 870°C or by heating in an atmosphere of H_2, CO, and N_2 to which controlled amounts of NH_3 and CH_4 are added. By these means both C and N are introduced. Although Fe does not react with molecular N_2 certain steels can be case-hardened by the action of ammonia at temperatures around 500°C. The Fe–N phase diagram is complex, and phases formed include:

α solid solution (about 1·1 atomic per cent N at 500°C).
γ phase (stable only above 600°C, at which temperature composition is approximately $Fe_{10}N$).
γ′ also f.c.c., Fe_4N.
ε h.c.p. (composition range extends from about Fe_3N to Fe_2N at low temperatures, but composition varies considerably with temperature).
ζ Fe_2N (orthorhombic, slightly deformed variant of ε).

1058

We mention these phases because they illustrate once again how the arrangement of metal atoms changes with increasing concentration of interstitial atoms and also because an outstanding feature of these nitrides is that with the exception of the γ phase (stable only at high temperatures and with rather low concentrations of N atoms) they have ordered arrangements of the N atoms in the octahedral interstices.[1]

(1) AC 1952 5 404.

Formula Index

Subject Index